Data Analysis Methods in Physical Oceanography

Data Analysis Methods in Physical Oceanography

Fourth Edition

Richard E. Thomson
Institute of Ocean Sciences, Fisheries and Oceans Canada,
Sidney, BC, Canada

William J. Emery
Aerospace Engineering Sciences, University of Colorado, Boulder, CO, United States

ELSEVIER

Elsevier
Radarweg 29, PO Box 211, 1000 AE Amsterdam, Netherlands
125 London Wall, London EC2Y 5AS, United Kingdom
50 Hampshire Street, 5th Floor, Cambridge, MA 02139, United States

ISBN: 978-0-323-91723-0

For information on all Elsevier Science publications visit our website at https://www.elsevier.com/books-and-journals

Publisher: Candice Janco
Acquisitions Editor: Maria Elekidou
Editorial Project Manager: Joshua Mearns
Production Project Manager: Kumar Anbazhagan
Cover Designer: Miles Hitchen

Typeset by TNQ Technologies

Dedication

Richard Thomson dedicates this book to his wife Irma, daughters Justine and Karen, and grandchildren Brenden and Nicholas.
Bill Emery dedicates this book to his wife Dora Emery, his children Alysse, Eric, and Micah, and to his grandchildren Margot and Elliot.

Contents

Preface

Numerous books have been written on data analysis methods in the physical and geophysical sciences over the past several decades. Most of these books are primarily directed toward the more theoretical aspects of data processing or narrowly focused on one particular topic. Unlike this book, few texts span the range from basic data sampling and statistical analysis to more modern techniques such as wavelet analysis, rotary spectral decomposition, Kalman filtering, extreme value analysis, self-organizing maps, fractals, and neural networks. Texts that also provide detailed information on the sensors and instruments that collect the data are even more rare. In writing this book, we saw a clear need for a practical reference volume for earth and ocean sciences that brings together established and modern processing techniques in remote and *in situ* observations under a single cover. This text is intended for students, established scientists, and marine engineers alike.

For the most part, graduate programs in oceanography have some form of methods course in which students learn about the measurement, calibration, processing, and interpretation of geophysical records. The classes are intended to give the students needed experience in both the logistics of data collection and the practical problems of data processing and analysis. Because class material generally is based on the experience of the faculty members giving the course, each class emphasizes different aspects of data collection and analysis. Formalism and presentation can differ widely. While it is valuable to learn from the first-hand experiences of the class instructor, it seemed to us important to have available a central reference text that could be used to provide some uniformity in the material being covered within the oceanographic community. This fourth Edition provides an extensive update on oceanographic instrumentation—including much expanded sections on satellite measurements, wavegliders and saildrones—and markedly expanded chapters on data processing methods that have become more widely available over the past decade. New material added to Edition Four include an Introduction to the International Thermodynamic Equation of Seawater-2010 (TEOS-10), an in-depth section on Extreme Value Analysis, a fully updated section on determining the Effective Degrees of Freedom when estimating statistical parameters, a list of data formats used for oceanographic and meteorological data, a link to modern software for Optimal Interpolation provided specifically for this book by Professor John Wilkin, and a two-chapter introductory segment on machine learning methods and neural networks.

Many of the data analysis techniques most useful to oceanographers can be found in books and journals covering a wide variety of topics. Much of the technical information on these techniques is detailed in texts on numerical methods, time series analysis, and statistical methods. In this book, we bring together as many as possible key data processing methods found in the literature, as well as adding new information on spatial and temporal data analysis techniques that were not readily available in older texts. For example, Chapter 1 provides a description of most of the instruments used in physical oceanography today, with updated information on satellites and satellite-derived products, as well as autonomous data collection platforms. This is not a straightforward task given the rapidly changing technology for both remote and *in situ* oceanic sensors, and the ever-accelerating rate of data collection and transmission. Our hope is that this book will provide instructional material for students in the marine sciences and serve as a general reference volume for those directly involved with all branches of geophysical and environmental research.

The broad scope and rapidly evolving nature of oceanographic sciences has meant that it has not been possible for us to fully detail all existing instrumentation or emerging data analysis methods. However, we believe that many of the methods and procedures outlined in this book will provide a basic understanding of the kinds of options available to the user for the interpretation of data sets. Our intention is to describe general statistical and analytical methods that will be sufficiently fundamental to maintain a high level of utility over the years.

Finally, we wish to reiterate that the analysis procedures discussed in this book apply to a wide readership in the geophysical and environmental sciences. As with oceanographers, this wider community of scientists and engineers should benefit from a central source of information that encompasses not only a description of the mathematical methods, but also considers some of the practical aspects of data collection and analyses. It is this synthesis between theoretical concepts and the logistical limitations of real-data measurement that is a primary goal of this text.

Richard E. Thomson, William J. Emery
North Saanich, British Columbia and Boulder, Colorado

Acknowledgments

Many people have contributed to this edition. Dudley Chelton (Oregon State University) and Alexander Rabinovich (P.P. Shirshov Institute of Oceanology and the Institute of Ocean Sciences) have generously provided input to several chapters. Dudley, along with Ismael Núñez-Riboni (Thünen-Institut für Seefischerei, Bremerhaven), helped immensely with the new section on Effective Degrees of Freedom in Chapter 3 (including careful reviews of the new material) and Alexander, along with Isaac Fine (Institute of Ocean Sciences, Fisheries and Oceans Canada), provided new analyses and figures that have significantly extended the topics covered in this book. Kurtis Anstey (University of Victoria) and Andreas Thurnherr (Lamont-Doherty Earth Observatory, Columbia University) generously provided plots showing the full-depth distributions of oceanic density structure and mean currents derived from recent GO-Ship repeat hydrography surveys in the Pacific and Atlantic oceans. John Wilkin (Rutgers University) graciously formulated "MatLab Live" scripts for the book, with several worked examples, for those wanting to become proficient in Optimal Interpolation as discussed in Chapter 4 and Appendix F. We thank Mirko Orlic, Mira Pasarić, and Iva Međugorac at the Geophysical Institute of the University of Zagreb for providing the sea level data for the Bay of Bakar that we used in the new section of Extreme Value Analysis (Chapter 5). Jadranka Šepić (University of Split) and Kejia Wang (Institute of Ocean Sciences) generously contributed to the analysis and graphics presented in Chapter 5.

We are very grateful to Olga Yakovenko (P.P. Shirshov Institute of Oceanology) for drafting high-quality figures for many sections of the book and to Igor Medvedev (also from the P.P Shirshov Institute of Oceanology) for his analysis and graphics regarding high-resolution spectra of the tides in Chapter 3. We thank Humfrey Melling and Michael Dempsey (Institute of Ocean Sciences) for providing us with important information on the special requirements for current meter moorings in the Arctic Ocean (Chapter 1). Similarly, we thank Richard Lumpkin (NOAA/AOML, Miami) for assisting our discussion on the global drifter programs and Franco Reseghetti (Italian National Agency for New Technologies, Energy and Sustainable Economic Development, Rome) for contributing to the XBT calibration discussion in Chapter 1. Igor Yashayaev (Bedford Institute of Oceanography) kindly provided exquisite plots of his updated Labrador Sea time series.

All of the acknowledgments from the previous editions of this book are still valid for Edition 4. We thank Andrew Bennett (Oregon State University, Corvallis) for reviewing the section on inverse methods in Chapter 4, Brenda Burd (Ecostat Research Inc., North Saanich) for reviewing the bootstrap method in Chapter 3, and Steve Mihály (Ocean Networks Canada, Victoria) for assisting with the section on self-organizing maps in Chapter 4. Roy Hourston and Maxim Krassovski of the Institute of Ocean Sciences helped considerably with various sections of this book, including sections on regime shifts and wavelet analysis in Chapter 5. The technical contributions to Chapter 1 from Tamás Juhász and David Spear of the Institute of Ocean Sciences are gratefully acknowledged. Patricia Kimber of Tango Design (Sidney) helped draft many of the figures.

Expert contributions to all editions—including the reports of errors and omissions in the previous editions—were also provided by Michael Foreman, Robie Macdonald, Diane Masson, Patrick Cummins, Joseph Linguanti, Ron Lindsay, Germaine Gatien, Steve Romaine, and Lucius Perreault (Institute of Ocean Sciences), Jason Middleton (University of New South Wales, Australia), Philip Woodworth (Permanent Service for Mean Sea Level, United Kingdom), William (Bill) Woodward (President and CEO of CLS America, Inc.), Laurence Breaker (California State University), Jo Suijlen (The Netherlands National Institute for Coastal and Marine Management), Guohong Fang (First Institute of Oceanography, China), Vlado Malačič (National Institute of Biology, Slovenia), Ųyvind Knutsen (University of Bergen, Norway), Parker MacCready (University of Washington), Andrew Slater (University of Colorado), David Dixon (Plymouth, United Kingdom), Drew Lucas (Scripps Institute of Oceanography), Wayne Martin (University of Washington), David Ciochetto (Dalhousie University), Alan Plueddemann (Woods Hole Oceanographic Institution), Fabien Durand (IRD/LEGOS, Toulouse), Jack Harlan

(NOAA, Boulder, Colorado), Denis Gilbert (The Maurice Lamontagne Institute), Ben Hamlington (University of Colorado), John Hunter (University of Tasmania), Irene Alonso (Instituto de Ciencias Marinas de Andalucia, Spain), Yonggang Liu (University of South Florida), Gary Borstad (ASL Environmental Sciences Ltd., Canada), Earl Davis and Bob Meldrum (Pacific Geosciences Centre, Sidney), and Mohammad Bahmanpour (University of Western Australia, Australia).

Lastly, we are grateful for the professional support we received from the personnel at Elsevier Science throughout this major technical undertaking. Particular thanks to Kumar Anbazhagan (Project Manager), Indhumathi Mani (Senior Copyrights Specialist), and the publisher and editorial team led by Candice Janco, Maria Elekidou, and Joshua Mearns.

Chapter 1

DATA ACQUISITION AND RECORDING

1.1 INTRODUCTION

Physical oceanography is an ever-evolving science in which the instruments, types of observations, and methods of analysis undergo continuous advancement and refinement. The changes that have occurred since we completed the 3rd Edition of this book over a decade ago continue to be impressive. Recent progress in oceanographic theory, instrumentation, sensors, platforms, software development, and data accessibility has led to significant advances in marine science and the way that findings are presented. The advent of high-speed digital computers and the linkages to global datasets through the internet have revolutionized measurement collection procedures and the manner in which data are shared, reduced and analyzed. In contrast to the situation that existed up to about the 1960s, the individual scientist is no longer able to be personally familiar with each data point and its contribution to his or her study. Instrumentation and data collection have generally moved out of direct application by the scientist and into the hands of skilled technicians and research assistants, who are becoming increasingly more specialized in the operation and maintenance of equipment. For example, oceanographers running open ocean glider programs typically require dedicated technicians to maintain the equipment, launch and recover the gliders and to process the data. New electronic instruments operate at data rates and storage capacities not possible with earlier mechanical devices, and they produce volumes of information that can only be handled by high-speed computers.

Most modern data collection systems transmit sensor output directly to computer-based data acquisition systems through satellite, cell phone or subsea cable links to shore stations or are stored in digital format on some type of electronic medium such as hard drives, flash cards, or optical disks. High-speed analog-to-digital converters and digital-signal-processors are now used to convert voltage or current signals from sensors to digital values. Increasing numbers of cabled observatories extending into the deep ocean from shore stations are now providing high-bandwidth data flows in near real time supported by previously impossible sustained levels of power and storage capacity. As funding for research vessels diminishes and existing fleets continue to age, open-ocean studies are gradually being assumed by satellites, drifting buoys, gliders, pop-up drifters, and long-term moorings. The days of limited power supply, insufficient data storage space, and weeks at sea on ships collecting routine survey data is becoming increasingly less common. Ships will still be needed but their role will be more focused on repeat surveys for climate research, on process-related studies (including those sponsored by wealthy individuals or major corporations) and on the deployment, servicing, and recovery of oceanographic and meteorological equipment, such as sensor packages incorporated in cabled observatory networks or satellite telemetering systems. All of these developments are moving physical oceanographers into analysts of what is becoming known as "big data". Faced with large volumes of information, the challenge to oceanographers is deciding how to approach these mega data accumulations and how to select the measurements and numerical simulations that are most relevant to the problems of interest. One of the goals of this book is to provide insight into the analyses of the ever-growing volume of oceanographic information in order to assist the practitioner in deciding where to invest his/her effort.

With the many technological advances taking place, it is important for marine scientists to be aware of both the capabilities and limitations of their sampling equipment. This requires a basic understanding of the sensors; the recording systems and the data processing tools. If these are known and the experiment is carefully planned, many problems commonly encountered during the processing stage can be avoided. We cannot overemphasize the need for thoughtful experimental planning and proper calibration of all oceanographic sensors. If instruments are not in near-optimal locations or the researcher is unsure of the values coming out of the instruments, then it will be difficult to believe the results gathered in the field. To be truly reliable, instruments should be calibrated on a regular basis at intervals determined by use and the susceptibility of the sensor to drift. More specifically, the output from all oceanic instruments such as thermometers, pressure sensors, dissolved oxygen probes, and fixed pathlength transmissometers drift with time and need to be calibrated before and after each field deployment. For example, the zero point for the early Paroscientific Digiquartz (0–10,000 psi) pressure sensors used in the Hawaii Ocean Time-series at station "Aloha" 100 km north of Honolulu drift

Data Analysis Methods in Physical Oceanography. https://doi.org/10.1016/B978-0-323-91723-0.00010-6

about 4 dbar (~ 4 m) in 3 years. As a consequence, the sensors are calibrated about every 6 months against a Paroscientific laboratory standard, which is recalibrated periodically at special calibration facilities in the United States (Lukas, 1994). A more recent examination of the long-term drift of Paroscientific pressure sensors by Polster et al. (2009), based on 118 seafloor pressure series having durations of 2 months to 9 years, yielded a mean *in situ* drift of only -0.88 ± 0.73 kPa/year, equivalent to -0.09 ± 0.07 m/year. Drift increased slightly with instrument depth. Even the most reliable platinum thermometers—the backbone of temperature measurement in marine sciences—can drift on the order of $0.001°C$ during a year. Our shipboard experience also shows that opportunistic, over-the-side field calibrations during oceanic surveys can be highly valuable to others in the science community, even if the calibrations are specific to one's own research program. As we discuss in the following chapters, there are a number of fundamental requirements to be considered when planning the collection of field records, including such basic considerations as the sampling interval, sampling duration, and sampling location.

The purpose of this chapter to review many of the standard instruments and measurement techniques used in physical oceanography in order to provide the reader with a common understanding of both the utility and limitations of the resulting measurements. The discussion is not intended to serve as a detailed "user's manual" nor as an "observer's handbook". Rather, our purpose is to describe the fundamentals of the instruments and measurement techniques in order to give some insight into the data they collect. An understanding of the basic observational concepts, and their limitations, is a prerequisite for the development of methods, techniques, and procedures used to analyze and interpret the data that are collected.

Rather than treat each measurement tool individually, we have attempted to group them into generic classes and to limit our discussion to common features of the particular instruments and associated techniques. Specific references to particular company's products and the quotation of manufacturer's engineering specifications have been avoided whenever possible. Instead, we refer to published material addressing the measurement systems or the data recorded by them. Those studies that compare measurements made by similar instruments are particularly valuable. On the other hand, there are companies whose products have become the "gold standard" against which other manufacturers are compared. Reliability and serviceability are critical factors in the choice of any instrument. The emphasis of each instrument review is to give the reader a general background in the collection of data using that particular instrument. For those readers interested in more detailed information regarding a specific instrument or measurement technique, we refer to the references at the end of the book, where we list the sources of the material quoted. We realize that, in terms of specific measurement systems, and their review, this text will become dated as new and better systems evolve. It is for that reason that we continue to update this book by bringing out new editions. Nevertheless, we hope that the general outline we present for accuracy, precision, and data coverage will serve as a useful guide to the employment of newer instruments and methods.

1.2 INTRODUCTION TO TEOS-10

The records collected by oceanic sensors need to be converted to practical quantities prior to processing and interpretation. The International Thermodynamic Equation of Seawater-2010 (TEOS-10) comprises the most recent suite of computer algorithms for calculating the fundamental properties of seawater, such as salinity, temperature and density. These algorithms supersede those provided by the Thermodynamic Equation of Seawater-1980 (EOS-80) that had been in use for the previous 3 decades. Central to TEOS-10 are revised formulations for salinity and temperature. Instead of Practical Salinity, S_p, and potential temperature, θ, used prior to 2010 (see Sections 1.3.5 and 1.4.2, respectively, for further details), oceanographers are now being encouraged to use Absolute Salinity, S_A, and Conservative Temperature, Θ (names are capitalized). These two variables, along with Sea Pressure, p, are the ingredients needed to derive the Gibbs function, $g(S_A, \Theta, p)$, which allows for the calculation of all thermodynamic properties of oceanic waters, including density, specific volume (inverse of density), enthalpy, entropy, dynamic height, sound speed, and chemical potentials, as well as the freezing temperature and the latent heats required in the transitions of state in melting and evaporation. The primary variables, S_A and Θ, are not available from the earlier International Equation of State-1980 but are considered essential for the accurate determination of "heat" in the ocean and for the consistent and accurate determination of air-sea and ice-sea heat fluxes. For example, Conservative Temperature, Θ, is defined to be proportional to potential enthalpy, which means that it provides a direct measure of the heat content per unit mass of seawater. According to Graham and McDougall (2013), the heat estimate based on Θ is two orders of magnitude more accurate than that based on potential temperature, θ.

The TEOS-10 computer software, TEOS-10 Manual and other documents are available online from www.TEOS-10.org. Introductory articles on TEOS-10 include "What every oceanographer needs to know about TEOS-10 (The TEOS-10 Primer)" by Pawlowicz (2010) and "Getting started with TEOS-10 and the Gibbs Sea Water (GSW) Oceanographic Toolbox" by McDougall and Barker (2011). Millero (2010) and Pawlowicz et al. (2012) provide historical accounts of how

TEOS-10 was developed. Millero (2010) is a personal account of the author's extensive involvement in studies of ocean thermodynamics over his career. Here, various aspects of TEOS-10 relevant to the collection and processing of oceanic data are woven into the various subsections of the book dealing with temperature, salinity, density and other oceanographic properties. Note that TEOS-10 uses t for *in situ* temperature rather than the more customary T. Throughout this text, t is used mainly for time so, rather than confuse matters, we continue to use T for temperature. The reader should be aware of this difference when using the TEOS-10 algorithms and texts. Algorithms in the TEOS-10 Tool Box have been given user-friendly names, like "gsw_SA_from_SP" (where gsw denotes the Gibbs Sea Water algorithm used to determine the Absolute Salinity from the Practical Salinity, itself obtained from seawater conductivity) and "gsw_CT_from_t" (the algorithm for the calculation of Conservative Temperature from *in situ* temperature, t). The input and output parameters of the GSW Oceanographic Toolbox are generally in units familiar to marine scientists, and can differ from the basic SI (*Système International*) units. The GSW Library is presently available in MATLAB, FORTRAN and C. Additional information is provided in the online literature specific to TEOS-10.

1.2.1 A Brief History

As noted above, the thermodynamic equations of state of seawater are essential for determining a wide range of oceanographic properties and for quantifying fluxes and exchange processes in the ocean. As with any formulation, the accuracies of these equations are only as good as the temperature, salinity and pressure data being utilized in developing the formulae. In the late 19th Century, the Danish physicist and oceanographer Martin Knudsen built a thermometer capable of measuring temperature in the deep ocean with a precision of $0.01°C$, which was impressive for its time (Modern oceanic thermometers have a precision better than $0.001°C$). Knudsen was also one of the first to provide a method for estimating the salinity from discrete samples collected in specially designed water bottles. Despite the fact that Knudsen's 1901 publication relating salinity, $S(‰)$, to measurements of Chlorinity, $Cl(‰)$,

$$S(‰) = 0.03 + 1.805 Cl(‰) \tag{1.1}$$

was based on only nine samples—one from each of the Red Sea, North Atlantic and North Sea, and six from the Baltic Sea—the formula (in parts per thousand, ‰) remained in use up until the 1970s. Other early work on seawater properties was conducted by Forch et al. (1902) and by Ekman (1908), who obtained high pressure measurements to depths of 600 bars ($\approx 5{,}850$ m).

The seawater relations obtained in the early part of the 20th Century were reformulated by Fofonoff et al. (1958) and Sweers (1971) and used for about 70 years, from about 1908 to 1980. From then until 2010, the Practical Salinity Scale-78 (PSS-78) (Lewis, 1980; Lewis and Perkin, 1978, 1981) was used for determining the Practical Salinity (S_p) and associated seawater properties (In this case, the subscript "p" refers to 'practical' and not to pressure). The random errors and bias of Knudsen's formula (1.1) toward the Baltic Sea continued to be overlooked or ignored, being replaced with a slightly modified version $S(‰) = 1.80655\,Cl(‰)$ in the 1970s at the time the Practical Salinity Scale was being developed. The EOS-80 algorithms that followed the development of PSS-78 combined S_p with *in situ* temperature T (now ITS-90) and water pressure, p, to calculate various seawater variables, including potential temperature (θ) and potential density (ρ_θ). Contours of potential density plotted on $S_p-\theta$ diagrams made it possible to analyze water-mass characteristics and spatial distributions.

EOS-80 was itself a major shift from the earlier standards developed in the early part of the 20th Century necessitated by the fact that, by the late 1970s, salinity was derived primarily by measuring the electrical conductivity of seawater as it passed through conductivity cells housed in modern Conductivity-Temperature-Depth (CTD) profilers. Salinity became synonymous with seawater conductivity. Bench salinometers and Standard Water samples are used to calibrate CTD records, while bottle samples are collected for field calibrations and to obtain samples for the determination of other oceanographic properties, such as dissolved oxygen and nutrient concentrations. Once PSS-78 was adopted, salinity went from being measured in parts per thousand or grams of solute per kilogram of water, as obtained from the chemical titration of water samples collected in bottle casts, to a dimensionless variable (S_p), sometimes (incorrectly) expressed as Practical Salinity Units (PSU). The requirement that values of S_p be presented without units was disconcerting to those of us who grew up with salt content being quantified in parts per thousand or in grams per kilogram. Millero (2010) expresses a similar sentiment.

In 1990, the International Practical Temperature Scale 1968 (IPTS-68) was replaced by the International Temperature Scale 1990 (ITS-90). The primary methods used to convert between the two temperature scales are Rusby's (1991) 8th order fit and the Saunders (1990) scaling

$$T_{68} = 1.00024\,T_{90}(°C) \tag{1.2}$$

According to the TEOS-10 manual (IOC et al., 2010), the two methods are interchangeable throughout the common oceanic temperature range of -2 to $40°C$, differing by less than the uncertainty in thermodynamic temperature (~ 1 mK, or $0.001°C$), or in the uncertainty in the practical applications of the IPTS-68 and ITS-90 scales (Pawlowicz, 2010). Differences between Rusby's and Saunders' formulae are also less than 0.03 mK ($0.03 \times 10^{-3°}C$) in the narrower temperature range -2 and $10°C$. Consequently, the oceanographic community is encouraged to use Saunders (1990) formula (1.2). One application of this formula is in the updated computer algorithm for the calculation of Practical Salinity (PSS-78) in terms of the conductivity ratio, K. Because the PSS-78 algorithms require T_{68} as the temperature argument, use of the PSS-78 algorithms with T_{90} data, requires the transformation to T_{68} using (1.2).

In 2005, the International Association of Physical Sciences of the Oceans (IAPSO) and the Scientific Committee on Oceanic Research (SCOR) established Working Group 127, chaired by Trevor McDougall of Australia, to update and improve EOS-80 (IOC et al., 2010). Because it is based on conductivity-derived measurements of salinity, the Practical Salinity Scale does not take into account contributions to water density from non-conducting solutes nor does it consider the spatially varying composition of seawater. EOS-80 also contained several polynomial expressions that were not fully mutually consistent and did not obey exactly the thermodynamic Maxwell cross-differentiation relations. EOS-80 is now replaced by TEOS-10 for examining the thermodynamic properties of seawater, ice and humid air. At the core of TEOS-10 is the calculation of the Gibbs free energy function, from which all of the thermodynamical properties of seawater, such as density, enthalpy, entropy, and internal energy are derived. Determining the Gibbs function and associated properties of seawater requires high-precision, *in situ* measurements of temperature and salinity from which are calculated new oceanographic variables, including Conservative Temperature (Θ) and Absolute Salinity (S_A). The authors of TEOS-10 emphasize in multiple written documents that salinity is still to be measured and archived using the old EOS-80 methodology (i.e., as Practical Salinity, S_p). What has changed is the way these data are used.

1.2.2 Sea Pressure (Depth) in TEOS-10

Pressure is the simplest of the three basic parameters required for TEOS-10. The new Sea Pressure, p, is defined as the Absolute Pressure, P, at the measurement depth, z, minus the Absolute Pressure of one standard atmosphere, $P_o \equiv 101.325$ Pascals (Pa), such that

$$p \equiv P - P_o \tag{1.3}$$

commonly expressed in decibars (1 dbar $= 10^4$ Pa). The use of decibars is most useful because it means that the Absolute Pressure and depth (z) in meters are roughly the same, because $P(\text{dbar}) \sim 1.025\, z(\text{m})$ in the open ocean. As an aside, we note that modern CTDs can typically resolve pressure equivalent to a centimeter (0.01 m or about 1 mbar), while bottom pressure recorders (BPRs) can now resolve variations in the overlying pressure to the order of 1 mm (10 Pa or 0.1 mbar). TEOS-10 publications also mention the common pressure variable arising during the calibration of sea-going instruments, termed gauge pressure, p^{gauge}, which is the Absolute Pressure minus the Absolute Pressure of the atmosphere at the time of the instrument's calibration, either in the laboratory or at sea. Because of the spatial and temporal variations in atmospheric pressure, Sea Pressure, p, is the preferred thermodynamic variable as it is unambiguously derived from Absolute Pressure, P.

1.2.3 Practical Salinity and Conductivity

Practical salinity, S_p, is defined on the Practical Salinity Scale of 1978 in terms of the conductivity ratio K_{15}—corresponding to the electrical conductivity of a sample at temperature $T_{68} = 15°C$ and pressure $p = 0$ dbar (Absolute Pressure $P = 101.325$ Pa)—divided by the conductivity of a standard potassium chloride (KCl) solution at this temperature and pressure (Lewis and Perkin, 1981; UNESCO, 1981, 1985; see Section 1.4.2 for further details). The mass fraction of KCl in the standard solution is 32.4356×10^{-3} (mass of KCl per mass of solution). Practical salinity S_p is, by definition, equal to 35 when $K_{15} = 1$ and is a unit-less quantity (sometimes written incorrectly as "psu"). When $K_{15} \neq 1$, S_p and K_{15} at $T_{68} = 15°C$, are related through the PSS-78 equation (Lewis and Perkin, 1981), given here as:

$$S_p = \sum_{i=0}^{5} a_i (K_{15})^{\frac{i}{2}}, K_{15} = \frac{C(S_p, T_{68} = 15°C, p = 0)}{C(35, T_{68} = 15°C, p = 0)} \tag{1.4}$$

where the coefficients, a_i ($i = 0-5$), have values of 0.0080, -0.1692, 25.3851, 14.0941, -7.021 and 2.7081, respectively. The sum of the six values $\sum a_i = 35$ and the equation is valid over the range $2 < S_p, < 42$. As a consequence of this definition, any oceanic water having a precisely known conductivity ratio of near unity at 15°C with the KCl solution is a secondary standard for routine calibration of oceanographic instruments. All sea waters having the same conductivity ratio have the same practical salinity. Chlorinity is to be regarded as a separate, independent variable in describing the properties of sea water.

For values of the sample temperature T_{68} different than from 15°C and *in situ* pressure, a more complicated relationship is required. The formulae for values of T_{68} in the range $-2°C < T < 35°C$ and $0 \leq p \leq 10,000$ dbar (Lewis and Perkin, 1981; UNESCO, 1985) are given in terms of the ratio

$$R_T = \frac{R}{r_T R_p}, \tag{1.5a}$$

where

$$R = \frac{C(S_p, T_{68}, p)}{C(35, T_{68} = 15°C, p)} \tag{1.5b}$$

is the conductivity ratio for the sample, r_T is a quadratic function of the *in situ* temperature (°C) and R_P is a function of the *in situ* pressure, temperature and R. Further details are found in Lewis and Perkin (1981) and in Section E.2 of IOC, SCOR and IAPSO (2010).

1.2.4 Absolute Salinity

Measuring water temperature to high precision is challenging but pales in comparison to measuring salinity to high precision. Basic to TEOS-10 is the need for precise values of the Absolute Salinity (S_A), which represents, to the best available accuracy, the mass fraction of dissolved solute found in Standard Seawater (SSW) having the same density (ρ) as that of the sample. [The term Absolute Salinity, S_A, is mentioned in Lewis and Perkin, 1981.] Standard Seawater is seawater with a specific composition, obtained from a reference material (IAPSO Standard Seawater) by adding pure distilled water or by removing it by evaporation (i.e., *not* by adding or removing solute). Most laboratory measurements of seawater properties are made using SSW. Absolute Salinity is also sometimes referred to as the "Density Salinity", S_A^{dens}, because it is the salinity value that yields the most accurate estimate of the density of a water sample when used in the TEOS-10 algorithms. Although S_A values are to be used to calculate all the thermodynamic properties of seawater in the major oceans using the new TEOS-10, it is the measured salinity, S_P, that is stored in archives, not S_A.

The TEOS-10 manuals provide a recipe for obtaining S_A — in SI units of grams of solute per kilogram of seawater (g/ kg) — from a Reference Salinity (S_R), which, in turn, is derived from "measured" values of the Practical Salinity (S_P) obtained primarily using modern water-column profiling systems or desktop salinometers. S_P is the measured version of salinity obtained from seawater conductivity from which the dependences upon temperature and pressure have been removed. Thus, we have the sequence, $S_P \rightarrow S_R \rightarrow S_A$, where the reference salinity, S_R, is the first best estimate of the Absolute Salinity. More formally, the goal is to obtain the most accurate possible values of the Absolute Salinity using the relationship,

$$S_A = S_R + \delta S_A, \tag{1.6}$$

where $S_R = constant \times S_P$ and where the "correction", $\delta S_A = \delta S_A(\phi, \lambda, p)$, is a function of the sample latitude (ϕ) in degrees north, longitude (λ) in degrees east (ranging from 0 to 360° E) and pressure, p. For seawater in the "Neptunian" range (i.e., $2 < S_P < 42$, $-2°C < T < 35°C$), a Reference Salinity (on the reference composition salinity scale in Chapter 2.4 of the TEOS-10 manual and Millero (2008)) can be estimated as

$$S_R = S_P \left(\frac{35.16504}{35.000}\right) g/kg \tag{1.7}$$

which is based on water samples collected in deep "blue-water" regions of the World Ocean, for example in the central North Pacific (e.g., Wright et al., 2011). The reference salinity estimates of the absolute salinity derived from relationship (1.4) is important for measurements requiring a precision better than 1 g/kg, or about 3% for typical offshore Reference Salinities in the range of 25−35 g/kg.

The term, δS_A, in (1.4) corrects for large-scale spatial variations in dissolved constituents and is commonly derived from mapped distributions based on bottle samples applicable to the specific region of the ocean or marginal sea where the measurements were conducted. As illustrated by Figure 1.1, this correction takes into account the added solutes resulting

FIGURE 1.1 (a) *Top panel*: Absolute Salinity Anomaly δS_A at $p = 2{,}000$ dbar; (b) *Bottom panel*: a south-to-north vertical section of Absolute Salinity Anomaly δS_A along $180°$ E in the Pacific Ocean. *From IOC, SCOR and IAPSO (2010).*

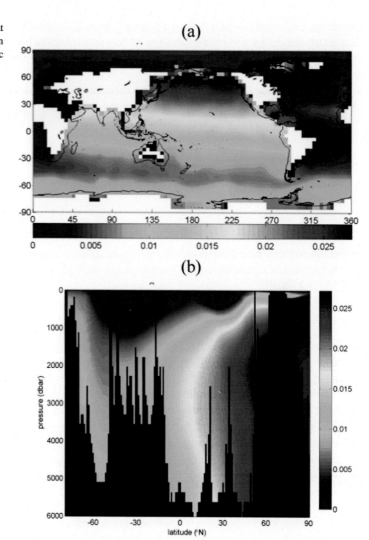

from the dissolution of calcium carbonate, $CaCO_3(s)$, and silicate, $SiO_2(s)$, in the deep waters, as well as the addition of carbon dioxide (CO_2), and the presence of nutrients like nitrate (NO_3) and phosphate (PO_4) from the oxidation of plant material. The mapped δS_A values arising from these added solutes are derived from the difference in density ($\Delta\rho$) between the measured densities of seawater samples compared with the densities calculated from the TEOS-10 equation of state for the same reference salinity, temperature, and pressure, using $\delta S_A = \Delta\rho/0.75179$ g/kg. The correction term can be as large as 0.02 g/kg in the open ocean, and as much as 0.09 g/kg in some coastal areas.

1.2.5 Simple Correction Factor Estimation

As one might expect, the simplest way to estimate the correction factor δS_A is to assume, in the absence of other information, that $\delta S_A = 0$. This is an acceptable approach when conducting work in coastal regions (e.g., on the continental shelf and inner coastal waters such as the Mediterranean and Red seas) and in shallow marginal seas that have had no direct anomaly measurements. Ignoring this correction is roughly equivalent to using the old PSS78/EOS80 approach, although even in this mode TEOS-10 provides estimates of more parameters, over a wider range in temperature, salinity, and pressure, with slightly better accuracy than the earlier methodology. Aside from simply setting the correction factor to zero, global atlases of δS_A exist for the open oceanic regions of the Atlantic, Pacific, Indian, and Southern oceans, and Baltic Sea.

The values of δS_A in the open ocean can also be estimated for waters at given longitudes, latitudes, and depths using correlations of δS_A with the concentration of $Si(OH)_4$ in the local waters (Figure 1.2). This dependence of δS_A on silicate reflects the fact that silicate affects the density of a seawater sample without significantly affecting its conductivity (i.e., its

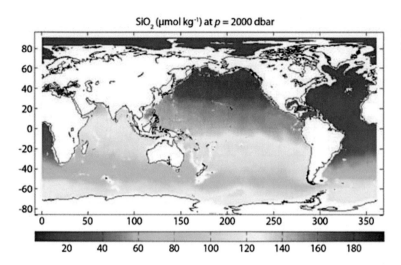

SiO$_2$ (µmol kg^{-1}) at p = 2000 dbar

FIGURE 1.2 Spatial distribution in the concentration of silicate, SiO$_2$ (µmol/kg), in the Pacific Ocean at a depth of 2,000 dbar. *From McDougall et al. (2009).*

Practical Salinity value). In practice, the silicate relationship explains about 60% of this effect and the remainder is due to the correlation of other composition anomalies (such as nitrate) with silicate. (This latter relation is possible because the composition variations arising from biogeochemical processes are themselves correlated.) In the McDougall et al. (2012) algorithm for the Baltic Sea, the region is treated separately, following the work of Millero and Kremling (1976) and Feistel et al. (2010a, b), because some rivers flowing into the Baltic are unusually high in calcium carbonate. Globally, McDougall et al. (2009) find that

$$\delta S_A \left(\frac{\text{g}}{\text{kg}}\right) = \frac{\Delta\rho}{0.75179 \left(\frac{\text{kg}}{\text{m}^3}\right)} \cong 98.24 \times 10^{-6}\, \text{SiO}_2 \left(\frac{\mu\text{mol}}{\text{kg}}\right) \cong 0.0054\, \frac{\text{g}}{\text{kg}} \tag{1.8}$$

Similar relationships are applicable for various regions of the ocean. For the Southern Ocean south of 30° S and for the Pacific, Indian and Atlantic Oceans for latitudes, λ, north of 30° S,

$$\text{Southern Ocean:}\ \delta S_A \left(\frac{\text{g}}{\text{kg}}\right) = 74.884 \times 10^{-6}\, \text{SiO}_2 \left(\frac{\mu\text{mol}}{\text{kg}}\right) \tag{1.9a}$$

$$\text{Pacific Ocean:}\ \delta S_A \left(\frac{\text{g}}{\text{kg}}\right) = 74.884 \times 10^{-6} \left(1 + 0.3622 \left[\frac{\lambda}{30°} + 1\right] \text{SiO}_2 \left(\frac{\mu\text{mol}}{\text{kg}}\right)\right) \tag{1.9b}$$

$$\text{Indian Ocean:}\ \delta S_A \left(\frac{\text{g}}{\text{kg}}\right) = 74.884 \times 10^{-6} \left(1 + 0.3861 \left[\frac{\lambda}{30°} + 1\right] \text{SiO}_2 \left(\frac{\mu\text{mol}}{\text{kg}}\right)\right) \tag{1.9c}$$

$$\text{Atlantic Ocean:}\ \delta S_A \left(\frac{\text{g}}{\text{kg}}\right) = 74.884 \times 10^{-6} \left(1 + 1.0028 \left[\frac{\lambda}{30°} + 1\right] \text{SiO}_2 \left(\frac{\mu\text{mol}}{\text{kg}}\right)\right) \tag{1.9d}$$

1.2.6 Less Simple Correction Factor Estimation

There are several additional ways to estimate the correction factor δS_A, aside from the two approaches noted above (Pawlowicz, 2010; updated 2013). In addition to simply assuming that $\delta S_A = 0$ or using a global atlas available for the Baltic and the open Atlantic, Pacific, Indian, and Southern oceans (McDougall et al., 2012), it is possible to use measurements of carbon system parameters. Specifically, if two measurements of dissolved inorganic carbon (DIC), total alkalinity (A_T), pH, or the fugacity (fCO$_2$), as well as measurements of silicate and nitrate are available, then δS_A can be estimated from a simple equation, linear in [A_T], [DIC], [NO$_3^-$], and [Si(OH)$_4$]. The coefficients for this equation—and the theoretical basis of the numerical model from which they are derived—are described in Pawlowicz et al. (2011). This model-based approach, which is applicable throughout the ocean and in numerical models that include biogeochemical processes, replicates the empirical (data-based) atlas approach to within ±0.004 g/kg, which is roughly the same as the observational error in the data used in the atlas. Alternatively, one can estimate the salinity anomaly by taking direct measurements of the density anomaly. The density anomaly, $\delta\rho_R$, is the difference between the true density and that computed using S_R as the salinity argument in TEOS-10 (i.e., ignoring the effects of salinity anomalies). Then, δS_A (g/kg) $\approx \delta\rho_R(\text{kg/m}^3)/0.7519$. This approach may be necessary to obtain accurate results in areas of unknown composition anomalies, although the global database of such measurements should improve over time.

In summary, the recipe for deriving S_A is:

(1) Calculate the Practical Salinity, S_P, using the electrical conductivity of water and PSS-78 (the Practical Salinity Scale-78), modified for the modern temperature standard (ITS-90);

(2) For seawater in the range $2 < S_P < 42$, $-2°C < T < 35°C$, calculate a first-order estimate of Absolute Salinity, S_A, using the Reference Salinity-Practical Salinity relationship

$$S_A \approx S_R = S_P\left(\frac{35.16504}{35.000}\right)\text{g}/\text{kg} \tag{1.10}$$

This relationship is needed for measurements that require a precision better than 1 g/kg and yields the best available approximation of the mass fraction of solute in Standard Seawater with the same conductivity as that of the data value;

(3) Apply a correction to (2) using

$$S_A = S_R + \delta S_A \tag{1.11}$$

where the salinity anomaly, δS_A, corrects for large-scale spatial variations in dissolved constituents applying values obtained either from detailed (mapped) distributions estimated from bottle samples applicable to the specific region of the World Ocean. Another way is to compute the difference $\Delta\rho$ between the measured density and the value calculated from TEOS-10 at the S_R-level, or from distributions of silicate, SiO_2. The authors of TEOS-10 emphasize the importance of stating the version number of the atlas used in the salinity correction, "as the size of the corrections may change as more direct density measurements are made in different parts of the ocean and the extrapolations are made more accurately". In publications, researchers should have a sentence similar to: "Salinities are reported on the TEOS-10 Absolute Salinity scale (IOC et al., 2010) with δS_A taken from version 3.0 of the McDougall et al. (2012) database" or modified as appropriate if another method for δS_A was used.

1.2.7 Conservative Temperature

In addition to a new salinity variable, TEOS-10 defines a new temperature variable, Conservative Temperature, Θ, which replaces potential temperature, θ, used in EOS-80. (Why one is in italics and the other is not, is not explained in the manuals.) Conservative Temperature is a function of Absolute Salinity, S_A, *in situ* temperature, T, and pressure, p, and is therefore similar to potential temperature in that the same concept is involved in their definitions. In both cases, we consider a seawater sample taken at an arbitrary pressure in the ocean being slowly brought to the ocean surface in an adiabatic and isohaline manner—i.e., with no loss or gain of heat and at constant salinity—until the pressure $p = 0$. The reduced compressive forces on the fluid parcel as it is raised in the water column causes its temperature to decrease, leading to a temperature at the end of the process called the potential temperature, θ. Similarly, the enthalpy at the end of this artificial experiment is defined to be the potential enthalpy, h_0, and Conservative Temperature, Θ, is the potential enthalpy divided by the fixed "heat capacity" $c_p^0 = 3991.86795711963\ \text{Jkg}^{-1}\text{K}^{-1}$.

The difference between potential temperature and Conservative Temperature ($\theta - \Theta$) can be as large as 1.4°C but is typically no more than $\pm0.1°C$ (Figure 1.3; from Figure A.17.1 of IOC et al., 2010). Note that a temperature difference of 0.1°C is roughly the difference between *in situ* temperature and potential temperatures for a pressure difference of 1,000 dbar (about 990 m depth), and it is approximately 40 times as large as the typical differences between T_{90} and T_{68} in the ocean. As noted in the Introduction, a major advantage of Conservative Temperature over potential temperature is that it represents the "heat content" of seawater more accurately. Once Absolute Salinity and Conservative Temperature are obtained for a dataset, TEOS-10 provides a suite of algorithms for calculating other properties of seawater such as density, potential density and various forms of geostrophic variable, including dynamic height anomaly. Instead of using the measured, conductivity-derived salinity, S_p, of PSS-78, the TEOS-10 algorithms are based on Absolute Salinity, S_A, and Conservative Temperature, Θ.

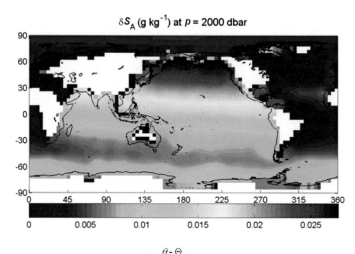

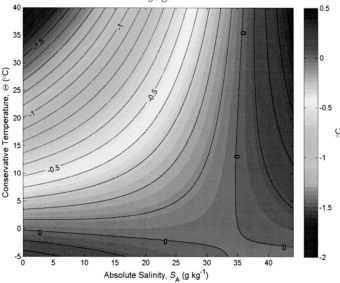

FIGURE 1.3 Differences between potential temperature and Conservative Temperature ($\theta - \Theta$). (a) *Top panel*: difference for the sea surface for the annually averaged atlas of Gouretski and Kolteermann (2004); and (b) *Bottom panel*: contours (°C) of ($\theta - \Theta$) versus Conservative Temperature and Absolute Salinity, S_A. *From IOC, SCOR and IAPSO (2010).*

Air-sea heat fluxes are exactly proportional to the flux of Conservative Temperature. Moreover, because Θ is almost a perfectly conservative variable (unlike potential temperature, θ), the meridional "heat" flux is accurately determined by the meridional flux of Θ. Also, the parameterized lateral diffusion of "heat" along neutral density planes based on potential temperature gradients can differ from that based on Conservative Temperature gradients by more than 1% (see Figure A.14.1 of IOC et al. (2010) which is reproduced below). For these reasons, Conservative Temperature Θ is the recommended temperature variable to be used in ocean analyses. Conservative Temperature Θ takes the place of potential temperature, θ, just as Absolute Salinity S_A takes the place of Practical Salinity, S_p.

In those cases where turbulent mixing processes are being examined either implicitly or explicitly, the appropriate temperature variable is Conservative Temperature, because it is designed to be a "heat" variable conserved under both adiabatic mixing and changes in depth (neither *in situ* nor potential temperature is conservative under these conditions). Thus, Conservative Temperature should be used for most applications in which the oceanic general circulation is being studied. This includes situations where salinity/temperature plots are used to determine the characteristics of water masses and their mixtures. However, the TEOS-10 manual issues a caveat: in cases where molecular diffusion or sensible conduction processes dominate (e.g., in studies of double-diffusion, or air-sea interaction), the appropriate temperature variable is still the *in situ* temperature. This is because fluxes driven by these mechanisms are proportional to gradients in *in situ* temperature, and not to gradients in heat content.

Because of large-scale variations in seawater composition, it is not possible to use TEOS-10 to draw simple curved isolines of potential density on a $S_p - \theta$ diagram. Instead, a given value of potential density defines an area on the $S_p - \theta$

diagram. Hence, for the analysis of ocean data using the TEOS-10 system, the user needs to switch from using the $S_P-\theta$ diagrams that were applicable to the EOS-80 system, to using $S_A-\Theta$ diagrams. It is on a $S_A-\Theta$ diagram that the isolines of potential density, ρ^Θ, can be drawn within TEOS-10. As the authors of the TEOS-10 instructions note, these calculations can be performed using the functions of the GSW Oceanographic Toolbox as follows: the observed variables S_A, T, p, together with longitude and latitude, are used to first determine Absolute Salinity S_A using equation "gsw_SA_from_SP". Then, Conservative Temperature, Θ, is calculated using "gsw_CT_from_t", where t denotes temperature T in file names. Oceanographic water masses are then analyzed on the $S_A-\Theta$ diagram (for example, by using "gsw_SA_CT_plot"), and potential density contours can be drawn on this $S_A-\Theta$ diagram using "gsw_rho(SA,CT,p_re)".

1.2.8 Gibbs Function

At the core of TEOS-10 is the Gibbs function, $g(S_A, T, p)$, calculated from the known values of Absolute Salinity, S_A, *in situ* temperature, T, and pressure, p. Once derived, the Gibbs function can be used to determine all of the thermodynamic properties of seawater. Historically known as Gibbs free energy (Gibbs, 1873), g combines enthalpy, h, and entropy, η, into a single relation,

$$g(S_A, T, p) = h - (T_0 + T)\eta \tag{1.12}$$

where $T_0 = 273.15$ K corresponds to $T = 0°C$ on the Celsius scale. Enthalpy $h = U + P \cdot V$ is the sum of the internal energy (U) and the product of Absolute Pressure from Eqn (1.3) and volume ($P \cdot V$). When a process occurs at constant pressure, the heat evolved (either released or absorbed) is equal to the change in enthalpy (i.e., $\Delta h = \Delta U + P\Delta V$). The second variable, entropy (η), is a qualitative measure of the degree to which the energy of atoms and molecules become more spread out in a process that can be defined in terms of statistical probabilities of a system or in terms of various thermodynamic variables. Quantitatively, entropy obeys the Second Law of Thermodynamics, which describes the changes in entropy with respect to the system and its surroundings, and the Third Law that characterizes the entropy of substances. Gibbs originally called the energy, g (or, G) the "available energy" in a system. His 1873 paper, "Graphical Methods in the Thermodynamics of Fluids," outlined how his equation could predict the behavior of systems when enthalpy and entropy are combined. Unfortunately, the term "free energy" has led to so much confusion that many researchers refer to it simply as the Gibbs energy. The "free" part of the name has its origins in steam-engine thermodynamics with its interest in converting heat into work: the change in Gibbs energy is the maximum amount of energy which can be "freed" from the system to perform useful work. Here, "useful" refers to work other than that which is associated with the expansion, ΔV, of the system. Much of the development of the Gibbs energy, g, at center of TEOS-10 is credited to Feistel (2003, 2008) who conducted his work while he was at the Leibniz Institute for Baltic Sea Research.

In TEOS-10, the Gibbs function of seawater is partitioned into the sum of a pure water component, $g^W(T, p)$, and a saline component, $g^S(S_A, T, p)$ (IAPWS-08 (International Association for the Properties of Water and Steam)), whereby

$$g(S_A, T, p) = g^W(T, p) + g^S(S_A, T, p) \tag{1.13}$$

TEOS-10 provides a set of algorithms in the Gibbs Sea Water (GSW) Oceanographic Toolbox from which the thermodynamic properties of seawater are then determined. Included in the toolbox are quantities such as density (ρ), specific volume ($1/\rho$), entropy, enthalpy, sound speed, buoyancy frequency, geostrophic, hydrostatic and thermal wind equations, and dynamic height anomaly. The seawater Gibbs function in the GSW Toolbox is expressed as a function of sea pressure (functionally equivalent to the use of Absolute Pressure P in the IAPWS Releases and in the SIA library); that is, g is a function of p, not p^{gauge} that was noted earlier in this section.

The Gibbs function for the pure liquid water component, $g^W(T, p)$, can be obtained from the IAPWS-95 Helmholtz function of a pure-water substance, which is valid from the freezing temperature, or from the sublimation temperature, up to 1,273 K. An alternative thermodynamic description of pure water in IAPWS-95, spanning the oceanic ranges of temperature (ITS-90°C) and pressure, is given by IAPWS-09 as

$$g^W(T, p) = g_u \sum_{j=0}^{7} \left[\sum_{k=0}^{6} \left(g_{jk} y^j z^k \right) \right] \tag{1.14}$$

while that for the combined polynomial and logarithmic form and the coefficients for the saline component of the Gibbs function

$$g^S(S_A, T, p) = g_u \sum_{j,k} \left\{ \left[g_{1,j,k} x^2 \cdot \ln(x) + \sum_{i>1} g_{i,j,k} x^i \right] y^j z^k \right\} \tag{1.15}$$

where, for both components,

$$g_u = 1\left(\frac{J}{kg}\right); x^2 = \frac{S_A}{40.188617}(\text{in g / kg}); y = \frac{T}{40}(\text{in °C}); z = P \times 10^{-8}(\text{in Pascals, Pa}) \tag{1.16}$$

The polynomial form and the coefficients for the pure water Gibbs function [from Feistel (2003) and IAPWS-09] are given in Appendix G of the TEOS-10 Manual, while the combined polynomial and logarithmic form and the coefficients for the saline part of the Gibbs function [from Feistel (2008) and IAPWS-08] are reproduced in Appendix H of the Manual. Pure water coefficients with $i = 0$ do not occur in the saline contribution.

The Gibbs function, g^W, obtained from the IAPWS-09 Gibbs function, is valid in the oceanographic ranges of temperatures and pressure, extending from less than the freezing temperature of seawater (at any pressure) to 40°C (specifically from $-[2.65 + 0.0743 (P_o + p) \text{ MPa}^{-1}]$°C to 40°C, and in the pressure range $0 < p < 10^4$ dbar). In practical terms, it is expected that oceanographers will use IAPWS-09 because it executes approximately two orders of magnitude faster than the IAPWS-95 code for pure water. Researchers concerned with temperatures between 40 and 80°C need to use the IAPWS-95 version of g^W, expressed in terms of absolute temperature (K) and Absolute Pressure (P), rather than the IAPWS-09 version.

The saline part of the Gibbs function is valid over the "Neptunian" oceanic ranges $0 < S_A < 42$ g/kg, -6.0°C $< T <$ 40°C, and $0 < p < 10^4$ dbar (roughly 9,000 m depth), although its thermal and colligative properties hold up to $T = 80$°C and $S_A = 120$ g/kg at the surface ($p = 0$). The GSW Toolbox, itself, is restricted to the oceanographic standard range in temperature and pressure. However, the validity of the results at $p = 0$ extends to Absolute Salinity values of up to mineral saturation concentrations (Marion et al., 2009). Specific volume (which is the pressure derivative of the Gibbs function) is presently an extrapolated quantity outside the "Neptunian" range given earlier for temperature and Absolute Salinity at $p = 0$, and has errors at this pressure of up to 3%.

Determination of the Gibbs function requires a reliable estimate of the absolute salinity of seawater, S_A. As described earlier, S_A is based upon the reference salinity S_R, which provides the best first estimate of S_A obtained from the practical salinity S_P. The correction factor, δS_A, arising from the added solutes in seawater and (in deep water) from the dissolution of $CaCO_3$ and SiO_2, CO_2, and nutrients like NO_3 and PO_4 from the oxidation of plant material. This correction, δS_A, is usually (but not always) positive due to the fact that added solutes are estimated from the differences between the measured densities of seawater samples compared with the densities calculated from the TEOS-10 equation of state ($\Delta\rho$) at the same reference salinity, temperature, and pressure, using $\delta S_A = \Delta\rho/0.75179$ g/kg. The values of δS_A in the ocean can be estimated for water at given longitudes, latitudes, and depths using correlations of δS_A and the concentration of $Si(OH)_4$ in the seawater. The S_A values can then be used to calculate all the thermodynamic properties of seawater in the major oceans using the new TEOS-10. It will be very useful to modelers examining the entropy and enthalpy of seawater.

1.2.9 Density

The density of seawater, $\rho(S_A, T, p)$, and its reciprocal, specific volume, $v(S_A, T, p)$, can be considered as the last of the building blocks of oceanic properties after salinity, temperature and pressure (Figure 1.4). In TEOS-10, this variable is obtained from the partial derivative of the Gibbs function with respect to pressure at constant Absolute Salinity, S_A, and temperature, T; specifically

$$\rho(S_A, T, p) = v(S_A, T, p)^{-1} = (g_P)^{-1} = \left(\partial g/\partial p|_{S_A,T}\right)^{-1} \tag{1.17}$$

where ρ has units of kg/m³. In the TEOS-10 algorithms, density is a function of Absolute Salinity rather than of Reference Salinity (S_R) or Practical Salinity (S_P). As noted in the TEOS-10 manual, this is quite important because Absolute Salinity in units of g/kg is numerically greater than Practical Salinity by between 0.165 and 0.195 g/kg in the open ocean. If Practical Salinity were inadvertently used as the salinity argument for the TESO-10 density algorithm, a marked difference in density of between 0.12 and 0.15 kg/m³ would result.

For many theoretical and modeling studies, it is convenient to regard density as a function of Conservative Temperature, Θ, rather than of *in situ* temperature T. Seawater density then has the functional form $\rho = \hat\rho(S_A, \Theta, p)$. The authors of TEOS-10 adopted the convention that when enthalpy h, specific volume v or density ρ are taken to be functions of potential temperature, θ, they are distinguished by an over-tilde, as in $\tilde v$ or $\tilde\rho$, and when they are taken to be functions of Conservative Temperatures, Θ, they are distinguished by a caret, as in $\hat v$ or $\hat\rho$. The computationally efficient expression for $\hat v(S_A, \Theta, p)$ involves 75 coefficients (Roquet et al., 2015) and is provided in the GSW computer software library as the function "gsw_specvol(SA,CT,p)"; $\hat\rho(S_A, \Theta, p)$ is obtained from "gsw_rho(SA,CT,p_ref)". Integration of $\hat v(S_A, \Theta, p)$ with respect to pressure yields a closed expression for enthalpy $\hat h(S_A, \Theta, p)$, which is available as the function "gsw_enthalpy(SA,CT,p)".

FIGURE 1.4 A sketch indicating how thermodynamic quantities such as density are calculated as functions of Absolute Salinity. Absolute Salinity is found by adding an estimate of the Absolute Salinity Anomaly δS_A to the Reference Salinity. *From IOC, SCOR and IAPSO (2010).*

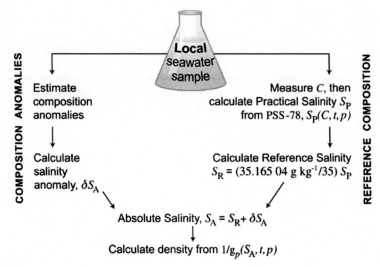

1.2.10 TEOS-10 and GSW in a Nutshell

To analyze oceanographic data using TEOS-10, the observed values of Practical Salinity, S_p, and *in situ* temperature, T, need to be converted into Absolute Salinity, S_A, and Conservative Temperature, Θ, as follows (keeping in mind that TEOS-10 uses "t" for *in situ* temperature):

Step 1: Calculate Absolute Salinity, S_A = gsw_SA_from_SP(SP, p, long, lat);
Step 2; Calculate Conservative Temperature, Θ = gsw_CT_from_t(SA, t, p);
Step 3: Having converted (S_p, t, p) to (S_A, Θ, p), use the Gibbs Sea Water (GSW) functions listed on page 14 of the TEOS-10 MANUAL to analyze the data. Use of these GSW functions ensures consistency between theoretical oceanography, observational oceanography and ocean modeling. (From McDougall and Barker, 2011: Getting started with TEOS-10 and the Gibbs Sea Water (GSW) Oceanographic Toolbox, 28 pp., SCOR/IAPSO WG127, ISBN 978-0-646-55621-5.)

1.3 TEMPERATURE

The measurement of temperature in the ocean uses conventional techniques except for deep observations, where hydrostatic pressures are high and there is a need to protect the sensing system from ambient depth/temperature changes arising from pressure rather than from temperature changes. Also important at great depths is the fact that temperature gradients are generally quite weak, so that it is necessary to have very accurate temperature measurements to capture these gradients and possible associated dynamic processes such as deep density currents and sill overflows.

Temperature is the easiest ocean property to measure accurately, with sensor resolutions better than 0.001°C for modern, more high-end thermistors. Some of the ways in which ocean temperature can be measured are:

1. Expansion of a liquid or a metal.
2. Differential expansion of two metals (bimetallic strip).
3. Vapor pressure of a liquid.
4. Thermocouples.
5. Change in electrical resistance (thermistors).
6. Infrared radiation from the sea surface (skin sea surface temperature).

In most of these sensing techniques, the temperature effect is very small and some form of amplification is necessary to make the temperature measurement detectable. Usually, the response is nearly linear with temperature so that only the first-order term in the calibration expansion is needed when converting the sensor measurement to temperature. However, in order to achieve high precision over large temperature ranges, second, third and even fourth order terms must sometimes be used to convert the measured variable to temperature. These points will be emphasized in the subsequent discussion.

1.3.1 *In Situ* Sea Surface Temperature (SST) Measurement

1.3.1.1 Mercury In-Glass Thermometers

Of the above methods, (1), (5), and (6) have been the most widely used in physical oceanography. Historically, the most common type of the liquid expansion sensor is the mercury-in-glass thermometer, invented by Daniel Gabriel Fahrenheit in Amsterdam in 1714. The thermometer is made up of a narrow diameter glass tube attached to a larger bulb reservoir that contains a lot more mercury than can be contained in the tube. Since it is the expansion in the tube that measures the temperature, a critical aspect of this device is that tube have very straight sides. Any deviation from linearity in the tube sides will introduce non-linearity into the temperature measurement making it very difficult to calibrate the thermometer.

To calibrate such a thermometer, the bulb is made to reach thermal equilibrium with a standard known temperature such as that of an ice/water point. This phase change takes place at a well-known and constant temperature. Another equilibrium point is the temperature of water vapor, which is another phase change point that takes place at a known constant temperature. One can assume linearity between these two points and mark off the temperature scale. In 1724 Fahrenheit created the Fahrenheit temperature scale which was the first standardized temperature scale to be used. It had the drawback that 0.0°F was not the freezing point of water. This was later corrected with the Celsius temperature scale. Originally known as the centigrade scale (derived from the Latin *centrum* which means 100 and *gradus* which means steps), the scale was renamed Celsius in 1948 to honor Swedish astronomer Anders Celsius (1701−44) who developed a similar temperature scale.

Archives of Sea Surface Temperature (SST) measurements date back more than 160 years and are mostly from measurements made at coastal stations. The earliest measurements were made for scientific exploration. Later, after the connection was made between SST patterns and ocean currents, many more SST measurements were made to influence navigational charts. By the twentieth century, it was clear that SST also influenced weather forecasts, which further motivated SST measurements. Most SST observations were not collected by scientific vessels but rather by voluntary observing ships (VOS), whose owners recognized that SST would contribute to safety at sea. Thus, historical SST observations are primarily along major shipping lines.

In their earliest oceanographic application, simple mercury thermometers were used to measure SST by filling a wooden bucket with surface sea water and inserting a mercury-in-glass thermometer into the bucket to measure the temperature of this surface sample. When sailing ships moved from wooden to canvas buckets for surface samples, the process introduced a negative bias to the SST measurements, as the canvas bucket samples were much more quickly cooled by the wind while sitting on the deck of the ship. Wooden buckets have much better insulating characteristics that protect the sea water sample from cooling by the wind (Figure 1.5). These problems were only discovered much later when people started closely examining the SST data (Figure 1.6).

The pronounced drop in SST between 1890 and 1900 in Figure 1.6 marks the time when canvas buckets were introduced for collecting surface water samples. These buckets had the advantage that they could be collapsed for storage, unlike the larger bulky wooden buckets shown in Figure 1.5. Still, the wooden buckets provided better insulation to protect the sample from the effects of wind cooling while the sample was on-deck. The increase in SST that starts around 1930 and hits a peak around 1940 reflects the transition from sailing ships to powered vessels. Along with this shift in propulsion, there was a shift in the collection of SST observations, which were now taken in the engine room of the ship as a measure of the sea water used to cool the engines. While this method made for a much easier SST collection, it introduced many differences in the nature of the SST data. First of all, the intake port where the sea water was being sampled was lower on the ship's hull, meaning it did not represent a real SST. Secondly, and more importantly, the engine room was generally a hot environment due to the heat of the engines. This subsequently led to a warm bias in the SSTs measured using the ship's intake. This is what caused the switch from a cold bias to a warm bias in the historical ship SST observations. These effects both ignore the random errors that were introduced in the data by humans incorrectly reading the temperature and transmitting the value by voice over the radio.

1.3.1.2 Drifting Buoy Thermistors

When satellite-tracked drifting buoys became a widely used and accepted platform, one of the first measurements added to the buoy was sea surface temperature. Because the shape of the buoy hull evolved rapidly in the early days of the drifting buoy platforms, it wasn't until a widely accepted buoy hull design was accepted that the observed temperature values could be considered as an accurate measure of the SST. In today's global drifter program, the buoys are equipped with thermistor SST sensors with an accuracy of ±0.1°C and a stability of 0.1°C/year (World Climate Research Program, 1988). These buoy SSTs were considered as the best possible *in situ* SST measurement for the calibration of satellite skin SSTs (Emery

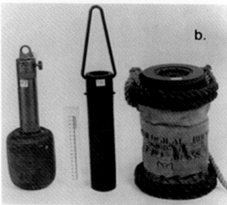

FIGURE 1.5   (a, b) A selection of surface water bucket samplers, including a wooden bucket in (a).

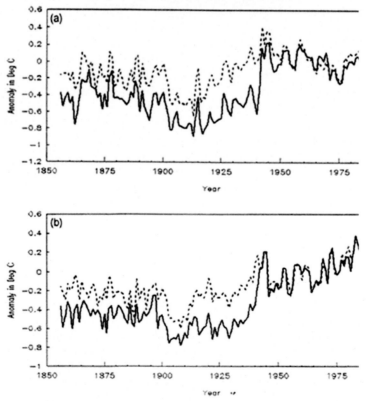

FIGURE 1.6   Time series of northern (a) and southern (b) hemisphere mean sea surface temperatures (SSTs) from 1850 to 1980. The dashed line is a correction to the canvas bucket SST measurements to account for wind cooling.

and Schluessel, 1989). It should be noted that the design of the buoys (Sybrandy et al., 2009) requires thermal insulation to be included to ensure that the solar heating at the top of the surface float does not impact the SST measurement. Moreover, the SST sensor should be accurate to better than 0.1°C for conditions when the inside of the buoy is 1.0°C warmer than the outside surface.

In reviewing SSTs from past drifting buoys, Kennedy (2014) finds errors or uncertainties of drifting buoy SSTs ranging from 0.12 to 0.67°C. The authors suggest separating the SST observational errors into random (∼0.6°C) and systematic (∼0.35°C) components. Modern drifting buoys are constructed in a common format and have their SST sensors located on the buoy, so they are at a nominal depth of between 10 and 20 cm. Designed in the early 1980s to study ocean currents as part of the global Surface Velocity Program (SVP), the buoys were next equipped to measure SST and barometric pressure for meteorological purposes. The buoys needed to be inexpensive, easy to deploy and reliable for at least 18 months of operation. This design became firmly standardized in 1991. Buoys equipped with both a barometric pressure sensor and an SST thermistor are called SVP-B drifters.

A typical SCP-B drifter is presented in Figure 1.7, which shows the location of the analog SST sensor, the digital High-Resolution SST (HRSST) sensor coupled with a hydrostatic pressure sensor and a barometric intake port for measuring atmospheric pressure. The strain gauge at the bottom is to provide some indication of whether the drogue element is still attached (or not). Another way of detecting the loss of a drogue, is by observing the daily displacements of the buoy and noting if there is a dramatic change in accelerations, indicating that the buoy has lost its drogue. The thick rubber connector at the bottom of the drifter serves as a hydraulic coupler that best accommodates the continued stress and strain experienced at the connection between the buoy and its drogue. The HRSST sensor is used to calibrate coincident satellite measurements, while the hydrostatic pressure sensor gives an indication of whether or not the buoy is submerged by waves or is floating on the sea surface.

The buoy in Figure 1.7 is a spherical buoy with a diameter of 40 cm and is made of high pressure molded polypropylene (Poli et al., 2019). A 12.5 m-long line (including the elastic hydraulic hose connected to the float), attached at the bottom of the buoy, connects to a "holey sock" drogue that is centered at 15 m depth. The holy sock drogue is 0.8 m in diameter and 6 m long (Lumpkin and Pazons, 2006). The SST is measured by two different sensors: an analog SST sensor with an accuracy of ±0.1°C and a new High-Resolution SST sensor (HRSST). Both are protected from solar radiation by a cap. The analog SST sensor is made from two cupronickel bolts of diameters 1.4 and 1.9 cm, protecting a 6 mm tube in which a thermistor is inserted. This is the common construction for SST sensors on all SVP buoys.

The HRSST sensor is composed of a thermistor inserted in a small stainless-steel needle that is 0.9 cm long and 0.12 cm in diameter. The thermistor has a resolution of 0.001°C (1 mK) and is accurate to ±0.01°C (LeMenn et al., 2019). The HRSST is enclosed along with a hydrostatic pressure sensor in a removable cylindrical housing that contains the electronics board for the two sensors. This cylindrical housing is needed to calibrate these two instruments in thermo-controlled baths but is removed when these modules are integrated into the buoys. The pressure sensor has a resolution of 0.05 dbar (≈ 0.05 m) and an accuracy of 0.05% over a range of 0−30 dbar.

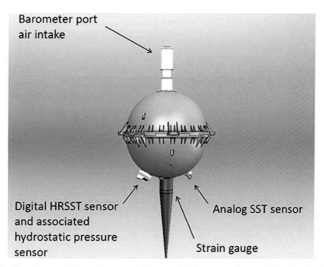

FIGURE 1.7   A Surface Velocity Program SVP-B buoy with its associated sensors. HRSST denotes the High-Resolution SST sensor.

As discussed by Sybrandy et al. (2009), the thermistor installation for the Global Drifter Program (GDP) buoys is very similar to that for the Surface Velocity Program. The GCP buoy is a 32 cm sphere tethered to a holy sock drogue at 15 m depth. The thermistor is thermally isolated from the inside of the float and is designed to react quickly to changes in SST. Thermal isolation prevents solar heating at the top of the surface float from influencing the SST measurements. The thermistor should be accurate to better than $\pm 0.1°C$ when the inside of the float is 1°C warmer than the sea surface. A thermistor fitting that reacts quickly to temperature changes also speeds up temperature calibration in the laboratory. One such option is the Betatherm assembly (part no. XP36K53D93), which includes a linearized thermistor composite within a stainless-steel fitting. Alternatively, an assembly can be constructed by potting a linearized thermistor composite within a suitable tubing connector (Figure 1.8). In their study of the errors in these SST measurements, LeMenn et al. (2019) developed equations for the theoretical response times of both the analog and HRSST sensors for two different water velocities (Table 1.1). The response time of both sensors is plotted as a function of water velocity in Figure 1.9.

The results in Figure 1.9 show the faster general response time for the HRSST sensor, which is one reason why it provides a better SST value for satellite calibration. It is also useful to consider the slower response time for the analog SST sensor as this is what is used for most SST calibrations. As reported by LeMenn et al. (2019), laboratory calibrations of these two types of SST sensors reveal a repeatability of 0.5 mK for the HRSST sensor (Table 1.2). The main source of uncertainty is due to the reproducibility of measurements that are impacted by the thermal inertia of the buoys. This results in the "expanded uncertainty."

The analog SST sensor was also calibrated over the temperature range of 5–35°C. The expanded SST results are shown for two sensors in Figure 1.10. It is possible to improve the accuracy and uncertainty of the SST measurements by calculating the coefficients of a straight line. This correction, however, introduces a linearity error. The expanded uncertainty of the analog SST sensors is found to be twice as large as the expanded uncertainties of the HRSST sensors. The analog SST sensors are also more sensitive to radiation effects such as solar heating. All of these errors cause the analog SST sensors to have larger systematic errors and measurement uncertainties (random error).

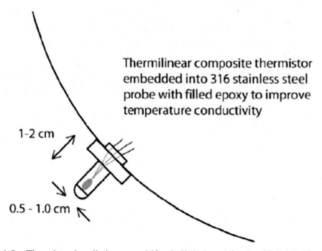

FIGURE 1.8 Thermistor installation on a drifter hull designed for the Global Drifter Program.

TABLE 1.1 Response Times of Buoy High Resolution SST (HRSST) Sensors for fast and slow water flow past the sensor.

	HRSST	HRSST	Analog SST	Analog SST
	(Slow)	(Fast)	(Slow)	(Fast)
Water Velocity (m/s)	0.001	1.0	0.001	1.0
Sensor diameter (cm)	0.15	0.15	1.4	1.4
Mass of the Sensor (g)	0.35	0.35	46.45	46.45
Response Time (ms)	77.5	2.5	57.1	20.1

After LeMenn et al. (2019).

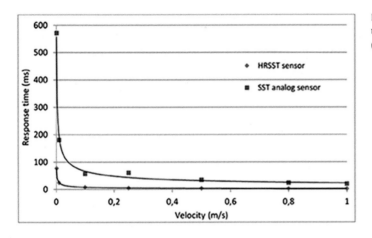

FIGURE 1.9 High Resolution SST (HRSST) sensor response time as a function of seawater velocity. *From LeMenn et al. (2019).*

TABLE 1.2 The uncertainty budget of High Resolution SSST (HRSST) sensors in milli-degrees Kelvin. Column numbers refer to specific sensor identifications.

Uncertainty budget of HRSST Measurements	N° Y17-07 (mK)	N° Y18-24 (mK)
Reference temperature (U_{ref})	0.9	0.9
Bath stability (U_{Bath})	0.3	0.3
Buoy HRSST reproducibility (S)	2.5	3.4
Buoy HRSST repeatability (S_{rep})	0.5	0.5
Expended uncertainty (U_c)	5.5	7.2

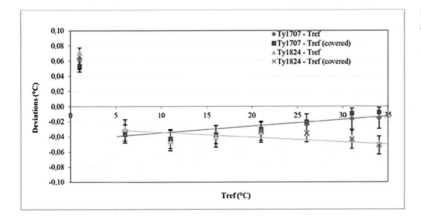

FIGURE 1.10 Deviations in the calibrations of two analog SST sensors.

1.3.1.3 Buoy and CTD Temperature Comparisons

Again, as reported in LeMenn (2019), two GDP drifters were deployed during an oceanographic cruise in the Mediterranean. An intercomparison between the two buoys and a CTD profiler (Sea Bird SBE 911) was set up. The CTD profiler was mounted beneath a rosette sampler, which was equipped with an additional reference thermometer (Sea Bird SBE 35) that had previously been calibrated to the International Temperature Scale of 1990 (ITS-90), which approximates the thermodynamic temperature scale. (A rosette consists of a carrousel holding up to thirty-six 10-liter sample bottles that can be "tripped" from the ship by sending an electric pulse down the conducting CTD support cable. The CTD is generally placed in the center at the lead end of the carrousel so that the sensors encounter water that is relatively undisturbed by the trailing rosette system.) The ITS-90 calibration uses 14 points ranging from 0.65 to 1357.77 K. The CTD was lowered to 15 m depth, which revealed that the upper first 4 m of the Sea were homogeneous and very close to 16.4°C. This made it possible to compare the buoy sensors with the CTD and the rosette temperatures. Between 4 and 5.5 m depth, there was a very strong temperature gradient of -1.25°C/m.

With the buoys tied to the CTD frame, five series of SBE35 SST measurements were collected, consisting of 29, 57, 113, 53 and 149 values. These were then averaged to give a temperature of 16.3968°C, with a standard deviation of 0.0057°C. The CTD was at 1.08 m depth and gave a temperature of 16.398°C. The buoy SST temperatures and differences relative to the CTD are given in Table 1.3.

The first two rows of this table are from the SST analog sensors, which first show the raw temperature and then the temperature corrected with the slopes and offsets from Figure 1.9. The corrected temperatures are much closer to the SBE35 and CTD temperatures. The HRSST values are well within the uncertainties of the CTD and SBE35 temperatures. Without any corrections, the analog SST sensors are within 0.05°C of the reference temperatures, while with corrections they are within the uncertainties of the reference temperatures. This is an important result as the analog sensors are typical of most SST sensors installed on drifters in the Global Drifter Program.

1.3.2 Satellite Sea Surface Temperature Measurement

The landmark paper by Minnett et al. (2019) reviews the "half-century" of satellite remote sensing of SST. Originally SST patterns were observed by single channel infrared images from polar orbiting weather satellites. These infrared channel images were created to provide the meteorologists with cloud imagery at night when the visible channels could not provide data. The infrared images revealed the complexity of SST patterns on the ocean surface. Because the single channel images were subject to spatial variations introduced by the atmospheric water vapor between the satellite and the sea surface, more quantitative satellite observations of SST needed to wait for the advent of satellite radiometers that allowed for a correction of the infrared radiances using multiple channel observations in the thermal infrared. At the same time, microwave radiometers were flown on polar-orbiting satellites that could also measure SST. Since then, SST has been observed from both infrared and microwave instruments on polar-orbiting satellites and from infrared radiometers on geostationary satellites.

The infrared imagery provided much greater spatial resolution than the microwave imagery but at the expense of being blocked by cloud cover and attenuated by atmospheric water vapor. It was this higher spatial resolution of the infrared imagery that made it possible to capture the complex space-time variability of the Gulf Stream (Legeckis, 1975). While infrared radiometers on geostationary satellites could image the Earth's surface every half hour, the weaker infrared signal at the 36,000 km orbit of these satellites resulted in reduced spatial resolution. It wasn't until later that better spatial resolution was possible at geostationary orbits. Subsequently, the frequent coverage of geostationary satellites made it possible to filter out moving cloud cover.

It is generally recognized that systematic time series measurement of SST began in the mid-19th century (Rayner et al., 2006). At about the time that observers began to collect quantitative infrared SST data from polar orbiting satellites, the open ocean drifting buoy program began to provide *in situ* SSTs that were better suited to act as fiducial SST measurements for the satellite infrared data. Prior to that, ship-based SST measurements were used for satellite infrared measurement calibrations, but it was clear that the water used to cool ship engines was collected much deeper in the water column. In addition, the temperature in the warm engine rooms in which the SST was being measured, resulted in a positive SST bias.

TABLE 1.3 Buoy SST measurements (°C) coincident with the SBE35 thermometer and the SBE911 CTD. The three right-hand columns are the differences in temperature (°C) between the various sensors and high-resolution SST data.

	Value transmitted	SST corrected	$T_{trans} - T_{ctd}$	$T_{trans} - T_{sbe35}$	$SST_{cor} - T_{sbe35}$
SST 58002	16.35	16.382	−0.048	−0.047	−0.014
SST 58019	16.35	16.389	−0.048	−0.047	−0.008
HRSST 58002	16.391	–	−0.007	−0.006	–
HRSST 58019	16.398	–	0.000	0.001	–

The first two lines show the results for the SST analog sensors; the last two lines show results for the HRSST sensors.

Over time, it became apparent that the infrared satellite sensors were responding to the radiation emitted from the very skin of the ocean's surface. This led the Group for High Resolution SST (GHRSST) science team to develop a set of definitions for SST (Figure 1.11). Note the nonlinear vertical scale in this figure, which allows for the 10 μm "thin" skin layer to be better depicted. Thus, the interface SST (SSTint) is always cooler than the temperature of the 10 μm thick skin layer (SSTskin) lying beneath. The skin layer represents the molecular layer that connects the turbulent atmosphere to the turbulent ocean. SSTskin is the temperature measured by an infrared radiometer operating at wavelengths between 3.7 and 12 μm. Below that is the subskin (SSTsubskin) layer formed at the base of the conductive laminar sub-layer, which for practical purposes can be approximated by the temperature measured by a microwave radiometer operating in the 6–11 GHz frequency range. Still deeper is the surface temperature at depth (SSTdepth), which represents the SSTs measured by buoys, ships, and other instruments. Finally, at the bottom, is the foundation temperature (SSTfnd), the temperature that is free from surface variability introduced by diurnal variations in solar radiation and wind stress at the sea surface. The value SSTfnd is, therefore, the temperature for which the heat gain from solar radiation absorption just exceeds the heat loss at the sea surface.

The SSTs derived from satellite radiance measurements are called "brightness temperatures" (BTs) to distinguish them from SSTs that would be measured by a radiometer located immediately above the sea surface. Differences between the two measurements are due to the effects of the intervening atmosphere on the SST measurements. These effects include cloud contamination and the attenuation of the SST signal by atmospheric water vapor. The first step in deriving SST from brightness temperatures involves confident identification of the satellite pixels that include radiance from clouds or aerosols that must be eliminated before clear sky atmospheric correction algorithms can be used (Minnett et al., 2019; Wick et al., 1992; Emery et al., 1994).

1.3.2.1 Clear Sky Conditions

Clear sky conditions are required for the computation of SST from infrared measurements. Therefore, the first task is to identify and remove cloud contaminated pixels. Identification of clouds in satellite infrared measurements usually consists of a binary test based on the Brightness Temperature (BT, also written as T_b) uniformity and then finding the BT minima by comparisons with lower resolution gap-free reference fields (Kilpatrick et al., 2001, 2015). These methods perform less well near SST fronts, at cloud edges, in the presence of small cirrus clouds and in the presence of low-level uniform stratus clouds, with temperatures similar to the SST values, particularly at high latitudes. The dependence on strict spatial uniformity often results in strong SST fronts being identified as clouds. There are differences in identifying clouds during the day and at night as channels, such as the 3.7 μm channel with its greater temperature sensitivity, can only be used at night due to the inclusion of reflected radiation during the day.

Early in the development of satellite infrared SST, it was found that differential absorption by atmospheric water vapor at the different thermal infrared wavelengths could be used to correct the pixels for the atmospheric water vapor absorption

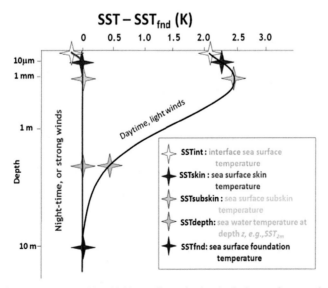

FIGURE 1.11 Cartoon of near-surface temperatures with a highly nonlinear depth-axis. Scales are for general guidance and are not to be used for derivations. *The SST temperature definitions are from Minnett and Kaiser-Weiss (2012).*

effects. Originally, only two thermal infrared channels were used to compute a corrected SST (McMillin and Crossby, 1984; Barton, 1995), such that

$$SST = aT_i + \gamma\left(T_i - T_j\right) + c \tag{1.18}$$

where T_i and T_j are the brightness temperatures measured in the two thermal infrared channels, c is an offset and γ is the differential absorption coefficient of the atmospheric water vapor (McMillin, 1975). This equation became known as the Multi-Channel SST (MCSST).

The role of diurnal variability in dictating the SST and the structure of the near-surface ocean is emphasized in Figure 1.11, where the marked difference in the near surface profiles is due to the absorption of solar radiation that takes place in the upper few meters of the ocean. This solar heating is particularly effective when the winds are weak and do not mix the upper ocean. Some of the first examples of diurnal heating were observed by Stommel et al. (1969), Halpern and Reed (1976), and Kaiser (1978). We note that, even when solar heating warms the shallow upper layer, the interface temperature and the skin temperature are both colder than the nominal subsurface SST. Thus, any satellite SST estimate will be cooler than the *in situ* SST used to validate it.

The identification and characterization of diurnal warming in all oceans prompted efforts to simulate this behavior numerically and to also estimate the impact of neglecting diurnal warming in large scale modeling systems. The consequences of not including diurnal warming in model simulations was studied in terms of heat budget errors in the Tropics (Clayson and Bogdanoff, 2013), the Mediterranean Sea (Marullo et al., 2016), and the North Sea (Fallman et al., 2017).

1.3.2.2 Infrared SST Observations

Historically, satellite observations of SST were from the infrared channels on weather satellites that had been created to provide meteorologists with cloud imagery during the nighttime. Later, the instrument was altered to improve the remote sensing of SST. The channels introduced into the Advanced Very High-Resolution Radiometer (AVHRR) to measure SST were designed to take advantage of atmospheric windows that allowed infrared radiation to pass through the air (Figure 1.12).

Note that the AVHRR channel 3 (night only 3.55–3.93 μm; Figure 1.12) responds to both thermally emitted radiation and reflected solar radiation. Thus, it could only be used at night to compute SST, when there is no reflected solar radiation present. As we remarked earlier, the advantage of this channel was its lower sensitivity to thermal radiation giving it the ability to respond to much higher surface temperatures. This channel became critical for the detection and mapping of forest fires. Channels 4 (10.3–11.3 μm) and 5 (11.5–12.5 μm) were used in the SST algorithm to correct for atmospheric water vapor attenuation of the emitted SST signal. These wavelengths correspond to the last iteration of the AVHRR suite of sensors. Observations from the AVHRR began in October 1978 and represent one of the longest time series of measurements in geoscience. Remarkably, there are still AVHRR sensors operating today, corresponding to a sequence of 43 years of SST measurements.

FIGURE 1.12   Atmospheric transmittance (%) for different wavelengths and AVHRR thermal infrared channels.

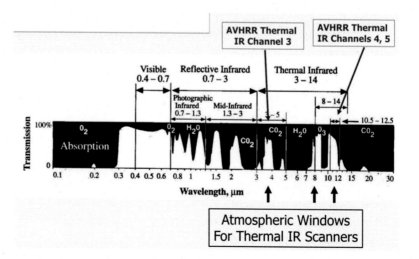

1.3.2.3 Cloud Clearing Process

As we stated previously, the process of computing SST requires the user to first clear cloud contaminated pixels from the computation to yield clear sky radiances. Most cloud identification procedures are based on the brightness temperature (BT) uniformity, the BT minima, comparisons with lower resolution cloud free pixels, and, finally, on a temporal filter that monitors the change in BT in a given pixel, realizing that clouds move around much more quickly than SST changes (Devittorio and Emery, 2002). All these cloud filtering methods are less effective over ocean thermal fronts and at cloud boundaries. The fact that cloud filters rely on strict spatial uniformity often results in identifying strong SST fronts as clouds. There are marked differences in cloud masking capabilities between daylight hours, when visible radiances are available, versus nighttime, when only the thermal infrared is available for cloud screening.

An alternative to the threshold cloud identification method in infrared imagery is the statistical approach developed by Merchant et al. (2005) based on the Bayes sampling theorem. In this method, joint probability distributions of the Brightness Temperatures (BTs) in two infrared channels are computed using radiative transfer simulations under clear sky conditions and under cloudy conditions by visual inspection of satellite imagery. Radiative transfer simulations from weather prediction models that characterize the atmospheric state are then used to give an expectation of clear-sky BTs. These are then compared with probability distributions of cloudy measurements and the measured BTs are assigned a probability of having been influenced by clouds. This approach has been widely used for several satellite missions such as the Advanced Along-Track Scanning Radiometer (AATSR) series (Merchant et al., 2012), the NOAA Geostationary Operational Environmental Satellite (GOES, Maturi et al., 2008) and Himawari-8 satellite (Kurihara et al., 2016).

Minnett et al. (2019) report another alternative to the threshold method—called the Alternating Decision Tree (ADT; Freund and Mason, 1999; Kilpatrick et al., 2019) — that was developed to avoid some of the earlier problems (Liu and Minnett, 2016; Liu et al., 2017). Instead of the standard decision tree approach, where a pixel was either cloudy or cloud-free, the ADT uses various tests to determine the likelihood of a pixel being identified as being clear or cloudy. Threshold values and the weights given to each test were determined by a Machine Learning algorithm (Hall et al., 2009) that was applied to a subset of matchup data for four conditions: nighttime, daytime, moderate sun-glint and strong sun-glint. When applied to MODIS and VIIRS (Visible and Infrared Imaging Radiometer Suite) data, the ADT algorithm improves on the decision-tree approach for all metrics considered (Kilpatrick et al., 2019).

As mentioned earlier, once a pixel is identified as cloud-free one still needs to correct for the atmospheric attenuation of the infrared SST signal due to the presence of water vapor in the atmosphere. Early in the development of SST remote sensing, it was found that the atmospheric attenuation would be different in the $3.5-4.1 \, \mu m$ channel and in the $9.5-12.5 \, \mu m$ channels and that this difference provided the potential for correcting for the intervening atmospheric water vapor (McMillin, 1975). The difference in the SSTs could be expressed by the Multi-Channel SST (MCSST) algorithm (Eqn 1.18). The coefficients depend on the atmospheric state at the time of observation (primarily the temperature and water vapor profiles in the atmosphere). Thus, they can be calculated from collocated, coincident *in situ* SST measurements ("matchups") or by using radiative transfer simulations. Generally, the matchup SST observations are taken from drifting buoys and are constrained to be within a 10 km spatial window and to be less than 1 h apart (McCllain, 1985; Minnett, 1991). This latter radiative transfer approach requires a large number of radiosonde observations to characterize the global atmosphere in order to be able to link the SST measurement with the proper atmospheric correction (Zavody et al., 1995).

While the MCSST was the main formulation for satellite infrared SST for many years, it is missing a correction for the effects of increasing path lengths that are a function of spacecraft attitude and scan angle. This limitation was later compensated for by adding a term $(\sec(\theta) - 1)$, where θ is the satellite zenith angle measured at the surface and scaled by an additional coefficient. The errors in the derived SST showed a regional dependence, indicating that the water vapor correction term was not adequate. In response, Walton et al. (1998) introduced the Non-Linear SST (NLSST) algorithm, which incorporated a non-linear combination of the top-of-atmosphere brightness temperatures (BTs) measured in the $10-12 \, \mu m$ wavelength interval, viz;

$$\mathrm{SST} = a_0 + a_1 T_{11} + a_2(T_{11} - T_{12})T_{\mathrm{sfc}} + a_3(\sec(\theta) - 1)(T_{11} - T_{12}) \qquad (1.19)$$

where T_{11} and T_{12} are the brightness temperatures measured at the $10-12 \, \mu m$ wavelengths and T_{sfc} is a "first guess" estimate of the SST in the area. This first guess is often computed as the MCSST in Eqn (1.19). In a very different study of atmospheric water vapor, Emery et al. (1990, 1994) examined atmospheric water vapor estimates from the Special Sensor Microwave Imager (SSM/I) and found that traditional water vapor estimates from the Fleet Numerical Oceanography Center (FNOC) matched well the SSM/I estimate for mid-latitudes, whereas for both tropical and higher latitudes the FNOC series were decidedly lower than the SSM/I water vapor estimates (Schluessel and Emery, 1989). It is likely that this is true for SST corrections as well, where the NLSST and even the MCSST work well at mid-latitudes but fail in the heavy water vapor laden tropics or in the dry atmospheres of the polar regions.

None of these SST algorithms we have listed have yet used the mid-infrared channels that are only valid at night. Thus, we can write a Visible Infrared Imaging Radiometer Suite (VIIRS) three-channel night-time algorithm as:

$$SST_{triple} = a_0 + a_1 T_{11} + a_2 (T_{3.7} - T_{12}) T_{sfc} + a_3 (\sec(\theta) - 1) \tag{1.20}$$

For the MODerate-resolution Imaging Spectrometer (MODIS), which has two narrow bands in the mid-infrared, the most accurate nighttime SST algorithm is:

$$SST_{3.95} = a_0 + a_1 T_{3.95} + a_2 (T_{3.95} - T_{4.95}) + a_3 (\sec(\theta) - 1) \tag{1.21}$$

There are two different methods for finding the coefficients in Eqns (1.18)–(1.21). The first is to use numerical simulations of the brightness temperature measurements using radiative transfer methods (a close approximation to skin SST) and the second is to matchup with collocated *in situ* SST measurement primarily from drifting buoys (not a skin SST). To better fit the conditions imposed by seasonal and regional variations in atmospheric water content, sets of co-efficients that depend on month of the year and latitude bands can be used. Because these SST algorithms involve the differences in infrared BTs from radiometers with large Noise Equivalent Temperature Differences (NE Δ Ts), the differences can amplify the errors in the SST algorithm. In locations, such as the polar regions, SSTs may be dominated by instrument and water vapor correction errors.

1.3.2.4 Microwave Sea Surface Temperatures

The first microwave radiometer specifically designed to, in part, measure sea surface temperature, was the Scanning Multi-channel Microwave Radiometer (SMMR) on the short-lived Seasat satellite. This instrument had a channel at 6.63 GHz for measuring SST (Figure 1.13).

At most microwave wavelengths, the atmosphere is essentially transparent and the radiation passes through to the satellite sensor unimpeded. This is the great benefit of using passive microwave sensing of SST but it is gained at the price of much lower spatial resolutions, as passive microwave spot sizes are very large compared to infrared pixel sizes. There are two ways of calculating SST from microwave data: (1) using radiative transfer models (RTMs) to find the parameter values that fit the microwave observations; and (2) applying a statistical approach that requires coincident *in situ* measurements. The RTM method requires instrument information (azimuth and incident angles, frequencies, polarizations) and in general retrieves the wind speed, the water vapor and cloud parameters at the same time as the SST. As reported by Minnett et al. (2019), differences between the simulated and observed Top of Atmosphere (TOA) BTs are generally due to errors in the RTM parameterizations or measurement errors (poor instrument calibration, channel crosstalk, etc.), which then require 'ad-hoc' corrections to the geophysical SST retrievals (Meissner and Wentz, 2012). The statistical approach can overcome some of these errors by incorporating them into the statistical model, which is developed using satellite TOA BT measurements collocated with *in situ* and model observations (Shibata, 2006; Pearson et al., 2018; Nielsen-Englyst et al., 2018).

Methods (1) and (2) have their respective benefits. The RTM approach ties the physics of the satellite BTs to the relevant parameters allowing for the inclusion of satellite instrument calibration errors. Their primary weakness is the

FIGURE 1.13 Relative sensitivity in the microwave portion of the electromagnetic spectrum. Also shown are the optimal channels for sensing various land and sea surface parameters.

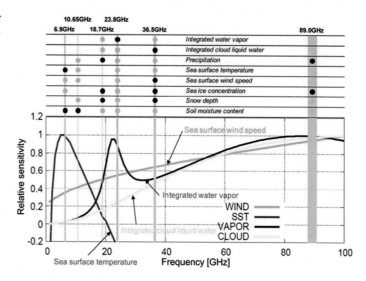

inability to account for errors in the RTM model that result in errors in the derived geophysical parameters. These RTM models can also be used to estimate instrument and other observable errors. The statistical approach is much simpler to apply and may result in more accurate SSTs as the instrument errors can be built into the retrievals. In either case it is important to first filter out as many of the external errors as possible. One source of error in the microwave, that is not present in the infrared, is the possibility of radio frequency interference (RFI) introduced by the operations of communication systems at the same or neighboring frequencies. Any pixels affected by RFI must be removed from the SST calculations. In the RTM retrievals the SST accuracy may be affected by extreme values in the other parameters being retrieved. Thus, cases where the SST retrievals are compromised by rain, strong winds, sea ice, sun glint and adjacent land, must be flagged and removed from the SST calculation.

While the passive microwave sensing of SST is attractive for its ability to penetrate cloud cover (the primary coverage problem for infrared SST) it does have the disadvantage that due to the weak nature of microwave emissions the signal must be integrated over large areas to be observable. This results in a much lower spatial resolution of the microwave SST retrievals. For example, the typical infrared spatial resolution is 1 km while the passive microwave spatial resolution is 25—30 km. Thus, stand-alone microwave SST retrievals do not have the spatial resolution needed to map the SST patterns. As a result, the microwave SSTs have generally been merged with the infrared SSTs to take best advantage of the optimal characteristics of both. One wonders if it makes sense to merge these two different SST products since the infrared sensors measure the SST emissions at the skin temperature while the passive microwave sensors measure the sub-skin temperature (Figure 1.11). A pictorial example of such a merging algorithm is given here in Figure 1.14.

In Figure 1.14, the following acronyms are used.

1. AVHRR, Advanced Very High Resolution Radiometer
2. VISSR, Visible Infrared Spin Scan Radiometer
3. VIIRS, Visible Infrared Imaging Radiometer Suite
4. TMI, TRMM (Tropical Rainfall Mapping Mission) Microwave Imager
5. SSM/I, Special Sensor Microwave Imager
6. OSCAR, Ocean Surface Current Analysis.

We note that, at the time this algorithm was developed, the only microwave instrument flying was the TMI, which was not optimal for sea surface temperature as its lowest frequency was 10.7 GHz, considerably higher than the 6.9 GHz channel that was optimal for observing SST. Still, for the period that it operated, the TMI was the only option for the computation of microwave SST. Another limitation of the TMI was that the TRMM was limited by its orbit to a coverage of ±40° of latitude since it was intended to study the rainfall conditions in the tropics.

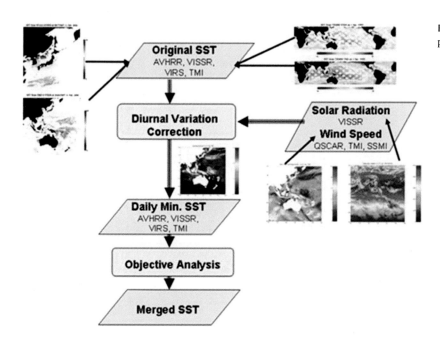

FIGURE 1.14 Scheme for merging infrared and passive microwave SST observations.

1.3.2.5 Along Track Scanning Radiometer Measurement

One instrument that provided a different approach to the atmospheric correction for the SST calculation was the Along Track Scanning Radiometer Measurement (ATSR) developed by the United Kingdom's Rutherford Appleton Laboratory, for deployment on the European Space Agency's (ESA's) series of polar-orbiting satellites. To account for the water vapor attenuation, this instrument measured the SST in multi-channels through two distinctly different atmospheric pathlengths by looking forward and then subsatellite on each scan (Figure 1.15). Note that with this scan geometry each spot on the surface is imaged twice by the radiometer with different atmospheric path lengths, thus making it possible to correct for the atmospheric attenuation of the SST signal. This required a stable and well-calibrated infrared radiometer. The required accuracies for ATSR derived SST are better than 0.3°C with a stability approaching 0.1°C/decade. To perform the required atmospheric corrections, the sensor views the Earth at nadir below the satellite and once at 55° from nadir.

The Along Track Scanning Radiometer (ATSR) also carries an exceptionally precise and stable on-board calibration system, with two reference targets designed for high uniformity and stability. These two black-body targets are maintained at temperatures near to the extremes of the Earth's temperatures (heated to 300 K, while the other floats to the outside ambient temperature estimated to be ~256 K) and these reference bodies are viewed during each scan cycle of the radiometer. These black bodies are cylinders with nonreflecting interior coatings, good insulation and a temperature monitoring system designed for high accuracy.

The ATSR has three channels at thermal infrared wavelengths of 3.7, 11 and 12 μm for the computation of SST. A reflected shortwave infrared channel at 1.6 μm is used to improve the detection of cloud cover by day. In addition, ATSR has three more visible and near infrared channels at 0.56, 0.67 and 0.87 μm which are used to map cloudy areas, and to measure solar radiation. These channels provide measurements from which land-cover properties can be extracted, such as the Normalized Difference Vegetation Index (NDVI). In this case, the ATSR scan geometry can help in mapping vegetation accuracy, particularly in the presence of excessive aerosols in the atmosphere. The ATSR was followed by an Advanced ATSR (AATSR) and the Sea and Land Surface Temperature Radiometer (SLSTR). In this last instrument, the angle view was switched to be behind the nadir scan rather than in front of it.

1.3.2.6 Visible Infrared Imaging Radiometer Suite Measurement

At present, the primary instrument for SST measurements is the Visible Infrared Imaging Radiometer Suite (VIIRS) that flies on the NOAA polar-orbiting satellites, which built on the experience gained from NASA's MODIS instruments. This included the sixteen detectors for each band and the grooved blackbody for infrared calibration. It did not, however, use the paddle-wheel scan mirror of MODIS and has instead a rotating telescope with a "half-angle mirror" to avoid image rotation across the swath that would otherwise occur due to the use of multiple detectors. The spatial resolution of the SST channels is 0.75 km at nadir and VIIRS has an innovative sampling scheme of aggregating rectangular pixels in the along-track

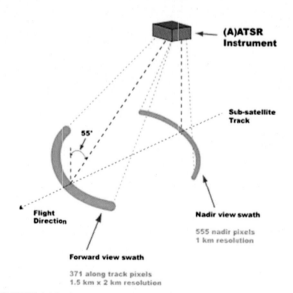

FIGURE 1.15 Scan geometry of the along track scanning radiometer.

direction to reduce the growth of pixel size across the scan away from nadir (Schueler et al., 2013; Gladkova et al., 2016). The infrared bands used for SST are the traditional 2 bands in the 10−12 μm interval and two bands in the 3.7−4.1 μm atmospheric window for use during night. VIIRS has a special higher resolution channel that samples at nadir with a 375 m resolution over the interval of 10.6−12.4 μm using the entire thermal infrared transmission window. While the absence of a split-window in this higher resolution band eliminates the capability of using the split window for a water vapor correction, it has been possible to use the atmospheric correction for the 750 m resolution bands applied to the higher resolution bands to map SST. This approach provides VIIRS with the capability of mapping submesoscale ocean features by their SST expressions.

1.3.2.7 Japanese SST Satellites

Japan launched the Global Change Observation Mission (GCOM) in 2012 with two satellites into the A-Train, which is an orbit with many satellites carrying complimentary payloads (L'Ecuyer and Jiang, 2010). These satellites carried the Advanced Microwave Scanning Radiometer-2 (AMSR-2), which has the same channels as AMSR-E (12 channel, 6 frequencies at 6.925, 10.65, 18.7, 23.8, 36.5 and 89.0 GHz, which are all dual polarized). In addition, AMSR-2 had a 7.3 GHz channel to mitigate Radio Frequency Interference (RFI) in the SST retrievals (Gentemann and Hilburn, 2015). An example of global SST derived from AMSR-2 measurements is shown here in Figure 1.16. Comparisons of the AMSR-2 SSTs with drifting and moored buoy SSTs yielded a mean difference of −0.49 K with a standard deviation of 1.63 K (Minnett et al., 2019).

Another Japanese satellite with SST observing capability was GCOM-C (climate) which carried the Second-generation Global Imager (SGLI), an imaging radiometer with channels in the visible and the infrared. The SGLI was made up of two instruments, the Visible and Near Infrared Radiometer (VNR) and the Infrared Scanning Radiometer (ISR). The ISR had two thermal infrared channels at 10.78 and 11.97 μm, which have a resolution of 250 m at nadir and a swath width of 1,400 km. The ISR SSTs are of very high quality and comparisons with buoy SSTs yield a mean difference of −0.063 K during the day and −0.181 K at night, with standard deviations of 0.333 and 0.619 K, respectively. The relatively high nighttime standard deviations are attributed to flaws in the night cloud clearing algorithm. This satellite flew in a sun-synchronous polar orbit consistent with the majority of the infrared sensing satellites.

1.3.2.8 Chinese SST Satellites

The responsibility for SST mapping in China rests with the China Meteorological Administration (CMA), who have developed a series of sun-synchronous polar orbiting satellites known as FengYun (FY), meaning "wind and cloud." These satellites carry the Multichannel Visible Infrared Scanning Radiometer (MVISR), which has 10 channels in the visible and infrared. The three thermal infrared channels are 3.75, 10.8 and 12.0 μm. Unfortunately, these channels had a very high Noise Equivalent temperatures (NE ΔT)—the change in the observable temperature due to instrument noise—of 0.4 K at

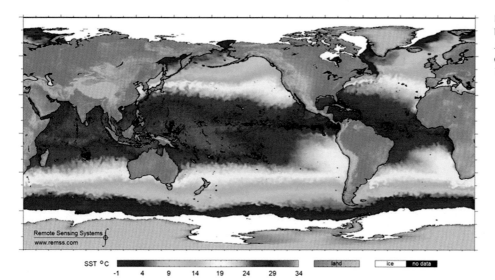

FIGURE 1.16 Global SST distribution derived by compositing AMSR2 measurements over 3 days ending 21 November 2018. *Credit: Wentz et al. (2014).*

SST °C   -1   4   9   14   19   24   29   34   land   ice   no data

300 K for the 3.75 μm channel and 0.22 K at 300 K for the 10.8 and 12.0 μm channels. This, combined with the difficulty of accessing satellite data from China, has led to very little use of these data for mapping SST temperature. The series was upgraded in 2008 with the Visible and InfraRed Radiometer (VIRR) and later by the Medium Resolution Spectral Imager-2 (MERSI-2). The satellites that carried these sensors were known as FY-3.

The Chinese State Oceanic Administration has sponsored a separate suite of polar orbiters known as the Hai Yang (meaning ocean, HY) series. These were smaller satellites launched in 2002 and 2007. They carried two imagers: the Coastal Zone Imager with its four visible channels, and the China Ocean Color and Temperature Scanner (COCTS), a ten-channel radiometer with eight channels in the visible and near infrared. There are also two in the thermal infrared (10.85 and 11.95 μm). SSTs computed from the latter compared with buoy SSTs gave a mean difference of 1.22 K and a standard deviation of 1.78 K. This series is expected to continue.

1.3.2.9 Geostationary Satellites

For many years the best resolution of the infrared channels on geostationary satellites was about 4 km, which, even with the increased temporal coverage to overcome cloud contamination, did not provide sufficient spatial resolution to map SST. This has changed in recent years and sensors in geostationary orbit are now much more capable. The Advanced Baseline Imager flying on the GOES-R series of geostationary satellites (and in a slightly modified form as the Japanese Advanced Himawari Imager (AHI)) has a spatial resolution in the thermal infrared of 2 km at nadir. These instruments have 16 spectral channels, five of which can be used for SST calculations: 3.85, 8.6, 10.45, 11.2 and 12.35 μm. These instruments provide full disk images every 10 min. Thus, with this improved infrared spatial resolution and the rapid refresh, this is a good data source to add to the computation of SST.

1.3.2.10 Other Satellite SST Sensors

There are a number of other satellites carrying sensors that could be used for SST observation such as the Thermal Infrared Sensor (TIS) on Landsat 8 and 9. While the TIS has the appropriate channels for the SST calculation and offers a superb spatial resolution of 100 m, the 15-day repeat cycle of Landsat satellites makes it inappropriate for mapping the mesoscale and smaller changes in space and time that are characteristic of SST features. Similarly, atmospheric sounding instruments like the AIRS (Atmosphere Infrared Sounder) have some appropriate SST channels but a very low spatial resolution $(13 \times 13 \text{ km}^2)$. The European Infrared Atmospheric Sounding Interferometer (IASI) has narrow channels in the spectral range appropriate for SST calculations. The IASI calibration, which consists of an internal black-body and a cold space view, has been found to be a very stable calibration system and hence IASI has been selected as a reference instrument by the Global Space-Based Inter-Calibration System (GSICS; Hewisonn et al., 2013). A comparison of best quality IASI SST_{skin} retrievals with subsurface SSTs from drifting buoys yielded a mean difference of -0.16 K and a standard deviation of 0.33 K.

While there are some additional satellite sensors that could be used for SST calculation, we have tried to summarize the most widely used to give the reader some idea of the common practices in mapping SST. We have also tried to provide the best estimates of the accuracies relative to drifting and moored buoy SST observations.

1.3.3 Subsurface Temperature Measurement

Mercury-in-glass thermometers were lowered into the ocean with hopes of measuring the temperature at great depths. Two effects were soon noticed. First, thermometer housings with insufficient strength succumbed to the great pressure in the ocean and were crushed. Second, the process of bringing an active thermometer through the oceanic vertical temperature gradient sufficiently altered the readings that it was not possible to accurately measure the deeper temperatures. An early solution to this problem was the development of min-max thermometers that were capable of retaining the minimum and maximum temperatures encountered over the descent and ascent of the thermometer. This type of thermometer was widely used on the British Challenger Expedition of 1873−76. These min-max thermometers are still used today to measure the extremes, such as at a mountain location.

The real breakthrough in thermometry was the development of the reversing thermometer, first introduced in London by Negretti and Zambra in 1874 (Sverdrup et al., 1942, p. 349). The reversing thermometer contains a mechanism (called a pigtail appendix) such that, when the thermometer is inverted, the mercury in the thermometer stem separates from the bulb reservoir and captures the temperature at the time of inversion. Subsequent temperature changes experienced by the thermometer have limited effects on the amount of mercury in the thermometer stem and can be accounted for when the temperature is read on board the observing ship. This "break-off mechanism" is based on the fact that more energy is required to create a gas-mercury interface (i.e., to break the mercury) than is needed to expand an interface that already exists. Thus, within the "pigtail" section of the reversing thermometer is a narrow region called the "break-off point", located near Appendix C in Figure 1.17, where the mercury will break when the thermometer is quickly inverted. Note in this figure that the thermometers on the left is protected, which means it is encased in glass and protected from the hydrostatic pressure of the ocean, while the unprotected thermometer on the right allows the seawater to flow in so that the ambient pressure can influence the temperature measured by the reversing thermometer.

Note also that each thermometer holds two thermometers. One is the reversing thermometer that captures the temperature at the depth the thermometer is inverted while the other "auxiliary" thermometer is used to compensate for the temperature on the vessel where the thermometers are being read. The accuracy of the reversing thermometer depends on the precision with which the mercury break occurs. In good reversing thermometers, this precision is better than 0.01°C. In standard mercury-in-glass thermometers, as well as in reversing thermometers, there are factors, other than the break point, which affect the precision of the temperature measurement. These are:

(a) Linearity in the expansion coefficient of the liquid.
(b) The constancy of the bulb volume.
(c) The uniformity of the capillary bore.
(d) The exposure of the thermometer stem to temperatures other than the bulb temperature.

Mercury expands in a near-linear manner with temperature. Therefore, it has been the liquid used in most high precision, liquid-glass thermometers. Other liquids such as alcohol and toluene are used in precision thermometers only for very low temperature applications where the higher viscosity of mercury is a limitation. Expansion linearity is critical in the construction of the thermometer scale, which would be difficult to engrave precisely if expansion were nonlinear. This nonlinearity becomes important in the mechanical bathythermograph (MBT) that we will discuss later, where a liquid in metal thermometer results in very nonlinear temperatures requiring that each MBT reader be calibrated for each individual instrument and cannot be transferred from one MBT to another.

In a mercury thermometer, the volume of the bulb is equivalent to about 6,000 stem-degrees Celsius. This is known as the "degree volume" and usually is considered to comprise the bulb plus the portion of the stem below the mark. If the thermometer is to retain its calibration, this volume must remain constant with a precision not commonly realized by the casual user. For a thermometer precision to be within ±0.01°C, the bulb volume must remain constant within one part in

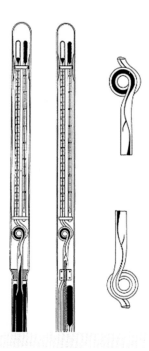

FIGURE 1.17   Richter reversing thermometers, protected (left) and unprotected (right). Also shown are expanded images of the "pigtail" appendices, shown before (upper) and after (lower) inversion of the instruments following sampling. The protected thermometer has a glass jacket to nullify the compression effects of water pressure; both instruments have scales for determining the main and auxiliary temperatures.

600,000. Glass does not have ideal mechanical properties and it is known to exhibit some plastic behavior and deform under sustained stress. Repeated exposure to high pressures may produce permanent deformation and a consequent shift in bulb volume. Therefore, precision can only be maintained by frequent laboratory calibration of the thermometer. Such shifts in bulb volume can be detected and corrected by the determination of the "ice point" (a slurry of water plus ice) which should be checked frequently if high accuracy is required. The procedure is obvious, but a few points should be considered. First, the ice should be made from distilled water and the water-ice mixture should also be made from distilled water. The container should be insulated and at least 70% of the bath in contact with the thermometer should be chopped ice. The thermometer should be immersed for five or more minutes during which time the ice-water mixture should be stirred continuously. The control temperature of the bath can be taken by an accurate thermometer of known reliability. Comparison with the temperature of the reversing thermometer, after the known calibration characteristics have been accounted for, will give an estimate of any offsets inherent in the use of the reversing thermometer in question.

The uniformity of the capillary bore is critical to the accuracy of the mercury thermometer. To maintain the linearity of the temperature scale it is necessary to have a uniform capillary as well as a linear response liquid element. Small variations in the capillary can occur because of small differences in cooling during its construction or to inhomogeneities in the glass. Errors resulting from the variations in capillary bore can be corrected through calibration at known temperatures. The resulting corrections, including any effect of the change in bulb volume, are known as "index corrections". These remain constant relative to the ice point and, once determined, can be corrected for a shift at the ice point by addition or subtraction of a constant amount. With proper calibration and maintenance, most of the mechanical defects in the thermometer can be accounted for. Reversing thermometers are then capable of accuracies of 0.01°C, as given earlier for the precision of the mercury breakpoint. This accuracy, of course, depends on the resolution of the temperature scale etched on the thermometer. For high accuracy in the typically weak vertical temperature gradients of the deep ocean, thermometers are etched with scale intervals between 0.1 and 0.2°C. Most reversing thermometers have scale intervals of 0.1°C.

The reliability and calibrated absolute accuracy of reversing thermometers continue to provide standard temperature measurements against which all forms of electronic sensors are compared and evaluated. In this role as a calibration standard, reversing thermometers continue to be widely used. In addition, some oceanographers still believe that standard hydrographic stations made with sample bottles and reversing thermometers, provide the only truly reliable data. For these reasons, we briefly describe some of the fundamental problems that occur when using reversing thermometers. An understanding of these errors may also prove helpful in evaluating the accuracy of reversing thermometer data that are archived in the historical data record. The primary malfunction that occurs with a reversing thermometer is a failure of the mercury to break at the correct position. This failure is caused by the presence of gas (a bubble) somewhere within the mercury column. Normally all thermometers contain some gas within the mercury. As long as the gas bubble has sufficient mercury compressing it, the bubble volume is negligible, but if the bubble gets into the upper part of the capillary tube, it expands and causes the mercury to break at the bubble rather than at the break-off point. The proper place for this resident gas is at the bulb end of mercury; for this reason, it is recommended that reversing thermometers always be stored and transported in the bulb-up (reservoir-down) position. Rough handling can be the cause of bubble formation higher up in the capillary tube. Bubbles lead to consistently offset temperatures and a record of the thermometer history can clearly indicate when such a malfunction has occurred. Again, the practice of renewing, or at least checking, the thermometer calibration is essential to ensuring accurate temperature measurements. As with most oceanographic equipment, a thermometer with a detailed history is much more valuable than a new one without some prior use information.

As discussed above, there are two basic types of reversing thermometers: (1) protected thermometers, which are encased completely in a glass jacket and not exposed to the pressure of the water column; and (2) unprotected thermometers, for which the glass jacket is open at one end so that the reservoir experiences the increase of pressure with ocean depth, leading to an apparent increase in the actual temperature. The increase in temperature with depth is due to the compression of the glass bulb, so that if the compressibility of the glass is known from the manufacturer, the pressure and hence the depth can be inferred from the temperature difference, $\Delta T = T_{\text{Unprotected}} - T_{\text{Protected}}$. The difference in thermometer readings, collected at the same depth, can be used to compute the depth of the temperature measurement to an accuracy of about 1% of the depth. This subject will be treated more completely in the section on depth/pressure measurement. We note that the 1% full scale accuracy for reversing thermometers is better than the accuracy of 2–3% normally expected from modern depth sounders but is much poorer than the 0.01% full-scale pressure accuracy expected from strain gauges used in most modern Conductivity-Temperature-Depth (CTD) probes.

Unless collected for a specific observational program or taken as calibrations for electronic measurement systems, reversing thermometer data are most commonly found in historical data archives. In such cases, the user is often unfamiliar with the precise history of the temperature data and thus cannot reconstruct the conditions under which the data were collected and edited. Under these conditions, one generally assumes that the errors are of two types; either they are large offsets (such as errors in reading the thermometer) which are readily identifiable by comparison with other regional historical data, or they are small random errors due to a variety of sources and difficult to identify or separate from real physical oceanic variability. Parallax errors, which are one of the main causes of reading errors, are greatly reduced through use of an eye-piece magnifier. Identification and editing of these errors depend on the problem being studied and will be discussed in a later section on data processing.

Another important aspect of traditional reversing thermometers is the very involved method for reading these thermometers. This procedure was developed to minimize the errors introduced when reading these analog devices and converting them to digital numbers. This procedure involves two persons to read all of the thermometers. They will each choose some type of magnifying unit to read the thermometers as the numbers are rather small (Figure 1.18).

The procedure is for each person to read all thermometers twice. The form is setup to feed into calculations to (a) correct the reversing thermometers for the auxiliary thermometer measurements and (b) to accurately compute the depth at which the sample was collected. It is important to link these measurements to the sample bottle it was collected with. The multiple readings by multiple people are intended to reduce any errors introduced by the readers and their interpretation of the thermometer scale divisions. Two people offer the potential of removing a bias that might be introduced by a single reader. Where there is an obvious disagreement in values for a particular thermometer, the thermometer may need to be re-read for a third time.

1.3.3.1 The Modern Digital Thermometer

New "digital reversing thermometers" have been created to transition the temperature measurement into the digital world. These have been created in a form factor that fits into the same receptacle as a standard mercury-in-glass reversing thermometer. As with the earlier thermometers, digital thermometers are activated when the thermometer rack reverses, locking the temperature at that depth into the device memory to be read out digitally at the surface. For this type of instrument, no correction is needed for the ambient temperature.

Digital thermometers are long-term stable platinum thermometers with a wide range of temperatures (-2.0 to $40.0°C$). This means that one can cover the ocean temperature range with a single thermometer rather than a set of mercury-in-glass thermometers. Digital thermometers have resolutions of $\pm0.0001°C$ for -2.0 to $9.99°C$ and $\pm0.001°C$ for $10.0-40.0°C$. The sampling frequency of the device is 1.25 Hz (0.8 s per sample). The accuracy is $\pm0.003°C$ and the instrument burst samples with a size of 16 samples. The overall stability of the temperature measurement is $\pm0.00025°C$ per month. The instrument weighs 170 g is 327 mm long and 20 mm in diameter. It will fit in a standard rosette thermometer rack. The unit displays the mean temperature value where reversed and the standard deviation of the burst sample. The thermometer has a glass and titanium housing, which can withstand the pressure at 10,000 m depth.

1.3.3.2 The Mechanical Bathythermograph

The mechanical bathythermograph (MBT) uses a liquid-in-metal thermometer to register temperature and a Bourdon tube sensor to measure pressure. The temperature sensing element is a fine copper tube nearly 17m in length filled with toluene (Figure 1.19). Temperature readings are recorded by a mechanical stylus, which scratches a thin line on a coated glass slide. Although this instrument has largely been replaced by the expendable bathythermograph (XBT), the historical archives contain a plethora of temperature profiles collected using this device. It is, therefore, worthwhile to describe the instrument and the variables it measures. Only the temperature measurement aspect of this device will be considered here; the pressure/depth recording capability will be addressed in a latter section.

FIGURE 1.18   Professor Jens Meincke (University of Hamburg, retired) reading reversing thermometers.

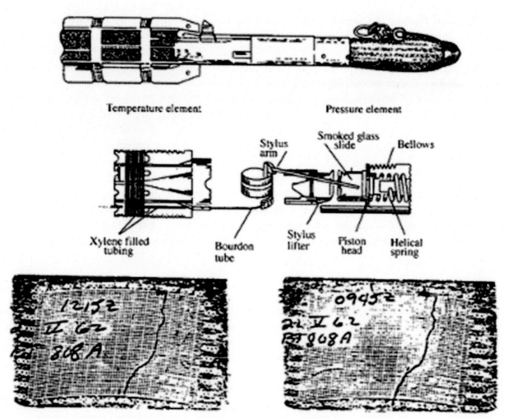

FIGURE 1.19   Mechanical Bathythermograph (MBT) instrument with a cutaway showing the internal components. At the bottom are two examples of the analog temperature profiles measured by an MBT.

There are numerous limitations to the MBT. To begin with, it is restricted to depths less than 300 m. While the MBT was intended to be used with the ship underway, it is only possible to use it successfully when the ship is traveling at no more than a few knots. At higher speeds, it becomes impossible to retrieve the MBT without the risk of hitting the instrument against the ship. Higher speeds also make it difficult to properly estimate the depth of the probe from the length of wire paid out from the winch. The temperature accuracy of the MBT is restricted by the inherent lower accuracy of the liquid-in-metal thermometer. Metal thermometers are also subject to permanent deformation. Since metal is more subject to changes at high temperatures than is glass it is possible to alter the performance of the MBT by continued exposure to higher temperatures (i.e., by leaving the probe out in the sun). The metal return spring of the temperature stylus is also a source of potential problems in that it is subject to hysteresis and creep. Hysteresis, in which the up-trace does not coincide with the down-trace, is especially prevalent when the temperature differences are small. Creep occurs when the metal is subjected to a constant loading for long periods. Thus, an MBT continuously used in the heat of the tropics may be found later to have a slight positive temperature error.

Most of the above errors can be detected and corrected for by frequent calibration of the MBT. Even with regular calibration, it is doubtful, however, that the stated precision of 0.1°F (0.06°C) can be attained. Here, the value is given in °F since most of the MBTs were produced with the Fahrenheit temperature scale. When considering MBT data from the historical data files, it should be realized that these data were entered into the files by hand. The usual method was to produce an enlarged black-and-white photograph of the temperature trace using the nonlinear calibration grid unique to each instrument. Temperature values were then read by eye off these photographs and entered into the data file at the corresponding depths. The usual procedure was to record temperatures for a fixed depth interval (i.e., 5 or 10 m) rather than to select out inflection points that best described the temperature profile. The primary weakness of this procedure is the ease with which incorrect values can enter the data file through misreading the temperature trace or incorrectly entering the measured value. Usually, these types of errors result in large differences with the neighboring values and can be easily identified by their apparent anomaly. Care should be taken, however, to remove such values before applying objective methods to search for smaller random errors. It is also possible that data entry errors can occur when entering date, time and position of the temperature profile and tests should be made to detect these errors.

Another source of errors in MBT profiles was the operation of the MBT winch. This unit (Figure 1.20) was usually mounted on the ship about midship, and it could not always be on the downwind side. Remember the goal of this instrument was to collect an upper ocean temperature profile with the ship underway. The winch had a single lever of operation, where one position activated the electrical motor to reel in the MBT probe and on the other side was the brake that would stop the cable from going out. In between was neutral, which was used to drop the probe. Problems arose when the probe was reeled in. The probe was known to "swim" from side to side when the suspending wire was under tension. Thus, when the probe neared the ship, it could be heard to bang against the side of the ship. The retrieval motor was very fast, and it was difficult to control the return speed. In addition, to activate the brake, the winch operator had to go through neutral and the probe would fall back into the water. Experienced operators learned to stop when the probe was near and then pull in the probe by hand.

1.3.3.3 Resistance Thermometers

Since the electrical resistance of metals (such as platinum, copper, or nickel) changes with temperature, these materials can be used as temperature sensors over a specific operating range. Platinum was first proposed by Sir William Siemens in 1871 as it is a noble metal and has the most stable resistance-temperature relationship over the largest temperature range. Nickel elements have a limited temperature range because the change in resistance per degree of temperature change becomes very non-linear for temperatures above 300°C. Copper has a very linear resistance-temperature relationship but unfortunately copper oxidizes at moderate temperatures and cannot be used for temperatures over 150°C.

The resistance R of most metals depends on temperature (T) and can be expressed as a polynomial

$$R = R_0\left(1 + aT + bT^2 + cT^3 + ...\right) \tag{1.22}$$

where a, b, and c are constants and R_0 is the resistance at $T = 0°C$. In practice, it is usually assumed that the response is linear over some limited temperature range and the proportionality can be given by the value of the coefficient α (called the temperature resistance coefficient). This coefficient can be written as

$$\alpha = (R_T - R_0) / (R_0 \cdot T) \tag{1.23}$$

where R_0 and R_T are the resistances of the sensor at 0 and T °C, respectively.

The most commonly used metals copper, platinum, and nickel have temperature coefficients, α, of 0.0043, 0.0039, and 0.0066 (°C)$^{-1}$, respectively. Of these, copper has the most linear response, but its resistance is low so that a thermal element would require many turns of fine wire and would consequently be expensive to produce. Nickel has a very high resistance but deviates sharply from linearity at higher temperatures. Platinum has a relatively high resistance level, is

FIGURE 1.20 Mechanical Bathythermograph (MBT) being operated onboard ship.

highly stable and has a relatively linear behavior. For these reasons, platinum resistance thermometers have become a standard by which the international scale of temperature is defined. Platinum thermometers are also widely used as laboratory calibration standards and have accuracies of 0.001°C.

The semiconductors form another class of resistive materials used for temperature measurements. These are mixtures of oxides of metals such as nickel, cobalt, and manganese that are molded at high pressure, followed by sintering (i.e., by heating to incipient fusion). The types of semiconductors used for oceanographic measurements are commonly called thermistors. Thermistors have the advantages that: (1) the temperature resistance coefficient of $-0.05(°C)^{-1}$ is about ten times as great as that for copper; and (2) the thermistors may be made with high resistance for a very small physical size.

The temperature coefficient, α, of thermistors is negative, which means that the resistance decreases as temperature increases. This temperature coefficient is not a constant except over very small temperature ranges; hence the change of resistance with temperature is not linear. Instead, the relationship between resistance and temperature is given by

$$R(T) = R_0 \exp\left[\beta\left(T^{-1} - T_0^{-1}\right)\right] \tag{1.24}$$

where $R_0 = R(T_0)$ is the conventional temperature coefficient of resistance, T and T_0 are absolute temperatures (K) with respective resistance values of $R(T)$ and R_0, and the constant β is determined by the energy required to generate and move the charge carriers responsible for electrical conduction. (As β increases, the material becomes more conducting.) Thus, we have a relationship whereby temperature T can be computed from the measurement of resistance $R(T)$.

Thermistors are generally produced using powdered metal oxides and the formulas and methods have progressed greatly over the past 20 years. They can now achieve temperature accuracies of $\pm0.1°C$ from 0 to 70°C, with excellent long-term stability. Thermistors come in many different styles, such as axial-leaded glass-encapsulated, glass-coated chips, epoxy-coated and surface-mounted. Another advantage of the thermistor is that the resistance change response to temperature change is very fast so that the instrument can quickly measure temperature changes in the presence of strong temperature gradients. Finally, a significant advantage to thermistors is the fact that they are relatively inexpensive and can be purchased in large quantities. These features have proven to be very important for the expendable bathythermograph, where the thermistor is discarded with every vertical profile measurement. The high accuracy and rapid response needed in a conductivity-temperature-depth (CTD) profiler is achieved by combining miniature and fast thermistors (30 ms response times) with a highly accurate, linear and slower platinum resistance thermometers. Both the thermistor and the platinum thermometer have excellent long-term stability.

1.3.3.4 The Expendable Bathythermograph

One of the most common uses of thermistors in oceanography is in expendable bathythermographs (XBTs). The XBT was developed to provide an upper ocean temperature profiling device that operated while the ship was underway. The crucial development was the concept of depth measurement using the elapsed time for a "freely-falling" probe with a well-known and constant fall-rate. To achieve "free-fall", independent of the ship's motion, the data transfer cable is constructed from fine copper wire with feed-spools in both the sensor probe and in the launching canister (Figure 1.21). The details of the depth measurement capability of the XBT will be discussed and evaluated in the section on depth/pressure measurements.

The XBT probes employ a thermistor placed in the nose of the probe as the temperature sensing element. According to the manufacturer (Sippican Corporation, now a subsidiary of Lockheed Martin Corporation, Marion, Massachusetts, U.S.A.), the accuracy of this system is $\pm0.1°C$. This figure is determined from the characteristics of a batch of semiconductor material that has known resistance-temperature $(R-T)$ properties. To yield a given resistance at a standard temperature, the individual thermistors are precision-ground, with the XBT probe thermistors ground to yield 5,000 Ω (here, Ω is the symbol for the unit of ohms) at 25°C (Georgi et al., 1980). If the major source of XBT probe-to-probe variability can be attributed to imprecise grinding, then a single-point calibration should suffice to reduce this variability in the resultant temperatures. Such a calibration was carried out by Georgi et al. (1980) both at sea and in the laboratory.

To evaluate the effects of random errors on the calibration procedure, twelve probes were calibrated repeatedly. The mean difference between the measured and bath temperatures was $\pm0.045°C$ with a standard deviation of 0.01°C. For the overall calibration comparison, 18 cases of probes (12 probes per case) were examined. Six cases of T7s (good to 800 m depth and vessel speeds up to 30 knots) and two cases of T6s (good to 500 m depth and at less than 15 knots) were purchased new from Sippican, while the remaining 10 cases of T4s (good to 500 m and up to 30 knots) were acquired from a large pool of XBT probes manufactured in 1970 for the U.S. Navy. The overall average standard deviation for the probes was 0.023°C, which reduces to 0.021°C when consideration is made for the inherent variability of the calibration procedure.

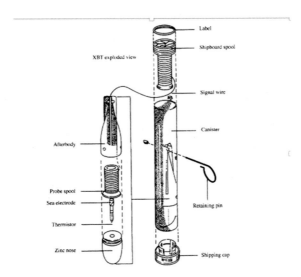

FIGURE 1.21 A cutaway diagram of the XBT showing the wire spools both on the probe (left) and on the canister that stays on the ship (right). The pin shown is pulled to release the probe.

A separate investigation was made of the resistance-temperature ($R-T$) relationship by studying the response characteristics for nine probes. The conclusion was that the $R-T$ differences ranged from $+0.011$ to $-0.014°C$, which means that the measured relationships were within $\pm0.014°C$ of the published relationship and that the calculation of new coefficients, following Steinhart and Hart (1968), is not warranted. Moreover, the final conclusions of Georgi et al. (1980) suggest an overall accuracy for XBT thermistors of $\pm0.06°C$ at the 95% confidence level and that the consistency between thermistors is sufficiently high that individual probe calibration is not needed for this accuracy level.

Another method of evaluating the performance of the XBT system is to compare XBT temperature profiles with those taken at the same time with a higher accuracy profiler such as a CTD system. Such comparisons are discussed by Heinmiller et al. (1983) for data collected in both the Atlantic and the Pacific using calibrated CTD systems. In these comparisons, it is always a problem to achieve true synopticity in the data collection since the XBT probe falls much faster than the recommended drop rate of around 1 m/s for a CTD probe. Most of the earlier comparisons between XBT and CTD profiles (Flierl and Robinson, 1977; Seaver and Kuleshov, 1982) were carried out using XBT temperature profiles collected between CTD stations separated by 30 km. For the probe, the sensor begins collecting data when the probe hits the water, and a seawater switch closes the circuit. The cap at the bottom of the launch tube is removed before deployment. For purposes of intercomparison, it is better for the XBT and CTD profiles to be collected as simultaneously as possible.

The primary error discussed by Heinmiller et al. (1983) is in the measurement of depth rather than temperature. There were, however, significant differences between temperatures measured at depths where the vertical temperature gradient was small, and the depth error should make little or no contribution. Here, the XBT temperatures were found to be systematically higher than those recorded by the CTD. Sample comparisons were divided by probe type and experiment. The T4 probes (as defined above) yielded a mean XBT-CTD difference of about $0.19°C$, while the T7s (defined above) had a lower mean temperature difference of $0.13°C$. Corresponding standard deviations of the temperature differences were $0.23°C$, for the T4s, and $0.11°C$ for the T7s. Taken together, these statistics suggest that XBTs are less accurate than the $\pm0.1°C$ given by the manufacturer and far less accurate than the $0.06°C$ reported by Georgi et al. (1980) from their calibrations.

From these divergent results, it is difficult to decide where the true XBT temperature accuracy lies. Because the Heinmiller et al. (1983) comparisons were made *in situ*, there are many sources of error that could contribute to the larger temperature differences. Even though most of the CTD casts were made with calibrated instruments, errors in operational procedures during collection and archival could add significant errors to the resultant data. Also, it is not easy to find segments of temperature profiles with no vertical temperature gradient and therefore it is difficult to ignore the effect of the depth measurement error on the temperature trace. It seems fair to conclude that the laboratory calibrations represent the ideal accuracy possible with the XBT system (i.e., better than $\pm0.1°C$). In the field, however, one must expect other influences that will reduce the accuracy of the XBT measurements and an overall accuracy slightly more than $\pm0.1°C$ is perhaps realistic. Some of the sources of these errors can be easily detected, such as an insulation failure in the copper wire, which results in single step offsets in the resulting temperature profile. Other possible temperature error sources are interference due to shipboard radio transmission (which shows up as high frequency noise in the vertical temperature profile) or problems with the recording system. Hopefully, these problems are detected before the data are archived in historical data files.

We comment that, until recently, most XBT data were digitized by hand. The disadvantage of this procedure is that chart paper recording doesn't fully realize the potential digital accuracy of the sensing system and that the opportunities for operator recording errors are considerable. Again, some care should be exercised in editing out these large errors, which usually result from the incorrect hand recording of temperature, date, time or position. It is becoming increasingly popular to use digital XBT recording systems, which improve the accuracy of the recording and eliminate the possibility of incorrectly entering the temperature trace. Such systems are described, for example, in Stegen et al. (1975) and Emery et al. (1986). Today, essentially all research XBT data are collected with digital systems, while the analog systems are predominantly used by various international navies.

The circuit diagram of the XBT probe is shown here in Figure 1.22. Supplying a constant current (about 30×10^{-6} Amp) through both the thermistor and resistor to ground allows voltages V_1 and V_2 to be measured relative to ground. Using $V = I \times R$ we can write

$$V_1 = I \cdot (R_{\text{thermistor}} + R_{\text{wire}} + R_{\text{water}}) \tag{1.25a}$$

$$V_2 = I \cdot \left(10^4 \text{ watts} + R_{\text{wire}} + R_{\text{water}}\right) \tag{1.25b}$$

The difference $V_2 - V_1 = I \cdot (10^4 \text{ W} - R_{\text{thermistor}})$, yields $R_{\text{thermistor}}$ and hence the ocean temperature once the calibration curve of the thermistor is applied. It is noteworthy that the individual thermistor is not calibrated but rather a subsample of say 200 out of 2,000 thermistors are calibrated and the average calibration is applied to all thermistors. The resistances of thermistors vary inversely with temperature.

The probe depth is calculated from the elapsed time. Because the probe lightens as its wire spools out, the depth is calculated from a second-degree polynomial:

$$z(\text{meters}) = 6.472t - 0.00216t^2 \tag{1.26}$$

where t is time in seconds (Sippican, 1994).

As mentioned earlier, the original recorders for the XBT were pressure sensitive chart paper with a pressure stylus that marked out the temperature as the paper rolled forward at the rate set by the time elapse and the depth given by Eqn (1.26). This analog system added to the inaccuracies of the XBT system and the Sippican Company then came out with a digital system. Soon, XBT users realized that they could build equivalent digital systems using a personal computer (PC) and a simple interface board. Thus, the digital data could be stored by the PC or plotted on an attached printer/plotter. These PC-based systems were a lot less expensive than the digital systems that were available from Sippican (Emery et al., 1986).

There were a variety of XBT probe launchers that were used depending on the setting. Perhaps the most widely used was the hand-held launcher (Figure 1.23), which gave the operator the freedom of moving the XBT launch to the downwind side of the ship. This was important as it was critical to keep the XBT wire away from the side of the ship. If the wire scraped the side of the ship, it was possible to scratch off the insulation dripped on the wire thus breaking the circuit and ending the temperature profile. This was a much greater danger if the probe was dropped on the windward side of the

FIGURE 1.22 Circuit diagram of an XBT probe.

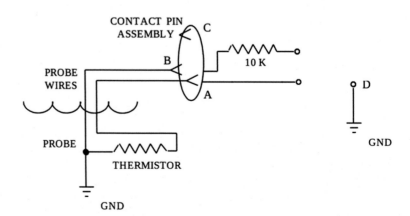

ship. Note in Figure 1.23 that the operator in the left picture is throwing the shipboard wire canister into the water before loading the new XBT unit. In the right picture, the operator is about the pull the pin that will release the XBT probe from the canister in the hand-held launcher.

The existence of a system that could easily measure the upper ocean temperature profile from a ship underway led to the establishment of "Ship of Opportunity Programs" (SOOPs), in which merchant vessels, such as the container vessel shown in the top panel of Figure 1.24, are equipped to collect XBT profiles (and other data) during their routine crossing of various ocean areas; a summary of the existing systems is presented in the bottom panel of Figure 1.24. SOOP programs were really enabled when the data collected became digital and there was no need to save paper output. The digital data

FIGURE 1.23 Two XBT launchers; vessel-mounted (left) and hand-held (right).

FIGURE 1.24 The World Meteorological Organization's (WMO's) Ship of Opportunity Program (SOOP). Top: The New Century 2, a 200 m long, 52,863 Gross Tonnage Vehicle Carrier that deploys XBTs in the Pacific Ocean. Bottom: Display of the components of a ship-of-opportunity XBT program showing (upper) a model of a merchant ship, (middle) components of the XBT that show the wire on the deck spool and within the clear probe and (bottom) a deck-mounted XBT launcher and PC based digital XBT recording system.

could then be transferred to shore via satellite data links and made available online to anyone interested in upper ocean temperature profiles from that part of the ocean. Initially, the XBT data collection was overseen by one of the ship's officers but it soon became apparent that the data collection could be automated. Multiple XBT launchers (Figure 1.25) were developed that could be activated from the ship's bridge or even automated to collect at specified times. GPS locations were coupled with the digital XBT data and all of it was then relayed to shore via a satellite link such as Iridium.

There are 49 "transects" that make up the global XBT sampling network (Figure 1.26). The lines in red are "high density" transects that are occupied at least 4 times per year and XBTs are deployed at approximately 25 km intervals along the ship's track. Frequently repeated transects are occupied approximately 18 times per year and XBTs are deployed at 100 km intervals. Previous "low density" transects (4 XBTs were deployed every 400 km along track) have been discontinued in favor of profiles from Argo profiling floats. Some XBT transects have been operating for more than 20 years. This XBT network requires collaboration between many organizations in many countries. The logistics and problems of the XBT network are unique but some aspects are common with other observational platforms such as surface drifters and Argo floats to be discussed later in this chapter. In recent years, 16,000−20,000 XBTs have been deployed each year. These upper ocean temperature profiles help determine seasonal and inter-annual fluctuations in the transport of mass and heat across the transects occupied.

1.3.3.5 Unmanned Aerial Vehicles

The advent of Unmanned Aerial Vehicles (UAVs) has dramatically expanded in recent years. Rules controlling the application of such platforms are changing such that they can be considered as a useful platform for many oceanographic applications. These planes have the advantage that they can fly low and slow over the ocean for very long periods of time making them ideal for the collection of space−time maps of oceanographic surface properties. At present, most sensors deployed on these planes have focused on surface expressions of oceanographic phenomena by looking at the optical and thermal infrared portions of the electromagnetic spectrum. Passive microwave sensors are being used for synoptic-scale

FIGURE 1.25   The "8 Shooter" multiple XBT launcher originating with NOAA's Atlantic Oceanographic and Meteorological Laboratory (AOML) in Miami, Florida.

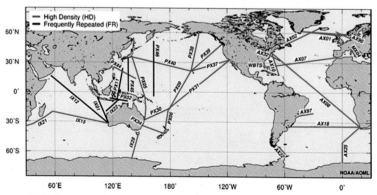

FIGURE 1.26   OceanObs09 recommended XBT transects (NOAA/AOML).

UAV applications and presently collect averaged information from the upper few cm of the ocean. Experimental programs are underway to develop packages that can be dropped from these aircraft to profile the upper portions of the ocean.

1.3.4 Salinity/Conductivity-Temperature-Depth (STD/CTD) Profilers

Resistance thermometers are widely used on continuous profilers designed to replace the earlier hydrographic profiles collected using a series of sampling bottles. The *in situ* electronic instruments continuously sample the water temperature, providing much higher resolution information on the ocean's vertical and horizontal temperature structure. Since density also depends on salinity, electronic sensors had to be developed to measure salinity *in situ* and were incorporated into the profiling system. As discussed by Baker (1981), an early electronic profiling system for temperature and salinity was described by Jacobsen (1948). The system was limited to 400 m depth and used separate supporting and data transfer cables. Next, a system called the STD (salinity-temperature-depth) profiler was developed by Hamon and Brown in the mid-1950s (Hamon, 1955; Hamon and Brown, 1958). The evolution of conductivity measurement, the basic parameter for the derivation of salinity, will be discussed in the section on salinity. This evolution led to the introduction of the conductivity-temperature-depth (CTD) profiling system (Brown, 1974). This change in name identified improvements, not only in the conductivity sensor, but also in the temperature sensing system designed to overcome the mismatch in the response times between the temperature and conductivity sensors. This mismatch often resulted in erroneous salinity spikes in the earlier STD systems (Dantzler, 1974). Most STD/CTD systems use a platinum resistance thermometer as one leg of an impedance bridge from which the temperature is determined.

An important development was made by Hamon and Brown (1958), whereby the sensing elements were all connected to oscillators that converted the measured variables to audio frequencies that could then be sent to the surface via a single conducting element in the profiler support cable. The outer cable sheath acted as both mechanical support and the return conductor. This data transfer method has subsequently been used on most electronic profiling systems, despite the ever-present concern for possible ground-fault problems. The early STDs were designed to operate to 1,000 m and had a temperature range of 0−30°C with an accuracy of ±0.15°C. Later STDs, such as the widely used Plessey Model 9040, had accuracies of 0.05°C with temperature ranges of −2 to +18°C or +15 to +35°C (range was switched automatically during a cast). Modern CTDs, such as the Sea Bird Electronics SBE 25*plus* and 911*plus* (Figure 1.27), the General Oceanics Idronaut Ocean Sciences 316 (modified after the EG&G Mark V), the Valeport Midas CTD, the Falmouth Scientific Integrated CTD profiler (FSI ICTD), the Applied Microsystems (AML) Plus v2 CTD, and the RBR (Branker) CTD, typically have accuracies of 0.001°C spanning a range of roughly −2 to +35°C. The glass-coated platinum alloy thermistor beads used in the SBE profilers are required to have a stability (drift) of less than 0.001°C within a range of −5 to +35°C. This compares with the drift of 0.001°C/month from earlier platinum resistance thermometers (Brown and Morrison, 1978; Hendry, 1993).

1.3.4.1 Dynamic Response of Temperature Sensors

Before considering more closely the problem of sensor response time for STD/CTD systems, it is worthwhile to review the general dynamic characteristics of temperature measuring systems. For example, no temperature sensor responds instantaneously to changes in the environment that it is measuring. If the environment temperature is changing, the sensor element lags in its response. This can be seen in the response of the combined platinum thermometer and miniature thermistor probe in Figure 1.28 (from Brown and Morrison, 1978). In this case, the response of the thermistor probe was designed to match the 25 ms response time of the conductivity cell. A simpler example is a reversing thermometer which, lowered through the water column, would at no time read the correct environment temperature until it had been stopped and was allowed to equilibrate with the surrounding water for some time. The time (K) that it takes the thermometer to respond to the temperature of a new environment is known as the response time or "time constant" of the sensor.

The time constant K is best defined by writing the heat transfer equation for the temperature sensor as

$$-\frac{dT}{dt} = \frac{1}{K}(T - T_w) \tag{1.27}$$

where T_w and T are the temperatures of the medium (water) and thermometer and t refers to the elapsed time. If we assume that the temperature change occurs rapidly as the sensor descends, the temperature response can be described by the integration of Eqn (1.27) from which:

$$\frac{(T - T_w)}{(T_0 - T_w)} = \frac{\Delta T}{\Delta T_0} = \exp\left(-\frac{t}{K}\right) \tag{1.28}$$

In this solution, T_0 refers to the temperature of the sensor before the temperature change and K is defined so that the ratio $\Delta T/\Delta T_0$ becomes e^{-1} (=0.368) when 63% of the temperature change, ΔT, has taken place. The time for the temperature sensor to reach 90% of the final temperature value can be calculated using $e^{-t/k} = 0.1$. A more complex case is when the temperature of the environment is changing at a constant rate; that is

$$T_w = T_1 + ct \tag{1.29}$$

where T_1 and c are constants. The temperature sensor then follows the same temperature change but lags behind so that

$$T - T_w = -cK \tag{1.30}$$

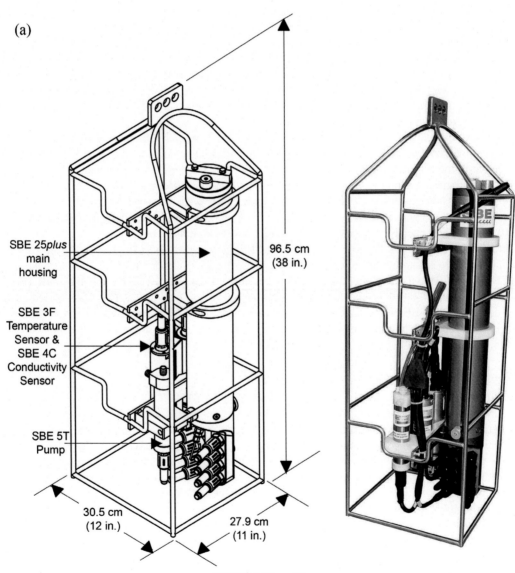

(a)

SBE 25*plus* main housing

SBE 3F Temperature Sensor & SBE 4C Conductivity Sensor

SBE 5T Pump

96.5 cm (38 in.)

30.5 cm (12 in.)

27.9 cm (11 in.)

FIGURE 1.27 cont'd.

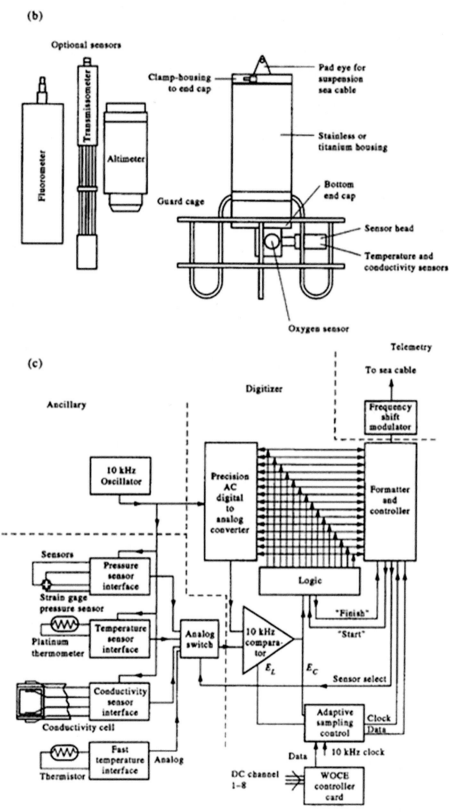

FIGURE 1.27 (a) Schematic (left) and photograph (right) of a Sea-Bird Scientific SBE 25*plus* Sealogger CTD. The SBE 25*plus* contains 8 analog inputs and 2 serial inputs for the integration of auxiliary sensors measuring dissolved oxygen, pH/ORP, nitrate, turbidity, PAR, beam transmittance, and a variety of fluorescence and scattering wavelengths. Analog (0-5V) and serial sensors from other manufacturers may also be integrated; (b) Schematic of General Oceanics MK3C/WOCE CTD and optional sensors; and (c) Schematic of electronics and sensors of General Oceanics MK3C/WOCE CTD. ORP, Oxidation-Reduction potential; DO, Dissolved oxygen. *(a) Courtesy: Janelle Hrycik, Sea-Bird Scientific, Inc (b,c) Courtesy: Dan Schaas and Mabel Gracia, General Oceanics.*

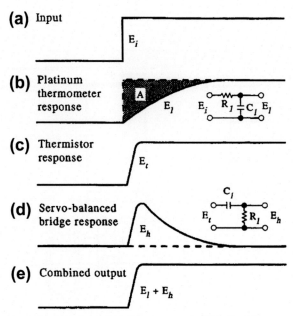

FIGURE 1.28   Combined output and response times of the resistance thermometer of a CTD (the different parts of the figure (a–e) are described to the right of each letter). *CTD*, conductivity-temperature-depth profiler.

The response times, as defined above, are given in Table 1.4 for various temperature sensing systems. Values refer to the time in seconds for the sensor to reach the specified percentage of its final value.

The ability of the sensor to attain its response level depends strongly on the speed at which the sensor moves through the medium. An example of the application of these response times is an estimate for the period of time a reversing thermometer must be allowed to "soak" in order to register the appropriate temperature. (Soaking allows the ship's motion to flush the sampling bottles and gives the thermometers time to settle to their final value.) Suppose we desired an accuracy of $\pm 0.01°C$ and that our reversing thermometer is initially $10°C$ warmer than the water. From Eqn (1.28), $0.01/10.0 = \exp(-t/K)$, so that for a 99% completed response, $t = 553$ s or 9.2 min. Thus, the standard recommended soaking period of 5 min (for a hydrographic cast) is set by thermometer limitations rather than by the imperfect flushing of the water sample bottles.

Suppose that the CTD thermistor listed in Table 1.4 is being used to profile in the thermocline, where the temperature change with depth is about $2°C/m$. To sense a change in temperature in the thermocline with a high-performance resolution of $0.001°C$ at the 90% response level for every meter of depth, the response equation requires that $\exp(-t/0.15) = 0.001/2.0$ from which we find $t = 1.14$ s. Thus, we have the usual recommendation for a CTD lowering rate of roughly 1 m/s. To achieve a 99% response, the CTD should be lowered at about 0.5 m/s.

The thermistors on the widely used SBE 911*plus* CTD are calibrated in a computer-controlled bath having a repeatable resolution of $0.0001°C$ and an accuracy of $0.0002°C$ (info@seabirdscientific.com). Calibrated sensor temperatures, $T\ (°C)$, are based on a third-order polynomial

$$T = 1 / \left\{ [a + b \ln(f_0 / f)] + c[\ln(f_0/f)]^2 + d[\ln(f_0/f)]^3 \right\} + T_{abs} \tag{1.31}$$

TABLE 1.4 Response times (in seconds) for various temperature sensors to reach 63%, 90% and 99% of their end points.

Device	$K_{63\%}$	$K_{90\%}$	$K_{99\%}$
Mechanical bathythermograph	0.13	0.30	0.60
SBE 911 plus CTD	0.065	0.15	0.30
Thermistor	0.04	0.08	0.16
Reversing thermometer	17.40	40.00	80.00

CTD = Sea-Bird SBE 911plus profiling Conductivity-Temperature-Depth system (standard lowering speed of 1 m/s).

where $a-d$ are constants, f is the CTD output frequency (in Hz), f_0 is an arbitrary scaling term historically set to the lowest frequency generated during the calibration procedure and set to $f_0 = 1,000$ Hz for ITS-90 temperatures and $T_{abs} = 273.15°$C is absolute temperature. The Sea-Bird calibration on 14 January 2022 for SBE 911*plus* CTD #5130 owned by the Institute of Ocean Sciences (Sidney, BC) yielded a residual (Bath-CTD) difference of $(0.1818182 \pm 6.867579) \times 10^{-5}$ °C with coefficients $a = 4.36024661 \times 10^{-3}$, $b = 6.40004881 \times 10^{-4}$, $c = 2.13647212 \times 10^{-5}$, and $d = 1.84972711 \times 10^{-6}$.

For completeness, we note that conductivity, C, in Siemens per meter (S/m) is calibrated in a bath using the relationship

$$C(S/m) = \left[a + bf^2 + cf^3 + df^4 \right] / \left[10(1 + \delta \cdot T + \varepsilon \cdot p) \right] \tag{1.32}$$

where for CTD #3500 calibrated on 16 February 2022, we have $a = -1.03816237 \times 10^1$, $b = 1.23686312$, $c = -1.14253601 \times 10^{-3}$, $d = 1.32314090 \times 10^{-4}$, $\delta = CT_{cor} = 3.2500 \times 10^{-6}$ and $\varepsilon = CP_{cor} = -9.5700 \times 10^{-8}$. The residual (Bath-CTD) difference was $(-2.44286 \pm 6.388643) \times 10^{-4}$ S/m. Similarly, the calibration of the pressure sensor on CTD #0506 on 29 January 2021 yielded a mean pressure difference of $-0.0002727273 \pm 0.239060998$ psia (1 PSIA $= 6.89475720$ kPa $= 0.689475720$ dbar).

1.3.5 Potential Temperature

The deeper one goes into the ocean, the greater the heating of the water caused by the compressive effect of hydrostatic pressure. The temperature for a parcel of water at depth is significantly higher than it would be in the absence of pressure effects. Potential temperature is the *in situ* temperature corrected for this internal heating caused by adiabatic compression as the parcel is transported to depth in the ocean. To a fairly reasonable approximation, the potential temperature defined as $\theta(p)$, or T_θ, is given in terms of the measured *in situ* temperature $T(p)$ as $\theta(p) = T(p) - F(R)$, where $F(R) \approx 0.1°$C/km is a function of the adiabatic temperature gradient R. The results can have important consequences for oceanographers studying water mass characteristics in the deep ocean. For example, measured ambient temperatures in the deeper waters of the 2,750 km long, maximum 6,669 m deep Middle America Trench off the west coast of Central America gradually increase with depth, whereas the calculated potential or Conservative temperature profiles reveal that the water column below about 4,000 m depth is typically "isothermal" to within 0.001°C.

The difference between the ambient temperature and θ increases slowly from zero at the ocean surface to about 0.5°C at 5,000 m depth (Table 1.5). For depth differences of approximately 100 m and temperatures less than 5°C, the difference between the two forms of temperature is roughly the absolute resolution ($\sim 0.01°$C) of most thermistors (Figure 1.29). Differences of this magnitude are significant in studies of deep ocean heating from hydrothermal venting or other heat sources where temperature anomalies of 10 millidegrees (0.010°C) are considered large. In fact, if the observed temperatures are not converted to potential temperature, it is impossible to calculate the anomalies correctly.

TABLE 1.5 Comparison of different forms of sigma-θ for the Western Pacific Ocean near Japan. Here, sigma, $\sigma_\theta = (\rho_\theta - 1,000)$, is determined relative to different pressure levels: the sea surface (σ_0), 2,000 dbar (σ_2), and 4,000 dbar (σ_4); ρ is in units of kg/m^3.

Depth (m)	In situ T (°C)	Potential T(0) (°C)	Salinity (psu)	σ_0	σ_2	σ_4
0	18.909	18.909	32.574	23.192	31.706	39.852
100	1.160	1.156	33.158	26.555	35.830	44.689
500	3.338	3.305	34.108	27.145	36.286	45.020
1,000	2.697	2.632	34.410	27.447	36.619	45.382
2,000	1.868	1.734	34.600	27.672	36.890	45.696
3,000	1.528	1.311	34.661	27.752	36.993	45.820
4,000	1.456	1.138	34.679	27.778	37.029	45.865
5,000	1.503	1.069	34.686	27.788	37.043	45.883
5,460	1.547	1.054	34.688	27.791	37.046	45.886

Columns 2 and 3 give the *in situ* and potential temperatures, respectively. Sigma units are kg/m^3. Practical Salinity was often written with no units; in TEOS-10 these psu values are in g/kg.
Data are from Talley et al. (1988).

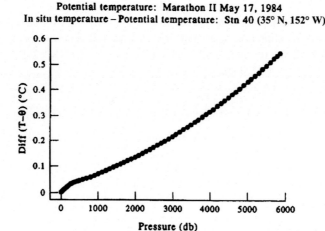

Potential temperature: Marathon II May 17, 1984
In situ temperature – Potential temperature: Stn 40 (35° N, 152° W)

FIGURE 1.29 Difference between the *in situ* temperature (*T*) recorded by a CTD versus the calculated potential temperature (*θ*) for a deep station in the North Pacific Ocean (35° N, 152° W). Below about 500 m, this curve is applicable to any region of the World Ocean. *CTD*, conductivity-temperature-depth profiler. *Data from Martin et al. (1987).*

1.4 SALINITY

It is the salt in the ocean that separates physical oceanography from other branches of fluid dynamics. Most oceanographers are familiar with the term "salinity" but many are not aware of its precise definition. Moreover, physical oceanographers often forget that salinity is a non-observable quantity and was traditionally defined by its relationship to a measurable parameter, "chlorinity". For the first half of the 20th century, chlorinity was measured by the chemical titration of a seawater sample. In 1899, the International Council for the Exploration of the Sea (ICES) established a commission, presided over by Professor M. Knudsen, to study the problems of determining salinity and density from seawater samples. In its report (Forch et al., 1902), the commission recommended that salinity be defined as follows: "The total amount of solid material in grams contained in 1 kg of seawater when all the carbonate has been converted to oxide, all the bromine and iodine replaced by chlorine, and all the organic material oxidized." The Thermodynamic Equation of Seawater-1980 (EOS-80) and the more recent Thermodynamic Equation of Seawater-2010 (TEOS-10) are, in large, methods for converting some measurable quantity (in this case the conductivity of water) to salt concentration.

Using the above definition, and available measurements of salinity, chlorinity (Cl), and density for a relatively small number of samples (a few hundred), the commission produced the empirical relationship (1.33)

$$S(\text{\textperthousand}) = 1.805Cl(\text{\textperthousand}) + 0.03 \tag{1.33}$$

known as Knudsen's equation and a set of tables referred to as Knudsen's tables. The symbol ‰ indicates "parts per thousand" (ppt) in analogy to percent (%) which is parts per hundred. In the Practical Salinity Scale, salinity is a unitless quantity but, because many oceanographers are uncomfortable with unitless values, salinity is sometimes written as "psu" for *practical salinity units*. (*Note*: Although we are aware that salinity is often written as a unitless variable, we are guilty of switching among the three formats in the text, especially, when discussing historical salinity measurements. We also find that including the unit *psu* helps eliminate ambiguity since it is clear that the variable under discussion is salinity, rather than some other nondimensional variable. The fact that TEOS-10 has switched back to unit defined variables is somewhat comforting!) It is also interesting to note that Knudsen himself considered using electrical conductivity (Knudsen, 1901) to measure salinity. However, due to the inadequacy of the apparatus available, or similar problems at the time, he decided that the chemical method was superior.

There are many different titration methods used to determine salinity but that most widely applied is the colorimetric titration of halides with silver nitrate ($AgNO_3$) using the visual end point provided by potassium chromate (K_2CrO_4), as described in Strickland and Parsons (1972). With a trained operator, this method is capable of an accuracy of ±0.02‰ in salinity using the empirical Knudsen relationship. For precise laboratory work, Cox (1963) reported on more sensitive techniques for determining the titration end point, which yielded a precision of 0.002‰ in chlorinity. Cox also describes an even more complex technique, used by the Standard Sea-water Service, which is capable of a precision of about ±0.0005‰ in chlorinity. It is fairly safe to say that these levels of precision are not typically obtained by the traditional titration method and that pre-conductivity salinities are generally no better than ±0.02‰ (±0.02 g/kg, or ±0.02 psu).

1.4.1 Salinity and Electrical Conductivity

In the early 1950s, technical improvements in the measurement of the electrical conductivity of seawater turned attention to using conductivity as a measure of salinity rather than the titration of chlorinity. Seawater conductivity depends on the ion content of the water and is therefore directly proportional to the salt content. The primary reason for moving away from titration methods was the development of reliable methods of making routine, accurate measurements of conductivity. The savings in time and labor has been exceptional. As noted earlier, the potential for using seawater conductivity as a measurement of salinity was first recognized by Knudsen (1901). Later papers explored further the relationship between conductivity, chlorinity, and salinity. A paper by Wenner et al. (1930) suggested that electrical conductivity was a more accurate measure of total salt content than of chlorinity alone. The authors' conclusion was based on data from the first conductive salinometer developed for the International Ice Patrol. This instrument used a set of six conductivity cells, controlled the sample temperature thermostatically and was capable of measurements with a precision of better than $\pm0.01\%_0$. With an experienced operator, the precision could be as high as $\pm0.003\%_0$ (Cox, 1963). The latter is a typical value for most modern conductive and inductive laboratory salinometers and is an order-of-magnitude improvement in the precision of salinity measurements over the older titration methods.

It is worth noting that the conductivity measured by either inductive or conductive laboratory salinometers, such as the widely used Guildline 8410A Portasal (portable) Salinometer, are relative measurements that are standardized by comparison with "standard seawater". As an outgrowth of the ICES commission on salinity, the reference, or standard seawater, was referred to as "Copenhagen Water" due to its earliest production by a group in Denmark. This standard water is produced by diluting a large sample of seawater, until it has a precise salinity of $35\%_0$ (Cox, 1963). Standard UNESCO seawater is now being produced by the "Standard Seawater Service" in England as well as at locations in the U.S.A. (e.g., the Woods Hole Oceanographic Institution).

Standard seawater is used as a comparison standard for each "run" of a set of salinity samples. To conserve standard water, it is customary to prepare a "secondary standard" with a constant salinity measured in reference to the standard seawater. A common procedure is to check the salinometer every 10−20 samples with the secondary standard and to use the primary standard every 50 or 100 samples. In all of these operations, it is essential to use proper procedures in "drawing the salinity sample" from the hydrographic water bottle into the sample bottle. Assuming that the hydrographic bottle remains well sealed on the upcast, two effects must be avoided: first, contamination by previous salinity samples (that have since evaporated leaving a salt residue that will increase salinity in the present bottle sample); and second, the possibility of evaporation of the present sample. The first problem is avoided by "rinsing" the salinity bottle and its cap two to three times with the sample water. Evaporation is avoided by using a screw cap with a gasket seal. A leaky *in situ* sample bottle will give values that are distorted by upper ocean values. For example, if salinity increases and dissolved oxygen decreases with depth, deep samples drawn from a leaky bottle will have anomalously low salinities and high oxygens.

Salinity samples are usually allowed to come to room temperature before being run on a laboratory bench salinometer. When running the salinity sample, one must be careful to avoid air bubbles and insure the proper flushing of the sample through the conductivity cell. Some bench salinometers correct for the marked influence of ambient temperature on the conductivity of the sample by controlling the sample temperature, while other salinometers merely measure the sample temperature in order to be able to compute the salinity from the conductivity and coincident temperature.

Another reason for the shift to conductivity measurements was the potential for *in situ* profiling of salinity. The development history of the STD/CTD profilers has been sketched out in Section 1.3.4 in terms of the development of continuous temperature profilers. The salinity sensing aspects of the instrument played an important role in the evolution of these profilers. The first STD (Hamon, 1955) used an electrode-type conductivity cell in which the resistance or conductivity of the seawater sample was measured and compared with that of a sample of standard seawater in the same cell. Fouling of the electrodes can be a problem with this type of sensor. Later designs (Hamon and Brown, 1958) used an inductive cell to sense conductivity. The inductive cell salinometer consists of two coaxial toroidal coils immersed in the seawater sample in a cell of fixed dimensions. An alternating current is passed through the primary coil, which then induces an electromagnetic force (EMF) and hence a current within the secondary coil. The EMF and current in the secondary coil are proportional to the conductivity (salinity) of the seawater sample. Again, the instrument is calibrated by measuring the conductivity of standard seawater in the same cell. The advantage of this type of cell is that there are no electrodes to become fouled. A widely used inductive type of STD was the Plessey model 9040, which claimed a salinity accuracy of $\pm0.03\%_0$. Precision was somewhat better, being between $0.01\%_0$ and $0.02\%_0$, depending on the resolution selected. Modern electrode-type cells measure the difference in voltage between conductivity elements at each end of the seawater passageway. With the conducting elements potted into the same material, this type of salinity sensor is less prone to contamination by biological fouling. At the same time, the response time of the conductive cell is longer than that of an inductive sensor, which leads to the problem of salinity spiking due to a mismatch with the temperature response.

The mismatch between the response times of the temperature and conductivity sensors is the primary problem with STD profilers. Spiking in the salinity record occurred because the salinity is computed from a temperature measured at a slightly different time than the conductivity measurement. Modern CTD systems record conductivity directly, rather than the salinity computed by the system's hardware, and have faster response thermal sensors. In addition, most modern CTD systems use electrodes rather than inductive salinity sensors. As shown in Figure 1.30, this sensor has a set of four parallel conductive elements that constitute a bridge circuit for the measurement of the current passed by the connecting seawater in the glass tube containing the conductivity elements. The voltage difference is measured between the conducting elements in the bridge circuit of the conductivity cell. The primary advantage of the conductive sensor is its greater accuracy and faster time response. Moreover, the mismatch problem is further mitigated in CTD systems, which use pumps to maintain a constant flow of water past the conductivity sensors. The flow rate is designed to minimize the mismatch between the conductivity and temperature sensors. In their discussion of the predecessor of the modern CTD, Fofonoff et al. (1974) give an overall salinity accuracy for this instrument of $\pm 0.003‰$. This accuracy estimate was based on comparisons with *in situ* reference samples whose salinities were determined with a laboratory salinometer also accurate to this level (Figure 1.31). Accuracies of this level are the same as the standard deviation of duplicate salinity samples run in the lab, demonstrating the high level of accuracy of CTD profilers.

1.4.1.1 A Comparison of Two CTDs

During sea trials in the North Atlantic, scientists at the Bedford Institute of Oceanography (Bedford, Nova Scotia) examined *in situ* temperature and salinity records from an EG&G Mark V CTD and a SBE 9 CTD (Hendry, 1993). The standards used for the comparisons were temperatures measured by SIS digital-reading reversing thermometers and salinity samples drawn from 10-l bottles on a Rosette sampler and analyzed using a Guildline Instruments Ltd. Autosal 8400A salinometer standardized with IAPSO Standard Water. Here, IAPSO stands for the International Association for the Physical Sciences for the Oceans.

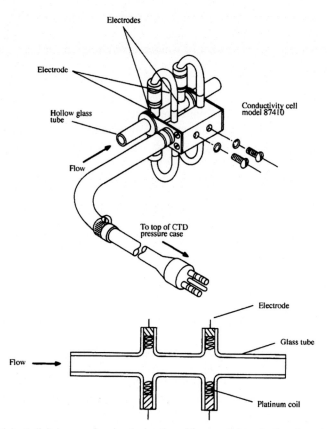

FIGURE 1.30 A Guildline conductivity (salinity) sensor showing the location of four parallel conductive elements inserted into the hallow glass tube. Conductivity is measured as the water flows through the glass tube. Cable plugs into the top of the CTD end plate on the pressure case. *CTD*, conductivity-temperature-depth profiler.

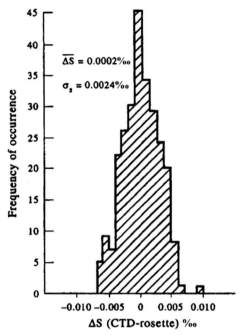

FIGURE 1.31 Histogram of salinity differences (in parts per thousand, ‰). Values used are the differences in salinity between salinity recorded by an early Neil-Brown CTD and deep-sea bottle samples taken from a Rosette sampler. $\overline{\Delta S}$ are the mean salinity differences and σ_s is the standard deviation. *Modified after Fofonoff et al. (1974).*

The Mark V was able to sample at 15.625 Hz and used two thermometers and a standard inductive salinity cell. The fast response (250 ms time constant) platinum thermometer was used to record the water temperature while the slower resistance thermometer, whose response time is more closely tuned to that of the conductivity cell, was used in the conversion of conductivity to salinity. Plots of the differences in temperature and salinity between the bottle samples and the CTD are presented in Figure 1.32. Using only the manufacturer's calibrations for all instruments, the Mark V CTD temperatures were lower than the reversing thermometer values by $0.0034 \pm 0.0023°C$ with no obvious dependence on depth (Figure 1.32a). In contrast, the Mark V salinity differences (Figure 1.32c) showed a significant trend with pressure, which may be related to the instrument used or a peculiarity of the cell. With pressure in decibar, regression of the data yields

$$\text{SalinityDiff}(\text{bottle} - \text{CTD}) = 0.00483 + 6.259 10^{-4} \text{ Pressure(CTD)} \tag{1.34}$$

with a squared correlation coefficient $r^2 = 0.84$. Removal of the trend gives salinity values accurate to about $\pm0.003‰$. Pressure errors of several decibars (several meters in depth) were noted.

The SBE 9 and SBE 25 sample at 24 Hz using a high-capacity pumping system and T−C duct to flush the conductivity cell at a known rate (e.g., 2.5 m/s pumping speed for a rate of 0.6−1.2/s). When on deck, the conductivity cell must be kept filled with distilled water. To allow for the proper alignment of the temperature and conductivity records (so that the computed salinity is related to the same parcel of water as the temperature), the instrument allows for a time shift of the conductivity channel relative to the temperature channel in the deck unit or in the system software Seasoft (module AlignCTD). In the Bedford study, the conductivity was shifted by 0.072 s earlier to align with temperature. (The deck unit was programmed to shift conductivity by one integral scan of 0.042 s and in the software the remaining 0.030 s.) Using the manufacturer's calibrations for all instruments, the SBE 9 CTD temperatures for the nine samples were higher than the reversing thermometer values by $0.0002 \pm 0.0024°C$ with only a moderate dependence on depth (Figure 1.32b). Salinity data from 30 samples collected over a 3,000-db depth range (Figure 1.32d) gave CTD salinities that were lower (fresher) than Autosal salinities by $0.005 \pm 0.002‰$, with no depth dependence. By comparison, the precision of a single bottle salinity measurement is $\pm0.0007‰$. Pressure errors were less than 1 dbar. Due to geometry changes and the slow degradation of the platinum on the electrode surfaces, the thermometer calibration is expected to drift by 2 m°C/year and the electronic circuitry by 3 m°C/year.

Based on the Bedford report, modern CTDs are accurate to approximately $\pm0.002°C$ in temperature, $\pm0.005‰$ in salinity and $<0.5\%$ of full-scale pressure in depth. The report provides some additional interesting reading on oceanic

FIGURE 1.32 CTD correction data for temperature (bottle–CTD) based on comparison of CTD data with *in situ* data from bottles attached to a Rosette sampler. (a) Temperature difference for the EG&G Mark V; (b) same as (a) but for the Sea-Bird SBE 9 CTD; (c) salinity difference for the EG&G Mark V; and (d) same as (c) but for the SBE 9 CTD. Regression curves are given for each calibration in terms of the pressure, *P*, in decibar; r^2 is the squared correlation coefficient. *Adapted from Hendry (1993).*

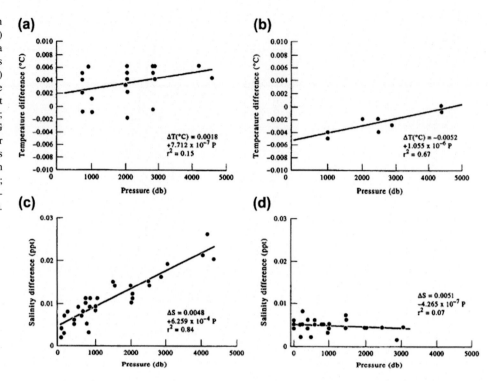

technology. To begin with, the investigators had considerable difficulty with erroneous triggering (misfiring) of bottles on the Rosette. Those of us who have endured this notorious "grounding" problem appreciate the difficulty of trying to decide if the bottle did or did not misfire and if the misfire registered on the shipboard deck unit. If the operator triggers the unit again after a misfire, the question arises as whether the new pulse fired the correct bottle or the next bottle in the sequence. Several of these misfires can lead to confusing data, especially in well-mixed regions of the ocean. It is good policy to keep track of the misfires for sorting out the data later.

According to repeat deep CTD-Rosette casts at Ocean Station "P" in the central northeast Pacific (50° N, 145° W) carried out by investigators at the Institute of Ocean Sciences (Sidney, British Columbia), there is good repeatability in the pumped SBE 911*plus* CTD systems. For example, variations referenced to common density surfaces at around 1,200 m depth during a June 2012 survey found temperature and salinity changes of 0.001°C and 0.0002 g/kg (‰) over a period of 4 h and 0.002°C and 0.0005 g/kg over a period of 2.5 days. Other years showed similarly small variations. No drift in the sensors was detected during post-cruise calibrations, and some of the change was attributed to slight changes in salinity values obtained from the Rosette bottle samples (Germaine Gatien and Steve Romaine, pers. comm., 2012).

1.4.2 The Practical Salinity Scale

As noted in Section 1.2, the Thermodynamic Equation of Seawater-2010 (TEOS-10), now recommended for oceanographic applications, is founded on the Practical Salinity Scale (PSS) devised in the 1970s. Because of this underlying foundation, it is useful to expand on the historical origins of the PSS. When using either chlorinity titration or the measurement of conductivity to compute salinity, one employs an empirical definition relating the observed variable to salinity. The need to move toward a salinity defined in terms of electrical conductivity, rather than chlorinity, led to the development of the Practical Salinity Scale in 1978 (Lewis and Perkin, 1978). Known as PSS-78, this defines salinity (*S*) in terms of the ratio K_{15} of the electrical conductivity of the seawater sample at the temperature of 15°C and a pressure of one standard atmosphere, to that of a potassium chloride (KCl) solution, in which the mass fraction of KCl is 32.4356×10^{-3}, at the same temperature and pressure. The K_{15} value exactly equal to 1 corresponds, by definition, to a practical salinity exactly equal to 35 psu (practical salinity units). Thus, the practical salinity is defined by the following equation:

$$S = 0.008 - 0.1692\, K_{15}^{1/2} + 25.3851\, K_{15} + 14.0941\, K_{15}^{3/2} - 7.0261\, K_{15}^{2} + 2.7081\, K_{15}^{5/2} \qquad (1.35)$$

This formulation was adopted by the UNESCO/ICES/SCOR/IAPSO Joint Panel on Oceanographic Tables and Standards, Sidney, B.C., Canada, Sept. 1—5, 1980 and endorsed separately by the International Association for the Physical Sciences of the Ocean (IAPSO) in December, the International Council for the Exploration of the Sea (ICES) in October, 1979, the Scientific Committee on Oceanic Research (SCOR) in September 1980 and the Intergovernmental Oceanographic Commission (IOC) of UNESCO in June, 1981. This equation is valid for a practical salinity S from 2‰ to 42‰.

In deriving this equation Lewis and Perkins (1978) concluded that while there was no unique solution to the "salinity problem" they suggested that any useful definition: (1) must be reproducible in any major laboratory throughout the world irrespective of the ionic content of local waters, (2) must be a conservative property, and (3) must allow density differences in any given water mass to be computed to acceptable limits. Earlier studies had already established that a "conductivity ratio" defined salinity scale is better than a "chlorinity" scale in terms of accurate density determination. In addition, Farland (1975) clearly demonstrated that in the hands of "average observers," titration is a much less precise procedure than is a conductivity measurement. To eliminate the ambiguity introduced by differing ionic makeups PSS-78 uses a salinity-conductivity ratio relationship; all waters of the same conductivity ratio have the same salinity regardless of ionic composition. It has been demonstrated (Millero, 1979) that the densities of water with the same absolute salinity are the same within experimental error. This is because the molal volumes of the various salts are not different enough to affect the density with the normal ranges of the ionic contents of seawater.

In light of the increased use of conductivity to measure salinity, and its more direct relationship to total salt content, a new definition of salinity had to be developed. As a first step in establishing the relationship between conductivity and salinity, Cox et al. (1967) examined this relationship in a variety of water samples from various geographic regions. These results were used in formulating new salinity tables (UNESCO, 1966) from which Wooster et al. (1969) derived a polynomial fit giving a new formula for salinity in terms of the conductivity ratio (K) at 15°C. The RMS deviation of this fit from the tabulated values was 0.002‰ in chlorinity for values greater than 15 and 0.005‰ for smaller values. It is worth noting that Cox et al. (1967) found that deep samples (>2,000 m depth) had a mean salinity, computed from chlorinity, that was 0.003‰ lower than that for conductivity.

As noted earlier, the "Practical Salinity Scale" or PSS-78 (Lewis, 1980) was quickly accepted by major oceanographic organizations and was recommended as the scale in which to report future salinity data (Lewis and Perkin, 1981). The primary objections to the earlier salinity definition of Wooster et al. (1969) were:

1. With salinity defined in terms of chlorinity it was independent of the different ionic ratios of seawater.
2. The mixtures of reference seawaters used to derive the relationship between chlorinity and conductivity ratio were nonreproducible.
3. The corresponding International Tables do not go below 10°C which makes them unsuitable for many *in situ* salinity measurements.

In the practical salinity scale, it is suggested that standard seawater should be a conductivity standard corresponding to, and having the same ionic content as, Copenhagen Water. The salinity of all other waters will be defined in terms of the conductivity ratio (R_{15} or C_{15} in the nomenclature of Lewis and Perkins) derived from a study of dilutions of standard seawater. This becomes then a practical salinity scale as distinct from an absolute salinity defined in terms of the total mass of salts per kilogram of solution.

A major problem in applying the practical salinity scale was its application to archived hydrographic data. As discussed by Lewis and Perkins (1981), the correction procedure for such data depends not only on the reduction formula used, but also on the calibration procedure used previously for the salinity instrument. Essentially, the correction procedure amounts to performing this calibration a second time using the differences, provided by Lewis and Perkins (1981) between the older salinity scale and PSS-78. Another alternative would be to return to the original raw data, if they have been saved, and to recompute the salinity according to PSS-78. From the discussion of Lewis and Perkins it is clear, however, that for salinities in the range of 33—37‰ differences of about ±0.01‰ (g/kg) can be anticipated between archived salinities and the corresponding values computed using PSS-78. This is about the same overall accuracy of modern CTD profilers.

It is interesting that the primary motivation for the development of PSS-78 came from people working in low salinity polar waters, where the UNESCO tables did not apply. In areas such as estuarine environments, which have very low salinities, and mid-ocean ridge regions with strong hydrothermal fluid venting, even PSS-78 is not adequate and there are still serious limitations to computing accurate salinities from conductivity measurements, even in the case of TEOS-10. There are several reasons for these limitations. First of all, the approach of relating specific conductance to salinity of total dissolved solids requires that the proportions of all the major ions in the natural electrolyte remain constant in time and space, and second, that the salinity expression represents all the dissolved solids in the fluid. Another factor to keep in mind is that the density calculated from the conductivity values is based on conducting ions in the fluid. If there are chemical components—or suspended particles—that contribute to the density of the fluid but not to the conductivity (i.e., nonconductive ions) then the density will be wrong. This is exactly the problem faced by McMannus et al. (1992) for deep CTD data from Crater Lake in Oregon. Here, silicic acid from hydrothermal venting in the south basin of the lake below 450 m depth did not contribute to the conductance but did alter the density. Without accounting for silicic acid the water column was weakly stratified; after accounting for it, the bottom waters became stratified. The different combinations of ions in hydrothermal fluids from mid-ocean ridges can seriously alter the salinity structure observed at the source during a normal deep CTD measurement. However, because of rapid entrainment of ambient bottom water, the ion mix becomes similar to that of normal seawater a few meters above the vent orifice. Typical ratios of ambient to hydrothermal fluid volumes are 7000:1. The vertical structure in buoyant plumes, such as that in Figure 1.33 taken a few meters above venting fields on Juan de Fuca Ridge in the northeast Pacific, presumably results from unstable conditions arising from turbulent mixing in the rising plume, not from sensor response problems.

Suspended particles can also contribute to increases in water density without affecting the "salinity". In recent studies, Thomson et al. (2010) and Thomson and Davis (2017) suggest that prolonged, along-axis episodic flows of 0.2 m/s and greater observed at depths of over 4,000 m at the bottom of the Middle America Trench off Central America are likely rotationally modified, auto-suspending turbidity currents initiated by tidal current resuspension of sediments above the 100-km long segment of the trench that shoals by 1,000 m to the southeast of their mooring site. Suspended particles in the turbidity currents are estimated to be about 0.0003—0.006% by volume. CTD measurements reveal that the potential density of the water is uniform in the lower 1,000 m of the trench and therefore unlikely to account for such strong bottom currents. Because of the strong density flows, the submersible *Alvin* was sometimes forced to hang onto bottom-mounted equipment to stop the vehicle from drifting down-trench while working on deep Ocean Drilling Program (ODP) borehole observatories 1253 and 1255 in the region (Earl Davis, pers. com., 2009). Results suggest that tidally induced turbidity currents may be common to steep, well-mixed regions of the deep ocean adjacent to sediment rich continental margins.

1.4.3 Nonconductive Methods

Efforts have been made to infer salinity directly from measurements of refractive index and density. Since the refractive index (n) varies with the temperature (T) and salinity (S) of a water sample (and with the wavelength of the illumination), measurements of n and T can be used to obtain *in situ* estimates of salinity. In order to achieve a salinity accuracy of ± 0.01 g/kg (‰), it is necessary to measure n to within 20×10^{-7} and to control temperature to within $\pm 0.005°C$. Some refractometers are capable of measuring n to 100×10^{-7}, leading to a salinity precision of 0.06 g/kg. Handheld

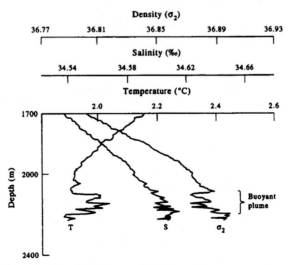

FIGURE 1.33   Vertical profiles of temperature, salinity (‰) and potential density collected from a CTD mounted on the submersible *Alvin*. Data collected during ascent away from the main hydrothermal vent field at Endeavour Ridge in the northeast Pacific (48° N, 129° W). Density is unstable over the depth range of the buoyant portion of the plume. *From Lupton et al. (1985).*

refractometers are simple and easy to use but yield salinity measurements no better than ± 0.2 g/kg. For higher sensitivity, interference methods can be used giving a precision for n of 5×10^{-7} corresponding to a salinity precision of ± 0.003 g/kg. This is a comparative interference technique and requires a reference seawater sample. Because it is a comparative method, knowledge of the exact temperature is not critical as long as both samples are observed at the same temperature. Direct measurements of seawater density can yield a precision of ± 0.008 in sigma-t (Kremling, 1972), which can be used to calculate salinities to within ± 0.02 g/kg. As the measurement of density is much more complicated than those of temperature and electrical conductivity, these latter quantities are usually observed and used to compute the *in situ* density.

1.4.4 Remote Sensing of Salinity

The measurement of sea surface salinity (SSS) from space was first attempted on Skylab (Lerner and Hollinger, 1977) using a 1.4 GHz microwave radiometer. Although many of the corrections needed to adjust for ambient atmospheric and oceanographic conditions were not well understood in the 1970s, the strong correlation between the sensor data (after correcting for other effects) and SSS were sufficiently encouraging to proceed with the technology. Spacecraft remote sensing of SSS using low-frequency microwave radiometry was first proposed by Swift and McIntosh (1983). The AMSR-E was placed in orbit on the Aqua satellite in 2002 but because of its limited ocean salinity measurement accuracy, it was not suitable for measuring the small surface salinity gradients of the open ocean. However, AMSR-E data taken over the Amazon River plume were originally used to demonstrate the feasibility of measuring ocean surface salinity with microwave radiometers from space (Reul et al., 2009; Klemas, 2011).

Two satellite missions are measuring (or have measured) SSS: the European Space Agency (ESA) Soil Moisture and Ocean Salinity (SMOS) mission launched in November 2009 and the NASA Aquarius mission launched in June 2011 (and ended June 8, 2015). Both missions were designed to last about 3 years but only SMOS continues as of July 2022. The satellites provide large-scale SSS measurements at accuracies of ± 0.5 g/kg for SMOS and ± 0.2 g/kg for Aquarius, at temporal scales of weeks to months over spatial scales of 50–100 km. The SMOS mission is measuring soil moisture over continental surfaces and surface salinity over the ocean, while the Aquarius mission observed SSS only. SMOS uses a unique interferometric antenna system to avoid the need for a very large antenna to measure the salinity signal. Aquarius uses a large reflector to sense the same low frequency microwave signals as SMOS.

The Microwave Imaging Radiometer with Aperture Synthesis on the SMOS satellite measures the passive microwave emission of the earth's surface (the brightness temperature, T_b) at a frequency of 1.400–1.427 GHz in the L-band, which is the band best suited to determining SSS. In this case, radiation downwelling from space at ~ 1.4 GHz impinges on the ocean surface and is partly absorbed within an ~ 1 cm thick upper layer and is partially reflected toward the atmosphere. Both moisture and salinity decrease the microwave radiation emitted from the earth's surface. SMOS collects data at 6 AM and 18 PM (local time) each day and the surface of the earth is fully covered every 3 days. Aquarius has an active microwave system coupled to the passive system, yielding a sensitivity that is a least an order of magnitude better than SMOS but over a larger spatial footprint and a longer 8-day repeat cycle for full coverage of the earth.

The polarized radiometric brightness temperature (the microwave radiation from the sea surface) used to determine SSS is given by $T_b = e \times T_s$ where T_s is the actual SST, and e is the emissivity of the sea surface, which itself is a function of SSS and T_s. The effects on T_b from ocean roughness and Faraday rotation must also be taken into consideration. Once the apparent brightness temperature of a body of water is measured, and where the thermodynamic SST is measured by other means, the salinity at the surface can be determined. For typical oceanic ranges of surface salinity and T_s, T_b has a brightness range of about 4–6 K at L-band frequencies. The sensitivity of T_b to changes in SSS versus T_s is greatest in warm water (0.7 K/(g/kg)) at 30°C) and least in cold water (0.3 K/(g/kg) at 0°C).

Salinity measurements from space are complicated by several factors, of which the most significant is the effect of sea surface roughness on microwave emission (Yueh et al., 2001). Sea surface roughness associated with wind waves, swell, and foam from breaking waves give rise to a change in the apparent brightness temperature of up to ~ 5 K. Because of the short timescales of ocean winds, accurate measurements of sea surface roughness must be simultaneous with the brightness measurements. These stringent requirements pose technical challenges for achieving the required radiometric accuracy and stability. Finally, the low frequency involved requires the use of very large antennas (or an innovative antenna solution such as was used in SMOS) to achieve a moderate spatial resolution on the ground. The brightness temperature at L-band for a space orbiting radiometer is also affected by galactic emissions ($\Delta T_b \approx 2$–8 K), atmospheric emissions ($\Delta T_b \approx 2.4$–2.8 K), and emission from water vapor and cloud liquid water (both of which have a small effect on ΔT_b). The influence of surface roughness, and hence wind, has led to a recent release of ocean surface wind speed calculated from Soil Moisture and Ocean Salinity (SMOS) measurements on 4 August 2021.

1.4.4.1 Satellite Microwave Soil Moisture and Ocean Salinity (SMOS) Technology

At the heart of SMOS is the Microwave Imaging Radiometer using Aperture Synthesis (MIRAS) which exploits the principle of interferometry to achieve the performance of a much larger antenna using interferometry of 69 Lightweight Cost-Effective Front-end (LICEF) receivers installed on 3 arms extending out from the central satellite housing (Figure 1.34).

The interferometric method measures the phase differences for the component receivers and using cross-correlations from all possible combinations of receiver pairs a two-dimensional "measurement image" is taken every 1.2 s. As the satellite moves along its orbital path, each observed area is seen under various viewing angles (Figure 1.35). To achieve the desired spatial resolution at L-band, a very large antenna would be required (the solution employed by Aquarius). For SMOS, however, the antenna aperture has been synthesized through a multitude of small antennas (receivers). MIRAS consists of a central structure with three deployable arms. Three of the "antennas" are also able to operate as accurate, highly stable Noise Injection Radiometers (NIR) and they are evenly distributed over the three arms and central structure. The overall antenna structure extends up to 8 m. By taking advantage of the interferometric technique, the receivers combine to give a spatial resolution similar to a filled-aperture antenna with the same overall dimensions. Because MIRAS uses the interferometric method, the raw image has areas of varying resolution.

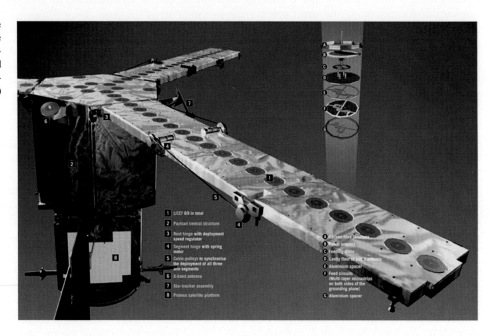

FIGURE 1.34 The Microwave Imaging Radiometer using Aperture Synthesis (MIRAS) instrument antenna components with an exploded view of a single Lightweight Cost-Effective Front-end (LICEF) component.

FIGURE 1.35 Soil moisture and ocean salinity (SMOS)-microwave imaging radiometer using aperture synthesis (MIRAS) field of view (FOV). *Image from CNES.*

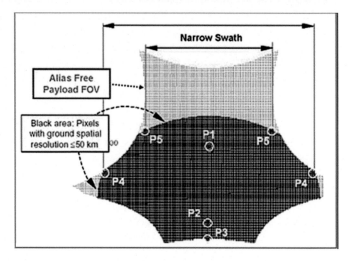

A snapshot of brightness temperature of the Field of View (FOV) is taken every 0.3 s with an average radiometric resolution of about 5 K over 200 K. Due to platform motion in orbit, each pixel is measured several times with different spatial and radiometric resolutions and incidence angles. The minimum FOV along-track dimension is about 800 km, which corresponds to 380 snapshots at 0.3 s each. Ground data processing is needed to account for all the observation variations in size, shape, angles, overlapping conditions, weighting and averaging schemes, resulting eventually in a brightness temperature (T_b) map for each snapshot as well as for the accumulated and averaged along-track observation incidences within a FOV. The repetitive measurements scheme of radiation, from varying footprints and incidences, has a similar summation effect on the retrieval of the overall signal as a Time Delay Integration (TDI) scheme for an optical imager.

The MIRAS is designed to operate in three modes. The full polarization mode has been the default mode since June 2010. The three modes are:

1. Dual-polarization mode, in which receivers are switched synchronously to either Horizontal (H) or Vertical (V) polarization.
2. Full polarimetric mode, in which segments of the array are switched according to a predefined sequence between H and V.
3. Calibration mode, in which measurements of the internal load, the noise diodes, or the so-called "fringe washing function" are determined.

Global comparisons between SMOS and Aquarius (Figure 1.36) show good general agreement but do have some detailed differences. Aquarius provides better pixel accuracy than SMOS, whereas SMOS provides more rapid revisit times and a greater spatial resolution. These satellite-only surface salinity maps should be compared with the surface salinity map based on all the *in situ* data collected over many years (Figure 1.37). While many of the large-scale features are similar, the

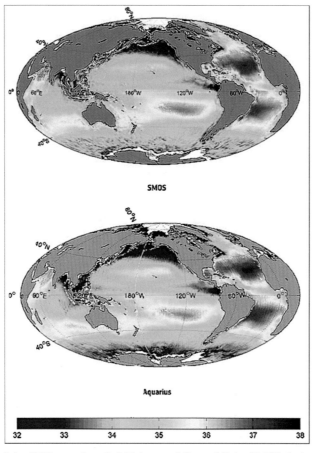

FIGURE 1.36 Global sea surface salinity (SSS) maps from Soil Moisture and Ocean Salinity (SMOS) (top) and Aquarius (bottom). *Image credits: IFREMER, ESR, ESA, NASA.*

FIGURE 1.37 Annual mean sea surface salinity (SSS) based on data from the World Ocean Atlas 2009. SSS values are in Practical Salinity Units (PSU), where PSU corresponds to g/kg).

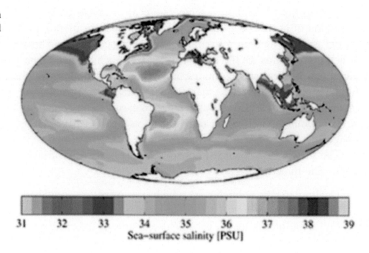

amplitudes are completely different. Some of the detailed features of the satellite-only fields do not even appear in the historical data map.

1.4.4.2 NASA's Aquarius Satellite

The Aquarius SSS instrument flew on the Argentinian SAC-D satellite mission and was a collaborative mission between the U.S. and Argentina. It was launched 10 June 2011, and ended on 8 June 2015, when the power and attitude control system of the spacecraft stopped operating. Aquarius was designed to measure and map SSS for at least 3 years with a spatial resolution of 150 km, repeating its global pattern every 7 days. The satellite flew in a sun-synchronous orbit 657 km above the Earth. The Aquarius instrument was jointly built by NASA's Jet Propulsion Lab (JPL) and NASA's Goddard Space Flight Center (GSFC). JPL managed Aquarius through the mission's commissioning phase and archived mission data. GSFC managed the mission's operations and routinely processed Aquarius data.

Aquarius was the primary instrument on the SAC-D spacecraft. It consisted of three passive microwave radiometers to detect the surface emission that was used to obtain salinity and an active scatterometer to measure the ocean waves that affect the precision of the salinity measurement. While SSS generally range from 32 to 37 psu (g/kg), the Aquarius sensor was able to detect changes in SSS as small as 0.2 psu.

1.4.5 Satellite Salinity Summary

In summary, calculation of the sea surface salinity (SSS) involves: (1) determination of the polarized radiometric brightness temperature, T_b, at the sea surface by correcting for ionospheric, atmospheric and extraterrestrial radiation; (2) correcting for sea surface roughness and SST; and (3) calculation of SSS from T_b. Radiative transfer models allow correction for up- and downwelling emission from the atmosphere, for atmospheric and ionospheric attenuation, and for Faraday rotation of the polarized microwave emissions as the radiation passes through the ionosphere. Downwelling galactic emissions are taken into account using maps provided by radio astronomers. With these corrections, and knowledge of the SST and roughness, salinity can be calculated from the brightness temperature, T_b.

1.5 DENSITY

It is the density of the ocean that exerts a controlling influence on the currents in the ocean but because it can't be observed directly, it has to be computed from other variables that are directly observable. That is why we have first discussed temperature and salinity, the two variables that, along with the pressure, are needed to compute seawater density. Density increases linearly with pressure, increases with increasing salinity but generally decreases with increasing temperature. Because the first two digits of seawater density, $\rho \sim 1{,}025 \text{ kg/m}^3$, never change, oceanographers often work with the quantity

$$\sigma_{S,T,p} = \rho - 1{,}000 \text{ kg/m}^3 \tag{1.36}$$

termed the "density anomaly", where S and T are the *in situ* salinity and temperature at the pressure, p, of the measurement At standard atmospheric pressure, the anomaly is known as "sigma-t" (σ_t), where

$$\sigma_t = \rho_{S,T,0} - 1,000 \text{ kg/m}^3 \tag{1.37}$$

Related to density is the coefficient of thermal expansion, α, which can be written as

$$\alpha = -\rho^{-1}(\partial\rho/\partial T)_{p,S}, \tag{1.38}$$

where $\partial\rho/\partial T$ is computed while holding constant the subscript variables p and S. In addition, the compressibility of seawater can be written as

$$\gamma = \rho^{-1}(\partial\rho/\partial T)_{T,S}. \tag{1.39}$$

1.5.1 Equation of State-1980

Before oceanographers were encouraged to move to the TEOS-10 definition of the equation of state introduced in Section 1.2, the 1980 Equation of State was widely used to define density. The equation uses T in °C and S from the Practical Salinity Scale, with pressure, p, in dbar (1 dbar $= 10^4$ Pa $= 10^4$ Nm^{-2}) and ρ in kg·m^{-3}. The ranges in the ocean are: $-2°C \leq T \leq 40°C$ and $0 \leq p \leq 10^5$ kPa (depth 0–10,000 m). Based on laboratory measurements, this formulation for density, ρ, is said to be accurate to at least 9×10^{-3} kg·m^{-3} (Talley et al., 2011). A graphical view of this relationship is presented in Figure 1.38, which shows the curved σ_t density lines as functions of temperature and salinity. Also shown are the temperature of maximum density and the freezing point. The shaded region represents 90% of the World Ocean, demonstrating that the range of ocean characteristics is relatively limited. The mean T, S and most common T, S are also indicated.

We note that the σ_t lines are more nonlinear with respect to T than to S. The maximum density meets the freezing line at $S = 24.7$ psu and, after crossing this density maximum, water becomes lighter toward the surface and eventually freezes over, while the deep basins of the ocean are filled with water of maximum density. Today it is relatively straightforward to calculate ocean variations using density directly. In the past, it was convenient to use the related variable of specific volume, defined as

$$\alpha = \rho^{-1} \text{ m}^3/\text{kg} \tag{1.40}$$

Again, it was most useful to subtract out the reference ocean value and deal with the "specific volume anomaly" defined as

$$\delta = \alpha_{S,T,p} - \alpha_{35,0,p} = \delta_s + \delta_T + \delta_{S.T} + \delta_{S,p} + \delta_{S,T,p} \tag{1.41}$$

In general, the last three terms are small enough that they can be neglected, leading to the definition of the "thermosteric anomaly," given by

$$\Delta_{S,T} = \delta_s + \delta_T + \delta_{S.T} \tag{1.42}$$

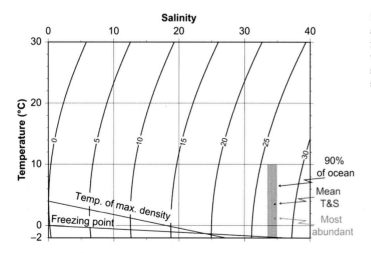

FIGURE 1.38 Density lines (curves) as functions of temperature and salinity. Also shown are the lines of maximum density and the freezing point. The shaded box represents 90% of the World Ocean while a single point corresponds to the mean temperature and salinity of the ocean. The most common values of T and S are also shown (Pickard and Emery (1982).)

1.5.2 Potential Density

In Section 1.3.5 on temperature, we introduced the concept of potential temperature as being the actual thermodynamic temperature corrected for the adiabatic temperature increase due to the greater pressure at depth. Technically, the potential density of a fluid parcel at pressure p is the density of that parcel if it was adiabatically brought to a reference pressure p_0, often the 1 bar (100 kPa) pressure at the sea surface. However, using the ocean surface as a reference level to calculate the potential density (ρ_θ) is not reasonable if one is comparing conditions at a very deep depth. In such cases, a deeper pressure surface is used, where for example, ρ_4, is the potential density at a depth of 4,000 m. The potential density ρ_4 is then the density that a parcel of water would have if it were raised (or lowered) adiabatically to the isobaric surface of 4,000 dbar ($\sim 4,000$ m). Regardless of the chosen reference level, it is assumed that the adiabatic movement to the reference pressure takes place with no change in salinity. Therefore, the potential density is calculated using the potential temperature and the *in situ* salinity.

A good example of the importance of potential temperature and potential density is given in Table 1.6, which presents measured values at depths greater than 1,400 m of *in situ* temperature, potential temperature, and densities σ_0, σ_4, and σ_{10}. Note that the temperature differences become significant below 5,000 m, while there is relatively little difference in salinity. The increase in temperature below 4,685 m depth makes the water column appear unstable, which is actually a consequence of adiabatic heating and not a real change of oceanic thermal conditions. Note also the slight increase even in the potential temperatures in the last four rows, which is also manifested in σ_0 but is corrected in σ_4 and σ_{10}.

The consequences of using different depth references are well demonstrated by Figure 1.39, where the top panel represents conditions using the sea surface as the reference and the lower panel using 4,000 m as the reference level. The water parcels denoted by the dots represent Mediterranean Water (saltier) and Nordic Seas Water(fresher) at their sill depths. The lines labeled (1) and (2) in the top panel have the same density at either end relative to the surface for the two sites, while they are different at the σ_4 level in the bottom panel. It should be noted that cold water is slightly more compressible than warm water.

In the case where two water parcels that have identical densities but very different T and S mix, the result is a new water mass with a different density (Figure 1.40). This process is known as cabbeling. During cabbeling, seawater almost always gets denser if it is slightly colder or slightly saltier. Medium-warm, medium-salty water can be denser than both fresher, colder water and saltier, warmer water because the equation of state of seawater is monotonic and non-linear. Because pure water is densest at about 4°C, cabbeling can also occur in fresh water. For example, a mixture of 1°C water with 6°C water that combines to form fresh water with a temperature of 4°C will be denser than either of the parent water masses. Ice is also less dense than water, so although ice floats in warm water, meltwater sinks in warm water (https://en.wikipedia.org/wiki/Cabbeling).

1.5.3 Neutral Density

Neutral density (γ^n) is a density variable introduced by Jackett and McDougal (1997). It is a function of three state variables (salinity, temperature and pressure) and the geographical location (longitude and latitude) and has the usual units

TABLE 1.6 Comparison of *in situ* temperature versus potential temperature and potential density relative to the sea surface (σ_0), 4,000 dbar (σ_4), and 10,000 dbar (σ_{10}) in the Mariana Trench in the western North Pacific.

Depth (m)	Salinity (psu)	Temperature (°C)	Potential temperature (°C)	σ_0 (kg m^{-3})	σ_4 (kg m^{-3})	σ_{10} (kg m^{-3})
1,487	34.597	2.800	2.695	27.591	45.514	69.495
2,590	34.660	1.730	1.544	27.734	45.777	69.903
3,488	34.680	1.500	1.230	27.773	45.849	70.015
4,685	34.697	1.431	1.028	27.800	45.898	70.090
5,585	34.699	1.526	1.004	27.803	45.904	70.099
6,484	34.599	1.658	1.005	27.803	45.904	70.099
9,940	34.700	2.266	1.007	27.804	45.904	70.099

Data are from the R/V T. Washington, 1978.

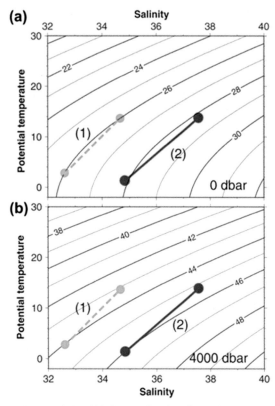

FIGURE 1.39 Potential density relative to (a) 0 dbar and (b) 4,000 dbar as a function of potential temperature (relative to 0 dbar) and *in situ* salinity. The points on each line in the top panel (a) have the same density, while in the lower panel (b) the densities are different at the ends of the lines. *From Talley et al. (2011).*

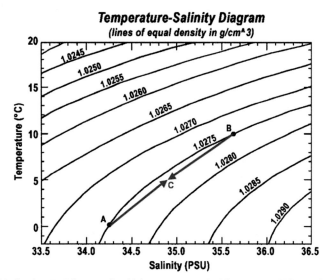

FIGURE 1.40 Visualization of cabbeling in a *T, S* diagram. Combining water mass A with water mass B in equal portions forms water mass C, which is denser than either A or B. https://en.wikipedia.org/wiki/Cabbeling.

of density (mass/volume). Isosurfaces of γ^n form density surfaces along which water masses flow in the deep ocean. This is equivalent to saying that water follows isentropic surfaces. Because these surfaces are difficult to define, we instead substitute isopycnal surfaces to trace the distributions of water masses. This is similar to the previously used method (Reid, 1994) that approximated neutral surfaces by linked sequences of potential density surfaces referred to a discrete set of reference pressures. If a water parcel is followed along a γ^n surface back to its starting point, the depth difference is about 10 m whereas if one uses potential density surfaces, the difference can be hundreds of meters.

1.5.4 Buoyancy and Static Stability

Because the ocean is strongly vertically stratified, the vertical gradient of density is almost always negative ($\partial\rho/\partial z < 0$) and water tends to remain at the depth level of its density. If a water parcel is displaced downward, it will experience a force pushing it upward to regain the depth consistent with its density. This buoyancy force was first discovered by Archimedes in 212 BCE, who stated that "Any object, wholly or partially immersed in a fluid, is buoyed up by a force equal to the weight of the fluid displaced by the object." We can write this buoyancy force in mathematical form as

$$F = \Delta V(\rho_1 - \rho_2) \tag{1.43}$$

where ΔV is the water parcel's volume, ρ_1 is its initial density and ρ_2 the density of the displaced parcel given by

$$\rho_2 = \rho + (\partial\rho/\partial p)_\theta \cdot \Delta p \tag{1.44}$$

where $\Delta p = -\rho g \Delta z$, from which we obtain the well know hydrostatic balance:

$$(\partial p/\partial z) = -\rho g \tag{1.45}$$

To understand the effects of static stability in the ocean, we consider a parcel of water that is displaced vertically in a fluid at rest and in hydrostatic equilibrium. Initially, the parcel has the same thermodynamic properties as the surrounding fluid that is at the same vertical level. Once displaced from its initial position, the parcel properties will typically be different from those at the new water level. The following assumptions apply to the parcel:

1. The parcel retains its identity and does not mix with the environment.
2. The parcel motion does not disturb the environment.
3. The pressure p of the parcel adjusts instantaneously to the ambient pressure of the fluid surrounding the parcel.
4. The parcel moves isentropically, so that its potential temperature remains constant.

The symbol for Static Stability, E, is defined by:

$$E = -\rho^{-1}(\partial\rho/\partial z) - g/c^2 \tag{1.46}$$

where c is the speed of sound in water. For this expression we have the following three ocean conditions: Stable: $E > 0$; Unstable: $E < 0$: and Neutral: $E = 0$. Note that, for a stable ocean for which $E > 0$,

$$(\partial\rho/\partial z) < (\partial\rho/\partial z)_{adiabatic} \tag{1.47}$$

As a consequence, a stable layer in the water column should have a vertical density lapse rate whose magnitude is larger (i.e., more negative) than the adiabatic lapse rate as defined earlier in the temperature section. Another way of writing E is

$$E = -\rho^{-1}\left[(\partial\rho/\partial z)_{water} - (\partial\rho/\partial z)_{parcel}\right] \tag{1.48}$$

Gill (1982) and McDougall (1987). In the upper kilometer of the ocean, the stability is high and the first term in (1.48) is much greater than the second. The first term represents the rate of vertical change of density in the water column while the second term is proportional to the compressibility of seawater, which is very small. Neglecting the second term, we can write the stability equation as

$$E \approx -\rho^{-1}\left[(\partial\rho/\partial z)_{water}\right] \tag{1.49}$$

This approximation is valid for $E > 50 \times 10^{-8}\,\text{m}^{-1}$. In the upper kilometer of the ocean ($z < 1{,}000$ m), $E = (50-1{,}000) \times 10^{-8}\,\text{m}^{-1}$, while in deep trenches ($z > 7{,}000$ m), $E \sim 1 \times 10^{-8}\,\text{m}^{-1}$.

1.5.5 The Stability or Brunt-Väisälä Frequency

The Brunt-Väisälä frequency, or buoyancy stability frequency, commonly written as $N(z)$, quantifies the importance of stability in determining internal oscillations within the water column. Basically, this frequency is that at which a fluid parcel will freely oscillate if vertically displaced. Consider a parcel of water that has a density ρ_0. This parcel is in an environment where the density is given by $\rho = \rho(z)$. If the parcel is displaced by a small vertical increment, z', and it maintains its original density (no thermal or haline change), so that its volume does not change, it will be subject to an extra gravitational force relative to its surroundings of:

$$\rho_0 \frac{\partial^2 z'}{\partial t^2} = -g\left[\rho(z) - \rho(z+z')\right] \tag{1.50}$$

where g is the gravitational acceleration. If we make the linear approximation

$$\rho(z+z') - \rho(z) = \frac{\partial \rho(z)}{\partial z} z' \tag{1.51}$$

and substitute (1.51) into (1.50), we obtain

$$\frac{\partial^2 z'}{\partial t^2} = \frac{g}{\rho_0} \frac{\partial \rho(z)}{\partial z} z' \tag{1.52}$$

This second-order differential equation has straightforward solutions of the form

$$z' = z'_o e^{it\sqrt{N^2}} \tag{1.53}$$

where $N(z)$, is given by

$$N(z) = \sqrt{-\frac{g}{\rho_o} \frac{d\rho}{dz}}, \text{or } N^2 = -gE \tag{1.54}$$

and where $\rho = \rho_\theta(z)$ is the potential density that depends upon both temperature and salinity. When perturbed vertically from its starting position, the water parcel experiences a vertical acceleration. If the acceleration is back toward the initial starting position, the stratification is said to be stable ($N^2 > 0$) and the parcel oscillates vertically with an angular frequency N. If the parcel is accelerated away from the initial position and the vertical stratification is unstable ($N^2 < 0$), vertical overturning or convection will soon ensue. The frequency N is the maximum frequency of internal waves in the ocean relative to the background stratification in the absence of a more slowing varying ambient current. Typical values of N are a few cycles per hour or around 0.005 radians per second (Figure 1.41). The periods of internal waves for the stability values given earlier are: 10 min for $E = 1,000 \times 10^{-8}$ m^{-1} and 6 h for $E = 1 \times 10^{-8}$ m^{-1}.

1.5.6 Dynamic Height

Dynamic height was created as an oceanic analog to geopotential height in the atmosphere and is only relevant when calculating and mapping geostrophic currents. For example, a dynamic meter (dyn·m) is defined as

$$D(\text{dyn} \cdot \text{m}) = gh/10 \ (\text{m}^2\text{s}^{-2}) \tag{1.55}$$

where h is in meters and the acceleration due to gravity, g, in m/s^2. (Note that, if we want work in units of dynamic centimeters (dyn·cm), we have to change the definition so that the denominator on the right side is 100 rather than 10 to balance the change in g. In addition, h must be in centimeters.) We can write pressure in terms of the dynamic height as

$$p = \int_{D_0}^{D} \rho dD \tag{1.56}$$

where D_0 is the dynamic height of the sea surface, which is the uppermost isobaric surface with sea level pressure zero. Conversely, we can define the dynamic height in terms of the pressure:

$$D = \int_{p_0}^{p} \alpha dp \tag{1.57}$$

where α = specific volume. Traditionally, D has been broken into two parts:

$$D = D_{35,0,p} + \Delta D, \tag{1.58}$$

where the first term represents the contribution of a standard ocean having a uniform salinity of 35‰ (35 psu) and uniform temperature of 0°C. The second term is the departure of the ocean from this standard ocean and is known as the dynamic height anomaly. Here, ΔD is the only variable between pressure values of p_0 and p. Hence, there are two ways to present

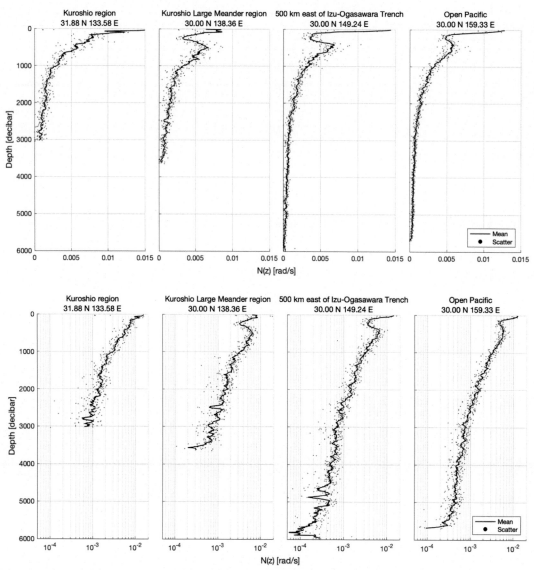

FIGURE 1.41   Examples of the stability frequency, $N(z)$, as a function of depth (z) from west to east along Line P02 ($\sim 30°$ N) in the North Pacific Ocean in 2022, beginning in the Kuroshio region. Dotted values are at every 8 m depth, corresponding to each ADCP-CTD depth bin, and the solid lines are smoothed values averaged over 48 m ($= 6 \times 8$ m bins). *Top panels*: Stability frequency profiles in radians/second; *Bottom panels*: as in the top panels but on a log-scale; i.e., log(rad/s), showing details in the bottom layer. Conversions are: 1 dbar $\approx$ 1.02 m; 1 rad/s $= 1/(2\pi)$ Hz $= 572.958$ cph (1 cph $= 0.0017453$ rad/s). *Analysis and plots are courtesy of Kurtis Anstey, University of Victoria.*

the relative pressure field of the ocean: (a) by constructing isobaric maps that give the pressure distribution on given level surfaces below the sea surface; or (b) by constructing topographic maps for given isobaric surfaces; i.e., maps showing lines of equal D on an isobaric surface beneath the sea surface.

We note that the dynamic height expressed by Eqn (1.57) is measured relative to the sea surface, which, except for small sea level changes associated with the atmospherically induced inverse barometer effect, is basically a level where pressure is uniform over large spatial scales. However, it is clear that dynamic height would be more useful if it had a depth-integrated value at the ocean surface that was measured relative to some deeper level that had no spatial pressure gradient. The conversion of dynamic height relative to the sea surface to the dynamic height oceanographers use to map ocean circulation takes place when a deeper layer is assumed to be the reference level and all the intervening layers are corrected so that the height difference is now expressed at the sea surface. Thus, in this traditional manner, geostrophic currents are referred to as "relative currents" because they are relative to the assumption that some deeper level provides a uniform, broad-scale reference level. This can be assumed to be the greatest depth of the density profiles (i.e., the observed T and S profiles)—a very limiting assumption—or it can be a level at which the current is measured directly and used to determine the offset for all the other depths. This procedure is well described in Pond and Pickard (1983) for two stations in the North Atlantic Current extending eastward from the terminus of the Gulf Steam. The tables for these two stations are repeated in Tables 1.7 through 1.8. Here, we assume that we have hydrographic station data down to 1,000 m in the northern hemisphere at two different stations, A and B, separated by 50 km along a line of latitude. We can write the relative geostrophic velocity, V_{rel}, between these two stations at level "j":

$$V_{rel,j} = \frac{10}{L \, 2\Omega \, \sin\phi} \left[\Delta D_{B,j} - \Delta D_{A,j} \right] \tag{1.59}$$

where L is in meters, $\Omega = 0.72921 \times 10^{-4}$ radians/s (0.04178 cph) is earth's angular rate of axial rotation, ϕ is the latitude, and the relative velocity, $V_{rel,j}$, is in m/s. The velocity is perpendicular to, and proportional to, the "dynamic height gradient" between stations A and B (Figure 1.42).

The previous equation can also be written in terms of the geopotential anomaly, $\Delta\Phi$, which has the benefit of preserving units, whereas the dynamic height anomaly mixes units requiring a change in coefficients to compensate for the change in the units for gravity. In terms of the geopotential, (1.59) can be written as

$$V_{rel,J} = \frac{1}{L \, 2\Omega \, \sin\phi} \left[\Delta\Phi_{B,j} - \Delta\Phi_{A,j} \right] \tag{1.60}$$

TABLE 1.7 Hydrographic data and the calculation of geopotential for Station A.

Station A 41°55′ N, 50°09′ W				Units of $10^{-8}\,m^3kg^{-1}$				Units of $m^3kg^{-1}\,Pa = m^2s^{-2}$		
Depth (m)	T (°C)	S (psu)	σ_t	Δ_{ST}	δ_{SP}	δ_{TP}	δ	Avg δ	Avg $\delta \times \Delta P$	$\Delta\Phi_{jA}$
0	5.99	33.71	26.56	148	0	0	148			6.638
								146	0.365	
25	6.00	33.78	26.61	144	0	0	144			6.273
								135	0.338	
50	10.30	34.86	26.81	125	0	1	126			5.935
								126	0.315	
75	10.30	34.88	26.83	123	0	2	125			5.620
								122	0.305	
100	10.10	34.92	26.89	117	0	2	119			5.315
								112	0.560	
150	10.25	35.17	27.06	101	0	3	104			4.755
								99	0.455	
200	8.85	35.03	27.19	89	0	4	93			4.300
								83	0.830	
300	6.85	34.93	27.41	68	0	5	73			3.470
								65	0.650	
400	5.55	34.93	27.58	52	0	5	57			2.820
								52	1.040	
600	4.55	34.95	27.71	39	0	7	46			1.780
								45	0.900	
800	4.25	34.95	27.74	37	0	8	45			0.880
								44	0.880	
1,000	3.90	34.95	27.78	33	0	10	43			0

TABLE 1.8 Hydrographic data and calculation of geopotential for Station B.

Station B 41°28′ N, 50°09′ W			Units of $10^{-8} \text{m}^3\text{kg}^{-1}$					Units of $\text{m}^3\text{kg}^{-1}\,\text{Pa} = \text{m}^2\text{s}^{-2}$		
Depth (m)	T (°C)	S (psu)	σ_t	Δ_{ST}	δ_{SP}	δ_{TP}	δ	Avg δ	Avg $\delta \times \Delta P$	$\Delta\Phi j_B$
0	13.04	35.62	26.88	118	0	0	118			7.894
								119	0.298	
25	13.09	35.63	26.88	118	0	1	119			7.596
								119	0.298	
50	13.07	34.63	26.88	118	0	1	119			7.298
								119	0.298	
75	13.05	35.64	26.89	117	0	2	119			7.000
								120	0.300	
100	13.05	35.62	26.88	118	0	3	121			6.700
								122	0.610	
150	13.00	35.61	26.88	118	0	4	122			6.090
								122	0.610	
200	12.65	35.54	26.90	116	0	5	121			5.480
								117	1.170	
300	11.30	35.36	27.02	105	0	7	112			4.310
								98	0.980	
400	8.30	35.09	27.32	76	0	7	83			3.330
								70	1.400	
600	5.20	34.93	27.61	49	0	8	57			1.930
								52	1.030	
800	4.20	34.92	27.73	38	0	8	46			0.900
								45	0.900	
1,000	4.20	34.97	27.77	34	0	10	44			0

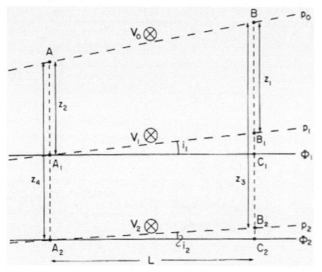

FIGURE 1.42 Schematic of geostrophic velocity, $V_{rel,j}$, calculated for level "j" between stations A and B for a deep horizontal reference level of 1,000 m. *Adapted from Pond and Pickard (1983).*

To illustrate how this calculation is performed, we use two tables from Pond and Pickard (1983) that demonstrate well how the actual dynamic heights are not correctly calculated until a reference level is assumed and applied. The last columns on the right in both Tables 1.7 and 1.8 show how the columns are inverted, with zero now being at 1,000 m. Here, the difference $[\Delta\Phi_B - \Delta\Phi_A]/g$ is an estimate of the height difference (dynamic topography) of isobaric surfaces at the two stations. If we take $g = 10$ ms^{-2}, this difference means that, between stations A and B, the dynamic surface only differs by 0.13 m relative to the zero level at 1,000 m ($\sim 10^4$ kPa) and the dynamic surface is 0.13 m higher at station B relative to station A. This emphasizes how small elevation differences in dynamic height can be responsible for very strong geostrophic ocean currents.

In Tables 1.7 and 1.8, the depth of 1,000 m was taken as a level of "no motion" which means there should be no horizontal pressure gradient between stations A and B at that level. In the second and third columns are the temperature and salinity from which the specific volume anomaly (σ_t) was calculated and entered in the fourth column. Columns 5−10 contain the dynamic equivalents, while the last column has been corrected for the assumption of no motion at 1,000 m. Notice how the dynamic anomaly (δ) is at a minimum at 1,000 m. Finally, we can calculate the geopotential height anomalies and enter them into the final column.

Once we have the "adjusted" geopotential anomalies, we can compute the differences between stations A and B and then compute the relative geostrophic velocities knowing the station latitudes and the Coriolis parameters (Table 1.9). The geostrophic current will be directed perpendicular to the line between stations A and B, with the higher value of dynamic height on the right in the northern hemisphere (on the left in the southern hemisphere). While this geostrophic current is not the only current present at this location, it is a significant component of the current that varies more slowly than do the wind driven currents. These are the same concepts that are applied to altimeter data in computing the geostrophic currents that they reflect. Instead of assuming a "level of no (or known) motion" with altimeter data, the procedure is to define a reference surface for the globe from which the altimeter data is corrected. As gravity models improve with the collection of satellite gravity data, the need for an arbitrary reference surface will be eliminated and altimeter data will be used to directly calculate geostrophic ocean currents.

It is important to keep in mind that estimating geostrophic currents using geopotential surface estimation involves many assumptions. To begin with, the method deals with a gradient, with no consideration of how the gradient was created. Thus, we are considering a balance of forces and not a Newtonian force and reaction. Next, the method assumes that there are no additional temporal changes to this gradient and that there are no frictional forces involved. Finally, we have assumed that

TABLE 1.9 Geopotential anomalies, $\Delta\Phi$, from Tables 1.7 and 1.8, and the calculated mean relative velocities, V_{rel}, between stations A and B at the various listed depths. Here, the separation distance $L = 50$ km, the mean of $\sin\phi = 0.665$. See Eqn 1.60 calculating V_{rel}.

Depth	$\Delta\Phi_B$	$\Delta\Phi_A$	$\Delta\Phi_B - \Delta\Phi_A$	V_{rel}
(m)	(m^{2}s^{-2})	(m^{2}s^{-2})	(m^{2}s^{-2})	(ms^{-1})
0	7.894	6.638	1.256	0.26
25	7.596	6.273	1.323	0.27
50	7.298	5.935	1.363	0.28
75	7.000	5.620	1.380	0.29
100	6.700	5.315	1.385	0.29
150	6.090	4.755	1.335	0.28
200	5.480	4.300	1.180	0.24
300	4.310	3.470	0.840	0.17
400	3.330	2.820	0.510	0.11
600	1.930	1.780	0.150	0.03
800	0.900	0.880	0.020	0.005
1,000	0	0	0	0

all of these motions are of a large enough spatial scale and are changing so slowly in time that the Coriolis force becomes a dominant factor. Geographic currents are then simply a balance between the pressure gradient force and the Coriolis force.

1.6 DEPTH OR PRESSURE

1.6.1 Hydrostatic Pressure

The depths recorded by profiling instruments are mainly derived from measured hydrostatic pressure, p. This is possible because of the almost linear relationship between hydrostatic pressure, $p = p(z)$, and geometric depth, z. The relationship is such that the "pressure expressed in decibars is nearly the same as the numerical value of the depth expressed in meters" (Sverdrup et al., 1942). The validity of this approximation can be seen in Table 1.10 in which we have compared values of hydrostatic pressure and geometric depth for a standard ocean. At depths shallower than 4,000 m, the difference is less than 2%. For many applications, this error is sufficiently small that it can be neglected and hydrostatic pressure values can be converted directly into geometric depth. The cause for the slight difference between pressure in decibars and depth in meters is found in the familiar hydrostatic relation,

$$p(z) = -g \int_z^0 \rho(z)dz \qquad (1.61)$$

Where $g = 9.81$ m/s^2 is the acceleration due to gravity and $\rho \approx 1,025$ kg/m^3 is the mean water density. Units of p are (kg/m^3) (m/s^2) (m) = (kg/m) (l/s^2). Also, p = force/area has units (N/m^2) = (Pa) = (10^{-5} bar). One Newton (N) = 1 kg m/s^2, so that $p \approx 1025 \cdot (9.81) \cdot z = 1.005525 \times 10^4 \cdot z$ (Pa) = $1.005525 \cdot z$ (dbar), where depth z is expressed in meters. A different value of density (or gravity) gives a slightly different p versus z relation. Based on the above formulation, z(m) $\approx p$(dbar)/$1.005525 = 0.99451 \cdot p$(dbar).

Certain techniques allow for continuous measurement of hydrostatic pressure while others can be carried out at discrete depths only. An example of the latter is the computation of "thermometric depth" using a combination of protected and unprotected reversing thermometers to sense the effects of pressure on the temperature reading. This is still considered one of the most accurate methods of determining hydrostatic pressure and is often used as an *in situ* calibration procedure for CTD profilers. Specifically, the pressure, p (in decibars, dbar), obtained from a CTD is related to the temperature difference, ΔT, between the protected and unprotected thermometers by $p \approx g\Delta T/k$, where the pressure constant for each

TABLE 1.10 Comparison of pressure (dbar) and depth (m) at standard oceanographic depths using the UNESCO algorithms. As a rough approximation, an increase in pressure of 1 dbar corresponds to an increase in depth of 0.99451 m.

Pressure (dbar)	Depth (m)	Difference (%)
0	0	0
100	99	1
200	198	1
300	297	1
500	495	1
1,000	990	1
1,500	1,483	1.1
2,000	1,975	1.3
3,000	2,956	1.5
4,000	3,932	1.7
5,000	4,904	1.9
6,000	5,872	2.1

Percent difference = (pressure in dbar − depth in meters)/pressure in dbar × 100%.

individual thermometer is $k \approx 0.1°C/(kg/cm^2)$. The details of this procedure are well described in Sverdrup et al. (1942, p. 350) but with a significant printing error (a missing plus sign in the second bracket), which is not corrected in Defant (1961, Vol. 1, p. 35). When correctly applied (see LaFond, 1951; Keyte, 1965), the thermometer technique is capable of yielding pressure measurements accurate to ±0.5% (Sverdrup, 1947). Most modern CTD systems claim a similar accuracy using strain-gauge sensors to directly measure pressure. The accuracy of early CTD pressure sensors was a function of the depth (pressure) itself and varied from 1.5 dbar in the upper 1,500 dbar to over 3.5 dbar below 3,500 dbar (Brown and Morrison, 1978). A test of a Sea-Bird SBE 9 CTD (Hendry, 1993) found pre- and post-cruise pressure calibration offsets of less than 1 dbar. Nonlinearity and hysteresis were less than 0.5 dbar over the full range of the sensor. A January 2021 calibration of the pressure sensor on a SBE 9*plus* CTD (SN 0506) by Sea-Bird Scientific in their Bellevue (Washington) facility yielded a difference between the applied and measured pressure of only -2.7273×10^{-4} pounds per square inch absolute (psia), equivalent to -1.8804×10^{-4} dbar, with a small standard deviation of ±0.2391 psia (equivalent to ±0.1648 dbar ≈ ±0.15 m). The calibration range was 14−9,951 psia (roughly 9.7−6,861 dbar) at a constant temperature of 21.2°C.

The mechanical bathythermograph (MBT) introduced earlier in this chapter measures pressure with a Bourdon tube sensor. The problem with this sensor is that the response of the tube to volume change is nonlinear and any alteration in tube shape or diameter will lead to changes in the pressure response. As a result of the nonlinear scaling of the MBT, pressure readout required a special optical reader (unique to the individual instrument) to read the scales; this read-out error added to the inaccuracies of the Bourdon tube, resulting in the limited accuracy of the MBT.

1.6.2 Free-Fall Velocity: The Expendable Bathythermograph (XBT)

Unlike the MBT, the more commonly used XBT, does not measure depth directly but rather infers it from the elapsed time of a "freely falling" probe. While this is a key element that makes such an expendable system feasible, it is also a possible source of error. In their study, Heinmiller et al. (1983) first corrected XBT profiles for systematic temperature errors and then compared the XBT profiles with corresponding CTD temperature profiles. In all cases, the XBT isotherm depths were less than the corresponding CTD isotherm depths for observations deeper than an intermediate depth (150 m for T4s and 400 m for T7s), with the largest differences at the bottom of each trace. Near the bottom of the XBT temperature profile, the difference errors exceeded the accepted limit of 2% error, with the deviations being far greater for the shallower T4 probes (Figure 1.43a). Added to this systematic error is an RMS depth error of approximately 10 m regardless of probe type (Figure 1.43b). Based on the data they analyzed, Heinmiller et al. (1983) provided a formula to correct for the systematic depth error. There are two primary sources of this depth error: first, the falling probe loses weight (and density) as the wire runs out of the probe supply spool, thus changing the fall rate; and second, frictional forces increase as the probe enters more dense waters. The increasing length of copper wire being paid out behind the probe must also add to the net drag on the probe.

The issue of XBT depth error, first reported by Flierl and Robinson (1977), has been extensively investigated by many groups (Georgi et al., 1980; Seaver and Kuleshov, 1982; Heinmiller et al., 1983; Green, 1984; Hanawa and Yoritaka, 1987; Roemmich and Cornuelle, 1987; Hanawa and Yoshikawa, 1991; Hanawa and Yasuda, 1991; Rual, 1991; Hallock and Teague, 1992; Thadathil et al., 2002; Kizu et al., 2010; Santos et al., 2018; Reseghetti et al., 2018) with varying results There is general agreement that the XBT probes fall faster than specified by the manufacturer and that some corrections are needed. Most of these assessments have been performed as a comparison with nearly coincident CTD profiles. The concentration assessments in the western Pacific Ocean show the interest in this problem in Japan, Australia, and Nouméa (New Caledonia).

A sample comparison between XBT and CTD temperature profiles (Figure 1.44a) shows the differences between the XBT and CTD temperature profiles as a function of depth. These profiles have not been corrected using the standard depth equation. The sample in Figure 1.44b has been depth corrected using the formulation given by Hanawa and Yoritaka (1987). Note the substantial changes in the shape of the difference profile and how the depth correction eliminates apparent minima in the differences. The overall magnitude of the differences has also been sharply reduced, demonstrating that many of the apparent temperature errors are, in reality, depth errors.

These XBT depth errors are known to be functions of depth since they depend on an incorrect fall-rate equation. This is clearly demonstrated in Figure 1.45, which gives the mean depth difference of a collection of 126 simultaneous temperature profiles, along with the standard deviation (shown as bars that represent the standard deviation on either side of the mean line) at the various depths. Also shown are the ±2%, or ±5 m limits, which are given as depth error bounds by the manufacturer. From this figure, it is clear that there is a bias with the XBT falling faster than specified by the fall-rate

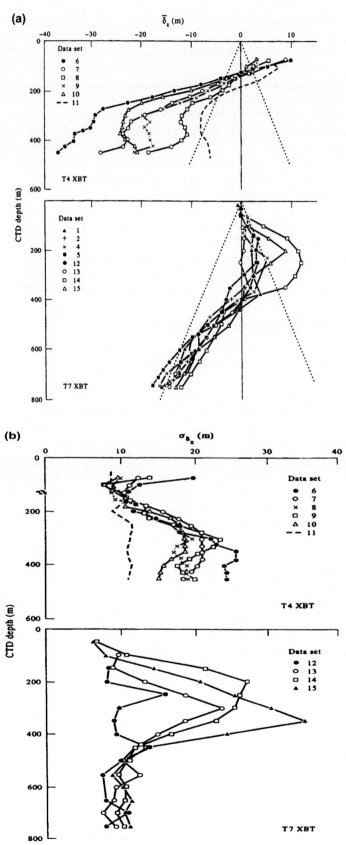

FIGURE 1.43 Vertical profiles of XBT-CTD depth differences for T4 and T7 XBTs for different data sets. (a) Mean values, $\bar{\delta}_z(m)$; (b) standard deviation $\overline{\sigma}_z(m)$.

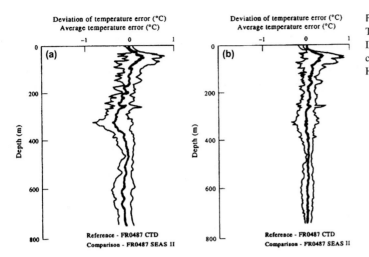

FIGURE 1.44 Average temperature error profiles (TXBT-TCTD) for XBT/CTD comparisons on FR0487 using the SEAS II XBT systems; center line gives the mean value. (a) Depth uncorrected; (b) depth corrected using the formulation given by Hanawa and Yoritaka (1987). *From IOC-888, Annex IV, p. 6.*

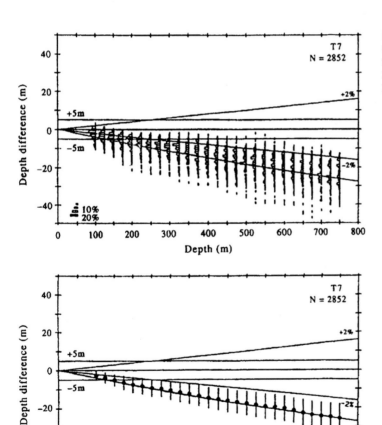

FIGURE 1.45 Depth error ($z_{CTD} - z_{XBT}$) for selected points on temperature profiles. *Upper panel*: XBT depths calculated using the Sippican Fall-Rate Equation (FRE); *Lower panel*: XBT depths calculated with the Hallock and Teague (1992) FRE.

equation, resulting in negative differences with the CTD profiles. The mean depth error of 26 m at 750 m depth translates into 3.5%, which obviously exceeds the manufacturer's specification.

Various investigators reduced these comparisons to new fall-rate equations (FREs), for which the depth (z) is given in terms of elapsed time, t, by

$$z = at - bt^2 \qquad (1.62)$$

and found that the coefficients were not very different (Figure 1.46). Along with the coefficients, this figure also shows contours of maximum deviations in depth relative to the revised equation of Hanawa and Yoshikawa (1991) for these different combinations of constants a and b in the fall-rate Eqn (1.61). Most of the errors in Figure 1.46 lie within the ± 10 m envelope of depth deviations, suggesting that it might be possible to develop a new fall-rate equation (i.e., new coefficients) that represents a universal solution to the fall-rate problem for XBT probes. An effort was made to develop this universal equation by reanalyzing existing XBT-CTD comparison profiles. This revised equation is

$$z = 6.733t - 0.00254t^2 \tag{1.63}$$

This revised fall-rate equation only applies to the T7 (roughly 700 m depth) XBT probes that were used in the comparisons. It was concluded that similar comparisons must be carried out for the other types of XBT probes.

In their study, Hallock and Teague (1992) found that the XBTs fall faster than specified by the Sippican fall rate equation. They subsequently developed a new fall-rate equation from both theoretical considerations and coincident XBT and CTD temperature profile comparisons in an area near Barbados. The Sippican fall-rate equation was found to underestimate the depth (z) by as much as 35 m at $z = 760$ m, leading them to formulate a new fall-rate equation:

$$z = 6.733t - 0.00238t^2 - 4.01 \tag{1.64}$$

which was valid for depths greater than 10 m. Hallock and Teague (1992) also found considerable probe-to-probe variability that was likely due to variations in the probe body changing the effective drag as the probe fell through the water column. To compare the XBT temperature profiles with CTD temperature profiles, the authors selected points on the profiles that were essentially the inflection points and made a comparison with the Sippican fall-rate equation (FRE) (Figure 1.47).

Hallock and Teague (1992) found that their FRE resulted in a residual RMS depth error of about 5.2 m. The RMS error is believed to be the result of probe-to-probe differences in the terminal velocity. This variation diminishes the relative importance of difference in fall-rate. Further investigation is required to determine the cause of this variability in probe terminal velocities. The authors also recommended further study of the near-surface transient problem for probe depths shallower than 10 m.

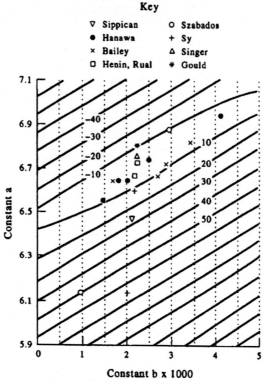

FIGURE 1.46   Fall-rate equation (FRE) coefficients for different XBT studies listed in the key. The different symbols refer to different XBT launch systems. *From IOC/INF-888.*

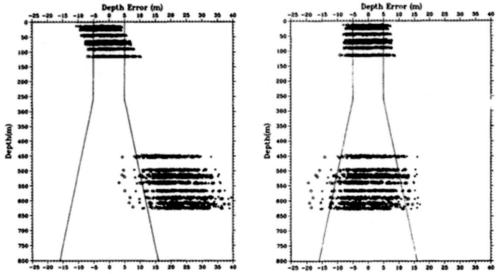

FIGURE 1.47 Depth error ($z_{CTD} - z_{XBT}$) for selected points on temperature profiles. *Left*: XBT depths calculated with the Sippican Fall-Rate Equation (FRE); *Right*: Improved XBT depths calculated with the Hallock and Teague (1992) FRE. Different symbols refer to different XBT launch systems.

A study by Kizu et al. (2010) examined the fall rate of XBT probes manufactured by Sippican versus those made in Japan by TSK. The investigators also examined T7 XBT profiles together with CTD temperature profiles in the area east of Japan. Results showed that the probes from the different manufacturers fell at rates that differed by about 3.5%. The Sippican probes fell slower than the standard equation by about 2.1%, while the TSK probes were faster by about 1.4%. Probes were also found to have different weights and other structural differences. Based on their study, Kizu et al. (2010) concluded that structural differences had a much greater influence on the fall rate than did the weight alone.

The study by Santos et al. (2018) explored the XBT FRE in the cold waters of the Southern Ocean. The authors examined 157 collocated XBT and CTD temperature profiles collected at three different locations in the region. The Sippican fall-rate equation (FRE) proved to perform better in the Southern Ocean, only overestimating the true depth by about 2%. The overall depth bias was positive which will lead to an overestimation of ocean heat content in the upper layer. While the Sippican FRE performed better in the Southern Ocean than it does elsewhere, corrections offered by other investigators still lead to improvements in the Southern Ocean. Further studies are needed to understand the thermal bias in the Southern Ocean.

One of the most comprehensive reviews of Sippican XBT performance was by Reseghetti et al. (2018). The investigators used both laboratory measurements and more than 350 XBT−CTD match ups from their field measurements and the World Ocean Database. In their laboratory tests, they examined the linear density of the copper wire, the probe weight in air, the diameter of the control hole and the external maximum diameter of the zinc nose. Measurements were based on a variety of probes from years 2002, 2003, 2007, 2008, 2010, 2014 and 2017. The results are presented in Table 1.11.

Reseghetti et al. (2018) found the probe weight values to be within the range provided by the manufacturer without any apparent correlation with the year of manufacture. They did find an unexpected gap between the zinc nose and the plastic cylinder in some of the probes, as well and variations in the lengths of their plastic fins. It is likely that these probe variations led to the depth differences observed by earlier investigators. They also tested the recording systems looking

TABLE 1.11 Wire linear density, nose diameter and hole diameter for T5 and T5/20 probes as measured in laboratory tests.

Parameter	No. of probes	Admitted range	Observed range	Mean value	Median value
Wire linear density (gm⁻¹)	55	–	0.1192−0.1214	0.1199 ± 0.0008	0.1198
Nose diameter (cm)	60	5.050−5.075	5.055−5.070	5.062 ± 0.005	5.060
Hole diameter (cm)	91	1.062−1.087	1.055−1.085	1.069 ± 0.008	1.070

From Reseghetti et al. (2018).

both at analog recording systems (ARS) and digital recording systems (DRS). The study used two different tester probes manufactured with high precision ($\sim 0.01\%$) resistors working at two different temperatures (12.20 and 26.75°C). Measurements were repeated in rooms at slightly different temperature conditions. They found that the different models of Sippican DRS yielded a variety of responses in the calibration curves, some being linear and some parabolic.

Turning to their own field data in the Mediterranean, Reseghetti et al. (2018) examined 55 pairs of XBT and CTD temperature profiles. They also examined the effect of launch height by dropping 7 of the XBTs from a higher platform. Unlike earlier studies, the authors did not find any significant dependence of fall-rate on weight differences. The differences in the zinc hole size were also found to be unimportant. They further explored the wire density and found no significant difference. To check the effect of the height of deployment, some probes were launched at even greater heights. Results showed that the depth uncertainty decreased with increasing height of launch. Many other studies have also found that the fall-rate, and hence depth uncertainty, increased with increasing temperature. Thus, it is important to consider the ocean temperature when making fall-rate comparisons. In their conclusions, the investigators state that:

1. The T5 probes examined had slightly more problems than other Sippican models;
2. The laboratory examinations have shown that wire linear density, hole diameter and length of the plastic tube have remained "reasonably" constant over the years. Probe weight, however, has shown larger variability that doesn't appear to dramatically affect the results. Electronic components of an XBT system influence the quality of the data. The circuit time constant (which is variable and linked to the recording system type) causes a delay in detecting the start of a thermal event (up to 2–3 m) causing errors in the uppermost part of the temperature profile. The results are irreversible changes in fall-rate and temperature reading;
3. The probe weight in air, the launch height and the size of the central hole appear to influence the fall-rate equation (FRE). In other words, short, heavier probes or those with a large hole diameter have a slightly slower fall rate, causing larger depth errors.
4. New fall-rate equation coefficients are recommended for the T5 probes, but it is cautioned that the observer take the geographic region and mean water temperature into account.

Finally, Bordone et al. (2020) examine XBT, ARGO Float and Ship-based CTD temperature profiles in the Mediterranean Sea to assess their compatibility. Ship of Opportunity XBT data were matched with ARGO float profiles. XBT temperature profiles deeper than 100 m agreed fairly well with ARGO profiles, with the XBT having a small positive bias (~ 0.05°C) and a standard deviation of 0.10°C. Side-by-side comparisons of ship CTD with ARGO profiles showed excellent agreement.

1.6.3 Echo Sounding

Acoustic depth sounders are now standard equipment on all classes and sizes of vessels. Marketed under a variety of names including echo sounder, fish finder, depth sounder, or depth indicator, the instruments all work on the same basic principle: The time it takes for an acoustic signal to make the round trip from a source to an acoustic reflector, such as the seafloor, is directly proportional to the distance traveled. Water supports the propagation of acoustic pressure waves because it is an elastic medium. The acoustic waves radiate spherically and travel with a speed $c(E, \rho)$ which depends on the elasticity (E) and density [$\rho = \rho(S, T, P)$] of the water. If the speed of sound is known at each time t along the sound path, then the distance d from the sound source to the seafloor is given in terms of the two-way travel time by

$$d = \frac{1}{2} \int_{t_t}^{t_r} c(t) dt \tag{1.65}$$

where

$$\Delta t = t_r - t_t = 2 \int_{t_t}^{t_r} [1/c(S, T, P; t)] dz(t) \tag{1.66}$$

is the time between transmission time (t_t) of the sound pulse and reception time (t_r) of the reflected pulse or echo. In practice, the values of c along the sound paths are not known and Eqn (1.65) must be approximated by

$$d = \frac{1}{2} \langle c \rangle \Delta t \tag{1.67}$$

where $\langle c \rangle$ is a mean sound speed over the pathlength, a value normally entered into the echo sounder during its calibration. The depth determined using the time delay is called a "sounding". In hydrography, a "reduced" sounding is one that is referenced to a particular datum. As noted by Watts and Rossby (1977), Eqn (1.66) is similar in form to the equation for dynamic height (geopotential anomaly), suggesting that travel time measurements from an inverted echo sounder (IES) can be used to measure geostrophic currents (cf. Section 1.5.5).

Because the bulk properties of water depend on the temperature (T), salinity (S), and pressure (P), sound speed also depends on these parameters through the relation

$$c = c_{0,35,0} + \Delta c_T + \Delta c_S + \Delta c_P + \Delta c_{S,T,P} \tag{1.68}$$

in which $c_{0,35,0} = 1449.22$ m/s (= the speed of sound at 0°C, 35 psu, and pressure $p = 0$, corresponding to water depth $z = 0$). The remaining terms are the first-order Taylor expansion corrections for temperature, salinity, and hydrostatic pressure; the final term, $\Delta c_{S,T,P}$, is a nonlinear corrective term incorporating the simultaneous variation of all three properties. A well-known set of values for this equation, having a stated experimental standard deviation of 0.29 m/s, is attributed to W. Wilson (Hill, 1962, p. 478). To a close approximation (Calder, 1975; MacPhee, 1976)

$$c(\text{m}/\text{s}) = 1449.2 + 4.6T - 0.055T^2 + 0.00029T^3 + (1.34 - 0.010T)(S - 35) + 0.016z \tag{1.69}$$

or (Mackenzie, 1981)

$$c(\text{m}/\text{s}) = 1448.96 + 4.591T - 5.304 \times 10^{-2}T^2 + 2.374 \times 10^{-4}T^3 + 1.340(S - 35)$$

$$+1.630 \times 10^{-2}z + 1.675 \times 10^{-7}z^2 - 1.025 \times 10^{-2}T(S - 35) - 7.139 \times 10^{-13}Tz^3 \tag{1.70}$$

where T is the temperature (°C) and S is the salinity (psu) measured at depth z (m). Accurate profiles of sound speed clearly require accurate measurements of temperature, which may not be available in advance. A commonly used oceanic approximation is the mean calibration speed $\langle c \rangle = 1,490$ m/s that is generally applied to ship's sounders. Note that the speed of sound increases with increasing temperature, salinity, and pressure, with temperature having by far the greatest effect (Figure 1.48a). For example, c increases by 1.3 m/s per 1 psu (g/kg) for a salinity range 34−35 psu (g/kg); increases by 4.5 m/s per 1.0°C for temperature (range 0−10°C); and increases by 1.6 m/s per 100 m depth. The depth capability of any sounder is limited by the power output of the transducer transmitting the sound pulses, by the sensitivity of the receiver listening for the echo returns, and by the capability of the instrument electronics and software to resolve the signal from noise. In modern sounders, pulse lengths typically range from 0.1 to 50 ms and a single transducer with a transmit/receive switching arrangement is used to both generate and receive the acoustic signals. The depth capability of an echo sounder is limited also by a number of important environmental factors. In general, sound waves are attenuated rapidly in water according to the relation:

$$\text{propagation loss (dB)} = 20 \log(r) + \alpha r/1,000 \tag{1.71}$$

where the propagation loss is measured in decibels (dB), r is the distance (or depth range in the case of depth sounders) in meters, and α (Figure 1.48b) is the attenuation coefficient (dB/km) in seawater as a function of frequency, temperature, and salinity (Urick, 1967). The first term in the equation accounts for geometrical spreading of the transmitted and received signal while the second encompasses scattering and absorption. Diffraction and refraction arising from density gradients have a minor effect on the attenuation compared with these other factors. The higher the frequency of the source the greater the attenuation due to absorption and the more limited the depth range (Figure 1.49). It is for this reason that most deep-sea sounders operate in the 1.0−50 kHz range. Even though high-frequency sounders can provide more precise depth resolution through shorter wavelengths and narrower beam widths, they cannot penetrate deeply enough to be of use for general soundings. However, they do have other important applications, including bioacoustical studies of the distribution and biomass of zooplankton, fish, and marine mammals. It was echo-sounder observations of tidally induced undular bores in the lee of a shallow sill in Knight Inlet (British Columbia) by Farmer and Smith (1980a, b) that helped rekindle interest in the observation and modeling of hydraulic jumps, internal waves, and soliton formation in a variety of oceanic settings (Apel et al., 1985; Farmer and Armi, 1999).

The output transducer converts electrical energy to sound energy and the receiving transducer converts sound vibrations to electrical energy. Loss of the acoustic signal through geometrical spreading is independent of frequency and results from the spherical spreading of wave fronts as a function of distance, while frequency-dependent absorption leads to the conversion of sound into heat through viscosity, thermal conductivity, and infra-molecular processes. Signal loss due to geometric wave front spreading follows a $1/r^2$ inverse square law. Scattering is caused by suspended particles, density

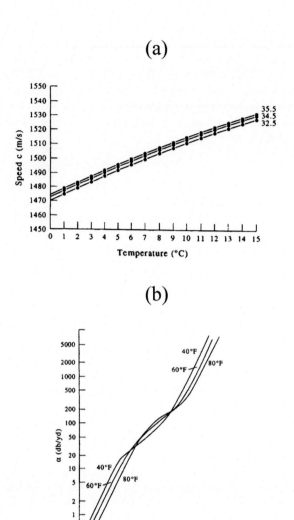

FIGURE 1.48 (a) Speed of sound as a function of temperature for different mean salinities (32.5, 34.5, and 35.5 psu) and fixed depth, $z = 1,500$ m; and (b) absorption coefficient in seawater at salinity 35 psu (g/kg) as a function of frequency for three temperatures ($40°F = 4.4°C$; $60°F = 15.6°C$; $80°F = 26.7°C$). Conversion factor; 1 dB/km = 1.0936 dB/kyard; 1 kHz = 1,000 cps; db ≡ dB. *After Urick (1967).*

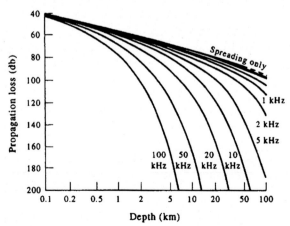

FIGURE 1.49 One-way sound attention (propagation loss, PL) in water as a function of sounder frequency (1 khz = 1,000 cps). Curves are derived using PL = 20log$r$ + αr/1,000, where r is in kilometers and α is taken from Figure 1.48 using the conversion to dB/km. Note that, in the figure, db denotes dB.

microstructure, and living organisms. In the upper 25 m of the water column, air bubbles from breaking waves and gas exchange processes are major acoustic scatterers. If I_0 is the intensity (e.g., power in watts) of the transducer and I_a is some reference intensity (nominally the output intensity in watts measured at 1 m distance from the transducer head) then the measured backscatter I_r is given by

$$I_r / I_a = b \cdot \exp(-\alpha r)/r^2 + A_n \tag{1.72}$$

where $b = I_0/I_a$ is the gain of the transducer, α is the inverse scale length for absorption of sound in water, $1/r^2$ gives the effect of geometric spreading over the distance r from which the sound is being returned, and A_n is a noise level. A return signal intensity reduction by a factor of two corresponds to a loss in intensity of -3 dB $[= 10 \log (1/2)]$ while a reduction by a factor of 100 corresponds to a loss of -20 dB $[= 10 \log (1/100)]$. Echo sounders are generally limited to depths of 10 km. The low (1−15 kHz) frequencies needed for these depths result in poor resolutions of only tens of meters. Since it takes roughly 13 s for a transducer "ping" to travel to 10 km and back, the recorded depth in deep water is not always an accurate measure of the depth beneath a moving ship. Better resolution is provided for depths less than 5 km using frequencies of 20−50 kHz. High-resolution sounders operate in a few hundred meters of water using frequencies of 30−300 kHz. Transducer beam width, side-lobe contamination, and side echoes ultimately reduce the resolution of any sounder.

1.6.3.1 Height Above the Bottom

In many oceanographic applications, the investigator is interested in real-time, highly accurate measurements of the altitude of his or her instrument package above the bottom. A real-time reporting echo sounder on the package will serve as an altimeter, enabling the operator to safely lower the sampling package, such as a CTD-Rosette system, to within 10 m of the bottom. Alternatively, the investigator can choose a cheaper route and attach a high-power omnidirectional "pinger" to the package. Rather than trying to measure the total depth of water, one uses the pinger on the package together with the ship's transducer (or hydrophone lowered over the side of the ship) to obtain the difference in depth (i.e., difference in time) between the signal, which has taken a direct route from the pinger and one which has been reflected from the bottom at an angle θ, where θ is roughly the angle the tow line makes with the bottom (Figure 1.50). The direct path (PB) takes a time t_d while the path reflected from the bottom (PAB) takes a total time t_r. The height, d, of the package above the bottom is

$$d = \frac{1}{2}c \cdot \Delta t \cdot \sin \theta = \frac{1}{2}c(t_r - t_d)\sin \theta \tag{1.73}$$

in which Δt is the time delay between the direct and reflected pings and c is the speed of sound in water. As the instrument package approaches the bottom, the two strongest analog traces on the depth recorder can be seen converging toward a "crossover" point. When the time taken to cover the direct and reflected paths are equal ($\Delta t = 0$), the package has

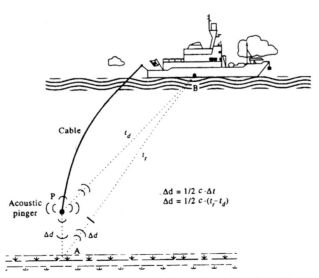

FIGURE 1.50 Schematic of how an acoustic pinger and the ship's sounder in the receive mode can be used to accurately determine proximity of a probe to the bottom. c is the speed of sound in water and t is time.

"hit" the bottom. The novice operator will be confused by the number of "false" bottom crossovers or wrap-around points when working in water that is deeper than integer multiples of the chosen sounder range. To avoid high levels of stress as each crossover is approached, the operator must know in advance how many false crossovers or wrap-arounds to expect before the instrument is truly in proximity to the bottom. For example, in a water of depth 3,230 m, a recorder chart set for a full-scale depth range of 0–750 m will register false bottoms when the pinger reaches water depths of 230, 980, and 1,730 m; i.e., 3,230 m − ($n \times$ 750 m), where n = 4, 3, 2. As far as the depth recorder is concerned, 3,230 m − (4 × 750 m) is the same as 230 m.

Analog sounder devices such as the PDR (Precision Depth Recorder) are the only real choice for this application since the analog trace provides a continuous visual record of how many crossovers have passed and how rapidly the final crossover is being approached. (The depth is first obtained from the ship's sounder, which can then be turned to "receive mode" only.) The two traces give a history of what has been happening so that it is easier to project in one's mind what to expect as the instrument nears the bottom. Problems arise if the package gets too far behind the vessel and the return echoes become lost in the ambient noise or if the bottom topography is very rugged and numerous spurious side echoes and shadows begin to appear. We recommend omnidirectional rather than strongly directional pingers so that, if the package streams away from the ship or twists with the current and cable, there is still some acoustic energy making its way to the ship's hull. To help avoid hitting the bottom, it is best to have the ship's sounder output turned off so that only the receive mode is working and the background noise is reduced; the operator can check on the total depth every so often by reconnecting the "transmit pulse". The ship's echo sounder correctly measures the height above bottom since it is programmed to divide any measured time delay by a factor of two to account for two-way travel times. The depth accuracy of the method improves as the package approaches the bottom because both the direct and reflected paths experience the same sound speed c. A value of c more closely tuned to deep water is applicable here since all that really counts is c near the seafloor in the region of study. With a little experience and a good clean signal, one can get accuracies of several meters above the bottom using 8–12 kHz sounders in several kilometers of water. Attaching both a pinger and an altimeter is clearly the preferable solution.

Note that the depth errors using the pinger method are negligible while the actual sounding depths from a depth sounder can be quite large. For example, if spatial differences in sound speed vary from 1,470 to 1,520 m/s over the sounding depth, then the maximum percentage depth errors are $\Delta c/c$ = 50/1,500 = 0.033 = 3.3%. In 4,000 m of water, this would amount to an error of 0.033 × 4,000 m ≈ 133 m.

1.6.4 Depth Measurement in Shallow Waters with Optical Satellite Imagery

Polar orbiting satellites, such as Landsat-8, Suomi National Polar-orbiting Partnership (SNPP) and Sentinel-3, all have sensor channels capable of making ocean color measurements. These channels can also be employed to map ocean bathymetry in shallow coastal waters having depths of 0–30 m, comparable to the depths accessible to airborne LiDAR—Light Detection and Ranging—measurements (see Section 1.6.5). This is an important application of satellite imagery, as coastal bottom topography is generally complex, consisting of substrates that change with time, such as corals, seagrass, and sand. In optically shallow waters, where the contribution of bottom reflection is not negligible, the light reflected in the ocean color bands carries information on the water depth, bottom albedo and water column optical properties. As such, the remote sensing reflectance, $R_{rs}(\lambda)$, has been used to generate bathymetry maps over optically shallow environments.

The algorithms to infer ocean depth from satellite channel reflectance can be separated into two general categories: empirical and physics-based. The empirical modules are based on statistical computations that link known depths and geolocated satellite pixels in one or several bands. This approach is fairly easy to implement, as long as accurate depth information and geolocation of multiple satellite channels are available. A weakness of the method is that it is restricted to regions where these two conditions are met. As a consequence, an empirical algorithm for one region cannot be generally applied to other areas of the global ocean. Physics-based systems utilize radiative transfer theory to estimate the radiances passing to the satellite sensors from the reflectance off the seafloor. In principle, this approach yields *in situ* bottom data for model tuning and therefore can be extended to the global coastal ocean. It must be remembered, however, that the shallow water radiative transfer equation is complex and difficult especially in the diverse waters of the shallow coastal ocean. According to earlier studies (Lee et al., 1998, 1999), the shallow water properties can be derived from hyperspectral reflectance measurements as a function of wavelength, using a spectral optimization algorithm (SOA). Use of a SOA is an effective method to estimate shallow water depth and it has been extensively evaluated and continuously refined. The SOA needs the many wavelength bands that are available with hyperspectral imagery, and leads to greater uncertainty when the number of bands is reduced, as it the case with multi-spectral sensors.

The polar-orbiting satellites mentioned above generally carry multi-spectral sensors. In their study, Wei et al. (2020) use reflectance data from the Landsat-8 Operational Land Imager (OLI), the Suomi NASA Preparatory Platform (SNPP), which carries the Visible Infrared Imaging Radiometer Suite (VIIRS), and the Sentinel-3A satellite, with its Ocean and Land Color instrument the OLCI. The OLI has four visible bands (including a blue band at 443 nm), the VIIRS has six visible bands (including one at 638 nm), and the OLCI has ten visible bands, with a purple band at 400 nm. Wei et al. (2020) developed an optimization approach for deriving depth maps in shallow coastal waters from the multi-spectral radiometers. Their approach uniquely takes advantage of the temporal variation in two satellite images over a relatively short period of time. This is a departure from standard approaches that retrieve depth from a single image. The time difference between the two images is short enough that the bottom albedo and water depth are assumed unchanged (after tide height correction). The two-image method is called the 2-SOA, versus the traditional 1-SOA method. The new method uses spectral optimization to reach an optimal solution for the water depth by searching for the minimum between the two observed and two modeled reflectance spectra. The workflow for this method is presented in Figure 1.51.

The study of Wei et al. (2020) adopted the hyperspectral optimization model of Lee et al. (1998, 1999) to describe the water-column optical properties and bottom albedo. This led to an expression for the reflectance arriving from just below the sea surface of

$$r_{rs} = r_{rs}^{dp}(\lambda)\left\{1 - \exp\left[-\left(\frac{1}{\cos\theta_w} + \frac{D_0(1 + D_1\mu(\lambda))^{0.5}}{\cos\theta_a}\right)k(\lambda)H\right]\right\} +$$

$$\frac{\rho(\lambda)}{\pi}\exp\left[-\left(\frac{1}{\cos\theta_w} + \frac{D_0'(1 + D_1'\mu(\lambda))^{0.5}}{\cos\theta_a}\right)k(\lambda)H\right] \tag{1.74}$$

where $r_{rs}(\lambda)$ is the sum of the contributions from the water column and from the bottom reflection. Here, $r_{rs}^{dp}(\lambda)$ refers to the reflectance at the surface in optically deep water, the parameter θ_a is the solar-zenith angle, θ_w is the subsurface solar-zenith angle, $k(\lambda)$ is an inherent optical property where $k(\lambda) = a(\lambda) + b_b(\lambda)$, H is the water depth to be solved for, and the values D_0, D_1, D_0' and D_1' are taken from Lee et al. (1999). Finally, $r_{rs}(\lambda)$ is propagated through the water surface to obtain a $R_{rs}(\lambda)$ spectrum (Lee et al., 1999).

$$R_{rs}(\lambda) = \frac{0.5r_{rs}(\lambda)}{1 - 1.5r_{rs}(\lambda)} \tag{1.75}$$

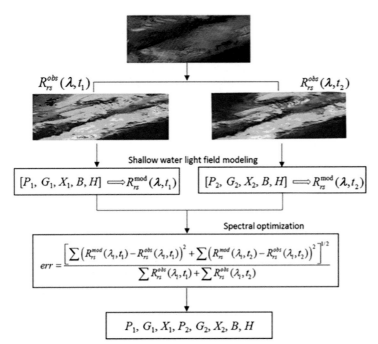

FIGURE 1.51 Schematic workflow of the two-spectrum optimization approach (2-SOA) to derive water depths from two multi-spectral satellite images. Here, R_{TS}^{obs} is the satellite observed reflectance, R_{TS}^{mod} is the modeled reflectance, P is the phytoplankton absorption coefficient at 443 nm, G is the CDM absorption coefficient at 443 nm, X is the particle backscattering coefficient at 443 nm, B is the normalized bottom albedo at 550 nm, H is the water depth, and err is the least squares residual error. The subscripts 1 and 2 refer to the two sequential images at times t_1 and t_2.

The reflectance spectrum is a function of five unknowns, one of which is the depth, H, that we wish to retrieve. The others are the adsorption coefficients listed earlier. The depth is found using an optimization procedure to minimize the expression

$$err = \frac{\left[\sum_i \left(R_{rs}^{mod}(\lambda_i) - R_{rs}^{obs}(\lambda_i)\right)^2\right]^{1/2}}{\sum_i R_{rs}^{obs}(\lambda_i)} \tag{1.76}$$

A similar cost function is commonly found in ocean color inversions to define the chlorophyll content. The standard approach is to use a single satellite image, which we label as 1-SOA. In contrast, the 2-SOA approach uses two images from slightly different times, where the factor to minimize is

$$err = \frac{\left[\sum_i \left(R_{rs}^{mod}(\lambda_i, t_1) - R_{rs}^{obs}(\lambda_i, t_1)\right)^2 + \sum_i \left(R_{rs}^{mod}(\lambda_i, t_2) - R_{rs}^{obs}(\lambda_i, t_2)\right)^2\right]^{1/2}}{\sum_i R_{rs}^{obs}(\lambda_i, t_1) + \sum_i R_{rs}^{obs}(\lambda_i, t_2)} \tag{1.77}$$

where t_1 and t_2 are the two different times that the images are taken. This optimization is solved with a MATLAB solver known as *fmincon* that includes a maximum of 2,000 iterations and a tolerance of 10^{-5}. Initially, the model was used to generate a hyperspectral data set (400−700 nm, for every 5 nm) over shallow water. Using this synthetic dataset to retrieve depth, it is possible to evaluate the performance of the optimization procedure before using real satellite data.

Turning to satellite data, Lee et al. (1999) used two images from the Landsat-8 Operational Land Imager (L8/OLI) 16 days apart, corresponding to the revisit period of L8. According to the NOAA tidal data, both images were collected during low-tide with the estimated tidal difference within ±0.33 m. Atmospheric corrections were those developed by Gordon and Wang (1994). The target area was the Olowalu Reef off Maui, Hawaii. Two S3A/OLCI and 2 SNPP/VIIRS images covered the Florida Keys and the Great Bahama Bank areas, respectively (Figure 1.52).

In their analysis, Lee et al. (1999) used the near-infrared (NIR) and shortwave infrared (SWIR) (865 and 2,201 nm) in combination to define aerosol type for correction of the imagery. The two S3A images for the Florida Keys were both captured at the same time of day keeping the tidal water level within ~0.3 m. In the Bahamas, the tidal level between the two images was <0.3 m. For SNPP/VIIRS, a combination of NIR+SWIR bands was used to define aerosol type for image correction. For S3A/OLCI, the aerosol type was chosen with the two NIR bands (779 and 865 nm; Gordon and Wang,

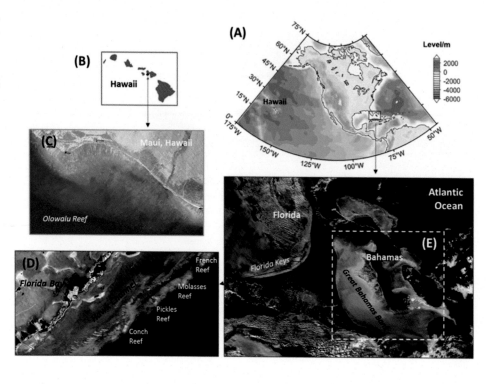

FIGURE 1.52 (a) Locations of study areas; (b) the study area off Maui, Hawaii, (c) a true-color image of the coral reefs off Maui, (d) a true-color image of the Florida Keys, and (e) a true-color image of the Bahamas (highlighted with dashed lines). *From Wei et al. (2020).*

1994). The retrieval accuracy values based on the synthetic approach of Wei et al. (2020) are presented in Table 1.12 for both the new 2-SOA model and the standard 1-SOA approaches. To evaluate the algorithm performance in estimating water depth, the authors calculated the bias $\varepsilon = median[(H - D)/D] \times 100\%$, where H and D refer to the estimated and known depth values, respectively. The authors also derived the absolute percentage difference, $|\varepsilon| = median|[(H - D)/D]| \times 100\%$ and the root-mean-square difference (RMSD) to assess the model uncertainty, with

$$RMSD = \sqrt{median[(H - D)^2}.$$

From these results, it is clear that the 2-SOA method can estimate water depths with significantly smaller errors that the traditional 1-SOA method. Among the three different satellite sensors simulated, the Landsat-8 Operational Land Imager (L8/OLI) benefited most from the 2-SOA, largely because it has the fewest number of wavelengths available for spectral optimization. The S3A/OLCI showed the smallest degree of improvement, again due to the fact that the OLCI has the largest number of bands.

The depths of the Olowalu Reef off the southwest coast of Maui (Hawaii) were computed with L8/OLI images. The results shown in Figure 1.53a are compared with a depth map from the Scanning Hydrographic Operational Airborne LiDAR Survey (SHOALS) in Figure 1.53b. The LiDAR results were spatially averaged. Pixels were within 30 m of the L8/OLI pixel locations. Both the L8 and LiDAR depth maps are very similar in their spatial distributions. Figure 1.53c presents the scatterplot and regression of the L8 depths on the LiDAR results. The majority of the points are grouped around the 45° line, indicating good correspondence between the two depth fields ($r^2 = 0.7$). The lack of complete agreement can, in part, be explained by the nature of the L8/OLI sensor, where in shallow waters of less than 5 m, the L8 depths generally overestimate the LiDAR and actual depths. Similarly, many of the L8 depth measurements in areas over 10 m deep are found to underestimate the LiDAR values. Despite these deviations from the LiDAR map, the outliers account for less than 2% of the data points.

A second location where the 2-SOA method was tested is the area east of the Florida Keys National Marine Sanctuary. The predominant bottom material is seagrass, with the addition of some coral barrier reefs. This study area is large enough that the investigators could use S3A/OLCI images with the algorithm. LIDAR images were, unfortunately, not available in this area to serve as ground truth. Instead, a rasterized version of the bathymetry map from the NOAA coastal relief model (CRM) was used for validation. The bathymetry maps from S3A and CRM are presented here in Figure 1.54. While the bathymetry maps in Figures 1.54a and b are similar, the regression line in Figure 1.54c does not show as high a correlation as with L8. Most of this area is shallower than 10 m. Both bathymetry maps do not appear to adequately separate land and water pixels in Florida Bay located near the barrier reefs. Part of the lack of agreement may be due to the CRM data themselves as they were generated by spatial interpolation based on historical low-tide hydrographic data. In the deeper water (>10 m), the S3A depths are biased low relative to the CRM values, which is likely due to the heterogeneity around the 15–30 m isobaths, where the CRM data have undergone spatial gridding, extrapolation and other spatial modifications that may have resulted in incorrect depths.

In a study that used the 2-SOA method, Lee et al. (2021) combined multi-band L8/OLI imagery with ICESat-2 LiDAR data to generate bathymetric maps for the Great Bahama Bank. The investigators argued that many satellite-image depth

TABLE 1.12 Retrieval accuracy of the estimated water depth (H) derived from the synthetic data with three ocean color satellites (L8/OLI, SNIPP/VIIRS and S3A/OLCI). The results from the two-spectrum optimization (2-SOA) and the standard approach (1-SOA) are provided for comparison of their performance.

		L8/OLI			SNPP/VIIS			S3A/OLCI		
		Coral	Seagrass	Sand	Coral	Seagrass	Sand	Coral	Seagrass	Sand
1-SOA	$|\varepsilon|$	42%	43%	21%	22%	31%	13%	22%	26%	10%
	ε	14%	13%	7%	6%	18%	2%	5%	14%	3%
	RMSD	9.3	9.5	6.0	8.8	9.1	4.6	8.3	8.5	4.1
2-SOA	$|\varepsilon|$	26%	28%	15%	14%	19%	11%	14%	16%	10%
	ε	1%	1%	3%	4%	9%	0%	1%	5%	2%
	RMSD	8.3	8.7	5.2	7.7	8.2	4.0	7.2	7.4	3.9

The root-mean-square difference (RMSD) is in units of meters.
Modified after Wei et al. (2020).

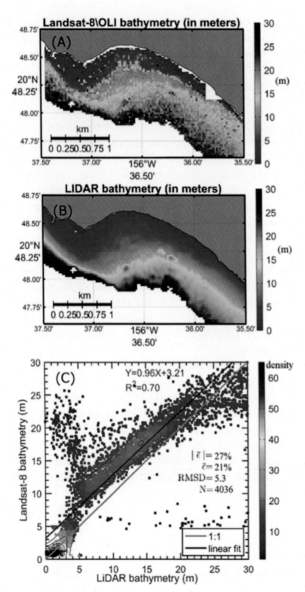

FIGURE 1.53   (a) Bathymetric map (m) of the Oluwalu Reef off southwest Maui derived from L8/OLI, (b) bathymetric map derived from LiDAR measurements, (c) scatterplot for the water depth from the Landsat-8 Operational Land Imager (L8/OLI) and LiDAR (colors indicate data density). The red solid line in (C) is the linear regression of L8 on LiDAR, where $Y = 0.96X + 3.21$, $r^2 (= R^2) = 0.7$. *From Wei et al. (2020).*

algorithms suffer from a lack of the ground truth data needed to establish the accuracy of the satellite depth observations. Consequently, LiDAR depths served as the fiducial depth data to combine with the depths obtained from the multi-spectral images. Because LiDAR only provides depth along the satellite track, the image can provide depths at this track and also fill in the rest of the target area (Figure 1.55).

The image-derived depths follow the method used by Wei et al. (2020) as reported earlier in this section and won't be repeated here. The significant difference is the use of the LiDAR depths to provide a confidence score for the multiband image derived depths. As in Wei et al. (2020), the Landsat-8 OLI image data are used. The LiDAR data are from ICESat-2, which was designed to provide altimetry of changes in polar ice cover and thickness. The instrument onboard is the Advanced Topographic Laser Altimeter System (ATLAS), a green (532 nm) wavelength, photon-counting laser altimeter with a 10 kHz pulse repetition rate. ATLAS uses photomultiplier tubes (PMTs) as detectors in the photon-counting mode so that a single photon reflected back to the receiver triggers a detection within the data acquisition system. This single-photon sensitive detection method provides an extremely high vertical resolution designed to measure small temporal changes in polar ice elevations. It also provides the sensitivity needed to measure the bottom depth of optically shallow ocean areas.

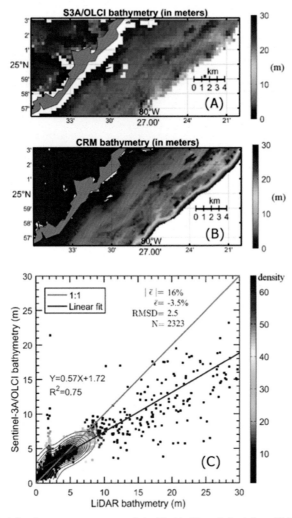

FIGURE 1.54 (a) Bathymetric map (m) for the seagrass area east of the Florida Keys derived from S3A/OLCI images with the 2-SOA method; (b) bathymetry from the NOAA CRM model; and (c) scatterplot of S3A against CRM matchup data that were averaged of 3 × 3 pixels. The red solid line is the linear regression $Y = 0.57X + 1.72$, r^2 (= R^2) = 0.75. *From Wei et al. (2020).*

The challenge in combining these data types is first to decide how to matchup the L8/OLI depths with those from ICESat-2. First, all of the L8 pixels with a dense cloud, were based on a Short Wave Infrared (SWIR) band having a 1,238 nm threshold, and were identified and removed from the dataset. In addition, the pixels of low-quality (poor atmospheric correction, land contaminated, strong glint contamination) were also removed. The ICESat-2 depths are part of the ATL03 data product and both L8/OLI depths and ATL03 depths include specific latitude and longitude data. Because the variation of depth (H) is negligible for short periods (after tidal correction), the time constraint for data matchup was set to ±2 weeks. To match H from the ICESat-2 with that from the OLI, the ICESat-2 track was first located in the OLI image. For ICESat-2 data, a L8 pixel was selected for the smallest separation distance. Because the along-track spatial resolution of ICESat-2 is 0.7 m, while the OLI spatial resolution is 30 m, there are many ICESat-2 observations within an OLI pixel. Thus, all of the ICESat-2 values within an OLI pixel were averaged together to form the H value for comparison with the L8 depth.

After several neural net studies to determine how the ICESat-2 and L8 data are related, Lee et al. (2021) produced a map having a 30 m spatial resolution over the Great Bahama Bank using the L8/OLI data, tuned with matchup data from the coincident ICESat-2 bathymetry (Figure 1.56). Later, LiDAR data were used to produce the confidence values that are presented in Figure 1.57, after being adjusted for tidal elevation differences.

Because the satellite-derived depths were computed by the fusion of the OLI image from Landsat-8 and LiDAR data (represented by the red dashed line in Figure 1.56), the investigators used another ICESat-2 data track within the image to obtain an independent depth estimate (dashed black line in Figure 1.57).

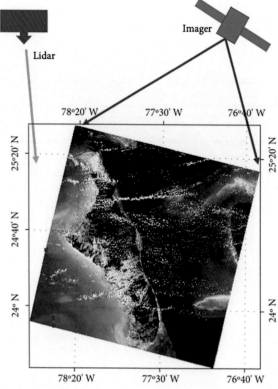

FIGURE 1.55 Example of collocated LiDAR and multiband image data. The dashed red line is the ICESat-2 track, while the background image is an OLI image from Landsat-8 (L8/OLI). *From Lee et al., (2021).*

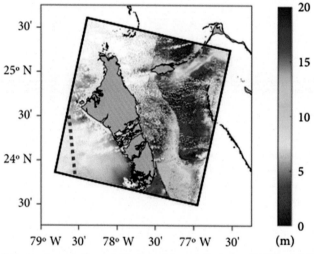

FIGURE 1.56 Bathymetric map for the image shown in Figure 1.55 obtained through the fusion of ICESat-2 LiDAR depths and Landsat-8 radiometric inferred depths. Gray indicates land while the white colors are areas of cloud or poor Landsat-8 Operational Land Imager (OLI) data values.

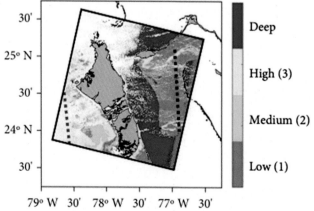

FIGURE 1.57 Confidence score map for the H_{imager} depths in Figure 1.56. The black dashed line is the ICESat-2 track on March 16, 2019, where H_{LiDAR} is used as an independent measurement to evaluate the confidence map.

1.6.5 Other Depth Sounding Methods

For the sake of completeness, several remote sensing, depth-sounding methods are introduced in the following sections.

1.6.5.1 LiDAR (Light Detection and Ranging) from Aircraft

Light detection and ranging (LiDAR) is an active electro-optical remote-sounding method that uses a pulsed laser system as a radiation source flown from the air (Figure 1.58) or, as discussed in previous sections regarding the Advanced Topographic Laser Altimeter System (ATLAS) on ICESat-2, from satellites. Research on bathymetric LiDAR technology began in the 1960s and was followed by hardware testing in early 1969, in an effort to collect ocean depths to serve nautical charting applications in hostile territories. The airborne sensor measures the distance to the surface of the ocean and to the seafloor based on the time interval between emission of the pulse and the reception of its reflections. Bathymetric LiDAR uses water-penetrating green light with a wavelength of 532 nm to determine seafloor and riverbed elevations, while topographic LiDAR use infrared light with wavelengths of 900–1,064 nm. The strength of LiDAR returns (Figure 1.59) varies with the composition of the surface object reflecting the transmitted light along the flight path. The reflective percentages are referred to as LiDAR intensity. The advantages of airborne hydrographic surveys are (1) consistent swath width, (2) efficient coverage in very shallow water (<5–10 m depth), (3) accurate depth measurement, and (4) the ability to combine the airborne survey with shipboard multibeam surveys for optimal results, efficiency and safety.

A typical airborne LiDAR unit consists of a pulsed laser transmitter, a receiver, and a signal analyzer–recorder. As the aircraft travels over the ground, LiDAR scans from side-to-side. While some pulses will be directly below at nadir, most

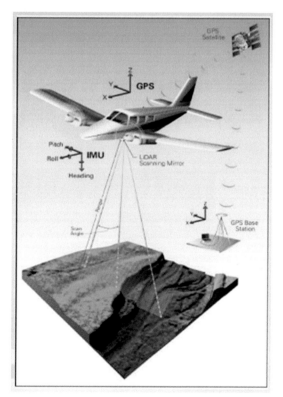

FIGURE 1.58 Cartoon showing LiDAR survey from a twin engine airplane. IMU (Inertial Measurement Unit) is part of the aircraft's inertial navigation system to measure the orientation of the aircraft using a combination of accelerometers and gyroscopes; GPS refers to the satellite-based Global Positioning System.

FIGURE 1.59 Reflected green light intensity from various objects during a coastal airborne LiDAR survey. Green LASER light has greatest depth penetration. Peak reflection is from the seafloor in the left and center panels but from the top of the rock in the right-hand panel. *(After Vasileios Alexandridis, Delft University of Technology, 2020)*

pulses are from off-nadir locations. Typically, linear LiDAR used in hydrography has a swath width of 1,000 m. New technologies, like Geiger LiDAR, have scan widths of 5,000 m, generating much wider footprints than traditional systems. Airborne LiDAR transmits over 160,000 light pulses per second, whereby each 1 m pixel coverage is based on about 15 pulses. Adjacent swath overlaps of 20% are used to prevent data gaps. The redundant data are also used to confirm that data collected on adjacent lines in opposite directions match. The technique is good to depths of a few tens of meters in coastal waters, where extinction coefficients are typically around 0.4–1.6 m^{-1}. The rapid spatial sampling capability of this technique makes it highly useful to hydrographers wanting to map shoals, rocks and, other navigational hazards. LiDAR topographic data are also important for generating coastal inundation maps for storm surge and tsunami inundation and runup. Several factors affect light intensity, including range and incident beam angle. Surface composition strongly influences light intensity. In practice, LiDAR systems are accurate to about 15 cm vertically and 40 cm horizontally.

There are two basic types of LiDAR: (a) Discrete LiDAR and (b) Full Waveform LiDAR. Figure 1.60 depicts the discrete LiDAR case, with pulses scanning the water column for three different coastal regions. The system records a series of returns as the pulse hits the ocean surface and then multiple reflectors within the water column. That is followed by a strong final pulse from the seafloor. The separate recorded returns constitute a "discrete return LiDAR". In this example, discrete LiDAR records each peak and deals separately with each return. In the case of Full Waveform LiDAR

FIGURE 1.60 Light signal return intensity from the ocean surface and the seafloor as a function of time along a survey line (full waveform case). https://gisgeography.com/lidar-light-detection-and-ranging/. *Source: Guenther (2007). Reproduced with permission from the American Society for Photogrammetry & Remote Sensing, asprs.org.*

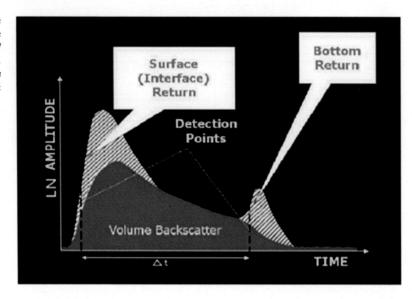

(Figure 1.60), the return signal is recorded as one continuous wave. The operator then considers the peaks as evidence of discrete reflectors. Even though full waveform data is more complicated, LiDAR is moving toward full waveform systems.

One of the main challenges for hydrographic systems is to determine the surface return (Figure 1.59). The system analyzes the signal-to-noise ratio (SNR) to identify the surface and bottom peaks of a waveform using a variety of complex algorithms that derive a depth confidence. Typically, the processors will examine the potential error for a sounding and the derived depth confidence, making sure these are all calculated for the return. The software will then make a decision to reject or accept the depth value. Ground-control by multibeam ship depth surveys can be used to compare with LiDAR depths on a daily basis. Cross lines are run perpendicular to the direction of the main survey lines and are typically collected on different days and different tidal ranges to confirm that cross line data matches the main survey data. GPS/IMU accuracies are important for direct georeferencing of the LiDAR and imagery data.

1.6.5.2 Synthetic Aperture Radar

One of the surprising aspects of synthetic aperture radar (SAR) is its ability to "see" shallow banks, ridges, and shoals in the coastal ocean. In this case, SAR does not measure the bottom topography directly but, instead, detects the distortion of the wave ripple field over the feature caused by deflection and/or acceleration of the ocean currents. For a discussion of this effect, the reader is referred to Robinson (1985). SAR is also being used to detect surface wakes generated by submarines and ships, and to delineate ocean fronts arising from changes in current shear and marked thermal gradients such as the "Wall of the Gulf Stream" (Belkin and O'Reilly, 2009; Williams et al., 2013). A sequence of airborne SAR images collected off the west coast of British Columbia in 1990 reveals atmospheric internal waves (resembling oceanic internal waves) with wavelengths of 1.2—2.3 km, periods of 3—8 min and group velocities of 3—6 ms^{-1} traveling counter to the 10 ms^{-1} surface winds (Thomson et al., 1992). Comparison of the observed waves with a simple three-layer density model of the troposphere suggests that the waves were low-mode oscillations trapped within a 0.75-km-thick temperature inversion overlying a 1-km-thick weakly stratified marine boundary layer (Vachon and Emery, 1988).

1.6.5.3 Satellite Gravity Measurement

The suggestion that there is a close connection between oceanic gravity anomalies and water depth was postulated as early as 1859 (Pratt, 1859, 1871). The idea of using measured gravity anomalies to estimate water depth began with Siemens (1876) who designed a gravity meter that obviated the need to spend hours for a single sounding (Vogt and Jung, 1991). It was not until the launching of the SEASAT radar altimeter in 1978 that this concept could be used as alternative to large-scale sounding measurements. Indeed, the first SEASAT-derived gravity map of the World Oceans (Haxby, 1985) closely resembles a bathymetric map with a horizontal resolution of about 50 km (Vogt and Jung, 1991). The idea of using satellite radar altimetry as a "bathymeter" is based on the good correlation observed between gravity anomalies and bathymetry in the 25—150 km wavelength radar band. Satellite bathymetry is especially valuable for sparsely sounded regions of the World Ocean such as the South Pacific, and in regions where depths are based on older soundings in which navigation errors are 10 km or more. As pointed out by Vogt and Jung, however, one-dimensional predictions cannot be accurately ground-truthed with a single ship survey track since the geoid measured along the track is partly a function of the off-track density distribution. (Geoid refers to a constant geopotential reference surface, such as long-term mean sea-level, which is everywhere normal to the earth's gravitational field.) A broad swath of shipborne data is needed to overlap the satellite track swath.

In addition to SEASAT, early satellites equipped with radar altimeters to map sea surface topography include GEOS-3, GEOSAT, ERS-1, and TOPEX/Poseidon. Modern satellites with high-resolution altimeters include Jason-1, Jason-2, Jason-3, Sentinel-6 and HY-2A,2B (China). Of the earlier satellites, only data from the U.S. Navy's GEOSAT have been processed to the accuracy and density of coverage needed to clearly resolve tectonic features in the marine gravity field on a global basis (Marks et al., 1993; EOS). This mission (which has been superseded by the Gravity Recovery And Climate Experiment, GRACE) was designed to map the marine geoid to a spatial resolution of 15 km. Detailed maps of the seafloor topography south of 30° S from the GEOSAT Geodetic Mission were declassified in 1992. These maps have a vertical RMS resolution of about 10—20 cm and, together with SEASAT data, have been used to delineate fracture zones, active and extinct mid-ocean ridges, and propagating rifts. Satellite altimetry from ERS-1 has been used to map the marine gravity field over the permanently ice-covered Arctic Ocean (Laxon and McAdoo, 1994). Future declassification of military status satellite data will lead to further analysis of the seafloor structure over the remaining portions of the World Ocean.

The launch of TOPEX/Poseidon satellite in 1992 with an accurate radar altimeter, resulted in important data sets for tidal analysis of global sea levels (e.g., Schrama and Ray, 1994; Ray, 1998; Cherniawsky et al., 2001). In 2002, the satellite was replaced in the same orbit by Jason-1 with a very accurate altimeter. A companion altimetry satellite, Jason-2, was launched in 2008. As we discuss later in more detail, satellite altimetry has proven invaluable for interpreting large-scale circulation previously deduced from dynamic height data. Satellite altimetry also is particularly well suited for examining the spatial structure, temporal variability, and propagation of mesoscale features in the ocean, including mesoscale eddies (Di Lorenzo et al., 2005), geostrophic surface currents, and Rossby waves (cf. Chelton and Schlax, 1996).

GRACE uses a highly accurate microwave ranging system to measure changes in the speed and distance between two identical spacecraft (nicknamed "Tom" and "Jerry") flying in a polar orbit about 220 km apart, 500 km above earth (Wikipedia, 2013). Launched on March 17, 2002, the ranging system on GRACE is able to detect separation changes as small as 10 μm (approximately one−tenth the width of a human hair) over the satellite separation distance. As they circle the globe 15 times a day, the satellites measure variations in earth's gravitational pull. When the first satellite passes over a region of slightly stronger gravity (corresponding to a positive gravity anomaly), it is pulled slightly ahead of the trailing satellite. This causes the distance between the satellites to increase. The first spacecraft then passes the anomaly, and slows down again while the following spacecraft accelerates, then decelerates over the same point. By measuring the constantly changing distance between the two satellites and combining that data with precise global positioning systems (GPS) positioning measurements, scientists can construct a detailed map of earth's gravity. This information, in turn, can be used to study a variety of global processes, including the rate at which mass is being added to the oceans from the land (Peltier, 2009).

1.7 SEA-LEVEL MEASUREMENT

The measurement of sea-level is one of the oldest forms of oceanic observation. Pytheus of Marseilles, who is reported to have circumnavigated Britain around 320 BCE, was one of the first to actually record the existence of tides and to note the close relationship between the time of high water and the transit of the moon. Nineteenth-century sea-level studies were related to vertical movements of the coastal boundaries in the belief that, averaged over time, the height of the mean sea level was related to movements of the land. More recent applications of sea-level measurements include the resolution of tidal constituents for coastal tidal height predictions, assisting in the prediction of El Niño/La Niña events in the Pacific, and determining the contributions of global sea level rise due to climate change. Tide gauge data are essential to studies of wind-generated storm surges, which can lead to devastating flooding of highly populated low-lying areas such as Bangladesh and the Eastern Seaboard of the United States. (In late October 2012, parts of the northeast coast of the US were inundated by a major storm surge during Hurricane Sandy. New York registered 10 m waves and a maximum 4.23 m storm surge). High storm surges from Category 5 hurricanes Irma and Maria in September 2017 were on the order of 10 m and mostly confined to the Caribbean and the Southeastern regions of the US. Tide gauges located along the perimeter of the Pacific Ocean and on Pacific islands are integral to the Pacific Tsunami Warning Center (PTWC) headquartered in Honolulu (Hawaii) and Palmer (Alaska) that alerts coastal residents to possible seismically generated waves associated with major underwater earthquakes and crustal displacements. The changing emphasis of digital tide gauge observations on tsunami warning and research, rather than for tidal analysis, means that the gauges must record at intervals of 6 min or shorter (preferably 1 min) and to be directly accessible online at all times.

In addition to measuring the vertical movement of the coastal land mass, long-term sea-level observations reflect variations in large-scale ocean circulation, surface wind stress, and oceanic volume. Because they provide a global-scale integrated measure of oceanic variability, long-term (>50 years) sea-level records from the global tide-gauge network provide some of the best information available on global climate change. Long-term trends in mean sea level are called *secular* changes while changes in mean sea level that occur throughout the World Ocean are known as *eustatic* changes. As described later in this section, changes in eustasy are associated with variations in land-based glaciation, variations in steric height arising from changes in global ocean temperature and salinity, fluctuations in the accumulation of oceanic sediments, and tectonic activity, such as the change in ocean volume and the shape of the ocean basins. Since coastal stations really measure the movement of the ocean relative to the land, land-based sea-level measurements are referred to as relative sea-level (RSL) measurements. Mean sea level is the long-term average sea level taken over periods of months or years. Datum levels used in hydrographic charts can be defined in many ways and generally differ from country to country (Thomson, 1981; Woodworth, 1991), but are generally referenced to some measure of low water, such as "lower low water, large tides". For geodetic purposes, mean sea-level needs to be measured over many years.

1.7.1 Specifics of Sea-Level Variability

As the previous discussion indicates, observed sea-level variations about some mean equilibrium, "datum" level can arise from four principal components: (1) *Short-term* temporal fluctuations in the height of the sea surface including those associated with wind waves and oceanic tides forced by the changing alignment of the sun, moon, and earth. In addition, there are changes due to atmospheric pressure (the isostatic "inverse barometer effect", corresponding to a ~ 1 cm *rise* in sea level for a 1 mb *drop* in atmospheric pressure), to wind-induced current setup/setdown along the coast (including Coriolis effects on the alongshore component of wind-generated currents), to changes in river runoff, and to changes in the large-scale ocean currents and gyres caused by fluctuations in the oceanic wind field and water mass redistribution. There is also a global non-isostatic, non-equilibrium atmospheric response of about 1 cm at periods of around 5 days due to forcing by the westward propagating Rossby-Haurwitz air pressure mode; (Luther, 1982; Sakazaki and Hamilton, 2020; Thomson and Fine, 2021; Ponte and Schindelegger, 2022) and a weaker response at 1.35 days due to the first mode atmospheric Kelvin wave (Sakazaki and Hamilton, 2020); (2) *Long-term* variations resulting from the slow precessions in the orbits of the earth and the moon, and long-term eustatic changes arising from variations in the mass of the ocean due to melting (or accumulation) of land-based ice in the major continental ice sheets that sit atop Antarctica and Greenland, changes in grounded polar ice caps, and changes in the smaller ice sheets and mountain glaciers. There are also major long-term variations in sea level due to *Steric effects*—slow variations in sea level arising from changes in ocean volume (i.e., density changes due to heating/cooling) without a change in ocean mass. At present, steric heights are generally increasing throughout much of the World Ocean as a result of global warming (expansion) of the upper ocean (IPCC, 2007, 2013, 2022). Diminishing the salt concentration of the upper ocean due to enhanced rainfall or riverine input has the same steric effect as thermal heating; (3) *Coastal subsidence* involving the lowering of the land brought about by reduction in the thickness (compaction) of unconsolidated coastal sediments, erosion, sediment deposition and with the withdrawal of fluids (water, oil, etc.) from the sediments; (4) Large-scale vertical crustal land movements that produce sea-level change through *tectonic processes* (mountain building) and ongoing *glacio-isostatic adjustment* arising from the continued viscoelastic response and rebound of the earth to melting of glaciers formed during the last ice age. In addition to the change in land level due to the unloading of the crust with the removal of glacial ice, sea levels are affected by the gravitational effects of continental ice sheets on the adjacent sea water. This spatially varying "sea-level fingerprinting" occurs because, as the ice disappears, relative sea levels fall at decreasing rates from highly glaciated areas to lesser glaciated areas (Mitrovica et al., 2001; Riva et al., 2010; U.S. National Research Council, 2012; James et al., 2021).

The principal semidiurnal (M_2) and diurnal (K_1) tidal constituents, with respective periods of 12.42 and 23.93 h, can be accurately resolved using a 15.3-day tidal record of hourly values. Further resolution of the important *spring-neap* cycle of the tides (the 15-day fortnightly cycle associated with the alignment of the sun, moon, and earth) and the tropic (declinational) cycle (the roughly 15-day cycle arising from the tilt of earth's axis relative to the moon's orbital plane) require a record length of roughly 29 days (or 0.98 lunar months; one lunar day = 25 h). For most practical purposes, this is the minimum length of record that is acceptable for construction of local tide tables. In fact, many countries maintain primary tide-gauge stations as reference locations for secondary (short-term) tide-gauge stations. Differences in the tide heights and times of high/low water are tabulated relative to the primary location. Accurate resolution of all 56 principal tidal constituents requires a record length of 365 days while an accurate measure of all components for long-term tidal applications requires a record of 18.6 years. The 18.6-year "Metonic" cycle or nutation is linked to the 5° tilt of the plane of the moon's orbit with respect to the plane of the earth's orbit and is the time it takes the line of intersection of these two planes to make one complete revolution (Thomson, 1981). Other tidal constituents include: The centimeter-scale Pole Tide (Chandler Effect) with a period of 14.3 months that arises from the Chandler Wobble in the instantaneous axis of the earth's rotation; an 8.8-year cycle associated with alterations in the eccentricity of the moon's orbit about the earth; and a 20,940-year cycle due to a wobble in the earth's orbit about the sun (precession of the equinox). Meteorological tides are caused by local atmospheric forcing and include large (~ 1 m) sea-level changes associated with storm surges that often flood low-lying areas (Murty, 1984). Their periodicities are related to changes in wind and atmospheric pressure.

In addition to the relatively small changes in sea level associated with the Metonic cycle and other orbital factors, there are a variety of major variations due to geological processes. Geological techniques such as coring of Greenland glaciers show a relatively rapid rise of sea-level from 18,000 years ago, when global sea levels were roughly 130 m lower than today. The rate of sea-level rise, which averaged 10 mm/year during the glacial—interglacial transition period, slowed dramatically about 8,000 years ago when the levels were 15 m below those of today. Present levels were reached roughly 4,000 years ago. Since that time, mean sea-level changes have consisted of oscillations of small amplitude (Barnett, 1983). However, there is now concern that global sea levels are rising at over 3 mm/year due to global warming through buildup of CO_2 and other "greenhouse gases" in the atmosphere. Some studies (e.g., Pfeffer et al., 2008; Grinsted et al., 2009) estimate that the mean rate during the 21st Century could be as high as 20 mm/year. Many long-term stations show definite long-term trends (Figure 1.53) that are most likely related to global climate change. There is presently a general increase of about 0.5—1 m per century for gauges located in geologically "stable" regions of the world. This could reach 2 m per century if some of the higher estimates are correct. It is now generally accepted that the long-term increase in global sea level is linked to melting of water locked in the polar ice sheets and northern hemisphere glaciers due to global warming.

Changes in the land-based ice cover alter both the ocean volume and the geodetic loading. However, it should be noted that the rate of rise varies considerably from location to location and can be strongly dependent on regional tectonic activity and the lingering effects of the continuing glacio-isostatic response (Peltier, 1990; Tushingham and Peltier, 1992; James et al., 2021). Investigators attempting to extract a possible climate change component from global sea-level records spend considerable effort generating spatially smoothed data sets consisting of a relatively small subset of the total tide gauge data available from the world archives (Figure 1.61).

Mean sea levels are usually computed from long series of hourly observations. Generally, a simple arithmetic average of hourly values is computed, but other methods, including the application of low-pass numerical filters to eliminate tides and storm surges, may be used before the means are computed. The average of all high and low water levels is called the mean tide height; it is close to, but not identical with, mean sea level. Monthly and annual mean sea-level series from a global network of stations are collected and published by the Permanent Service for Mean Sea level (PSMSL) in England, together with details of gauge location and the definitions of the datums to which the measurements are referred. Data at PSMSL are held for 2,167 "Metric" file stations of which 1,384 have had their data adjusted to a tide gauge benchmark datum to form the Revised Local Reference (RLR) data set. This datum is approximately 7,000 m below mean sea level, with the arbitrary choice made to avoid negative numbers in the resulting RLR monthly and annual mean values. Of these stations, 133 have data from before 1900. Most of these stations are in the northern hemisphere so that careful analysis is necessary to avoid geographic bias in their interpretation. Amsterdam has the longest tide-gauge record in the world but the oldest data that satisfy the selection criteria of the PSMSL are from Brest, starting in 1807 (Figure 1.61). For many stations, the PSMSL Website (www.psmsl.org) provides links to other sources of sea level and land level (GPS) information. The site also has a considerable amount of background information on making sea-level measurements and analyzing the data. Recent developments at PSMSL have been described by Holgate et al. (2013).

Since the 1990s, tide gauge records have been augmented by satellite altimetry measurements of global sea level. The long-term trend in the spatially averaged altimetry data (Figure 1.62) supports arguments for accelerated eustatic sea-level rise compared to the beginning of the twentieth century. According to satellite records, global sea levels are presently rising at a mean rate of about 3.4 ± 0.4 mm/year.

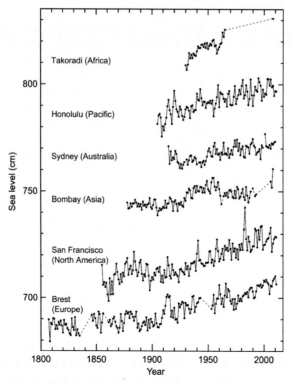

FIGURE 1.61 Annual mean sea-level values for the longest records for each continent. Data are Revised Local Reference (RLR) records from the Permanent Service for Mean Sea-level at the Bidston Observatory in Merseyside, United Kingdom. Each record has been given an arbitrary offset for presentation purposes. The Takoradi record was truncated in 1965 when major problems with the gauge were reported. *Courtesy of Philip Woodworth, Permanent Service for Mean Sea Level, National Oceanography Centre, UK.*

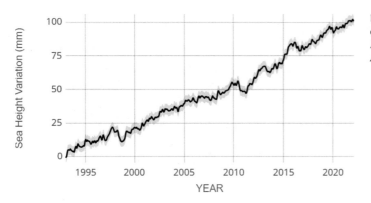

FIGURE 1.62 Global mean sea level rise based solely on satellite data (1993−2022). Mean rate of rise =3.4 ± 0.4 mm/yr. *Data source credit: Physical Oceanography Distributed Active Archive Center (PO.DAAC). JPL/NASA.*

1.7.2 Tide and Pressure Gauges

Although pressure and acoustic gauges are becoming increasingly more popular around the world, many sea-level measurements are still made using a float gauge, in which the float rises and falls with the water level (Figure 1.63). Modern recording systems replace the analog pen with a digital recording system that allows for unattended, real-time data acquisition, control, and communication. Data are recorded on flash memory with removable media support using SD cards, MMC cards, and USB thumb drives. Measurements can be event-driven or independently scheduled, with sample intervals ranging from 1 s to a day. Loggers can have a wide variety of built-in functions such as min/max and averaging over specified periods. Many of the digital recording systems have been equipped with telemetering systems that send sea-level heights and time via satellite transmitters (e.g., GOES, INMARSAT, METEOSAT, INSAT, IRIDIUM), modem (Cell, CDPD (Cellular Digital Packet Data), telephone landline), radio (UHF/VHF, spread spectrum), or other serial communication devices. Direct and reliable communication links are critical for tsunami warning systems.

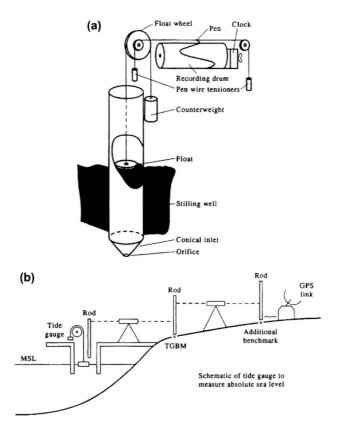

FIGURE 1.63 (a) A basic float stilling-well gauge used to measure water levels on the coast; (b) schematic of tide-gauge station with the gauge, network of benchmarks and advanced geodetic link. *TGBM*, tide gauge benchmark. *Source: Woodworth (1991). Reproduced and modified with permission from the Coastal Education and Research Foundation, Inc.*

The most important aspect of this type of sea-level measurement is the installation of the float stilling well. The nature of this installation will determine the frequency response of the float system and will help damp out unwanted high-frequency oscillations due to surface gravity waves. Stilling wells can have their inlet at the bottom of the well or use a pipe inlet connected to the lower part of the well. Both designs damp out the high-frequency sea-level changes. Maintenance of the sea-level gauge ensures that the water inlet orifice is kept clear of obstructions from silt, sand, or marine organisms. Also, in areas of strong stratification such as rivers or estuaries, the water in the stilling well can be of a different density than the water surrounding it. When installing such a measurement system, it is important to provide adequate protection from contamination of the stilling well and from damage to the recorder. A potential hazard in all harbor gauge installations is damage from ship traffic or contamination of the float response from ship wakes.

Proper installation of coastal sea-level gauges requires that they be surveyed into a legal benchmark so that measured changes will be known relative to a known land elevation (Figure 1.63b). When properly tied to a benchmark height, changes in gauge height relative to land can be taken into account when computing the mean sea-level. This tie-in with the local benchmark datum is done by running the level back to the nearest available geodetic datum. Modern three-dimensional satellite-based GPS and long-baseline telemetry systems now make it possible to accurately determine the vertical movement of a tide gauge relative to the geoid. Satellite altimeter measurements initiated in the early 1990s provide global measurements of the relative sea surface, which can then be compared with the conventional sea-level measurements. The importance of the altimeter sea level records will continue to grow as the measurement accuracy and record length continue to improve. The usual test of a sea-level instrument (called the Van de Casteele test) involves operating the instrument over a full tidal cycle and comparing the results against simultaneous measurements made with a manual procedure. This procedure only shows if the recording device is operating properly so that a separate test of the stilling well is needed to accurately measure the response of the float. Other than mechanical problems and poorly documented repositioning errors, timing errors are one of the major sources of error in sea-level records. Sea-level gauges have either mechanical or electronic clocks, which must be periodically checked to ensure that there is no significant drift in the timing of the mechanism. When possible, these checks should be made weekly. Depending on the specific instrument, well-maintained sea-level recorders are capable of measurements accurate to within a several millimeters.

Another type of sea-level measuring device is the pneumatic or bubbler gauge (Figure 1.64). This system links changes in the hydrostatic pressure at the outlet point of the bubbles to variations in sea level. Like other pressure sensing gauges, this gauge measures the combination of sea-level height and atmospheric pressure. As a consequence, most bubbler gauges operate in the differential mode whereby the recorded value is the difference between the measured pressure and the atmospheric pressure. While these instruments are somewhat less accurate than float gauges, they are useful in installations where a float gauge would be subject to either damage from ship traffic or strongly influenced by wave motion. In a study in Tasmania in the 1970s, Australian technicians reported that the plastic pressure tubes leading from the electronics package to the ocean were constantly being destroyed by curious wombats.

Sea-level heights are recorded in a variety of formats. Graphical records must be digitized and care taken to record only values properly resolved by the instrument. Differences in recording scale will lead to variations in the resolution of the

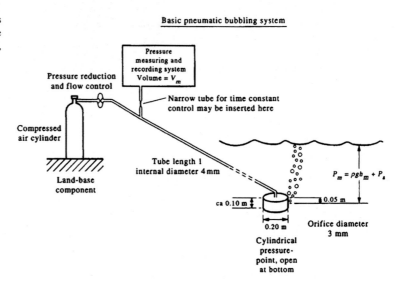

FIGURE 1.64 The pneumatic or bubble gauge. This system links changes in the hydrostatic pressure, P_m, at the outlet point of the bubbles to variations in sea-level, h_m, water density, ρ, and atmospheric pressure, P_a.

gauge thereby limiting the accuracy of the digitized data. Modern gauges eliminate this possible problem by recording digital data. (For most tsunami work, digital sampling intervals of less than 6 min—and preferably 1 min or less—are highly recommended.) During digitization of analog data, it is important to edit out any of the obvious errors due to pen-ink problems or mechanical failures in the advance mechanism. Also, when long-term sea-level variations are of interest, one must be careful to filter out high-frequency fluctuations due to waves and seiches. The choice of time reference is important in creating a sea-level time series. The usual convention is to use the local time at the location of the tide gauge, which can then be referenced to Greenwich Mean Time (GMT, now called Universal Time Coordonné, UTC). The sea-level record should also contain some information about the reference height datum on which the sea-level heights are based. Digital recording systems are subject to clock errors and care should be taken to correct for these errors when the digital records are examined.

In recent years, sensitive and accurate pressure sensors have been developed for measurement of deep-sea tides and trans-oceanic tsunamis where fluctuations on the order of 1 mm need to be detected in depths of thousands of meters. At first, these sensors were largely based on the "Vibraton" built by United Control Corporation, which measured pressure by changes in the frequency of oscillation of a wire under tension. This frequency change was measured to an accuracy of 6×10^{-4} Hz and led to a sea-level accuracy of 0.8 mm (Snodgrass, 1968). To maintain this high level of accuracy, it was necessary to correct for temperature effects to a resolution of $0.001°C$. When Vibraton sensors ceased to be commercially available they were replaced by resonating quartz crystal transducers (Wimbush, 1977), which are now the standard for measurement in both deep sea and coastal pressure gauge recorders. Manufactured by Paroscientific (Paros, 1976), these sensors have a depth sensitivity of 1×10^{-4} dbars (approximately 10^{-4} m or 0.1 mm) for both shallow and deep-sea measurements. Most modern pressure gauges used in coastal- and deep-sea tidal and tsunami measurements make use of these types of sensors. In addition, Aanderaa commonly used pressure sensors manufactured by the Finnish company Vaisala in their water level gauges; these sensors have a resolution of 0.01 dbars (0.01 hPa) over a range of 500–1,100 hPa. (Pressure gauges used in coastal waters are often known as water level gauges.) Temperature correction is required to maintain accurate depth measurements. Wearn and Baker (1980) report measurements made by such quartz sensors from year-long moorings in the Southern Ocean. Unfortunately, instabilities in the quartz sensors lead to sensor drifts, which limited the use of the sensors in long-term, deep pressure measurements. The use of dual pressure sensors helps to correct for drift since each pressure sensor will have somewhat different drift characteristics but will produce similar responses to higher frequency oceanic variability. A technical report by Paroscientific that compared 10-min data from a bottom pressure recorder (BPR) in 7,000 m of water with a surface nano-barometer that was measuring ambient atmospheric pressure (an instrument that is 700 times more accurate) shows that the BPR-tracked changes in barometric pressure with a resolution of 0.25 Pa or about 0.025 mm of equivalent water depth (Schaad, 2009; Technical Note, Paroscientific, Inc.). Oceanographic instruments can now resolve pressure variations to a fraction of a millimeter at full ocean depth, corresponding to parts per billion in water depth. Tides, infragravity waves, tsunamis, meteotsunamis—tsunami-like oscillations generated by atmospheric pressure pulses and disturbances (Monserrat et al., 2006; Vilibić et al., 2021)—and seasonal variations in sea level are readily resolved at this precision. However, sensor drift of roughly 1 cm/year precludes long-term accuracy measurements to this level.

1.7.2.1 DART and Cabled Observatory BPRs

Paroscientific BPRs form the backbone of the DART (Deep-ocean Assessment and Reporting of Tsunamis) buoy system operated as part of the Tsunami Warning Centers (TWCs) in the Pacific, Atlantic and Indian oceans. The most extensive warning facilities are the NOAA Pacific Tsunami Warning Center (PTWC) in Hawaii and the U.S. National Tsunami Warning Center in Palmer, Alaska, which has primary coverage for the U.S., Canada, the Caribbean and the Gulf of Mexico. BPRs also form the backbone of the regional-scale tsunami and infrared gravity wave array of the Ocean Networks Canada (formerly NEPTUNE Canada) cabled observatory off the coast of British Columbia (maximum 2,700 m depth) (www.neptunecanada.com) and the seafloor monitoring program at Axial Seamount (an active submarine volcano at ~1,500 m depth on Juan de Fuca Ridge) as part of the U.S. cabled Ocean Observatories Initiative (OOI) off the west coast of Oregon (Fox, 2016; Nooner and Chadwick Jr., 2016; Fine et al., 2020).

DART II became operational in 2005 (Green, 2006). Although most DART buoys are located around the Pacific Rim (Figure 1.65), several buoys are now located in the Caribbean, off the east coast of the United States, and in the Indian Ocean to the west of Sumatra (Indonesia), the site of the devastating magnitude 9.0 Sumatra-Andaman earthquake and tsunami of 26 December 2004. Tsunami wave heights recorded by a bottom-anchored seafloor BPR (Figure 1.66) are sent in real time by an acoustic modem link to a nearby surface mooring, which then transmits the data via satellite to a shore site in the United States (Gonzalez et al., 1998). The BPR collects temperature and pressure at 15 s intervals. The pressure

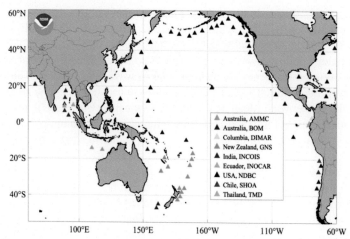

FIGURE 1.65 Locations of DART buoys in the World Ocean and the countries responsible for maintaining the systems. *From NDBC US National Data Buoy Center (NOAA). June 2022.*

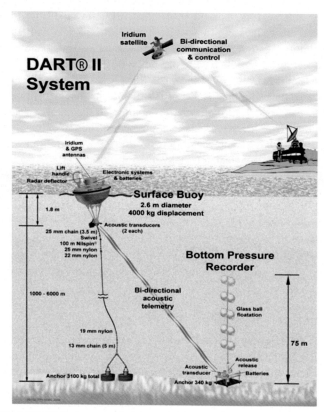

FIGURE 1.66 The components of the DART II System presently in operation in the ocean. Note the separate moorings for the buoy and the acoustic link. *From NDBC US National Data Buoy Center (NOAA).*

data are corrected for temperature and the pressure converted to an equivalent sea surface height (height of the ocean surface above the seafloor) using a constant conversion of 670 mm/psia (pounds per square inch absolute).

The system has two data reporting modes: standard and event. The system operates routinely in standard mode, in which four spot-values of 15 s data are sent at 15 min intervals at scheduled transmission times. When the internal detection software identifies an event, the system begins event mode transmissions (Mofjeld, 2009). In the event mode, 15 s values are transmitted during the initial few minutes, followed by 1 min averages. Event mode messages also contain the time of the initial occurrence of the event. The system returns to standard transmission after 4 h of 1 min real-time transmissions if no further events are detected. Two-way communication via the Iridium commercial satellite communications system (Meinig et al., 2005) allows the Tsunami Warning Centers to set stations in event mode in anticipation of possible tsunamis or retrieve the high-resolution (15 s intervals) data in 1-h blocks for detailed analysis. The DART buoys have two independent and redundant communications systems. The National Data Buoy Center (NDBC) distributes the data from both transmitters under separate transmitter identifiers. NDBC receives the data from the DART II systems, formats the data into bulletins grouped by ocean basin (see the NDBC − DART GTS Bulletin Transmitter List, for a listing of the bulletin headers used for each transmitted identifier), and then delivers them to the National Weather Service Telecommunications Gateway (NWSTG) that then distributes the data in real time to the TWCs via NWS communications and nationally and internationally via the Global Telecommunications System. A comprehensive appraisal of DART is provided by Mungov et al. (2013) and Rabinovich and Eblé (2015).

The system maintains a measurement accuracy within ±1.0 cm of the observed tides deployed in similar depths off Hawaii within 100 nautical miles of an operational DART station using a standard tide model for tidal adjustments between stations. Measurement resolution is 0.25 mm in water depths of 1,000−6,000 m. For 15 s data spanning a duration of 1 h, the time from a request for data to the receipt of data at the TWC must be less than 15 min. In event mode, the time from the start of the event to receipt of the data at the TWC's server must be less than 3 min.

The BPRs deployed within the NEPTUNE-Canada and VENUS cabled observatory networks of Ocean Networks Canada (ONC) are being used to study tsunamis, seafloor loading, tides, infragravity waves, seismic waves, and other ocean and earth phenomena off the west coast of Canada (Barnes et al., 2008). A low power high-precision signal period counter developed jointly by the Pacific Geoscience Centre and Bennest Enterprises (E. Davis, R. Meldrum, J. Bennest, pers. Com., Sidney, BC, 2013) is coupled to standard Paroscientific quartz pressure sensors to provide a resolution of better than 0.04 mm in 4,000 m of water at a sampling rate of 1 Hz. Beginning with the 2009-Samoan tsunami, the 2010-Chilean tsunami and the 2011 Tohoku tsunami originating off Japan, most tsunamis originating from submarine earthquakes having magnitudes greater than 8.0 have been clearly recorded by the bottom-mounted NEPTUNE-Canada "tsunami array" of BPRs moored in depths of up to 2,700 m off the coast of Vancouver Island, by open-ocean DART stations located nearby, and by tide gauges on the British Columbia−Washington coast (Thomson et al., 2009, 2013; Rabinovich et al., 2012; Wang et al., 2020). These high quality, far-field tsunami records provide a comprehensive analysis of the events up to 11,000 km from the source area.

1.7.3 Satellite Altimetry

Conventional sea-level measurement systems are limited by the need for a fixed platform installation. As a result, they are only possible from coastal or island stations where they can be referenced to the land boundary. Unfortunately, there are large segments of the world's ocean without islands, so that the best hope for long-term global-scale sea-level measurements lies with satellite-borne radar altimetry. Early studies (Huang et al., 1978) using GEOS-3 altimeter data with its fairly low precision of 20−30 cm, demonstrated the value of such data for estimating the variability of the sea surface from repeated passes of the satellite radar. In this case, the difference between repeated collinear satellite passes eliminates the unknown contribution of the earth's geoid to the radar altimeter measurement. This same technique was employed by Cheney et al. (1983), using 1,000 orbits of high quality SEASAT radar altimeter data.

With a known precision of 5−8 cm, the early radar data provided some of the first large-scale maps of mesoscale variability in the world's ocean. Satellite altimetry is now actively pursued by physical oceanographers interested in ocean circulation problems, including the formation and propagation of mesoscale eddies, and major current systems. The experience with GEOS-3 and SEASAT altimeter data demonstrated the great potential of these systems, which now provide sufficient accuracy to allow the specification of the mean ocean circulation related to the ocean surface topography. As with earlier applications, the primary concern is with the contribution of the earth's geoid to the satellite altimeter measurements. The geoid is known to have variations with space scales similar to the scales of sea-level fluctuations associated with the mean and mesoscale ocean circulation. In addition, satellite altimetry data must be corrected for: (1) variations in satellite orbit; (2) atmospheric effects, requiring knowledge of the intervening atmospheric temperature and water vapor profiles; and (3) sea state, which affects the shape of the reflected radar waveforms. Altimeters onboard TOPEX/Poseidon and a follow-on series of satellites (combined with more exact processing methods) allow more precise determination of sea-level variations, to a level of about 2 cm, while assimilation of radar altimeter data into numerical ocean models has paved the way to improved mapping of mean sea level and of the geoid. More recent altimeter missions, such as the Surface Water Ocean Topography Mission (SWOT) launched on 16 December 2022, also allow higher-resolution description of the ocean mesoscale variability. Developed jointly by NASA and Centre National D'Etudes Spatiales (CNES), with contributions from the Canadian Space Agency (CSA) and United Kingdom Space Agency, SWOT is the first satellite mission that will observe the heights of most surface water on Earth, including that in the oceans, lakes, rivers, and reservoirs.

Marked headway has been made in the area of satellite altimetry due to the successful deployment of a number of spaceborne altimeters. The first to generate a lot of new data was the GEOSAT satellite first launched in 1985 by the U.S. Navy in an effort to more precisely map the influences of the geoid on missile tracks. After an 18-month "geodetic mission" the Navy was convinced by Dr. Jim Mitchell and others to put the satellite into an "exact repeat orbit" in November, 1986, using the same orbit as the previous SEASAT satellite (Tapley et al., 1982). The altimeter data from this orbit had already been made public and thus the classified altimeter data from the geodetic mapping mission were already compromised for this orbit. By having the satellite operating in this orbit, scientists would be able to collect and analyze data on the ocean's height variability. Fortunately, the GEOSAT altimeter continued to function into 1989 providing almost three full years of repeated altimetry measurements. In addition, the navy has now released the "crossover" data from the geodetic mission. In this mission, the track did not repeat but the crossovers between ascending and descending tracks provided valuable information on ocean height variability. Thus, it is possible to combine data from the earlier crossovers and repeat orbits from the "exact repeat mission" to form a nearly 5-year time series of sea surface height variations. It should be stressed that without a detailed knowledge of the earth's geoid it is not possible to compute absolute currents and the main area of investigation provided by the GEOSAT data was in studying the ocean's height variability.

Considerable experience was gained in computing the various corrections that are needed for satellite altimeter data (Chelton, 1988). These include the ionospheric correction, the dry tropospheric correction, and the wet tropospheric correction. Added to these are the errors due to electromagnetic bias, antenna mispointing, antenna gain calibration, the inverse barometer effect, ocean/earth tides, and precise orbit determination. Since the GEOSAT satellite did not carry a radiometer to compute tropospheric water vapor, other operational satellite sensors were used to compute the atmospheric moisture to correct the altimeter pathlength (Emery et al., 1989a). Many experiments were conducted to better understand the electromagnetic bias correction (Born et al., 1982; Hayne and Hancock, 1982). Other corrections can be routinely computed from available sources, including the dry troposphere correction, which requires knowledge of the atmospheric pressure (Chelton, 1988).

GEOSAT data have been used to map both the large-scale and smaller scale regional circulations of the ocean. Miller and Cheney (1990) used GEOSAT data to monitor the meridional transport of warm surface water in the tropical Pacific during an El Niño event. Combining crossover and colinear data, the authors constructed a continuous time series of sea-level changes on a 2° × 1° grid in the Pacific between 20° N and 20° S for the 4-year period from 1985 to 1989. They concluded that the 1986−87 El Niño was a low-frequency modulation of the normal seasonal sea-level cycle and that a buildup of sea level in the western Pacific was not required as a precursor to an El Niño event. A similar analysis of colinear GEOSAT data for the tropical Atlantic (Arnault et al., 1990) showed good agreement between the satellite-sensed sea-level changes and those measured *in situ* using dynamic height methods. Using GEOSAT sea-level residuals computed from a 2-year mean, Vazquez et al. (1990) examined the behavior of the Gulf Stream downstream of Cape Hatteras. Comparisons with NOAA infrared satellite imagery show a fair agreement with the Gulf Stream path, but with some sea-level deviation maps not showing a clear location of the main stream. In the same geographic, region Born et al. (1987) used a combination of GEOSAT altimetry and airborne XBT data to map geoid profiles as the difference between the altimetric sea level and the baroclinic dynamic height. Many other oceanographers have used GEOSAT data to study a great variety of oceanographic circulation systems.

In the summer of 1992, the long awaited TOPEX-Poseidon altimetric satellite was launched. Carrying two altimeters (one French and one U.S.) with a single antenna, TOPEX/Poseidon marked a significant step forward in altimetric remote sensing. The NASA altimeter was a dual-frequency altimeter, which was able to compensate for the influence of ionospheric changes. The French altimeter was the first solid-state instrument to be deployed in space. In addition, there was a boresight microwave radiometer (TOPEX Microwave Radiometer, TMR) to provide real time atmospheric water vapor measurements for the computation of wet troposphere corrections for the onboard altimeters. The resulting combination of data provided altimeter heights accurate to ±10 cm, giving important data for tidal analysis of global sea levels and other large-scale processes (e.g., Schrama and Ray, 1994; Ray, 1998; Di Lorenzo et al., 2005; Foreman et al., 1998; Cherniawsky et al., 2001). A truly joint project, the satellite was built in the U.S. and launched by the French *Ariane* launch vehicle. In early 2002, the altimetric satellite Jason-1 (launched in December 2001, carrying only the French digital altimeter) was placed into the TOPEX/Poseidon (TP) orbit and the older satellite shifted to an orbit midway between that of the Jason-1 orbital paths after a 6 month intercalibration period. During this time Jason-1 followed TP by 70 s in time. This phase of the mission confirmed that the altimeter record started by TP could be continued by Jason-1.

Jason-1 had a repeat orbital period of 10 days and was capable of high-precision ocean altimetry measurements from satellite to the ocean surface of a few centimeters. A companion altimetric satellite OSTM/Jason-2—where OSTM stands for Ocean Surface Topography Mission—was launched in June 2008 and is now in a circular, non-sun-synchronous orbit at an inclination of 66° to Earth's equator which allows it to survey 95% of Earth's ice-free ocean every 10 days. In 2009,

Jason-1 was moved to the opposite side of Earth from the Jason-2, and now flies over the same region of the ocean that Jason-2 flew over 5 days earlier. The ground tracks of Jason-1 fall mid-way between those of Jason-2, which are about 315 km apart at the equator. The follow-on, Jason-3, was launched January 17, 2016, again as cooperation between France and the U.S. It continues to fly in the same orbit as Topex/Poseidon. This combined mission provides twice the number of measurements of the ocean's surface for accurate observations of sea surface height variation that include changes in global sea level, the velocity of ocean currents, including those associated with mesoscale eddies, and changes in ocean heat storage (thermosteric effect). The combined mission helps pave the way for a future ocean altimeter mission that would collect much more detailed data with its single instrument than the two Jason satellites now do together.

Altimetry data are available through the French Web site AVISO (Archiving, Validation and Interpretation of Satellite Oceanographic data). Several data sets are available, including the mean sea surface height anomaly (msla) acquired from the Ssalto/Duacs project. AVISO presently distributes satellite altimetry data from TOPEX/Poseidon, Jason-1, ERS-1 and ERS-2, EnviSat, Jason 2, Jason-3, Sentinel-6 and DORIS (precise orbit determination and positioning; a radio system deployed by France/CNES) products. Altimetry is also provided by Cryosat-2 with an altimeter (Siral) working in an interferometric mode, with a high orbit inclination of 92° to satisfy scientific needs for observing the polar regions and ice sheets, and with an orbit non-sun-synchronous (commonly used for remote sensing satellites). The Chinese altimeter satellites HY-2A, 2B have 14-day orbits.

Figure 1.67 is an example of what can be achieved using data downloaded from AVISO. Here, we have compared the alongshore surface currents derived from altimetry with co-located alongshore currents observed at 35 m depth at long-term current meter mooring site A1 in 500 m of water off the west coast of Vancouver Island, British Columbia (cf., Thomson and Ware, 1996). The satellite node (at 48.5° N, 126.3° W) and the mooring location are within 7 km of one another. In general, the 10-day altimetry data reproduces the seasonal cycle of the alongshore currents over the continental slope but cannot reproduce the energetic higher frequency motions associated with local, as well as remote, wind forcing (cf. Hickey et al., 2003; Connolly et al., 2014).

1.7.3.1 Accuracy of Altimeter Sea Level Heights

Modern satellite altimeters are capable of measuring the height of the sub-satellite sea surface to centimeter accuracy. This accuracy is due to a combination of a number of effects. First there is the need for a very accurate knowledge and control of the satellite's orbit. This is accomplished by a number of redundant orbital altitude measurements systems (Figure 1.68). The method to accurately calculate the exact satellite orbit is known as "precise orbit determination" or POD. The tracking system used in the POD consists of (a) an onboard laser reflector that is used to range from 10 to 20 laser ranging stations over the Earth's surface and (b) The French DORIS (Doppler Orbitography and Radiopositiong Integrated by Satellite) system that communicates with 60 ground stations, again well distributed over the Earth's surface. The DORIS receiver measures the Doppler shift of microwave signals to support the POD. In addition, NASA provides a GPS receiver onboard the spacecraft to give precise and continuous tracking of the satellite position by monitoring the range and timing signals

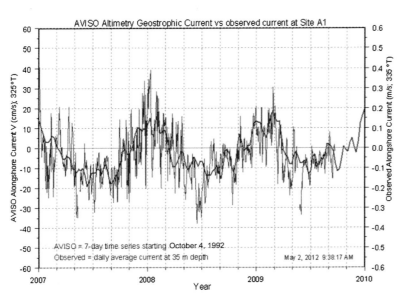

FIGURE 1.67 Comparison of the alongshore component of geostrophic current velocity derived from satellite altimetry (blue line) and the corresponding alongshore current observed by a current meter moored at 35 m depth (red line). The geostrophic currents are computed at 7-day time steps by AVISO; the current meter data are daily mean values for a nearby long-term mooring site located in 500 m of water on the continental slope off southwest Vancouver Island, British Columbia. The altimeter products were produced by SSALTO/DUACS, where DUACS (Data Unification and Altimeter Combination System) is part of the CNES/CLS multi-mission ground segment (SSALTO) providing satellite altimeter sea level products and distributed by AVISO (http://www.aviso.oceanobs.com/duacs/).

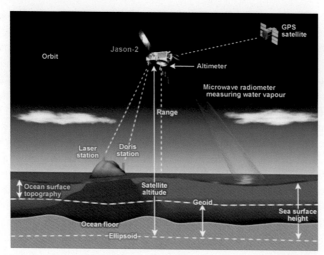

FIGURE 1.68 Measurement systems used to calculate the altimeter satellite orbit including the laser ranging system, the DORIS system and the GPS receiver.

from up to 12 GPS satellites at the same time. POD combines all of these measurements with accurate models of the forces (e.g., gravity, aerodynamic drag) that govern the satellite's motion. The result is an orbit accurate to about 2 cm.

The next component of measuring the ocean's height is the range measurement from the satellite to the Earth's surface. This is done with a radar altimeter that emits microwave pulses at frequencies of 13.6 and 5.3 GHz. These pulses are bounced off the sea surface back to the satellite which is used to calculate the distance from the satellite. The U.S.–French altimeter satellites carry the Poseidon-3 solid state altimeters. The signal must be corrected for a number of factors, but the most important correction is for atmospheric water vapor between the satellite and the sea surface. For this correction the satellite has a bore-sighted passive microwave radiometer known as the Advanced Microwave Radiometer (AMR). It measures microwave emissions in 23.8, 18.7 and 34 GHz channels that are used to calculate the radar path delay caused by atmospheric water vapor. Other corrections are the ionospheric correction, the sea-state bias correction, tidal correction and the inverse barometer correction. After all of these corrections are made the altimeter individual measurement is accurate to less than 2 cm, which applies to a few-kilometer diameter spot directly beneath the satellite. If one averaged the few-hundred thousand measurements collected by the satellite in the time it takes to cover the global ocean (10 days), global mean sea level can be determined with a precision of several millimeters.

To verify and validate the accuracy of the satellite altimeter heights both NASA and CNES (French Centre national d'etudes spatiales) operate verification sites along the ground track of the satellite. The CNES validation site is on the French island of Corsica in the Mediterranean Sea, while the NASA verification site is on an oil platform known as Harvest (Figure 1.69), located at the eastern entrance to the Santa Barbara Channel, off the coast of central California. Both sites are

FIGURE 1.69 Harvest oil platform used as an altimeter validation site located at the eastern end of the Santa Barbara Channel, California.

equipped with accurate tracking systems (e.g., GPS) that enable accurate surveying of the stations into the same Earth frame that is used in the POD orbit calculations. They are also equipped with tide gauges (mechanical and laser) to directly measure water level beneath the satellite on a satellite overpass. Every 10 days the satellite passes directly overhead providing two independent observations of sea level to compare with the satellite measurement. These measurements have verified the sea height altimeter accuracy of slightly less than 2 cm.

1.7.3.2 Global Sea Level Rise

One of the important contributions that altimeter measurements have made recently is providing better measurements of global sea level rise (Figure 1.70). Before altimeter data were available, sea level rise was based solely on coastal sea level stations, generally referred to as tide gauges. Global mean sea level has risen steadily since the start of the longest running tide gauge recording at Brest (France) in 1807 (Permanent Service for Mean Sea Level (PSMSL), National Oceanography Centre, Liverpool, UK).

There is a marked increase in slope at about 2006, after which global mean sea level rose by 3.6 mm per year up to 2015, at rate which was 2.5 times the average rate of 1.4 mm per year throughout most of the twentieth century. By 2100, sea global sea level will likely increase by at least 0.3 m relative to the year 2000, regardless of whether or not there is a leveling off of greenhouse gas emissions. This net additional one-third of a meter in sea level rise will result in major flooding events even under nominal conditions. When strong storms hit, many coastal areas will be flooded by the higher combined water levels. Across the globe, sea level rise is more intense in some regions than others, with many of the highest increases on eastern seaboards (Figure 1.71). Some of the higher values are in the Gulf of Mexico, where storm surges from hurricanes regularly devastate U.S. coastal communities. The fact that the distribution in Figure 1.71 is mostly blue indicates that most locations have already experienced ongoing sea level rise and that there are only a very few areas that have seen a drop in sea level, specifically Finland and northern Canada, where the land continues to uplift due to rebound of the earth's mantle from the melting of ice contained in glaciers from the last glaciation.

There is a variety of forecasts for future sea level rise. If we consider only the effect of thermal expansion due to warming of the atmosphere (Figure 1.72), we find a considerable spread in these forecasts. The average of these forecasts still predicts an increase in sea level of about 1 m above the 2020 global average by the end of the century. The global sea level rise due to thermal expansion alone will lead to the flooding of major coastal cities (Figure 1.73).

Although coastal cities in Asia will be the most impacted by rising sea levels, cities in Oceania, Africa, Europe and the Americas will also be affected. In fact, Manhattan and Miami will be at least partially under water. Most coastal cities are densely populated but the increased frequency of flooding will occur slowly enough to allow people time to relocate. Problems will first occur during the storm surges mentioned earlier. In Oceania, many of the low islands and atolls will be submerged and no longer inhabitable. In the U.S., 40% of the population lives in coastal areas that will be influenced by sea

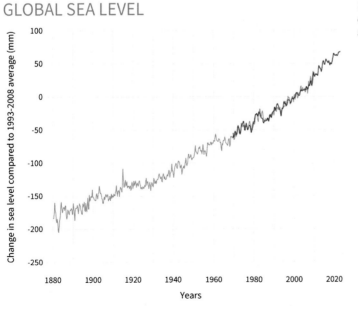

GLOBAL SEA LEVEL

FIGURE 1.70   Global mean sea level rise (mm) 1880–2020. The dark line is based on added tide gauge data and after 1985 satellite altimeter data. *Courtesy of NOAA/Climate.*

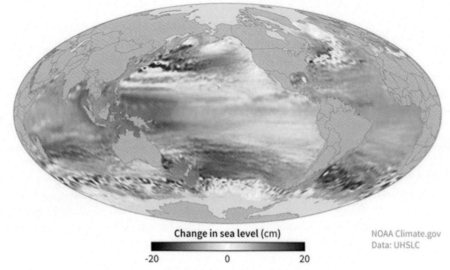

Change in sea level (cm)

-20 0 20

NOAA Climate.gov
Data: UHSLC

FIGURE 1.71 Global sea level change from 1993 to 2019 (cm).

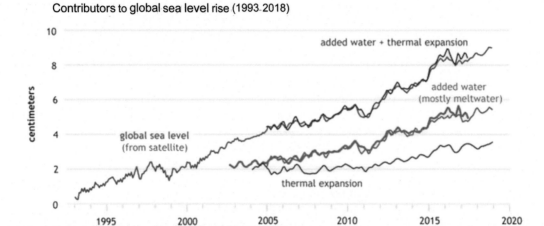

Contributors to global sea level rise (1993-2018)

FIGURE 1.72 Independent estimates of the contributions of thermal expansion (red line) and added mass mainly from glacial melt (blue) to the net change in the global mean sea level rise derived from satellite altimeter measurements from 1993 to 2018. The purple line (the summation of the estimated thermal and mass contributions) closely agrees with the satellite record (black line). *Modified after Lindsey (2022) and adapted from the "State of the Climate in 2018", NOAA Climate.gov.*

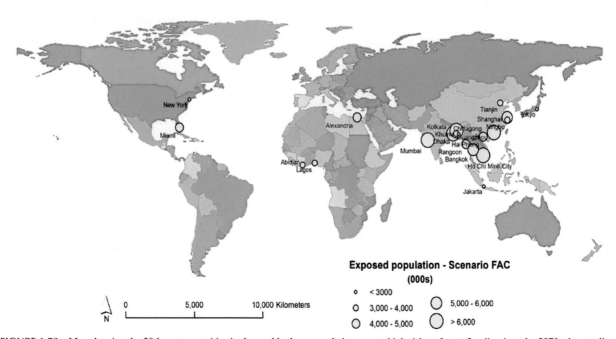

FIGURE 1.73 Map showing the 20 largest port cities in the world whose populations are at high risk to future flooding into the 2070s due to climate change. Estimates take into account global sea level rise, storm surge enhancement, local subsidence and socio-economic factors. FAC denotes Future All Cities. *Modified after Hanson et al. (2011).*

level rise. In Miami, the authorities are already constructing dikes to prevent flooding. In Louisiana, dikes are now protecting the areas around New Orleans. At some point, however, the ocean will prevail, and all these preventive measures will prove inadequate. Once again, we note that the present rate of sea level rise of around 3.6 mm/year is more than twice the rate of 1.4 mm/year that existed throughout most of the 20th century.

Of the factors causing sea levels to rise, the greatest is the addition of new volumes of water through the melting of land-based glaciers and ice sheets due to global warming. This melt water contribution is much larger than that caused by the thermal expansion of the ocean water. These two effects combined account for the majority of the sea level rise. Viewing these increases as forecasts (Figure 1.74) for the end of the 21st Century, we obtain a range of global sea level increase from a low of a few centimeters to an extreme value of about 2.5 m.

As evident in Figure 1.75a, the extreme of 2.5 m global sea level rise will lead to seawater inundation for many coastal areas of the United States by the end of the century. Much higher sea levels are also expected for Canada by 2100 (James et al., 2021), which will be further augmented under an enhanced climate scenario that features an additional 65 cm of sea-level rise sourced from West Antarctica being added to the median IPCC (2013) RCP8.5 projections (Figure 1.75b).

1.7.4 Inverted Echo Sounder

As noted in Section 1.5, accurate depth measurements using acoustic sounders require corrections on the order of $\pm 1\%$ for variations in sound speed introduced by changes in oceanic density. Rossby (1969) suggested that this effect could be used to advantage since it provided a way to measure variations in travel times of acoustic pulses sent from the sea floor due to changes in the depth of the thermocline. Moreover, the fact that travel times are integrated measurements means that they effectively filter out all but the fundamental mode of any vertical oscillations. This idea led to the development of the IES in which the round-trip travel time of regularly spaced 10 kHz acoustic pulses from the seafloor are now used to determine temporal variability in the integrated density structure of the ocean. The IES has been widely used in studies of the Gulf Stream, where its records are interpreted in terms of thermocline depth, heat content, and dynamic height (Rossby, 1969; Watts and Rossby, 1977). It has also been used in the equatorial Pacific and Atlantic although interpretation of the data is more uncertain because of a lack of repeated deep CTD casts to determine density variability (Chiswell et al., 1988).

Tidal period variability and large changes caused by El Niño-Southern Oscillation events are potentially serious problems in the interpretation of echo sounding data. In particular, the CTD data are needed to convert time series of acoustic travel time Δt between two depth levels (z_1, z_2) to a time series of dynamic height ΔD integrated over the pressure range p_1 to p_2 with an accuracy of $\pm 0.01-0.04$ dynamic meters. The obvious similarity between these two parameters (Watts and Rossby, 1977) can be seen from the relations

$$\Delta t_{z1/z2} = 2 \int_{z_1}^{z_2} [1 / c(S,T,p)]dz \tag{1.78}$$

and

$$\Delta D_{p2/p1} = \int_{p_1}^{p_2} [1/\rho(S,T,p) - 1/\rho(35,0,p)]dp\, 10^3 \int_{p_1}^{p_2} \delta\, dp \tag{1.79}$$

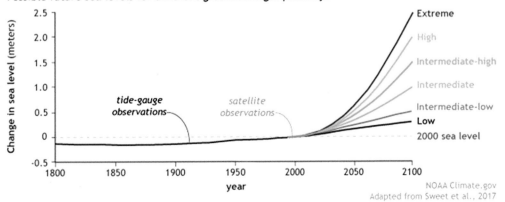

FIGURE 1.74 Possible future global sea level for different greenhouse gas conditions. *Courtesy of NOAA/Climate.*

FIGURE 1.75 (a) Flooded areas for the continental United States (excluding Alaska) under the extreme scenario of a 2.5 m global sea level rise by the year 2100; (b) projected relative sea-level at 2100 for the coast of Canada (relative to average conditions from 1986 to 2005). For the enhanced climate scenario, which features an additional 65 cm of sea-level rise sourced from West Antarctica and added to the median of IPCC RCP8.5 (IPCC, 2007, 2013). *(a) Courtesy NOAA/Climate. (b) From James et al. (2021).*

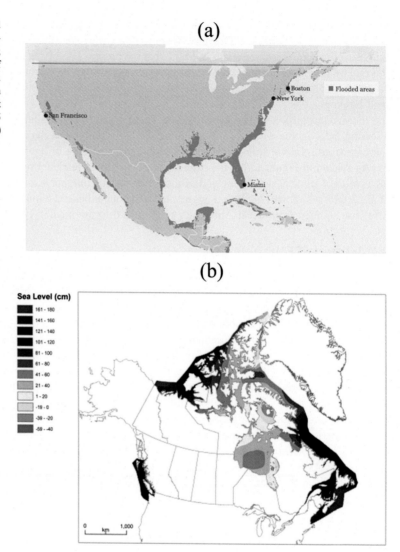

where ρ is the water density, c is the speed of sound (which is dependent on the water density through specified salinity S, temperature T, and pressure p), and z = depth (positive downward). Finally, δ, defined as

$$\delta = \alpha(S,T,p) - \alpha(35,0,p) = 1/\rho(S,T,p) - 1/\rho(35,0,p) \tag{1.80}$$

is the specific volume anomaly. In these expressions, we use SI units with depth in meters, density in kilogram per meter cubed, pressure in decibars, and dynamic height in dynamic meters (1 dyn·m = 10 m²/s²).

Chiswell et al. (1988) compare time series of dynamic height from an IES with sea-level height ($z_1 = -\eta$) from a pressure sensor located 70 km away on Palmyra Island in the central equatorial Pacific. The spectra for the dynamic height variations determined from the IES closely resembled those from the pressure gauge. Significant coherence was found between the two signals at the 99.9% level of significance. Although, in principle, varying mixtures of vertical internal modes could produce a frequency dependence in the conversion of IES to dynamic height, the effect was not significant over the year-long data series. Wimbush et al. (1990) discussed moorings in 4,325 m of water 72 km west of the subsurface pressure gauge in the Palmyra Lagoon (5°53′ N, 162°05′ W). The IES was set to 1/2-h sampling, each sample consisting of 20 pulses 10 s apart. Outliers were eliminated and the median value taken as representative of the acoustic travel time. According to Wimbush et al., a conventional IES without a pressure sensor adequately records synoptic-scale dynamic height oscillations with 20—100-day periods. Chiswell (1992) discussed 14-month records from five IESs deployed in February 1991 in a 50 km array in 4,780 m of water near 23° N, 158° W north of Hawaii. The CTD and acoustic Doppler current profiler (ADCP) data collected during monthly surveys at the array site provided sufficient density data to calibrate the IES data in terms of dynamic height and geostrophic currents.

Wimbush et al. (1990) used the response method of Munk and Cartwright (1966) to determine the diurnal and semidiurnal tides, and filtered the data with a 40-h Gaussian low-pass filter to examine the residual (nontidal) motions. Chiswell has attempted to resolve the tidal motions through 36 h burst sampling of the density structure from 3-hourly CTD profiles. IES deployments show that there is a linear relationship between dynamic height and travel time (Figure 1.76), with the calibration slope dependent on the particular T-S properties of the region. In this case, we can link variations in ΔD (for depths shallower than the reference level p_{ref} used in the dynamic height calculation) to the acoustic travel time Δt_{ref}

$$\Delta D = m\Delta t_{ref} \tag{1.81}$$

where total acoustic travel time to the bottom is

$$\Delta t = \Delta t_{ref} + \gamma H_2 \tag{1.82}$$

in which $\gamma = 2/c_b$ and c_b is the average sound speed (assumed constant) between p_{ref} and the bottom. The depth H_2 is the depth range between the seafloor (pressure $= p_b$) and the reference pressure level, p_{ref}. Solving yields,

$$\Delta D = m\left[\Delta t - \left(\gamma / \rho_b' g'\right)p_b\right] \tag{1.83}$$

where gravity and bottom density are scaled as $g' = 0.1g$ and $\rho_b' = 10^{-3}\rho_b$, respectively. For oscillations in density having periods longer than about 20 days, the second term on the right-hand side of Eqn (1.83) may be dropped, whereby

$$\Delta D = m\Delta t \tag{1.84}$$

Wimbush et al. (1990) find $m = -70$ dyn·m/s to convert Δt to ΔD, while Chiswell (1992) finds $m = -57.8$ dyn·m/s for Δt defined for $z = 0-4{,}500$ m and ΔD at 100 m referenced to 1,000 m. The high squared correlation coefficient, $r^2 = 0.93$, is based on 186 shallow ($<1{,}000$ m) and 17 deep ($>4{,}500$ m) CTD casts. The error in the slope using the deep casts is 4 dyn·m/s, with an RMS deviation of 0.017 dyn·m/s; for the shallow casts, the mean is 0.1 dyn·m/s with a standard deviation of 0.029 dyn·m/s. The travel times for the subtropical Pacific moorings of Chiswell correlate better with dynamic height measured below 100 m than with surface dynamic heights. This is because large variations in the temperature and salinity relation in the upper 100 m affect dynamic height more than they affect acoustic travel time (Chiswell et al., 1988). The tidal range of 0.08 dyn·m is relatively large compared with the seasonal range of 0.25 dyn·m and illustrates the need for detailed CTD sampling. Geostrophic currents have been derived from the array using the time series of dynamic height created from the multiple IES moorings. Aliasing of the records by high-frequency motions and a lack of CTD data to the depth of the IES remain as problems for this method.

Today, researchers can purchase commercial IES combined with an optional Paroscientific pressure sensor (the instrument is designated a "PIES"). A combined IES, data logger, and acoustic release with both pressure and Aanderaa current velocity sensors is called a CPIES. PIES is a long-life sensor logging unit that transmits a wideband acoustic pulse to accurately measure the average sound velocity through a column of water from the seabed to the sea surface and back

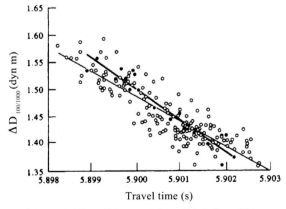

FIGURE 1.76 Dynamic height at 100 m relative to 1,000 m ($\Delta D_{100/1,000}$) from 186 shallow CTD casts plotted against corresponding travel time measured by IESs (open circles). Thin line is the least squares fit. Thick line and solid circles give $\Delta D_{100/1,000}$ calculated from 17 deep casts plotted against the corresponding travel time from 4,500 m to the surface, $T_{0/4,500}$ ($r^2 = 0.93$ and slope $= -57.8$ dyn·m/s). *From Chiswell (1992).*

again. Units are commonly used to examine tides, currents, and long-term, *in situ*, changes in the thermal properties of the ocean along with coincident variations in barotropic pressure. The pressure sensor provides an accurate measurement of depth (distance to the surface). The sampling interval of PIES can be configured serially before deployment and also via its internal acoustic telemetry link. This telemetry link also allows recorded data to be transmitted to surface at data rates ranging from 100 to 6,000 bits per second. PIES can be free-fall deployed. When the study is complete, its integrated acoustic release enables it to be commanded to disconnect from its tripod stand and return to the surface under its own buoyancy ready for collection by the surface vessel. Results from IESs can be found in Meinen and Watts (2000), Meinen et al. (2004), and Li et al. (2009). Numerous agencies (including NOAA/AMOL (Atlantic Oceanographic and Meteorological Laboratory)) have acquired these instruments from the University of Rhode Island (URI), the only known manufacturer of these systems.

1.7.5 Wave Height and Direction

Any discussion of sea-level would be incomplete without some mention of surface gravity wave measurements. Methods include: a capacitance staff which measures the change in capacitance of a conductor as the air—water interface moves up and down with passage of the waves; an upward-looking, high-frequency acoustic sounder or ADCP with a vertical-pointing transducer, which can be used to examine both the surface elevation and the associated orbital currents; a fixed graduated-staff attached to a drill platform, stuck in the sand or otherwise attached to the seafloor; satellite altimetry; a bottom-mounted pressure gauge with rapid sampling time; a shipborne Tucker wave-recorder system; and wave-riding buoys with three-axis accelerometers. For brevity, we limit our presentation to the directional waverider since it represents a reliable off-the-shelf technology that is widely used. Commercial units include the Datawell Directional Waverider Mark III and Directional Waverider 4 and the TRIAXYS Directional Wave Buoy. Both companies offer a solar-powered unit.

The Datawell directional waverider is a spherical, 0.9 m diameter buoy for measuring wave height, wave direction and water temperature. The buoy contains a heave-pitch-roll sensor, a three-axis fluxgate compass and two fixed "x" and "y" accelerometers. The directional (x, y, z) displacements in the buoy frame of reference are based on digital integration of the horizontal (x, y) and vertical (z) accelerations. Horizontal motions rather than wave slope are measured by this system. Vertical motions are measured by an accelerometer placed on a gravity-stabilized platform. The platform consists of a disk, which is suspended in a fluid within a plastic sphere placed at the bottom of the buoy. Accelerations are derived from the electrical coupling between a fixed coil on the sphere and a coil on the platform. A fluxgate compass is used to convert displacements from the buoy frame of reference to true earth coordinates. The operation of the solar-powered three-axis accelerator TRIAXYS Directional Wave Buoy is similar to that of the Datawell buoy.

Displacement records are internally filtered at a high-frequency cutoff of 0.6 Hz. Onboard data reduction computes energy density, the prevailing wave direction, and the directional spread of the waves. Frequency resolution is around 0.01 Hz for waves in the range 0.033—1.0 Hz (periods of 30—1 s). Transmission of data is optional through a high-frequency (HF) transmitter with a range of 50 km, through Iridium and/or Argos satellite links, or via the Global System for Mobile Communications (GSM) internet. Waverider buoys allow for GPS monitoring and tracking through the HF link. The buoy will measure heave in the range ±20 m with 0.5% wave height resolution for wave periods of 1.6—30 s in the moored configuration. The direction range is 0—360° with a resolution of 1.5°. The manufacturers of the TRIAXYS buoy (AXYS Technologies, Inc.) claim better battery life because of the use of solid-state accelerometers, and a more linear response to heave as a function of frequency. Successful deployments in water depths of up to 300 m have been undertaken. Upon activation, the satellite system is also capable of secondary telemetry should the primary system be unavailable, as well as for a WatchMan500 controller alarm should the buoy drift too far out of position. If the buoy were to move beyond a defined radius, the end user will receive a series of alarm messages. The instrument collects, processes, and logs wave and SST data on the buoy, which is then transmitted via VHF, cellular, or satellite telemetry to a base station hosting the AXYS Data Management System. The TRIAXYS Waves plug-in data display allows for full presentation and archiving of data along with diagnostic utilities.

A crucial aspect of collecting reliable, long-term wave data is the mooring configuration. If not designed and moored correctly, there is little chance the mooring will survive the constant stresses of the wave motions. As illustrated in Figure 1.77, the recommended configuration consists of a single-point vertical mooring with two standard rubber shock cords and heavy bottom chain. This arrangement ensures sufficient symmetrical horizontal buoy response for small motions at low frequencies while the low stiffness of the rubber cords allows the waverider to follow waves up to 40 m high. Current velocities can be up to 2.5 m/s, depending on water depth (Datawell bv, 1992).

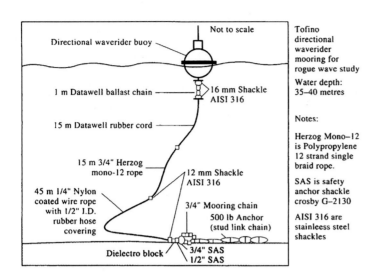

Not to scale

Directional waverider buoy

1 m Datawell ballast chain

16 mm Shackle
AISI 316

15 m Datawell rubber cord

15 m 3/4" Herzog
mono-12 rope

12 mm Shackle
AISI 316

45 m 1/4" Nylon
coated wire rope
with 1/2" I.D.
rubber hose
covering

3/4" Mooring chain

500 lb Anchor
(stud link chain)

Dielectro block

3/4" SAS
1/2" SAS

Tofino directional waverider mooring for rogue wave study

Water depth: 35–40 metres

Notes:

Herzog Mono–12 is Polypropylene 12 strand single braid rope.

SAS is safety anchor shackle crosby G–2130

AISI 316 are stainleess steel shackles

FIGURE 1.77 Morning configuration for a Datawell Directional Waverider buoy on a shallow continental shelf. *Modified after Datawell bv, 1992; Courtesy T. Juhász and R. Kashino.*

In addition to surface moored systems, several companies make it possible to measure directional waves using surface mounted electromagnetic current meters (S4ADW) and bottom-mounted Acoustic Doppler Current Profilers (ADCPs) equipped with an additional upward looking central transducer or with specialized internal software for processing wave motions. Among these are the Teledyne Workhorse Waves Array and the Rowe Technologies SeaWAVE, which measures waves from relatively shallow depths using 300, 600 or 1,200 kHz ADCPs. Wave cutoffs around 1−2 s depend on deployment depth (e.g., 2.5−14 m for 1,200 kHz and 10−80 m for 300 kHz products). The 600 kHz Aanderaa (Xylem) SeaGuard II DCP uses internal software to extract wave height and period information when measuring current velocity. The instrument outputs statistical products such as significant wave height, peak wave period, and mean direction over a specified integration time. Wave height is determined by a vertical beam that measures the distance to the surface using the echo from short pulses and simple peak estimation algorithms of internal software in the case of the 600 kHz SeaGuard II. The wave direction is found by cross correlating the along-beam velocity estimates and the wave height measurement from the vertical beam. Wave measurements are typically available for seafloor-mounted instruments but recent improvements permit the instrument to be mounted also on rotating subsurface buoys. EIVA, a Danish Marine supply and development company, offers the ToughBoy Panchax 1.2 which comes in four different models with ADCP frequencies of 300, 600 and 1,200 kHz like its U.S. counterparts. The choice of model depends on which ADCP one wishes to include in the solution—or if the user wishes to purchase the wave buoy without an ADCP. This model comes with a 600 kHz ADCP.

1.8 EULERIAN CURRENTS

The development of reliable, self-recording current meters is one of the major technological advances of modern oceanography. These sturdy, comparatively lightweight instruments are, in part, a by-product of the rapid improvement in electronic recording systems, which make it possible to record large volumes of digital data at high sampling rates (Baker, 1981). Although they can be used in moored, vessel-mounted, or tethered profiling modes, most current meters are used in time-series measurement of current speed and direction at fixed locations. (Such fixed-location measurements are called Eulerian measurements after the Swiss mathematician Leonhard Euler (1707−83) who first formulated the equations for fluid motion in a fixed frame of reference.) The development of reliable mooring technology and improved technical capability of Remotely Operated Vehicles (ROVs) have also played a major role in advancing the use of moored current meters and other fixed platform applications. Acoustic release technology, which has proven so critical to oceanic research from moored platforms, will be discussed at the end of this section.

Most commercially available current meters have sufficient internal power and data storage to be moored for several months to several years. With the exception of current meters incorporated within real-time cabled observatory networks, which supply power through cables connected directly to a shore-based facility, the instrument's longevity in the ocean clearly depends on the selected sampling rate, the data storage capacity, the battery life, and the ambient water temperature. Greater power can be obtained from lithium batteries than from more conventional alkaline batteries but the user typically faces numerous transportation-of-hazardous goods regulations and operational concerns with lithium batteries. Operating times for all types of batteries decreases with increasing water temperature.

Despite their sophistication, most current meters are made to withstand a fair amount of abuse during deployment and recovery operations. Typical "off-the-shelf" current meters (and releases) can be deployed to depths of 1,000–2,000 m, and many manufacturers fabricate deep versions of their products with heavy-duty pressure cases and connectors for deployments to depths of 6,000 m or more, corresponding to the depths within major oceanic trenches such as the Middle America Trench in the eastern Pacific off Central America, the Japan Trench off the east coast of Japan and the Sunda Trench adjacent to the south coast of Java. Most modern current meters also allow for the addition of ancillary sensors for concurrent measurement of temperature, conductivity (salinity), water clarity (light attenuation and turbidity), pressure, dissolved oxygen (DO) and other scalars. Instrument failure is less likely than in the past but is certainly not uncommon. Leakage through connector ports, seals, and cables are major causes for failures. Damaged (or missing) O-rings, saltwater corrosion, or poorly tightened fasteners are also causes for instrument damage and data loss. Corrosion sometimes occurs as a result of electrical faults, the use of combined dissimilar metals, or crevice corrosion in stainless steel (as with some acoustic releases and instrument frames and hardware). Instruments are typically much more reliable than in the past, due, in part, to reduced size, and the use of integrated surface mounts, rather than discrete components, in the electronics. Battery life, rather than memory capacity, has become the limiting factor for self-contained instrument deployments, but can be extended using external battery packs.

Current meters differ in their type of speed and direction sensors, and in the way they internally process and record data. Although most oceanographers would prefer to work with the scalar components u, v of the horizontal current velocity vector, $\mathbf{u} = (u, v)$, current meters can directly measure only the speed ($|\mathbf{u}|$) and direction (θ) of the horizontal flow. (For now, we ignore the vertical velocity component, w.) It is because of this constraint that most current meter editing and analysis programs historically work with speed and direction. From a practical point of view, both the (u, v) and the $(|\mathbf{u}|, \theta)$ representations have their advantages, despite the difficulties with the discontinuity in directional values at the ends of the interval of 0–360°. [To avoid confusion, researchers need to be aware of the difference between oceanic and meteorological convention when it involves moving fluids. For historical reasons, likely linked to the days of sailing ships, meteorologists specify the direction the wind is from (i.e., from which direction the wind is blowing into the sails), whereas oceanographers specify the direction toward which the water is going (i.e., the direction in which the ship is drifting).]

Speed sensors can be of two types: *mechanical sensors*, which measure the current-induced spin of a rotor or paddle wheel; and *nonmechanical sensors*, which measure the current-induced change in: (1) a known electromagnetic field; (2) the difference in acoustic transmission times along a fixed acoustic path; or (3) the Doppler shift in frequency (or corresponding shift in phase) in gated acoustic pulses reflected off backscatters considered to be drifting passively in the water column. Current meters can also be classified as "single-point" instruments, which measure the flow in a small region of the water column in the immediate vicinity of the instrument, or as "profiling" instruments, which measure the flow over a range of ensonified regions at a distance from the instrument. Despite these fundamental differences, all current meters have certain basic components that include speed sensors, a compass to determine orientation relative to the earth, built-in data processing algorithms, a digital storage device, and a source of power, such as a battery pack or shore-based cable connection. Possible speed sensors include:

1. Propellers (with or without ducts).
2. Savonius rotors.
3. Acoustic detectors (sound propagation time or Doppler frequency shift).
4. Electromagnetic sensors (induced magnetic field).
5. Platinum resistors (flow-induced cooling).

Flow direction relative to the axes of the current meter is usually sensed using a separate vane or by configuring the speed sensors along two or three orthogonal axes. In all current meters, the absolute orientation of the instrument, relative to the earth's magnetic field, is determined by an internal compass. At polar latitudes where the horizontal component of the earth's magnetic field is weak, the measurement of absolute current requires that the meter be positioned rigidly in a known orientation. Direction resolution depends on the type of compass used in the measurement, e.g., clamped potentiometer for the earlier Aanderaa Recording Current Meters (RCMs), optical disk for Marsh-McBirney electromagnetic current meters (ECMs), and flux gate (Hall effect) compasses for the EG&G Vector Measuring Current Meter (VMCM), the InterOcean S4 current meter, and the Teledyne-RDI, Nortek, and SonTek acoustic current meters (ACMs). For each deployment, compass direction must be corrected for the local deviation of the earth's magnetic field before the velocity data are converted to true north–south and east–west components. The accuracy, precision, and reliability of a particular current meter are functions of the specific sensor configuration and the kind of processing applied to the data. Rather than comment on all the many possible variations, we will discuss a few of the more generic and successful configurations.

The problems and procedures associated with the use of these instruments, and the analysis of the resultant data, are sufficiently similar that the discussion should be instructive to cover the use of instruments not specifically mentioned. Current meter technology has advanced considerably over the past several decades and several of the companies mentioned in earlier editions of this book are no longer manufacturing current meters for ocean science. Most manufacturers of ACMs and ECMs are still in operation; the high accuracy and durability provided by the single-point ECMs and ACMs, and profiling ADCPs built by these companies, have enabled them to dominate marine research and industry applications. However, much of the data in historical archives are from older types of current meters. For this reason, we have retained parts of the sections on current meters published in previous editions of this book. For further details, the reader is directed to the Third Edition of this book.

1.8.1 Early Current Meter Technology

One of the earliest forms of current measurement was the tilt of a weighted line lowered from a ship. The time it took an object dropped alongside a vessel to travel the known length of the ship also provided a measure of the surface flow. (The term "knot" is from the use by Dutch sailors of a knotted line to measure the speed of their sailing vessel.) Although we like to think of the current meter as a recent innovation, the Ekman current meter was in use as early as the 1930s (Ekman, 1932). Although many different mechanical current meters were built in those days (see Sverdrup et al., 1942), few worked and most scientists went back to the Ekman meter. To measure the current, the instrument was lowered over the side of the ship to a specific depth, started by a "messenger" (a weight that is slid down the line), and then allowed several minutes before being stopped by a second messenger. The current speed for each time increment was determined by reading a dial that recorded the number of revolutions of an impeller turned by the current. A table was used to convert impeller revolutions to current speed. Current direction was determined from the distribution of copper balls that fell into a compass tray below the meter at a fall rate that was a function of the rotation of the propellor. A profile from 10 to 100 m typically took about 30 min. Obvious problems with this instrument included low accuracy in speed and direction, limited endurance, and the need to work from a ship or other stationary platform.

One of the first commercial current meters was the self-contained Geodyne 850 current meter built in the United States in the 1960s. The Geodyne was a large and bulky, vertically standing unit with a small direction vane and a four-cup Savonius rotor. Burst sampling was permissible in the range of 60−660 s. The Nerpic CMDR current meter built in France in the 1960s was a torpedo-like device that oriented itself with the current flow and used an impellor-type rotor to measure current speed. In the original versions, data were recorded on punched paper tape. The Kaijo Denki current meter built in Japan in the 1970s was one of the earliest types of instruments that used differential acoustical travel time between two transmitter-receiver pairs to measure flow velocity.

In our opinion, the age of the modern current meter started with the Aanderaa RCM (Recording Current Meter) developed by Ivar Aanderaa in Norway in the early 1960s under the sponsorship of the North Atlantic Treaty Organization (NATO). The fact that many of these internally recording current meters remained in widespread use until very recently attests to the instrument's durability. There was a time when many oceanographers considered the Aanderaa RCM4 (Figure 1.78) and its deep (>2,000 m) counterpart, the RCM5, the workhorses of physical oceanography. It certainly was the most common and reliable current meter of its time. For this reason, there have been more studies, intercomparisons, and soul searching with this instrument than with any other type of meter.

1.8.2 Rotor-Type Current Meters

1.8.2.1 The RCM Series of Current Meters

The Geodyne and RCM4 current meters were the first current meters to use a Savonius rotor to measure current speed. This rotor consists of six axisymmetric, curved blades enclosed in a vertical housing, which is oriented normally to the direction of flow (Figure 1.78). Data collected by the RCM4 were originally recorded on a small 1/4-inch reel-to-reel magnetic tape, with allowable sampling rate settings of 3.75×2^N min (e.g., 3.75, 7.5, 15, 30, 60 min) where N (= 0, 1, 2, ...) is an integer. Speed is obtained from the number of rotor revolutions for the entire sample interval while direction is the single direction recorded at the end of the sample period. Thus, speed is based on the average value for the recording interval, whereas direction involves a single measurement over the interval. In the past, the number of revolutions per recorded data "count" was varied by changing the entire rotor counter module. Later RCMs allowed the investigator to set the number of revolutions per count (e.g., 2^M revolutions per count, where $M = 1, 2, 3, ...$) so that the speed range of the instrument can be adjusted for the flow conditions. The direction vane of the RCM4 is rigidly affixed to the pressure case containing the data logger. The unit is then inserted in the mooring line and the entire current meter allowed to orient in the direction of

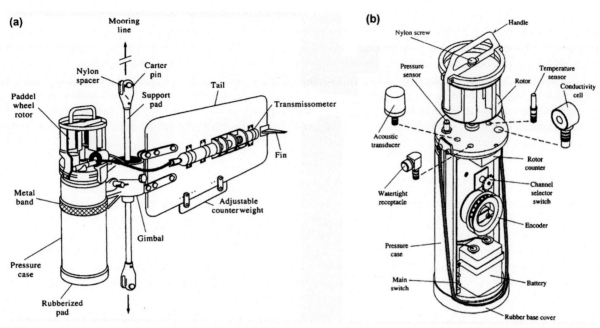

FIGURE 1.78 (a) Aanderaa RCM4 current meter; (b) exploded view of the encoder side of the Aanderaa RCM4 current meter. The reverse side contains a reel-to-tape system for recording the data from the different channels. The recorder unit is attached to a directional vane. *Courtesy Gail Gabel.*

the current. Although the RCM does not average internally, vector-averaged currents can be obtained through post-processing of the data (Thomson et al., 1985).

Part of the reason for the popularity of the RCM series of current meters was their reliability, comparatively low cost, and relatively simple operation. Both calibration and maintenance of the instruments could be performed by individuals with fairly limited electronics expertise. In recent years, many of the newer types of current meters, such as ECMs and ACMs, have advanced to the point that they require considerable electronics expertise and advanced computer diagnostic's capability on the part of the user. Moreover, modern high maintenance instruments often need to be shipped back to the manufacturer for warranty-supported calibration. Another attractive feature of the RCM was the easy addition of sensors for measuring temperature, conductivity, and pressure (depth). The Aanderaa RCM7 introduced in the late 1980s could be purchased with standard temperature (range, -2 to $+35°C$), expanded temperature (i.e., over a narrower range, such as $0-10°C$), conductivity (for salinity), and total pressure. The $0-5$ V output from a Sea Tech transmissometer for measuring water clarity was also readily incorporated in the instrument package. Thus, there was the potential to collect a wide range of parameters other than just currents alone.

To convert the dated raw data from RCMs to physical units (i.e., speed, direction), calibration constants are needed for the individual sensors. For most parameters the calibration values are found for each meter separately as a quadratic fit to the calibration data. However, this is not the case for the speed parameter for which a general curve can be used for all rotors, if currents are typically greater than 10 cm/s. Directions also are handled somewhat differently in that no formula is derived from the calibration data but rather a simple lookup table is developed for the calibration data from which the compass readings can be converted directly into degrees from true or magnetic north (taking into account the slow change in local magnetic declination over time). As a check on the timing accuracy of a data series, the analyst needs to calculate the number of observed values that should have been recorded during the measurement period. This should equal the number of records in the raw data file. If not, then the data have timing errors, or missing value, which must be corrected before processing can continue. The number of recorded values, less 1 ($= N - 1$) multiplied by the sampling interval, Δt, should equal the record length, T; i.e., correct timing in a record requires $T = (N - 1)\Delta t$. A similar calculation should actually be applied to all time series collected by oceanic instruments.

1.8.2.2 The Vector Averaging Current Meter

As discussed by Baker (1981), one of the important data reduction techniques in oceanography was the introduction of the "burst sampling" scheme (Richardson et al., 1963) whereby short samples of densely packed data are interspersed with longer periods without sampling. In continuous mode, the average current speed and instantaneous direction are recorded once per sampling interval. In burst mode, a rapid series of speed and direction measurements are averaged over a short segment of the sampling interval.

In vector-average mode, the instrument uses speed and direction to calculate the horizontal and vertical components of the absolute velocity during the burst. The instrument then separately averages each component internally to provide a single value of the velocity vector for each burst. If enough is known about the spectrum of the flow variability, the burst samples can be used to adequately estimate the total energy in the various frequency bands. This procedure greatly reduces the amount of recording space needed to sample the currents. The vector-averaging current meter (VACM) introduced in the 1970s used both burst sampling and internal processing to compute the vector-average components of the current for each sampling period. Current speed was obtained using a Savonius rotor similar to that on the RCMs but direction was from a small vane that was free to rotate relative to the chassis of the current meter. Vectors were computed for every eight revolutions of the rotor and averaged over periods of from 4 to 15 min, depending on the selected sampling interval.

1.8.2.3 Problems with the Savonius Rotor

A principal shortcoming of Aanderaa RCM current meters was their inability to record currents accurately in regions affected by surface wave motions. (Here, we are using the past tense, assuming that most researchers no longer use these types of instruments.) The problem with the Savonius rotor response is that it is omnidirectional and therefore responds excessively to oscillatory wave action. An intercomparison experiment using a mooring array shown schematically in Figure 1.79a demonstrated the differences between Savonius rotor measurements and those made with an electromagnetic current meter (see Section 1.8.3 for a discussion of electromagnetic current meters). Even under moderate wave conditions, the near-surface moored RCM4 can have its speeds increased by a factor of two through wave pumping (Figure 1.79b). The effect of wave pumping on the Savonius rotor significantly increases the spectral energy at both low and high frequencies (Figure 1.79c). Hence, the instrument was best suited to moorings supported with subsurface floats but was not suitable for moorings beneath surface buoys or in the upper ocean wave regime. Unlike the earlier Aanderaa current meters, VACMs provided accurate measurements when deployed in near-surface wave fields and from surface-following moorings (Halpern, 1978). In a comparison between Aanderaa and VACM measurements, Saunders (1976) concluded that "the Aanderaa instrument, excellent though it is on subsurface moorings, is not designed, nor should it be used, where wave frequency fluctuations are a significant fraction of the signal." In this, and a later paper, Saunders (1980) pointed out that the contamination of the Aanderaa measurements in near-surface applications was due also to a lag in the response of the direction vane to oscillatory flow.

In 1991, Aanderaa Instruments began manufacturing the vector-averaging RCM7 with a paddle-wheel rotor and internal solid state, E-prom modular memory. In earlier versions of the RCM7, the paddle-wheel rotor was partially shielded by a semicircle baffle, which was intended to reduce wave-induced "pumping". This was subsequently abandoned since the baffle was found to shed small-scale eddies, which interfere with the response of the paddle wheel in other operations. Field tests indicated that the vector-averaging RCM7 had only slightly better wave-region performance than the earlier RCMs (Figure 1.79c) and the overall improvements were marginal for most applications. A problem with the electronic memory of the RCMs was that data were lost if the instrument flooded, as it often did when the instrument was hit by fishnet or tug boat lines. This was not the case for the 1/4-inch magnetic tapes used in the old RCM4s. Thomson (1977) reports finding a long-lost RCM4 that had lain on the bottom of Johnstone Strait, British Columbia for over 3 years. Although the metal components and circuit boards had turned to mush, the salt-encrusted tape contained a full record of error-free data.

Another problem common to all Savonius rotor current meters is that bearing friction results in a fairly high threshold of the rotor response and an improper response of the rotor to low current speeds. For the Aanderaa RCM4/5 current meters, this threshold level is about 2 cm/s and current measurements taken in quiescent portions of the ocean often had many missing values where the currents were too low to turn the rotors during the sampling interval. An attempt in the 1990s to use mechanical RCM7s to measure dense, oxygen-rich water intruding over the shallow sill at the head of Effingham Inlet, a fjord with anoxic bottom water on the west coast of Vancouver Island, British Columbia, failed to generate more than a few rotor turns over a period of many months, despite evidence of intrusive events in the coincident-moored CTD time series. In contrast, a 300 kHz acoustic Doppler current meter (ADCM) moored at the same location and depth the following year, yielded numerous weak, ≤ 1 cm/s, inflow events over the same months, consistent with the CTD time series. (Similarly, high-frequency 300 and 2,000 kHz ACMs, moored in the inlet starting in 2010, were able to measure the surge-like intrusion of oxygen-rich water, bottom sediment resuspension, and zooplankton disruption caused by the arrival of the March 2011 Tohoku tsunami that originated over 7,000 km to the west; Thomson et al., 2013.)

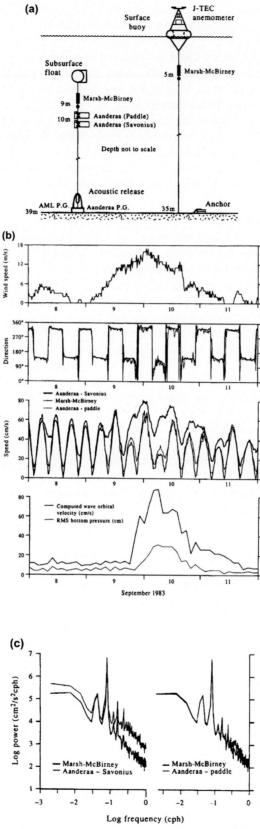

FIGURE 1.79 (a) Mooring arrangement for comparison of current speed and direction from Aanderaa RCM4 (Savonius rotor) and RCM7 (paddle wheel) current meters and Marsh-McBirney (Electromagnetic) current meters moored at 10 m depth during September 1983 in an oceanic wave zone (Hecate Strait, British Columbia); (b) winds were measured using a J-Tec vortex-shedding anemometer. In moderate wind-wave conditions, a surface or near-surface moored RCM4 with Savonius rotor can have its speeds increased by a factor of two through wave pumping. The paddle-wheel RCM7 behaves somewhat better; (c) power spectra for current measurements in (a). *Adapted from Woodward et al. (1990).*

According to manufacturer specifications, the response of the RCMs is linear for current speeds between 2.5 and 250 cm/s so that once the rotor is turning it has acceptable response characteristics. In this range, accuracy is given as 1 cm/s, or 2% of the speed, whichever is greater. Accuracies for the other associated sensors are $\pm 1\%$ for pressure, $\pm 0.3°C$ for temperature, and ± 0.05 g/kg for salinity. All of these accuracies are really "relative values and regular calibration is required to ensure accurate and reliable measurements".

As described by Pillsbury et al. (1974), calibration of the RCM4 compass is important because more compass failures occurred for a set of instruments than all other sensor failures combined. From calibration work reported by Gould (1973), it is clear that there is a significant departure from linearity in most RCM4 compasses. The magnitude of the nonlinearity errors (approximately 1% of the scalar mean speed per degree of compass nonlinearity) means that many of the residual velocity values observed in the ocean could be introduced by a nonlinearity of $1°$ or $2°$ in the direction sensor.

1.8.2.4 Vector Measuring Current Meter (VMCM)

To circumvent the nonlinear response problems of the RCM4, Weller and Davis (1980) developed the VMCM, which used two orthogonal propeller current sensors with an accurate cosine response. This instrument produced negligible rectification and therefore accurately measured mean flow in the presence of unsteady oscillating flow. In laboratory tests, the VMCM performed well in the presence of combined mean plus oscillatory flow as compared with poorer performances by Savonius rotor/vane systems and by electromagnetic and acoustic sensors. The open fan-type rotors of the VMCM were highly susceptible to fouling by small filaments of weed and other debris. Hogg and Frye (2007) considered VMCMs and VACMs as their long-standing benchmark current meters.

1.8.3 Nonmechanical Current Meters

Nonmechanical current meters use electromagnetic or acoustic sensors to determine the current velocity relative to the current meter. Once the flow relative to the current meter is determined, the absolute velocity in earth coordinates is found using direction from a built-in magnetic compass. Single-point current meters measure the flow velocity in the immediate vicinity of the instrument, whereas profiling ACMs (addressed in the following section) provide flow measurements at specified distances from the transducer head designed to vertically profile the horizontal current.

1.8.3.1 ACM: Differential Travel Time

Differential travel time ACMs measure the current-induced differences in the time delay of short, high-frequency (megahertz) sound pulses transmitted between an acoustic source and receiver separated by a fixed distance, L, or alternatively, the phase shift in continuously transmitted high-frequency sound signals over the path-length. In both cases, the transducer and receiver are combined into one source-receiver unit. The greater the speed of the current component in the direction of sound propagation, the shorter the pulse travel time and the greater the phase shift between sequences of pulse, and vice versa. For instance, suppose that the speed of sound in the absence of any current has a value c. The times for sound to travel simultaneously in opposite directions from two combined transducer–receiver pairs in the presence of an along-axis current of speed v are: $t_1 = L/(c + v)$ for transducer–receiver pair No. 1 and $t_2 = L/(c - v)$ for transducer–receiver pair No. 2. The velocity component along the transducer axis is therefore

$$v = L(t_2 - t_1) / (2t_1 t_2) \tag{1.85}$$

A three-axis current meter determines the three-dimensional velocity by simultaneously measuring time differences along three orthogonal axes. This technology does not depend upon the presence of acoustic scatterers in the water for measuring currents and can, therefore, be relied upon to measure velocity in clear water where scatterer-dependent Doppler current meters (ADCMs) may fail to give a continuous signal. The instruments also operate close to the water surface and to the seafloor where ADCMs may give spurious acoustic reflections.

The shortest sampling interval for ACMs is determined by the need to sample twice for the time that an eddy, having the characteristic scale, typically $L \sim 10$ cm, of the sensor pathlengths, is advected through the sensor volume. Assuming the "frozen-field" hypothesis (in which changes in turbulent eddies are small compared to the time it takes for the eddies to be carried past a point at flow velocity, v), the smallest eddy that can be resolved by the sensor has a transit time L/v, which requires a minimum sample interval $\Delta t = 2v/L$ (Williams 3rd, 2004). Failure to take this into account means that the spatially resolved, but under sampled, eddies will contribute energy to a lower wavenumber part of the spectrum, thereby aliasing the true spectrum. The effects of eddies smaller than the sensor volume are smoothed because the contribution on one side of an eddy is canceled by that of the other side of the eddy. In theory, it is expected that the turbulence spectrum for high wavenumbers would begin to decrease faster than the slope of $-5/3$ for the inertial subrange (e.g., Phillips, 1966) for wavenumbers equivalent to the reciprocal of the acoustic pathlength. However, observations find that the spectrum continues to fall at $-5/3$ at least out to the scale of the beam width and possibly beyond (Williams 3rd, 2004).

Examples of commercial ACMs include the three-directional axis Modular Acoustic Velocity Sensor (MAVS 5)-ACM available from NOBSKA (Woods Hole, USA) and the three-directional axis ACM-Plus available from Falmouth Scientific, Inc. (FSI) (Falmouth, USA). The former measures time differences and the latter phase changes of high frequency pulses. Earlier versions of ACMs are the SimTronix UCM 40, the Neil Brown ACM current meters and the MAVS-2 and 3. Because of the rapid ($\approx 1{,}500$ m/s) propagation of sound in water, these current meters are capable of high-frequency sampling and processing, with typical data rates of 25 Hz and higher. The instruments also can provide estimates of the sound velocity, c, along the two paths of length L (~ 10 cm) between the sensors. More specifically, $c = 2L/t$, where $t = t_1 t_2/(t_1 + t_2)$ is the effective time of propagation. Manufacturer specifications for the MAVS 5 and FSI ACM-Plus vector-averaging current meters using four acoustic paths are listed in Table 1.13.

NOBSKA was formed in 1997 to commercialize the MAVS Current Meter designed, with federal funding, by the Woods Hole Oceanographic Institution to address the need for low velocity, bottom boundary layer measurements. The product has evolved into a general-purpose current meter that retains its ability to measure in low flow, clear water environments. MAVS-4 and 5 have been developed for the U.S. Ocean Observatories Initiative (OOI). Additional sensors include temperature, conductivity, and pressure. Sampling is possible in burst mode (programmed for timed sampling), externally triggered measurement, and in continuous mode. Research results based on MAVS can be found in (Garcia-Berdeal et al., 2006; Johnston et al., 2006; U.S. Geological Survey, 2017; Polzin et al., 2021). Similar specifications apply to the FSI ACMs, which also includes options for turbidity, dissolved oxygen, optical backscatter and other 5-volt DC input instrumentation.

Because of their sophisticated technology, ACMs are often difficult to operate and maintain without dedicated technical support. For example, biofouling of the transducers can be a problem on any long-term mooring in the euphotic (near-surface light influenced) zone. The instruments also must undergo frequent recalibration due to possible sensor misalignment and changes in the physical dimensions of the transducer–receiver pairs. As discussed by Weller and Davis (1980), this is a particular weakness of this type of ACM, which has proved difficult to calibrate due to drifts in the zero level and in the amplifier gain. In one comparison, they found that the background electrical noise of the ACM had the

TABLE 1.13 Manufacturer's specifications for two commercially available vector-averaging Acoustic Current Meters (ACMs) using differences in time (phase) lag for high-frequency acoustic pulses transmitted between pairs of orthogonal send-and-receive sensors.

	NOBSKA MAVS 5	Falmouth Scientific, Inc. (FSI) ACM Plus
Speed (range)	0–200 cm/s (standard)	0–600 cm/s
Speed (accuracy)	0.3 cm/s or 1% of the range, whichever is greater	$\leq 1\%$ (Full Scale) ± 0.5 cm/s
Speed (resolution)	0.03 cm/s for range 0–200 cm/s	0.01 cm/s
Depth range	Standard = 2,000 m; optional = 6,000 m	Standard = 300 and 3,000 m; optional = 7,000 m
Compass direction	Accuracy 2°; resolution 1°, with response time of 0.05 s	Accuracy 2°; resolution 0.01° (3-axis magnetometer)
Sampling rate	10 Hz in earth (east, north, up) velocity coordinates; 15 Hz in instrument coordinates	10 Hz (max)
Allowable tilt	0–45° a true cosine tilt response up to ±20°	0–30°; accuracy 0.5°; resolution 0.01°
Temperature	Accuracy 0.1°C; standard resolution 0.03°C	Accuracy ≤ 0.1°C; resolution ≤ 0.01°C
Conductivity	Accuracy 0.2 mS/cm; resolution 0.02 mS/cm	Accuracy 0.01 mS/cm; resolution 0.001 mS/cm; stability 0.0005/month
Pressure	Accuracy 0.5% Full Scale (FS); resolution 0.024% FS depth	Accuracy 0.1% Full Scale (FS); resolution 0.01% FS depth; stability 0.01%/month
Speed of sound	Range of 1,350–1,600 m/s and accuracy of ± 5 m/s	

same level as the signal. As they point out, these problems are with the system electronics and have obviously proven to be solvable. Similar problems were encountered by Kuhn et al. (1980) in their intercomparison test using an early prototype model. More recent studies (Trivett et al., 1991; Polzin et al., 2021) have indicated problems with vortex shedding from the instrument support structure.

1.8.3.2 Electromagnetic Current Meters (ECMs)

ECMs such as the Marsh-McBirney 512 and the InterOcean S4 use the fact that an oceanic current behaves as a moving electrical conductor. As a result, when an ocean current flows through a magnetic field generated within the instrument, an electromotive force is induced, which, according to Faraday's law of electromagnetic induction, is directly proportional to the speed of the ocean current and at right angles to both the magnetic field and the direction of the current. In general, the magnetic field may be that of the earth or the one generated by an electric current flowing through appropriately shaped coils (Figure 1.80). In marine sciences, two-axis ECMs with an internal compass are used to determine the horizontal components of the flow velocity referenced to earth coordinates. The electrical voltage induced between the electrodes by the water motion through the magnetic field produced by the instrument gives the oceanic flow components relative to the instrument axes while the internal compass determines the orientation of the axes relative to the horizontal component of earth's magnetic field. ECMs such as the S4 measure the electrical potential generated across two pairs of exposed metal (titanium) electrodes located on opposite sides of the equatorial plane on the surface of a plastic sphere (Figure 1.81). The

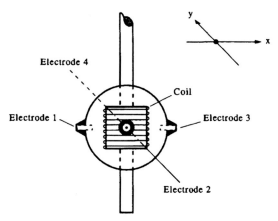

FIGURE 1.80 Principle of the electromagnetic current meter. The instrument measures the electromotive force on an electric charge (the oceanic flow) moving through the magnetic field generated by the coil. This produces a voltage potential at right angles to both the magnetic induction field and the direction of flow.

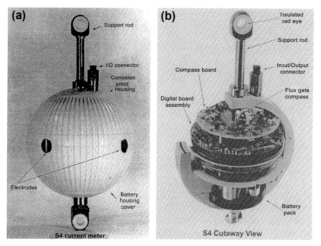

FIGURE 1.81 InterOcean S4 electromagnetic current meter, (a) view of the instrument showing electrodes; (b) cut-away view of the electronics. The spherical hull has a diameter of 25 cm and the instrument weighs 1.5 kg in water. *Courtesy, Mark Geneau, InterOcean.*

electrodes form orthogonal (x, y) axes that detect changes in the induced electrical potential generated by the ocean current. The induced voltage potential (or EMF) **E** is found by Faraday's Law through the cross product,

$$\mathbf{E} = \int_0^\infty \mathbf{v} \times \mathbf{B} dL \qquad (1.86)$$

where **v**, is the velocity of the flow past the electrodes, **B** is the strength of the applied magnetic field supplied by a battery-driven coil oriented along the vertical axis of the instrument, and L is the distance from the center of the coil. The component of the magnetic field directed vertically past the electrodes and the current flow parallel to the x-axis generates a voltage along the horizontal y-axis that is directly proportional to the strength of the water flow (see Section 1.8.6 for further details). The electric current induced by the voltage potential can be measured directly and converted to components of the flow velocity using laboratory calibration factors. Alternatively, a gain-controlled amplifier can be used to maintain a constant DC voltage at the logical output. The feedback current needed to maintain the electric current is directly proportional to the flow speed. InterOcean's S4 ECM specifications are provided in Table 1.14 along with those for the MAVS 5 copied from Table 1.13 (Marsh-McBirney no longer makes an oceanic ECM).

The standard S4A and deep S4AH ECMs have fast response platinum temperature sensors, inductive flow-through conductivity sensors, and high-resolution depth sensors. Additional add-on sensors include optical backscatter (OBS; suspended solids), transmissometer (turbidity meter), dissolved oxygen, pH and redox. Standard memory is 32 megabytes (MB), which is expandable to 256 MB. The S4A also provides for two separate autonomous sampling cycles. This multitasking allows the instrument to collect data that normally requires several instruments running different cycles. Each "mode" allows for complete control of the sampling interval, averaging time (continuous or burst sampling), and parameters to record. All of the sensors available on the S4 can be programmed to sample in either or both of the chosen sampling modes. Data can be averaged over regular intervals of a few seconds to tens of minutes or set to burst sample with a specified number of samples per burst at a given sampling interval. In addition, one can set the number of times the velocity is sampled compared with conductivity and temperature. There is also an adaptive sampling option, which makes it possible to program the instrument to only save to memory those events exceeding a preset threshold. This feature in the S4 allows burst-mode recording of data over extended periods previously not possible due to memory limitations with

TABLE 1.14 Manufacturer's specifications for the InterOcean S4 Electromagnetic Current Meter (ACM). The specifications for the MAVS 5 from Table 1.1 are included for comparison.

	InterOcean ECM	NOBSKA MAVS 5
Speed (range)	Standard 0–350 cm/s; expand to 0–750 cm/s; or reduce for higher resolution	Standard 0–200 cm/s.
Speed (accuracy)	±2% of reading ±1 cm/s	0.3 cm/s or 1% of the range, whichever is greater
Speed (resolution)	Standard 0.2 cm/s (range 0–350 cm/s); varies for range (e.g., 0.06 cm/s for a 0–100 cm/s range)	0.03 cm/s for range 0–200 cm/s
Depth range	Standard = 1,000 m; optional = 6,000 m	Standard = 2,000 m; optional = 6,000 m
Compass direction	Accuracy ±2° (within tilts of 25°); resolution 0.5°	Accuracy ±2°; resolution 1°, with response time of 0.05 s
Sampling rate	S4A has a rate of 2 Hz; S4AH has a rate of 5 Hz	10 Hz in earth (east, north, up) velocity coordinates; 15 Hz in instrument coordinates
Allowable tilt	Cosine tilt response up to ±25°	0–45° a true cosine tilt response up to ±20°
Temperature	Accuracy 0.2°C; resolution 0.05°C (thermistor or platinum)	Standard accuracy 0.1°C; resolution 0.03°C
Conductivity	Accuracy 0.2 mS/cm; resolution 0.1 mS/cm	Accuracy 0.2 mS/cm; resolution 0.02 mS/cm
Pressure	Accuracy 0.25% Full Scale (FS); resolution 1 dBar	Accuracy 0.5% Full Scale (FS); resolution 0.024% FS depth

normal burst-mode recording. This has particular application for current and wave data recording where the user is only interested in recording high level occurrence events. The surface of the S4 housing is grooved to maintain a turbulent boundary layer and prevent flow separation at higher speeds.

1.8.4 Acoustic Doppler Current Meters (ADCMs)

ADCMs are nonmechanical current meters that measure flow speed and direction by transmitting high-frequency sound waves along geometrically configured transducer axes and then determining the Doppler frequency shift, Δf, (or, equivalent the phase shift) of the return signals reflected from assemblages of "drifters" that are assumed to be moving with the velocity of the ensonified water volume. In a sense, the instrument "whistles" at a known frequency, f, and listens for changes in the frequency (or phase) of the echo. This technology makes it possible to measure the current velocity within a meter or so of the instrument, as well as to profile the velocity as a function of distance from the transducers. Profiling instruments, known as ADCPs (Acoustic Doppler Current Profilers) or acoustic Doppler profilers (ADPs), have become key components of modern oceanographic research (Figure 1.82). Velocity measurements are now recorded by ADCPs on fixed moorings, installed within through-hull fittings for underway shipboard (SADCP) sampling, or lowered by a winch (LADCP) for full water column profiling. Lowered ADCPs are typically combined with a CTD-Rosette (bottle) system so that the velocity measurements are recorded at the same time as the ambient water properties (Fischer and Visbeck, 1993). Studies using ADCPs range from the centimeter-scales of turbulent dissipation (e.g., Polzin et al., 2002; Thurnherr et al., 2014; Klymak et al., 2012), to regional-scale circulation in semi-enclosed basins (Hosokawa and Okura, 2022), to mesoscale eddies and continental shelf circulation (Li et al., 2022; Dalbosco et al., 2020), to the basin-scales of open ocean circulation (Zhang et al., 2017). Moreover, it soon became apparent after the introduction of commercial ADCPs that the acoustic backscatter signal used to measure the currents could also be used to estimate zooplankton biomass concentrations (Flagg and Smith, 1989; Burd and Thomson, 1993) and to detect the presence of hydrothermal plumes (Thomson et al., 1991, 1992). In a recent study, Cutter Jr. et al. (2022) used a combination of moored echosounders and ADCPs to examine Krill biomass and flux on the continental shelf of the northern Antarctic Peninsula.

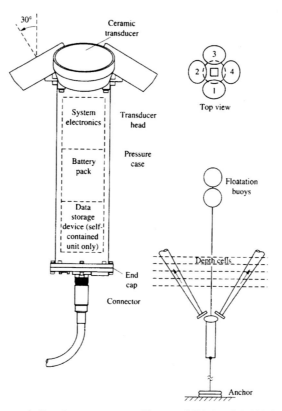

FIGURE 1.82 A direct reading 150 kHz acoustic Doppler current meter with external Teledyne-RD-232 link manufactured by RD Instruments. Side view shows three of the four ceramic transducers. Each transducer is oriented at 30° to the axis of the instrument. The pressure case holds the system electronics and echo-sounder power boards. *From RD Instruments manual (1989).*

The Doppler acoustic technique relies on the fact that: (1) sound is reflected and/or scattered when it encounters marked changes in density; and (2) the frequency of the reflected sound is increased (decreased) in direct proportion to the rate at which the reflectors are approaching (or receding from) the instrument along the acoustic path (Figure 1.83). Principle (2) is responsible for the shift to lower frequency in the whistle from a passing train and is used by astronomers to measure the rate at which stars and galaxies are moving relative to the earth. The commonly observed "red shift" of starlight reveals that most distant galaxies are receding from the earth at an accelerating rate due to the expansion of the universe. Reflectors ensonified by ADCMs include "clouds" of planktonic organisms such as euphausiids, copepods, and jellies, as well as pteropods, fish (with and without swim bladders), suspended particles, and discontinuities in water density. Buoyant wastewater plumes from coastal sewage outfalls and buoyant hydrothermal plumes rising from seafloor spreading regions are two common examples of density discontinuities that can be detected acoustically.

1.8.4.1 Narrowband versus Broadband ADCPs

There are two types of ADCPs in use: "narrowband" ADCPs, which determine flow velocity based on the measured Doppler frequency shift, Δf, in single return pulses received by the transducers, and "broadband" ADCPs, which derive flow velocity from temporal differences in the delay (or advance), Δt, between a series of received pulses. The time difference, Δt, in a propagating sound pulse and its associated phase change, $\Delta \phi = (\Delta t / \Delta T) \cdot 360°$, are mathematically equivalent to the Doppler frequency shift used in narrowband instruments (here, ΔT is the fixed time between the transmitted sound pulses). For both types of ADCP, the instruments are programmed to receive return signals within a narrow temporal window determined by the travel time to-and-from scatterers at a given "bin" distance, R, from the transducer (Figure 1.84). Rather than determining the frequency change, Δf, arising from the water motion relative to the instrument, broadband technology emits a series of pulses, carried within a longer pulse, and then determines the shift in phase arising from the time delay (or advance) between successive wave forms. Specifically, the offsets in time between adjacent sound pulses transmitted by the instrument are decreased (increased) in direct proportion to the rate at which the reflectors are approaching (or receding from) the instrument along the acoustic path. The phase changes between a series of coded pulses provide a more robust estimate of the water velocity and standard deviation (Teledyne-RD Instruments, 2011).

Unlike the single-point current meters discussed in the previous sections, which measure current time series at a fixed depth at, or within about a meter or so of, the instrument, both narrow and broadband profiling ADCMs (ADCPs) provide time series of the flow averaged over a suite of ensonified depth bins centered at known distances from the transducers. ADCPs are like having a stack of current meters, albeit with the assumption that the flow is homogeneous over each of the individual volumes (cells) of water being sampled acoustically (Figure 1.84). The transducers used to measure the currents

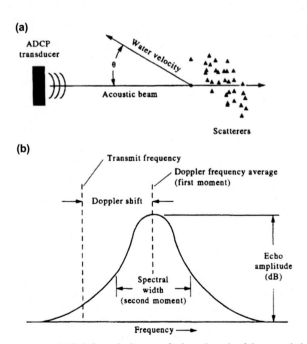

FIGURE 1.83 Principles of ADCP measurement. (a) Relative velocity, $v \cos\theta$, along the axis of the acoustic beam between the backscatterers and the transducer head; (b) auto-spectrum of returned acoustic signal showing the Doppler frequency shift for a given bin (RD Instruments (1989).)

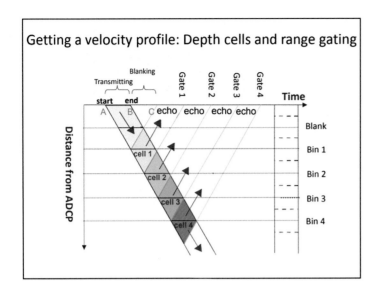

FIGURE 1.84 Distance (*R*) from the ADCP transducer versus the time (*t*) of arrival of the reflected pulse. Downward arrows denote the transmitted pulse and upward arrows the reflected pulses from each cell (bin) located at the cell's distance from the transducers. The gates open to allow the reflected pulses to be recorded by the transducers based on the travel time for the known speed of sound. *Modified after "Acoustic Doppler Current Profiler (ADCP): Principles of Operation and Setup, Christian Mohn & Martin White, SMARTSkills Workshop for Vessel Users and Researchers, Marine Institute, Galway 29th April 2016.*

are tilted by 20−30° relative to the vertical axis of the instrument, which means that the volume of water being ensonified within each cell increases with distance from the transducer head. A common arrangement consists of four transducers in a Janus-type configuration with 60° angle separations between positions on the transducer head. (Janus was the Roman god who looked both forward and backward at the same time). Commercial ADCMs—which includes both single-point and profiling instruments—are built by Aanderaa Instruments (now part of Xylem Inc.), Teledyne-RD Instruments, Nortek USA/Nortek International, and Son Tek/YSI. We have worked with ACMs from all four manufacturers and found them all to be quality products with very similar flow accuracy and resolution. There are, however, differences in instrument reliability, range of products and customer service that the investigator should address before purchasing a particular instrument.

Because it has been around longer, the Teledyne-RD Instruments (formerly RDI) ADCP has been the focus of numerous comparisons and analyses (e.g., Pettigrew and Irish, 1986; Pettigrew et al., 1986, 2005; Flagg and Smith, 1989; Schott and Leaman, 1991; Burd and Thomson, 2015; Layton et al., 2018; Thomson and Speer, 2020; Nagano et al., 2022). In recent years, Nortek, Aanderaa and FSI, have expanded into the acoustic instrument market with products for measuring three-component current velocity (profiling and single-point), as well as water levels (tides), surface waves, ice thickness, and vessel position (navigation by Doppler velocity log). Nortek lists over two dozen separate acoustic devices on its website. Teledyne-RD Instruments and Nortek make self-contained internally recording units, direct reading units, and vessel-mounted instruments, and both have adopted broadband technology (while maintaining narrow-band technology) for determining current velocity. Aanderaa Instruments presently focuses on a single 600 kHz frequency ADCM, while Teledyne-RDI and Nortek instrumentation is available at selected frequencies within the 75−1,200 kHz range. The M9 ADCP built by SonTek for underway profiling, has nine downward-looking transducers, with two pairs of four transducers each having Janus-type configurations with 25° upward slant angles operating at 3.0 and 1.0 MHz, the ninth is a vertical-looking 0.5 MHz transducer used as an echosounder to measure the water depth. The choice of frequency is dependent on the particular application, the amount of power available, and the required profiling range, which, in turn, depends on the frequency-dependent rate of sound absorption in seawater (See Table 1.16). For example Camilli et al. (2012) used a 1.2 KHz ADCP mounted on a ROV on 31 May 2010 to quantify well-flow rate from the *Deepwater Horizon* "Macondo Well" prior to removal of its broken riser. Two beams were looking horizontally and the other at 60°, pointed toward the riser and sampling at 5 Hz from a distance of 4−7 m.

Nortek also makes high quality ADCMs, specifically the 0.4−2.0 MHz Aquadopp for high-resolution single-point and medium range profiling measurements (e.g., Grimaldi et al., 2022) and the 55−1,000 kHz Signature series for longer range profiler measurements. The lower frequency Signature 55 (55 kHz) ADCM claims that it has a nominal range of 1,000 m compared to the range of around 500 m for the Teledyne-RDI 75 kHz Sentinal ADCP. Nortek claims a range of 600 m for their 75 kHz Signature instrument, although the actual range depends on the presence of adequate scatterers in the water column. The 600 kHz Aanderaa SeaGuard II ACM can accommodate a wide range of optional oceanic sensors including high resolution temperature, conductivity, waves and tides, turbidity, and dissolved oxygen (via the Optode DO concentration sensor).

Because acoustic profiling current meters are well suited to oceanographic applications, we will consider this instrument in some detail. Table 1.15 lists several popular instruments presently in operation. We have intentionally not compared instruments having similar frequencies and applications, but rather have selected a range of instruments to provide an indication of the types of profilers on the market. Nortek and Teledyne-RDI also make long-range 5-beam ADCPs with a central beam pointing directly along the axis of the instrument to measure backscatter intensity from turbulence and from zooplankton and other biomass in the overlying water column (Gargett et al., 2004; Togneri et al., 2017).

As noted above, narrowband ADCMs measure currents by determining the relative frequency change, Δf, of backscattered echoes from a single transmit pulse (Gordon, 1996), while Broadband ADCP technology measures the ocean current by determining the phase shifts ("time dilation") $\Delta \phi$ of backscattered echoes from a series of multiple transmitted pulses (Teledyne-RD Instruments, 2011). The Aanderaa Seaguard II Doppler Current Meter (DCM) and Nortek Signature ADP profilers operate at similar frequencies to those offered by Teledyne-RDI ADCPs and with comparable numbers of bins. As with other ADCPs, the Nortek profilers typically have a ring of four separate transducer beams to determine the 3-D flow. The use of four transducers provides redundancy in the flow calculations. For example, Teledyne-RDI ADCPs use the redundancy to provide an error estimation for the calculated flow. Instantaneous flow estimates that have widely different values can be ignored when calculating an ensemble average over some specified averaging time. A report on an intercomparison between a 614 kHz Broadband ADCP and two 607 kHz DCMs moored in 11.5 m of water in Øresund, Denmark has been prepared by the Danish Hydraulic Institute (Rørbaek, 1994).

As noted above, standard ADCPs employ four separate transducers oriented in a Janus configuration with beams pointing at an angle of 20−30° to the plane of the transducers. The smaller angle of 20° enables the ADCP to record values closer to "hard" reflectors, such as the ocean surface or the seafloor. The Aanderaa 2-MHz RCM11 and SeaGuard II have two orthogonal transducer paths measuring the Doppler or phase shift in the horizontal plane only, so that the vertical component of velocity is not measured. For simplicity, consider the case of a narrow-band ADCP with a transmit pulse having a fixed length of a few milliseconds. Here, the Doppler frequency shift, Δf, of the backscattered signal is proportional to the component of relative velocity, $v \cos \theta$, along the axis of the acoustic beam between the backscatterers and the transducer head (Figure 1.83a). For a given source frequency, f, and bin k (depth range $= D_k$) we find

$$v_k = \frac{\frac{1}{2}\left(\frac{\Delta f_k}{f}\right)c}{\cos(\theta_k)} \tag{1.87}$$

where v_k is the relative current velocity for bin k at depth D_k, θ_k is the angle between the relative velocity vector and the line between the scatters and the ADCP beam, and c is the speed of sound at the transducer. The factor of ½ appears in (1.87) because the Doppler shift occurs twice; once during the transmit pulse when, for example, the sound waves are trying to catch up with receding scatterers, and again as the receding scatters reflect the sound waves back to the instrument. The ADCP first determines the current velocity relative to the instrument by combining the observed values of frequency change, or phase change, in the case of broadband instruments, along the axes of each of the acoustic beams (the instrument can only "see" along the *axis* of a given transducer, not across it; Figure 1.83a). Note that, as θ_k in Eqn (1.87) approaches 90° (current at right angles to the beam path), the water velocity would need to approach infinity in order for the ADCP to register a Doppler (phase) shift.

Velocities measured relative to the ADCP are converted internally to absolute velocity components in east−west and north−south coordinates, called "earth" coordinates, using the azimuth measurements from an internal magnetic compass. Calibration of the ADCP compass as a function of azimuth (from 0° to 360°) is presumed to have taken place prior to deployment, taking into account spatial variations in the deployment region. Specific difficulties arise in the high Arctic, where the internal compass (such as the three-axis flux gate compass used in the Teledyne-RDI Workhorse ADCP) may not be able to reliably measure the weak horizontal component of the earth's magnetic field. In such cases, current meters are often deployed in a torsion-free configuration, which allows for pitch and roll but do not rotate. It is then possible to calibrate the instrument compass by comparing the orientation of the observed tidal ellipses with those from numerical tidal models. This approach works best in channels and straits that have well-defined tidal flows but is less successful in less restricted areas. This approach also assumes that the mooring does not change orientation during the deployment. It is also possible to mount a more accurate compass and data logger in a stand-alone pressure case to give a better estimate of real heading and to track changes in mooring orientation (Humfrey Melling and Michael Dempsey, Institute of Ocean Sciences, Fisheries and Oceans Canada, pers. Comm., 2022).

TABLE 1.15 Manufacturer's specifications for three commercially available broadband Doppler Current Meters: Aanderaa (Xylem) SeaGard II DCP, Nortek Signature 100 Long-range CM, (with 70–120 kHz optional center beam for zooplankton sampling), and Teledyne-RD Instruments Profiling Long-ranger Acoustic Current Meter (ACM).

	Aanderaa Seagard II (4 beams at 25° slants)	Nortek Signature 100 (4 beams at 20° slants; Optional center beam)	Teledyne-RDI Workhorse (4 beams at 20° slants)
Frequency (kHz)	600	100	300
Wavelength (cm)	0.25	1.5	0.50
Profile range (m)	30–70	300–400	116 (BB); 155 (NB)
No. cells (max)	150	200	256 (max)
Cell size (m)	0.5–5	3–15	1–8
Echo intensity (dynamic range)	>50 dB (resolution < 0.01 dB)	70 dB (resolution 0.5 dB)	80 dB (precision 1.5 dB)
Speed (range)	0–400 cm/s	0–250 (or 0–500) cm/s	0–500 cm/s; 20 m/s (maximum)
Speed (accuracy)	0.3 cm/s or ±1% of reading	1% of reading ±0.5 cm/s	0.5% of reading ±0.5 cm/s
Speed (resolution)	0.1 cm/s	0.1 cm/s	0.1 cm/s
Depth rating (m)	300, 3,000, 4,500, 6,000	1,500 (6,000)	Max 1,000
Compass direction	Accuracy < 0.5° (RMS) ±3.5°; resolution 0.1°.	Accuracy ±2° for tilt <30°; resolution 0.01°	±2° @ 60° magnetic dip angle (resolution 0.01°)
Ping rate	10 Hz	1 Hz	2 Hz
Allowable tilt	±90°.	Full 3D	±15°
Temperature (embedded)	Accuracy 0.05°C; resolution 0.001°C	Accuracy 0.1°C; resolution 0.01°C	Accuracy 0.04°C; resolution 0.01°C
Conductivity (optional)	Accuracy 0.005 S/m, resolution 0.0002 S/m	NA	NA
Pressure	Accuracy ±0.02% FS; resolution <0.0001% FS	Accuracy 0.1% FS; resolution better than 0.002% FS	Accuracy 0.1% FS; max drift 0.25%

BB, Broad Band; *NB*, Narrow Band.

1.8.4.2 Operational Factors

The relative frequency shift, Δf_k, or corresponding phase shift, $\Delta \phi_k$, for bin D_k, are derived using the observed frequency (or change in phase) of the returning echo (Figure 1.83b). To calculate the shifts in frequency and phase caused by the moving scatterers, the ADCP first estimates the autocovariance function, $C(\tau)$, of the return echo using an internal hardware processing module. The slope of $C(\tau)$ as a function of time lag, τ, is then related to the frequency change due to the movement of the scatterer region during the time that it was ensonified by the transmit pulse. Similarly, the time lags (or leads) obtained from the correlations between coded pulses, combined with estimates of the local sound speed, provide a determination of changes in distance and, hence, velocity between the ADCP and the cloud of scatterers. Because of inherent noise in the instrumentation and the environment, as well as the distortion of the backscattered signal due to differences in acoustic responses of the possible targets, the returned signal will have a finite spectral shape centered about the mean Doppler shifted frequency (Figure 1.83b). The spectral width (SW) of this signal has the form $SW = 500/D$, where D is the bin thickness in meters, and is a direct measure of the uncertainty of the velocity estimate due to the finite pulse length, turbulence, and nonuniformity in scattering velocity. In the case of the standard Teledyne-RDI ADCP, depth cell lengths, D, can range from 1 to 32 m but are usually set at 4—8 m. Depth cell size for profiling ACMs ranges from 0.12 m for the higher frequencies to 32 m for the lower frequencies. Each acoustic beam of the Teledyne-RDI ADCP has a width of 2—4° (at the −3 dB or half-power point of the transducer beam pattern) so that the "footprint" over which the acoustic averaging is performed is fairly small. Nortek profilers have corresponding beam widths of 3.7° for the 400-kHz system to 1.7° for the 2.0 MHz system; the vertical beam has a frequency of 600 kHz and beam width of 1.7°. At a distance of 300 m, the footprint of a Teledyne-RDI ADCP has a radius of 5—10 m. However, the horizontal separation between beams is roughly equal to the distance to the depth cell so that the assumption of horizontal uniformity of the current velocity is not always valid, especially for those cells farthest from the transducers.

Sidelobes of the transducer acoustic pattern can limit the reliability of the data. For the 4-beam 30°-angle ADCP, measurements taken over the last 15% $[\approx (1 - \cos 30°)]$ of the full-scale depth range are not valid if the ocean surface (or seafloor) are within the range of an upward (or downward) looking instrument. In general, the maximum range, R_{max}, of acceptable data for a vertically oriented ADCP within proximity to a "hard" reflecting surface such as the sea surface or sea floor is given by $R_{max} \approx H \cdot \cos \varphi$, where H is the distance from the ADCP to the reflecting surface and φ is the angle that the transducers make with the instrument axis (for a 20° instrument, only 6% of the range $[\approx (1 - \cos 20°)]$ is lost near the sea surface or seafloor). For vessel-mounted systems working in areas of rough or rapidly sloping bottom topography, a more practical estimate is $R_{max} \approx H \cdot (\cos(\varphi) - \alpha)$, where $\alpha \approx 0.05$ is a correction factor that accounts for differences in water depth during short (<10 min) ensemble averaging periods.

The higher the frequency, the shorter the distance an acoustic sounder can penetrate the water, but the greater the instrument's ability to resolve velocity structure according to the signal wavelength (Table 1.16). The 75- and 150 kHz units are mainly used for surveys over depth ranges of 0—500 m while higher frequencies such as the 600 and 1,200 kHz units are favored for examining flow velocity in shallow water of 25—50 m depth. Profilers have an approximate range of 60—90 m for the low-frequency 400-kHz system to 4—10 m for the high-frequency 2.0 MHz system. As noted above, most ADCPs employ four separate transducers each pointing at an angle of 20—30° to the plane of the transducers. Since only the current speeds along each of the beam axes can be estimated, trigonometric functions must be applied to the velocities to transform them into horizontal and vertical velocity components. The instrument provides one estimate of the horizontal velocity and two

TABLE 1.16 Teledyne-RD instruments acoustic wavelengths (λ), depth ranges (m) and acoustic absorption (α) for different transducer frequencies for the low power (1) and high power (2) settings. N = number of 1-s acoustic pings per sample ensemble.

Freq. (kHz)	λ (mm)	Depth range (m)		Standard deviation (cm/s)			α_1 (dB/m)	α_2 (dB/m)
		Low	High	$N = 15$	$N = 30$	$N = 60$		
76.8	20	400	700	6.72	4.75	3.36	0.025	0.0221
153.6	10	240	400	3.36	2.38	1.68	0.039	0.0395
307.2	5	120	240	1.68	1.19	0.84	0.062	0.0726
614.4	2.5	60	60	0.84	0.59	0.42	0.139	0.1884
1228.8	1.25	25	25	0.42	0.30	0.21	0.440	0.6466

From RD Instruments (1989).

independent estimates of the vertical velocity. The Teledyne-RDI ADCP senses the Doppler frequency (narrowband) or phase shift (broadband) in each 1 s acoustic "ping" by looking at the time-delayed gated signal returning from distinct "bins" (also depth cells or distance ranges) from the transducer along each of the four-beam axes. The resultant speed estimates are then converted within the instrument to common bin positions centered at $D_o + M \times D - D/2$ m ($M = 1, 2, ..., 8$ to a maximum of 128 bins or cells) along the central axes normal to the plane of the transducers. Here, D is the cell width and D_o is a constant blanking length (see below). The three-transducer Nortek profilers also have a maximum of $M = 128$ cells. Since the different time delays t_k of each pulse correspond to different distances D_k from the transducers, the instrument provides estimates of the horizontal (u, v) and vertical (w) components of velocity averaged over adjoining depth ranges (or depth bins). As illustrated by Figure 1.85, the averaging consists of a linear weighting over twice the bin length, $D = z_{k+1} - z_k = c(t_{k+1} - t_k)$, where c is the sound speed. For the 4 m bin length selected in Figure 1.85, the triangular weighted average is over 8 m. The depth range of a particular bin covers the distance:

$$from: \text{blankdepth} + (\text{binnumber}) \times (\text{binlength}) - (\text{binlength})/2$$

$$to: \text{blankdepth} + (\text{binnumber}) \times (\text{binlength}) + (\text{binlength})/2$$

A 4 m blanking is applied to the beginning of the beam to eliminate nonlinear effects near the transducer. The minimum length of the blank is frequency dependent but a larger value can be selected by the user. (For the Nortek Aquadopp profilers, the minimum blanking ranges from 1 m for the 0.4-MHz unit to 0.05 m for the 2.0-MHz unit.) For the particular setup shown in Figure 1.85, there are 15 1-s pings for each 20 s ensemble; bottom tracking is turned on every four pings.

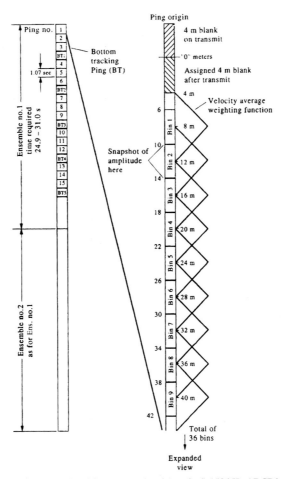

FIGURE 1.85 Allocation of depth bins and machine overhead for a narrow band (standard) 150 kHz ADCP having a bin length of 4 m, a blanking range of 4 m and a depth range of 36 bins. The instrument obtained 15 1 s pings for each 20 s ensemble and used the remaining time for internal processing and data transmission up an electrical cable. The information on the right is an expansion of the bin allocation for the first ping. A triangular weighting is used to determine the velocity for each bin. Similar results apply to the remaining pings for each of ensemble. A 4 m blanking is applied to the beginning of the beam to eliminate nonlinear effects the transducer. *Courtesy, George Chase, Institute of Ocean Sciences.*

This option, together with machine processing "overhead" and time for transmission up the tow cable, uses up a segment of the total time available for each ensemble-averaging period.

The maximum range of the standard (single–transmit pulse) ADCP depends on the depth at which the strength of the return signal drops to the noise level. Depending on the rate of energy loss and heat dissipation, the instrument is generally capable of measuring current velocity to a range R(in *meters*) $= 250(300/f)$, where f is the frequency in kilohertz (kHz). The velocities (and backscatter intensity, which we discuss later in the section) from a series of pings are averaged to form an "ensemble" record. This saves on storage space in memory, reduces the amount of processing, and improves the error estimate for the velocity record. Each acoustic ping lasts about 1–10 ms and 10 or more separate pings, together with an equal number of compass readings, are typically used to calculate an ensemble-averaged velocity estimate for each recorded increment of time in the time series. The random error of the horizontal velocity (in meters per second) for each ensemble is given as

$$\sigma(\mathrm{m}/\mathrm{s}) = 1.6 \times 10^2 / \left(fDN^{1/2}\right) \tag{1.88}$$

where N is the number of individual 1 s pings per ensemble and D is the bin length in meters. For example, a 30 s ensemble-averaging period chosen during the instrument setup procedure, generally allows for about 20 pings plus 10 s of processing time. This overhead time is inherent to the system and must be taken into account when determining the error estimates. As indicated in Table 1.16, the standard deviation of the vertical and horizontal velocity estimates for this case is about 3 cm/s for $D = 8$ m and a 150 kHz transducer. The greater the number of pings used in a given ensemble, the greater the accuracy of the velocity estimate, with $\sigma \approx N^{-1/2}$. Tilt sensors are used to calculate changes in the orientation of the transducer axis and to ensure that data are binned into correct depth ranges. These sensors are limited to $\pm 20°$ for the ADCPs and $\pm 30°$ for the Nortek ADPs for highest accuracy. For greater tilts, the velocity components are determined much less accurately. Only three of the four beams of the ADCPs are needed for each three-dimensional velocity calculation. The built-in redundancy provides for an "error velocity" estimate for each ensemble velocity, which involves subtracting the two independent estimates of the vertical velocity component for each ping. When the two vertical velocity estimates agree closely, the horizontal velocity components are most likely correct. In addition to the reliability check, the fourth beam serves as a backup should one of the transducers fail. Another measure provided by the ADCP is the "percent good" which is the percentage of pings that exceed the signal-to-noise threshold. Normally, the percent good rapidly falls below 50% at some depth and stays below that level. In practical terms, there usually is little difference in the data for assigned values of 25%, 50%, or 75% good.

We note that the low power settings in Table 1.16 are for self-contained units, while high power is for either self-contained or externally powered units. The standard deviations for velocity at a given frequency are for ensemble averages of N pings per ensemble, a depth cell size (bin length and length of transmit pulse) of 8 m and a 30° beam angle orientation. For a 20° angle, one must multiply N values by 1.5; for other depth cell sizes D (m) multiply values by 8/D. The values α_1 and α_2 are different published estimates of the absorption of sound at 4°C, 35 g/kg and atmospheric pressure. At high frequencies, the range of transducers is limited by nonlinear dynamics (cavitation) and heat dissipation so that the ranges at high and low power output are the same.

The only way to improve velocity measurement accuracy with the standard single-pulse narrow-band ADCP is to lengthen the transmit pulse. A longer transmit pulse extends the length of the autocorrelation function and increases the number of lag values that can be used in the calculation of velocity. Since bin length is proportional to a loss of depth resolution (see Table 1.17 for a list of goals and trade-offs. Here, SADCP stands for a hull-mounted shipboard ADCP). By transmitting a series of short pulses, the Broadband ADCPs circumvent these problems. Because of the multiple transmit pulses, the Broadband ADCP is capable of much better velocity resolution and higher vertical resolution. The time between

TABLE 1.17 Requirements for Acoustic Doppler Current Profiler (ADCP) setup and the consequences (trade-offs).

Requirement	ADCP setup	Trade-offs
Improve depth (bin) resolution	Smaller depth bins	Reduced profiling range
Reduce noise in depth bins	Larger depth bins	Reduced vertical resolution
Reduce noise in SADCP data	Time average profiles	Lower horizontal resolution
Improve profiling range	Use narrowband mode	Increased random noise

pulses sets the correlation lags available for velocity computation, while pulse length governs the size of the depth cells, as in the standard unit. Moreover, velocity is determined from differences in the arrival times (or phase) of successive pulses. By increasing the effective bandwidth of the received signal by two orders of magnitude, the Broadband ADCP can reduce the variance of the velocity measurement by as much as two orders of magnitude. The ADCP systems provided by Teledyne-RDI and Nortek Instruments offer "real-time" computer screen display for at-sea operations. The standard narrow-band ADCP uses a data acquisition system, that is no longer supported by Teledyne-RD Instruments, to display output of velocity components, beam-averaged backscatter intensity, percent good, and other ship related parameters such as heading and pitch and roll.

1.8.4.3 Shipboard ADCPs (SADCPs)

Doppler current meters were originally designed for determining vessel speed and, therefore, included sensors (heading, pitch, roll, and yaw) for measuring currents from a moving platform. The same instrumentation is needed for in-line moorings, which can move laterally and vertically with changes in the strength and direction of the current flowing past the mooring components. These ancillary data are used to correct the measured velocities. In order to determine the true current velocity in "earth coordinates" from a moving vessel, ADCPs are capable of measuring the velocity of the instrument over the seafloor, provided that bottom is within range of the transducers and the bottom reflection exceeds the background noise level. A separate bin is used for this bottom tracking. The bottom tracking mode is usually turned on for a fraction of the total sampling time, uses a longer pulse length and provides a more accurate estimate of relative velocity than other bins. In their study of decade-long current measurements at the mouth of Tokyo Bay, collected using a through-hull 300 kHz Teledyne-RDI SADCP mounted on a commercial ferry, Hosokawa and Okura (2022) report bottom-tracked current velocity accuracies of a few millimeters per second with a standard deviation of 1.0 cm/s. Modern shipboard positioning systems that use multiple carrier frequencies, such as GPS (U.S.), Galileo (Europe), GLONASS (Russia) and BDS (Chinese BeiDou navigation satellite system), are accurate to better than ± 1 m. Accuracies of centimeter scale are achievable when (a) combined with differential systems and (b) when convergence times (the time required for carrier-phase ambiguities encountered by the satellite arrays are fully resolved) are of the order of tens of minutes. GPS Standard Positioning Service Performance Standard used in combination with augmentation systems, or with other satellite systems such as Galileo, can resolve horizontal and vertical coordinates to around 20 cm (Kiliszek et al., 2022; Langley, 2021). It is the augmentation systems that improves the accuracy of military applications of GPS. A ± 10 m instantaneous accuracy means that estimates of the ship speed taken at time increments of, say, $10-100$ s can have errors as high as $10-100$ cm/s, which are generally comparable to the kinds of current speeds we are trying to measure. Augmented Differential GPS (DGPS), which relies on error corrections transmitted from fixed land-based reference stations for which satellite positioning and timing errors have been calculated, or using multiple systems such as GPS and Galileo, is accurate to better than ± 1 m (to ± 0.1 m in some implementations). Shipboard systems working in these modes can be used to determine absolute currents to an accuracy of roughly $\pm 1-10$ cm/s by subtracting the accurately determined ship's velocity over the ground from relative currents measured by the ADCP (see note at the end of this section). The use of SADCP measurements to study warm and cold-core eddies in the Kuroshio Extension is presented in Li et al. (2022).

DGPS uses a network of fixed, ground-based reference stations to broadcast the difference between the positions indicated by the satellite systems and the known fixed positions. These stations broadcast the difference between the measured satellite pseudo-ranges and actual (internally computed) pseudo-ranges, whereby the receiver stations may correct their pseudo-ranges by the same amount. The digital correction signal is typically broadcast locally over ground-based transmitters of shorter range. A similar system that transmits corrections from orbiting satellites instead of ground-based transmitters is called a Wide-Area DGPS (WADGPS) or Satellite-Based Augmentation System.

There are several factors that limit the accuracy of ADPs: (1) The accuracy of the frequency shift measurement used to obtain the relative velocity. This estimate is conducted by software within the instrument and strongly depends on the signal/noise ratio and the velocity distribution among the scatters; (2) the size of the footprint and the homogeneity of the flow field. For example, at a distance of 300 m from the transducers of the Teledyne-RDI Sentinal ADCP, the spatial separation between sampling volumes for opposite beams is 300 m so that they are seeing different parts of the water column, which may have different velocities; and (3) the actual passiveness of the drifters, i.e., how representative are they, in aggregate, of the *in situ* current? (Many species of zooplankton are active swimmers and vertical migrators.) In the shipboard system, the ADCP can track the bottom and obtain absolute velocity, provided the acoustic beam ranges to the bottom. Once out of range of the bottom, only the velocity relative to the ship or some level of no motion can be measured. As noted above, standard navigational positioning systems like GPS without the highly accurate (<1 m) differential mode or multiple system support, cannot be used to obtain ship velocity since the accuracy of the standard mode yields ship speed accuracies that are, at best, comparable to the absolute current speeds being measured. Erroneous velocity and backscatter data are commonly obtained from shipboard ADCP measurements due to vessel motions in moderate to heavy seas. In addition to exposure of the transducer head, the acoustic signal is strongly attenuated by air bubbles under the ship's hull or through the upper portion of the water column. Much better data is collected from a ship "running" with the seas than one lying in the trough or hove to in heavy seas. Our experience is that data collected in moderate to heavy seas are often unreliable and need to be carefully scrutinized. In deep water, zooplankton aggregations can lead to the formation of "false bottoms" in which the instrument mistakes the high reflectivity from the scattering layer as the seafloor.

A further note on satellite-based navigation measurements. There are currently a variety of chart datums that are used in the setup menu of a shipboard system and one must be sure to select that datum, that matches the chart being used for navigation. The general default datum is WGS-84 (World Geodetic System, 1984), which applies to any region of the world. An older, commonly used datum in the eastern North Pacific and western North Atlantic is NAD-27 (North American Datum, 1927); it was subsequently replaced by NAD-83 (North American Datum, 1983). Other datums are WGS-72, Australian, Tokyo, European, and Alaska/Canada. *Selective Availability* is the name given by the United States Department of Defense for degradation of the GPS satellite constellation accuracy for civilian use. When disabled (as it was during the 1st Gulf War involving Iraq), GPS accuracy increased by about a factor of 10. (At present, GPS accuracy has increased to such a high level that, even with Selective Availability, civil use of differential GPS is just as accurate as military GPS.) The modern GEBCO (General Bathymetric Charts of the Oceans) gridded bathymetric data set (GEBCO_2021 Grid) provides a global terrain model for the ocean and land. Data are in meters on a 15 arc-second interval grid (1 arc-sec = 1/3,600 of a degree of latitude or longitude $\sim$ 30 m at the equator). Depths are accompanied by a Type Identifier (TID) grid that gives information on the sources used to build the GEBCO_2021 grid. GEBCO's global elevation models are generated by the assimilation of heterogeneous data types assuming all are referred to mean sea level. However, care is required in that in some shallow water areas, the grids include data from sources having a vertical datum other than mean sea level. It is incumbent on the user to ensure that chart datums for different countries are referred to the same tidal level. For example, in Canada, Higher High Water Mean Tide (HHWMT) levels are based on the predicted astronomical tides derived from tidal elevation data for a given site, whereas HHAWMT in the United States is based on the observed tidal elevations at the site and therefore include the contributions from meteorological tides and other non-astronomical factors. As an example, at Point Atkinson near Vancouver in the shared waters of the southern Strait of Georgia between British Columbia and Washington State, HHWMT (Canada) is 1.30 m above Mean Sea Level (MSL) while MHHW (U.S.) is 1.18 m above MSL. (MSL is 0.18-0.19 m above the Canadian geodetic datum CVD2013).

1.8.4.4 Lowered Acoustic Doppler Current Profilers (LADCPs)

Unlike the case for rigid moorings affixed to the seafloor, ADCP measurements of ocean currents from moving platforms can be subject to considerable uncertainly due to the fact that the platform speeds are often two orders of magnitude greater than the ocean current speeds O(0.1 m/s) being measured. Even buoy-supported moorings undergo "watch-circles" due to wave and flow-induced stresses on the mooring components. Removing platform motions, including those arising from pitch and roll, requires high sampling rates and sophisticated data extraction algorithms. Care is also needed to ensure that nearby metallic structures (such as ships) don't induce compass biases. Because the goal is often to examine geostrophic flows or other non-tidal motions, an effort to remove the dominant M_2 and K_1 tidal current constituents using global or regional tidal models is sometimes warranted. Beginning in the 1990s, self-contained ADCPs (often with one transducer array looking up and one looking down) with frequencies of 300 kHz or higher have been strapped to Rosette-CTD packages to collect current velocity along with the scalar properties such as salinity, temperature, dissolved oxygen and nutrients (Thurnherr et al., 2014). To accurately record the vertical velocity recorded by the LADCP, the measurements must be accurate to several mm/s during a typical CTD lowering speed of around 1−2 m/s (Thurnherr, 2011). The vertical velocity due to the CTD/LADCP being lowered or raised through the water column can be obtained, after some careful processing (Thurnherr, 2011) from the time derivative of the CTD pressure ($w_{CTD} = \partial p / \partial t$). To help eliminate the bias in the LADCP attitude (i.e., heading, pitch and roll) introduced by the ship's motion, Thurnherr et al. (2017) incorporate independent magnetometer/accelerometer sensors (microchips manufactured by STMicroelectronics) recording at 100 Hz that help avoid distortion of the ADCP compass by magnetic interference from the ship's hull and equipment. An example of currents obtained using a combination of shipboard and lowered Broadband ADCPs (i.e., SADCPs and LADCPs) is shown in Figure 1.86 for repeat line A22 in the western North Atlantic. Data used in the figure were obtained during the spring of 2021 during a GO-SHIP survey by researchers aboard the R/V *Thomas G. Thompson*. The line begins near the shelf break off Columbia (South America), extends northward across the Caribbean Sea, through Anegada Passage, north to Bermuda, and then northwestward to Woods Hole, Massachusetts. The SADCP consisted of Teledyne-RDI 150 and 38 kHz Ocean Surveyors, while the LADCP consisted of a SeaBird CTD and two self-contained 300 kHz Teledyne-RDI Workhorse Monitors with Deep Sea Power and Light rechargeable 48 V batteries and cabling. The upward looking Workhorse was fitted with a custom self-recording accelerometer/magnetometer package (IMP). Differences in the horizontal currents in the upper ocean measured by the shipboard ADCP and the lowered ADCPs were generally small, with an

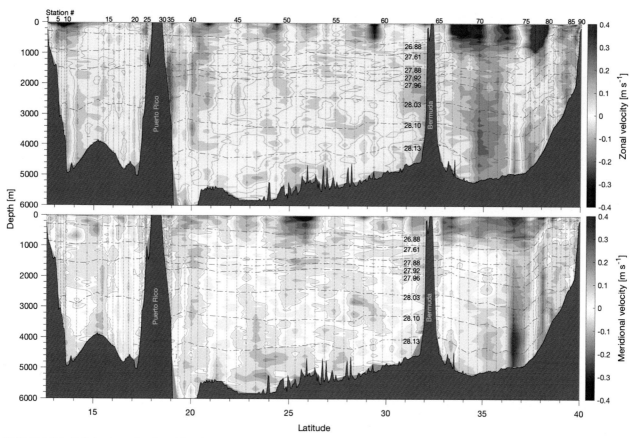

FIGURE 1.86 Full depth zonal and meridional components of current velocity as functions of latitude along GO-SHIP Line A22 in the western North Atlantic in the spring of 2021. Currents in red (blue) are to the east (west) in the top panel and to the north (south) in the bottom panel. Measured velocities are constrained by GPS, SADCP, CTD, and bottom tracking data, and smoothed for vertical scales <50 m to filter out instrumental noise. Also shown are the zero-velocity contours (thin gray lines), and neutral density contours (dot-dash gray lines) calculated using a histogram equalization algorithm. The cruise occupied 90 stations, spanning the Caribbean Current, the Gulf Stream, and the North Atlantic Deep Western Boundary Current. *Data are from the US GO-SHIP program; analysis and plotting are courtesy of Kurtis Anstey, University of Victoria.*

overall RMS difference of 4.6 cm/s (2021 GO-SHIP A22 LADCP Post-Cruise QC Report; A.M. Thurnherr, 12 October 2022; for details on the data and their processing, contact athurnherr@mailbox.org.)

Lowered ADCP measurements in high latitude regions, like the Arctic Ocean, typically encounter a host of problems, including poor compass response, weak currents in deeper water and a paucity of scatterers. The performance is improved in shallow water environments, where bottom tracking can be used to resolve ADCP rotation. Thurnherr (2011, 2017) overcame the compass limitation by collecting the LADCP profiles in conjunction with external accelerometer/magnetometer measurements. Researchers at the Woods Hole Oceanographic Institution (WHOI) used a laser ring gyro as a substitute for a compass. They improved the instrument heading, but still had issues with clear water.

1.8.4.5 Deep-Sea Observations Using an ADCP in Low Scatter Environments

One of the prime concerns with ADCMs is the possible low signal-to-noise (S/N) level due to a lack of scatterers in the water column. Recent studies by Thomson et al. (2012) and Thomson and Davis (2017) provide some confidence that even a marginal S/N level can often yield high quality time-series data. In November 2005, a 2 MHz single-point Nortek Aquadopp ACM was deployed in an upward-looking configuration in 4,386 m of water near the middle of the Middle America Trench off Costa Rica, roughly 0.72 km northwest of Ocean Drilling Program Borehole Site 1253. The ACM recorded three-dimensional current velocity, three-beam acoustic backscatter intensity, pressure, and temperature at an elevation of 21 m above the bottom (mab) every 15 min based on a 1 Hz sampling rate and 2 min burst-averaging period. Currents were measured in earth coordinates based on an ensonified water volume within a radius of a few meters of the three acoustic transducers. The instrument was recovered by the DSV *Alvin* from the R.V. *Atlantis* in February 2009 and, with the exception of two missing data points, functioned flawlessly until its battery failed on 21 April 2007. According to the data, sensor resolutions were ±0.005 m/s for velocity, $\pm 0.01°C$ for temperature, ± 2 counts for backscatter, and ± 0.013 m for depth. The depth resolution was roughly 0.0003% of the full-scale pressure based on a background density of 1,027.744 kg/m^3 from CTD data; the backscatter resolution was ~ 0.90 dB based on Norteks specification of 0.40–0.47 dB/count (Lohrmann, 2001).

Linear interpolation was used to fill in the two missing 15 min values that occurred part way through the 523-day velocity time series. The current direction was then corrected for local magnetic declination and the current vectors rotated from a north and east reference frame to a principal component reference frame in which horizontal velocity components u, v are in the northeast (cross trench; 45° T) and northwest (along trench; 315° T) directions, respectively. As indicated in Figure 3 in the Thomson et al. (2012) study, horizontal currents were strongest in the along-trench direction and reached 15 min average speeds of up to 0.3 m/s toward the northwest. Vertical velocities were also strong and commonly exceeded the instrument resolution of 0.005 m/s. The acoustic backscatter intensity had a range of 25 counts (~ 11 dB) superimposed on a background value of about 40 counts, with no suggestion of an acoustic noise threshold. Backscatter from the three beams was nearly identical.

1.8.4.6 Acoustic Backscatter

Although originally designed to measure currents, the ADCP has become a highly useful tool for investigating the distribution and abundance of zooplankton in the ocean. In particular, the intensity of backscattered sound waves for each depth bin—actually a "snapshot" of the intensity at a distance of two-thirds the way along the bin (Figure 1.85)—can be used to estimate the integrated mass of the backscatters over the "footprint" volume (width and thickness) of the original acoustic beams (Flagg and Smith, 1989; Deines, 1999; Cochrane et al., 2003). As with velocity, the instrument compensates for apparent changes in bin depth due to instrument tilt and roll. Calculation of the backscatter anomaly caused by plankton or other elements in the water column, including bottom sediments resuspended by internal solibores (Thomson and Speer, 2020) or as turbidity currents (Hill and Lintern, 2022), requires an understanding of the various factors causing dispersion and attenuation of the sound waves in water. Proper calibration of the acoustic signal as a function of acoustic range is essential for correct interpretation of the ADCP backscatter data. The measured backscatter intensity (also energy or amplitude squared) I_r is given by

$$I_r / I_a = b \cdot \exp(-2\alpha_i z) / z^2 + A_n \tag{1.89}$$

where $b = I_0/I_a$ is the transducer gain, I_0 is the intensity of the ADCP transducer output, I_a is a reference intensity, α_i, is the absorption coefficient for water (cf. Table 1.16; $i = 1, 2$), $1/z^2$ is the effect of geometric beam spreading over the range z, and A_n is the relative noise level. The factor b arises because the ADCP does not record output intensity from the transducers, only relative intensity. The acoustic volume scattering strength, S_v, of the ADCP is then given by the logarithm of Eqn (1.89) as

$$S_v = 10 \log(I_r / I_0) - 10 \log(b) \tag{1.90}$$

where the first term is the absolute acoustic scattering strength of the ADCP and the second term is an unknown additive constant. Since the later term is unknown, a measure of the relative volume scattering strength, S'_v, referenced to some standard calibration region, S_{vc}, can be determined as $S'_v = S_v - S_{vc}$ (Thomson et al., 1991, 1992; Burd and Thomson, 1994, 2012, 2015). Thomson et al. (1992) use a vertically towed vehicle and are therefore able to calibrate their data relative to the near-uniform backscatter reference layer at intermediate depths (1,000–1,500 m) in the northeast Pacific. The full sonar equation for the volume cross-scattering cross section, σ_b, from which we derive $S_v = \log(\sigma_b)$, can be found in Urick (1967). Equations for estimating target strength (TS; dB/m^2) and volume backscattering strength (S_v; dB/m) can also be obtained from the data by converting the recorded squared backscatter amplitudes (I, dB) resulting from processing of the raw data (cf. Cochrane et al., 2003; Cutter Jr. et al., 2022).

A more formal approach to determining the acoustic backscatter anomalies, $I'(t; z)$, induced by scatterers in the water column, is to first remove the effects, $I_o(t; z)$, of acoustic spreading, along-beam absorption, temporal (t) variations in the power output of the ADCP transducers and other physical factors. Following Deines (1999), variations in the backscatter intensity anomaly, $I'(t; z)$, for each beam and bin, z, due to short-term changes in backscatter distributions within the water column have the form

$$I'(t; z) = I(t; z) - I_o(t; z) \tag{1.91a}$$

where

$$I_o = k_c^{-1} \left[C + 10 \log(TR^2) - L_{DBM} - P_{DBW} + 2\alpha R - k_c Er \right] \tag{1.91b}$$

defines the spatial and temporal variability in the background backscatter intensity associated with basic operations of the ADCP. Here, C is a constant, T is the water temperature at the ADCP in degrees Kelvin (K), $R = R(z)$ is the slant range in meters to the acoustic bin, $L_{DBM} = 10\log$(pulse-length in meters), $P_{DBW} = 10\log$(transmit power in Watts), α is the absorption coefficient of sea water (dB/m), k_c is the transducer sensitivity (dB/count), and Er is the reference noise level (counts) for a specific transducer. During any given observation period, the terms in (1.91b) are either constant or vary slowly over a time scale, t^*, which is much longer than the recording period under investigation (i.e., $I_o(t;z) \sim I_o(t^*;z)$, where $t^* >> t$). The term L_{DBM} is generally constant for all three (or four) beams and changes in backscatter intensity due to variations in water temperature, $T(K) = T(°C) + 273.16$, which are typically small. The maximum changes during internal wave events examined by Thomson and Spear (2020) at the bottom of the southern Strait of Georgia in British Columbia were less than 1°C. The measured power output, P_{DBW}, for a moored ADCPs undergoes gradual linear decrease over a deployment period, while background values of α can be assumed to have changed slowly over a seasonal time scale. Resuspension of bottom sediments by strong currents, such as in the Strait of Georgia study, might affect the acoustic absorption (α) but these affects are difficult to determine. Determination of the noise level, Er, is also difficult but can be assumed to remain constant.

The transducer sensitivity, k_c, is provided by the manufacturer and should be nearly the same for each of the transducers on the ADCP so that the individual values can be replaced by the mean of the four values. In the Thomson and Spear (2020) study, the two 600 kHz, four-beam Teledyne-RDI Workhorse ADCPs had mean sensitivity values of $k_c \sim 0.434$ (± 0.018) and 0.390 (± 0.003) dB/count. This result, together with the observation that temporal variations in the background intensity (1.51b) for each ADCP were dominated by a linear trend in the power output, P_{DBW}, enabled the authors to compute the beam-averaged, 1-min interval backscatter intensity $\overline{I}(t;z) = \frac{1}{4}\sum I'(t;z)$ as

$$\overline{I'}(t;z) \sim \overline{I}(t;z) - \overline{I}_o(t;z) \tag{1.92}$$

where overbars denote a beam-average for each time step and bin. The term, $\overline{I}_o$, is a strongly varying function of depth and a slowly varying linear function of time, whereby the anomalies, $\overline{I'}$, could be obtained from the 1 min beam-averaged observations, $\overline{I}(t;z)$ by subtracting the calculated mean and linear trend in backscatter intensity for each of the bins collected by each ADCP. The mean, beam-averaged vertical backscatter intensity structure for the two upward-looking ADCPs at sites GVRD-N and S using this simple approach is presented in Figure 1.87 along with the standard deviations of the anomaly records, $\overline{I'}$, as functions of depth. As discussed, the depth-dependent changes in the variance structure may be related to zooplankton and fish aggregations, as well as to current-induced sediment suspension.

Teledyne-RDI ADCMs do not measure directly the input or output of the acoustic backscatter intensity but rather the voltage from the so-called Automatic Gain Control (AGC), which is an internal adjustment, positive feedback circuit in the output device that attempts to keep the transducer output power constant. The average compensation voltage in the AGC is recorded and can be used to estimate the relative backscatter intensity. By incorporating a user exit program, the ensemble average AGC for each of the beams for each bin can also be recorded. As we will discuss later, this is proportional to the biomass (density × cross section) of the scatterers. The instrument also measures temperature—which it needs to calculate response correctly—and, in the case of ADCPs, the percent good, which is a measure of the number of reasonable pings per ensemble.

The speed of sound, c_s, in water varies with temperature, salinity, and depth but is generally around 1,500 m/s. Therefore, sound oscillations of frequency $f = 150$ kHz (a common frequency used on shipboard systems and moored systems) have a wavelength, $\lambda = c_s/f$, of about 1 cm. Using the standard rule of thumb that the acoustic wave detects objects of about one-quarter of its wavelength, objects greater than 2.5 mm will reflect sound, while objects less than this will scatter the sound. The proportion of the sound beam transmitted, reflected, or scattered by the object is influenced by small contrasts in compressibility and density between the water and the features of the object. Organisms with a bony skeleton, scaly integument, and air bladder reflect/scatter more sound than an organism made up mostly of protoplasm such as salps and jellyfish (Flagg and Smith, 1989). Similarly, organisms that are aggregated into patches or layers return more scattered sound energy per unit volume (i.e., have a greater volume scattering strength) than uniform distributions of the same organisms.

A major problem with using the ADCP for plankton studies is common to all bioacoustical measurements; namely, determining the species composition and size distribution of the animals contributing to the acoustic backscatter. Invariably, *in situ* sampling using net tows is needed to calibrate the acoustic signal. If the ADCP is incorporated in the net system, the package has the advantage that the volume flow through each net can be determined accurately using the ADCP-measured velocity (Burd and Thomson, 1993). An attempt to calibrate the ADCP against net samples was conducted by Flagg and Smith (1989) who also pointed out problems with the response of the shipboard system to temperature

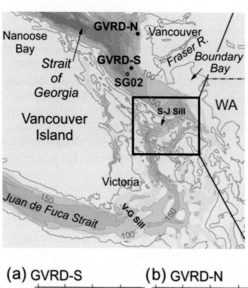

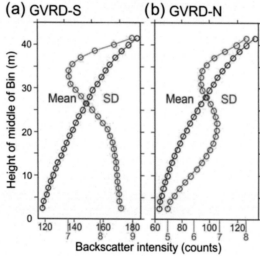

FIGURE 1.87 Record-average mean (blue) and standard deviation (red) of the beam-averaged, 1-min acoustic backscatter anomaly records as a function of height (z) above the bottom from the ADCP moorings in the southern Strait of Georgia (see the top panel for locations): (a) GVRD-S; and (b) GVRD-N. The bin nearest the transducers (bin 1) had anomalously low acoustic levels and was omitted. Also omitted are the last two bins (29 and 30), which were contaminated by transducer side-lobe effects. *From Thomson and Spear (2020).*

fluctuations in the ADCP electronics. A more recent "calibration" (Burd and Thomson, 2012, 2019) compares volume scattering strength for a 153-kHz ADCP against 197 coincident, mixed-species zooplankton samples collected over six summers using a Tucker trawl net system towed to depths of up to 3,000 m in the northeast Pacific. Results show that the acoustic backscatter data from the single-frequency ADCP mounted near the opening of the towed net system accounts for 84% of the variance in total net biomass, despite the extensive mix of faunal types, depth range, and broad spatial and temporal extent of the study.

1.8.5 Comparisons of Current Meters

As noted earlier, a major problem with the Savonius rotor is contamination of speed measurements by mooring motions (Gould and Sambuco, 1975). The contamination of the rotor speed is caused primarily by vertical motion or "rotor pumping" as the mooring moves up and down under wave action. In effect, the speed overestimates of the rotor result from its ability to accelerate about three times faster than it decelerates. Pettigrew et al. (1986) summarize studies on the ability of VMCMs and VACMs in laboratory tests to accurately measure horizontal flow in the presence of surface waves. For wave orbital velocities, W, of the same magnitude as the steady towing speed, U, of the current meter through the water (i.e., $W/U \approx 1$), the accuracy of the VACM depends on the ratio W/U. The percentage error increases as the ratio W/U increases and substantial overestimation of the true speed occurs for $W/U > 0.5$. The results for the VMCM differ significantly from those of the VACM. In particular, the VMCM underestimates the true velocity by as much as 30% for $W/U \approx 1$, while for $W/U > 2$, speed errors do not appear to be strongly dependent on either W/U or on the relative orientation of the mean and wave current motions. For $W/U < 1/3$, the VMCM was within 2% of the actual speed. While vector averaging can reduce the effect of vertical motion on the recorded currents by smoothing out the short-term oscillatory flow, the basic sensor response is not well tuned to conditions in the wave zone or those for surface moorings. Intercomparisons of conventional current meters (Quadfasel and Schott, 1979; Halpern et al., 1981; Beardsley et al., 1981) have shown that VACM speeds are only slightly higher on surface moorings than on subsurface moorings and that contamination by mooring motion was only important for higher frequencies (>1 cph). At frequencies above 3–4 cph, ocean current spectra computed from VACM current meters did not flatten (i.e., not decrease with frequency) as much as spectra from other rotor equipped current meters. Near the surface this is due to horizontal motion of the mooring (Zenk et al., 1980), which is rectified by the Savonius rotor; at greater depths, the surface float motion contributes to vertical motion, which aliases the rotor speed due to rotor pumping. Further details can be found in Weller and Davis (1980), Mero (1982), Beardsley (1987), and Hogg and Frye (2007).

Another problem with the Savonius rotor is that it does not have a cosine response to variations in the angle of attack of the flow due to interference of the support posts. In a study of rotor contamination, Pearson et al. (1981) conclude that Savonius rotor measurements, made from a mooring with a float 18 m below the sea surface, were not seriously contaminated by surface wave-induced mooring motion. In sharp contrast, Woodward et al. (1990) compared a standard Savonius rotor with a paddle-wheel rotor designed for wave-field applications, and an ECM. The electromagnetic speed sensors appeared to perform well in the near-surface wave field while the standard Savonius rotor was severely contaminated by wave-induced currents (Figure 1.88).

Field comparisons (Halpern et al., 1981) demonstrated that above the thermocline (5–27 m depth) the VMCM, the VACM, and ACM all produced similar results for frequencies below 0.3 cph, regardless of mooring type. Above 4 cph, it was recommended that the VACM be used with a spar buoy surface float while both the VMCM and the ACM could be used with surface-following floats such as a donut buoy. In general, better-quality measurements were made at depths from subsurface moorings than from surface moorings, indicating that even the VMCM data were contaminated somewhat by mooring motion.

The processing of current meter data is specific to the type of meter being used. It is interesting to read in current meter comparisons such as Beardsley et al. (1981) or Kuhn et al. (1980), the variety of processing procedures required to produce compatible data for the intercomparison of observations from different current meters. An important part of the data processing is the application of the instrument-specific calibration values to render measurements in terms of engineering units. In this regard, it is also important to have both a pre- and post-experiment calibration of the instrument to detect any serious changes in the equipment that might have occurred during the measurement period.

One of the earliest comparisons between a bottom-mounted ADCP and conventional mechanical current meters was conducted in 133 m of water near the shelf-break off northern California in 1982 (Pettigrew and Irish, 1983; Pettigrew et al., 1986). The 90-day time series of horizontal currents from a prototype upward-looking 308-kHz ADCP using a 4 m bin length was compared with currents from a nearby (~300 m) string of VACMs and VMCMs. Despite the fact that only two of the beams could be used and the instrument had a 10° list, results show striking agreement between the two sets of data (Table 1.18). Mean differences between corresponding acoustic and mechanical current meters were typically less than 0.5 cm/s while RMS differences were about 2 cm/s. Since acoustic currents were based on two beams tilted at 20° to the vertical, the relatively poor correlation at 10 m depth probably resulted from rotor pumping and overspeeding of the VACM rather than side-lobe contamination of the ADCP which would occur in the upper 6% of the depth range. Similar results were obtained by Schott (1986). A comparison in the moored performance of the earlier versions (2008–09) of the Aanderaa SeaGuard RCM and the Teledyne RDI Doppler volume sampler (DVS) is presented for the Scotian shelf and slope of the North Atlantic by Drozdowski and Greenan (2013), along with comparisons between the Aanderaa RCM8 mechanical (paddle wheel rotor) current meter, the Aanderaa RCM11 acoustic current meter, and 307-kHz Teledyne RDI ADCPs. The RMS of the speed difference between concurrent instrument combinations was 1.0–1.6 cm/s, or about 3–6% of the upper limit of speeds observed at the mooring sites.

In 2002, the Institute of Ocean Sciences (Sidney, British Columbia) conducted a 4 month comparison of three, single-point ACMs (the Nortek Aquadopp, the Sontek Argonaut MD, and the Aanderaa RCM11) and one single-point ECM (the InterOcean S4) against Aanderaa RCM5 and RCM8 paddle-wheel current meters. Because they had been used in many studies prior to 2002, the half-shielded paddle-wheel RCMs provided the standard against which the other instruments were compared. Instrument pairs were roughly 2 m apart vertically and moored at roughly 2,200 m depth at three separate locations within the axial valley of Endeavour Ridge in the northeast Pacific where currents are typically less than 10 cm/s. The results were never published and problems were experienced with all of the nonmechanical current meters. The

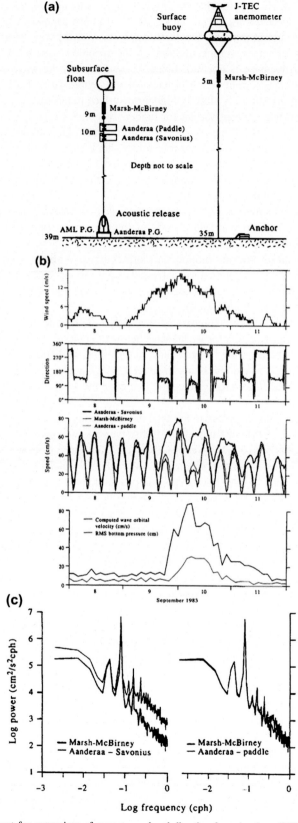

FIGURE 1.88 (a) Mooring arrangement for comparison of current speed and direction from Aanderaa RCM4 (Savonius rotor) and RCM7 (paddle wheel) current meters and Marsh-McBirney (Electromagnetic) current meters moored at 10 m depth during September 1983 in an oceanic wave zone (Hecate Strait, British Columbia); (b) winds were measured using a J-Tec vortex-shedding anemometer. In moderate wind-wave conditions, a surface or near-surface moored RCM4 with Savonius rotor can have its speeds increased by a factor of two through wave pumping. The paddle-wheel RCM7 behaves somewhat better; (c) power spectra for current measurements in (a). *Adapted from Woodward et al. (1990).*

TABLE 1.18 Comparison of hourly time series of longshore currents over a 90-day period between a 308-kHz ADCP and conventional current meters moored off Northern California. The flow velocity in the acoustic bin of the ADCP closest to the depth of the conventional current meter was chosen for the comparison.

Depth (m)	Moored current meter	Correlation coefficient, r	Speed difference (cm/s)	
			Mean	RMS
10	VACM	0.94	−3.7	8.1
20	VMCM	0.97	0.8	4.6
35	VMCM	0.98	0.2	2.7
55	VMCM	0.98	0.0	2.4
70	VMCM	0.98	0.3	2.2
90	VMCM	0.98	1.0	2.2
110	VMCM	0.98	0.5	1.9
120	VACM	0.97	−0.1	2.0

Results are found using the two-beam solution for the ADCP. *VACM*, Vector averaging current meter; *VMCM*, vector measuring current meter. Adapted from Pettigrew and Irish (1986).

principal finding was that the nonmechanical current meters were capable of measuring the very weak flows in the valley whereas rotor friction caused the RCM5/8s to erroneously record zero speeds for extended periods of time, for as much as a half semidiurnal tidal period. Subsequent experience with the nonmechanical instruments shows that quality, reliability, and durability have greatly improved since the time of this intercomparison. The results of this intercomparison helped in the design for the four long-term moorings presently installed at depths of around 2,200 m within the axial valley of Endeavour Ridge as part of the Ocean Networks Canada cabled observatory (see Figure 1.91).

1.8.6 Electromagnetic Submarine Cable Measurements

As discussed in Section 1.8.3.2 regarding single-point electromagnetic current meters, the dynamo interaction of moving, conducting seawater (salt ions) within the earth's stationary magnetic field induces electric currents in the ocean. In 1832, Faraday tried to measure the flow of the Thames River using large electrodes positioned on either side but was unsuccessful because his galvanometers were not sensitive enough to record a change in the electric current induced by the river in the presence of earth's magnetic field. Following the Second World War, the principle was used successfully to estimate the flow along the English Channel by measuring the difference in potential between electrodes on either side of the channel using a telegraph cable for one of the electrodes and the current induced by the vertical component of the earth's magnetic field. These "motional" electric fields produce a spatially smoothed measure of the water velocity at subinertial periods (periods longer than $1/f = 11.964$ h/sin(latitude)). For a given point on the seafloor, the electric fields are proportional to the vertically averaged, seawater-conductivity weighted water velocity averaged over a horizontal radius of a few water depths (Chave and Luther, 1990). Technologies that measure the horizontal electric field (HEF) yield direct observations of the barotropic transport in the overlying water column. Electric field measurements of transport are obtained from abandoned submarine communication cables or from self-contained bottom recorders. For a submarine cable, the motional HEF is integrated along the entire cable length.

1.8.6.1 Basic Electrodynamic Theory

According to theory (Sanford, 1971; Chave and Luther, 1990; Chave et al., 1992), the horizontal velocity vector field $\mathbf{v}^*$ is related to the HEF $\mathbf{E}_h$ by

$$\mathbf{E}_h = F_z \mathbf{k} \times \mathbf{v}^* \tag{1.92a}$$

(sensor in a reference frame fixed to the seafloor), or

$$\mathbf{E}_h = -F_z \mathbf{k} \times (\mathbf{V} - \mathbf{v}^*) \tag{1.92b}$$

(sensor moving relative to seafloor), where F_z is the local vertical component of the geomagnetic field, **k** is a unit vector in the upward vertical direction, **V** is the vector sum of the horizontal velocities of the ocean relative to the earth and the sensor relative to the ocean, and

$$\mathbf{v}^* = C \int_{-H}^{0} \sigma(z')\mathbf{v}_h(z')dz' \Big/ \int_{-H}^{0} \sigma(z')dz' \tag{1.93}$$

is the scaled (by the constant C) horizontal water velocity. The water velocity is averaged vertically over the water column of thickness H and weighted by the seawater conductivity, $\sigma(z)$. Eqn (1.93) reduces to the scaled barotropic velocity, $C\mathbf{v}$ when either the conductivity profile or the horizontal velocity is depth-independent. In the northern hemisphere, where **F** points into the earth, the north electric field is proportional to the west component of velocity while the east electric field is proportional to its north component. Neglecting the noise, we can solve Eqn (1.92a) to obtain

$$\mathbf{v}^* = -\mathbf{k} \times \mathbf{E}_h / F_z \tag{1.94}$$

Because **F** is known to one part in 10^4 for the entire globe, measurement of $\mathbf{E}_h$ yields the horizontal flow field.

Measurement of the HEF is entirely passive, being based on naturally occurring fields, and hence has low power requirements and is nonintrusive. Motional electromagnetic devices may be used in a Eulerian configuration (bottom recorders or submarine cables) or a Lagrangian configuration (surface drifter, subsurface float, or towed fish). Eqn (1.92b) shows that a relative velocity estimate is possible by measuring the HEF from a moving platform. On many instances, lack of a specific knowledge of $\mathbf{v}^*$ is not a critical limitation since it is independent of depth by Eqn (1.93). The moving frame of reference Eqn (1.92b) is exploited by vertical profilers such as the electromagnetic velocity profiler and the expendable current profiler produced by Sippican. Horizontal profiles of the HEF can be obtained from a towed instrument and used with precise navigation to yield estimates of $\mathbf{v}^*$ and the surface water velocity. The original form of such a towed instrument is the geomagnetic electrokinetograph (GEK) of von Arx (1950).

1.8.6.2 Land and Satellite-Based Magnetometer Observations

Because oceanic motions induce electric currents and associated magnetic fields, spaceborne and ground-based magnetometers should, theoretically, be capable of monitoring integrated information on seawater temperature and salinity, tidal current transport and the strength of the radial component of the geomagnetic field. Tidal magnetic field amplitudes and phases have been extracted from magnetometer measurements in the past but due to uncertainties caused by a variety of factors, the characteristics and temporal variability of the signals are not well known (Minami, 2017). Petereit et al. (2022) use 10-year magnetometer time series from three coastal island stations (Ascension Island, the Crozet Archipelago, and San Juan on Puerto Rico) to characterize seasonal and long-term trends in oceanic variability, with a focus on the M_2 tidal motions. The analyses reveal: (1) trends in signal amplitude changes of up to ≈ 1 nT and phase changes of order $O(10°)$; (2) at least 4 years of data are needed to obtain reliable amplitude and phase values with the extraction methods used; and (3) signal phases are a less dependent on the chosen extraction method than signal amplitudes. Seasonal magnetic field signal variations at the M_2 period deviate up to 25% from the annual mean.

Ocean-dynamo signals induced by tides are the only ocean-dynamo motions that have been observed from space. The first successful magnetic signal extraction of the diurnal principal lunar tide (M_2) from CHAMP data was achieved by Tyler et al. (2003). Additional tides (N_2, O_1) have also been extracted successfully (Sabaka et al., 2016; Grayver and Olsen, 2019). (Here, CHAMP = Challenging Minisatellite Payload, was a German satellite launched July 15, 2000 from Plesetsk, Russia and was used for 10 years for atmospheric and ionospheric research, as well as other geoscientific applications, such as GPS radio occultation; Wikipedia, 2022.)

1.8.6.3 Fiber Optic Interferometry Using Transoceanic Submarine Cables

The sensitivity of optical fibers to environmental perturbations indicates that submarine cables can be used to monitor seismic disturbances, tsunamis and ocean currents. Observational methods that exploit this sensitivity have shown that phase changes in light transmitted by the existing global network of submarine cables could be used as seafloor sensors. Zhan et al. (2021) further show that changes in environmental perturbations in light polarization can also be used for sensing physical processes over submarine cables up to 10,500 km in length. The difficulty is that, in both the phase and polarization approaches, measured changes are integrated over the entire length of the cable. Although the detection of earthquakes and ocean waves has been successfully demonstrated, measuring the integrated perturbations over the full extent of the cable has two major drawbacks: (1) the noise floor of the measurements is set by the background environmental noise summed over the entire cable length, which clearly limits the detection of smaller environmental perturbations; and (2) because the cable is effectively a single sensor, signals from multiple cables are required to triangulate the source locations of physical events, such as earthquakes and tsunamis. Marra et al. (2022) show a way to overcome the first limitation and substantially improve on the second by using optical interferometry to detect environmentally induced optical phase changes over individual sections of the submarine cable, rather than its entire length. The cable sections correspond to one or more spans between the one hundred or so repeaters used for the amplification of the optical signal along the cable. Using this approach, the investigators were able to detect earthquakes, microseisms, and ocean currents at resolutions of less than 100 km along the cable. The substantially higher sensitivity achieved by this interferometry-based method enables the detection of signals that could not be observed when measuring the cumulative optical perturbation across the entire cable. Moreover, by performing spanwise measurements at repeater intervals along the cable, they showed that a single cable is sufficient to identify the epicenters of tele-seismic earthquakes. The environmentally induced optical perturbations in each cable section are measured independently, effectively converting a submarine cable into an array of fiber-based sensors. The number of sensors can be as high as the number of repeater-to-repeater cable spans. For an intercontinental link, such as the 5,860-km-long intercontinental submarine optical fiber link between the west coast of the United Kingdom and the east coast of Canada used in the Marra et al. (2022) study, there are 128 spans (Figure 1.89).

The UK-Canada link consists of two cables: a 248-km-long cable from Southport (UK) to Dublin (Ireland), and a 5,612-km-long cable from Dublin to Halifax. The total of 128 optical repeaters installed on the link correspond to an average span between repeaters of 46 km. The experimental setup in the Marra et al. study allowed the investigator to simultaneously measure loop-back signals from up to 12 selected repeaters at a time. The viability of the technique was demonstrated through the detection of two earthquakes on multiple submarine cable spans; the Northern Peru moment magnitude (*Mw*) 7.5 earthquake of 28 November 2021 and the Flores Sea 7.3 earthquake on 14 December 2021. The Peru earthquake was detected on six of the nine sections of the cable that were tested. The technique also recorded the tidal currents over each of the different segments, with the amplitude of the signal pattern in the Irish Sea well correlated with the known neap-spring cycle in the region. The observed signals are thought to arise from "strumming" of exposed cable sections or by pressure changes induced by the tidal currents flowing past the cable. Observed periodic signals on the section close to the southeastern coast of Ireland were less pronounced and were accompanied by a strong microseismic component in the 0.07−0.15 Hz frequency range that were likely caused by the coupling of ocean wave energy to the shallow seafloor.

On the fourth section examined on the cable, extending from the end of the continental shelf to the region of the Mid-Atlantic Ridge (MAR), the system detected excess noise between 0.5 and 1.5 Hz, with a 12 h repeating pattern indicative of tidal currents in the MAR area. The authors also observed a large increase in the detected noise in the cable where the cable approaches and crosses the MAR. The recordings in this case were attributed to the interaction of currents with the rapidly changing MAR topography, where seamounts rise from 3,740 m to 2,330 m depth, with the cable possibly being

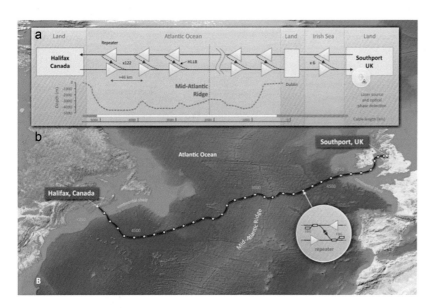

FIGURE 1.89 Map and details of the UK-Canada submarine link. (a) Illustration of the high-loss loop back (HLLB) architecture and cable bathymetry; (b) map of the UK-Canada submarine cable. The actual number of repeaters (128) was reduced in the figure for illustration purposes. Each repeater has a pair of fiber Bragg grating (FBG) reflectors. *From Marra et al. (2022).*

suspended between irregularities of the seafloor and thus more subject to current-induced movement. The MAR region of the cable was found to be a major contributor to the integrated round-trip noise over the entire Halifax-Southport cable. There was also a high level of correlation of the periodic signals observed on the shallow water sections of the cable with the tidal current velocity and high correlation between the signals in the 0.1−0.5 Hz range and the wave heights measured at the M2 buoy in the Irish Sea. Hurricane Larry was identified as the likely source of the dispersive microseisms in the 0.05−0.1 Hz frequency range between 13 and 17 September 2017.

According to the authors, the technique has the potential to transform Earth-monitoring capabilities, as the currently greatly under-sampled regions of the seafloor could be instrumented with thousands of interferometry-based sensors without modifying the existing submarine telecommunication infrastructure. Marra et al. (2022) further suggest that the more compact and lower cost narrow-linewidth telecommunications lasers could be used with limited or no loss of sensitivity. This feature makes the technique highly scalable toward potential conversion of the existing subsea cable infrastructure into a global network of environmental sensors. By converting submarine cables into arrays of environmental sensors, a large network of hundreds or thousands of permanent and real-time seafloor sensors could be implemented without modification of the existing subsea infrastructure. This has the potential to transform understanding of both shallow and deep processes inside Earth. The ability to record seismic phases in the middle of ocean basins could enable the imaging of previously obscured structures such as mid-ocean ridges and oceanic fault zones, advancing our understanding of the processes that underlie the generation of oceanic crust and the mechanisms by which oceanic plates are hydrated. The cable sensitivity to water currents could be explored to improve our understanding of the thermohaline circulation, including the proposed slowing down of ocean currents due to rising global temperatures. Furthermore, the joint detection of deep currents and seafloor pressure changes caused by the passing of ocean waves implies strong potential for tsunami sensing. Lastly, while more research is needed to characterize long-term measurement drifts, the sensitivity of the optical cables to temperature could be explored for climate change research.

For a recent summary and proposal for the use of underwater telecommunications cables for climate-scale research, the reader is directed to Howe et al. (2019). In this study, the authors outline the SMART subsea cables initiative (Science Monitoring And Reliable Telecommunications), which would "piggyback" on the power and communications infrastructure of a million kilometers of undersea fiber optic cables and thousands of repeaters, creating the potential for seafloor-based global ocean observing at a modest incremental cost. Initial sensors would measure temperature, pressure, currents and seismic acceleration.

1.8.7 Other Methods of Current Measurement

There are numerous other ways to make Eulerian current measurements though not all have been successfully commercialized. For example, prior to the ADCP, scientists in Japan used towed electrodes at the ocean surface (the GEK) to routinely monitor the currents off the east coast of Japan. One of us (WJE) has taken similar GEK measurements in Hawaiian waters. Goldstein et al. (1989) report on the use of SAR to measure surface currents from the phase-delay maps of aircraft-borne radar.

1.8.7.1 High Frequency Coastal Radar

High Frequency (HF) radar has become a wide spread technique for mapping surface currents along coastlines (Paduan and Washburn, 2013; Roarty et al., 2019). This method, which benefits from the fact that ocean waters are electrically conducting, uses short-baseline radars with carrier frequencies ranging from 5 to 45 MHz to sense the current-induced Doppler shift of Bragg backscatter from wind-generated waves having half the wavelength, λ, of the radar signal. The Doppler frequency shift yields the component of the current velocity along radial lines extending seaward from the radar shore station. For two radar stations installed along a shoreline, or on opposite sides of a channel (or strait), the spatial footprints of the radial velocities partially intersect, making it possible to compute a true horizontal surface current vector within the overlapping area. The higher the frequency of the radar the higher the spatial resolution of the surface currents. This coverage is typically limited to the near shore region or to the central area between opposing radar stations. Often these HF radars are operated at two different frequencies with that at 25 MHz giving a 2 km spatial resolution in a region extending about 50−60 km offshore. The 12 MHz band yields a 6 km spatial resolution over an area extending 150 km from shore. Because the transmitters are briefly turned off immediately after transmission, there is a blanking region of about 2 km in the vicinity of the array.

One of the most widely used HF radar systems is CODAR (Coastal Ocean Dynamics Applications Radar), which is marketed by Ocean Sensors Ltd. under the name "SeaSonde". These radars can also provide information on significant wave height, as well as wave direction and period along the radials emanating out from the shore stations to a range of about 3 km from the coast. There is an extensive array of CODAR stations along the west coast of the United States and results from the radar systems can be found in Kim et al. (2011). A comprehensive comparison between CODAR measured surface currents and surface current estimates from satellite altimetry has been carried out by Roesler et al. (2013).

Masson (1996) used a 12 MHz SeaSonde CODAR array to examine the effect of tidal currents on wind waves off the southern end of the Queen Charlotte Islands (now officially called Haida Gwaii) in northern British Columbia and Cummings et al. (2022) used a 5 MHz SeaSonde to measure winter currents through Hecate Strait, separating Haida Gwaii from the mainland coast of British Columbia (one station was located at Sandspit Airport on Haida Gwaii and the other on the eastern side of the strait near the lighthouse on Bonilla Island; Figure 1.90a). The system was sensitive to Bragg scattering from surface gravity waves with wavelengths $\lambda = 0.5c/f = 30$ m, where frequency $f = 5$ MHz and $c = 3 \times 10^8$ m s^{-1} is the speed of light. Each radar unit derived the currents from the Doppler shift of the Bragg waves along the radar line-of-sight based on the deep-water approximation to the dispersion relation for gravity waves, typically assumed to be valid for depths greater than $\lambda/2$. The measured current is a weighted vertical average of the near-surface current and is dependent on the radar frequency and the vertical distribution of the Eulerian current through the Bragg wave. Simple theoretical models of wind-driven flow yield an effective measurement depth of $1.3-2.5$ m for HF radar operating near 5 MHz (Chavanne, 2018). In addition to the Eulerian current, the measured flow likely includes a contribution from the Stokes drift, although there is uncertainty regarding its magnitude. The gridded 5-km resolution radar currents recorded by Cummins et al. (2022) were compared to historical current meter observations and to the output from a numerical model, the Coastal Ice Ocean Prediction System for the West Coast of Canada (CIOPS-W), a component of the pan-Canadian operational ocean modeling system. Tidal currents derived from the radar data are in reasonable agreement with

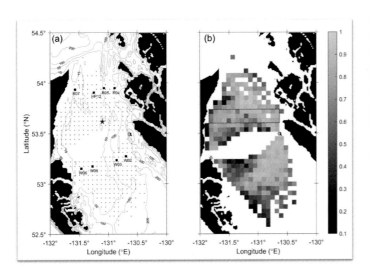

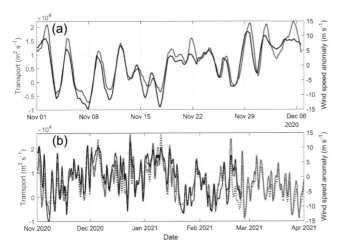

FIGURE 1.90 *Top two panels*: (a) Plan view of Hecate Strait (northern British Columbia) including topographic contours for the 15, 50, 100, 200 and 300 m isobaths, locations of the HF radars (cyan diamonds), grid for the surface currents (blue dots), current meter moorings (solid squares), and location of the North Hecate buoy, C46183 (red pentagram); (b) Fraction of hourly data returned over the HF radar grid for the period Nov. 1, 2020–Mar. 31, 2021; *Bottom two panels*: (a) northward transport for November 2020 based on the *y*-component of velocity averaged over the three rows of grid points enclosed by the dotted lines; (b) as for (a) but for the period November 2020 to April 2021. *From Cummins et al. (2022).*

historical current meter observations and can be used to describe surface current variability at subtidal time scales. Specifically, the main semidiurnal (M_2) tidal currents in Hecate Strait obtained from the 5 MHz HF radar and current meter records reveal differences of roughly 5.5 ± 2.6 cm/s, with radar values lower than the typical 0.25–0.50 m/s current meter values (Figure 1.90b). The northward transport through the strait computed from the CODAR and current meter data show nearly identical, low-frequency, wind-driven variability.

Halverson et al. (2017) provide estimates of the factors affecting the accuracy of 25 MHz HF radar measurements of the Fraser River plume in the southern Strait of Georgia. The *working range*—the distance to the farthest radial velocity solution along a fixed bearing—was observed to vary with surface water conductivity, sea state, and tides, which account for 34%, 16%, and 14% (or less) of the total variance in the currents, respectively. The working range was found to increase nearly linearly with conductivity, while the dependence on wind speed was linear at low wind speeds but peaks and subsequently decreases at higher speeds. The effect of tides on the working range was unclear; however, the range leads sea level by about 1.5 h, which the authors suggest implies that tidal currents are likely responsible. Errors are also expected from the multiple signal classification (MUSIC) algorithm used by the manufacturer for direction finding. While conductivity variations had the largest impact of the measurements, they are not considered a limiting factor for the ranges of 30 km of the radar, which can still be achieved under low-salinity conditions (e.g., Absolute salinity $S_A \sim 15$ g/kg). Because conductivity cannot approach freshwater values in the ocean, a highly conductive medium will always be present. The salinity in the Fraser River plume under freshet is lower than is found in most coastal regions, except for perhaps the Baltic Sea, or the near-field region of major river plumes, such as that of the Mississippi River off the Gulf Coast of the U.S. and the Amazon River off the Atlantic coast of Brazil. This means that salinity will not limit the working range of most HF radar installations. Instead, the limiting factor for most systems, at least at 25 MHz in fetch-limited waters, is sea state. In the southern Strait of Georgia, observed winds during the 1.6-year study exceeded 6.5 m/s—the minimum speed required to maximize working range—only 25% of the time. Thus, the range was limited by sea state 75% of the time. Tidal variations in the working range affected the working range through several mechanisms: first, tides modulate sea state; second, tides modulate the river discharge and therefore the salinity of the river plume.

1.8.7.2 Acoustic Correlation

Other techniques, such as the correlation sonar and acoustic "scintillation" flow measurements use pattern recognition and cross-correlation methods, respectively, to determine the current over a volume of ensonified water (Farmer et al., 1987; Lemon and Farmer, 1990). The acoustic scintillation method determines the flow in a turbulent medium by comparing the combined spatial and temporal variability of forward-scattered sound along two closely spaced parallel acoustic paths separated by a distance, Δx. Assuming that the turbulent field does not change significantly during the time it takes the fluid to travel between the two paths, the pattern of amplitude and phase fluctuations at the downstream receiver will, for some time lag, Δt, closely resemble that of the upstream receiver. Examination of the time delay in the peak of the covariance function for the two signals gives Δt, which then determines the mean velocity $v = \Delta x / \Delta t$ normal to the two acoustic paths. The technique has been used successfully to measure the horizontal flow in tidal channels and rivers, as well as the vertical velocity of a buoyant-plume rising from a deep-sea hydrothermal vent in the northeast Pacific (Lemon et al., 1996; Xu and Di Iorio, 2011; Di Iorio et al., 2012).

1.8.7.3 Displacement of Oceanic Features

Numerous papers have discussed the computation of surface currents from the displacements of patterns of SST in thermal AVHRR imagery. In the maximum cross-correlation (MCC) method, the cross correlation between successive satellite images is used to map the displacements due to the advection of the SST pattern (Emery et al., 1986, 1992). Wu (1991, 1993) has advanced a "relaxation labeling method" for computing sea surface velocity from sequential time-lapsed images. The method attempts to address two major deficiencies with the MCC method, namely: (1) the MCC approach is strictly statistical and does not exploit *a priori* knowledge of the physical problem; and (2) pattern deformation and rotation, as well as image noise, can introduce significant error into MCC vector estimates. The latter problem was addressed by Emery et al. (1992) who showed that rotation can be resolved using large search windows. (We note that the correlation method mentioned in the previous section is a form of feature displacement method. It is the "frozen" spatial structure of the dominant eddies rising within the turbulent plumes that enable the technology to determine the vertical velocity of buoyant plume.)

1.8.7.4 Other Doppler Current Measurements

Measurement of the vertical velocity, volume flux, and expansion rate of black smokers rising from hydrothermal vents can now be made using acoustic Doppler backscatter time series from bottom-mounted acoustic systems such as Cabled

Observatory Vent Imaging Sonar (COVIS) (Jackson et al., 2003). In their field study, Xu et al. (2013) connected the 400-kHz COVIS to the NEPTUNE-Canada (Ocean Networks Canada) Cabled Observatory to measure the flow velocity over a 10 m segment of a buoyant near-bottom plume at the Main Endeavour Field on the Juan de Fuca Ridge in the northeast Pacific. Analysis is based on the covariance method of Jackson et al. (2003) in which the velocity component, v_r, in the direction of the acoustic line of sight is given by

$$v_r = \frac{c\Delta f}{2f} \tag{1.95a}$$

where

$$\Delta f = \frac{1}{2\pi\Delta t} angle\left[\sum_{n=1}^{N}\int_{t=0}^{T}E(t)E^*(t+\Delta t)dt\right] \tag{1.95b}$$

is the Doppler frequency shift from acoustic signals backscattered from particles and turbulence in the plume, c is the sound speed, and $f \sim 400$ kHz is the sonar frequency in the Doppler mode. The angle operator in Eqn (1.95b) calculates the phase angle in radians of a complex number. $E(t)$ is a demodulated complex signal corresponding to a given azimuthal beam and a given ping, whose amplitude and phase at the time are related to the amplitude of the acoustic backscatter and its phase shift relative to the transmitted pulses (* denotes the complex conjugate). The integral in Eqn (1.95b) estimates the autocorrelation function at the time lag, Δt. A rectangular window with length $T_w = 1$ ms is used to truncate the received signal. A summation over $N_p = 40$ pings at each elevation angle of the tiltable acoustic transducer reduces the uncertainty in the measurement caused by turbulence and background noise. The standard deviation, v_{std}, of v_r is calculated over the 40 pings and is used as a metric for uncertainty in the Doppler measurements. Over the roughly 1 month proof-of-concept study, the method yielded temporal variations of the plume vertical volume flux of approximately $2-5$ m³/s, centerline vertical velocities in the range $0.11-0.24$ m/s, and plume radius expansion rates of $0.082-0.21$ m per meter of vertical rise (Xu et al., 2013). For a detailed discussion of the instrumentation and its application to observations of deep seafloor hydrothermal discharge and the interaction of buoyant plumes with the ambient currents, the reader is directed to Xu et al. (2021) and the references therein. The study also includes information on the U.S. Ocean Observatories Initiative (OOI) cabled seafloor network and research being conducted at the "Ashes" vent field at 1,540 m depth within the caldera of the active underwater Axial Volcano off the coast of Oregon.

1.8.7.5 Wire-Crawling Profilers

The Cyclesonde (van Leer et al., 1974; Baker, 1981) was an early water column profiler consisting of a RCM4 current meter affixed to a buoyancy-driven platform, which made repeated automatic round trips between the surface and some specified depth (typically <500 m) along a taut-wire mooring. Vertical cycling of the instrument was controlled by changing the density of the instrument package by a few percent using an inflatable bladder. Depending on the prescribed sampling interval and the duration of each round-trip (or depth of water sampled), the instrument provided time series of currents, temperature, and salinity over periods of weeks to months at depths of every 10 m or so through the water column (Stacey et al., 1988; Webb and Pond, 1986). The Cyclesonde technology has been superseded by the McLane Moored Profiler (MMP; Toole et al., 2017; McLane Research Laboratories, Inc., 2022), which uses a drive motor to provide the ascent/descent of the CTD/ACM sampling system along a fixed wire moored to the bottom (Figure 1.91). The instrument weighs 71 kg in air but is nearly neutrally buoyant in water. Standard profiling speeds are 0.25 m/s, with 0.1 and 0.33 m/s motor options, and is able to reach depths of $30-6,000$ m. Data are collected continuously along the profile and existing packages allow for a broad suite of sensors, including Sea Bird Electronics and RBR CTDs, Falmouth Scientific, Nortek and Nobska acoustic current meters (ACMs), Aanderaa Optode dissolved oxygen probes, Seapoint fluorometer and turbidity meters, ProOceanus CO_2 or CH_4 sensors and Biospherical photosynthetically active radiation (PAR) sensor (www.mclanelabs.com).

1.8.8 Mooring Logistics

In terms of accuracy and reliability, current meter data from surface and subsurface moorings cannot be divorced from the mooring itself. While many common mooring procedures are available, there is no single accepted technique nor is there agreement on the subsequent behavior of the mooring while in the water. Surface moorings with their flotation on the wavy surface of the ocean will behave differently than subsurface moorings over which the buoyancy is distributed vertically

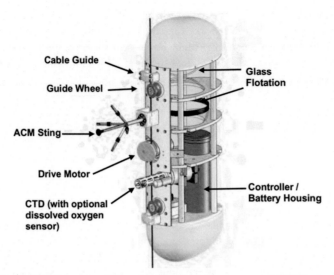

FIGURE 1.91   McLane Moored Profiler (MMP) carrying CTD and dissolved oxygen sensors, and MAVS acoustic current meter (ACM). *From the McLane Research Laboratories, Inc. website.*

along the mooring line as in Figure 1.92. The mooring in Figure 1.92 consists of paired, real-time recording single-point Nortek Aquadopp 3,000 2 MHz ADCMs and SBE 37 MicroCAT CTDs at 5, 50, 125, and 200 m above bottom (mab), and an upward-looking Teledyne-RDI 75-kHz Long Ranger ADCP housed in a 45-inch Flotation Technologies syntactic foam float at 250 mab. Results show that the ADCP is providing current velocity and acoustic backscatter intensity records for all 128 4 m bins over a depth range of 512 m above the top of the mooring, or ∼1,000 mab. There are also Flotation Technologies CF-12 12 inch clamp-on football floats distributed along the length of the mooring to support the mooring and the cable when the system is on the surface during recovery.

The standard procedure in mooring deployment is to stream-out the mooring elements from the ship, starting with the topmost buoyancy element. The instruments and spacer lines are then attached on deck at designated positions along the line and the mooring elements allowed to float away on the surface from the deployment vessel (line payout on the deck is temporarily "stopped off" to allow technicians and crew to attach each element, putting considerable stress on all elements streaming through the water). The anchor—often several rail wheels or a large concrete block—is held for last with the attached acoustic release (see Section 1.8.9) and generally supported from the ship with two ropes or with a mechanical quick-release hock affixed to the top of the acoustic release. After the anchor/release unit is lifted from the deck to over the water, the second rope suspends the anchor, and the rest of the mooring continues to stream away from the ship. The ship is maneuvered into the desired geographic deployment position (taking into account the trajectory the mooring is likely to follow as it falls to the bottom), where the second support line is cut, or the quick-release activated by a deck-hand pulling on a tethered line. The anchor then descends into the sea carrying the mooring elements with it. In this way the mooring will be located as closely as possible to the desired deployment position. For short moorings that need to be deployed very accurately near a particular feature on the seafloor (such as a hydrothermal vent), the entire mooring is assembled on deck and then lifted by crane over the water by a hock attached to the top buoyancy element. Once the ship has maneuvered into position, the mooring, which is hanging vertically from the crane, is released anchor-first for its descent to the seafloor. One of us (RET) has also deployed a 250 m long, multiple-element moorings in 2,000 m of water by first loading the anchor (a concrete block) and attached acoustic release on a Zodiac and then slowly motoring away from the ship, while towing the ship-assembled mooring behind. Once we reached the desired mooring site, the anchoring unit was pushed over the side of the Zodiac and the entire mooring allowed to fall freely to the bottom. (Note: this procedure is not generally recommended because of safety concerns and was nearly vetoed by the ship's Captain at the time. There is also the lonely feeling one gets sitting on a small rubber boat far from the research ship in the middle of the open Pacific Ocean.)

For the case of subsurface moorings, the addition of pressure sensors to most current meters gives confidence on the instrument depth (especially over steep or complicated bottom topography) and helps characterize mooring motion and determine its effect on the measured currents. Variations in the depth of the sensor can be calculated from the pressure fluctuations and used to estimate the depth and position of the moored instruments as a function of time. Also, models of mooring behavior have been developed which enable the user to predetermine line tensions and mooring motions based on the cross-sectional areas of the mooring components and estimates of the horizontal current profile. For example, the

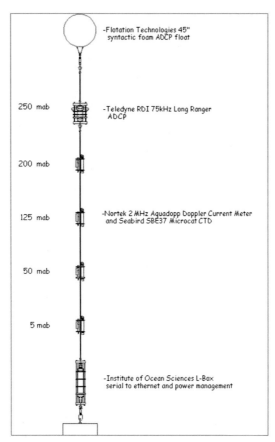

FIGURE 1.92   Schematic of a bottom-anchored mooring located at around 2,200 m depth within the axial valley of Endeavour Ridge, northeast Pacific. The mooring is connected by cable to the Ocean Networks Canada cabled observatory, which supplies power and transmits 1-s data from the five oceanographic instruments to a Node, and hence by fiber optic cable to shore. The mooring release consists of a pull-pin that is operated by a remotely operated vehicle (ROV). Elevations are in meters above bottom (mab). *Mooring design and image courtesy of David Spear, Tamás Juhász, and Lucius Perreault, Institute of Ocean Sciences.*

subsurface mooring program, SSMOOR, distributed by Cable Dynamics and Mooring Systems in Woods Hole (Berteaux, 1990, 1991), uses a finite element technique to integrate the differential equilibrium equations for cables subjected to steady state currents. Factors taken into consideration include: the mooring wire (or rope) diameter, weight in water, and modulus of elasticity; and the shapes, cross sections, drag coefficients, weights, and centers of buoyancy of the recording instruments. Up to 10 current speeds can be specified for the current profile and as many as 20 instruments inserted in the anchoring line. A mooring design and dynamics program developed by Richard Dewey at Ocean Networks Canada (University of Victoria) that combines the Berteaux model with the DOS MoorDesign model formulated by the Pacific Marine Environmental Laboratory (NOAA, Seattle) is available at http://canuck.seos.uvic.ca/rkd/mooring/moordyn.php.

Mooring motions are largest when surface floats are used. For surface moorings in deep water, the length of the mooring line creates a relatively large "watch circle" that the surface float can occupy. This will add apparent horizontal motion to the attached current meters while, at depth, the surface wave and wind-driven fluctuations translate into mainly vertical oscillations of the mooring elements. Some intercomparison experiments have tried to use a variety of mooring types to test the effects of moorings alone. Zenk et al. (1980) compare VACM measurements from a taut-line surface mooring with a single-line spar buoy float and a more rigid two-line, H-shaped mooring. As expected, the H-shaped mooring was more stable and the other two exhibited much stronger oscillations. The current meters on the rigid H-mooring registered the greater current oscillations since the meters on the other, less restricted, moorings moved with the flow, reducing the measured current speed.

In their current meter comparison, Halpern et al. (1981) discuss four different types of mooring buoyancy; three surface and one subsurface. The surface floats were: a toroid, a spar-buoy, and a torpedo-shaped float. They found that rotor pumping was much greater under the toroid than under the spar buoy and that the effect of rotor pumping on the resulting current spectra was significant at frequencies above 4 cph. While this was true for near-surface current meters, they also found that for deeper instruments the spar buoy float transmitted larger variations to the deeper meters making it a poor candidate for flotation in deep water current measurements. They found that both the VMCM and the ACM are less affected by the surface motions of a toroidal buoy. In a different comparison, Beardsley et al. (1981) tested an Aanderaa current meter suspended from a surface spar buoy, and found a significant reduction in the contamination of the measured signal by wave effects due to both currents and orbital motion with the spar buoy. Even with this flotation system, however, the Aanderaa current meter continued to register high current speeds compared with other sensors.

In an overall review of the recent history of current meter measurement, Boicourt (1982) makes the interesting observation that "results from current measurement studies are independent of the quality of the data". In making this claim, he remarks that often the required results are only qualitative, placing less rigorous demands on the accuracy of the measurements. He also points out that present knowledge of the high-frequency performance of most flow sensors is inadequate to allow definitive analysis of the current measuring system. In this regard, he states that ACMs and ECMs, with their fast velocity response sensors, hold great promise for overcoming the fundamental problems with mechanical current sensing systems, as observations have now shown. Finally, he calls for added research in defining the high-frequency behavior of common current meters.

Fieldwork by the Bedford Institute of Oceanography on Georges Bank in the western Atlantic has revealed another unwelcome problem with moored rotor-type current meters. Comparisons between currents measured by a subsurface array of Aanderaa current meters on the bank and a shipboard ADCP indicated that current speeds from the moored array were 20–30% lower than concurrent speeds from the profiler. To test the notion that the under-speed estimates were due to high-frequency mooring vibration caused by vortex shedding from the spherical floatation elements, an accelerometer was built into one of the subsurface moorings. Accelerations measured by this device confirmed that the current meters were being subjected to high-frequency side-to-side motions. Under certain flow conditions, the amplitudes of the horizontal excursions were as large as 0.5 m at periods of 3 s. Tests confirmed that the spherical buoyancy packages were the source of the motions. By enclosing the spherically shaped buoyancy elements in more streamlined torpedo-shaped packages, the mooring line displacements were reduced to about 10% of what they were for the original configuration. Excellent agreement was found between the current meter and vessel-mounted ADCP current records.

In certain areas of the world (e.g., Georges Bank off the east coast of North America), the survivability of a mooring can have more to do with fishing activity than to environmental conditions. Also, in the early days of deep-sea moorings, the Scripps Institution of Oceanography lost equipment on surface moorings to theft and vandalism. Preventing mooring and data loss in such regions can be difficult and expensive. For fishery oceanography studies the dilemma is that, to be of use, the measurements must be obtained in areas where they are most vulnerable to fishnet fouling and fish-line entanglement. Damage to nets equates to lost fishing time and damaged or lost instrumentation. Aside from providing detailed information on the mooring locations in printed material handed out to commercial fishermen, or published in "Notices to Mariners", or provided on designated Web sites, fish processing companies and coast guard, the scientist may need to resort to closely spaced "guard buoys" in an attempt to keep fishermen and shipping traffic from subsurface moorings. Our experience is that a limited array of only three or so coast guard-approved buoys more than 0.5 km from the mooring is inadequate, and that certain operators will even use the buoys to guide their operations, thereby increasing the chance of damage. The other concern is whether to "reward" or "punish" those that damage or dislodge moorings. Again, our experience on the west coast of Canada is that the researcher wants to encourage commercial fishers to return the mooring components caught in their nets rather than causing them to throw them back into the sea and not report their locations. (The question of whether to levy library fines for late book returns comes to mind.) Even with modern navigational coast guard tracking, it is almost impossible to link the loss of a mooring to a particular vessel. As no rational commercial mariner will go out of his/her way to uproot a mooring, it is better to thank the operator for the return of mooring components than to lose the equipment and its data.

1.8.8.1 Mooring Logistics in the High Arctic

The conversion from velocity relative to the current meter to velocity in (x, y, z) earth coordinates requires a 360° calibration of the internal compass with respect to the local magnetic field. Consistent with all current meter deployments, it is best practice to calibrate the instrument compass at a latitude close to that of the instrument deployment. As noted earlier in this section, difficulties arise at high latitudes because of the weak horizontal component of the earth's magnetic field. In the High Arctic, only calibrations relative to earth's magnetic field are possible. Not possible is the part of the calibration that involves ferromagnetism within the battery pack—end caps of the individual cells, electronical connections between cells spot-welded to the end caps—and nearby (<50 cm) mooring components. The need to minimize deviations from correct zenith-pointing is accomplished by an "intelligent" design of the mooring. Because of the weak horizontal magnetic field, mooring depression by current drag does not matter as much as flow-induced tilting of the instrument and internal compass. To minimize drag, it is necessary to use floats with a high buoyancy-to-drag ratio and by ensuring that the mooring has a large as possible righting moment, made possible by raising the center of buoyancy of the mooring package and lowering the center of mass. Shaped syntactic floats, such as the M40, with more-or-less collocated centers of mass and buoyancy are impractical unless a tilt-stabilizing float is placed above the instrument. Internal compasses also tend to have nonlinear direction responses. For example, the non-linearity of the Teledyne-RDI Workhorse fluxgate compass (attributable to ferromagnetism) is typically between $\pm 15°$ and $\pm 45°$ before calibration in the Beaufort Sea region, where the horizontal component (H) of earth's magnetic field is about $7,000-8,000$ nT (nano Tesla, $= 10^{-9}$ T), for a vertical component at about $82-83°$ to the horizontal (the full magnetic strength in the region is over $55,000$ nT). The amplitude of residual non-linearity, after calibration, typically ranges between $1°$ and $10°$; the mean over all azimuth is usually not zero. The non-linearity and bias are measured and applied as corrections to the headings that the ADCP records.

Because of the compass limitations—the likelihood of a compass malfunction increases as the horizontal component, H, of earth's magnet field decreases, becoming unreliable at about $H = 5,500$ nT—current meters moored in the High Arctic are often deployed in a torsion-free configuration. This is when "twist-less" moorings, which allow for pitch and roll but do not twist, become essential. Canadian researchers working in the High Arctic have used "twist less" moorings for several decades (Humfrey Melling and Michael Dempsey, Fisheries and Oceans Canada, pers. com., 2022). These produce good results but with a number of limitations. Chief among these is that the (fixed) heading of the instrument depends on the local tidal current ellipse, the actual heading of which must be determined from a reliable tidal model. In well-mapped, relatively strait channels with strong tidal currents, this is not a problem, but is less robust in regions of weak flow. Moreover, the generally stronger semi-diurnal tidal motions are of little use in the Arctic because of the proximity of the semi-diurnal inertial latitude (75° N). Wind forced, near-circular, counterclockwise rotating near-inertial oscillations "contaminate" motions in the semi-diurnal band and thicken the boundary layer of the clockwise rotary tidal flows (barotropic tidal models such as WebTide lack boundary layers). Best results are obtained using the weaker diurnal tides. However, diurnal tidal currents in the Arctic are generally even weaker than the weak semi-diurnal tidal flows. Moreover, it is difficult to conduct long-term "twist-less" moorings, and even more difficult to deploy them. As a result, the investigator is generally limited in the use of ADCPs to relatively shallow water. Researchers working in the Canadian Arctic have used 75-kHz instruments in water as deep as 450 m and 600-kHz instruments in water of about 40 m depth. Using ADCPs closer to the surface than 30 m is ill-advised in waters prone to drifting ice.

1.8.9 Acoustic Transponder Releases

An acoustic transponder release ("acoustic release") is a remotely controlled motorized linkage device that connects a bottom anchor (often an expendable set of used train wheels or specially designed concrete block) to the recoverable elements of a mooring (Figure 1.93). Modern acoustic releases are critical for the reliable recovery of mooring equipment that has been deployed from ships or placed on the bottom by Remotely Operated Vehicles (ROVs) (Heinmiller, 1968). Operation of the release from a ship requires a deck unit specially built for the particular type of release and a transducer for

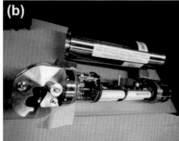

FIGURE 1.93 (a) A deep-sea mooring using an IXBLUE SAS Oceano 2500S-Universal AR861 acoustic release (attached to a blue rope) being prepared for deployment off the deck of the Canadian Coast Guard Research Ship "John P. Tully". The yellow MF45 Series float manufactured by Flotation Technologies supports an upward-looking Teledyne-RDI LR75 kHz ADCP; a Sea Bird SBE 37 CTD has been attached to the release. Two train wheels form the anchor; (b) details of an IXBLUE SAS Oceano 2500S-Universal AR861 acoustic release open on the bench. The hock release arm is on the left end and the acoustic transponder on the right end. The central part shows the electronic circuit boards and the battery packs. The stainless-steel deep pressure case appears in the background. *Photos courtesy of David Spear, Institute of Ocean Sciences.*

acoustic interrogation of the release. A ship's sounder can be used in place of the hand-held transducer provided it is of compatible frequency and has a wide beam. This is useful since it allows the technician to "talk" to the release from the ship's laboratory rather than by lowering a transducer over the side of the ship. However, for "acoustically noisy" ships, lowering a transducer over the side is often the only way to communicate with the release. More advanced releases enable the user to measure the *slant range* from the ship to the release based on the two-way time delay. By taking into account the slant of the acoustic path, the user can determine the coordinates of the release. Long-life (e.g., 3-year) acoustic "pingers" built into the releases also are used to locate the depth and position of moorings using triangulation procedures. This is particularly useful for those moorings that fail to surface on command and must be dredged from the ship using a long line and grappling hook. In some acoustic releases, a rough estimate of the orientation of the release can be obtained remotely through changes in ping rate. For example, in the case of the 1090D InterOcean acoustic releases used by the Institute of Ocean Sciences, a doubled ping rate means that it is lying on its side rather being upright in the water column.

Acoustic releases made from stainless steel and/or titanium construction for most offshore oceanic research applications are manufactured by EdgeTech (formerly ORE Offshore), InterOcean Systems LLC, Teledyne-Benthos, iXBlue SAS Oceano Technologies, and Sonardyne. Marine industry publications also mention Subsea Sonics, UTC, Unique Group, Marine Electronics, Desert Star System, and Mitcham Industries, which appear to be mainly used for shallow water environmental studies. Catalogs of available marine products are available on-line for Sonardyne, iXblue (Oeano Technologies). Table 1.19 provides a list of deep-sea acoustic release transponders manufactured by the main players in the market. Most companies also have releases with shallower depth ratings (500–3,000 m) for more coastal applications. For example, the Sonardyne RT 6–1,000 is rated to 1,000 m, the Edge Tech Port MFE to 3,500 m and the Teledyne-Bethos R2K to 2,000 m depth.

As a side note, the acoustic releases used successfully in a 3-year study of the deep currents in the Middle America Trench (Thomson et al., 2010; Thomson and Davis, 2017), were 35-year old Oceano RT181 releases that were still "chirping" when returned to the laboratory many months after recovery.

Most modern releases use separate "load" (or "arm") and "release" codes so that the release can be remotely opened for instrument recovery. Some releases also provide a release code, such as "release confirmed", that signals to the operator on the ship or launch that the mooring has released and should be expected to surface in a time appropriate for the depth and net buoyancy of the mooring elements. Some releases also have a pinging mode, which allows the user to follow its rise using the ship's echo sounder (commonly 12 kHz). Releases typically can operate *in situ* for 2 years on alkaline batteries with another year in reserve. Using a lithium battery pack can extend their service life to 10 years. Thorough maintenance of the releases is an effective way to reduce mooring loss. As a result of improved technology and maintenance, the reliability of acoustic releases has improved to the point where they now function nearly 100% of the time.

As it is sometimes difficult to predict precisely where a mooring will surface, the time immediately after the release code has been sent can be quite tense. Spotting the mooring from the ship can be a real challenge, especially in rough sea conditions and reduced visibility. Attachment of a pressure-rated radio beacon and flashing light to the top float of the mooring can aide considerably with the recovery operation. In addition, floats and other mooring elements are generally colored yellow or bright orange to make them easier to spot when they are on the surface. If the mooring fails to surface after an appropriate time, a search can be initiated assuming that the mooring has surfaced and has not been spotted. Past experience has demonstrated the wisdom of having a dual release system with two acoustic releases side-by-side in a parallel harness often from different release manufacturers. A triangular bridle at the top of the dual system connects the

TABLE 1.19 Deep-sea acoustic release transponders available from the major manufacturers in support of scientific research.

	EdgeTech	InterOcean	iXblue Oceano	Sonardyne	Teledyne Benthos
Headquarters	USA	USA	France	UK	USA
Model ID	8242XS	1090ED	R5	RT 6-6000	R12K
Depth rating (m)	6,000	8,000	6,000	6,000	12,000
Release load (kg)	5,500	4,600	2,500	1,700	4,540
Battery life (yrs)	2.4/9	1	5	2.7	2
Weight (kg) air/water	36/28	11.2/8.1	25/19	20/15	33/25

releases to a single point in the mooring line while a spreader bar connects the bottom of the package to a single attachment point on the anchor chain. If one release fails to open, the second can be triggered, allowing both releases to rise to the surface. The extra cost can help avoid the need to dredge for the mooring if a single release should fail. (Sonardyne offers a parallel release kit with a heavy-duty release frame.) Dredging is a last (and often unrewarding) resort that can be extremely harmful to the mooring hardware, leading to severe damage to the current meters and other instruments on the mooring line. In addition, there is a correct dredging procedure and oceanographers new to the field should talk to more experienced colleagues for guidance. Safety issues include tangled mooring lines that need to be guided past the ship's propeller and carefully handled on deck. Tangled lines that are under load from the mooring components and/or still-attached anchor can be especially dangerous.

Moorings are sometimes inadvertently "hit" by fishing gear, towlines, ships, or submerged vessels. Improperly designed mooring components can also cause moorings to break apart. To facilitate the protection and recovery of fragmented moorings, technicians at the Institute of Ocean Sciences (IOS) in British Columbia have devised a protocol for mooring loss prevention, including: (1) the reduction in corrosion of mooring components by applying protective surface coatings, isolating dissimilar metals, using marine grade materials, and installing sacrificial anodes such as zinc blocks; (2) the use of distributed floatation such that any portion of the mooring remains positively buoyant should there be an unexpected break in the mooring line; (3) selecting mooring locations that minimize potential loss and aid recovery, while maintaining the integrity of the science program. In addition to the scientific program, consideration must be given to marine traffic, the nature of the seafloor, and proximity of submarine hazards such as transmission cables; (4) the use of simple acoustic beacons ("pingers") to mark the two ends of a mooring section. These can be recovered should the pinging mode of the acoustic release fail when the mooring arrives on the surface. In addition to providing recovery backup, the two pingers can be used to confirm that the mooring is still in position and intact prior to recovery. Pingers used at IOS transmit a 27-kHz frequency pulse, with 1 s-repetition rate, at 1/4 W output. They are mounted on an "in-line" strength member and have a conservative 2.5-year continuous operation life. They are simple, salt water activated, and have a limited range of roughly 0.5−1.0 nautical miles, depending on environmental conditions; and (5) equipping each mooring with a variety of locating beacons, for example VHF radio beacons, Argos Platform transmit terminal (PTT) satellite beacons, and XF xenon flashing lights. Iridium beacons serve the same purpose as Argos but with 2-way communication capability. Because of their expense and service fees, beacons are placed on the most valuable portion of the mooring only. As an example, IOS uses a Model 265 Beacon made locally by Oceanetic Measurement Ltd. The beacons are submersible to 1,500 m depth, have either a pressure or saltwater switch, and a service life of about 1 month. The beacon signals when a mooring has risen to the surface and periodically updates a track of its latitude and longitude within 1-km accuracy, to 300 m highest accuracy. The PTT is a "dumb" transmitter so that the number and quality of the position fixes are determined by the current satellite constellation. The commonly used VHF and XF units are manufactured by Novatech, a branch of Metocean Ltd. In addition, Xeos Ltd. provides VHF, Argos, XF and Iridium beacons, with modern electronics and low power consumption.

1.9 LAGRANGIAN CURRENT MEASUREMENTS

A fundamental goal of physical oceanography is to provide a first-order description of the global ocean circulation and its spatial and temporal variability. The idea of following individual parcels of water (the Lagrangian perspective) is attractive because it permits investigation of a range of processes taking place within a tagged volume of water. Named after Joseph L. Lagrange (1736−1811), a French mathematician noted for his early work on fluid dynamics and tides, Lagrangian descriptions of flow can be used to investigate a broad suite of biophysical and geochemical processes, ranging from the dispersion of substances discharged into the ocean from a point source—such as the radioactive emissions from the March 2011 Fukushima Dia-ichi nuclear incident (Aoyama et al., 2016; Smith et al., 2017) − to the seaward advection of nutrients and associated semi-enclosed marine ecosystems by mesoscale eddies originating over continental margins (Whitney and Robert, 2002). Early Lagrangian measurements consisted of tracking some form of tracer, such as a surface float or dye patch. While giving vivid displays of water motions over short periods of time, these techniques demanded considerable onsite effort on the part of the investigator. Initial technical advances were made more rapidly in the development of moored current meters, which yielded a strictly Eulerian picture of the current. However, improvements in tracking systems and buoy technology since the 1970s, have made it possible to follow unattended surface and subsurface drifters for periods of many months to several years. Satellite-tracked surface buoys and acoustically tracked, neutrally buoyant SOFAR (Sound Fixing And Ranging) sub-surface floats (also known as "Swallow" floats; John Swallow was the first to track sub-surface, neutrally buoyant floats) have been able to provide reliable, long-term, quasi-Lagrangian trajectories for many different parts of the world. (The trajectories are called quasi-Lagrangian because the drifters have a small "slip" of the order of 1−3 cm/s—which varies with platform and drogue—relative to the advective flow and because they do not move on true density surfaces. Surface drifters, for example, move on a two-dimensional plane rather than a three-dimensional density surface.) More recently, the more than 3,000 active vertically profiling floats maintained since 1999 as part of the international Argo program are providing oceanographers with an unprecedented volume of data on the temperature, current velocity, and (to a lower extent) the salinity of the World Ocean. By December 2012, these drifters had yielded more than one million temperature and salinity profiles to depths of 2,000 m. By March 2022, the number of profiling Argo drifters providing oceanographic data on the global ocean had increased to 3,975 (www.aoml.noaa.gov/phod/argo/).

Remotely tracked drifters provide a convenient and relatively inexpensive tool for investigating ocean variability without continued direct involvement of the investigator. In the case of the satellite-tracked buoys, the scientist can now access the positions of drifters or collect data from ancillary sensors on the buoys. The number of possible satellite positional fixes varies with latitude for the older buoys tracked by the ARGOS system (Table 1.20). Recently, buoy positions are calculated by onboard GPS receivers and relayed via the Iridium satellite communication system. Time delays between the time that the data are collected by the spacecraft and the time they are available to the user is typically less than a few hours (Figure 1.94). Time delays are even shorter for the newer system using GPS. This feature makes the drifters useful for tracking objects in near real-time such as oil spills. Oceanic platforms and satellite data transmission systems have become so reliable that both moored and drifting platforms are now used for the collection of a variety of oceanic and meteorological data, including sea surface temperature (SST), sea surface pressure (SSP), wind velocity, dissolved oxygen concentration, fluorescence, and mixed layer temperature. A new era of oceanographic data collection is in progress with less direct dependence on ships and more emphasis on data collection from autonomous platforms and satellites.

1.9.1 Drift Cards and Bottles

Until the advent of modern tracking techniques, estimates of Lagrangian currents were obtained by seeding the ocean surface with marked waterproof cards or sealed bottles and determining where these "drifters" came ashore. The cards or bottles contained a note requesting that the finder contact the appropriate addressee with the time and location of recovery. To improve the chances of notification, a small token reward was usually offered (one Australian group gave out boomerangs). Although drift cards and bottles provide a relatively low-cost approach to Lagrangian measurements, they have major limitations. Because they float near the surface of the ocean, the movements of the cards and bottles are strongly affected by wind and wave-induced motions. In fact, much of what these types of drifter measure is wind and wave-induced drift ("windage") rather than underlying ocean currents (see Section 1.9.1.1). Moreover, even if the recovery rate was fairly high (1% is considered excellent for most drift card studies), the drifters provide, at best, an estimate of the lower bound of the mean current averaged over the time from deployment to recovery. Unless the card/bottle was recovered at sea, the scientist could never know if the drifter had recently washed ashore or had been lying on the beach for some time. In addition, the drifter provided no information on the current patterns between the deployment and recovery points.

TABLE 1.20 The mean number of satellite passes per day (24 h) for the Argos System for different numbers of satellites in view.

Transmitter latitude (°North and South)	With two satellites	With three satellites	With four satellites
0	7	10	14
15	8	12	16
30	9	13	18
45	11	16	22
55	16	24	32
65	22	33	44
75	28	42	56
90	28	42	56

There were seven satellites that carried the Argos system at this time. Because the number of passes scales linearly, the tabulated values can be extrapolated to give the mean number of passes for the seven-satellite system.
Table courtesy of Bill Woodward, President and CEO of CLS America, Inc. (2013).

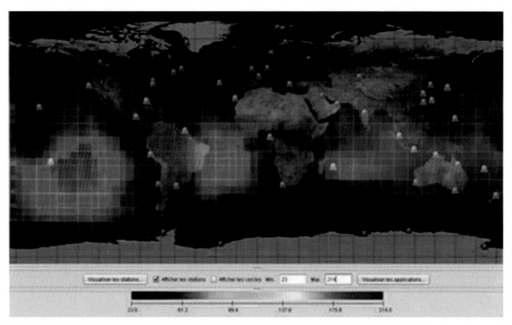

FIGURE 1.94 Argos data disposal time (data throughput time) in minutes for May 2013. The time between the satellite observation and the time that the data are processed and available from Service Argos. The data timeliness is a function of several variables, including the number of satellites in the system and the number of antennae which are connected and receiving Argos data in real time. *Courtesy of Bill Woodward, President and CEO of CLS America, Inc. (2013).*

1.9.1.1 Windage

Estimating windage is important when determining the drift of surface objects and has proven to be particularly important for predicting the trans-oceanic drift of the estimated 1.5 million metric tons of surface debris from the March 11, 2011-Tohoku-oki tsunami in the North Pacific (Cummins et al., 2011). The velocity, $\mathbf{U}$, of objects subject to windage can be separated into two components, $\mathbf{U} = \mathbf{U}_{water} + \mathbf{U}_{wind}$, where $\mathbf{U}_{water}$ is the horizontal velocity of the water at the ocean surface and $\mathbf{U}_{wind}$ is the velocity of the object relative to the water (windage). Windage results from the direct forcing on an object by the wind blowing over the sea surface. Assuming a balance of wind force and ocean drag, the wind-driven component of velocity of an object can be estimated as

$$\mathbf{U}_{wind} = k\sqrt{\frac{A}{B}}\,\mathbf{U}_{10} \tag{1.96}$$

where $\mathbf{U}_{10}$ is the wind velocity at the standard height of 10 m above the sea surface, A and B are, respectively, the surface area of the object above and below the water line, and $k \sim 0.025$ is a constant drag coefficient. According to this relation, an object with equal surface area above and below the water line ($A/B = 1$) drifts relative to the water surface at 2.5% of the wind speed at 10 m elevation. This drift is in the direction of the wind, which may be different than the direction of the prevailing surface current. An object with high windage will be transported across the ocean more rapidly than an object with low windage.

1.9.2 Evolution of the Modern Drifting Buoy

The evolution to the drifting buoys now being used in the Global Drifter Program (GDP) took place over about a decade. The most important development was that of the Random Access Measurement System (RAMS) that was launched on the last of the NIMBUS satellites. RAMS made it possible to geolocate a drifting platform on the Earth's surface with a reasonable accuracy of ±5 km, while also collecting a modest volume of data transmitted from the platform. The importance of RAMS is that it could perform these tasks for multiple platforms (up to eight) at the same time. This capability made it possible to routinely collect information from buoys on the surface of the ocean. The data were stored on the NIMBUS satellite and downlinked once per 108-min orbit. In response, the oceanography community began developing buoys to collect surface current information (Lagrangian drifters) along with sea surface temperature and atmospheric pressure. A wide variety of hull shapes were initially developed (e.g., Figure 1.95), each with its own specific characteristics.

FIGURE 1.95   Two types of first-generation drifting buoys on the deck of a ship. *W.J. Emery photo.*

The white and gray buoy in Figure 1.95 is the one first used by the National Oceanic and Atmospheric Administration (NOAA) in the United States. It had the disadvantage that the buoy beneath the large float was subject to considerable sheer stress that caused the buoy to break and fail. The reddish colored buoys were developed by Dennis Kirwan and Gerry McNally. They were very robust, but expensive to build. The flanges located around the hull were there to keep the buoy from bobbing up and down in the wave field. Another first-generation drifting buoy that was commercially available is presented in Figure 1.96. While it had the same general design of the NOAA buoys that failed, the Polar Science buoys were made of aluminum and therefore did not break apart beneath the float like the plastic NOAA buoys. Also shown in Figure 1.96 is a rolled-up "window shade" drogue that was commonly used in first-generation buoys. The problem with

FIGURE 1.96   A Polar Science drifting buoy with a wrapped-up window-shade drogue prior to deployment. *W.J. Emery photo.*

window-shade drogues is that they would "sail" in the current (rise up and allow the current to flow underneath). This causes the drogue to lose a considerable fraction of the external fluid force, as the current flows under, rather than at the drogue, thereby failing to couple the buoy to the ocean current.

Although launching any one of these drifters was an easy process—one just had to have a sufficient number of people to pick up the unit and toss it over the side of the ship—the prices in those times meant that it was equivalent to tossing a new Chevy Chevette into the ocean. On rare occasions, it was possible to recover a buoy after a long time at sea. One of the drifters in Figure 1.96 was retrieved after 2 years in the North Pacific Ocean; It no longer had its window shade drogue but did manage to acquire a drogue of barnacles (Figure 1.97).

The advent of the Surface Drifter Program of the World Ocean Circulation Experiment (WOCE) required a standardization of the drifting buoys. Peter Niiler was instrumental in unifying the drifter platform and drogue. The WOCE drifter in Figure 1.98 shows the relative sizes of the surface buoy and the "holey-sock" drogue that enables the currents to pull the unit around the ocean. A weak point of the drifter was the coupling between the surface float and the line leading to the drogue. The line experienced repeated shock loads from surface wave motions and was known to fail due to fatigue. To avoid this problem, the float-drogue coupling was made out of thick rubber "radiator hose" and a smaller intermediate buoy was added to further reduce the shock loading on the line.

1.9.3 Modern Drifters

Lagrangian drifters can be separated into three basic types: (1) Surface drifters having a surface buoy that has a short underwater appendage or is tethered to a subsurface drogue at some specified depth (typically less than 30 m, with 15 m the accepted standard drogue depth); (2) Subsurface, neutrally buoyant floats that are designed to remain on fixed subsurface density surfaces; and (3) Pop-up floats that cycle between subsurface density surfaces (where they remain for some fixed period of time) and the ocean surface. The early RAMS satellite-tracking system noted earlier was replaced in 1978 by the French Argos system (Collecte Localisation Satellite, CLS) carried onboard U.S. NOAA polar-orbiting weather satellites. As reported by Krauss and Käse (1984), this twin satellite system is capable of positional accuracies better than ±2 km. Location quality depends on a number of factors, such as the quality attributed to the position of a PTT (Platform Transmit Terminal) as per one of seven location classes (LC). When ≥4 location messages are received, the upper bounds for an error estimate are 250 m (LC3), 500 m (LC2), 1,500 m (LC1), or >1,500 m (LC0). For cases with <4 messages, no error estimate is given by Argos, but it is more than the 1.5 km and is likely closer to the ±2 km stated earlier. For most drifting buoys, a position is determined by fewer than four contacts. The number of fixes per day is a function of latitude and the number of satellites available. Higher accuracy is possible when the platform's position is fixed over periods longer than 7 min during two successive satellite passes. The drifters themselves cost about $2,000 and are considered expendable. Typical tracking costs (in 2013) were of the order of $2,500 per year for full tracking (no positional fixes omitted) and one-third of this for the one-third-duty cycle permitted by Service Argos (i.e., full-time tracking for 8 or 24 h followed by no

FIGURE 1.97 A Polar Science drifter recovered after 2 years floating in the North Pacific Ocean. *W.J. Emery photo.*

FIGURE 1.98 The WOCE drifter with its holey-sock drogue and connecting rubber lines.

The WOCE/TOGA
Lagrangian Drifter

tracking for 16 or 48 h, respectively; cf. Bograd et al., 1999a). Studies of surface and subsurface currents using Argos-tracked drifters can be found throughout the literature. Examples for the North Pacific involving mesoscale eddies, basin-scale North Pacific trajectories, inertial and tidal currents, mean surface currents, diurnal shelf waves, and anticyclonic eddies have been presented by Thomson et al. (1990, 1997, 1998), Bograd et al. (1999b), Rabinovich and Thomson (2001), and Rabinovich et al. (2002). Fratantoni (2001) provides a detailed discussion of the surface circulation throughout the North Atlantic as observed by satellite-tracked drifters during the 1990s. Results also show that there remain many "unvisited" areas of the Atlantic Ocean.

The development of low-cost and accurate GPS receivers in 2000 began a transition to buoys equipped with GPS receivers that transmitted both position and data via the Iridium satellite communication system. This transition from Argos to GPS positions transmitted via Iridium is clearly evident in Figure 1.99 (Rick Lumpkin, personnel communication, 2022). Iridium's unique constellation of communications satellites (Figure 1.100) makes it possible to relay data from platforms anywhere on the Earth's surface. Originally conceived as a method of relaying data from Earth platforms at high latitudes, where geostationary communication satellites fail, Iridium has expanded to be a system that works for all latitudes. The idea is that an Earth platform will be able to "see" and communicate with at least one, if not more than one, Iridium satellites at any given time. Another advantage of the Iridium data transfer is that more than 32 bytes (the Argos data limit) can be transferred, opening up drifting buoys as platforms that can carry many more sensors and relay their data in real time. Almost all drifting buoys are now using GPS locations and Iridium data transfer.

Examples of drogue configurations for satellite-tracked drifters are presented in Figure 1.101a along with the design for the standard holey-sock (World Ocean Circulation Experiment) WOCE/TOGA near-surface velocity drifter (Figure 1.101b). The purpose of the drogue is to reduce slippage between the drifter package and the water. It should be noted that early studies demonstrated that the window shade "sailed" with the current, allowing the current to flow under the drogue. The parachute, while effective, was prone to collapse in weak currents and then would not redeploy. Another drogue, called the "triton" drogue, was difficult and expensive to construct and failed to perform any better than the holey-sock, which was subsequently adopted as the global standard design for surface drifters.

The surface float contains the transmitter, temperature and atmospheric pressure sensors, along with a drogue sensor (strain gauge) and instrument electronics (Figure 1.101c). The purpose of the subsurface buoy is to reduce the "snap

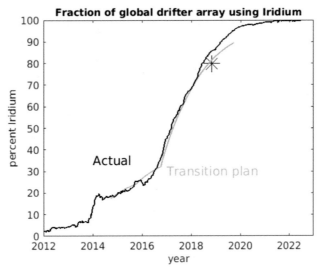

FIGURE 1.99 The transition of drifting buoy locations and data transfer from Argos to Iridium satellite systems. The blue line shows the planned transition percentage; the solid black line is the actual percentages. *Courtesy Richard Lumpkin, Director, GDP program, NOAA/AOML.*

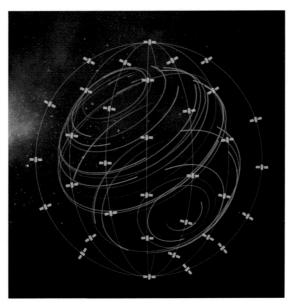

FIGURE 1.100 The Iridium constellation of communication satellites covering the globe. *Courtesy Richard Lumpkin, Director, GDP program, NOAA/AOML.*

loading" on the drogue and cable by absorbing some of the shock from surface wave motions. In the case of the WOCE/TOGA holey-sock drifter, the ratio of the drogue cross-sectional area to the cross-sectional area of the other drifter components (such as the wire tether and subsurface float) is about 45:1, a relatively high drag-area ratio for typical drogues.

The WOCE/TOGA drifter has further evolved into the now standard drifter (Figure 1.102) of the Global Drifter Program (GDP), an international program to keep ∼ 1,300 drifting buoys repeatedly deployed in most areas of the ocean.

All GDP drifters are equipped with holey-sock drogues centered at 15 m below the surface. The holes in the socks give rise to uniform distributions of the pressure and viscous forces over the surface of the drogue (Peimani et al., 2022). There is an effort to maintain a global drifter coverage but, as evident in the most recent drifter map shown in the next section, the coverage is not uniform. The buoys are rapidly advected away by the strong currents in the equatorial and western boundary current regions; the Southern Ocean is almost completely bare due to the strong Circumpolar Current.

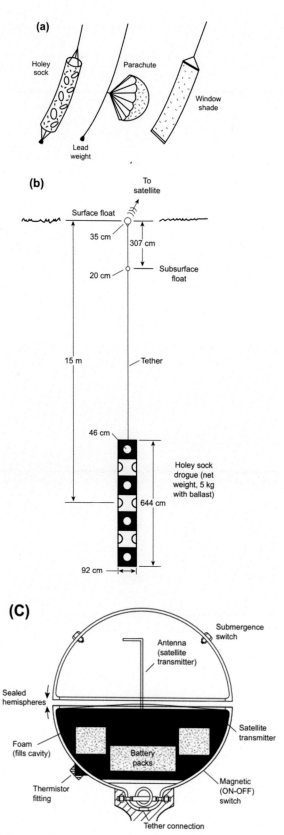

FIGURE 1.101 (a) Examples of the basic drogue designs for satellite-tracked drifters: holey-sock drogue; parachute drogue; and window-shade drogue; (b) schematic of the standard WOCE/TOGA holey-sock surface drifter showing pattern of holes in cloth panels; (c) cut-away view of foam-filled Plexiglass shell and PTT (satellite transmitter) used for the surface buoy. When complete, the surface float has an excess buoyancy greater than 7 kg. *(a) Niiler et al. (1987). (b and c) From Sybrandy and Niiler (1990).*

FIGURE 1.102 Dr. Richard (Rick) Lumpkin, Director of the Physical Oceanography Division of the Atlantic Oceanographic and Meteorological Laboratory, holding a GDP drifter with its holey sock drogue prior to launch.

1.9.4 The Global Drifter Program

The successful use of drifters as observational platforms led to the establishment of the Global Drifter Program (GDP), a principal component of the Global Surface Drifting Buoy Array, which in the United States is part of NOAA's Global Ocean Observing System (GOOS). The objectives of the GDP are to:

1. Maintain a global 5° × 5° array of 1,250 satellite-tracked surface drifting buoys to meet the need for an accurate and globally dense set of *in situ* observations of upper mixed layer currents, sea surface temperature, salinity, atmospheric pressure, and wind velocity. (Although salinity measurements were included in the original plans, they now appear to have been temporarily abandoned.)
2. Provide a data processing system for scientific use of these data. At present, the data support short-term (seasonal to interannual) climate predictions as well as climate research and monitoring.

In the United States, the GDP is administered by NOAA's Atlantic Oceanographic and Meteorological Laboratory (AOML), which coordinates deployments, processes the data, archives the data, maintains META files describing each drifter deployed, develops and distributes data-based products and updates the GDP Web site: https://www.aoml.noaa.gov/phod/gdp/index.php. AOML coordinates the drifter industry, upgrades the technology, purchases drifters, and develops enhanced data sets. NOAA's GDP is part of an international program involving over 27 countries with major contributions, in addition to the United States, from Argentina, Australia, Brazil, Canada, France, Germany, Iceland, India, Italy, Republic of Korea, Mexico, New Zealand, South Africa, Spain, and the United Kingdom. The United States, France and the UK are the three leading contributors. The Lagrangian Drifter Laboratory at the Scripps Institution (La Jolla, California) undertakes drifter engineering technology, improves on existing designs, develops new drifters, manages the real-time data stream, including posting drifter data to the Global Telecommunication System. Scripps also helps to supervise the industry, purchase and fabricate drifters and develops enhanced data sets.

The drifting buoy configuration for the GDP is presented in Figure 1.103. Although there are a number of manufacturers of these drifters—so that individual buoys can look somewhat different—the basic configuration is the same as that in Figure 1.103. Deployment of the buoys is carried out from research vessels, volunteer observing ships or from the air. Because the GDP objective is to maintain a global array of ~ 1,300 buoys operating simultaneously, there is a need for repeated annual deployments to fill in gaps that arise from buoy failures or to cover areas from which buoys have been rapidly advected away.

As noted previously, the GDP Drifter Data Assembly Center has the goal of assembling and providing uniform quality control of barometric pressure, SST and surface velocity measurements to help improve climate prediction models, which require accurate estimates of SST to initialize their ocean component. In addition to supporting numerical modeling, the drifters provide SST for the calibration and validation of satellite infrared SSTs which, in turn, allow for greater spatial coverage for climate prediction models. The status of the GDP drifter array of 1,300 buoys as of late October 2021 is shown in Figure 1.104. The red dots in this figure denote buoys that measure SST only (a total of 383) and the blue dots denote those that also measure sea-level barometric pressure (892). A few buoys also measure surface salinity. Another

FIGURE 1.103   Launching a holey-sock drifter from the stern of a vessel. *From NOAA's Global Drifter Program website.*

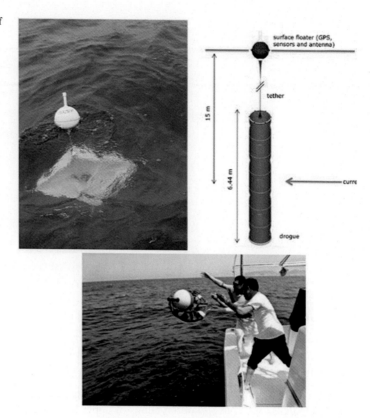

FIGURE 1.104 Status of the Global Drifter Array as of October 25, 2021. *NOAA Global Drifter Program.*

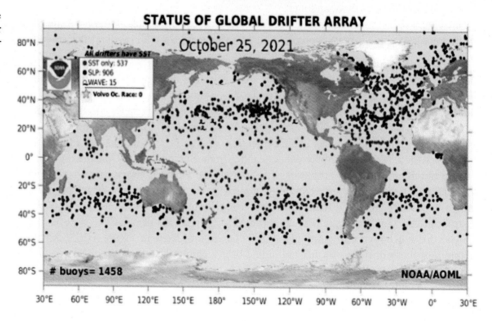

version of this map (not shown) reveals that the majority of the drifters have been provided by the U.S., followed by France. Both surface water temperature and atmospheric pressure measurements are critical for weather and climate models. A different view of the data distribution is given in Figure 1.105, which shows the growth of the GDP since 1988. The plot reveals how many buoys measure SST, and how many sense atmospheric pressure, and wind. The SST sensing buoys constitute a large majority of the total number of buoy days; atmospheric pressure sensing platforms began in 1994 and continue to increase along with the SST. The addition of wind sensors began in 1997 and peaked in 2002. Salinity

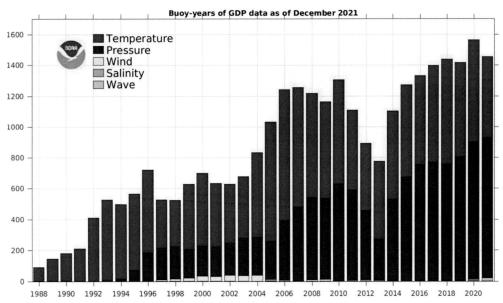

FIGURE 1.105 Growth of the Global Drifter Program drifter database in terms of sensor years of data for the period 1988–2021. *NOAA Global Drifter Program.*

sensors were added in 2010 and their numbers steadily increased through 2012 but appear to have ended in April 2013. Many of these additional measurements may now reappear since the data transfer through Iridium can be much greater than through Argos.

An informative application of drifter data is to map surface currents and the distribution of eddy kinetic energy (EKE). The EKE distribution in Figure 1.106 clearly highlights regions with strong zonal currents and strong poleward flowing western boundary currents, such as the Kuroshio, Gulf Stream, and East Australian Current. Many of these same areas correspond to areas of high EKE with some additional areas of high EKE, including the Brazil–Malvinas Current confluence off the south-central east coast of South America. The EKE is substantially lower in the Antarctic Circumpolar Current where strong zonal currents do not exhibit the same high degree of mesoscale variability. Yu et al. (2019) separated the temporal variability in EKE into specific frequency bands that reveal, among other findings, a resonance of diurnal motions near latitude 30° (the turning latitude in both hemispheres) and enhanced semidiurnal tidal currents at high latitudes in the Pacific Ocean. Intensified kinetic energy is also present at 30° latitudes in Figure 1.106.

Because surface drifters closely track the two-dimensional flow of the ocean surface, they are ideal for studying the dispersion of surface particles, such as fish larvae, and buoyant pollutants, such as oil spills. Drifter observations of dispersion can also be used to quantify the effects of mesoscale variability on the mean transport (Davis, 1991). Lumpkin et al. (2002) compared the Lagrangian and Eulerian length scales from surface drifters and altimetry, and found that they were proportionally related only in the most energetic part of the ocean (such as the Gulf Stream), where Lagrangian particles are advected around eddies at timescales much shorter than the Eulerian timescale. Here, "proportionally related" means that the two scales had different magnitudes but varied in a similar manner. Where the currents are weaker, the two scales were completely unrelated. Following Davis (1983) and Middleton (1985), Yu et al. (2019) note that, for turbulent flows that are isotropic, stationary, and nondivergent, the relationship between Lagrangian and Eulerian velocity spectra depends on a parameter $\alpha = U_{rms} \cdot (TE/LE)$, where U_{rms} is the root-mean-square velocity and TE and LE are, respectively, the temporal and spatial decorrelation scales of the motions. For small α (the "fixed float" regime), the velocity data provided by drifters behaves like that from fixed moorings, whereby the Eulerian and Lagrangian velocity spectra are similar. For moderate to large values of α, the Eulerian spatial variability is aliased into the temporal Lagrangian variability, whereby the Lagrangian and Eulerian spectra will differ.

Bauer et al. (2002) separated mean and eddy drifter velocities in the tropical Pacific using optimized bicubic splines and found that eddy diffusivity in this region is strongly anisotropic: zonal diffusion is up to seven times greater than meridional diffusion due to the trapping of water parcels in coherent features such as equatorial and tropical instability waves. Two years later, Zurbas and Oh (2004) mapped the global surface diffusivity using a method designed to avoid the contamination of their eddy statistics by the time-varying mean flow shear. Their maps of apparent diffusivity (Figure 1.107) reveal enhanced values in zonal bands near 30° latitude in both the southern and northern hemispheres, similar to results of Yu et al. (2019). Zurbas and Oh suggested that these features were due to meandering, eddy-rich

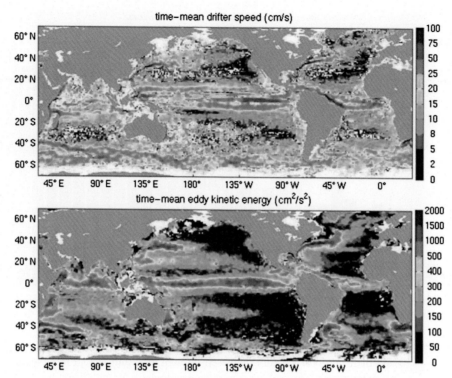

FIGURE 1.106 Time-averaged speed (top) and eddy kinetic energy (bottom) calculated from drifter observations. Values are shown at 1° spatial resolution. *Courtesy of Rick Lumpkin, Director, GDP program, NOAA/AOML.*

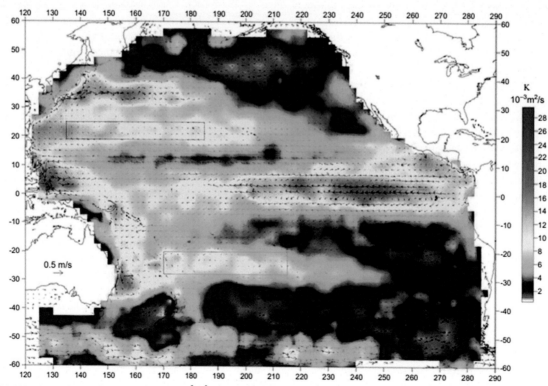

FIGURE 1.107 Map of the lateral diffusivity, K (10^{-3} m^2/s), in the Pacific Ocean, combined with the mean current vectors in the surface layer, derived from satellite-tracked drifters. *From Zurbas and Oh (2004).*

eastward currents in the North Atlantic (Azores Current) and the North Pacific (North Subtropical Countercurrent) and to the westward drift of Agulhas retroflection eddies in the South Atlantic. In the South Pacific, they attributed high diffusivity values to the presence of a South Subtropical Countercurrent.

Some oceanic regions have been well sampled during parts of the year but poorly sampled at other times due to infrequent batch buoy deployments from research vessels or volunteer observing ships. In areas with strong seasonal fluctuations, such as the tropical Atlantic, this produces biased estimates of the mean flow when they are averaged in bins. Lumpkin (2003) addressed this problem by dividing the Lagrangian time series into spatial bins and, within each bin, decomposing the time series into a time-mean, annual and semiannual harmonics, and a residual component. This decomposition was performed using a Gauss-Markov method, familiar in the oceanographic literature for its use in box inverse models (e.g., Ganachaud and Wunsch, 2000). This approach can resolve the amplitudes and phases of seasonal fluctuations throughout most of the tropical Atlantic, with the present density of drifter observations (Lumpkin and Garraffo, 2005).

An important application of drifter data is its synthesis with satellite altimetry. Altimetry is excellent at capturing the variations of sea-level height associated with geostrophic velocity anomalies but cannot yet map absolute sea level with sufficient accuracy to resolve time-mean currents. In addition, the sea level anomaly fails to account for the significant ageostrophic components of current variations, such as inertial effects that may account for differences in drifter-derived and altimetry-derived eddy kinetic energy (EKE) on either side of the Gulf Steam front (Fratantoni, 2001). Moreover, gridded altimetric data tend to smooth the observations, which results in systematic underestimates of the EKE. Niiler et al. (2003) describe a method to synthesize Ekman current-removed drifter speeds with gridded altimetric velocity anomalies in regions where they are significantly correlated. This method uses concurrent drifter and altimetry velocities to calibrate altimetry, making its amplitude consistent with the *in situ* drifter observations while using the time series from altimetry to correct for biased drifter sampling of the mesoscale to interannual variations. Using this method applied to the global drifter data set, Niiler et al. (2004) produced a map of the absolute sea level for the period 1992—2002 (Figure 1.108). For a list of publications linked to the Global Drifter Program, the reader is directed to the bibliography at https://www.aoml.noaa.gov/phod/gdp/bib/bibliography_chronological.php.

1.9.5 Processing Satellite-Tracked Drifter Data

Positional data obtained through satellite tracking needs to be carefully examined for erroneous locations and the time of drogue loss. In fact, one of the main problems with surface drifters, aside from the need for accurate positioning, is knowing if and when the drogue has fallen off. Strain sensors are often installed to sense drogue attachment, but they are

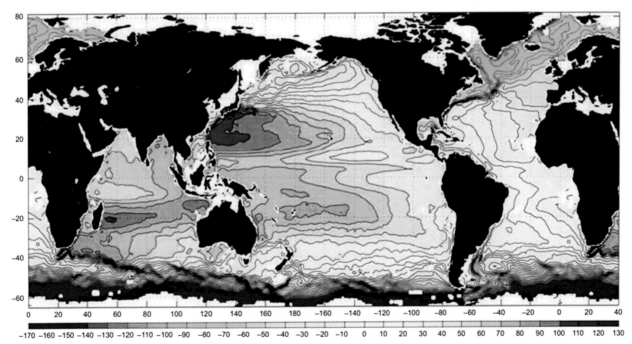

FIGURE 1.108 Long-term absolute mean (1992—2002) sea level from satellite altimetry and drifter trajectories. Elevations are in cm. *From Niiler et al. (2004).*

frequently unreliable. The tether linkage between the surface buoy and the drogue is the major engineering problem in designing robust and long-life drifters. Because of this problem, drifters often have a subsurface float to help absorb the snap loading on the drogue caused by surface waves and also ensure that the surface element is not constantly submerged in rough weather. An abrupt and sustained order-of-magnitude increase in the velocity and its variance derived from first differences of the edited positional data can be considered as evidence for drogue loss. The cubic spline routine in most software analysis packages works well for positional data provided the sampling interval is only a few hours (Bograd et al., 1999a). Although it is not recommended, the user can obtain the velocity components (u, v) directly from spline co-efficients for the positional data.

1.9.6 Drifter Response

As with all Lagrangian tracers, it is difficult to know how accurately a drifter is coupled to the water and what effects external forces on the drifter's hull might have on its performance. In most applications, the coupling between the buoy and the water is greatly improved by the drogue. Studies by Dahlen and Chhabra (1983) have determined that a holey-sock drogue is more efficient than either the window shade or the parachute drogue. This shape is easy to deploy and was selected for the standard drifter used in the WOCE program (Sybrandy and Niiler, 1990) and later in the Global Drifter Program (GDP).

In addition to the disagreements about which type of drogue is best, early field studies of Kirwan et al. (1978) reported that the wind-drag correction formula, given by Kirwan et al. (1975), is much too large for periods of high wind. The subsequent conclusion was that drifter velocities uncorrected for wind drag are better indicators of the true prevailing surface currents than are those corrected for the influence of wind drag on the buoy hull. In this context, it should be recognized that Lagrangian drifting buoys respond to the integrated drag forces including the forces on the drogue and the direct forcing on the hull. The driving forces in the water column consist of a superposition of geostrophic currents plus wind- and tide-generated currents. To evaluate the role of wind forcing on drifter trajectories, McNally (1981) compared monthly mean drifter trajectories with the flow lines for mean monthly winds computed from Fleet Numerical Weather Central's (now the Fleet Numerical Meteorology and Oceanography Center, FNMOC, Monterey, California) synoptic wind analysis. He found that the large-scale, coherent surface flow followed isobars of sea-level pressure and was 20−30° to the right of the surface wind in the North Pacific (Figure 1.109). Overall buoy speeds were 1.5% of the geostrophic wind speed during periods of strong atmospheric forcing (fall, winter, and spring). In the summer, mesoscale ocean circulation features, unrelated to the local wind, tended to determine the buoy trajectories.

McNally (1981) also compared trajectories among buoys with drogues at 30 m depth, buoys with drogues at 120 m, and buoys without drogues. He found that drifters drogued below 100 m depth behaved very differently from ones drogued at 30 m, but that those with drogues at 30 m and those without drogues behaved similarly. Using the record from a drogue tension sensor (drogue on−off sensor), McNally found that it was not possible to detect from the trajectory alone when a buoy had lost its drogue. This result suggests a lack of vertical current shear in the upper 30 m, where the flow apparently responds more directly to wind-driven currents than to baroclinic geostrophic flow. The result was supported by the poor correlation between mean seasonal dynamic height maps and the tracks of near-surface drifters reported by McNally et al. (1983). McNally (1981) also described an annual increase by a factor of 5 of the wind speed in the North Pacific, while the drifter speeds increased by a factor of 3.5, somewhat surprising considering that the wind stress that drives the currents is proportional to the square of the wind speed. During this same time the mean seasonal dynamic height amplitude changed only slightly. (It should be cautioned that studies by McNally and Kirwan were based on the large red buoys shown in Figure 1.110, which can be expected to behave differently than the standard drifters of the GDP.)

Emery et al. (1985) confirmed the lack of agreement between drifter tracks and the synoptic geostrophic current estimates, as well as the high correlation between drifter displacements and the geostrophic wind speed and direction (Figure 1.110). In a rather complex analysis of the wind-driven current derived from drifter trajectories, Kirwan et al. (1979) concluded that, while the drifter response is best described by a two-parameter linear system (consistent with the driving of the buoy by wave-driven Stokes drift), a combination of Ekman current plus Stokes drift also adequately described the resulting trajectories. Calculations by Emery et al. (1985), based on the nominal hull size, suggest that the Stokes drift component is relatively small and that the current in the surface Ekman layer is the primary driving mechanism for the mean drifter motion. That the angle this current makes to the wind is less than the 45° predicted by Ekman (1905), is expected since the real ocean conditions never seem to meet the steady wind conditions for Ekman's derivation. McNally (1981) found an average angle of 30° while Kirwan et al. (1979) reported an angle of 15°, both to the right of the wind for the northern hemisphere. A vector-regression approach to relating wind-stress forcing to the surface current response can be found in Cummins et al. (2022) (see also Chapter 5, Section 5.11).

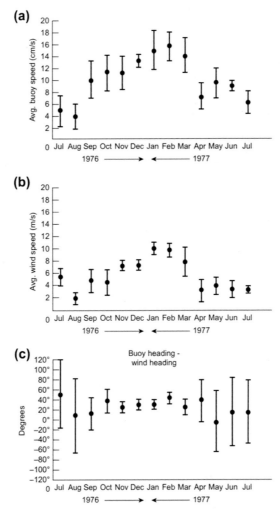

FIGURE 1.109 Monthly average wind and buoy speeds over the North Pacific from June 1976 through July 1977. (a) Monthly average drifter speeds; (b) monthly average wind speeds; (c) monthly average difference angle between wind direction and drifter direction. Vertical bars denote ±1 standard deviation (McNally (1981).)

A search for the elusive Ekman spiral was conducted from November 20, 1991, to February 29, 1992, by Krauss (1993) using 10 satellite drifters drogued at five different levels within well-mixed homogeneous water of 80 m depth in the North Sea midway between England and Norway. The holey-sock drogues used in the study were 10 m long and centered at 5 m depth intervals from 7.5 to 27.5 m (Figure 1.111a). Results for the first 4 weeks of drift, when the drifters were relatively close together, revealed a clockwise turning and decay of the apparent wind stress with depth as required by Ekman-layer theory (Figure 1.111b). Here, the apparent wind stress is derived from the fluctuations in current velocity shear measured by the satellite-tracked drifters. Sea surface slopes needed to complete the calculations are from a numerical model. (The current field in the study region is a superposition of barotropic currents due to sea level variations and Ekman currents.) The observed amplitude decay of 0.90 and deflection of 10° near the surface, followed by an angle of 41.6° at 25 m depth, appear to be in close agreement with theory (the apparent wind is toward 0° at the surface and should give rise to an Ekman current at depth that is 45° to the right of the apparent wind). However, in general, classical Ekman theory was unable to fully describe the observed deflection of the apparent wind (and Ekman current) to the right of the wind and its decay with depth. To be consistent with Ekman's theory, an eddy viscosity of 10^3 cgs units would be needed, which is well beyond the norm. Yet, as noted by Krauss, "... the deflections are a strong indication that some type of Ekman spiral dominates within the upper 30 m". In a related study, Lenn and Chereskin (2009), used repeated high-resolution profiles from a shipboard ADCP to examine the Ekman spiral and transport in Drake Passage in the Southern Ocean. Based on data from 156 transects between South America and the Antarctic Peninsula from September 1999 to October 2006, the authors show that the mean Ekman spiral penetrates to roughly 100 m depth in the unstratified surface waters and is compressed vertically

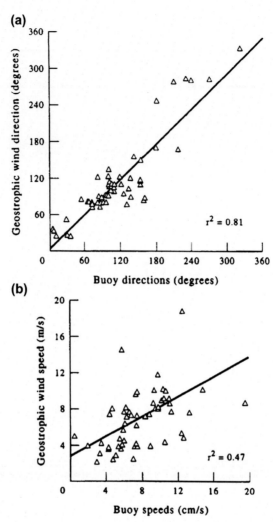

FIGURE 1.110 (a) Comparison between monthly mean buoy and geostrophic wind directions; (b) comparison between monthly mean buoy and geostrophic wind speeds. *From Emery et al. (1985).*

relative to theoretical predictions based on a constant eddy viscosity. The amplitude of the current was found to decay more rapidly than it rotated anticyclonically with depth. Fluctuations in the upper ocean stratification associated with diurnal buoyancy fluxes were thought to contribute to the Ekman spiral compression and the nonparallel shear—stress relation in the passage (see also Yoshikawa et al., 2007).

In an earlier study, McNally and White (1985) examined wind-driven flow in the upper 90 m using a set of buoys drogued at different depths. They found a sharp change in buoy behavior when the drogue entered the deepening surface mixed layer. This response was characterized by a sudden increase in the amplitude of near-inertial motions with a downwind drifter velocity component three times that of the crosswind component. They also found that 80—90% of the observed crosswind component could be explained by an Ekman slab model. The large downwind response leads to surface currents, calculated from the buoy displacements, that are greater than 0° but less than 45° (about 30°) to the right of the wind. This behavior was true for all buoys with drogue depths greater than the upper mixed layer thickness; once in the mixed layer, all buoys behaved the same regardless of the drogue depth.

In summary, the question of the relative coupling of drogued and undrogued drifting buoys to the water is still not completely resolved. Drifters measure currents, but which components of the flow dominate the buoy trajectories is still a topic of debate. Based on the recent literature, it appears that shallow, open-ocean drifters with drogue depths less than about 50 m are driven mainly by the wind-forced surface frictional Ekman layer, whereas deep drifters with drogue depths exceeding 100 m are more related to geostrophic currents. The likely percentage of contribution by these two current types depends on the type of drogue system that is used. A problem with trying to measure the deeper currents is that deeper

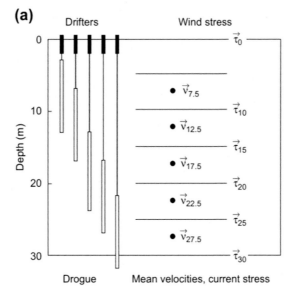

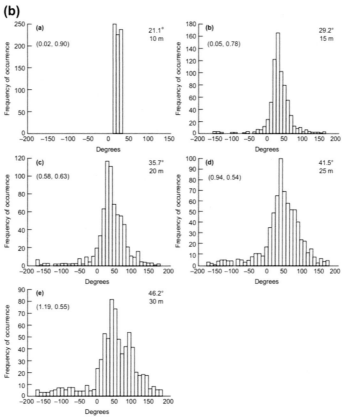

FIGURE 1.111 Test of Ekman's 1905 theory. The clockwise turning and decay of the apparent wind stress τ_D at depth D (m) relative to the observed surface wind stress. The apparent wind stress is derived from the current velocity shear dv_D/dz measured by satellite-tracked drifters (a) drogued at different depths during homogeneous winter conditions; (b) histogram of the relative angle (in degrees) between the surface wind-stress vector and the calculated apparent wind-stress as a function of depth (surface wind minus apparent wind). Linear regression values (α, β) give apparent wind-stress as a function of surface wind stress. Offset results from the different time scales of the winds and the currents. Mean values given in upper right corner of figure. *Courtesy W. Krauss (1994).*

drogue systems tend to fail sooner, and it is difficult to access quantitatively the role of the drogue in the buoy trajectories. Drogue loss due to wave loading and mechanical decoupling of the surface buoy and the drogue is still the main technical problem to extending drifter life. In more confined coastal waters, windage on the drifter hull and grounding are major factors affecting Lagrangian studies. The relatively long-lived drogues required to maintain a hull-to-drogue cross-section of around 40:1 leads to an increased likelihood of the drifter being temporarily or permanently grounded.

There are other problems with drifters worth noting. In addition to drogue loss and errors in positioning and data transmission, the transmitters submerge in heavy weather and lose contact with the passing satellite. Low drag ratios lead to poor flow response characteristics and, because of the time between Service Argos satellite passes in the past, there was generally inadequate sampling of tidal and near-inertial motions (especially at low latitudes or for the 1 day on-two days off duty cycles) leading to aliasing errors. At present, the satellite positional sampling rate is not generally a problem for drifters now being tracked by GPS. Regardless of the tracking method, however, drifters still have an uncanny tendency to go aground and to concentrate in areas of surface convergence.

1.9.7 Applications of GDP Drifter Data

The profusion of relatively uniform drifter data over the past several decades has made the study of their trajectories an attractive topic for research. For example, Lumpkin and Pazos (2005) use the GDP data to estimate the surface currents within the North Atlantic. After a comprehensive discussion of the drifters now used in the GDP, the authors argue that surface velocity drifters provide important, albeit imperfect, pseudo-Lagrangian observations at the depth of the drogue element. The data are pseudo-Lagrangian because the water parcels can oscillate up and down, whereas the drifter stays at the ocean surface; results are "imperfect" because of the slip of the drifter elements relative to the surrounding water. Slip—the difference in the horizontal component of the drifter motion relative to the lateral motion of currents averaged over the drogue depth—is caused by direct wind forcing on the surface float, drag on the float and tether induced by the wind-driven shear in the upper ocean currents, and by surface gravity wave rectification (Niiler et al., 1987; Geyer, 1989). The resulting velocity observations (Figure 1.112) are therefore a combination of slip, plus the directly wind-driven flow in the upper mixed layer, plus the 15 m deep eddy and gyre-scale currents.

A related study was published by Lumpkin and Johnson (2013), in which the authors calculate global near-surface currents from GDP data on a $0.5° \times 0.5°$ latitude-longitude grid using a new methodology. Data considered at each grid location lie within a centered bin of specified area, with a shape defined by the variance ellipse of the current fluctuations within that bin. The time-mean current, its annual harmonic, semiannual harmonic, correlation with the Southern Oscillation Index (SOI), spatial gradients and residuals are estimated, along with formal error bars for each component. The temporal mean resolves the major surface current systems of the World Ocean, while the variance shows the enhanced eddy kinetic energy in the western boundary current systems, in equatorial regions and within the Antarctic Circumpolar Current. Lumpkin and Johnson also identify three large "eddy deserts", two in the Pacific and one in the Atlantic. Seasonal variations reveal phenomena, such as the gyre-scale shifts in the convergence centers of the subtropical gyres, as well as the seasonal evolution of currents and eddies in the western tropical Pacific Ocean. Results of the study are available as a monthly climatology. The number of drogued drifter observations used in the study is shown in Figure 1.113, measured in units of drifter days per square latitude-longitude degree; Figure 1.114 presents the mean current speeds and streamlines from this analysis.

The current speeds in Figure 1.114 are from unsmoothed drifter data (to better indicate local current magnitudes), while the streamlines are from spatially smoothed velocities to better reflect the large-scale currents, such as the western boundary currents. The smoothing is applied only to bins with mean speeds <25 cm/s so that strong currents are not oversmoothed, while spurious localized convergences and divergence in regions of weak velocity are reduced. Lumpkin and Johnson (2013) also demonstrated their ability to determine seasonal changes in the fields depicted in Figure 1.114. As illustrated by Figure 1.115, there are dramatic changes in the Indian Ocean, as conditions switch from the Northeast Monsoon to the Southwest Monsoon. Changes are also readily apparent in the North Pacific and North Atlantic currents. In the North Pacific, the flow transitions from a dual gyre system in February to primarily meridional flow in August. In the North Atlantic, the single gyre in February gives way to meridional flow in August. Lumpkin and Johnson (2013) also computed errors estimates and correlations with the Southern Oscillation Index (SOI) (Figure 1.116).

The errors in Figure 1.116 indicate where currents are well resolved by the data and where more data are needed. It is noteworthy that the errors in Figure 1.116a attain their highest values in the equatorial regions and in the Southern Ocean, where strong currents quickly transport buoys out of the areas. The signal-to-noise ratio for the time-mean current (Figure 1.116b) is highest in the strong current regions and smallest in the weak current regions, such as the centers of the subtropical gyres. Seasonal variations are well-resolved (Figure 1.116d) throughout the tropical Pacific, along the path of

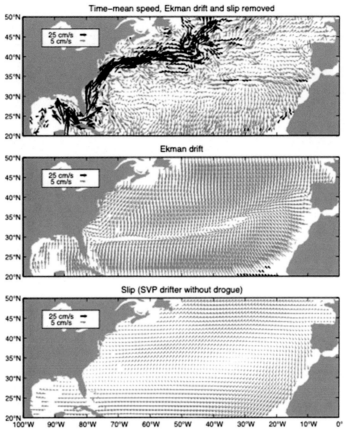

FIGURE 1.112 *Top*: Time-mean speed of all Surface Velocity Program (SVP) drifters in the subtropical North Atlantic, October 1989 to April 2004, with Ekman drift and slip removed. A separate scale (bold arrows) is used for speeds exceeding 10 cm/s. *Middle*: Time-mean Ekman drift (Ralph and Niiler, 1999; Niiler, 2001), showing the wind-driven convergence that forces the subtropical gyre (also see Figure 15 of Rio and Hernandez, 2003). *Bottom*: Time-mean slip of undrogued SVP drifters, using the parameterizations of Niiler and Paduan (1995) and Pazan and Niiler (2001). Drogued drifters experience a slip that is smaller by an order of magnitude. The total time-mean drift of undrogued drifters is given by the sum of the three panels.

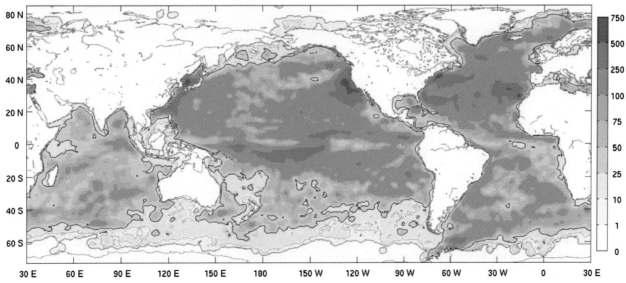

FIGURE 1.113 Number of drogued drifter observation used to map files at each binned grid point (in drifter days per square spatial degree). Thresholds for estimating a mean field (0.8 drifter days per square degree, red lines), a seasonal cycle (7 days per square degree, orange lines), and a Southern Oscillation Index (SOI) regression (29 days per square degree, black line) are contoured. *From Lumpkin and Johnson (2013).*

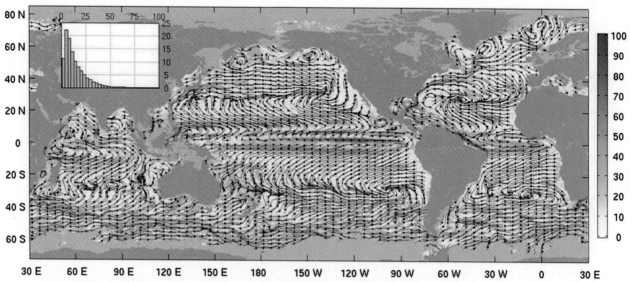

FIGURE 1.114 Mean currents speeds (color legend in cm/s) from near-surface drifter data along with streamlines (black lines). Streamlines are calculated from spatially smoothed currents to indicate the flow direction and qualitatively illustrate large-scale circulation features, including surface divergence regions. Light gray areas have fewer than 10 drifter days per bin (0.8 per square degree). In addition, only bins with mean current speeds statistically different from zero at one standard error of the mean, are shaded. Inset (top left) shows histograms of mean current speed (cm/s; horizontal axis is from 0 to 100 at 3.125 cm/s per bin interval) versus number of bins (in kilo-bins; vertical axis, from 0 to 25,000 bins). *From Lumpkin and Johnson (2013).*

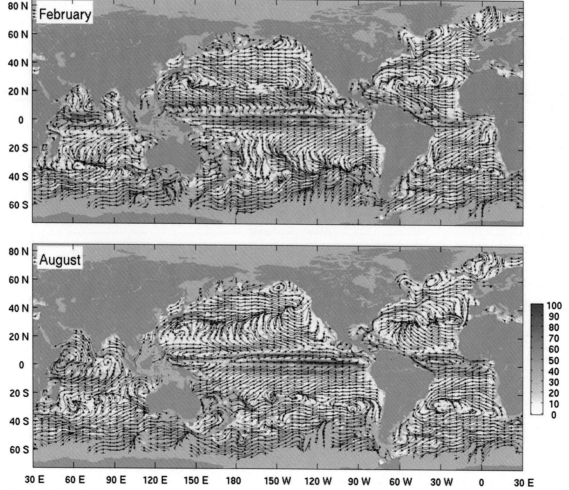

FIGURE 1.115 Near-surface ocean currents from surface drifter data. Details follow Figure 1.114, except the results presented here are climatological means for: February (top) and August (bottom). Light gray areas have fewer than 90 drifter days per bin (7 per square degree). *From Lumpkin and Johnson (2013).*

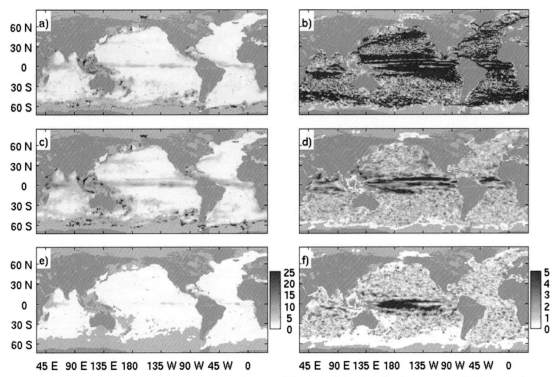

FIGURE 1.116 *Left column*: (a) absolute error (square root of the sum of the squared errors) in the time-mean speed, (c) speed of seasonal current variations, (e) current correlation with the Southern Oscillation Index (SOI) (cm/s, color bar adjusted). *Right column*: (b) signal-to-noise ratio, i.e., magnitude of the coefficient (or square root of the sum of the squared coefficients) divided by the error estimates for (b) the absolute speed, (d) speed of the seasonal current variations, and (f) SOI-correlated currents (color bar to be used for all but (e)).

the North Equatorial Counter Current (NECC) in the central and western tropical Atlantic, and off the equator in the tropical Indian basins. Seasonal variations are more marginally resolved in the Gulf of Guinea and the southern tropical Atlantic and equatorial Indian Ocean (where observations are sparse).

The next major step in the analysis of GDP data was published by Laurindo et al. (2017). As a follow on to Lumpkin and Johnson (2013), the study improves on the analytical procedure by correcting the slip bias of undrogued drifters so that they may be included in the overall statistics, thus recovering about half of the GDP dataset. They also introduce a new method of decomposing Lagrangian data into mean, seasonal and eddy components, with a focus on reducing the smoothing of spatial gradients inherent in data binning methods. Laurindo et al. (2017) further examined the sensitivity of the results to the selection of method parameters and evaluate the method performance relative to other data, such as altimeter-derived geostrophic currents.

An indication of the number of drifter days used in the Laurindo et al. (2017) slip analysis of undrogued drifters is apparent in the data distributions in Figure 1.117. The density of data from drogued drifters is, as usual, greatest in the traditional deployment sites, such as the western North Atlantic and both the western and eastern North Pacific, the tropical Pacific, and the Sea of Japan (also East Sea). What is surprising is the sparse area of drogued buoys in the South Pacific, particularly in the western part. The undrogued buoys are prevalent in the centers of the subtropical gyres, known to be convergence regions of wind driven material. These undrogued buoys tend to be older buoys that have lost their drogues and subsequently moved into the interior parts of the central gyres. While Ekman convergence plays a role in this concentration, Beron-Vera (2016) demonstrated that the main mechanism driving the accumulation of undrogued drifters at large-scale convergence zones is the combined action of wind and ocean current on the finite-sized floating objects. The new formulation of slip derived by Laurindo et al. (2017) for the undrogued buoys used the 10 m wind from the European Center for Medium Range Weather Forecasts (ECMWF), which were scaled for downwind slip in a manner similar to the of Pazan and Niiler (2001). The wind was linearly interpolated to the buoy position and the slip was calculated for both drogues and undrouged buoys. The slip was calculated using 6-h drogued and undrogued drifter observations selected within $4° \times 4°$ spatial bins centered at the grid points of a $1° \times 1°$ global grid. Only bins with more than 300 data values were considered and results for which the slip exceeded 3 standard deviations from the mean of all bins were considered as outliers and excluded. The downwind slip coefficients are shown in Figure 1.118.

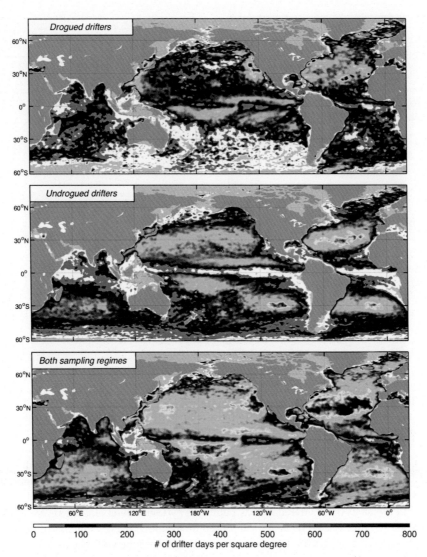

FIGURE 1.117   Number of drifter observation days per square degree for the period between February 1979 and June 2015 including drogued (top), undrogued (middle) and combined (bottom). The color bar for the number of drifter days is shown at the bottom. *From Laurindo et al. (2017).*

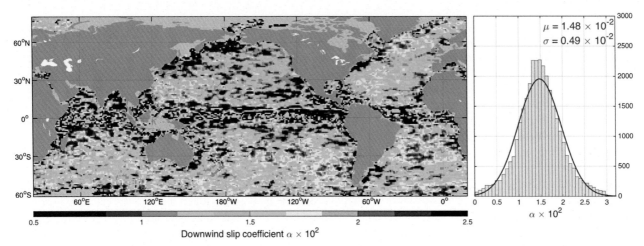

FIGURE 1.118   Downwind slip coefficient for undrogued GDP drifters. The histogram shows the slip coefficient distribution; the red line is a Gaussian fit to the distribution. *From Laurindo et al. (2017).*

The slip coefficients obtained by Laurindo et al. (2017) are consistent with earlier studies including 0.97×10^{-2} for the Pacific and North Atlantic oceans (Pazan and Niiler, 2001), 0.66×10^{-2} in the eastern Mediterranean Sea (Poulain et al., 2009), and 1.64×10^{-2} in the equatorial Atlantic and in the Indian Ocean (Perez et al., 2014; Peng et al., 2015). The spatial pattern in Figure 1.118 would not be observed in a purely random field and is most likely due to different geophysical conditions. Mean velocities computed from uncorrected and undrogued drifters differ on the order of 0.1 m/s due to the slip of the undrogued buoys. In general, individual velocities can be twice those of the drogued buoy velocities.

Following Lumpkin and Johnson (2013) and Laurindo et al. (2017) low-pass filtered the slip corrected 6-h drifter trajectory velocities using a 5th degree Butterworth filter with a cutoff period 1.5 times the local inertial period or 5 days, whichever is shorter, to remove tidal and near-inertial variability. The resultant velocities were then linearly interpolated to daily values, which reduces the amount of correlated data without significantly compromising data coverage in sparse regions. Data subsets of the zonal and meridional velocities, u and v, were then selected within circular spatial bins centered on grid points of a $0.25° \times 0.25°$ global grid. The bins have a radius of $1°$ longitude, meaning that they overlap each other by $0.75°$ in the zonal direction and that their area decreases poleward. This use of overlapping bins on a fixed Eulerian grid and the subsequent latitudinal dependence of their area results in an increase of the spatial resolution of the pseudo-Eulerian maps and reflects the poleward reduction of the Rossby radius of deformation (Lumpkin and Johnson, 2013).

With each bin, the Cartesian (u, v) velocity observations are treated as data series, $V(x, y; t)$, dependent on horizontal (x, y) and temporal (t) coordinates, and expanded in the form

$$V(x, y; t) = \langle V \rangle + V^*(x, y) + V^S(x, y; t) + V^e(x, y; t) \tag{1.97}$$

where $\langle V \rangle$ is an ensemble average, $V^*(x, y)$ denotes horizontal variations in the mean field, $V^S(x, y; t)$ represents seasonal variations, and $V^e(x, y; t)$ are the residual (eddy) fluctuations. To estimate the horizontal variations of the mean field, previous studies have fitted 2-D functions to the binned drifter data (Bauer et al., 1998; Johnson, 2001; Lumpkin and Johnson, 2013; Peng et al., 2015). Although this approach improves the definition of horizontal velocity gradients relative to straight bin-averaging (Frantantoni, 2001; Jakobsen et al., 2003; Reverdiin et al., 2003; Zhurbas et al., 2014), the resulting pseudo-Eulerian mean velocities are still very smooth when compared against directly observed Eulerian current fields. To further reduce this smoothing, Laurindo et al. (2017) used 1-D functions to model the drifter velocities.

The 1-D approach is based on the assumption that horizontal variations of the temporal-mean velocity field are highly anisotropic, with larger scales along the mean velocity isolines than across them (Huang et al., 2007). The advantage of 1-D over 2-D functions lies in the fact that their fitting requires the determination of a smaller number of coefficients, making the fit less prone to estimation errors due to numerical instability, while at the same time allowing the use of more complex functions to model the mean horizontal gradients. Summary maps of the resulting global mean pseudo-Eulerian currents derived by Laurindo et al. (2017) are presented in Figure 1.119.

1.9.8 Other Types of Surface Drifters

Before leaving this topic, it is appropriate to mention that, while the satellite-tracked buoys are perhaps the most widely used type of surface follower for open waters, there are other buoy tracking methods being used in more confined coastal waters. A common method is to follow the surface buoy using ship's radar or radar from a nearby land-based station. More expensive buoys are instrumented with both radar reflectors and transponders to improve the tracking. The accuracies of such systems all depend on the ability of the radar to locate the platform and also on the navigational accuracy of the ship. Fixes at several near-simultaneous locations are needed to triangulate the position of the drifter accurately. Data recording techniques vary from hand plotting on the radar screen to photographing the screen continuously for subsequent digital analysis. These techniques are labor intensive when compared with satellite data transmission, which provides direct digital data output.

In addition to radar, several other types of buoy tracking systems have been developed. Older drifters relied on radio wave navigation techniques such as LORAN or NAVSTAR (early satellite navigation). For example, the subsurface drogued NAVocean and Candel Industries Sea Rover-3 Loran-C drifters had built-in Loran-C tracking systems that can both store and transmit the positional data to a nearby ship within a range of 25–50 km. Absolute positional accuracy in coastal regions was around 200 m but diminished offshore with decreased Loran-C accuracy. However, relative positional errors are considerably smaller. Based on time-delay transmission data from three regional Loran-C transmitters, Woodward and Crawford (1992) estimated relative position errors of a few tens of meters and drift speed uncertainty of ± 2 cm/s for drifters deployed off the west coast of Canada. Once it is out of range of the ship, the Loran-C drifter could be lost unless it was also equipped with a satellite transmission system. Meteor-burst communication is a well-known technique that makes use of the high degree of ionization of the troposphere by the continuous meteor bombardment of

FIGURE 1.119 Global maps of the mean pseudo-Eulerian zonal (a) and meridional (b) velocities and mean speeds (c) calculated from GDP drifter data. The curly vectors in (c) are streamlines calculated using the data depicted in (a) and (b) to indicate the general direction of the large-scale circulation. *From Laurindo et al. (2017).*

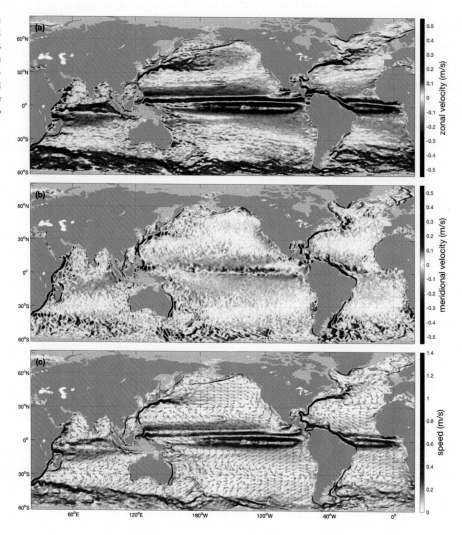

the earth. A signal sent from a coastal master station skips from the ionosphere and is received and then retransmitted by the buoy up to several thousand kilometers from the source. Since the return signal is highly directional, it gives the distance and direction of the buoy from the master station. Buoys can also be positioned using VHF radio transmission via direction and range.

As mentioned earlier, the introduction of small, low-cost GPS receivers such as SPOT Globalstar now makes it possible for buoy platforms to position themselves continuously to within several tens of meters. Provision for differential GPS (using a surveyed land-based shore station) has improved the accuracy to an order of a few meters. Data are then relayed via satellite to provide a higher resolution buoy trajectory than is presently possible with Argos tracking buoys. Given the high positioning rate possible for GPS systems, it is the spatial accuracy of the fixes that limits the accuracy of the velocity measurements. Japanese scientists have claimed recently to have used a highly accurate mini-drifter GPS-Argos tracking system to study nearshore currents based on 10−20 min positional data (http://www.argos-system.org/web/en/55-news.php?item=511).

Surface drifters used in coastal waters include: (1) the Surface Circulation Tracker (SCT) designed and built by Tamás Juhász and distributed through Oceanetic.com (http://www.oceanetic.com/oceanetic-surface-circulation tracker.html) and tracked using GPS and SPOT-Globalstar satellite communications (https://www.findmespot.com/en-us/products-services/spot-trace and https://www.globalstar.com/en-ca/products/spot-for-business/spot-trace). The user defines the position reporting from a choice of 5, 10, 30, or 60 min. The present monthly tracking costs in U.S. dollars are either $9.95 or $12.50, with activation fees of either $30 (for a 12-month term) or $30 + $35 (for monthly). Position accuracy is within 7.8 m at the 95% confidence level (pers. com., Roy Hourston, Institute of Ocean Sciences, Sidney, BC); (2) Xeos OSKER (https://xeostech.com/osker). The drifters use the GPS positioning system and Iridium satellite communication, with a positional reporting default of 10 min that can be redefined by the user. The positional resolution is unknown; and (3) MetOcean CODE-Davis https://metocean.com/products/code-davis-drifter/. These drifters also use a GPS system, with either Iridium or ARGOS-based satellite telemetry. The default position reporting decreases after deployment, but can be redefined by the user (e.g., to a constant 10 min). The position resolution is unknown. Figure 1.120 shows a SCT drifter and three selected tracks recorded for the coastal waters of British Columbia (Hourston et al., 2021).

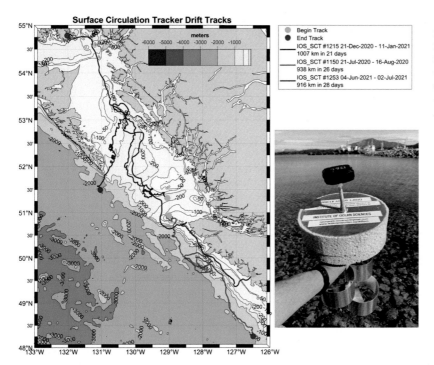

FIGURE 1.120 Oceanetic SCT drifter and surface trajectories for three of the many drifters launched off the west coast of British Columbia in 2020 and 2021. The black box is a GPS-Globalstar transmitter. The full drogue length is 0.39 m, with center at 0.3 m. *Courtesy Roy Hourston and Charles Hannah, Institute of Ocean Sciences. cf. Hourston et al. (2021).*

1.10 SUBSURFACE FLOATS

New technological advances in subsurface, neutrally buoyant float design have improved interpretation and understanding of deep ocean circulation in the same way that surface drifters have improved research in the shallow ocean. In their earliest form (Swallow, 1955), subsurface quasi-Lagrangian drifters were followed on the surface by a ship. Later SOFAR floats took advantage of the small absorption of low-frequency sound emitted by the buoys and trapped in the sound channel (the sound velocity minimum layer) located at intermediate depths in the ocean. The positions of these subsurface floats were triangulated from the signal reception at SOFAR coastal listening stations. The development of the autonomous SOFAR float, which was tracked from listening stations moored in the sound channel (Rossby and Webb, 1970), removed the burden of ship tracking and made the SOFAR float a practical tool for the tracking of subsurface water movements. Although positional accuracies of SOFAR floats depend on both the tracking and float-transponder systems, the location accuracy of ±1 km given by Rossby and Webb (1970) is a representative value. In this case, neutrally buoyant SOFAR floats have a positional accuracy that is markedly lower than that of satellite-tracked drifting surface buoys. Using high-power 250 Hz sound sources, the early SOFAR floats are credited with the discovery of mesoscale variability in the deep ocean and for pioneering our understanding of Lagrangian eddy statistics (Freeland et al., 1975). The familiar "spaghetti-diagram" (Figure 1.121) is characteristic of the type of eddy-like variability measured by SOFAR floats deployed in the upper ocean sound channel (Richardson, 1993).

SOFAR floats transmit low-frequency sound pulses, which are tracked from shore-listening positions or from specially moored "autonomous" listening stations. The need to generate low-frequency sound means that the floats are long (8 m) and heavy (430 kg), making them expensive to build and difficult to handle. In addition, stable sound sources were expensive to be installed on an expendable platform. Because greater expense is involved in sending sound signals than receiving them, a newer type of float called the RAFOS (SOFAR spelled backwards) float has been developed by Thomas

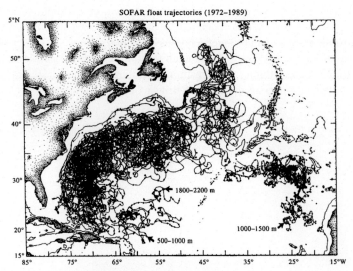

FIGURE 1.121 "Spaghetti-diagram" of all SOFAR float tracks from 1972 to 1989, excluding data from the POLYMODE Local Dynamics Experiment. Ticks on tracks denote daily fixes. Short gaps have been filled by linear interpretation. Plots are characteristic of the type of eddy-like variability measured by SOFAR floats deployed in the upper ocean sound channel. *Courtesy, Phillip Richardson (pers. comm., 1994).*

Rossby in which the buoys listen for, rather than transmit, the sound pulses (Figure 1.122). In this configuration, the float acts as a drifting acoustic listening station that senses signals emanating from moored sound sources especially deployed for the experiment (Rossby et al., 1986). The positions of RAFOS floats in a particular area are then determined through triangulation from the known positions of the moored source stations. A typical moored sound source, which broadcasts for 80 s every 2 days at a frequency of 260 Hz, has a range of 2,000 km and an average lifetime of 3 years (*WOCE Notes*, June 3, 1991).

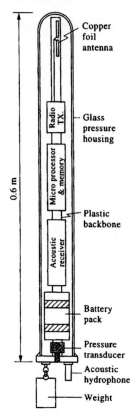

FIGURE 1.122 Schematic of a RAFOS float. *Courtesy, Thomas Rossby, University of Rhode Island.*

Because RAFOS floats are much less expensive to construct than SOFAR floats (since they are not a sound source), RAFOS floats are considered expendable. The data processed and stored by each RAFOS buoy as it drifts within the moored listening array must eventually be transmitted to shore via the Argos system, the Iridium satellite constellation, or other satellite link. To do this, the RAFOS float must come to the surface periodically to transmit its trajectory information. After "uplinking" its data, the buoy again descends to its programmed depth (density surface) and continues to collect trajectory data. The cycle is repeated until the batteries run out.

The need for deep ocean drifters that are independent of acoustic tracking networks has led to the development of the "pop-up" float. The float is primarily a satellite PTT and a ballast device that periodically comes to the surface and transmits its location data and "health" status (an update on its battery voltage and other parameters) to a satellite tracking system (Davis et al., 1992). The only known points on the buoy trajectory are those obtained when the buoy is on the surface. As with the RAFOS buoys, the pop-up drifter sinks to its prescribed depth level after transmitting its data to the satellite system and continues its advection with the deep currents. The advantage of such floats is that they can be designed to survive for a considerable time using limited power consumption. Assuming that deep mean currents are relatively weak, the pop-up float is an effective tool for delineating the spatial pattern of the deep flow, which up to now has not been possible over large areas. The Autonomous Lagrangian Circulation Explorer (ALACE) described by Davis et al. (1992) and the Argo floats presently being maintained by the international oceanographic community (basically, modern versions of the earlier ALACE floats) drift at a preset depth (typically, 2,000 m) for a set period, for example 10 days, then rise to the surface for about a day to transmit their position to the satellite. Water property profiles are obtained on the upward leg of the cycle. The drifter then returns to a prescribed depth, which is maintained by pumping fluid to an external bladder that changes its volume and hence its buoyancy (see Section 1.9.7.1 for further details).

As with the surface drifter data, the real problem in interpreting SOFAR/RAFOS float data is their fundamental "quasi-Lagrangian" nature (Riser, 1982). From a comparison of the theoretical displacements of true Lagrangian particles in simple periodic ocean current regimes with the displacements of real quasi-Lagrangian floats, Riser concludes that the planetary scale (Rossby wave) flows in his model contribute more significantly to the dispersion of 700 m depth SOFAR floats than do motions associated with near-inertial oscillations or internal waves of tidal period. Based on these model speculations, he suggests that while a quasi-Lagrangian drifter will not always behave as a Lagrangian particle, it will, nevertheless, provide a representative trajectory for periods of weeks to months. For his Rossby wave plus internal wave model, Riser derived a correlation timescale of about 100 days. He also suggests that the residence of some floats in the small-scale (25 km) features, in which they were deployed, provides some justification for his conclusions.

For pop-up floats, problems in the interpretation of the positional data arise from: (1) interruptions in the deep trajectory every time the drifter surfaces; (2) uncertainty in the actual float position between satellite fixes; and (3) contamination of the deep velocity record by motions of the float when it is on the surface or during ascent and descent. An essential requirement in the accurate determination of the subsurface current is to find the exact latitude/longitude coordinates of the buoy when it first breaks the ocean surface and when it first begins to resink. The ability to interpolate satellite fixes to these times is determined by the nature of the surface flow and the number satellite fixes. ALACE and Argo floats ascend more rapidly than they descend and spend little time at the surface. In a trial of an ALACE float to 1 km depth, the drifter spent 0.3 h in the upper 150 m, and 4 h between 150 and 950 m depth. Thus, according to Davis et al. (1992), most of the error was from vertical velocity shear at depths of 150 m and deeper, below the surface wind-driven layer. As an example, ALACE drifter 653 was launched at 2004 UTC on June 22, 1997, over the northern end of Loihi Seamount (18°56.229′ N, 155°15.347′ W). The float was equipped with temperature, conductivity, and pressure sensors, and ballasted to float at roughly 1,350 dbar ($\approx$ 1,335 m), coincident with the core of the hydrothermal plume emanating from the seamount. The float was programmed to rise to the ocean surface every 15 days, spend roughly 24 h reporting its position and sensor data through Service Argos, and then submerge to its predetermined depth. Satellite positioning in the region had an accuracy of roughly 300 m at the 95% level of confidence, with generally seven to eight fixes per day at tropical latitudes. Location data provided averages of the mid-depth current at 15-day intervals and estimates of the surface currents at roughly 3-h intervals. Because the study was mainly interested in the response of the drifter trajectory to topographic features, no attempt was made to correct the observed drift for shear-induced velocity errors (cf. Thomson and Freeland, 1999 for further details).

1.10.1 Profiling Argo Pop-Up Floats

Argo floats are pop-up drifters with high-resolution sensors that profile the upper portion of the ice-free global ocean (Figure 1.123a). Over a typical cycle, the float starts at the surface and descends to a depth of 1,000 m (the "parking depth"), where it rests. After 9 days at the parking depth, the float descends to a depth of 2,000 m, where it turns on its

FIGURE 1.123 (a) Cutaway of a SOLO-II Argo drifter built by MRV Systems (USA); and (b) the standard 10-day Argo cycling mode. The float first descends to a target depth of 1,000 m to begin drifting. After 9.5 days, the float then descends to 2,000 m from which it begins its temperature and salinity profile. After a 10-day submergence, the float surfaces to transmit data to a satellite. *From www.aoml.noaa.gov/phod/argo/.*

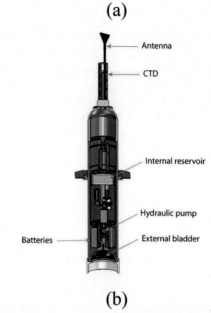

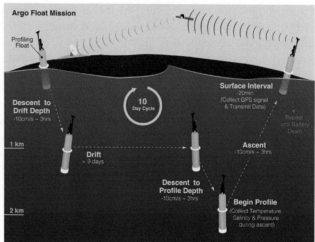

sampling equipment. The float then rises to the surface, providing profiles of the water temperature and conductivity (salinity) as it ascends (Figure 1.123b). Additional sensors, such as dissolved oxygen, pH and nitrate have also been incorporated into some of the floats. The drifter then remains on the surface long enough to transmit its position and profile data through the Argos or Iridium satellite systems. In an evolution similar to the drifting buoys, most Argo floats now use GPS for location and transmit their data via Iridium satellites. The drifter subsequently returns to the parking depth to start another cycle. A typical complete cycle has a duration of 10 days. Modern Argo floats are capable of making about 150 roundtrips to depths of 2 km over a lifetime of about 5 years.

For pop-up drifters, the current velocity at the parking depth is determined from time-varying changes in the positional data broadcast to satellites when the float is on the surface. This limits the temporal resolution of velocity time series to periods longer than at least twice the subsurface drift duration. The deployment of Argo drifters began in year 2000, supported by 31 different nations. By November 2007, Argo had achieved its initial goal of 3,000 floats and by November 29, 2012, there were 3,619 active floats in the World Ocean gathering profiles at a rate of one about every 4 min—equal to 360 profiles a day or about 11,000 every month. By November 4, 2012, Argo had collected its one-millionth vertical profile. At present, Argo is providing over 400 profiles per month from over 4,000 floats. This is an impressive achievement considering that since the beginning of deep-sea oceanography in the late nineteenth century, ship-based observations have gathered just over half a million temperature and salinity profiles to a depth of 1,000 m and only 200,000 to a depth of 2,000 m (Gilbert, 2012). For Argo to be maintained at the 3,000-float level, nations need to provide about 800 floats per year. Even with the 3,000 floats, additional floats are needed because some areas of the ocean are overpopulated while others have gaps that need to be filled. Argo strives to maintain an average distance of 300 km between floats and works to provide a quality real-time data system that delivers 90% of profiles to users via two global data centers within 24 h. A delayed mode quality control system (DMQC) has been established and 60% of all eligible profiles have had DMQC applied. Float reliability has improved each year. Argo is now a major contributor to the World Climate Research Program's Climate Variability and Predictability Experiment (CLIVAR) project and to the Global Ocean Data Assimilation Experiment (GODAE). The Argo array is part of the Global Climate Observing System/Global Ocean Observing System (GCOS/GOOS). While continuing its core mission of monitoring ocean temperature and salinity, Argo is extending into ice-covered and shallower areas and partners are adding measurements of ocean biogeochemistry.

1.10.1.1 Technical Information

In addition to their sensors, Argo floats are comprised of three subsystems: (1) Hydraulics, which control the buoyancy adjustment through an inflatable external bladder, allowing the float to surface and dive; (2) microprocessors, which manage function control and scheduling; and (3) a data transmission system that controls communication with passing satellites. The floats have an approximate weight of 25 kg, a maximum operating depth of 2,000 m, and crush depth of 2,600 m. The models presently in use are the ARVOR (the predecessor of the PROVOR float) built by NKE-Instrumentation (France) in close collaboration with IFREMER, the APEX float produced by Teledyne-Webb Research Corporation (USA), the SOLO-II float designed and built by Scripps Institution of Oceanography (USA), the S2A float produced by MRV Systems (USA), who bought the rights to the SOLO-II and manufactures it under the rebranded name of the S2A float and NAVIS built by Sea-Bird (USA). A smaller number of NOVA, NEMO and HM2000 floats are also manufactured. The Sea-Bird SBE temperature and salinity sensors are now used almost exclusively on core Argo floats and the SBE is the only approved CTD. Falmouth Scientific Inc. (FSI; USA) sensors were also used at the beginning of the program and the RBR (Canada) CTD is presently in the pilot approval phase for the project. The temperature data are accurate to a few millidegrees over the float lifetime (see the Argo Website for a discussion of salinity and oxygen data accuracy.)

The autonomous battery-powered Argo floats are designed to be neutrally buoyant at their parking depth. At this pressure, the floats have a density equal to the ambient pressure and a compressibility that is less than that of seawater. The three types of floats operate in a similar fashion but differ slightly in their design characteristics. At 10-day intervals (a typical float duty cycle), the internal pumps direct fluid into an external bladder, causing the float to rise to the ocean surface over a period of about 6 h. A series of roughly 200 temperature-salinity-pressure measurements is obtained during the ascent and stored internally in the float. Satellite systems (GPS) determine the drifter position and retrieve the data from the float once it reaches the surface. The bladder then deflates, returning the fluid to the float, and the float returns to its original density level. Floats are designed to make about 150 complete cycles. As noted above, the present standard Argo mode consists of a "park and profile" cycle during which the float descends to a target depth of 1,000 m to begin its drift (Figure 1.123). After 9.5 days, the float then descends to 2,000 m from which it begins its water property profiling (Figure 1.124). Starting in 2010, 70% of the floats were profiling to depths greater than 1,500 m. Another 20% profiled between depths of 1,000 and 1,500 m.

To ensure error-free data reception and location in all weather conditions, the float must spend between 6 and 12 h at the ocean surface. Positions are accurate to roughly 100 m depending on the number of satellites within range and the geometry of their distribution. Until recently, Argo used GPS positioning and data communication through both Service Argos and Iridium satellites. However, Iridium is an attractive option to Service Argos as it allows more detailed profiles to be transmitted within a shorter time at the surface. Iridium also allows for two-way communication. By 2010, 250 floats had been deployed with Iridium antennas. As we discussed in Section 1.9.3, only a small number of Argo drifters now transmit through Système Argos, and that number continues to shrink. A majority of the floats in the Argo array use the Global Positioning System to establish float positions and Iridium to transmit their data. Since 2013, a majority of the deployed floats use Iridium communications and this will slowly replace the Argos-tracked floats as they reach the end of their life.

When on the surface, Argo floats are affected by wind, currents, and other factors. There also is potentially significant contamination by current shear during the slow ascent and descent of the float cycle. A study of velocity errors for the oceanic region southwest of Japan by Ichikawa et al. (2001) has shown that surface drift and vertical shear can cause Argo drift velocities at the 1,000 m parking depth to be overestimated by as much as 10−25%. However, the analysis of Ichikawa et al. (2001) invokes a model for the interaction between a profiling float and a model shear. The model shear used is representative of an area close to the Kuroshio, a Western Boundary Current region with high current shear. Much smaller shears of order 10% or less can be expected in less dynamic regions of the ocean. Katsumatai and Yoshinari (2010) used global Argo float data (excluding marginal seas and ice-covered areas) from early 2000 to early 2010 to estimate average currents at the 1,000 dbar depth level at a $1°$ spatial resolution. The average flow field had an estimated uncertainty of 0.03 ms^{-1}, with higher uncertainty (>0.03 ms^{-1}, in both the zonal and meridional directions) near the Equator and in the Southern Ocean. The array bias (i.e., the bias due to the horizontal gradient in the density of the float data), was generally negligible, with an average magnitude of 0.007 ms^{-1} outside the equatorial region, only becoming relatively large (>0.01 ms^{-1}) near coastal regions. The measurement uncertainty was assumed to be spatially uniform and included

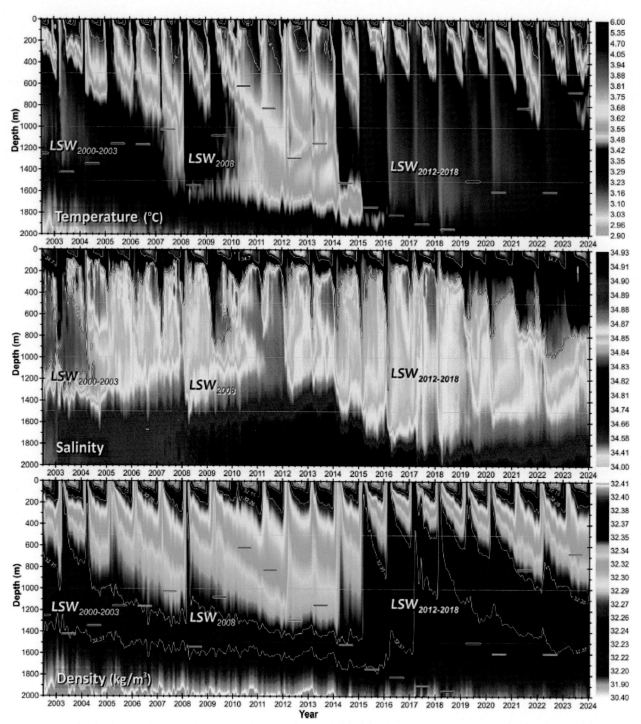

FIGURE 1.124 Central Labrador Sea (North Atlantic Ocean) temperature, salinity and density time series spanning the period 2002-2023 based on fully quality-controlled and calibrated Argo float and ship-based observations. Subscripts denote the Labrador Sea Water year classes. Dots at 500, 1000 and 1500 m depth correspond to the time bins having observations. Short horizontal lines indicate the winter convection depths. *Figure generated courtesy of Igor Yashayaev, Fisheries and Oceans Canada, Bedford Institute of Oceanography (2024). For further information see Yashayaev (2024), Yashayaev & Loder (2017), Yashayaev & Loder (2016) and Yashayaev & Loder (2009).*

errors due to Argos positioning, internal clock drift, unknown surface drift before submerging and after resurfacing, and unknown drifts during ascent and descent between the surface and the parking depth. The overall uncertainty was not sensitive to the assumed value of the measurement uncertainty.

1.11 SURFACE DISPLACEMENTS IN SATELLITE IMAGERY

Well-navigated (geographically located) sequential satellite images can be used as "pseudo-drifters" to infer surface currents. The assumption is that the entire displacement of surface features seen in the imagery is caused by surface current advection. This displacement estimate method (called the maximum cross correlation or MCC) was applied successfully to sea ice displacements by Ninnis et al. (1986). Later, the same approach, was applied to infrared images of SST by Emery et al. (1986). The patterns and velocities of the SST-inferred currents were confirmed by the drifts of shallow (5 m drogue) drifters and by a CTD survey. Later studies (Tokamamkian et al., 1990; Kelly and Strub, 1992) confirmed the utility of this method in tracking the surface displacements in different current regimes. When applied to the Gulf Stream (Emery et al., 1992), the MCC method reveals both the prevailing flow and meanders. A numerical model of the Gulf Stream, used to evaluate the reliability of the MCC currents, found that, for images more than 24 h apart, noise in this strong flow regime begins to severely distort the surface advection pattern.

The MCC method can also be applied to other surface features such as chlorophyll and sediment patterns mapped by ocean color sensors. It is possible to combine ocean color tracking with infrared image tracking (Crocker et al., 2007). Infrared features are influenced by heating and cooling, in addition to surface advection, while surface chlorophyll patterns respond to *in situ* biological activity. Because these two features reflect the same advective patterns (assuming similar advective characteristics for temperature and color), the differences in calculated surface vectors should reflect differences in surface responses. Thus, by combining both color and SST it is possible to produce a unique surface flow pattern that corrects for heating/cooling and primary biological production.

Another use of the MCC method is its application to SAR imagery. In this case, surface slicks result in the suppression of the SAR backscatter so that, provided that wind speeds are not too high to mask the oceanic features, the MCC method can be used to track the movements of the slick present in the surface layer (Qazi et al., 2013). Unlike infrared and ocean color images, which can be separated by as much as 24 h, the SAR images perform best in the MCC application when they are 30 min apart, as was the case with ENVISAT and ERS-2 satellites until mid-2009 (Qazi et al., 2013). The MCC-SAR surface currents are much more highly resolved than those computed from infrared or ocean color. In addition, the MCC-SAR currents can be computed much closer to the shoreline due to the higher spatial resolution of the SAR image.

1.12 AUTONOMOUS PLATFORMS AND UNDERWATER VEHICLES

1.12.1 Wave Gliders

Wave gliders were invented in 2005 by Roger Hine, the CEO of Liquid Robotics, to enable new types of ocean observations that did not require costly deep-water moorings or expensive ship operations. The idea was to harvest the abundant energy in ocean waves using a propulsion system that didn't require a propeller that creates noise at frequencies that conflict with whale communication (Hine et al., 2009). Solar panels would provide the battery power for the vehicle's control electronics, measurement systems and communications. Although wave gliders presently have just sufficient speed to make headway against low-velocity flow regions of the ocean, the buoyancy-derived propulsion provides a marked increase in range and duration compared to vehicles operating with electrically driven propellers, thereby extending ocean surveys from hours to months, and over ranges of thousands of kilometers. The propulsion system is a submerged "glider" that is tethered to the surface float. The glider rectifies the wave action to provide thrust to the tether and hence the surface vehicle (Figure 1.125). This system is purely mechanical and there is no electrical power generated in the glider propulsion system. Wave gliders can carry a variety of sensors including wind speed and direction, air temperature, near-surface sea temperature, and temperature along the vertical tether, along with pressure. In past deployments, a wave glider was launched on the US west coast and traveled to Hawaii and then on to Australia and separately to Japan (http://liquid.com/pacx/pacific-crossing.html).

The exact separation distance between the float and the subsurface glider is a critical design element and some experimentation was required to define the optimal separation distance. A typical forward speed of the surface platform is 0.8 m/s in a typical sea with 1 m high waves. The glider's forward speed depends upon the amplitude of the surface waves, the overall buoyancy force provided by the float and the glider's weight.

The wave glider system has been designed to withstand extreme sea conditions as tested by a wave glider's interaction with Hurricane Flossie in 2007, where it demonstrated the ability to operate in 2.2 m seas and 21 m/s winds (Sea State 5). The wave glider system exhibits a natural "sea anchor" character that helps it to weather abnormally strong weather conditions. As the sea surface becomes more turbulent, the Wave Glider tends to "dig in" to the waves and the float dives through the middle of the larger waves. A result is a remarkable robustness in the face of strong and adverse sea condition. The opposite extreme of very calm seas represents a greater challenge to the successful operation of the Wave Glider. Without significant wave energy to harvest, the Wave Glider is not able to maintain course and will not able to maintain station. Fortunately, the ocean is rarely so calm that there is no wave motion. The Wave Glider has been designed to be able to make significant forward progress even in very mild sea conditions (wave heights of tens of cm). Even in these very

(a)

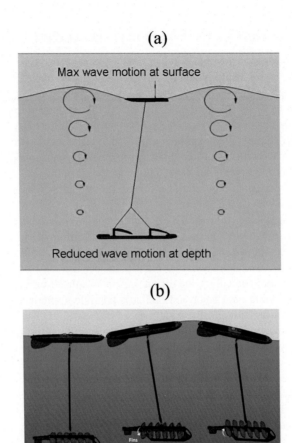

(b)

FIGURE 1.125 (a) The two-part Wave Glider part vehicle with a surface float connected to a submerged glider via a flexible tether. Surface waves pull the submerged glider up from below as the wave crest passes the float. (b) The glider's wing converts a portion of the wave-powered vertical motion into forward motion. The magnitude of the glider's forward propulsive force is proportional to the difference in the ocean wave amplitudes at the surface and at the depth of the glider. https://www.liquid-robotics.com/wave-glider/how-it-works/.

calm conditions, the Wave Glider is able to maintain a forward speed of 0.13—0.26 m/s. This speed is adequate to allow the platform to station keep against typical ocean surface currents.

To supply the Wave Glider with electrical power to run its navigation and control systems, provide communications and run the science instruments, a system of 665 Watt-hours of rechargeable lithium-ion batteries is used. This battery subsystem is composed of seven smart battery packs that are electrically isolated from each other. Only two batteries are in use at any given time and each battery has a separate discharging and monitoring circuitry. The Wave Glider's navigation, control and communications systems require only 0.7 W of average continuous power. The longest Wave Glider mission duration without a battery recharge is 23 days, which will vary with the specific science payload. The Wave Glider carries two photovoltaic solar panels that can deliver up to 43 W of peak power each. This amount of power is influenced by the mission latitude, the number of daylight hours and the angle of incidence at which the solar rays impinge upon the solar panels. Cloud cover and other weather conditions will also influence the charging capability of the PV panels. While just a few Watts is sufficient for many science payloads, such as cameras and passive receivers, there are other instruments, like radar, that require considerable energy, which can be a problem, particularly when operating in cloudy polar regions. Liquid Robotics is exploring ways to harness wave energy to power these devices.

For navigation, the Wave Glider uses a 12-channel GPS receiver and also carries a tilt-compensated magnetic compass with three-axis accelerometers. Some floats also carry a water speed sensor to enable short-term dead reckoning. Typical positional accuracy is better than 3 m. The Wave Glider navigates autonomously to preprogrammed waypoints and to keep station. The platform has demonstrated the ability to hold station in the open ocean for a long duration, with a "watch-circle" of only 25 m (Figure 1.126). This watch-circle is much better than long, deep-ocean moorings where circle diameters are on the order of the ocean depth. The communications and control systems are described in Figure 1.127.

Wave Gliders are controlled via a simple web-based command and control interface (Figure 1.127). Each Wave Glider vehicle communicates with the shore-based web server using an Iridium modem messaging session which is then put onto the internet for access by Wave Glider operators and all interested science parties. These communication sessions take

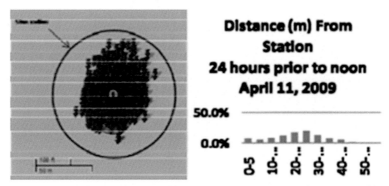

FIGURE 1.126 Wave Glider as a virtual buoy, maintaining a watch-circle of radius 12 m under typical sea state conditions. Unlike the case for a surface mooring, the size of the watch circle is independent of depth. The communication and control systems are presented in Figure 1.127.

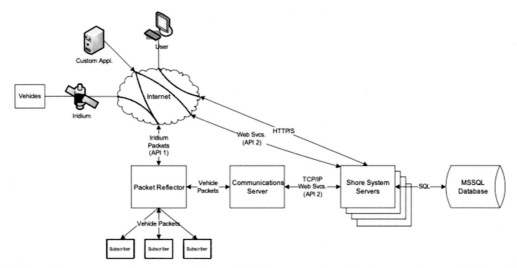

FIGURE 1.127 Wave Glider communications and control scheme. An Iridium satcom link provides the command-and-control channel for all deployed vehicles. A web interface is dedicated to an individual Wave Glider communications server, which allows operators to control the vehicle and for multiple subscribers to receive Wave Glider data.

place at configurable intervals, typically every 5 or 15 min. The web-based interface can be accessed on a computer or a mobile device. Wave Gliders also carry short-range, high-bandwidth radio modems that are used primarily to communicate with surface support vessels during ocean trials. The Wave Glider float is shaped like a boat and is highly modular being able to accommodate a variety of payloads (Figure 1.128).

All of the command and control, communications and navigation electronics are contained in the core electronics module located at the center of the float. The payload instrumentation can be housed in either the forward or aft water-tight compartments. They can be connected to the communications system for data relay. The fore and aft compartments are then covered with solar panels. A wide variety of payloads have been accommodated by the Wave Glider. These are: passive hydrophones and towed hydrophone arrays, marine weather stations, still cameras, video cameras, acoustic Doppler current profilers (ADCPs) and CTD sensors. The Wave Glider is a platform that can be used for a wide variety of ocean applications (tsunami and seismic monitoring, fish and marine mammal monitoring, offshore energy, and marine defense applications).

1.12.1.1 Autonomous Tracking of an Oceanic Thermal Front

The goal of thermal-front tracking project (Zhang et al., 2019) was to enable the Wave Glider to autonomously recognize and delineate oceanic thermal features that have strong surface expressions and are known to have biological significance. The method was used to track and map an upwelling event in Monterey Bay (California) over a 39-h period. In earlier efforts, this group used an Autonomous Underwater Vehicle (AUV) to detect and track an upwelling front. Each time the AUV crossed the front, the vehicle would make a turn at an oblique angle to recross the front, thus zigzagging along the front to map its structure. In one study, Zhang et al. (2019) mapped an upwelling event in Monterey Bay over a 5.5-day period. To apply a similar analysis to the Wave Glider, they would need to look for an upwelling signature in sea surface

FIGURE 1.128 The Modular Wave Glider float. *Source: Liquid Robotics, a Boeing Company (https://www.liquid-robotics.com/wave-glider/how-it-works/).*

temperature (SST). In its usual mode of operation, the Wave Glider is programmed to reach set waypoints. In this application, it is necessary to have the Wave Glider adapt by detecting the SST front and then changing its behavior in response. In this way, the Wave Glider can autonomously recognize and track the SST signature of the upwelling front.

This group's previous work with an AUV took advantage of the yo-yo (vertical) trajectory (see Slocum glider description) to measure the vertical temperature difference in the water column. Because Wave Gliders are restricted to measurements at the ocean's surface, they use the strong SST gradient between the cold upwelled water and the warmer water on the seaward side of the upwelling to indicate where upwelling has occurred. Thus, the Wave Glider will detect the upwelling front as an SST gradient whether going from warm to cold or from cold to warm SST. The temperature gradient is calculated by differentiation, which has the effect of amplifying small scale temperature variations which are not indicative of the upwelling front. Therefore, to prevent the small-scale variations from aliasing the temperature gradient, the raw temperature is first low-pass filtered with a sliding window of constant span. The span of the window is set wide enough to filter out the small-scale variations, yet narrower than the width of the upwelling frontal zone. The horizontal temperature gradient was then calculated as the difference between two low-pass-filtered SSTs.

The upwelling SST front-tracking method of Zhang et al. (2019) was modeled on their experience with the AUV data. Consider the following tracking steps:

1. *Front detection*: Assume that the Wave Glider has crossed from cold upwelled water to warmer offshore water and that the SST gradient is greater than a set gradient threshold. A "distance accumulator" then starts to grow. Whenever the gradient falls below the threshold, the accumulator resets to zero. (Setting of the threshold is based on earlier data.) When the distance accumulator exceeds a distance threshold, the vehicle determines that it has encountered a consistently high-gradient zone corresponding to the upwelling SST front.
2. *Zigzag tracking*: Once the SST front is detected, the Wave Glider first continues driving into the warm water for a specific duration to be sure that it has crossed the frontal zone. It then turns an oblique angle to go back to the colder upwelled water, as illustrated in Figure 1.129. On the way back to the colder upwelled water, if the absolute value of the SST gradient (the gradient is now negative) is greater than the threshold value the distance accumulator again starts to grow. When this accumulator exceeds the preset threshold, the vehicle determines that it has once again found the front. As with the warm water, the Wave Glider keeps going a bit farther to ensure that it has crossed the front and then turns back toward the warmer offshore water (Figure 1.129).
3. *Sweeps*: The Wave Glider repeats this cycle zigzagging across the frontal zone. The zigzag tracks alternate between northward and southward sweeps as shown in Figure 1.129. The operation is terminated when the mission duration has elapsed or when the Wave Glider operator ends the operation. All of the boundaries and thresholds are set using prior knowledge based on previous observations

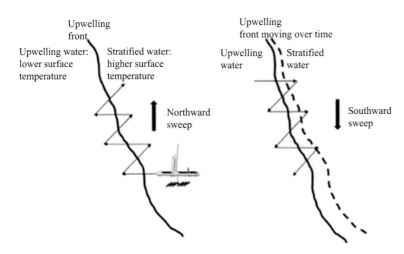

FIGURE 1.129 A Wave Glider conducting SST (upwelling) front-tracking on a zigzag track. *From Zhang et al. (2019).*

To test this method, an experiment was conducted off of Monterey Bay during a typical upwelling wind blowing from the northwest along the coast. Monterey Bay is unique in that the northeast Bay is sheltered from this wind by a high mountain range, which forms an "upwelling shadow." As a consequence, upwelling is restricted to the southern portion of the bay. In Figure 1.130, the zigzag tracks of a Monterey Bay Aquarium Wave Glider are overlain on two satellite SST images.

The solid black lines in Figure 1.130 are the Wave Glider tracks for the appropriate day and the gray tracks represent the second day. The expressions of the upwelling are very different on each of these days and the Wave Glider tracked very different SST fronts. The overall Wave Glider SST tracking is summarized in Figure 1.131, where the Wave Glider tracks have been colored to reflect the raw SST values. This particular front tracking had a duration of 39 h.

The track in Figure 1.131 shows how the Wave Glider's zigzag track is designed to cross the SST front at near to normal angles. This is a demonstration of how an autonomous platform can be preprogrammed to map a feature like an upwelling SST front.

1.12.1.2 A Wave Glider Approach to Persistent Ocean Observation and Research

The study by Daniel et al. (2011) in Monterey Bay was a demonstration project to explore the utility of the Wave Glider platform for a wide variety of ocean measurements, including CTDs and ADCPs. Earlier studies had shown the viability of the Wave Glider carrying a weather station, fluorometers, and the ability to use acoustic telemetry to communicate with undersea and bottom mounted instruments. In addition, the Wave Glider was found to be an excellent platform for the study of marine mammals due to the absence of a propellor propulsion that interferes with mammal underwater

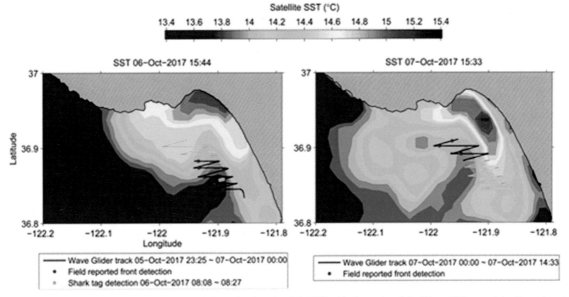

FIGURE 1.130 Satellite SST images (color bar at top) from October 6 and 7, 2017 with the tracks of the Wave Glider overlain. Front detections are marked by the red dots. Shark tag detections are marked by the green squares. Time is in UTC. *From Zhang et al. (2019).*

FIGURE 1.131 Summary of front tracking study in October 2017. Color on the Wave Glider track denotes the measured temperature (°C) at 0.4 m depth. *From Zhang et al. (2019).*

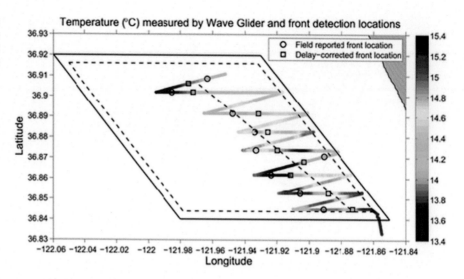

communication. All of these applications are summarized in Table 1.21. This comprehensive list makes it clear that a platform such as the Wave Glider can carry out the majority of measurements required to research upper ocean physical oceanography and marine meteorology. The Wave Glider has also been used to tow bodies from which were suspended vertical chains of thermistors and CTDs. These chains have obvious length limitations given that the drag on the chain can overcome the forward movement of the wave glider itself.

In the Monterey Bay experiment, a wave glider was instrumented with a Sea Bird Electronics (SBE) 37-SI un-pumped MicroCAT (CTD) mounted on the glider element at a depth of 6 m. The mounting of the CTD on the glider element is shown in Figure 1.132. Temperature and conductivity were recorded at 2 min intervals for the 12 day duration of the experiment. The CTD data and GPS times/positions were transferred to the onshore Wave Glider Management System at 5 min intervals via the vehicle's Iridium satellite link. At the same time, instrumented moored buoys in the bay measured temperature and conductivity profiles. These were correlated with the wave glider measurements when they were in the vicinity. The buoys measure conductivity and temperature at multiple depths including 1 and 10 m. Only 5% of the Wave Glider measurement possibilities were within 1.9 km of the buoy M1 and 4.1 km of buoy M2, while 10% are within 2.6 km of M1 and 4.9 km of M2. The overall correlations are presented in Table 1.22 for both buoy M1 and M2. If we restrict the correlations to the closest 5%, we obtain the much higher correlations of Table 1.23.

Graphical presentations of the temperature (Figure 1.133a) suggest that there are processes taking place at the surface for both moorings that are not captured by the Wave Glider measurements. These processes likely include atmospheric heating effects that produce higher surface temperatures that are attenuated at the 6 m depth of the Wave Glider records and the 10 m depth of the buoy records. Note that in Figure 1.133a, RF refers to the wave glider that for this experiment was named the Red Flash (RF). The conductivity time series (Figure 1.133b) also shows peaks of the surface buoy conductivities that greatly exceed any of the subsurface conductivities. There is slightly better agreement between the RF (glider) conductivities and the M2 10 m conductivity. At the end of the time series, there is some form of unusual coastal event that caused a disruption in the overall conductivity time series.

1.12.2 Saildrones

A somewhat later entry into the world of autonomous ocean platforms is the Saildrone (https://www.saildrone.com) created by Richard Jenkins and Dylan Owens, who launched their first test platform from San Francisco Bay on 1 October, 2013. The platform was preprogrammed to sail to Hawaii, and was subsequently picked up north of Oahu by the investigators. The "wind-powered autonomous surface vehicle" made the 2,248 nautical mile (4,165 km) trip half-way across the Pacific Ocean in 34 days. The voyage included a storm, with gale-force winds followed by 2 weeks of almost no wind. During the storm, the Saildrone reported speeds of up to 16 mph (∼26 km/h) and angles of 75° as it surfed down the backside of breaking waves, waves that could have destroyed the vehicle if it had caught the wave at the wrong angle.

Saildrones have a narrow hull stabilized by two outriggers, one on each side. Its "sail" is a 20-foot (6.1 m) tall, solid carbon-fiber wing (Figure 1.134a and b). Extending from the back of the wing, about halfway between the top and bottom, is a tail. (Jenkins says that is a little trick he stole from the Wright Brothers.) Above the water line, the first Saildrone was painted "safety orange" and labeled "OCEAN RESEARCH IN PROGRESS." Below the water line, the hull is black with bottom paint and the drone is labeled "Honey Badger." This Honey Badger has made the trip from California to Hawaii

TABLE 1.21 Wave Glider study goals, platform operations and result status for operations in Monterey Bay (California).

Science need	Parameters/Phenomenon	Measurements/Instruments required	Wave Glider capability	Status
Climate change	Ocean: CO_2 T Atm: CO_2-T, CH_4, H_2O_2, pH	CO_2, flux most useful—PMEL buoy based sensor available	Compatible with this sensor, cost-effective time series measurements over large areas	Collaborating with PMEL
Polar melting -> S increase, sea level rise	S—Surface vertical profiles	CTD—horizontal profiles, vertical profiles also needed for traditional dynamics	Excellent platform for horizontal profiles; need solution for vertical profile	Horizontal profiles completed, analysis promising
Geostrophic circulation (surface translational motion)	Dynamic calculation from density field, derived from S, T, Z	Vertical CTD profiles, coupled with sea-surface height or approximations (level of no motion, etc.)	Requires vertical profiling apparatus	Profiling concepts being considered
Thermohaline circulation ("ocean conveyor")	T, S, versus Z; tracer distribution (O_2, D_2O, ...)	CTD with winch and poss. O_2, CO_2, iodine-128, etc. sensors in winch package	Requires vertical profiling apparatus	Profiling concepts being considered
Thermal structure/ thermocline depth (ASW, climate change)	T versus Z. (i.e., BT, XBTX)	BT, CTD, XBT/XCTD can be deployed, but payload capacity limits duration	Compatible with XBT/XCTD but vertical profiling apparatus is most desirable for cost-effective time series measurements over large areas	Profiling concepts being considered
Primary productivity— fisheries, ecosystems	Chl a (fluorimeter spectrophotometers) nutrients, O_2, CO_2, optical clarity, turbidity	Cell counters, nutrients— autoanalyzer beam attenuation, volume scattering, absorption	Compatible with most of these sensors, cost-effective time series measurements over large areas	Some instruments already demonstrated
Marine mammal activity, behavior, distribution	Passive acoustics	Hydrophone and processing	Compatible with these sensors, cost-effective time series measurements over large areas	Preliminary results published (Wiggins et al., 2010)
Surface circulation	$V(x, y)$	ADCP or current profiler	Compatible with this sensor, cost-effective time series measurements over large areas	ADCP capability is now available
Surface waves—safety, meteorological forecasts, coastal protection	H, L, O (directional spectrum)	ADCP, motion sensors	Compatible with these sensors, cost-effective time series measurements over large areas	ADCP capability is now available
Pollution—offshore, meristic (littoral), estuarine	Nutrients, heavy metals, color dissolved organic matter, etc.	Fluorometer, optical backscatter, wet chemistry *in situ* sensors	Compatible with many of these sensors, cost-effective time series measurements over large areas	Some capabilities deployed other in development
Meteorological climate forecasting, ENSO, etc. forecasting	Surface currents, air: T, P, precipitation, wind, RH	ADCP, navigation derived current, atmospheric sensor	Compatible with these sensors, cost-effective time series measurements over large areas	Demonstrated
Safety ashore	Tsunamis	Bottom pressure	Effective relay for seafloor sensor	Demonstrated (Manley and Willcox, 2010)

From Daniel et al. (2011)

FIGURE 1.132 Mounting position of the Sea Bird Electronics SBE 37-SI CTD on the glider portion of a Wave Glider. *From Daniel et al. (2011).*

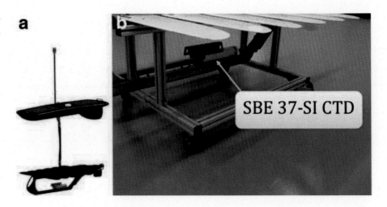

TABLE 1.22 Correlations between the wave glider CTD and the M1, M2 buoy CTDs for the total data set collected in Monterey Bay (California).

Buoy ID: depth	Temperature	Conductivity	Salinity
M1: 0 m	0.83	0.80	0.78
M1: 10 m	0.85	0.32	0.73
M2: 0 m	0.75	0.75	0.63
M2: 10 m	0.78	0.78	0.56

From Daniel et al. (2011).

TABLE 1.23 Correlations between the wave glider CTD and M1, M2 buoy CTD for the closest 5% of the data collected in Monterey Bay (California).

Buoy ID: depth	Temperature	Conductivity	Salinity	Max distance (km) from buoy
M1: 0 m	0.93	0.87	0.97	1.35
M1: 10 m	0.96	0.93	0.99	1.34
M2: 0 m	0.95	0.96	0.98	4.36
M2: 10 m	0.94	0.95	0.98	4.40

Note: There is a marked increase in correlation when only the closest wave glider data are compared. Also, the buoy 10 m depth values correlate slightly better than do the surface values.
From Daniel et al. (2011).

undamaged and has proven the sea worthiness of the craft and verified its ability to operate unmanned in the open ocean. The operation of the saildrone is explained in Figure 1.135 and associated text, taken from: https://www.wired.com/2014/02/saildrone/.

The components of the saildrone are shown in Figure 1.135 and are as follows:

1. *The Wing*: As the wind passes over it, the wing produces thrust. That force is concentrated on its axis of rotation, preventing the wing from spinning wildly.
2. *The Tail*: A little tab at the back of the tail can be set to the left or right, causing the wing to rotate a few degrees and maintain an efficient angle of attack.
3. *The Counterweight Positioned at the end of a spar*: it adjusts the wing's equilibrium, so that its center of gravity is balanced, allowing it to rotate as needed.
4. *The Rudder*: While in theory it's possible to operate the Saildrone by using only the sail, it's more efficient to use a rudder to point the boat where the operator wants it to go.

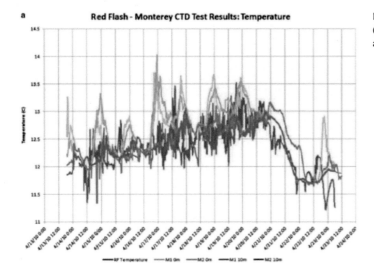

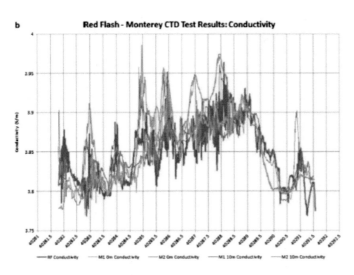

FIGURE 1.133   Time series from the Wave Glider "Red Flash" (RF; in red) and moored buoys M1 and M2. (a) Temperature (°C); and (b) conductivity (S/m).

5. *The Autopilot:* GPS provides information on data and location. That's all Saildrone needs to know. Navigation instructions reach the autopilot via satellite.
6. *The Keel:* If the Saildrone gets knocked over, it will right itself because of the keel's weighting. Its steep angle sheds debris like kelp and lost fishing nets.

The Saildrone has its genesis in the background of Richard Jenkins who grew up in a sailing town on the southern coast of England. At 17 he was working as a draftsman at a boatyard near his hometown of Lymington when he came across a "land yacht" with wheels instead of a keel and a hard wing instead of a flexible sail. He decided to rebuild the yacht, which he managed to do in 6 months after which he made his first attempt at a new speed record for a wind-driven vehicle. Over time, Jenkins eventually built a Mark 4 version of the "boat" and moved to the U.S. for a speed trial at a dry lakebed just south of Las Vegas. It was there that he piloted the wind-powered sail yacht, called Greenbird, to a speed of 126 mph (203 km/h), setting a new wind-powered vehicle land speed record.

The wing that Jenkins had used for his record-breaking runs had a tail. It is not an obvious design since even the wing powered sailboats do not have tails. Once Jenkins started reaching airplane-like speeds in the Greenbird, he realized he needed control surfaces that worked less like a sailboat and more like an airplane. The tab on a saildrone's wing makes it pull left or right. Unlike a normal sail, which has to be shifted each time the boat changes direction, the saildrone's wing is always correctly angled into the wind. To optimize this angle, the tail on the wing further aligns the wing and the wind direction to refine the input of the wind in providing "thrust" on the vehicle. It was after he moved to San Francisco to work

(a)

(b)

FIGURE 1.134   The Saildrone (a) as seen from the side, with its wing and tail; and (b) as seen from above showing the outriggers for stability.

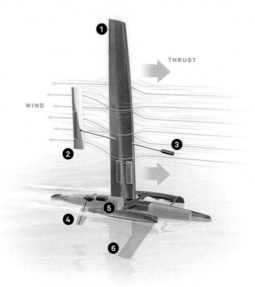

FIGURE 1.135   Components of the Saildrone and its operation (see details that follow).

for Google cofounders Larry Page and Sergey Brin that he realized that the wing he had developed for Greenbird would be perfect for an oceangoing drone. The drone would need only three moving parts: the tab on the tail of the wing, the free rotating wing itself and the rudder. Only two of these parts, the tab and the rudder, would need power to control them. He knew that a few off-the-shelf solar panels would provide more than enough power for these controls. Interestingly, Dylan Owens was working at the same shipyard and shared the idea of building an autonomous ocean drone.

The voyage of the Honey Badger from California to Hawaii was the demonstration that Jenkins and Owens needed to get their company off the ground. After this first successful trip, the Honey Badger was sent on a trip down and around the Antarctic and back up to the western equatorial Pacific. Early funding came from Wendy Schmidt, wife of Google executive chair Eric Schmidt who was impressed with Jenkins's resourcefulness.

Since these early days, the Saildrone has been used in a great many applications and continues to expand its horizons. One important application is the study of hurricanes—somewhere that the Saildrones can go and research vessels can't. Scientists at the National Oceanic and Atmospheric Administration (NOAA) are sending saildrones equipped with meteorological sensors into hurricanes with 120 mph (193 km/h) winds and 50 foot (15 m) waves. An important parameter to study is the rapid intensification of the maximum sustained winds of a tropical cyclone, defined as an increase of 35 mph (56 km/h) in 24 h. This intensification poses a significant threat to coastal communities that need to be evacuated before the intensification events occur. Almost all Category 4 and 5 hurricanes undergo rapid intensification at some point in their lifetimes. A Saildrone sailed into the middle of the Category 4 hurricane named Sam and brought back data and video of what it was like to be at sea in the middle of a Category 4 hurricane. Here, it encountered 125 mph ($\sim$200 km/h) winds and 50 foot (15 m) waves. As global warming and climate change progresses this rapid intensification will likely go to 70 mph ($\sim$113 km/h) winds over 24 h.

Climate change is another target suited to Saildrone capabilities. In mid-December 2021, three Saildrones were launched from Newport, Rhode Island to travel along the Gulf Stream collecting oceanographic data in the winter season, which is difficult for typical research vessel operations. The goal of this measurement program was to gather data needed to improve medium and long-range weather forecasting and involved participation by scientists from the European Center for Medium Scale Weather Forecasting. The project was funded by a $1 million grant from the philanthropic arm of Google to study Climate Change.

Saildrones come in small, medium and large sizes capable of carrying a variety of keel, deck and mast mounted payloads powered by solar energy or, with the larger platforms, a diesel supplement. The company has built over 100 of its smallest 23 foot (7 m) Explorer model, which are intended for long endurance ocean data gathering and measurements (fisheries and weather) because they rely only on wind for propulsion (averaging 3 knots) and solar panels on the wing to charge batteries for cameras, ocean/atmosphere sensors and computers. Explorers have a range of about 15,000 nautical miles (27,780 km) and can operate over a year in conditions as adverse as the Arctic and the Antarctic. Honey Badger, the company's first drone, circumnavigated the Antarctic and returned to Hawaii. The next size up is the Voyager which is 33 feet (10 m) long and adds a diesel-electric engine to power additional sensors including radar. (Solar panels are inadequate to supply the power to operate radar.) The endurance of the Voyager is reduced to 180 days, but the diesel engine increases the vehicle speed to 5–10 knots ($\sim$9–19 km/h). Finally, the 72 foot (22 m) Surveyor has more diesel capacity, a larger sail for 5-10 knot speeds. As with the Voyager the endurance is limited to 180 days (limited primarily by diesel fuel) but the platform can carry many more sensors.

Another application is Saildrone's work with the U.S. Coast Guard Southern Command on potential maritime domain awareness (MDA) for drug interdiction. The company offered acoustic, visual and infrared solutions, paired with computer vision and machine learning, but they soon learned that Coast Guard had no existing database of sea-level images. "Without millions of images of ships or boats, you can't train a machine learning database", Jenkins explained. The company then deployed a fleet of 30 Explorers to collect surface images of ships, boats and other platforms. The Saildrone company may have the most comprehensive set of images of surface ships in the world. This can be used with Artificial Intelligence (AI) methods to identify the suspicious vessels. Saildrones can be dispatched to detect and locate suspect ships smuggling drugs. The U.S. Coast Guard has recently identified illegal and unregulated fishing as a bigger problem than drug trafficking. In October 2020, Saildrone carried out a successful 30-day demonstration of MDA capabilities for the U.S. Coast Guard off the coast of Hawaii. Saildrone data transferred in real-time by satellite demonstrated the ability of the Saildrone to provide the detection and monitoring capability sought by the U.S. Coast Guard.

1.12.2.1 Exploring the Pacific Arctic Seasonal Ice Zone with Saildrone USVs

High-quality, *in situ* measurement of marine variables is needed in the seasonal ice zone (SIZ) to better understand polar weather, climate and ecosystems. To provide such observations, five uncrewed surface vehicles (USVs) Saildrones were deployed in the Chuckchi and Beaufort Seas following the 2019 retreat of sea ice (Chiodi et al., 2021). The vehicles were instrumented to measure surface oceanic and atmospheric parameters needed to estimate air-sea fluxes of heat, momentum and carbon dioxide. The Saildrones were launched from Unalaska, Alaska in early May 2019 and sailed northward through Bering Strait in early June 2019. To complete the mission, the drones needed to sample in the SIZ and return to Unalaska before the availability of solar power was diminished to the point where the drone could no longer function. Navigation and operational information were relayed to the Saildrone Command Center in near real time (a few minutes delay). Four cameras mounted on the wing provided images that took up to 30 min to be transferred. The cameras were oriented upward to view the sky, downward to view the vehicle's hull (Figure 1.136), and horizontally for views fore and aft of the wing. The downward looking camera confirmed those times when the saildrone was in close proximity to sea ice.

Remote sensing information on sea ice cover is collected from passive and active microwave sensor and can be provided to users via information providers such as the U.S. National Ice Center (NIC). While it was important for the saildrones to operate close to sea ice, it is not possible for them to sail through the ice as it is for ships with ice breaking and ice strengthened capabilities. Thus, while the saildrones were preprogrammed for fixed waypoints, it became necessary to vector them away from their course if there was an ice floe in the way. The focus to this study was to collect data that could be used to improve weather and climate prediction models.

The Saildrones used in the Chiodi et al. (2021) study had ~7 m long hulls, rigid wing heights of 5 m above the water line and keel depths of 2.5 m. Each drone was equipped with a Rototronic HC_2-S_3 mounted at 2.3 m height on the wing that measured air temperature and relative humidity, a Sea Bird Electronics SBE-37 Microcat at 0.5 m depth that measured conductivity and temperature, a Vailsala PTB-210 Barometer on the hull (0.2 m height) that measured sea level pressure, and a Gill model 15990-PK-020 anemometer atop the wing (5.2 m height) that measured wind velocity at 10 Hz. These wind measurements, coupled with the onboard inertial measurement unit (VectorNav model VN-300) and GPS, allowed for georeferenced wind velocity to be calculated onboard and telemetered via the Global Telecom System and onshore data repositories as 1 min averages. Higher frequency sampled data were stored onboard for later detailed analysis. Two of the drones were equipped with Autonomous Surface Vehicle CO_2 sensors capable of measuring pCO_2 in both the air and water. Along with the wind speed, the pCO_2 observations made it possible to estimate the air-sea flux of carbon dioxide. These same drones also had instruments to measure solar irradiance, longwave radiation, ocean skin temperature, ocean color (Chl-a, CDOM), dissolved oxygen, pH and ocean current speed and direction (ADCP). All but one of the drones carried the three cameras described earlier. All of the data collected in this experiment are hosted at NOAA's Pacific Marine Environmental Laboratory in its Science Data Integration Group.

To enable comparisons between the *in situ* saildrone measurements and other observations of sea ice information, the study defined two sea ice related quantities. The first is called In Situ Sea Ice (ISSI) and is defined as "1" during periods when ice was visible in the downward or horizontally oriented cameras and "0" when ice was not visible. The resulting time series allows for the comparison of ice presence with coincident time series of surface temperature, salinity, and other variables. The second quantity is called "Ice-Blocked Vehicle (IBV)" and indicates points where the drone is blocked by the ice after transiting from open water. Three hours is used as the comparison period.

FIGURE 1.136 A birds-eye look showing the Saildrone wing in sea ice.

The European agency for meteorological satellites (EUMETSAT) provides daily gridded sea-ice concentration maps based on the Advanced Scanning Microwave Radiometer (AMSR-2), which will be referred to as EA2. EA2 daily estimates were available starting September 2016 on a 10-km horizontal grid. Daily EA2 updates were used for planning during the Saildrone mission. The investigators binned the ice concentrations in 1% bins and considered the boundary of the ice extent as those grid cells with >10% ice concentration in keeping with the boundary as defined by the National Ice Center (NIC). The NIC 10% contour was drawn by NIC analysts using a variety of satellite information (Figure 1.137).

The minimum ice extent in the Chiodi et al. (2021) study occurred on 15 September 2019. The return of ice to the study region in 2019 was also slower than in the previous 2 years. For those Saildrone observations taken while sea ice was visible in the saildrone camera images, the most common EA2 sea-ice estimate is still 0%. There were 8.3 h of measurements made with sea-ice visible in the saildrone camera images, when the EA2 value was 15% ice concentration. It should be noted that frequently the NIC and EA2 ice extents did not agree.

Sea-ice is known to influence the surface properties of the seawater that surrounds it. Thus, knowledge of the marine surface variables can provide useful information regarding the presence of sea-ice, which may prove helpful in navigation in polar seas. An examination of 1 min averages of the surface variables shows why surface marine observations may not prove useful for this purpose. The approach of a single Saildrone to the ice edge provides a useful example (Figure 1.138).

The Saildrone used in the Chiodi et al. (2021) was blocked by the sea-ice as indicated by the sharp drop in vehicle speed from roughly 0.8 m/s to 0.3 m/s (labeled as IBV in Figure 1.138). As a confirmation, there was a camera image taken at this time that showed the drone in contact with the ice. This first drop in speed happened at 00:40 UTC, while the drop in water density due to the change in salinity only occurs at 00:42 UTC. Between 00:40 and 00:42 UTC the water temperature just gradually decreases. Thus, it is clear that the density drop is caused by the salinity change when the drone is firmly in the sea-ice. The delay in the surface parameters suggests that they should not be used as indicators of the ice extent. An examination of other surface variables, such as near-surface air temperature and humidity, also failed to clearly indicate when the drone first was blocked by the sea-ice. Thus, it was concluded that surface variables are not good indicators of the location of the ice edge.

1.12.2.2 Uncrewed Surface Vehicle (USV) Survey of Walleye Pollock

In 2020, the developing COVID-19 pandemic disrupted fisheries surveys to a large extent. Many surveys were canceled, including those for walleye Pollock (*Gadus chalcogrammus*) on the Chukchi shelf of the eastern Bering Sea (EBS), the largest fishery in the U.S. This posed a significant challenge for fisheries management, which relies on timely and consistent abundance estimates of fish stocks to support management decisions. In the EBS, the largest single species fishery results in landings valued at $1.4 billion dollars. The ship-based acoustic-trawl and bottom trawl surveys were

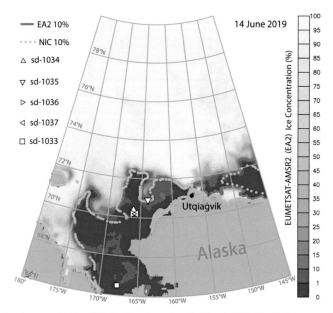

FIGURE 1.137 Sea-ice concentration image with the 10% ice boundary from EA2 and NIC. The Saildrone positions are indicated by the triangles. *From Chiodi et al. (2021).*

FIGURE 1.138 One-minute averaged data collected just before and after a Saildrone was blocked by sea-ice (IBV) on 18 July 2019. (a) The georeferenced vehicle speed, (b) water density, (c) salinity and (d) ocean surface temperature. The IBV block is inferred from the dramatic block in speed at 00:40 UTC. *From Chiodi et al. (2021).*

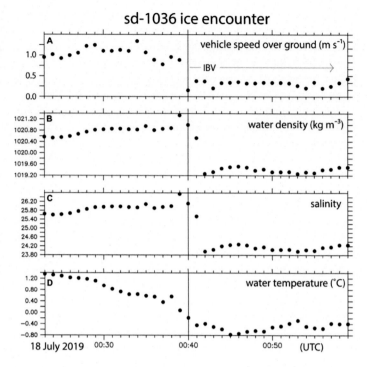

canceled due to the risk of the virus spreading through the ship's crew. In response, De Robertis et al. (2021) proposed using USVs equipped with calibrated echosounders to conduct a USV-based acoustic survey to mitigate the loss of information resulting from the canceled ship surveys.

The primary limitation of USVs for acoustic fish abundance surveys is their inability to perform simultaneous biological sampling to identify specific acoustic targets. Acoustic signals are presently insufficient to reliably distinguish between fish species and size classes in most operational conditions. During ship-based surveys, trawl samples are usually used to identify species and size/age compositions. In environments where a single species of fish predominates (as does Pollock), acoustic signals alone may be converted into estimates of the Pollock abundance even without the confirmation of a species by a coincident trawl sample. Based on previous research, the authors believed they could proceed with their USV acoustic surveys to estimate Pollock abundance. To conduct the surveys, three Saildrones were chartered and equipped with calibrated echosounders.

The USVs surveyed the same 3.5×10^5 km^2 area that previous surveys had covered but transects were spaced farther apart due to time constraints. The USVs measured acoustic backscatter, but population biomass is used in the stock assessment model. As a result, the USV backscatter values had to be converted to biomass using an empirical relationship between Pollock backscatter and biomass derived from previous ship surveys. Finally, the uncertainty introduced by the wider spaced sampling and the backscatter-to-biomass conversion was established using Monte Carlo simulations. The USVs followed a shortened survey plan in case an abbreviated ship survey became possible. The survey consisted of 14 transects spaced 74 km apart for a total length of 4,727 km (Figure 1.139).

Due to COVID-19 travel restrictions, the Saildrones could not be launched from Alaska and it was decided to have the drones make the transit to the study area themselves. The drones began their survey operations on July 4 and finished on August 20, 2020. The drones then sailed back to Alameda, California where the raw acoustic data were downloaded and sent by overnight courier to the interested scientists for further analysis. The USVs were equipped with Simrad WBT mini split-beam echosounders using an ES38-18/200-18CR transducer gimbal mounted on the keel at a depth of 1.9 m. The echosounders continuously transmitted 0.512 ms narrowband signals every 3 s. Although previous backscatter observations were collected down to 1,000 m, only a very small amount of Pollock backscatter was measured >300 m. Thus, 300 m was selected as a cut off for the USV echosounder measurements. Raw echosounder backscatter was recorded on flash memory and data summaries were transmitted over the Iridium satellite link. The echosounders were calibrated before and after the surveys in San Francisco Bay on 12−14 May and 28 October 2020. These calibrations were carried out at warmer water temperatures than occurred at the surveys, which increased the possible uncertainties introduced by this temperature difference. The mean gain from the two calibrations was used to correct the echosounder data.

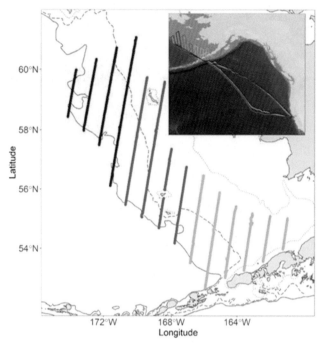

FIGURE 1.139 Uncrewed surface vehicle (USV) tracks along the survey transects on the Chukchi shelf. Each USV (each has a different color) surveyed a series of transects spaced 74 km apart starting from the south. The 50, 100 and 200 m depth contours are also shown. The inset shows where the survey was located, and the paths followed by the drones to and from their launch site in California. *From De Robertis et al. (2021).*

The acoustic data were processed using Echoview v11.0 software. Backscatter showing evidence of attenuation of the transmitted signal by bubbles at high sea states were removed from the record. The data were processed by four independent analysts, who on previous ship surveys visually inspected acoustic records to identify Pollock. In most cases, acoustic backscatter consistent with fish concentrations in the lower one-third to two-thirds of the water column is dominated by 1-year plus (1+) walleye Pollock, while at the deepest depths the backscatter is generally attributed to adult Pollock. In some areas, Pollock co-occur with diffuse backscatter caused by other scatters, which makes the identification of Pollock much more difficult. A filter was applied that eliminated these weak targets as non-Pollock.

The Saildrone navigational course depends on wind speed and direction as the Saildrone must "tack" back and forth to sail upwind and drift when becalmed. Therefore, the distance traveled on a pre-planned track depends very much on the wind. This was accounted for by averaging the USV observations into 0.5 nm (0.926 km) along-track elementary distance sampling units (EDSUs) defined along the pre-planned transects. More than 50 "ping" backscatter intervals were associated with each EDSU at low speeds, with the drone traveling upwind. On average, 7.6 ping intervals were associated with each EDSU. The USV transducers had much wider beam widths than had been traditionally used. Thus, the USVs are expected to detect less backscatter near the sea bottom. This bias depends on the depth, the beam width, and the degree to which the fish aggregate near the seafloor. As a result, the USV measurements were corrected for the difference in beamwidth, which increased total Pollock backscatter by 3.2%.

The USVs completed the survey without problems. The wind speeds averaged 5.4 ± 2.5 m/s with 95% of the winds <9.7 m/s. Vertical Pollock distributions (Figure 1.140) were consistent with previous surveys.

In general, there was a shallow backscatter layer that was not attributed to Pollock and a near-seafloor layer of backscatter that was attributed to age 1+ Pollock (Figure 1.140a). There were cases where backscatter was attributed almost entirely to age 1+ pollock (Figure 1.140b), while in others, scattering from sources other than Pollock filled most of the water column (Figure 1.140c). The strong backscatter from fish schools near the surface (<40 m in Figure 1.140a–c) typically represents the presence of age-0 Pollock. The diffuse scatters just beneath these layers are poorly characterized, but these layers have in general a different frequency response from Pollock, and previous studies have shown that age 1+ Pollock are scarce in these layers. Previous studies of backscatter and trawls have suggested that they are linearly related to each other and that we can use this relationship to convert acoustic backscatter to biomass (Figure 1.141).

All of the plots in Figure 1.141 clearly show a very tight correlation between acoustic backscatter and Pollock biomass. Thus, this relationship can be used to compute Pollock biomass from the USV acoustic surveys to be used in the Pollock

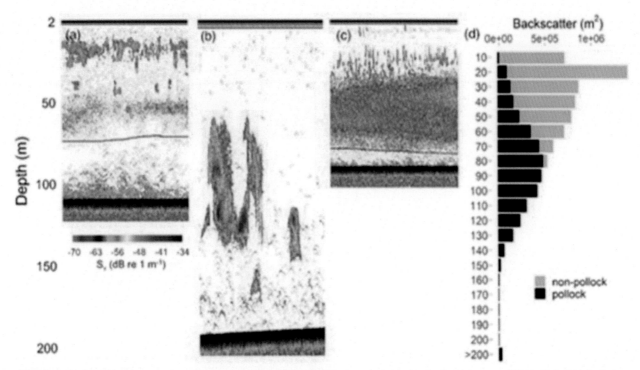

FIGURE 1.140   Example echograms from the uncrewed surface vehicle (USV) survey in the Bering Sea in the summer of 2020: (a) Pollock in the lower one-third of the water column, (b) a high-density Pollock aggregation with few other scatters, (c) a challenging case where other scatterers dominate the water column and only backscatter close to the bottom can be attributed to Pollock, (d) stacked bars depicting the survey-wide daytime depth distribution of Pollock backscatter. *From De Robertis et al. (2021).*

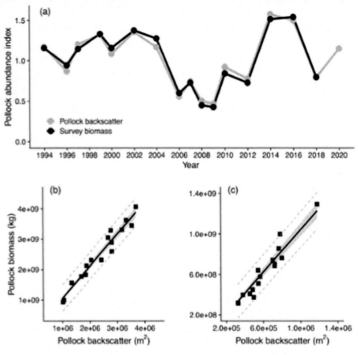

FIGURE 1.141   Comparison of Pollock backscatter and coincident biomass from trawl measurements from 1994 to 2018 in the Bering Sea. The 2020 backscatter survey is from the USV measurements. (a) Time series of Pollock backscatter and biomass for all of the earlier surveys. (b) Regression plot for measurements >3 m above the bottom; and (c) regression for points 0.5−3 m above the bottom layers. *From De Robertis et al. (2021).*

abundance models. The biomass estimated for 2020 was 3.62×10^9 kg of Pollock, which represents an increase of 44.7% over the previous year (Figure 1.142).

Fish are known to react differently to approaching survey vessels, which typically biases surveys lower. Given their smaller size and shape and their lack of motor propulsion, USVs are likely to elicit less reaction by fish and are therefore likely to be more representative of the true abundance of Pollock. Another issue is the difference of the transducer depth of the USV compared with the survey vessel. USVs have their transducers at 2 m while the survey vessels have theirs at 9 m. This increases the potential for bad weather influencing the USV backscatter measurements and the need to filter out erroneous signals. We further note that the USV is not able to carry the larger (and therefore heavier) echosounders needed to achieve narrow beamwidths. This effect needs to be compensated for as well. The conclusion is that while USVs are able to conduct useful acoustic surveys, they aren't able to carry out trawls nor is their transducer beamwidth optimal for Pollock surveys.

1.12.3 Unmanned Aerial Vehicles (UAVs) or Drones

Drones have become a part of daily life as they have found applications in the military, weather forecasting, mapping, farming, architecture, archeology, and oceanography. This list is by no means exhaustive and there are many uses of UAVs that we may have missed. When one mentions drones to most people, they think of electrically powered, small multi-blade copters that can hover and fly and primarily carry still and video cameras that record onboard on memory cards. These systems are so abundant that one can buy them in toy departments and hobby shops. These aircraft are ubiquitous throughout the world and many people have developed the skill needed to fly them. There are even organized races and photo contests using these platforms.

There are, of course, fixed wing UAVs that are used for reconnaissance and other data collection activities. These range from modest size aircraft that can be launched with a catapult and land on a wire. to normal plane size drones that take off and land on a runway. The larger military drones can carry missiles that can be used to remotely strike an enemy. Meteorological drones have been instrumented to do the atmospheric profiling that is usually assigned to weather balloons. Up to a few years ago, the rules governing the civil operation of drones were very restrictive, but they have been substantially relaxed and drones are performing a lot more tasks than in the past. Delivery services like UPS, FedEx and Amazon are investigating having drones deliver packages to customers. It is clear that new drones and new drone applications will continue to be defined in the near future. They are now an accepted part of modern life.

1.12.3.1 Estimating the Body Mass of Free-Living Whales Using Aerial Photogrammetry and 3D Volumetrics

One of the more unique applications of a drone to oceanography was a study by researchers from the Aarhus Institute of Advanced Studies (AIAS) in Denmark, in collaboration with Woods Hole Oceanographic Institution (WHOI) in the U.S., to estimate the body mass of Right Whales (Christensen et al., 2019). The mass of an animal influences its metabolic rate and food requirements along with its growth, fasting and endurance, thermoregulation, foraging capability, home and range size, locomotion and cost of transport. Thus, body mass is a critical component in studies of whale physiology and biogenics.

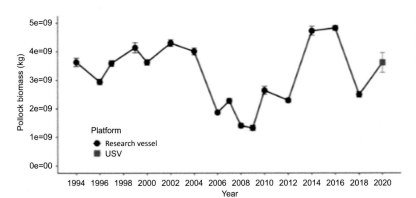

FIGURE 1.142 The abundance of Bering Sea Pollock time series with the 2020 estimate value added. Error bars indicate the ± 1 standard deviation of the estimate based on Monte Carlo simulations.

Baleen Whales are the largest animals on the planet, ranging in size from the 6.5 m long, 3,500 kg Pygmy Right Whale to the 33 m long and 190,000 kg Blue Whale. Their body size reduces their mass-specific metabolic rate and allows them to store large amounts of energy reserves to survive periods of low resource availability, which in turn enables them to undertake long distance migrations to exploit spatially and temporally clustered food resources. The large size of Baleen Whales is also correlated with foraging efficiency, diving capacity and relative cost of reproduction. Although size is a fundamental characteristic, it is seldom included in studies of Baleen Whales due to the difficulty of directly measuring size in the field.

Most data on Baleen Whale body mass are derived from measurements of dead animals, either caught in whaling operations or from accidental fisheries catch or beach strandings. Working with dead animals has limitations in that samples are not collected for science but rather for harvesting. By-catch and stranding samples are also not designed for study. While there have been studies to model body mass as a function of length, most of these studies have not considered the condition of the body. Attempts to account for body condition either rely on body girth data, which cannot be obtained from free-living whales, or by assuming a circular cross-sectional body shape when converting body width to girth.

The Danish and U.S. investigators used unmanned aerial-vehicle (UAV) photogrammetry, along with historical catch records, to estimate body mass. In addition to a camera, the UAV carried a Lightware SF11/C laser range finder mounted on the back of the UAV. The accuracy of the range finder was estimated to be 0.7 cm. The photos were used to estimate body length, width (lateral distance) and height (dorso-central distance) of free-living Southern Right Whales (Figure 1.144). The investigators then developed a model to predict the body volume of the whales, which incorporated both their size (body length) and body condition (width and height). To test this model, they predicted the body volume of eight North Pacific Right Whales caught in scientific whaling operations for which the body mass was measured. From these predictions, they subsequently derived a body volume-to-mass conversion factor, which could then be used to calculate the body mass of the whales measured by aerial photogrammetry. In addition to being able to estimate body volume and mass, the method made it possible to create an accurate 3-D model of the whales being photographed. A drone image of a Southern Right Whale and her calf is presented in Figure 1.143.

Few measurements of body mass exist for Right Whales. Most of these were collected as part of scientific studies between 1961 and 1968 that included the collection of 10 whales caught during their summer feeding season. To establish the relationship between body volume and mass for Right Whales, the width and height data from this sample were used to estimate the girth of the whales using the complete elliptic integral of the second kind formula to calculate the circumference of an ellipse (Figure 1.144). Linear models were then used to model the body volume as a function of its measured body length and estimated girths.

For their whale study, Christensen et al. (2019) collected aerial photographs of the dorsal surface and lateral side of 102 Southern Right Whales in Argentina between 4 August and 3 November 2018. The photographs were all taken by a UAV (multi-blade copter) at altitudes ranging from 19.4 to 38.7 m (mean = 27.2 m, standard deviation, SD = 3.32 m). After removing duplicates and animals with missing range finder data, a total of 86 whales remained. Of these, 48 were calves, seven were juveniles and 31 lactating females. The relationship between body length and the height-to-width ratio is summarized in Figure 1.145.

The body length of the whales did not influence their HW ratio. In addition, their relative body condition did not influence the body shape of juveniles or lactating females. The body shape of calves, however, changed significantly as a function of body condition as the calves grew. In their study, Christensen et al. (2019) computed body volume estimates for Southern Right Whales ranging from 1.45 to 55.56 m³. The estimated girths of the 86 whales, calculated from the width and height data, ranged from 2.3 to 9.22 m across the pectoral fin, from 2.76 to 9.28 m across the umbilicus and from 1.69 to 6.81 m across the anus. There was a strong correlation between the body volume of the Southern Right Whales and their

FIGURE 1.143 Drone image of a Southern Right Whale and her calf. An example of how these drone images were converted to measurements of length, width and height is presented in Figure 1.144.

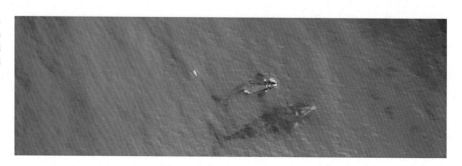

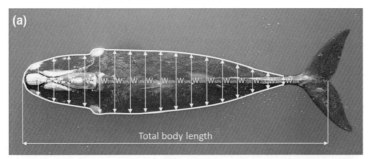

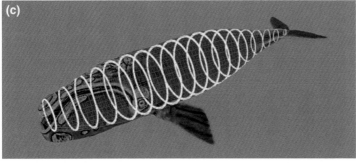

FIGURE 1.144 (a) Aerial photograph of the dorsal surface of a Southern Right Whale used to measure body length (BL) and width at 5% increments along the body axis, from 5% to 85% of the body length from the rostrum (white arrows). (b) Lateral side of the same whale used to extract body height (dorso-ventral distance) along the same measurement sites. The white solid lines indicate the location of the predicted girths, at 25%, 50% and 72% of the BL from the rostrum, along the body axis (dotted white line). (c) A 3D model of the whale used to estimate body volume. The cross-sectional ellipses illustrate the variation in the height-width ratio across the whale's body.

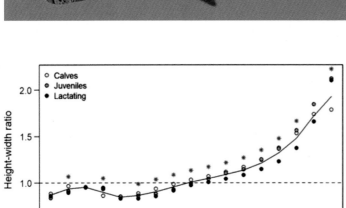

FIGURE 1.145 The body shape of Southern Right Whales measured as height-width ratios (HWs) across the body from 5% to 85% body length from the rostrum. The different reproductive classes are indicated, and the solid black line represents the average HW-ratio of all classes. The dashed line indicates a ratio of 1:1, equivalent to a circular body shape. Significant differences in HW ratios between reproductive classes for each measurement site are indicated with red asterisks. *From Christensen et al. (2019).*

body length ($r^2 = 0.996$). The predicted body volume of the dead North Pacific Right Whales varied between 34.82 and 114.05 m^3, with a mean of 77.35 m^3 and a standard deviation of 24.62 m^3. The estimated volume-to-mass conversion factor, or density, was 754.63 kg/m^3 (standard deviation = 50.03 kg/m^3). Using this conversion, the predicted body mass of the free-living Southern Right Whales in Argentina ranged between 1,092 and 41,928 kg (Figure 1.146).

Although the true body mass of the Southern Right Whales measured in the Christensen et al. (2019) study was unknown, the predicted estimates are consistent with existing measured body mass data, and earlier derived length-to-mass

FIGURE 1.146 Predicted body mass (M) and volume of Southern Right Whales (circular points, $n = 86$) and North Pacific Right Whales (black crosses, $N = 8$, Omura et al., 1969) as a function of body length (BL). The different age classes are indicated. The lines represent the predicted body mass (left axis) and the body volume (right axis) from body length. *From Christensen et al. (2019).*

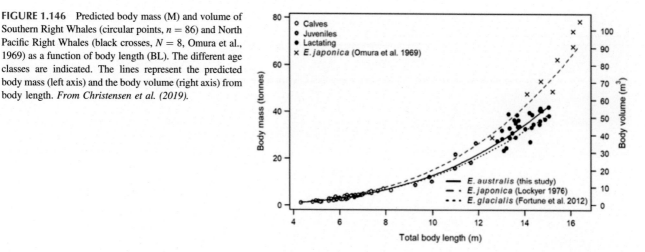

models based on actual measurements of whales in the North Atlantic. Few studies have tried to describe the cross-sectional body shape of whales. By photographing both their dorsal surface and their lateral side these investigators were able to show that the HW ratio of southern right whales varied considerably across their bodies. For juveniles and adults, the HW ratio remained constant across body length and condition, meaning that once the HW ratios are known, only data on body length and width are needed to accurately calculate the body volume and hence the body mass for these reproductive classes. The calves however vary greatly with age and no one model fits the HW to BL relationships.

1.12.4 Autonomous Underwater Vehicles (AUVs)

Autonomous underwater vehicles (AUVs) and Remotely Operated Vehicles (ROVs) provide sensor platforms for measuring oceanic water properties. ROVs such as the two-body 5,000 m depth Remotely Operated Platform for Ocean Sciences (ROPOS, operated by The Canadian Scientific Submersible Facility) and the 6,500 m depth Jason/Medea (designed by the Woods Hole Oceanographic Institution, Deep Submergence Laboratory) are highly maneuverable, un-manned tethered submersibles (decoupled from surface motion by intermediary controller package) that are "flown" from a ship to safely study and instrument many features of the World Ocean including hydrothermal venting systems at seafloor spreading regions of the deep ocean. They can be instrumented with CTDs, water sample carousels, high resolution still and video cameras, as well as robotic tools such as drills and mechanical manipulators. In contrast to the tethered ROVs, AUVs (including Gliders) are free of any surface vessel and therefore capable of conducting marine work during poor sea conditions at greater through-the-water speeds. The downside of all these devices is that they are expensive to purchase, highly expensive to operate, and labor intensive, requiring dedicated technical support, ship time, and maintenance. Although many research groups have purchased AUVs for oceanic surveys, most platforms are used by the military and the ocean industry. The Autosub 6,000 built by Underwater Systems Laboratory at the National Oceanography Centre (Southampton, United Kingdom) is a battery powered unit that supports magnetometer, turbidity, CTD, and electro-magnetic EH sensors. It also has a 3 m altitude collision avoidance system and a 6,000 m design limit. REMUS AUVs (REMUS 100, 600, and 6,000) developed originally by the Woods Hole Oceanographic Institution and then transferred to HYDROID (now a subsidiary of Kongsberg Marine) are also capable of extensive marine research.

Underwater gliders use small changes in buoyancy, similar to Argo drifters, in conjunction with wings that convert vertical fall to forward motion, allowing the instrument to move horizontally at speeds of around 1 knot (0.5 m/s) with low power consumption. The concept of a glider with a buoyancy engine powered by a heat exchanger was introduced to the oceanographic community by Henry Stommel in a 1989 article in *Oceanography*. In the article, Stommel proposed the use of a glider, called *Slocum*, developed with research engineer Doug Webb (Figure 1.147). The glider name was for Joshua Slocum, who made the first solo circumnavigation of the globe by sailboat. Stommel and Webb proposed harnessing energy from the thermal gradient between deep ocean water (2−4°C) and surface water (near-atmospheric temperature) to achieve globe-circling range constrained only by battery power on board for communication, sensors, and navigational computers. By 2005, not only had a working thermal-powered glider (*Slocum Thermal*) been demonstrated by Webb Research, but they and other institutions had introduced battery-powered gliders with impressive duration and efficiency, far exceeding that of traditional survey-class AUVs.

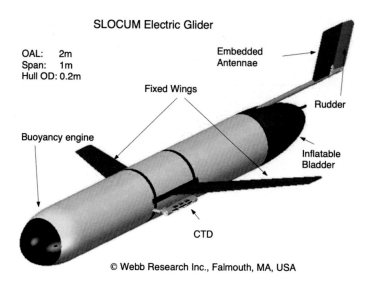

SLOCUM Electric Glider

OAL: 2m
Span: 1m
Hull OD: 0.2m

Embedded Antennae

Fixed Wings

Rudder

Buoyancy engine

Inflatable Bladder

CTD

© Webb Research Inc., Falmouth, MA, USA

FIGURE 1.147 Rendering of a Slocum glider built by Webb Research Inc., Falmouth, Massachusetts (U.S.A.).

Starting in the early 1960s, Autonomous Underwater Vehicles were developed to carry out measurements within the water column that were difficult to perform from surface platforms. These AUVs have many different forms and applications. We focus our review on the active and well-developed program at the Woods Hole Oceanographic Institution (WHOI), which hosts a wide variety of AUVs. In essence, AUVs are programmable, robotic vehicles that can drift, drive or glide through the ocean without real-time control by human operators. Some AUVs are programmed to report periodically to operators either directly or through satellite links or underwater acoustic beacons. These communications can provide commands to alter the programming of an AUV. Some AUVs can also make decisions on their own, changing their mission profile based on some environmental conditions they observe through their onboard sensors.

The schematic diagram of the Slocum glider presented in Figure 1.147 shows the important elements of the design. The "buoyancy engine" in the nose that controls the vertical movement of the platform also gives it a forward motion by changing the "angle of attack" of the wings. A rudder controls the direction of this forward motion. The CTD sensor is mounted under the starboard wing. (Sea Bird Scientific makes a special CTD unit for the Slocum glider.) Although the vehicle is designed for various depths, most Slocum gliders are limited to 200 m depth. Similarly designed Sea gliders and Spray gliders can operate down to 1,000 and 1,500 m, respectively. The Slocum glider has a weight in air of approximately 50 kg and its total volume change capacity is between 0.5% and 1% of its total displacement. Horizontal speeds relative to the surrounding water are typically about 35 cm/s. Slocum Gliders are capable of moving to preset locations and depths.

The long-range and duration capabilities of Slocum gliders have made them ideal for subsurface sampling on a regional scale. The platform consumes only 240 mW of power while sampling continuously at ½ Hz and a single Alkaline C-cell can operate the CTD continuously for 37 h. One Rutgers University glider, named the "Scarlet Knight", crossed the entire North Atlantic, leaving New Jersey on 27 April 2009 and arriving in France 221 days later (Figure 1.148). The glider called home over 1,000 times (via satellite) to report its location and relay data, made 11,000 dives and 11,000 ascents, and traveled 2,200 km vertically while covering the 7,409.6 km distance between start and end.

Remote Environmental Monitoring Units (REMUS) are low-cost AUVs designed by WHOI to operate with a simple laptop computer. Initially conceived for coastal monitoring, this torpedo shaped platform is now used for a wide variety of studies operating at a wide range of ocean depths. REMUS is propellor driven and uses fins for steering and diving. After entering the water, REMUS uses acoustic navigation to independently survey an area, while sensors inside sample and record data. Each REMUS has a control computer that performs the data collection as well as the platform command and control functions. During the U.S. military Operation "Iraqi Freedom" in 2003, the U.S. Navy used REMUS vehicles to detect mines in the Persian Gulf harbor of Um Qasr. Navy officers said that REMUS could do the work of 12−16 human divers and that they were immune to long exposure to cold water temperatures, murky water, sharks or hunger. Another REMUS was specifically adapted to survey New York City's Delaware River Aqueduct for leaks.

There is now a whole family of REMUS vehicles. REMUS 100 is a compact, lightweight AUV designed for coastal environments having depths of up to 100 m. As with other REMUS platforms, it can be configured for a wide variety of instruments. REMUS 600 is the most versatile member of the REMUS family, and its modular design enables an even

FIGURE 1.148 The track of the "Scarlet Knight" Slocum Glider that crossed the Atlantic Ocean in the spring of 2009, beginning on the east coast of the U.S. and arriving on the coast of France 7 months later.

greater variety of instruments and sensors to be hosted on the platform. It can operate down to 600 m depth and for durations of up to 70 h at speeds of up to 5 knots and a range of 286 nautical miles (530 km). The REMUS 3,000 is similar in size and construction to the REMUS 600 but made out of titanium, allowing the vehicle to descend to much greater ocean depths and carry advanced sensors for underwater mapping and imaging. REMUS 6,000 is an entirely new shape and can operate anywhere from 25 to 6,000 m depth. The REMUS Tunnel Inspection Vehicle is a customized vehicle built in 2004 to inspect a section of the Delaware Aqueduct. The REMUS SharkCam is a specially adapted REMUS 100 vehicle equipped with video cameras, navigational and science instrumentation that enables it to locate and track marine animals such as the Great White Shark. The REMUS TurtleCam follows turtles that have an attached transponder, but is otherwise similar to the SharkCam.

Another WHOI AUV is SENTRY (Figure 1.149), which is part of the National Deep Submergence Facility (NDSF). SENTRY is capable of operating down to 6,000 m depth and carries a suite of sensors for generating bathymetric, sidescan-subbottom and magnetic maps of the seafloor. The vehicle can also collect digital bottom photographs in a variety of deep-sea terrains, such a mid-ocean ridges, deep-sea hydrothermal vents or cold seeps at ocean margins. SENTRY is also uniquely capable of operating in extreme terrain such as underwater volcanoes. SENTRY's navigation system uses a Doppler velocity log and an inertial navigation system aided by acoustic systems. The acoustic system also provides communications that can be used to call up the vehicle's state and sensor status, in addition to re-tasking the vehicle's operations. SENTRY can also be used to locate and quantify hydrothermal fluxes and operate in tandem with deep-dive submersibles, such as *Alvin*.

The Woods Hole Oceanographic Institution's latest AUV is named "Orpheus" and is a new class of AUVs capable of reaching and operating in the ocean's greatest depths. Two identical Orpheus AUVs were built in 2018 and named Orpheus and Eurydice after the famous pair from Greek mythology who adventured through the depths of Hades. Access to depths

FIGURE 1.149 The Woods Hole Oceanographic Institution (WHOI) AUV "SENTRY".

below 6,000 m has been strongly limited by the great pressures that make operations difficult. With the loss of the Nereus at hadal depth, the deepest parts of the ocean have been inaccessible with AUVs. To return to operations at these depths, engineers from WHOI and NASA's Jet Propulsion Lab (JPL) collaborated to create a fleet of hadal AUVs that are small, lightweight and with a modular design based on the "cubesat" concept of a common core that can then be added to and modified to suit any particular mission. The AUV incorporates control and mapping software developed by JPL that enables the platform to reconfigure its objective while operating. The fixed directional thrusters (Figure 1.150) and compact shape make Orpheus nimble, allowing it to explore near the sea floor, land to collect samples and lift off again to continue its mission.

Unlike other vehicles designed to work at hadal depths, Orpheus is designed to be widely produced, simple to operate, and deployable from small ships of opportunity rather than being limited to large research vessels. This design philosophy enables access to the Orpheus platform to a much broader range of scientists. It also is an example of the requirements for autonomous systems that will explore great depths on other planets. Orpheus also carries a unique piece of heritage. In 2012, when James Cameron descended into the Challenger Deep in the Mariana Trench, he piloted Deepsea Challenger, a submersible that was constructed mainly of syntactic foam—a rigid flotation material composed of microscopic glass spheres embedded in epoxy resin. The syntactic foam used in Orpheus and its twin came from spare material produced for the Deepsea Challenger. Orpheus was first tested in September 2018 and in September 2019 from the WHOI-operated research vessel [*Neil Armstrong*]. Each test extended the operation to greater depths.

1.12.5 Instrumenting Marine Mammals

Initially, marine mammals were simply tagged to record their history, but it was then realized that they could be instrumented to collect oceanographic data. This was particularly useful in that marine mammals spend considerable time in polar regions that are largely under-sampled. It should be noted that the instruments attached to the seals are shed when the seal molts after returning from their annual migration. So, the seal is not saddled with these attachments for life.

Between March 1997 and March 1999, Boehlert et al. (2001) tracked and studied nine instrumented seals in the eastern North Pacific. The six female and three male seals carried time-temperature-depth (TTDR) recorders and ARGOS platform terminal transmitters (Figure 1.151). The ARGOS Platform Transmit Terminals (PTTs), Model ST-6 (Teleonics, Mesa, Arizona) transmitted every 34 s while the animal was at the surface. This provides the time and location to go with the temperature-depth pairs from the TTDR. Due to the low power transmission and the small antenna coupled with the limited surface time, each surface period will not result in a geolocation. Instead, positions were calculated with the ARGOS system using auxiliary location processing, wherein lower quality locations are still calculated and reported.

Temperature and depth were measured with a Mk 3 data recording system (Wildlife Computers, Seattle, Washington). This instrument has a temperature resolution of 0.1°C and an accuracy of ±0.5°C. The minimum recorded temperature was 4.8°C, although one animal showed truncation at 5°C. Each TTDR was laboratory calibrated before deployment and two of them exhibited nonlinear calibration curves and were therefore excluded from the experiment. The TTDR data were recorded in memory every 30 s and retrieved after the animals returned to the rookery months later. Examples of their

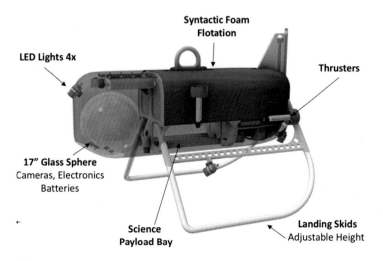

FIGURE 1.150   The Woods Hole Oceanographic Institution (WHOI) Orpheus hadal deep AUV.

LED Lights 4x

Syntactic Foam Flotation

Thrusters

17" Glass Sphere
Cameras, Electronics
Batteries

Science
Payload Bay

Landing Skids
Adjustable Height

FIGURE 1.151   Adult female elephant seal with ARGOS satellite transmitter and Mk 7 TTDR attached. Past studies have reported that only the head is out of water when a seal surfaces to breathe. *From Boehlert et al. (2001).*

tracks in the northeast Pacific are presented in Figure 1.152. Females tracked in 1998−99 show a very different set of tracks from those of the males in 1997−98. Some of the male return tracks are missing data, with only the track from "Moose" being complete in both directions. Typical vertical dive profiles are shown in Figure 1.153.

Location accuracy was a concern in the Boehlert et al. (2001) study. The number of ARGOS positions per day ranged from 1.2 to 4.3, with an average of 2.5 positions per day. Filtering these data resulted in a loss of 5.8% of the positions.

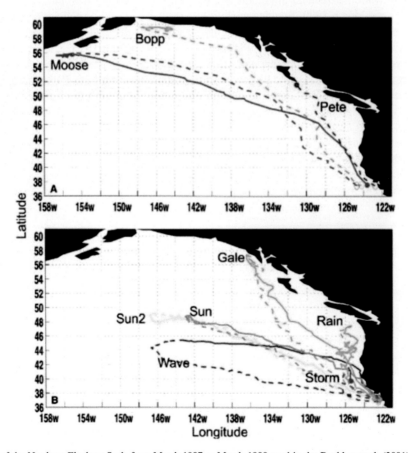

FIGURE 1.152   Tracks of the Northern Elephant Seals from March 1997 to March 1999 used in the Boehlert et al. (2001) study. Outward tracks are dashed lines and return tracks are solid lines. Only those tracks with both TTDR and ARGOS data are shown. *Upper*: males tracked in 1997; *Lower*: Females tracked in 1998−99. Note the apparent differences in foraging areas. *From Boehlert et al. (2001).*

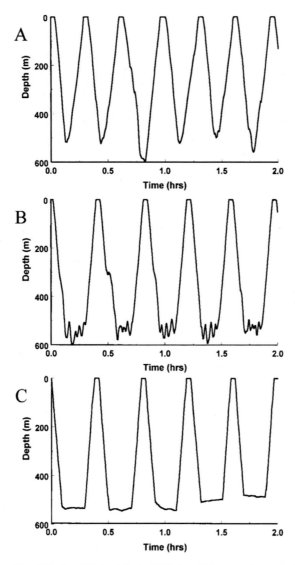

FIGURE 1.153 Typical vertical dive profiles of Northern Elephant Seals. (a) V-shaped dives are characteristic of seals transiting to foraging grounds; (b) dives with bottom time having vertical excursions are pelagic foraging dives; and (c) dives with flat-bottom times are foraging or transit dives along the continental shelf or slope. *From Boehlert et al. (2001).*

There were marked differences between the males and females in this study. The females had ARGOS fixes on average 2.3 times as frequently as the males. This probably represents a difference in behavior at the surface affecting PTT performance. For both males and females, the difference was linearly related to the time between adjacent fixes, with the slopes of the regression lines being 7.26 km/day for females and 6.53 km/day for males. These are considered to be consistent with the errors in the ARGOS system itself. The regularity and depth of the dives of these instrumented animals provides an excellent source of surface and subsurface temperature information (Figure 1.154).

The Marine Mammals Exploring the Oceans Pole to Pole (MEOP) consortium (Treasure et al., 2017) was set up to study the polar regions of the ocean using instrumented marine mammals. Since 2003, instrumented animals have been gathering exceptionally large data sets consisting of oceanographic CTD casts (>500,000 profiles) that are now freely available to the scientific community through the MEOP data portal (http://meop.net). The MEOP consortium is made up of international collaborators that use marine animals to collect biological and physical oceanographic data. To this end, the conductivity-temperature-depth satellite relay data loggers (CTD-SRDLs, Figure 1.155) were developed. The CTD-SRDLs are built at the Sea Mammal Research Unity at the University of St. Andrews in the UK. The sensor head consists of a pressure transducer, a platinum resistance thermometer and an inductive cell for measuring conductivity. The temperature

FIGURE 1.154 Dive profiles (solid lines) and thermal structure from a time-temperature-depth recorder on a Northern Elephant Seal from a series of dives over a 9.6 h period on 16 March 1998. The seal was located at 52.58° N, 144.63° W. Note the regularity of the dives.

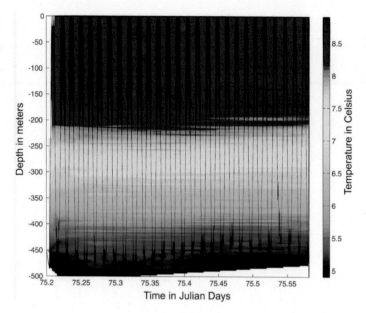

FIGURE 1.155 A CTD-SRDL with hardware components labeled. The unit is housed in solid epoxy rated to either 500 or 2,000 m depth. Standard sensors include a wet/dry saltwater switch, temperature and conductivity. The unit has a PC interface and is powered by a battery. The data is transferred via ARGOS in a compressed form using between 10 and 25 data points per profile. Sample location is also provided by ARGOS.

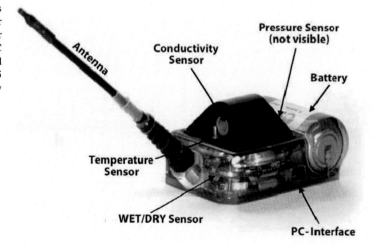

and conductivity sensors have precisions (repeatability) of 0.005°C and 0.005 mS cm^{-1}, respectively. The units are noninvasive and are glued to the animal's fur so that they fall off when the animal molts.

The majority of the animals instrumented as part of MEOP have been various types of seals but some CTD-SRDLs have been deployed on turtles, which also must surface to breathe and thus provide opportunities to use ARGOS to geolocate and transfer data. As the program has grown since 2003, it has collected a considerable amount of data in polar regions (Figure 1.156). This concentration of CTD profiles in the polar regions has greatly expanded coverage of these logistically difficult to visit regions. These data have aided physical oceanographers to constrain models in such critically under sampled regions. The marine animals' profiles have been particularly useful when merged with ARGO profiles and previously collected ship data. Seal-derived data have played an instrumental role in developing our understanding of the Antarctic Shelf Circulation and the formation of Antarctic Bottom Water (AABW). In 2011, seal-borne CTD-SRDLs were central to solving a 30+ year old mystery regarding AABW in the Weddell-Enderby Basin. Observations of very high salinity shelf water were linked to a new source of AABW off Cape Darnley, East Antarctica. Analysis of an additional 2 years of seal-profile data made it possible to show that Prydz Bay, located just east of Cape Darnley, provides a secondary contribution to AABW. It was also found that recent melting of nearby ice shelves partially suppressed the AABW formation, demonstrating a sensitivity of AABW formation to global warming.

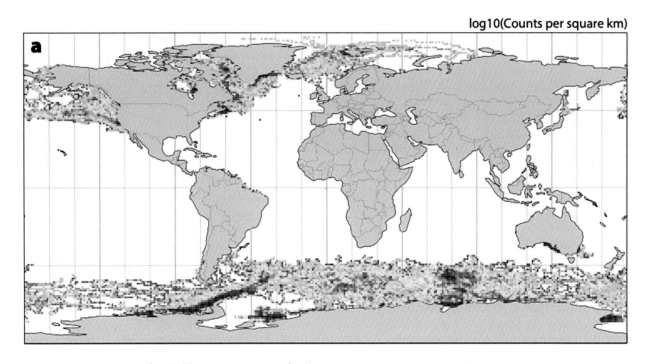

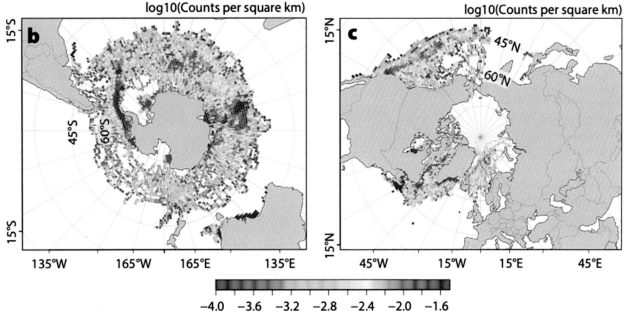

FIGURE 1.156 Data density distribution for CTD profiles from the Marine Mammals Exploring the Oceans Pole to Pole (MEOP) program showing (a) the entire globe and, separately, the (b) southern and (c) northern hemispheres.

1.13 WIND MEASUREMENTS

The surface wind stress is one of the primary mechanisms driving ocean variability over a broad range of frequencies and spatial scales, ranging from high frequency internal waves and turbulence in the ocean interior to the large-scale Antarctic Circumpolar Current. Turbulent mechanical wind mixing is a major factor in the deepening of the surface mixed layer, while passing storm fronts are responsible for the generation of near-inertial currents, coastally trapped waves, storm surge, the large-scale gyral circulation of the open ocean, and numerous other oceanic processes. Coastal upwelling and baroclinic undercurrents, such as the California Undercurrent, forced mainly by the alongshore pressure gradients along eastern boundary regions of the World Ocean, also respond to changes in the wind forcing (cf. Connolly et al., 2014; Vélez-Belchi et al., 2021). It is therefore not surprising to find a section on wind data in an oceanographic text. Moreover, we can state with some confidence that most of the scientific assessment of wind data over the ocean has been done by oceanographers searching for the best way to define the meteorological forcing field for oceanic processes. This is especially true of investigators working on upper ocean dynamics and numerical modelers who require synoptic or climatological winds to drive their circulation models. It is not the intent of this book to discuss in detail the many types of available wind sensors and to evaluate their performance, as is done with the oceanographic sensors. Instead, we will briefly review the types of wind data available for ocean regions and make some general statements about the usefulness and reliability of these data.

Every day, the World Meteorological Organization (WMO) provides key atmospheric, oceanic and land data from "well over 10,000 manned and automatic surface weather stations, 1,000 upper-air stations, 7,000 ships, 100 moored and 1,000 drifting buoys, hundreds of weather radars and 3.000 specially equipped commercial aircraft". To this the WMO can add roughly 30 meteorological and 200 research satellites as part of the global network for meteorological, hydrological and other geophysical observations. Observations are quality-controlled, based on technical standards defined by the WMO Instruments and Methods of Observation Programme (IMOP), and made available to all countries through the WMO Information System (WIS). Open-ocean wind data are of four basic types: (1) six-hourly geostrophic wind data computed from measured distributions of atmospheric sea surface pressure over the ocean; (2) directly measured wind data from ships and moored platforms (typically provided at hourly intervals); (3) 12-hourly wind speed and direction inferred from scatterometers flown on selected polar-orbiting satellites; and (4) hourly to six-hourly reanalysis wind velocity data derived from a blend of observations and numerical models. Wind speeds are also available from altimeters and passive microwave. In addition to directly observed winds, multiple governmental and non-governmental agencies around the world use observations and atmospheric models to generate nowcasts, medium range (3−7 day) forecasts, and extended range (monthly to seasonal) forecasts of precipitation, atmospheric pressure, wind velocity, and other parameters for various sectors of the global ocean. For example, up to 1-week wind forecasts are provided for the contiguous waters of North America by the National Centers for Environmental Prediction (National Weather Service) in the United States and the Meteorological Service of Canada (Environment and Climate Change Canada). Regional (and, in many cases, global) medium range meteorological forecasts are provided by the US Weather Service (USWS), the European Centre for Medium-range Weather Forecasts (ECMWF) headquartered in the United Kingdom, the Japan Meteorological Agency (JMA) in Tokyo, the Australian Bureau of Meteorology in Melbourne, the India Meteorological Department in Puna, the China Meteorological Administration (CMA) in Beijing and the South African Weather Service (SAWS) in Pretoria. A complete list of international weather bureaus can be found at the Wikipedia site (https://en.wikipedia.org/wiki/List_of_meteorology_institutions).

Atmospheric pressure maps over the ocean are prepared from combinations of data recorded by ships at sea, from moored or drifting platforms, and from ocean island stations. Analysis procedures have changed over the years with early efforts depending on the subjective change as hand-contouring of the available data. More recently, there has been a shift to computer-generated "objective analysis" of the atmospheric pressure data. Because they are derived from synoptic weather networks, the pressure data are originally computed at six-hourly intervals (00, 06, 12, and 18 UTC). While some work has been done to correct barometer readings from ships to compensate for installation position relative to sea level, no systematic study has been undertaken to test or edit these data or analyses. However, in general, sea-level pressure patterns appear to be quite smooth, suggesting that the data are generally reliable. Objective analysis smooths the data and suppresses any noise that might be present.

It is not a simple process to conformally map a given atmospheric pressure distribution into a surface wind field. While the computation of the geostrophic wind velocity from the spatial gradients of atmospheric pressure is fairly straightforward, it is more difficult to extrapolate the geostrophic wind field through the sea-surface boundary layer. The primary problem is our imperfect knowledge of the oceanic boundary layer and the manner in which it transfers momentum from the wind to the ocean surface. While most scientists have agreed on the drag coefficient for low wind speeds (<5 m/s), there continues to be some disagreement on the appropriate coefficient for higher wind speeds and wave-current conditions (cf. Kara et al., 2007). Added to this is a lack of understanding of boundary layer dynamics and how planetary vorticity affects this layer. This leads to a lack of agreement on the backing effect and the resulting angle one needs to apply between the geostrophic wind vector and the surface wind vector. Thus, wind stress computations have required the *a priori* selection of the wind stress formulae for the transformation of geostrophic winds into surface wind stresses. The application of these stress calculations will therefore always depend on the selected wind stress relation and any derived oceanographic inferences are always subject to this limitation.

Anemometers installed on ships, moored buoys, or island stations provide another source of open-ocean wind data. The ship and buoy records are subject to problems arising from measuring the wind around structures and relative to a moving platform, which is itself being affected by the wind. These effects are difficult to estimate and even more difficult to detect once the data have been recorded or transmitted. Many of the earlier ship-wind data in climatological archives are based on wind estimates made by the ship's officers from their evaluation of the local sea state. (The Beaufort Scale was designed for the days of sailing vessels and uses the observed wave field to estimate the wind speed.) Analysis of the ship-reported winds from the Pacific (Wyrtki and Meyers, 1975) has demonstrated that, with some editing and smoothing, these subjective data can yield useful estimates of the distribution of wind over the equatorial Pacific. Barnett (1983) has used objective analysis on these same data to produce an even more filtered set of wind observations for this region. Following a slightly different approach, Busalacchi and O'Brien (1981) reanalyzed the ship wind-data to fill in spatial gaps before applying the wind fields to oceanographic model studies.

There are now a large number of moored meteorological buoys in the World Ocean that can be found at http://www.ndbc.noaa/gov. The buoys typically have dual sensors (in case of instrument failures or damage) that provide hourly measurements of wind speed and direction, atmospheric pressure, air and water temperature, significant wave height and peak wave frequency. (Significant wave height is the average height of the highest 1/3 of all the measured waves over some specific period, typically 20 min.) The data are noisy, have numerous spikes, and generally require considerable effort to "clean up". Meteorological buoys and their sensors are also subject to damage or loss during extreme wave conditions. Because such conditions generally occur in winter when it is difficult to service the platforms from ships, there are often gaps in the data series.

Included in other widely used sets of wind data are the synoptic wind fields produced by the Fleet Numerical Meteorological and Ocean Center (FNMOC; https://www.metoc.navy.mil/fnmoc/fnmoc.html)—formerly the Fleet Numerical Oceanography Center (FNOC)—in Monterey, California. These analyses use not only ship, buoy, and island reports but also winds inferred from satellite-borne scatterometers and the tracking of clouds in sequences of visible and infrared satellite imagery. The latter technique uses the infrared image to estimate the temperature and, therefore, infer the elevation of the cloud mass being followed. By examining sequences of satellite images, specific cloud forms can be followed, and the corresponding wind speed and direction computed for the altitude of the cloud temperature. As might be expected, this procedure is dependent not only on the accuracies of the satellite sensors but also on the interpretive skills of the operator. As a consequence, no real quantitative levels of accuracy can be attached to these data. Comparison between the FNMOC winds and coincident winds measured from an open-ocean buoy (Friehe and Pazan, 1978) showed excellent agreement in speed and direction over a period of 60 days. Although this single-point comparison is too limited to establish any uncertainty values for the FNMOC wind fields, the comparison provides some confirmation of the validity of techniques used to derive the FNMOC winds.

A wind product for the Pacific Ocean similar to the FNMOC winds is generated by the National Marine Fisheries Service (NMFS) in Monterey, California (Holl and Mendenhall, 1972; Bakun, 1973). In this product, the geostrophic "gradient" winds are first computed at a $3° \times 3°$ latitude−longitude grid spacing from spatial gradients in the 6-hourly synoptic atmospheric pressure fields at the 500 or 800 mb surfaces. To obtain the surface-wind vectors in the frictional atmospheric boundary layer, the magnitudes of the calculated geostrophic wind vectors are reduced by a factor of 0.7 and the wind vectors rotated (backed) by 15°; here, "backed" refers to a counterclockwise motion in the northern hemisphere and a clockwise rotation in the southern hemisphere. (Some of the original work on this method can be traced to Fofonoff, 1960.)

Winds computed by the NMFS were compared with winds measured from moored buoys off the coast of British Columbia during the summers of 1979 and 1980 (Figure 1.157). In this comparison, Thomson (1983) concluded that winds computed from atmospheric pressure provided an accurate representation of the oceanic winds for timescales longer than several days but failed to accurately resolve short-term wind reversals associated with transient weather systems. Computed winds also tended to underestimate percentages of low and high wind speed. Similar results were reported by Marsden (1987) for the northeast Pacific (including Ocean Weather Station P) and by Macklin et al. (1993) for the rugged coast of western Alaska. The poor correlation of observed and computed winds at short timescales is thought to be due to the coarse ($3° \times 3°$) spacing, the low (6 h) temporal resolution of the pressure field and the strong influence of orographic effects in mountainous coastal regimes. In Thomson's study, peak-computed winds were roughly 20° to the right of the observed peak inner-shelf winds, suggesting that the computed winds were representative of more offshore conditions or that the 15° correction for frictional effects was too small. Spectra of observed winds were found to be dominated by motions at much larger wavelengths than were found in the computed values. The NMFS winds were found to contain a significant 24-h sea-breeze component in the inner shelf observed winds but not in the records farther offshore. Based on spectral comparisons it was concluded that the NMFS winds closely represented the actual winds for periods longer than 2 days (frequencies less than 0.02 cph) and only marginally matched actual winds for periods shorter than 2 days.

The Comprehensive Ocean-Atmosphere Data Set (COADS) and the international version (ICOADS) are cooperative efforts between the National Center for Atmospheric Research (NCAR, USA) and the National Oceanic and Atmospheric Administration (NOAA, USA) to generate observed wind fields for the global ocean. COADS consists of global marine

FIGURE 1.157 Comparison of observed and calculated oceanic winds for the period May to September 1980 on the west coast of Vancouver Island. Inset shows location of the moored buoys for 1979 and 1980 (triangles) and location (solid square) of grid point (49° N, 127° W) for the geostrophic winds. Observed winds are from anemometers on moored buoys; calculated winds are the 6-hourly geostrophic winds provided by the National Marine Fisheries Service (NMFS) in Monterey, California. *From Thomson (1983).*

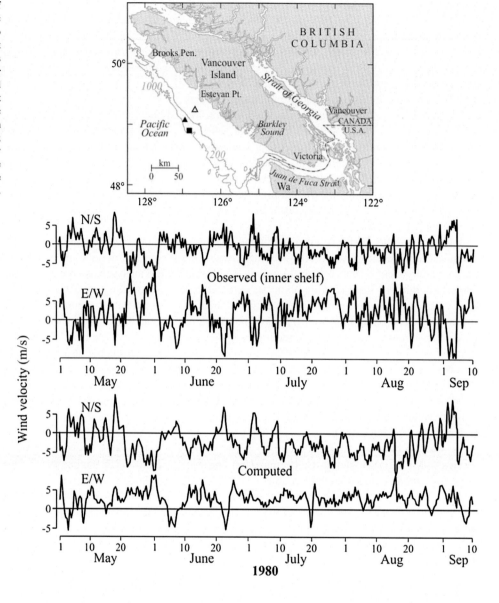

surface observations for the period 1784–1997 and monthly summary statistics of these observations for 1800–1997. The observations are primarily from ships (merchant, ocean research, fishing, and naval), moored platforms, and drifting buoys. Other sources of real-time and delayed mode data are included. Because coastal oceanographic studies rely on accurate nearshore wind data, Cherniawsky and Crawford (1996) compared monthly mean wind speeds and directions from buoys off the west coast of Canada to those from COADS for the period 1987–92. Differences between the 2° × 2° COADS and buoy winds were mainly due to inconsistencies in the ship recording methods. The effect of large ocean waves on buoy wind measurements was also a potential source of measurement error for winds greater than 7–10 m/s; above this range, buoy winds may be underestimated.

Koráčin and Dorman (2001) used wind data generated by a 9-km grid, University of Washington Mesoscale Model 5 (MM5 regional model) to examine topographic influences on the summer marine boundary layer (MBL) along the California coast in June 1996. The modeled winds, which include orographic effects on the wind fields, and coastal buoy winds were deemed to be sufficiently similar for the authors to be confident in the MM5 results. The wind structure near coastal capes appears to be typically composed of an upstream convergence zone (compression bulge) and a downstream supercritical divergent flow (expansion fan) followed by a "deceleration zone". This flow structure undergoes marked diurnal variability, which affects the local wind divergence field and cloud formation. The authors further concluded that the overall MBL structure (including winds) is governed primarily by topography in the inner coastal zone lying within 100 km of shore. In a more recent study, Tinis et al. (2006) compared MM5 winds observed at coastal meteorological buoys from British Columbia to northern California in order to assess their suitability for use in regional biological ocean modeling (Pitcher et al., 2010). Two 3 month study periods from 2003 were chosen: summer (July to September), which is most important for the growth of toxic algae off the Washington State coast, and fall (October to December) when downwelling favorable wind events force the onshore movement of potentially toxin-contaminated shelf water. Wind speeds determined by the MM5 model ranged from 81% to 101% of observed wind speeds. Mean winds were well modeled in summer but were, on average, 35° clockwise in the fall compared to buoy winds. Winds were strongest in the diurnal and 2−5 day bands in both seasons; spectral coherence between the model and observed winds in both of these frequency bands were highest (0.66−0.93) for the coast of Washington and northern Vancouver Island. In some near-shore regions, modeled winds were insufficiently accurate to represent the observed winds.

As the previous discussion indicates, a primary caution when using coastal wind data is that winds often need to be corrected for local orographic effects especially along mountainous coasts (Macklin et al., 1993). This is also true of winds generated by orographically sensitive regional numerical models such as the University of Washington MM5 winds. If the measured wind data are to be considered representative of the coastal ocean region, the wind sensor must be unobstructed along the direction of the wind. If not ideally situated, the measured wind data can still be used if the directional data are weighted to account for the bias due to local wind channeling by the topography. Marsden (1987) found good agreement between measured and calculated winds at the rugged but exposed anemometer site at Cape St. James on the central British Columbia coast, but relatively poor agreement for these winds at the protected coastal station at Tofino Airport 300 km to the south of the Cape.

Buoy wind measurements must be properly designed. The anemometer should be located on the windward side of the buoy tower so that it "sees" an undisturbed airflow. To ensure this aspect many buoy operators use a large wind vane on the buoy mast to align the buoy hull with the wind. Cup or propellor anemometers are primarily used for moored buoy applications. Such anemometers will measure the correct wind speed regardless of the buoy's orientation. The wind speed uses a generator or a pulse generator to create a signal to record while the wind direction is measured using either a potentiometer or a vertical tail to detect the wind direction (Weller et al., 1990). The threshold for these wind measurement sensors is about 1 m/s, which is needed to overcome the mechanical friction in the system. The anemometer surface elements (cups, propeller and vanes) and their bearings, tend to wear and corrode with time requiring constant maintenance to keep the buoy wind measurement accurate and functioning. A sonic anemometer can overcome these limitations and will also provide measurements with a very fine resolution. The primary disadvantage of these sensors is that the supporting structure for the transducers distorts the wind flow, and the sonic pulses are influenced by precipitation.

1.13.1 Satellite Measurement

Other than moored buoy measurements, as discussed in the preceding section, measuring the wind vector (wind speed and direction) over the ocean can be very challenging. Traditional anemometer winds from ships at sea are influenced not only by the ship motion but also by the supporting structure of the anemometer. Also, there are large portions of the ocean that are not frequently visited by normal ocean going ship traffic. Thus, satellite-based ocean wind measurements are an attractive way to measure the wind.

The Seasat scatterometer was the first spaceborne instrument to demonstrate the capability of a radar scatterometer to observe the wind vector over the entire World Ocean. The term "scatterometry" was coined by Richard Moore for the measurement of the wind vector from the backscatter of a radar signal. Previously, navy ships had been using radar backscatter to detect submarine periscopes and other nearby vessels. In November 1963, Moore received a phone call from Peter Badgley of NASA who wanted to pool together various instruments to study the Earth on a very large scale. Moore made a presentation to the National Academy of Sciences in Washington D.C. stating that they could develop a scatterometer to measure the ocean wind vector at his lab in Kansas. He made a similar presentation to the National Academy in 1965 in San Diego, where the Department of Defense agreed to incorporate a scatterometer on a small manned orbital platform to monitor the Soviet Union and China. This manned platform never got off the ground. Moore's next project was to build a 15 GHz scatterometer in conjunction with NASA. The U.S. Navy had already demonstrated that backscattering could measure winds up to 15 knots. Badgley and Moore, along with Willard Pierson, thought they could measure much higher wind speeds. Linwood Jones supervised the project at NASA's Langley facility. Together with Moore and Pierson, Jones flew the scatterometer on a NASA C-130 aircraft over the Gulf of Mexico. The scatterometer had its first spaceflight going up on Skylab in 1973. These data were compared with winds measured from ship radars to confirm that the orbital measurements could accurately measure the wind vector. These developments culminated in the deployment of a Ku-Band scatterometer on Seasat in 1978.

As reported by Grantham et al. (1976), the Seasat scatterometer (SASS) was a 14.6 GHz scatterometer using 4 fan-beam antennas (Figure 1.158) to measure wind speed and direction over a 1,000 km wide swath with a ground resolution cell size of 50 km × 50 km. This scatterometer was considered to be accurate for wind speeds from 4 to 24 m/s for the Seasat orbit of 790 km altitude, 108° inclination and 0.001 eccentricity. The antennas were dual-polarized and together formed a starlike pattern of illumination on the Earth (Figure 1.158). The 14.6 GHz signal switched sequentially through the four antenna-polarization combinations taking 1.89 s for each, resulting in a total of 7.56 s to complete the switching sequence. The received RF pulses backscattered from the ocean surface were processed for each Doppler channel to yield the mean ocean scattering coefficient and its error due to communication noise. A number of return pulses are averaged to obtain a mean backscatter. Additional errors, such as system biases, were removed using coincident measurements taken on an aircraft calibration under flight.

The brief operation of Seasat cut short the important global mapping of the SASS. The operation did, however, demonstrate that global wind vectors over the ocean could be monitored from space using a scatterometer. Even this limited wind vector data set was used to improve weather forecasts (Peteherych et al., 1988). These investigators examined six cases for the west coast of North America to determine what improvements would be gained by adding SASS wind vectors to the usual forecast information. They found that in all six cases that the weather forecasts improved when the SASS winds were included. As might be expected, wind forecasts improved in general but specifically in areas where the SASS observations were the only data available.

In a slightly different application, Freilich and Chelton (1986) computed the wavenumber spectra of Pacific Ocean winds from the SASS data. To do this, the Pacific was broken into four regions: the first two covering the South Pacific and the second two the North Pacific (Figure 1.159). As it was not possible to simply apply fast-Fourier transform (FFT) techniques to the SASS data (which were restricted to swaths with gaps in between), the authors first objectively mapped the irregularly spaced and gappy data to a regular grid to which they could then apply standard 2-D FFT methods. To apply optimal interpolation, Freilcih and Chelton initially assumed that the wavenumber spectra of the wind field would be isotropic. Thus, the covariance functions corresponding to these spectra were obtained with an inverse Fourier transform, which gave damped cosine functions. These positive definite approximations to the assumed covariance were then used to optimally interpolate the SASS vector winds.

A benefit to the optimal interpolation method is that it provides an estimate of the error of each interpolated field. This is a mapping error and does not address any instrumental or other sampling errors that may be present. As might be expected, the mapping error was found to be small within the SASS swaths but significantly large in the gaps between the swaths. Because the SASS swaths were rather narrow, the wavenumber spectra were restricted to one-dimensional auto- and cross-spectra in the along-track direction. Spectra were computed for each individual swath that contained a minimum of 70 wind vectors. The wind vectors were transformed into zonal (positive toward the east) and meridional (positive toward the north)

FIGURE 1.158   Seasat scatterometer antennas.

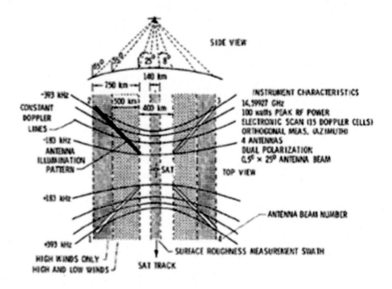

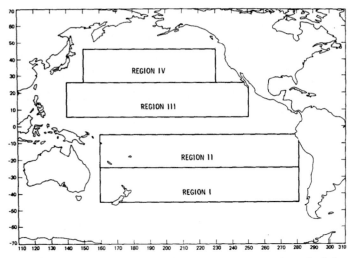

FIGURE 1.159 Geographic regions used to computer wind wavenumber spectra. *From Freilich and Chelton (1986).*

component speeds. The one-dimensional spectra for each study region are showed in Figure 1.160 along with an aircraft scatterometer spectrum presented by Nadstrom et al. (1984). The power laws that go with these curves are given in Table 1.24 (Freilich and Chelton, 1986).

The averaged spectra are nearly linear when plotted in log-log space, indicating that the vector winds follow a simple power law dependence on wavenumber, written as k^{-b}. Spectra from the midlatitude regions (regions I and IV) have nearly identical power-law dependencies with $b = 2.20 \pm 0.08$ for region I and $b = 2.21 \pm 0.10$ for region IV. Normalized spectra from the tropical regions also share similar power-law dependencies, with $b = 1.76 \pm 0.04$ for region II and $b = 1.85 \pm 0.06$ for region III. The difference between spectral slopes between the tropics and midlatitudes is statistically significant.

In a scatterometer record, the wind speed is proportional to the measured scattering cross section. The wind direction is found by determining the angle that is most consistent with the backscatter observed from multiple angles. In roughly 5 min, a satellite in low polar orbit will move far enough to view a particular spot on the ocean from angles spanning 90°.

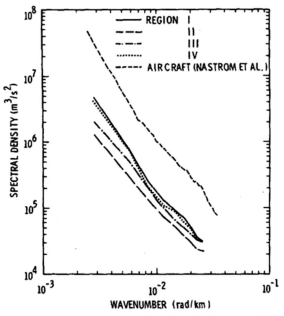

FIGURE 1.160 Log-log plots of averaged kinetic energy spectra (unnormalized) from the four geographic regions in Figure 1.159 along with a kinetic energy spectrum measured by Nastrom and Jasperson (1984). The abscissa is wavenumber in units of rad/km and the ordinate is spectral density in units of m^3/s^2. *From Freilich and Chelton (1986).*

TABLE 1.24 Power law exponents and averaged variances for the kinetic energy spectra for each geographic region. Uncertainties are standard deviations. Power laws are of the form as k^{-b}. See Figure 1.159 for locations of the four regions.

Geographic region	Normalized power law exponent (b)	Unnormalized power law exponent (b)	Total variance ($m^2 s^{-2}$)
I	2.20 ± 0.08	2.26 ± 0.09	17.7
II	1.76 ± 0.04	1.86 ± 0.05	5.6
III	1.85 ± 0.06	1.97 ± 0.07	8.8
IV	2.21 ± 0.10	2.21 ± 0.10	15.3

After Freilich and Chelton (1986).

The mathematical function describing the fit of the observed backscatter (as a function of the wind direction) has multiple minima (solutions) and it is necessary to observe the spot from many angles. SASS achieved this with multiple fan antennas (Figure 1.158). This angle ambiguity is most severe at low wind speeds (less than 7 m/s) but once the wind exceeds 8 m/s, it becomes much easier to resolve this ambiguity.

The success of the Seasat scatterometer prompted many follow-on instruments. The next scatterometers to fly in space were on the European Remote Sensing satellites, ERS-1 and ERS-2. This scatterometer was configured with three "stick" antennas covering a swath width of 500 km (Figure 1.161). The specifics of the ERS scatterometer are given here in Table 1.25.

The conversion of radar cross section (σ^0) observations into wind measurements uses a mathematical model, which defines the relationship between σ^0, wind speed and wind direction knowing the incidence angle of the scatterometer pulse and polarizations This model is based on previous measurements made prior to launch with a series of aircraft flights equipped with airborne scatterometers. The scatterometer measurements are made continuously over a 500 km wide swath so that wind speed and direction can be observed at points that are separated by 25 km. The scatterometer is equipped with a calibrator designed to ensure highly stable measurements. The instrument is also equipped to compensate for the Doppler shift introduced by the motion of the spacecraft and the Earth's rotation. The next development in satellite scatterometers was the NASA scatterometer known as NSCAT (Table 1.26).

Developed as a follow on to the SASS flown on Seasat, NSCAT first flew on the Japanese Advanced Earth Observing Satellite (ADEOS) launched in August 1996. NSCAT had six antennas and operated at a frequency of 14 GHz. The antennas scanned two 600 km bands of the ocean that are separated by 330 km gaps. The resolution within these bands was

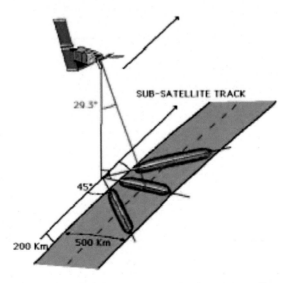

FIGURE 1.161   Geometry of the European Remote Sensing (ERS) satellite scatterometer.

TABLE 1.25 European Remote Sensing (ERS) satellite scatterometer specifications. SD denotes the standard deviation.

Swath width	500 km
Spatial resolution	50 km
Grid spacing	25 km
Wind speed	0.5–30 m/s
Accuracy	SD: 2 m/s
	Bias: 0.3 m/s
Wind direction	0–360°
Accuracy	SD: 20°
	Bias: 0.8°

TABLE 1.26 Satellite scatterometer history and development.

Background	Duration	Spatial resolution	Grid spacing	Scan character	Frequency
SeaSat-A scatterometer	1978/7/7–1978/10/10	50 km	100 km	Two sided Double swath	Ku band (14.6 GHz)
ERS-1 scatterometer	1991/7–1997/5/21	50 km	50 km	One sided Single swath	C band (5.3 GHz)
ERS-2 scatterometer	1997/5/21–2011/7	50 km	50 km	One sided Single swath	C band (5.3 GHz)
NSCAT	1996/9/15–1997/6/30	25 km	25 km	Two sided Double swath	Ku band (13.995 GHz)
Seawinds on QuikSCAT	1999/7/19–2009/11/23	25 km	12.5 km	Conical scan One wide swath	Ku band (13.4 GHz)
Seawinds on ADEOS II	2002/12–2003/10	25 km	12.5 km	Conical scan One wide swath	Ku band (13.4 GHz)
ASCAT-A	2006/10–present	50 km	12.5 km	Two sided Double swath	C band (5.3 GHz)
ASCAT-B	2012/10/29–present	50 km	12.5 km	Two sided Double swath	C band (5.3 GHz)
OCEANSAT2	2009/9/23–2014	25 km	25 km	Conical scan One wide swath	Ku band (13.5 GHz)
HY-2A	2011/9–present	25 km	25 km	Conical scan One wide swath	Ku band (13.256 GHz)
ISS RapidSCAT	2014/9/20–2016/8	25 km	12.5 km	Conical scan One wide swath	Ku band (13.4 GHz)

25 km and the satellite covered 90% of the ice-free ocean every 2 days under clear and cloudy conditions. The ADEOS satellites flew at 800 km altitude, with an inclination of 98.6° and an orbital period of 101 min and a repeat period of 41 days. Unfortunately, this satellite failed in July 1997 after sustained structural damage to the solar panel, ending the operation of NSCAT. The next major development in scatterometry was the SeaWinds instrument, which NASA built to replace NSCAT. SeaWinds first flew on QuikSCAT, a satellite built in a hurry to replace the NSCAT that flew only briefly on ADEOS. The satellite was built by Ball Aerospace using a common satellite bus they had available for many missions (Figure 1.162).

FIGURE 1.162   The QuikSCAT satellite.

QuikSCAT was launched on 19 June 1999, slightly fewer than 2 years after the failure of ADEOS. This rapid response satellite continued to operate until 2 October 2018. The SeaWinds instrument suffered an antenna rotation bearing failure on 23 November 2009, but the satellite continued to operate until it ran out of fuel almost a decade later. SeaWinds took a very different approach to measuring backscatter than any of the earlier scatterometers. The earlier instruments all used "fan-beam" measurements, whereas SeaWinds used a "pencil-beam" with a rotating dish antenna of about 1 m in diameter. The rotating dish has two center feeds and rotates at a rate of 18 revolutions per minute. The elevation or "look angles" of the two antennas are 40° and 46° relative to nadir. At the satellite orbital altitude of 803 km, the incidence angles of the pencil beams are 46° and 55°, respectively. The two beams are electrically polarized in the horizontal (perpendicular to the incidence plane) direction for the inner or 40° beam. The 46° outer beam is vertically polarized. The antenna beamwidths produce a surface footprint of approximately 26 km × 36 km (Figure 1.163).

One major advantage of the conical scanning SeaWinds is that it allows for a smaller instrument that can be easily accommodated on smaller spacecraft such as QuikSCAT. Another advantage of SeaWinds is that, because it measures at a constant incidence angle, there is no "nadir gap" in the swath coverage as there was for the fan beam scatterometers. Thus, pencil beam scatterometers offer an important improvement in terms of Earth coverage.

A consequence of the 18 rpm rotation of the dish is that each point within the 700 km swath is viewed from four different azimuth angles, twice by the outer beam looking forward and then aft, and twice by the inner beam in the same

FIGURE 1.163   SeaWinds measurement geometry.

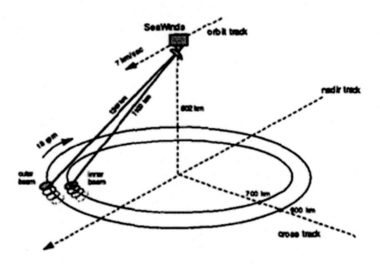

fashion. At the outside edge of the swath, the ocean is viewed twice by the outer beam only. Note that, unlike the fan-beam scatterometers, the azimuth angle "mix" of the radar cross-section (σ^0) measurements used in the wind retrieval is not constant but varies from nadir out to the edge of the swath. Thus, the wind retrieval performance of SeaWinds is observed to vary as a function of the distance from the nadir track. The optimal resolution is when the azimuth differences are near 90°. The radar has a 110-watt pulse at a frequency of 13.4 GHz and the overall instrument weighs 200 kg. The σ^0 measurements are continuous over a 1,800 km swath as the satellite progresses on its orbit. This wide orbit is such that the satellite covers 90% of the World Ocean within 24 h. Wind speeds between 3 and 30 m/s are estimated to have speed and direction accuracies of ~2 m/s and 20°, respectively.

The relative long life of QuikSCAT (designed for a 2-year mission) provided ocean vector wind products for a variety of important applications. One of these was in hurricane forecasting. Specifically, the U.S. Hurricane Forecast Office became very dependent on QuikSCAT winds for their forecasts. An example presented here is an image of the winds associated with Hurricane Katrina (Figure 1.164).

SeaWinds data were used for many other applications. One such study involved the mega-urbanization in Beijing, China that showed that the physical area of the city quadrupled between 2000 and 2009. Taking advantage of the fact that human structures produces stronger backscatter than soil or vegetation, the larger and taller the buildings are the greater the backscatter (Figure 1.165).

Another interesting application of QuikSCAT data was to the study of sea ice and icebergs, which was pioneered by David Long and his students at Brigham Young University (BYU). QuikSCAT identified an iceberg the size of Rhode Island known as B10A in its very first pass over Antarctica. B10A is about 38 km × 77 km and extends 90 m above the

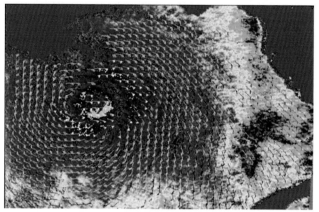

FIGURE 1.164 QuikSCAT imaged winds of Hurricane Katrina in the Gulf of Mexico in August 2005. Highest wind speeds are purple, and winds weaken outward from the eye. Barbs show wind direction; white barbs indicate heavy rainfall. *Image credit: NASA/JPL.*

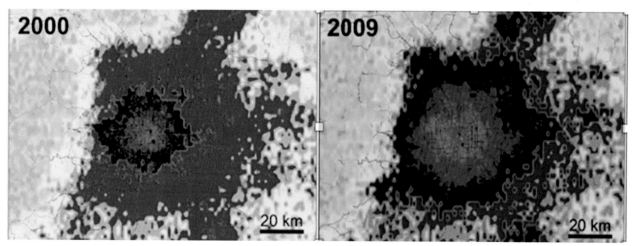

FIGURE 1.165 Data from QuikSCAT showing the changing extent of Beijing between 2000 and 2009 through changes to its infrastructure. *Image credit: NASA/JPL.*

water line. Its keel likely reaches to a depth of 300 m. The iceberg has posed a threat to shipping in the area but eventually broke into smaller chunks and melted. The BYU group has also studied sea ice around Antarctica. The greatest benefit of QuikSCAT was its ability to map the ocean vector winds over the ocean on a daily basis (Figure 1.166).

QuikSCAT launched in June 1999 and remained fully operational until November 2009 when the primary SeaWinds instrument antenna stopped rotating due to a bearing failure in the spin mechanism. The instrument continued to operate in a non-scanning mode until October 2018. Another SeaWinds instrument was supplied to Japan for inclusion on ADEOS-2, which unfortunately also failed after less than a year of operation. Still, the success of the initial Seasat SASS and the subsequent SeaWinds on QuikSCAT led to the launch and operation of a great many other scatterometers (Figure 1.167).

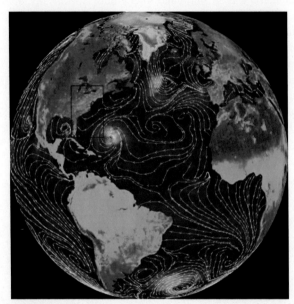

FIGURE 1.166 QuikSCAT winds on 20 September 1999. Orange areas are the strongest winds and blue areas are the relatively lightest winds. *Image credit: NASA.*

The virtual scatterometer constellation

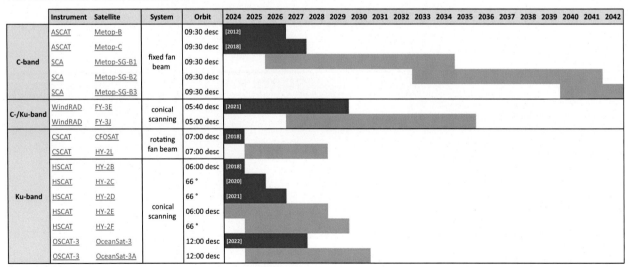

FIGURE 1.167 Overview of satellite missions with scatterometer instruments for global ocean wind vector observations. *Image credit: Committee on Earth Observation Satellites (CEOS).*

Wind speeds can be sensed by a number of active and passive microwave satellite instruments. The Special Sensor Microwave Imager (SSM/I) could measure wind speed over the ocean quite accurately. Satellite altimeters were also able to measure ocean wind speeds. These sensors, however, do not yield the important ocean wind vectors that were observed by the scatterometer. The U.S. Navy wanted a passive microwave instrument that could sense the ocean wind vector, which led to the creation of the Windsat instrument flying on the Coriolis satellite. Windsat is a polarimetric microwave radiometer designed to demonstrate the retrieval of wind speed and direction from the fully polarimetric measurements. Windsat uses a conically scanned 1.83 m offset parabolic reflector (Figure 1.168) with multiple feeds for the three fully polarized and two dual polarized channels (Table 1.27).

The 6.8 GHz channel is used to correct for sea surface temperature and the 23.8 GHz channel to correct for atmospheric moisture. Thus, wind speed and direction are computed from the 10.7, 18.7 and 37 GHz channels. The wind speed uses the 6.8 and 10.7 GHz channels while the wind direction uses the anisotropic feature of the 37 GHz channel together with the 10.7 GHz channel and the third and fourth stokes parameters of the 18.7 GHz channel. Wind direction is only possible in areas without precipitation and for wind speeds larger than 6−8 m/s (recall that the scatterometer covers the range from 3 to 30 m/s). Figure 1.169 is an example of the wind speeds and direction produced by the Japan Meteorological Agency for the region northeast of Japan.

Even before the advent of satellite wind measurements there was a need for wind observations over the ocean. A technique was developed to track clouds in sequential satellite imagery to infer cloud drift winds (CDW). These clouds could be identified in visible and infrared imagery and later in water vapor imagery. This meant that the CDW could be computed well before the microwave instruments were developed. The one big difference is that these CDW were not

FIGURE 1.168   Windsat Instrument with its large reflector and multiple feed horns.

TABLE 1.27 Windsat frequencies and specifications. *IFOV*, instantaneous field of view; NEΔτ(1), noise equivalent differential temperature (in Kelvin).

Freq (GHz)	Channels	Band width (MHz)	τ (ms)	NEΔτ(1)	Earth incidence Angle (deg)	IFOV (km)
6.8	V, H	125	5.00	0.48	53.5	40 × 60
10.7	V, H ±45, L, R	300	3.50	0.37	49.9	25 × 38
18.7	V, H ±45, L, R	750	2.00	0.39	55.3	16 × 27
23.8	V, H	500	1.48	0.55	53.0	12 × 20
37.0	V, H ±45, L, R	2,000	1.00	0.45	53.0	8 × 13

FIGURE 1.169 Windsat wind speeds and directions (wind barbs). The contours are of atmospheric pressure. *Image credit: Japan Meteorological Agency.*

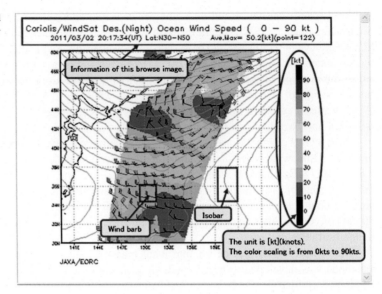

always at the sea surface and therefore did not represent wind stress over the ocean, in contrast to present day microwave instruments. Fortunately, the temperatures in the infrared clouds could be used to infer the altitude at which the CDW were observed so that the CDW could be incorporated into numerical forecast models.

The passive microwave sensors on the U.S. Defense Meteorological Satellite Program (DMSP) satellites, called the Special Sensor Microwave Imager (SSM/I), were able to sense wind speed but not direction (Figure 1.170). The SSM/I is a seven-channel four-frequency, linearly polarized microwave radar operating in a sun-synchronous orbit at an altitude of 860 km. Three of the four channels (19.3, 37.0, and 85.5 GHz) are dual polarized while the 22.2 GHz channel is only vertically polarized, for a total of seven channels. The nearly 1,400 km swath of the conically scanned SSM/I produces complete coverage between 89.833 S to 89.833 N every 3 days per satellite (Halpern et al., 1993). There were usually at least two SSM/I satellites in operation. While the spatial resolution is poor due to the sensing capabilities at the microwave frequencies, algorithms have been developed that appear to produce reliable estimates of wind speed over the open ocean (Wentz et al., 1986; Gooberlet et al., 1990; Halpern et al., 1993). Wind speed accuracies are about ±2 m/s for the range of speeds between 3 and 25 m/s under rain-free conditions. Because the emissivity of land is very different from that of water, the SSM/I cannot be used to estimate wind speed within 100 km of land. Similarly, surface wind speed within 200 km of

FIGURE 1.170 Global annual mean of the SSMI (Special Sensor Microwave Imager) surface wind speed for 1991. *Credit: JAXA/EORC.*

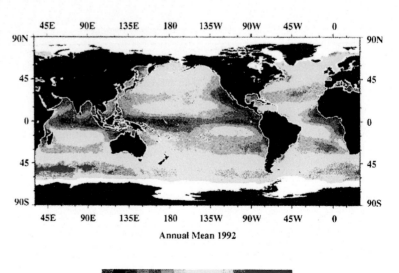

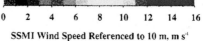

SSMI Wind Speed Referenced to 10 m, m s⁻¹

the ice edge cannot be computed from SSM/I data. However, wind speeds computed from the SSM/I compare reasonably well with open-ocean winds (Emery et al., 1994). Waliser and Gautier (1993) find that in the central and eastern equatorial Pacific, SSM/I wind-speed comparisons were well within the accuracies specified for the SSM/I. Biases (buoy − SSM/I) wind speeds were generally less than 1 m/s and RMS differences were less than 2 m/s. However, in the western equatorial Pacific, biases were generally greater than 1−3 m/s and RMS differences closer to 2−3 m/s. According to Waliser and Gautier "... there are still some difficulties to overcome in understanding the influences that local synoptic conditions (e.g., clouds/rainfall), and even background atmospheric and oceanic climatology effects, have on the retrieval of ocean-surface wind speeds from spaceborne sensors".

There are, have been and will be, many other satellite sensors operated in the microwave band that are able to sense wind speed such as: the Advanced Microwave Radiometer on the NASA/CNES Surface Water Ocean Topography (SWOT) mission, the Advanced Microwave Sounding Unit (AMSU) flying on the NASA/AQUA satellite and on NOAA polar orbiters, the Advanced Technology Microwave Sounder (ATMS) ON NOAA's Suomi National Polar-orbiting Partnership (SNPP) satellite, the GPM Microwave Imager (GMI) on the NASA/JAXA GPM satellite, the Advanced Microwave Radiometer (AMR) flying on the Jason 1, 2, 3 altimeter satellites of NASA/CNES, the MicroWave Radiometer on Argentina's Microwave Radiometer flying on the SAC-D satellite, the Microwave Imaging Radiometer using Aperture Synthesis (MIRAS) on ESA's Soil Moisture Ocean Salinity (SMOS) satellite, the Microwave Imaging/Sounding Radiometer (MTVZA-GY) flying on the Russian Meteor M and M1 meteorological polar-orbiting satellites, the MicroWave Radiation Imager (MWRI) flying on the Chinese FY-3 weather satellite, the MicroWave Temperature Sounder on the Chinese FY-3 weather satellite series, the OSCAT scatterometer on the ISROm (India) Oceansat-2 mission, the PolSCAT that flies on the NASA C-130 aircraft, SMAP that is an L-band scatterometer flying on NASA'a Soil Moisture Active Passive (SMAP) mission, the Scanning Multichannel Microwave Radiometer (SMMR) that first flew on Seasat and then the engineering model was upgraded to fly on the NIMBUS 7 mission, and the Tropical Rainfall Measurement Mission (TRMM) from NASA/JAXA.

In addition all satellite altimeters are able to measure wind speed by the shape of the waveform. Added to this is the possibility of using Synthetic Aperture Radars (SARs) to measure wind speed. As a result, there is a large volume of data available to map wind speed. Unfortunately, only scatterometer and specially equipped passive sensors are able to measure wind direction. In some cases, a SAR image can be used to infer wind direction. So, there are now and have recently been, a large number of satellite sensors that can measure wind speed and a fair number that can observe wind direction as well. Hopefully, there will be additional scatterometers flying in space in the future.

1.13.2 Reanalysis Meteorological Data

Many oceanographers make use of spatially gridded meteorological time series generated by a variety of governmental organizations throughout the world. Open access provision of reanalysis records began with the 40-year dataset (1957−96) presented by the US National Centers for Environmental Prediction (NCEP)/National Center for Atmospheric Research (NCAR) (Kalnay et al., 1996; Kistler et al., 2001). This was followed by further NCEP reanalysis products (Kanamitsu et al., 2002), the Climate Forecast System Reanalysis (Schneider et al., 2014), and the Modern-Era Retrospective analysis for Research and Applications Version 2 (Rienecker et al., 2011). Early descriptions of the weekly SST data can be found in Reynolds et al. (2002). At present, the US National Centers for Environmental Protection (NCEP) provides gridded meteorological fields from three main datasets: (1) the global, 2.5° × 2.5°, NCEP/NCAR 6-hourly Reanalysis-1 data from 1948 to present; (2) the global, 2.5° × 2.5°, NCEP/DOE 6-hourly Reanalysis-2 atmospheric fields from 1979 to present (the satellite era) at 2.5° spatial resolution; and (3) 32-km resolution, 3-hourly North American Regional Reanalysis (NARR) data, which is similar to (1) and (2) but with more snow, ice and precipitation products. Surface and near-surface (0.995 sigma level) reanalysis data include wind velocity, pressure, air temperature and relative humidity. https://www.ncdc.noaa.gov/data-access/model-data/model-datasets/reanalysis-1-reanalysis-2. In addition to the 6-hourly times series, reanalysis also include daily and monthly versions of the data series. There is also a caveat: "Please read problem list before using the data."

High quality global atmospheric reanalysis products are also available through the ERA5−Fifth generation of the European Centre for Medium Range Forecasts (ECMWF) (e.g., Hersbach et al., 2020), accessed through the Copernicus Climate Change Service Climate Data Store (CDS), the Japan Reanalysis (JRA-55) meteorological data and by regional reanalysis by Australia, Canada, China, India and others. ERA5 is based on a high-resolution forecast system with ¼ degree resolution and 1-h sampling from 1979 to present. An analysis of large-scale atmospheric pressure variations based on the global ERA-5 data is presented in Sakazaki and Hamilton (2020).

Gridded reanalysis fields are classified into three categories which define the relative contributions from the observations and model used to derive the variable. Wind velocity and SST are classified as type-A variables since they rely most heavily on the data and are considered the most reliable. Type-B variables, such as surface air temperature and relative humidity, are influenced by both observation and model, and are less reliable. The least reliable type-C data, which includes surface heat fluxes, are derived solely from the model. The main data sets used in all mixed layer depth models (e.g., Turner and Kraus, 1967; Gaspar, 1988; Thomson and Fine, 2009) are the short and long-wave radiation fluxes, the latent heat flux, and the sensible heat flux. Bulk models also require time series of precipitation and wind speed.

Reanalysis data have been subjected to several quality reviews. For example, the study by Kalnay et al. (1996) shows that monthly and annual mean heat fluxes from reanalysis agree favorably on a global scale with those derived from observational data sets; the study by Ladd and Bond (2002) indicates that the reanalysis shortwave radiation flux for the vicinity of Station "P" in the central northeast Pacific has a positive bias of roughly 20 W/m^2. Figure 1.171 provides a comparison between hourly winds observed at meteorological buoy C46206 moored on La Perouse Bank off the southwest coast of Vancouver Island and corresponding 3-hourly winds from a nearby NARR site for the period 29 April to 2 September 2012. The observed and reanalysis winds are in close agreement.

FIGURE 1.171 Comparison of observed and North American Regional Reanalysis (NARR) winds for the period April 29 to September 2, 2012, for the west coast of Vancouver Island, British Columbia. Wind vectors have been defined in terms of their northward (top two panels) and eastward components (bottom two panels). Inset shows location of the moored meteorological buoy C46206 (triangle) and location (solid square) of the grid point 49.0° N, 126.1° W for the reanalysis winds. Observed winds are from dual anemometers on the moored buoy; calculated winds are NCAR/NOAA. *Figure courtesy, Roy Hourston, Institute of Ocean Sciences (2013).*

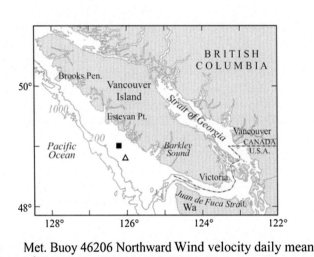

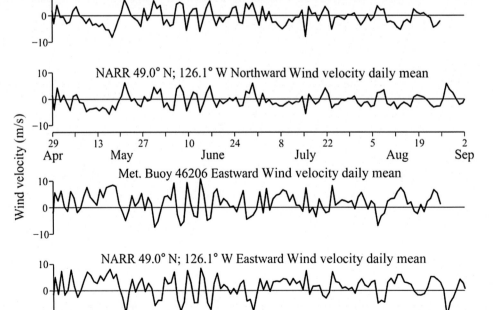

Some regions also benefit from dense local meteorological networks. For example, we have used a unique set of high-resolution 1 min sampled air pressure and wind velocity records from the Victoria School-Based Weather Station Network (herein, the "VS-network") of 171 meteorological stations located mainly in the southern part of Vancouver Island and on the Gulf Islands (www.victoriaweather.ca; Weaver and Wiebe, 2006). Additional atmospheric observations are provided at 1-min sampling by four USA Automated Surface Observing System (ASOS; https://mesonet.agron.iastate.edu/) weather stations (Bellingham, Friday Harbor, Everett and La Push) and three NOAA stations (Port Angeles) in Washington State.

Reanalysis products are likely to be least accurate in coastal regions due to limited spatial resolution and local effects such as hilly or mountainous terrain. In a study of a major storm surge event in the southern Strait of Georgia within the Salish Sea on the west coast of Canada (Fine and Thomson, 2021), the authors needed to force the numerical model with the best available coastal wind and pressure fields. The meteorological products with the highest spatial resolution (2.5 km) were the regional 1-h gridded data from the Environment Canada High Resolution Deterministic Prediction System (HRDPS) and available for the area since 2014. Gridded wind velocities and air pressure were also obtained from the NARR and ERA5 1-hourly reanalysis. To determine the quality of the different atmospheric forcing fields, the reanalysis wind velocity data were compared to hourly winds observed over the period 1 January 2014 to 31 December 2019 at nine meteorological buoys along the coast of Vancouver Island and Washington State (Figure 1.172). Results of the vector regression analysis are shown in Table 1.28).

The next step was to validate the storm surge model and to examine the effect of using different atmospheric forcing fields. To do this, the authors chose the historically high storm surge of December 20, 2018, in the Salish Sea (separating British Columbia from Washington State) as a test case. Model runs were undertaken with NARR, ERA5 and HRDPS forcing and modeled storm surge heights and timing compared with the storm extracted from 21 tide coastal gauge records. The main results are:

1. The model results using the HRDPS forcing outperformed that of the two other reanalysis datasets by providing the best fit (in a least squares sense) to the observed surge at almost all tide gauge stations (not shown).
2. The model results using the ERA5 forcing provided results that were very close to those of the much higher resolution HRDPS (Table 1.28). The differences between model runs were less than the differences between each run and the observations. Thus, the ERA5 dataset is acceptable for the retrospective storm surge modeling.
3. Model runs with the NARR forcing were unacceptable for modeling inside the Salish Sea because of the large spatial grid scales of the first and second generation NCEP reanalysis fields (Kistler et al., 2001).

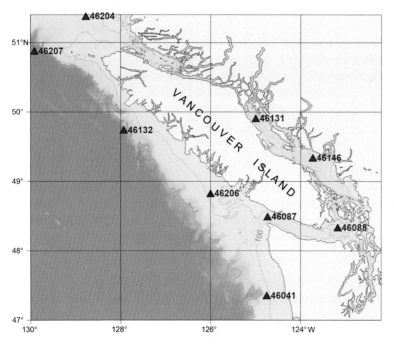

FIGURE 1.172 Meteorological buoys on the west coast of Vancouver Island (British Columbia) and northern Washington State used in the correlation analysis with nearby reanalysis data (see Table 1.28). *Courtesy Isaac Fine, Institute of Ocean Sciences.*

TABLE 1.28 Regression of the hourly wind velocity observed at nine meteorological buoys off the coast of Vancouver Island (British Columbia) (see Figure 1.172) versus hourly wind velocity from the nearest grid site for three reanalysis datasets covering the period 2014–19. HRDPS denotes winds from the Canadian High Deterministic Resolution Prediction System, ERA5 is the European Reanalysis data and NARR denotes the North American Regional Reanalysis from the U.S. National Centers for Environmental Information; $\rho_{1,2}$ = correlation coefficient, θ (Theta) is the angle (in °) between the wind vectors and skill is the observed variance in the U, V horizontal wind velocity components explained by a particular reanalysis dataset. Skill = 1 indicates perfect skill (all variance accounted for).

Buoy	HRDPS			ERA5			NARR		
	$\rho_{1,2}$	Theta	Skill	$\rho_{1,2}$	Theta	Skill	$\rho_{1,2}$	Theta	Skill
46041	0.92	8.5	0.79	0.94	8.3	0.83	0.89	2.7	0.79
46087	0.87	4.7	0.74	0.86	0.4	0.70	0.75	1.9	0.50
46206	0.86	7.0	0.66	0.88	5.2	0.72	0.79	−2.8	0.63
46088	0.85	4.3	0.69	0.85	0.8	0.73	0.57	−2.6	0.26
46146	0.82	0.0	0.62	0.83	−1.8	0.64	0.59	19.4	0.28
46131	0.84	4.0	0.66	0.83	−5.9	0.53	0.72	−13.0	0.41
46132	0.90	17.1	0.70	0.93	16.2	0.77	0.88	8.6	0.74
46207	0.87	6.9	0.71	0.89	7.0	0.77	0.84	2.1	0.69
46204	0.92	−2.0	0.84	0.95	−3.3	0.90	0.89	−9.4	0.77

Data courtesy of Isaac Fine, Institute of Ocean Sciences.

1.14 PRECIPITATION

Precipitation is one of the most difficult and challenging measurements to make over the ocean. Simple rain gauges installed on ships are invariably affected by the pitch and roll of the ship, by salt spray and by wind flow over the ship's hull and superstructure. The short space and time scales of precipitation make it difficult to interpret point measurements. Rain gauges have two conflicting requirements that make use on shipboard difficult. First the gauge needs to be installed away from the ship influences, such as salt spray, which calls for positioning as high as possible on a mast. However, this conflicts directly with the second requirement, which calls for the regular maintenance of the gauge by ship's personnel. Few systematic studies have been made of precipitation measurements taken from ships, and little effort is made today to instrument ships to routinely observe rainfall over the ocean. A 25-year time series from Ocean Station "P" (Figure 1.173) in the northeast Pacific is one of a few in the open ocean (most others were taken at Ocean Weather Stations similar to Station P). Specialized rain gauges have been developed for use at sea, but they are easily damaged or stolen when mounted on buoys on the ocean surface.

One new technique is to infer rainfall from variations in the upper-ocean acoustic noise. While it may seem a bit confusing to interpret ocean upper-layer acoustic noise both in terms of rainfall and wind, the frequency signatures of the two noise-generating mechanisms are sufficiently different as to be distinguishable. The unique characteristics of the sounds produced by different kinds of rainfall make it possible to use Acoustic Rain Gauges (ARGs) to identify and measure raindrop size, fall rate, and other properties of rainfall over the ocean. This area of the ocean surface contributing to the sound recorded by the ARG increases as the gauge is placed deeper in the ocean. Because rainfall can vary markedly over short distances, this spatially integrated rain measurement is considered preferable to a point measurement. Heavy rain can increase noise levels by up to 35 dB over frequencies ranging from roughly 1 kHz to greater than 50 kHz. Extreme rain events produce loud signals that can reach as much as 50 dB above the background noise. Individual raindrops create underwater sound in two ways. The first sound is made by the impact of the raindrop hitting the ocean surface. Following the initial impact, sound can radiate from air bubbles trapped under water during the splash. These bubbles generally produce the louder sound. Raindrops of different sizes produce different sound intensity and frequency. Small raindrops (0.8–1.2 mm diameter) are surprisingly loud because they generate bubbles with every splash. Sound from these drops has frequencies between 13 and 25 kHz. Medium raindrops (1.2–2.0 mm) do not generate bubbles and are therefore

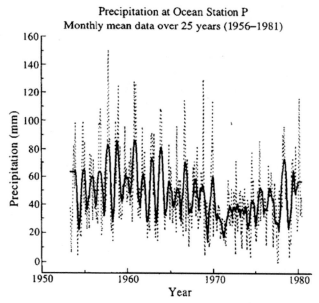

FIGURE 1.173   A 25 year time-series (1956–81) of precipitation collected from Canadian Weather Ships at Ocean Station PAPA (50° N, 145° W) in the northeast Pacific. Solid line is from use of a Savitsky-Golay smoother (order = 13 months). *Courtesy: Isaac Fine, Institute of Ocean Sciences.*

remarkably quiet. Large (2.0–3.5 mm) and very large (>3.5 mm) raindrops trap larger bubbles, which can produce frequencies as low as 1 kHz. These measurements can only be made from a mooring away from ship influences.

Satellite techniques that relate the area of cold clouds to surface rainfall have had some success (Joyce and Arkin, 1997). However, infrared (IR) techniques underestimate warm rain, and lead to frequent false readings for certain anvil and thick cirrus clouds with cold IR brightness temperatures. A commonly used rainfall infrared product is the Global Precipitation Index (GPI); an algorithm that was developed by Joyce and Arkin (1997). The GPI overestimates rainfall from the large areas of cold clouds that form over the Western Pacific Ocean (Liu et al., 2007) and underestimates certain other kinds of rainfall, but appears to do quite well in other selected regimes.

The 1987 launch of the Special Sensor Microwave Imager (SSM/I) on one of the Defense Meteorological Satellite Program satellites provided a new opportunity to infer precipitation from microwave satellite measurements. While a precipitation algorithm was developed prior to the launch (Hollinger, 1989), later studies have improved upon this algorithm to formulate better retrievals of precipitation over both land and ocean. In the list of "environmental products" for the SSM/I, the "precipitation over water" field shows a 25 km resolution, a range of 0–80 mm/h, an absolute accuracy of 5 mm/h for quantization levels of 0, 5, 10, 15, 20, and ≥25 mm/h. This algorithm utilized both the 85.5 and 37 GHz SSM/I channels, thus limiting the spatial resolution to the 25 km spot sizes of the 37 GHz channel.

A study by Spencer et al. (1989) employed only the two different polarizations of the 85.5 GHz channel, thus allowing the resolution to improve to the 12.5 km per spot size of this channel. This algorithm was compared with 15 min rain gauge data from a squall system in the southeast United States (Spencer et al., 1989). The 0.01 inch (0.039 mm) rain gauge data were found to correlate well ($r^2 = 0.7$) with the SSM/I polarization corrected 85.5 GHz brightness temperatures. This correlation is surprisingly high considering the difference in the sampling characteristics of the SSM/I versus the rain gauge data. Portions of a rain system adjacent to the squall line were found to have little or no scattering signature in either the 85.5 or the 37 GHz SSM/I data due likely to the lack of an ice phase presence in the target area. This appears to be a limitation of the passive microwave methods to discern warm rain over land. Microwave sensors and precipitation radar have been used increasingly in recent years, with the potential to improve precipitation estimates from the surface and from space. Unlike IR, these techniques directly sense precipitation particles rather than cloud tops. However, significant difficulties remain. Rainfall retrieved from the Tropical Rainfall Measuring Mission (TRMM) precipitation radar suffers from uncertain attenuation correction, problems over complex terrain, and the limit of minimum detectable signal (Iguchi et al., 2000). Microwave retrievals over the ocean are thought to rival radar retrievals for accuracy, but retrievals over land are compromised because of variations of the surface emissivity (Spencer et al., 1989; Kummerow et al., 2001). The Tropical Rainfall Measuring Mission has led to a significant advancement in the quantification of moderate to intense rainfall over the ocean. Despite this success, current rain measuring sensors lack sufficient sensitivity and retrieval difficulties to detect and estimate light rainfall, especially over subtropical and high latitude oceans. Among various space-borne sensors, CloudSat provides superior retrieval of light rainfall and drizzle. By complementing rain estimates from CloudSat and precipitation radar aboard TRMM, it has been determined that the quasi-global (60° S to 60° N latitude) mean oceanic rain rate is about 3.05 mm/day, considerably larger than that obtained from any individual sensor product. Together with careful consideration of scaling issues, rainfall estimates from TRMM PR, TRMM TMI, AMSR-E, MHS, IR, and CloudSat CPR sensors have been analyzed. Results show that the highest agreement among sensors in measuring the frequency and amount of rain occurs in the zone between 20° S and 20° N. However, toward higher latitudes and within the subtropical high-pressure regions, a majority of the sensors miss a significant fraction of the rainfall. This underestimation can exceed 50% of the total rain volume. The dual frequency precipitation radar (DPR), with Ka/Ku-bands and a multi-channel passive microwave radiometer on the Global Precipitation Mission (GPM) "core" satellite extends the TRMM capability to measure light rain over 65° S to 65° N, creating an unprecedented opportunity to improve global quantification and properties of precipitation.

The TRMM satellite, launched in 1997, used both passive and active microwave instruments to measure rainfall in the tropics. It demonstrated the importance of making observations in a non-sun-synchronous orbit at different times of day interlaced with polar orbiting sensors at fixed times of day to improve near real-time monitoring of hurricanes and give accurate estimates of total rainfall accumulation. The TRMM satellite was designed to collect 3 years of data and it operated for 17 years until it ran out of fuel to stabilize its orbit and it dropped out of orbit. Building on the success of TRMM, the Global Precipitation Mission of the U.S. (NASA) and Japan (JAXA) extended the GPM core satellite carried instruments similar to TRMM but in a polar orbit as part of the GPM satellite constellation. In addition to the GPM core satellite this constellation incorporates the SuomiNPP satellite, the NOAA 18/19/20 satellites, the MetOp A/B/C satellites the DMSP F17/F18 satellites, the GCOM-W1 satellite and the Megha-Tropiques satellite (Figure 1.174).

The GPM core satellite carries the first space-borne Ku/Ka-band Dual-frequency Precipitation Radar (DPR) and a multi-channel GPM Microwave Imager (GMI) as shown in Figure 1.175. The DPR instrument has a Ka-band precipitation radar (KiPR) that operates at 35.5 GHz and a Ku-band precipitation radar (KuPR) operating at 13.6 GHz. Together these channels make it possible for the DPR to provide three-dimensional measurements of precipitation structure and characteristics. The DPR began operation collecting data over a swath of 125 and 245 km for the Ka and Ku band radars, respectively. Since May 2018 both radars have swaths of 245 km. As compared with the TRMM data the DPR is more sensitive to light rain rates and snowfall. In addition, by combining the overlapping Ka/Ku-bands it is possible to derive new information about the particle drop size distributions for moderate rainfall intensities. The DPR is also expected to give further insight into how precipitation processes are affected by human activities.

The GMI is a conically scanning multi-channel microwave radiometer (Figure 1.175) with a swath width of 885 km and having 13 channels between 10 and 183 GHz (Table 1.29).

The components of the GMI are the: (a) instrument spacecraft structure; (b) spin mechanism assembly; (c) slip-ring assembly; (d) instrument base structure; (e) hot load tray (keeps the sun off the hot load); (f) feed horns; (g) hot load

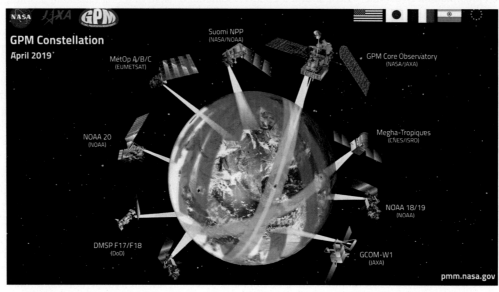

FIGURE 1.174   The Global Precipitation Mission (GPM) constellation of precipitation observing satellites.

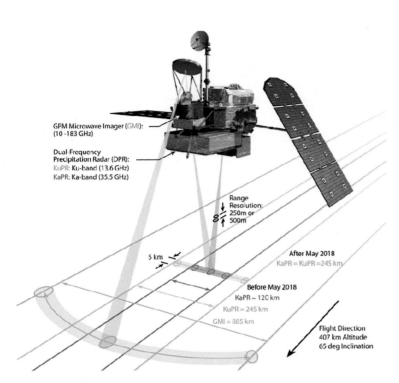

GPM Microwave Imager (GMI):
(10 -183 GHz)

Dual-Frequency
Precipitation Radar (DPR):
KuPR: Ku-band (13.6 GHz)
KaPR: Ka-band (35.5 GHz)

Range
Resolution:
250m or
500m

5 km

After May 2018
KaPR = KuPR = 245 km

Before May 2018
KaPR = 120 km
KuPR = 245 km
GMI = 885 km

Flight Direction
407 km Altitude
65 deg Inclination

FIGURE 1.175 The Global Precipitation Mission (GPM) Core Satellite showing the two main instruments of the Dual-frequency Precipitation Radar (DPR) and multi-channel GPM Microwave Imager (GMI).

TABLE 1.29 Band specifics for the Global Precipitation Mission (GPM) Microwave Imager (GMI).

No.	Central freq (GHz)	Bandwidth (MHz)	Polarization	NE$\Delta\tau$ (K)	IFOV km × km	Pixel km × km
1	10.65	100	V	0.96	19 × 32	12 × 13.4
2	10.65	100	H	0.96	19 × 32	12 × 13.4
3	18.7	200	V	0.84	11 × 18	6 × 13.4
4	18.7	200	H	0.84	11 × 18	6 × 13.4
5	23.8	400	V	1.05	9.2 × 15	6 × 13.4
6	36.5	1,000	V	0.65	8.6 × 14	6 × 13.4
7	36.5	1,000	H	0.65	8.6 × 14	6 × 13.4
8	89.0	6,000	V	0.57	4.4 × 7.2	3.0 × 13.4
9	89.0	6,000	H	0.57	4.4 × 7.2	3.0 × 13.4
10	166.5	4,000	V	1.5	4.4 × 7.2	3.0 × 13.4
11	166.5	4,000	H	1.5	4.4 × 7.2	3.0 × 13.4
12	183 ± 3	2,000	V	1.5	4.4 × 7.2	3.0 × 13.4
13	183 ± 7	2,000	V	1.5	4.4 × 7.2	3.0 × 13.4

IFOV, Instantaneous Field of View; *NE$\Delta\tau$*, noise equivalent differential temperature (in Kelvin).
NASA Global Precipitation Measurement Mission.

(GMI calibration uses the temperature of the hot load reference, the cold sky reflector and the noise diodes); (h) de-spin assembly; (i) cold sky reflector; (j) launch locks; (k) RF receivers; (l) wave-guides, noise diodes, mixer-preamplifiers, RF coaxial cables; (m) sunshade and struts; and (n) main reflector (and its support structure), and the instrument control assembly (ICA).

NASA provides an interesting pictorial to envision the applications of the various GMI channels (Figure 1.176).

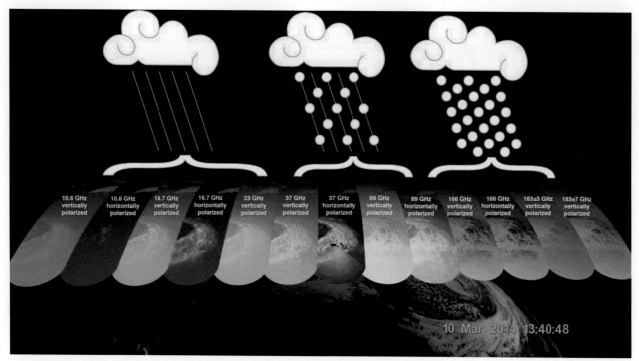

FIGURE 1.176 Applications of the 13 GMI (GPM Microwave Imager) channels. *NASA Global Precipitation Measurement Mission.*

In this depiction, the 5 channels on the left are sensitive to heavy and moderate rainfall while the four in the middle sense mixtures of precipitation with snow and ice within clouds. Such mixtures are generally the result of ice or snow melting into rain as it falls. Finally, the four channels on the right sense both water vapor and snowfall. Overall, the frequencies get larger as you go to the right. As seen in Table 1.29, most channels are both vertically and horizontally polarized.

In the GPM constellation, the GMI data are used as a reference standard for all of the other instruments in the constellation. The data from the GPM core satellite and the partner satellites are merged together into a single global precipitation dataset called IMERG, which is updated every 3 h. A half-hourly image is shown in Figure 1.177.

As of 2020, NASA and JAXA have collected 22 years of rain and snowfall data from space. Interested scientists can access these data from IMERG (Integrated Multi-satellite Retrievals for GPM) and it is now possible to merge the new GPM data with the earlier TRMM precipitation data. Applications of these data include ecology, water and agriculture, energy, natural disasters, health and, most obviously, weather. We will consider only a few examples.

1.14.1 Hurricane Irma

At 1 pm EDT (1700 UTC) on September 5, 2017, the radar on the GPM core satellite captured a 3-D view of the heat engine inside the category-5 hurricane "Irma" (Figure 1.178). Under the central ring of clouds that circles the eye, water that had evaporated from the sea surface condenses, releasing heat that powers the circling winds of the hurricane. The GPM radar was able to estimate how much water is falling as precipitation inside the hurricane, which gives a guide as to how much energy is being released inside the hurricane's central heat engine. This energy then radiates out into the outer bands of the hurricane from the central core as it weakens. It is the warm sea surface temperatures that feed energy into hurricanes. This is the reason that hurricanes tend to follow the pattern of maximum sea surface temperature as the pattern moves.

1.14.2 Hurricane Harvey

Hurricane "Harvey" deposited a large volume of rain on Texas, causing major flooding and creating a general disaster. At the time of Figure 1.179, the storm was over Austin and Houston, and later extended further into eastern Texas. During this

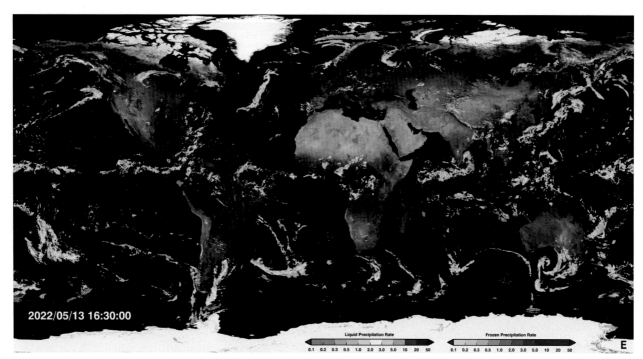

FIGURE 1.177 GPM image of global precipitation from a constellation of satellites. *NASA Global Precipitation Measurement Mission.*

FIGURE 1.178 A 3-D view of Hurricane Irma from the radar on the core GPM satellite.

period, the storm was slowly meandering southeastward at only 4 km/h. This image depicts rain rates as measured by the GMI and the GPM DPR radar. GPM shows that the rain pattern is highly asymmetric, with the majority of the rain being north and east of the hurricane center. A broad area of modest rain can be seen stretching from Galveston Bay to just north of Houston. Within this band are embedded pockets of heavy rain (red areas). The GPM measured rain rate in the heavy precipitation areas was estimated to be 96 mm/h. Since at this time Harvey still extended out over the Gulf of Mexico, the storm was able to draw in a continuous supply of heat from the ocean and warm moist air from the atmosphere to sustain the large amounts of rain it was producing.

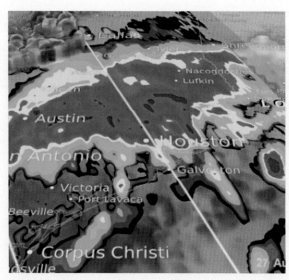

FIGURE 1.179 Rainfall in Texas as seen by the GMI on the GPM core satellite at 11:45 UTC, 27 August 2017.

1.14.3 Hurricane Zeta

As Hurricane "Zeta" moved toward landfall on the U.S. Gulf Coast, NASA monitored the storm with a variety of Earth-observing satellite instruments (Figure 1.180) and stood ready to aid affected communities with satellite data and analyses. Zeta followed a path similar to that of the earlier Hurricane Delta, which crossed the Yucatan Peninsula and then made its way across the Gulf of Mexico going onshore at the Louisiana coast as a category 2 hurricane on 9 October 2020.

1.14.4 Atmospheric River Brings Heavy Rainfall to Central California

"Atmospheric rivers" are long, narrow bands of water vapor that cause heavy precipitation when they encounter land. Using the GPM constellation of satellites, NASA was able to document an atmospheric river event that brought heavy rain to the U.S. West Coast in January 2021 (Figure 1.181). Image documentation was at half-hourly steps from 25 to 29 January. Authorities estimated that much of central California received over 2 inches (50 mm) of rainfall during this event, with areas south of San Francisco, particularly the Big Sur region, receiving up to 14 inches (356 mm) at high elevations.

FIGURE 1.180 Hurricane Zeta in the Gulf of Mexico 28 October 2020. When Zeta made landfall on the northern Gulf Coast of the United States, it became the 7th named storm to do so in the record-breaking hurricane season of 2020, following a path taken earlier by the Tropical Storm Cristobal.

FIGURE 1.181 Satellite image of the "atmospheric river" that was soaking Central California on 29 January 2021. Blue color indicates regions with high rain rates. *Source: https://gpm.nasa.gov/applications/weather/.*

1.15 CHEMICAL TRACERS

Oceanographers use a variety of chemical substances to track diffusive and advective processes in the ocean. Understanding these processes leads to insight into the oceanic uptake of anthropogenically generated heat and CO_2, variations in open ocean circulation, and changes in the global climate and marine ecosystems. Chemical tracers are also used to validate ocean circulation models. Tracers used in this context include tritium, chlorofluorocarbons (CFCs), natural and bomb-produced radiocarbon, as well as dissolved oxygen, silicate, phosphate, isotopes of organic and inorganic carbon compounds, and some noble gases (e.g., helium and argon). Information from chemical tracers provides information that cannot be obtained from temperature-salinity alone, so that determining a model's ability to capture chemical tracer patterns is important. Natural chemical tracers such as isotopes of carbon, argon, and oxygen are useful for examining old water masses, such as those in the North Pacific and the Circumpolar Deep Water. Transient tracers, such as tritium, chlorofluorocarbons, and bomb-produced carbon-14, ^{14}C, are best suited for analyzing circulation over decadal timescales, including thermocline ventilation, the renewal of Antarctic Intermediate Water, and the ventilation pathways of North Atlantic Deep Water and Antarctic Bottom Water. Tracer model studies have helped to reveal inadequacies in the model representation of certain water mass formation processes, for example, convection, downslope flows, and deep ocean currents.

Chemical tracers can be divided into two primary categories: *conservative* tracers such as salt and helium whose concentrations are affected only by mixing and diffusion processes in the marine environment; and *nonconservative* tracers such as dissolved oxygen, silicate, iron, and manganese whose concentrations are modified by chemical and biological processes, as well as by mixing and diffusion. The *conventional* tracers, temperature, salinity, dissolved oxygen and nutrients (nitrate, phosphate and silicate), have been used since the days of Wüst (1935) and Defant (1936) to study ocean circulation. More recently, *radioactive* tracers such as radiocarbon (^{14}C) and tritium (^3H) also are being used to study oceanic motions and water mass distribution. The observed concentrations of those substances, which enter from the atmosphere, must first be corrected for natural radioactive decay and estimates made of these substance's atmospheric distributions prior to their entering the ocean. If these radioactive materials decay to a stable daughter isotope, the ratio of the radioactive element to the stable product can be used to determine the time that the tracer was last exposed to the atmosphere. *Transient tracers*, which we will consider separately, are chemicals added to the ocean by anthropogenic sources in a short time span over a limited spatial region. Most transient tracers presently in use are radioactive. What is important to the physical oceanographer is that chemical substances that enter the ocean from the atmosphere or through the seafloor provide valuable information on a wide spectrum of oceanographic processes ranging from the ventilation of the bottom water masses, to the rate of isopycnal and diapycnal (cross-isopycnal) mixing and diffusion, to the downstream evolution of effluent plumes emanating from hydrothermal vent sites.

Until recently, many of these parameters required the collection and post-cast analysis of water bottle samples using large samples, which are then subsampled and analyzed by various types of chemical procedures. There are excellent reference books presently available that describe in detail these methods and their associated problems (e.g., Grasshoff et al., 1983; Parsons et al., 1984). The book by Grasshof et al. (1983) also contains an excellent section on water samples and their application to chemical analyses. There are important concerns for the reliability of the chemical measurements regarding contamination of the sampling bottle or the subsampling procedure. Also, the volumes required for different chemical analyses vary greatly. A list of sample volumes for chemical observations as part of WOCE can be found in Volume I of the WOCE Implementation Plan (WOCE, 1988). It is certain that the collection of these volumes will include both presently available "off-the-shelf" samplers, sampling systems newly developed by private companies and sampling units designed and built by scientists." In any case, the precision and accuracy of these measurements depend, in part, on the sampling technique used.

Modern chemical "sniffers" (or chemical pumps) are being developed that allow for *in situ* analysis of samples (Lupton et al., 1993). The requirement for *in situ* chemical sampling of hydrothermal vents leads to the development of the submersible chemical analyzer (SCANNER) for analyses of Mn and Fe, the submersible system used to assess vented emissions (SUAVE) for Mn, Fe, Si, H_2S, and one of PO_4 or Cl. This requirement also lead to the development of the zero-angle photon spectrometer (ZAPS) for detecting dissolved Mn concentration ratio to ambient seawater concentrations (≤ 1 nmol/l) (Lilley et al., 1995). The SCANNER and SUAVE systems comprise online colorimetric chemical detectors while ZAPS is a fiber-optic spectrometer, which combines solid-state chemistry with photomultiplier tube detection to make flow through *in situ* chemical measurements. Recent publications on the use of modern chemical sensors at hydrothermal vents can be found in the American Geophysical Union monograph "Mid-Ocean Ridges: Hydrothermal interactions between the lithosphere and oceans" (German et al., 2004) and the RIDGE 2000, Special Issue of *Oceanography* (Fornari et al., 2012).

For many chemical measurements, no single set of procedures applies so that groups, or individual scientists, must be responsible for their own data quality. It is impossible to evaluate after-the-fact the influences of sampling technique, sample history (storage, etc.) and analysis technique. It is therefore more difficult to attach levels of accuracy to these diverse methods. In this text, we will make some general comments regarding potential problems for each of the important parameters. For a more extensive discussion of chemical tracers, the reader is referred to Broecker and Peng (1982) and Charnock et al. (1988). A review on the contribution to the ocean carbon cycle of chemoautotrophic production, as well as secondary production and respiration from meso-zooplankton and micro-nekton below 400 m depth, is provided in Burd and Thomson (2022).

1.15.1 Conventional Tracers

1.15.1.1 Temperature and Salinity

If it were not for large-scale geographical differences in heat and buoyancy fluxes through the ocean surface from the overlying atmosphere, ocean temperatures and salinity would be nearly homogeneous, disrupted only by input from geothermal heating through the seafloor (Warren, 1970; Jenkins et al., 1978; Reid, 1982). In fact, below 1,500 m depth the salinity range throughout the World Ocean is only about 0.5 g/kg despite the regular deep-water formation at high latitudes (Warren, 1983). Temperature, salinity, and density distributions enable us to identify different water masses and track the movement of these water masses in the World Oceans.

Atlases of temperature and salinity for the Atlantic Ocean were produced by Wüst (1935) and Defant (1936) using data from the 1925–27 *Meteor Expedition.* These maps help define the depths of vertical mixing and upwelling in the upper ocean and reveal the extent of ventilation of deep and intermediate waters by sinking of cold, high salinity, high-density water from the Southern Ocean and the Labrador Sea.

Updated atlases for the Atlantic were presented in Fuglister (1960) and Worthington (1976). Similar maps for the Pacific Ocean were produced by Reid (1965) and Barkley (1968). Reid's atlas included distributions of dissolved oxygen and phosphate/phosphorous. An atlas of water properties for the North Pacific, was presented by Dodimead et al. (1963) and Favorite et al. (1976). Wyrtki (1971) provided conventional tracer data for the Indian Ocean obtained from the International Indian Ocean Expedition. An atlas of the Bering Sea is provided by Sayles et al. (1979). A summary of the global water mass distribution can be found in Emery and Meincke (1986). Surveys conducted during the World Ocean Circulation Experiment (1991–97) provide updated maps of conventional tracer distributions in the global ocean. As discussed in Section 1.10.1, the international Argo program is presently yielding unprecedented volumes of high-resolution temperature and salinity data within the upper 2,000 m of the World Ocean. Experimentalists and numerical modelers alike have enthusiastically endorsed these global datasets.

1.15.1.2 Dissolved Oxygen

Along with temperature and salinity, dissolved oxygen concentration is considered one of the primary scalar properties needed to characterize the physical attributes of marine and freshwater environments. Although it is not usually a conservative quantity, dissolved oxygen serves as a valuable tracer for mixing and ventilation throughout the water column and is a key index of water quality in regions of strong biological oxygen demand. This demand may arise from animal respiration, bacteria-driven decay, or nonorganic chemical processes (the discharge of pulp-mill effluent into the marine environment places a heavy burden on oxygen levels). Dissolved oxygen is widely used by physical oceanographers to delineate water-mass distributions, to estimate the timing and intensity of coastal upwelling processes and to establish the occurrence of deep-water renewal events in coastal fjords. In a study of the North Pacific, Reid and Mantyla (1978) found that dissolved oxygen gives the clearest signal of the subarctic cyclonic gyre in the deep ocean. Profiles of dissolved oxygen collected in the spring of 2014 immediately before and after a pronounced cooling event recorded at ~4,500 m depth in North Pond—a small (8 km × 15 km) basement depression on the western flank of the mid-Atlantic Ridge near ~23° N—helped support the idea that the events are intrusions of highly diluted Antarctic Bottom Water originating from the Southern Ocean (Becker et al., 2021).

The apparent oxygen utilization (AOU) is the difference between the possible saturated oxygen content at a given pressure and temperature, and the actually observed oxygen content (Figure 1.182). Specifically. AOU = [O_2 solubility] − [O_2 observed] at the *in situ* temperature and pressure. Below the euphotic zone, this parameter provides an approximate measure of biological demand due to respiration and decay. It also is commonly used to trace water-mass movement and to determine the "age" (defined as the time away from exposure to the surface source) of oceanic water masses. Use of AOU suggests that the intermediate waters of the northeast Pacific have an age of several thousand years and are among the oldest (last to be ventilated) waters of the World Ocean. Mantyla and Reid (1983) arrived at similar conclusions based on global distributions of potential temperature, salinity, oxygen and silicate. A more complete discussion of this parameter can be found in Chapter 3 of Broecker and Peng (1982).

The "core-layer" method introduced by Wüst (1935) identified water masses, and their boundaries, on the basis of maxima or minima in temperature, salinity and dissolved oxygen content. In the ocean, dissolved oxygen levels are high

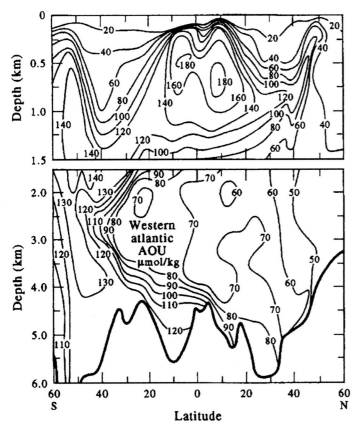

FIGURE 1.182 Vertical section of Apparent Oxygen Utilization (AOU) in mol/kg for the western basin of the Atlantic Ocean. The section is broken at 1,500 m depth. *Figure 3.9 from GEOSECS program, Broecker and Peng (1982).*

near the surface where they contact the atmosphere but rapidly diminish to a minimum near 500−1,000 m depth due to the decay of upper-ocean detritus. Oxygen values again increase with depth toward the bottom. Wyrtki (1962) discusses the relationship between the observed subsurface oxygen minimum in the North Pacific and the general circulation of the ocean, suggesting that it is a balance between upward advection, downward diffusion and *in situ* biological/chemical consumption. Miyake and Saruhashi (1967) argued that the effect of horizontal advection has a much greater effect on dissolved oxygen distributions than horizontal diffusion and biological consumption. In certain deep regions of the ocean, such as the Weddell Enderby Basin off Antarctica, the consumption of oxygen is below the detectable limit of the data so that oxygen may serve as a conservative chemical tracer (Edmond et al., 1979). Within coastal regions, narrow fjords often have one or more shallow cross-channel sills that greatly limit the exchange of offshore oceanic water with the bottom water in deep adjoining basins. The suboxic to anoxic conditions found in these inner basins of inlets are indicative of weak vertical mixing (stagnant bottom water) and infrequent dense water intrusions. In many of these deep anoxic basins, high levels of dissolved hydrogen sulfide (H_2S) can develop. Examples include Koljöfjord and Byfjord in western Sweden (Hansson et al., 2013) and Effingham Inlet on the west coast of Canada (Dallimore et al., 2005; Patterson et al., 2007, 2013).

When water bottle sampling was the only method for oceanographic profiling, the measurement of dissolved oxygen was only slightly more cumbersome and time consuming than the measurement of temperature and salinity. The advent of the modern CTD with its rapid temperature and conductivity responses left oxygen sampling behind. Thus, despite the importance of dissolved oxygen distributions to our understanding of chemical processes and biological consumption in the ocean, dissolved oxygen is far less widely observed than temperature or salinity. At present, there are three principal methods for measurement of dissolved oxygen: (1) water bottle sampling followed by chemical "pickling" and endpoint titration using the Winkler method (Strickland and Parsons, 1968; Hichman, 1978); (2) electronic sampling using a membrane covered polarographic "Clark" cell (Langdon, 1984); and (3) electronic sampling using luminescence-quenching technology (Thomson et al., 1988; Tengberg et al., 2006). The primary problems with standard water-bottle sampling of dissolved oxygen are the potential for sample contamination by the ambient air when the subsampling is carried out on deck, poor sampling procedure (such as inadequate rinsing of the sample bottles), and the oxidization effects caused by sunlight on the sample. Thus, laboratory procedures call for the immediate fixing of the solution after it is drawn from the water bottle by the addition of manganese chloride and alkaline iodide. During the pickling stage of the Winkler method, the dissolved oxygen in the sample oxidizes Mn(II) to Mn(III) in alkaline solution to form a precipitate MnO_2. This is followed by oxidation of added I− by the Mn(III) in acidic solution. The resultant I_2 is titrated with thiosulfate solution using starch as an endpoint indicator. After the sample is chemically "fixed", the precipitate that forms can be allowed to settle for 10−20 min. At this stage, samples may be stored in a dark environment for up to 12 h before they need to be titrated. Parsons et al. (1984) give the precision of their recommended spectrophotometric method as $\pm 0.064/N$ (mg/l), where N is the number of replicate subsamples processed. The Winkler method is accurate to 1% provided the chemical analysis methods are rigorously applied. Another measure is the percentage saturation, which is the ratio of dissolved oxygen in the water to the amount of oxygen the water could hold at that temperature, salinity and pressure. Saturation curves closely follow those for dissolved oxygen.

In situ electronic dissolved oxygen sensors have been developed for use with profiling systems such as CTDs or as ancillary sensors attached to single-point current meters. All existing sensors use a version of the Clark cell, which operates on the basis of electro-reduction of molecular oxygen at a cathode, or are based on the ability of certain substances to quench fluorescence from a blue light source. The fluorescent substance is embedded in a gas permeable material that is exposed to the surrounding water. When used in a polarographic mode, the electric current supplied by the cathode in a Clark cell is proportional to the oxygen concentration in the surrounding fluid. To lessen the sensitivity of the device to turbulent fluctuations in the fluid, the electrode is covered with an electrolyte and membrane. Oxygen must diffuse down gradient through the membrane into the electrolyte before it can be reduced at the surface of the cathode. There are a number of drawbacks with the present systems. First of all, the diffusion of oxygen through the boundary layer near the surface of the probe is slow, limiting the response time of the cell to several minutes. Also, the electrochemical reaction within the cell consumes oxygen and stirring may be required to maintain the correct external oxygen concentration. Changes in the structure of the cell—due to alterations in the diffusion characteristics of the membrane as a result of temperature, mechanical stress and biofouling and to deterioration of the electrolyte and surfaces—require that the cell be recalibrated every several hours. The need for frequent re-calibration limits the use of the polarographic technique for profiling and mooring applications. Langdon (1984) uses a pulse technique to reduce the calibration drift. This improves long-term stability, but time constants are still the order of minutes.

The Yellow Springs Instruments and Beckman polarographic dissolved oxygen sensors (Brown and Morrison, 1978) sense the oxygen content by the current in an electrode membrane combined with a thermistor for membrane temperature correction. The current through this membrane depends on the dissolved oxygen in the water and the temperature of the membrane. Samples of both membrane current and temperature are averaged every 1.024 s giving a resolution of 0.5 MA (microamps) with an accuracy of ± 2 MA over a range of 0–25 MA. These *in situ* sensors have yet to be critically evaluated with reference to well-tested and approved methods. There are concerns with changes in the membrane over the period of operations and problems with calibration. Nevertheless, as measurement technology improves, an *in situ* oxygen sensor will be a high priority in that it saves considerable processing time and avoids errors possible with shipboard processing.

The Aanderaa Data Instruments (AADI) Oxygen Optode employs fluorescence quenching to obtain rapid and stable measurements of dissolved oxygen in a wide range of marine conditions (Tengberg et al., 2006). Optodes are now incorporated in profiling and moored CTD platforms and in single-point current meters. Although the use of fluorescence quenching for oxygen determination has been known since the 1930s (Kautsky, 1939) and widely used for *in vivo* measurement of the partial pressure of oxygen in blood (Peterson et al., 1984), the first application in oceanography was not reported until 1988 (Thomson et al., 1988). As with the modern Optode sensor, this prototype fluorescence-based dissolved oxygen sensor operated on the principle that the fluorescence intensity of an externally light-excited fluorophore is attenuated or "quenched" in direct relation to the concentration of dissolved oxygen in an ambient fluid (Figure 1.183a). Optimum results were obtained using a high-intensity blue-light source (wavelength of 450–500 nm) since this is the wavelength that most readily excites the known fluorophores. Results from a 6-day time-series record of dissolved oxygen concentration from a moored instrument in Saanich Inlet in 1987 suggested that the technique can be

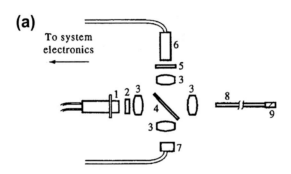

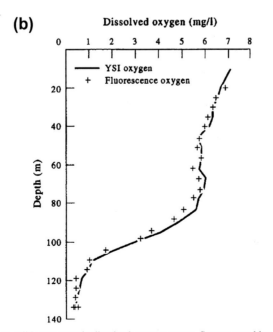

FIGURE 1.183   (a) Schematic of the first solid-state oceanic dissolved oxygen sensor. System uses blue light from (1) to excite a fluorophore in the sensor tip (9). The concentration of dissolved oxygen in the ambient fluid sensed by (6) is proportional to the degree of quenching of blue light fluoresced by the chemical-doped sensor; (b) simultaneous profiles of oxygen in Saanich Inlet. *YSI*, YSI dissolved oxygen sensor. *From Thomson et al. (1988).*

used to build a rapid (<1 s) response profiling sensor (Figure 1.183b) with long-term stability and high (<0.1 ml/l) sensitivity. The fact that the oxygen spectrum closely resembled the temperature spectrum for the entire frequency band up to a period of 2 h suggests that the oxygen data were at least as stable as the thermistor on the Aanderaa RCM4 current meter that was used in the moored study. Since no blue-light source was available at the time, the prototype device relied on a high-power white-light source and a car battery to drive the system. Modern luminescence quenching sensors employ the needed blue-light source and a chemically stable probe capable of withstanding the rigors of shipboard operations and high hydrostatic pressures. The lack of a commercial blue-light source with sufficient power (≈ 1 MW) to produce a strong fluorescence response is no longer an impediment to future improvements in this technology.

The AADI Optode for measuring dissolved oxygen became commercially available in 2002. According to the company's Web site, instruments have been used on autonomous Argo floats (e.g., Joos et al., 2003; Johnson et al., 2010) and gliders (Nicholoson et al., 2008), long-term monitoring in coastal environments with high bio-fouling (Martini et al., 2007), on coastal buoys (Jannasch et al., 2008), on Ferry box systems (Hydes et al., 2009), on profiling CTD instruments down to 6,000 m, and in chemical sensor networks (Johnson et al., 2007). The manufacturer specifies an operating range of 0−500 MM (0−16 ml/l), a resolution of 1 MM (micromolar), an accuracy of better than ±8 MM (or 5% of the value, whichever is greater), and a 63% response time of 25 s (a response time of 8 s is also listed but it is not clear if this is the capability of the technology or the sensor).

1.15.1.3 Nutrients

Nutrients such as nitrate, nitrite, phosphate, and silicate are among the "old guard" of oceanic properties obtained on standard oceanographic cruises. One need only examine the early technical reports published by oceanographic institutions to appreciate the considerable effort that went into the collection of these data on a routine basis. Oceanographers are again beginning to use these data on a routine basis to understand the distribution and evolution of water masses. However, there are a number of problems with nutrient collection that need to be heeded. To begin with, the data must be collected in duplicate (preferably triplicate) in small 10 mm vials and frozen immediately after the samples are drawn using a "quick freeze" device or alcohol bath. This is to prevent chemical and biological transformations of the sample while it is waiting to be processed. Careful rinsing of the nutrient vials is required as the samples are being drawn. Silicate must be collected using plastic rather than glass vials to prevent contamination by the glass silicate. Plastic caps must not be placed on too tightly and some space must be left in the vials for expansion of the fluid during freezing. Nutrient sample analysis is labor-intensive, time-consuming work. Although storage time can be extended to several weeks, we strongly recommend that nutrients be processed as soon as possible after collection, preferably on board the research ship using an autoanalyzer. With individual parameter techniques this is less likely to be possible than with more recent automated methods, which have been developed to handle most nutrients (Grasshof et al., 1983). These automated systems, which use colorimetric detection for the final measurement, need to be carefully standardized and maintained. Under these conditions, they are capable of providing high quality nutrient measurements on a rapid throughput basis.

Profiles of nutrients and dissolved oxygen for the North Pacific are presented in Figure 1.184. As first reported by Redfield (1958), the concentrations of nitrate, phosphate, and oxygen are closely linked except near source or sink regions of the water column. A weaker relationship exists between these variables and silicate. Nitrite only occurs in significant amounts near the sea surface where it is associated with phytoplankton activity in the photic zone and in the detritus layer just below the seasonal depth of the mixed layer. Although the linear relationships between these parameters vary from region to region, the reason for the strong correlations is readily explained. Within the photic zone, phytoplankton fix nitrogen, carbon and other materials using sunlight as an energy source and chlorophyll as a catalyst. In regions of high phytoplankton activity such as mid-latitudes in summer, the upper layers of the ocean are supersaturated in oxygen and depleted in nutrients. That is, there are sources and sinks for oxygen and nutrients. However, below the photic zone, bacterial decay and dissolution of detritus raining downward from the upper ocean leads to chemical transformations of oxidized products. This, in turn, leads to a reduction of oxygen compounds and a corresponding one-to-one release of nitrate, phosphate and silicate. This linear relation would prevail throughout the ocean below the photic zone if were not for other sources and sinks for these chemicals. For example, we now know that silicate enters the ocean through resuspension of bottom sediments and from hydrothermal fluids vents from mid-ocean ridge systems (Talley and Joyce, 1992). Chemosynthetic production by bacteria in hydrothermal plumes is also a source/sink region as the analog to photosynthetic processes in the upper ocean.

It is generally thought that limitations in upper ocean nutrients, especially nitrates, combined with zooplankton predation (grazing) and turbulent mixing processes control primary (phytoplankton) productivity in the ocean. It is generally accepted that other nutrients such as the aeolian supply of iron compounds might ultimately control productivity in areas

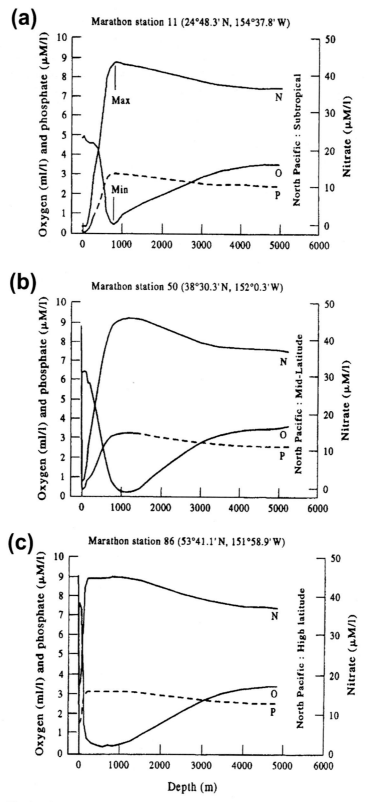

FIGURE 1.184   Plots of nitrate (N), phosphate (P) and dissolved oxygen (O) for the North Pacific. (a) Station 11 at 24°48.3′ N; 154°37.8′ W; (b) Station 50 at 38°30.3′ N; 152°00.3′ W; (c) Station 86 at 53°41.1′ N; 151°58.9′ W. *Data from Martin et al. (1987).*

such as the equatorial and subarctic regions of the Pacific and the Southern Ocean where nitrate concentrations are high year-round but spring and fall blooms do not occur (Chisholm and Morel, 1991). These high nutrient, low chlorophyll (HNLP) regions have become the focus of increasing numbers of multi-disciplinary studies (Hamme et al., 2010; Parsons, 2012). From a climate change perspective, it is important to understand how to increase productivity in the upper ocean in order to increase sequestering of atmospheric CO_2 into the deep ocean, which holds about 50 times more CO_2 than the atmosphere. This higher CO_2 concentration in the deep ocean is due to the "biological pump" whereby phytoplankton use photosynthesis to fix CO_2 at the ocean surface into organic carbon. When the plankton die, they sink into the deep ocean, where the organic carbon is consumed by other organisms and respired back to CO_2 to form dissolved carbonate.

In a classic paper, Redfield (1958) suggested that organisms both respond to and modify their external environments. His premise was that the nitrate of the ocean and the oxygen of the atmosphere are determined by the biochemical cycle and not conditions imposed on the organisms through factors beyond their control. Support for his thesis was derived from the fact that the well-defined nitrogen, phosphorous, carbon, and oxygen compositions of plankton in the upper ocean were almost identical to the concentrations of these elements regenerated from chemical processes in the 95% of the ocean that lies below the autotrophic zone. As pointed out by Redfield, the synthesis of organic material by phytoplankton leads to oceanic changes in concentrations of phosphorous, nitrogen, and carbon in the ratio 1:16:106. During heterotrophic oxidation and remineralization of this biogenic material (i.e., decomposition of these organisms), the observed ratios are 1:15:105. Thus, for every phosphorous atom that is used by phytoplankton during photosynthesis in the euphotic zone, exactly 16 nitrogen atoms and 106 carbon atoms are used up. Alternatively, for every phosphorous atom that is liberated during decomposition in the deep ocean, exactly 15 nitrogen and 105 carbon atoms are liberated. The oxidation of these atoms during photosynthesis requires about 276 oxygen atoms while during decomposition 235 oxygen atoms are withdrawn from the water column for each atom of phosphorous that is added. If this process were simply one way, the primary nutrients would soon be completely depleted from the upper ocean. That is why life-supporting replenishment of depleted nutrients to the upper ocean through upwelling and vertical diffusion of deeper nutrient rich waters is such an important process to the planet. Bruland et al. (1991) give a modern version of the Redfield ratios based on phytoplankton collected under bloom conditions as: C:N:P:Fe:Zn:Cu,Mn,Ni,Cd = 106:16:1:0.005:0.002:0.0004 (see also Martin and Knauer, 1973).

According to the above ratios, the formation of organic matter by phytoplankton, in the surface autotrophic zone, leads to the withdrawal of carbonate, nitrate, and phosphate from the water column. Oxygen is released as part of photosynthesis and the upper few meters of the ocean can be supersaturated in oxygen at highly productive times of the year. When the plants die and sink into the deeper ocean, decomposition by oxidation returns these compounds back to the seawater. Thus, increases in carbonate, nitrate, and phosphate concentrations below the euphotic zone are accompanied by a corresponding decrease in oxygen levels. This process leads to a rapid increase in nitrate and phosphate and a corresponding rapid decrease in oxygen within the upper kilometer or so of the ocean (Figure 1.184). Nitrate and phosphate reach subsurface maxima at intermediate depths and then begin to decrease slowly with depth to the seafloor. Oxygen, on the other hand, falls to a mid-depth minimum (the oxygen minimum layer) before starting to increase slowly with depth toward the seafloor. In the upper zone, the balance of chemicals is altered considerably by biological activity while near the coast the balance is altered by runoff, which provides a different ratio of nutrients. However, below the surface layer, the changes occur in the manner suggested by the Redfield ratios (Redfield et al., 1963). Note that the concentration of silicate is almost like that of the other nutrients, except that it doesn't reach a maximum at mid-depth and becomes more decoupled from the accompanying oxygen curve. This suggests a source function for silicate in the deep ocean. Indeed, there are two sources: resuspension and dissolution of siliceous material from rocks and other inorganic material on the seafloor and the injection of silicates into the ocean from hydrothermal venting along mid-ocean ridges and other magmatic source regions in the deep ocean.

The fact that carbon and oxygen concentrations greatly exceed the levels required by plankton while those of phosphorous and nitrogen were identical to those observed on average in the ocean (carbon is at least 10 times that needed for photosynthesis), prompted Redfield to suggest that phosphate and nitrate are limiting factors to oceanic primary productivity. It is thought that nitrate (NO_3) is the primary limiting factor although phosphorous limitation is still important in certain coastal areas. Airborne iron is also thought to be a limiting nutrient for primary productivity in the open ocean. Evidence for this is based on the year-round absence of phytoplankton blooms in the subarctic Pacific, equatorial Pacific and Southern Ocean despite the high near-surface concentrations of nitrate and phosphate. In these areas, autotrophic processes fail to exploit NO_3 and PO_4. The idea is that iron, or some other mineral, limits growth, which is not the case in areas served by aeolian transport from the land. Unfortunately, it is not yet possible to sort out the effects of iron limitations from grazing by herbivorous zooplankton or from physical mixing in the surface layer which prevents stratification from confining the animals to a thin upper layer. The first open-ocean iron fertilization experiment (Ironex 1), conducted in

October 1993 within the equatorial Pacific near the Galapagos Islands, showed that iron enrichment with $FeSO_4$ could dramatically increase surface productivity (Coale et al., 1998). Using sulfur hexafluoride to track the 64 km^2 iron-enriched area containing 443 kg of iron sulfate solution, scientists found that the rate of growth and total mass of phytoplankton doubled over a period of 3 days. However, the iron soon precipitated out of solution as ultra-fine particles and sank, causing a sharp decrease in productivity levels. Since then, there have been numerous international iron fertilization experiments to examine carbon sequestering in the ocean. A recent experiment (LOHAFEX) was conducted in March 2009 in the southwestern Atlantic by Indian and German scientists onboard the research vessel *Polarstern*. Over a period of two and a half months, scientists fertilized a 300 km^2 cyclonic mesoscale eddy using six metric tons of dissolved iron and followed the effects of the fertilization on the plankton continuously for 39 days. As in previous studies, the iron stimulated the growth of plankton, which doubled their biomass within the first 2 weeks by taking up carbon dioxide from the water. However, further plankton growth, and hence further drawdown of CO_2 in the upper ocean was prevented by increased grazing pressure from small crustacean zooplankton (copepods). Algal species, which regularly generate blooms in coastal regions including the Antarctic, were most heavily grazed (Wikipedia, 2013). At present, the question of iron enrichment and ocean productivity remains unresolved (Boyd et al., 2007).

1.15.1.4 Silicate

The oceanic distribution of many elements is determined by their involvement in the biochemical cycle. Nitrate and phosphate are associated with the labile tissue and protoplasm of surface plankton whereas silicate and alkalinity are linked to the refractory hard parts of the organisms such as the silica shells of diatoms. The term "silicate" applies to dissolved reactive silicate (monosilicic acid, $Si(OH)_4$; Iler, 1979) measured from water samples. Since most of the silicate undergoes dissolution in the water column rather than the seafloor (Edmond et al., 1979), its distribution serves as tracer for water-mass mixing and advection. A north-south plot of silicate concentration in the Atlantic Ocean (from 80° S to 60° N), with the derived movement of principal water masses, is presented in Meckler et al. (2013). The advantage of silica over carbonates or other compounds is that siliceous sediments are found only in well-defined areas associated with surface upwelling and their distribution is not particularly dependent on depth. According to Edmond et al. (1979) the average flux of dissolved silica from the sediments to the deep ocean is about 3 $\mu mol/cm^2/year$, which is sufficient to make it a useful tracer of deep-sea flow. Large fluxes are observed in the Weddell-Enderby Basin off Antarctica and in the northern Indian Ocean. In the Meckler et al. (2013) study, a 550,000-year opal sediment core formed by the siliceous diatom shells originating from the Northwest Africa upwelling system, is used to show degassing of CO_2 into the atmosphere off Antarctica at the time of major stadial events (relatively short cooling events, such as the Younger Dryas, that interrupt the warming leading out of a major glacial period).

In the extreme northeast Pacific (northeast of 45° N, 160° W), silicate concentration increases with depth to the bottom while in the equatorial Pacific no anomalies are observed despite the presence of opaline deposits. The increased silicate with depth in the northeast Pacific appears to be associated, in part, with westward advection of dissolved siliceous sediments deposited over the continental margin of the wind-induced upwelling domain that extends from British Columbia to Baja California along the west coast of North America. As noted by Edmond et al. (1979), the existence of silica sources at the seafloor makes it impossible to use global correlations with the extensive silicate distribution data to determine the distribution of other variables such as trace metals. Historical data, together with transect data collected along 47° N (Talley et al., 1988), further suggest that both the intermediate silica maximum in the depth range 2,000–2,400 m and the near-bottom silica maximum in Cascadia Basin to the east of the Juan de Fuca Ridge (Figure 1.185) may be due, in part, to hydrothermal venting of high silicate waters (Talley and Joyce, 1992). The silica in the hydrothermal plumes emanating from the vents originates as silicates stripped from the crustal rocks by the high-temperature hydrothermal fluids. Other factors include vertical flux divergence of settling silicate particles, dissolution from opaline bottom sediments, and up-slope injection from the bottom boundary layer.

Macdonald et al. (1986) point out that improper thawing of frozen silicate samples can result in a significant and variable negative bias in seawater determination of silicate. The problem arises from conversion of reactive silicate to a nonreactive, polymetric form in the frozen sample. This polymerization need not affect accuracy for frozen samples provided that sufficient thawing time is allowed for depolymerization to the reactive form. To control bias, the analyst must adjust the length of time between thawing and analysis. The appropriate "waiting time" varies according to the salinity of the sample, the silicate concentration and the length of time the sample was frozen. Waiting time increases with the time that the samples were frozen and with silicate/salinity ratio. For example, deep silicate samples collected from the northeast Pacific (salinity $\approx$ 35 g/kg; silicate $\approx$ 180 $\mu mol/l$) and stored for one to 2 months must be thawed for about 8 h before processing. This increases to 24 h for samples stored for more than five to six months. Macdonald et al. (1986) conclude

FIGURE 1.185 The meridional distribution of silica (micromoles per liter) in the North Pacific along approximately 152° W (Hawaiian region to Kodiak Island, Alaska). Mid-depth maximum values in excess of 180 µmol/l are emphasized. *From Talley et al. (1991) and Talley and Joyce (1992).*

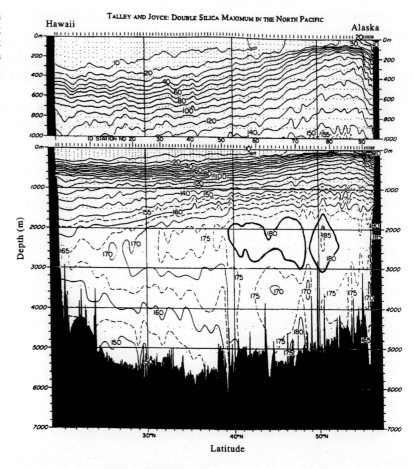

that "If the objectives of sampling can accept a 5% negative bias and a slight loss of precision, then freezing is a simple method for storing a wide range of samples. However, samples should be analyzed as soon as practicable."

1.15.1.5 pH

The concern for increased acidification of the World Ocean—arising from reduced availability of carbonate ions (CO_3^{2-}) due to increased anthropogenic input of atmospheric CO_2 emissions (Feely et al., 2012; Franco et al., 2021)—has led to the need for accurate and reliable aquatic pH sensors. A primary concern is that changes in pH change the aragonite saturation state, Ω_{arag}, which is of importance for calcifying organisms. In thermodynamic equilibrium, $\Omega_{arag} > 1$ gives rise to the precipitation of aragonite while for $\Omega_{arag} < 1$, seawater is corrosive to calcium carbonate so that, in the absence of biologically mediated protective mechanisms, dissolution will begin (Fabry et al., 2008; Espinosa, 2012). Marine organisms have been observed to reduce their calcification rates in undersaturated waters (Feely et al., 2004; Hoegh-Guldberg et al., 2007; Doney et al., 2009).

In aqueous solutions, pH is defined in terms of the proton (hydrogen ion, H^+) activity,

$$pH = -\log a_{H^+} \tag{1.98a}$$

where the activity of a substance is proportional to the concentration

$$a_{H^+} = \gamma_{H^+}{}^{[H^+]} \tag{1.98b}$$

The activity coefficient, γ_{H^+}, varies with both total ionic strength (i.e., salinity) and the concentration of other ions in the fluid (Robie Macdonald, Institute of Ocean Sciences, Sidney, BC, pers. Comm., 2012). This "activity" depends on is how many protons there are in the solution, including interactions with all the other ions in the solution. Activity is usually lower than concentration, but not always; a number of ionic interactions enhance activity, particularly for acid-base reactions. Both the old litmus paper test (colored acid-base indicator) and modern glass electrodes respond to "activity" (apparent concentration) rather than to absolute concentration. The historic gravimetric measurements generally reported concentrations, rather than activities. Most modern pH probes measure activity, but activity is defined operationally and is dependent on both the method used and the analyte solution; a glass electrode placed into a glacial freshwater pond measures something quite different from a spectrophotometric measurement, or even a measurement using that same electrode in the ocean (Robie Macdonald, pers. Com., Fisheries and Oceans Canada, Institute of Ocean Sciences, Sidney, BC, 2012). There are presently two primary methods for measuring ocean pH: electrochemical and spectrophotometric (see *Guide to Best Practices for Ocean CO₂ Measurements* by Dickson et al., 2007, PICES Special Publication 3). Most *in situ* pH probes use electrochemical methods based on a glass combination electrode. In these probes (Figure 1.186), the liquid junction potential (essentially a "blank" signal resulting from the construction of the electrode and its interaction with the sample) is either minimized or very accurately defined, allowing for extremely precise pH measurements. However, when the electrode is placed in seawater, the high salinity changes the liquid junction potential, destabilizing the electrical signal and resulting in prohibitively long analysis times (In many situations, the electrode never attains a stable reading, although some threshold level of change is accepted as a stable measurement). In addition, glass electrodes respond not only to hydrogen ions but also to other positively charged ions such as sodium (Na^+), which occurs at extremely high concentrations in seawater, creating a substantial interference in the pH measurement. Most efforts to adapt electrochemical pH measurements for practical seawater applications have focused on ways to minimize the sodium interference and eliminate the liquid junction potential.

The difficulties with electrochemical pH seawater measurements have led to the resurrection of older pH measurement methods based on dye molecules that change color in acids versus bases (the principle behind litmus paper). For example, *meta*-cresol purple, currently the most commonly used dye in seawater pH measurements, is yellow in its protonated (acidic) form and purple in its deprotonated (basic) form. To determine the pH of a solution, a small amount of the dye is added, and the absorbance measured at the maximum wavelengths (yellow and purple) of the two dye forms. The ratio of those absorbance values gives the precise pH of the solution. Because pK_a, the pH at which half the dye molecules are protonated and half are not, is about eight for *meta*-cresol purple (the same as for seawater), it is generally a good indicator for seawater analyses. However, other dyes are sometimes used for some special conditions, such as in estuaries or polar waters. While this method is very quick and easy to perform, its accuracy is dependent on the quality of the spectrophotometer used and the accuracy to which the pK_a of the dye has been determined for the sample type, taking into account the effects of salinity, temperature, and interfering ions.

Because pH is dependent on the analytical method used, careful and accurate calibration is critical. Standard buffer solutions of organic compounds of very specific and well-characterized pK_a are generally used as standards for pH analyses. The most commonly used buffers are those produced by the US National Institute for Standards and Technology (NIST buffers) in purified water at pHs of 4, 7, and 10 (these are generally dyed pink, yellow, and blue, respectively). Because NBS standards are prepared in dilute solution, they are inappropriate for calibrating seawater pH analyses. For the

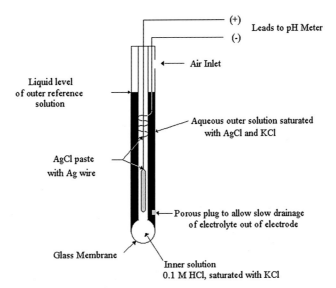

FIGURE 1.186 Schematic of a combination glass pH electrode. The liquid junction potential arises across the porous plug separating the exterior reference solution from the sample. *David Reckhow, University of Massachusetts at Amherst, U.S.A.*

spectrophotometric method, the dye stability constants that match the samples will not match the buffers, and in electrochemical analyses, using buffers that do not match the samples introduces an additional liquid junction potential and destabilizes the measurements. Although recipes for seawater standard buffers have been available for a couple decades, they have not been reproducible between laboratories. Truly standardized seawater buffers (produced by Andrew Dickson at Scripps Oceanographic Institution, who also produces standards for seawater alkalinity and total inorganic carbon) have only recently become available, and they have not yet been fully tested and implemented.

With respect to what buffers are used to calibrate the analysis, several different pH scales have been proposed for environmental analyses, based on how they've been calibrated. For seawater analyses, the appropriate scale to use depends on whether the dyes or electrodes are calibrated in NaCl or artificial seawater solutions, termed the "free" versus "total" hydrogen ion scales. In studies involving only high-salinity seawater samples, the "total" scale is preferred, but in studies of more varied environments such as estuaries or polar waters influenced by sea ice melt, the "free" scale can be preferable. Converting between the scales requires knowledge of or assumptions about the behavior of sulfate and fluoride ions in the samples.

Biological processes in closed systems can cause large changes in pH. Most of these issues are not important, if one is just trying to keep your swimming pool, aquarium, or possibly even fish farm in balance. However, when trying to look at interannual or climate change variations, including anthropogenic ocean acidification, extremely precise and accurate analyses are required in order to confidently identify the relatively small changes that might be occurring. For example, mixed layer pH is thought to be changing at about 0.002/year. The standard electrochemical method for seawater has an optimum precision of 0.004, while the spectrophotometric method is at about 0.001. In comparison, typical low price pH meters intended for use in low-salinity waters have a precision of about 0.01. As yet, successful use of a pH probe for long-term deployment on a mooring has not been reported. However, a number of manufacturers are now claiming to have pH sensors with sufficient stability for long-term deployment in seawater with precision adequate for climate studies. Their claims just have not yet been confirmed by independent deployments. In summary, it's safe to say that while we have made excellent progress on the precision of pH analyses in seawater, accuracy is still an unknown. However, even to get the necessary precision in seawater analyses requires specialized methods and equipment; the pool kit or the cheap hand-held pH probe are not going to give meaningful pH numbers in seawater.

1.15.1.6 Light Attenuation and Scattering

The light energy in a fluid is attenuated by the combined effects of absorption and scattering. In the ocean, absorption involves a conversion of light into other forms of energy such as heat; scattering involves the redirection of light by water molecules, dissolved solids and suspended material without the loss of total energy. Transmissometers are optical instruments that measure the clarity of water by measuring the fraction of light energy lost from a collimated light beam as it passes along a known pathlength (Figure 1.187). Attenuation results from the combined effects of absorption and shallow-angle Rayleigh (forward) scattering of the light beam by impurities and fine particles in the water. Water that is completely free of impurities is optically pure. Nephelometers (or turbidity meters) measure scattered light and respond primarily to the first-order effects of particle concentrations and size. Depending on manufacturer, commercially available nephelometers examine scattered light in the range from 90° to 165° to the axis of the light beam. Most instruments use IR light with a wavelength of 660 nm. Because light at this wavelength is rapidly absorbed in water (63% attenuation every 5 cm), there is little contamination of the source beam due to sunlight except within the top meter or so of the water column.

The intensity $I(r)$ of a light beam of wavelength λ traveling a distance r through a fluid suspension attenuates as

$$I(r) = I_0 \exp(-cr) \tag{1.99}$$

FIGURE 1.187 Exploded view of a Sea-Tech transmissometer. Red light of wavelength 660 nm passes from the light-emitting diode (LED) to the sensor over a fixed pathlength of 0.25 m.

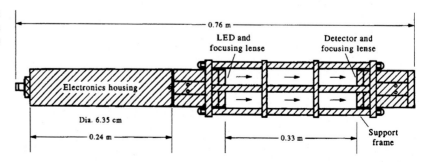

where I_0 is the initial intensity at $r = 0$ and $c = c(\lambda)$ is the rate of attenuation per unit distance. Attenuation of the light source occurs through removal or redirection of light beam energy by scattering and absorption. In the ocean, visible long-wave radiation (red) is absorbed more than visible short-wave radiation (blue and green) and what energy is left at long wavelengths undergoes less scattering than at short wavelengths. As a consequence, the ocean appears blue to blue-green when viewed from above. The exact color response depends on the scattering and absorption characteristics of the materials in the water including the dissolved versus the suspended phase—factors that are used to advantage in remote sensing techniques. For a fixed monochromatic light source, the clarity of the water, measured relative to distilled uncontaminated water, provides a quantitative estimate of the mass or volume concentration of suspended particles. Such material can originate from a variety of sources including terrigenous sediment carried into the coastal ocean by runoff, from current-induced resuspension of material in the benthic layer, or from detectable concentrations of plankton.

The "Secchi disk" is one of the simplest and earliest methods for measuring light attenuation in the upper layer of the ocean. A typical Secchi disk consists of a flat, 30-cm diameter white plate that is lowered on a marked line (suspended from the disk center) over the side of the ship. The depth at which the disk can no longer be seen from the ship is a measure of the amount of surface light that reaches a given depth and can be used to obtain a single integrated estimate of the extinction coefficient, $c(\lambda)$. The disk is still in use today. For example, Dodson (1990) used Secchi disk data from a series of lakes in Europe and the U.S.A. that suggest a direct relationship between the depth of day—night (diel) migration of zooplankton and the amount of light penetrating the epihelion. In this case, the zooplankton minimize mortality from visually feeding fish and maximize grazing rate. Despite its simplicity, there are a number of problems with this technique, notwithstanding the fact that it fails to give a measure of the water clarity as a function of depth and is limited to near-surface waters. In addition, the visibility of the disk will depend on the amount of light at the ocean surface (and type of light through cloud cover), on the roughness of the ocean surface, and the eyesight of the observer. Today, oceanographers rely on transmissometers and nephelometers to determine the clarity of the water as a function of depth.

A typical transmissometer consists of a constant intensity, single-frequency light source and receiving lens separated by a fixed pathlength, r_0. The attenuation coefficient in units of per meter is then found from the natural logarithm relation

$$c = -(1/r_0)\ln(I/I_0) \qquad (1.100)$$

in which r_0 is measured in meters, and I/I_0 is the ratio of the light intensity at the receiver versus that transmitted by the red (660 nm) LED. This choice of light wavelength is useful because it eliminates attenuation from dissolved organic substances consisting mainly of humic acids or "yellow matter" (also called "gelbstoff"; Jerlov, 1976). The Sea Tech transmissometer (Bartz et al., 1978) has an accuracy of $\pm0.5\%$ and a small ($<1.03°$ or 0.018 radians) receiver acceptance angle that minimizes the complication of the collector receiving specious forward-scattering light. To obtain absolute values, the source and lens must be calibrated in distilled water and air since scatter can affect the results. As an example, a 0.25 m pathlength transmissometer which has a calibration value of $I_0 = 94.6\%$ in clean water and reading of 89.1% in the ocean corresponds to a light attenuation coefficient

$$c = -4\ln(I/I_0) = -4\ln(0.891/0.946) = 0.240\text{m}^{-1} \qquad (1.101)$$

Values of c in the ocean range from around 0.15/m for relatively clear offshore water for concentrations of particles as low as 100 µg/1 to around 21/m for turbid coastal water with particle concentrations of 140 mg/l (Sea Tech user's manual). In studies of hydrothermal venting, measurement of water clarity is often one of the best methods to determine the location and intensity of the plume (Baker and Massoth, 1987; Thomson et al., 1992; Figure 1.188).

Problems with the transmissometer technique are: (1) drift in the intensity of the light source with time; (2) clouding of the lens by organic and inorganic material which affect the *in situ* calibration of the instrument; and (3) scattering, rather than absorption, of the light. If we ignore the influence of dissolved substances, the attenuation coefficient, c, depends on the concentration of the suspended material but also on the size, shape, and index of refraction of the material (Baker and Lavelle, 1984). Thus, a linear relationship between c and particle concentration C such that

$$c = \alpha C + \alpha_o \qquad (1.102)$$

only occurs when the effects of size, shape and index of refraction are negligible or mutually compensating; here, α_o denotes the offset in c at zero particle concentration. After concentration, particle size is the next most important variable effecting clarity. Accurate estimates of concentration therefore require calibration in terms of the distribution of particle sizes and shapes in suspension as for example in Baker and Lavelle (1984). Laboratory results demonstrate that calibrations of beam transmissometer data in terms of particle mass or volume concentration are acutely sensitive to the size distribution of the particle population under study. There is also a trend of decreasing calibration slopes from environments where large

FIGURE 1.188 Cross sections of temperature anomaly (°C) and light attenuation coefficient (1/m) for the "meg-aplume" observed near the hydrothermal main site on the Cleft Segment of Juan de Fuca Ridge in the northeast Pacific in September 1986. The temperature anomaly gives temperature over the plume depth relative to the observed background temperature. Dotted line shows σ_θ surfaces and solid line the saw tooth track of the towed CTD path. *From Baker et al. (1989).*

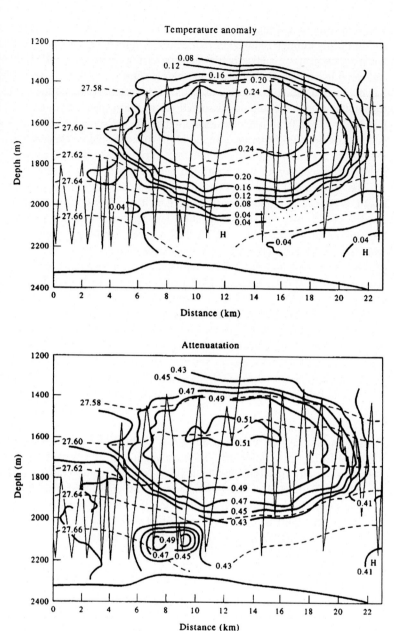

particles are rare (deep ocean) to those where they are common (shallow estuaries and coastal waters). Theoretical atten-uation curves agree more with observations when the natural particles are treated as disks rather than as spheres as in Mie scattering theory. The need for field calibration is stressed.

The results of Baker and Lavelle (1984) can be summarized as follows: (1) calibration of beam transmissometers is acutely sensitive to the size distribution of the particle population under study; (2) theoretical calculations based on Mie scattering theory and size distributions measured by a Coulter counter agree when attenuation for glass spheres is observed but underestimate the attenuation of natural particles when these particles are assigned an effective optical diameter equal to their equivalent spherical diameter deduced from particle volume measurements; (3) treating particles as disks expands their effective optical diameter and increases the theoretical attenuation slope close to the observed values; (4) there is a need to collect samples along with the transmissometer measurements, especially where the particle environment is nonhomogeneous.

Transmissometers are best used for measuring the optical clarity of relatively clear water whereas nephelometers are most suitable for measuring suspended particles in highly turbid waters. In murky waters, nephelometers have superior linearity over transmissometers while transmissometers are more sensitive at low concentrations. "Turbidity" or cloudiness of the water is a relative, not an absolute term. It is an apparent optical property depending on characteristics of the scattering particles, external lighting conditions and the instrument used. Turbidity is measured in nephelometer units referenced to a turbidity standard or in Formazin Turbidity Units (FTUs) derived from diluted concentrations of 4000-FTU formazin, a murky white suspension that can be purchased commercially. Since turbidity is a relative measure, manufacturers recommend that calibration involve the use of suspended matter from the waters to be monitored. This is not an easy task if one is working in a deep or highly variable regime.

1.15.1.7 Oxygen Isotope: $\delta^{18}O$

The ratio of oxygen isotope 18 to oxygen isotope 16 in water is fractionated by differences in weight. The lighter element ^{16}O is more easily evaporated than ^{18}O and is therefore a measure of temperature; the higher the temperature the greater the $H_2^{18}O/H_2^{16}O$ ratio. In contrast to the variability in the surface ocean, average $H_2^{18}O/H_2^{16}O$ ratios for the deep ocean (>500 m depth) vary by less than 1%. This ratio (in percent) is expressed in conventional delta "δ" notation as

$$\delta^{18}O(\%) = \left(R_{std} / R_{sample} - 1\right) \times 10^{10} \tag{1.103}$$

where $R = H_2^{18}O/H_2^{16}O$ is the ratio of the two main isotopes of oxygen and the subscript "std" refers to Standard Mean Ocean Water. The low variability in $\delta^{18}O$ values in waters in the deep sea has led to widespread use of oxygen isotopes as a paleothermometric indicator. These methods assume relatively little variation (about 1%) in the $\delta^{18}O$ values of deep ocean water over geological time. The $\delta^{18}O$ values of carbonate, silica, and phosphate precipitated by both living and fossil marine organisms, such as foramininferans, radiolarians, coccolithophorids, diatoms, and barnacles, have been used to estimate temperatures of the water in which the organism lived based on temperature-dependent equilibria between the oxygen in the water and the biomineralized phase of interest. The $\delta^{18}O$ values vary in space and time in different regions of the ocean. For example, shallow continental shelves are influenced by freshwater input, particularly at high latitudes. Thus, oxygen removed from seawater by organisms should reflect oceanic conditions at the time. Salinity and ^{18}O content are related in most ocean waters with similar processes influencing both in tandem.

According to Kipphut (1990), the $H_2^{18}O/H_2^{16}O$ ratio in seawater in the Gulf of Alaska shows only slight variation except near those coastal margins where there is significant input of freshwater from melting of large glaciers ($\delta^{18}O \approx -23\%$) and runoff from coastal precipitation ($\delta^{18}O \approx -10\%$). Precipitation is generally depleted in the heavier isotopes of oxygen because of isotopic fractionation processes, which occur during evaporation and condensation. Because the fractionation processes are temperature dependent, precipitation at higher latitudes and elevations shows progressively lower $H_2^{18}O/H_2^{16}O$ ratios. The ratio is a conservative property of water and when combined with salinity may be useful in determining distinct components of water masses. The isotope data south of Alaska suggest that the coastal waters in southwestern Alaska are derived from a combination of glacier melt and runoff from as far eastward as south-central Alaska. If we add the freshwater added by runoff from the large rivers of northern British Columbia, the Alaska Coastal Current (Royer, 1981; Schumacher and Reed, 1986) is a continuous feature flowing more than 1,500 km from the southern Alaska Panhandle to Unimak Pass at the beginning of the Aleutian Island chain. Information regarding the important application of $\delta^{18}O$ observed in cores to the study of paleoclimate change is found in Chapter 6 of the 2007 report of the Intergovernmental Panel on Climate Change (Jansen et al., 2007).

1.15.1.8 Helium-3; Helium/Heat Ratio

Helium-3 (^3He) is an inert and stable isotope of helium whose residence time of about 4,000 years in the ocean makes it a useful tracer for oceanic mixing times and deepsea circulation. There are two main sources in the ocean. In the upper mixed layer and thermocline, ^3He is produced by the β-decay of anthropogenic tritium; in the deep ocean, ^3He originates with mantle degassing of primordial helium from mid-ocean ridge hydrothermal vents. Anderson (1993) also argues that ^3He and neon from hotspot magmas and gases may reflect an extraterrestrial origin; specifically, subduction of ancient pelagic sediments rich in solar ^3He and neon originate with interplanetary dust particles now being recycled at oceanic hotspots. [For counterarguments see Hiyagon (1994) and Craig (1994).] The distinct isotopic ratio of mantle helium ($^3He/^4He = 10^{-5}$) versus a ratio of 10^{-6} for atmospheric helium makes $^3He/^4He$ a useful tracer in the ocean. In a classic paper, Lupton and Craig (1981) showed that the $^3He/^4He$ ratio in the 2,500 m deep core of the hydrothermal plume emanating from the East Pacific Rise at 15° S in the Pacific Ocean was 50 times higher than the ratio of atmospheric helium. The helium plume could be traced more than 2,000 km westward from the venting region on the crest of the mid-ocean ridge (Figures 1.189 and 1.190). To quote the authors, "In magnitude, scale, and striking asymmetry, this plume is one of the most remarkable features of the deep ocean, resembling a volcanic cloud injected into a steady east wind". Helium-3 is now used extensively

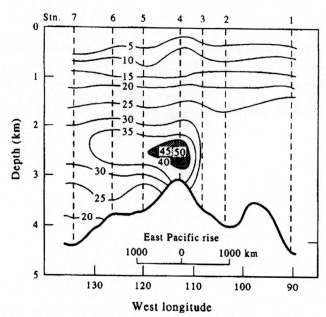

FIGURE 1.189   Cross section of $\delta(^3\text{He})$ over the East Pacific Rise at 15° S. The level of neutral plume buoyancy, as determined by the core depth of the ^{3}He plume, is about 400 m above the ridge crest. The ratio is defined as $\delta(^3\text{He}) = (R/R_{\text{ATM}} - 1) \times 100$ where $R = {}^3\text{He}/{}^4\text{He}$ and $R_{\text{ATM}} = 1.40 \times 10^{-6}$. *From Lupton and Craig (1981).*

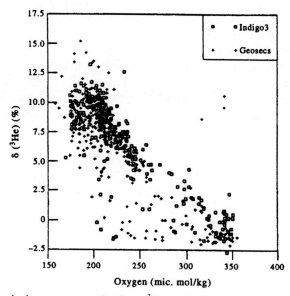

FIGURE 1.190   Correlation between dissolved oxygen concentration O_2 and ^{3}He in the Southern Ocean indicates that the distributions of these tracers in the region of the World Ocean are mainly determined by ventilation processes. Combination of GEOSECS and INDIGO-3 data. *From Jamous et al. (1992).*

as tracer for hydrothermal plumes in active spreading regions such as the Juan de Fuca Ridge in the northeast Pacific and the East Pacific Rise in the South Pacific.

Data collected during GEOSECS indicates that the deep Pacific is the oceanic region most enriched in ^{3}He with a mean ratio concentration δ^3He value of 17% compared with 10% in the Indian Ocean, 7% in the Southern Ocean and 2% in the Atlantic (Jamous et al., 1992). The core of the plume at the East Pacific Rise has a value of 50%. (Here, $\delta^3\text{He}(\%) = (R/R_a - 1) \times 100$, where $R = {}^3\text{He}/{}^2\text{He}$ is the isotopic ratio of the sample and R_a is the atmospheric ratio.) The differences in concentration relate directly to the differences in hydrothermal input and inversely to the degree of deep-water ventilation. For example, there is a considerably greater hydrothermal activity in the Pacific than in the Atlantic while the Atlantic deep water is highly ventilated compared with the Pacific. Similarly, the values of δ^3He (≈ 28) at the bottom of the Black Sea reflect the presence of a strong source at the seafloor. In contrast, the strong correlation between dissolved oxygen concentration and ^{3}He in the Southern Ocean (Figure 1.190), indicates that the distributions of these tracers in this region of the world ocean are mainly determined by ventilation processes.

Early vent-fluid samples taken from hydrothermal systems on the Galapagos Rift and at $21°$ N on the East Pacific Rise were found to have nearly equal ratios of ^{3}He to heat despite the considerable geographical separation of the sites and widely different fluid exit temperatures ($\approx$ 20 and $350°C$, respectively). Here, "heat" is the excess amount of heat (in calories or joules) added to the ambient water by geothermal processes. By combining independent estimates of the mantle flux of ^{3}He within the ocean with the observed ratio ^{3}He/Heat $\approx 0.5 \times 10^{-12}$ cm^3 STP/cal, Jenkins et al. (1978) calculated a global oceanic hydrothermal heat flux of 4.9×10^{19} cal/year. An examination of the ^{3}He/Heat ratios in the 20-km wide megaplume observed in August 1986 on Juan de Fuca Ridge (Lupton et al., 1989) has shown that the ratios can vary by as much as an order of magnitude and that heat fluxes based on ^{3}He measurements must be taken with caution. Specifically, the ratio ^{3}He/Heat was found to vary with height within the megaplume formed during the hydrothermal event. The megaplume had lower helium values and five times the temperature anomaly as the near-bottom chronic venting regime. Since helium is extracted from the magma by the circulating fluids in the hydrothermal system, the relatively low ratios of ^{3}He/Heat in the megaplume presumably resulted from relatively high water-to-rock ratios and the youth of the hydro-thermal fluid prior to its injection into the overlying ocean. Lupton et al. (1989) suggest that a value of $\approx 2 \times 10^{-12}$ cm^3 STP He/cal may be a reasonable estimate for the average ^{3}He/Heat signature of fluids vented into the oceans by mid-ocean ridge hydrothermal systems.

1.15.2 Transient Chemical Tracers

"Transient tracers" are anthropogenic compounds that are injected into the ocean over spatially limited regions within well-defined periods of time. The time "window" makes these compounds especially well-suited to studies of upper ocean mixing and deep-sea ventilation. Transient tracers are commonly used to constrain solutions of global "box" models used to investigate climate-scale carbon dioxide fluxes within coupled atmosphere-ocean systems (Broecker and Peng, 1982; Sarmiento et al., 1988), and in generalized inverse models incorporating both data and ocean dynamics to determine oceanic flow structure (see Bennett, 1992). The timed release into the ocean may take place over a few hours, as in the case of rhodamine dye, or last longer than a century, as in the case of chlorofluorocarbons (CFCs). Injection of certain tracers, such as radiocarbon (^{14}C) greatly augments the natural distributions of these chemicals while for others, such as CFCs and tritium (^{3}H), the tracer is superimposed on an almost nonexistent background concentration. Because of the slow advection and mixing processes in the ocean, as well as the extensive research needed to measure the tracer distributions, most tracers are used in the study of seasonal to decadal scale oceanic variability. For all transient tracers, studies are limited by imperfect knowledge of the surface boundary conditions during water-mass formation. This is especially true of tracers entering from the atmosphere. Tritium and radiocarbon are radioactive isotopes whose observed concentrations must first be corrected for natural radioactive decay. Both tracers have widespread use in descriptive studies and large-scale nu-merical modeling of ventilation of the deep ocean and the transformation of water masses over periods of decades. Our main purpose in this section is to provide a brief outline of the types of studies possible with transient tracers. Only results for the main tracers will be presented; secondary tracers such as krypton-85 and argon-39 are not discussed.

1.15.2.1 Tritium

During the late 1950s and early 1960s, large amounts of bomb-produced radiocarbon (^{14}C), strontium (^{90}Sr) and tritium (^{3}H) were released into the stratosphere during aboveground testing of thermonuclear weapons (Figure 1.191a and b). Of these, "bomb" tritium (the heaviest isotope of hydrogen) has an extensive database and is measurable to high precision and sensitivity. Tritium is incorporated directly in water molecules as HTO so that it is a true water-mass tracer. Most of the tritium was produced by tests conducted in the northern hemisphere and was eventually deposited onto the earth's surface north of $15°$ N (Weiss and Roether, 1980; Broecker et al., 1986). Deposition into the oceans is through vapor diffusion and rainfall at a ratio of roughly 2:1 according to observational data. A study by Lipps and Hemler (1992) suggests that the ratio varies according to the type of rainfall. The large fronts across the Pacific and Atlantic oceans at subtropical latitudes impede lateral mixing and the southward transport of tritium. As a result, tritium with a half-life of 12.43 years serves as a useful tracer for water motions on timescales of decades. It is most useful when combined with measurements of its stable, inert daughter product ^{3}He. This combination helps determine the age of tritium entering the ocean and provides additional information on the distribution of tritium in the atmosphere before it entered the ocean (Jenkins, 1988). Most large-scale studies are based on the extensive tritium data collected in the North Pacific during the Geochemical Ocean Sections Study (GEOSECS: 1972−74) and Long Lines (1983−85). Roughly 0.3 l of seawater are required for the measurement of tritium by beta-decay counting.

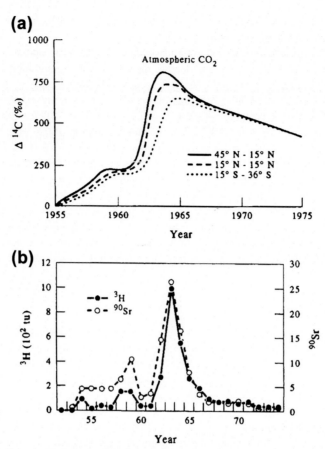

FIGURE 1.191   Time series of bomb-produced elements released into the stratosphere during above-ground testing of thermonuclear weapons during the late 1950s and early 1960s: (a) radiocarbon (^{14}C) and (b) strontium (^{90}Sr) from measurements of atmospheric carbon dioxide and tritium (^3H) based on rain at Valencia Ireland. *Adapted from Quay et al. (1983) and Broecker and Peng (1982).*

Tritium in natural waters is expressed in "tritium units" (TU), which is the abundance ratio $^3H/^1H \times 10^{18}$. The ratio abundance corresponds to 7.09 disintegrations per min per kg of water. To remove the effect of normal radioactive decay from a data series, the tritium concentrations are corrected to a common reference of January 1, 1981. Thus, TU81N is the ratio of $^3H/^1H$ a sample would have as of 1981/01/01. The measurement error for "decay-corrected" data is 0.05 TU or 3.5%, whichever is greater (Van Scoy et al., 1991). Water having values less than 0.2TU81 are considered to reflect cosmogenic background levels or arise from dilution by mixing of bomb tritium. The fact that decay-corrected tritium is a conservative quantity that was added to a selected area of the World Ocean in a relatively short period of time (Figure 1.192) makes it attractive as an oceanic tracer. Changes in the spatial distribution of tritium with time provide a measure of horizontal advection while depth penetrations on isopycnals that do not outcrop to the atmosphere are indicative of cross-isopycnal (diapycnal) mixing. Fine (1985) uses upper ocean tritium data from the GEOSECS program to show that there is a net transport of 5×10^6 m^3/s in the upper 300 m from the Pacific to the Indian Ocean through the Indonesian Archipelago. This contrasts with values of 1.7×10^6 m^3/s obtained using hydrographic data (Wyrtki, 1961) and $5-14 \times 10^6$ m^3/s from salt and mass balances (Godfrey and Golding, 1981; Piola and Gordon, 1984; Gordon, 1986). Gargett et al. (1986) have examined the 9 year record of tritium from Ocean Station P (50° N, 145° W) in the northeast Pacific. Results suggest that the observed vertical distribution of tritium in this region is determined mainly through advection along isopycnals rather than by isopycnal or diapycnal diffusion in the density range of maximum vertical tritium gradient. Tritium data studied by Van Scoy et al. (1991) show evidence for wind-driven circulation to the depth of the dissolved oxygen minimum near 1,000 m depth ($\sigma_t = 27.40$) in subpolar regions of the North Pacific. The authors conclude that, after 2 decades of mixing, advection along isopycnal surfaces appears to be the dominant process influencing the distribution of tritium in the North Pacific and that cross-isopycnal mixing in the subpolar region is important for ventilating the non-outcropping isopycnals. According to Van Scoy et al. (1991), tritium has penetrated on average 100 m deeper into the ocean during the 10 years between the GEOSECS and Long Lines surveys. Depletions of tritium in the

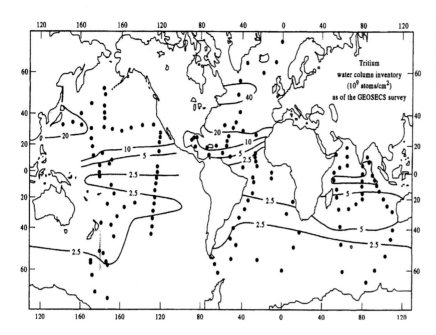

FIGURE 1.192 Decay-corrected tritium (TU81) water column inventories over the World Oceans based on results obtained as part of the GEOSECS program and NAGS expedition. *Adapted from Broecker et al. (1986).*

upper ocean are seen in the tropics and at high southern latitudes. Moreover, the above-background tritium levels observed on non-outcropping isopycnals surfaces in the North Pacific indicate that ventilation is still taking place despite the absence of deep convective mixing in this region. In the Atlantic Ocean, deep convection is the dominant mechanism for the invasion of surface waters into the deep ocean (Figure 1.193).

Tritium data are used to constrain circulation models for the World Ocean. For example, tritium records combined with a three-box model of the Japan Sea (a.k.a., the East Sea) —a comparatively isolated oceanic region with a mean depth of 1,350 m—have yielded overturn times for the deep water of 100 years and overall residence times of 1,000 years (Watanabe et al., 1991). Similar estimates for this region based on the same box-model constrained by ^{226}Ra and ^{14}C data yielded a turnover time of 300—500 years for deep water and 600—1,300 years for the residence time (Harada and Tsunogai, 1986). Applications to larger oceanic basins are generally less successful. Memery and Wunsch (1990) found that the tritium data did not strongly constrain their circulation model for the North Atlantic and that large errors ($\approx 20\%$) in the input of tritium at the surface can be accommodated by relatively minor changes in the model circulation. According to Wunsch (1988), "Any uncertainty in the transient tracer boundary conditions and sparse interior ocean temporal coverage greatly weakens the ability of such tracers to constrain the ocean circulation". Although the authors still believe in the usefulness of tritium records, they suggest that chlorofluorocarbons will improve modeling capability since the atmospheric concentration of these compounds remains relatively high despite the 1988 Montreal Accord and are better known than for tritium.

Jenkins (1988) describes the use of the tritium-^{3}He age, which takes advantage of the radioactive clock of ^{3}He and the long timescale of tritium to measure the elapsed time since the Helium gas was in equilibrium with the atmosphere. Time scales for which this combined tracer is useful are 0.1—10 years.

1.15.2.2 Radiocarbon

Carbon-14 (^{14}C) dating requires prior knowledge of long-term variations in the ^{14}C/^{12}C ratio in the atmosphere. Because of the difficulties in separating radiocarbon produced from thermonuclear devices and cosmic rays, bomb-generated radiocarbon is a less useful tracer of upper ocean processes than is tritium. The problem of using radiocarbon data collected prior to 1958 together with tritium measurements to establish the prenuclear levels of radiocarbon is discussed by Broecker and Peng (1982). Once the prenuclear surface-water cosmic radiocarbon concentration is known for each locality, water column inventories for bomb-radiocarbon can be obtained from the depth profiles of ^{14}C/^{12}C, ^{3}H, and $\sum CO_2$ concentration obtained as part of the GEOSECS, NORPAX, and TTO programs (Broecker et al., 1985). Bomb-produced radiocarbon is delivered through a nearly irreversible process from the atmosphere to the ocean so that it is possible to estimate the amount of this isotope that has entered any given region of the ocean. As a result of this production, levels of ^{14}CO_2$ increased by about a factor of two in the northern hemisphere during the late 1950s and early 1960s. Measurement of

FIGURE 1.193   Cross sections of tritium (TU81N) concentration in the western Atlantic Ocean (a) GEOSECS 1972; (b) TTO 1981. Results suggest that the observed vertical distribution of tritium in this region is determined mainly through advection along isopycals, rather than by isopynal or diapycnal diffusion in the density range of maximum vertical tritium gradient. *(a and b) From Östlund and Rooth (1990).*

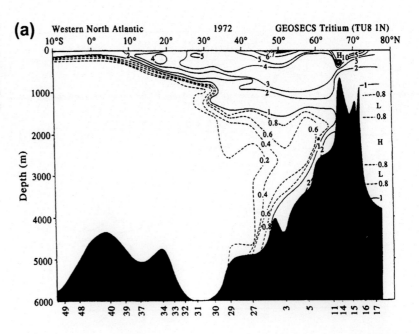

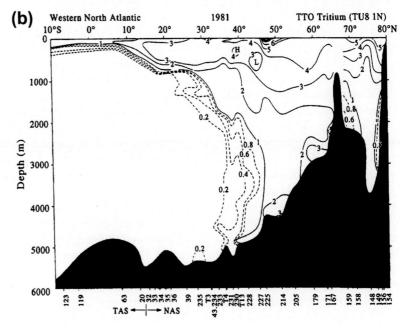

radiocarbon by beta decay requires 200–250 l of seawater to give the desired accuracy of 3–4 parts per thousand (ppt). Age resolution is 25–30 years for abyssal oceanic conditions for which the introduction of bomb-radiocarbon effects remain negligible. Radiocarbon has a half-life of 5,680 years and decays at a fixed rate of 1% every 83 years. A rapid onboard technique for measuring radiocarbon using an accelerator mass spectrometer is described by Bard et al. (1988). This technique decreases the sample size by 2,000 compared with that using the standard β-counting method.

By convention, radiocarbon assays are expressed as $\Delta^{14}C$, which is the deviation in parts per 1,000 (ppt) of the $^{14}C/^{12}C$ ratio from that of a hypothetical wood standard with $\delta^{13}C = {}^{13}C/^{12}C = -25$ ppt and corrected from the actual $\delta^{13}C$ values (around 0 ppt for seawater to exactly -25 ppt to compare with the wood standard). The standard is a way to compare the observed ratio of carbon isotopes to the atmospheric value prior to the industrial revolution starting about 1850. The quantity of ^{14}C in a sample of seawater is proportional to the actual uncorrected $\delta^{14}C = {}^{14}C/^{12}C$ ratio $(1 + 0.001\delta^{14}C)$. More precisely

$$\Delta^{14}C = \delta^{14}C - 2(\delta^{13}C + 25)(1 + \delta^{14}C / 1,000) \qquad (1.104a)$$

where

$$\delta^{14}C = 1,000\left[\left(\left(^{14}C/C\right)_{sample} - \left(^{14}C/C\right)_{standard}\right) / \left(^{14}C/C\right)_{standard}\right] \qquad (1.104b)$$

Pre-bomb $\Delta^{14}C$ values from corals collected in the early 1950s average around -50 (± 5) ppt (Druffel, 1989). Thus, any $\Delta^{14}C$ value above -50 ppt will indicate the presence of anthropogenic radiocarbon, mainly produced by the atmospheric nuclear testing in the early 1960s. The determination of inventories for bomb-produced radiocarbon in the ocean is much more complex than for bomb tritium. The reason is that the amount of natural tritium in the sea is negligible compared with the amount of bomb-produced tritium. In the case of radiocarbon, the delivery of isotopes to the ocean requires a better knowledge of wind speeds over the ocean and of the wind speed dependence of the CO_2 exchange rate.

The concentration of ^{14}C in the ocean is influenced by several processes. For example, bottom water formation in the Weddell Sea and the North Atlantic provides a direct input of surface water ^{14}C (Figure 1.194). Additional input of ^{14}C to the deep sea can occur by transport along isopycnals, by vertical mixing in the main oceanic thermocline, by lateral mixing of water masses and by upwelling in coastal and equatorial regions. Addition of CO_2 and ^{14}C comes from the dissolution of carbonate skeletons and the oxidation of organic materials from sinking particles. Stuiver et al. (1982) use radiocarbon data from GEOSECS to estimate abyssal ($>1,500$ m) waters replacement times for the Pacific, Atlantic and Indian Oceans of 510, 275, and 250 years, respectively. The deep waters of the entire World Ocean are replaced on average every 500 years. Östlund and Rooth (1990) found a relative decrease in the difference in $\Delta^{14}C$ between the surface and the northerly abyssal layers of the North Atlantic of 25–30%. If this were due to vertical diffusivity a high value of 10 cm^2/s would be required based on a scale depth of 1 km and 10 years between surveys. This is a factor of 10 too large so that high latitude injection processes must be responsible for the observed evolution below 1,000 m depth. Measurements of $\Delta^{14}C$ from seawater and organisms from the Pacific coast of Baja California (Druffel and Williams, 1991) revealed the effects of coastal upwelling and bottom-feeding habits. Dilution of nearshore waters by upwelling accounts for reduced radioactive carbon levels observed near the coast while feeding on sediment-derived carbon explains the reduced levels of ^{14}C in sampled organisms relative to dissolved inorganic carbon in the water column. Broecker et al. (1991) have addressed the concerns about the accuracy of ventilation flux estimates for the deep Atlantic due to temporal changes in the $^{14}C/C$ ratio for atmospheric CO_2. Despite the fact that $\Delta^{14}C$ values have declined from about 10 to 20 ppt over the past 300 years due to changes in the solar wind and the addition of $\Delta^{14}C$-free CO_2 to the atmosphere from fossil fuel burning, temporal effects have been considerably buffered in the ocean and errors in radiocarbon ages are too low by only 10–15%. The reason is that the northern and southern source waters for the Atlantic deep water have $\Delta^{14}C$ ratios (and hence relative time variability) considerably lower than the atmospheric ratio. For further details on the methodology, the reader is again referred to Chapter 6 of the 2007 IPCC report (Jansen et al., 2007).

1.15.2.3 Chlorofluorocarbons

Chlorofluorocarbons (CFCs) are a group of volatile anthropogenic compounds that until the 1988 Montreal Protocol found increasingly widespread use in aerosol propellants, plastic foam blowing agents, refrigerants and solvents. Also known as chlorofluoromethanes (CFMs) and "Freons" (a Dupont trade name), most of these chemicals eventually find their way into the atmosphere where they play a primary role in the destruction of stratospheric ozone. The two primary compounds CFC-12 or F-12 (CF_2Cl_2) and CFC-11 or F-11 ($CFCl_3$) have respective lifetimes in the troposphere of 111 and 74 years. Although more than 90% of production and release of F-11 and F-12 takes place in the northern hemisphere, the meridional distributions of these compounds in the global troposphere are relatively uniform due to the high stability of the compounds and the rapid mixing that occurs in the lower atmosphere. The source function at the ocean surface differs by only about 7% from the northern hemisphere to the southern hemisphere (Bullister, 1989). During the period 1930–75 the ratio F-11/F-12 in the atmosphere and ocean surface increased with greater uses of these chemicals (Figure 1.195). The regulation of CFC use in spray cans in the U.S.A. during the late 1970s decreased the rate of CFC-11 increase so that the ratio F-11/F-12 ratio in the atmosphere has remained nearly constant. As a consequence, measurements of the ratio provide information on when a particular water mass was last in contact with the atmosphere. In shelf waters the CFCs concentration is determined by rates of mixed layer entrainment, gas exchange and mixing with source water. At a removal rate of about 1% per year from the atmosphere by stratospheric photolysis, CFCs will serve as ocean tracers well into the next century.

Because they are chemically inert in seawater, Chlorofluorocarbons are used to examine gas exchange between the atmosphere and ocean, ocean ventilation and mixing on decadal scales. The limit of detection of F-11 and F-12 in seawater volumes as small as 30 ml is better than 5×10^{-15} mol/kg seawater (Bullister and Weiss, 1988), or roughly three orders of magnitude higher than near-surface concentrations in the ocean. Modern techniques allow for the processing of CFCs at

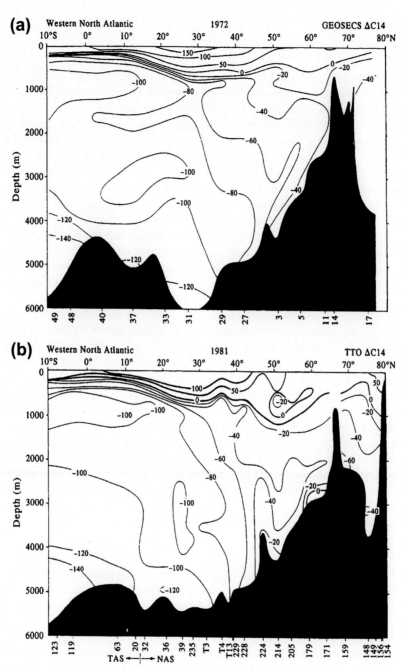

FIGURE 1.194 Cross sections of radiocarbon concentrations in the western Atlantic Ocean (a) GEOSECS 1972; (b) TTO 1981. Note that significant changes occur mainly in the deep waters north of 40° N. *From Östlund and Rooth (1990).*

sea with processing times of the order of hours. Gammon et al. (1982) examined the vertical distribution of CFCs at two offshore sites in the northeast Pacific. Using a one-dimensional vertical diffusion/advection model driven by an exponential surface source term, they obtained a characteristic depth penetration of 120–140 m. For a Gulf of Alaska station at 50° N, 140° W, vertical profiles of F-11 and F-12 gave consistent vertical diffusivities of order 1 cm^2/s and an upwelling velocity of 12–14 m/year. Woods (1985) used CFCs to estimate the transit time and mixing of Labrador Sea Water from its northern source region to the equator along the western Atlantic Ocean boundary. In a related study, Wallace and Lazier (1988) used CFCs and a simple convection model to examine recently renewed Labrador Sea Water formed by deep convection to depths greater than 1,500 m following a severe winter in the North Atlantic. Their observed CFC levels of 60% saturation with respect to contemporary atmospheric concentrations suggest that deep convection took place too

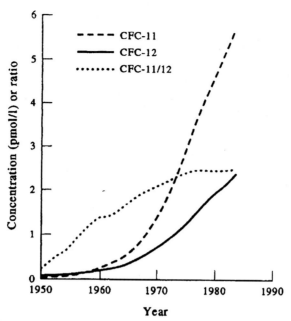

FIGURE 1.195 CFC-12 and CFC-11 concentrations in the upper ocean for $T = -1°C$ and $S = 34.3$ psu (g/kg) as function of time. *From Trumbore et al. (1991).*

rapidly for air-sea gas exchange to bring CFC levels to equilibrium. Trumbore et al. (1991) have used CFCs collected in 1984 to examine recent deep-water ventilation and bottom water formation near the continental shelf in the Ross Sea in the Antarctic Ocean. Using CFC data combined with conventional (temperature, salinity, dissolved oxygen, and nutrient) tracer data in a time-dependent convection model they estimate shelf-water resident times of about 3 years for the Ross Sea. At the other end of the globe, Schlosser et al. (1991) use hydrographic and CFC data to suggest that formation of Greenland Sea Deep Water decreased in the 1980s. The dissolved F-12 concentration in Figure 1.196 illustrates several aspects of the circulation in the North Atlantic. In particular, we note the core of the Labrador Sea Water mentioned earlier in this section, the presence of a lens of Mediterranean outflow water ("Meddy") at 22° N and the core of high CFC over the equator which is thought to be a longitudinal extension of flow from the western boundary near Brazil (Bullister, 1989).

1.15.2.4 Radon-222

Radon (^{222}Rn) is a chemically inert gas with a radioactive half-life of 3.825 days. It occurs naturally as a radionuclide of the ^{238}U series and is injected into the atmosphere by volcanic eruptions. The gas has proven particularly useful at timescales of a few days to weeks for examining the rate of gas transfer between the atmosphere and the ocean surface (Peng et al., 1979), in studies of water column mixing rates (Sarmiento et al., 1976), and for estimating the heat and chemical fluxes from hydro-thermal venting at mid-ocean ridges (Rosenberg et al., 1988; Kadko et al., 1990).

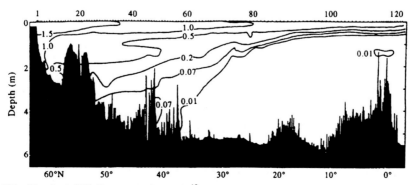

FIGURE 1.196 Dissolved CFC-12 concentrations (10^{-12} mol/kg) along a North Atlantic section. *From Bullister (1989).*

The new application of ^{222}Rn studies to hydrothermal venting regions has been especially successful (Rosenberg et al., 1988). In this case, it is assumed that there is a constant flux of radon into the effluent plume that typically rises several hundred meters above the venting region at depths of 2–3 km on the ridge axis. Typical venting regions have scales of 100 m and are spaced at several kilometers along the ridge axis. Waters exiting from black smokers can be up to 400°C. At steady state, the amount of radon lost to radioactive decay at some point in the laterally spreading nonbuoyant plume is balanced by a supply of radon from the venting region. To obtain the total heat (or chemical species) issuing from the venting region, the observer first uses a submersible or towed sensor package to measure the ratio of radon to heat (or species) anomaly, ^{222}Rn/ΔT, in the plume near the vent orifice—before the radon in the plume has a chance to disperse or age. The observer then uses a towed sensor package to map the total inventory of radon in the spreading plume (Figure 1.197). Taking into account the effect of cold-water entrainment on the rising plume at Endeavour Ridge in the northeast Pacific (47°47′ N, 129°06′ W), Rosenberg et al. (1988) found an initial radon/ΔT value of 0.03 dpm (disintegrations per minute—the standard unit of measurement for radioactive materials) or 55 atoms per joule. They then used hydrocast bottle data to estimate the standing crop of radon above 2,100 m depth as ^{222}Rn(Total) = 8×10^{12} dpm. At steady-state, hydrothermal venting must be adding this much radon to the system so that the total heat emanating from the vents is.

$$^{222}\text{Rn(Total)} / \left(^{222}\text{Rn} / \Delta T\right) = 3(\pm 2) \times 10^9 \text{ watts} \tag{1.105}$$

which compares with estimates based on direct measurements of the total heat content anomaly of the plume in combination with local currents (Baker and Massoth, 1986, 1987; Baker et al., 1995). Gendron et al. (1993) have used ^{222}Rn to examine time variability in hydrothermal venting on the Cleft Segment of Juan de Fuca Ridge and to estimate the age of the plume as a function of location relative to the known vent sites. They found that the hydrothermal flux decreased from 2.2 ± 0.3 GW in 1990 to 1.2 ± 0.2 GW in 1991 (1 GW = 10^9 W).

The estimates using radon-222 in the ocean are complicated by the fact that radon concentration is a function of both radioactive decay and dilution with ambient seawater. Similar estimates can be made using ^{3}He to heat ratios combined with the total inventory of ^{3}He in the ocean. The result (Jenkins, 1978) is a global hydrothermal heat flux of 4.9×10^{19} cal/year. Baker and Lupton (1990) have used the ^{3}He/heat ratio as a possible indicator of magmatic/tectonic activity at ridge segments. The change from a ratio of 4.4×10^{-12} cm^3 STP/cal immediately following the megaplume eruption at Cleft segment to 1.3×10^{-12} cm^3 STP/cal 2 years later suggests that high ratios may be indicative of venting created or profoundly perturbed by a magmatic-tectonic event, while lower values may typify systems at equilibrium.

FIGURE 1.197 Apparent age of the neutrally buoyant plume on the isopycnal surface $\sigma_\theta = 27.68$ (roughly 2,100 m depth) at the Cleft Segment of Juan de Fuca Ridge in the northeast Pacific. Distribution based on Radon-222 data for September 1990. Depths in meters. Plume rises from the hydrothermal vent depth of 2,280 m to approximately 2,100 m. *From Gendron et al. (1993).*

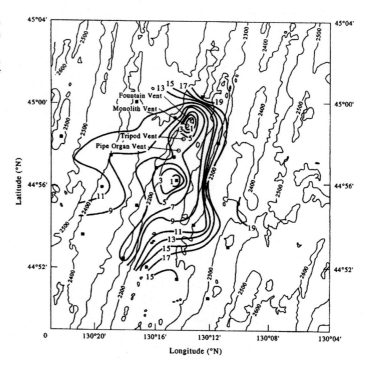

1.15.2.5 Sulfur Hexafluoride

In certain instances, there is a distinct advantage to a controlled and localized release of a chemical into the environment. Prefluorinated tracers such as sulfur hexachloride (SF_6) and perfluorodecalin (PFD) are among the new generation of deliberately released tracers used to measure mixing and diffusion rates in the ocean. These substances are particularly good at examining vertical mixing. Their appeal is that they are readily detectable conservative tracers that have no significant effect on the environment and no toxicity. A thorough description of the use of these tracers as well as rhodamine dyes can be found in Watson and Ledwell (1988). In the case of rhodamine dyes, the detection limit by fluorometers is set by the background fluorescence of natural substances in water, which is about one part in 10^{12} in the deep ocean, although Lane et al. (1984) and J.M. Suijlen (pers. comm., 2002) suggest that this limit can be increased to one part in 10^{16}. For SF_6 the background limit is set by dissolution from the atmosphere where the compound is present at one to two parts in 10^{12} by volume. Surface values in the ocean are roughly 5×10^{-17} and diminish to zero in deep water. The instrumental detection for SF_6 is limited to about 1/10 of the near-surface value (Watson and Liddicoat, 1985). PFD has no measurable background level in the ocean and is limited by instrumental detection to about one part in 10^{16}. For a release of 1 metric ton ($\equiv 1,000$ kg) at a given density level in an experiment, these detection limits translate to maximum horizontal scales of 100 km for rhodamine dye, 1,000 km for PFDs and basin scales for SF_6. Lifetimes for the tracers range from months to about a year. Despite their usefulness, the long-term prognosis for SF_6 and PFDs is limited as industrial injection of SF_6 into the atmosphere and medical use of PFDs will eventually increase background levels and take away from their ability to serve as tracers.

Rhodamine dye is used mainly in coastal studies and can be as effective as SF_6 as a conservative tracer for studying a broad range of mixing processes near the sea surface provided one uses two rhodamines with different photolytic decay rates (Suijlen and Buyse, 1994; Upstill-Goddard et al., 2001). SF_6 has been used successfully in WOCE. The North Atlantic Tracer Release Experiment (NATRE) was a large-scale WOCE-related study using SF_6 to examine the stirring and diapycnal mixing in the pycnocline of the North Atlantic. In May 1992, 139 kg of sulfur hexafluoride was released on the isopycnal surface 26.75 kg/m^3 (310 dbar) along with eight SOFAR floats and six pop-up drifters in the eastern subtropical Atlantic near 25.7° N, 28.3° W. To sample the tracer, investigators towed a vertical array of 20 integrating samplers at 0.5 m/s through the patch. A prototype 18-chamber sampler at the center of the array obtained a lateral resolution of about 360 m. The average profile increased from an RMS thickness of 6.8 m after 14 days to an RMS thickness of about 45 m by April 1993, yielding a diapycnal eddy diffusivity of 0.1–0.2 cm^2/s ($= 0.1–0.2 \times 10^{-4}$ m^2/s). To be successful, experiments like NATRE require the tracer to be injected on a constant density surface rather than a constant depth. Internal wave oscillations and other vertical motions would broaden the tracer concentration more than necessary if it were released at a constant depth. Care must be taken during injection to ensure the tracer's buoyancy is correct and that the turbulent wake of the injection apparatus is not excessive. A recent large-scale study of diapycnal mixing near 1,500 m depth in the Antarctic Circumpolar Current using trifluoromethyl sulfur pentafluoride (CF_3SF_5) is presented by Ledwell et al. (2011). Released to the west of Drake Passage, the tracer indicates that the diapycnal diffusivity, averaged over 1 year and over tens of thousands of square kilometers, is $(1.3 \pm 0.2) \times 10^{-5}$ m^2/s. The authors report that turbulent diapycnal mixing of this intensity is characteristic of the mid-latitude ocean interior, where the energy for mixing is thought to originate with internal wave breaking. Results support evidence that diapycnal mixing in the interior mid-depth ocean is weak and is likely too small to dictate the mid-depth meridional overturning circulation of the ocean.

1.15.2.6 Strontium-90

Most of the radioactive strontium released to the environment was from the atmospheric testing of nuclear bombs from 1945 to 1980. The nuclear weapons tests injected radioactive material into the stratosphere, which resulted in extensive dispersal of radioactive strontium and other radionuclides. The Chernobyl nuclear power plant incident in the Ukraine in April 1986 also caused the release of strontium-90 into the atmosphere. Because the half-life of strontium-90 decay is around 29 years, some strontium-90 reached the upper atmosphere and was subsequently transported around the world. Radioactive strontium release also occurs from the operation and maintenance of nuclear power plants, but these values are negligible compared to the levels associated with the atmospheric testing of nuclear weapons and the Chernobyl incident.

The distribution of bomb-produced ^{90}Sr in the ocean is quite similar to that of tritium. However as pointed out by Toggweiler and Trumbore (1985), ^{90}Sr has the virtue that the ratio ^{90}Sr/Ca incorporated into coral skeletons has the same value as this ratio in seawater. Corals average out seasonal variations in the ^{90}Sr content of seawater so that annual bands provide a time-averaged measure of the amount of strontium in the water. The results of Toggweiler and Trumbore (1985) revealed that waters move into the Indian Ocean via passages through the Indonesian Archipelago. In addition, the data suggested that there is a large-scale transport of water between the temperate and tropical North Pacific. Strontium-90 in the surface waters of the North Pacific has decreased steadily from 0.923×10^{-12} Ci/L sea water in the early 1960s to present day levels of approximately 0.081×10^{-12} Ci/L sea water, where 1 Ci (1 Curie) $= 3.7 \times 10^{10}$ disintegrations per second. The total amount of strontium-90 released to the atmosphere from weapons testing (1945–80) was 1.6×10^7 Ci (or 6×10^{17} Bequel).

1.16 OCEANOGRAPHIC DATA FORMATS

Physical oceanographic data are archived in a number of different formats. These include character format, native format, packed binary, or one of several "scientific" data formats. The purpose of the scientific data formats is to provide additional information to better describe the contents of the dataset. It is important for users of these archived data to know what format the data are in and how the data can be retrieved. Usually, a comprehensive description of the dataset is given along with some level of software that enables the user to read the data and incorporate the values to subsequent computations.

In discussing common archival formats, it is critical to ensure that the computer terminology is clear. The smallest quantity of information in a data file is called a "bit." Each bit may be "on" or "off", as represented by a 0 or a 1. Computers store both text and numbers as sequences of bits. A sequence of 8-bits is called a *byte* and it often used to describe a character of text. An ordered sequence of bytes is called a *word*. Generally, workstations and supercomputers used by the oceanographic community use 32- or 64-bit words. If bytes refer to characters, then a 32-bit word would contain 4 characters and a 64-bit word could contain 8 characters. A computer word used for storing floating-point numeric values consists of three segments: a sign bit, a characteristic (biased exponent) and a mantissa. An integer is represented by two segments: a sign bit and a sequence of bits. A 32-bit word can store floating point numbers to a precision of six-to-seven decimal places, while a 64-bit word can store numbers with a precision of thirteen-to-fourteen decimal places.

A workstation with 32-bit precision can use and store data in the 64-bit mode by declaring the values in double precision in FORTRAN and double-*int* and long *int* in "C." Most computers use numbers where the most significant bytes are ordered from left to right (called "big-endian" form). Other systems use the "little-endian" form and reverse the byte order. This order characteristic of a computer can result in difficulties when dealing with binary data created on different machines but there is usually software provided to convert between the two forms. In the early days of computers, datasets were limited in size based on what a 7- or 9-track magnetic tape could hold. Today, there are many different forms of storage media available, and the sizes of datasets are in terms of kilobytes (KB) — a kilobyte refers to 1,024 bytes and not to 1,000 bytes — mega-bytes (MB, 10^6), giga-bytes (GB, 10^9), tera-bytes (TB, 10^{12}) and peta-bytes (PB, 10^{15}).

1.16.1 Character Format

Generally, this is the most convenient form for the user. The most common and widely used character set is that based on the ASCII standard. The other character set is EBCDIC, which is an IBM standard. ASCII and EBCDIC are both 8-bit character sets. It is possible to convert from one character set to another using software tools (or "filters" in UNIX jargon). The advantage of using character data is that it can be read by a human being or through standard computer input/output statements in either the FORTRAN or C programming languages. Character data are ideal for "small" datasets that need to be read on a variety of computers. This becomes a problem when the dataset is large because the number of computer instructions needed to read character data is considerable and character data take up a large amount of space on computer external media, such as discs, tapes or other portable memory. For example, the numbers "9", "679.43", $-0.123456E+05$ require 1, 6, and 13 bytes respectively, to store and manipulate in ASCII. These numbers can be archived much more concisely using packed binary representation.

1.16.2 Native Format

Computers from different manufacturers may use different schemes to represent both numbers and characters. Formats that are internal to a given type of computer are commonly referred to as the "native format." As mentioned earlier, characters are generally stored in ASCII, although some machines will use EBCDIC. Numeric values may be in an IEEE format standard or a vendor specific representation (e.g., Cray, IBM). Reading and writing data in native format is fast if one is on the appropriate computer and no conversion is required. In addition, no precision is lost. However, reading one machine's native format on a different company's computer, which uses a different native format, can be very slow because of the complex conversion algorithm that must be used. Another drawback is that each numeric value archived in native form requires the full word length of the computer (e.g., 32 or 64 bits). Thus, each value is stored with full machine precision even if it uses more bits than necessary for the known accuracy of the values.

1.16.3 Packed Binary Format

Datasets in physical oceanography can be quite large (satellite images, float and drifter data, high resolution ADCP time series, etc.), therefore it is best to minimize the amount of information that needs to be archived to be able to retrieve the data. This optimization of information using a minimum number of bits is accomplished using something called packed binary. This is a very efficient way of archiving data, and it is independent of computer representation. This means that it will work across platforms. Packing data values is done by expressing integer and most often floating-point values in sequences of bits sufficient to capture the required precision of the data. For example, the floating-point number "3.1" requires 32- or 64-bit in native format or 24-bit (three ASCII or EBCDIC characters) in character format. However, in packed binary it requires only 5 bits. In general, binary files make the best use of the storage space and it is still possible to retrieve the ASCII equivalent. In some cases, however, the binary representation will not be exactly the same as the ASCII file. We need to recognize that to a computer all data is binary as that is the only format that a computer can deal with, and it is the computer system firmware that creates the bits and bytes.

To understand packed binary, we first need to review the binary representation of data. Suppose we want to store the number 4,000,000,000 (4 billion). In the usual ASCII system, we would write 400000000 using a total of 10 ASCII characters (or in this case, 10 bytes), where each byte has 8 bits, or $2^8 = 256$ possible values. In the binary system, the storage of 4 billion values requires roughly $2^{32} = (2^8)^4$ bits, whereby the number 4 billion can be stored in only 4 bytes. While binary formats are very efficient, they are not intuitively easy to read and work with. Packed binary usually means that more (if not all available) bit combinations are used to encode values, while unpacked means that some bit combinations remain unused either to improve readability or to make certain calculations easier. The basic message is that unpacked data require more storage space.

Packed binary numbers are stored sequentially and retrieved from a bit stream. There are many software packages to pack and unpack binary bit steams. Although software must be used to convert the bit groups to a machine's readable internal format, this conversion is considerably faster than converting a character data set. An additional benefit to packed binary data is that it enables network data transfer due to the far smaller volume.

1.16.4 Scientific Data Formats

There is a plethora of "standard" scientific data formats that are basically designed to provide as much information about the data as possible within the data format itself. Often these science formats are dictated by the computer architecture. We list many of these formats and provide a brief description of each.

- GRIB (Grid in Binary) is a World Meteorological Organization (WMO) standard data format, which is an efficient method for transmitting and archiving large volumes of two-dimensional meteorological and oceanographic data. GRIB was first created in 1985 and has undergone many updates. The latest version (GRIB2), declared operational in 2001, is widely used by all global weather information systems. GRIB2 is more flexible that the original GRIB and can handle radar and satellite data. It also allows for greater data compression.
- CDF (Common Data Format) was initially developed around 1980 by the NASA Goddard Space Flight Center (GSFC) as an interface and a toolkit for archival and access to multidimensional data on a VAX computer using VMS FORTRAN. Over the years, CDF has evolved into a machine-independent standard and is used by different groups (both NASA and others) for storing both space and Earth science data. CDF is a "self-describing" data format for the storage of scalar and multidimensional data in a platform and discipline independent way. The format allows users to easily manage the collected arrays of data. Built into CDF is support for data compression (gZip, RLE, Huffman) and automatic uncompressing and checksum. CDF provides support for large (>2 Gbytes) files and has a library that supports CDF data processing. CDF interacts well with higher-level programming packages such as IDL and MATLAB.
- NetCDF (network CDF) is a commonly used data format developed in 1986 by the Unidata program of UCAR in Boulder, Colorado. The format uses CDF as a starting point, with the goal to make the format machine independent so that it could be used by users across the network to access and store data. Strangely, netCDF is not compatible with the NASA CDF. NetCDF provides a single common interface to the data, which is implemented on top of an architecture-independent representation of the data. A netCDF file includes information about the data it contains. A netCDF file can be accessed and utilized by computers independent of the way they store integers, characters and floating-point numbers. Small subsets of large datasets in various formats may be accessed efficiently through netCDF interfaces, even from remote servers. Data may be appended to a properly structured netCDF file without having to recopy the dataset or redefine its structure. A netCDF file may be shared between writers and readers of the same

netCDF file. Unidata provides continued support of all earlier and present forms of netCDF software. For more information see http://www.unidata.ucar.edu/software/netcdf/.

- HDF (Hierarchical Data Format) is a general, extensible scientific data exchange format created by the National Center for Supercomputing Applications (NCSA) at the University of Illinois at Urbana-Champaign. There are two main compatible versions: HDF4 and HDF5. HDF emphasizes a single common data format, on which many interfaces can be built. A netCDF interface to HDF4 is provided but there is no support for mixing HDF and netCDF structures. In other words, HDF4 software can read HDF and netCDF but can only write in HDF4. Both HDF4 and HDF5 are more flexible than netCDF but also are more complicated. HDF is often used to archive and transmit images.

- BUFR (Binary Universal Format Representation) is another WMO (see GRIB above) standard data format for the representation of meteorological and oceanographic data. Although it can be used for any kind of data, BUFR's primary use is for observational data from measurement stations such as radiosonde profiles sea surface temperature, air temperature, and other variables. It was designed to reduce redundancy for efficient transmission over the Global Telecommunications System (GTS) and to reduce the time needed to decode the information. CSV (Comma-separated/variables) is typically used with ASCII tables and is common in applications like Excel.GIF (Graphics Interchange Format), a copyrighted format widely used on webpages and limited to 256 colors (called the GIF palette), although there are many different color mapping tables. GML (Geography Markup Language) is the XML (see below) grammar defined by the Open Geospatial Consortium (OGC) to express geographical features. Noteworthy for providing a noncommercial, logical, ASCII replacement for shapefiles and for coverages. JPEG (Joint photographic experts' group compression scheme) is an image format based on Scheme, a widely used image format.

- KML (Google's keyhole markup language). A metadata "wrapper" around local or remote images; can also contain vector drawing instructions like a shapefile.PNG (Portable Network Graphics). A public domain format to replace JPEG and GIF (which are copyrighted by the originators).TIF, TIFF (Tagged Image Format). This format can hold images, but the specification also allows data rasters to be stored (for example, floating point numbers or integers); both forms can include internal geo-referencing tags. When the georeferencing is internal, then the name GeoTIFF is used. When the file contains numerical data, it is called a "raster."

- XML (Extensible Markup Language). A standard that provides a set of rules for encoding documents in a format that is both human-readable and machine-readable. It is widely used for the representation of arbitrary data structures (websites).ZIP (Data compression and archiving format). Widely used format, originally developed by the company PKZIP.

1.16.5 Merging of HDF and netCDF

A release of netCDF in 2005 has an Applied Programming Interface (API) that has been extended and implemented on top of the HDF5 data format. NetCDF users will be able to create HDF5 files with benefits not available with the netCDF format alone, such as much larger files and multiple unlimited dimensions. This new release also supports access to older netCDF files. A combined software library will include the best features of both netCDF and HDF5, while taking advantage of their individual strengths.

This is by no means a comprehensive list of formats used in physical oceanography as there are many more and, it is safe to assume, more will arise in the near future. Why so many formats? The answer is both historical and practical. Historically, many agencies developed their own internal format standards for data archival. This was during a period when scientists spent a lot of time with their own data. As our ability to observe expanded, datasets became much larger and groups began to share data. As a consequence, data formats became standards between the participating members of a group that shared that data. Soon, it became apparent that a common format would greatly facilitate the ability for researchers to work together. Members of science organizations devoted considerable effort to data matters and soon people became interested in creating common or standard formats. Still, there has been no unifying organization that guides the use of specific standards. The result has been a proliferation of "standard" formats. Added to this is the constant desire to improve these standard formats, which has led to new releases and efforts to combine various formats. Another driver of standard formats is changes in technology, which expands the need for new kinds of formats. The advent of satellite data increased the need to handle large image files and the explosion of meteorological observations generated a need to standardize read/write formats.

Glossary

AABW Antarctic Bottom Water
AADI Aanderaa Instruments Inc. (Norway)

AATSR Advanced Along-Track Scanning Radiometer

ABI Advanced Baseline Imager

ACC Antarctic Circumpolar Current

ACM Acoustic Current Meter

ADCM Acoustic Doppler Current Meter

ADCP Acoustic Doppler Current Profiler

ADEOS A Japanese satellite

ADT Alternating Decision Tree

AGC Automatic Gain Control

AHI Japanese Advanced Himawari Imager

AIAS Aarhus Institute of Advanced Studies in Denmark

AIRS Atmospheric Infrared Sounder AMR (Advanced Microwave Radiometer)

AMSR-2,E Advanced Microwave Scanning Radiometer

AMSU Advanced Microwave Sounding Unit (NOAA, USA)

AOML Atlantic Oceanographic and Meteorological Laboratory (USA)

AOU Apparent oxygen utilization

ARG Acoustic rain gauge

ARGO International program using profiling/drifting floats

ARGOS Advanced Research and Global Observation Satellite (France)

ASCAT Scatterometer on the European Metop satellites

ATLAS Advanced Topographic Laser Altimeter System

ATMS Advanced Technology Microwave Sounder (NOAA, USA)

AUV Autonomous Underwater Vehicle

AVHRR Advanced Very High Resolution Radiometer

AVISO Archiving, Validation and Interpretation of Satellite Oceanographic Data (France)

BCE Before the common era

BL Body length

BPR Bottom pressure recorders

Bragg grating A microstructure within the core of an optical fiber comprising a periodic modulation of the refractive index of the underlying glass material.

BT Brightness Temperature (T_b)

CDPD Cellular Digital Packet Data

CDS Climate Data Store (Copernicus Climate Change Service)

CDW Cloud drift winds

CFC Chlorofluorocarbon

CHAMP CHAllenging Minisatellite Payload

CLIVAR Climate Variability and Predictability Experiment

CLS Collecte Localisation Satellites (France)

CMA China Meteorological Administration

CNES Centre national d'études spatiales; French National Centre for Space Studies

COADS Comprehensive Ocean-Atmosphere Data Set

CODAR Coastal Ocean Dynamics Applications Radar

Conservative temperature Potential enthalpy divided by the fixed heat capacity

cph Cycles per hour

CTD Conductivity-Temperature-Depth profiler

DART Deep-ocean Assessment and Reporting of Tsunamis (NOAA, USA)

DBAR Decibar (pressure)

DGPS Differential GPS; WADGPS: Wide area differential GPS

DMOC Delayed mode quality control system

DOE Department of Energy (USA)

DORIS Doppler Orbitography and Radiopositioning Integrated by Satellite

DPR Dual frequency precipitation radar

DUACS Data Unification and Altimeter Combination System

EBS Eastern Bering Sea

ECM Electromagnetic Current Meter

ECMWF European Centre for Medium-range Weather Forecasts

EDSU Elementary Distance Sampling Unit

EKE Eddy kinetic energy

Enthalpy Thermodynamic quantity equivalent to the total heat content of a system
ENVISAT Large European polar orbiting satellite
EOS-80 Thermodynamic Equation of Seawater-1980
ERA-5 Fifth generation of the ECMWF analysis
ESR Earth and Space Research (ESA)
FNOC Fleet Numerical Oceanography Center (USA)
FOV Field of View (satellite sensor)
FY FengYun (wind and cloud)
g Grams (*g*: acceleration due to gravity)
GCOM Global Change Observation Mission
GCOS Global Climate Observing System
GDP Global Drifter Program
GEBCO General Bathymetric Charts of the Oceans
GEK Geomagnetic electrokinetograph
Geoid The hypothetical shape of the earth, coinciding with mean sea level and its imagined extension under (or over) land areas.
GEOSAT GEOdetic SATellite, an early altimeter satellite of the US Navy
GEOSECS Geochemical Ocean Sections Study
GHRSST Group for High Resolution SST
Glider Subsurface vehicle that uses changes in buoyancy and wings for forward motion.
Global Drifter Program International program to deploy and monitor satellite-tracked drogued drifting buoys
GMI GPM microwave imager
GMT Greenwich Mean Time ("Zulu"), now UTC for "Universal Time Coordinated" or "Universel Temps Coordonné"; also Coordinated Universal Time
GODAE Global Ocean Data Assimilation Experiment
GOES Geostationary Operational Environmental Satellite (NOAA)
GPI Global Precipitation Index
GPM Global Precipitation Mission
GPS Global positioning system
GRACE Gravity Recovery and Climate Experiment
GSW Gibbs Sea Water Oceanographic Toolbox
GTS Global telecommunication system
HEF Horizontal electric field
HF radar High Frequency Radar
HHWMT Higher High Water Mean Tide; nearly equivalent to MHHW = Mean Higher High Water as used in the US
HRDPS High Resolution Deterministic Prediction System (Environment Canada)
HRSST High Resolution SST
IAPSO International Association of Physical Sciences of the Oceans (UNESCO)
IAPWS-08, 09, 95 The saline component of the Gibbs function of seawater
IASI Infrared Atmospheric Sounding Interferometer (Europe)
ICES International Council for the Exploration of the Sea
ICESat-2 NASA polar satellite
IES Inverted echo sounder
IFREMER French National Institute for Ocean Science
IMOP Instruments and Methods of Observations Program
INMARSAT International Maritime Satellite Organization (non-profit for marine satellite communications)
INSAT Indian National Satellite System
IOC Intergovernmental Oceanographic Commission (UNESCO)
IPCC International Panel on Climate Change (United Nations)
IPSS-78 Practical Salinity Scale-1978
IPTS-68 International Practical Temperature Scale-1968
IRIDIUM US based global satellite communications using non geostationary satellites
ISR Infrared Scanning Radiometer
ITS-90 International Temperature Scale of 1990
JAXA Japan Aerospace Exploration Agency
JMA Japan Meteorological Agency
JPL Jet Propulsion Laboratory (NASA)
JRA-55 Japan Reanalysis data
LADCP Lowered Acoustic Doppler Current Profiler (ADCP)

LiDAR Light Detection and Ranging

m/s Meters/second

MAR Mid-Atlantic Ridge

MAVS Modular Acoustic Velocity Sensor

MBL Marine Boundary Layer

MBT Mechanical bathythermograph

MCC Maximum Cross Correlation

MCSST Multi-Channel SST

MEOP Marine Mammals Exploring the Ocean Pole to Pole

META data A set of data that describes and gives information about other data.

MIRAS Microwave Imaging Radiometer using Aperture Synthesis (sensor on SMOS)

mK Milli-degrees Kelvin

MMC MultiMedia SD card

MODIS MODerate resolution Imaging Spectrometer

ms Millisecond

MTVZA-GY Russian Microwave Imaging/Sounding Radiometer

MUSIC Algorithm used for frequency estimation and radio direction finding

MVISR Muultichannel Visible Infrared Scanning Radiometer

MWRI Microwave Radiation Imager

m°C Milli degrees Celsius

NARR North American Regional Reanalysis (USA)

NATO North Atlantic Treaty Organization

NCAR National Center for Atmospheric Research (USA)

NCEP US National Center for Environmental Prediction

NDBC US National Data Buoy Center

NE $\Delta\tau$ Noise Equivalent Temperature Difference

NECC North Equatorial Counter Current

Neptunian oceanic range $0 < S_A < 42$ g/kg, $-6.0°$ C $< T < 40°$ C, and $0 < p < 10^4$ dbar

NIC National Ice Center

NIMBUS NASA satellites series to test new instruments in space

NIR Noise Injection Radiometer

NLSST Non-Linear SST

NMFS National Marine Fisheries Service (USA)

NOAA National Oceanic and Atmospheric Administration (USA)

NSCAT NASA scatterometer

NWASTG National Weather Service Telecommunications Gateway (USA)

ODP Ocean Drilling Program

OLCI Ocean and Land Color Instrument

OLI Operational Land Imager (Landsat 8)

Orpheus Woods Hole Oceanographic Institution (WHOI) AUV

OSCAR Ocean Surface Current Analysis

OSCAT Scatterometer on the ISRO (Indian) satellite

OSTM Ocean Surface Topography Mission (NASA)

PIES/CPIES Pressure Inverted Echo Sounder/Current and Pressure Inverted Echo Sounder

PO.DAAC Physical Oceanography Distributed Active Archive Center (NASA JPL)

POD Precision orbit determination

PSIA Pounds per Square Inch Absolute; 1 PSI $= 6.89475720$ kPa $= 0.689475720$ dbar

PSMSL Permanent Service for Mean Sea Level (National Oceanography Centre, Liverpool, England)

PSU Practical salinity unit

PTT Platform transmit terminal

PTWC Pacific Tsunami Warning Center (centers in Hawaii and Alaska)

QuikSCAT Dedicated scatterometer satellite

RAMS Random Access Measurement System

RapidSCAT Scatterometer installed on the space station

RCM Recording Current Meter

REMUS Remote Environmental Monitoring Unit (WHOI AUV)

RFI Radio Frequency Interference

RLR Revised Local Reference

ROPOS Remotely Operated Platform for Ocean Sciences (Canada)

ROV Remotely Operated Vehicle

RSS Remote Sensing Systems

RTM Radiative Transfer Model

S/N Signal-to-noise level

S3A Sentinel 3A Satellite (Europe)

S4ADW Surface mounted electromagnetic current meter

SAC-D Argentinian Satellite

SADCP Through-hull Shipboard ADCP

Saildrone Autonomous surface vehicles driven by wind

Salp Barrel-shaped, planktic tunicate

SAR Synthetic Aperture Radar

SASS Seasat scatterometer (USA)

Satellite altimetry An instrument to monitor the relative ocean height

SAWS South African Weather Service

SBE Sea Bird Electronics, Inc. (USA)

SCANNER Submersible chemical analyzer

SCOR Scientific Committee on Oceanic Research (UNESCO)

SD SanDisk digital storage cards

Seawinds NASA scatterometer on QuikSCAT

SENTRY Woods Hole Oceanographic Institution (WHOI) AUV

SIA Seawater-Ice-Air library for TEOS-10 subroutines for evaluating a wide range of thermodynamic properties of pure water (using IAPWS-95).

SMAP Soil Moisture Active/Passive mission

SMART Science Monitoring And Reliable Telecommunications

SMOS Soil Moisture and Ocean Salinity mission

SNPP Suomi NASA Preparatory Platform

SOA Spectrum Optimization Approach

SOFAR float Sound Fixing and Ranging Floats; subsurface floats that listen to acoustic signals from moored beacons and position by triangulation

SOI Southern Oscillation Index

SPOT-Globalstar Satellite communications system

SRDL Satellite relay data logger

SSALTO French Salto multimission ground segment

SSS Sea Surface Salinity

SST Sea Surface Temperature

SSW Standard seawater

STD Salinity-Temperature-Depth profiler

SUAVE Submersible system to assess vented emissions

SVP Surface Velocity Program; global surface drifter program

SVP-B Global drifters equipped with barometric pressure and temperature sensors

SWOT Surface Water Ocean Topography mission (NASA/CNES)

TDI Time Delay Integration

TEOS-10 Thermodynamic Equation of Seawater-2010

TGBM Tide gauge bench mark

TIS Thermal Infrared Sensor (Landsat)

TMI TRMM Microwave Imager

TOA Top of Atmosphere

TOGA Tropical Ocean Global Atmosphere

TOPEX/Poseidon Satellite altimeter mission. (NASA/CNES)

TRIAXYS Directional Wave Buoy (AXYS Technologies Inc., Canada)

TU Tritium Units

UAV Unmanned Aerial Vehicle (drone)

UHF Ultra high frequency (300 MHz−3 GHz)

UNESCO United Nations Educational, Scientific and Cultural Organization

URL University of Rhode Island (USA)

USV Uncrewed surface vehicle

USWS US Weather Service

VACM Vector Averaging Current Meter

VHF Very high frequency (30 − 300 MHz)

VIIRS Visible Infrared Imaging Radiometer Suite

VMCM Vector Measuring Current Meter

VNR Visible and Near Infrared Radiometer

VOS Volunteer Observing Ships

Wave glider Autonomous surface vehicles driven by wave motion

WebTide Barotropic tidal model

WGS-84 World Geodetic Survey 1984

WHOI Woods Hole Oceanographic Institution (USA)

Windsat Satellite designed to measure the wind vector with passive microwave radiation

WIS WMO Information System

WMO World Meteorological Organization (UNESCO)

WOCE World Ocean Circulation Experiment

XBT Expendable Bathythermograph; ship based expendable temperature profiler

ZAPS Zero-angle photon spectrometer

Chapter 2

DATA PROCESSING AND PRESENTATION

2.1 INTRODUCTION

Most instruments neither measure oceanographic properties directly nor store the related engineering or geophysical parameters that the investigator eventually wants from the recorded data. For example, the temperature and conductivity data generated by Sea-Bird 911 conductivity-temperature-depth (CTD) instruments originate as measurements of the electrical resistivity of the sensors in response to their external environment. In turn, the sensors form the variable element in separate Wien bridge oscillator modules whose frequencies of oscillation change in response to the changes in electrical resistance. These changes are converted to an analog voltage, which is then converted to a digital signal before being converted to temperature and conductivity using calibration against an accurately known standard in the laboratory. Pressure (depth) is determined from the electrical output from a mechanical strain gauge transducer, with temperature compensation. In these pressure sensors, voltages are obtained from pressure-induced changes in the dimension of a flexural-vibrating, load-sensitive resonator.

This progression from sensor electrical response, to the response of an oscillator circuit, to oceanic parameter is further complicated by the fact that all measurement systems alter their characteristics with time and therefore require repeated calibration to define the relationship between the measured and/or stored values and the geophysical quantities of interest. The usefulness of any individual observation depends strongly on the care with which the calibration and subsequent data processing are carried out. Data processing consists of using calibration information to convert instrument values to engineering units and then using specific formulas to produce the geophysical data. As an example, calibration coefficients are used to convert voltages collected in the different channels of a CTD to salinity, temperature, and depth. Salinity is a function of conductivity, temperature, and pressure, while the depth derived from CTD profiles is a function of pressure and temperature. These can then be used to derive such quantities as potential or Conservative temperature (the pressure compression-corrected temperature) and steric height (the vertically integrated specific volume anomaly derived from the density structure).

Once the data are collected, further processing is required to check for errors and to remove erroneous values. In the case of temporal measurements, for example, a necessary first step is to check for timing errors. Such errors can arise from problems with the recorder's clock, including gradual changes in the sampling interval, Δt, or erroneous skipping of digital samples during the recording stage. As noted in Section 1.8.2.1, if N is the number of samples collected, then $(N-1)\Delta t$ should equal the total length of the record, T. This points to the obvious need to keep accurate records of the exact start and end times of the data record and, where possible, inserting "time stamps" at regularly spaced intervals throughout the data series. If a preliminary check shows that $T \neq (N-1)\Delta t$, the investigator needs to conduct a search for possible missing records. Abrupt simultaneous changes in recorded values on all channels often point to times of missing or likely incorrect data. Gradual changes in the clock sampling rate ("clock drift") are more of a problem and one has often to assume some sort of linear change in Δt over the recording period. Marked shifts in clock speed are generally easily determined. When either the start or the end time is in doubt, the investigator must rely on other techniques to determine the reliability of the sampling clock and sampling rate. Oceanographers are fortunate to have the astronomical tides as a highly reliable clock. Specifically, in the case of moored time series records obtained in regions with reasonably strong tidal motions, one can check that the amplitude ratios among the normally dominant K_1, O_1 (diurnal) and M_2, S_2 (semidiurnal) tidal constituents (Table 2.1) are consistent with previous observations. If they are not, there may be problems with the clock (or calibration of the signal amplitude). If the phases of the tidal constituents are known from previous observations in the region, these can be compared with phases from the suspect instrument. For diurnal motions, each 1-hour error in timing corresponds to a phase change of $15°$ per hour; for semidiurnal motions the phase change is $30°$ per hour. Large discrepancies suggest timing problems with the data. We recommend using M_2 for the semidiurnal variations and O_1 for the diurnal variations. Although the K_1 constituent is typically predominant relative to the O_1 constituent, it is likely to be "contaminated" by daily (24 h) atmospheric processes, such as the diurnal sea-breeze.

Data Analysis Methods in Physical Oceanography. https://doi.org/10.1016/B978-0-323-91723-0.00008-8

TABLE 2.1 Frequencies (cycles per hour, cph) and periods (hours) for the Major Diurnal (O_1, K_1) and Major Semidiurnal (M_2, S_2) Tidal Constituents.

Tidal constituent	O_1	K_1	M_2	S_2
Frequency (cph)	0.03873065	0.04178075	0.08051140	0.08333333
Period (hours)	25.8193446	23.934467	12.420601	12.000000

In the case of the high data collection and transmission rates needed for cabled ocean observatories, there is a tendency to insert too many time stamps into the data stream. The finite time it takes to insert the time stamp can lead to disruptions in the data stream and a corresponding loss of cadence in the information flow.

Two types of errors must be considered in the editing stage: (1) large "accidental" errors or spikes" that result from equipment failure, power surges, or other major data flow disruptions (including plankton, such as salps and small jellyfish, which partially impede the flow of seawater past or through the sensor) and (2) small random errors or "noise" that arise from changes in the sensor configuration, electrical and environmental noise, and unresolved environmental variability. The noise can be treated using statistical methods, while elimination of the larger errors generally requires the use of more subjective evaluation procedures. Data summary diagrams or histogram distributions are useful in identifying the large errors as sharp deviations from the general population, while the treatment of the smaller random errors requires knowledge of the population density function for the data. It is often assumed that random errors are statistically independent and have a normal (Gaussian) probability distribution. A summary diagram can help the investigator evaluate editing programs that "automatically" remove data points having magnitudes that exceed the record mean value by some integer multiple of the record standard deviation. For example, the editing procedure might be asked to eliminate data values for which, $x - \bar{x} >$ 3σ, where $\bar{x}$ and σ are the mean and standard deviation of x, respectively. This is wrought with pitfalls, especially if one is dealing with highly variable or episodic systems. By not directly examining the data points in conjunction with adjacent values, the investigator can never be certain that he or she is not throwing away reliable values. For example, during the strong 1983–1984 El Niño, water temperatures at intermediate depths along Line P in the northeast Pacific exceeded the historical mean temperature by 10 standard deviations (10σ). Had there not been other evidence for basin-wide oceanic heating during this period, there would have been a tendency to dispense with these "abnormal" values. Another classic example of questioning the validity of one's data is when German oceanographer Georg Wüst calculated the fluxes of water properties in the South Atlantic but omitted the horizontal heat flux, which is the easiest flux to calculate. When one of Wüst's former students was contacted, he said that Wüst had found that the heat flux was equatorward and that he considered that to be counterintuitive. Subsequent research proved the calculation to be correct.

In July 2009, technicians collecting high-resolution CTD data in the nearly landlocked Belize Inlet on the central coast of mainland British Columbia observed anomalous, small-scale (1–10 m) steplike structures in the temperature and salinity data from 70 to 210 m depth at several inner sites in the basin. The initial response on the deck of the ship was that these features were due to an instrument malfunction and that the CTD should be replaced with the backup system. However, on closer examination, the observed structures were recognized as thermohaline staircases, which turned out to be the first observation of double-diffusive features within a coastal fjord (Spear and Thomson, 2012). Salt-fingering staircases of $\sim$10 m thickness were present between 70 and 140 m depth (the temperature and salinity minima were at 150 m depth), while diffusive convection staircases of $\sim$1 m thickness were present between 160 and 210 m depth. A global event of this kind occurred when the NASA Solar Backscatter Ultraviolet instrument continued to record extremely low levels over the Antarctic in the 1980s. These data were considered to be "bad" until, years later, Dr. Susan Solomon of the Massachusetts Institute of Technology made direct measurements over Antarctica and discovered the ozone hole in the overlying atmosphere.

2.2 BASIC SAMPLING REQUIREMENTS

A primary concern in most observational work is the *accuracy* of the measurement device, a common performance statistic defining how well the instrument measures the actual value. Absolute accuracy requires frequent instrument calibration to detect and correct for any shifts in response. The inconvenience (and frequently the cost) of regular calibration often causes the scientist to substitute instrument *precision* as the measurement capability of an instrument. Unlike absolute accuracy, precision is a relative term and simply represents the ability of the instrument to repeat the observation without deviation. Absolute accuracy further requires that the observation be consistent in magnitude with some universally accepted reference standard. In most cases, the user must be satisfied with having good precision and repeatability of the measurement rather than having absolute measurement accuracy. Any instrument that fails to maintain its precision, fails to provide data that can be handled in any meaningful statistical fashion. The best instruments are those that provide both high precision and defensible absolute accuracy. It is sometimes advantageous to measure simultaneously the same variable with more than one reliable instrument. However, if the instruments have the same precision but not the same absolute accuracy, we are reminded of the saying that "a man with one watch knows the time, a man with two watches does not".

Digital instrument *resolution* is measured in bits, where a resolution of N bits means that the full range of the sensor is partitioned into 2^N equal segments ($N = 1, 2...$). For example, eight-bit resolution means that the specified full-scale range of the sensor, say $V = 10$ V, is divided into $2^8 = 256$ increments, with a bit resolution of $V/256 = 0.039$ V. Whether the instrument can actually measure to a resolution of $V/2^N$ units is another matter. The sensor range can always be divided into an increasing number of smaller increments but eventually one reaches a point where the value of each bit is buried in the noise level of the sensor and is no longer significant. In summary, accuracy is how well an instrument measures the true value, precision is how closely repeated measurements are to one another, and resolution is how well the instrument can resolve actual differences between measurements.

2.2.1 Sampling Interval

Assuming the instrument selected can produce reliable and useful data, the next highest priority sampling requirement is that the measurements be collected often enough in space and time to resolve the phenomena of interest. For example, in the days when oceanographers were only interested in the mean stratification of the World Ocean, water property profiles from discrete-level hydrographic (bottle) casts were adequate to resolve the general vertical density structure. On the other hand, these same discrete-level profiles failed to resolve the detailed structure associated with interleaving and mixing processes, such as those associated with shear-induced turbulence and thermohaline staircases (salt fingering and diffusive convection), which are now resolved by the rapid vertical sampling provided by modern acoustic Doppler current profilers and conductivity-temperature-depth (CTD) probes. The need for higher resolution assumes that the oceanographer has some prior knowledge of the process of interest. Often this prior knowledge has been collected with instruments incapable of resolving the true variability and may, therefore, only be suggested by highly aliased (distorted) data collected using earlier techniques. In addition, laboratory and theoretical studies may provide information on the scales that must be resolved by the measurement system.

For discrete digital data $x(t_i)$ measured at times t_i, the choice of the sampling increment Δt (or Δx in the case of spatial measurements) is the quantity of importance. In essence, we want to sample often enough that we can pick out the highest frequency component of interest in the time series but not oversample so that we fill up the data storage file, exceed the response time of the sensor, use up all the battery power, or become swamped with unnecessary data. In the case of real-time cabled observatories, it is also possible to sample so rapidly (hundreds of times per second) that inserting the essential time stamps in the data string can disrupt the cadence of the record. We might also want to sample at irregular intervals to avoid built-in bias in our sampling scheme. If the sampling interval is too large to resolve higher frequency components, it becomes necessary to suppress these components during sampling using a sensor whose response is limited to frequencies equal to that of the sampling frequency. As we discuss in our section on processing satellite-tracked drifter data, these lessons are often learned too late—after the buoys have been cast adrift in the sea.

The important aspect to keep in mind is that, for a given sampling interval Δt, the highest frequency we can hope to resolve is the *Nyquist* (or *folding*) *frequency*, f_{Nq}, defined as

$$f_{Nq} = 1/(2\Delta t) \tag{2.1}$$

(Note that the subscript Nq stands for "Nyquist" and is used throughout the book so that there is no confusion with our use of N for the buoyancy or Brunt-Väisälä frequency, $N(z)$, derived from the vertical density structure, or for N representing the number of data samples; $N = 1, 2, ...$) We cannot resolve any frequency higher than f_{Nq}. For example, if we sample every 10 h, the highest frequency we can hope to see in the data is $f_{Nq} = 0.05$ cph (cycles per hour). Eqn (2.1) states the obvious—that it takes at least two sampling intervals (or three data points) to resolve a sinusoidal-type oscillation with period $1/f_{Nq}$ (Figure 2.1). In practice, we need to contend with noise and sampling errors so that it takes something like three or more sampling increments (i.e., $\geq$ four data points) to accurately determine the highest observable frequency. Thus, f_{Nq} is an upper limit. The highest frequency we can resolve for a sampling of $\Delta t = 10$ h in Figure 2.1 is closer to $1/(3\Delta t) \approx 0.033$ cph. (Replacing Δt with Δx in the case of spatial sampling increments allows us to interpret these limitations in terms of the highest wavenumber (*Nyquist wavenumber*) the data are able to resolve.)

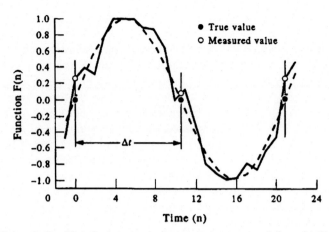

FIGURE 2.1   Plot of the function F(n) = sin $(2\pi n/20 + \phi)$ where time is given by the integer $n = -1, 0, ..., 24$. The period $2\Delta t = 1/f_{Nq}$ is 20 units and ϕ is a random phase with a small magnitude in the range ± 0.1 radians. Open circles denote measured points and solid points the curve F(n). Noise makes it necessary to use more than three data values to accurately define the oscillation period.

An important consequence of Eqn (2.1) is the problem *of aliasing*. In particular, if there is energy at frequencies $f > f_{Nq}$—which we obviously cannot resolve because of the Δt we picked—this energy gets folded back into the range of frequencies, $f < f_{Nq}$, which we are attempting to resolve (hence, the alternate name "folding frequency" for f_{Nq}). This unresolved energy does not disappear but gets redistributed within the frequency range of interest. To make matters worse, the folded-back energy is disguised (or aliased) within frequency components different from those of its origin. We cannot distinguish this folded-back energy from that which actually belongs to the lower frequencies. Thus, we end up with erroneous (aliased) estimates of the spectral energy variance over the resolvable range of frequencies. An example of highly aliased data would be current meter data collected using 13-h sampling in a region dominated by strong semidiurnal (12.42-h period) tidal currents. More will be said on this topic in Chapter 5. One of the most common examples is a television or movie clip of a wheel that appears to rotate backwards as the vehicle moves forward. This effect is a consequence of the mismatch between the rate (frequency) at which the wheel rotates and the rate at which images of the wheel are sampled by the video capture device. Because this effect is a function of the vehicles speed, one can often can see the backwards motion slow down and then even reverse to forward as the wheel rotation slows down.

As a general rule, the investigator should plan a measurement program based on the frequencies and wavenumbers (estimated from the corresponding periods and wavelengths) of the parameters of interest over the study domain. This requirement may then dictate the selection of the measurement tool or technique. If the instrument cannot sample rapidly enough to resolve the frequencies of concern it should not be used. It should be emphasized that the Nyquist frequency concept applies to both time and space and the Nyquist wavenumber is a valid means of determining the fundamental wavelength that must be sampled.

2.2.2 Sampling Duration

The next concern is that one samples long enough to establish a statistically significant determination of the process being studied or to capture the occurrences of episodic events, such as underwater volcanic eruptions (e.g., Chadwick et al., 2013; Nooner and Chadwick, 2016) or the formation of near-bottom "solibores" − rotor-like internal bores that are followed by trains of high-amplitude internal waves (Henyey and Hoering, 1997; Hosegood and van Haren, 2004; Thomson and Spear, 2020). For time-series measurements, this amounts to a requirement that the data be collected over a period sufficiently long that repeated cycles of the phenomenon are observed. This also applies to spatial sampling where statistical considerations require a large enough sample to define multiple cycles of the process being studied. Again, the requirement places basic limitations on the instrument selected for use. If the equipment cannot continuously collect the data needed for the length of time required to resolve repeated cycles of the process, it is not well suited to the measurement required.

Consider the duration, T, of a record sampled at time step Δt. The longer the record, the better we are able to resolve different frequency components in the data. (In the case of spatially separated data, Δx, wavenumber resolution increases with increased spatial coverage, L, of the data.) It is the total record length $T = N\Delta t$ (or $L = N\Delta x$ for spatial data) obtained for N data samples that: (1) determines the lowest frequency *(the fundamental frequency)*

$$f_0 = 1/(N\Delta t) = 1/T \qquad (2.2)$$

that can be extracted from the time-series record; (2) determines the frequency resolution or minimum difference in frequency $\Delta f = |f_2 - f_1| = 1/(N\Delta t)$ that can be resolved between adjoining frequency components, f_1 and f_2 (Figure 2.2); and (3) determines the amount of band averaging (averaging of adjacent frequency bands) that can be applied to enhance the statistical significance of individual spectral estimates. In Figure 2.2, the two separate waveforms of equal amplitude but different frequency produce a single spectrum. The two frequencies are well resolved for $\Delta f = 2/(N\Delta t)$ and $3/(2N\Delta t)$, just resolved for $\Delta f = 1/(N\Delta t)$, and not resolved for $\Delta f = 1/(2N\Delta t)$.

In theory, we should be able to resolve all frequency components, f, in the frequency range $f_0 \leq f \leq f_{Nq}$, where f_{Nq} and f_0 are defined by Eqns (2.1) and (2.2), respectively. Herein lies a classic sampling problem. In order to resolve the frequencies of interest in a time series, we need to sample for a long enough time (T large) so that f_0 covers the low end of the frequency spectrum and Δf is small (frequency resolution is high). At the same time, we would like to sample sufficiently rapidly (Δt small) so that f_{Nq} extends beyond all frequency components with significant spectral energy. Unfortunately, the longer and more rapidly we want to sample, the more data we need to collect and store, the more power we need to provide, and the more time, effort, and money we need to put into the sensor design and sampling program. There is also the question of whether the sensor can actually measure accurately at the high frequency end of the sampling interval of interest.

Our ability to resolve frequency components follows from Rayleigh's criterion for the resolution of adjacent spectral peaks in light that has been shone onto a diffraction grating. It states that two adjacent frequency components are just resolved when the peaks of the spectra are separated by frequency difference $\Delta f = f_0 = 1/(N\Delta t)$ (Figure 2.2). For example, to separate the spectral peak associated with the lunar–solar semidiurnal tidal component M_2 (frequency = 0.08051 cph) from that of the solar semidiurnal tidal component S_2 (0.08333 cph), for which $\Delta f = 0.00282$ cph, requires $N = 355$ data points at a sampling interval $\Delta t = 1$ h or $N = 71$ data points at $\Delta t = 5$ h. Similarly, a total of 328 data values at 1 h sampling is needed to separate the two main diurnal constituents K_1 and O_1, for which $\Delta f = 0.00305$ cph (cf. Table 2.1). Note that since f_{Nq} is the highest frequency we can measure and f_0 is the limit of our frequency resolution, then

$$f_{Nq}/f_0 = (1/2\Delta t)/(1/N\Delta t) = N/2 \tag{2.3}$$

is the maximum number of Fourier components that we can hope to estimate in any analysis.

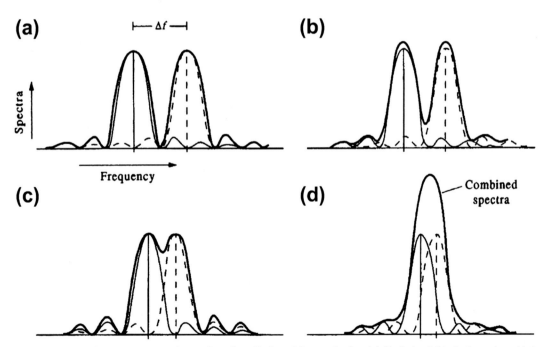

FIGURE 2.2 Spectral peaks of two separate waveforms of equal amplitude and frequencies f_1 and f_2 (dashed and thin line) together with the calculated spectrum (solid line). (a) and (b) are well-resolved spectra; (c) just resolved spectra; and (d) not resolved. Thick solid line is the total spectrum for two underlying signals with slightly different peak frequencies.

2.2.3 Sampling Accuracy

According to the two previous sections, we need to sample long enough and often enough if we hope to resolve the range of scales of interest in the variables being measured. It is intuitively obvious that we also need to sample as accurately as possible—with the degree of recording accuracy determined by the response characteristics of the sensors, the number of bits per data record (or parameter value) needed to raise measurement values above background noise, and the volume of data we can live with. There is no use attempting to sample the high end of the spectrum if the instrument cannot respond sufficiently rapidly or accurately to resolve changes in the parameter being measured. A tell-tale sign that an instrument has reached its limit of resolution is a flattening (i.e., "whitening") of the high-frequency end of the power spectrum; the frequency at which the spectrum of a measured parameter begins to flatten out as a function of increasing frequency typically marks the point where the accuracy of the instrument measurements is beginning to fall below the noise threshold, or "noise floor". In addition, there are several approaches to this aspect of data sampling including the brute-force approach in which we measure as often as we can at the degree of accuracy available and then improve the statistical reliability of each data record through postsurvey averaging, smoothing, and other manipulation. This is the case for observations provided through shore-powered, fiber-optic, cabled observatories such as the ALOHA observatory located 100 km north of the island of Oahu (Hawaii), the Monterey Accelerated Research System (MARS) in Monterey Canyon, California, the Ocean Networks Canada (ONC) cabled observatory systems — formerly identified as the Victoria Experimental Network Under the Sea (VENUS), and North-East Pacific Time Series Underwater Networked Experiments (NEPTUNE) — extending from the Strait of Georgia to the continental margin and Cascadia Basin out to the Juan de Fuca Ridge off the west coast of British Columbia, the Ocean Observatories Initiative (OOI) Regional Scale Nodes off the coasts of Oregon and Washington in the Pacific Northwest of the United States (including the underwater volcano sites on Axial Seamount), the Dense Ocean-floor Network System for Earthquakes and Tsunamis off the east coast of Japan, and the Hellenic Integrated Marine Inland Water Observing, Forecasting and offshore Technology System (HIMIOFoTS) off Pylos on the southwest coast of Greece. Data can be sampled as rapidly as possible and the data processing left to the post-acquisition stage at the onshore data management facility.

2.2.4 Burst Sampling vs Continuous Sampling

Regularly spaced, digital time series can be obtained in two different ways. The most common approach is to use a *continuous sampling mode*, in which the data are sampled at equally spaced intervals $t_k = t_0 + k\Delta t$ from the start time t_0, where k is a positive integer. Regardless of whether the equally spaced data have undergone internal averaging or decimation using algorithms built into the machine, the output to the data storage file is a series of individual samples at times t_k. (Here, "decimation" is used in the loose sense of removing every nth data point, where n is any positive integer, and not in the sense of the ancient Roman punishment of putting to death one in 10 soldiers in a legion guilty of mutiny or other crime.) Alternatively, we can use a *burst sampling mode*, in which rapid sampling is undertaken over a relatively short time interval, Δt_B, or "burst" embedded within each regularly spaced time interval, Δt. That is, the data are sampled at high frequency for a short duration starting (or ending) at times t_k for which the burst duration $\Delta t_B \ll \Delta t$. The instrument "rests" between bursts. There are advantages to the burst sampling scheme, especially in noisy (high frequency) environments where it may be necessary to average out the noise to get at the frequencies of interest. Burst sampling works especially well when there is a "spectral gap" between fluctuations at the high and low ends of the spectrum. As an example, there is typically a spectral gap between surface gravity waves in the open ocean (periods of 1−20 s) and the 12.4-hourly motions that characterize semidiurnal tidal motions. Thus, if we wanted to measure surface tidal currents using the burst-mode option for our current meter, we could set the sampling to a 2-min burst every hour; this option would smooth out the high-frequency wave effects but provide sufficient numbers of velocity measurements to resolve the tidal motions. Burst sampling enables us to filter out the high-frequency noise and obtain an improved estimate of the variability hidden underneath the high-frequency fluctuations. In addition, we can examine the high-frequency variability by scrutinizing the burst sampled data. If we were to sample rapidly enough, we could estimate the surface gravity wave energy spectrum. Many oceanographic instruments use (or have provision for) a burst sampling data collection mode.

A "duty cycle" was sometimes used in the past to collect positional data from Service Argos satellite-tracked drifters as a cost-saving form of burst sampling. In this case, all positional data within a 24 h period (about 10 satellite fixes) were collected only every third day. Tracking costs paid to Service Argos were reduced by a factor of three using the duty cycle. Unfortunately, problems arose when the length of each burst was too short to resolve energetic motions with periods comparable to the burst sample length. In the case of satellite-tracked drifters poleward of tropical latitudes, these problems are associated with highly energetic inertial motions whose periods $T = 1/(2\Omega|\sin\theta|)$ are comparable to the 24 h duration of the burst sample (here, $\Omega = 0.1161 \times 10^{-4}$ cycles per second is the earth's rate of rotation and $\theta \equiv$ latitude). Beginning in 1992, it became possible to improve resolution of high-frequency motions using a 1/3-duty cycle of 8 h "on" followed by 16 h "off". According to Bograd et al. (1999), even better resolution of high-frequency mid-latitude motions could be obtained using a duty cycle of 16 h "on" followed by 32 h "off".

A duty cycle is used in the Deep-ocean Assessment and Reporting of Tsunamis (DART) buoys moored in various regions of the World Ocean as part of enhanced tsunami warning systems. To save battery life and reduce data storage needs, the bottom pressure recorders (BPRs) in the DART systems, report time-averaged pressure ($\cong$ water depth) every 15 min to orbiting satellites through an acoustic link to a nearby surface buoy. When the built-in algorithm detects an anomalous change in bottom pressure due to the arrival of the leading wave of a tsunami, the instrument switches into "event mode". The instrument then transmits bottom pressure data every 15 s for several minutes followed by 1-min averaged data for the next 4 h (González et al., 1998; Bernard et al., 2001; González et al., 2005; Titov et al., 2005; Mungov et al., 2013; Rabinovich and Eblé, 2015). At present, DART buoys switch to event mode if there is a threshold change of 3 cm in equivalent water depth for earthquake magnitudes greater than 7.0 and epicenter distances of over 600 km. Problems arise when the leading wave form varies too slowly to be detected by the algorithm or if large waves continue to arrive well after the 4 h cutoff. For example, several DART buoys in the northeast Pacific failed to capture the leading tsunami wave from the September 2009, $M_w = 8.1$ Samoa earthquake, or to detect the slowly varying trough that formed the lead wave from the magnitude 8.8, February 2010 earthquake off the coast of Chile (Rabinovich et al., 2013). Vertical acceleration of the seafloor associated with seismic waves (mainly Rayleigh waves moving along the water−bottom interface) can also trigger false tsunami responses (Mofjeld et al., 2001). Because of the duty cycle, only those few buoys providing continuous 15 s internal recording can provide data for the duration of major tsunami events, which typically have frequency-dependent, e-folding decay timescales of around a day (Rabinovich et al., 2013).

Major storms can also trigger tsunami-like signals in bottom pressure recorders (BPRs). For example, the bottom-mounted BPR at around 1,500 m depth within the caldera of Axial Seamount off the coast of Oregon in the U.S. initially appeared to record waves from the Rat Islands tsunami of November 17, 2003 in the Aleutian Islands (Fine et al., 2020). However, close examination of the time series and wavelet diagram for the event found that the background noise was extremely high because of a major storm on November 17 off the Oregon-Washington coast. Meteorological observations from nearby stations revealed high sea conditions around this time, with a maximum significant wind wave height of 11.3 m in the morning of November 17 at Meteorological buoy 46005 located roughly 80 km to the west of Axial Seamount. Significant wave heights of 8−9 m were recorded later in the day at more coastal buoys 46029 and 46050. Also, the data and numerical simulations of the Rat Islands event showed that most of the energy went to the south toward the Hawaiian Islands, with little energy propagating to the east, where open ocean tsunami amplitudes were predicted to be only a few millimeters. Such small waves are impossible to detect during times of high background noise, as occurred at Axial Seamount at this time. The event was characterized as "non-detectable" and was not included in the study.

2.2.5 Regularly vs Irregularly Sampled Data

In certain respects, an irregular sampling in time or non-equidistant placement of instruments can be more effective than a more esthetically appealing uniform sampling. For example, unequal spacing permits a more statistically reliable resolution of oceanic spatial variability by increasing the number of quasi-independent estimates of the dominant wavelengths (wavenumbers). Since oceanographers are almost always faced with having fewer instruments than they require to resolve oceanic features, irregular spacing can also be used to increase the overall spatial coverage (fundamental wavenumber) while maintaining the small-scale instrument separation for Nyquist wavenumber estimates. The main concern is the lack of redundancy should certain key instruments fail, as so often seems to happen. In this case, a quasi-regular spacing between locations is better. Prior knowledge of the scales of variability to expect is a definite plus in any experimental array design, although this may preclude discovering any behavior that was previously unknown.

In a sense, the quasi-logarithmic vertical spacing historically adopted by oceanographers for water bottle (hydrographic) sampling—specifically 0, 10, 20, 30, 50, 75, 100, 125, 150, 175, 200 m, etc.—represents a "spectral window" adaptation to the known physical-chemical structure of the ocean. Highest resolution is required in the upper layer, where vertical changes in most oceanic variables are most rapid. Added to this was the reality that the wire supporting the sample bottles could not carry all of the bottles at the same time for a "deep cast", even though the deeper bottles are more widely spaced and, hence, fewer in number. As a result, the practice was to divide the sampling into two or three sections each covering a certain depth interval and carrying the appropriate number of sample bottles. For this reason, earlier hydrographic stations must be considered to be samples over a longer time period during which time the ship drifted. Consequently, the bottle samples integrated over space as well as time.

Similarly, an uneven horizontal arrangement of observations increases the number of quasi-independent estimates of the horizontal wavenumber spectrum. Digital data are most often sampled (or subsampled) at regularly spaced time increments. Aside from the usual human propensity for order, the need for regularly spaced data derives from the fact that most analysis methods have been developed for regular-spaced, gap-free data series. In practice, the investigator is frequently faced with irregularly sampled data in space and time. One of the motivations for writing this book was to make sure that data analysts in physical oceanography are aware of alternative methods to analyze such irregular data. Although digital data do not necessarily need to be sampled at regularly spaced time increments to give meaningful results, some form of interpolation between values may eventually be required. Since interpolation involves a methodology for estimating unknown values from known data, it can lead to its own set of problems.

2.2.6 Independent Realizations

As we review the different instruments and methods, the reader should keep in mind the three basic concerns with respect to observations: accuracy/precision; resolution (spatial and temporal); and statistical significance (statistical sampling theory). A fundamental consideration in ensuring the statistical significance of a set of measurements is the need for independent realizations. If repeated measurements of a process are strongly correlated, they provide no new information and do not contribute to the statistical significance of the measurements. Often a subjective decision must be made on the question of statistical independence. While this concept has a formal definition, in practice it is often difficult to judge. A simple guide is that any suite of measurements that is highly correlated (in time or space) cannot be independent. At the same time, a group or sequence of measurements that is totally uncorrelated must be independent. In the case of no correlation between each sample or realization, the number of "degrees of freedom" is defined by the total number of independent measurements; for the case of perfect correlation, the redundancy of the data values reduces the degrees of freedom to unity for a scalar quantity and to two for a vector quantity. The degree of correlation within the data set provides a way of estimating the number of degrees of freedom within a given suite of observations. While more precise methods are presented in later chapters of this text, a simple linear relation between degrees of freedom and correlation often gives the practitioner a way to proceed without developing complex mathematical constructs.

As we discuss in detail in Chapters 3 and 5, all of these sampling recommendations have statistical foundations and the guiding rules of probability and estimation can be carefully applied to determine the sampling requirements and dictate the appropriate measurement system. At the same time, these very statistical methods can be applied to existing data in order to better evaluate their ability to measure phenomena of interest. These comments are made to assist the reader in evaluating the potential of a particular instrument (or method) for the measurement of some desired variable.

2.3 CALIBRATION

Before data records can be examined for errors and further reduced for analysis, they must first be converted to meaningful physical units. The integer format generally used in the past to save storage space and to conduct onboard instrument data processing is not amenable to simple visual examination. Binary and American Standard Code for Information Interchange (ASCII) formats are the two most common ways to store the raw data (cf. Section 1.16), with the storage space required for the more basic Binary format being about 20% less than for the integer values of ASCII format. Conversion of the raw data requires the appropriate calibration coefficients for each sensor. These constants relate recorded values to known values of the measurement parameter. The accuracy of the data then depends on the reliability of the calibration procedure as well as on the performance of the instrument and the number of individual measurements used to generate each output value. Very precise instruments with poor calibrations will produce incorrect, error-prone data. Common practice is to fit the set of calibration values by least squares quadratic expressions, yielding either functional (mathematical) or empirical relations between the recorded values and the appropriate physical values. This simplifies the postprocessing since the raw data can readily be passed through the calibration formula to yield observations in the correct units. We emphasize that the editing and calibration work should always be performed on *copies* of the original data; never work directly on the raw, unedited data.

In some cases, the calibration data do not lend themselves to description in terms of polynomial expressions. An example is the direction channel in the older Aanderaa RCM current meter data, for which the calibration data consists of a table relating the recorded direction in raw 10-byte integer format (range, 0–1,024) to the corresponding direction in degrees from the compass calibration (Pillsbury et al., 1974). Some thought should be given to producing calibration "functions" that best represent the calibration data. With the availability of modern computing facilities, it is no more burdensome to build the calibration into a table than it is to convert it to a mathematical expression. Most important, however, is the need to ensure that the calibration accurately represents the performance range and characteristics of the instrument. Unquestioned acceptance of the manufacturer's calibration values is not always recommended for the processing of newly collected data. Instead, individual laboratory and/or field calibration may be needed for each instrument. In some instances, this is not possible (for example, in the case of expendable bathythermograph (XBT) probes which come prepackaged and ready for deployment) and some overall average calibration relation must be developed for the measurement system regardless of individual sensors. The accuracy value supplied by the manufacturer of XBTs is based on the calibration of 200 out of 2000 thermistors used.

Some instruments are individually calibrated before and after each experiment to determine if changes in the sensor unit had occurred during its operation. In fact, most manufacturers of oceanic equipment ask that their equipment be sent back for regular scheduled calibration. Although these calibrations can, for the most part, be done by the purchaser themselves, the companies have both the equipment and the expertise to do the work efficiently, and to their specifications. For example, Sea-Bird Electronics will calibrate not only their sensors (temperature, conductivity, pressure, and oxygen) but also third-party add-ons such as transmissometers and fluorometers. These companies also have the ability to repair the instruments should the calibration reveal a problem with one of the sensors.

The conversion to geophysical units must take both pre- and post-calibrations into account. Often the pre- and post-calibration are averaged together or used to define a calibration trend line, which can then be used to transform the instrument engineering units to the appropriate geophysical units. Sometimes a post-calibration reveals a serious instrument malfunction and the data record must be examined to find the place where the failure occurred. Data after this point are eliminated (or modified to account for the instrumental problems) and the post-calibration information is not used in the conversion to geophysical values. Even if the instrument continues to function in a reasonable manner, the calibration history of the instrument is important to producing accurate geophysical measurements from the instrument.

Because each instrument may use a somewhat different procedure to encode and record data, it is not possible to discuss all the techniques employed. We therefore have outlined a general procedure only. Appendix A provides a list of the many physical units used today in physical oceanography. Although there have been many efforts to standardize these units, one must still be prepared to work with data in nonstandard units. This may be particularly true in the case of older historical data collected before the introduction of acceptable international units. These standard units also are included in Appendix A. To give an example of different units for the same quantity, pressure can be expressed as Pascals (Pa)—including hPa (hundreds of Pascals) or kPa (thousands of Pascals), bars (including mbars, where 1 mbar = 1 hPa), psi (pounds per square inch), mm of Mercury (mmHg, or torr), or meters of water (m H_2O). Oceanographers typically measure pressure as hPa (or mbar), while physicians taking a patient's systolic and diastolic blood pressure measure in mmHg.

2.4 INTERPOLATION

Data gaps or "holes" are a problem fundamental to many geophysical data records. This is particularly true of physical oceanographic data which are typically collected by instruments subjected to harsh environmental conditions. Gappy data are frequently the consequence of uneven or irregular sampling (in time and/or space), or they may result from the removal of erroneous values during editing, from sporadic recording system failures, or from scheduled or unscheduled system shutdowns (as, for example, in the case of cabled observatories). An example of gaps in spatial data is the one that occurs in infrared sea surface temperatures introduced by clouds in infrared images. Infrequent data gaps, having a limited duration relative to strongly energetic periods of interest, are generally of minor concern, unless one is interested in short-term episodic events rather than stationary periodic phenomena. Major difficulties arise if the length of the holes exceeds a significant fraction (1/3—1/2) of the signal of interest and the overall data loss rises beyond 20—30% (Sturges, 1983). Gaps have a greater effect on weak signals than on strong signals and the adverse effects of the gaps increase most rapidly for the smallest percentages of data lost. While some useful computational techniques have been developed for unevenly spaced data (Meisel, 1978, 1979) and there are even some advantages to having a range of Nyquist frequencies within a given data set (Press et al., 1992), most analysis methods require data values that are regularly spaced in time and/or space. As a consequence, it is generally necessary to use an interpolation procedure to create the required regular set of continuous data values as part of the data processing. Noteworthy exceptions are tidal analysis programs used to derive the tidal constituents of scalar and vector time series (Foreman, 1977, 1978; updated 2004; Pawlowicz et al., 2002). Because they are based on least squares analyses for sine and cosine Fourier components at specified frequencies, these programs have no difficulty working with gappy time series. How accurately the programs determine the various tidal constituents is another matter and depends strongly on the quality of the data. The problem of interpolation and smoothing is discussed in more detail in Chapter 3.

2.5 MAP PROJECTIONS

One neglected aspect of mapping oceanographic variables is the selection of an appropriate map projection. A wide variety of projections has been used in the past. The nature of the analysis, its scale, and geographic region of interest dictate the type of map projection to use (Bowditch, 1977). Polar studies generally use a conic or other polar projection to avoid distortion of zonal variations near the poles. An example of a simple conic projection for the Northern Hemisphere is given in Figure 2.3. In this case, the cone is tangent at a single latitude (called a standard parallel), which can be selected by changing the angle of the cone (Figure 2.3a). The resulting latitude–longitude scales are different around each point and the projection is said to be nonconformal (Figure 2.3b). Here, nonconformal (=non-orthomorphic) means that the map does not conserve shape and angular relationships. A conformal conic projection is the Lambert conformal projection, which cuts the Earth at two latitudes. In this projection, the spacing of latitude lines is altered so that the distortion is the same as along meridians. This is the most widely used conic projection for navigation because straight lines nearly correspond to great circle routes. A variation of this mapping is the "modified Lambert conformal projection". This projection amounts to selecting the top standard parallel very near the pole, thus closing off the top of the map. Such a conic projection is conformal over most of its domain. Mention should also be made of the "polar stereographic projection" that is favored by meteorologists. Presumably, the advantages of this projection are its ability to cover an entire hemisphere, and its low distortion at temperate latitudes.

At mid and low latitudes, it is common to use some form of Mercator projection, which accounts for the meridional change in the Earth radius by a change in the length of the zonal axis. Mercator maps are conformal in the sense that distortions in latitude and longitude are similar. The most common of these is the transverse Mercator or cylindrical projection (Figure 2.4). As the name implies, it amounts to projecting the Earth's surface onto a cylinder which is tangent at the equator (equatorial cylindrical). This type of projection, by definition, cannot include the poles. A variant of this is called the oblique Mercator projection, corresponding to a cylinder that is tangent to the Earth along a line tilted with respect to the equator. Unlike the equatorial cylindrical, this oblique projection can represent the poles (Figure 2.5a). This

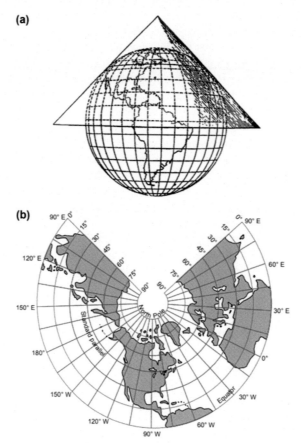

(a)

(b)

FIGURE 2.3   Example of a simple conic projection for the Northern Hemisphere. The single tangent cone in (a) is used to create the map in (b). *From Bowditch (1977).*

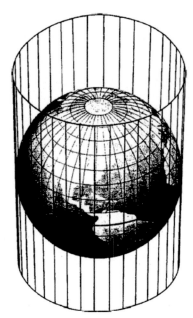 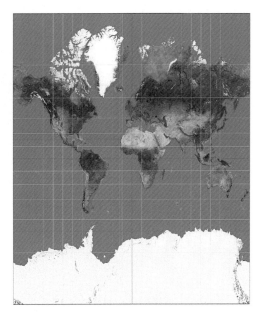

FIGURE 2.4 The transverse Mercator or cylindrical projection showing the projection of the globe (left side) onto a two-dimensional grid (right side). *From Bowditch (1977) and Wikipedia (2024).*

form of Mercator projection also has a conformal character, with equal distortions in lines of latitude and longitude (Figure 2.5b). The most familiar Mercator mapping is the universal transverse Mercator grid, which is a military grid using the equatorial cylindrical projection. Another popular midlatitude projection is the rectangular, or equal-area projection, which is a cylindrical projection with uniform spacing between lines of latitude and lines of longitude. In applications where actual Earth distortion is not important, this type of equal-area projection is often used. Whereas Mercator projections are useful for plotting vectors, equal-area projections are useful for representing scalar properties. For studies of limited areas, special projections may be developed, such as the azimuthal projection, which consists of a projection onto a flat plane tangent to the Earth at a single point. This is also called a gnomonic projection. Stereographic projects perform similar projections; however, where gnomonic projections use the center of the Earth as the origin, stereographic projections use a point on the surface of the Earth.

The effects of map projection on mapped oceanographic properties should always be considered. Often the distortion is unimportant because only the distribution relative to the provided geography (land boundaries) is important. In other cases, such as plots of Lagrangian drifter trajectories, it is important to compare maps using the same projection from which it should be possible to roughly estimate velocities along the paths. Variations in map projections can also introduce unwanted variations in the displays of properties.

2.6 DATA PRESENTATION

2.6.1 Introduction

The analysis of most oceanographic records necessitates some form of "first-look" visual display. Even the editing and processing of data typically requires a display stage, as, for example, in the exact determination of the start and end of a time series, or in the interactive removal and interpolation of data spikes and other erroneous values. A useful axiom is, "when in doubt, look at the data". In order to look at the data, we need specific display procedures. A single set of display procedures for all applications is not possible since different oceanographic data sets require different attributes. Often, the development of a new display method may be the substance of a particular research project. For instance, the advent of satellite oceanography has greatly increased the need for interactive graphics display and digital image analysis. The development of interactive edge and gradient detection software for determining the location of ocean features remains near the forefront of satellite-based Earth observation studies (Belkin and O'Reilly, 2009; Belkin et al., 2009; Williams et al., 2013). Our discussion begins with traditional types of data and analysis product presentations. These were developed as oceanographers sought ways to depict the ocean they were observing. The earliest shipboard measurements consisted of temperatures taken at the sea surface and soundings of the ocean bottom. These data were

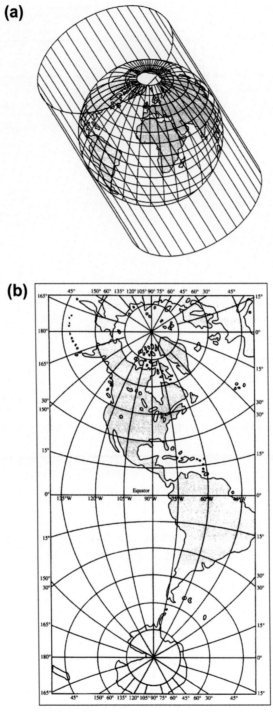

FIGURE 2.5 An oblique Mercator or oblique cylindrical projection that includes the poles. The cylinder in (a) is used to generate the transverse Mercator map of the western hemisphere in (b). *From Bowditch (1977).*

most appropriately plotted on maps to represent their geographical variability. The data were then contoured by hand to provide a smooth picture of the variable's distribution over the survey region. Examples of historical interest are the meridional sections of salinity from the eastern and western basins of the North Atlantic based on data collected during the German *Meteor* Expedition of 1925–1927 (Figure 2.6; Spiess, 1928). The water property maps from this expedition were among the first to indicate the meridional movements of water masses in the Atlantic Ocean. Figure 2.7 shows a more recent cross-section and color presentation of roughly the same south-north section but based on Absolute Salinity

Latitude

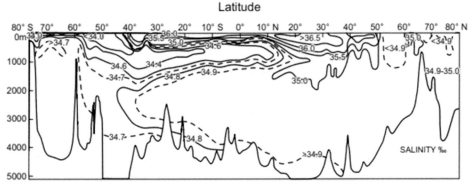

FIGURE 2.6 Latitudinal section of salinity (‰; ppt) in the western basin of the Atlantic Ocean drawn by hand over a century ago. *After Spiess (1928).*

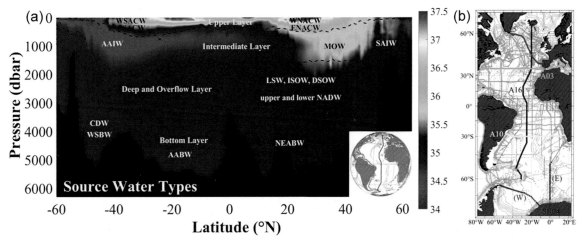

FIGURE 2.7 (a) Distributions of water masses in the Atlantic Ocean based on absolute salinity (g/kg) collected during a north-south A16 section in 2013 (see inset on lower right). The dashed lines denote the boundary of the four vertical layers defined by the neutral density levels (see Section 1.2 for definitions); and (b) the five WOCE/GO-SHIP sections that were selected to represent the vertical distribution of the main water masses. *From Liu and Tanhua (2021).*

(g/kg) collected during a 2013 survey (Liu and Tanhua, 2021). The water mass distributions in Figures 2.6 and 2.7—mapped nearly a century apart—are similar but bottom topography is now better resolved and water masses more clearly defined.

Vertical profiling capability makes it possible to map quantities on different types of horizontal surfaces. Usually, specific depth levels are chosen to characterize spatial variability within certain layers. The near-vertical homogeneity of the deeper layers means that fewer surfaces need to be mapped to describe the lower part of the water column. Closer to the ocean surface, additional layers may be required to properly represent the strong horizontal gradients. In their study of Atlantic Ocean water masses (e.g., Figure 2.7), Liu and Tanhua (2021) used a total of four primary neutral density layers.

The realization by oceanographers of the importance of both along- and cross-isopycnal processes has led to the practice of displaying water properties on specific isopycnal, or neutral density, surfaces. Because these surfaces do not usually coincide with constant depth levels, the depth of the isopycnal or neutral density surface also is sometimes plotted. Isopycnal surfaces are chosen to characterize the upper and lower layers separately. Often, processes not obvious in a horizontal depth plot are clearly shown on selected isopycnal surfaces. This practice is especially useful in tracking the lateral distribution of tracer properties such as the deep and intermediate depth silicate maxima in the North Pacific (Talley and Joyce, 1992) or the spreading of hydrothermal plumes that have risen to a density surface corresponding to their level of neutral buoyancy (Feely et al., 1994). Plots of potential temperature (θ) vs salinity (the $\theta-S$ relationship) or θ vs potential density ($\sigma-\theta$) have proven to be particularly useful in defining the maximum height of rise of hydrothermal plumes formed over venting sites along midocean ridges (Figure 2.8).

A plot relating one property to another is of considerable value in oceanography. Known as a "characteristic diagram" the most common is that relating temperature and salinity called the *TS* diagram when using EOS-80, or Θ-S_A diagram when using TEOS-10 (see Chapter 1). Originally defined with temperature and salinity values obtained from the same

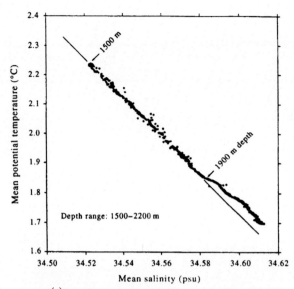

FIGURE 2.8   Plot of mean potential temperature $\left(\overline{\theta}\right)$ vs mean salinity $\left(\overline{S}\right)$ for depths of 1,500–2,200 m on the Endeavour Segment of Juan de Fuca Ridge in the northeast Pacific. The least squares linear fit covers the depth range 1,500–1,900 m, where $\overline{\theta} = -6.563\cdot\overline{S} + 228.795°C$. The abrupt change in the $\overline{\theta} - \overline{S}$ relationship at 1,900 m depth marks the maximum height of rise of local hydrothermal vent plumes. *From Thomson et al. (1992).*

sample bottles, the *TS* relationship was used to detect incorrect bottle samples and to define oceanic water masses (Figure 2.9; *Top panel*). *TS* plots have been shown to provide consistent relationships over large horizontal areas (Helland-Hansen, 1918) and have been the focus of studies on the formation of water masses (McDougall, 1985a,b; Liu and Tanhua, 2021). In the top panel of Figure 2.9, the mean *TS* relationship is formulated as the average of *S* over intervals of *T*. Depth values have been included and represent a range of *Z* values spanning the many possible depths at which a single *TS* pair is observed. Thus, it is not possible to define a unique mean *T, S, Z* relationship for a collection of different profiles.

Characteristic diagrams are not limited to temperature and salinity. Various combinations of scalar quantities including temperature, salinity, dissolved oxygen, silicate, nitrate, phosphate, alkalinity, and/or derived biogeochemical quantities have been used. For data processed using the TEOS-10 algorithms, the user plots Conservative Temperature (CT; Θ) versus Absolute Salinity (S_A) instead of potential temperature versus Practical Salinity. Within TEOS-10, the "Θ - S_A" diagram becomes the new "θ - S" diagram. Moreover, as with EOS-80, identification of water mass types is not limited to Θ and S_A. For example, the Source Water Types (SWTs) identified in Figure 2.9 (*Bottom panel*) for the Atlantic Ocean are derived from the biased-adjusted Global Ocean Data Analysis Project version 2 (GLODAPv2) data product that takes into account conservative temperature and absolute salinity, as well as non-conservative properties (oxygen, silicate, phosphate and nitrate). The distributions of these water masses are investigated with the use of the optimum multi-parameter (OMP) method (Liu and Tanhua, 2021).

Except for some minor changes, plots of vertical profiles, vertical sections, horizontal maps, and time series continue to serve as the primary display techniques for physical oceanographers. The development of electronic instruments, with their rapid sampling capabilities and the growing use of high-volume satellite data, may have changed how marine scientists display certain data but most of the basic display formats remain the same. Today, a computer is programmed to carry out both the required computations and to plot the results. Image formats, which are common with satellite data, require further sophisticated interactive processing to produce images with accurate geographical correspondence. Despite this, the combination of vertical sections and horizontal maps continues to provide most investigators with the requisite geometrical display capability.

2.6.2 Vertical Profiles

As oceanographic sampling became more detailed and sophisticated, vertical profiling of water properties became possible and new data displays were required. Of immediate interest were simple vertical profiles of temperature and salinity, such as those shown in Figure 2.10. These property profiles, based on a limited number of sample bottles suspended from the hydrographic wire at standard hydrographic depths, originally served to both depict the vertical stratification of the measured parameter and to detect any sampling bottles that had not functioned properly. The "bad" data points could then

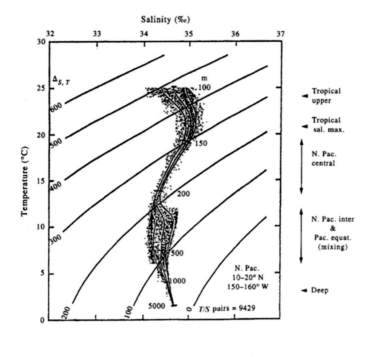

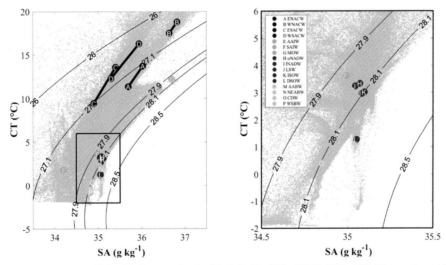

FIGURE 2.9 *Upper*: Mean temperature-salinity ($T-S$) curves for the North Pacific ($10°-20°$ N; $150°-160°$ W), together with isolines of the corresponding thermosteric density anomaly, $\Delta_{S,T}$ (from Pickard and Emery (1992)); *Bottom*: (a) Θ-S_A diagram of all data based on the GLODAPv2 data product (gray dots), representing 16 main source water types (SWTs) for the Atlantic Ocean; (b) expanded panel for values in the box in (a). The colored dots with letters A–D show the upper and lower boundaries of the central waters, and E–P denote the mean values of other SWTs [see Figures 2.18 and 2.22 for the properties of the Labrador Sea Water]. *From Liu and Tanhua (2021).*

either be excluded or discarded entirely from the data set. Leakage of the watertight seals, failure of the bottle to trip, and damage against the side of the ship are the major causes of sample loss. Leakage problems can be especially difficult to detect.

Vertical profiles obtained from ships, buoys, aircraft, AUVs and other platforms provide a convenient way to display oceanic structure. One must be careful in selecting the appropriate scales for the vertical and horizontal property axes. The vertical axis may change scale or vary nonlinearly to account for the marked changes in the upper ocean compared with the relative homogeneity of the lower layers. The property axis needs to have a fine enough scale so as to define the small vertical gradients in the deeper layer without the upper layer going off-scale. When considering a variety of different vertical profiles together (Figure 2.10), a common property scale is an advantage, although consideration must be given to

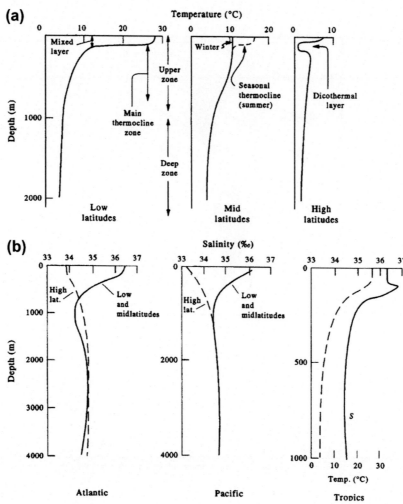

FIGURE 2.10 Vertical profiles. (a) Temperature (°C) profiles for tropical (low) latitudes, midlatitudes, and polar (high) latitudes in the Pacific Ocean. (b) Salinity (ppt) profiles for the Atlantic, Pacific, and tropical oceans for different latitudes. The dicothermal layer in (a) is formed from intense winter cooling followed by summer warming to shallower depths. Both salinity (solid line) and temperature (dashed line) are plotted for the tropics in (b). *From Pickard and Emery (1992).*

the strong dependence of vertical property profiles on latitude and season. In Figure 2.11, it was possible to truncate the series at 100 m depth as almost all of the seasonal variability occurred above that depth.

The development and regular use of continuous, high-resolution, electronic profiling systems such as CTD/Rosette packages, often combined with lowered high-frequency acoustic Doppler current profilers (LADCPs), now provides oceanic fine structure and turbulent dissipation information previously not resolvable with standard hydrographic casts (Waterhouse et al., 2014; Thurnherr et al., 2015). Profiles from standard bottle casts required smooth interpolation between observed depths so that structures finer in scale than the smallest vertical sampling separation were missed. Vertical profiles from modern CTD/Rosette systems are of such high resolution that they are generally either vertically averaged or sub-sampled to reduce the large volume of data to a manageable level for display. For example, with the rapid (≈ 24 Hz) sampling rates of modern CTD systems, parameters such as temperature and salinity, which are generated approximately every 0.01 m, are not presentable in a plot of reasonable size. For plotting purposes, the data are typically averaged to create files with sampling increments of 1 m or larger.

Studies of fine-scale (centimeter scale) variability require the display of full CTD and LADCP resolution and will generally be limited to selected portions of the vertical profile. These portions are chosen to reflect the part of the water column that is of greatest concern for the study. Full-resolution CTD/LADCP profiles reveal fine-scale structure in T, S, and velocity shear and can be used to study mixing processes such as interleaving and double diffusion. Expressions of these processes are also apparent in full-resolution TS diagrams using CTD data. One must be careful, however, not to confuse instrument noise (e.g., those due to vibrations or "strumming" of the support cable caused by vortex shedding)

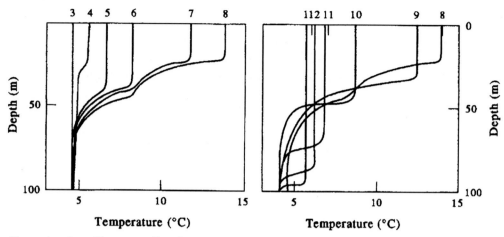

FIGURE 2.11 Time series of monthly mean profiles of upper ocean temperature at Ocean Weather Station "P", northeast Pacific (50° N, 145° W). Numbers denote the months of the year. *From Pickard and Emery (1992).*

with fine-scale oceanic structure. Analyses should be used where possible to separate the instrument noise from the wave number, band-limited signal of mixing processes. Often, computer programs for processing CTD data contain a series of different display options that can be used to manipulate the stored high-resolution digital data. The abundance of raw CTD digital data, and the variety of *in situ* calibration procedures, makes it difficult to interpret and analyze CTD records using a universal format. This is a fundamental problem in assembling a historical file of CTD observations. Hopefully, the statistics of CTD data that have been smoothed to a resolution comparable to that of traditional bottle casts are sufficiently homogeneous to be treated as updates to the hydrographic station data file. The increasingly wide use of combined CTD/ ADCP and rosette profiling systems has led to a dramatic decrease in the number of standard bottle casts.

2.6.3 Vertical Sections

The data collected from a research vessel at a series of CTD/rosette water property stations or from a CTD package carried by autonomous platforms cycling vertically through the water column while advancing horizontally underway at depth, may be represented as vertical section plots. Here, the discretely sampled data are entered into a two-dimensional (2D) vertical section at the sample depths and then contoured to produce the vertical structure along the section (Figure 2.12). Two things need to be considered in this presentation. First, the depth of the ocean, relative to the horizontal distance of most sections, is very small and vertical exaggeration is required to form readable sections. Second, the stratification can be separated roughly into two near-uniform layers with a strong density gradient layer (the pycnocline) sandwiched between. This two-layer system led early German oceanographers to introduce the terms "troposphere" and "stratosphere" (Wüst, 1935; Defant, 1936), which they described as the warm and cold water spheres of the ocean, respectively. Introduced in analogy to the atmospheric vertical structure, this nomenclature has subsequently not been widely used in oceanography. The consequence of this natural vertical stratification, however, is that vertical sections are sometimes best displayed in two parts, a shallow upper layer, with an expanded scale to show the considerable detail normally found in the upper ocean, and a deeper layer with a much more compressed vertical resolution due to the smaller vertical structural variability. Figure 2.13 shows a more recent plot of the same region as Figure 2.12 that now also includes details of dissolved oxygen and the primary nutrients associated with the temperature and salinity distributions. Note that the northward extension of the Antarctic Bottom Water in Figure 2.12 appears to stop where the section crosses from the western to eastern side of the Mid Atlantic Ridge. In reality, the bottom water continues northward on the western side of the ridge (cf. Becker et al., 2021) and the apparent "stoppage" is due to the section line crossing the ridge, from west to east, near the equator.

Vertical sections are a way to display vertically profiled data collected regionally along the track of a research vessel or taken from more extended crossings of an ocean basin (usually, meridionally or zonally). Marked vertical exaggeration is necessary to make oceanic structure visible in these sections. A basic assumption in any vertical section is that the structure being mapped has a persistence scale longer than the time required to collect the section data. Depending on the type of data collected at each station, the length of the section, and the speed of the vessel, shipboard collection times can run from a few days to a few weeks. Thus, only phenomena with timescales longer than these periods are properly resolved by the vertical sections. Recognizing this fact leads to a trade-off between spatial resolution (between-station spacing) and the

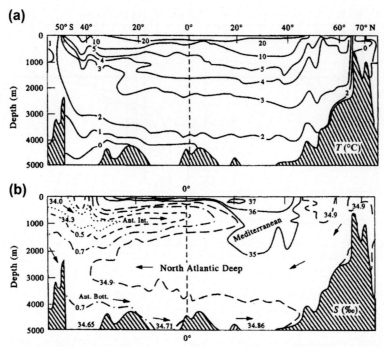

FIGURE 2.12 Latitudinal (meridional) cross-sections of (a) in situ temperature and (b) salinity for the Atlantic Ocean. Arrows denote direction of water mass movement based on the distribution of properties. Ant. Bott., Atlantic Bottom Water; Ant. Int., Antarctic Intermediate Water. *From Pickard and Emery (1992).*

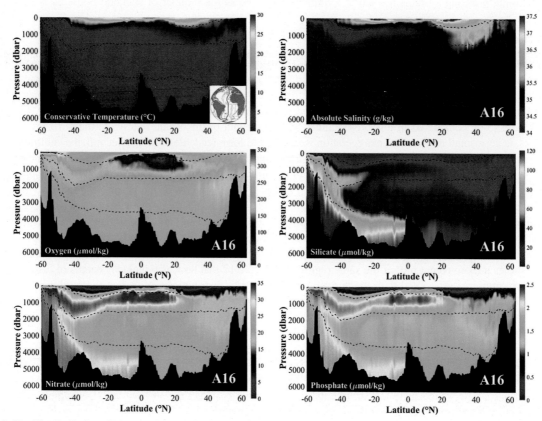

FIGURE 2.13 The distribution of water properties in the Atlantic Ocean along repeat-section A16 in 2013 as characterized by the Optimum Multi-Parameter (OMP) method. The dashed lines show the neutral densities of 27.10, 27.90 and 28.10 kg m^{-3}. See inset in the upper left panel for the transit location. *From Liu and Tanhua (2021).*

time to complete the section. Sampling time decreases as the number of profiles decreases and the samples taken approach a true *synoptic* representation (samples collected at the same time). Airborne surveys using expendable probes such as airborne XBTs from fixed-wing aircraft and helicopters yield much more synoptic information but are limited in the type of measurement that can be made and by the depth range of a given measurement. Although aircraft often have hourly charge-out rates that are similar to ships and generally are more cost-effective than ships on a per datum basis, operation of aircraft is usually the domain of the military or coast guard. Satellite images are collected in seconds and minutes and can, therefore, be considered as truly synoptic in nature.

Fewer sample profiles mean wider spacing between stations and reduced resolution of smaller, short-term variations. There is a real danger of short timescale or space-scale variability aliasing quasi-synoptic, low-resolution vertical sections. Thus, the data collection scheme must be designed to either resolve or eliminate (by filtering) scales of oceanic variability shorter than those being studied. With the ever-increasing interest in ocean climate, and at a time when the importance of mesoscale oceanic circulation features has been recognized, investigators should give serious consideration to their intended sampling program to optimize the future usefulness of the data collected.

Traditional bottle hydrographic casts were intended to resolve the slowly changing background patterns of the property distributions associated with the mean "steady-state" circulation. As a result, station spacings were usually too large to adequately resolve mesoscale features. In addition, bottle casts require long station times leading to relatively long total elapsed times for each section. The fact that these data have provided a meaningful picture of the ocean suggests that there is a strong component of the oceanic property distributions related to the steady-state circulation. For these reasons, vertical sections based on traditional bottle-cast station data provide useful definitions of the meridional and zonal distributions of individual water masses (Figure 2.12).

The importance of mesoscale oceanic variability has prompted many oceanographers to decrease their sample spacing. Electronic profiling systems, such as the CTD and CTD/Rosette, require less time per profile than standard bottle casts so that the total elapsed time per section has been reduced over the years, consistent with the need for greater spatial resolution. Still, most oceanographic sections are far from being synoptic owing to the low speeds of ships and some consideration must be given to the definition of which time/space scales are actually being resolved by the measurements. For example, suppose we wish to survey a 1,000-km zonal oceanic section and collect a meager 20 salinity–temperature profiles to 2,000 m depth along the way. At an average speed of 12 knots, steaming time alone will amount to about 2 days. Each CTD rosette station takes about 2 hours. Thus, our survey time would range from 3 to 4 days, which is just marginally synoptic by most oceanographers' standards. In more protected coastal waters, the advent of self-contained CTD systems has made it possible to use Hovercraft, large inflatables, or other small craft to move rapidly between stations. For example, scientists in the STRATOGEM (Strait of Georgia Ecosystem Modelling) project used Canadian Coast Guard hovercraft and large car ferries, with as many as 14 daily scheduled sailings, for their studies in the Strait of Georgia, British Columbia (Pawlowicz et al., 2007; Halverson and Pawlowicz, 2008; Riche, 2011).

Expendable profiling systems such as the XBT make it possible to reduce sampling time by allowing profile collection from a moving ship. Ships also can be fitted with an acoustic current profiling system (SADCP), which allows for the measurement of ocean currents in the upper few hundred meters of the water column while the ship is underway. The depth of measurement is determined by frequency and is about 500 m for the commonly used Teledyne-RDI 150-kHz transducers. Most modern oceanographic vessels also have Shipboard ASCII Interrogation Loop (SAIL) systems for rapid (≈ 1 min) sampling of the near-surface temperature and salinity at the intake for the ship's engine cooling system. SAIL data are typically collected a few meters below the ship's waterline. Oceanographic sensor arrays towed in a saw-tooth pattern behind the ship provide another technique for detailed sampling of the water column. This method has seen wide application in studying near-surface fronts, turbulent microstructure, and hydrothermal venting. These technological improvements have lowered the sample time and increased the vertical resolution. Unfortunately, the instruments typically require considerable technical support and processing of the data can be highly labor intensive. Determining the locations of the data values also requires integration of the oceanic instrumentation with a reliable Global Positioning System.

As referred to earlier, it is common practice when plotting sections to divide the vertical axis into two parts, with the upper portion greatly expanded to display the larger changes of the upper layer. The smaller (closer spacing between lines) contour interval used in the upper part may be greater than that used for the weaker vertical gradients of the deeper layer. It is important, however, to maintain a constant contour interval within each layer to faithfully represent the gradients. In regions with particularly weak vertical gradients, additional contours may be added by introducing a change in line weight, or by changing line type (dots, dashes, etc.) to distinguish the added line from the other contours. All contours must be clearly labeled. Color is often very effective in distinguishing gradients represented by the contours. While it is common practice to use shades of red to indicate warm regions, and shades of blue for cold, there is no recommended or universal color coding for properties such as salinity, dissolved oxygen, or nutrients. The color atlas of water properties for the Pacific Ocean published by Reid (1965) provides a useful color scheme.

In sections derived from bottle samples, individual data points are usually indicated by a dot or by the actual data value. In addition, the station number is indicated in the margin above or below the profile. Stations collected with CTDs usually have the station position indicated but no longer have dots or sample values for individual data points. Because of the high vertical resolution, only the contours are plotted.

The horizontal axis usually represents distance along the section and many sections have a small inset map showing the section location. Alternatively, the reader is referred to another map, which shows all section locations. Since many sections are taken along parallels of latitude or meridians of longitude, it is customary to include the appropriate latitude or longitude scale at the top or bottom of each section (Figure 2.14). Even when a section only approximates zonal or meridional lines, estimates of the latitude or longitude are frequently included in the x-axis label to help orient the reader. Station labels should also be added to the axis.

A unique problem encountered when plotting deep vertical sections of density is the need to have different pressure reference levels for the density determination to account for the dependence of seawater compressibility on temperature. Since water temperature generally decreases with pressure (greater depths), artificially low densities will be calculated at the greatest depths when using the surface pressure as a reference (Lynn and Reid, 1968; Reid and Lynn, 1971). When one wants to resolve the deep density structure, and at the same time display the upper layer, different reference levels are used for different depth intervals. As shown in Figure 2.14, the resulting section has discontinuities in the density contours as the reference level changes.

A final comment about vertical sections concerns the representation of bottom topography. The required vertical exaggeration makes it necessary to represent the bottom topography on an exaggerated scale. This often produces steep-looking islands and bottom relief. There is a temptation to ignore bottom structure, but as oceanographers become more aware of the importance of bottom topography in dictating certain aspects of the circulation and turbulent dissipation, it is useful to include some representation of the bottom structure in the sections.

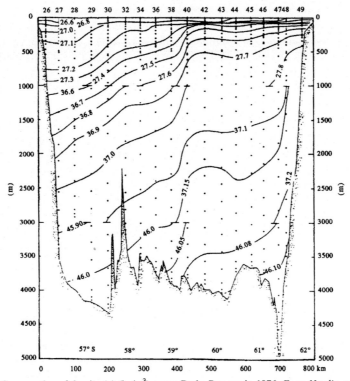

FIGURE 2.14 Cross-section of density (σ) (kg/m^3) across Drake Passage in 1976. *From Nowlin and Clifford (1982).*

2.6.4 Horizontal (Plan-View) Maps

In the introduction, we mentioned the early mapping of ocean surface properties and bottom depths. Following established traditions in map making, these early maps were as much works of art as they were representations of oceanographic information. The collection of hydrographic profiles later made it possible to depict property distributions at different levels of the water column (Figure 2.15). As with vertical sections, the question of sample time vs horizontal resolution needs to be addressed, especially where maps cover large portions of an ocean basin. Instead of the days to weeks needed to collect data along a single short section, it may take weeks, months, and even years to obtain the required data covering large geographical regions. Often, horizontal maps consist of a collection of sections designed to define either the zonal/meridional structure or cross-shore structure for near-coastal regions. In most cases, the data presented on a map are contoured with the assumption that the map corresponds to a stationary property distribution. For continental shelf regions, data used in a single map should cover a time period that is less than the approximately 10-day e-folding timescale of mesoscale eddies. In this context, the "e-folding time" is the time for the mesoscale currents to decay to $1/e^1 = 0.368$ of their peak values. Satellite derived observations provide widespread coverage of the surface global ocean at relatively short time scales but still need to be averaged over space to highlight regions of highest variability, as with the mean eddy kinetic energy (EKE) of surface currents mapped by altimetry observations in Figure 2.16.

Much of what we know about the overall structure of the ocean, particularly the deep ocean, has been inferred from large-scale maps of water properties. A presentation developed by Wüst (1935) to better display the horizontal variations of particular water masses is based on Wüst's *core layer* method. Using vertical property profiles, vertical sections, and characteristic (one property vs. another property) diagrams, Wüst defined a core layer as a property extremum and then traced the distribution of properties along the surface defined by these extrema. Because each core layer is not strictly horizontal, it is first necessary to present a map showing the depth of the core layer in question. Properties such as

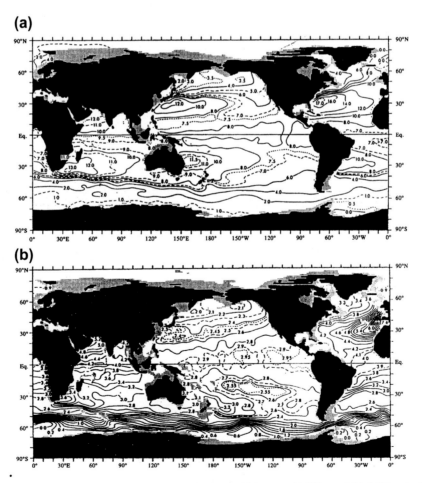

FIGURE 2.15 Horizontal maps of annual mean potential temperature in the World Ocean at (a) 500 m and (b) 1,000 m depth. *From Levitus (1982).*

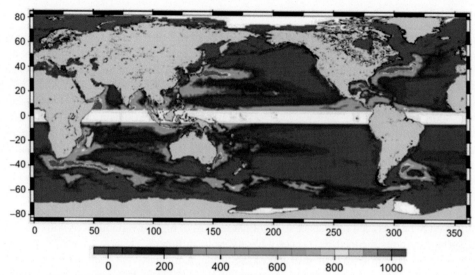

FIGURE 2.16 Global distribution of the climatological mean sea surface EKE (cm^2/s^2) derived from satellite altimetry during the period 1993–2011. The equatorial regions are blank because the Coriolis parameter is too close to zero for the accurate estimation of geostrophic velocities from altimetric sea surface heights. *From Ducet et al. (2000) and Bibarboure et al. (2011); adapted from "Western Boundary Currents" Chapter 13 by S. Imawaki, A.S. Bower, L. Beal, and B. Qiu.*

temperature, salinity, oxygen, and nutrients also can be plotted along these layers in addition to the percentage of the appropriate water mass defined from the characteristic diagrams. A similar presentation is the plotting of properties on selected density surfaces. This practice originated with Montgomery (1938) who argued that advection and mixing would occur most easily along surfaces of constant entropy. Because these isentropic surfaces are difficult to determine, Montgomery suggested that surfaces of constant potential density would be close approximations in the lower layers and that sigma-*t* would be appropriate for the upper layers. Known as *isentropic analysis* because of its thermodynamic reasoning, this technique led to the practice of presenting horizontal maps on sigma-*t* or sigma-*θ* (potential density) surfaces. While it may be difficult to visualize the shape of the density surfaces, this type of format is often better at revealing property gradients. As with the core layer method, preparing maps on density surfaces includes the plotting of characteristic property diagrams to identify the best set of density surfaces. Inherent in this type of presentation is the assumption that diapycnal (cross-isopycnal) mixing does not occur. Sometimes steric surfaces or surfaces of thermosteric anomaly are chosen for plotting rather than density.

The definition and construction of contour lines on horizontal maps has evolved in recent years from a subjective hand-drawn procedure to a more objective procedure carried out by a computer. Hand analyses usually appear quite smooth but it is impossible to adequately define the smoothing process applied to the data since it varies with user experience and prejudice. Only if the same person contoured all the data, is it possible to compare map results directly. Differences produced by subjective contouring are less severe for many long-term and stationary processes, which are likely to be well represented regardless of subjective preference. Shorter term and smaller space-scale variations, however, will be treated differently by each analyst and it will be impossible to compare results. In this regard, we note that weather maps used in 6-hourly weather forecasts are, in part, still drawn by hand since this allows for needed subjective decisions based on the accumulated experience of the meteorologist. Hand contouring by physical oceanographers connected the analyst much more closely to the data generating the contours. This was realistic when the volume of data was small and each point could be individually considered by the analyst. However, this is no longer possible with the large volumes generated by electronic sampling instruments. Objective analysis and other computer-based mapping procedures have been developed to carry out the horizontal mapping and contouring. Some of these methods are presented individually in later sections of this, and later, chapters. Since there is such a wide selection of mapping methods, it is not possible to discuss each individually. However, the reader is cautioned in applying any specific mapping routine to ensure that any implicit assumptions are satisfied by the data being mapped. The character of the result needs to be anticipated so that the consequences of the mapping procedure can be evaluated. For example, the mapping procedure called "objective analysis" or "optimum interpolation", is inherently a smoothing operation. As a consequence, the output gridded data may be smoothed over a horizontal length scale greater than the scale of interest in the study. One must decide how best to retain the variability of interest and still have a definable mapping procedure for irregularly spaced data.

2.6.5 Characteristic or Property vs Property Diagrams

As noted in the introduction to this section, it is useful in many oceanographic applications to relate two simultaneously observed variables. Helland-Hansen (1918) first suggested the utility of plotting temperature (*T*) against salinity (*S*). He found that *TS* diagrams were similar over large areas of the ocean and remained constant in time at many locations. An early application of the *TS* diagram was the testing and editing of newly acquired hydrographic bottle data. When compared with existing *TS* curves for a particular region, *TS* curves from newly collected data quickly revealed erroneous samples which could then be corrected or eliminated. Similar characteristic diagrams were developed for other ocean properties. Many of these, however, were not conservative and could not be expected to exhibit the constancy of the *TS* relationship.

As originally conceived, characteristic diagrams such as the *TS* plots were straightforward to construct. Pairs of property values from the same water bottle sample constituted a point on the characteristic plot. The connected points formed the *TS* curve for the station. Each *TS* curve represented an individual oceanographic station and similarities between stations were judged by comparing their *TS* curves. These traditional *TS* curves exhibit a unique relationship between *T*, *S*, and *Z* (the depth of the sample). What stays constant is the *T* verses *S* relationship, not the correspondence with *Z*. As internal waves, eddies, and other unresolved dynamical features move through a region, the depth of the density structure changes. In response, the paired *TS* value moves up and down along the *TS* curve, thus maintaining the water mass structure. This argument does not hold for near-surface layers where the water is being modified by wind mixing and heat and buoyancy fluxes with the atmosphere, or in frontal zones where the water mass is being modified by turbulent diapycnal mixing and interleaving with intruding water types. Figure 2.17 shows the progression in the *TS* relationship along the west coast of North America beginning with Pacific Equatorial Water in the south, which mixes with Pacific Subarctic Upper Water in the north as currents move the water mass poleward toward Alaska. Aside from the equatorial water, plotted *TS* curves span upper continental slope CTD stations from Newport (Oregon) to Shumagin Island (Aleutian Islands, Alaska).

Although *TS* diagrams remain in widespread use, Θ-S_A diagrams have been shown to improve representativeness of water mass characteristics. In Figure 2.18, we show the water mass properties (both conservative and non-conservative)

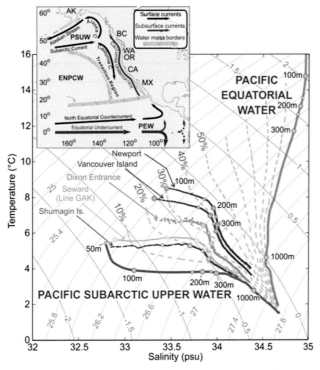

FIGURE 2.17 Average summer (April–October) temperature-salinity (*TS*) curves and associated specific density, σ_t ($\sigma_t = \sigma(T,S,0) = \rho - 1,000$) for the west coast of North America; ρ is the *in situ* water density in kg/m^3 adjusted to water pressure, $p = 0$, at the ocean surface (solid gray curves). The inset shows the locations of the main water masses and prevailing currents. Thick blue and red curves are characteristic curves for the Pacific Subarctic Upper Water (PSUW) and Pacific Equatorial Water (PEW), respectively. Intervening dashed gray curves give the percentage of PEW at 10% increments assuming that mixing between the PSUW and PEW occurs along surfaces of constant σ_t. Selected depths along the *TS* curves are also shown; near-surface *T* and *S* values are ignored. Thin green curves are contours of constant *spiciness* (high-temperature, high-salinity water). Locations of the specific *TS* curves are provided in Figure 2 of Thomson and Krassovski (2010).

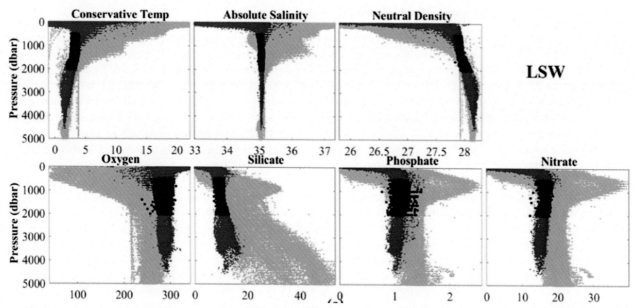

FIGURE 2.18 Characteristic water properties (in blue) used to define Labrador Sea Water (LSW) versus pressure (1 dbar ≈ 1.005525 m), conservative temperature (°C), absolute salinity (g/kg), neutral density (kg/m³), dissolved oxygen (μmol/kg) and nutrients (μmol/kg). *From Liu and Tanhua (2021).*

used by Liu and Tanhua (2021) to define the Labrador Sea Water in the North Atlantic, with Conservative Temperature and Absolute Salinity used in place of Potential temperature and Practical Salinity. As noted with respect to the bottom panels in Figure 2.9, plots are derived from the Global Ocean Data Analysis Project version 2 (GLODAPv2) data product and use the Optimal Multi-Parameter (OMP) method.

Temporal oceanic variability has important consequences for the calculation of mean *TS* diagrams where *TS* pairs, from a number of different bottle or CTD casts, are averaged together to define the *TS* relationship for a given area or lapsed time interval. Perhaps the easiest way to present this information is in the form of a time series plot (Figure 2.19), where the open circles represent monthly averaged *TS* values for three different lagoon locations off the coast of Queensland, Australia.

The traditional *TS* curves presented in Figures 2.9 and 2.20 are part of a family of curves relating measured variables such as temperature and salinity to density (sigma-*t*) or thermosteric anomaly ($\Delta_{S,T}$). The curvature of the lines in these figures is due to the nonlinear nature of the ocean's equation of state. In a traditional single-cast *TS* diagram, the stability of the water column, represented by the *TS* curve, can be easily evaluated. Unless one is in an unstable region, density should always increase with depth along the *TS* curve. Furthermore, the analysis of *TS* curves can shed important light on the advective and mixing processes generating these characteristic diagrams. We note that the thermosteric anomaly, $\Delta_{S,T}$, is used for *TS* curves rather than specific volume anomaly, $\delta_{S,T}$, since the pressure term that is included in $\delta_{S,T}$ has been found to be negligible for hydrostatic computation and can be approximated by $\Delta_{S,T}$, which lacks the pressure term.

Introduced by Montgomery (1958), Figure 2.20 presents a volumetric census of the water mass with the corresponding *TS* properties. The analyst must decide the vertical and horizontal extent of a given water mass and assign to it certain *TS* properties. From this information, the volume of the water mass can be estimated and entered on the *TS* diagram (Figure 2.20). The border values correspond to sums across *T* and *S* values. Worthington (1981) used this procedure, and a three-dimensional plotting routine, to produce a volumetric *TS* diagram for the deep waters of the World Ocean (Figure 2.16). The distinct peak in Figure 2.21 corresponds to common deep water which fills most of the deeper parts of the Pacific. Sayles et al. (1979) used the method to produce a good descriptive analysis of Bering Sea Water. This type of diagram has been made possible with the development of computer graphics techniques, which greatly enhance our ability to display and visualize data.

In a highly site-specific application of *TS* curves, McDuff (1988) has examined the effects of different source salinities on the thermal anomalies produced by buoyant hydrothermal plumes rising from midocean ridges. In potential temperature—salinity (θ−S) space, the shapes of the θ−S curves strongly depend on the salinity of the source waters and lead to markedly different thermal anomalies as a function of height above the vent site.

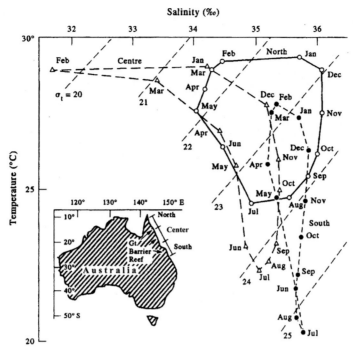

FIGURE 2.19 Monthly mean temperature–salinity pairs for surface water samples over a year in the lagoon waters of the Great Barrier Reef (see inset). The solid line applies to the "North" location in the insert; the long-dashed lines apply to the "Center" location and the short-dashed lines to the "South" region. *From Pickard and Emery (1992).*

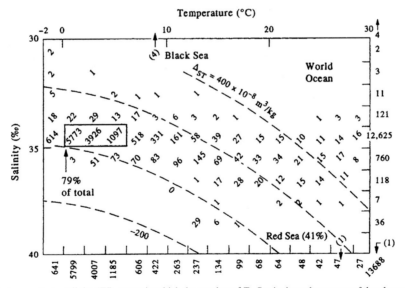

FIGURE 2.20 Volumetric temperature–salinity (*TS*) curves in which the number of *T*–*S* pairs in each segment of the plot can be calculated. The values in the rectangular box apply to 79% of the World Ocean. *From Montgomery (1958).*

2.6.6 Histograms

As oceanographic sampling matured, the concept of a stationary ocean has given way to the notion of a highly variable system requiring frequent sampling in space and time. Data display has graduated from a purely pictorial presentation to statistical representations. A plot format, related to fundamental statistical concepts of sampling and probability, is the histogram or frequency-of-occurrence diagram. This fundamental descriptive diagram presents information on how often a certain value occurred in any set of sample values (Figure 2.22). As we discuss in the section on basic statistics, there is no set rule for the construction of histograms and the selection of a sample variable interval (called "bin size") is somewhat

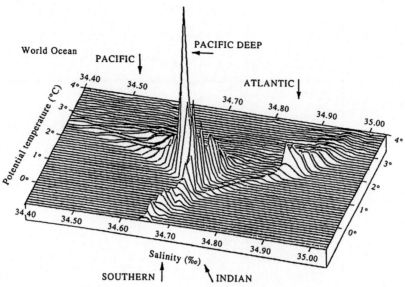

FIGURE 2.21 Three-dimensional volumetric *TS* diagram for the deep waters of the World Ocean. The distinct peak corresponds to common deep water, which fills most of the abyssal parts of the Pacific Ocean. *From Pickard and Emery (1992).*

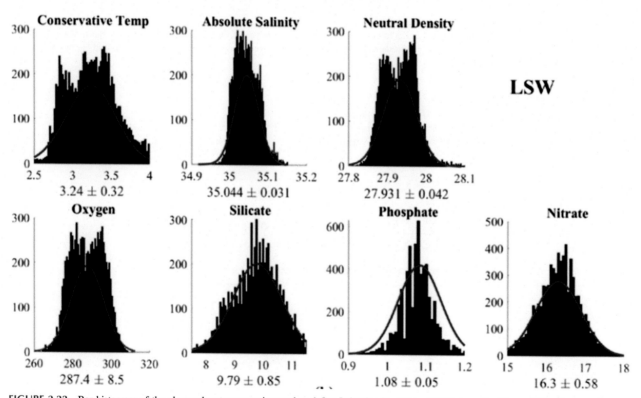

FIGURE 2.22 Bar histogram of the observed water properties used to define Labrador Sea Water (LSW) (see also Figure 2.18). Conservative temperature (°C), absolute salinity (g/kg), neutral density (kg/m³), dissolved oxygen and nutrients (μmol/kg). The Gaussian fit (red curve) shows the mean value and spread in values. *From Liu and Tanhua (2021).*

arbitrary. This choice of bin size determines the smoothness of the presentation, with the caveat that an appropriately wide interval is needed to ensure statistically meaningful frequency-of-occurrence values.

2.6.7 Time Series Presentation

The graphical presentation of time series records is of particular importance in oceanography. Early requirements were generated by shore-based measurements of sea-level heights, sea surface temperature, and other relevant parameters. As ship traffic increased, the need for offshore beacons led to the establishment of light or pilot ships, which also served as platforms for offshore data collection. Some of the early studies, made by geographers in the emerging field of physical oceanography, were carried out from light ships. The time series of wind, waves, surface currents, and surface temperature collected from these vessels needed to be displayed as a function of time. Later, dedicated research vessels such as weather ships were used as "anchored" platforms to observe currents and water properties as time series. Today, many time series data are collected by moored instruments which record internally or send their data back to a shore station via satellites, radio telemetry, and electrical or fiber-optic bottom cables. The need for real-time data acquisition for operational oceanography and meteorology has created an increased interest in new methods of telemetering data. The development of bottom-mounted acoustical modem systems, cable observatories, such as the Monterey Accelerated Research Systems and Ocean Networks Canada, and satellite data collection systems, such as Service Argos and Iridium, have opened new possibilities for the transmission of oceanographic data to shore stations and for the transmission of operational commands back to the offshore modules.

The simplest way to present time series information is to plot a scalar variable against time. The timescale depends on the data series to be plotted and may range in intervals from seconds to years. Scalar quantities can be displayed as time series plots on single or multiple vertical axes. Two-dimensional vector quantities (e.g., horizontal currents or surface winds) can be plotted as time series of speed (V) and direction (θ)—the two horizontal scalars actually measured by current meters and anemometers and more immediately available from the instrument—or as time series of two orthogonal components of velocity that need to be specified. Scalar time series of the u (x-direction) and v (y-direction) components of velocity are presented in Figure 2.23 along with simultaneously collected values of water temperature. Note that it is common practice in oceanography to rotate the horizontal u,v velocity components in order to align them with the dominant geographic or topographic orientation of the study region. This is especially true near coastal boundaries. Although there is no established convention on axes orientation, the horizontal orthogonal axes are commonly specified with x and y as the cross-shore and longshore directions, respectively. Alternatively, x and y are used for the across- and along-isobath directions. Over the continental shelf, the terms across-shelf and along-shelf are often used in place of across-shore and alongshore. The vertical (z) component of current can also be plotted, although a separate vertical scale is generally needed because of the much weaker speeds compared to the horizontal speeds. We note that oceanographic convention has vectors of current (and wind) pointing in the direction the flow is *going* whereas meteorological convention has wind vectors pointing in the direction the wind is *from*. To avoid confusion, marine scientists are advised to use the oceanographic convention. Plots of the current velocity components derived from acoustic Doppler Current Profiler (ADCP) measurements are especially appealing since they show the flow as functions of depth and time.

In Figure 2.24, we show the daily mean alongshore component of the current at mooring site A1 located in 500 m of water on the continental slope 60 km off southwest Vancouver Island, British Columbia. Current velocity at the site in 2012 was measured at 4 m bin intervals by a bottom-moored, upward-looking 75-kHz ADCP with 10 min sampling. Appended above the oceanic record are the daily-mean values of the alongshore component of wind stress from four reanalyses (NARR-1) grid points, ranging from San Diego (California) in the south to Estevan Point (southern Vancouver

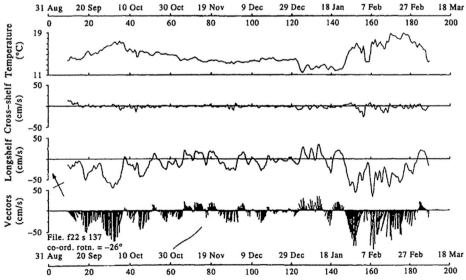

FIGURE 2.23 Time series of the low-pass filtered u (across-shelf, x) and v (along-shelf, y) components of velocity, together with the simultaneously collected values of temperature (T) for the east coast of Australia immediately south of Sydney, August 31, 1983 to March 18, 1984. The axes for the stick vectors are rotated by $-26°$ from North so that "up" is in the alongshore direction. The current meter was at 137 m depth in a total water depth of 212 m. Time in Julian days as well as calendar days. *From Freeland et al. (1985), © Copyright CSIRO Australia.*

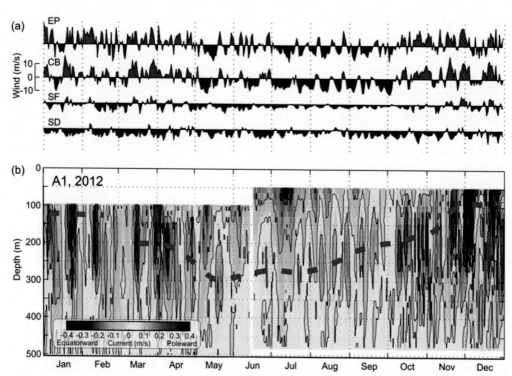

FIGURE 2.24 Time series of daily wind and current velocity during 2012. (a) Alongshore wind velocity at 10 m elevation at four North American Regional Reanalysis (NARR-1) locations along the west coast of North America (EP: Estevan Point, British Columbia; CB: Cape Blanco, Oregon; SF: San Francisco, California; SD: San Diego, California); (b) alongshore current velocity as a function of depth at mooring A1 off southwest Vancouver Island from an upward-looking 75 kHz ADCP moored at 500 m depth on the slope. Red and blue denote poleward and equatorward flows, respectively. The dashed line in 2b shows the estimated "core depth" of the poleward flowing California Undercurrent. *From Thomson and Krassovski (2015).*

Island) in the north. The strong poleward current centered between 150 and 300 m depth represents the California Undercurrent, whose temporal variability at A1 is most highly correlated with variations in the alongshore winds off Cape Blanco on the Oregon Coast, rather than with nearby winds off the island (Thomson and Krassovski, 2015).

Another common approach is the use of the "stick plot" (Figure 2.25; also Figure 2.23) where each "stick" (i.e., vector) corresponds to a measured speed and direction at the specified time. The length of the stick is scaled to the current speed. Direction may be relative to true north (pointed upward on the page) or the coordinate system may be rotated to align the axes with the dominant geographic or topographic boundaries (oceanographers in the Southern Hemisphere use true south instead of true north). The stick plot presentation is ideal for displaying directional variations of the measured currents. Rotational oscillations, due to the tides and inertial currents, are clearly represented. The once popular, but now less often used progressive vector diagram (PVD), can also be used to plot vector velocity time series (Figure 2.26). In this case, the time-integrated displacements along each of two horizontal orthogonal directions (x, y) are calculated from the corresponding velocity components (u,v) such that $(x,y) = (x_0,y_0) + \sum (u_i,v_i)\Delta t_i$ (where times Δt_i are for observations $i = 1, 2, ...$) to give "pseudo" downstream displacements of a parcel of water from its origin (x_0, y_0). The plot gives the vector sum of the individual current vectors plotting them head to tail for the period of interest. Distance in kilometers may be used in place of Earth coordinates, although there are distinct advantages to using the Mercator projection. Residual or long-term vector-mean currents are readily apparent in the PVD and rotational behavior also is well represented. The signature rotational motions of inertial and tidal currents can be easily distinguished in this type of diagram. The drawback is that the plots give the impression of a Lagrangian measurement with time (i.e., one that appears to be following the path of a fluid parcel) when in reality the data are from a single point (an Eulerian measurement) that cannot take into account the inevitable spatial inhomogeneities in the flow field; measurements at one location cannot inform us of the downstream trajectory of water parcels once they had crossed the recording location unless the flow is uniform in space.

Yet another type of time series plot consists of a series of vertical profiles at the same locations as functions of time (Figure 2.27). The time series plot has a vertical axis, much like a vertical profile, but with time replacing the horizontal distance axis. Similarly, a time series of horizontal transects along a repeated survey line is like a horizontal map but with time replacing one of the spatial axes. Property values from different time-depth (t, z) or time-distance (t, x) pairs are then

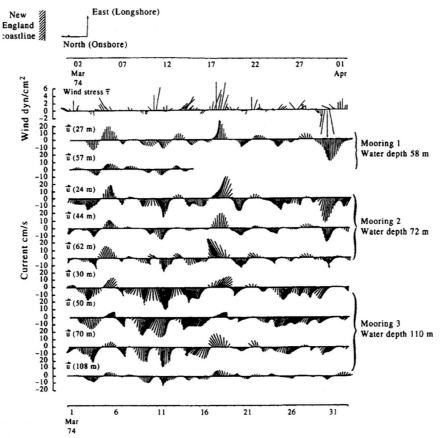

FIGURE 2.25   Vector (stick) plots of low-pass filtered wind stress and subtidal currents at different depths measured along the East Coast of the United States about 100 km west of Nantucket Shoals. East (up) is alongshore and north is cross-shore. Brackets give the current meter depth (m). *Source: Warren and Carl (1980). Fig. 7.11 (p. 221), © 1980 Massachusetts Institute of Technology, by permission of The MIT Press.*

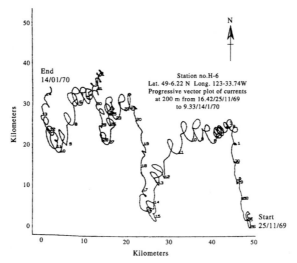

FIGURE 2.26   Progressive Vector Diagram (PVD) constructed using the east—west and north—south components of velocity for currents measured every 10 min for a period of 50 days at a depth of 200 m in the Strait of Georgia, British Columbia. Plotted positions correspond to horizontal displacements of the water that would occur if the flow near the mooring location was the same as that at the derived location. *From Tabata and Stickland (1972).*

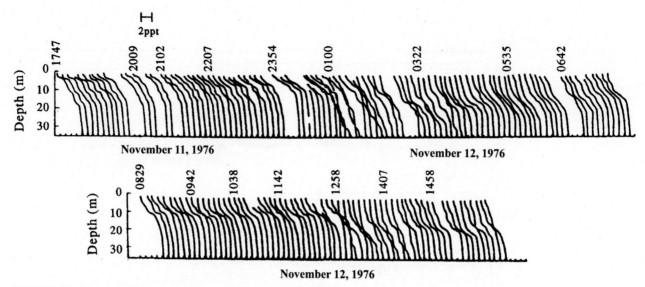

FIGURE 2.27   Time series of salinity profiles ("waterfall plot") taken in a highly stratified fjord. The effects of large internal waves can be seen around 0100 and 1300 on 12 November. The salinity scale is shown at the top of the figure. *From Farmer and Smith (1980).*

contoured to produce time series plots, which look very similar to horizontal maps. This type of presentation, called a Hovmöller diagram, is useful in depicting temporal signals that have a pronounced vertical or horizontal structure, such as seasonal heating and cooling in the upper ocean, Kelvin-Rossby wave propagation in the equatorial Pacific (Figure 2.28) or cumulative coastal wind-driven upwelling along the west coast of North America (Figure 2.29). Other temporal changes due to vertical layering (e.g., from river a plume) are well represented by this type of plot. Figures 2.28 and 2.29 display marked interannual variability. Figure 2.29 also indicates that this variability is significantly different north and south of about 40° N (near the Oregon-California border).

2.6.8 Other Forms of Graphical Presentation

Plotting oceanographic data has gone from a manpower-intensive process to one primarily carried out by computers. Computer graphics have provided oceanographers with a variety of new presentation formats. For example, all the data display formats previously discussed can now be carried out by computer systems. Much of the investigator's time is spent ensuring that computer programs are developed, not only for the analysis of the data but also for the presentation of results. These steps are often combined, as in the case of objective mapping of irregularly spaced data. In this case, an objective interpolation scheme is used to map a horizontal flow or property field. Contouring of the output objective map is then done by the computer. Frequently, both the smoothing provided by objective analysis, and the computer contouring, can be performed by existing software. Sometimes problems with these programs arise, such as continuing to contour over land or the restriction to certain contour intervals. Both of these problems must be overcome in the computer routine or the data altered in some way to avoid the problems. For example, "masks" can be applied to avoid contouring over land surfaces.

2.6.8.1 Three-Dimensional Displays

In addition to computer mapping, the computer makes it possible to explore other presentations not possible in hand analyses. Three-dimensional plotting is one of the more obvious examples of improved data display possible with computers. For example, Figure 2.30 shows a three-dimensional plot of coastal bottom topography and a 2D projection (contour map) of the same field. One main advantage of the three-dimensional plot is the geometrical interpretation given to the plot. We can more clearly see both the sign and the relative magnitudes of the dominant features. A further benefit of this form of presentation is the ability to present views of the data display from different angles and perspectives. For example, the topography in Figure 2.30 can be rotated to emphasize the different canyons that cut across the continental slope. Any analysis, which outputs a variable as a function of two others can benefit from a three-dimensional display. A well-known oceanic example is the Garrett-Munk spectrum for internal wave variability in the ocean (Figure 2.31) in which spectral amplitude based on observational data is plotted as a function of vertical wave number (m) and wave frequency (ω). The diagram tells the

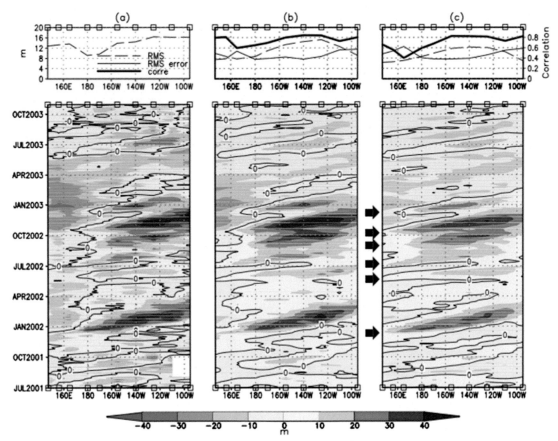

FIGURE 2.28 Variations in thermocline depth anomalies (m) along the equator in the Pacific Ocean between July 2001 and October 2003 for (a) the NOAA TAO (Tropical Atmosphere Ocean) data, (b) contributions of Kelvin and Rossby waves for the first three baroclinic modes, and (c) the Kelvin wave contribution. Upper panels indicate the RMS (dashed line), RMS error (solid line), and correlation (thick solid line) between model and observations. Open squares in the top panels indicate the position of the TAO mooring buoys along the equator (E). The black arrows in (c) show the starts of downwelling Kelvin waves. *From Mosquera-Vásquez et al. (2013).*

observer what kind of spectral shape to expect from a specific type of profiling method. for sections of the given spatial and temporal sampling intervals.

In Figure 2.32, color is used to denote the directional spectra ($D_{\eta\eta}$) of wave heights, η, during a 20 min burst sample of surface waves at Martha's Vineyard Coastal Observatory's (MVCO's) Air—Sea Interaction Tower. The tower is located about 3 km south of Martha's Vineyard, Massachusetts, in approximately 16 m of water. The spectral peaks and directions are markedly different for the swell and the wind waves, which is typical for this site. The mean wind is from the southwest.

2.6.8.2 Taylor Diagrams

Taylor (2001) introduced a method for graphically summarizing the statistical relationship between two related fields or sets of parameters. These "Taylor diagrams" are particularly useful for evaluating the relative skill of different numerical models against observations. In this case, the similarity between the modeled and observed fields is quantified in terms of their correlation coefficients (r), the degree of variation as measured by their standard deviations, and the centered root-mean-square (RMS) differences between the models and the observations. The mean values of the fields are removed before computing their higher order statistics so that the diagrams do not contain any bias information. Only the centered pattern errors are applied.

An example of a Taylor diagram is presented in Figure 2.33. The purpose of the diagram is to show the relative skill with which eight climate models are capable of simulating the global pattern of annual mean precipitation (Taylor, 2005). The red dots with letters A—H show where the eight models are positioned within the 2D fields represented by three sets of curved lines for the three statistical parameters: (1) the circular lines radiating out from the origin (0, 0) and terminating on the x- and y-axes span the range of standard deviations (in millimeters/day) for the model simulations and for the data; the

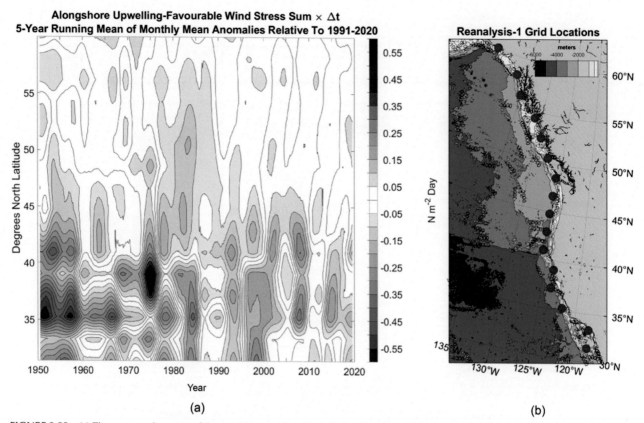

FIGURE 2.29 (a) Five-year running mean of the monthly anomalies of cumulative (time-integrated), upwelling-favorable (i.e., alongshore) wind stress for (b) selected Reanalysis-1 sites (red dots) from Mexico to Alaska off the west coast of North America. Areas in red denote periods with strengthened upwelling-favorable winds relative to the period 1991−2020. Units are wind-stress$\times \Delta t$ in (N/m^2)·days. *Courtesy Roy Hourston, Institute of Ocean Sciences.*

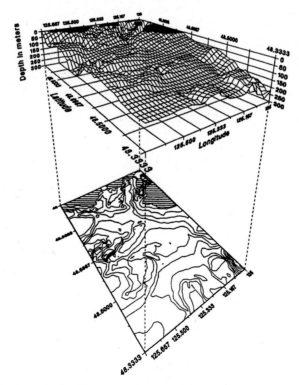

FIGURE 2.30 Three-dimensional plot of water depth at 20-m contour intervals off the southwest coast of Vancouver Island. The bottom plot is the 2D projection of the topography. *Courtesy Gary Hamilton.*

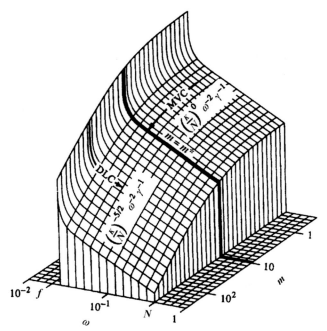

FIGURE 2.31 Garrett—Munk energy spectrum for oceanic internal waves based on different types of observations. Spectral amplitude (arbitrary units) is plotted against m (the vertical wave number in cycles per meter) and ω (the wave frequency in cycles per hour). Here, m^* is the wave number bandwidth, κ is the horizontal wave number, $N=N(z)$ is the buoyancy frequency, f is the Coriolis parameter, and $\gamma = \left(1 - f^2/\omega^2\right)^{1/2}$. MVC, moored vertical coherence and DLC, dropped lag coherence between vertically separated measurements. *From Garrett and Munk (1979).*

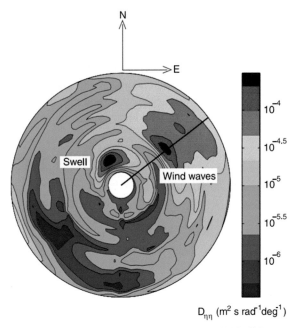

$$D_{\eta\eta} \ (m^2 \ s \ rad^{-1}deg^{-1})$$

FIGURE 2.32 Directional wave spectrum ($D_{\eta\eta}$) from a 20-min sampling burst on 8 October 2003 off the east coast of the United States showing distinct peaks due to swell and wind waves with amplitudes, η. The line pointing toward 54° true compass bearing (i.e., to the northeast) shows the wind direction toward which the wind is blowing. During this burst, the swell were propagating toward the north-northwest and wind waves toward the northeast. *From Gerbi et al. (2009).*

standard deviation for the single set of observations is shown by the dashed circular line (ending in "observed"); (2) the spokes emanating outward from origin span the full range (0—1) of possible correlation values that one might expect to find between the models and the observations; and (3) the last set of curved lines that span outward from where the dashed line of the observations intersects the x-axis denotes the centered RMS difference between the simulated and observed

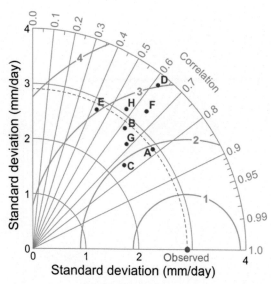

FIGURE 2.33 A Taylor diagram showing a statistical comparison of eight climate model estimates (models A through H) vs observations of the global pattern of mean annual precipitation. The dashed line labeled "observed" denotes the standard deviation of the observations. Green curves denote the RMS differences between the observations and the models. *From Taylor (2005).*

values of the annual mean precipitation. The model which combines the highest correlation with the lowest RMS difference with the data (i.e., lies closest to the dashed line) would be considered as a possible "best performance" model. If the standard deviation of the model also resembles that of the observations, then the model is clearly performing very well.

In this example, Taylor (2005) considers model F, which has a pattern correlation of $r \sim 0.65$ and a centered RMS value of about 2.6 mm/day (based on the curved lines that radiate out from the dashed point on the *x*-axis), as a good candidate. The standard deviation of the simulated rainfall field is about 3.3 mm/day which is slightly greater than the standard deviation of 2.9 mm/day of the observations. Models A, B, and E lie closest to the dashed line (i.e., have standard deviations similar to the data) but only model A has a relatively low RMS error (roughly 2 mm/day) and only model A (and C) account for over 50% of the variance ($r > 0.71$) between the model and the data. Taylor (2001) notes that the diagram can be extended into a second quadrant to allow for negative correlation coefficients and that the statistics can be normalized by dividing both the RMS difference and standard deviations of the model results by the standard deviation of the observations.

2.6.8.3 The Age of Color

As with many of the figures presented in this book, the extended use of color in journal papers, online publications and scientific reports provides major improvements in data and data analysis presentations. Previously, most color presentations were restricted to atlas and report presentations and were not available in journal articles. New printing procedures and online access have made color more affordable and much wider use is being made of color displays. One area of recent study where color display has played a major role is in the presentation of satellite and topographic images. Here, the use of false color enables the investigator to expand the dynamic range of the usual gray shades so that they are more easily recognizable by eye. False color is also used to enhance certain features such as sea surface temperature patterns and fronts inferred from infrared satellite images. The enhancements, and pseudo-color images, may be produced using a strictly defined function or may be developed in the interactive mode in which the analyst can produce a pleasing display. One important consideration in any manipulation of satellite images is to have each image registered to a ground map, which is generally called "image navigation". This navigation procedure (Emery et al., 1989b) can be carried out using satellite ephemeris data (orbital parameters) to correct for Earth curvature and rotation. Timing and spacecraft attitude errors often require the image to be "remapped" to fit the map projection exactly. An alternative method of image correction is to use a series of ground control points (GCPs) to warp the image to fit the GCP. GCPs are usually features such as bays or promontories that stand out in both the satellite image and the base map. In using GCP navigation a primary correction is made assuming a circular orbit and applying the mean satellite orbital parameters. It is possible to use these GCPs and the MCC (Chapter 1) in reverse to find the final image corrections after the orbital method is applied (Emery et al., 2003).

Access to digital image processing has greatly increased the investigator's capability to present and display data. Conventional data may be plotted in map form and overlain on a satellite image to show correspondence. This is possible since most image systems have one or more graphics overlay planes. Another form of presentation, partly motivated by satellite imagery, is the time sequence presentation of maps or images. Called "scene animation", this format produces a movie-style output which can be conveniently recorded on video. With widespread home use of video recorder systems and the ability of modern laptop computers to animate imagery, this form of data visualization is readily accessible to most people. A problem with this type of display has been the inability to publish videos or film loops. Many publishers now allow videos to be embedded in a published article. The move to open access publication has greatly enabled this form of publication. ADCP data from two or more nearby mooring sites also makes it possible to generate three-dimensional movies (using software like QuickTime, VLC Media Player, or AVI player) of the currents, winds, and other properties simultaneously. With the greatly enhanced video capacity of modern cell phones, this type of display may soon become commonplace on these devices.

Digital image manipulation has also changed the way oceanographers approach data display. Using an interactive system, the scientist-operator not only can change the brightness scale assignment (enhancement) but also can alter the orientation, the size (zoom in, zoom out), and the overall location of the output scene using a joystick, trackball, or mouse (digital tablet and cursor). With an interactive system, the three-dimensional display can be shifted and rotated to view all sides of the output. This allows the user to visualize areas hidden behind prominent features. Some of the most powerful applications have been developed by marine geologists and hydrographers whose display software and graphical information systems are used for navigation, marine resource mapping, and marine research (Figure 2.34).

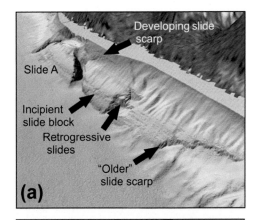

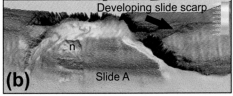

FIGURE 2.34 Perspective views of a mid-Holocene slide and surrounding area in Douglas Channel, British Columbia, using Fledermaus™ software. (a) Oblique view looking to the north showing several unstable features adjacent to the slide; and (b) oblique view looking to the east, showing several smaller slides at "n" at the top and backslope of the main slide. The slide volume of 60 million m³ was used to study landslide-generated tsunamis in the coastal region (Thomson et al., 2012). The color code gives water depth in meters. *From Conway et al. (2012).*

As more oceanographers become involved with digital image processing and pseudo-color displays, there should be an increase in the variety of data and result presentations. These will not only add new information to each plot but also make the presentation of the information more interesting and "colorful". The old adage of a picture being worth a thousand words is often true in oceanography and the interests of the investigators are best served when their results can be displayed in some interesting graphical or image form.

GLOSSARY

ADCP Acoustic Doppler Current Profiler
ASCII American Standard Code for Information Interchange
BPR Bottom Pressure Recorder
cph cycles per hour
CTD Conductivity-Temperature-Depth profiler

DART Deep-ocean Assessment and Reporting of Tsunamis (NOAA, USA)
GCP Ground control points
GLODAPv2 Global Ocean Data Analysis Project version 2
HIMIOFoTS Hellenic Integrated Marine Inland Water Observing, Forecasting and offshore Technology System (Greece)
LADCP Lowered Acoustic Doppler Current Profiler
MARS Monterey Accelerated Research System (USA)
NEPTUNE North-East Pacific Time Series Underwater Networks Experiments (Canada)
OMP Optimum multi-parameter
ONC Ocean Networks Canada (University of Victoria)
Pa pascals (pressure)
RCM Recording Current Meter
RMS Root mean square
SADCP Ship-mounted Acoustic Doppler Current Profiler
SAIL Shipboard ASCII Integration Loop (underway salinity and temperature sampling)
Service Argos Satellite positioning system (France)
STRATOGEM Strait of Georgia Ecosystem Modelling project (University of British Columbia, Canada)
VENUS Victoria Experimental Network Under the Sea (ONC, University of Victoria)

Chapter 3

STATISTICAL METHODS AND ERROR HANDLING

3.1 INTRODUCTION

This chapter provides a review of some of the basic statistical concepts and terminology used in processing data. The investigator needs this information to deal properly with the specific techniques used to edit and analyze oceanographic data. Our review is intended to establish a common level of understanding by the readers, not to provide a summary of all available procedures.

In the past, all collected data were processed and reduced by hand so that the individual scientist had an opportunity to become personally familiar with each data value. During this manual reduction of data, the investigator took into account important information regarding the particular instrument used and was able to determine which data were "better" in the sense that they had been collected and processed correctly. Within the limits of the observing systems, an accurate description of the data could be achieved without the application of statistical procedures. Individual intuition and familiarity with shipboard procedures took precedence in this type of data processing and analyses were made on comparatively few data values. In such investigations, the question of statistical reliability was seldom raised and it was assumed that individual data points were a valid representation of the parameter being measured. The data values were considered to be "correct."

For the most part, the advent of the computer and electronic data collection methods has meant that a knowledge of statistical methods has become essential to any reliable interpretation of results. Circumstances still exist, however, for which physical oceanographers still assign considerable weight to the quality of individual measurements. This is certainly true of water sample data, such as dissolved oxygen, nutrients, and chemical tracers collected from bottle and Rosette casts. In these cases, the established methods of data reduction, including familiarity with the data and knowledge of previous work in a particular region, still produce valuable descriptions of oceanic features and phenomena with a spatial resolution not possible with statistical techniques. However, for those more accustomed to having data collected and/or delivered on high density storage media, such as CD-ROM, optical disc, USB flash drive (also, thumb drive or key drive), or portable hard drive, statistical methods are essential to determining the value of the data and to decide how much of it can be considered useful for the intended analysis. This statistical approach arises from the fundamental complexity of the ocean, a multivariate system with many degrees of freedom in which nonlinear dynamics and sampling limitations make it difficult to separate scales of variability.

A fundamental problem with a statistical approach to data reduction is the fact that the ocean is not a stationary environment in which we can make repeated measurements. By "stationary" we mean a physical system whose statistical properties remain unchanged with time. In order to make sense of observations, oceanographers are forced to make some rather strong assumptions about the processes under investigation. Basic to these assumptions are the concepts of randomness and the consequent laws of probability. Since each oceanographic measurement can be considered as a superposition of the desired signal and unwanted noise (due to measurement errors and unresolved geophysical variability), the assumption of random behavior is often applied to both the signal and the noise. We must consider not only the statistical character of the signal and noise contributions individually but also the fact that the signal and the noise can interact with each other. Only through the application of the concept of probability can we make the assumptions required to reduce this complex set of variables to a workable subset. Our brief summary of statistics will emphasize concepts pertinent to the analysis of random variables, such as probability density functions (PDFs) and statistical moments (mean, variance, etc.). A brief glossary of statistical terms can be found in Appendix B.

Data Analysis Methods in Physical Oceanography. https://doi.org/10.1016/B978-0-323-91723-0.00006-4

3.2 SAMPLE DISTRIBUTIONS

Fundamental to any form of data analysis is the realization that we are usually working with a limited set (or sample) of random events drawn from a much larger population. We use samples to make estimates of the true statistical properties of the population. Historically, studies in physical oceanography were dependent on too few data points to allow for statistical inference and individual samples were considered representative of the true ocean. Often, an estimate of the population distribution is made from the sample set by using the relative frequency distribution, or histogram, of the measured data points. There is no fixed rule on how such a histogram is constructed in terms of ideal bin interval or number of bins. Generally, the more data there are, the greater the number of bins used in the histogram. Bins should be selected so that the majority of the measurements do not fall on the bin boundaries. Because the area of a histogram bin is proportional to the fraction of the total number of measurements in that interval, it represents the probability that an individual sample value will lie within that interval (Figure 3.1).

The most basic descriptive parameter for any set of measurements is the sample mean. The mean is generally taken over the duration of a time series (time average) or over an ensemble of measurements (ensemble mean) collected under similar conditions (Table 3.1). If the sample has N data values, $x_1, x_2, \ldots, x_N$, the sample mean is calculated as

$$\bar{x} = \frac{1}{N} \sum_{i=1}^{N} x_i \tag{3.1}$$

The sample mean is an unbiased estimate of the true population mean, μ. Here, an "unbiased" estimator is one for which the expected value, $E[x]$, of the estimator is equal to the parameter being estimated. In this case, $E[x] = \mu$ for which $\bar{x}$ is an unbiased estimator. The sample mean locates the center of mass of the data distribution such that

$$\sum_{i=1}^{N} (x_i - \bar{x}) = 0 \tag{3.2}$$

FIGURE 3.1 Histogram giving the percentage occurrences for the times of satellite position fixes during a 24-h day. Data are for satellite-tracked surface drifter #4851 deployed in the northeast Pacific Ocean from 10 December 1992 to 28 February 1993. During this 90-day period, the satellite receiver on the drifter was in the continuous receive mode.

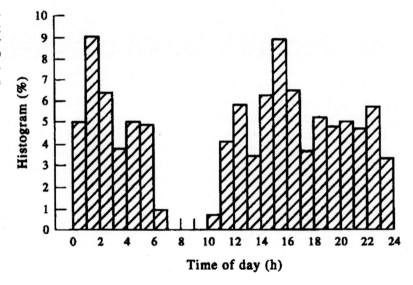

TABLE 3.1 Statistical values for the data set $x = \{x_i, i = 1, \ldots, 9\} = \{-3, -1, 0, 2, 5, 7, 11, 12, 12\}$.

Mean, $\bar{x}$	Biased variance, s'^2	Unbiased variance, s^2	Standard deviation, s	Range	Median	Mode
5.00	30.22	34.00	5.83	15	5	12

that is, the sample mean splits the data so that there is an equal weighting of negative and positive values of the fluctuation, $x' = x_i - \bar{x}$, about the mean value, $\bar{x}$. The weighted sample mean is the general case of Eqn (3.1) and is defined as

$$\bar{x} = \frac{1}{N} \sum_{i=1}^{N} f_i x_i = \sum_{i=1}^{N} (f_i / N) x_i \tag{3.3}$$

where f_i/N is the relative frequency of occurrence of the ith value for the particular experiment or observational data set. In Eqn (3.1), $f_i = 1$ for all i.

The sample mean values give the center of mass of a data distribution but not its width or how broadly the sample values are distributed. To determine how the data are spread about the mean, we need a measure of the sample variability or *deviation*, which is expressed in terms of the positive square root of the sample *variance*. For the data used in Eqn (3.1), the *sample variance* is the average of the square of the sample deviations from the sample mean, expressed as

$$s'^2 = \frac{1}{N} \sum_{i=1}^{N} (x_i - \bar{x})^2 \tag{3.4}$$

The *sample standard deviation*, $s' = \sqrt{s'^2}$, the positive square root of Eqn (3.4), is a measure of the typical difference of a data value from the mean value of all the data points. In general, these differ from the corresponding true *population variance*, σ^2, and the *population standard deviation, σ*. As defined by Eqn (3.4), the sample variance, s'^2, is a biased estimate of the true population variance, σ^2. An unbiased estimator of the population variance, s, is obtained from

$$s^2 = \frac{1}{(N-1)} \sum_{i=1}^{N} (x_i - \bar{x})^2 \tag{3.5a}$$

$$= \frac{1}{(N-1)} \left[\sum_{i=1}^{N} (x_i)^2 - \frac{1}{N} \left(\sum_{i=1}^{N} x_i \right)^2 \right] \tag{3.5b}$$

where the denominator $N - 1$ expresses the fact that at least two values are needed to define a sample variance and standard deviation, s. The use of the estimators s versus s' is often a matter of debate among oceanographers, although it should be noted that the difference between the two values decreases as the sample size increases. Only for relatively small samples ($N < 30$) is the difference significant. Because s' has a smaller mean-square (MS) error than s (suggesting a lower error than might be the case) and because s is an unbiased estimator when the population mean is known *a priori*, we recommend the use of Eqns (3.5a) and (3.5b). However, a word of caution: if your hypothesis depends on the difference between s and s', then you have ventured onto shaky statistical ground supported by questionable data. We further note that the expanded relation Eqn (3.5b) is a more efficient computational formulation than Eqn (3.5a) in that it allows one to obtain s^2 from a single pass through the data. If the sample mean must be calculated first, two passes through the same data set are required rather than one, which is computationally less efficient when dealing with large data sets.

Other statistical values of importance are the range, mode, and median of a data distribution (Table 3.1). The *range* is the spread or absolute difference between the endpoint values of the data set, while the *mode* is the value of the distribution that occurs most often. For example, the data sequence 2, 4, 4, 6, 4, 7 has a range of $|2 - 7| = 5$ and a mode of 4. The *median* is the middle value in a set of numbers arranged according to magnitude (the data sequence $-1, 0, 2, 3, 5, 6, 7$ has a median of 3). If there is an even number of data points, the median value is chosen midway between the two candidates for the central value. Two other measures, *skewness* (the third moment of the distribution and degree of asymmetry of the data about the mean) and *kurtosis* (a nondimensional number measuring the flatness or peakedness of a distribution) are less often used in oceanography.

As we discuss more thoroughly later in this chapter, the shapes of many sample distributions can be approximated by a *normal* (also called a *bell* or *Gaussian*) distribution. A convenient aspect of a normal population distribution is that we can apply the following empirical "rule of thumb" to the data:

$$\begin{cases} \mu \pm \sigma, \text{ spans approximately 68\% of the measurements} \\ \mu \pm 2\sigma, \text{ spans approximately 95\% of the measurements} \\ \mu \pm 3\sigma, \text{ spans 99\% of the measurements} \end{cases} \tag{3.6}$$

The percentages are represented by the areas under the normal distribution curve spanned by each of the limits (Figure 3.2). We emphasize that the above limits apply only to normal distributions of random variables.

FIGURE 3.2 Plot of the normal (Gaussian) probability density function, $f(x)$, for mean μ and standard deviation σ of the random variable, X. Numerical values denote the relative areas for the different limit ranges. *After Harriett and Murphy (1975).*

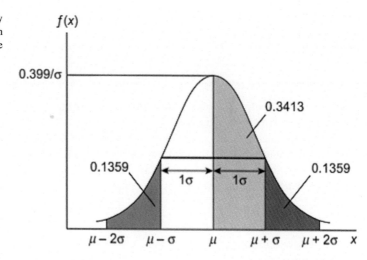

3.3 PROBABILITY

Most data collected by oceanographers are made up of samples taken from a larger unknown population. If we view these samples as random events of a statistical process, then we are faced with an element of uncertainty: "What are the chances that a certain event occurred or will occur based on our sample?" or "How likely is it that a given sample is truly representative of a certain population distribution?" (The last question might be asked of political pollsters who use small sample sizes to make sweeping statements about the opinions of the populace as a whole). We need to find the best procedures for inferring the population distribution from the sample distribution and to have measures that specify the goodness of the inference. Probability theory provides the foundation for this type of analysis. In effect, it enables us to find a value between 0 and 1, which tells us just how likely is a particular event or sequence of events. A probability is a proportional measure of the occurrence of an event. If the event has a probability of zero, then it is impossible; if it has a probability of unity, then it is certain to occur. Probability theory as we know it today was initiated by Pascal and Fermat in the seventeenth century through their interest in games of chance. In the eighteenth century, Jacob and Nicholas Bernoulli continued to study the theory of chance in gaming applications, while Gauss and Laplace extended the theory to the social sciences and actuarial mathematics. Well-known names like R.A. Fisher, J. Neyman, and E.S. Pearson are associated with the proliferation of statistical techniques developed in the twentieth century. The *frequentist* statistical techniques developed by these authors—which is the focus of this section—define probability in terms of the outcomes of a large number of near identical trials. The alternative *Bayesian* statistical technique, named after Thomas Bayes, an eighteenth-century theologian and mathematician, views probability in terms of conditional probability, in which the likelihood of a particular result or event depends on some prior probability, which is then updated when new information is obtained. Whereas mathematicians prior to Bayes reasoned their way from chance to relative frequency, Bayes presented a mathematical foundation for statistical inference progressing from relative frequency to chance (Diaconis and Skyrms, 2018).

The *probability mass function*, $f(x) = P(X = x)$ — also known as the *probability density function* (PDF) — gives the relative frequency of occurrence of each possible value of a random variable, X. In the discrete case, the function specifies the point probabilities $f(x_i) = P(X = x_i)$ and assumes nonzero values only at points $X = x_i$ ($i = 1, 2,...$). One of the most common examples of a discrete probability mass function is the sum of the dots obtained from the roll of a pair of dice (Table 3.2). (Originally, this would have been called a pair of "die", but modern English usage has "dice" as both a singular and plural noun.) According to probability theory, the dice player is most likely to roll a 7 (highest probability mass function value) and least likely to roll a 2 or 12 (lowest probability mass function value). The dice example reveals two of the fundamental properties of all probability functions: (1) $0 \le f(x_i)$; and (2) $\sum f(x_i) = 1$, where the summation is over all possible values of x. For the case of a continuous variable, the above fundamental properties become: (1) $0 \le f(x)$; and (2) $\int f(x)dx = 1$ where the integration is over all x in the range $(-\infty, \infty)$.

To further illustrate the concept of probability, consider N independent trials, each of which has the same probability of "success" p and probability of "failure" $q = 1 - p$. The probability of success or failure is unity; $p + q = 1$. Such trials

TABLE 3.2 The discrete probability mass function and cumulative probability functions for the sum of the dots (variable X) obtained by tossing a pair of dice.

Sum of dots (X)	Frequency of occurrence	Relative frequency	Probability mass function, $f(x) = P(X = x)$	Cumulative probability function $F(x) = P(X \leq x)$
2	1	1/36	$P(x = 2) = 1/36$	$F(2) = P(X \leq 2) = 1/36$
3	2	2/36	$P(x = 3) = 2/36$	$F(3) = P(X \leq 3) = 3/36$
4	3	3/36	$P(x = 4) = 3/36$	$F(4) = P(X \leq 4) = 6/36$
5	4	4/36	$P(x = 5) = 4/36$	$F(5) = P(X \leq 5) = 10/36$
6	5	5/36	$P(x = 6) = 5/36$	$F(6) = P(X \leq 6) = 15/36$
7	6	6/36	$P(x = 7) = 6/36$	$F(7) = P(X \leq 7) = 21/36$
8	5	5/36	$P(x = 8) = 5/36$	$F(8) = P(X \leq 8) = 26/36$
9	4	4/36	$P(x = 9) = 4/36$	$F(9) = P(X \leq 9) = 30/36$
10	3	3/36	$P(x = 10) = 3/36$	$F(10) = P(X \leq 10) = 33/36$
11	2	2/36	$P(x = 11) = 2/36$	$F(11) = P(X \leq 11) = 35/36$
12	1	1/36	$P(x = 12) = 1/36$	$F(12) = P(X \leq 12) = 1$
Total sum	36	1.00		1.00

involve binomial distributions for which the outcomes can be only one of two events: for example, a tossed coin will produce a head or a tail; an XBT will work or it will not work. If X represents the number of successes that occur in the N trials, then X is said to be a discrete random variable having parameters (N, p). The term "Bernoulli trial" is sometimes used for X. The probability mass function that gives the relative frequency of occurrence of each value of the random variable X having parameters (N, p) is the binomial distribution,

$$p(x) = \binom{N}{x} p^x (1 - p)^{N-x}, x = 0, 1, \ldots, N \tag{3.7a}$$

where the expression

$$\binom{N}{x} = \binom{N}{N - x} = {}_N C_x \equiv \frac{N!}{[(N - x)! x!]} \tag{3.7b}$$

is the number of different *combinations* of groups of x objects that can be chosen from a total set of N objects without regard to order (i.e., abc is the same combination as cba, or any other combination). The number of different combinations of x objects is always fewer than the number of *permutations*, ${}_N\tilde{P}_x$, of x objects $[{}_N\tilde{P}_x \equiv N!/(N - x)!]$, written here with a tilde ($\sim$) to distinguish it from probability, P. In the case of permutations, different ordering of the same objects counts for a different permutation (i.e., ab is different than ba). As an example, the number of possible different batting orders (permutations) a coach can create among the first four hitters on a nine-person baseball team is $9!/(9 - 4)! = 9!/5! = 3,024$. Each different ordering of the four players counts as a permutation. In contrast, the number of different groups of ball players a coach can put in the first four lead-off batting positions without regard to any particular batting order among the groups of four players is $9!/[(9 - 4)! 4!] = 9!/5! 4! = 126$. The numbers

$$\binom{N}{x}$$

are often called *binomial coefficients* since they appear as coefficients in the expansion of the binomial expression $(a + b)^N$ given by the binomial theorem:

$$(a + b)^N = \sum_{k=0}^{N} \binom{N}{k} a^k b^{N-k} \tag{3.8}$$

The summed probability mass function

$$f(a \leq x \leq b) = \sum_{i=a}^{b} f(x_i) \qquad (3.9)$$

for variable X over a specified range of values (a, b) can be demonstrated by a simple oceanographic example. Suppose there is a probability, $(1 - p)$, that a current meter will fail when moored in the ocean and that the failure is independent from current meter to current meter. Assume that a particular string of single-point meters will successfully measure the expected flow structure if at least 50% of the meters on the string remain operative. For example, a two-instrument string used to measure the barotropic flow will be successful if one current meter remains operative while a four-instrument string used to resolve the low mode, baroclinic flow will be successful if at least 2 instruments remain operative. We then ask: "For what values of p is a 4-instrument array preferable to a 2-instrument array?" Since each current meter is assumed to fail or function independently of the other meters, it follows that the number of functioning current meters is a binomial random variable. The probability that a 4-instrument mooring is successful is then

$$P(2 \leq x \leq 4) = \sum_{k=2}^{4} \binom{4}{k} p^k (1-p)^{4-k}$$

$$= \binom{4}{2} p^2 (1-p)^2 + \binom{4}{3} p^3 (1-p)^1 + \binom{4}{4} p^4 (1-p)^0$$

$$= 6p^2 (1-p)^2 + 4p^3 (1-p)^1 + p^4$$

Similarly, the probability that a 2-instrument array is successful is

$$P(1 \leq x \leq 2) = \sum_{k=1}^{2} \binom{2}{k} p^k (1-p)^{2-k}$$

$$= 2p(1-p) + p^2$$

From these two relations, we find that the 4-instrument string is more likely to succeed in its application than the 2-instrument array when

$$6p^2 (1-p)^2 + 4p^3 (1-p)^1 + p^4 \geq 2p(1-p) + p^2$$

or, after some factoring and simplification, when

$$p(p-1)^2 (3p-2) \geq 0$$

for which we require $3p - 2 \geq 0$, or $p \geq 2/3$. When compared to the 2-instrument array, the 4-instrument array is more likely to do its intended job when the probability, p, that a given instrument works is $p \geq 2/3$. The 2-instrument array is more likely to succeed when $p \leq 2/3$.

In the previous example, specification of the probability p requires information on the rate of failure of the current meters. This, in turn, requires information on success rates from previous mooring programs that used these particular instruments. When examining these success rates, we would likely make the fundamental assumption, applicable to most of the data sets we collect, that each sample in our set of observations is an independent realization drawn from a random distribution. Individual events in this distribution cannot be predicted with certainty but their relative frequency of occurrence, for a long series of repeated trials (samples)—following Bernoulli's Law of Large Numbers—is often remarkably stable. We further remark that the binomial distribution discussed in the context of current meters is only one type of PDF. Other distribution functions will be discussed later in the chapter.

3.3.1 Cumulative Probability Functions

The probability mass function yields the probability of a specific event or probability of a range of events. From this function we can derive the *cumulative probability function, $F(x)$*—also called the cumulative distribution function, cumulative mass function, and probability distribution function—defined as that fraction of the total number of possible outcomes X (a random variable), which are less than a specific value x (a number). Thus, the distribution function is the probability that $X \leq x$, or

$$F(x) = P(X \leq x)$$

$$= \sum_{all\ X \leq x} f(x_i), -\infty < x < \infty \tag{3.10a}$$

(discrete random variable, X_i)

$$= \int_{-\infty}^{x} f(x)dx, -\infty < x < \infty \tag{3.10b}$$

(continuous random variable, X)

The discrete cumulative distribution function for tossing a pair of fair dice (Table 3.2) is plotted in Figure 3.3. Because the probabilities P and f are limited to the range 0 and 1, we have $F(-\infty) = 0$ and $F(\infty) = 1$. In addition, the distribution function $F(x)$ is a nondecreasing function of x, such that $F(x_1) \leq F(x_2)$ for $x_1 < x_2$, where $F(x)$ is continuous from the right (Table 3.2).

It follows that, for the case of a continuous function, the derivative of the distribution function F with respect to the sample parameter, x

$$f(x) = \frac{dF(x)}{dx} \tag{3.11}$$

recovers the PDF, f. As noted earlier, the PDF has the property that its integral over all values is unity

$$\int_{-\infty}^{\infty} f(x)dx = F(\infty) - F(-\infty) = 1$$

In the limit $dx \to 0$ the fraction of outcomes for which x lies in the interval $x < x' < x + dx$ is equal to $f(x')dx$, the probability for this interval. The random variables being considered here are continuous so that the PDF can be defined by Eqn (3.11). Variables with distribution functions that contain discontinuities, such as the steps in Figure 3.3, are called discrete variables. A random variable is considered discrete if it assumes only a countable number of values. In most oceanographic sampling, measurements can take on an infinity of values along a given scale and the measurements are best considered as continuous random variables. The function $F(x)$ for a continuous random variable X is itself continuous and

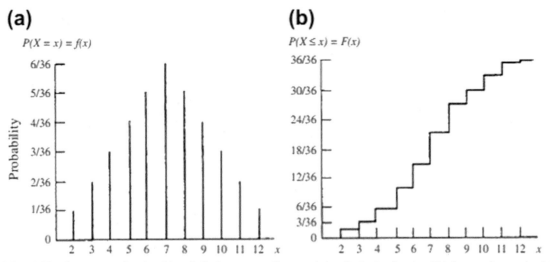

FIGURE 3.3 (a) The discrete mass function $f(x)$ and (b) the corresponding cumulative distribution function $F(x)$ from tossing a pair of dice (see Table 3.2). *After Harnett and Murphy (1975).*

appears as a smooth curve. Similarly, the PDF for a continuous random variable X is continuous and can be used to evaluate the probability that X falls within some interval $[a, b]$ as

$$P(a \leq X \leq b) = \int_a^b f(x)dx \qquad (3.12)$$

3.4 MOMENTS AND EXPECTED VALUES

The discussion in the previous section allows us to determine the probability of a single event or experiment, or describe the probability of a set of outcomes for a specific random variable. However, the discussion is not concise enough to describe fully the probability distributions for specific data sets. The situation is similar to Section 3.2 in which we started with a set of observed values. In addition to presenting the individual values, we seek properties of the data, such as the sample mean and variance to help us characterize the structure of our observations. In the case of probability distributions, we are not dealing with the *observed* mean and variance but with the *expected* mean and variance obtained from an infinite number of realizations of the random variable under consideration.

Before discussing some common PDFs, we need to review the computation of the parameters used to describe these functions. These parameters are, in general, called "moments" by analogy to mechanical systems where moments describe the distribution of forces relative to some reference point. The statistical concept of degrees of freedom is also inherited from the terminology of physical mechanical systems where the number of degrees of freedom specifies the motion possible within the physical constraints of the mechanical system and its distribution of forces. As noted earlier, the population mean, μ, and standard deviation, σ, define the first and second moments that describe the center and spread (distribution about the center) of the probability function. In general, these parameters do not uniquely define the PDF since many different PDFs can have the same mean and standard deviation. However, in the case of the Gaussian distribution, the PDF is completely described by μ and σ. In defining moments, we must be careful to distinguish between moments taken about the origin and moments taken about the mean (central moments).

When discussing moments, it is useful to introduce the concept of expected value. This concept is analogous to the notion of weighted functions. For a discrete random variable, X, with a probability function $P(x)$ (the discrete analogue to the continuous PDF), the expected value of X is written as $E[X]$ and is equivalent to the arithmetic mean, μ, of the probability distribution. In particular, we can write the expected value for a discrete PDF as

$$E[X] = \sum_{i=1}^N x_i f(x_i) = \mu \qquad (3.13)$$

where μ is the population mean introduced in Section 3.2. The probability function $P(x)$ serves as a weighting function similar to the function f_i/N in Eqn (3.3). The difference is that f_i/N is the relative frequency for a single set of experimental samples whereas $f(x)$ is the expected relative frequency for an infinite number of samples from repeated trials of the experiment. The expected value, $E[X]$, for the sample which includes X, is the sample mean, $\bar{x}$. Similarly, the variance of the random variable X is the expected value of $(X - \mu)^2$, or

$$V[X] = E[(X - \mu)^2] = \sum_{i=1}^N (x_i - \mu)^2 f(x_i) = \sigma^2 \qquad (3.14)$$

In the case of a continuous random variable, X, with PDF $f(x)$, the expected value is

$$E[X] = \int_{-\infty}^\infty x f(x)dx \qquad (3.15)$$

while for any function $g(X)$ with a PDF $f(x)$, the expected value can be written as

$$E[g(X)] = \int_{-\infty}^\infty g(x)f(x)dx \text{(continuous variable)} \qquad (3.16a)$$

$$= \sum_{i=1}^N g(x_i)f(x_i) \text{(discreate case)}. \qquad (3.16b)$$

Some useful properties of expected values for random variables are:

1. For c = constant; $E[c] = c, V[c] = 0$;
2. $E[cg(X)] = cE[g(X)], V[cg(X)] = c^2V[g(X)]$;
3. $E[g_1(X) \pm g_2(X) \pm \ldots] = E[g_1(X)] \pm E[g_2(X)] \pm \ldots$;
4. $V[g(X)] = E[(g(X) - \mu)^2] = E[g(X)^2] - \mu^2$, (variance about the mean);
5. $E[g_1 g_2] = E[g_1]E[g_2]$;
6. $V[g_1 \pm g_2] = V[g_1] + V[g_2] \pm 2C[g_1, g_2]$.

Property (6) introduces the *covariance function* of two variables, C, defined as

$$C[g_1, g_2] = E[g_1 g_2] - E[g_1]E[g_2] \tag{3.17}$$

where, using property (5), $C = 0$ when g_1 and g_2 are independent random variances. Using properties (1)–(3), we find that $E[Y]$ for the linear relation $Y = a + bX$ can be expanded to

$$E[Y] = E[a + bX] = a + bE[X]$$

while from (1) and (6) we find

$$V[Y] = V[a + bX] = b^2 V[X]$$

At this point, we remark that averages, expressed as expected values, $E[X]$, apply to ensemble averages of many (read, infinite) repeated samples. This means that each sample is considered to be drawn from an infinite ensemble of identical statistical processes varying under exactly the same conditions. In practice, we do not have an infinite number of repeated samples taken under identical conditions but rather time (or space) records having limited temporal (or spatial) extent. In using time or space averages as representative of ensemble averages, we are assuming that our records are *ergodic*. This implies that averages over an infinite ensemble can be replaced by an average over a single, infinitely long time series. According to a proof by Birkhoff (1931), an ergodic dynamical system is one in which the average of a measurable function, $x(t)$, over all probability space (p) equals the limiting time average,

$$\int x \, dp = \lim_{n \to \infty} \left\{ \frac{1}{n} [x(t_1) + \ldots + x(t_n)] \right\} \tag{3.18a}$$

or, equivalently,

$$\langle x(t) \rangle = \lim_{T \to \infty} \frac{1}{2T} \int_{-T}^{T} x(t) \, dt = \lim_{N \to \infty} \frac{1}{N} \sum_{k=1}^{N} x_k(t) \tag{3.18b}$$

where $\langle x(t) \rangle$ denotes an ensemble average over points, x_k, or an ensemble average of replications of the time series, $x(t)$. The time average of a process at a particular geographical point is equal to the average at a particular time (t) over an ensemble of points, x_k, or an ensemble average of replications of the time series at x. For natural hazards, this allows replacing an estimate of the source or hazard statistics at a particular location, where there is limited knowledge throughout time, with the statistics of an ensemble of known source or hazard variables over a broad region (or even globally).

An ergodic process is not to be confused with a stationary process for which the PDF of $X(t)$ is independent of time. In reality, time/space series can be considered stationary if major shifts in the statistical characteristics of the series occur over intervals that are long compared to the averaging interval so that the space/time records remain homogeneous (exhibit the same general behavior) throughout the selected averaging interval. A data record that is quiescent during the first half of the record and then exhibits large irregular oscillations during the second half of the record is not stationary.

3.4.1 Unbiased Estimators and Moments

As we stated earlier, $\bar{x}$ and s^2, as defined by Eqns (3.3) and (3.5a, 3.5b) are unbiased estimators of the true population mean, μ, and variance, σ^2. That is, the expected values of x and $(x - \bar{x})^2$ are equal to μ and σ^2, respectively. To illustrate, we first prove that $E[\bar{x}] = \mu$. To do this, we write the expected value as the normalized sum of all $\bar{x}$ values, whereby

$$E[x] = E\left[\frac{1}{N} \sum_{i=1}^{N} x_i \right] = \frac{1}{N} \sum_{i=1}^{N} E[x_i] = \frac{1}{N} \sum_{i=1}^{N} \mu = \mu$$

as required. Next, we demonstrate that $E[s^2] = \sigma^2$. We again use the appropriate definitions and write

$$
\begin{aligned}
E\left[s^2\right] &= E\left[\frac{1}{N-1}\sum_{i=1}^{N}(x_i - \bar{x})^2\right] \\
&= E\left[\frac{1}{N-1}\left\{\sum_{i=1}^{N}\left[(x_i - \mu)^2 - N(\bar{x} - \mu)^2\right]\right\}\right] \\
&= \frac{1}{N-1}\left\{\sum_{i=1}^{N}E\left[(x_i - \mu)^2\right] - NE\left[(\bar{x} - \mu)^2\right]\right\} \\
&= \frac{1}{N-1}\left\{\sum_{i=1}^{N}(\sigma^2) - N\frac{\sigma^2}{N}\right\} = \frac{\sigma^2}{N-1}(N-1) = \sigma^2
\end{aligned}
$$

where we have used the relations $x_i - \bar{x} = (x_i - \mu) - (\bar{x} - \mu)$, $E[(x_i - \mu)^2] = V[x_i] = \sigma^2$ (the variance of an individual trial) and $E[(x_i - \mu)^2] = V[x_i] = \sigma^2/N$ (the variance of the sample mean relative to the population mean, μ). The last expression derives from the central limit theorem discussed in Section 3.6.

Returning to the discussion of statistical moments, we define the ith moment of the random variable X, taken about the origin, as

$$
E[X^i] = \mu_i \tag{3.19}
$$

Thus, the first moment about the origin ($i = 1$) is the population mean, $\mu = \mu_1$. Similarly, we can define the ith moment of X taken about the mean (called the ith central moment of X) as

$$
E[(X - \mu)^i] = \mu_i \tag{3.20}
$$

The population variance, σ^2, is the second ($i = 2$) central moment, μ_2.

3.5 COMMON PDFS

The purpose of this section is to provide examples of three common PDFs. The first is the uniform PDF given by

$$
\begin{aligned}
f(x) &= \frac{1}{x_2 - x_1}, x_1 \leq x \leq x_2 \\
&= 0, \text{ otherwise}
\end{aligned} \tag{3.21}
$$

Figure 3.4 which is the intended PDF of random numbers generated by most computers and handheld calculators. The function is usually scaled between 0 and 1. The cumulative density function $F(x)$ given by Eqns (3.10a) and (3.10b) has the form

$$
\begin{aligned}
F(x) &= 0, x < x_1 \\
&= \frac{x - x_1}{x_2 - x_1}, x_1 \leq x \leq x_2 \\
&= 1, x \geq x_2
\end{aligned} \tag{3.22}
$$

while the mean and standard deviation of Eqn (3.21) are given by $\mu = (x_2 + x_1)/2$ and $\sigma = (x_2 - x_1)/(2\sqrt{3})$.

Perhaps the most familiar and widely used PDF is the normal (or Gaussian) density function:

$$
f(x) = \frac{e^{\left[-(x-\mu)^2/2\sigma^2\right]}}{\sigma\sqrt{2\pi}}; \sigma > 0, -\infty < \mu < \infty, -\infty < x < \infty \tag{3.23}
$$

where the parameter σ represents the standard deviation (or spread) of the random variable X about its mean value μ (Figure 3.2). For convenience, Eqn (3.23) is often written in shorthand notion as $N(\mu, \sigma^2)$. The height of the density

(a)

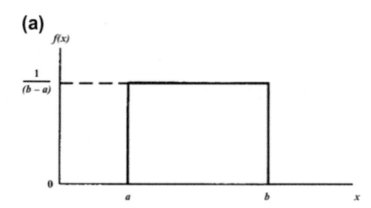

(b)

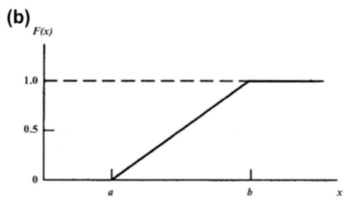

FIGURE 3.4   Uniform probability density distribution functions: (a) the probability density function, $f(x)$; and (b) the corresponding cumulative probability distribution function, $F(x)$. *From Bendat and Piersol (1986).*

function at $x = \mu$ is $0.399/\sigma$. The cumulative probability distribution of a normally distributed random variable, X, lying in the interval a to b is given by the integral (Eqn 3.12),

$$P(a \leq X \leq b) = \frac{\int_a^b e^{\left[-\frac{(x-\mu)^2}{2\sigma^2}\right]}}{\sigma\sqrt{2\pi}}\, dx \qquad (3.24)$$

which is the area under the normal curve between a and b. Since a closed form of this integral does not exist, it must be evaluated by approximate methods, often involving the use of tables of areas. We have included a table of curve areas in Appendix D, Table D.1. The normal distribution is symmetric with respect to μ so that areas need to be tabulated only on one side of the mean. For example, $P(\mu \leq x \leq \mu + 1\sigma) = 0.3413$ so by symmetry $P(\mu - 1\sigma \leq x \leq \mu + 1\sigma) = 2(0.3413) = 0.6826$. The latter is the value used in the rule of thumb estimates for the range of the standard deviation, σ. For the normal distribution, the tabulated values represent the area between the mean and a point z, where z is the distance from the mean measured in standard deviations. This leads to the familiar transform for a normal random variable X given by

$$Z = \frac{X - \mu}{\sigma} \qquad (3.25)$$

called the "standardized normal variable". The variable Z gives the distances of points measured from the mean of the normal random variable in terms of the standard deviation of the normal random variable, X (Figure 3.5). The standard normal variable Z is normally distributed with a mean of zero (0) and a standard deviation of unity (1). Thus, if X is described by the function $N(\mu, \sigma^2)$ then Z is described by the function $N(0, 1)$.

The third continuous PDF is the gamma density function, which applies to random variables, which are always nonnegative, thus producing distributions that are skewed to the right. The gamma PDF is

$$f(x) = \frac{x^{\sigma-1} e^{-\frac{x}{\beta}}}{\beta^{\alpha}\Gamma(\alpha)} \quad \alpha, \beta > 0;\ 0 \leq x \leq \infty \qquad (3.26)$$

$$= 0, \text{ elsewhere}$$

FIGURE 3.5 Distribution $f(z)$ for the standardized normal random variable, $Z = (X - \mu)/\sigma$ (cf. Figure 3.2). *After Harriett and Murphy (1975).*

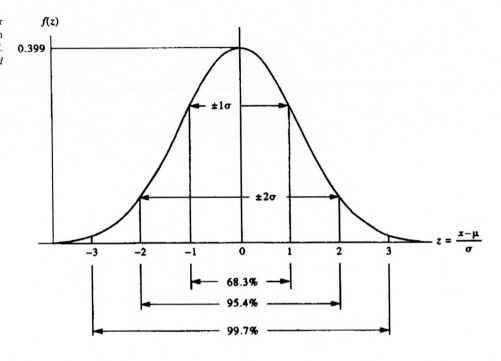

where σ and β are parameters of the distribution and $\Gamma(\alpha)$ is the gamma function

$$\Gamma(\alpha) = \int_0^{\infty} x^{\alpha-1} e^{-x} dx \tag{3.27}$$

For any integer n,

$$\Gamma(n) = 1 \cdot 2 \cdot 3 \cdot \ldots \cdot (n-1) = (n-1)! \tag{3.28}$$

while for a continuous variable α

$$\Gamma(\alpha) = (\alpha-1)\Gamma(\alpha-1), \text{ for } \alpha \geq 1 \tag{3.29}$$

where $\Gamma(1) = 1$. Plots of the gamma PDF for $\beta = 1$ and three values of the parameter α are presented in Figure 3.6. Because it is again impossible to define a closed form of the integral of the PDF in Eqn (3.26), tables are used to evaluate probabilities from the PDF. One particularly important gamma density function has a PDF with $\alpha = \nu/2$ and $\beta = 2$. This is the *chi-square random distribution* (written as χ_ν^2 and pronounced "ki square") with ν degrees of freedom (Appendix D,

FIGURE 3.6 Plots of the gamma function for various values of the parameter α ($\beta = 1$).

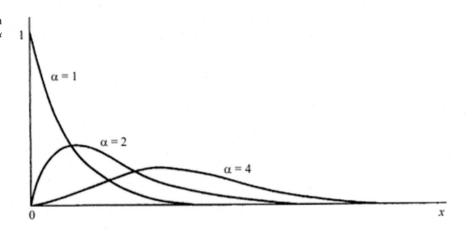

Table D.2). The chi-square distribution gets its name from the fact that it involves the square of normally distributed random variables, as we will explain shortly. Up to this point, we have dealt with a single random variable X and its standard normalized equivalent, $Z = (X - \mu)/\sigma$. We now wish to investigate the combined properties of more than one standardized independent normal variable. For example, we might want to investigate the distributions of temperature differences between reversing thermometers and a CTD (conductivity-temperature-depth) thermistor for a suite of CTD versus reversing thermometer intercomparisons taken at the same location. Each cast is considered to produce a temperature difference distribution x_k with a mean μ_k and a variance σ_k^2. The set of standardized independent normal variables Z_k formed from the casts is assumed to yield ν independent standardized normal variables $Z_1, Z_2, ..., Z_\nu$. The new random variable formed from the sum of the squares of the variables $Z_1, Z_2, ..., Z_\nu$ is the chi-square variable, χ_ν^2,

$$\chi_\nu^2 = Z_1^2 + Z_2^2 + ... + Z_\nu^2 \tag{3.30}$$

having ν degrees of freedom. For the case of our temperature comparison, this represents the square of the deviations for each cast about the mean. Properties of the distribution are

$$\text{Mean} = E[\chi_\nu^2] = \nu \tag{3.31a}$$

$$\text{Variance} = E\left[(\chi_\nu^2 - \nu)^2\right] = 2\nu \tag{3.31b}$$

We will make considerable use of the function χ_ν^2 in our discussion concerning confidence intervals for spectral estimates.

It bears repeating that PDFs are really just models for real populations whose distributions we do not know. In many applications, it is not important that our PDF be a precise description of the true population since we are mainly concerned with the statistics of the distributions as provided by the probability statements from the model. It is not, in general, a simple problem to select the right PDF for a given data set. Two suggestions are worth mentioning: (1) use available theoretical considerations regarding the process that generated the data; and (2) use the data sample to compute a frequency histogram and select the PDF that best fits the histogram. Once the PDF is selected, it can be used to compute statistical estimates of the true population parameters.

We also keep in mind that our statistics are computed from, and thus are functions of, other random variables and are, therefore, themselves random variables. For example, consider sample variables $X_1, X_2, ..., X_N$ from a normal population with mean μ and variance σ^2, then

$$\overline{X} = \frac{1}{N} \sum_{i=1}^{N} X_i \tag{3.32}$$

is normally distributed with mean μ and variance σ^2/N. From this it follows that

$$Z = \frac{\overline{X} - \mu}{\sigma_x} = \frac{\overline{X} - \mu}{\sigma/\sqrt{N}} = \sqrt{N}\,\frac{\overline{X} - \mu}{\sigma} \tag{3.33}$$

has a standard normal distribution $N(0, 1)$ with zero mean and variance of unity. Using the same sample, $X_1, X_2, ..., X_N$, we find that

$$\frac{1}{\sigma^2} \sum_{i=1}^{N} (X_i - \overline{X})^2 = \frac{(N-1)s^2}{\sigma^2} = \chi_\nu^2 \tag{3.34}$$

has a chi-square distribution (χ_ν^2) with $\nu = (N - 1)$ degrees of freedom. (Only $N - 1$, not N, degrees of freedom are available since the estimator requires use of the mean, which reduces the degrees of freedom by one.) Here, the sample standard deviation, s, is an unbiased estimate of σ. Thus, if s^2 is the variance of $N - 1$ random samples drawn from a normal population with variance σ^2, then the variable $(N-1)s^2/\sigma^2$ has the same distribution as a χ_ν^2 variable with $\nu = N-1$ degrees of freedom.

We also can use $(X - \overline{X})/(s/\sqrt{N})$ as an estimate of the standard normal statistic, $(X - \mu)/(\sigma/\sqrt{N})$. The continuous sample statistic $(X - \overline{X})/(s/\sqrt{N})$ has a PDF known as the *Student's t-distribution* (Appendix D, Table D.3) with $(N - 1)$ degrees of freedom. The name derives from the Irish statistician W.S. Gossett who was one of the first to work on the statistic. Because his employer would not allow employees to publish their research, Gossett published his results under the name "Student" in 1908. Mathematically, the random variable t is defined as a standardized normal variable, Z, divided by the square root of an independently distributed chi-square variable divided by its degrees of freedom; viz, $t = Z / \sqrt{(\chi_\nu^2/\nu)}$. Thus, one can safely use the normal distribution for samples $\nu > 30$, but for smaller samples one must use the t-distribution. In other words, the normal distribution can be relied upon to give a good approximation to the t-distribution only for $\nu > 30$. The t-distribution is much more "robust" than the chi-square distribution.

The above relations for statistics computed from a normal population are important for two reasons:

1. Often, the data or the measurement errors can be assumed to have population distributions with normal PDFs;
2. One is working with averages that themselves are normally distributed regardless of the PDF of the original data. This statement is a version of the well-known *central limit theorem*.

3.6 CENTRAL LIMIT THEOREM

Let $X_1, X_2, ..., X_i, ...$ be a sequence of independent random variables with $E[X_i] = \mu_i$ and $V[X_i] = \sigma_i^2$. Define a new random variable $X = X_1 + X_2 + ... + X_N$. Then, as N becomes large, the standard normalized variable

$$Z_N = \frac{\left(X - \sum_{i=1}^{N} \mu_i \right)}{\left(\sum_{i=1}^{N} \sigma_i^2 \right)^{1/2}} \tag{3.35}$$

takes on a normal distribution regardless of the distribution of the original population variable from which the sample was drawn. The fact that the X_i values may have any kind of distribution, and yet the sum X may be approximated by a normally distributed random variable, is the basic reason for the importance of the normal distribution in probability theory. For example, X might represent the summation of fresh water added to an estuary from a large number of rivers and streams, each with its own particular form of annual variability. In this case, the sum of the rivers and stream inputs would result in a normal distribution in Z_N for the annual input of fresh water. Alternatively, the variable X, representing the success or failure of an AXBT launch, may be represented as the sum of the following independent binomially distributed random variables (a variable that can only take on one of two possible values)

$$X_i = 1 \text{ if the } i\text{th cast is a success}$$
$$= 0 \text{ if the } i\text{th cast is a failure}$$

with $X = X_1 + X_2 + ... + X_N$. For this random variable, $E[X] = Np$ and $V[X] = Np(1 - p)$. For large N, it can be shown that the variable $(X - E[X]) / \sqrt{V[X]}$ closely resembles the normal distribution, $N(0, 1)$.

A special form of the central limit theorem may be stated as: the distribution of mean values calculated from a suite of random samples X_i ($X_{i,1}, X_{i,2}, ...$) taken from a discrete or continuous population having the same mean μ and variance σ^2 approaches the normal distribution with mean μ and variance σ^2/N as N goes to infinity. Consequently, the distribution of the arithmetic mean

$$\overline{X} = \frac{1}{N} \sum_{i=1}^{N} X_i \tag{3.36}$$

is asymptotically normal with mean μ and variance σ^2/N when N is large. Ideally, we would like $N \to \infty$ but, for practical purposes, $N \geq 30$ will generally ensure that the population of X is normally distributed. This certainly appears to be the consensus in the literature we have examined. When N is small, the shape of the sample distribution will depend mainly on the shape of the parent population. However, as N becomes larger, the shape of the mean of the sampling distribution becomes increasingly more like that of a normal distribution no matter what the shape of the parent population (Figure 3.7). In many instances, the normality assumption for the sampling distribution for $\overline{X}$ is reasonably accurate for $N > 4$ and quite accurate for $N > 10$ (e.g., Harriett and Murphy, 1975; Bendat and Piersol, 1986).

The central limit theorem has important implications for we often deal with average values in time or space. For example, current meter systems average over some time interval, allowing us to invoke the central limit theorem and assume normal statistics for the resulting data values. Similarly, data from high-resolution CTD systems are generally vertically averaged (or averaged over some set of cycles in time), thus approaching a normal PDF for the data averages, via the central limit theorem. An added benefit of this theorem is that the variance of the averages is reduced by the factor N,

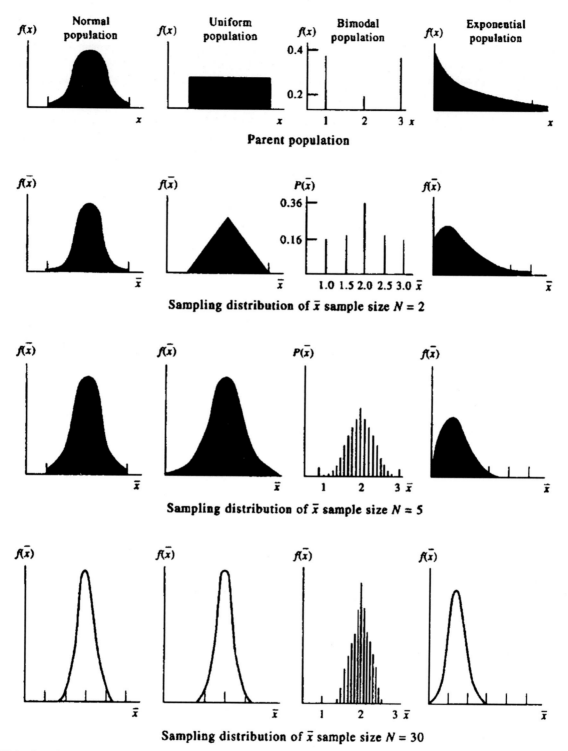

FIGURE 3.7　Sampling distribution of the mean, $\bar{x}$, for different types of population distributions for increasing sample size, $N = 2$, 5, and 30. The shape of the sampling distribution becomes increasingly more like that of a normal distribution regardless of the shape of the parent population.

the number of samples averaged. The theorem essentially states that individual terms in the sum contribute a negligible amount to the variation of the sum, and that it is not likely that any one value makes a large contribution to the sum. Errors of measurements certainly have this characteristic. The final error is the sum of many small contributions none of which contributes very much to the total error. Note that the sample standard error is an unbiased estimate (again in the sense that the expected value is equal to the population parameter being estimated) even though the component sample standard deviation is not.

To further illustrate the use of the central limit theorem, consider a set of independent measurements of a process whose probability distribution is unknown. Through previous experimentation, the distribution of this process was estimated to have a mean of 7 and a variance of 120. If $\bar{x}$ denotes the mean of the sample measurements, we want to find the number of measurements, N, required to a give a probability

$$P(4 \leq \bar{x} \leq 10) = 0.866$$

where 4 and 10 are the chosen problem limits. Here, we use the central limit theorem to argue that, while we do not know the exact distribution of the specific variable, we do know that mean values are normally distributed. Using the standard variable, $Z = (x - \mu)/(\sigma/\sqrt{N})$, substituting $\bar{x}$ for x, and then using the fact that $\sigma = \sqrt{120} = 2\sqrt{30}$, we can then write the probability function as

$$
\begin{aligned}
P(4 \leq \bar{x} \leq 10) &= P\left[\frac{(4 - \mu)\sqrt{N}}{\sigma} < Z < \frac{(10 - \mu)\sqrt{N}}{\sigma}\right] \\
&= P\left[\frac{(4 - 7)\sqrt{N}}{2\sqrt{30}} < Z < \frac{(10 - 7)\sqrt{N}}{2\sqrt{30}}\right] \\
&= P\left[\frac{-3\sqrt{N}}{2\sqrt{30}} < Z < \frac{3\sqrt{N}}{2\sqrt{30}}\right] \\
&= 2P\left[Z < \frac{3\sqrt{N}}{2\sqrt{30}}\right] - 1 = 0.866
\end{aligned}
$$

where we have used the fact that $P[0 \leq Z] = 1$, from which we find

$$P\left[Z < \frac{3\sqrt{N}}{2\sqrt{30}}\right] = 0.933$$

Assuming that we are dealing with a normal distribution, we can look up the value 0.933 in a table to find the value of the integrand to which this cumulative probability corresponds. In this case, $3/2\sqrt{N/30} = 1.5$, so that $N = 30$.

3.7 ESTIMATION

In most oceanographic applications, the population parameters are unknown and must be estimated from a sample. Faced with this estimation problem, the objective of statistical analysis is twofold: to present criteria that allow us to determine how well a given sample represents the population parameter; and to provide methods for estimating these parameters. An *estimator* is a random variable used to provide *estimates* of population parameters. "Good" estimators are those that satisfy a number of important criteria: (1) have average values that equal the parameter being estimated (*unbiasedness* property); (2) have relatively small variance (*efficiency* property); and (3) approach asymptotically the value of the population parameter as the sample size increases (*consistency* property). We have already introduced the concept of estimator bias in discussing variance and standard deviation. Formally, an estimate $\widehat{\theta}$ of a parameter θ (here, as usual, the hat symbol (^) indicates an estimate), is an unbiased estimate provided that $E\left[\widehat{\theta}\right] = \theta$; otherwise, it is a biased estimate with a bias $B = E\left[\widehat{\theta}\right] - \theta$. An unbiased estimator is any estimate whose average value over all possible random samples is equal to the population parameter being estimated. An example of an unbiased estimator is the mean of the noise in an acoustic current meter record created by turbulent fluctuations in the velocity of sound speed in water; an example of a biased estimator is the linear slope of a sea-level record in the presence of a long-term trend (a slow change in the average value). Other examples of unbiased estimators are $\bar{x}$ for $\widehat{\theta}$, μ for $E\left[\widehat{\theta}\right]$, and $\frac{\sigma^2}{N}$ for σ_{θ}^2. The mean-square (MS) error of our estimate $\widehat{\theta}$ is

$$E\left[\left(\widehat{\theta} - \theta\right)^2\right] = V\left[\widehat{\theta}\right] + B^2 \qquad (3.37)$$

The most efficient estimator (property 2) is the estimator with the smallest MS error. Since it is possible to obtain more than one unbiased estimator for the same target parameter, θ, we define the efficiency of an estimator as the ratio of the variances of the two estimators. For example, if we have two unbiased estimates $\hat{\theta}_1$ and $\hat{\theta}_2$, we can compute the relative efficiency of these two estimates as

$$\text{Efficiency} = V\left[\hat{\theta}_2\right] / V\left[\hat{\theta}_1\right] \tag{3.38}$$

where $V\left[\hat{\theta}_1\right]$ and $V\left[\hat{\theta}_2\right]$ are the variances of the estimators. A low value of the ratio would suggest that $V\left[\hat{\theta}_2\right]$ is more efficient while a high value would indicate that $V\left[\hat{\theta}_1\right]$ is more efficient. As an example, consider the efficiency of two familiar estimators of the mean of a normal distribution. In particular, let $\hat{\theta}_1$ be the median value and $\hat{\theta}_2$ be the sample mean. The variance of the sample median is $V\left[\hat{\theta}_1\right] = \left(1.2533^2\sigma^2/N\right)$ while the sample mean has a variance $V\left[\hat{\theta}_2\right] = \sigma^2/N$. Thus, the efficiency is

$$\begin{aligned}\text{Efficiency} &= V\left[\hat{\theta}_2\right]/V\left[\hat{\theta}_1\right] \\ &= \left(\sigma^2/N\right)/\left(1.2533^2\sigma^2/N\right) \\ &= 0.6366\end{aligned}$$

Therefore, the variability of the sample mean is 63.7% of the variability of the sample median, which indicates that the sample *mean* is a more efficient estimator than the sample *median*.

As a second example, consider the sample variances s'^2 and s^2 given by Eqns (3.4) and (3.5a, 3.5b), respectively. The efficiency of these two sample variances is the ratio of s'^2 to s^2, namely,

$$\frac{1/N \sum\limits_{i=1}^{N} (x_i - \bar{x})^2}{1/(N-1) \sum\limits_{i=1}^{N} (x_i - \bar{x})^2} = \frac{N-1}{N} < 1$$

which indicates that s'^2 is a more efficient statistic than s^2.

We can view the difference $\hat{\theta} - \theta$ as the distance between the population "target" value and our estimate. Because this difference is also a random variable, we can ask probability-related questions, such as: "What is the probability

$$P\left(-b < \left(\hat{\theta} - \theta\right) < b\right)$$

for some range $(-b, b)$?" It is common practice to express b as some multiple of the sample standard deviation of σ_θ (e.g., $b = k\sigma_\theta$, $k > 1$). A widely used value is $k = 2$, corresponding to two standard deviations. Here, we can apply an important result known as *Tchebysheff's theorem*, which states that for any random variable Y, for $k \geq 1$:

$$P(|Y - \mu| < k\sigma) \geq 1 - \frac{1}{k^2} \tag{3.39a}$$

or,

$$P(|Y - \mu| \geq k\sigma) \leq \frac{1}{k^2} \tag{3.39b}$$

where $\mu = E[Y]$ and $\sigma^2 = V[Y]$. Applying this to the problem at hand, we find that for $k = 2$, $P\left(|\hat{\theta} - \theta| < 2\sigma_\theta\right) \geq 1 - 1/(2)^2 = 0.75$. Therefore, most random variables occurring in nature can be found within two standard deviations ($\pm 2\sigma$) of the mean with a probability of 0.75. Note that the probability statement (Eqn 3.39a) indicates that the probability is greater than or equal to the value of $1 - 1/k^2$ for any type of distribution. We can therefore, expect somewhat more than 75% of the values to lie with the range $(-2\sigma, 2\sigma)$. In fact, this is generally a conservative estimate. If we assume that oceanographic measurements are typically normally distributed, we obtain $P(|Y - \mu| < 2\sigma) = 0.95$, so that 95% of the observations lie within $\pm 2\sigma$. This is an important conclusion in terms of data editing methods that use criteria designed to select erroneous values from data samples based on probabilities.

3.8 CONFIDENCE INTERVALS

An important application of interval estimates for probability distribution functions is the formulation of *confidence intervals* for parameter estimates. These intervals define the degree of certainty that a given quantity, θ, will fall between specified lower and upper bounds θ_L, θ_U, respectively, of the parameter estimates. The confidence interval (θ_L, θ_U) associated with a particular confidence statement is usually written as

$$P(\theta_L < \theta < \theta_U) = 1-\alpha, 0 < \alpha < 1 \tag{3.40}$$

where α is called the *level of significance* (or confidence coefficient) for the confidence statement and $(1 - \alpha)100$ is the percent significance level for the variable θ. (The terms confidence coefficient, significance level, confidence level, and confidence are commonly used interchangeably.) A typical value for α is 0.05, which means that 95% of the cumulative area under the probability curve (Eqn 3.40) is contained between the points θ_L and θ_U (Figure 3.8). For both symmetric and asymmetric probability distributions, each of the two points cuts off $\alpha/2$ of the total area under the distribution curve, leaving a total area under the curve of $1 - \alpha$; θ_L cuts off the left-hand part of the distribution tail and θ_U cuts off the right-hand part of the tail.

If θ_L and θ_U are derived from the true value of the variable θ (such as the population mean, μ), then the probability interval is fixed. However, where estimates determined from a sample (for example, the sample mean $\bar{x}$) are being used to determine the variable value, θ, the probability interval will vary from sample to sample because of changes in the sample mean and standard deviation. Thus, we must inquire about the probability that the true value of θ will fall within the intervals generated by each of the given sample estimates. The statement that $P(\theta_L < \theta < \theta_U)$ does not mean that the population variable θ has a probability of $P = 1 - \alpha$ of falling in the sample interval (θ_L, θ_U), in the sense that θ was behaving like a sample. The population variable is a fixed quantity. Once the interval is picked, the population variable θ is either in the interval or it is not (probability 1 or 0). For the sample data, the interval may shift depending on the mean and variance of the particular sample that was selected from the population. We should, therefore, interpret (Eqn 3.40) to mean that there is a probability P that the specified random sample interval (θ_L, θ_U) contains the true population variable θ a total of $(1 - \alpha)100\%$ of the time. That is, $(1 - \alpha)$ is the fraction of the time that the true variable value, θ, is contained by the sample interval (θ_L, θ_U).

In general, we need a quantity, called a *pivotal quantity*, that is, a function of the sample estimator $\hat{\theta}$ and the unknown variable θ, where θ is the only unknown. The pivotal quantity must have a PDF that does not depend on θ. For large samples $(N \geq 30)$ of unbiased point estimators, the standard normal distribution $Z = \left(\hat{\theta} - \theta\right)\big/\sigma_\theta$ is a pivotal quantity. In fact, it is common to express the confidence interval in terms of Z. For example, consider the statistic $\hat{\theta}$ with $E\left[\hat{\theta}\right] = \theta$ and $V\left[\hat{\theta}\right] = \sigma_\theta^2$; find the $100(1 - \alpha)\%$ confidence interval. To do this, we first define

$$P\left(-Z_{\alpha/2} < Z < Z_{\alpha/2}\right) = 1-\alpha \tag{3.41}$$

and then use the above relation $Z = \left(\hat{\theta} - \theta\right)\big/\sigma_{\hat{\theta}}$ to obtain

$$P\left(\hat{\theta} - Z_{\alpha/2}\sigma_{\hat{\theta}} < \theta < \hat{\theta} + Z_{\alpha/2}\sigma_{\hat{\theta}}\right) = 1-\alpha \tag{3.42}$$

FIGURE 3.8 Location of the limits $(\theta_L, \theta_U) = (-Z_{\alpha/2}, +Z_{\alpha/2})$ for a normal probability distribution. For $\alpha = 0.05$, the cumulative area $1 - \alpha$ corresponds to the 95% interval for the distribution.

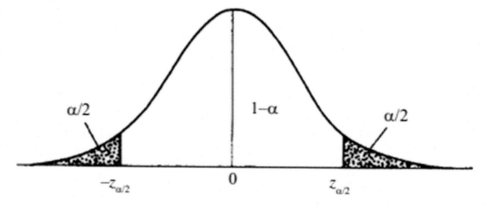

This formula can be used for large samples to compute the confidence interval for θ once α is selected. Again, the significance level, $1 - \alpha$, refers to the probability that the population parameter θ will be bracketed by the given confidence interval. The meaning of these limits is shown graphically in Figure 3.8 for a normal population. We remark that if the population standard deviation σ is known it should be used in Eqn (3.42) so that $\sigma_{\hat{\theta}} = \sigma$; if not, the sample standard deviation s can be used with little loss in accuracy if the sample size is sufficiently large; i.e., $N > 30$, although quoted values in the literature can sometimes be as high as 50 or more.

The three types of confidence intervals commonly used in oceanography are listed below. Specific usage depends on whether we are interested in the mean or the variance of the quantity being estimated.

3.8.1 Confidence Interval for μ (σ Known)

When the population standard deviation, σ, is known and the parent population is normally distributed ($N > 30$), the $100(1 - \alpha)$ percent confidence interval for the population mean is given by the symmetrical distribution for the standardized normal distribution, Z:

$$\bar{x} - Z_{\alpha/2}\frac{\sigma}{\sqrt{N}} < \mu < \bar{x} + Z_{\alpha/2}\frac{\sigma}{\sqrt{N}} \tag{3.43}$$

As an example, we wish to find the 95% confidence interval ($\alpha = 0.05$) for μ, given the sample mean $\bar{x}$ and known normally distributed population variance, σ^2. Suppose that a thermistor installed at the entrance to the ship's engine cooling water intake takes samples every second for 20 s (so, $N = 20$) and yields a mean ensemble temperature $\bar{x} = 12.7°C$ for the particular burst. Further, suppose that the water is isothermal and that the only source of variability is instrument noise, which we know from previous calibration in the laboratory has a known noise level $\sigma = 0.5°C$. Since we want the 95% confidence interval, the appropriate values of Z for the normal distribution are $Z_{\alpha/2} = 1.96$ and $-Z_{\alpha/2} = -1.96$ (see value for Z for $\alpha/2 = 0.9750$ in Appendix D, Table D.1). Substituting these values into Eqn (3.43) along with $N = 20$, $\sigma = 0.5°C$, and $\bar{x} = 12.7°C$ we find

$$\left[12.7 - (1.96)0.5 / \sqrt{20}\right]°C < \mu < \left[12.7 + (1.96)0.5\sqrt{20}\right]°C$$

so that

$$12.48°C < \mu < 12.92°C$$

Based on our 20 data values, there is a 95% probability that the true mean temperature of the water will be bracketed by the interval (12.48°C, 12.92°C) derived from the interval

$$\left(\bar{x} - Z_{\alpha/2}\frac{\sigma}{\sqrt{N}}, \bar{x} + Z_{\alpha/2}\frac{\sigma}{\sqrt{N}}\right)$$

3.8.2 Confidence Interval for μ (σ Unknown)

In most real circumstances, σ is not known and we must resort to the use of the sample standard deviation, s. Similarly, for small samples ($N < 30$), we cannot use the above technique but must introduce a formalism that works for any sample size and distribution, as long as the departures from normality are not excessive. Under these conditions, we resort to the variable $t = (\bar{x} - \mu)/(s/\sqrt{N})$, which has a student's t-distribution with $\nu = (N - 1)$ degrees of freedom. Derivation of the $100(1 - \alpha)\%$ confidence interval follows the same procedure used for the symmetrically distributed normal distribution, except that we must modify the limits. In this case

$$P\left[-t_{\alpha/2,\nu} < (\bar{x} - \mu)/\frac{s}{\sqrt{N}} < t_{\alpha/2,\nu}\right] = 1 - \alpha \tag{3.44}$$

This is easily arranged to give the $100(1 - \alpha)\%$ confidence interval for μ

$$\bar{x} - t_{\alpha/2,\nu}\frac{s}{\sqrt{N}} < \mu < \bar{x} + t_{\alpha/2,\nu}\frac{s}{\sqrt{N}} \tag{3.45}$$

Note the similarity between Eqn (3.45) and the form Eqn (3.42) obtained for μ when σ is known. We return to our previous example of ship injection temperature and this time assume that $s = 0.5°C$ is a measured quantity obtained by subtracting the mean value $\bar{x} = 12.7°C$ from the series of 20 measurements. Turning to Appendix D (Table D.3A) for the cumulative t-distribution, we look for values of $F(t)$ under the column for the 95% confidence interval ($\alpha = 0.05$) for which $F(t) = 1 - \alpha/2 = 0.975$. Using the fact that $\nu = (N - 1) = 19$, we find $t_{\alpha/2,\nu} = t_{0.025,19} = 2.093$. Substituting these values into Eqn (3.45) yields

$$\left[12.7 - 2.093\left(0.5/\sqrt{20}\right)\right]°C < \mu < \left[12.7 + 2.093\left(0.5/\sqrt{20}\right)\right]°C$$

$$12.47°C < \mu < 12.93°C$$

Thus, there is a 95% probability that the interval (12.47°C, 12.93°C) will bracket the true mean temperature, μ. Because of the large sample size, this result is only slightly different than the result obtained for the normal distribution in the previous example when σ was known *a priori*.

3.8.3 Confidence Interval for σ^2

Under certain circumstances, we are more interested in the confidence interval for the signal variance than the signal mean. For example, to determine the reliability of a spectral peak in a spectral density distribution (or spectrum), we need to know the confidence intervals for the population variance, σ^2, based on our sample variance, s^2. To do this, we seek a new pivotal quantity. Recall from Eqn (3.34) that for N samples of a variable x_i from a normal population, the expression

$$\frac{1}{\sigma^2}\sum_{i=1}^{N}(x_i - \bar{x})^2 = \frac{(N-1)s^2}{\sigma^2} \tag{3.46}$$

is a χ^2 variable with $(N - 1)$ degrees of freedom. Using this as a pivotal quantity, we can find the upper and lower bounds χ_U^2 and χ_L^2 for which

$$P\left[\chi_L^2 < \frac{(N-1)s^2}{\sigma^2} < \chi_U^2\right] = 1 - \alpha \tag{3.47}$$

or, upon rearranging terms,

$$P\left[\frac{(N-1)s^2}{\chi_U^2} < \sigma^2 < \frac{(N-1)s^2}{\chi_L^2}\right] = 1 - \alpha \tag{3.48}$$

Note that χ^2 is a skewed function (Figure 3.9), which means that the upper and lower bounds in Eqn (3.48) are asymmetric; the point $1 - \alpha/2$ rather than $-\alpha/2$ determines the point that cuts off $\alpha/2$ of the area at the lower end of the chi-square distribution.

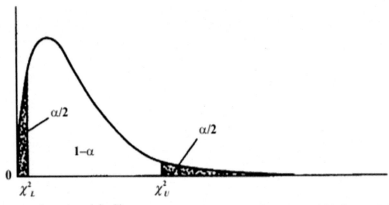

FIGURE 3.9   Location of the limits $(\theta_L, \theta_U) = (\chi_L^2, \chi_U^2)$ for a normal probability distribution. For $\alpha = 0.05$, the cumulative area $1 - \alpha$ corresponds to the 95% confidence interval for the distribution. χ^2 is a skewed function, so that the upper and lower bounds are asymmetrically distributed.

From Eqn (3.48) we obtain the well-known $100(1 - \alpha)\%$ confidence interval for the variance σ^2 when sampled from a normal population

$$\frac{(N-1)s^2}{\chi^2_{\alpha/2,\nu}} < \sigma^2 < \frac{(N-1)s^2}{\chi^2_{1-\alpha/2,\nu}} \tag{3.49}$$

where the subscripts $\alpha/2$ and $1 - \alpha/2$ characterize the endpoint values of the confidence interval and $\nu = (N-1)$ gives the degrees of freedom of the chi-square distribution. The smaller value of $\chi^2 \left(\chi^2_L = \chi^2_{1-\alpha/2,\nu} \right)$ appears in the denominator of the upper endpoint for σ^2 while the larger value of $\chi^2 \left(\chi^2_U = \chi^2_{\alpha/2,\nu} \right)$ appears in the denominator of the lower endpoint for σ^2. As an example, suppose that we have $\nu = 10$ in a spectral estimate of the eastward component of current velocity and that the background variance of the spectrum near a distinct spectral peak is $s^2 = 10$ cm^2/s^2. What is the 95% confidence interval for the variance? How big would the peak have to be to stand out statistically from the background level? (Details on spectral estimation are presented in Chapter 5). In this case, $\alpha/2 = 0.025$ and $1 - \alpha/2 = 0.975$.

Turning to the cumulative distribution $F(\chi^2)$ for $\nu = N - 1 = 9$ degrees of freedom in Appendix D (Table D.2), we find that $\chi^2_{0.025,9} = 19.02$ for cumulative integral $F(\alpha/2 = 0.025)$ and that $\chi^2_{0.95,9} = 2.70$ for cumulative integral $F(1 - \alpha/2 = 0.975)$. Thus, $P\left(2.70 < \chi^2_{\nu=9} < 19.02\right) = 1 - \alpha = 0.95$. Substituting $N - 1 = 9$, $s^2 = 10$ cm^2/s^2, $\chi^2_{\alpha/2,9} = 19.02$ for the value that cuts off $\alpha/2$ of the upper end area under the curve and $\chi^2_{1-\alpha/2,9} = 2.70$ for the value that cuts off $1 - \alpha/2$ of the lower end area of the curve, Eqn (3.49) then yields

$$9\left(10 \text{ cm}^2 / \text{s}^2\right)/19.02 < \sigma^2 < 9\left(10 \text{ cm}^2 / \text{s}^2\right)/2.70$$

$$4.7 \text{ cm}^2/\text{s}^2 < \sigma^2 < 33.3 \text{ cm}^2/\text{s}^2$$

Thus, the true background variance will lie between 4.7 and 33.3 cm^2/s^2. If a spectral peak is found to have a greater range than these limits, it represents a statistically significant departure from background spectral density levels.

In most instances, spectral densities are presented in terms of the log (logarithm to the base 10) of the spectral density function versus frequency or log-frequency (see Chapter 5). Dividing through by s^2 in Eqn (3.49) and taking the log, yields

$$\log(N-1) - \log\left(\chi^2_{\alpha/2,\nu}\right) < \log\left(\frac{\sigma^2}{s^2}\right) < \log(N-1) - \log\left(\chi^2_{1-\alpha/2,\nu}\right) \tag{3.50a}$$

or, upon subtracting log $(N-1)$ and rearranging, the inequality becomes

$$\log\left(\chi^2_{1-\alpha/2,\nu}\right) < \log(N-1) - \log(\sigma^2/s^2) < \log\left(\chi^2_{\alpha/2,\nu}\right) \tag{3.50b}$$

The range, R, of the variance is then

$$R = \log\left(\chi^2_{\alpha/2,\nu}\right) - \log\left(\chi^2_{1-\alpha/2,\nu}\right) \tag{3.51}$$

while the pivot point, p_0, of the interval is

$$p_0 = \log(N-1) - \log(\sigma^2/s^2) \tag{3.52}$$

If we assume that the measured background value of s^2 is a good approximation to σ^2, so that $\sigma^2/s^2 = 1$, then $p_0 = \log(N - 1)$. The ranges between the maximum value and p_0, and the minimal value and p_0, are $\log\left(\chi^2_{\alpha/2,\nu}\right) - p_0$ and $\log\left(\chi^2_{1-\alpha/2,\nu}\right) - p_0$, respectively, with (3.50b) taking the form

$$\log\left(\chi^2_{1-\alpha/2,\nu}\right) < \log(N-1) < \log\left(\chi^2_{\alpha/2,\nu}\right) \tag{3.53}$$

Returning to our previous example for the 95% confidence interval, we find that

$$\log(2.70) < \log(9) < \log(19.02)$$

$$0.43 < 0.95 < 1.28$$

giving a range $= 0.848$ with the pivot point at $p_0 = 0.95$.

3.8.4 Spurious Significance and the Bonferroni Adjustment

The previous sections dealt with the confidence intervals for the mean and standard deviation of a single data series. Now, suppose that we want to compare the statistical values of pairs of measurements obtained from the same or different physical process. Inferential statistics, such as a multiple-sample t-test, in which many pairs of samples are compared, is meant to inform the investigator of the probability that the statistical values of the samples are significantly different at a specified level of significance. Consider the example of the ship's cooling intake temperatures discussed in Section 3.8.1 and suppose the ship has thermistors in separate intakes at the same depth on both the port and starboard sides of the vessel. One side records a mean temperature of $\bar{x}_1 = 12.4 \pm 0.3°C$ and the other a mean temperature of $\bar{x}_2 = 14.5 \pm 0.6°C$; both values are based on a series of 20 simultaneous measurements at each intake. Is the observed difference of $2.1°C$ in mean temperatures due to random variations in the measurements or is the difference significant at, say, the $1 - \alpha = 1 - 0.05 = 0.95$ (95%) confidence level? To examine this question, we follow the t-test calculation used in the previous section. Again using $t_{\alpha/2,\nu} = t_{0.025,19} = 2.093$ we obtain

$$\left[12.4 - 2.093 \left(0.3 / \sqrt{20} \right) \right]°C < \bar{x}_1 < \left[12.7 + 2.093 \left(0.3 / \sqrt{20} \right) \right]°C$$

$$\left[14.5 - 2.093 \left(0.6 / \sqrt{20} \right) \right]°C < \bar{x}_2 < \left[14.5 + 2.093 \left(0.6 / \sqrt{20} \right) \right]°C$$

whereby

$$1.9143°C < |\bar{x}_2 - \bar{x}_1| < 2.2872°C$$

indicating that the observed difference of $2.1°C$ likely occurred by chance at the 95% confidence level. At this level of certainty, the two thermistors were measuring in the same body of water.

Now suppose that a total of 50 such comparisons are available for the time that the ship was transiting between ports. Based on general statistics for the $\alpha = 0.05$ level, we would expect that roughly 5% of the differences $|\bar{x}_2 - \bar{x}_1|$ in the observed (ostensibly independent) means to exceed the range for chance occurrences. In other words, there is the possibility of obtaining roughly $0.05 \times 50 = 2.5$ (i.e., around 2−3) spurious values that would indicate that the differences in water temperature from one side of the ship to the other were due to real differences in water temperature. Clearly, the greater the number of independent t-tests, the greater the probability that one or more spuriously significant differences will be found.

For statistically independent means, the "family-wise" probability, P_{fw}, of one or more spuriously significant t-tests—and therefore the greater the likelihood of committing a Type-I error (see Section 3.14)—is determined as

$$P_{fw} = 1 - (1 - \alpha)^m \tag{3.54}$$

where m is the number of comparisons. For $m = 1$ comparisons and α set to 0.05, the probability of at least one Type I error is 5% (as noted above), while for 50 comparisons, the chance of at least one Type I error is $1 - (1 - 0.05)^{50} = 0.92$ (92%). [In statistical hypothesis testing, a "Type I error", or "false positive" finding or conclusion, is the mistaken rejection of an actually true *null hypothesis*—an hypothesis one is trying to nullify.] For non-independent means, probabilities are even higher, where for 10 independent comparisons we obtain 40% spurious values while for 10 non-independent comparisons this increases to obtain 60% spurious values (Cochran and Cox, 1992).

Using the above, we can calculate the probability that a researcher will commit a Type I statistical error by mistakenly rejecting an actually true null hypothesis but, because we cannot determine which of the significant differences in a large group of comparisons are spurious, the interpretation of studies involving multiple comparisons becomes nearly impossible. This is where the *Bonferroni adjustment* comes into play. Named after an Italian mathematician, Carlo Emilio Bonferroni (1892−1960), who worked on probability theory, the Bonferroni adjustment was introduced to reduce the likelihood of a Type I statistical error. A common version of the Bonferroni adjustment is to specify $\alpha_{adjusted} = \alpha/m$, where α is the overall experimental value, m is the number of comparisons made and $\alpha_{adjusted}$ is the adjusted level at which each of the comparisons must test for significance. Thus, for the 50 temperature comparisons mentioned above, we have $\alpha_{adjusted} = \frac{0.05}{50} = 0.001$, corresponding to a $p \leq 0.005$ value of roughly $3.5°C$ for 50 degrees of freedom (see values in Table D.3A). Replacing 2.093 (~ 2.1)$°C$ with $3.5°C$ in the previous example, we find that $1.8395°C < |\bar{x}_2 - \bar{x}_1| < 2.3605°C$, so that the observed difference of $2.1°C$ for this particular comparison almost certainly occurred by chance. Replacing α with $\alpha_{adjusted}$ in the above calculations helps reaffirm that the difference of $2.1°C$ is not statistically significant. Had the difference been, for want of an example, $2.3°C$, we would have accepted the value as significant based on the calculation using Eqn (3.45) but not significant after the Bonferroni adjustment that increased the difference to over $2.36°C$.

As with all statistical inferences, it is best to err on the side of caution. In many cases, use of the Bonferroni adjustment can be used to confirm nullification of the null hypothesis and not to publish as significant results that simply may have occurred by chance.

3.8.5 Goodness-of-Fit Test

The goodness-of-fit statistical model is, as the model states, the best fit to data obtained as outcomes of an experiment. Measures of goodness-of-fit typically summarize the differences between observed values and the values expected from the model in question. When the set of outcomes for an experiment is limited to two outcomes (such as success or failure, on or off, and so on), the appropriate test statistic for the distribution is the binomial variable. However, when more than two outcomes are possible, the preferred statistic is the chi-square variable. In addition to providing confidence intervals for spectral estimates and other measurement parameters, the chi-square variable is used to test how closely the observed frequency distribution of a given parameter corresponds to the expected frequency distribution for the parameter. The expected frequencies represent the average number of values expected to fall in each frequency interval based on some theoretical probability distribution, such as a normal distribution. The observed frequency distribution represents a sample of values drawn from some probability distribution. What we want to know is whether the observed and expected frequencies are similar enough for us to conclude that they are drawn from the same probability function (the "null hypothesis"). The test for this similarity using the chi-square variable is called a "goodness-of-fit" test.

Consider a sample of N observations from a random variable X with observed PDF $p_0(X)$. Let the N observations be grouped into K intervals (or categories) called *class intervals*, whose graphical distribution forms a frequency histogram (Bendat and Piersol, 1986). The actual number of observed values that fall within the ith class interval is denoted by f_i, and is called the *observed frequency* in the ith class. The number of observed values that we would expect to fall within the ith class interval if the observations really followed the theoretical probability distribution, $p(X)$, is denoted F_i, and is called the *expected frequency* in the ith class interval. The difference between the observed frequency and the expected frequency for each class interval is given by $f_i - F_i$. The total discrepancy for all class intervals between the expected and observed distributions is measured by the sample statistic

$$X^2 = \sum_{i=1}^{K} \left[(f_i - F_i)^2 / F_i \right] \tag{3.55}$$

where division by F_i transforms the sum of the squares into the chi-square-type variable, χ^2.

The number of degrees of freedom, ν, for the variable X^2 is equal to K minus the number of different independent linear restrictions imposed on the observations. As discussed by Bendat and Piersol (1986), one degree of freedom is lost through the restriction that, if $K - 1$ class intervals are determined, the Kth class interval follows automatically. If the expected theoretical density function is normally distributed then the mean and variance must be computed to allow comparison of the observed and expected distributions. This results in the loss of two additional degrees of freedom. Consequently, if the chi-square goodness-of-fit test is used to test for normality of the data, the true number of degrees of freedom for X^2 is $\nu = K - 3$.

Eqn (3.55) measures the goodness-of-fit between f_i and F_i as follows: when the fit is good (that is, f_i and F_i are generally close), then the numerator of Eqn (3.55) will be small and the hence the value of X^2 will be low. On the other hand, if f_i and F_i are not close, the numerator of Eqn (3.55) will be comparatively large and so the value of X^2 will be large. Thus, the critical region for the test statistic X^2 will always be in the upper tail of the chi-square function (Figure 3.9) because we wish to reject the null hypothesis, whenever the difference between f_i and F_i, is large. More specifically, the region of acceptance of the null hypothesis (see Section 3.14) is

$$X^2 \leq \chi^2_{\alpha;\nu} \tag{3.56}$$

where the value of $\chi^2_{\alpha;\nu}$ is available from Appendix D (Table D.2). If X^2 is less than or equal to $\chi^2_{\alpha;\nu}$, the hypothesis that $P(X) = P_0(X)$ is accepted at the α level of significance (i.e., there is a $100\alpha\%$ chance that we are wrong in accepting the null hypothesis that our data are drawn from a normal distribution; or $100(1 - \alpha)\%$ chance that we are right in accepting the null hypothesis regarding the normal distribution). However, if the sample value X^2 is greater than $\chi^2_{\alpha;\nu}$, the hypothesis is rejected at the level of significance, i.e., there is $100(1 - \alpha)\%$ chance that we are wrong in accepting the null hypothesis that the date is drawn from a normal distribution.

As an example, suppose our analysis involves 15 class intervals and that the fit between the 15 estimates of f_i and F_i (where F_i is normally distributed) yields $X^2 = 23.1$. From tables for the cumulative chi-square distribution, $F(X) = P\left(X^2 > \chi^2_{\alpha;\nu}\right)$, we find that $P(X^2 > 21.03) = 0.05$ for $\nu = K - 3 = 12$ degrees of freedom. Thus, at the $\alpha = 0.05$ level of significance (95% certainty level), we cannot accept the null hypothesis that the observed values came from the same distribution as the expected values. In this case, our chances of being wrong are less than $100\alpha\%$.

Chi-square tests for normality are typically performed using a constant interval width. Nonuniform distributions will yield different expected frequency distributions from one class interval to the next. Bendat and Piersol recommend a class interval width of $\Delta x \approx 0.4s$, where s is the standard deviation of the sample data. A further requirement is that the expected frequencies in all class intervals be sufficiently large that X^2 in Eqn (3.55) is an acceptable approximation to $\chi^2_{\alpha;\nu}$. A common recommendation is that $F_i > 3$ in all class intervals. When testing for normality, where the expected frequencies diminish on the tails of the distribution, this requirement is attained by letting the first and last intervals extend to $-\infty$ and to $+\infty$, respectively, so that $F_1, F_2, ..., F_K > 3$.

As an example of a goodness-of-fit test, we consider a sample of $N = 200$ surface gravity wave heights measured every 0.78 s by a Datawell waverider buoy deployed off the west coast of Canada during the winter of 1993–94 (Table 3.3). The wave record spans a period of 2.59 min and corresponds to a time of extreme (5 m high) storm-generated waves. According to one school of thought (e.g., Phillips et al., 1993), extreme wave events in the ocean are part of a Gaussian process and the occurrence of maximum wave heights is related in a linear manner to the statistical distribution of the surrounding wave field. If this is true, then the heights of high-wave events relative to the background seas should follow a normal frequency distribution. To test this at the $\alpha = 0.05$ significance level, $K = 10$ class intervals for the observed wave heights were fitted to a Gaussian probability distribution. The steps in the goodness-of-fit test are as follows:

TABLE 3.3 Wave heights (mm) during a period of anomalously high waves as measured by a Datawell Waverider buoy deployed in 30 m depth on the inner continental shelf of Vancouver Island, British Columbia.

4,636	4,840	4,901	4,950	4,980	5,014	5,034	5,060	5,095	5,130
4,698	4,842	4,904	4,954	4,986	5,014	5,037	5,066	5,095	5,135
4,702	4,848	4,907	4,955	4,991	5,015	5,037	5,066	5,096	5,135
4,731	4,854	4,907	4,956	4,994	5,017	5,038	5,069	5,102	5,145
4,743	4,856	4,908	4,956	4,996	5,020	5,039	5,069	5,103	5,155
4,745	4,867	4,914	4,956	4,996	5,020	5,040	5,071	5,104	5,157
4,747	4,867	4,916	4,959	4,996	5,021	5,040	5,072	5,104	5,164
4,749	4,870	4,917	4,960	4,997	5,023	5,044	5,073	5,104	5,165
4,773	4,870	4,923	4,961	4,998	5,024	5,045	5,074	5,106	5,166
4,785	4,874	4,925	4,963	5,003	5,025	5,045	5,074	5,110	5,171
4,793	4,876	4,934	4,964	5,006	5,025	5,047	5,074	5,111	5,175
4,814	4,877	4,935	4,964	5,006	5,025	5,048	5,078	5,115	5,176
4,817	4,883	4,937	4,966	5,006	5,025	5,050	5,079	5,119	5,177
4,818	4,885	4,939	4,966	5,006	5,028	5,051	5,080	5,119	5,181
4,823	4,886	4,940	4,970	5,006	5,029	5,052	5,081	5,120	5,196
4,824	4,892	4,941	4,971	5,010	5,029	5,053	5,086	5,121	5,198
4,828	4,896	4,942	4,972	5,011	5,029	5,057	5,089	5,122	5,201
4,829	4,897	4,942	4,974	5,011	5,030	5,058	5,091	5,123	5,210
4,830	4,898	4,943	4,977	5,012	5,031	5,059	5,093	5,125	5,252
4,840	4,899	4,944	4,979	5,012	5,032	5,059	5,094	5,127	5,299

The original $N = 200$ data values have been rank ordered, starting with the lowest value in the upper left corner. The upper bounds of the K-class intervals have been underlined.
Courtesy, Diane Masson, Institute of Ocean Sciences, Sidney, B.C.

1. Specify the class interval width Δx and list the upper limit of the standardized values, $Z_{\alpha/2}$, of the normal distribution that correspond to these values (as in Table 3.4; the lower bound for the first rank is $-\infty$). Commonly Δx is assumed to span 0.4 standard deviations, s, such that $\Delta x \approx 0.4s$; here we use $\Delta x \approx 0.5s$. For $\Delta x = 0.4s$, the values of $Z_{\alpha/2}$, we want are $(\ldots, -2.4, -2.0, \ldots, 2.0, 2.4, \ldots)$ while for $\Delta x = 0.5s$, the values are $(\ldots, -2.5, -2.0, \ldots, 2.0, 2.5, \ldots)$.

2. Determine the finite upper and lower bounds for $Z_{\alpha/2}$ from the requirement that $F_i > 3$. Since $F_i = NP_i$ (where $N = 200$ and P_i is the normal probability distribution for the ith interval), we require $P > 3/N = 0.015$. From tables of the standardized normal density function, we find that $P > 0.015$ implies a lower bound $Z_{\alpha/2} = -2.0$, and an upper bound $Z_{\alpha/2} = +2.0$. Note that for a larger sample, say $N = 2,000$, we have $P > 3/2,000 = 0.0015$ and the bounds become ± 2.8 for the interval $\Delta x = 0.4s$ and ± 2.5 for the interval $\Delta x = 0.5s$.

3. Calculate the expected upper limit, $x = sZ_{\alpha/2} + \bar{x}(\text{mm})$, for the class intervals and mark this limit on the data table (Table 3.3). For each upper bound, $Z_{\alpha/2}$, in Table 3.4, find the corresponding probability density value. Note that these values apply to intervals so, for example, $P(-2.0 < x < -1.5) = 0.0668 - 0.0228 = 0.044$; $P(2.0 < x < \infty) = 0.0228$.

4. Using the value of P, find the expected frequency values $F_i = NP_i$. The observed frequency f_i is found from Table 3.3 by counting the actual number of wave heights lying between the marks made in step 3. Complete the table and calculate X^2. Compare to $\chi^2_{\alpha;\nu}$.

In the above example, $X^2 = 10.02$ and there are $\nu = 7$ degrees of freedom. From Appendix D (Table D.2), we find

$$P\left(X^2 > \chi^2_{\alpha,\nu}\right) = P\left(X^2 > \chi^2_{0.05,7}\right) = 14.07.$$ Thus, at the $\alpha = 0.05$ level of significance (95% significance level), we can accept the null hypothesis that the large wave heights measured by the waverider buoy had a Gaussian (normal) distribution in time and space.

3.9 SELECTING THE SAMPLE SIZE

It is not possible to determine the required sample size N for a given confidence interval until a measure of the data variability, the population standard deviation, σ, is known. This is because the variability of $\overline{X}$ depends on the variability of X. Since we do not usually know *a priori* the population standard deviation (the value for the true population), we use the best estimate available, the sample standard deviation, s. We also need to know the frequency content of the data variable so that we can ensure that the N values we use in our calculations are statistically independent samples. As a simple example, consider a normally distributed, continuous random variable, Y, with the units of meters. We wish to find the

TABLE 3.4 Calculation table for goodness of fit test for the data in Table 3.3.

Class interval	Upper limit of data interval						$\left(\frac{F_i - f_i}{F_i}\right)^2$
	$Z_{\alpha/2}$	$x = sZ_{\alpha/2} + \bar{x}$	P_i	$F_i = NP_i$	f_i	$F_i - f_i$	
1	-2.0	4767.4	0.0228	4.6	8	3.4	2.51
2	-1.5	4825.0	0.0440	8.8	8	0.8	0.07
3	-1.0	4882.5	0.0919	18.4	16	2.4	0.31
4	-0.5	4940.1	0.1498	30.0	23	7.0	1.63
5	0	4997.6	0.1915	38.3	33	5.3	0.73
6	0.5	5055.2	0.1915	38.3	48	9.7	2.46
7	1.0	5112.7	0.1498	30.0	35	5.0	0.83
8	1.5	5170.2	0.0919	18.4	18	0.4	0.01
9	2.0	5227.8	0.0440	8.8	9	0.2	0.00
10	∞	∞	0.0228	4.6	2	2.6	1.47
Totals			1.0000	200	200		10.02

The number of intervals has been determined using an interval width $\Delta x = 0.5s$ with $Z_{\alpha/2}$ in units of 0.5 and requiring that $F_i > 3$. Here, $N = 200$, $\bar{x}$(mean) $= 4997.6$ mm, s (standard deviation) $= 115.1$ mm, and ν (degrees of freedom) $= k - 3 = 7$.

average of the sample and want it to be accurate to within ± 5 m. Since we know that approximately 95% of the sample means of a normally distributed random variable will lie within $\pm 2\sigma_Y$ of the true mean, μ, we require that $2\sigma_Y = 5$ m. Using the central limit theorem for the mean, we can estimate σ_Y by

$$\hat{\sigma}_Y = \sigma/\sqrt{N}$$

so that $2\hat{\sigma}_Y = 2\sigma/\sqrt{N} = 5$ m, whereby $N = 4\sigma^2/25$ (assuming that the N observations are statistically independent). If σ is known, we can easily find N.

When we do not know σ, we are forced to use an estimate from an earlier sample within the range of measurements. If we know the sample range, we can apply the empirical rule for normal distributions that the range is approximately 4σ and take one-fourth the range as our estimate of σ. Suppose our range in the above example is 84 m. Then, $\sigma = 21$ m and

$$N = 4\sigma^2/25 = (4)(21 \text{ m})^2/(25 \text{ m}^2) = 70.56 \approx 71$$

This means that, for a sample of $N = 71$ statistically independent values, we would be 95% sure (probability $= 0.95$) that our estimate of the mean value would lie within $\pm 2\sigma_Y = \pm 5$ m of the true mean.

One method for selecting the sample size for relatively large samples is based on Tchebysheff's theorem known as the "weak law of large numbers". Let $f(x)$ be a density function with mean μ and variance σ^2, and let $\bar{x}_N$ be the sample mean of a random sample of size N from $f(x)$. Let ε and δ be any two specified numbers satisfying $\varepsilon > 0$ and $0 < \delta < 1$. If N is any integer greater than $(\sigma^2/\varepsilon^2)\delta$ then

$$P[-\varepsilon < \bar{x}_N - \mu < \varepsilon] \geq 1-\delta \tag{3.57}$$

To show the validity of condition (Eqn 3.57), we use Tchebysheff's inequality

$$P[g(x) \geq k] \geq \frac{E[g(x)]}{k} \tag{3.58}$$

for every $k > 0$, random variable x, and nonnegative function $g(x)$. An equivalent formula is

$$P[g(x) < k] \geq 1 - \frac{E[g(x)]}{k} \tag{3.59}$$

Let, $g(x) = [(\bar{x}_N - \mu) < \varepsilon]^2$ and $k = \varepsilon^2$ then

$$P[-\varepsilon < (\bar{x}_N - \mu) < \varepsilon] = P[|\bar{x}_N - \mu| < \varepsilon]$$
$$= P[|\bar{x}_N - \mu|^2 < \varepsilon^2] \geq 1 - \frac{E[(\bar{x}_N - \mu)^2]}{\varepsilon^2} \tag{3.60}$$
$$= 1 - \frac{\sigma^2}{N\varepsilon^2} \geq 1-\delta$$

For $\delta > \sigma^2/N\varepsilon^2$ or $N > \sigma^2/\delta\varepsilon^2$, the latter expression becomes

$$P[|\bar{x}_N - \mu| < \varepsilon] \geq 1-\delta. \tag{3.61}$$

We illustrate the use of the above relations by considering a distribution with an unknown mean and variance $\sigma^2 = 1$. How large a sample must be taken in order that the probability will be at least 0.95 that the sample mean, $\bar{x}_N$, will lie within 0.5 of the true population mean? Given are: $\sigma^2 = 1$ and $\varepsilon = 0.5$. Rearranging the inequality (Eqn 3.61)

$$\delta \geq 1 - P[|\bar{x}_N - \mu| < 0.5] = 1 - 0.95 = 0.05$$

Substituting into the relation $N > (\sigma^2/\delta\varepsilon^2) = 1/(0.05)(0.5)^2$ shows that we require $N \geq 80$ independent samples.

3.10 CONFIDENCE INTERVALS FOR ALTIMETER-BIAS ESTIMATES

As an example of how to estimate confidence limits and sample size, consider an oceanographic altimetric satellite, where the altimeter is to be calibrated by repeated passes over a spot on the earth where surface-based measurements provide a precise, independent measure of the sea surface elevation. A typical reference site is an offshore oil platform having sea-level gauges and a location system, such as the ubiquitous multi-satellite global positioning system (GPS). For the TOPEX/POSEIDON satellite, one such reference location is the oil-rig "Harvest Platform" in the Santa Barbara channel off southern California (Christensen et al., 1994). Each pass over the reference site provides a measurement of the satellite altimeter bias, which is used to compute an average bias after repeated calibration observations, x. This bias is just the difference between the height measured by the altimeter and the height measured independently by the *in situ* measurements at the reference site. If we assume that the observation errors are normally distributed with a mean of zero, then the uncertainty of the true mean bias,

$$\sigma_b = Zs_b/\sqrt{N}$$

where $Z = \frac{\bar{x}-\mu}{\sigma}$ is the standard normal distribution (3.25), s_b is the standard deviation of the measurements, and N is the number of measurements (i.e., the number of calibration passes over the reference site).

Suppose we are required to know the true mean of the altimeter bias to within 2 cm, and that we estimate the uncertainty of the individual measurements to be 3 cm. We then ask: "What is the number of independent measurements required to give a bias of 2 cm at the 90%, 95%, and 99% confidence intervals?" Using the above formulation for the standard error, we find

$$N = \left(Z_{\alpha/2}s_b/\sigma_b\right)^2 \tag{3.62}$$

from which we can compute the required sample size. As before, the parameter α refers to the chosen significance level which, in the present case, corresponds to $\alpha = 0.10, 0.05$, and 0.01. Now $\sigma_b = 2$ cm (required) and $s_b = 3$ cm (estimated), so that we can use the standard normal table for $Z_{\alpha/2} = N(0, 1)$ in Appendix D (Table D.1) to obtain the values shown in Table 3.5. If we require the true mean bias to be 1.5 cm instead of 2.0 cm, the values in Table 3.5 become those in Table 3.6. Note that for the 90% confidence level, we find values of Z in Table D.1 corresponding to $1-\alpha/2 = 0.950$; similarly for the 95 and 99%, we find Z for tabulated values of 0.975 and 0.995, respectively.

Lastly, suppose the satellite is in a 10-day repeat orbit so that we can only collect a reference measurement every 10 days at our ground site; we are given 240 days to collect reference observations. What confidence intervals can be achieved for both of the above cases if we assume that only 50% of the calibration measurements are successful and that the 10-day observations are statistically independent? Now, since we have only one calibration measurement every 10 days for 50% of 240 days we have

$$c = (0.5)(240 \text{ days})(1 \text{ measurement} / 10 \text{ days}) = 12 \text{ measurements}$$

Referring to the previous two tables, we see that for the first case (Table 3.5), where the mean bias was required to be 2.0 cm, we can achieve the 95% interval with 12 measurements; for the case where the mean bias is restricted to 1.5 cm (Table 3.6), only the 90% confidence interval is possible. Note that the values in the last column on the right are rounded-up, integer versions of the calculated values of N.

TABLE 3.5 Calculation of the number of satellite altimeter required to attain a given level of confidence in elevation using the relation (3.62) for $\sigma_b = 2$ cm and $s_b = 3$ cm.

Confidence level (α)	Standard normal value ($Z_{\alpha/2}$)	Exact number of observations (N)	Actual number of observations
90%	1.645	6.089	7
95%	1.960	8.644	9
99%	2.576	14.931	15

TABLE 3.6 Calculation of the number of satellite altimeter observations needed for a given level of confidence in the level elevation using Eqn (3.62) for $\sigma_b = 1.5$ cm and $s_b = 3$ cm.

Confidence level (α)	Standard normal value (Z_{α})	Exact number of observations (N)	Actual number of observations
90%	1.645	10.82	11
95%	1.960	15.37	16
99%	2.576	26.54	27

3.11 ESTIMATION METHODS

Now that we have introduced methods to calculate confidence intervals for our estimates of the population statistics μ and σ^2, we need procedures to estimate these quantities themselves. There are many different methods we could use but space does not allow us to discuss them all. We first introduce a very general technique, known as Minimum Variance Unbiased Estimation (MVUE), and then later discuss a popular method called the Maximum Likelihood Method, which leads to MVUE estimators.

Before introducing the MVUE procedure, we need to define two terms: *sufficiency* and *likelihood*. Let $y_1, y_2, ..., y_N$ be a random sample from a probability distribution with an unknown statistical parameter, θ (mean, variance, etc.). The statistic $U = g(y_1, y_2, ..., y_N)$ is said to be sufficient for θ (e.g., the population mean) if the conditional distribution $y_1, y_2, ..., y_N$, given U, does not depend on θ. In other words, once the form of U is known, no other combination of samples $y_1, y_2, ..., y_N$, provides additional information about θ. This tells us how to check if our statistic is sufficient but does not tell us how to compute the statistic.

To define likelihood, we again let $y_1, y_2, ..., y_N$ be sample observations of random variables $Y_1, Y_2, ..., Y_N$. For continuous variables, the likelihood $L(y_1, y_2, ..., y_N)$ is the joint probability density $f(y_1, y_2, ..., y_N)$ evaluated at the observations, y_i. Assuming that the variables Y_i are statistically independent

$$L(y_1, y_2, ..., y_N) = f(y_1, y_2, ..., y_N) = f(y_1)f(y_2)...f(y_N) \tag{3.63}$$

where $f(y_i)$, $i = 1, 2, ..., N$, is the PDF for the random variable Y_i.

As an oceanographic example, consider a record of daily average current velocities, V_i ($i = 1, ..., N$), obtained using a single current meter moored for a period of 1 month ($N = 30$ days). Show that the monthly mean velocity, V, is a sufficient statistic for the population mean if the variance is known (in this case, estimated from the range of current values). Since the daily velocities are average values of shorter-term current velocity measurements (e.g., 10 min values), we can invoke the central limit theorem to conclude that the daily velocities are normally distributed. Hence, the PDF for each daily value can be written as

$$f(V_i) = \frac{1}{\sigma(2\pi)^{1/2}} \exp\left[\frac{-1}{2\sigma^2}(V_i - \mu)^2\right]$$

We can write the likelihood L of the sample as

$$L = f(V_1, V_2, ..., V_{30}) = f(V_1)f(V_2)...f(V_{30})$$

$$= \frac{1}{\sigma(2\pi)^{\frac{1}{2}}} \exp\left[-\frac{1}{2\sigma^2}(V_1 - \mu)^2\right] \times \frac{1}{\sigma(2\pi)^{\frac{1}{2}}} \exp\left[-\frac{1}{2\sigma^2}(V_2 - \mu)^2\right] \times ... \times \frac{1}{\sigma(2\pi)^{1/2}} \exp\left[-\frac{1}{2\sigma^2}(V_{30} - \mu)^2\right]$$

$$= \frac{1}{\left[\sigma(2\pi)^{\frac{1}{2}}\right]^{30}} \exp\left[-\frac{1}{2\sigma^2} \sum_{i=1}^{30}(V_i - \mu)^2\right]$$

Because σ is known from our range of current velocities, then L is a function of V and μ, only. Hence, V is a sufficient statistic for the population mean, μ.

3.11.1 Minimum Variance Unbiased Estimation

For random variables $Y_1, Y_2, ..., Y_N$, with PDF, $f(y)$, and unknown parameter, θ, let one set of sample observations be $(x_1, x_2, ..., x_N)$ and another be $(y_1, y_2, ..., y_N)$. The ratio of the likelihoods of these two sets of observations can be written as

$$L(x_1, x_2, ..., x_N) / L(y_1, y_2, ..., y_N) \tag{3.64}$$

In general, this ratio will not be a function of θ if, and only if, there is a function $g(x_1, x_2, ..., x_N)$ such that $g(x_1, x_2, ..., x_N) = g(y_1, y_2, ..., y_N)$ for all choices of x and y. If such a function can be found, it is the minimum sufficient statistic for θ. Any unbiased estimator that is a function of a minimal sufficient statistic will be a Minimum Variance Unbiased Estimation (MVUE); this means that it will possess the smallest possible variance among the unbiased estimators.

We illustrate what is meant with an example. Again, let $x_1, x_2, ..., x_N$ be a random sample from a normal population with the unknown parameters μ and σ^2. We want to find the MVUE of μ and σ^2. In this case, the likelihood ratio Eqn (3.64) can be written as

$$
\begin{aligned}
\frac{L(x_1, x_2, ..., x_N)}{L(y_1, y_2, ..., y_N)} &= \frac{f(x_1, x_2, ..., x_N)}{f(y_1, y_2, ..., y_N)} \\
&= \frac{\frac{1}{\sigma\sqrt{2\pi}}\exp\left[-\left(\frac{1}{2\sigma^2}\right)\sum_{i=1}^{N}(x_i - \mu)^2\right]}{\frac{1}{\sigma\sqrt{2\pi}}\exp\left[-\left(\frac{1}{2\sigma^2}\right)\sum_{i=1}^{N}(y_i - \mu)^2\right]} \\
&= \exp\left\{-\left(\frac{1}{2\sigma^2}\right)\left[\sum_{i=1}^{N}(x_i - \mu)^2 - \sum_{i=1}^{N}(y_i - \mu)^2\right]\right\} \\
&= \exp\left\{-\left(\frac{1}{2\sigma^2}\right)\left[\left(\sum_{i=1}^{N}x_i^2 - \sum_{i=1}^{N}y_i^2\right) - 2\mu\left(\sum_{i=1}^{N}x_i - \sum_{i=1}^{N}y_i\right)\right]\right\}
\end{aligned}
\tag{3.65}
$$

For ratio (3.65) to be independent of μ, we must have

$$
\sum_{i=1}^{N}x_i = \sum_{i=1}^{N}y_i
\tag{3.66}
$$

Similarly, for the ratio to be independent of σ^2, requires Eqn (3.66) as well as

$$
\sum_{i=1}^{N}x_i^2 = \sum_{i=1}^{N}y_i^2
\tag{3.67}
$$

Thus, both $\sum x_i$ and $\sum x_i^2$ are minimum sufficient statistics for μ and σ^2. Because $\bar{x}$ is an unbiased estimate of μ

$$
s^2 = \frac{1}{N-1}\sum_{i=1}^{N}(x_i - \bar{x})^2
\tag{3.68}
$$

is an unbiased estimate of σ^2. Since both $\bar{x}$ and s^2 are functions of the minimal sufficient statistics

$$
\sum_{i=1}^{N}x_i \text{ and } \sum_{i=1}^{N}x_i^2
$$

as expressed by Eqns (3.66) and (3.67), they also are MVUEs for μ and σ^2.

3.11.2 Maximum Likelihood

The procedure introduced earlier to compute the MVUE is complicated by the fact that one must find some function of the minimal sufficient statistic that gives the sought-after target parameter. Finding this function is generally a matter of trial and error. A more sophisticated procedure, the maximum likelihood method, often leads to the MVUE. Thus, an estimate that yields minimum variance equates to an estimate that has maximum likelihood of being correct.

The formal statement of this method is quite simple. Choose as estimates those parameter values that maximize the likelihood $L(y_1, y_2, ..., y_N)$. A simple example using discrete variables helps to illustrate the logic in the maximum likelihood method. This example is intended to give the reader an intuitive sense of the maximum likelihood method that can be more formerly defined and applied to continuous functions and PDFs. Assume we have a bag containing three marbles. The marbles can be black or white. We randomly sample two of the three and find that they are both black. What is the best estimate of the total number of black marbles in the bag? If there are actually two black and one white in the bag, the probability, p, of sampling two black marbles is from Eqn (3.7a)

$$
p(r = 2\ black) = \binom{N}{r}p^r(1-p)^{N-r} = \binom{3}{2}\left(\frac{2}{3}\right)^2\left(\frac{1}{3}\right)^1 = \frac{4}{9}
$$

where the binomial expression (for the number of combinations of samples taken r at a time from a total of N samples) is

$$\binom{N}{r} = N! \, / \, [r!(N-r)!] \tag{3.69}$$

Now, if there are actually three black marbles in the bag, the probability of sampling two black marbles is obviously $p(r = 2 \; black) = 1$. On this basis, it seems reasonable to choose 3 as the estimate of the number of black marbles in the bag in order to maximize the probability of the observed sample.

A more complex example can be used to illustrate the application of this method to our estimates of the mean, μ, and variance, σ^2, for a normal population. Again, let $y_1, y_2, \ldots, y_N$ be a random sample from a normal population with parameters μ and σ^2. We want to find the maximum likelihood estimators of μ and σ^2. To find the maximum likelihood, L, we need to write the joint PDF of the independent observations $y_1, y_2, \ldots, y_N$

$$
\begin{aligned}
L &= f(y_1, y_2, \ldots, y_N) \; = f(y_1)f(y_2)\ldots f(y_N) \\[6pt]
&= \frac{1}{\sigma(2\pi)^{\frac{1}{2}}} \exp\left[-\frac{1}{2\sigma^2}(y_1 - \mu)^2\right] \times \frac{1}{\sigma(2\pi)^{\frac{1}{2}}} \exp\left[-\frac{1}{2\sigma^2}(y_2 - \mu)^2\right] \times \ldots \\[6pt]
&\quad \times \frac{1}{\sigma(2\pi)^{\frac{1}{2}}} \exp\left[-\frac{1}{2\sigma^2}(y_N - \mu)^2\right] \\[6pt]
&= \frac{1}{[\sigma(2\pi)^{\frac{1}{2}}]^N} \exp\left[-\frac{1}{2\sigma^2}\sum_{i=1}^{N}(y_i - \mu)^2\right]
\end{aligned}
\tag{3.70}
$$

We simplify this expression by taking the natural logarithm, $\ln(L)$, which we then differentiate to find the maximum. Specifically, we begin with

$$\ln(L) = -\frac{N}{2}\ln(\sigma^2) - \frac{N}{2}\ln(2\pi) - \frac{\sum_{i=1}^{N}(y_i - \mu)^2}{2\sigma^2} \tag{3.71}$$

Taking derivatives of Eqn (3.71) with respect to μ and σ^2, we find

$$\frac{\partial[\ln(L)]}{\partial\mu} = \sum_{i=1}^{N}\left[\left(\frac{(y_i - \mu)}{\sigma^2}\right)\right] \tag{3.72a}$$

$$\frac{\partial[\ln(L)]}{\partial\sigma^2} = -\frac{N}{2\sigma^2} + \sum_{i=1}^{N}\left[\left(\frac{(y_i - \mu)^2}{2\sigma^4}\right)\right] \tag{3.72b}$$

Setting Eqns (3.72a) and (3.72b) to zero and solving yields the required estimates of μ and σ^2. Beginning with Eqn (3.72a), $\sum_{i=1}^{N}\left[\left(\frac{(y_i - \mu)}{\sigma^2}\right)\right] = 0$ requires that

$$\mu = \frac{1}{N}\sum_{i=1}^{N}y_i = \bar{y} \tag{3.73}$$

Substituting $\bar{y}$ into Eqn (3.72b) and setting the equation to zero, yields

$$-N/\sigma^2 + \sum_{i=1}^{N}(y_i - \bar{y})^2 \, / \, \sigma^4 = 0 \tag{3.74a}$$

whereby the estimate for σ^2 becomes

$$\hat{\sigma}^2 = \sum_{i=1}^{N}(y_i - \bar{y})^2 \, / \, N = (s')^2 \tag{3.74b}$$

Thus, $\bar{y}$ and $(s')^2$ are the maximum likelihood estimators of μ and σ^2. Although $\bar{y}$ is an unbiased estimate of μ, s'^2 is not an unbiased estimate of σ^2, as noted at the beginning of the chapter. However, s'^2 can easily be adjusted to the unbiased estimator

$$s^2 = \frac{1}{N-1} \sum_{i=1}^{N} (y_i - \bar{y})^2 \tag{3.75}$$

Because the maximum likelihood method has widespread application, we present another simple example to illustrate its use. Let $y_1, y_2, \ldots, y_N$, be a random sample taken from a uniform probability distribution $f(y_i) = 1/\theta = $ constant, $0 \le y_i \le \theta$, and $i = 1, 2, \ldots, N$. We want to find the maximum likelihood estimate of θ. Again, we write the likelihood, L, as the joint probability function

$$\begin{aligned} L = f(y_1, y_2, \ldots, y_N) &= f(y_1)f(y_2)\ldots f(y_N) \\ &= (1/\theta)(1/\theta)\ldots(1/\theta) = (1/\theta)^N \end{aligned} \tag{3.76}$$

In this case, L is a monotonically decreasing function of θ and nowhere is $\frac{dL}{d\theta} = -N\left(\frac{1}{\theta}\right)^{N+1}$ equal to zero. As a consequence, L must be greater than, or equal to, the largest sample value, y_N. L is, therefore, not an unbiased estimator of θ but can be adjusted using

$$\theta = \frac{(N+1)}{N} y_N \tag{3.77}$$

which is an unbiased estimate. We note that if any statistic U can be shown to be a sufficient statistic for estimating θ, then the maximum likelihood estimator is always some function of U. If this maximum likelihood estimate can be found, and then adjusted to be unbiased, the result will generally be an MVUE.

To demonstrate the application of the maximum likelihood approach, assume that a random sample of size N is selected from the normal distribution (Eqn 3.23) with μ and σ^2 as the mean and variance for each x (where we assume that the x_i values are independent). We ask: if $\bar{\theta} = (\theta_1, \theta_2) = (\mu, \sigma^2)$ is the parameter space for the PDF $f(x_1, x_2, \ldots, x_N)$, then what is the likelihood function? Also, find the maximum likelihood estimator $\widehat{\theta}_1$ of θ_1, which maximizes the likelihood function and find the maximum likelihood estimator $\widehat{\theta}_2$, which maximizes the likelihood function θ_2. We first write the PDF as

$$f(\bar{x}, \bar{\theta}) = \frac{1}{[\sigma^2(2\pi)]^{N/2}} \exp\left[\frac{1}{2\sigma^2} \sum_{i=1}^{N} (x_i - \mu)^2\right]$$

$$L(\bar{x}, \bar{\theta}) = \prod_{i=1}^{N} \left\{ \frac{1}{(2\pi\sigma^2)^{1/2}} \exp\left[\frac{-(x_i - \mu)^2}{\sigma^2}\right] \right\}$$

where L is the likelihood function written in terms of the product, Π, of the exponential. Taking the natural log of the above expression with respect to our estimated parameter, θ_1, and setting it equal to zero to find the maximum, we obtain

$$\ln(L) = \frac{-N}{2} \ln(\sigma^2) - \frac{N}{2} \ln(2\pi) - \frac{\sum_{i=1}^{N} (x_i - \mu)^2}{2\sigma^2}$$

where $\sigma > 0$ and $-\infty < \mu < \infty$. The derivative of this function with respect to θ_1 (which is μ) is

$$\frac{\partial L}{\partial \mu} = \frac{-1}{2\sigma^2} \sum_{i=1}^{N} (x_i - \mu)(-2) = \frac{1}{\sigma^2} \sum_{i=1}^{N} (x_i - \mu) = 0$$

so that our estimate of μ is

$$\widehat{\mu} = \bar{x} = \frac{1}{N} \sum_{i-1}^{N} x_i$$

Furthermore, the maximum likelihood estimator of θ_2 (which is σ^2) is given by

$$\frac{\partial L}{\partial \sigma^2} = \frac{-N}{2\sigma^2} - \frac{(-1)}{2\sigma^4} \sum_{i=1}^{N} (x_i - \mu)^2 = \frac{1}{2\sigma^2} \left[\frac{1}{\sigma^2} \sum_{i=1}^{N} (x_i - \mu)^2 - N \right] = 0$$

which yields the estimator

$$\hat{\sigma}^2 = \frac{1}{N} \sum_{i=1}^{N} (x_i - \hat{\mu})^2$$

For a normally distributed oceanographic data set, we can readily obtain maximum likelihood estimates of the mean and variance of the data. However, the real value of this technique is for variables that are not normally distributed. For example, if we examine spectral energy computed from current velocities, the spectral values typically have a chi-square distribution rather than a normal distribution. If we follow the maximum likelihood procedure, we find that the spectral values have a mean of ν, the number of degrees of freedom, and a variance of 2ν. These are the maximum likelihood estimators for the mean and variance. This example can be used as a pattern for applying the maximum likelihood method to a particular sample. In particular, we first determine the appropriate PDF for the sample values. We then find the joint likelihood function, take the natural logs, and then differentiate with respect to the parameter of interest. Setting this derivative equal to zero to find the maximum subsequently yields the value of the parameter being sought.

3.12 LINEAR ESTIMATION (REGRESSION)

Linear regression is one of a number of statistical procedures that fall under the general heading of linear estimation. Since linear regression is widely treated in the literature and is available in many software packages, our primary purpose here is to establish a common vocabulary for all readers. In our previous discussion and examples, we assumed that the random variables $Y_1, Y_2,...,Y_N$ were independent (in a probabilistic sense) and identically distributed, which implies that $E[Y_i] = \mu$ is a constant. Often this is not the case and the expected value $E[Y_i]$ of the variable is a function of some other parameter. We now consider the values y of a random variable, Y, called the dependent variable, whose values are a function of one or more *nonrandom* variables $x_1, x_2,...,x_N$, called independent variables (in a mathematical, rather than probabilistic sense).

If we model our random variable as

$$y = E[y] + \varepsilon = b_0 + b_1 x + \varepsilon \tag{3.78}$$

we invariably find that the points, y, are scattered about the regression line $E[y] = b_0 + b_1 x$. The random variable ε in the right-hand term of Eqn (3.78) gives the departure from linearity and has a specific PDF with a mean value $\bar{\varepsilon} = 0$. In other words, we can think of y as having a deterministic part, $E[y]$, and a random part, ε, that is randomly distributed about the regression line. By definition, simple linear regression is limited to finding the coefficients b_0 and b_1. If N independent variables $(x_1, x_2,...,x_N)$ are involved in the variability of each value y, we must deal with *multiple linear regression*. In this case, Eqn (3.78) becomes

$$y = b_0 + b_1 x_1 + b_2 x_2 + ... + b_N x_N + \varepsilon \tag{3.79}$$

3.12.1 Method of Least Squares

One of the most powerful techniques for fitting a dependent model parameter, y, to independent (observed input) variable x_i ($i = 1, 2, ..., N$) is the *method of least squares*. We apply the method in terms of linear estimation and will later readdress the topic in terms of more general statistical models. (Note: by "linear" we mean linear in the parameters $b_0, b_1, ..., b_N$. Thus, $y = b_0 + b_1 x_i + \varepsilon$ is linear but $y = b_0 + \sin(b_1 x_1) + \varepsilon$ is not.) We begin with the simplest case, that of fitting a straight line to a set of points using the "best" coefficients b_0, b_1 (Figure 3.10). In a sense, the least squares procedure does what we do (and often quite well) by eye—it minimizes the vertical deviations (residuals) of data points from the fitted line. The plots in Figure 3.10 are for two separate data sets; $y = y(x)$ and $z = z(x)$. Let

$$y_i = \hat{y}_i + \varepsilon_i \tag{3.80}$$

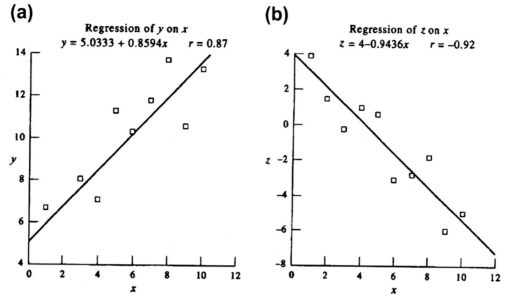

FIGURE 3.10 Straight-line (linear regression) fits to the sets of points in Table 3.7 using the "best" fit coefficients b_0, b_1. (a) Regression of y on x, for which $(b_0, b_1) = (5.0333, 0.8594)$; and (b) regression of z on x, for which $(b_0, b_1) = (4.0, -0.9436)$. r is the correlation coefficient.

where

$$\widehat{y}_i = b_0 + b_1 x_i \tag{3.81}$$

is the estimator for the deterministic portion of the data and ε is the residual or error. To find the coefficients b_0, b_1 we need to minimize the sum of the squared errors (SSE), where SSE is the total variance that is not explained (accounted for) by our linear regression model given by Eqns (3.80) and (3.81)

$$\text{SSE} = \sum_{i=1}^{N} \varepsilon_i^2 = \sum_{i=1}^{N} (y_i - \widehat{y}_i)^2 = \sum_{i=1}^{N} [y_i - (b_0 + b_1 x_i)]^2 \tag{3.82a}$$

$$= \text{SST} - \text{SSR} \tag{3.82b}$$

in which

$$\text{SST} = \sum_{i-1}^{N} (y_i - \bar{y})^2 \text{ and SSR} = \sum_{i-1}^{N} (\widehat{y}_i - \bar{y})^2 \tag{3.82c}$$

Here, sum of squares total (SST) is the variance in the data, and sum of squares regression (SSR) is the amount of variance explained by our regression model. Minimization amounts to finding those coefficients that minimize the unexplained variance (SSE). Taking the partial derivatives of Eqn (3.82a) with respect to b_0 and b_1 and setting the resultant values equal to zero, the minimization conditions are

$$\frac{\partial \text{SSE}}{\partial b_0} = 0; \frac{\partial \text{SSE}}{\partial b_1} = 0 \tag{3.83}$$

Substituting Eqn (3.82a) into Eqn (3.73), we have for b_0

$$\frac{\partial \text{SSE}}{\partial b_0} = \frac{\partial}{\partial b_0} \left\{ \sum_{i=1}^{N} [y_i - (b_0 + b_1 x_i)]^2 \right\} = -2 \sum_{i=1}^{N} [y_i - (b_0 + b_1 x_i)]$$

$$= -2 \left(\sum_{i=1}^{N} y_i - N b_0 - b_1 \sum_{i=1}^{N} x_i \right) = 0 \tag{3.84a}$$

Now for b_1

$$\frac{\partial \text{SSE}}{\partial b_1} = \frac{\partial}{\partial b_1} \left\{ \sum_{i=1}^{N} [y_i - (b_0 + b_1 x_i)]^2 \right\}$$

$$= -2 \sum_{i=1}^{N} (x_i)[y_i - (b_0 + b_1 x_i)] \tag{3.84b}$$

$$= -2 \left(\sum_{i=1}^{N} x_i y_i - b_0 \sum_{i=1}^{N} x_i - b_1 \sum_{i=1}^{N} x_i^2 \right) = 0$$

Once the mean values of y and x are calculated, these least squares equations can be solved simultaneously to find an estimate of the coefficient b_1 (the slope of the regression line); this is then used to obtain an estimate of the second coefficient, b_0 (the intercept of the regression line). In particular

$$b_1 = \frac{\sum_{i=1}^{N}(x_i - \bar{x})(y_i - \bar{y})}{\sum_{i=1}^{N}(x_i - \bar{x})^2} = \frac{\left[N \sum_{i=1}^{N} x_i y_i - \sum_{i=1}^{N} x_i \sum_{i=1}^{N} y_i \right]}{\left[N \sum_{i=1}^{N} x_i^2 - \left(\sum_{i=1}^{N} x_i \right)^2 \right]} \tag{3.85a}$$

$$b_0 = \bar{y} - b_1 \bar{x} \tag{3.85b}$$

Several features of the regression values are worth noting. First, if we substitute the intercept $b_0 = \bar{y} - b_1 \bar{x}$ into the line $\hat{y} = b_0 + b_1 x$, we obtain

$$\hat{y} = \bar{y} = b_1(x - \bar{x})$$

As a result, whenever $x = \bar{x}$, we have $\hat{y} = \bar{y}$. This means: (1) that the regression line always passes through the point $(\bar{x}, \bar{y})$, the centroid of the distribution, and (2) that because the operation $\partial \text{SSE}/\partial b_0 = 0$ minimizes the error $\sum \varepsilon_i = 0$, the regression line not only goes through the point of averages $(\bar{x}, \bar{y})$ but it also splits the scatter of the observed points so that the positive residuals (where the regression line passes below the true point) always cancel exactly the negative residuals (where the line passes above the true point). The sample regression line is therefore an unbiased estimate of the population regression line. To summarize the linear regression procedure, we note that:

1. For each selected x (independent variable) there is a distribution of y (or z in the case of the examples shown in Figure 3.10) from which the sample (dependent variable) is drawn at random.
2. The population of y corresponding to a selected x has a mean μ that lies on the straight line $\mu = b_0 + b_1 x$, where b_0 and b_1 are regression parameters.
3. For each population, the standard deviation of y about its mean, $b_0 + b_1 x$, has the same value ($s_{xy} = s_\varepsilon$; $y = b_0 + b_1 x + \varepsilon$). Note that ε is a random variable drawn from a normal population with $\mu = 0$ and $s = s_{xy}$.

As indicated by the above analysis, y is the sum of a random component ε and a deterministic component x; the deterministic part determines the mean values of the y population samples, with one distribution of y for each distribution of x (x_i; $i = 1, 2, ..., N$) that we pick. The mean values of y lie on the straight line, $\mu = b_0 + b_1 x$, which is the population regression line. The regression parameter b_0 is the y mean, where $x = 0$, and b_1 is the slope of the regression line. The random part, ε, is independent of x and y. To compute the regression parameters, we need values of $N, \bar{x}, \bar{y}, \sum x^2, \sum y^2$, and $\sum xy$. Earlier, we discussed the computational shortcuts to compute $\sum x^2$ and $\sum y^2$ without first computing the means of x and y. The same can be accomplished for xy using

$$\sum(x - \bar{x})(y - \bar{y}) = \sum(xy) - \left(\sum x \sum y \right) / N$$

As examples of linear regression, consider the data sets in Table 3.7 for dependent variables y_i and z_i, which are both functions of the same independent variable x_i (for example, y_i, could be the eastward and z_i the northward component of velocity as functions of time, x_i). We compute the regression coefficients b_0, b_1 plus the sample variance s^2 and the percent of explained variance (100 SSR/SST) for each data set.

TABLE 3.7 Values for dependent variables y_i, z_i as function of x_i.

x_i	y_i	$\widehat{y}_i$	z_i	$\widehat{z}_i$
1.0	6.7	5.9	3.9	3.1
2.0	4.7	6.8	1.5	2.1
3.0	8.1	7.6	−0.2	1.2
4.0	7.1	8.5	1.0	0.2
5.0	11.3	9.4	0.6	−0.7
6.0	10.5	10.2	−3.1	−1.7
7.0	11.8	11.1	−2.8	−2.6
8.0	13.7	11.9	−1.8	−3.6
9.0	10.6	12.8	−6.0	−4.5
10.0	13.3	13.7	−5.0	−5.4

SST(y) = 80.64; SSR(y) = 61.11; SSE(y) = 19.53

SST(z) = 86.39; SSR(z) = 73.46; SSE(z) = 12.93

The estimated values $\widehat{y}$ and $\widehat{z}$ are derived from the linear regression analysis. Formulae at the bottom of the table are the sum of squares total (SST), sum of squares regression (SSR), and the sum of the squared errors (SSE) to be derived in our regression analysis for $N = 10$.

To estimate the regression parameters, we must first compute the means of the three series

$$\bar{x} = 5.50; \bar{y} = 9.78; \bar{z} = −1.19$$

We then use the means to calculate the sums in Eqns (3.82a)−(3.82c)

$$\sum_{i=1}^{10} (x_i - \bar{x})^2 = 82.50; \sum_{i=1}^{10} (x_i - \bar{x})(y_i - \bar{y}) = 71.00; \sum_{i=1}^{10} (x_i - \bar{x})(z_i - \bar{z}) = −77.85$$

$$\text{SST}(y) = \sum_{i=1}^{10} (y_i - \bar{y})^2 = 80.64$$

$$\text{SST}(z) = \sum_{i=1}^{10} (z_i - \bar{z})^2 = 86.36$$

For the regression of y on x $(\widehat{y} = b_0 + b_1 x)$ we find

$$b_0 = 5.05; b_1 = 0.861; s^2 = 2.44$$

$$\text{SSR}(y) / \text{SST}(y) = 61.11 / 80.64 = 0.758 \, (= 75.8\%)$$

while for the regression of z on x $(\widehat{z} = b_0 + b_1 x)$, we have

$$b_0 = 4.00; b_1 = −0.94; s^2 = 1.62$$

$$\text{SSR}(z) / \text{SST}(z) = 73.46 / 86.36 = 0.850 \, (= 85.0\%)$$

The ratio SSR/SST (variance explained/total variance) is a measure of the goodness of fit of the regression curves called the *coefficient of determination*, r^2 (*squared correlation coefficient*). If the regression line fits perfectly all the sample values, all residuals would be zero. In turn, SSE = 0 and SSR/SST = r^2 = 1. As the fit becomes increasingly less representative of the data points, r^2 decreases toward a possible minimum of zero. In the above example, both ratios, SSR/SST, exceed 0.75 (75%), indicating that the linear regression values explain a considerable fraction of the signal variance.

3.12.2 Standard Error of the Estimate

The measure of the absolute magnitude of the goodness of fit of our estimate, $\widehat{y}$, is the standard error of the estimate, s_ε (= s_{yx}), defined as

$$s_\varepsilon = [\text{SSE}/(N{-}2)]^{1/2}$$
$$= \left[\frac{1}{N{-}2}\sum_{i=1}^{N}(y_i - \widehat{y})^2\right]^{1/2} \tag{3.86}$$

The number of degrees of freedom, $N - 2$, for s_ε is based on the fact that two parameters, b_0 and b_1, are needed for any linear regression estimate. If s_ε is from a normal distribution then approximately 68.3% of the observations will fall within $\pm 1s_\varepsilon$ units of the regression line, 95.4% will fall within $\pm 2s_\varepsilon$ units of the line and 99.7% will fall within $\pm 3s_\varepsilon$ units of this line. For the examples of Table 3.7 (which includes estimates $\widehat{y}$ and $\widehat{z}$):

$$\text{Variable } y\text{: } s_\varepsilon = [\text{SSE}(y)/(N{-}2)]^{1/2} = (19.53/8)^{1/2} = 1.56$$

$$\text{Variable } z\text{: } s_\varepsilon = [\text{SSE}(z)/(N{-}2)]^{1/2} = (12.93/8)^{1/2} = 1.27$$

As a result, the $\pm 2s_\varepsilon$ ranges are $\pm 2(1.56)$ and $\pm 2(1.27)$, respectively.

The standard error for the estimate of the y-intercept, $\widehat{b}_0$, of the regression line is given as

$$s_{b_0} = s_\varepsilon \left(\frac{1}{N} + \frac{\bar{x}^2}{\sum_{i=1}^{N}(x_i - \bar{x})^2}\right)^{1/2} \tag{3.87a}$$

and the standard error for the slope, $\widehat{b}_1$, of the regression line as

$$s_{b_1} = \frac{s_\varepsilon}{\left[\sum(x_i - \bar{x})^2\right]^{1/2}} \tag{3.87b}$$

For small samples ($N < 30$), we can write the 90% confidence interval for the true value of the regression slope, b_1, as

$$\widehat{b}_1 - t_{0.05}s_{b_1} \leq b_1 \leq \widehat{b}_1 + t_{0.05}s_{b_1} \tag{3.88}$$

where $t_{0.05}$ is the Student-t statistic for $\alpha/2 = 0.05$ and $N - 2$ degrees of freedom. Turning to the regression line itself, it is useful to know how the confidence intervals for the regression estimates of: (1) $\bar{y}$ (the mean of the dependent variable, y), and (2) y_i (a specific value of the dependent variable, y) vary for given values of the independent variable, x_i. Specifically, the confidence intervals for $\bar{y}$ and y_i are given by:

$$\bar{y} \pm t_{\alpha/2,N-2}s_\varepsilon\sqrt{\bar{x}_i}; \quad \text{estimate of } \bar{y} \text{ given } x_i$$
$$y_i \pm t_{\alpha/2,N-2}s_\varepsilon\sqrt{1 + \bar{x}_i}; \quad \text{estimate of } y_i \text{ given } x_i \tag{3.89a}$$

respectively, where

$$\bar{x}_i = \frac{1}{N} + \frac{(x_i - \bar{x})^2}{\sum_{i=1}^{N}(x_i - \bar{x})^2} \tag{3.89b}$$

Because of the dependence on x_i, these confidence limits would appear as hyperbolae in regression diagrams, such as Figure 3.10. The hyperbolae, when plotted, would represent confidence belts for the different significance levels. Note the increasing uncertainty of making predictions for y for values of x far removed from the mean value, $\bar{x}$. Because the lines indicate that y must be within the confidence belt, higher significance levels have narrower belts. For all significance levels, estimates of $\bar{y}$ and y_i get worse as we move away from $\bar{x}$. Remember that these confidence belts are for the regression line itself and not for the individual points. Hence, in the case of the 95% confidence interval, if repeated samples of y_i are taken of the same size and the same fixed value of x, then 95% of the confidence intervals, constructed for the mean value of y and x, will contain the true value of the mean of y and x. If only one prediction is made for x, then the probability that the calculated interval will contain the true value is 95%.

3.12.3 Multivariate Regression

To extend the regression procedure to multivariate regression, it is best to formulate our linear estimation model in matrix terms. Suppose our model is of the form

$$Y = b_0 + b_1 X_1 + b_2 X_2 + \ldots + b_k X_k + \varepsilon \tag{3.90}$$

and that we make N independent (probabilistic) observations $y_1, y_2, \ldots, y_N$ of Y. This means that we can write

$$y_i = b_0 + b_1 x_{i1} + b_2 x_{i2} + \ldots + b_k x_{ik} + \varepsilon_i \tag{3.91}$$

where x_{ik} is the kth independent variable for the ith observation. Writing this in matrix form we have

$$\mathbf{Y} = \begin{pmatrix} y_1 \\ y_2 \\ \ldots \\ \ldots \\ y_N \end{pmatrix}, \mathbf{X} = \begin{pmatrix} x_{10} & x_{11} & \cdots & x_{1k} \\ x_{20} & x_{21} & \cdots & x_{2k} \\ \cdots & \cdots & \cdots & \cdots \\ \cdots & \cdots & \cdots & \cdots \\ x_{N0} & x_{N1} & \cdots & x_{Nk} \end{pmatrix}$$

$$\mathbf{B} = \begin{pmatrix} b_0 \\ b_1 \\ \ldots \\ \ldots \\ b_k \end{pmatrix}, \mathbf{E} = \begin{pmatrix} \varepsilon_1 \\ \varepsilon_2 \\ \ldots \\ \ldots \\ \varepsilon_N \end{pmatrix} \tag{3.92}$$

where the boldface capital letters denote matrices. Using Eqn (3.92), we can represent the N equations relating y_i the independent variable x_{ij} as

$$\mathbf{Y} = \mathbf{B} \cdot \mathbf{X} + \mathbf{E} \tag{3.93}$$

If we restrict our analysis to the first two coefficients, Eqn (3.93) reduces to the simple straight-line fit model (Eqn 3.81). In this case, the matrices for N observations become

$$\mathbf{Y} = \begin{pmatrix} y_1 \\ y_2 \\ \ldots \\ \ldots \\ y_N \end{pmatrix}, \quad \mathbf{X} = \begin{pmatrix} x_{10} & x_{11} & \cdots & x_{1N} \\ x_{20} & x_{21} & \cdots & x_{2N} \\ \cdots & \cdots & \cdots & \cdots \\ \cdots & \cdots & \cdots & \cdots \\ x_{N0} & x_{N1} & \cdots & x_{NN} \end{pmatrix}$$

$$\mathbf{B} = \begin{pmatrix} b_0 \\ b_1 \end{pmatrix}, \quad \mathbf{E} = \begin{pmatrix} \varepsilon_1 \\ \varepsilon_2 \\ \ldots \\ \ldots \\ \varepsilon_N \end{pmatrix} \tag{3.94}$$

Using these N observations in Eqn (3.93), the least squares equations are

$$
Nb_0 + b_1 \sum_{i=1}^{N} x_i = \sum_{i=1}^{N} y_i
$$

$$
b_0 \sum_{i=1}^{N} x_i + b_1 \sum_{i=1}^{N} x_i^2 = \sum_{i=1}^{N} x_i y_i
$$

(3.95)

which we can solve for b_0 and b_1. We can generalize the procedure further by realizing that for $x_{i0} = 1$ in Eqn (3.91), we have

$$
\mathbf{X}' \cdot \mathbf{X} = \begin{pmatrix} 1 & \cdots & \cdots & 1 \\ \cdots & \cdots & \cdots & \cdots \\ x_1 & \cdots & \cdots & x_N \end{pmatrix} \begin{pmatrix} 1 & x_i \\ \cdots & \cdots \\ 1 & x_N \end{pmatrix} = \begin{pmatrix} N & \sum x_i \\ \cdots & \cdots \\ \sum x_i & \sum x_i^2 \end{pmatrix}
$$

(3.96)

where $\mathbf{X}'$ is the transpose of the matrix $\mathbf{X}$ and, the sums are from 1 to N, and

$$
\mathbf{X}' \cdot \mathbf{Y} = \begin{pmatrix} \sum_{i=1}^{N} y_i \\ \sum_{i=1}^{N} x_i y_i \end{pmatrix}
$$

(3.97)

The least squares equations can then be expressed as

$$
\left(\mathbf{X}' \cdot \mathbf{X} \right) \cdot \mathbf{B} = \mathbf{X}' \cdot \mathbf{Y}
$$

(3.98)

where

$$
\mathbf{B} = \begin{pmatrix} b_0 \\ b_1 \end{pmatrix}
$$

(3.99)

Solving the above equations for $\mathbf{B}$, we obtain

$$
\mathbf{B} = \left(\mathbf{X}' \cdot \mathbf{X} \right)^{-1} \left(\mathbf{X}' \cdot \mathbf{Y} \right)
$$

(3.100)

3.12.4 A Computational Example of Matrix Regression

Because linear regression is widely used in oceanography, we will illustrate its use by a simple example. Suppose we want to fit a line to the data pairs consisting of the independent variable x_i and the dependent variable y_i given in Table 3.8. From these we find

$$
\sum_{i=1}^{N} x_i = 0, \ \sum_{i=1}^{N} y_i = 5, \ \sum_{i=1}^{N} x_i y_i = 7, \ \sum_{i=1}^{N} x_i^2 = 10
$$

Substituting into Eqns (3.85a) and (3.85b), we have

$$
b_1 = \frac{\left[N \sum_{i=1}^{N} x_i y_i - \sum_{i=1}^{N} x_i \sum_{i=1}^{N} y_i \right]}{\left[N \sum_{i=1}^{N} x_i^2 - \left(\sum_{i=1}^{N} x_i \right)^2 \right]}
$$

$$
= \frac{[(5)(7) - (0)(5)]}{[(5)(10) - 0]} = 0.7
$$

$$
b_0 = \bar{y} - b_1 \bar{x} = 5/5 - (0.7)(0) = 1
$$

TABLE 3.8 Data values used in least squares linear fit of a two-coefficient regression model, $y_i = F(x_i)$.

Data		Solution values	
x_i	y_i	$(x_i)(y_i)$	x_i^2
−2	0	0	4
−1	0	0	1
0	1	0	0
1	1	1	1
2	3	6	4

This same problem can be put in matrix form (see the previous section)

$$
\mathbf{Y} = \begin{pmatrix} 0 \\ 0 \\ 1 \\ 1 \\ 3 \end{pmatrix}, \mathbf{X} = \begin{pmatrix} 1 & -2 \\ 1 & -1 \\ 1 & 0 \\ 1 & 1 \\ 1 & 2 \end{pmatrix}
$$

$$
\mathbf{X}' \cdot \mathbf{X} = \begin{pmatrix} 5 & 0 \\ 0 & 10 \end{pmatrix}, \mathbf{X}' \cdot \mathbf{Y} = \begin{pmatrix} 5 \\ 7 \end{pmatrix}, (\mathbf{X}' \cdot \mathbf{X})^{-1} = \begin{pmatrix} \frac{1}{5} & 0 \\ 0 & \frac{1}{10} \end{pmatrix}
$$

$$
\mathbf{B} = (\mathbf{X}' \cdot \mathbf{X})^{-1}(\mathbf{X}' \cdot \mathbf{Y}) = \begin{pmatrix} \frac{1}{5} & 0 \\ 0 & \frac{1}{10} \end{pmatrix} \begin{pmatrix} 5 \\ 7 \end{pmatrix} = \begin{pmatrix} 1 \\ 0.7 \end{pmatrix}
$$

so that by Eqn (3.99), $b_0 = 1$ and $b_1 = 0.7$.

An important property of the simple straight-line least-square estimators we have just derived is that b_0 and b_1 are unbiased estimates of their true parameter values. We have assumed that $E[\varepsilon] = 0$ and that $V[\varepsilon] = \sigma^2$; thus, the error variance is independent of x and $V[Y] = V[\varepsilon] = \sigma^2$. Since σ^2 is usually unknown, we estimate it using the sample variance (Eqns 3.5a and 3.5b) given by

$$
s^2 = \frac{1}{N-1} \sum_{i=1}^{N} (y_i - \bar{y})^2 \tag{3.101}
$$

However, if we use the output values, $\hat{y}_i$, from the least squares to estimate $\varepsilon_i(Y) = y_i - \hat{y}$, we must write Eqn (3.101) as

$$
s^2 = \frac{1}{N-2} \sum_{i=1}^{N} (y_i - \hat{y}_i)^2 = \frac{1}{N-2} \text{SSE} \tag{3.102}
$$

where SSE, given by Eqn (3.82a), represents the sum of the squares of the errors and the $N-2$ corresponds to the fact that two parameters, b_0 and b_1, are needed in the model. In matrix notation we can write the SSE as

$$
\text{SSE} = \mathbf{Y}' \cdot \mathbf{Y} - (\mathbf{B}' \cdot \mathbf{X}') \cdot \mathbf{Y} \tag{3.103}
$$

Using this with the previous numerical example, we write Eqn (3.103) as

$$(0 \quad 0 \quad 1 \quad 1 \quad 3) \begin{pmatrix} 0 \\ 0 \\ 1 \\ 1 \\ 3 \end{pmatrix} - (1 \quad 0.7) \begin{pmatrix} 1 & 1 & 1 & 1 & 1 \\ -2 & -1 & 0 & 1 & 2 \end{pmatrix} \begin{pmatrix} 0 \\ 0 \\ 1 \\ 1 \\ 3 \end{pmatrix}$$

$$= 11 - (1 \quad 0.7) \begin{pmatrix} 5 \\ 7 \end{pmatrix} = 11 - 9.9 = 1.1$$

Because $s^2 = \text{SSE}/(N - 2)$, we have $s^2 = 1.1/(3) = 0.367$ as our estimator of σ^2.

3.12.5 Polynomial Curve Fitting with Least Squares

The use of least-squares fitting is not limited to the straight-line regression model discussed thus far. In general, we can write our linear model as any polynomial of the form

$$Y = b_0 + b_1 x + b_2 x^2 + \ldots + b_N x^N + \varepsilon \tag{3.104}$$

The procedure is the same as with the straight-line case except that now the **X** matrix has $N + 1$ columns. Thus, the least-squares fit will have $N + 1$ linear equations with $N + 1$ unknowns, $b_0, b_1, \ldots, b_N$. These equations are called the *normal equations*.

3.12.6 Relationship Between Least Squares and Maximum Likelihood

As discussed earlier, the maximum likelihood estimator is one that maximizes the likelihood of sampling a given parameter. In general, if we have a sample x_i from a population with the PDF $f(x_i, \theta)$, where θ is the parameter of interest, the maximum likelihood estimator $L(\theta)$ is the product of the individual independent probabilities

$$L(\theta) = f(x_1, \theta) f(x_2, \theta) \ldots f(x_N, \theta) \tag{3.105}$$

If the errors all come from a normal distribution, this becomes from Eqn (3.70)

$$L(\theta) = \frac{\exp\left[-\sum_{i=1}^{N} \frac{(x_i - \theta)^2}{2\sigma^2} \right]}{\sigma^N (2\pi)^{\frac{N}{2}}} \tag{3.106}$$

When Eqn (3.70) is maximized, it leads to the least-squares estimate

$$\widehat{\theta} = \frac{1}{N} \sum_{i=1}^{N} x_i = \bar{x}$$

In other words, the least-squares estimate of the mean of θ can be derived from a normal distribution using the maximum likelihood criterion. This value is found to be the average of the independent variable x.

3.13 RELATIONSHIP BETWEEN REGRESSION AND CORRELATION

The subject of correlation will be considered in more detail when we examine time series analysis methods. Our intension, here, is simply to introduce the concept in general statistical terms and relate it to the simple regression model just discussed. As with regression, correlation relates two variables but unlike regression it is measured without estimation of the population regression line.

The *correlation coefficient, r* (also commonly denoted as R), is a way of determining how well two (or more) variables covary in time or space. (Note: the correlation coefficient is not to be confused with the *correlation function*, ρ, discussed later in this chapter.) For two random variables x ($x_1, x_2, \ldots, x_N$) and y ($y_1, y_2, \ldots, y_N$) the correlation coefficient, r, can be written as

$$r = \frac{1}{N-1} \frac{\sum\limits_{i=1}^{N}(x_i - \bar{x})(y_i - \bar{y})}{s_x s_y} = \frac{C_{xy}}{s_x s_y} \tag{3.107}$$

where

$$C_{xy} = \frac{1}{N-1} \sum_{i=1}^{N}[(x_i - \bar{x})(y_i - \bar{y})] \tag{3.108}$$

is the *covariance* of x and y, and s_x and s_y are the standard deviations for the two data records as defined by Eqns (3.5a) and (3.5b). We note two important properties of r:

1. r is a dimensionless quantity since the units of the numerator and the denominator are the same. In contrast, the covariance function has the units of x and y;
2. the value of r lies between -1 and $+1$ because it is normalized by the product of the standard deviations of both variables.

For $r = \pm 1$, the data points (x_i, y_i) lie along a straight line and the samples are said to have a perfect correlation (plus (+) for "in-phase" fluctuations and minus ($-$) for $180°$ "out-of-phase" fluctuations). For $r \approx 0$, the points are scattered randomly on the graph and there is little or no relationship between the variables. The variables x_i, y_i in Eqns (3.107) and (3.108) could be samples from two different, independent random variables or they could represent the independent (input) and dependent (output) variables of an estimation model. Alternatively, they could be samples from the same variable. Known as an *autocorrelation*, the latter is usually computed for increasing lag or shifts in the starting value for one of the time series. A lag of "m" means that the first m values of one of the series, say the x series, are removed prior to the calculation so that x_{m+1} becomes the new x_1, and so on.

Some authors prefer to use r^2 (the coefficient of determination discussed in Section 3.12.1 in the context of straight-line regression) rather than r (the correlation coefficient) because the squared value can be used to construct a significance level for r^2 in terms of a hypothesis test when the true correlation squared is zero. Writing

$$C_{xy}^2 / \left(s_x s_y\right)^2 = \text{SSR} / \text{SST} = r^2 \tag{3.109}$$

we see that r^2 = variance explained/total variance, as stated earlier. A value $r = 0.75$ means that a linear regression of y on x explains $r^2(\%) = 56.25\%$ of the total sample variance. Our approach is to use r to obtain the sign of the correlation and to use r^2 to estimate the joint variances. It is worth noting that a moderate value $r = 0.25$, for example, might seem significant (perhaps when comparing a current velocity record against a coincident wind stress record) until it is realized that it means that the variance in the wind stress only accounts for roughly $r^2(\%) = 6.25\%$ of the variance in current velocity.

3.13.1 The Effects of Random Errors on Correlation

Before discussing the relationship between r and our simple regression model, it is important to realize that sampling errors in x_i and y_i can only cause r to decrease. This can be shown by writing our two variables as a combination of true values (α_i, β_i) and random errors (δ_i, ε_i). In particular

$$\begin{aligned} x_i &= \alpha_i + \delta_i \\ y_i &= \beta_i + \varepsilon_i \end{aligned} \tag{3.110}$$

Using Eqns (3.107) and (3.109), we can write the correlation between x_i and y_i as

$$r_{xy} = \frac{s_\alpha s_\beta r_{\alpha\beta} + s_\beta s_\delta r_{\beta\delta} + s_\alpha s_\varepsilon r_{\alpha\varepsilon} + s_\delta s_\varepsilon r_{\delta\varepsilon}}{s_x s_y} \tag{3.111}$$

where for convenience we have dropped the index i. Because the random errors δ and ε are assumed to be independent of each other and of the variables α and β, we know that

$$r_{\beta\delta} = r_{\alpha\varepsilon} = r_{\delta\varepsilon} = 0$$

so that Eqn (3.111) becomes

$$r_{xy} = \frac{s_\alpha s_\beta}{s_x s_y} r_{\alpha\beta} \tag{3.112}$$

This result means that the ratio between the product of the true standard deviations (s_α, s_β) to the product of the measured variable (s_x, s_y) determines the magnitude of the computed correlation coefficient (r_{xy}) relative to the true value ($r_{\alpha\beta}$).

To determine Eqn (3.112), we expand the variances of x and y as

$$\left(s_x^2, s_y^2 \right) = \frac{1}{N-1} \sum_{i=1}^{N} \left[(x_i - \bar{x})^2, (y_i - \bar{y})^2 \right]$$

where, as usual, $\bar{x}, \bar{y}$ are the average values for samples x_i, y_i, respectively. Expanding the numerator into its component terms through Eqn (3.109), and using the fact that the errors are independent of one another, and of x and y, yields

$$\sum_{i=1}^{N} (x_i - \bar{x})^2 = \sum_{i=1}^{N} \left[(\alpha_i - \bar{\alpha})^2 + \delta_i^2 \right]$$

$$\sum_{i=1}^{N} (y_i - \bar{y})^2 = \sum_{i=1}^{N} \left[(\beta_i - \bar{\beta})^2 + \varepsilon_i^2 \right]$$

Dividing through by $(N - 1)$ and using the definitions for standard deviation, we find

$$s_x^2 = s_\alpha^2 + \frac{\sum_{i=1}^{N} \delta_i^2}{N-1}; \quad s_y^2 = s_\beta^2 + \frac{\sum_{i=1}^{N} \varepsilon_i^2}{N-1} \tag{3.113}$$

Because the second terms in each of the above expressions can never be negative ($N > 1$), the observed variances s_x^2 and s_y^2 are always greater than the corresponding true variances. Applying this result to Eqn (3.112), we see that the calculated correlation, r_{xy}, derived from the observations is always smaller than the true correlation, $r_{\alpha\beta}$. Because of random errors, the correlation coefficient computed from the observations will be smaller than (or, at best, equal to) the true correlation coefficient.

3.13.2 The Maximum Likelihood Correlation Estimator

Returning to the relationship between correlation and regression, we note the maximum likelihood estimator of the correlation coefficient is, by Eqn (3.107)

$$r = \left[\sum_{i=1}^{N} (x_i - \bar{x})(y_i - \bar{y}) \right] \bigg/ \left[\sum_{i=1}^{N} (x_i - \bar{x})^2 \sum_{i=1}^{N} (y_i - \bar{y})^2 \right]^{1/2} \tag{3.114}$$

for a bivariate normal population (x_i, y_i). We can expand this using Eqn (3.108) to derive

$$r = \frac{N \sum_{i=1}^{N} (x_i y_i) - \left(\sum_{i=1}^{N} x_i \right) \left(\sum_{i=1}^{N} y_i \right)}{\left\{ \left[N \sum_{i=1}^{N} x_i^2 - \left(\sum_{i=1}^{N} x_i \right)^2 \right] \left[N \sum_{i=1}^{N} y_i^2 - \left(\sum_{i=1}^{N} y_i \right)^2 \right] \right\}^{1/2}} \tag{3.115}$$

Note that the numerator in Eqn (3.115) is similar to the numerator of the estimator for b_1 in Eqn (3.85a).

For the case where the regression line passes through the origin in Eqn (3.85b), we have $b_1 = 0$ and our model is

$$\hat{y}_i = \hat{b}_1 x_i$$

and we can rewrite Eqn (3.85a) as

$$\widehat{b}_1 = \frac{\sum\limits_{i=1}^{N}(x_iy_i)}{\sum\limits_{i=1}^{N}x_i^2} = \frac{rs_y}{s_x}; \text{ or, } r = \frac{\widehat{b}_1 s_x}{s_y} \tag{3.116}$$

Thus, r can be computed from $\widehat{b}_1$ and vice versa if the standard deviations of the sample variance x and y are known. Also, using the relationship between $\widehat{b}_1$ and r we can write the variance of the parameter estimate in Eqn (3.116) as

$$s^2 = \frac{1}{N-2}\sum\limits_{i=1}^{N}(y_i - \overline{y})^2 = \frac{1}{N-2}\text{SSE} \tag{3.117}$$

We can use this result to better understand the relationship between correlation and regression by writing the ratio of the regression variance in Eqn (3.117) to the sample variance for y alone; for large N, this becomes

$$\frac{s^2}{s_y^2} = \frac{(N-1)(1-r^2)}{N-2} \approx \left(1-r^2\right) \tag{3.118}$$

Thus, for N large, r^2 is that portion of the variance of y that can be attributed to its regression on x while $(1 - r^2)$ is that portion of y's variance that is independent of x. Earlier it was noted that a computationally efficient way to calculate the variance was to use Eqn (3.5b) which required only a single pass through the data sample. A similar saving can be gained in computing the covariance by expanding the product

$$\sum\limits_{i=1}^{N}(x_i - \overline{x})(y_i - \overline{y}) = \sum\limits_{i=1}^{N}(x_iy_i) - \frac{1}{N}\left(\sum\limits_{i=1}^{N}x_i\right)\left(\sum\limits_{i=1}^{N}y_i\right) \tag{3.119}$$

3.13.3 Correlation and Regression: Cause and Effect

A point worth stressing is that a high correlation coefficient or a "good" fit of a regression curve $y = y(x)$ to a set of observations x, does not imply that x is "causing" y. Nor does it imply that x will provide a good predictor for y in the future. For example, the number of sockeye salmon returning to the Fraser River of British Columbia each fall from the North Pacific Ocean is often highly correlated with the mean fall sea surface temperature (SST) at Amphitrite Point on the southwest coast of Vancouver Island. No one believes that the fish are responding directly to the temperature at this point, but rather that temperature is a proxy variable for the real factor (or combination of factors) influencing the homeward migration of the fish. Of course, we are not saying that one should not draw inferences or conclusions from correlation or regression analysis (for example, SSTs often have large spatial and temporal correlation scales so that temperature may, indeed, be the main driving variable) but only that caution is advised when seeking cause-and-effect relationships between variables. We further remark that there is little point in drawing any type of line through the data unless the scatter about the line is appreciably less than the overall spread of the observations. There is a tendency to fit trend lines to data with large variability and scatter even if a trend is not justified on statistical grounds. If $|r| < 0.5$, it may be unreasonable to fit a line for predictive purposes.

There is another important aspect of regression-correlation analysis that is worth stressing: although the value of the correlation coefficient or coefficient of determination does *not* depend on which variable (x or y) is designated as the independent variable and which is designated as the dependent variable, this distinction *is* very important when it comes to regression analysis. The regression coefficients a, b for the conditional distribution of y given x ($y = a_1 + b_1 x$) are different than those for the conditional distribution of x given y ($x = a_2 + b_2 y$). In general, $a_1 \neq -a_2/b_2$ and $b_1 \neq 1/b_2$ and so that the regression lines are different. In the first case, we are solving for the line shown in Figure 3.11a, while in the second case we are solving for the line in Figure 3.11b.

As an example, consider the broken lines in Figure 3.11c which show the two different linear regression lines for the regression of the observed cross-channel sea-level differences $y = \Delta\eta_c$, as measured by coastal tide gauges, and the calculated cross-channel sea-level difference $x = \Delta\eta_m$ obtained using concurrent current meter data from cross-channel moorings. The term $\Delta\eta_c$ is simply the difference in the mean sea level from one side of the 25-km-wide channel to the other, while $\Delta\eta_m$ is calculated from the current meter records assuming that the time-averaged along-channel flow is in geostrophic balance (Labrecque et al., 1994). The dotted line is the regression $\Delta\eta_c = a_1 + b_1\Delta\eta_m$, while the dashed line is the regression $\Delta\eta_m = a_2 + b_2\Delta\eta_c$, with $b_1 \neq b_2$. The correlation coefficient $r = 0.69$ is the same for the two regressions. The solid line in Figure 3.11c is the so-called *neutral* regression line for the two parameters (Garrett and Petrie, 1981) and might seem the line of choice since it is not obvious which parameter should be the independent parameter and which should be the dependent parameter. Neutral regression is equivalent to minimizing the sum of the square distances from the regression line (Figure 3.11d).

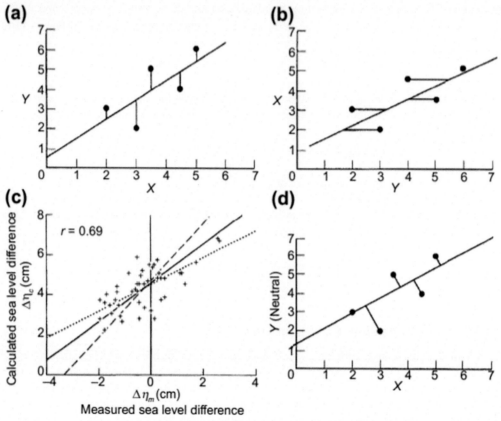

FIGURE 3.11   Straight-line regressions (a) y on x, and (b) x on y showing the "direction" along which the variance is minimized. (c) Scatter plot of $\Delta\eta_c$ versus $\Delta\eta_m$ for a cross-section of the 22-km-wide Juan de Fuca Strait separating Vancouver Island from Washington State. Plots give the regression of the observed cross-channel sea level differences $y = \Delta\eta_c$, as measured by coastal tide gauges, and the calculated cross-channel sea-level difference, $x = \Delta\eta_m$, obtained using concurrent current meter data from cross-channel moorings. The solid slopping line in plot (c) gives the bisector regression fit to the data (slope and 95% confidence level = 0.96 ± 0.37); the dotted line (slope = 0.66 ± 0.14) and the dashed line (slope = 1.40 ± 0.32) are the standard slopes for $\Delta\eta_c$ versus $\Delta\eta_m$ and $\Delta\eta_m$ versus $\Delta\eta_c$, respectively. Here, $r = 0.69$. (d) The "direction" along which the variance for the data points in (a) and (b) is minimized. *(c) is From Labrecque et al. (1994).*

In fisheries research, neutral regression is known as *geometric mean functional regression* (GMFR) and is commonly used to relate fish body proportions when there is no clear basis to select dependent and independent variables (Sprent and Dolby, 1980). For two variables with zero means, the slope estimator, b, is given by the square roots of the variance ratios

$$b_{yx} = \text{sgn}(s_{xy})\left[\frac{\sum_{i=1}^{N}(y_i - \bar{y})^2}{\sum_{i=1}^{N}(x_i - \bar{x})^2}\right]^{1/2} \quad ; \text{ regression } \widehat{y}_i = \widehat{b}_{yx}x_i$$

$$(3.120)$$

$$b_{xy} = \text{sgn}(s_{xy})\left[\frac{\sum_{i=1}^{N}(x_i - \bar{x})^2}{\sum_{i=1}^{N}(y_i - \bar{y})^2}\right]^{1/2} \quad ; \text{ regression } \widehat{x}_i = \widehat{b}_{xy}y_i$$

where $\text{sgn}(s_{xy})$ is the sign of the covariance function $s_{xy} = \sum(x_i - \bar{x})(y_i - \bar{y})$ and $b_{yx} = 1/b_{xy}$ as required ($\text{sgn}x = +1$ for $x > 0$, $= -1$ for $x < 0$). Note that the slope b_{yx} lies midway between the slopes b_1 and b_2

$$
b_1 = \frac{\sum\limits_{i=1}^{N}(x_i - \bar{x})(y_i - \bar{y})}{\sum\limits_{i=1}^{N}(x_i - \bar{x})^2}; \quad \text{regression line } \widehat{y}_i = a_1 + \widehat{b}_1 x_i
$$

$$
b_2 = \frac{\sum\limits_{i=1}^{N}(x_i - \bar{x})(y_i - \bar{y})}{\sum\limits_{i=1}^{N}(y_i - \bar{y})^2}; \quad \text{regression line } \widehat{x}_i = a_2 + \widehat{b}_2 y_i
$$

(3.121)

given by Eqn (3.85a) for standard regression analyses (Figure 3.11a). The GMFR is then the geometric mean slope of the least-squares regression coefficient for the regression slope of y on x and the regression of x on y; $b_{yx} = (b_1/b_2)^{1/2}$. Because the slope from the GMFR is simply a ratio of variances, it is "transparent" to the determination of correlation coefficients or coefficients of determination. It is these correlations, not the slope of the line, that test the strength of the linear relationship between the two variables. Moreover, none of the standard linear regression models reduces to the GMFR slope estimate except under unlikely circumstances. According to Sprent and Dolby (1980), *ad hoc* use of the GMFR is not recommended when there are errors in both variables. The GMFR model, though appealing, rests on shaky statistical ground and its use remains controversial.

3.14 HYPOTHESIS TESTING

Statistical inference takes one of two forms. Either we make estimates of population variables, as we have done thus far, or we test hypotheses about the implications of these variables. Statistical inference, in which the investigator chooses between two conflicting hypotheses about the value of a particular population variable, is known as *hypothesis testing*.

Hypothesis testing follows scientific methodology from whose nomenclature the terms are borrowed. The investigator forms a "hypothesis," collects some sample data and uses a statistical construct to either reject or accept the original hypothesis. The basic elements of a statistical test are: (1) the *null hypothesis*, H_o (the hypothesis to be tested); (2) the alternate hypothesis, H_a; (3) the test statistic to be used; and (4) the region of rejection of the hypothesis. The active components of a statistical test are the test statistic and the associated rejection region, with the latter specifying the values of the test statistic for which the null hypothesis is rejected. We emphasize the point that "pure" hypothesis testing originated from early work in which the null hypothesis corresponded to an idea or theory about a population variable that the scientist hoped *would be rejected*. "Null" in this case means incorrect and invalid so that we could call it the "invalid hypothesis." In other words, the null hypothesis specified those values of the population variable, which it was thought did *not* represent the true value of the variable. (We touched briefly on the null hypothesis in Section 3.8.4 when discussing confidence intervals.) This is a form of negative thinking and is the reason that many of us would rather think in terms of the *alternate hypothesis* in which we specify those values of the variable that we hope will hold true (the "valid" hypothesis). Oceanographers are generally a positive group, so if one has a hypothesis, research is typically conducted in order to prove the hypothesis is true not false. Regardless of which hypothesis is chosen, it is important to remember that the true population value under consideration must either lie in the test set covered by H_o or in the set covered by H_a. There are no other choices.

We restrict consideration of hypothesis testing to large samples ($N > 30$). In hypothesis testing, two types of errors are possible. In a Type-1 error, the null hypothesis H_o is rejected when it is true. The probability of this type of error is denoted by α. Type-2 errors occur when H_o is accepted when it is false (i.e., H_a is true). The probability of Type-2 errors is written as β. In Table 3.9, the probability P (accept $H_o | H_o$ is true) $= 1 - \alpha$ corresponds to the $100(1 - \alpha)\%$ confidence interval. Alternatively, the probability $P(\text{reject } H_o | H_o \text{ is false}) = 1 - \beta$ is the power of the statistical test since it indicates the ability of the test to determine when the null hypothesis is false and H_o should be rejected.

For a parameter θ based on a random sample $x_1, x_2, \ldots, x_N$, we want to test various values of θ using the estimate $\widehat{\theta}$ as a test statistic. This estimator is assumed to have an approximately normal sampling distribution. For a specified value of $\widehat{\theta}$ ($= \theta_0$), we want to test the hypothesis, H_o, that $\widehat{\theta}$ ($= \theta_0$) (written H_o: $\theta = \theta_0$) with the alternate hypothesis, H_a, that $\widehat{\theta} > \theta_0$ (written H_a: $\theta > \theta_0$). An efficient test statistic for our assumed normal distribution is the standard normal Z defined as

TABLE 3.9 The four possible decision outcomes in hypothesis testing and the probability of each decision outcome in a test hypothesis.

		Possible situation	
Action	Accept H$_o$	H$_o$ is true Correct confidence level $1 - \alpha$	H$_o$ is false Incorrect decision; (Type-2 error); β
	Reject H$_o$	Incorrect decision (Type-1 error); α	Correct decision; power of the test $1 - \beta$
	Sum	1.00	1.00

$$Z = \frac{\left(\widehat{\theta} - \theta\right)}{\widehat{\sigma}_{\widehat{\theta}}} \tag{3.122}$$

where $\widehat{\sigma}_{\widehat{\theta}}$ is the standard deviation of the approximately normal sampling distribution of $\widehat{\theta}$, which can then be computed from the sample. For this test statistic, the null hypothesis (H$_0$: $\theta = \theta_0$) is rejected for $Z > Z_\alpha$ where α is the probability of a Type-1 error. Graphically, this rejection region is depicted as the shaded portion in Figure 3.12a, which is called an "upper-tail" test. Similarly, a "lower-tail" test would have the shaded rejection region starting at $-Z_\alpha$ and corresponds to $Z < -Z_\alpha$ and $\theta < \theta_0$ (Figure 3.12b). A two-tailed test (Figure 3.12c) is one for which the null hypothesis rejection region is $|Z| > Z_{\alpha/2}$ and $\theta \neq \theta_0$. The decision of which test alternative to use should be based on the form of the alternate hypothesis. If one is interested in parameter values greater than θ_0, an upper-tail test is used; for values less than θ_0, a lower-tail test is appropriate. If one is interested in any change from θ_0, it is best to use a two-tailed test. The following is an example for which a two-tailed test is appropriate.

Suppose that daily averaged currents for some mooring locations are available for the same month from two different years (e.g., January 2022 and January 2023). We wish to test the hypothesis that the monthly means of the alongshore component of the flow, V, for these two different years are the same. If the daily averages are computed from hourly observations, we invoke the central limit theorem and conclude that our sampling distributions are normally distributed. Taking each month as having 31 days, we satisfy the condition of a large sample ($N > 30$) and can use the procedure outlined above. Suppose we observe that for January 2022 the mean and standard deviation of the observed current is $V_{2022} = 23 \pm 3$ cm/s while for January 2023 we find a monthly mean speed $V_{2023} = 20 \pm 2$ cm/s (here, the standard deviations are obtained from the signal variances). We now wish to test the null hypothesis that the true (as opposed to our sampled) monthly mean current speeds were statistically the same for the two separate years. We use the two-tailed test to detect any deviations from equality. In this example, the *point estimator* used to detect any difference between the monthly mean records calculated from daily observed values is the sample mean difference, $\widehat{\theta} - \theta_0 = V_{2022} - V_{2023}$. Our test statistic Eqn (3.122) is

$$Z = \frac{(V_{2022} - V_{2023})}{\left[s^2_{2022}/N_{2022} + s^2_{2023}/N_{2023}\right]^{1/2}}$$

which yields

$$Z = \frac{(23 - 20)}{[9/31 + 4/31]^{1/2}} = 4.63$$

To determine if the above result falls in the rejection region, $Z > Z_\alpha$, we need to select the significance level α for Type-1 errors. For the 95% significance level, $\alpha = 0.05$ and $\alpha/2 = 0.025$. From the standard normal table (Appendix D, Table D.1) $Z_{0.025} = 1.96$, where the range $(-1.96, +1.96)$ encompasses 95% of the values of the area of the Z normal distribution. Our test value $Z = 4.63$ is greater than 1.96 so that it falls within the rejection region, and we must reject the hypothesis that the monthly mean current speeds are the same for both years. In most oceanographic applications hypothesis testing is limited to the null hypothesis and thus Type-1 errors are most appropriate. We will not consider here the implementation of Type-2 errors that lead to the acceptance of an alternate hypothesis as described in Table 3.9.

(a)

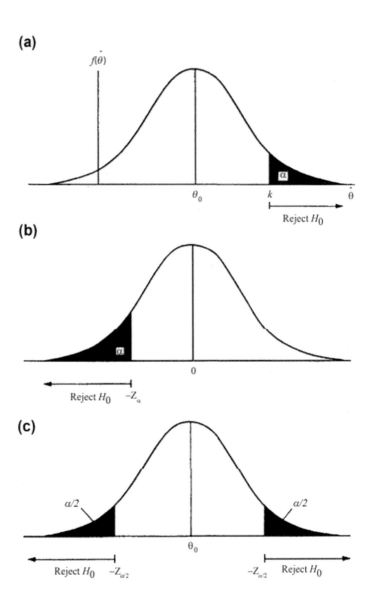

(b)

(c)

Turning again to satellite altimetry for an example, we note that the altimeter height bias, H_{bias}, discussed earlier in Section 3.10, is one of the error sources that contributes to the overall error "budget" of altimetric height measurements. Suppose that we wish to know if the *overall* height error H_T in the absence of the bias error, H_{bias}, is less than some specified amount, H_ε. We first set up the null hypothesis (H_0: $H_T - H_{bias} < H_\varepsilon$) that the overall height error in the absence of any bias is less than H_ε. At this point, we must also select a significance level for our test. A significance level of $1 - \alpha$ means that we do not want to make a mistake and reject the null hypothesis more than $\alpha(100)\%$ of the time. We begin by defining our hypothesis limit, H_T, as

$$H_T = H_\varepsilon + \frac{Z_\alpha s_b}{\sqrt{N}} \tag{3.123}$$

where the standard normal distribution Z_α for the bias error is given by Eqn (3.122) and s_b is the standard error (uncertainty) in our measurements. If the mean error of our measurements is greater than $H_T - H_{bias}$, then we reject H_0 and conclude that the height error in the absence of any bias error is greater than H_ε with a probability α of being wrong.

Suppose we set $H_e = 13$ cm and consider $N = 9$ consecutive statistically independent satellite measurements in which each measurement is assumed to have an uncertainty of $s_b = 3$ cm. If the observed height error is $H_T = 15$ cm, do we accept or reject the null hypothesis for the probability level $\alpha = 0.10$? What about the cases for $\alpha = 0.05$ and $\alpha = 0.01$? Given our hypothesis limit $H_e = 13$ cm and the fact that $N = 9$ and $s_b = 3$ cm (so, $s_b/\sqrt{N} = 1$ cm), we can write Eqn (3.123) as $H_T = 13 + Z_\alpha$ cm. According to the results of Table 3.10, this means that we can accept the null hypothesis that the overall error is less than 13 cm at the 5% and 10% probability levels but not at the 1% probability level (corresponding to the 95%, 90%, and 99% significance levels, respectively).

3.14.1 Significance Levels and Confidence Intervals for Correlation

One useful application of null hypothesis testing is the development of significance levels for the correlation coefficient, r. If we take the null hypothesis as $r = r_o$, where r_o is some estimate of the correlation coefficient, we can determine the rejection region in terms of r for a chosen significance level, α, for different degrees of freedom $(N - 2)$. A list of such values is given in the table in Appendix E. In that table, the correlation coefficient r for the 95% and 99% significance levels (also called the 5% and 1% levels depending on whether or not one is judging a population parameter or testing a hypothesis) are presented as functions of the number of degrees of freedom.

For example, the differences among a sample of 20 pairs of (x, y) values with a correlation coefficient, r, less than 0.444 and $N - 2 = 18$ degrees of freedom would not be significantly different from zero at the 95% confidence level. It is interesting to note that, because of the close relationship between r and the regression coefficient b_1 of these pairs of values, we could have developed the table for r values using a test of the null hypothesis for b_1.

The procedure for finding confidence intervals for the correlation coefficient r is to first transform it into the standard normal variable Z_r as

$$Z_r = \frac{1}{2}[\ln(1+r) - \ln(1-r)] \tag{3.124}$$

which has the standard error

$$\sigma_z = \frac{1}{(N-3)^{1/2}} \tag{3.125}$$

independent of the value of the correlation. The appropriate confidence interval is then

$$Z_r - Z_{\alpha/2}\sigma_z < Z < Z_r + Z_{\alpha/2}\sigma_z \tag{3.126}$$

which can be transformed back into values of r using Eqn (3.124).

Before leaving the subject of correlations we want to stress that correlations are merely statistical constructs and, while we have some mathematical guidelines as to the statistical reliability of these values, we cannot replace common sense and physical insight with our statistical calculations. It is entirely possible that our statistics will deceive us if we do not apply them carefully. We again emphasize that a high correlation can reveal either a close relationship between two variables or their simultaneous dependence on a third variable. It is also possible that a high correlation may be due to complete coincidence and have no causal relationship behind it. The basic question that needs to be asked is "does it make sense?" A classic example (Snedecor and Cochran, 1967) is the high negative correlation (-0.98) between the annual birthrate in Great Britain and the annual production of pig iron in the United States for the years 1875–1920. This high correlation is statistically significant for the available $N - 2 = 43$ degrees of freedom, but the likelihood of a direct relationship between these two variables is very low.

TABLE 3.10 Test results for the null hypothesis that the overall error H_T of satellite altimetry data in the absence of a bias error (H_e) is less than 13 cm (assuming a normal error distribution).

Significance level, α	Standard normal distribution, Z_α	Total error height, H_T	Decision
0.01	2.575	15.575 cm	Reject H_o
0.05	1.960	14.960 cm	Accept H_o
0.10	1.645	14.645 cm	Accept H_o

3.14.2 Analysis of Variance and the *F*-Distribution

Most of the statistical tests we have presented to this point are designed to test for differences between two populations. In certain circumstances, we may wish to investigate the differences among three or more populations simultaneously rather than attempt the arduous task of examining all possible pairs. For example, we might want to compare the mean lifetimes of drifters sold by several different manufacturers to see if there is a difference in survivability for similar environmental conditions; or, we might want to look for significant differences among temperature or salinity data measured simultaneously during an intercomparison of several different commercially available CTDs. The *analysis of variance* (ANOVA) is a method for performing simultaneous tests on data sets drawn from different populations. In essence, ANOVA is a test between the amount of variation in the data that can be attributed to chance and that which can be attributed to specific causes and effects. If the amount of shared variability *between* samples is small relative to the shared variability *within* samples, then the null hypothesis H_o—that the variability occurred by chance—cannot be rejected. If, on the other hand, the ratio of these variations is sufficiently large, we can reject H_o. "Sufficiently large," in this case, is determined by the ratio of two continuous χ^2 probability distributions. This ratio is known as the *F-distribution*.

To examine this subject further, we need several definitions. Suppose we have samples from a total of *J* populations and that a given sample consists of N_j values. In ANOVA, the *J* samples are called *J* "treatments," a term that stems from early applications of the method to agricultural problems where soils were "treated" with different kinds of fertilizer and the statistical results compared. In the one-factor ANOVA model, the values y_{ij} for a particular treatment (input), x_j, differ from some common background value, μ, because of random effects; that is

$$y_{ij} = \mu + x_j + \varepsilon_{ij}; j = 1, 2, ..., J$$
$$i = 1, 2, ..., N_J$$

(3.127)

where the outcome y_{ij} is made up of a common (grand average) effect (μ), plus a treatment effect (x_j), and a random component, ε_{ij}. The grand mean, μ, and the treatment effects, x_j, are assumed to be constants while the errors, ε_{ij}, are independent, normally distributed, variables with zero mean and a common variance, σ^2, for all populations. The null hypothesis for this one-factor model is that the treatments have zero effect. That is, H_o: $x_j = 0$ ($j = 1, 2,..., J$) or, equivalently, H_o: $\mu_1 = \mu_2 = ... = \mu_J$ (i.e., there is no difference between the populations aside from that due to random errors). The alternative hypothesis is that some of the treatments have a nonzero effect. Note that "treatment" can refer to any basic parameter we wish to compare such as buoy design, power supply, or CTD manufacturer. To test the null hypotheses, we consider samples of size N_j from each of the *J* populations. For each of these samples, we calculate the mean value $\bar{y}_j (j = 1, 2,..., J)$. The grand mean for all the data is denoted as $\bar{y}$.

As an example, suppose we want to intercompare the temperature records from three types of CTDs placed in the same temperature bath under identical sampling conditions. Four countries take part in the intercomparison and each brings the same three types of CTD. The results of the test are reproduced in Table 3.11.

If H_o is true, $\mu_1 = \mu_2 = \mu_3$ and the measured differences between $\bar{y}_1, \bar{y}_2$, and $\bar{y}_3$ in Table 3.11 can be attributed to random processes.

The treatment effects for the CTD example are given by

$$x_1 = \bar{y}_1 - \bar{y} = -0.002°C$$
$$x_2 = \bar{y}_2 - \bar{y} = +0.001°C$$
$$x_3 = \bar{y}_3 - \bar{y} = 0.0°C$$

TABLE 3.11 Temperatures in °C measured by three makes of CTD in the same calibration tank.

Measurement (*i*)	CTD type 1 Sample *j* = 1	CTD type 2 Sample *j* = 2	CTD type 3 Sample *j* = 3
1	15.001	15.004	15.002
2	14.999	15.002	15.003
3	15.000	15.001	15.000
4	14.998	15.004	15.002
Mean $\bar{y}$(°C)	15.000 $= \bar{y}_1$	15.003 $= \bar{y}_2$	15.002 $= \bar{y}_3$

Four instruments of each type are used in the test. The grand mean for the data from all three instruments is $\bar{y} = 15.002°C$.

where $\bar{y} = (\bar{y}_1 + \bar{y}_2 + \bar{y}_3)/3$. The ANOVA test involves determining whether the estimated values of x_j are large enough to convince us that H_o is not true. Whenever H_o is true, we can expect that the variability between the J means is the same as the variability within each sample (the only source of variability is the random effects, ε_{ij}). However, if the treatment effects are not all zero, then the variability between samples should be larger than the variability within the samples.

The variation within the J samples is found by first summing the squared deviations of y_{ij} about the mean value $\bar{y}_j$ for each sample, namely

$$\sum_{i=1}^{N} (y_{ij} - \bar{y}_j)^2 ; j = 1, 2, 3$$

where N_j is the number of measurements in each sample. If we then sum this variation over all J samples, we obtain the *sum of squares within* (SSW).

$$\text{Sum of squares within: SSW} = \sum_{j=1}^{J} \left(\sum_{i=1}^{N} (y_{ij} - \bar{y}_j)^2 \right) \tag{3.128}$$

Note that the sample lengths, N_j, need not be the same since the summation for each sample uses only the mean for that particular sample. Next, we will need the amount of variation between the samples (SSB). This is obtained by taking the squared deviation of the mean of the Jth sample, $\bar{y}_j$, and the grand mean, $\bar{y}$. This deviation must then be weighted by the number of observations in the Jth sample. The overall sum is given by

$$\text{Sum of squares between SSB} = \sum_{j=1}^{J} N_j (\bar{y}_j - \bar{y})^2 \tag{3.129}$$

To compare the variability within samples to the variability between samples, we need to divide each sum by its respective number of degrees of freedom, just as we did with other variance expressions, such as s^2. For SSB, the degrees of freedom (DOF) = $J - 1$ while for SSW

$$\text{DOF} = \left(\sum_{j=1}^{J} N_j \right) - J$$

The MS values are then:

$$\text{Mean-square between MSB} = \frac{\text{SSB}}{J-1} \tag{3.130a}$$

$$\text{Mean-square within MSW} = \frac{\text{SSW}}{\left(\sum_{j=1}^{J} N_j \right) - J} \tag{3.130b}$$

In the above example, $J - 1 = 2$ and $\sum N_j - J = 9$. The calculated values of mean-square between (MSB) and mean-square within (MSW) for our CTD example are given in Table 3.12. Specifically,

$$\text{SSW} = \sum_{i=1}^{4} (y_{i1} - \bar{y}_1)^2 + \sum_{i=1}^{4} (y_{i2} - \bar{y}_2)^2 + \sum_{i=1}^{4} (y_{i3} - \bar{y}_3)^2$$

$$\text{SSB} = N_1 (\bar{y}_1 - \bar{y})^2 + N_2 (\bar{y}_2 - \bar{y})^2 + N_3 (\bar{y}_3 - \bar{y})^2 + N_4 (\bar{y}_4 - \bar{y})^2$$

where the total N_j (for $j = 1, ..., 4$) = 12. To determine if the ratio of MSB to MSW is large enough to reject the null hypothesis, we use the F-distribution for $J - 1$ and

$$\left(\sum_{j=1}^{J} N_j \right) - J$$

TABLE 3.12 Calculated values of sum of squares and mean-square values for the CTD temperature intercomparison. DOF denotes the number of degrees of freedom.

Type of variation	Sum of squares (°C²)	DOF	Mean-square (°C²)
Between samples (type of CTD)	20×10^{-6}	2	10×10^{-6}
Within samples (all CTDs)	18×10^{-6}	9	2×10^{-6}
Total	38×10^{-6}	11	(Ratio = 5.0)

degrees of freedom.

Named after R. A. Fisher, who first studied it in 1924, the F-distribution is defined in terms of the ratio of two independent χ^2 variables divided by their respective degrees of freedom. If X_1 is a χ^2 variable with ν_1 degrees of freedom and X_2 is another χ^2 variable with ν_2 degrees of freedom, then the random variable

$$F(\nu_1, \nu_2) = \frac{X_1/\nu_1}{X_2/\nu_2} \tag{3.131}$$

is a nonnegative chi-square variable with ν_1 degrees of freedom in the numerator and ν_2 degrees of freedom in the denominator. (The distribution is asymmetrical and can never be negative because it is the ratio of two squared terms.) If $J = 2$, as in the CTD example above, the F-test is equivalent to a one-sided t-test. There is no upper limit to F, which like the χ^2-distribution is skewed to the right. Tables are used to list the critical values of $P(F > F_\alpha)$ for selected degrees of freedom ν_1 and ν_2 for the two most commonly used significance levels, $\alpha = 0.05$ and $\alpha = 0.01$. In ANOVA, the values of SSB and SSW follow χ^2-distributions. Therefore, if we let $X_1 = $ SSB and $X_2 = $ SSW, then

$$F\left(J-1, \sum_{j=1}^{J} N_j - J\right) = \frac{[\text{SSB}/(J-1)]}{\text{SSW}/(\sum N_j - J)} = \frac{\text{MSB}}{\text{MSW}} \tag{3.132}$$

When MSB is large relative to MSW, F will be large and we can justifiably reject the null hypothesis that the different CTDs (different treatment effects) measure the same temperature within the accuracy of the instruments. For our CTD intercomparison (Table 3.12), we have MSB/MSW = 5.0, $\nu_1 = 2$ and $\nu_2 = 9$. Using the values for the F-distribution for 2 and 9 degrees of freedom from Appendix D, Table D.4a, we find $F_\alpha(2, 9) = 4.26$ for $\alpha = 0.05$ (95% confidence level) and $F_\alpha(2, 9) = 8.02$ for $\alpha = 0.01$ (99% confidence level). Since, $F = 5.0$ in our example, we conclude that a difference exists among the different makes of CTD at the 95% confidence level, but not at the 99% confidence level.

3.15 EFFECTIVE DEGREES OF FREEDOM

Up to this point, we have assumed that we are dealing with random variables such that the N values in a given sample are statistically independent. For example, in calculating the unbiased standard deviation for N data points, we assume there are $N - 1$ degrees of freedom. (We use $N - 1$ rather than N since we need a minimum of two values to calculate the standard deviation of a sample). Similarly, in Sections 3.8 and 3.10, we specify confidence limits in terms of the number of samples rather than the "true" number of degrees of freedom to be used to calculate the sample statistics. This distinction arises because, with the exception of samples generated by white-noise processes, consecutive data values are unlikely to be independent. Contributions from low-frequency components and persistent narrow band oscillations can lead to a high degree of serial correlation within measured data sets or between data sets. As a consequence, the record length, N, is not generally representative of the actual number of degrees of freedom for the statistical quantity of interest, such as the mean, variance or cross-correlation, computed from a given geophysical record.

The most common examples of highly coherent narrow-band signals are the tides and tidal currents, which possess strong temporal and spatial coherence. Similar considerations also apply to most climate-scale processes such as those affected by El Niño-La Niña variability in the Pacific Ocean and the volume of Labrador Sea Water in the Atlantic Ocean. If our statistics are to have a rigorous basis, we are forced to find the *effective number of degrees of freedom*, $N^*(< N)$ derived using lagged autocorrelations and cross-correlations of the data series being analyzed. Determination of the data segments to which to apply the lagged correlation estimates also needs to be addressed. The value of N^* for a sample estimate, $\widehat{\theta}$, of a probabilistic parameter depends on the particular statistic, θ, under consideration and should, therefore, be denoted with explicit indication of the statistic to which it applies (i.e., $N^* \equiv N^*(\theta)$). The procedure for determining $N^*\left(\widehat{\theta}\right)$ for a sample involves the derivation of an expression for the variance of the sample statistic for the case of the N sample observations. A formal expression for $N^*\left(\widehat{\theta}\right)$ is then obtained by comparing this variance with the variance in the case for which the observations are statistically independent (i.e., serially uncorrelated). We emphasize that the general derivation of $N^*(\theta)$ is not akin, for example, to determining the speed of light in a vacuum; $N^*(\theta)$ does not have a singular value but is a probabilistic entity with a "fuzzy" distribution.

The limitations to estimating the Pearson correlation coefficients (correlations) between the data series, $x(t)$, and the series, $y(t)$, as functions of variable, t, can be summarized as follows:

1. Accurate assessment of the statistical significance of the correlation between two data series (for example input $x(t)$ and response $y(t)$) requires the use of the effective number of degrees of freedom, N^*, which is generally markedly smaller than the total number of observations, N, in each data set.
2. The accuracy of derived regression coefficients increases as N^* increases.
3. The accuracy of the regression coefficients in multivariate linear regression decreases as the number of inputs M increases (measurement error is added).
4. The accuracy increases as the model skill increases and decreases as the input parameters become more correlated.
5. Detrending time series when determining the effective degrees of freedom is not recommended as it has the effect of removing low frequency contributions, which can lead to an "artificial" increase in N^*.

The above considerations emphasize the need for careful selection of the input data and the careful evaluation of the characteristics of these data. As pointed out by Davis (1977), a fundamental part of this selection process is the determination of the space and timescales to be studied. The methods used to extract this fundamental scale information from the input data can range from spectral and cross-spectral analyses (see Chapter 5) to a filtering of the data using preselected windows (Chapter 6). (Performing filtering in the time domain rather than in the frequency domain is often less complicated). The filtering process has the goal of eliminating scales that are not expected to contribute to the true correlation coefficients but which will add artificial correlation due to instrument and sampling errors.

Once the space and/or timescales are determined, selected or set by the filtering, the next step is the selection of the input series to use in the estimate. At this stage, a dilemma arises between limiting the effects of errors while, at the same time, including as many as possible of the available uncorrelated input variables in order to increase the degrees of freedom. Davis (1977) recommends using dynamical considerations to make this selection and shows how the data required for proper statistical estimation are generally those required to make the dynamical system well posed. However, he also mentions that, in general, the dynamics of most processes are not well enough understood and that specification data are not known with certainty. Nevertheless, some quantitative understanding of the physical system can serve as a useful guide to the selection of estimation data.

3.15.1 Basic Concepts

The effects of coherent (nonrandom) processes within data series leads to the question of data redundancy in multivariate linear regression and the need to estimate the effective number of degrees of freedom in the data prior to deriving confidence intervals and other statistics. Our general regression model is

$$\widehat{y}(t_i) = \sum_{m=1}^{M} b_m x_m(t_i); i = 1, \dots, N \tag{3.133}$$

where the x_m represents M observed parameters or data series measured at times t_i. The b_m are M linear regression coefficients relating the independent variables $x_m(t_i)$ to the model estimates, $\widehat{y}(t_i)$. Here, the x_m observations can be measurements of different physical quantities or of the same quantity measured at different times or locations. The case $m = 1$ corresponds to the simple linear regression model, $\widehat{y}(t_i) = bx(t_i)$.

The estimate $\widehat{y}(t_i)$ differs from the true parameter by an error $\varepsilon_i(t_i) = \widehat{y}(t_i) - y(t_i) = \widehat{y}_i - y_i$ ($i = 1, \dots, N$). Following our earlier discussion, we assume that this error is randomly distributed and is therefore uncorrelated with the input data $x_m(t_i)$. To find the best estimate, we apply the method of least squares to minimize the mean-square (MS) error, given as

$$\overline{\varepsilon^2} = \sum_{i=1}^{M} \left(\sum_{m=1}^{M} b_i b_m \overline{x_i x_m} \right) - 2 \sum_{m=1}^{M} b_m \overline{x_m y} + \overline{y^2} \tag{3.134}$$

In this case, the over bars represent *ensemble averages* derived from the M records, each of N-duration. To assist us in our minimization, we invoke the Gauss−Markov theorem, which states that the estimator, given by Eqn (3.133), with the smallest MS error is that with coefficients

$$b_m = \sum_{j=1}^{M} \left[\left\{ \overline{x_m x_j} \right\}^{-1} \overline{x_j y} \right] \tag{3.135}$$

where $\left\{ \overline{x_m x_j} \right\}^{-1}$ is the (m, j) element of the inverse of the $M \times M$ cross-covariance matrix of the input variables (note: $\left\{ \overline{x_m x_j} \right\}^{-1} \neq 1 / \overline{x_m x_j}$). This MS product matrix is always positive definite unless one of the input variables x_m can be expressed as an exact linear combination of the other input values. The presence of random measurement errors in all input data makes this "degeneracy" highly unlikely. It should be noted, however, that it is the partial correlation between inputs that increases the uncertainty in our estimator by decreasing the number of degrees of freedom through a reduction in the independence of the input parameters. We can write the minimum least-square error ε_o^2 as

$$\overline{\varepsilon_o^2} = \overline{y^2} - \sum_{m=1}^{M} \left[\sum_{j=1}^{M} \overline{x_j y} \left\{ \overline{x_m x_j} \right\}^{-1} \overline{x_j y} \right] \tag{3.136}$$

At this point, we introduce a measure of the reliability of our estimate called the *Skill* of the model. This skill is defined as the fraction of the actual parameter variance explained by our linear statistical estimator; thus

$$Skill = \left\{ \sum_{i=1}^{M} \left[\sum_{j=1}^{M} \left[\overline{x_j y} \cdot \left\{ \overline{x_j x_i} \right\}^{-1} \cdot \overline{x_i y} \right] \right] \right\} / \overline{y^2} \tag{3.137}$$

The skill value ranges from no skill ($Skill = 0$) to perfect skill ($Skill = 1$). We note that for the case ($M = 1$), *Skill* is the square of the correlation between x_j and y.

The fundamental trade-off for any linear estimation model is that, while one wants to use as many independent input variables as possible to avoid interdependence among the estimates of the dependent variable, each new input contributes random measurement errors that degrade the overall estimate. As pointed out by Davis (1977) the best criterion for selecting the input data parameters is to use *a priori* theoretical considerations. If this is not possible, some effort should be made to select those inputs which contribute most to the estimation skill. This procedure is referred to as "screening". It is difficult to quantify the effects of screening on the skill of a regression model as some input variables will improve the skill by random chance, rather than because of true predictive skill (Dudley Chelton, pers. comm., 2022).

The conflicting requirements of limiting M (the number of observed variables) and including all candidate input variables is a dilemma. In considering this dilemma, Chelton (1983) concludes that the only way to reduce the error limits of the estimated regression coefficients is to increase the "effective degrees of freedom N^*." This can be done only by increasing the sample size of the input variable (i.e., using a longer time series, N) or by high-passing the data to eliminate contributions from unresolved, and generally coherent, low-frequency components. Detrending the data sets, which is necessary to reduce bias and to better approximate the true value of a statistic, is similar in its effect on the value of N^* of removing very low frequency components in the data and is not recommended in the determination of N^* (Núñez-Riboni et al., 2023). Because we are generally forced to deal with relatively short data records in which ensemble averages are replaced by sample averages over time or space, we need procedures to evaluate the effective degrees of freedom.

In the case of real data, the skill given by Eqn (3.137) is generally derived in terms of temporal or spatial averages rather than ensemble averages. If we assume for a moment that the x_k input data are serially uncorrelated (i.e., we can expand the data series into orthogonal functions), the sample estimate of the skill can be written as

$$Skill = \frac{\sum_{i=1}^{M} \left[\sum_{j=1}^{M} \left(\frac{\overline{x_i x_j}^2}{\overline{x_i^2}} \right) \right]}{\overline{y^2}} \tag{3.138}$$

where overbars denote temporal or spatial averages (Davis, 1976). A consequence of the summations in Eqn (3.138), and the reason this expression is included here, is that the skill increases with increases in the number, M, of variables used in the model.

3.15.2 A Simple Method for Estimating N^*

As we show in the sections that follow, calculation of the effective degrees of freedom can be complex and, in some cases, it is difficult to derive a robust quantitative estimate. We therefore begin with a simple (albeit, fundamentally flawed) approach for estimating the effective degrees of freedom. If the correlation coefficient, $r_{xy} = r$ between two variables (data series x and y) is unity ($|r| = 1$) then the variables are perfectly correlated and there are only 2 degrees of freedom, corresponding to the fact that their relationship can be written in terms of two parameters (e.g., phase and amplitude). If instead, the two variables have a correlation of $r = 0$, they are serially uncorrelated and can be considered as independent variables. Under the latter conditions, there are as many degrees of freedom as there are data values, N. In general, $|r|$ lies somewhere between 0 and 1. Under these circumstances, one can use the correlation coefficient to derive an approximate, zeroth order estimate of N^*. The rational is that the correlation coefficient represents a ratio of the independent to dependent values of the two variables. However, because fully dependent variables have a correlation of unity, corresponding to the minimum of degrees of freedom, it is clear that we cannot simply use the correlation coefficient to estimate the effective degrees of freedom, N^*. Instead, we need to subtract the correlation from unity (i.e., $1 - |r|$) and multiply this difference by the number of data values, N, in each series. The value

$$N^* \approx (1 - |r|)N \tag{3.139}$$

provides a rough estimate of the effective degrees of freedom. While it is much less accurate than the procedures outlined in the sections that follow, this method gives a "ball-park" estimate of N^* for the data sets being correlated. Thus, for $|r| = 0.25$, $N^* \approx 0.75\,N$, indicating that there is a high degree of independence between the two data series. (Note: We do not recommend using the method in a submission to a scientific journal).

3.15.3 Estimating N^* Using the Integral Time Scale

A reliable estimate of N^* for the mean $\mu = \bar{x}$ of a given data series (variable), x, is obtained using its autocovariance, $C_{xx}(j) = E[(x(i) - \bar{x})(x(i+j) - \bar{x})]$ as a function of lag, j. To do this, we first find the integral timescale, $T(\mu)$, for the mean of the data record as defined by

$$T(\mu) = \frac{\Delta t}{C_{xx}(0)} \sum_{j=1}^{N-1} \left[C_{xx}(0) + 2\left(\frac{N-j}{N}\right)C_{xx}(j) \right] \text{ (discrete case)} \tag{3.140a}$$

$$= \frac{1}{C_{xx}(0)} \int_{-\infty}^{\infty} C_{xx}(\tau)d\tau \text{ (continuous case)} \tag{3.140b}$$

where, as usual, N is the total number of observations and Δt is the sampling interval. The factor "2" in Eqn (3.140a) accounts for the need to include both positive and negative values of the lag j in the summation and integral functions, taking into account the fact that autocovariance functions are symmetrical, so that only the positive lags are considered. The contribution from lag $j = 0$ is included as a separate term. Once the integral timescale is known, the effective number of degrees of freedom for the sample mean, $\hat{\theta} = \hat{\mu}$, is derived as

$$N^*(\hat{\mu}) = \frac{N\Delta t}{T(\mu)} \tag{3.141}$$

where $N\Delta t$ is the total length (i.e., duration or length) of the record. If, for example, $N = 120$, $\Delta t = 1$ h, and $T = 10$ h, then $N^*(\hat{\mu}) = 12$ (i.e. $\ll N$).

In the case $\hat{\theta} = \hat{\sigma}^2$, the sample variance, we use the square of the integral time scale, while taking into account the need to retain a linear weighting function $(N - j)/N$, whereby

$$T(\sigma^2) = \frac{\Delta t}{(C_{xx}(0))^2} \sum_{j=1}^{N-1} \left[(C_{xx}(0))^2 + 2\left(\frac{N-j}{N}\right)(C_{xx}(j))^2 \right] \text{ (discrete)} \tag{3.142a}$$

$$= \frac{1}{(C_{xx}(0))^2} \int_{-\infty}^{\infty} (C_{xx}(\tau))^2 d\tau \text{ (continuous)} \tag{3.142b}$$

and so

$$N^*\left(\widehat{\sigma}^2\right) = \frac{N\Delta t}{T(\sigma^2)} \tag{3.142c}$$

For the case of $N^*\left(\widehat{r}_{xy}\right)$ for the cross-correlation, r_{xy} (or autocorrelation r_{xx}), the corresponding expressions are (cf. Section S3 of the Supplement of Nunez-Riboni et al., 2023; Chelton, 1983)

$$T\left(r_{xy}\right) = \frac{\Delta t}{C_{xx}(0)C_{yy}(0)} \sum_{j=1}^{N-1}\left\{ \left[C_{xx}(0)C_{yy}(0)\right] + 2\left[C_{xx}(j)C_{yy}(j) + C_{xy}(j)C_{yx}(j)\right]\right\} \tag{3.143a}$$

$$= \frac{1}{C_{xx}(0)C_{yy}(0)} \int_{-\infty}^{\infty} \left[C_{xx}(\tau)C_{yy}(\tau) + C_{xy}(\tau)C_{yx}(\tau)\right] d\tau \,(\text{continuous}) \tag{3.143b}$$

whereby

$$N^*\left(\widehat{r}_{xy}\right) = \frac{N\Delta t}{T\left(r_{xy}\right)} \tag{3.143c}$$

We note that the integral time scales listed above are fixed quantities (based on true lagged correlations), so that the only adjustable quantity in the integrals is the numerator, $N\Delta t$. As a consequence, N^* can only be increased by increasing the record length. In other words, for a fixed sample interval Δt, N^* can only be increased by increasing the number of samples, N.

To find the autocovariance functions required in the above expressions, we let $i = j\Delta\tau$ be the jth lag ($j = 0, 1, \ldots$), then

$$C_{xx}(j) = \frac{1}{N - (1 + j)} \sum_{i=1}^{N-j}\left[(x_i - \bar{x})(x_{i+j} - \bar{x})\right]; j = 0, \ldots, N_{max} \tag{3.144a}$$

$$C_{xx}(0) = \frac{1}{N-1} \sum_{i=1}^{N} (x_i - \bar{x})^2 = s_x^2 \tag{3.144b}$$

where $C_{xx}(0)$ is the variance, s_x^2, of the full data series. In both Eqns (3.144a) and (3.144b), the data start with the first value for $i = 1$; N_{max} is the maximum number of reasonable lag values, starting at zero lag and going to N_{max} ($<N$) that can be calculated before the summation becomes erratic. In theory, we would like $C_{xx}(j) \to 0$ as $j \to N$. In reality, however, the data series will contain low-frequency components, which will cause the autocovariance function to oscillate about zero or asymptote toward a nonzero value. It is also evident that the statistical significance of the summation becomes meaningless at large lag due to the fact that the statistic is based on fewer and fewer values as the lag becomes large. For example, at a lag $j = (N - 3)$ there are only four values that go into the summation and these are derived from neighboring points that are likely to be highly correlated.

We can picture the integral timescale using Eqn (3.140b). Letting $C(j) = C_{xx}(j)$, and writing

$$T \cdot C(0) = \int_{-\infty}^{\infty} C(\tau) d\tau \tag{3.145}$$

we find that the area under the curve $C(\tau)$ for both positive and negative lag values, τ, is equated to the rectangular region $T \cdot C(0)$ (Figure 3.13). In essence, we take a reasonable portion of the curve $C(\tau)$, obtain its area and divide the integral (sum) by its value, $C(0)$, at zero lag. An example of the autocovariance function and the integral timescale derived from it are shown in Figure 3.14 for satellite-tracked drifter data in the North Pacific.

3.15.4 The General Formulation for N^* for the Skill of a Regression Model

Following Bartlet (1946), Davis (1978), and Chelton (1983, 1984), we can expand the previously defined skill estimate into a true skill (S_T) plus an artificial skill (S_A) such that

$$Skill = S_T + S_A \tag{3.146}$$

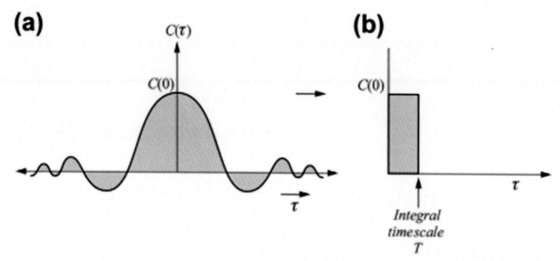

FIGURE 3.13   Definition of the integral timescale, T. The area under the correlation curve $C(\tau)$ versus τ in (a) is equated to the rectangular region $T \cdot C(0)$ in (b). In practice, only a portion of the curve $C(\tau)$ contributes significantly to the area in (a).

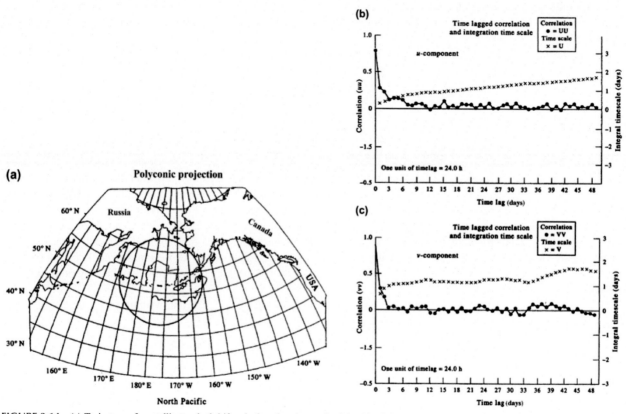

FIGURE 3.14   (a) Trajectory of a satellite-tracked drifter deployed to the south of the Aleutian Islands in the northeast Pacific and covering the period November 13, 1991 to July 30, 1993 based on a 6-hourly sampling interval; (b, c) autocovariance functions and corresponding integral timescales for zonal (u) and meridional (v) velocities of the satellite-tracked drifter. *Courtesy of Adrian Dolling.*

The artificial skill, S_A, arises from errors in the estimates and can be calculated by evaluating the skill in Eqn (3.146) at a very long time (or space) lag where no real skill is expected. At this point, there is no true estimation skill and $Skill = S_A$.

Davis (1976) derived an appropriate expression for the expected (mean) value of this artificial skill, which relates it to the effective number of degrees of freedom, N^*

$$\overline{S}_A = \sum_{m=1}^{M} \left(N_m^*\right)^{-1} \tag{3.147}$$

where N_m^* is the effective degrees of freedom associated with the sample estimate of the covariance between the output y and the M inputs, x_m, of the model. Under the conditions that S_T (the true skill) is not large, that the record length N is long compared to the autocovariance scales of y and the M time series x_m, and that the N_m^* are the same for all x_m, we can write N^* in the general form given by Chelton (1983) as

$$N^* = \frac{N}{\sum\limits_{j=-\infty}^{\infty} \left\{ \left[C_{xx}(j)C_{yy}(j) + C_{xy}(j)C_{yx}(j) \right] / \left[C_{xx}(0)C_{yy}(0) \right] \right\}} \tag{3.148a}$$

$$= \frac{N}{\sum\limits_{j=-\infty}^{\infty} \left[\left(\rho_{xx}(j)\rho_{yy}(j) + \rho_{xy}(j)\rho_{yx}(j) \right) \right]} \tag{3.148b}$$

where

$$\rho_{\eta\eta}(j) = C_{\eta\eta}(j) / C_{\eta\eta}(0) = C_{\eta\eta}(j) / s_\eta^2 \tag{3.149a}$$

is the normalized autocovariance function (i.e., the autocorrelation function) at lag j for any variable η (with variance s_η^2), and

$$C_{\eta\eta}(j) = E[(\eta(i) - \overline{\eta})(\eta(i+j) - \overline{\eta})] \tag{3.149b}$$

Expression (3.148a) includes the cross-covariances and expression (3.148b) the cross-correlations between y and the M time series x_k (e.g., $C_{xy}(j)$ and $\rho_{xy}(j)$), both of which are assumed to be normally distributed, and is not limited to cases where the skill is small. Although the cross-covariances (and cross-correlations) are generally much smaller than the autocovariances (autocorrelations) values that appear in the denominator, this is not always the case. Given the high computational speeds of modern computers, there is little reason to avoid calculating them (just to be sure). Note also that $C_{yx}(j) = C_{xy}(-j)$, and correspondingly, $\rho_{yx}(j) = \rho_{xy}(-j)$.

In principle, the effective degrees of freedom, N^*, as expressed by (3.148a) and (3.148b) can be used with standard tables to find the selected significance levels for the *Skill*. In the ideal case, when all input variables are neither cross- nor serially-correlated (and therefore independent), the effective number of degrees of freedom is N, the sample size. Typically, however, input data series are serially or spatially correlated and $N^* < N$. The larger the time/space correlation scales in Eqns (3.148a) and (3.148b), the smaller the value of N^*, so that it is the large scale, low-frequency components of the input data that lead to a decrease in the number of independent values in the data series. In practice, the true auto- and cross-correlations in (3.148a) and (3.148b) are not known. Substitution of sample estimates of the correlations can, and often does, lead to spurious results.

3.15.4.1 Picking the Maximum Lag J for the Correlation Functions

As noted above, the true auto- and cross-covariances (correlations) in (3.148a) and (3.148b) are not known and the computation of N^* requires the substitution of sample estimates over a finite range of lags $j(= -J, ..., J)$ for the correlations in Eqns (3.148a, 3.148b) and (3.149a, 3.149b). One option for choosing the finite range of lags is to specify J as the lag at which the autocorrelation function, ρ_{xx}, decreases to $\rho_{xx}(j) = 0.5$. Alternatively, one can pick J as that value of lag j at which the summation (integral) of the autocovariance (autocorrelation) function reaches a steady plateau before starting to oscillate with increasing lag. Alternatively, we can, as a last resort, set $J = N^*$, where N^* is derived using the integral time scale given byEqn (3.140a).

3.15.5 The Long-Lag Artificial Skill Method

The problem of estimating N^* was revisited in the Supplementary Material of Núñez-Riboni et al. (2023), which examines the correlation among 1,790 environmental time series related to ecosystems and their associated environmental (climatic) and anthropogenic driving mechanisms. The study used 402 ecological time series from the Living Planet Database (LPD, 2021), 35 atmospheric and oceanic time series, including the El-Niño-Southern Oscillation (ENSO) Index, the Pacific and

Atlantic oscillations (PDO and NAO) and the volume of Labrador Sea Water, and 1,162 time series of human-related activities from "Our World in Data" (OWID, 2021). The ecological time series includes population sizes of terrestrial, freshwater and marine vertebrates, while the human-related data include time series of air pollution, demographic growth and energy and food production.

This broad range of yearly time series can usually be well characterized as having "red" spectra, with dependence on frequency, f, of the form

$$S(f) \approx f^{-\lambda} \tag{3.150}$$

where the exponent, λ, is derived from Monto Carlo simulations of the spectra for each data set resolved at steps of 0.25 and found to lie in the range $0 \leq \lambda \leq 5$ for most of observed times series. White-noise processes (e.g., natural disasters) have $\lambda \to 0$, while anthropogenic series can have $\lambda \to 8$. For statistical analyses, multiple versions of each annual time series $x(t)$, were generated using Monto Carlo simulations. To do this, all of the time series were first Fourier analyzed to obtain their spectral amplitudes $a(f_j) = \sqrt{S(f_j)\Delta f}$ for each frequency component, $f_j = j/(N\Delta t)$, $j = 1, ..., N$ at annual time steps, Δt, and bandwidth $\Delta f = 1/(N\Delta t)$. The phase, $\phi(f_j)$, of each Fourier component was then drawn randomly from a uniform probability distribution $0 \leq \phi \leq \pi$ to generate a large number of Monto Carlo simulations of the spectral contributions to each data series. The spectra were then fitted to the spectra (3.150) to derive λ at steps of 0.25 and covering a broad range $0 \leq \lambda \leq 12$.

The spectral distributions were then inverse Fourier transformed to provide 5,000 pairs of simulated series $x(t_i)$ and $y(t_i)$, with total durations N for the different series ranging from 10 to 110 years. Based on the formulation presented in Núñez-Riboni et al. (2023), the effective number of degrees of freedom for the climatic-scale data series is

$$N^* = 2N \frac{(k_2 - k_1 + 1)}{\sum\limits_{j=k_1}^{k_2} \left[\left(N_j \cdot \widehat{\rho}_{xy}^2(j) \right) + \left(N_{-j} \cdot \widehat{\rho}_{xy}^2(-j) \right) \right]} \tag{3.151}$$

where the cross-correlation between two time series $x(t_i) = x_i$ and $y(t_i) = y_i$ $(i = 1, ..., N)$ is given by

$$\widehat{\rho}_{xy}(j) = \left(\frac{N}{N-j} \right) \frac{\sum\limits_{i=1}^{N-k} \left[(x_i - \bar{x})(y_{i+j} - \bar{y}) \right]}{\left\{ \sum\limits_{i=1}^{N} \left[(x_i - \bar{x}) \right]^2 \sum\limits_{i-1}^{N} \left[(y_i - \bar{y}) \right]^2 \right\}^{1/2}} \tag{3.152}$$

The lag times $(t_j = j)$ that range from a lower value of k_1 to a maximum lag k_2 $(k_2 > k_1)$ should include only the long lags at which the analyst is confident there is no true skill in the relation between x and y. The mean values (denoted by the overbars) on $\bar{x}$ and $\bar{y}$ in (3.152) are calculated in the usual manner using all N values of x and y. As shown by Figure 3.15 for the particular cases of $N = 20$, 50 and 100, the number of effective degrees of freedom as a fraction of the total record length, N, falls off with the increasing autocorrelation time scale of the time series, as characterized by the parameter λ (see Figure 2 of Núñez-Riboni et al., 2023). As one would expect, the confidence level for significance of the correlation functions between x and y becomes increasing more stringent going from the 90 to the 95 and then to the 99% confidence interval (corresponding to $\alpha = 0.10$, 0.05 and 0.01, respectively).

For the case of the cross correlation between two time series (i.e., $M = 1$ in (3.137)), the skill is

$$Skill(k) = \widehat{\rho}_{xy}^2(k) \tag{3.153}$$

Based on 500 Monto Carlo (MC) simulations, Núñez-Riboni et al. (2023) generated a matrix with coordinates k_1 versus k_2 (Figure 3.16) that presents the differences between the average of $Skill(k) \cdot N(k)$ for the uncorrelated data series x,y and the average of $Skill(k) \cdot \widetilde{N}(k)$ for the correlated versions $\widetilde{x}, \widetilde{y}$, with the latter cross correlation at lag $k = 0$ specified, for the purpose of illustration, to be $\rho_{xy}(0) = 0.6$ (i.e., not zero). More specifically, the authors determined values of $\Phi = \overline{Skill(k)\widetilde{N}(k)} - \overline{Skill(k)N(k)}$, corresponding to the differences averaged over the subset of lags between k_1 and k_2 for the correlated time series $\widetilde{x}, \widetilde{y}$ and those averaged between lags $k_1 = 0$ and $k_2 = 80\%$ for the original uncorrelated time series x, y. The goal was to determine the lag indices k_1 and k_2 that yielded values of Φ derived using the Artificial Skill Method (ASM) that were closest to zero. The analysis considered choices of k_1 and k_2 that included a range of lags near zero for which a nonzero correlation was imposed for the simulations.

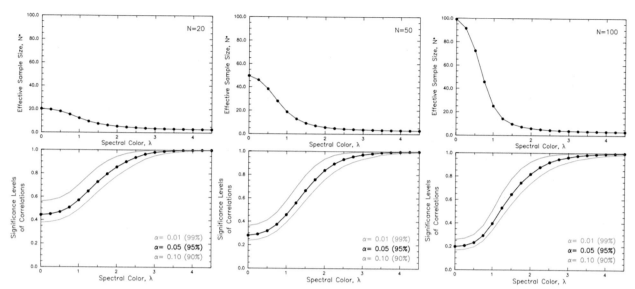

FIGURE 3.15 *Upper panels*: The variation in N^* (the effective number of degrees of freedom) as a function of the spectral exponent, λ, derived from correlations between Monto Carlo simulated environmental time series at annual sampling over record lengths (N) of 20, 50 and 100 years. *Bottom Panels*: The corresponding minimum correlation levels for the correlation significance (α) between time series x and y for different values of α as functions of λ. *From Núñez-Riboni et al. (2023).*

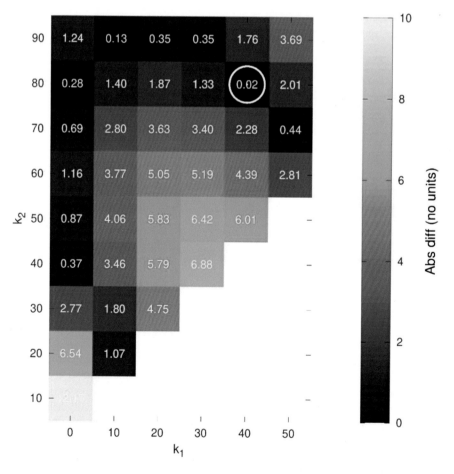

FIGURE 3.16 Average absolute differences between the product $Skill(k)\widetilde{N}(k)$ averaged between k_1 and k_2 from Monto Carlo simulations of 500 correlated time series $\widetilde{x}$, $\widetilde{y}$ and the product $Skill(k)N(k)$ averaged between $k_1 = 0$ and $k_2 = 80\%$ for the original uncorrelated time series x, y. The circled value combines lags k_1 and k_2 yielding results for the expected value $E\left[\rho_{\widetilde{xy}}(0)\right] > 0$ closest to that for the idealized case of uncorrelated time series. $E\left[\rho_{xy}(0)\right] = 0$. *From Núñez-Riboni et al. (2023).*

As per the Monto Carlo simulation results in Figure 3.16, Núñez-Riboni et al. (2023), recommend setting the lower and upper bounds as

$$k_1 = 0.4 \, N \text{ and } k_2 = 0.8 \, N \tag{3.154}$$

where k_1 is more narrowly defined than k_2. This assumes that any apparent skill in the lagged cross correlations between x and y within this lag range is purely artificial. The upper end, k_2, of the range is imposed to avoid inclusion of very long lags for which the cross correlations often become noisy because of the small number of paired values of x and y. The lower end, k_1, of the range is a subjective choice of the analyst and should be chosen based on knowledge or intuition regarding the range of lags for which the true skill is nonzero (Ismael Núñez-Riboni, pers. comm., 2023). For example, Dhage and Strub (2016) set $k_1 = 0.2 \, N$, arguing that there was no true skill in the correlation at lags longer than $0.2 \, N$, while Chelton and Risien (2020) used the more conservative approach of setting $k_1 = 0.4 \, N$. Unless N^* is small, the estimate of the critical value of the cross correlation is not highly sensitive to errors in the estimate of N^* (see Figure S1 of Núñez-Riboni et al. (2023)), and, therefore, to any errors that arise from the choice of k_1 used to estimate N^*.

Once the effective number of degrees of freedom, N^*, has been determined, it can be used to estimate the statistical significance of the cross-correlation, r, of a sample as

$$r^2(\alpha) = \frac{t^2\left(\frac{\alpha}{2}; N^*-2\right)}{t^2\left(\frac{\alpha}{2}; N^*-2\right) + (N^*-2)} \tag{3.155}$$

where $t\left(\frac{\alpha}{2}; N^*-2\right)$ is the score at probability $\frac{\alpha}{2}$ of the t-distribution with N^*-2 degrees of freedom. This expression is not defined for $N^* \leq 2$, while for $N^* \gg 3$, the t-distribution can be replaced with an asymptotically equivalent χ^2 distribution with 1 degree of freedom (see Figure S1 of Núñez-Riboni et al., 2023), whereby

$$r^2(\alpha) \approx \frac{\chi_\alpha^2}{(\chi_\alpha^2 + N^*)} \approx \frac{\chi_\alpha^2}{N^*} \tag{3.156}$$

3.15.6 The Pyper and Peterman Method

In their investigations of fisheries data, Pyper and Peterman (1998a, b) approximate N^* as

$$N^* = \frac{N}{\sum\limits_{j=-J}^{J} \widehat{\rho}_{xx}(j)\widehat{\rho}_{yy}(j)} \tag{3.157}$$

where $\widehat{\rho}_{xx}(j)$ and $\widehat{\rho}_{yy}(j)$ are the sample estimates of the autocorrelations at lag times $j\Delta t$ and for an upper limit, J, of the summation index. This differs from the full expression (3.148) in that: (1) it omits the product of the two lagged cross-correlations; (2) the doubly infinite summation is truncated at lags $\pm J\Delta t$; and (3) the true autocorrelations in (3.148) are replaced with sample estimates in (3.157). The auto-correlations are obtained from

$$\widehat{\rho}_{xx}(j) = \left(\frac{N}{N-j}\right) \frac{\sum\limits_{i=1}^{N-k}\left[(x_i - \bar{x})(x_{i+j} - \bar{x})\right]}{\sum\limits_{i=1}^{N}\left[(x_i - \bar{x})^2\right]} \tag{3.158}$$

Pyper and Peterman recommend using $J = N/5$, corresponding to the first 20% of each data set. A comparison of this method with the Artificial Skill Method of Núñez-Riboni et al. (2023) finds that this short range for J is too small, especially when N is small. Here, we suggest that researchers find J using the autocorrelation function and the integral time scale, T. Restricting correlation estimates to short lag durations (e.g., the first 20% of the data record) typically leads to an overestimation of N^*, while use of long lags (e.g., more than 80% of the data records) leads to more spread-out and less statistically reliable estimates of N^*.

Nunez-Riboni et al. (2023) found that both the Artificial Skill Method summarized in Section 3.15.5 and the Pyper and Peterson Method summarized above perform better without detrending, with generally poor performance for large λ when a trend is removed. Linear trends contribute to the low-frequency variability. Detrending decreases autocorrelation and therefore increases N^*. As a consequence, the critical value decreases and both the number of significant correlations and the error rate deviation increase.

The Núñez-Riboni et al. (2023) results stress that, of the two methods considered in their study, the Artificial Skill Method, rather than the Pyper and Pearson method, more accurately reproduces prescribed error rates, $e(\alpha)$, for the wide range of spectral colors representative of climatic, ecological and anthropogenic time series. With this goal, the authors characterized roughly 1,800 observational records in different categories of spectral colors, including climate variability, abundance of vertebrate species, and pollution. Specific focus was on time series with annual sampling over data records of at least 40 years, which are particularly relevant for climate studies. The methodology advocated in this study provides a simple and realistic assessment of the significance of sample estimates of cross correlation for time series with any sample interval and record length.

Other methods of estimating N^* include the cross-validation method of Michaelsen (1987) and the spectral method of Ebisuzaki (1997) (cf. Núñez-Riboni et al., 2023). It appears that these are the methods that are most commonly used in atmospheric sciences but that are rarely used in oceanography.

3.15.7 Trend Estimates

As noted in the previous sections, oceanographic variability comprises a combination of random and nonrandom processes. Consequently, there is invariably a nonzero correlation between values in the series that must be taken into account when the investigator tallies up the true number of independent samples or degrees of freedom that apply to the system being studied. This number is important when it comes to determining the confidence limits of linear regression slopes and parameter estimates. As an example, consider the confidence limits on the slope of the least squares linear regression $\widehat{y} = b_0 + b_1 x$ (where, again, $\widehat{y}$ denotes an estimator for the function y). From Eqn (3.45), the limits are

$$\pm \left(s_\varepsilon t_{\alpha/2,\nu} \right) / \left[(N-1)^{1/2} s_x \right] \tag{3.159a}$$

Or, in terms of the estimator β_1 for b_1

$$b_1 - \frac{\left(s_\varepsilon t_{\alpha/2,\nu} \right)}{(N-1)^{1/2} s_x} < \beta_1 < b_1 + \frac{\left(s_\varepsilon t_{\alpha/2,\nu} \right)}{(N-1)^{1/2} s_x} \tag{3.159b}$$

where $\nu = N - 2$ is the number of degrees of freedom for the student's t-distribution at the $100(1 - \alpha)\%$ confidence level, and the standard error of the estimate, s_ε, is given by

$$s_\varepsilon = \left[\frac{1}{N-2} \sum_{i=1}^{N} (y_i - \widehat{y})^2 \right]^{1/2} = \left[\frac{1}{N-2} \text{SSE} \right]^{1/2} \tag{3.160}$$

The standard deviation for the x variable, s_x, is given by

$$s_x = \left[\frac{1}{N-1} \sum_{i=1}^{N} (x_i - \bar{x})^2 \right]^{1/2} \tag{3.161a}$$

or,

$$(N-1)^{1/2} s_x = \left[\sum_{i=1}^{N} (x_i - \bar{x})^2 \right]^{1/2} \tag{3.161b}$$

The question is: what do we use for the number of degrees of freedom if the N samples in our series are not statistically independent? The reason we ask this question is that the characteristic amplitudes of the fluctuations s_ε and s_x are calculated using all N values in our data series when, in reality, we should be using some form of *effective* number of degrees of freedom N^* ($< N$), which takes into account the degree of correlation that exists between data points (see the previous sections).

Suppose we decide to err on the conservative side by agreeing to work with that value of N^*, which makes the confidence limits $\pm \left(s_\varepsilon t_{\frac{\alpha}{2},\nu} \right) / \left[(N^*-1) s_x^2 \right]^{1/2}$ as small as justifiably possible. This means that, when we estimate the confidence limits for a regression slope for a given confidence coefficient, α, we know that we have probably been too cautious and that the confidence limits on the slope probably bracket those that we derive.

We begin by keeping s_ε as it is. If there are high frequency (possibly random) fluctuations superimposed on coherent low-frequency motions, retaining the high-frequency variability adds to the magnitude of s_ε. Had we low-pass filtered the data first and recomputed se based on the true number of data points in our low-pass filtered record, we would expect s_ε to be somewhat smaller. By using s_ε as it is, we are assuming that it is a fixed quantity no matter how we subsample or filter the data (s_ε = constant). We do the same with s_x but now replace $N - 1$ with $N^* - 1$, where $N^* < N$. This increases the magnitude of the confidence limits. All that remains is to assume that the number of degrees of freedom for the t-distribution are given by the effective number of degrees of freedom $\nu = N^* - 2$. This statistic has a larger value than for $\nu = N - 2$ so that, again, we are overestimating the magnitude of the confidence interval. This confidence interval is then given by

$$\pm \left(s_\varepsilon t_{\frac{\alpha}{2},\nu} \right) / \left[(N^* - 1) s_x^2 \right]^{1/2} \tag{3.162a}$$

that is,

$$b_1 - \frac{\left(s_\varepsilon t_{\alpha/2,\nu} \right)}{(N^* - 1)^{1/2} s_x} < \beta_1 < b_1 + \frac{\left(s_\varepsilon t_{\alpha/2,\nu} \right)}{(N^* - 1)^{1/2} s_x} \tag{3.162b}$$

with $\nu = N^* - 2$.

3.16 EDITING AND DESPIKING: THE NATURE OF ERRORS

A major concern in processing oceanographic data is how to distinguish the true oceanic signal from measurement "errors" or other erroneous values. There are two very different types of measurement errors that can affect data. *Random errors*, usually equated with "noise," have random probability distributions and are generally small compared to the signal. Random errors are associated with inaccuracies in the measurement system or with real variability that is not resolved by the measurement system. The well-accepted statistical techniques for estimating the effects of such random errors are based largely on the statistics of a random population (see previous sections on statistics). Other errors that strongly influence data analysis are *accidental* errors. These errors are not representative of the true population and occur as a result of undetected instrument failures, misreading of scales, incorrect recording of data, and other human failings. In the following discussion, we will handle these two error types in reverse order since the large accidental errors must be removed first before techniques can be applied to treat the "statistical" (random) errors.

One example of a large accidental error would be assigning an incorrect geographic location to an oceanographic measurement, which then transfers the observations to a region with which they have no direct relationship. Some of these errors, such as oceanographic stations on land, are easily detected, while others are less obvious. Another example of such errors would be biases in a group of measurements due to the application of incorrect sensor calibrations or undetected instrument malfunctions. An all-too-common error occurs during preparation of moored instruments when the operator inadvertently sets the start time using local time (or daylight-saving time) while writing in the log that the time in UTC (Coordinated Universal Time). This typically happens when the person is rushed or is trying to work in rough seas on a rolling ship. Those working with the data will incorrectly interpret the time error as a phase shift.

A major goal of data processing is to remove any errors in order to make the data set as self-consistent as possible. If we know the history of the data, meaning the details of its collection and reduction, we may be in a better position to understand the sources of these errors. If we have received the data from another source, or are looking at archived data, we may not have available the necessary details on the "pedigree" of the data and may have to come to some rather arbitrary decisions regarding its reliability. Considering the widespread use of computer-linked data banks, this is not a trivial problem. The question is how to ensure the necessary quality control yet ensure rapid dissemination and accessibility to data files.

3.16.1 Identifying and Removing Errors

There are two important axioms to follow when dealing with large erroneous values or "spikes":

1. To identify the large errors, it is often necessary to examine all of the data in visual form in order to get a "feel" for the data;
2. When large errors are encountered, it is usually best to eliminate them altogether rather than try to "correct" them and incorporate them back into the data set.

Of course, care must be taken not to reject important data points just because they do not fit either the previous data structure or one's preconceived notion of the process. A good example is the determination of heat transport in the South

Atlantic. Bennett (1976) suggested that the oceanic heat transport in this ocean is directed toward the equator, contrary to the widely accepted notion that oceanic heat transports are generally poleward. Stommel (personal communication) noted that, in his tabulation of property fluxes for the South Atlantic, Wüst (1957) conspicuously left out the flux of heat while treating other less easily computed transports such as those of nutrients and oxygen. Through an exchange of letters with a former student of Wüst's, Stommel learned that the heat content calculation indeed showed that heat is transported equatorward. Wüst considered this to be the wrong direction and the results were not published along with the other flux values. The point of this story is to illustrate the way in which our prejudice can lead us to reject significant results. In such cases, there is no hard rule as to how this decision is made and a great deal of subjectivity will always be inherent in this level of data interpretation. As for the heat flux in the Southern Ocean, present estimates show it is poleward but with a high degree of uncertainty. Moreover, mesoscale eddies contribute a significant fraction of the poleward flux (Volkov et al., 2010; Foppert et al., 2017; Gutierrez-Villanueva et al., 2020), an aspect of the circulation that earlier could not be resolved by observations and numerical models.

The need to examine all the data to detect errors presents a difficult task because of the large numbers of values and the difficulty of looking at unprocessed data. In this case, it is more important to think of ways in which we can present the data so as to ask and answer the questions regarding consistency of the measurements. A compact overview of all the data is the best solution. This presentation may be as simple as a scatter diagram of the observations versus some independent variable, or a scatter diagram relating two concurrently measured parameters. While scatter diagrams cannot be used to resolve visually individual points, they do reveal groupings of points that relate to the physical processes expressed by the data. As an example, consider a temperature–salinity scatter diagram (Figure 3.17) computed using a large number of hydrographic data collected from bottle casts. Here, the groups of dots labeled "*a*," "*b*" refer to different water masses present in the 5° square 35–40° N, 15–20° W, where the data were collected. The data labeled "*c*" clearly represent a distinct water mass since the points lie along a line divergent from the rest of the scatter values. If we look at other similar *TS* scatter plots, we recognize that this line is consistent with the *TS* relationship from a corresponding square at this same longitude but south of the equator. Thus, it is likely that the latitude recorded was incorrect and that these data are simply

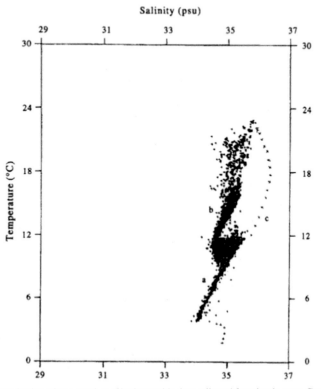

FIGURE 3.17 TS relationship computed using a large number of hydrographic data collected from bottle casts. Groups labeled "a," "b" refer to different water masses present in the 5° square (35–40° N, 15–20° W) where the data were collected. The data labeled "c" clearly represent a distinct water mass since the points lie along a line divergent from the rest of the scatter values.

misplaced. We correct this by eliminating the points "c" from our square. However, we cannot be sufficiently confident of our assumption to add the points to the other square even though the data coverage there is not very good.

Often it is not possible to develop a simple summary presentation of all the data. In the case of current meter data, a time series presentation is the most appropriate way of looking at the data. As noted by Pillsbury et al. (1974), error detection using this technique is very time consuming. They note that this procedure can be used successfully for speed, pressure, salinity, and temperature but not for direction, which varies widely. This is due to the fact that direction is limited to the range 0–360° and shows no extreme values. Because of the wrap-around ("2π discontinuity") problem, in which 0° = 360° (or, alternatively, −180° = +180°), direction records tend to be very "spiky," especially in regions of strong tidal flow. A scatter diagram of speed versus direction can be used to detect systematic errors between the speed and direction sensors and to pinpoint those times when the current speed is below the threshold recording level of the instrument. This would be displayed by the direction readings at speeds below threshold and would be easier to identify on the scatter plot than in the individual time series. The only way around the problem with the direction channel is to transform the recorded time series of speed and direction (U, θ) to orthogonal components of velocity (u, v). In particular, separate plots of the east–west (u) and the north–south (v) velocity components (or alongshore and cross-shore components for data collected near the coast) quickly reveal any erroneous values in the data (Figure 3.18).

To minimize possible errors when deploying current meters: (1) Inspect the O-ring for cuts or nicks and do not trap loose wiring under the ring seat when closing the case. Leakage of small amounts of water to the bottom of the instrument case can cause electrical malfunctions when the instrument tilts: (2) It is essential to hand-record accurate times for the first and last data records. Make sure the time zone is recorded. Record the time the instrument enters the water on deployment and leaves the water on recovery. More problems can be linked to poor bookkeeping than any other cause.

We remark that the two items mentioned above apply to all moored instruments. Modern acoustic current meters have been known to have O-ring leaks and subsequent electrical failures, and the problem of recording time zone continues to be a problem despite the best efforts of technical protocols. Other problems include measurement errors due to low numbers of acoustic scatterers (poorly delineated Doppler frequency shift), gradual deterioration of acoustic transducers and their power output due to long-term effects of high pressures and damage from rough handling procedures on-deck, and insufficient battery power for the experiment duration and sampling rate.

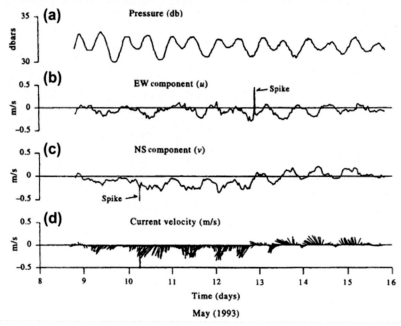

FIGURE 3.18   A Plot of hourly data obtained from an Aanderra RCM4 current meter moored at 30 m-depth data in 250 m of water near the entrance to Juan de Fuca Strait (48° 3.30′ N, 125° 18.80′ W) during the period May 8–16, 1993. (a) Ambient pressure (instrument depth in meters); (b) east-west (u) component of velocity (m/s); (c) north-south (v) component of velocity (m/s); (d) velocity stick vector (m/s). Erroneous current velocity values ("spikes") stand out in the (u, v) records. Flow consisted of moderate tidal currents superimposed on a surface estuarine outflow that weakened around May 13.

A standard method for isolating large errors is to compute a histogram of the sample values. This amounts to completing step 1 in a goodness of fit calculation since a histogram is nothing more than a diagram showing the frequencies of occurrence of sample values. While this is a very straightforward procedure some thought must go into selecting the parameter intervals, or bins, over which the sample frequencies are calculated. If the bins are too large, the histogram will not resolve the character of the sample probability distribution function (PDF) and the effects of large error values will be suppressed by being grouped with more commonly occurring values. On the other hand, if the bins are too narrow, individual values take on more influence and the resulting distribution will not appear smooth. This makes it difficult to "see" the real shape of the distribution.

The use of a histogram in locating large errors is that it readily identifies the number of widely differing values that occur and shows whether these divergent values fit into the assumed PDF for the assumed variable. In other words, we can not only see how many values ("outliers") differ widely from the mean values, but also determine if the number of large values in the sample is consistent with the expected distribution of large values for the population. Thus, we have an added guideline for deciding whether the sample values should be retained or eliminated for subsequent analysis. Both PDFs and histograms use visual means of detecting large error values. It is possible to use more automated and objective techniques, such as eliminating all values that exceed a specified standard deviation (e.g., $\pm 3\sigma$). However, these approaches have the weakness that they must first consider all data points, including the extreme values, as valid in order to determine decision levels for selecting or rejecting data. Here, we could use an iterative process in which the values outside the accepted range are omitted from each subsequent recalculation of the mean and standard deviation, until the remaining data have near constant statistics with each new iteration. Large errors, which are usually easy to spot using visual editing techniques, should be removed before proceeding to a more objective step involving the detection of less obvious random deviations. An objective technique for identifying outlier values is to compute a function, which selects extremes of the population, such as the first derivative of the measured variable with respect to an independent parameter. An example would be a time series of temperature measured from a line in a satellite image. After the extreme gradients are identified in the first derivative calculation, there is still the question of how widely the extremes should be allowed to differ from the rest of the population and whether a value should be considered as an error value or as simply as a maximum (or minimum) of the process being observed.

In making such a decision, it is necessary to have an estimate of the variability of the process. As discussed above, the dispersion (spread) of the population distribution is best represented by the variance or the standard deviation. If we are dealing with a normal population, we know that the standard deviation specifies the spread of the distribution and that 68% of the population values lie within $\mu \pm \sigma$ while 95% of these values are in the interval $\mu \pm 2\sigma$. Beyond $\mu \pm 3\sigma$ there is only 0.26% of the total frequency of occurrence, leaving 99.74% within this interval. Thus, it is again a matter of probabilities and significance level; and we must choose at what level we will reject deviations from the mean as errors. If we choose to discard all measurements beyond 2σ, we will have retained 95% of the sample population as our new sample population for which we will repeat out estimation of the statistics. This suggests that we will make our statistical estimate twice; first to decide what data to retain, and second to make statistical inferences about the behavior exhibited by the revised sample data. It is customary to use a much coarser subsampling interval, or to use broadly smoothed data, to compute the initial sample standard deviations for the purposes of editing the data. For our *TS* curve example (Figure 3.17), we might initially have used a computational interval of 1 or 2°C to compute a standard deviation for the first-stage editing and then have used the newly defined sample population (original sample minus large deviations >2°C) to recalculate the mean and standard deviation with a resolution of 0.1°C, closer to the measurement accuracy for reversing thermometers. In statistical analysis, we should not expect to exceed the inherent accuracy and resolution of our data. Modern computing facilities, and even pocket calculators, make it tempting to work with many decimal places despite the fact that higher place values are not at all representative of the ability of the instrument to make the measurement.

A form of two-step editing is used in the routine processing of CTD data, which is typically sampled at ≈ 25 samples per second per channel (≈ 25 Hz/channel). Because these instruments produce many more data than we are generally capable of examining, both smoothing and editing procedures are often built into the routine processing programs. The steps involved with processing calibrated CTD data are commonly as follows:

1. Write the data to a file for display on a computer screen using an interactive editing program written for the particular data set.
2. Examine all data for a given set of parameters by displaying the data simultaneously on a computer monitor; as a consistency check, it is important to know if large errors in one parameter, such as temperature, are associated with some real feature in another parameter, such as salinity.
3. With the cursor, eliminate erroneous values collected near the ocean surface, where the probe rises in and out of the water with the roll of the ship.

4. Using the file in (3), calculate the pressure gradient versus depth for the data and eliminate those data values for which the depth is decreasing with time for a down-cast and increasing with time for an up-cast (wave action eliminator).
5. Using the file of (4), produce a hard-copy or computer screen plot of the entire profile plus an expanded version for the upper ocean (say 0−300 m depth).
6. On the copy, "flag" erroneous values and irregularities in all data channels.
7. Use the interactive screen display to eliminate "bad" data identified in (5). If gaps between data points are small, linearly interpolate between adjacent values.
8. Smooth the edited file by averaging values over a specified depth range. Except for studies of turbulence or other fine-scale features, 1-m averaged files are typically generated for profile data and 1-s averaged files for time series data.

Because of improved CTD technology and data storage capacities in recent years, step (8) is often conducted first.

Fofonoff et al. (1974) used a 1/2-s average (15 scans) to smooth the measured pressure series. From this smoothed set, a 1/10th decibar pressure series was generated. Even with the smoothing, the pressure was oversampled, with roughly two observations for each pressure interval. The goal of this computation was to produce a uniform pressure series that could be used to generate profiles of T and S with depth. Processing routines could be added that first sorted out spurious extreme T and S values, based on a running mean standard deviation, and which ensured that the pressure series was monotonically increasing. This would correct for small variations in the depth of the probe due to ship motion or strong current shear. Also, in making these editing decisions we should always keep in mind the instrument characteristics and not discard data well within the noise level of the measurement system.

When editing newly collected data, we should always consider what is already known from similar, or related measurements in order to detect obvious errors. A typical example is the use of TS curves to evaluate the performance of sample bottles in a hydro-cast. Since TS curves are known to remain relatively stationary for many areas, previously sampled TS curves for an area can be used to locate data points that may have been caused by the erroneous performance of a water sampler; these are generally due to inadvertent bottle "trips" in which the sampler likely closed before or after the desired depth was achieved. Prior TS curves also have served as a means of interpolating a particular hydrocast or perhaps providing salinities to match measured temperatures. This approach is limited, however, to those areas and those parameters for which a sufficient number of existing observations are available to define the mean state and variability. In many areas, and for many parameters, information is too limited for existing data to be of any real use in evaluating the quality of new measurements. As a matter of curiosity, it would be interesting to determine the numbers of deep hydrocast data that were unknowingly collected at hydrothermal venting sites and discarded because they were "erroneous." Anomalously high temperatures would be difficult to justify if one did not know about hydrothermal circulation and associated buoyant plumes. Similar comments apply to "anomalous" CTD profiles obtained within thermohaline staircases (double-diffusive features). As noted in Section 2.1, salt-fingering and diffusive convection generate small scale (∼1−10 m) vertical structure that would appear to be highly erroneous if measured for the first time.

In contrast to large accidental errors, which lead to large offsets or systematic biases, random errors are generally small and normally distributed. These errors often are the result of inaccuracies in the instrumentation or data collection procedures and therefore represent the limit of our ability to measure the desired variable. Added to this is our inability to completely resolve the inherent variability in a particular parameter. This too may be a limitation of our instrument or of our sampling scheme. In either case, when we cannot directly measure a scale of oceanic variability that contributes to the alias of our measurement, the variability will form part of the uncertainty in the final calculated value.

The theory of random errors is well established (Scarborough, 1966). The fundamental approach is to treat the errors as random numbers with a normal PDF. Basic to this assumption is that positive and negative errors of the same size occur in about equal number and tend to cancel each other. This suggests that the appropriate way to treat data containing random errors is in terms of mean-square (MS) errors and root-mean-square (rms) values. Another fundamental assumption is that the probability of an error occurring depends inversely on its magnitude; thus, small errors are more frequent than large ones. Following the first of these two assumptions, the PDF of the random errors might be written as

$$p(\varepsilon_x) = f\left(\varepsilon_x^2\right) \tag{3.163}$$

where p is the PDF of the errors ε_x. The second characteristic requires that the probability decreases with increasing ε_x, so we can write for any real constant, k

$$p(\varepsilon_x) = C_o \exp\left(-k^2\varepsilon_x^2\right) \tag{3.164}$$

Using the fact that the integral under the curve of any PDF is unity, we solve for C_o and obtain

$$p(\varepsilon_x) = \frac{k}{\sqrt{\pi}} \exp\left(-k^2 \varepsilon_x^2\right) \tag{3.165}$$

This expression is known as the probability equation or the error equation. A graph of the function gives the normal or Gaussian probability curve. The term k is a constant called the *index of precision* and sets both the amplitude and the width of the normal curve. As k increases, the normal curve becomes narrower and the errors get smaller, making the measurement more precise. (This description applies only for small random errors and not to systematic errors.)

3.16.2 Propagation of Error

Suppose we have a quantity, F, which is calculated from a combination of a number (N) of independently observed variables. For example, F might be oceanic heat transport computed from independent velocity and temperature profiles, x_i ($i = 1, 2$). We can estimate the combined random error of F as the SSE of the individual variables provided that the errors are independent of the variables and that they are all normally distributed. As a simple example, let F be a linear combination of our measurement variables, x_i

$$F = a_1 x_1 + a_2 x_2 + \ldots + a_N x_N \tag{3.166}$$

where $a_1, \ldots, a_N$ are constants. The inverse of the squared error or *index of precision* (H) of F can be written

$$\frac{1}{H^2} = \frac{a_1^2}{h_1^2} + \frac{a_2^2}{h_2^2} + \ldots + \frac{a_N^2}{h_N^2} = \sum_{i=1}^{N} \left(a_i^2 / h_i^2\right) \tag{3.167}$$

where h_i is the error for the ith measurement, x_i.

A more generalized formula for error calculations for arbitrary F for which the contributing variables are uncorrelated is

$$\begin{aligned}
\frac{1}{H^2} &= \frac{(\partial F / \partial x_1)^2}{h_1^2} + \frac{(\partial F / \partial x_2)^2}{h_2^2} + \ldots + \frac{(\partial F / \partial x_N)^2}{h_N^2} \\
&= \sum_{i=1}^{N} \left[(\partial F / \partial x_i)^2 / h_i^2\right]
\end{aligned} \tag{3.168}$$

where partial derivatives $\partial F / \partial x_i$ are obtained from Taylor expansions of the function F in terms of the independent variables x_i. Specifically, $\partial F / \partial x_i = a_i$. To convert this expression to one in terms of relative errors, we use the fact that

$$\frac{1}{h^2} = \frac{r_e^2}{\rho_e^2} \tag{3.169}$$

where r_e is the corresponding relative error and $\rho_e = 0.4769$ is a constant obtained from the error Eqn (3.165). Using this definition, we can write our final error as

$$R_e = \left[(\partial F / \partial x_1)^2 r_1^2 + (\partial F / \partial x_2)^2 r_2^2 + \ldots + (\partial F / \partial x_N)^2 r_N^2\right]^{1/2} \tag{3.170}$$

In this form, R_e is really only the RMS error that describes the equivalent combined error in the equation of interest. This Taylor expansion of the contributing error terms is known as the *propagation of errors formula*. It is limited to small errors and uncorrelated independent variables.

Because these principles apply only to small random errors, it is necessary to use some data editing procedure to remove any large errors or biases in the measurements before using this formula. By using a mean-square (MS) formulation, we take advantage of the fact that small random errors can be expected to often cancel each other, resulting in a far smaller MS error than would result if the measurement errors were simply added regardless of sign to yield a maximum "worst case error." The primary application of Eqn (3.170) is in determining the overall error in a quantity derived from a number of component variables all with measurement errors. This is a situation common to many oceanographic problems.

A more complicated propagation of error formula is needed if there is a nonzero correlation between the independent variables, x. In this case, we must also retain the covariance terms in any Taylor expansion of the small error terms. For example, the density of water, ρ, is a function of both temperature T and salinity S so that the errors (variances) in density σ_ρ^2 can be related to the measurement errors in temperature σ_T^2 and salinity σ_S^2 by

$$\sigma_\rho^2 = (\partial\rho/\partial T)^2\sigma_T^2 + (\partial\rho/\partial S)^2\sigma_S^2 + 2[(\partial\rho/\partial T) \cdot (\partial\rho/\partial S)]C(T,S) \tag{3.171}$$

where $C(T, S)$ is the covariance between temperature and salinity fluctuations. Only when $C(T, S) = 0$ do we get the result in Eqn (3.170). An example of a detailed error calculation for the measurement of flow through trawl nets towed at various angles through the water column is given in Burd and Thomson (1993).

3.16.3 Dealing with Numbers: The Statistics of Roundoff

Because we must represent all measurements in discrete digital form, we are forced to deal with the consequences of numerical roundoff, or truncation. The problem results from the limitations of digital computing machines. For example, the irrational fraction 1/3 is represented in the computer as the decimal equivalent 0.3333...3, with an obvious round-off effect. This may not seem to be a problem for most applications since most computers carry a minimum of eight decimal places at single precision. The large number of arithmetic operations carried out in a problem lasting for only a few seconds of computer processing time can, however, lead to large errors due to roundoff and truncation errors. The case of greatest concern is when two nearly identical numbers are subtracted, requiring proper representation to the smallest possible digit. Such differences can easily occur unknowingly in a complicated computational problem. Rather than discuss procedures for estimating this roundoff error, we will discuss the nature of the problem and emphasize the need to avoid roundoff.

General floating-point values (decimal numbers) in a computer follow closely the so-called "scientific notation" and are represented as a mantissa (to the right of the decimal point) and an exponent (the associated power of 10). For example, in a three-digit system, the number 64.282 would be represented as 0.643×10^2 where the roundoff is accomplished by adding five in the thousands' decimal place and then truncating after the third digit. This process of rounding off results in a slight bias because it always rounds up when there is a 5 in the least significant digit. A way to overcome this rounding-up bias is to use the last digit retained, $k + 1$, to determine whether to round the kth digit up or down. This rule, which leads to the least possible error, is to roundup if the next to the last digit retained is odd and to leave the value unchanged when it is even, the latter having an overall effect of rounding down. This procedure can be summarized as follows. When rounding a number to k decimals:

1. if the $k + 1$ decimal is 0, 2, 4, 6, 8 then the k decimal is unchanged (in effect, rounding down);
2. if the $k + 1$ decimal is 1, 3, 5, 7, 9 then the k decimal is increased by 1 (rounding up).

This system of rounding-off will result in errors that are generally less than 0.5×10^{-k} and maximum roundoff errors of 0.6×10^{-k}. In most applications, the effect of this roundoff bias is too small to justify the added numerical manipulation required to implement this even—odd roundoff scheme.

In computing systems, floating-point numbers are handled in a binary representation having 24 bits (word length is 32 bits but 8 bits are used for the exponent), which results in seven significant decimal digits. Called *single precision*, this level of accuracy is adequate for many computations. For those problems with repeated calculations, and the subsequent high probability of differencing two nearly identical numbers, a *double-precision* representation is used which has 56 binary bits leading to 16 significant decimal digits. Roundoff, in the case of double precision, results in very small biases, which can be completely ignored for most applications. Another approach to the problem of roundoff errors is to consider them to be random variables. In this way, statistical methods can be applied to better understand the effects of roundoff errors. Consider the roundoff of a single number x; for this number, all numbers occurring in the interval $x_0 - 1/2 < x < x_0 + 1/2$ (measured in units of the last digit) become that number. Thus, the roundoff has a uniform probability distribution in the last digit. We can write the corresponding PDF $f(x)$ for x as

$$f(x) = \begin{cases} 1 & \left(x_0 - \dfrac{1}{2}, x_0 + \dfrac{1}{2}\right) \\ 0 & \text{elsewhere} \end{cases} \tag{3.172}$$

and note that

$$\int_{-\infty}^{\infty} f(x)dx = 1 \tag{3.173}$$

The most common measures of a PDF are its first two moments, the mean and the variance. The mean of $f(x)$ in Eqn (3.172) is x_0 and the variance is

$$V[f(x)] = \sigma^2 = \int\limits_{x_0-1/2}^{x_0+1/2} [x - x_0]^2 f(x)dx = \int\limits_{-1/2}^{+1/2} x'^2 dx' = \frac{1}{12} \tag{3.174}$$

Experimental tests have verified the uniform distribution of roundoff in computer systems. In fact, computers generate random numbers by using the overflow value of the mantissa. We can represent roundoff as an additive random error (ε) superimposed on the true variable (x). In this case, we can write the computer representation of our variable (which we assume is free from measurement and sampling errors) as $x + \varepsilon$. For a floating-point number system, it is better to use

$$x(1 + \varepsilon); |\varepsilon| < \frac{1}{2}(10^{-2}) \tag{3.175}$$

for the variable with roundoff error ε. This formulation has the effect of focusing attention on the consequences of roundoff for every application in which it appears. For example, the product

$$x_1(1 + \varepsilon_1)x_2(1 + \varepsilon_2) = x_1 x_2(1 + \varepsilon_1 + \varepsilon_2 + \varepsilon_1\varepsilon_2) \tag{3.176}$$

demonstrates how roundoff propagates during multiplication. Generally, the product $\varepsilon_1\varepsilon_2$ is sufficiently small to be ignored. However, in the above multiplication we must include the roundoff for this operation, whereby Eqn (3.176) becomes

$$x_1(1 + \varepsilon_1)x_2(1 + \varepsilon_2) = x_3(1 + \varepsilon_3) = x_1 x_2(1 + \varepsilon_1 + \varepsilon_2 + \varepsilon)$$

$$|\varepsilon| < \frac{1}{2}(10^{-2}); \varepsilon_3 = \varepsilon_1 + \varepsilon_2 + \varepsilon \tag{3.177}$$

Similar error propagation results are found for other arithmetical operations. We can extend this to a generalized product

$$y_1(1 + \varepsilon_1)y_2(1 + \varepsilon_2)...y_N(1 + \varepsilon_N) \tag{3.178}$$

which becomes

$$y_1 y_2...y_N[1 + (\varepsilon_1 + \varepsilon_2 + ... + \varepsilon_N)] \tag{3.179}$$

By the central limit theorem, the distribution of a multitude (M) of sums of N independent random numbers—in this case, $\bar{\varepsilon}_m = \sum_{i=1}^{N} \varepsilon_i$ ($m = 1,...,M$) for the roundoff errors—approaches a normal distribution with mean $\mu_{\bar{\varepsilon}}$ and variance $\sigma_{\bar{\varepsilon}}^2$ as M becomes large. The effect for the other operations, such as a sum of variables, y, is much the same; therefore, while individual roundoffs are from a uniform distribution, the result of many arithmetic roundoff operations tends toward a normal distribution. This also can be demonstrated experimentally.

As stated earlier, we will generally ignore roundoff as a source of error in the processing and analysis of oceanographic data. The above discussion has been presented here to make the reader aware of potential problems and provide some familiarity with the problems of using computing systems. In most data applications, the effects of roundoff error are small enough to be ignored. Only in the case of recursive calculations, where each computation depends on the previous one, do we anticipate large roundoff errors. This is usually a problem for numerical modelers who must deal with the repeated manipulation of computer-generated "data." In cases where roundoff errors are of some consequence, statistical methods can be used in which the errors can be treated as variables from a normal population.

3.16.4 Gauss–Markov Theorem

The term *Gauss–Markov process* is often used to model certain kinds of random variability in oceanography. To understand the assumptions behind this process, consider the standard linear regression model, $y = \alpha + \beta x + \varepsilon$, developed in the previous sections. As before, α, β are regression coefficients, x is a deterministic variable and ε is a random variable. According to the Gauss–Markov theorem, the estimators α, β can be found from least squares analysis and are the *best linear unbiased estimators* for the model with the following conditions on ε:

1. The random variable ε is independent of the independent variable, x;
2. ε has a mean of zero; that is $E[\varepsilon] = 0$;
3. Errors ε_j and ε_k associated with any two points in the population are independent of one another; the covariance between any two errors is zero; $C[\varepsilon_j, \varepsilon_k] = 0, j \neq k$;

4. ε has a finite variance $\sigma_\varepsilon^2 \neq 0$.

The estimators are *unbiased* because their expected values equal the population values (given conditions 1 and 2) and they are *best* in the sense that they are efficient (if conditions 3 and 4 hold true), the variance of the least-squares estimators being smaller than any other linear unbiased estimator. A further assumption that is often made is that the errors, ε, are normally distributed. In this case, the estimators of α, β, and μ using the least-squares requirements are identical to the estimators resulting from the use of maximum-likelihood estimation. This assumption, combined with the four previous assumptions, provide the rationale for the least-squares procedure.

3.17 INTERPOLATION: FILLING THE DATA GAPS

Most analysis procedures used in the physical sciences are designed for comparatively long and densely sampled series with equally spaced measurements in time or space. The wealth of information on time series analysis primarily applies to regularly spaced and abundant observations. There are two main reasons for this: (1) the mathematical necessity for long, equally spaced data for the derivation of statistically reliable estimates from modern analytical techniques; and (2) the fact that most modern measurement systems both collect and store data in digital format. Spectral estimates, for example, improve with increased duration of the data series in the sense that one is able to cover an increasing range of the dominant frequency constituents that make up the record. Digital sampling systems are considerably more economical than analogue recording systems in that they cut down on storage space, power consumption and postprocessing effort.

3.17.1 Equally and Unequally Spaced Data

Electronic systems now provide data at regularly spaced sampling increments. This includes data from autonomously recording instruments as well as data from instruments integrated into cabled observatory networks connected to a shore station. Equipment failure of any type generally leads to either data *gaps* or a premature termination of the record. The failure of electronic data logging systems—which in the case of cabled observatories also includes the hardware linking the instrument to the shore station—is but one source of gappy records in physical oceanography. Because of their very nature, shipborne measurements are a source of gappy records. Oceanographic research vessels are expensive platforms to operate and must be used in an optimal fashion. As a consequence, it is often impossible to collect observations in time or space of sufficient regularity and spacing to resolve the phenomenon of interest. Efforts are usually made to space measurements as evenly as possible but, for a variety of reasons, station spacings are often considerably greater than desired. Diminishing science budgets are having major impacts on maintaining historical sampling schedules for existing climate-based time series. Weather conditions, as well as ship and equipment problems, almost invariably lead to unwanted gaps in the data set. Sometimes equipment failures are not detected until the data are examined in the laboratory. In addition, editing out errors produces unwanted gaps in the data record.

The gap problem is even more severe when one is analyzing historical data or data collected from "platforms of opportunity." Historical data are a collection of many different sampling programs all of which had different goals and therefore very different sampling requirements. By their very nature, such collections of data will necessarily be irregularly spaced and variable in terms of accuracy and reliability. Further editing, dictated by the goals of the historical data analysis project, will add new gaps to the set of existing data series.

Monitoring stations, ships of opportunity, and satellite measurements frequently produce data series that are unevenly spaced. The geographic distribution of monitoring stations (e.g., sea-level stations on Pacific Ocean islands, Deep-ocean Assessment and Reporting of Tsunamis (DART) tsunami recording stations, and meteorological buoy stations) is far from uniform in terms of the spacing between stations. Thus, while the data series collected at each station, may themselves consist of evenly and densely spaced measurements in time, the space intervals between stations will be highly irregular. Open ocean buoys and current meter moorings also fit this classification of densely and evenly spaced temporal observations at widely and often irregularly spaced locations. Here again, any failure in the recording system (digital or analogue), whether minor or catastrophic, will lead to gaps in the time series record. Often these gaps are quite large since unplanned recovery efforts are required to correct the problem. Such a correction effort assumes that the telemetering of data is available which, with the exception of data sent through satellites or through cabled observatory networks, is not widely available. Failures of onboard recording systems must wait until the scheduled servicing of the instrument, which may then result in relatively large data gaps.

At the other end of the sampling spectrum, satellite observing systems provide dense and evenly spaced spatial measurements that are often irregular in time. A familiar source of temporal gaps in infrared image series is cloud cover.

Both occasional and persistent cloud cover can interrupt a sequence of images collected to study changes of sea surface temperature. The effects of cloud cover apply also to satellite remote sensing in the optical bands. In addition to the cloud-cover problem, there are often problems with the onboard satellite sensing systems or associated with the ground receiving station that led to gaps in time series of image data. Microwave sensing of the surface is not sensitive to cloud attenuation but it is subject to sensor and ground-recording failure problems. Satellite radar data collected at the ocean surface are distorted by winds above a certain threshold speed. Off the Atlantic coast of North America, where there are strong oceanic features associated with the Gulf Stream, the wind threshold speed is about 10 m/s whereas off the Pacific coast, where oceanic frontal structure is more subdued, the wind speed threshold is about 3 m/s (Williams et al., 2013).

Platforms of opportunity (usually merchant ships) produce uniquely irregular sets of measurements. Most merchant ships repeat the same course with minor adjustments for local weather conditions and season. A seasonal shift in course is generally seen at higher latitudes to take advantage of great circle routes during times of better weather. A return to lower latitudes is seen in winter as the ships avoid problems with strong storms. Added to the seasonal track changes is the nature of the daily sampling procedure. Usually, the ship takes measurements at some specified time interval which, due to variations in ship track, ship speed and weather conditions, may be at very different positions from sailing to sailing. Thus, the merchant ship data will be irregular in both space and time. Systems that operate continuously from ships of opportunity (e.g., injection SST) overcome this problem. These continuous measurements, however, are still subject to variations in ship track.

The net result of all these measurement problems is that oceanographers are often faced with short records of unequally spaced data. Even if the records are long, they are often gappy in time or space. It is, therefore, necessary to interpolate these data to produce series of evenly spaced measurements. While some analysis procedures, such as least-squares harmonic analysis favored in tidal analysis (cf., Foreman et al.,1995 (updated 2009); Pawlowicz et al., 2002), apply directly to uneven or gappy data, it is more often the case that irregularly spaced data are interpolated to yield evenly spaced, regular data. These interpolated records can then be analyzed with familiar methods of time series analysis.

Interpolation also may be required with evenly spaced data if the subject dynamics apply to smaller space/timescales than are resolved by the measurements. Thus, the data points that are interpolated produce another set of regularly spaced points with a finer resolution. Many interpolation procedures have been developed that only apply to evenly spaced data.

3.17.2 Interpolation Methods

Interpolation techniques are needed for both irregularly spaced and evenly spaced data series. Before deciding which interpolation method is most effective, we need to consider the particular application. A series of appropriate questions regarding the selection of the best interpolation procedures are:

1. What samples (original data series, derivatives, etc.) should we use?
2. What class of interpolation function (linear, higher-order polynomial, cubic spline, etc.) best satisfies the dynamical restrictions of the analysis?
3. What mathematical criteria (exact data point matching, least-squares fit, continuity of slopes, etc.) do we use to derive the interpolated values?
4. Where do we apply these criteria?

Answers to these questions serve as guides to the selection of a unique interpolation procedure.

3.17.2.1 Linear Interpolation

The type of interpolation scheme to be employed depends on how many data points we want our interpolation curve (polynomial) to pass through (i.e., to "fit"). Increasing the number of points we want the curve to pass through, increases the order of the polynomial we need to do the fitting. The most straightforward and widely used interpolation procedure is that of *linear interpolation*. This consists of connecting a straight line between two data points and choosing interpolated values at the appropriate positions along that line. For a data series $y(x)$, this linear procedure can be written as

$$
\begin{aligned}
y(x) &= y(a) + \frac{x-a}{b-a}[y(b) - y(a)] \\
&= \frac{(b-x)y(a) + (x-a)y(b)}{b-a}
\end{aligned}
\tag{3.180}
$$

where $x_{\text{start}} = a$ and $x_{\text{end}} = b$ are the times (or positions) of the data collection at the start and end of the sampling increment being interpolated, and x represents the corresponding time (or position) of the desired interpolated value within the interval $[a, b]$. This is the customary procedure for interpolating between values in most tables. The same formula can be applied to *extrapolation* (extending the data beyond the domain of the observations) where the point x lies beyond the interval $[a, b]$. Eqn (3.180) is a special case of the Lagrange polynomial interpolation formula discussed in the next section.

3.17.2.2 Polynomial Interpolation

If we wish to interpolate between more than two points simultaneously, we need to use higher-order polynomials than the first-order polynomial (straight line) used in the previous section. For example, through three points we can find a unique polynomial of degree 2 (a quadratic); through four points, a unique polynomial of degree 3 (a cubic), and so on. The two methods described below are computationally robust in the sense that they yield reasonable results at most points. Polynomial interpolation techniques, such as Vandermonde's and Newton's methods are awkward to program and suffer from problems with roundoff error.

3.17.2.3 Lagrange's Method

The Lagrange polynomial interpolation formula is a method for finding an interpolating polynomial $y(x)$ of degree N which passes through all of the available data points at the same time, (x_i, y_i); $i = 1, 2, ..., N + 1$. The general form for this polynomial, of which linear interpolation is a special case, is given as

$$y(x) = a_0 + a_1 x + a_2 x^2 + ... + a_N x^N = \sum_{k=0}^{N} a_k x^k$$

$$= \sum_{i=1}^{N+1} \left[y_i \left(\Pi_{\substack{k=1 \\ k \neq i}}^{N+1} \left(\frac{x - x_k}{x_i - x_k} \right) \right) \right]$$

(3.181)

where Π is the product function. Note that in the product function, the ith term—corresponding to the particular data point, x_i, in the denominator—is not included when calculating the product for the term involving x_i. Even though k ranges from 1 to $N + 1$, Π uses only N terms and the final polynomial is of order N, as required.

The goal of the Lagrange interpolation method is to find an Nth degree polynomial which is constrained to pass through the original $N + 1$ data points and which yields a "reasonable" interpolated value for any position x located anywhere between the original data points. To see that the polynomial passes through the original data points, note that the ith product function, Π_i, defined for the data point x_i in the denominator is constructed in such a way that $\Pi_i(x_j; x_i) = \delta_{ij}$ whenever $x = x_j$ is one of the data values (δ_{ij} is the Kronecker delta function). This means that $\Pi_i(x_j; x_i) = 0$ for all x_j except for the specific value $x = x_i$ found in the original data series that matches the term in the denominator. In the latter case, $\Pi_i(x_j; x_i) = 1$ and $y_i \Pi_i(x_j; x_i) = y_j$.

The general polynomial we seek is constructed as a sum of the product functions in Eqn (3.181) which can be expanded to give

$$y(x) = \sum_{i=1}^{N+1} y_i [Q_i(x) / Q_i(x_i)]$$

(3.182)

in which

$$Q_i(x) = (x - x_1)(x - x_2)...(x - x_{i-1})(x - x_{i+1})...(x - x_{N+1})$$

(3.183)

is the product of all the factors except the ith one. For any x, Eqn (3.182) can be expanded to give the interpolating polynomial

$$y(x) = y_1 \frac{(x - x_2)(x - x_3)...(x - x_{N+1})}{(x_1 - x_2)(x_1 - x_3)...(x_1 - x_{N+1})} + y_2 \frac{(x - x_1)(x - x_3)...(x - x_{N+1})}{(x_2 - x_1)(x_2 - x_3)...(x_2 - x_{N+1})} + ...$$
$$+ y_{N+1} \frac{(x - x_1)(x - x_2)...(x - x_N)}{(x_{N+1} - x_1)(x_{N+1} - x_3)...(x_{N+1} - x_N)}$$

(3.184)

Note that, for the original data points, $x = x_i$, the polynomial yields the correct output value $y(x_i) = y_i$, as required.

In the Lagrange interpolation method, the calculation is based on all the known data values. If the user wants to add new data to the series, the whole calculation must be repeated from the start. Although the above formula can be applied directly, programming improvements exist that should be taken into account (Press et al., 1992). Use of Neville's algorithm for constructing the interpolating polynomial is more efficient and allows for an estimate of the errors resulting from the curve fit.

As an example of this interpolation method, consider four points (x_i, y_i), $i = 1, ..., 4$ given as (0, 2), (1, 2), (2, 0) and (3, 0) through which we wish to fit a (cubic) polynomial. Substituting these values into Eqn (3.183), we obtain

$$y(x) = 2\frac{(x-1)(x-2)(x-3)}{(0-1)(0-2)(0-3)} + 2\frac{(x-0)(x-2)(x-3)}{(1-0)(1-2)(1-3)} + 0 + 0$$

$$= \frac{2}{3}x^3 - 3x^2 + \frac{7}{3}x + 2$$

The resulting third-order curve is plotted in Figure 3.19.

3.17.2.4 Spline Interpolation

In recent years, the method that has received the widest general acceptance is the spline interpolation method. Splines, unlike other polynomial interpolations, such as the Lagrange polynomial interpolation formula, apply to a series of segments of the data record rather than the entire data series. This leads to the obvious question to ask in selecting the proper interpolation procedure: Do we want a single, high-order polynomial for the interpolation over the entire domain, or would it be better to use a sequence of lower-order polynomials for short segments and sum them over the domain of interest? This integration is inherently a smoothing operation but one must be careful of discontinuities, or sharp corners, where the segments join together. Spline functions are designed to overcome such discontinuities, at least for the lower-order derivatives. It is because discontinuities are allowed in higher-order derivatives that splines are so effective locally. Constraints placed on the interpolated series in one region have only very small effects on regions far removed. As a result, splines are more effective at fitting nonanalytic distributions characteristic of real data. The term "spline" derives from the flexible drafting tool used by naval architects to draw piecewise continuous curves.

Splines have other favorable properties such as good convergence, highly accurate derivative approximation, and good stability in the presence of roundoff errors. Splines represent a middle ground between a purely analytical description and numerical finite difference methods, which breaks the domain into the smallest possible intervals. The piecewise approximation philosophy represented by splines has given rise to finite element numerical methods.

With spline interpolation, we approximate the interpolation function, $y(x)$, over the interval $[a, b]$ by dividing the interval into subregions with the requirement that there be continuity of the function at the joints. We can define a spline function, $y(x)$, of degree N with values at the joints. $a = u_0 \leq u_1 \leq u_2 ... \leq u_N = b$ and having the properties:

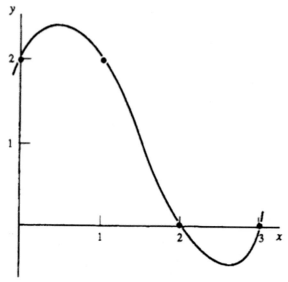

FIGURE 3.19 Use of Lagrange's method to fit a third order (cubic) polynomial through the data points (x_i, y) given by (0, 2), (1, 2), (2, 0), and (3, 0).

1. In each interval $u_{i-1} \leq x \leq u_i$ ($i = 1, ..., m$), the function $y(x)$ is a polynomial of degree not greater than N;
2. At each interior joint, $y(x)$ and its first $N - 1$ derivatives are continuous.

The spline function in widest use is the cubic spline ($N = 3$). To give the reader familiarity with the spline interpolation technique, we will develop the cubic spline equations and work through a simple example. Consider a data series with elements (x_i, y_i), $i = 1, ..., N$. Since we are working with a cubic spline interpolation, the first two derivatives $y'(x)$ and $y''(x)$ of the interpolation function, $y(x)$, can be defined for each of the points x_i while the third derivatives $y'''(x)$ will be a constant for all x. Here, the prime symbol denotes differentiation with respect to the independent variable x. We write the spline function in the form

$$y(x) = f_i(x); x_i \leq x \leq x_{i+1}, i = 1, ..., N-1 \tag{3.185}$$

and specify the following conditions at the junctions of the segments:

1. Continuity of the spline function:

$$
\begin{aligned}
f_i(x_i) &= y(x_i) = y_i, i = 1, 2, ..., N-1; \\
f_{i-1}(x_i) &= y(x_i) = y_i, i = 2, 3, ..., N;
\end{aligned}
\tag{3.186a}
$$

2. Continuity of the slope (first derivative):

$$f'_{i-1}(x_i) = f'_i(x_i), i = 1, 2, ..., N-1 \tag{3.186b}$$

3. Continuity of second derivative:

$$f''_{i-1}(x_i) = f''_i(x_i), i = 1, 2, ..., N-1 \tag{3.186c}$$

Because $y'''(x) = $ constant, $y''(x)$ must be linear, so that

$$
\begin{aligned}
f''_i(x_i) &= y''_i \frac{(x_{i+1} - x)}{(x_{i+1} - x_i)} \\
&= y''_{i+1} \frac{(x - x_i)}{(x_{i+1} - x_i)}
\end{aligned}
\tag{3.187}
$$

Integrating twice and selecting integration constants to satisfy the conditions, Eqns (3.186a) and (3.186b) on $f_i(x_i)$ and $f_{i-1}(x_i)$ gives

$$
\begin{aligned}
f_i(x_i) = {} & y_i \frac{(x_{i+1} - x)}{(x_{i+1} - x_i)} + y_{i+1} \frac{(x - x_i)}{(x_{i+1} - x_i)} - \frac{(x_{i+1} - x_i)^2}{6} y''_i \left\{ \frac{(x_{i+1} - x)}{(x_{i+1} - x_i)} - \left[\frac{(x_{i+1} - x)}{(x_{i+1} - x_i)} \right]^3 \right\} \\
& - \frac{(x_{i+1} - x_i)^2}{6} y''_{i+1} \left\{ \frac{(x - x_i)}{(x_{i+1} - x_i)} - \left[\frac{(x - x_i)}{(x_{i+1} - x_i)} \right]^3 \right\}
\end{aligned}
\tag{3.188}
$$

which uniquely satisfies the continuity condition for the second derivative but not, in general, for the first derivative (slope). To ensure continuity of the slope at the seams, we expand Eqn (3.187) by differentiation to get

$$f'_i(x_i) = \frac{(y_{i+1} - y_i)}{(x_{i+1} - x_i)} - \frac{(x_{i+1} - x_i)}{6} (2y''_i + y''_{i+1}) \tag{3.189a}$$

$$f'_{i-1}(x_i) = \frac{(y_i - y_{i-1})}{(x_{i+1} - x_i)} - \frac{(x_{i+1} - x_i)}{6} (y''_{i-1} + 2y''_i) \tag{3.189b}$$

We then set Eqns (3.189a) and (3.189b) to be equal in order to satisfy slope continuity Eqn (3.186b), whereby

$$(x_i - x_{i-1})y''_{i-1} + 2[(x_{i+1} - x_{i-1})]y''_i + (x_{i+1} - x_i)y''_{i+1}$$

$$= 6\frac{(y_{i+1} - y_i)}{x_{i+1} - x_i} - \frac{(y_i - y_{i-1})}{x_i - x_{i-1}}, i = 2, ..., N-1 \tag{3.190}$$

which must be satisfied at $N - 2$ points by the N unknown quantities, y''_i. We require two more conditions on the y''_i that we get by specifying conditions at the end points x_1 and x_N of the data sequence. After specifying these end values, we have $N - 2$ unknowns, which we find by solving the $N - 2$ equations. There are two main ways of specifying the end points: (1) we set one or both of the second derivatives, y''_i and y''_N at the end points to be zero (this is termed the *natural cubic spline*) so that the interpolating function has zero curvature at one or both boundaries; or (2) we set either y''_i and y''_N to values derived from Eqns (3.189a) and (3.189b) in order that the first derivatives of the interpolating function, y'_i, take on specified values at one or both of the termination boundaries.

As a general example, we consider the spline solution for six evenly spaced points with the data interval $h = x_{i+1} - x_i$ and function d_i defined in terms of y_i, as

$$d_i = \frac{(y_{i+1} - 2y_i + y_{i-1})}{2h^2} \tag{3.191}$$

We can write Eqn (3.188) for these six equally spaced points in matrix form as

$$\begin{pmatrix} 4 & 1 & 1 & 0 \\ 1 & 4 & 1 & 0 \\ 0 & 1 & 4 & 1 \\ 0 & 0 & 1 & 4 \end{pmatrix} \begin{pmatrix} y''_2 \\ y''_3 \\ y''_4 \\ y''_5 \end{pmatrix} = \begin{pmatrix} 12d_2 - \dfrac{y''_1}{h} \\ 12d_3 \\ 12d_4 \\ 12d_5 - \dfrac{y''_6}{h} \end{pmatrix} \tag{3.192}$$

If we want to specify y'_i rather than y''_i, we need an equation relating both. If the end conditions are not known, the simplest choice is $y''_i = 0$ (the *natural spline* noted above). Another, and smoother choice (in the sense of less inflection or curvature at the interpolated point) is $y''_1 = 0.05y''_2$. Although spline interpolation is a global, rather than a local, curve (altering a y''_i or an end condition affects the overall spline), the dominant diagonal terms in Eqn (3.192) cause the effects to rapidly decrease as the distance from the altered point increases.

We should point out the method of splines offers no advantage over polynomial interpolation when applied to either the approximation of well-behaved mathematical functions or to curve fitting when the experimental data are dense. "Dense" means that the number of data points in a subregion is more than an order of magnitude larger than the number of inflection points in the fitted curve and that there are no abrupt changes in the second derivative. The advantage of splines is their inherent smoothness when dealing with sparse data.

As a numerical example of spline fitting, we consider the six-point fitting of the points represented in Eqn (3.192) for the 11 data points in Table 3.13. Using a general polynomial fit yields the curve in Figure 3.20. Here, all but the last of the first six points lie on a near straight line. Due to this single point, the polynomial curve oscillates with an amplitude that does not decrease. In contrast, the spline amplitude (Figure 3.21) for the same 11 values reduces each cycle by a factor of 3.

Often the first or second derivatives of the interpolated function are important. In Figure 3.20, we see that fitting a polynomial to sparse data can result in large, unrealistic changes in the second derivatives. The spline fit to the same points (Figure 3.21) using the end-point conditions $y''_1 = y''_N = 0$ demonstrates the smoothness of the spline interpolation. In essence, the spline method sacrifices higher-order continuity to achieve second derivative smoothness.

Spline interpolation is generally accomplished by computer routines that operate on the data set in question. Computer routines solve for the spline functions by solving the equation

$$\sum_{i=1}^{N} [(g(x_i) - y_i)/\delta y_i]^2 = S \tag{3.193}$$

where $g(x_i)$ is composed of cubic parabolas

$$g(x) = a_i + b_i(x - x_i) + c_i(x - x_i)^2 + d_i(x - x_i)^3 \tag{3.194}$$

TABLE 3.13 Data pairs (x_i, y_i) used for interpolation schemes in Figures 3.20 and 3.21

i	x_i	y_i
1	0	16
2	14	19
3	27	36
4	33	48
5	41	53
6	48	90
7	62	119
8	74	120
9	89	96
10	99	71
11	114	36

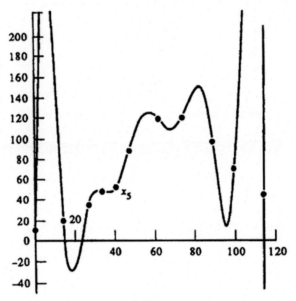

FIGURE 3.20   A general six-point polynomial fit to the data values in Table 3.13. Due to a single point, the polynomial curve oscillates with an amplitude that does not decrease with x.

for the interval $x_i \leq x \leq x_{i+1}$. The terms δy_i are positive numbers that control the amount of smoothing at each point; the larger is δy_i, the more closely the spline fits at each data point. A good choice for δy_i is the standard deviation of the data values.

The S term also controls smoothing, resulting in more smoothing when S increases. As S gets smaller, smoothing decreases and the splines fit the data points more closely. When $S = 0$, the data points are fitted exactly by the interpolating spline functions. A recommended value of S is $N/2$, where N is the number of data points. An even smoother interpolation can be achieved using splines under tension. Tension is introduced into the spline procedure to eliminate extraneous inflection points. An iterative procedure is usually used to select the best level for the tension parameter.

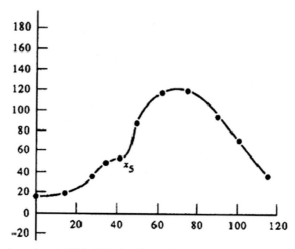

FIGURE 3.21   Cubic spline fit to the data values in Table 3.13. Amplitude of each cycle is reduced by a factor of three compared to Figure 3.20.

3.17.3 Interpolating Gappy Records: Practical Examples

Gaps or "holes" occur frequently in geophysical data series. Gaps in a stationary time series are, of course, analogous to gaps in a homogeneous spatial distribution. Small gaps are of little concern and linear interpolation is recommended for filling the gaps. If the gaps are large (of the size of the integral time or space scale), it is generally better to work with the existing short data segments than to "make up" data by pushing interpolation schemes beyond their accepted limitations. For the gray area between these two extremes, one wants to know how large the data loss can be and still permit reasonable use of standard interpolation techniques and processing methods. The problem of gappy data in oceanography was addressed by Thompson (1971) who suggested that a random sampling of data points might be an optimally efficient approach. Further insight into the problem of missing data can found in Davis and Regier (1977) and Bretherton and McWilliams (1980). In this section, we present two examples of how to deal with gappy data. One is a straightforward analysis by Sturges (1983), who used monthly tide gauge data to investigate what happens to spectral estimates when one punches holes in the data set. The other is a practical guide to the interpolation of satellite-tracked Lagrangian drifter data with its inherently irregular time steps.

3.17.3.1 Interpolating Gappy Records for Time Series Analysis

Sturges (1983) used a Monte Carlo technique to poke holes at random in a known time series of monthly mean sea level. The original record had a "red" spectrum, which fell off as f^{-3} at high frequencies (f) and contained a single major spectral peak at a period of 12 months. A total of 120 months of data were used in the analysis. The idea was to reconstruct the gappy series using a cubic spline interpolation method and see how closely the spectrum from the interpolated time series resembled that of the original time series. Data loss was limited to less than 30% of the record length and, for any individual experiment, the holes were all the same length. However, different hole lengths were used in successive runs. The only stipulation was that the length of the data segment before the next gap be at least as long as the gap itself. The program was not allowed to eliminate the first and last data points.

Cross-spectra were computed between the original time series and the interpolated gappy series. For a specified hole size, holes were generated randomly in the data series, the cross-spectra computed and the entire process repeated 1,000 times. The magnitudes of the resulting cross-spectra provided estimates of how much power was lost or gained during the interpolation while the corresponding phases was interpreted as the error introduced by the interpolation process (Figure 3.22). Several important conclusions arise from Sturges' analysis:

1. Gaps have a more adverse effect on weak spectral components (spectral peaks) than on strong ones embedded in the same background spectrum;
2. The phase can be estimated to roughly 10° uncertainty at the 99% confidence level for data losses of up to 30% for a strong spectral signal; the requirement is that the gaps are kept to about 1/3 (0.33) of the period of the signal being examined. If the gaps increase to 1/2 (0.5) of the period, maintenance of a 10° phase uncertainty at the 99% level

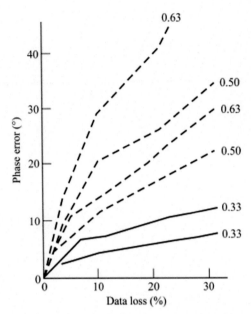

FIGURE 3.22 Absolute phase error in degrees (°) expressed as a function of the data lost (%) between the original sea level time series and the same series but with randomly generated holes that have been filled in using a cubic spline fit. Each line shows the ratio D/T, where D is the length of the random gap and T is the spectral period of interest; for example, the value 0.5 means that the holes were four units (months) long and that the period was 8 units long. Results are shown for the 90% and 99% confidence limits (lower and upper curves for each ratio value, respectively). *From Sturges (1983).*

requires that the data loss is less than 5% (in Figure 3.22, the x-axis indicates the percentage of data removed then filled with a cubic spline algorithm when examining the phase errors along the y-axis);

3. Although correlation functions can be computed for gappy data, it is much more difficult to compute the cross-correlation function for these data.

According to Sturges' analysis, the adverse effects of gaps depend on the length of gaps relative to the length of the data set and on the magnitudes of the dominant spectral components in the signal.

3.17.3.2 Interpolating Satellite-Tracked Positional Data

The analysis of positional (latitude, longitude) time series collected through the Service Argos and Iridium satellite-tracking systems illustrates some of the problems that may arise with standard interpolation procedures. Because the times that polar-orbiting satellites pass over an oceanic region change through the day and because drifters move relative to the orbits of the satellite, the times between satellite fixes are irregular. At midlatitudes, times between locational fixes for Service Argos can range from less than an hour to as long as 10 h. Typical average times between fixes are around 2—3 h (Thomson et al., 1997). The challenge is to generate regularly spaced time series of latitude (x) and longitude (y) from which one can derive regularly spaced time series of drifter zonal velocity ($u = \Delta x/\Delta t$) and meridional velocity ($v = \Delta y/\Delta t$). This challenge is especially problematic where a "duty cycle" has been programmed into the drifter transmitter to reduce the number (and cost) of transmissions to the passing satellites. A commonly used duty cycle, consisting of 1 day of continuous transmission followed by 2 days of no transmission, results in large data gaps that can make calculation of mean currents difficult in regions having strong currents in the inertial and tidal frequency bands. The duty cycle of 8 h continuous transmission followed by 16 h of silence is superior for midlatitude regions with strong inertial or tidal frequency variability.

Because of strong inertial motions in the upper layer of the open ocean and strong tidal motions over continental margins, sampling intervals of 3—4 h, or less, are preferable. A typical time step of 6 h used in many analyses of satellite-tracked drifters is inadequate to resolve inertial motions except in regions equatorward of 30° latitude where the inertial period $T = 1/f_{inertial}$ exceeds 24 h (at 50° latitude, $T \approx 16.5$ h; see *Coriolis frequency*). To generate time series at a reasonably short time step, say 3 h, we need to interpolate between irregularly spaced data points. To do this, we use a cubic spline interpolation for each of the positional records. After the correct start and end times for the oceanic portion of the record have been determined, the first step in the process is to remove any erroneous points from the "raw" data by calculating speeds over adjacent time steps, t_i; e.g., $u_i = (x_{i+1} - x_i)/(t_{i+1} - t_i)$. One then omits any unrealistic velocity values that exceed some threshold value (say 5 m/s). This "edited" record needs to undergo further editing by averaging successive data positions for which the time step Δt is less than an hour. The reason for this is quite simple: Because positional accuracies Δx and Δy are about 350 m approximately 63% of the time, velocity errors are roughly $\Delta x/\Delta t >$ 0.1 m/s when $\Delta t < 1$ h. Such error values are comparable to mean ocean currents and need to be eliminated from the records. Drifters located using GPS transmitters have much smaller positional errors and, therefore, better velocity resolution. The time series also need to be examined for drogue-on, drogue-off. If a reliable strain sensor is built into the drogue system, it can be used to determine if and when the drogue fell off. Otherwise, one needs to calculate the speed-squared from the raw data and look for sudden major "jumps" in speed that signal loss of the drogue (Figure 3.23). We recommend this approach for all modern-day drifters since strain gauge sensors appear to be unreliable. At the time that the third this edition of this book was being written, drogue loss and not battery or transmitter failure, was the primary cause of drifter "failure" in the open ocean. Argo drifters are not drogued, but drift at depth where the vertical current shear is generally weak, thereby reducing slippage and associated velocity error.

Provided there are more than about six accurate satellite fixes per day, the edited positional records can be interpolated to regularly spaced 3-h time series using a cubic spline interpolation algorithm. In general, the spline curve will be well behaved and the fit will resemble the kind of curve one would draw through the data by eye. Inertial and tidal loops in the trajectory will be fairly well resolved. Spurious results will occur where data gaps are too large to properly condition the spline interpolation algorithm. Assuming that the spline interpolation of positions looks reasonable, the next step is to calculate the velocity components from the rate of change of position. It is tempting to equate the coefficient for the linear term in the cubic spline interpolation to the "instantaneous" velocity at any location along the drifter trajectory. That would be a mistake. Although trajectories can look quite smooth, curvatures can be large and resulting velocities unrealistic. In fact, use of the spline coefficients to calculate instantaneous velocity components leads to an increase in the kinetic energy of the motions. The reader can verify this by artificially generating a continuous time series of position consisting of a linear trend and time varying inertial motions. The artificial position record is then decimated to 3-hourly values and a cubic spline interpolation scheme applied. Using instantaneous velocity values at the 3-hourly time steps derived from the interpolation, one finds that the kinetic energy in most frequency bands is increased relative to the original record. The recommended procedure is to calculate the two horizontal velocity components (u, v) from the central differences between three consecutive values of the 3-hourly positional data. From first differences, the velocity components at each point "i" are then: $u_i = (x_{i+1} - x_i)/(t_{i+1} - t_i)$ and $v_i = (y_{i+1} - y_i)/(t_{i+1} - t_i)$ for simple two-point differences or for the recommended centered values, $u_i = (x_{i+1} - x_{i-1})/(t_{i+1} - t_{i-1})$ and $v_i = (y_{i+1} - y_{i-1})/(t_{i+1} - t_{i-1})$. In summary, for those oceanic regions subject to pronounced inertial and tidal frequency motions, we have recommended the use of cubic spline interpolation to generate 2−4-hourly time series for position but simple linear interpolation of positional data to generate the corresponding time series for velocity. The interpolation requires more than six to eight satellite fixes through the day to be successful.

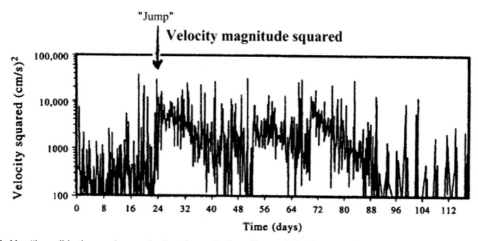

FIGURE 3.23 Sudden "jumps" in the speed-squared values from edited satellite-tracked drifter velocity data collected in the North Pacific near 50° N between 133 and 142° W longitude during the period September 4 to December 30, 1990. The "jump" indicates rapid acceleration of the drifter following probable loss of its drogue.

Trajectories with data gaps that are long relative to the local inertial period require special consideration. For gaps associated with a transmitter duty cycle of 8 h "on" followed by 16 h "off," we can obtain accurate daily mean positional values by least-squares fitting a time-varying continuous function to successive segments of the irregular data and then averaging the resulting function over successive 24-h periods. This filtering process is as follows (see Bograd et al., 1999):

1. Use least squares to fit a specified function, $\xi(t)$, to several (N) successive 8-h days of zonal (or meridional) trajectory data. The general model has the form $\xi(t) = a + bt + ct^2 + dt^3 + a_1.\sin(2\pi f_1 t + \phi_1) + a_2.\sin(2\pi f_2 t + \phi_2)$ where a, b, c, d, a_1, ϕ_1, a_2 and ϕ_2 are the unknown coefficients, f_1 is the local Coriolis frequency and f_2 the semidiurnal frequency (~ 0.081 cph). The phases ϕ_1, ϕ_2 for the two frequencies will vary from segment to segment. We suggest that four to 5 days ($N = 4$ or 5) of data be used for each segment fit. Shorter segments will have too few data for an accurate least-squares fit; longer segments will result in too much smoothing of the intermittent inertial and tidal motions;

2. Repeat the least-squares operation for each segment of length N days, shifting forward in time by 1 day after each set of coefficients is determined. This yields one estimate for the first day $\xi_1 = \xi\,(t = t_1)$, two estimates for the second day, ξ_2, three estimates for the third day and four estimates for all other days until near the end of the record when the number of estimates again falls to unity for the last record. Average all the values in each daily segment for each of the multiple curves $\xi_i(t)$ ($i = 1, \ldots,$ up to N) to get the average daily latitude $\xi_x(t)$ and longitude $\xi_y(t)$;

3. The pairs of coefficients a_1, ϕ_1 and a_2, ϕ_2 can be used to give rough reconstructions of the inertial and semidiurnal tidal motions, respectively. However, expect the phases to fluctuate considerably from segment to segment due to natural variability in the phases of the motions and from contamination by adjacent frequency bands.

For the duty cycle consisting of 1 day "on" followed by 2 days "off," the model is less useful (except at equatorial latitudes) and requires a much longer data segment (say 12 days instead of four) for each least-squares analysis.

3.17.3.3 Interpolation Records from Nearby Stations

Provided that the spatial scales of the processes being examined are large compared to the separation between sampling sites, short gaps in the time series at one location can be filled using an identical type of time series from a nearby location. For example, missing hourly tide heights at one coastal tide gauge station can be filled using hourly tide heights from an adjacent station further along the coast. To do this, we first use coincident data segments to determine the relative amplitudes and phases of the time series at the two locations.

A simple cross-correlation analysis can be used to determine the peak time lag between the series while the relative amplitudes can be obtained from the ratio of the standard deviations of the two series. Gaps in one time series (series 1) are then filled by applying the appropriate time lag and amplitude factor to the uninterrupted data series (series 2). Because tide gauges are generally in relatively protected embayments, each of which has its own particular frequency response characteristics, it is not possible to apply the tidal constituents for one site to an adjacent site even if the tidal constituents in the offshore region are nearly identical. A more sophisticated approach would be to first obtain the complex transfer function $H_{12}(\omega) = |H_{12}(\omega)|\exp[i\phi_{12}(\omega)]$ as a function of frequency ω for the two coincident time series. The missing time series values at site 1 could then be filled using the amplitudes $|H_{12}(\omega)|$ and phase differences $\phi_{12}(\omega)$ of the transfer function applied to the uninterrupted data series.

3.18 COVARIANCE AND THE COVARIANCE MATRIX

Covariance, like variance, is a measure of variability and has the units of the variables being examined. For two variables, the covariance is a measure of the joint variation about a common mean. When extended to a multivariate population, the relevant statistic is the covariance matrix. As we shall see, it is equivalent to what will be introduced later as the "mean product matrix." The covariance and covariance matrix are the fundamental concepts behind the spatial analysis techniques discussed in the next chapter.

3.18.1 Covariance and Structure Functions

The covariance $C(Y_1, Y_2)$, also written as $\text{cov}[Y_1, Y_2]$, between variables Y_1, Y_2 is

$$C(Y_1, Y_2) = E[(Y_1 - \mu_1)(Y_2 - \mu_2)] \tag{3.195}$$

where $\mu_1 = E[Y_1]$ and $\mu_2 = E[Y_2]$. A positive covariance indicates that Y_2 and Y_1 increase and decrease together while a negative covariance has Y_2 decreasing as Y_1 increases, and vice versa. We can expand Eqn (3.195) into a more convenient computational form

$$C(Y_1, Y_2) = E[Y_1 Y_2] - E[Y_1]E[Y_2] \tag{3.196}$$

Note, that if Y_1, Y_2 are independent random variables, then $C[Y_1, Y_2] = 0$.

For a two-dimensional isotropic velocity field, $u_i(\mathbf{y})$, the covariance tensor $C(r)$, also called the *structure function* from earlier studies of turbulence, takes the form

$$\begin{aligned} C_{ij}(r) &= \langle u_i(\mathbf{y})u_j(\mathbf{y}+\mathbf{r}) \rangle \\ &= \sigma^2 \frac{[f(r) - g(r)]r_i r_j}{r^2} + g(r)\delta_{ij} \end{aligned} \tag{3.197}$$

where $i (= 1, 2)$ and $j (= 1, 2)$ correspond to the two possible horizontal components of the velocity field, $\langle \cdot \rangle$ denotes an ensemble average, $r \equiv |\mathbf{r}|$, $\mathbf{y} = (y_1, y_2)$ is the position vector, $f(r)$ and $g(r)$ are, respectively, the one-dimensional longitudinal and transverse correlation functions, and $\sigma^2 = \langle u_i(\mathbf{y})^2 \rangle$. The longitudinal and transverse correlation functions are

$$f(r) = \langle u_L(\mathbf{y})u_L(\mathbf{y}+\mathbf{r}) \rangle \tag{3.198a}$$

$$g(r) = \langle u_P(\mathbf{y})u_P(\mathbf{y}+\mathbf{r}) \rangle \tag{3.198b}$$

where $u_L(\mathbf{y})$ and $u_P(\mathbf{y})$ are the velocity fluctuations parallel and perpendicular to $\mathbf{r} = (r_1, r_2)$. The velocity fluctuations are normalized so that the correlations equal unity at $|\mathbf{r}| = 0$. If the two-dimensional flow field is horizontally nondivergent, homogenous and isotropic, then $C_{ij}(r) = 0$ and

$$g(r) = \frac{d}{dr}[rf(r)] \tag{3.199}$$

Freeland et al. (1975) have used Eqn (3.199) to test for two-dimensional, nondivergent, homogenous, and isotropic low-frequency velocity structure in SOFAR (SOunding Fixing And Ranging) float data collected in the North Atlantic. Stacey et al. (1988) used this relation to test for similar flow structure in the Strait of Georgia, British Columbia. Although close to the error limits in certain cases, the observed structure is generally consistent with horizontal, nondivergent, homogeneous, and isotropic flow (Figure 3.24). The dotted lines in Figure 3.24 are the analytical functions

$$f(r) = (1 + br)e^{-br} \tag{3.200a}$$

$$g(r) = (1 + br - b^2 r^2)e^{-br} \tag{3.200b}$$

3.18.2 A Computational Example

If Y_1, Y_2 have a joint PDF

$$f(y_1, y_2) = \begin{cases} 2y_1, & 0 \le y_1 \le 1; 0 \le y_2 \le 1 \\ 0, & \text{elsewhere} \end{cases} \tag{3.201}$$

what is the covariance of Y_1, Y_2? We first write the expected value of Y_1, Y_2 as

$$\begin{aligned} E[Y_1 Y_2] &= \int_0^1 \int_0^1 y_1 y_2 f(y_1, y_2) dy_1 dy_2 = \int_0^1 \int_0^1 y_1 y_2 (2y_1) dy_1 dy_2 \\ &= \int_0^1 \frac{1}{3} y_2 \left(2y_1^3\right)\Big|_0^1 dy_2 = \int_0^1 \frac{2}{3} y_2 dy_2 = \frac{2}{3} \frac{y_2^2}{2}\Big|_0^1 = \frac{1}{3} \end{aligned}$$

Recall that, for discrete variables

$$E[g(Y_1, ..., Y_k)] = \sum_y ... \sum_y g(y_1, ..., y_k) P(y_1, ..., y_k)$$

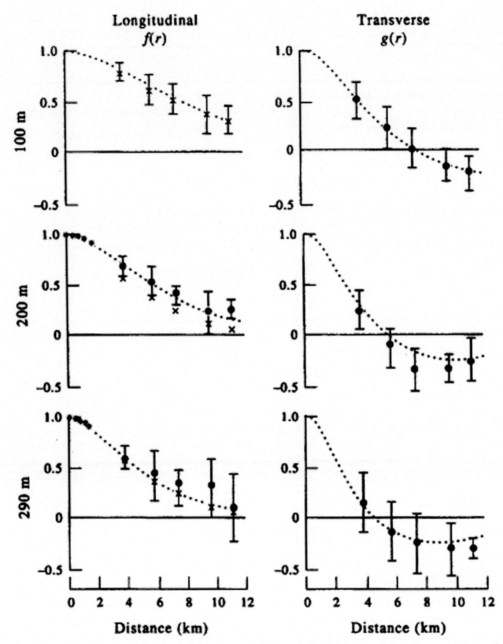

FIGURE 3.24   Longitudinal and transverse correlations at 100, 200, and 280/290 m depths. The dots are measured average values and error bars are the standard deviations. The mean and trend were removed from each time series before calculation of the correlations. The crosses are predicted values of $f(r)$ calculated using Eqn (3.199) by drawing straight line segments between the average values of $g(r)$ and integrating under the curve. *From Stacey et al. (1988).*

or for continuous variables

$$E[g(Y_1, \ldots, Y_k)] = \int_{y_k} \ldots \int_{y_1} g(y_1, \ldots, y_k) f(y_1, \ldots, y_k) dy_1 \ldots dy_k$$

For this example, we find $E[Y_1 Y_2] = 1/3$. Now

$$E[Y_1] = \int_0^1 \int_0^1 y_1(2y_1) dy_1 dy_2 = \int_0^1 \frac{2}{3} y_1^3 \Big|_0^1 dy_2 = \frac{2}{3} y_2 \Big|_0^1 = \frac{2}{3}$$

and $E[Y_2] = 1/2$, so that $C[Y_1Y_2] = E[Y_1Y_2] - \mu_1\mu_2 = 1/3 - (2/3)(1/2) = 0$. Therefore, Y_1 and Y_2 are independent. Of course, we could have anticipated this result since $f(y_1, y_2)$ in Eqn (3.201) is independent of y_2.

3.18.3 Multivariate Distributions

In the case of multivariate distributions, the covariance becomes the *covariance matrix*. If we have n measurements (samples) of N variables (Y), we can describe this as N random variables having a joint N-dimensional PDF

$$f_{1,2,...,N}(Y_1, Y_2, ..., Y_N) \tag{3.202}$$

If the random variables, Y, are mutually independent, the joint PDF can be factored in the usual way as

$$f_{1,2,...,N}(Y_1, Y_2, ..., Y_N) = f_1(Y_1)f_2(Y_2)...f_N(Y_N) \tag{3.203}$$

An important multivariate PDF is the multivariate normal PDF,

$$f_Y(Y) = \left(\frac{1}{(2\pi)^{\frac{N}{2}}|\mathbf{W}|^{\frac{1}{2}}}\right)\exp\left[-\frac{1}{2}(\mathbf{Y}-\boldsymbol{\mu})^T\mathbf{W}^{-1}(\mathbf{Y}-\boldsymbol{\mu})\right]$$

where $\mathbf{Y}^T = (Y_1, Y_2,..., Y_N)$, mean values $\boldsymbol{\mu}^T = (\mu_1, \mu_2,..., \mu_N)$, are row vectors and $\mathbf{W}^{-1}$ is the inverse of the covariance matrix $\mathbf{W}$

$$\mathbf{W} = \begin{pmatrix} \sigma_1^2 & \sigma_1\sigma_2\rho_{12} & \sigma_1\sigma_3\rho_{13} & ... & \sigma_1\sigma_N\rho_{1N} \\ \sigma_2\sigma_1\rho_{12} & \sigma_2^2 & \sigma_2\sigma_3\rho_{23} & ... & \sigma_2\sigma_N\rho_{2N} \\ ... & ... & ... & ... & ... \\ \sigma_N\sigma_1\rho_{N1} & \sigma_N\sigma_2\rho_{N2} & ... & ... & \sigma_N^2 \end{pmatrix} \tag{3.204}$$

where $\rho_{ij} = C_{ij}/\sigma_i\sigma_j$ is the correlation function; we can also write this as

$$\mathbf{W} = \begin{pmatrix} V[Y_1] & C[Y_1Y_2] & C[Y_1Y_3] & ... & C[Y_1Y_N] \\ C[Y_2Y_1] & V[Y_2] & C[Y_2Y_3] & ... & C[Y_2Y_N] \\ ... & ... & ... & ... & ... \\ C[Y_NY_1] & C[Y_NY_2] & C[Y_NY_3] & ... & V[Y_NY_N] \end{pmatrix} \tag{3.205}$$

Note that the covariance $C[Y_iY_j] = C[Y_jY_i]$ and, therefore, $\mathbf{W}$ is symmetric ($\mathbf{W} = \mathbf{W}^T$). In addition, $\mathbf{W}$ is positive semidefinite; that is, $|\mathbf{W}|$ and all its principal minors are nonnegative. Another way to show this is

$$V[\boldsymbol{\lambda}^T\mathbf{Y}] = E[\boldsymbol{\lambda}^T(\mathbf{Y}-\boldsymbol{\mu})(\mathbf{Y}-\boldsymbol{\mu})^T\boldsymbol{\lambda}] = \boldsymbol{\lambda}^T|\mathbf{W}| \tag{3.206}$$

which will always be nonnegative for any λ.

3.19 THE BOOTSTRAP AND JACKKNIFE METHODS

Many data series in the natural sciences are nonreproducible and the researcher is left with only one set of observations with which to work. With only one realization of a series, it is impossible to compare it with a related series to determine if they are drawn from the same, or from different, populations. There are numerous oceanographic examples, including tsunami oscillations recorded by a coastal tide gauge, a single seasonal cycle of monthly mean currents at a mooring location, and a trend in long-term temperature data from a climate monitoring station. Marine biologists face similar limitations when analyzing groups of animal species caught in nets or bottom grab samples. The problem is that empirical observations are prone to error and any interpretation of an event must be devised based on statistical measures of the probability of the event. A fundamental measure for testing the validity of any property of a data set is its variance. Parametric statistical models have been developed that help the investigator decide the degree of faith to be placed in a given statistic. However, data and model are often nonlinear so that it is not usually possible to find an analytical expression for model variance in terms of the data variance.

The *parametric* statistical methods presented in the previous sections were institutionalized long before the time of modern digital computers when use of analytical expressions greatly simplified the laborious hand calculation of statistical properties. During the past few decades, *nonparametric* statistical methods have been developed to take advantage of the increasing computational efficiencies of computers. An advantage of the new methods is that they permit investigations of the statistical properties of a sample, which do not conform to a specific analytical model. Equally importantly, they can be applied to small data sets while still providing a reliable estimation of confidence limits on the statistic of interest. "Bootstrapping" and "jackknifing" are two of the more commonly used methods that could not be used effectively until the invention of the digital computer. Both are resampling techniques in which artificial data sets are generated by selection of points from an original set of data. Specifically, we start with a single realization of an "experiment" and from that one set of experimental data we create a multitude of new artificial realizations of the experiment without having to repeat the observations. These realizations are then used to estimate the reliability of the particular statistic of interest, with the underlying assumption that the sample data are representative of the entire population.

In the bootstrapping method, random samples selected during the resampling process are replaced before each new sample is created. As a consequence, any data value has the possibility of being drawn many times. The name bootstrap arises from the expression "to lift oneself up by one's bootstraps." In jackknifing, artificial data sets are created by selectively and systematically removing samples from the original data set. The statistics of interest are recalculated for each resulting truncated data set and the variability among the artificial samples used to describe the variability of these statistics. "Cross-validation" is an older technique. The idea is to split the data into two parts and set one part aside. Curves are fitted to the first part and then tested against values in the second part. Cross-validation consists of determining how well the fitted curves predict the values in the portion of data set aside. The data can be randomly split in many ways and many times in order to obtain the needed statistical reliability. For additional information on this technique, the reader is referred to Efron and Gong (1983).

3.19.1 Bootstrap Method

Introduced by Efron in 1977 (Diaconis and Efron, 1983), bootstrapping provides freedom from two limiting factors that have constrained statistical theory since its beginning: (1) the assumption of normal (Gaussian) data distributions; and (2) the focus on statistical measures whose theoretical properties can be analyzed mathematically. As with other nonparametric methods, bootstrapping is insensitive to assumptions made with respect to the statistical properties of the data and does not need an analytical expression for the connection between model and data statistical properties. Resampling techniques are based on the idea that we can repeat a particular experiment by constructing multiple data sets from the one measured data set. Application of the resampling procedure must be modeled on a testable hypothesis so that the resulting probability can be used to accept or reject the null hypothesis. The methods can be applied just as well to any statistic, simple or complicated. A *bootstrap sample* is a "copy" of the original data that may contain a certain value (datum, x_n) more than once, once, or not at all (i.e., the number of occurrences of x_n lies between 0 and N, where N is the number of independent data points). Introductions into the bootstrapping procedure can be found in Efron and Gong (1983), Diaconis and Efron (1983), and Tichelaar and Ruff (1989). Nemec and Brinkhurst (1988) apply the method to testing the statistical significance of biological species cluster analysis for which there are duplicate or triplicate samples for each location. Connolly et al. (2009) use bootstrapping methodology to test three different models for species abundance distributions of Indo-Pacific corals and reef fishes.

Suppose that we have N values of a scalar or vector variable, x_n ($1 \leq n \leq N$), whose statistical properties we wish to investigate in relation to another variable. This could be a univariate variable, such as sea-level height $x_n = \eta(t_n)$ at a single location over a period of N time steps, t_n, or the structure of the first mode empirical orthogonal function $\phi_1(x_n)$ as a function of location, x_n. Alternatively, we could be dealing with a bivariate variable (x_{1n}, x_{2n}) such as water temperature versus dissolved oxygen content from a series of vertical profiles. Results apply to any other set of measurements whose statistics we wish to determine. We may want to compare means and standard deviations (variances) of different records to see if they are significantly different. Alternatively, we might want to place confidence limits on the slope of a line derived using a standard least-squares fit to our bivariate data (x_{1n}, x_{2n}), or, determine how much confidence we can have in the coefficients we obtained from the least-squares fit of an annual cycle to a single set of 12 monthly mean current records from a mooring location. Note that if there is a high degree of correlation among the N data values, the N are not statistically independent samples and we are faced with the usual problem of dealing with an effective number of degrees of freedom N^* for the data set.

The procedure is to equate each of our N independent data points with a number produced by a random number generator. We can do this by assigning each of the data values to separate uniform-width bins lying along the line $(-1, +1)$, or $(0, 1)$, depending on the random number generator being used. For N values, there will be N uniform-width bins on the line and each bin will be equated with one of the N data values (Figure 3.25). The bin width is $2/N$ if the line -1 to $+1$ is used. A random number generator such as a Monte Carlo scheme is used to randomly select sequences of N bins

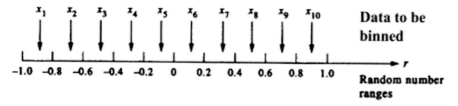

Data to be binned

Random number ranges

FIGURE 3.25 The assignment (binning) of observed data values x_n ($n = 1, ..., 10$) to 10 range values of the random number, r_k ($k = 1, ..., 10$). For each bootstrap sample of 10 values, 10 random numbers are selected and located according bin range. The datum values x_n assigned to each range are then used to form the bootstrap sample.

corresponding to the multiple bootstrap samples. Suppose that the random number generator picks a number, r, from the range $-1 \leq r \leq 1$. If this number falls into the range of bin k, corresponding to the range $[2(k - 1)/N] - 1 \leq r_k \leq (2k/N) - 1$, for $k = 1, ..., N$, then the data value x_k assigned to bin k is taken to be one of the samples we need to make up our bootstrap data set. In Figure 3.25, there are 10 data values and 10 corresponding random number segments of length 0.2, with datum value x_1 assigned to the range -1.0 to -0.8, x_2 assigned to -0.8 to -0.6, and so on. Since bootstrapping works with replacement, it is quite possible that the random number generator will come up with the same bin several times, or not at all. The first N data values from our resampling constitute the first bootstrap sample. The process is then repeated again and again until hundreds or thousands of bootstrap samples have been generated. Diaconis and Efron (1983) discuss making a US billion bootstrap samples. They also take another approach. Instead of generating one bootstrap sample at a time by equating bins along the real line $(-1, 1)$ with N samples, they generate all the needed multiple copies of all the N data values (say one million copies of each of the original data values or data points) and place them all in a rotating "lotto" bin. They then reach in and pull out all the requisite number of N-value bootstrap samples from the shuffled points, being careful to throw each data point back into the bin before selecting the next value. This requires some sort of label for each value in the bin based on a random selection process that can identify a data point that has been selected.

Although bootstrapping has yet to find widespread application in the marine sciences, there are several noteworthy examples in the literature. Enfield and Cid (1990) examined the stationarity of different groupings of El Niño recurrence rates based on the chronology of Quinn et al. (1987). For example, group 1 consisted of all strong (S) and very strong (VS) events for the period 1525–1983, while groups 4 and 5 consisted of S/VS events for times of high and low solar activity for this period. Groups 6–10 contained different samples of intensities for the modern period of 1803–1987. Maximum likelihood estimation was used to fit a two-parameter Weibull distribution $f(t)$ to each sample group,

$$f(t) = \left(\beta t^{\beta-1} / \tau^\beta\right)\exp\left[-(t/\tau)^\beta\right] \tag{3.207}$$

where β and τ are, respectively, the shape (peakedness) and timescale (RMS return interval) parameters, and t is the random variable for the return interval. For each group, only a single distribution could be fitted. To derive estimates of the mean and standard deviations of the parameters for each group, 500 bootstrap samples were generated and the Weibull parameters obtained for each sample. As indicated by Figure 3.26, this number of samples provides good convergence to the mean value for the Weibull distribution fit for each group. The distribution of El Niño return events for bootstrap samples for all intensities for the "early modern" period 1803–91 is shown in Figure 3.27 along with its corresponding Weibull distribution. Enfield and Cid (1990) use the resampling analysis to show that, for the groups associated with times of low solar activity and those associated with times of high solar activity, there is comparatively little overlap between the bootstrap-derived frequency histograms and mean return timescales, τ (years) (Figure 3.28). These results suggest that there is a statistical difference in the return times for the two groups and that return times are nonstationary.

Much of the present evidence for global warming is based on Northern Hemispheric annual surface air temperature records over the past 100 years (Jones et al., 1986; Hansen and Lebedeff, 1987; Gruza et al., 1988; Mann et al., 1998; IPCC, 2007). Interest in the reliability of the means and trends of these records (labeled H, J, and G) prompted Eisner and Tsonis (1991) to examine differences in means and trends of pairs of these records for the three global mean temperature curves. The data sets have been constructed using different averaging methods and different observational data bases. Data set H contains only observations from land stations whereas data set J uses both land-and ship-based observations. Averages for set H are derived using equal area boxes over the globe whereas data set G is constructed by visual inspection of anomalies from sea-level temperature analyses. The usual assumption is that these time series are representative of the same population, a result that appears to be supported by the statistically significant correlation $r > 0.79$ among the different curves. As pointed out by Eisner and Tsonis, however, the presence of trends in these data means that the linear cross-correlation coefficient may not be a reliable measure of the covariability of the records. Two questions can be addressed using the bootstrapping method: (1) are the three versions of the temperature records significantly different that

FIGURE 3.26 Estimation of the population mean Weibull distribution parameters (mean return time in years) using the bootstrap method for El Niño events taking place during times of low solar activity for the period 1525−1983; τ is the return time and σ its standard deviation. *From Enfield and Cid (1990).*

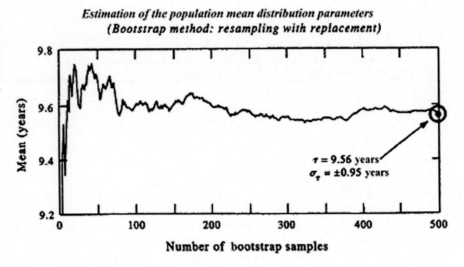

FIGURE 3.27 Histogram of El Niño return times for all events between 1803 and 1987 (group #7) derived using the bootstrapping resampling technique. The solid curve is the Weibull distribution fitted to the histogram. The modal and mean return intervals (3.3 and 3.8 years, respectively) are the derived from the Maximum Likelihood Estimation (MLE) estimates of the population parameters. *From Enfield and Cid (1990).*

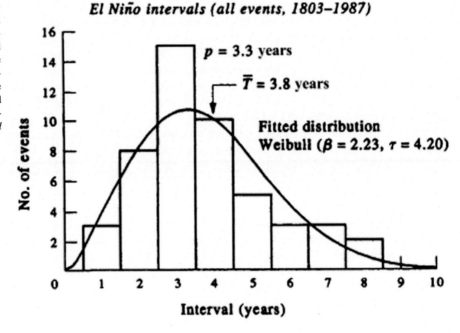

we can say they are not drawn from the same population? (The null hypothesis is false.) and (2) are the trends in the three records sufficiently alike that they are measuring a true rise in global temperature?

Because of the strong linear correlation in the records, the authors work with difference records. A difference record is constructed by subtracting the annual (mean removed) departure record of one data set from the annual departure record of another data set. Although not zero, the cross-correlation for the difference records is considerably less than those for the original departure records, showing that differencing is a form of high-pass filtering that effectively reduces biasing from the trends. The average difference for all 97 years of data used in the analyses (the difference record H-J relative to the years 1951−80) is −0.05°C, indicating that the hemispheric temperatures of Jones et al. (1986) are slightly warmer than those of Hansen and Lebedeff (1987). Similar results were obtained for H-G and J-G. To see if these differences are

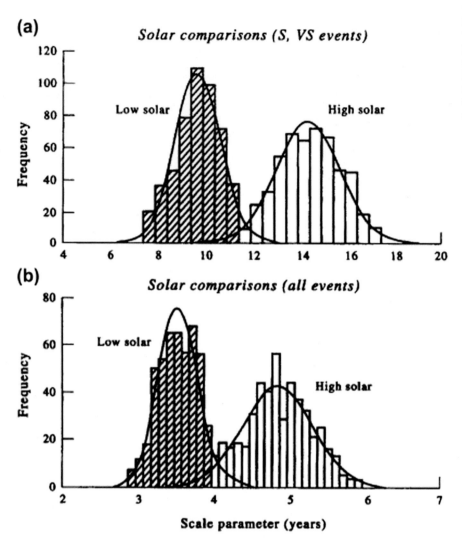

FIGURE 3.28 Histograms and fitted Weibull distributions obtained using the bootstrapping method. Plots show the occurrences of El Niño events for the times of low and high solar flare activity for (a) strong and very strong El Niño events, only; (b) all El Niño events. *From Enfield and Cid (1990).*

statistically significant, 10,000 bootstrap samples of the difference records were generated. The results (Figure 3.29a) suggest that all three hemispheric temperature records exhibit significantly different nonzero means. The overlap in the distributions is quite minimal. The same process was then used to examine the trends in the difference records. For the H-J record, the trend is +0.15°C/century so that the trend of Hansen and Lebedeff is greater than that of Jones et al. As indicated in Figure 3.29b, the long-term trends were distinct. On the basis of these results, the authors were forced to conclude that at least two of the data sets do not represent the true population (i.e., the truth). More generally, the results bring into question the confidence one can have that the long-term temperature trends obtained from these particular data are representative of trends over hemispheric or global scales.

Biological oceanographers often have difficult sampling problems that can be addressed by bootstrap methods. For example, the biologist may want to use cluster analyses of animal abundance for different locations to see if species distributions differ statistically from one sampling location (or time) to the next. Cluster analyses of ecological data use dendrograms—linkage rules which group samples according to the relative similarity of total species composition—to determine if the organisms in one group of samples have been drawn from the same or different statistical assemblages of those of another group of samples. Provided there are, at least, replicates for most samples, bootstrapping can be used to derive tests for statistical significance of similarity linkages in cluster analyses (Burd and Thomson, 1994). In a more recent

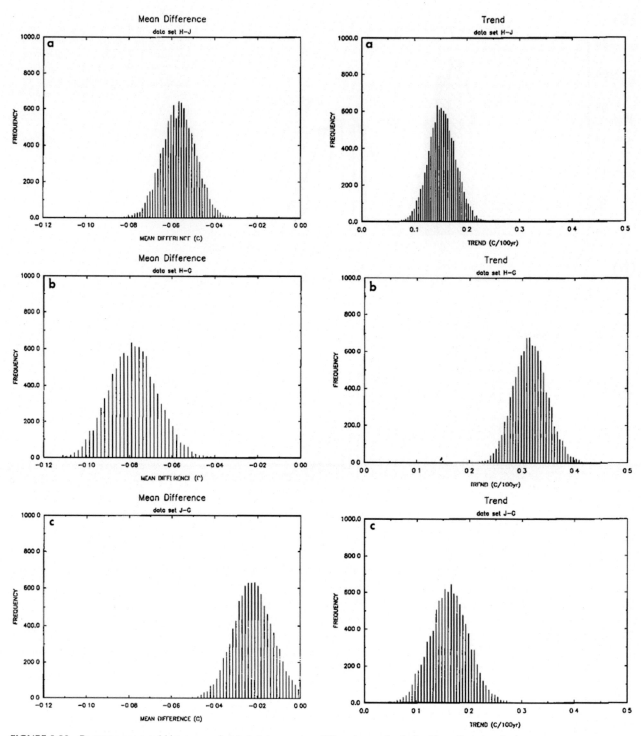

FIGURE 3.29 Bootstrap-generated histograms of global air temperature difference records obtained by subtracting the temperature records of Jones et al. (1986) (J), Hansen and Lebedeff (1987) (H), and Gruza et al. (1988) (G). (a) Frequency distributions of the mean differences plotted for 104 bootstrap samples. The *x*-axis (ordinate) gives the number of times the bootstrap mean fell into a given interval. All three distributions are located to the left of zero mean difference. (b) Same as (a), but for slope (trend) of the temperature difference curves. All three distributions are separated from zero indicating significant differences between long-term surface temperature trends given by each of the three data sets. *From Eisner and Tsonis (1991).*

study, Connolly et al. (2009) derive and use several alternative bootstrap analyses to test the ability of three different species abundance models to characterize ecological assemblages of Indo-Pacific corals and reef fishes. For further information on this aspect of bootstrapping, the reader is referred to Nemec and Brinkhurst (1988). Finally, in this section, we note that it is possible to vary the bootstrap size by selecting samples smaller than N, the original size of the data set, to compare various estimator distributions obtained from different sample sizes. This allows one to observe the effects of varying sample size on sample estimator distributions and statistical power.

3.19.2 Jackknife Method

Several other methods are similar in concept to bootstrapping but differ significantly in detail. The idea, in each case, is to generate artificial data sets and assess the variability of a statistic from its variability over all the sets of artificial data. The methods differ in the way they generate the artificial data. Jackknifing differs from bootstrapping in that data points are not replaced prior to each resampling. This technique was first proposed by Maurice Quenouille in 1949 and developed by John Tukey in the 1950s. The name "jackknife" was used by Tukey to suggest an all-purpose statistical tool.

A jackknife resample is obtained by deleting a fixed number of data points (j) from the original set of N data points. For each resample, a different group of j values is removed so that each resample consists of a distinct collection of data values. In the "delete-j" jackknife sample, there will be $k = N - j$ samples in each new truncated data set. The total number of new artificial data that can be generated is

$$\binom{N}{j}$$

which the reader will recognize as $_N P_j = N!/(N - j)!$, the number of permutations of N objects taken j at a time. Consider the simple delete-1 jackknife. In this case, there are $N - 1$ samples per artificial data set and a total of $_N P_j = N$ new data sets that can be created by systematically removing one value at a time. As illustrated by Figure 3.30, an original data set of four data values will yield a total of four distinct delete-1 jackknife samples, each of size three (3), which

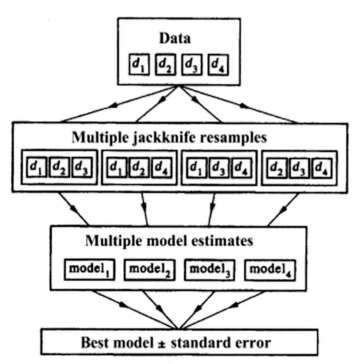

FIGURE 3.30 Schematic representation of the jackknife. The original data vector has four components (samples), labeled d1 −d4. The data are resampled by deleting a fixed number of components (here, one) from the original data to form multiple jackknife resamples (in this case, four). Each resample defines a model estimate. The multiple model estimates are then combined to a best model and its standard deviation. *From Tichelaar and Ruff (1989).*

can then be used to examine various statistics of the original data set. The sample average of the data derived by deleting the ith datum, denoted by the subscript (i), is

$$\bar{x}_{(i)} = \frac{N\bar{x} - x_i}{N-1} = \frac{1}{N-1} \sum_{\substack{j=1 \\ j \neq i}}^{N} x_j \tag{3.208}$$

where

$$\bar{x} = \frac{1}{N} \sum_{i=1}^{N} x_i$$

is the mean found using all the original data. The average of the N jackknife averages, $\bar{x}_{(i)}$, is

$$\bar{x}^* = \frac{1}{N} \sum_{i=1}^{N} \bar{x}_{(i)} = \bar{x} \tag{3.209}$$

The last result, namely that the mean of all the jackknife samples is identical to the mean of the original data set, is easily obtained using Eqn (3.208). The estimator for the standard deviation, σ_j, of the delete-1 jackknife is

$$\sigma_j = \sum_{i=1}^{N} \left[(\bar{x}_{(i)} - \bar{x}^*)^2 \right]^{1/2} \tag{3.210a}$$

$$= \frac{1}{N-1} \sum_{i=1}^{N} \left[(x_i - \bar{x})^2 \right]^{1/2} \tag{3.210b}$$

where Eqn (3.210b) is the usual expression for the standard deviation of N data values. Our expression differs slightly from that of Efron and Gong (1983), who use a denominator of $1/[(N-1)N]$ instead of $1/(N-1)^2$ in their definition of variance. The advantage of Eqn (3.210a) is that it can be generalized for finding the standard deviation of any estimator θ that can be derived for the original data. In particular, if θ is a scalar, we simply replace $x_{(i)}$ with $\theta_{(i)}$ and x^* with θ^* where $\theta_{(i)}$ is an estimator for θ obtained for the data set with the ith value removed. Although the jackknife requires fewer calculations than the bootstrap, it is less flexible and at times less dependable (Efron and Gong, 1983). In general, there are N jackknife samples for the delete-1 jackknife as compared with

$$_{2N-1}P_N = \binom{2N-1}{N}$$

bootstrap points.

Our example of jackknifing is from Tichelaar and Ruff (1989), who generated $N = 20$ unequally spaced data values y_i that follow the relation $y_i = cx_i + \varepsilon_i$ ($c = 1.5$, exactly), where ε_i is a noise component drawn from a "white" spectral distribution with a normalized standard deviation of 1.5 and mean of zero. The least squares estimator for the standard deviation of the slope is

$$\hat{\sigma} = \sum_{i=1}^{N} \left[(y_i - \hat{c}x_i)^2 \right] / \left[(N-1) \sum_{i=1}^{N} x_i^2 \right] \tag{3.211}$$

where $\hat{c} = \sum_{i=1}^{N}(y_i x_i) / \sum_{i=1}^{N}(x_i^2)$. Two jackknife estimators were used: (1) the delete-1 jackknife, for which the artificial sample sizes are $N - 1 = 19$; and (2) the delete-half $(N/2)$ jackknife for which the sample sizes are $N - N/2 = 10$. In both cases, the jackknife resamples had equal weighting in the analysis. For the delete-half jackknife, a Monte Carlo determination of 100 subsamples was used since the total samples $_{20}P_{10} = 20!/10!$ is very large ($\sim 6.7 \times 10^{11}$). The results are presented in Figure 3.31. The last panel gives the corresponding result for the bootstrap estimate of the slope using 100 bootstrap samples. Results showed that the bootstrap standard error of the slope was slightly lower than those for both jackknifing estimates.

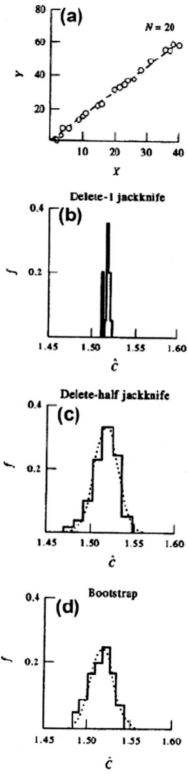

FIGURE 3.31   Use of the jackknife technique to estimate the reliability of a linear regression line. (a) A least-squares fit through the noisy data, for which the estimated slope $\hat{c} = 1.518 \pm 0.0138$ (± 1 standard error); (b) the normalized frequency of occurrence distribution, f, for the delete-1 jackknife which yields $\hat{c} = 1.518 \pm 0.0136$; (c) as in (b) but for the delete-half jackknife for which $\hat{c} = 1.517 \pm 0.0141$; (d) the corresponding bootstrapping estimate, for which $\hat{c} = 1.517 \pm 0.0132$. Note the subtle difference in distribution between (c) and (d). The dashed line is analytical distribution of $\hat{c}$. *From Tichelaar and Ruff (1989).*

3.20 EXTREME VALUE ANALYSIS

Recent destructive events, such as coastal flooding from storm surges (Needham et al., 2015; Kohno et al., 2018), the formation of "blobs" of anomalously warm surface water in the northeastern and southeastern regions of the Pacific Ocean (Peterson et al., 2015; Di Lorenzo and Mantua, 2016), the bleaching of equatorial reef systems (Hughes et al., 2017) and the number and intensity of wild fires around the world (Hirach and Koren, 2021), have made the concept of "extreme events" and their link to climate-induced changes increasingly more common and familiar (IPCC, 6th Assessment Report, 2021). The December 26, 2004 Sumatra tsunami, which killed over 320,000 people in the Indian Ocean region (Rabinovich et al., 2019), and the early May 2008 storm surge from Cyclone Nargis that was responsible for over 138,000 deaths in Myanmar and adjoining countries (Kohno et al., 2018), provide stark reminders of the impact that extreme events can have on people living in or visiting coastal communities around the globe. The desire to understand the science and statistical formulation behind the estimation of future extreme events prompted us to include a section on Extreme Value Analysis (EVA) in this edition of the book.

Extreme value analysis is used to (a) determine the probability distributions and frequencies of occurrence of extreme events as functions of their "size", and (b) to estimate the event size (*return level*) for a given "waiting time" (*return period*) between extreme occurrences. Here, "size" is used as a catch-all term covering amplitude, variance, duration, number of events, and other metrics. Extreme values are positioned at the tail ends of probability distributions and, by definition, deviate significantly from the expected or "normal" outcomes of "every-day" statistics (Figure 3.32). The word "significantly" in this case generally refers to outcomes that exceed a specified threshold, such as a high percentile level (e.g., the 99th percentile), corresponding to a large number of standard deviations relative to the mean, median, average variance or other measure. Our goal in this section is to lay the mathematical foundation for this branch of statistical analysis and then to provide worked examples applicable to physical oceanography.

Extreme value analysis has its roots in the work by the English statistician L.H.C. Tippett in the early part of the 20th Century. Tippett (1902−85) was employed by the "British Cotton Industry Research Association" and was mainly interested in strengthening cotton thread (Wikipedia, 2023). Foundational work was also provided by Wilfried Fritz Pareto (1848−1923), an Italian civil engineer who made important contributions to the study of income distribution and in the analysis of personal choices. He was the first to show that personal income follows a power-law probability distribution (now named after him) that characterized his observation that 80% of the wealth in Italy belonged to about 20% of the population. Important contributions to extreme value analysis were further provided by the German statistician Emil Gumbel (1891−1966), a pioneer in the application of extreme value theory to climate and hydrology studies, by Swedish engineer Waloddi Weibull (1887−1979), known for his work on the statistical treatment of fatigue, strength, and lifetime of materials, and by the French mathematician Maurice Fréchet (1878−1973), who made major contributions to probability and statistics, as well as to mathematics.

The data examined by extreme analyses are instrumentally recorded or simulated using numerical models, preferably models that have been calibrated against observations or systematically compared ("benchmarked") against other models. Once the extreme values have been determined—using the *Block Maximum* (BM) method or *Peaks over Threshold* (POT) method—they can be used to derive empirical (data-based) estimates of the frequencies of occurrence and return levels for events of given return periods. Although empirical estimates are important elements of extreme analysis, it is not possible

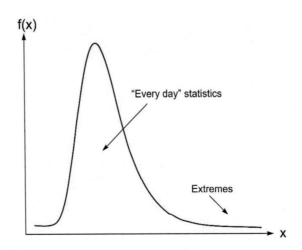

FIGURE 3.32 Extreme statistics deal with events located in the tails of probability distributions, $f(x)$, as functions of size, x. "Every-day" statistics deal with events in the main body of the distribution.

to make projections based on the data alone; determination of statistical properties for predictive purposes requires the application of one or more probability distribution functions.

Records of extreme occurrences are fitted to *Probability Density Functions* (PDFs) and their integrated counterparts, the *Cumulative Distribution Function* (CDF) and the *Complementary Cumulative Distribution Function* (CCDF). A PDF (also, *Probability Distribution Function* or *Probability Mass Function*, PMF, for discrete random variables) is the likelihood that an outcome drawn from variable X will equal a specified size, x, while a CDF is the likelihood that the outcome will not exceed x. CCDFs, also termed "Survival Functions", are the probability that the size of an event will occur beyond size, x, and are "complementary" in the sense that $CCDF(x) = 1 - CDF(x)$. Distribution models that closely reproduce the data make it possible to estimate the likelihood a particular extreme size and to estimate the inter-event time (return period) between future events of a given size. Determining the probability that an event of a specific size will take place within a given time span is a crucial step in the assessment of risk from natural hazards. For example, a 50-year tide gauge record can be used to quantify the water level expected to be exceeded, on average, once every 100 years (that is, an event that has a 1% chance of being exceeded in any given year). Studies of future risk also need to consider whether a process is stationary or non-stationary. For example, amplification of the intensity and frequency of future coastal flooding events during ongoing global sea level rise can differ markedly with region and amplitude of the storm surge (Ghanbari et al., 2019). Buchanan et al. (2017) find that, for the 0.5 m rise in global sea level that is predicted to occur well before the end of the 21st century, the 0.2%, 1% and 10% annual chance floods are expected to recur, respectively, 814, 335, and 108 times as often at Seattle on the west coast of the United States, but 4, 16 and 148 times as often at Charleston on the east coast of the US. The reason for this sharp contrast between the two coasts is the differences in their responses to meteorological and hydrodynamic processes, which subsequently determines the shapes of the associated extreme probability distributions. Regions exposed to tropical cyclones, such as those along the Gulf and Atlantic coasts, tend to have frequency distributions that tail-off relatively slowly. Conversely, sites along the Pacific coast, which are limited by the steeper coastal slopes leading to the seabed and fewer barrier beaches, tend to have frequency distributions that tail-off more quickly. Thus, some regions can expect disproportionate amplification of higher frequency events, leading to a greater number of historically *precedented floods*, while others face amplification of lower frequency events and particularly fast-growing risks of historically *unprecedented floods*.

The study of extreme events is a complex and ever-evolving subject that cannot be fully covered in one segment of this book. Consequently, our primary purpose is to present the basic concepts of Extreme Value Analysis and to introduce some of the research topics within the context of physical oceanography. Several oceanographic examples are presented to assist in understanding the approaches. For a more in-depth coverage of extreme data analysis, investigators are directed to the books and journal publications cited in the text.

3.20.1 Basics of Probability Distribution Functions

Because extreme value analysis deals with the tail-like ends of probability distribution functions, we begin with a condensed recap of some of the background material on basic probability theory abstracted from earlier parts of this chapter. Consider a random variable, X, which takes on the values x (which, in the discrete case, would consist of values $x_1, x_2, x_3, \ldots x_n$.) As noted at the beginning of this chapter, the likelihood that a sample selected from the variable, X, will assume a value equal to x, is the Probability Density Function (PDF),

$$f(x) = P(X = x) \tag{3.212}$$

which, in the continuous case. satisfies the conditions,

$$f(x) \geq 0, \text{ and } \int_{-\infty}^{\infty} f(x)dx = 1. \tag{3.213}$$

The Cumulative Distribution Function (CDF) is obtained from $f(x)$ as,

$$F(x) = P(X \leq x) = \int_{-\infty}^{x} f(x)dx, \tag{3.214}$$

and corresponds to the probability that *no* values sampled from X will exceed x, or equivalently, that $F(x)$ is the probability that all values sampled will be less than or equal to x. The probability function $F(x; \alpha) = P(X \leq x | X > \alpha)$ is the *Conditional Probability* that a value from X is less than or equal to x, for x greater than α. As this is a book about data, we deal mainly with discrete formulations. Continuous mathematical expressions for $f(x)$ and $F(x)$ can be found in the literature, with summation operations replaced by integrals. The reader is also directed to previous sections on probability presented in this chapter.

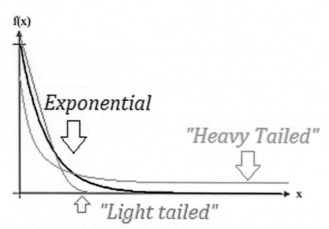

FIGURE 3.33 Heavy, light and exponential tails of extreme probability density functions, $f(x)$.

Because $F(x)$ is the sum of all outcomes $X \leq x$, it makes sense that there is a complementary function that is the sum of all outcomes $X > x$. This leads to the Complementary Cumulative Distribution Function (CCDF), or Survival Function, $\widetilde{F}(x)$ — sometimes written as $\Phi(x)$ (e.g., Kagan, 2002) — defined as

$$\widetilde{F}(x) = P(X > x) = 1 - F(x) = \int_x^\infty f(x)dx \tag{3.215}$$

the probability that the sampled values will exceed x. Thus, under the assumption of stationarity (i.e., that return periods for a given event size remain the same over time), a value $\widetilde{F}_P(x) = 0.1$ for $x = x_o$ means that there is a 10% probability that outcomes from the random variable, X, will exceed x_o in 1 time unit, or that it will likely take at least 10 time-units before the level x_o is once again exceeded. For $x = x_{max}$, $F(x) = 1$, as the probabilities of all events have been summed up, while $\widetilde{F}(x) = 0$, as no larger events can occur.

3.20.1.1 Tails of Extreme Probability Distributions

As illustrated in Figure 3.32, extreme value distributions have distinct tails. More specifically, extreme probability distribution functions are described as either heavy- or light-tailed, or, equivalently, as heavy- or thin-tailed. Both descriptions are related to the extent of the function's fall-off relative to the exponential (Figure 3.33). Heavy-tailed distributions, such as the *Fréchet* distribution discussed later in this section, have power-law decays that fall off slower than exponential functions (i.e., are not exponentially bounded) and therefore have more *mass* in their tails (i.e., are "heavier") than the exponential distribution. Distributions, such as the *Gumbel* distribution, have probability functions that decay faster than an exponential distribution and so are light-tailed (have less *mass* in their tails). In many applications, it is the right-hand side of the distribution that is of interest; however, distributions may also have a heavy left tail, or both tails may be heavy. Heavy-tailed distributions are further divided into three subclasses: the fat-tailed, the long-tailed and the sub-exponential distributions. All long-tailed distributions are heavy-tailed, but the converse is not true, as it is possible to construct heavy-tailed distributions that are not long-tailed. In effect, all commonly used heavy-tailed distributions belong to the sub-exponential class. Fat tails are related to the severity of the event and to the frequency of occurrence. For example, tsunamis generated by major earthquakes (with moment magnitudes, $M_w > 7.6$) are heavy-tailed and follow a *Pareto* power-law distribution (Geist et al., 2009). Such events are rare and can have destructive impacts. Thin-tailed distributions have upper bounds on their size and tend to have less extensive impacts.

3.20.2 Probability Distribution Functions for Extreme Values

There are two basic families of probability distribution functions for analyzing extreme datasets: (1) the family of *Generalized Extreme Value* (GEV) distributions; and (2) the set of *Generalized Pareto* (GP) distributions. As summarized by Table 3.14, determination of the generalized probability distributions for the analysis of extreme outcomes arising from random variable, X, are dependent on at least two of three parameters: the location parameter, μ; the scale parameter, s; and the shape parameter, ξ. For both distribution families, $\mu, s \geq 0$ and real over the range $-\infty \leq x \leq +\infty$, also written as $x \in (-\infty, \infty)$, where $\in$ signifies that the independent values, x, are elements within the range $(-\infty, \infty)$. The shape parameter, ξ, is real, and often specified as $\xi \geq 0$, which can lead to some confusion because (as discussed below) the sign of ξ is typically used to characterize sub-families of the GEV distributions. Unlike the mean of a record, $\mu = \bar{x}$, that is

TABLE 3.14 Probability distributions for extreme values x of the variable X. Here, $\xi \equiv$ shape parameter; $s \equiv$ scale parameter; $\mu \equiv$ location parameter. Typically: $\xi \in (0, \infty)$; $s \in (0, \infty)$. Note that ξ is always positive and that it is the way it appears in the exponents that changes.

	Range	$f(x) = $ PDF; $F(x) = $ CDF
Generalized Extreme Value (GEV)	$x \in (-\infty, \infty); \mu, s > 0;$ $1 + \frac{\xi(x-\mu)}{s} > 0; \xi \neq 0$	$f(x; \mu, s, \xi) = \frac{1}{s}\left[1 + \xi\left(\frac{x-\mu}{s}\right)\right]^{-\left(\frac{1}{\xi}+1\right)}$ $\times \exp\left\{-\left[1 + \xi\left(\frac{x-\mu}{s}\right)\right]^{-\frac{1}{\xi}}\right\}$ $F(x; \mu, s, \xi) = \exp\left\{-\left[1 + \xi\left(\frac{x-\mu}{s}\right)\right]^{-\frac{1}{\xi}}\right\}$
Gumbel Distribution (GEV for $\xi = 0$)	$x \in (-\infty, \infty); \mu, s > 0;$	$f(x; \mu, s) = \frac{1}{s}\exp\left\{-\left[\left(\frac{x-\mu}{s}\right) + \exp\left[-\left(\frac{x-\mu}{s}\right)\right]\right]\right\}$ $F(x; s, \mu) = \exp\left\{-\exp\left[-\left(\frac{x-\mu}{s}\right)\right]\right\}$
Weibull Distribution $\xi \rightarrow -\xi < 0$; this is for 3-parameters; set $\mu = 0$ for the 2-parameter version.	$x \geq 0; \mu, s, \xi > 0;$ $f(x) = 0$ for $x < 0$	$f(x; \mu, s, \xi) = \left(\frac{\xi}{s}\right)\left(\frac{x-\mu}{s}\right)^{-(1-\xi)} \times \exp\left[-\left(\frac{x-\mu}{s}\right)^{\xi}\right]$ $F(x; \mu, s, \xi) = 1 - \exp\left[-\left(\frac{x-\mu}{s}\right)^{\xi}\right]$
	$x \in (-\infty, \infty); s > 0$ $\mu = x_{min} = $ x-minimum $(x > x_{min});$	$f(x; s, \xi) = \frac{\xi}{s}\left(\frac{x-x_{min}}{s}\right)^{-(1+\xi)}\exp\left[-\left(\frac{x-x_{min}}{s}\right)^{-\frac{1}{\xi}}\right]$ $F(x; s, \xi) = \exp\left[-\left(\frac{x-x_{min}}{s}\right)^{-\frac{1}{\xi}}\right]$
Generalized Pareto Distribution (GPD)		$f(x) = \frac{1}{s}\left(1 + \xi\frac{x}{s}\right)^{-\left(\frac{1}{\xi}+1\right)}$ $F(x) = 1 - (1 + \xi x)^{-1/\xi}$ for $\xi \neq 0$ $F(x) = 1 - \exp(-x)$ for $\xi = 0$
Pareto Distribution (PD) (Type I)	$x \in (x_{min}, \infty); \alpha = \frac{1}{\xi} > 0;$ $s = x_{min};$ real $f(x)$ & $F(x) = 0$ if $x < x_{min}$	$f(x) = \alpha\frac{(x_{min})^{\alpha}}{x^{(\alpha+1)}}$ $F(x) = 1 - \left(\frac{x_{min}}{x}\right)^{\alpha}$

often used in the mathematical notation for a normal (Gaussian) probability distribution, the location parameter, μ, is not the record mean, although it does represent a form of central value for the distribution. Similarly, the scale parameter, s, is not the standard deviation, typically written as σ in normal distributions, although it plays a similar role in that it governs the size of the deviations relative to μ in extreme value data. For example, the location parameter used in studies of coastal flooding corresponds to a local sea level reference level (such as the local mean tide or chart datum) and the scale parameter denotes the variability in the maxima of water levels caused by a combination of tides and storm surges. [There appears to be no universally accepted notation for the distribution parameters. Although μ (or, u) is widely used for the location parameter, the shape parameter, ξ, is also written as β, k, λ, or $1/\alpha$, while the scale parameter, s, is sometimes presented as σ, λ or α. The reader is also cautioned that most software packages use $\log(x)$ for the natural logarithm of x, instead of $\ln(x)$, and $\log10(x)$ for the base-10 logarithms. In contrast, written texts typically use $\log(x)$ and $\ln(x)$ for base-10 and natural logarithms, respectively.]

The shape parameter, ξ, governs the fall-off in the tails of GEV and GP distributions, and each of these two principal families of extreme distributions has a set of sub-families whose characteristics are determined by the sign of ξ. The three sub-family types within the overarching GEV distribution are: Type I ($\xi = 0$), the light-tailed *Gumbel* sub-family of distributions, in which the tail tapers off more quickly than an exponential; Type II ($\xi > 0$), the heavy-tailed and un-bounded *Fréchet* sub-family of distributions, in which the tail tapers off less quickly than an exponential; and Type III ($\xi < 0$), the bounded *Weibull* sub-family of distributions, with a finite right-end point $\left(\hat{\mu} - \hat{s}/\hat{\xi}\right)$, where a hat ($\wedge$) is used to denote an estimated value. Geist et al. (2009) also note that *Gumbel* distributions fall-off more quickly than a power-law, while *Fréchet* distributions taper off as a power law as x becomes large. Higher positive values of ξ result in heavier tails. (The three types of Generalized Pareto distributions follow similar ξ-dependent patterns.) Of the three types of GEV distributions, the Gumbel distribution (Gumbel, 1958) is commonly used in natural hazard studies as it has an infinite right-hand tail but corresponds to a tapered power-law distribution for large values. On the other hand, the literature is replete with studies that favor one distribution over another and claims that previously used distributions are not the best choice. We further note that, although the sign of the shape parameter is used to distinguish between the three types of distributions, the sign of ξ for *Weibull* density functions sometimes is treated as strictly positive, $\xi \geq 0$, and then appears in the mathematical exponent as $-\xi$; instead of terms like e^{ξ} (where $\xi < 0$), many authors (cf. Hosking and Wallis, 1987) use terms like $e^{-\xi}$, where $\xi > 0$). In Table 3.14, we have mainly followed the common approach of keeping $\xi \geq 0$.

3.20.2.1 The Generalized Extreme Value (GEV) Distribution

If the distribution function for maximum values in a random sample of size N converges to a distribution function as N tends to infinity, then that function must be a GEV distribution. It is also true that this statement and other results of extreme value theory are valid even under general dependence conditions (Coles, 2001). The probability density function, $f(x; \mu, s, \xi)$, and the associated cumulative distribution function and complementary cumulative distribution function for a GEV distribution for size values $x \in (-\infty, \infty)$ are (Table 3.14),

$$f(x; \mu, s, \xi) = \frac{1}{s}\left[1 + \xi\left(\frac{x-\mu}{s}\right)\right]^{-\left(1+\frac{1}{\xi}\right)} \exp\left\{-\left[1 + \xi\left(\frac{x-\mu}{s}\right)\right]^{-\frac{1}{\xi}}\right\}; \quad \mu, s > 0; \xi \neq 0; \tag{3.216a}$$

$$1 + \frac{\xi(x-\mu)}{s} > 0;$$

$$F(x; \mu, s, \xi) = \exp\left\{-\left[1 + \xi\left(\frac{x-\mu}{s}\right)\right]^{-\frac{1}{\xi}}\right\} \tag{3.216b}$$

$$\widetilde{F}(x; \mu, s, \xi) = 1 - F(x; \mu, s, \xi). \tag{3.216c}$$

As noted above, there are three sub-families of GEV distributions determined by the sign of the shape parameter, ξ. The Type I case ($\xi = 0$) corresponds to the *Gumbel* distribution, the Type II case ($\xi > 0$) to the *Fréchet* distribution, where $x > \mu - s/\xi$ (the x-domain has a lower bound), and the Type III case ($\xi < 0$) to the *Weibull* distribution, for which $x < \mu - s/\xi$ (so that the x-domain has an upper bound). The definition (3.216c) is generic and, as it applies to all distribution functions, it is not repeated for probability distribution functions presented in the subsections that follow.

There exist numerous methods for determining the parameters μ, s, ξ from the extreme data. A well-established choice is the Maximum Likelihood Estimation (MLE) method (see Section 3.11), which involves maximizing the *log-likelihood function*, $l(x; \mu, s, \xi)$, with respect to each of the desired parameters; e.g., by setting $\partial/\partial\xi(l(x; \mu, s, \xi)) = 0$ to find the values for ξ. The parameters and their confidence intervals are estimated from the data using this function or other methods using software in statistical packages such as those in MATLAB and R. According to Coles (2001), Maximum Likelihood Estimators are regular (i.e., have the usual asymptotic properties) for $\xi > -0.5$, are generally attainable (without the standard asymptotic behavior) for $-1 < \xi < -0.5$, but not likely attainable for $\xi < -1$.

3.20.2.2 The Gumbel Distribution

The Gumbel probability distribution $f(X = x; \mu, s)$ and cumulative distribution function $F(X \leq x; \mu, s)$ correspond to Type-I GEV distributions (i.e., $\xi = 0$) for $-\infty < x < \infty$, $s > 0$, μ, s real. As per Table 3.14,

$$f(x; \mu, s) = \frac{1}{s}\exp\left\{-\left[\left(\frac{x-\mu}{s}\right) + \exp\left[\exp\left(-\left(\frac{x-\mu}{s}\right)\right)\right]\right]\right\} \tag{3.217a}$$

$$F(x; \mu, s) = \exp\left\{-\exp\left[-\left(\frac{x-\mu}{s}\right)\right]\right\} \tag{3.17b}$$

where the parameters μ and s are estimated using numerical optimization software in statistics packages that minimize the log-likelihood function,

$$l(x; \mu, s) = -N\cdot\ln(s) - \sum_{i=1}^{N}\left(\frac{x_i-\mu}{s}\right) - \sum_{i=1}^{N}\left[\exp\left(-\left(\frac{x_i-\mu}{s}\right)\right)\right] \tag{3.218}$$

(Coles, 2001). Several examples of Gumbel distributions are provided in Figure 3.34a and b. Rough estimates of μ and s are also obtained by substituting the mean and standard deviation of the extreme data record into the characteristic properties of the Gumbel functions, yielding,

$$\widehat{s} = 6^{1/2}(\text{extreme record standard deviation}/\pi) \tag{3.219a}$$

$$\widehat{\mu} = \text{record mean} - \widehat{s}\cdot\gamma, \tag{3.219b}$$

where $\gamma = \lim_{N\to\infty}\sum_{k=1}^{N}\left(\frac{1}{k}\right) + \ln(N) = 0.577215...$ is the Euler-Mascheroni constant. Approximate error estimates, $\Delta\widehat{\mu}$ and $\Delta\widehat{s}$, for the two parameters in (3.219) are found using

$$(\Delta\widehat{\mu}, \Delta\widehat{s}) = (\widehat{\mu}, \widehat{s})/\sqrt{N} \tag{3.220}$$

The Gumbel distribution is one of the most widely used distributions for climate modeling and has been used to examine the highest annual water levels on the coast and to estimate the probabilities of maxima occurrences

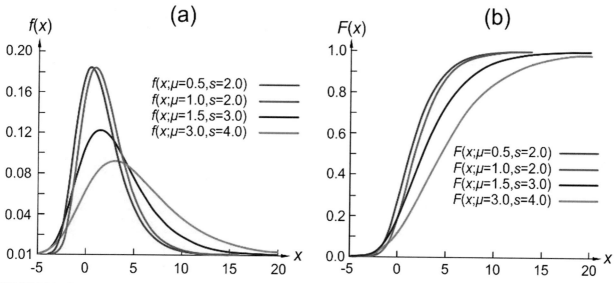

FIGURE 3.34 Gumbel distributions. (a) The probability distributions $f(x) = P(X = x)$ and (b) the corresponding cumulative distribution functions $F(X \leq x)$ for Gumbel distributions for a selected range of location (μ) and scale (s) parameters. *Adapted from Wikipedia (2023); https://www.wikipedia.org/.*

(e.g., Vitousek et al., 2017). Hourly values above a specified threshold, such as the 99th percentile, ensure that exceedance size describes extreme events (Buchanan et al., 2017). Among 20 different extreme-value analysis methods, Wahl et al. (2017) find that the Gumbel distribution gives the highest estimate of 100-year return water levels for the northeast Pacific. The Gumbel probability distribution is also used to analyze monthly and annual maximum values of daily rainfall and river discharge volume (Ritzema, 1994) and to describe droughts (Burke et al., 2010). It is further used to predict the waiting times between extreme events such as earthquakes, floods and other natural disasters. On the other hand, some studies have downplayed the applicability of the Gumbel distribution. In particular, Buchanan et al. (2017) suggest that the dependence on the Gumbel distribution in the fifth report of the Intergovernmental Panel on Climate Change (IPCC, 2014) was misleading as it is invariant to flood levels and does not capture the distinct effects of sea level rise on flooding in areas with heavy and thin-tailed flood frequency distributions. Use of a Generalized Pareto distribution was preferred.

3.20.2.3 The Weibull Distribution

The Weibull probability distribution has two main formulations: the three-parameter version $f(x; \mu, s, \xi)$ and the two-parameter version $f(x; s, \xi)$ obtained by setting $\mu = 0$ in the three-parameter version (Figure 3.35a and b). For variables, X, representing the "time-to-failure" of a machine part or materials, μ characterizes "the initial time-to-failure". The three-parameter Weibull distribution and cumulative distribution functions for size, x, are for $\mu, s, \xi > 0$ (Table 3.14),

$$f(x; \mu, s, \xi) = \left(\frac{\xi}{s}\right)\left(\frac{x-\mu}{s}\right)^{(\xi-1)} \exp\left[-\left(\frac{x-\mu}{s}\right)^{\xi}\right], x \geq \mu;$$

$$= 0, x < \mu; \tag{3.221a}$$

$$F(x; \mu, s, \xi) = 1 - \exp\left[-\left(\frac{x-\mu}{s}\right)^{\xi}\right], x \geq \mu$$

$$= 0, x < \mu \tag{3.221b}$$

(As noted earlier, the Weibull distribution formally corresponds to the Type-III GEV distribution ($\xi < 0$), but in keeping with most authors, we choose to specify $\xi > 0$ and instead change the sign of the shape parameter in the exponents of the GEV).

The structure of $f(x; \mu, s, \xi)$ changes markedly with ξ. For $0 < \xi < 1, f(x)$ tends to infinity as x approaches 0 from above and is strictly decreasing with increasing x; for $\xi = 1, f(x)$ tends to $1/s$ as as x approaches 0 from above and is again strictly decreasing. The main difference is for $\xi > 1$, in which case, $f(x)$ approaches 0 for decreasing x, then increases until its mode and then decreases again (see Wikipedia, 2023).

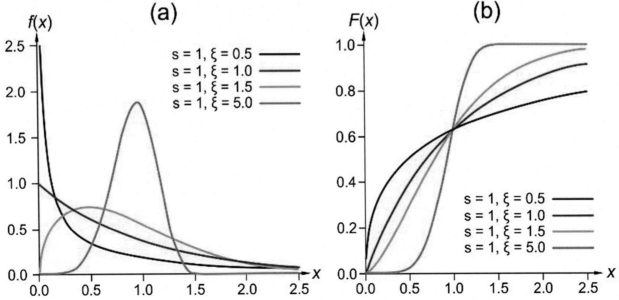

FIGURE 3.35 Weibull distributions. (a) The probability distributions $f(x) = P(X = x)$ and (b) the corresponding cumulative distribution functions $F(X \leq x)$ for two-parameter Weibull distributions for scale parameter $s = 1$ and a selected range of shape parameters, ξ. *Adapted from Wikipedia (2023).*

For the two-parameter case, $\mu = 0$, estimates of the shape parameter, ξ and scale parameter, s, are obtained by solving (numerically or graphically) the pair of nonlinear equations

$$\frac{1}{\hat{\xi}} = \frac{\sum_{i=1}^{N}\left[(x_i)^{\hat{\xi}}\ln(x_i)\right]}{\sum_{i=1}^{N}(x_i)^{\hat{\xi}}} - \frac{1}{N}\sum_{i=1}^{N}\ln(x_i) \tag{3.222a}$$

$$\hat{s} = \left(\frac{1}{N}\sum_{i=1}^{N}(x_i)^{\hat{\xi}}\right)^{1/\hat{\xi}}. \tag{3.222b}$$

Wikipedia (2023; https://www.wikipedia.org/) also presents a method for determining $\hat{s}$ and $\hat{\xi}$ after ranking of the N-highest extreme values from $x_1 > x_2 > \ldots > x_N$.

Regardless of the approach, the scale parameters need to be found either numerically or by graphing the two functions and determining where they intersect. The Weibull distribution is used in hydrology to examine extremes such as the annual maximum 1-day rainfalls and river discharges (Ritzema, 1994).

3.20.2.4 The Fréchet (Inverse Weibull) Distribution

The Fréchet distribution corresponds to the Type-II GEV distribution (for $\xi > 0$), with location parameter $\mu = x_{min}$ (the minimum extreme x-value), $x > x_{min},\; ;\; s > 0$;

$$f(x;\mu,s,\xi) = \frac{\xi}{s}\left(\frac{x - x_{min}}{s}\right)^{-(1+\xi)}\exp\left[-\left(\frac{x - x_{min}}{s}\right)^{-\xi}\right] \tag{3.223a}$$

$$F(x;\mu,s,\xi) = \exp\left[-\left(\frac{x - x_{min}}{s}\right)^{-\xi}\right] \tag{3.223b}$$

with the 2-parameter Fréchet (F) distribution (Figure 3.36a and b) related to the two-parameter Weibull (W) distribution by

$$f_F(x;s,\xi) = -f_W(x;s,-\xi) \tag{3.223c}$$

The Maximum Likelihood Estimators for the distribution parameters are found by minimizing the log-likelihood function for the distribution, yielding:

$$\frac{N}{\hat{\xi}} - \sum_{i=1}^{N}\ln(x_i) + \frac{N\sum_{i=1}^{N}\left[x_i^{-\hat{\xi}}\ln(x_i)\right]}{\sum_{i=1}^{N}\left(x_i^{-\hat{\xi}}\right)} = 0 \tag{3.224a}$$

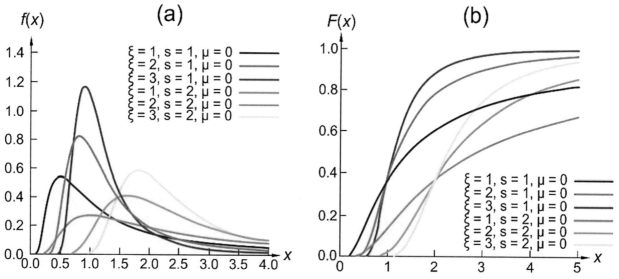

FIGURE 3.36 Fréchet distributions. (a) The probability distributions $f(x) = P(X = x)$ and (b) the corresponding cumulative distribution functions $F(X \leq x)$ for two-parameter Fréchet distributions for $\mu = x_{min} = 0$ and a selected range of scale (s) and shape (ξ), parameters. *Adapted from Wikipedia (2023); https://www.wikipedia.org/.*

$$\widehat{s}^{-1} = \frac{N}{\sum_{i=1}^{N}\left(x_i^{-\widehat{\xi}}\right)} \tag{3.224b}$$

The Fréchet distribution is used in hydrology to examine extremes, such as the annual maximum 1-day rainfall and river discharge.

3.20.2.5 Generalized Pareto Distribution

Generalized Pareto Distributions (GPDs) are a family of continuous probability functions used to model the tails of extreme probability distributions (Figure 3.37a and b). As with the Generalized Extreme Value (GEV) distributions, GPDs are characterized by location parameter, μ, scale parameter, s, and shape parameter, ξ. The probability density and cumulative distribution functions for GPDs as functions of size, x, are;

$$f(x;\mu,s,\xi) = P(X = x) = \frac{1}{s}\left(1 + \xi\left(\frac{x-\mu}{s}\right)\right)^{-\frac{1+\xi}{\xi}}; x \geq \mu; \xi > 0 \tag{3.225a}$$

$$\mu \leq x \leq \left(\mu - \frac{s}{\xi}\right); \xi < 0$$

$$= \exp\left(-\left(\frac{x-\mu}{s}\right)\right); \xi = 0 \tag{3.225b}$$

$$F(x;\mu,s,\xi) = P(X \leq x) = 1 - \left(1 + \xi\left(\frac{x-\mu}{s}\right)\right)^{-\frac{1}{\xi}}; x \geq \mu; \xi > 0 \tag{3.225c}$$

$$= 1 - \exp\left(-\left(\frac{x-\mu}{s}\right)\right); \xi = 0. \tag{3.225d}$$

GPDs are widely used to model extreme data with power-law dependance, including earthquakes, tsunami run-up, storm surge, and tsunami wave heights (cf. Geist, 2009). Wahl et al. (2017) report that the Generalized Pareto Distribution, with a 99th percentile for a threshold value, is the preferred approach for the analysis of extreme coastal sea levels. The approach almost never produced the highest or lowest return water levels of the broad range of distributions considered.

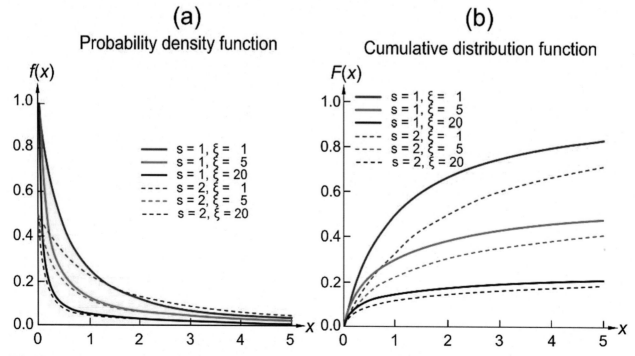

FIGURE 3.37   Generalized Pareto distributions. (a) The probability distributions $f(x) = P(X = x)$ and (b) the corresponding cumulative distribution functions $F(X \leq x)$ for Generalized Pareto Distributions for a selected range of slope (s) and scale (ξ), parameters for location parameter $\mu = 0$. *Adapted from Wikipedia (2023); https://www.wikipedia.org/.*

3.20.2.6 Pareto Distribution (Type 1)

The Pareto Distribution is a special case of the Generalized Pareto Distribution, in which the location parameter $\mu = x_{min} \geq 0$ is a minimum value of x (and the threshold level above which all data can be considered extreme) and the scale parameter $s = \xi x_{min}$ is proportional to the shape parameter, ξ. For convenience, we define $\alpha = 1/\xi$ and write the probability density and cumulative distribution functions for Type I Pareto distributions (Table 3.14) as simple power law functions:

$$f(x) = \left(\frac{\alpha}{x}\right)\left(\frac{x_{min}}{x}\right)^{\alpha}; x \geq x_{min}; \alpha = 1/\xi > 0 \tag{3.226a}$$

$$F(x) = 1 - \left(\frac{x_{min}}{x}\right)^{\alpha}; F(x) = 0, x \leq x_{min} \tag{3.226b}$$

The shape parameter, ξ, can be calculated from the data series using Maximum Likelihood Estimation (e.g., Kagan, 2002) as

$$\xi = \frac{1}{N}\sum_{i=1}^{N}\ln\left(\frac{x_i}{x_{min}}\right) \pm \Delta\xi \tag{3.227}$$

where, N is the total number of values in the extreme data series and we have added an estimate of the error using,

$$\Delta\xi = \xi/\sqrt{N} \tag{3.228}$$

(Wikipedia, 2021; Pareto Distribution; https://www.wikipedia.org/). An obvious advantage of the Type 1 model is that the shape parameter is readily derived from the data and doesn't, as is normally the case, require computation through graphical or numerical iteration techniques. [For modified Pareto distributions having a soft taper of the roll-off parameterized by a corner value x_c, numerical MLE techniques may be required (Geist et al., 2009).] Several examples of Pareto distributions are provided in Figure 3.37a and b. Note that some authors (e.g., Kagan, 2002; Geist et al., 2009) use the symbol $\beta = 1/\xi$ for the power-law exponent, while Wikipedia (2021) uses $\alpha = 1/\xi$.

The survival function—the likelihood that the observed values will exceed x at least once, on average—is from (3.226b)

$$\widetilde{F}_P(x) = \left(\frac{x_{min}}{x}\right)^{\alpha}; x \geq x_{min} \tag{3.229}$$

We provide an example of this simple power-law function later in this section.

3.20.3 Return Periods and Levels

Once a probability distribution function and its parameters have been derived for an extreme data series, the distribution can be used to estimate the return periods (between-event times) for outcomes of specified sizes and to estimate the return levels (events sizes) for specified return intervals of interest. Empirical estimates of return periods and expected numbers of events within a specified recurrence interval are also readily available directly from the data.

3.20.3.1 Empirical Estimates of Return Periods and Counting Rates

For a series of extreme values, we can obtain relationships between the size of an event (ranked according to its size) and its recurrence interval (return period), τ. In this approach, the data series x is first ranked in descending order of size, from the largest value ($rank = 1$), to the lowest value ($rank = M$), where M is the number of exceedances (extremes) in the dataset. For a dataset having N units (e.g., M values spanning N years of data), the formulation for recurrence interval, τ, and frequency of occurrence (counting rate), $\lambda_r = (1/\tau_r)$, for rank $(r) = 1, ..., M$, is

$$\tau_r = \frac{N+1}{rank}, \lambda_r = \frac{rank}{N+1} \tag{3.230}$$

Each value of rank is associated with an extreme value x, thus providing a one-to-one correspondence between data size and τ. As an example, Table 3.15 presents the $M = 25$ ranked values of winter storm surge heights for the 40-year period 1979 to 2018 recorded by the permanent tide gauge at Point Atkinson in the southern Strait of Georgia near Vancouver, British Columbia. Storm surge heights were extracted from the hourly gauge records by removing the tides. Data are then used to calculate the return periods, τ_r, and counting rates, λ_r, using a duration of $N+1 = 40$ years. In this example τ_r ranges from 40 years to $40/25 = 1.6$ years and λ_r from $1/40 = 0.025$ cycles (events) per year to $25/40 = 0.525$ events per year. Plots of τ_r and λ_r versus the size x for each ranked value (Figure 3.38) give estimates of the return period and corresponding frequency of occurrence as functions of storm surge height.

TABLE 3.15 Storm surge amplitudes (m) for Point Atkinson (49° 20.25′ N; 123° 22.42′ W) in the southern Strait of Georgia from 1979 to 2018. Surge values are ranked according to their height, and return period calculated as $\tau_r = \frac{N+1}{rank}$, where $N+1$ is the number of years spanned by the record. The last four columns give the probability of at least one event exceeding the specific height for the given return period. Multiplication by 100 gives the likelihood of exceedance as a percentage.

Rank	Height (m)	Return period (τ_r, years)	1-year return	5-year return	25-year return	50-year return
1	1.020	40.00	0.0247	0.1175	0.4647	0.7135
2	0.962	20.00	0.0488	0.2212	0.7135	0.9179
3	0.945	13.33	0.0723	0.3127	0.8466	0.9765
4	0.918	10.00	0.0952	0.3935	0.9179	0.9933
5	0.907	8.00	0.1175	0.4647	0.9561	0.9981
6	0.907	6.67	0.1393	0.5276	0.9765	0.9994
7	0.893	5.71	0.1605	0.5831	0.9874	0.9998
8	0.891	5.00	0.1813	0.6321	0.9933	0.9999
9	0.883	4.44	0.2015	0.6753	0.9964	~1.0000
10	0.875	4.00	0.2212	0.7135	0.9981	~1.0000
11	0.869	3.64	0.2404	0.7472	0.9990	~1.0000
12	0.863	3.33	0.2592	0.7769	0.9994	~1.0000
13	0.862	3.08	0.2775	0.8031	0.9997	~1.0000
14	0.856	2.86	0.2953	0.8262	0.9998	~1.0000
15	0.848	2.67	0.3127	0.8466	0.9999	~1.0000
16	0.838	2.50	0.3297	0.8647	~1.0000	~1.0000
17	0.836	2.35	0.3462	0.8806	~1.0000	~1.0000
18	0.824	2.22	0.3624	0.8946	~1.0000	~1.0000
19	0.824	2.11	0.3781	0.9070	~1.0000	~1.0000
20	0.823	2.00	0.3935	0.9179	~1.0000	~1.0000
21	0.819	1.90	0.4084	0.9276	~1.0000	~1.0000
22	0.814	1.82	0.4231	0.9361	~1.0000	~1.0000
23	0.812	1.74	0.4373	0.9436	~1.0000	~1.0000
24	0.806	1.67	0.4512	0.9502	~1.0000	~1.0000
25	0.801	1.60	0.4647	0.9561	~1.0000	~1.0000

3.20.3.2 The Poisson Probability Distribution

For stationary random processes, the probability of exactly m independent events occurring within the specified time interval, t, for occurrences having a return period, τ_r, and counting rate, $\lambda_r = 1/\tau_r$, is given by the *Poisson* distribution

$$f(t; m) = \frac{(t/\tau_r)^m}{m!}e^{-t/\tau_r} = \frac{(\lambda_r t)^m}{m!}e^{-\lambda_r t} \tag{3.231}$$

The immediate question is: *How do we obtain the return period, τ_r, and counting rate, λ_r, for determining $f(t; m)$, and therefore, the probability of m independent events taking place?* The obvious approach is to use the empirical values available through (3.230) as the distribution parameters. Once determined, the values can be substituted into (3.231) to obtain the likelihood of a single event in the specified time interval. For example, the probability of *exactly one event* occurring in a 1-year interval for a process having a return period of 5 years (or counting rate $\lambda_r = 0.2$ year^{-1}) is $f(1; 1) = \frac{(1/5)^1}{1!}e^{-\frac{1}{5}} = 16.37\%$. For one occurrence in 5 years, $f(5; 1) = \frac{(5/5)^1}{1!}e^{-\frac{5}{5}} = 36.79\%$, and in 50 years

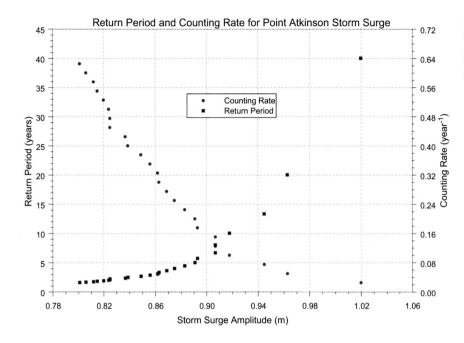

FIGURE 3.38 Return period and counting rates for storm surge events at Point Atkinson (49° 20.25′ N; 123° 22.42′ W) in the southern Strait of Georgia from 1979 to 2018 *See data in Table 3.15.*

$f(50;1) = \frac{(50/5)^1}{1!}e^{-\frac{50}{5}} = 0.05\%$. Thus, the likelihood of recording exactly one event over a time period equal to the return period (here, 5 years) is around 37%, while for much longer wait times, such as 50 years, it is almost zero. The latter applies because it is highly likely that there will be many more than just one event over the 50-year period, which easily verified by setting $m \geq 2$ in (3.231). In particular, for $m = 2$ and $\tau_r = 5$ years, $f(50;2) = 0.23\%$, which is five times greater than for a single event; for $m = 10$, the probability jumps to $f(50;10) = 12.51\%$, or 250 times more likely than one event.

We note that an event size, such as its amplitude or duration, with a return period, τ_r, does not mean that an event of that particular size is guaranteed to occur every τ_r years or that only one event of that size will only happen every τ_r years. It is possible that an event of that particular size occurs once or more, or never at all, over the return period. To determine the likelihood that an event will exceed the number of occurrences m for a given return period (as in Table 3.15), we turn to the complementary cumulative distribution function (CCDF), $\widetilde{F}(t;m)$. This function represents the probability that, in the time t, there will be additional events whose size exceeds the m events having return periods, τ_r, (or equivalently, that the time interval t_m for the process to reach the mth event will actually be less than t) and is expressed as

$$\widetilde{F}(t;m_{t_m} \geq m) = \widetilde{F}(t;t_m \leq t) = 1 - e^{-t/\tau_r}\sum_{k=0}^{m-1}\left[\frac{(t/\tau_r)^k}{k!}\right] \tag{3.232}$$

(Geist et al., 2009). For the special case of one or more stronger events over the time interval, t, exceeding the events with return periods, τ_r, Eqn (3.232) reduces to the commonly applied expression

$$\widetilde{F}(t;m_{t_m} = 1) = 1 - e^{-\frac{t}{\tau_r}} = 1 - e^{-\lambda_r t}. \tag{3.233}$$

If, instead, we are interested in non-exceedance occurrences over a given time interval, t, we use the cumulative distribution function (CDF),

$$F(t;m_{t_m} = 1) = e^{-t/\tau_r} = e^{-\lambda_r t}. \tag{3.234}$$

For a time period, t, equal to the return period, τ_r, both (3.233) and (3.234) yield fixed values: $\widetilde{F}(\tau_r;1) = 63.21\%$ and $F(\tau_r;1) = 1 - \widetilde{F}(\tau_r;1) = 36.79\%$. Because this holds for all cases for which $t = \tau_r$, the probability of exceedance (non-exceedance) within the time interval is independent of the return period and equal to 63.21% (36.79%) not 100% (0%), as one might have been expected.

A highly useful aspect of (3.233) is that the probability of at least one extreme event over the time, t, is directly determined by the counting rate, λ_r, for events of a specified size

$$\lambda_r \approx \frac{m_e}{T} = \frac{rank}{N+1} \tag{3.235}$$

where m_e is the number of extreme events of a particular size over the record length, $T = N\Delta t$ (in the same time units as t), and Δt is the sampling time step. Thus, for example, for $m_e = 2$ events in $T = 100$ years, $\lambda_r = 0.02$ year^{-1}, and the process has a return period $\tau_r = 1/\lambda_r = 50$ years. Then, the probability of more than one occurrence of events of that particular size in, say, 10 years is

$$\widetilde{F}(10; 2) = 1 - e^{-0.02 \times 10} = 18.13\% \tag{3.236}$$

For observations with gaps, the estimation of λ_r needs to take into account the missing data within each block of time (e.g., within each month or each year for monthly and annual data series, respectively). This leads to the adjusted counting rate

$$\lambda_r = \frac{1}{M} \sum_{i=1}^{M} \frac{M_i}{p_i} \tag{3.237}$$

(Ferriera and Guedes Soares, 1998; Caries, 2011) where M_i is the number of data peaks over the threshold in the ith time interval (e.g., a year), and $p_i = m_i / m_{total}$ is an adjustment for missing data within each time interval, for which m_i is the number of actual observations available in the ith time interval and m_{total} is the maximum possible number of observations in that time interval. For example, for a time series of annual extremes derived using daily values, $m_i \leq m_{total}$, and $m_{total} = 365$ (or, 366 during leap years).

3.20.3.3 The m-Observation Return Interval for BM GEV Distributions

Assume that the statistics for extreme variable X follow a Poisson probability distribution. Then, because the GEV function $F(x)$ is invertible [i.e., $F^{-1}(F(x)) = x$], the observed level, x_m, that will be exceeded once during m observations on average (e.g., over m months or m years) is

$$x_m = F^{-1}(1 - p_m) = \mu - \frac{s}{\xi}\left\{1 - [-\ln(1 - p_m)]^{-\xi}\right\} \text{ for } \xi \neq 0 \tag{3.238a}$$

$$x_m = F^{-1}(1 - p_m) = \mu - s \cdot \ln\{-\ln(1 - p_m)\} \text{ for } \xi = 0 \tag{3.238b}$$

in which $\ln(x)$ denotes the natural logarithm and $F(x_m) = 1 - p_m = 1 - 1/m$. Once μ, s, ξ have been estimated from the data, (3.238) can be used to determine, x_m, the **m-observation return level**, as a function of the **m-observation return period**, $1/p_m$. Where the data are annual values, the level x_m is expected to be exceeded by the annual maximum in any given year with the probability p_m; the m-observation return level then becomes the **m-year return level**, and the m-observation return period becomes the **m-year return interval**.

In Figure 3.39, we provide an example of a GEV distribution and its interpretation in terms of return period. In this example, the value $x_1 = 30$ occurs at an annual probability, $p(X = 30) = 0.01$, whereby the return period for x_{30} becomes $1/p = m = 100$ years. Similarly, the return value $x_2 = 40$, has a return period of 10,000 years.

Suppose we want to determine the 100-year return levels for extreme surface water temperatures ($x \to T$) at a mid-latitude site for which the existing data yields the following mean and 95% confidence intervals (shown in brackets) for the Generalized Extreme Value (GEV) distribution: $\mu = 11.52$ (10.86, 12.24)°C, $s = 1.50$ (1.03, 1.92)°C and $\xi = -0.30$ (−0.63, −0.03). Plugging these values into the GEV formulation (3.238a), we find that, for $p_m = 0.01$ (the 100-year return period), the temperature $T_{100} = 15.26$ (12.40, 20.49)°C. For the Gumbel distribution, a special case of the GEV distribution for which the shape parameter $\xi = 0$ (3.238b), the data give $\mu = 11.29$ (10.72, 11.93)°C and $s = 1.33$ (1.00, 1.65)°C, which yields $T_{100} = 18.42$ (15.60, 21.07)°C. The values from the Gumbel distribution are significantly higher than for the GEV distribution for both the mean and 95% confidence interval.

3.20.3.4 The m-Observation Return Interval for POT Distributions

For Generalized Pareto distributions for extreme value series derived using the Peaks Over Threshold method, the m-observation return levels are

$$x_m = \mu + \frac{s}{\xi}\left[(m \cdot p_\mu)^\xi - 1\right], \text{ for } \xi \neq 0, x > \mu \tag{3.239a}$$

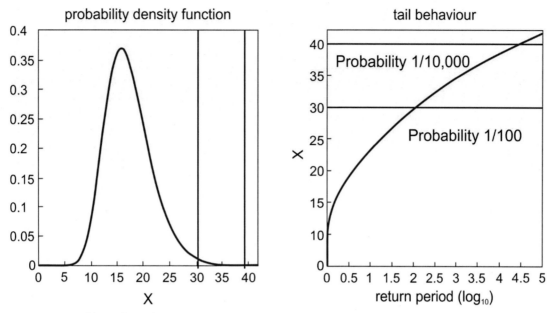

FIGURE 3.39 (a) Plot of $f(x) = P(X = x)$ for a Generalized Extreme Value (GEV) distribution that has been fitted to unitless geophysical data, x; (b) log plot of the return times derived for two values of x ($x_1 = 30$ and $x_2 = 40$) of the GEV (vertical lines in (a)), with the resulting values, x, obtained for probabilities $P(X = x_1) = 0.01$ and $P(X = x_2) = 0.0001$ and corresponding return times in log format.

$$x_m = \mu + s \cdot \ln(m \cdot p_\mu) \text{ if } \xi = 0, x > \mu \tag{3.239b}$$

(Coles, 2001; Bommier, 2014), and where $p_\mu = P\{X > \mu\}$ is the probability of exceedance for a high threshold level, μ. To estimate the M-year return level for a Generalized Pareto distribution, let m_{yr} be the number of observations per year, then the total number of observations in M years is $m = M \times m_{yr}$. The M-year return level subsequently becomes

$$x_M = \mu + \frac{s}{\xi}\left[(Mm_{yr}p_\mu)^\xi - 1\right], \text{ for } \xi \neq 0, x > \mu \tag{3.240a}$$

$$x_M = \mu + s \cdot \ln(Mm_{yr}p_\mu) \text{ if } \xi = 0, x > \mu \tag{3.240b}$$

Determination of the M-year return level requires estimation of the probability p_μ, and two parameters, s and ξ (which we can obtain using the maximum likelihood method). If we assume that the exceedances above the threshold μ are rare events, p_μ is expected to follow a Poisson distribution. As noted above, the Poisson distribution is characterized by the parameter, λ, representing the mean of the threshold exceedances per unit time, in this case, per year. Then, p_μ can be estimated as

$$p_\mu = \lambda/m_{yr}, \tag{3.241}$$

where an unbiased estimate of λ is given by

$$\widehat{\lambda} = n_\mu/N, \tag{3.242}$$

in which n_μ is the number of exceedances over the selected threshold μ in a record of N years, where $m_{yr}N = $ total number of observations in the record. Substituting the expressions for p_μ in terms of $\widehat{\lambda}$ in (3.241), we obtain

$$x_M = \mu + \frac{s}{\xi}\left[\left(\widehat{\lambda}M\right)^\xi - 1\right], \text{ for } \xi \neq 0, x > \mu \tag{3.243a}$$

$$= \mu + s \cdot \ln\left(\widehat{\lambda}M\right), \text{ for } \xi = 0, x > \mu \tag{3.243b}$$

for the M-year return level. As an example, Coles (2001) calculates $p_\mu = 0.00867$ based on $n_\mu = 152$ daily rainfall exceedances over a threshold of 30 mm in southwest England for a complete set of $N = 17{,}531$ daily observations over the period 1914–62.

3.20.3.5 Pareto Distribution: The Special Type I Case

For a Type I Pareto distribution, $\mu = \frac{\varsigma}{\xi} = x_{min}$ so that (3.243a) becomes

$$x_M = x_{min}\left(\widehat{\lambda}M\right)^{\xi} \text{ for } \xi \neq 0, x > \mu \tag{3.244}$$

where, from (3.242), $\widehat{\lambda} = $ (number occurrences above the minimum)/(number of values in the data set). Alternatively, we can also take the natural logarithm of $\widetilde{F}_P(x) = \left(\frac{x_{min}}{x}\right)^{\alpha}$, to obtain $\ln\left[\widetilde{F}_P(x)\right] = \alpha \cdot [\ln(x_{min}) - \ln(x)]$. Solving for $\ln(x)$ and then taking the exponential (the inverse), we derive the m-observation return interval

$$x_m = x_{min}\left[1/\widetilde{F}_P(x_m)\right]^{\xi} \text{ for } \xi \neq 0, x > \mu \tag{3.245}$$

where $\widetilde{F}_P(x_m) \approx 1/m$. To convert the m-observation return period to the M-year return period, we set $m = \widehat{\lambda}M$, so that $1/\widetilde{F}_P(x_M) \approx \widehat{\lambda}M$, and as for (3.244)

$$x_M = x_{min}\left[\widehat{\lambda}M\right]^{\xi} \text{ for } \xi \neq 0, x > \mu \tag{3.246}$$

The recent study of major storm surge in the Salish Sea on the southwest coast of Canada discussed in Section 3.20.3.2, revealed 25 events spanning 40 years in tide gauge data from the Point Atkinson light-station (Table 3.15). The of log(rank) of the storm surge amplitudes ($rank = 1-25$) versus log(amplitude) presented in Figure 3.38 shows a power-law distribution with an elbow-like change in slope at $rank = 15$, corresponding to a storm surge height $x = 0.849$ m. In this case, the slope in the log-log plot was a near straight line for $rank \geq 15$, yielding a shape parameter for the highest storm surge amplitudes ($rank = 1, ..., 15$) of $\xi = 0.05721$, which can then be substituted into (3.246) along with $x_{min} = 0.849$ m and $\widehat{\lambda} = 0.375$ (15 events/40 years). For the $M = 100$-year return level, we find $x_M = 0.849 \times [0.375 \times 100]^{0.05721} = 1.045$ m.

3.20.4 Extreme Data

The advantages of the Peaks Over Threshold (POT) method are that it considers much more of the data than the Block Maximum (BM) method (which focuses on one value per block) and yields estimators that are not overly influenced by the smaller sizes. The disadvantages of the POT method are that it depends on the threshold choice and may require a declustering of groups of maxima to ensure that each peak included in the extreme series is an independent occurrence. For example, in studies of extreme storm surge heights, it is customary to allow for 3 days between clusters to ensure that they are linked to separate weather systems. If the clusters are due to separate events, the maximum peak in the cluster is selected as representative of the event size and the other peaks associated with the storm are masked out. This need to address temporal dependence in a dataset is more of an issue in threshold exceedance models, for which the user can either decluster or, alternatively, adjust the threshold level.

As illustrated by Figure 3.40, each of the two main types of extreme data is commonly linked to a particular family of distributions. Although there is no fixed rule, data obtained using the BM method are typically fitted to *Generalized Extreme Value* (GEV) distributions, while data from the POT method are fitted to *Generalized Pareto* (GP) distributions. For POT data, the analysis may involve application of two separate distributions: one for the number of events expected to occur within a specified return period and a second for the size of the exceedance values for a specified return period. The BM method is used in a wide variety of determinations, including the peak annual water levels in harbors (Zhai et al., 2019), the amplitude of (hurricane-induced) storm surge at coastal tide gauges (Fang et al., 2021) and extreme significant wave heights at monthly and annual time scales (Rueda et al., 2016). Similarly, the POT method has been used to examine a range of processes, such as the number of extreme flooding events along the coastline of the United States (Buchanan et al., 2017; Ghanbari et al., 2018), the maximum amplitudes of propagating tsunamis recorded by bottom pressure sensors in the open eastern Pacific Ocean (Fine et al., 2020), and the extent of tsunami run-up along a coastline (Geist, 2009). A global perspective on the coastal impact and adaptation to extreme sea levels (based on 510 individual tide gauges) is provided by Wahl et al. (2017), who used Gumbel and GEV distributions to examine annual and monthly maxima, and GP distributions for peaks over threshold analyses.

Although the BM method is generally applied to just one exceedance value per block, the method can also be applied to more than one peak value per block. This leads to the hybrid *r-largest order statistics* method mentioned at the beginning of this section. For instance, Wahl et al. (2017) used GEV distributions to examine the peak annual sea level values, as well as the second to tenth largest annual values, for coastal sites around the globe, giving the probability distributions labels

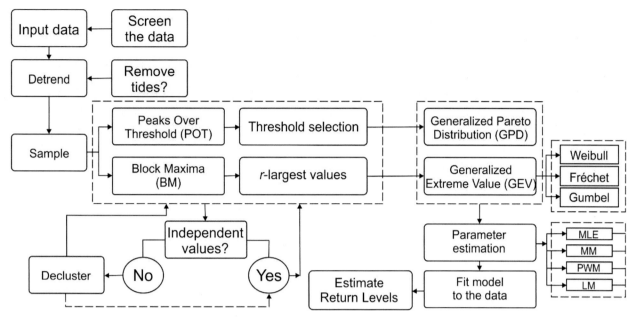

FIGURE 3.40 Flow diagram for processing and analyzing extreme data. *Modified after and Arns et al. (2013) and Fang et al. (2021).*

such as GEV-$r2$, ..., GEV-$r10$. Generalized Pareto distributions were also applied to the extreme time series using thresholds between the 98th and 99.75th percentiles, with declustering time differences of 3 days. Similarly, Fang et al. (2021) selected exceedance values ranging from $r = 1$ to 10 values/year in a study of typhon-induced storm surge on the coast of China. They too examined results based on selected thresholds ranging from the 98th to the 99.25th percentiles in increments of 0.25.

3.20.4.1 Choosing the Threshold Level for POT Data

The choice of threshold level in the Peaks over Threshold method is not straightforward and remains a difficult topic (Davison and Huser, 2015; Cavanaugh et al., 2015). As with many statistical applications, there is a trade-off between retaining too many or too few data points. A high threshold level reduces bias in the distribution, because it satisfies the requirement of convergence in extreme value theory, but increases the variance of the estimated probability distribution parameters owing to a paucity of data points. On the other hand, a low threshold decreases the variance of the estimators because of the greater number of values, but increases the bias at the extreme end of the probability distribution. In their study of storm surges on the coast of China, Fang et al. (2021) compared results from the BM approach with those from the POT approach by picking percentile thresholds that gave sample sizes that matched the ones used in the r-largest analysis (i.e., 99.88% corresponds to $r = 1$ value/yr, on average, and 99.44% to $r = 10$ values/yr, on average). The comparisons indicate that the Block Maximum method (for the annual maxima, in particular) misses some important extremes, which can lead to an underestimation of extreme sea levels.

The choice of threshold values can be quite nuanced. For example, the United States National Oceanic and Atmospheric Administration (NOAA) identifies three "official" coastal flood thresholds (Ghanbari et al., 2018): *minor flooding* (exceedance over a minor flood threshold) that causes minimal damage accompanied by a public threat and inconvenience; *moderate flooding* (exceedance over a moderate flood threshold) that causes considerable damage to private and commercial property; and *major flooding* (exceedance over a major flood threshold) that can result in extensive loss of property and life. Sweet et al. (2018) specify 0.5, 0.8 and 1.2 m above the local tidal range as minor, moderate and extreme levels, respectively, for NOAA tide gauge sites operated on the contiguous US.

Graphical methods are generally used to determine the threshold level. These include: (1) The Mean Residual Life Plot (Davidson and Smith, 1990); (2) the Parameter Stability Plot (Boomier, 2014); (3) the Dispersion Index (Ribatet, 2006); and (4) The Multiple-Threshold Model (Wadsworth and Tawn, 2012). Although there have been various attempts to automate threshold selection (e.g., Dupuis, 1998; Thompson et al., 2009; Solari and Losada, 2012), graphical methods—supplemented with consideration of goodness of fit (GOF), such as the use of Quantile-Quantile (Q-Q) plots or GOF tests (Davison and Huser, 2015)—remain the most common approach (e.g., Coles, 2001; Solari and Losada, 2012).

Instead of selecting a single threshold, some authors have proposed averaging the results obtained from a series of thresholds (e.g., Beguería, 2005).

Graphical methods have several drawbacks. To begin with, it is not possible to quantify the uncertainty associated with a particular threshold selection or its impact on the uncertainty of high return period quantiles (Solari et al., 2017). Secondly, graphical methods cannot be automated, resulting in a degree of subjectivity to extreme value analysis. The first issue can be partially solved by first defining the POT data series and then fitting the Generalized Pareto Distribution (GPD) using methods that can estimate the location parameter. As further outlined in the next section, a threshold μ_o, is established and only the peaks where $x_{peak} \geq \mu_o$ are retained for analysis. Then, the location parameter of the GPD, or threshold, μ, is estimated from the data where $x_{peak} \geq \mu$. With respect to the second issue, Solari et al. (2017) discuss methods for automatic threshold estimation for POT applications. The methods enable the user to quantify both the uncertainty of a threshold and its impact on the uncertainty of high return-period quantiles. The authors claim that the method is easily implementable and therefore accessible to a wide audience.

3.20.4.2 Threshold Levels for POT Data

The Parameter Stability Plot is a common graphical method for determining the threshold levels for Pareto distributions. Suppose the exceedance of a high threshold level, μ_o, follows a Generalized Pareto Distribution with parameters s_{μ_o} and ξ. Then, for any threshold $\mu > \mu_o$, exceedance values will follow a GPD with shape parameter $\xi_\mu = \xi$ and scale parameter $s_\mu = s_{\mu_o} + \xi(\mu - \mu_o)$. Let $s' = s_\mu - \xi_\mu \mu$, which no longer depends on μ because μ_o is already a high threshold value. Then for x_{max}, the maximum observed value, the plots $\{(\mu, s'); \mu < x_{max}\}$ and $\{(\mu, \xi_\mu); \mu < x_{max}\}$ will show s' and ξ_μ as constant for all $\mu > \mu_o$, provided that μ_o is a suitably high threshold value for the asymptotic approximation. The threshold is chosen for that value where the shape and scale parameters remain constant. The data presented in Table 3.15, for storm surges at Point Atkinson, provide a useful example of the choice of $\mu_o = 0.859$ m as the cutoff.

Another graphical approach for Pareto distributions is to plot the threshold level, μ, versus the mean threshold excess

$$\left\{\left(\mu, \frac{1}{m_\mu}\sum_{i=1}^{m_\mu}(x_i - \mu)\right); \mu < x_{max}\right\} \tag{3.247}$$

where the x_i are the m_μ observed values that exceed μ. For a range of thresholds, μ, we seek the value μ_o above which the plot becomes linear. A straight-line segment indicates that the assumption of a GPD distribution is likely valid, as in the storm surge example provided in Figure 3.38.

3.20.4.3 Software Used to Compute Parameters for Extreme Value Distributions

Caires (2011) lists several computer software products that can be used to calculate the parameters for fitting extreme value distributions. These include MATLAB, C++ with a MATLAB Graphical Users Interface; R, S-Plus and R; S-Plus and Fortran Python, and Excel.

3.20.5 A Recipe for Generating and Processing Extreme Value Records

Once the method for determining extreme values has been selected, there are a number of steps in the fabrication of a dataset (cf. Arns et al., 2013; Wahl et al., 2017; Fang et al., 2021). The steps assume that the extreme events under study are samples from independent random processes and that events occurring in one time period are independent of those in other time periods. A rough *rule of thumb* is that maximum m-observation return periods be extrapolated to approximately four times the length of the observational record (Pugh and Woodworth, 2014). On the other hand, analyses for broad-scale impact and adaptation studies also need to examine low-probability, high impact, events. For example, Wahl et al. (2017) note that a 1-in-10,000-year flood design standard has been adopted for the most densely populated areas of the Netherlands and is also used in other countries to protect critical coastal infrastructure such as nuclear power plants.

3.20.5.1 Basic Steps in Data Preparation

As outlined in Figure 3.40, the main steps in preparing extreme data are:

1. *Screening*: As with any data set, it's necessary to eliminate values that are invalid outliers originating with instrument malfunction, calibration errors, data transmission, or other factors. Even where log books or other notes on the data collection are available, it is not always straightforward to identify and screen out erroneous extreme values.

2. *Trend removal*: In many circumstances, it is necessary to remove the long-term trend from the original data prior to determination of the extremes. If the trend is affecting the extremes (for example, where rising global sea levels are amplifying storm surge heights), the process is *nonstationary* and contributions associated with the trend must be taken into account. In the Block Method, detrending involves subtracting the mean of the data record on a block-by-block (e.g., year-by-year) basis. Removal of the mean for each block ensures that the extremes are true extremes, and measured relative to a separately varying background level (cf. Muis et al., 2016; Wahl et al., 2017).

3. *De-tiding*: For those cases in which barotropic or baroclinic tides partially mask extremes in a time series, harmonic analysis methods (cf. Chapter 5) can be used to remove the stationary component of the tidal variations, i.e., the predictable component of the record that is phase-locked to the luni-solar tidal forcing. Where feasible, removal of the astronomical tidal constituents should include removal of the 18.61-year nodal cycle caused by the slow, 360° *nutation* ("rocking motion") of the moon's orbital plane about the earth. Any residual tidal motions—such as those due to baroclinic semi-diurnal tides or diurnal shelf waves that are not phased-locked with the astronomical forcing—can be removed by applying a low-pass filter with a 40-h cut-off. Accurate removal of the tides requires a record with a minimum sampling rate of once per hour.

4. *Extracting extreme values*: Application of the POT method, requires the choice of a threshold that defines the minimum acceptable extreme value for the particular process. This can be tricky: if the threshold is too low, the fitted probability function will be distorted by the large number of low values; if too high, the fitted function could be poorly constrained because of the limited number of data values. Use of the BM also requires some thought. For example, to study extreme storm surge, the background trend can be derived by averaging the data on an annual basis, which, at mid-latitudes, means focusing on the period from late fall to early spring of the next year, rather than averaging over a calendar year.

5. *Declustering*: To ensure that the extreme values in a data series are statistically independent, the separation of peaks in a cluster of peaks may be necessary. Extreme values that occur close together could be part of the same event and therefore not independent samples. In studies of storm surge heights, a time difference, $\Delta t \geq 3$ days, between extreme values seems reasonable (Arns et al. 2013; Wahl et al. 2017; Feng and Tsimplis, 2014). This scale is consistent with the times between extratropical weather systems. In general, the investigator first specifies a threshold value and locates all clusters exceeding the threshold. A minimum separation, Δt_{min}, between clusters is set and the boundary between adjoining clusters defined at times for which $\Delta t > \Delta t_{min}$. The maximum exceedance in each cluster is then chosen to represent the value for that cluster.

6. *Choice of distribution*: The time variations in event size obtained through the BM and POT methods are typically quantified using the Generalized Extreme Value (GEV) distribution and the Generalized Pareto Distribution (GPD), respectively. Which particular distribution is preferable, depends on the characteristics of the data and there is no loss of generality in using different forms of distribution for the same data series. Plots of the fitted distributions versus a histogram of the extreme values reveal how well a particular distribution represents the observations.

7. *Parameter estimation*: Fitting of extreme data to a particular distribution requires estimation of the defining parameters: location (μ), scale (s) and shape (ξ). A range of estimation methods exists, including: L-Moments, Probability Weighted Moments (PWM), least squares, Method of Moments, and Maximum Likelihood Estimation (MLE). As the uncertainty arising from the choice of parameter estimation method is small compared to other key uncertainties (Wahl et al., 2017), the MLE method is often the favored choice. Table 3.16 presents a comparison of the 100-year return levels for significant wave heights ($H_{1/3}$) in the North Sea derived using the BM and POT approaches, to which were applied two different methods for calculating the distribution parameters. The PWM approach gave slightly higher projections than the MLE approach for both the BM and POT analyses. Caires (2011) concludes that these results support the Hosking et al. (1985) recommendation to always use the PWM method for GPD or GEV estimation for relatively short data sets "with not too heavy-tailed distributions."

Removal of the astronomical tides in Step 3 can vary from simply removing the leading diurnal and semi-diurnal constituents, to removing as many constituents as possible based on the length of the data series. For example, in their study of extreme storm surge on the west coast of Canada, Zhai et al. (2019) fitted 45 astronomical constituents and 24 of the more important shallow-water constituents to the observations using the tidal analysis package, "T_Tide" (Pawlowicz et al., 2002), which includes nodal corrections based on the latitude of the tide gauge site. Eliot (2010), Menéndez and Woodworth (2010) and Talke et al. (2018) have shown that the 18.61-year nodal cycle affects the estimation of extreme storm tides and needs to be removed. Following removal of the tides from the tide gauge records, spectral analysis can be used to determine whether most of the tidal energy has been removed from the time series. If there is still tidal energy in the semi-diurnal and diurnal bands due to baroclinic tides that are not phase-locked to the astronomical tides, this energy can be removed using low-pass filters.

TABLE 3.16 Comparison of the 100-year return level of significant wave heights, $H_{1/3}$, derived using observations at the Schiermonnikoog Nord Buoy (53.596° N, 06.167° W) moored in 19 m of water off the Netherlands from 1970 to 2002. Data were processed with the Peaks Over Threshold (POT) and Block Maximum (BM) methods and fitted to GEV and GPD distribution using two different parameter estimation methods: the Probability Weighted Moments (PWM) and the Maximum Likelihood Estimation (MLE) methods. The PMW approach performs better than the MLE for small to moderate sample sizes. The 95% confidence intervals are from an adjusted percentile boot strap method.

	POT/GPD PWM	POT/GPD MLE	BM/GEV PWM	BM/GEV MLE
No. values	119	119	21	21
μ (m) location	8.27	8.27	10.52 (9.85, 11.24)	10.57 (8.52, 11.50)
ξ shape	−0.08 (−0.30, 0.16)	−0.15 (−0.33, 0.03)	−0.35 (−0.71, −0.03)	−0.38 (−1.15, −0.09)
s (m) scale	1.40 (1.04, 1.82)	1.49 (1.11, 1.93)	1.53 (1.06, 1.95)	1.46 (0.59, 2.46)
$H_{1/3}$ (m) 100-year	15.23 (12.98, 18.65)	14.36 (12.81, 18.57)	14.04 (12.78, 15.67)	13.72 (10.57, 16.76)

Modified after Caires (2007).

3.20.6 Worked Examples of Extreme Value Analysis

This section presents examples of extreme value analysis for one POT and one BM oceanic data series based on the methods outline in previous sections. The examples are also meant to illustrate the methods and the effort that is sometimes required to generate extreme value series.

3.20.6.1 Peak Over Threshold Method (Deep Sea Tsunamis)

The Gutenberg-Richter relationship published in the middle of the twentieth century (Gutenberg and Richter, 1944) found that the magnitudes of seismically generated earthquakes follow a power-law distribution over much of the earthquake intensity range. Except for the extreme end of the distribution, where the relationship fails due to physical limitations in the spatial extent of subduction zone failure regions (Geist, 2009), such power-law dependencies indicate the absence of a dominant scale of variability. A similar finding holds for seismically generated tsunamis (Fine et al., 2020), which serve as an example for the peaks-over-threshold method for extreme events. Fine et al. (2020) examines a 32-year record of tsunami waves in the northeast Pacific Ocean recorded by DART bottom pressure recorders maintained in the caldera of Axial Seamount—an active underwater volcano located at 1,500 m depth off the coast of Oregon (Chadwick et al., 2012). The advantage of using bottom pressures records, rather than coastal tide gauge records, to study tsunamis is that large segments of the sea-level displacement spectrum arising from surface air pressure loading, such that due to the inverse barometric effect, are compensated for, making it easier to extract small tsunami signals. Data stored internally in offshore DART systems also have high sampling rates (≤1 min), high vertical resolutions (better than 1 mm equivalent depth) and, unlike tide gauge data, remain uncontaminated by reflected tsunami waves for several hours after the passage of the first tsunami waves. Prior to 1992, the sampling rate was around 1 min but has since been increased to 15 s.

3.20.6.1.1 Data Preparation

1. The first step in the Fine et al. (2020) tsunami study was to construct a near continuous 32.4-year time series of bottom pressure for the period 1986−2018. To accomplish this, the authors used data from 35 individual Bottom Pressure Recorder (BPR) deployments at Axial Seamount (46° N, 130° W) from 1986 to 2016, and augmented these data with time series from four BPRs moored on the Cleft Segment of Juan de Fuca Ridge (100 km to the south of the seamount) and more recent BPR records from 2016 to 2018 from the Ocean Observatory Initiative (OOI) Cabled Array (Cowles et al., 2010; Kelley et al., 2014). Bottom pressure records from the Cleft Segment that were coincident with those at Axial Seamount were used to fill gaps in the seamount record.

2. Each tsunami detected in the seamount record was linked to a specific tsunamigenic earthquake. To establish this link, Fine et al. (2020) compared the observed arrival times of the tsunami waves at the seamount to the expected arrival times based on numerical simulations of the wave travel from the source region. This ensured that there was a direct correspondence between the earthquake and the tsunami waves at Axial Seamount. One event, the weak tsunami-like waves originally thought to have originated from an earthquake near the Rat Islands in the Aleutian Islands (Alaska) on 17 November 2003, was deemed to be storm-induced and omitted from the analysis. In total, the authors identified 41 tsunamis in the seamount record. Only tsunamis generated by earthquakes with seismic moments $M_w > 7.0$ were intense enough to be generate "significant" bottom pressure fluctuations.

3. The next step was to determine (a) the peak tsunami wave amplitude, x_{peak}, and (b) the root-mean-square (rms) amplitude, x_{rms}, for the first 12 h of a tsunami record. The 12-h period provides a time series of "uncontaminated" tsunami motions, recorded prior to the arrival of tsunami waves reflected from the coast. (The full datasets are listed in the Supplemental Material of Fine et al., 2020).

4. The Peaks Over Threshold time series created in steps 1–3 was examined for clustering—as could occur for tsunamis generated by strong aftershocks—to ensure that the tsunami events were statistically independent and that there was, at most, one tsunami per year. Here, we examine only the x_{rms} record because it is considered a better indicator of overall tsunami energy than the peak wave. No declustering was needed for either of the two forms of tsunami record.

5. Once the x_{rms} data series were obtained, the values were ranked in order of descending amplitude, from the highest ($rank = 1$) to the lowest ($rank = 41$). A plot of log($rank$) versus log(x_{rms}) (Figure 3.41) shows a marked change in slope at $rank = 25$, indicating a threshold $\mu = 1.5$ mm for extreme values. Further examination of the data reveals that the amplitude for $rank = 25$ can also be eliminated (there are too many small values around 1.5 mm) and the threshold set to $\mu = x_{min} = 1.59$ mm. This yielded 24 tsunami events over a time span of 29 years (taking into

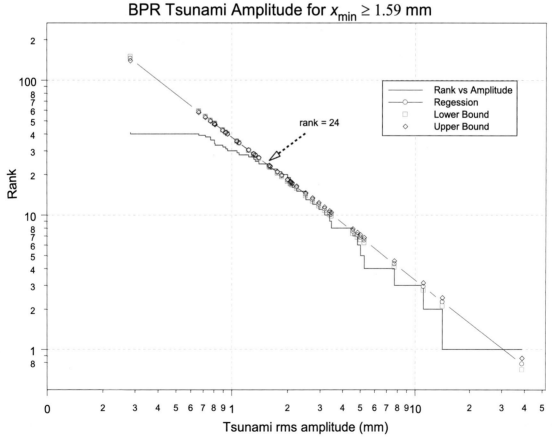

FIGURE 3.41 Log($rank$) versus log(x_{rms}) for root-mean-square tsunami amplitudes x_{rms} from bottom pressure records at 1,500 m depth at Axial Seamount in the northeast Pacific. Data are plotted in horizontal steps rather than as isolated points. The regression line was derived using values with $rank \leq 24$, corresponding to the extreme portion of the dataset having a clear power-law dependence. Regression values for the 95% confidence intervals on the slope are included.

consideration missing segments in the 32-year time span). The result was 24 tsunamis over a period of 29 years, ranging from a maximum rms-amplitude $x_{max} = 38.73$ mm to a minimum $x_{min} = 1.59$ mm (Table 3.17).

3.20.6.1.2 Analysis of Extreme Tsunami Data

As indicated by the linear least-squares regression line in Figure 3.41, the extreme value portion of the tsunami amplitude data (i.e., for $rank \leq 24$ and amplitudes $x_{rms} \geq x_{min}$), has a nearly perfect correlation coefficient (r), with $r^2 = 0.986$. Consequently, tsunami waves measured at Axial Seamount have a pronounced power-law dependence of the form

$$rank(x) = a(x_{rms})^b \qquad (3.248a)$$

$$\log rank(x) = \log(a) + b \cdot \log(x_{rms}) \qquad (3.248b)$$

where, from least squares analysis, $a = 37.7746$ (36.2994, 39.3097) and $b = -1.0615$ ($-1.0888, -1.0347$); here, the numbers in brackets denote mean ± 1 standard deviation. The fact that there is no saturation of the distribution curves plotted in Figure 3.42a and b, indicates that the tsunami wave energy did not reach a physical limit, consistent with the absence of a spatial limit to the earthquake source regions generating the tsunamis.

3.20.6.1.3 Interpreting the Zig-Zag Distribution

The reason for plotting the distributions in Figure 3.42a and b as zigzags (horizontal steps), rather than as scatter plots, is to emphasize the probabilistic nature of the distributions. The horizontal lines in the zigzags illustrate that the size for a specific rank has a range of possible values. For example, in the case of $rank = 1$, possible values lie in the range $14.26 < x \leq 38.73$ mm. Similarly, the size at $rank = 2$ has a probable value in the range $11.18 < x \leq 14.26$ mm, and so on, up to

TABLE 3.17 Values of the root-mean-square (rms) amplitudes, x_{rms} (mm), of tsunami waves versus *rank* of the signal for 24 tsunamis recorded at Axial Seamount in the northeast Pacific from 1986 to 2018; rms values are based on the first 12 h of the tsunami record.

Rank	1	2	3	4	5	6	7	8	9	10	11	12
RMS	38.73	14.21	11.11	7.71	5.25	5.01	4.82	4.55	3.46	3.38	3.17	2.95
Rank	13	14	15	16	17	18	19	20	21	22	23	24
RMS	2.73	2.51	2.51	2.24	2.12	2.09	2.00	2.07	1.86	1.76	1.61	1.59

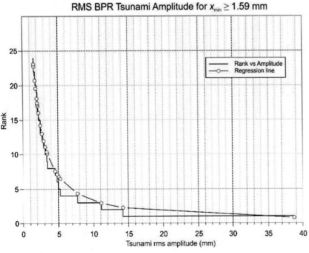
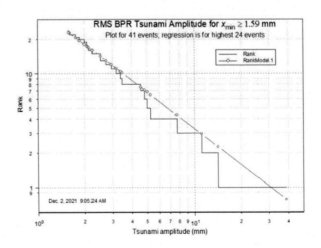

FIGURE 3.42 Plots of the *rank* of the root-mean-square tsunami amplitude, x_{rms}, versus x_{rms}, for the two versions of formula (3.20.27) for bottom pressure records at 1,500 m depth at Axial Seamount in the northeast Pacific: (a) Stepwise linear $rank \sim function(x_{rms})$; and (b) $\log(rank) \sim \log(x_{rms})$. The data are plotted in horizontal steps rather than as isolated points. The regression line was derived using values with $rank \leq 24$, corresponding to the extreme portion of the dataset having a clear power-law dependence.

rank = 24, where the possible size range is $0 < x \leq 1.67$ mm. Each range of values is a measure of the uncertainty in the estimate for a specific return period. That is, for a given return period, a different outcome than the one observed is possible.

3.20.6.1.4 Generalized Pareto Distributions

The power-law structure evident in the Axial Seamount data for *rank* ≤ 24 supports modeling the bottom pressure record, x_{rms}, as a Generalized Pareto Distribution, where from (3.225)

$$f(x_{rms}) = \frac{1}{s}\left(1 + \xi\left(\frac{x_{rms} - \mu}{s}\right)\right)^{-\frac{1+\xi}{\xi}}$$

(3.249)

for the shape parameter $\xi \neq 0$ and location parameter $\mu \equiv x_{min} = 1.59$ mm. Using MATLAB, we obtain the following mean and 95% confidence intervals for estimates of ξ and the scale parameter, s:

$$\widehat{\xi} = 0.1938(-0.1637, 0.5512)$$

(3.250a)

$$\widehat{s} = 4.2798\ (2.5169, 7.2775)\ \text{mm}$$

(3.250b)

Figure 3.43 presents a plot of the function $f(x_{rms})$ derived from Eqn (3.249) along with a histogram of the data, x_{rms}. The curvature of this function closely resembles that obtained using the least squares method in Figure 3.42. However, as indicated by the histogram of x_{rms}, the GPD distribution is weakly constrained by the maximum value, leading to the broad confidence intervals in (3.250) and therefore considerable uncertainty in return level estimates for large return intervals. This is clearly evident when we use the above parameters to calculate the $M = 100$-year return level using (3.243a),

$$x_{rms}(M = 100) = \mu + \frac{\widehat{s}}{\widehat{\xi}}\left[\left(\widehat{\lambda}M\right)^{\widehat{\xi}} - 1\right]$$

(3.251a)

$$= 31.47(9.50, 138.97)\ \text{mm}$$

(3.251b)

where the counting rate $\widehat{\lambda} = 24/29$ equates to 24 events exceeding x_{min} over the span of 29 years of extreme data. Clearly, this is not overly informative, especially considering that the broad range of the 95% confidence intervals and the result that the mean value of 31.37 mm for the 100-year return level is considerably less than the observed value of 38.73 mm over the 34-year record.

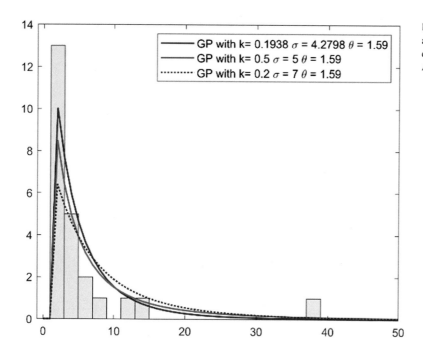

FIGURE 3.43 Generalized Probability Distribution and histogram for tsunami amplitudes (x_{rms}, mm) derived from bottom pressure records at the summit of Axial Seamount.

3.20.6.1.5 Pareto Distribution

We now turn to the less complex Pareto Distribution to model the bottom pressure data. For this model, the probability density function (PDF) and the cumulative distribution function (CDF) for the root-mean-square tsunami amplitude, x_{rms}, are

$$f(x_{rms}) = \xi \frac{(x_{min})^{1/\xi}}{x_{rms}^{(1+1/\xi)}}; x_{rms} \geq x_{min} = 1.59 \text{ mm} \tag{3.252a}$$

$$F(x_{rms}) = 1 - \left(\frac{x_{min}}{x_{rms}}\right)^{1/\xi} x_{rms} \geq 1.59 \text{ mm}$$
$$= 0; x_{rms} < 1.59 \text{ mm} \tag{3.252b}$$

where the shape parameter ξ is derived from the observations using Eqn (3.227), for which

$$\widehat{\xi} = \frac{1}{24}\sum_{n=1}^{24} \ln\left(\frac{x_{rms,n}}{1.59}\right) = 0.807 \pm 0.165. \tag{3.253}$$

with the error estimate $\Delta\widehat{\xi} = \frac{\widehat{\xi}}{\sqrt{N}} = \frac{0.807}{\sqrt{24}}$. The survival function, $\widetilde{F}(x)$—i.e., the probability that at least one occurrence exceeds x (for $x \geq 1.59$ mm)—is, from Eqns (3.215) and (3.252b),

$$\widetilde{F}(x_{rms}) = \left(\frac{1.59 \text{ mm}}{x_{rms}}\right)^{1/\xi}. \tag{3.254}$$

Plots of the three distribution functions versus x_{rms} are presented in Figure 3.44. For $x_{rms} = x_{min}$, $\widetilde{F}(x_{rms}) = 1$, indicating that all values exceed the minimal value, while the probability that all values will, for example, exceed 20 mm is $\widetilde{F}(x_{rms} > 20 \text{ mm}) \cong 0.05$. [Note that Fine et al. (2020) use β for the slope parameter and the term in the summation in Eqn (5) should be "$\ln(R_i/R_{min})$".].

3.20.6.1.6 Return Level for a 100-Year Event

From (3.243a), the mean and 95% confidence interval for the $M = 100$-year return values for x_{rms} (M) is derived from

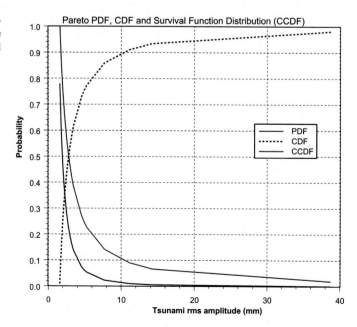

FIGURE 3.44 The probability distribution function (PDF), cumulative distribution function (CDF) and complementary cumulative distribution function (CCDF) for the Pareto Distribution for Axial Seamount tsunamis.

$$x_{rms}(M = 100) = x_{min}\left(\widehat{\lambda}M\right)^{\widehat{\xi}}$$

$$= 56.11(27.08, 116.28) \text{ mm} \tag{3.255}$$

where, as previously, $\widehat{\lambda} = 24/29$. These estimates for the return level seem more reasonable than those derived from the Generalized Pareto Distribution. Repeating the calculation for the $M = 50$-year return level, we obtain the mean $x_{rms}(50) = 32.07$ mm, which is lower than the peak observed value of 38.73 mm for the empirical time estimate, $t_r = 29$ years. However, there is no inconsistency here as the Pareto Distribution is based on all of the extreme data and also because the peak value in the model can, according to the zig-zag plot, have a range of values between 14.21 and 38.73 mm for $t_r = 29$ years. A return level of 32.07 mm for a return period of 50 years is statistically consistent with the data.

Wahl et al. (2017) report similar ranges in uncertainty in their study of 510 global tide gauge records. In their case, the 95% uncertainty ranges across different extreme value models for the 100-year events (events with a 1% exceedance probability in any given year) extend from less than 10 cm at many sites along the US west coast, South America, the Mediterranean, and parts of Australia, to more than 1 m along the US east coast, East Asia and northern Europe.

3.20.6.2 Block Maximum Method (Mediterranean Sea Levels)

A common application of the Block Maximum Method is the construction of an annual series of peak values. Here, we apply the method to hourly tide gauge data from the town of Bakar (45.30° N; 14.53° E; Permanent Service for Mean Sea Level, 2021) near the head of the 4.6-km long Bakar Bay on the Adriatic coast of Croatia. As the northward extension of the much larger Rijeka Bay, Bakar Bay is considered to be relatively well isolated from sea level oscillations originating in the Adriatic (Šepić et al., 2008, 2022; Vilibić et al., 2017). After Trieste, Bakar has the longest operating tide-gauge station in the Adriatic region and one of the longest and most complete sea level records within the Mediterranean. The record has been used to study seiches in Bakar Bay generated by tidal, wind and air pressure forcing, as well as to examine seasonal variability arising from air-sea interactions and interannual-to-interdecadal variability due to the combined effects of global sea-level rise and regional tectonic motions.

3.20.6.2.1 The Data

The Bakar tide gauge was first installed in 1929 but measurements were disrupted by the Second World War. A near continuous analogue hourly record of sea levels is now available from the beginning of 1956 to the present. The analogue data recorded from 1956 to 1982 were hand digitized by Professor Orlić's group at the University of Zagreb in 2003 and output to four decimal places. Data from 1983 to 2002 were next digitized by the Croatian Hydrographic Service in Split and rounded to whole number values under the knowledge that the data were only accurate to 1 cm. Subsequent data from 2003 to present were again digitized at the University of Zagreb. Our analysis covers the 1-cm resolution, 64-year sea level time series from 1 January 1956 to 31 December 2019 provided by Dr. Šepić (University of Split) through Professor Orlić's group. The data are measured relative to the geoid and have several small gaps, amounting to roughly 0.28% of the total record. Much of 1983 is missing along with some winter values in 1959, 1974 and 1980. Although tides are relatively high for this part of the Mediterranean (maximum range ~ 1.2 m), they have not been removed for this analysis. Astronomical tides are not the dominant sea level signal for the purposes of this study.

3.20.6.2.2 Extreme Values

Close examination of the data shows that all major peaks in sea level—with the sole exceptions of single late spring events on 6 June 2002 and 28 May 2007—occurred between late fall and early spring (Figure 3.45). As a consequence, each annual block can be defined as beginning on July 1 and ending on June 30 of the following year. This shortens the number of years in the extreme record from 64 to 63. The next step was to calculate the mean sea level for each block by averaging the hourly water elevations from 1 July to 30 June for each year, and then subtracting the means from the original record to produce annual anomalies of sea level height. As indicated by Figure 3.46, the annual mean varies by O(10) cm from year-to-year and underwent a sudden offset increase in year 2000 (1 July 2000 to 30 June 2001). The reason for this increase in the background sea level height is unknown and not germane to the present discussion.

From the record of annual sea level anomalies, we determined the $r1$, $r2$ and $r3$-highest extreme values for each block (year). Figure 3.47 shows the $r1$ time series, the series with the highest values. As illustrated by the histogram in Figure 3.45, we found that most peak values over the 63 years occurred between October and March, with maximum

FIGURE 3.45 Histogram of the monthly number of extreme sea level occurrences at Bakar, Croatia (45.30° N; 14.53° E), from 1 January 1956 to 31 December 2019. With the exception of single values in May and June, peak water levels were mainly recorded from October to March of each year.

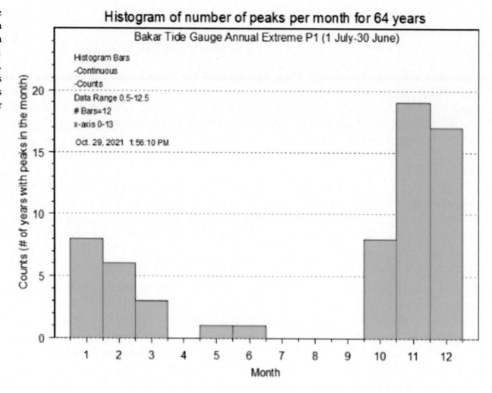

FIGURE 3.46 Time series of the annual mean sea level anomaly (cm) for Bakar, Croatia, for the period 1 January 1956 to 31 December 2019. Annual values span the period 1 July to 20 June of the following year.

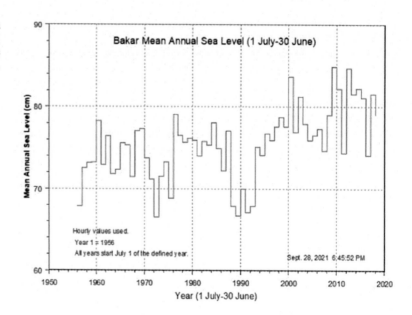

numbers in November (19 of 63) and December (17 of 63). Most missing values occurred in winter, so it is possible that the peaks in some years, such as 1983, were not the maximum values for that year. A more complete formulation of the Bakar data would involve filling the gaps using transfer functions with adjoining tide gauge stations or with the long-term tide gauge record for Trieste (Italy) on the opposite side of the Adriatic.

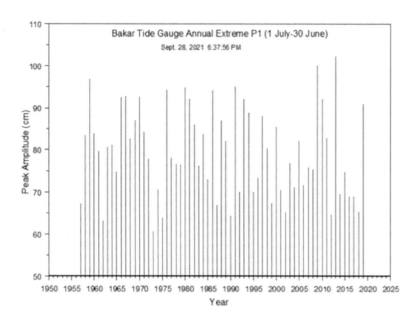

FIGURE 3.47 Time series of the sea level anomaly (cm) for Bakar, Croatia, for the period 1 January 1956 to 31 December 2019.

3.20.6.2.3 Probability Distributions

The Bakar sea-level anomaly data $(r1-r3)$ were first processed as Gumbel distributions (see Eqn 3.17):

$$f(x; \mu, s) = \frac{1}{s} \exp\left\{ -\left[\left(\frac{x - \mu}{s} \right) + \exp\left[\exp\left(-\left(\frac{x - \mu}{s} \right) \right) \right] \right] \right\}; \tag{3.256a}$$

$$F(x; \mu, s) = \exp\left\{ -\exp\left[-\left(\frac{x - \mu}{s} \right) \right] \right\} \tag{3.256b}$$

with the location, μ, and scale, s, parameters for the three r-highest extreme sea level series derived using Maximum Likelihood Estimation (MLE) software. Values for the two parameters and the three r-extremes are listed in Table 3.18; the corresponding Gumbel distributions for the highest $(r1)$ peak, $f(x; s_1, \mu_1)$, $F(x; s_1, \mu_1)$ and $\widetilde{F}(x; s_1, \mu_1)$ are plotted in Figure 3.48. Unlike the Pareto Distribution presented in the previous section, the Gumbel (G) distribution (for which $\xi = 0$), reaches a maximum for $x = \mu = 74.498$ cm and trails then off as an exponential rate for increasing x. For comparison, Table 3.18 also includes the empirically derived values of μ and s derived from the mean and standard deviation of the sea level anomaly data, where from (3.219):

$$\widehat{s} \approx 6^{1/2}(\text{standard deviation} / \pi) = 6^{1/2}(10.585 \text{ cm} / \pi) = 8.2519 \text{ cm} \tag{3.257a}$$

$$\widehat{\mu} \approx \text{mean} - \widehat{s} \cdot \gamma = 79.6492 - 8.2519 * 0.5772 = 74.8861 \text{ cm} \tag{3.257b}$$

$$(\Delta\widehat{\mu}, \Delta\widehat{s}) = \pm (\widehat{\mu}, \widehat{s}) / \sqrt{N} = (\pm 9.435, \pm 1.040) \text{ cm} \tag{3.257c}$$

It is apparent that, for this data set, the empirical estimates of the mean parameter from (3.257) agree closely with those obtained for the $r1$ extreme sea levels using the MLE method (Table 3.18). On the other hand, the 95% confidence intervals for the empirical estimates are significantly greater than for those from the MLE analyses, especially the location parameter, μ, whose range is a factor of four times greater than those based on MLE.

For completeness, we also computed the probability mass functions, $f(x; \mu_i, s_i, \xi_i)$, for the Generalized Extreme Value (GEV) Distribution, the Gumbel distribution and a three-parameter Weibull distribution for the three highest peak series $(r_i, i = 1,2,3)$. Results are provided in Table 3.19 and plots of the distributions presented in Figure 3.49a–c, together with the histogram for each data series. As Figure 3.49 indicates, the distributions are similar and there is little to choose between the functions (Figure 3.50). The survival functions, $P(X.x) = \widetilde{F}(x)$, for the three highest peaks are shown in Figure 3.20.

TABLE 3.18 Gumbel distribution parameters for the Bakar sea-level time series, spanning the 63- year period from 1 January 1956 to 31 December 2019, for the three highest annual peaks, r1, r2 and r3 derived using the Maximum Likelihood Estimation (MLE); μ = location parameter and s = scale parameter. For tabulation purposes, the 95% levels have been rounded to two decimal places. The last row contains the empirical estimates of the parameters obtained using the record mean and standard deviation.

	Gumbel r1 (peak)		Gumbel r2 (2nd highest)		Gumbel r3 (3rd highest)	
	μ_1 (cm)	s_1 (cm)	μ_2 (cm)	s_2 (cm)	μ_3 (cm)	s_3 (cm)
MLE	74.498	9.283	67.696	7.133	62.421	5.568
	72.07, 76.92	7.67, 11.24	65.83, 69.56	5.92, 8.60	60.97, 63.87	4.58, 6.77
Empirical	74.886	8.252	—	—	—	—
	65.45, 84.32	7.21, 9.29				

FIGURE 3.48 Plots of the Gumbel Probability Distribution Function $f(x;\mu_1,s_1)$, Cumulative Distribution Function $F(x;\mu_1,s_1)$ and Complementary Cumulative Distribution Function $\widetilde{F}(x;\mu_1,s_1)$ for the highest (r1) peak Bakar sea-level height anomaly (x) for location parameter, μ_1, and scale parameter, s_1, obtained using the MLE method. Data span the period for the period 1956–2019.

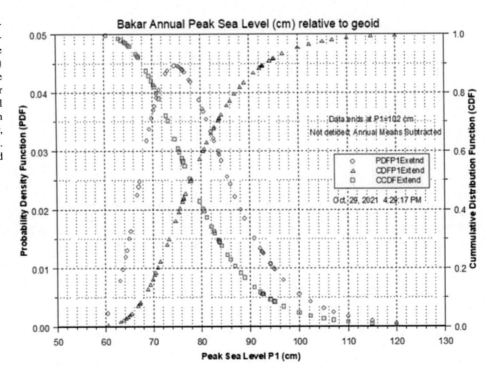

3.20.6.2.4 Return Levels for a 100-Year Event

The parameter values in the above tables can now be used in the formulae in Section 3.20.3 to obtain the return levels for sea level anomaly values for specific probabilities p_m = 1/(return period in years). For a 100-year return period (p_{100} = 0.01), the mean value and 95% confidence interval for the peak sea level anomaly series (r1) for the Gumbel (G) distribution derived using the MLE method (Table 3.18) is

$$x_G(100) = \mu_1 - s_1 \cdot \ln[-\ln(1 - p_{100})]$$
$$= 117.20 \text{ cm } (106.66 \text{ cm, } 127.61 \text{ cm}),$$

(3.258)

For a 200-year event (p_m = 0.005), and the mean $x_G(200)$ = 123.659 cm with 95% confidence interval (112.69 cm, 136.45 cm). The fact that there is little difference between the 100 and 200-year return levels is due to the paucity of high extreme values and the resulting light tail of the distribution.

TABLE 3.19 Parameters derived for the three annual extreme value time series (r1–r3) of sea level anomalies at Bakar (Croatia). GEV denotes the Generalized Extreme Value distribution and Weibull the three-parameter Weibull distribution. μ = location parameter, s = scale parameter, and ξ the shape parameter. In each case, the mean of the parameter (top value) is accompanied by the 95% confidence interval (lower values). Values in the row r1 are to be compared with those in the r1 column of Table 3.18. Time series used in the analysis are from 1 January 1956 to 31 December 2019.

	GEV			Weibull		
	μ (cm)	s (cm)	ξ	Shift (cm)	s (cm)	ξ
r1	75.837	10.122	−0.261	57.970	24.494	2.196
	72.94, 78.74	8.15, 12.57	−0.49, −0.03	51.73, 64.21	16.71, 32.28	1.26, 3.13
r2	68.428	7.434	−0.187	53.031	20.871	2.512
	66.39, 70.47	6.13, 9.02	−0.35, −0.03	47.65, 58.41	14.51, 27.23	1.53, 3.49
r3	62.694	5.765	−0.090	54.084	12.933	1.753
	61.00, 64.39	4.62, 7.20	−0.35, 0.17	52.40, 55.77	10.03, 15.83	1.22, 2.29

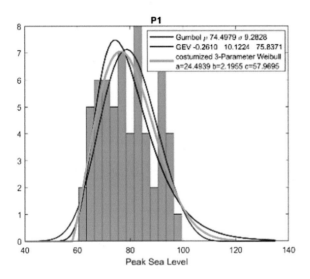

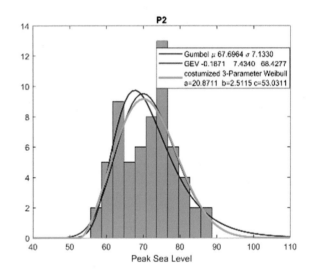

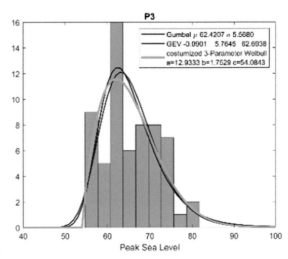

FIGURE 3.49 Plots of the Generalized Extreme Value (GEV), Gumbel and 3-parameter Weibull Probability Distribution Functions $f(x; \mu, s, \xi)$, for the Bakar (Croatia) sea level anomaly heights (x) for the three highest peak extreme time series r1–r3 (plots P1-P3, respectively) for location parameter, μ, scale parameter, s, and shape parameter, ξ. Data span the period for the period 1956 to 2019.

FIGURE 3.50 Survival functions $P(X \geq x) = \tilde{F}(x)$ for the three highest peaks. $r1$, $r2$ and $r3$ in the Bakar Bay (Croatia) extreme sea level anomaly data.

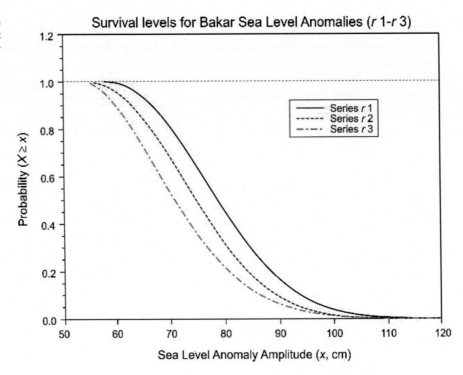

The corresponding 100-year return levels for the Generalized Extreme Value (*GEV*) and Weibull (*W*) distribution parameters for series $r1$ (Table 3.19) are obtained using

$$x_{GEV} = \mu - \frac{s}{\xi}\left\{1 - \left[-\ln(1 - p_m)\right]^{-\xi}\right\} \text{ for } \xi \neq 0 \tag{3.259}$$

$$x_W = \mu - \frac{s}{\xi}\ln\left\{-\ln(1 - p_m)\right\} \text{ for } \xi \neq 0. \tag{3.260}$$

Results are listed in Table 3.20. As these results indicate, the projected 100-year return levels are similar for all three types of distribution functions, although the upper bound of the GEV distribution of nearly 500 cm seems highly unlikely. Moreover, in all cases, the situation in 100 years will have changed and the data used to derive the distributions will no longer apply (Figure 3.51).

3.20.7 Nonstationary Extreme Value Analysis

Processes that have extreme outcomes can be directly affected by changing background conditions and other external factors. The case of future coastal flooding during rising global sea levels is a prime example. Not only is there considerable uncertainty regarding projected global sea level rise during the 21st Century but the rate of sea level rise differs regionally due to differences in tectonically induced uplift/subsidence of the land (James et al., 2021), the compaction of sediment and the removal of ground water for irrigation purposes (Mazzotti et al., 2008), and lingering glacio-isostatic

TABLE 3.20 The 100-year return levels (cm) for the Bakar sea-level anomaly data obtained using the distribution parameters for the Gumbel, GEV and Weibull probability functions; emp. denotes values derived from the empirically derived parameters in Table 3.18. Values in brackets give the 95% confidence intervals.

	Gumbel (MLE)	Gumbel (emp.)	GEV (MLE)	Weibull (MLE)
100-year (cm)	117.20 (106.66, 127.61)	112.85 (98.62, 127.06)	114.72 (89.65, 497.87)	109.28 (76.29, 182.06)

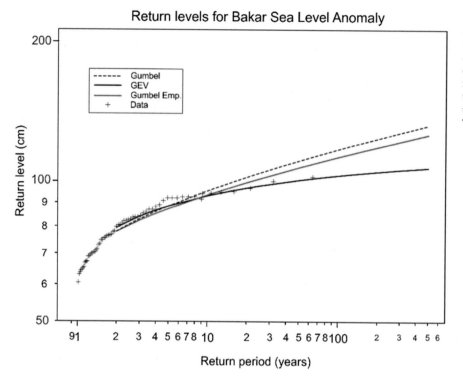

FIGURE 3.51 Sea level return levels (cm) as a function of return period periods (years) for the detided Bakar sea-level anomaly data (denoted by +s) and the corresponding Gumbel and GEV probability functions fitted to the data. The plot for the Gumbel Emp distribution was derived using the empirically fitted parameters in Table 3.18.

rebound effects from the last glaciation (Peltier, 2004). Changes in precipitation, river discharge, surface winds and ocean circulation also affect coastal flooding. As a result of these changes, the parameters μ, s, ξ that characterize GEV and GP distributions may vary with time, mean sea level, and other covariates.

The incorporation of nonstationary statistics in the estimation of probability distribution parameters has been examined through time-dependent parameters of GEV distributions (Boettle et al., 2013; Menéndez and Woodworth, 2010; Obeysekera et al., 2013; Salas and Obeysekera, 2014) and GP distributions (Méndez et al., 2006; Kyselý et al., 2010). Several studies also address the uncertainty in numerical projections of future sea level rise (SLR). For example, Ghanbari et al. (2019) suggest that, because extreme sea level data are correlated with mean sea level (Tebaldi et al., 2012), nonstationary should be tackled in terms of changing mean sea level rather than temporal goalposts, such as the year 2050 or 2100. Alternatively, Hunter (2012) addresses the impact of global sea level rise on infrastructure allowance (the height or other aspect of coastal infrastructure that needs to be altered in order to cope with climate change) but assumes that the expected frequency of flooding from storm surge under different sea level rise scenarios remains unchanged relative to mean sea level at the time.

Extreme value analysis deals with the tails of distribution functions. However, as global sea levels continue to rise, exceedance thresholds applied to present-day floods are likely to be exceeded more frequently during normal high tides in the future. Outcomes in the upper tails of present-day distributions will become more common and, therefore, no longer rare (Figure 3.52). As a consequence, it becomes necessary to take into account both the extremes and the changes in the background conditions (Stephens et al., 2018). Ghanbari et al. (2019) developed a statistically coherent nonstationary mixture-probability model (nonstationary Mixture Normal-GPD probability model) that considers changes in mean sea level as the covariate. A six-parameter model, consisting of three parameters for the Normal (Gaussian) background distribution and three for the Generalized Pareto Distribution (GPD) model, is considered, in which the daily maximum in sea level typically follows a Normal probability distribution for sea level elevation, z, below the threshold ($z < \mu$) and a Generalized Probability Distribution for data above the threshold ($z \geq \mu$). The spliced conditional Cumulative Distribution Function of the mixture model is defined as

$$F(z|\mu, s, \xi; \overline{\mu}, \sigma, \varphi) = (1 - \varphi) \frac{N(z|\overline{\mu}, \sigma)}{N(z|\mu, s)} \text{ if } z < \mu \qquad (3.261a)$$

$$= (1 - \varphi) + \varphi F_{GPD}(\mu, s, \xi) \text{ if } z \geq \mu \qquad (3.261b)$$

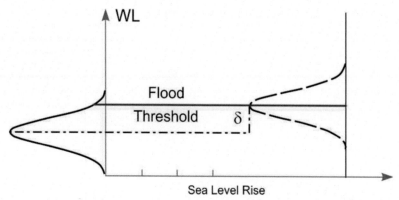

FIGURE 3.52 Schematic of the change in the water level (WL) probability distribution following a δ increase in mean sea level. The distribution function has the same structure but is offset to higher water levels. *From Ghanbari et al. (2019).*

(MacDonald et al., 2011) where F_{GPD} is the Generalized Pareto Distribution

$$F_{GPD}(x; \mu, s, \xi) = \left\{ 1 - \left[1 + \xi \left(\frac{z - \mu}{s} \right) \right]^{-\frac{1}{\xi}} \right\} \text{ if } \xi \neq 0 \qquad (3.262a)$$

$$= \left\{ 1 - \exp \left[- \left(\frac{z - \mu}{s} \right) \right] \right\} \text{ if } \xi = 0 \qquad (3.262b)$$

In the above formulation, $\bar{\mu}$ and σ denote the mean and standard deviation of sea level for the Normal Distribution and φ is the probability of independent exceedances over the threshold, μ. Variable φ is the ratio of the number of clusters above the threshold to the total number of observations (Coles, 2001). As illustrated by Figure 3.52, the approach assumes that the rising sea levels will shift the existing sea level distribution toward higher water levels without distortion of the distribution. Thus, no additional covariate dependency was assumed for the scale parameter of the Normal component or with the scale and shape parameters of the GPD component. Nonstationary in the background component of the distribution is addressed by changing the location parameter (i.e., the mean) of the Normal component as

$$\mu = \mu(\delta) = \mu_o + \delta \qquad (3.263)$$

where μ_o represents the estimated Normal distribution location parameter computed from the historical daily maximum sea level relative to a specific reference year. Similarly, Hunter (2012) finds that for an increase, $\delta = \Delta z$, in mean sea level, z, the expected number of exceedances, N, in extreme coastal flooding events over the return period, τ,

$$N = \exp \left[- \left(\frac{z - \mu}{s} \right) \right] \qquad (3.264)$$

is increased by a factor $\exp \left(\frac{\delta}{s} \right)$, where s is the shape parameter for a Gumbel distribution function.

Ghanbari et al. (2019) considers the coastal flood frequency and amplification factor,

$$AF = \frac{\tau_o(z)}{\tau_\delta(z)} \qquad (3.265)$$

in which $\tau_o(z)$ is the existing return period of water level, z, and $\tau_\delta(z)$ is the return period of water level under the δ increase in mean sea level. Results were applied to the Contiguous US coastal stations with long-term tide gauge observations. Buchanan et al. (2017) argue that the use of the Gumbel distribution for sea level rise scenarios in Intergovernmental Panel on Climate Change (IPCC) Fifth Assessment Report (AR5l Church et al., 2013)—used because of its exponential relationship between level and frequency of flooding—is not the optimal choice as it implies that amplification of flood frequency is invariant across different levels of flooding. For example, the distribution assumes that extreme events like a 500-year flood increase by the same magnitude as less extreme events. The authors suggest that the Generalized Pareto distribution is preferable for flood risk assessments for the emerging nonstationary climate. The authors further examine how the shape parameter, which determines the tail of the distribution, affects flood frequency distributions. Flood frequency with $\xi > 0$

are heavy tailed with a relatively high frequency of extreme flood levels. Conversely, flood frequency curves with $\xi < 0$ are thin-tailed and have an upper bound of extreme flood levels; $\xi = 0$ is identical to the Gumbel distribution. For further research on nonstationary extreme value analysis the reader is directed to Coles (2001), Salas and Obeysekera (2014), and Arns et al. (2017).

For Glossary, please refer to the Appendix B section at the end of the book.

Chapter 4

THE SPATIAL ANALYSES OF DATA FIELDS

4.1 INTRODUCTION

A fundamental problem in oceanography is how best to represent spatially distributed data (or statistical products computed from these data) in such a way that dynamical processes or their effects can best be visualized. As in most aspects of observational analysis, there has been a dramatic change in the approach to this problem due to the increased abundance of digital data and our ability to process them. Prior to the use of digital computers, data displays were constructed by hand and "contouring" was an acquired skill of the descriptive analyst. Hand contouring is still practiced by a few scientists today although, more likely, the data points being contoured are averaged values produced by a computer. In other applications, the computer not only performs the averaging but also uses objective statistical techniques to produce both the gridded values and the associated contours. This can lead to some strange results, as computers will contour values even when it is not justified by the data values or their distribution.

The purpose of this section is to review data techniques and procedures designed to reduce spatially distributed data to a level that can be visualized easily by the analyst. We will discuss methods that address both spatial fields and time series of spatial fields since these are the primary modes of data distribution encountered by the investigator. Our focus is on the more widely used techniques, which we present in a practical fashion, stressing the application of the method for interpretive applications.

4.2 TRADITIONAL BLOCK AND AREA AVERAGING

A common older method for deriving a gridded set of data is simply to average the available data over an arbitrarily selected rectangular grid. This averaging grid can lie along any chosen surface but is most often constructed in the horizontal or vertical plane. Because the grid is often chosen for convenience, without any consideration to the sampling coverage, it can lead to an unequal distribution of samples per grid "box". For example, because distance in longitude varies as the cosine of the latitude, the practice of gridding data by 5 or 10° squares in latitude and longitude may lead to increasingly greater spatial areas to be covered at low latitudes. Although this can be overcome somewhat by converting to distances using the central latitude of the box (Poulain and Niiler, 1989), it is easy to see that inhomogeneity in the sampling coverage can quickly nullify any of the useful assumptions made earlier about the Gaussian nature of sample populations or, at least, about the set of means computed from these samples. This is less of a problem with satellite-tracked drifter data since satellite ground tracks converge with increasing latitude, allowing the data density in boxes of fixed longitude length to remain nearly constant.

With markedly different data coverage between sample regions, we cannot always compare the values computed in these squares at the same level of statistical confidence. To help attain statistical uniformity, one must be careful to consider properly the amount of data being included in such averages and be able to evaluate possible effects of the variable data coverage on the mapped results. Each value should be associated with a sample size indicating how many data points, N, went into the computed mean. This will not dictate the spatial or temporal distributions of the sample data field but will at least provide a sample size parameter, which can be used to evaluate the mean and standard deviation at each point. While the standard deviation of each grid sample is composed of both spatial and temporal fluctuations (within the time period of the grid sample), it does give an estimate of the inherent variability associated with the computed mean value.

Despite problems with nonuniform data coverage, it has proven worthwhile to produce maps or cross-sections with simple grid-averaging methods since they frequently represent the best spatial resolution possible with the existing data coverage. The approach is certainly simple and straightforward. Besides, the data coverage often does not justify more complex and computer-intensive data reduction techniques. Specialized block-averaging techniques have been designed to improve the resolution of the corresponding data by taking into account the nature of the overall observed global variability and by trying to maximize the coverage appropriately. For example, averaging areas in offshore regions are frequently

Data Analysis Methods in Physical Oceanography. https://doi.org/10.1016/B978-0-323-91723-0.00004-0

selected which have narrow meridional extent and wide zonal extent, taking advantage of the generally stronger meridional gradients observed in the ocean. Thus, an averaging area covering 2° latitude by 10° longitude may be used to better resolve the meridional gradients that dominate the open ocean (Wyrtki and Meyers, 1975). This same idea may be adapted to more limited regions, such as continental margins, if the general oceanographic conditions are known. If so, the data can be averaged accordingly, providing improved resolution perpendicular to strong frontal features. For example, the averaging areas in the Benguela Current System off southwest Africa or the California Current System off the west coast of North America would be narrower in offshore directions at right angles to the coast than in the alongshore direction. A further extension of this type of grid selection would be to base the entire averaging area selection on the data coverage. This is difficult to formalize objectively because it requires the subjective selection of the averaging scheme by an individual. However, it is possible in this way to improve resolution without a substantial increase in sampling (Emery, 1983).

In general, all block or area-averaging techniques make the assumption that the data being considered in each grid box are statistically homogeneous and isotropic over the region of study. Under these assumptions, area sample size can be based strictly on the amount of data coverage (number of data values) rather than on having to know details about processes represented by the data. Statistical homogeneity does not require that all the data were collected by the same instrument having the same sampling characteristics. Thus, grid-square averaging can include data from many different instruments, which generally have the same error limits.

One must be careful when averaging different kinds of measurements, even if they are of the same parameter. It is very tempting, for example, to average mechanical bathythermograph (MBT) temperatures with newer expendable bathythermograph (XBT) temperatures to produce temperature maps at specific depths. Before doing so, it is worth remembering that XBT data are likely to be accurate to 0.1°C, as reported earlier (at least in temperature), while MBT data are decidedly less accurate and less reliable. Another marked difference between the two instruments is their relative vertical coverage. While most MBTs stopped at 250 m depth, XBTs are good to 500−1,800 m, depending on the type of probe. Thus, temperature profiles from MBTs can be expected to be different from those collected with XBTs. Any mix of the two will necessarily degrade the average to the quality of the MBT data and bias averages to shallow (<300 m) depths. In some applications, the level of degraded accuracy will be more than adequate, and it is only necessary to state assumptions clearly and be aware of the intended application when assembling the data from these different instruments. Also, one can expect distinct discontinuities as the data make the transition from a mix of measurements at shallower levels to strictly XBT data at greater depth. This same argument holds when mixing XBT temperature profiles with much more accurate but less plentiful CTD profiles.

Other important practical concerns in forming block averages have to do with the usual uneven geographic location of oceanographic measurements. Consider the global distribution of all autumn oceanographic research measurements up to 1970 of the most common oceanographic observation, temperature profiles (Figure 4.1). It is surprising how frequently these observations lie along meridians of latitude or parallels of longitude. This makes it difficult to assign the data to any particular 5 or 10° square when the border of the square coincides with integer values of latitude or longitude. When the latter occurs, the investigator must decide to which square the borders will be assigned to and be consistent in carrying this definition through the calculation of the mean values.

As illustrated by Figure 4.1, data coverage can be highly nonuniform. In this example, some areas were frequently sampled while others were seldom (or never) occupied. There was, and still is, a distinct concentration of measurements off the west coast of the U.S. near the location of Scripps Institution of Oceanography and there is an abundance of observations around Japan reflecting that country's interest in oceanographic research. Notice also the concentration of observations along ship routes such as the transits to and from Hawaii. Many of the ship tracks in the North Pacific follow "great circle routes", providing the ship with the shortest travel distance over the globe. Such nonuniformity in data coverage is a primary factor in considering the representativeness of simple block averages. It certainly brings into question the assumptions of homogeneity (spatially uniform sampling distribution) and isotropy (uniform sampling regardless of direction) since the sample distribution varies greatly with location and may often have a preferred orientation. The situation becomes even more severe when one examines the quality of the data in the individual casts represented by the dots in Figure 4.1. In order to establish a truly consistent data set in terms of the quality of the observations (i.e., the depth of the cast, the number of samples, the availability of oxygen and nutrients, and so on), it is generally necessary to reject many of the available hydrographic casts. The situation is changing with the advent of major oceanographic initiatives such as Argo, which has a dedicated objective to "... provide a quantitative description of the changing state of the upper ocean and the patterns of ocean climate variability from months to decades, including heat and freshwater storage and transport." (http://www.argo.ucsd.edu/About_Argo.html).

The question of data coverage depends on the kind of scientific questions the data set is being used to address. For studies not requiring high-quality hydrographic or CTD profile stations, a greater number of observations is available,

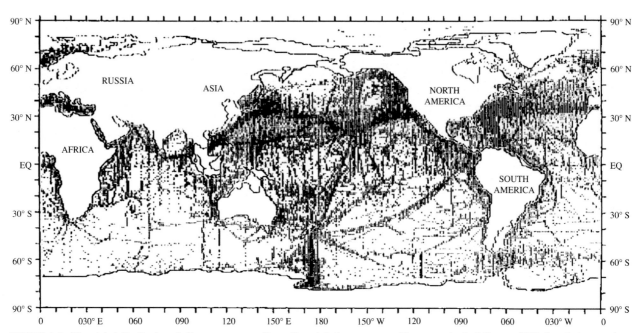

FIGURE 4.1 The global distribution of all temperature profiles collected during oceanographic surveys in the fall up to 1970. Sample is most dense along major shipping routes.

while for more restrictive studies requiring a higher accuracy, far fewer vertical casts would match the requirements. This is also true for other types of historical data but is less true of newly collected data. However, even now, one must ensure that all observations have a similar level of accuracy and reliability. Variations in equipment performance, such as sensor response or failure, must be compensated for in order to keep the observations consistent. Also, changes in instrument calibration need to be taken into account over the duration of a sampling program. For example, transmissometer lenses frequently become coated with a biotic film that reduces the amount of light passing between the source and receiver lenses. A nonlinear, time-dependent calibration is needed to correct for this effect.

Despite the potential problems with the block-averaging approach to data presentation, much information can be provided by careful consideration of the data rather than the use of more objective statistical methods to judge data quality. The shift to statistical methods represents a transition from the traditional oceanographic efforts of the early part of the twentieth century when considerable importance was given to every measurement value. In those days, individual scientists were personally responsible for the collection, processing and quality of their data. Then, it was a simple task to differentiate between "correct" and "incorrect" samples without having to resort to statistical methods to indicate how well the environment had been observed. In addition, earlier investigations were primarily concerned with defining the mean state of the ocean. Temporal variability was sometimes estimated but was otherwise ignored in order to emphasize the mean spatial field. With today's large volumes of data, it is no longer possible to "hand check" each data value. A good example is provided by satellite-sensed information, which generally consists of large groupings of data that are usually treated as a collection of individual data values.

In anticipation of our discussion of filtering in Chapters 5 and 6, we point out that block averaging corresponds to the application of a box-car-shaped filter to the data series. This type of filter has several negative characteristics such as a slow filter roll off and large side lobes, which can distort the information in the original data series.

4.3 COMPUTER CONTOURING

Most computer contouring programs today satisfy the basic concepts of contouring, which are (Banks and Henry, 1996):

1. Contours are faithful to the real data
2. Contours appear to have been drawn by hand
3. Handle faults properly
4. Are correct in three dimensions
5. Function in either time or space
6. Be fast and economical

As the volume of data continues to increase, the need for a rapid way to visualize the data also increases. No longer can the analyst take the time to carefully consider each data point and draw contours by hand. If they do, it will prove impossible to address the plethora of data that is available for analysis. Instead, investigators must turn to digital computers as the way to produce visual representations of the measurement data.

4.3.1 Interpolation: The Basic Problem of Contouring

In Chapter 3, we introduced interpolation and discussed the implications of various approaches. Interpolation is the essence of computer contouring. When data are contoured by hand, the brain performs the interpolation, taking into account the values around the contour line. While this is a relatively easy task, the number of contours that would need to be drawn for the large volume of data available today would be prohibitively time consuming. Thus, we would like to turn this task over to the computer. It is, however, difficult for computer software to emulate the subjective and aesthetic aspects of human contouring. So, there is a tradeoff between speed and aesthetics.

Bad data complicate the contouring process because they interrupt the flow and destroy the continuity of the contour. A computer contouring routine should be able proceed across such faults. This is called vertical interpolation. Any interpolation calculates the value of a surface between two known (observed) values. By hand, contouring is a bit of a lost art as computer contouring has taken over. There are two basic approaches to computer contouring: (a) indirect or gridding techniques; and (b) direct or triangulation methods.

4.3.2 Indirect Method (Gridding)

In this method, the original data values are used to generate a new set of values on a regular grid. This new set of values replace the original data values and are the ones that will be contoured. This approach simplifies the contouring by producing a regular grid for the interpolation. The grid can be oriented in any direction and the size of the gridded field defines the area that will be contoured.

There are four steps to gridding:

1. Select an origin for the data and decide on the grid size. Each corner is a "grid node."
2. Sort the original data by their x-y locations.
3. Determine the data values to assign to a grid box.
4. Create a set of values at the grid nodes by interpolating from the nearest neighbor observed values.

The last two steps have used numerous schemes to produce maps that are both correct and aesthetically pleasing. To demonstrate the differences that can occur, we present three different contour maps (Figures 4.2–4.4) that are based on the same data values.

The fact that these differently contoured fields are based on the same original data emphasizes the need to choose the interpolation method whose resultant contour field best represents the measurements. If there are five different ways to select the neighbor values that are to be used and ten different ways to interpolate the contours, then the analyst can be faced with 50 different representations of the contoured field. While it is easy to ensure that the contours will represent the grid points, it is not clear that the contours will faithfully represent the original observed data.

4.3.3 Indirect Method (Triangulation)

Triangulation was the first method used to generate computer-generated contour maps. This was a consequence of surveyors using triangulation when creating topographic maps. This form of direct contouring interpolates values from a pattern that is derived from the original data. The pattern preserves the values of the original measurement data. Triangulation also requires the selection of neighbors, but this selection is completed when the triangulation is done. Triangulation has the following characteristics:

1. The original data set of n points has been represented by a set of triangles;
2. These triangles have data points at each vertex, are as equilateral as possible and the contours do not change with orientation. These triangles are called *Delaunay* triangles;
3. For large data sets, each point will have an average of six neighbors. If the points are not located randomly, some points will have more than six neighbors;
4. Contours are not calculated outside of the domain of the observed data.

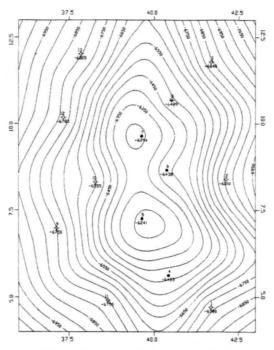

FIGURE 4.2 Contours picked from six nearest neighbors using a piece-wise least-squares fit. Note that, what is important here is the pattern, not the contour values.

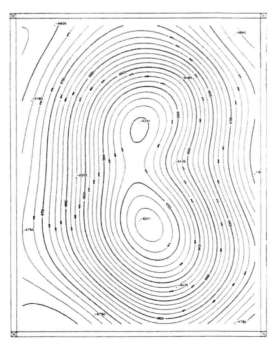

FIGURE 4.3 Contours using six nearest neighbors and adaptive fitting, in which the fitting curve is generated by iteratively adjusting the "active" contour points to better fit the data points. As with Figure 4.2, it is the pattern, not the contoured values that is important here.

Examples of triangulation are presented here in Figures 4.5 and 4.6.

As suggested by the figures, the smoothness controls the number of triangles used in the interpolation. (Here, smoothing is a procedure applied to the entire triangulation mesh, as with Lagrangian smoothing). The basic shape of the contoured field does not change. Unlike the indirect method, the triangulation method is faithful to all of the observed

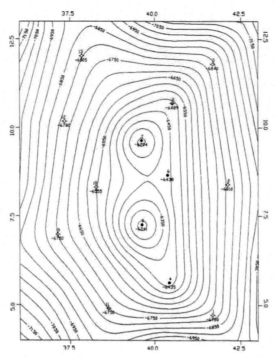

FIGURE 4.4   Contours using six nearest neighbors with linear projected slopes.

points. The gridding method is more flexible but often does not honor the original observed values (i.e., does not meet all of the given data values in the contouring).

4.4 OPTIMUM INTERPOLATION

Optimum interpolation (OI) is a method in which the investigator applies the statistics of the data field to be interpolated to dictate the interpolation and, hence, the contour system. OI is part of a more general analysis technique known as *objective analysis,* which is an estimation procedure that can be specified mathematically. The form of OI most widely used in the physical sciences is that of least-squares optimal interpolation, more appropriately referred to as *Gauss–Markov smoothing*, which is essentially an application of the linear estimation (smoothing) techniques discussed in Chapter 3. Because it is generally used to map spatially nonuniform data to a regularly spaced set of gridded values, Gauss–Markov smoothing might best be called "Gauss–Markov mapping". The basis for the technique is the Gauss–Markov theorem, which was first introduced by Gandin (1965) to provide a systematic procedure for the production of gridded maps of meteorological parameters. If the covariance function used in the Gauss–Markov mapping is the covariance of the data field (as opposed to a more *ad hoc* covariance function based on historical information, as is often the case), then Gauss–Markov smoothing is optimal in the sense that it minimizes the mean square error of the objective estimates. A similar technique, called "Kriging" after a South African engineer Daniel G. Krige, was developed in mining engineering (see Section 4.5). Oceanographic applications of OI are provided by Bretherton et al. (1976), Freeland and Gould (1976), Bretherton and McWilliams (1980), Hiller and Käse (1983), Bennett (1992), and others.

The two fundamental assumptions in optimal interpolation are that the statistics of the subject data field are stationary (unchanging over the sample period of each map) and homogeneous (the same characteristics over the entire data field). A further assumption often made to simplify the analysis is that the statistics of the second moment, or covariance function, are isotropic (the same structure in all directions). Bretherton et al. (1976) point out that if these statistical characteristics are known or can be estimated for some existing data field (such as a climatology based on historical data), they can be used to design optimum measurement arrays to sample the field. Because the optimal estimator is linear and consists of a weighted sum of all the observations within a specified range of each grid point, the objective mapping procedure produces a smoothed version of the original data field that will tend to underestimate the true field. In other words, if an observation point happens to coincide with an optimally interpolated grid point, the observed value and interpolated value will probably not be equal due to the presence of noise in the data. The degree of smoothing is determined by the characteristics

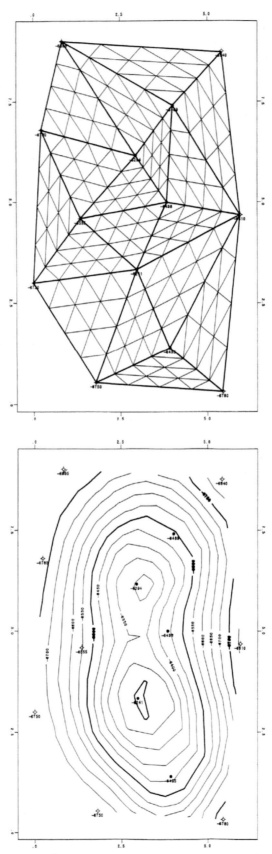

FIGURE 4.5 Triangulation with a smoothness of 4 ($4^2 = 16$ mesh values). As noted previously, it is the pattern that is of interest in this figure, not the contour values.

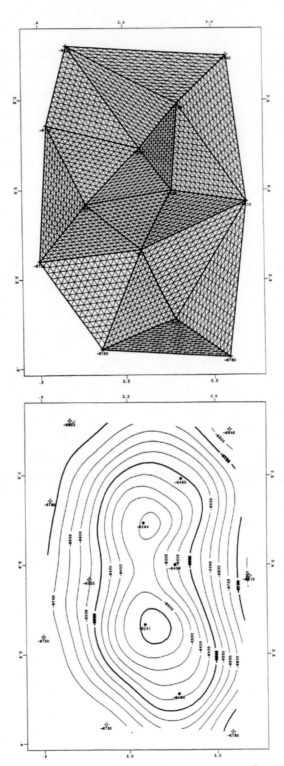

FIGURE 4.6 Triangulation with a smoothness of 16 ($16^2 = 256$), which smooths with 4^2 more mesh values than in Figure 4.5.

of the signal and error covariance functions used in the mapping and increases with increasing spatial scales for a specified covariance function.

The optimal gridding process is complicated by several factors, including the fact that the ability of the interpolation procedure to generate representative maps decreases as the number of measurements being used decreases. In physical oceanography, observations are often very sparse. As indicated in Figure 4.1, oceanographic observations are generally not evenly spaced and are frequently few in number. This inhomogeneity is characteristic of many regions of the ocean and drives the need for objective techniques for mapping the variables of interest.

The general problem is to compute an estimate $\widehat{D}(x, t)$ of the scalar variable $D(\mathbf{x}, t)$ at a position $\mathbf{x} = (x, y)$ from irregularly spaced and inexact observations $d(\mathbf{x}_n, t)$ at a limited number of data positions $\mathbf{x}_n$ ($n = 1, 2, ..., N$). These observations are "inexact" in that they are subject to various types of errors, such as instrumental noise and environmental variability, and to "errors" associated with the least-squares fit analysis used in the optimum interpolation procedure. Implementation of the procedure requires *a priori* knowledge of the variable's covariance function, $C(\mathbf{r})$, and uncorrelated error variance, ε, where $\mathbf{r}$ is the spatial separation between positions. For isotropic processes, $C(\mathbf{r}) \rightarrow C(r)$, where $r = |\mathbf{r}|$. Although specification of the covariance matrix should be founded on the observed structure of oceanic variables, selection of the mathematical form of the covariance matrix is hardly an "objective" process even with reliable data (cf. Denman and Freeland, 1985). In addition to the assumptions of stationarity, homogeneity, and isotropy, an important constraint on the chosen covariance matrix is that it must be positive definite (no negative eigenvalues). Bretherton et al. (1976) report that objective estimates computed from nonpositive-definite matrices are not optimal and the mapping results are poor. In fact, nonpositive-definite covariance functions can yield objective estimates with negative expected squared errors. One way to ensure that the covariance matrix is positive definite is to fit a function that results in a positive definite covariance matrix to the sample covariance matrix calculated from the data (Hiller and Käse, 1983). This results in a continuous mathematical expression to be used in the data weighting procedure. In attempting to specify a covariance function for data collected in continental shelf waters, Denman and Freeland (1985) further required that $\partial^2 C / \partial x^2$ and $\partial^2 C / \partial y^2$ be continuous at $r = 0$ (to ensure a continuously differentiable process) and that the variance spectrum, $S(k)$, derived from the transform of $C(\mathbf{r})$ be integrable and nonnegative for all wavenumbers, $\mathbf{k}$ (to ensure a realizable stochastic random process).

Calculation of the covariance matrix requires that the mean and "trend" be removed from the data (the trend is not necessarily linear). In three-dimensional space, this amounts to the removal of a planar or curvilinear surface. For example, the mean density structure in an upwelling domain is a curved surface, which is shallow over the outer shelf and deepens seaward. Calculation of the density covariance matrix for such a region first involves removal of the curved mean density surface (Denman and Freeland, 1985). Failure to remove the mean and trend would not alter the fact that our estimates are optimal, but it would redistribute variability from unresolved larger scales throughout the wavenumber space occupied by the data. The procedure would then map features that have been influenced by the trend and mean.

As discussed later in the Chapter 5 on time series, there are many ways to estimate the trend. If ample good-quality historical data exist, the trend can be estimated from these data and can then be subtracted from the data being investigated. If historical data are not available, or the historical coverage is inadequate, then the trend must be computed from the sample data set itself. Numerous methods exist for calculating the trend and all require some type of functional fit to the existing data using a least-squares method. These functions can range from straight lines to complex higher-order polynomials and associated nonlinear functions. We note that, although many oceanographic data fields do not satisfy the conditions of stationarity, homogeneity, and isotropy, their anomaly fields do. In the case of anomaly fields, the trend and mean have already been removed. Gandin (1965) reports that it may be possible to estimate the covariance matrix from existing historical data. This is more often the case in meteorology than in oceanography. In most oceanographic applications, the analyst must estimate the covariance matrix from the data set being studied. Regardless of the approach used, the practice of dealing with anomalies rather than with the observations themselves can make the optimum interpolation procedure considerably more accurate.

In the following, we present a brief outline of objective mapping procedures. The interested reader is referred to Gandin (1965) and Bretherton et al. (1976) for further details. As noted previously, we consider the problem of constructing a gridded map of the scalar variable $D(\mathbf{x}, t)$ from an irregularly spaced set of scalar measurements $d(\mathbf{x}, t)$ at positions $\mathbf{x}$ and times t. The notation $\mathbf{x}$ refers to a suite of measurement sites, x_n ($n = 1, 2, ...$), each with distinct (x, y) coordinates. We use the term "variable" to mean directly measured oceanic variables as well as calculated variables, such as the density or stream function derived from the observations. Thus, the data $d(\mathbf{x}, t)$ may consist of measurements of the particular variable we are trying to map, or they may consist of some other variables that are related to D in a linear way. The former case gives

$$d(\mathbf{x}, t) = D(\mathbf{x}, t) + \varepsilon(\mathbf{x}) \tag{4.1}$$

where the ε are zero-mean measurement errors that are not correlated with the gridded variable D or its anomaly. In the latter case

$$d(\mathbf{x},t) = F[D(\mathbf{x},t)] + \varepsilon(\mathbf{x}) \tag{4.2}$$

in which F is a linear functional that acts on the function D in a linear fashion to give a scalar (Bennett, 1992). For example, if $D(\mathbf{x}, t) = \Psi(\mathbf{x}, t)$ is the stream function, then the data could be current meter measurements of the zonal velocity field, $u(\mathbf{x}, t) = F[\Psi(\mathbf{x}, t)]$, where

$$d(\mathbf{x},t) = u(\mathbf{x},t) + \varepsilon(\mathbf{x}) = -\frac{\partial \Psi(\mathbf{x})}{\partial y} + \varepsilon(\mathbf{x}) \tag{4.3}$$

and $\partial\Psi/\partial y$ is the gradient of the stream function in the meridional direction.

To generalize the objective mapping problem, we assume that mean values have *not* been removed from the original data prior to the analysis. If we consider the objective mapping for a single "snapshot" in time (thereby dropping the time index, t), we can write linear estimates $\widehat{D}(\mathbf{x})$ of $D(\mathbf{x})$ as the summation over a weighted set of the measurements d_i ($i = 1$, ..., N)

$$\widehat{D}(\mathbf{x}) = \overline{D}(\mathbf{x}) + \sum_{i=1}^{N} b_i(d_i - \overline{d}) \tag{4.4}$$

where the overbar denotes an expected value (mean), $d_i = d(\mathbf{x}) = d(x_i)$, and $1 \le i \le N$ is shorthand notation for the data values, and the $b_i = b(\mathbf{x}) = b(x_i)$ are, as yet unspecified, weighting coefficients at the data points, x_i. The selection of the N data values is made by restricting these values to some finite area about the grid point. Often called "Cressman Weights", the b_i depends only on the distance between the grid point and the location of the observation and not on the observation itself. The search radius selected defines the length-scale over which an observation is used in the interpolation to a specified grid point. This search parameter is chosen by the user and can be made to vary in space depending on the data coverage and the inherent horizontal physical scales.

Estimates of the parameters b_i in Eqn (4.4) are found in the usual way by minimizing the mean square variance of the error $e(\mathbf{x})^2$ between the measured variable, D, and the linear estimate, $\widehat{D}$ at the data location. In particular, by minimizing

$$\overline{e(\mathbf{x})^2} = \overline{[D(\mathbf{x}) - \widehat{D}(\mathbf{x})]^2} \tag{4.5}$$

with respect to the b_i. On substitution of (4.4) in (4.5), we obtain

$$\overline{e(\mathbf{x})^2} = \overline{[D(\mathbf{x}) - \overline{D}(\mathbf{x})]^2} + \sum_{i=1}^{N}\sum_{j=1}^{N} b_i b_j \overline{(d_i - \overline{d})(d_j - \overline{d})} - 2\sum_{i=1}^{N} b_i \overline{(d_i - \overline{d})(D - \overline{D})} \tag{4.6}$$

Note, that if the mean has been removed, we can set $\overline{D}(\mathbf{x}) = \overline{d}(\mathbf{x}) = 0$ in (4.6).

The mean square difference in Eqn (4.6) is minimized when

$$b_i = \sum_{j=1}^{N} \left\{ \left[(d_i - \overline{d})(d_j - \overline{d}) \right]^{-1} (d_j - \overline{d})(D - \overline{D}) \right\} \tag{4.7}$$

To calculate the weighting coefficients in (4.7), and therefore the grid-value estimates in (4.4), we need to compute the covariance matrix by averaging over all possible pairs of data taken at points x_i, x_j; the corresponding covariance matrix is

$$\overline{(d_i - \overline{d})(d_j - \overline{d})} = \overline{(d(x_i) - \overline{d})(d(x_j) - \overline{d})} \tag{4.8}$$

We do the same for the interpolated value

$$\overline{(d_i - \overline{d})(D_j - \overline{D})} = \overline{(d(x_i) - \overline{d})(d(x_k) - \overline{D})} \tag{4.9}$$

where x_k is the location vector for the grid point estimate $\widehat{D}(x_k)$. This is a key step in optimum interpolation whereby the computation of the covariance function at the grid value $\widehat{D}(x_k)$ depends only on the distances between the measurement locations and the positions of the grid values. This is done by inferring the grid value covariance using the data covariance only as a function of the distance between the grid point and the data value location. We should also note that because the covariance is for the anomaly (here, the difference from the mean) it could also be considered as the "error covariance."

In general, we need a series of measurements at each location so that we can obtain statistically reliable expected values for the elements of the covariance matrices in (4.8) and (4.9). The expected values in the above relations could be computed as ensemble averages over spatially distributed sets of measurements. Typically, however, we have only one set of measurements for the specified locations x_i, x_j. As a consequence, we need to assume that, for the region of study, the data statistics are homogeneous, stationary and isotropic. If these conditions are met, the covariance matrix for the data distribution (for example, sea surface temperature) depends only on the distance r between data values, where $r = |x_j - x_i|$. Thus, we have elements i, j, of the covariance matrix given by

$$\overline{(d_i - \overline{d})(d_j - \overline{d})} = C(|x_j - x_i|) + \overline{\varepsilon^2}$$
$$\overline{(d_j - \overline{d})(D_j - \overline{D})} = C(|x_j - x_k|) + \overline{\varepsilon^2} \tag{4.10}$$

where $C(|r|) = \overline{d(x)d(x+r)}$ is the data covariance matrix as a function of the separation distance and the mean square error $\varepsilon(\mathbf{x})^2$ implies that this estimate is not exact and that there is some error in the estimation of the correlation function from the data. This is referred to as the "mean-square mapping error" of the interpolation. We note that this is not the same error in (4.6) that we minimize to solve for the weights in (4.7). The matrix can now be calculated by forming pairs of observed data values separated into bins according to the distance between sample sites, x_i. These are then averaged over the number of pairs that have the same separation distance to yield the product matrix

$$\overline{(d_i - \overline{d})(d_j - \overline{d})}$$

This computation requires us to define some "bin interval" for the separation distances so that we can group the product values together. To ensure that the resulting covariance matrix meets the condition of being positive definite, a smooth function satisfying this requirement can be fitted to the computed raw covariance function. This fitted covariance function is used for

$$\overline{(d_i - \overline{d})(D - \overline{D})}$$

and to calculate the terms

$$\left[\overline{(d_i - \overline{d})(d_j - \overline{d})}\right]^{-1}$$

of the inverse matrix. The weights b_i are then computed from (4.7). It is a simple process to then compute the optimal grid value estimates from (4.4). Note that, for the case where the data can provide no help in the estimate of D (that is, for $\varepsilon(\mathbf{x}) \to \infty$), then $b_i = 0$ and the only reasonable estimate is $\widehat{D}(\mathbf{x}) = \overline{D}$, the mean value. Similarly, if the data are error free (such that $\varepsilon(\mathbf{x}) \to 0$), then $\widehat{D}(x_i) = D(x_i)$ for all x_i ($i = 1, ..., N$). In other words, the estimated value and the measured data are identical at the measurement sites (within the limits of the inherent noise in the data values) and the estimator simply interpolates between the observations.

As with other interpolation methods, optimum interpolation is subject to sampling inhomogeneities. Consequently, for regions with very few observations, optimum interpolation can return a discontinuous field. The method also assumes, sometimes incorrectly, that all measurement values have the same error variance since the weighting calculation is based on distance only.

The critical step in the objective mapping procedure is the computation of the covariance matrix. We have described a straightforward procedure to estimate the covariance matrix from the sample data. As with the estimate of the mean or overall trend, it is often possible to use an existing set of historical data to compute the covariance matrix. This is frequently the case in meteorological applications where long series of historical data are available. In oceanography, however, the covariance matrix typically must be computed from the sample data. Where historical data are available, it is important to recognize that using these data to estimate the covariance matrix for use with more recently collected data is tantamount to assuming that the statistics have remained stationary since the time that the historical data were collected.

Bretherton et al. (1976) suggest that objective analysis can be used to compute the covariance matrix. In this case, they start with an assumed covariance function, $\widehat{F}$, which is then compared with a covariance function computed from data with a fixed distance x_o. The difference between the model $\widehat{F}$ and the real F computed from the data is minimized by repeated iteration.

To this point, we have presented optimum interpolation as it applies to scalar fields. We can also apply optimal (Gauss–Markov) interpolation to vector fields. One approach is to examine each scalar velocity component separately so that for n velocity vectors we have $2n$ velocity components

$$d_r = u_1(\mathbf{x}_r); \quad d_{r+n} = u_2(\mathbf{x}_r) \tag{4.11}$$

where u_1 and u_2 (or, alternatively, u and v) are the x, y velocity components at x_r. If the velocity field is non-divergent, we can introduce a scalar stream function, $\Psi(\mathbf{x})$, such that

$$u_1 = -\frac{\partial \Psi}{\partial y}; u_2 = \frac{\partial \Psi}{\partial x} \tag{4.12}$$

and apply scalar methods to Ψ.

Once the optimal interpolation has been executed, there is a need to return to Eqn (4.6) to compute the actual error associated with each optimal interpolation. To this end, we note that we now have the interpolated data from (4.4). Thus, we can use $\widehat{D}$ computed from (4.4) as the value for D in (4.6). The product in the last term of (4.6) is computed from the covariance in (4.10). In this way, it is possible to compute the error associated with each optimally interpolated value. Frequently, this error field is plotted for a specific threshold level, typically 50% of the interpolated values in the mapped field (see following examples). It is important to retain this error estimate as part of the optimal interpolation since it enables the investigator to assess the statistical significance of individual gridded values.

4.4.1 Optimum Interpolation: The Wilkin Code

As authors of this book, we have been frequently pressured to include processing code in the text, but we have resisted because of the great variety of coding languages and procedures that are available. We also believed that any material that we included would soon be out of date by the time this edition of this book appeared. After some thought, we decided to make an exception in the case of Optimum Interpolation (OI). This procedure has been difficult for many to formulate and so we decided to include OI code in this latest edition of this text. Appendix F provides advanced OI code written by Dr. John Wilkin of Rutgers University (New Jersey, USA) that one of the authors (WJE) has used in the past. In Appendix F, the reader will find MATLAB code that shows the user how to compute the autocorrelation function needed to map a Sea Surface Temperature (SST) field. Examples of the OI code applied to the SST are presented along with the inherent mapping error. We clearly hope that readers will find this code useful. Any questions should be directed to Dr. Wilkin who has kindly included his email link in the Appendix. We ask that users thank Dr. Wilkin for the use of his code.

4.4.2 Objective Mapping: Examples

An example of optimum interpolation applied to a single oceanographic survey is provided by the results of Hiller and Käse (1983). The data are from a CTD survey grid occupied in the North Atlantic about midway between the Azores and the Canary Islands (Figure 4.7). At each CTD station, the geopotential anomaly at 25 db (dBar) relative to the anomaly at 1,500 db (written 25/1,500 db) was calculated and selected as the variable to be mapped. The two-dimensional correlation function for these data is shown in three-dimensional perspective in Figure 4.8a. A series of different correlation functions were examined and an isotropic, Gaussian function that was positive definite, was selected as the best fit (Figure 4.8b).

FIGURE 4.7 Locations of CTD stations taken in the North Atlantic between the Azores and the Canary Islands in spring 1982 (experiment POSEIDON 86, Hiller and Käse, 1983). Also shown are locations of current profile (P) and hydrocast (W) stations.

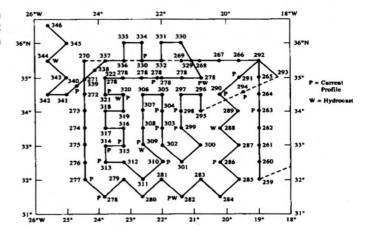

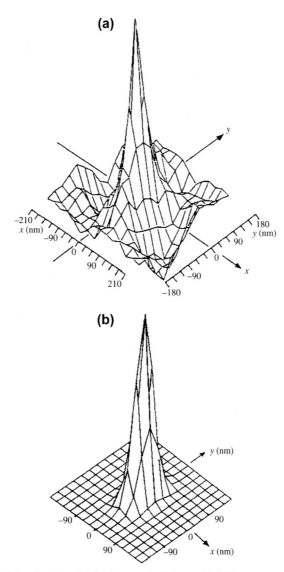

FIGURE 4.8 The two-dimensional correlation function $C(\mathbf{r})$ for the geopotential anomaly field at a pressure of 25 dbar referenced to 1,500 dbar (25/1,500 dbar) for the data collected at stations shown in Figure 4.7 (1 dbar $= 1\ m^2/s^2$). Here, $\mathbf{r} = (x, y)$, where x, y are the eastward and northward co-ordinates, respectively. Distances are in nautical miles. (a) The "raw" values of $C(\mathbf{r})$ based on the observations; (b) a model of the correlation function fitted to (a). *From Hiller and Käse (1983).*

Using this covariance function, the authors obtained the objectively mapped 25/1,500 db geopotential anomaly shown in Figure 4.9a. Removal of a linear trend gives the objective map shown in Figure 4.9b and the associated RMS error field shown in Figure 4.9c. Only near the outside boundaries of the data domain does the RMS error increase to around 50% of the geopotential anomaly field (Figure 4.9b).

As an example of objective mapping applied to a vector field, Hiller and Käse (1983) examined a limited number of satellite-tracked drifter trajectories that coincided with the CTD survey in space and time. Velocity vectors based on daily averages of low-passed finite difference velocities are shown in Figure 4.10a. Rather than compute a covariance function for this relatively small sample, the covariance function from the analysis of the 25/1,500 db geopotential anomaly was used. Also, an assumed error level, $\overline{\epsilon^2}$, was used rather than a computed estimate from the small sample. With the isotropic correlation scale estimated to be 75 km, the objective mapping produces the vector field in Figure 4.10b. The stippled area in this figure corresponds to the region where the error variance exceeds 50% of the total variance. Due to the paucity of data, the area of statistically significant vector mapping is quite limited. Nevertheless, the resulting vectors are consistent with the geopotential height map in Figure 4.9a.

FIGURE 4.9 Objective analysis of the geopotential anomaly field 25/1,500 dbar (m²/s²) using the lower of the two correlation functions in Eqn (4.10). (a) The approximate center of the frontal band in this region of the ocean is marked by the 13.5 dbar pressure isoline; (b) same as (a) but after subtraction of the linear spatial trend; (c) objectives analysis of the residual mesoscale perturbation field 25/1,500 dbar after removal of the composite mean field. *After Hiller and Käse (1983).*

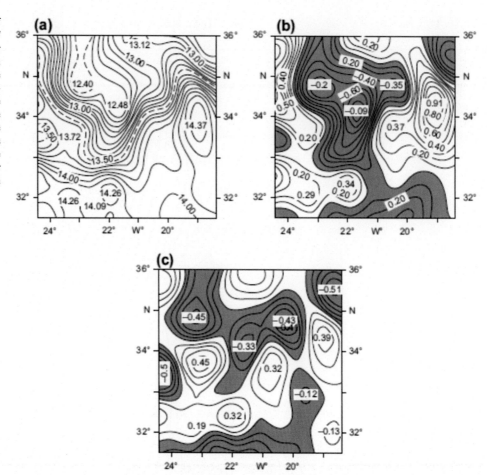

Another example is provided by McWilliams (1976) who used dynamic height relative to 1,500 m depth plus deep float velocities at 1,500 m to estimate the stream-function field. The isotropic covariance function for the random fluctuations in stream function $\Psi' = \Psi - \overline{\Psi}$ at 1,500 m depth was

$$
\begin{aligned}
C(r) &= \overline{\Psi'(\mathbf{x}, z, t)\Psi'(\mathbf{x} + \mathbf{r}, z, t)} \\
&= \overline{\Psi'^2}\left(1 - \varepsilon^2\right)\left(1 - \gamma^2 r^2\right)\exp\left(-\frac{1}{2}\delta^2 r^2\right)
\end{aligned}
\tag{4.13}
$$

where $\mathbf{r}$ is a horizontal separation vector, $r = |\mathbf{r}|$, ε is an estimate of relative measurement noise ($0 \leq \varepsilon \leq 1$), and γ^{-1}, δ^{-1} are decorrelation length scales found by fitting Eqn (4.13) to prior data. Denman and Freeland (1985) discuss the merits of five different covariance functions fitted to geopotential height data collected over a period of 3 years off the west coast of Vancouver Island. As discussed in Chapter 1, the widely used global sea surface temperature (SST) data generated by NOAA (Reynolds and Smith, 1994) are based on optimal interpolation on a 1° × 1° grid. The weekly and monthly analyses use 7 days of *in situ* (ship and buoy) and satellite SST records. Error statistics show that the SST rms data errors from ships are almost twice as large as the data errors from buoys or satellites, and that the average *e*-folding spatial scales for the error are 850 km in the zonal direction and 615 km in the meridional direction. The analysis also includes a preliminary step that uses Poisson's equation to correct any satellite biases relative to the *in situ* data. Reynolds and Smith (1994) demonstrate the importance of this correction using data following the 1991 eruptions of Mt. Pinatubo. For other examples, the reader is referred to Bennett (1992).

As a final point, we remark that the requirement of isotropy is easily relaxed by using direction-dependent covariance matrices, $C(r_1, r_2)$, whose spatial structure depends on two orthogonal spatial coordinates, r_1 and r_2 (with $r_2 \geq r_1$). For

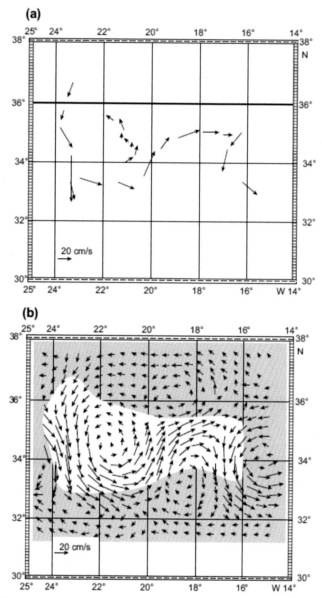

FIGURE 4.10 Analysis of the velocity field for the current profile collected in the grid in Figure 4.7. (a) The input velocity field; (b) objective analysis of the input velocity field with correlation scale $\lambda = 200$ km assumed noise variance of 30% of the total variance of the field. This approach treats mesoscale variability on scales less than 200 km as noise, which is smoothed out. In the shaded area, the error variance exceeds 50% of the total variance. *From Hiller and Käse (1983).*

example, the map of light attenuation coefficient at 20 m depth obtained from transmissometer profiles off the west coast of Vancouver Island (Figure 4.11) uses an exponentially decaying, elliptically shaped covariance matrix

$$C(r_1, r_2) = \exp\left[-a\Delta x^2 - b\Delta y^2 - c\Delta x\Delta y\right] \tag{4.14a}$$

where

$$a = \frac{1}{2}\left\{[\cos(\pi\phi/180)/r_1]^2 + [\sin(\pi\phi/180)/r_2]^2\right\}$$

$$b = \frac{1}{2}\left\{[\sin(\pi\phi/180)/r_1]^2 + [\cos(\pi\phi/180)/r_2]^2\right\} \tag{4.14b}$$

$$c = \cos(\pi\phi/180)\sin(\pi\phi/180)\left[r_2^2 - r_1^2\right]/(r_1 r_2)^2$$

FIGURE 4.11 Objective analysis map of light attenuation coefficient (per meter) at 20-m depth on the west coast of Vancouver Island obtained from transmissometer profiles. The covariance function $C(r_1, r_2)$ given by the ellipse is assumed to decay exponentially with distance with the longshore correlation scale $r_2 = 50$ km and cross-shore correlation scale $r_1 = 25$ km. Here, (r_1, r_2) is written as (R_1, R_2) in the figure.

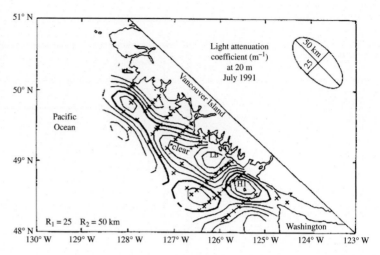

Here, Δx and Δy are, respectively, the eastward and northward distances from the grid point to the data point, and ϕ is the orientation angle (in degrees) of the coastline measured counterclockwise from north. In this case, it is assumed that the alongshore correlation scale, r_2, is twice the across-shore correlation scale, r_1. The idea here is that, like water-depth changes, alongshore variations in coastal water properties such as temperature, salinity, geopotential height, and log-transformed phytoplankton chlorophyll-a pigment concentration occur over longer length scales than across-shore variations.

In their introduction to optimal interpolation, Barth et al. (2008; http://modb.oce.ulg.ac.be/wiki) construct an artificial data field based on a spatially inhomogeneous sampling density (Figure 4.12). The white area in this figure corresponds to a physical barrier in the original data field (the colored region). Figure 4.13 presents the actual data values from the original data field for the data locations specified in Figure 4.12.

A comparison between the interpolated data field (colored map) in Figure 4.12 and the original data field in Figure 4.13, which was used to generate the interpolated map, clearly reveals that much of the structure in the original field is not captured by the gridded sample data. It is also clear that this data sample has some very different spatial characteristics in that there is a regular grid just above the barrier with very dense sampling, while much of the remaining data field is relatively sparsely sampled, with rather large regions not sampled at all. Moreover, the real data values would have errors associated with them, including instrumental errors, which may be random or have a bias. Representative errors arise when the observations do not precisely coincide with what we are attempting to determine. For example, Barth et al. (2008)

FIGURE 4.12 Optimal interpolation (colored region) of a data field and the sampling points (dots) of the original data field. The white region denotes a physical barrier. *From Barth et al. (2010).*

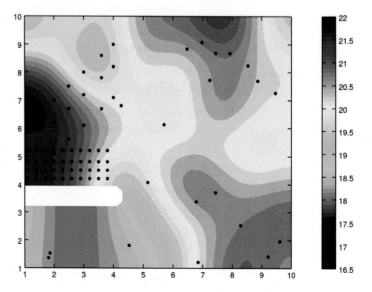

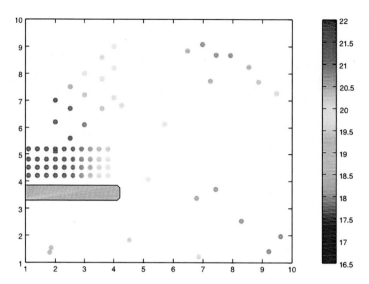

FIGURE 4.13   Actual values of the original field at the data locations shown in Figure 4.12. *From Barth et al. (2010).*

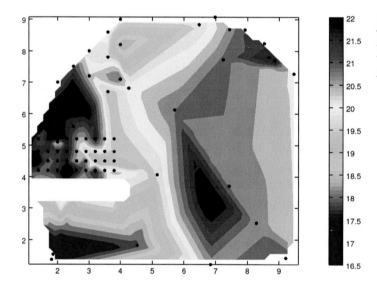

FIGURE 4.14   Linear interpolation of the data in Figure 4.12. This distribution was obtained by decomposing the domain into triangles, where the vertices are the location of the data points. Within each triangle, the value is interpolated linearly. *Source: Barth et al. (2010).*

remark that the investigator might desire a monthly average whereas the measurements being used are instantaneous or averaged over a short period of time. In addition, not all measurements are likely to have been collected at the same time. Added to these possible sources of error are a list of other errors, including human error, transmission or recording errors, and instrumental malfunctions.

To demonstrate a comparison with the earlier direct interpolation methods, we generated a linear interpolation of the data in Figure 4.13 using triangulation. Results are presented in Figure 4.14.

The distribution in Figure 4.14 looks quite different from that in Figure 4.12, which has been optimally interpolated. The smoothness of Figure 4.12 is much more attractive than the angular distribution in Figure 4.14, which we obtained using linear interpolation.

4.5 KRIGING

Kriging is another optimal interpolation technique and originated with the Master's thesis of Danie G. Krige who pioneered the distance-weighted average gold grades at the Witwatersrand reef complex mine in South Africa. The method was developed to improve mapping of the topography of a geographic region that had unevenly spaced data points, with the ultimate goal of estimating topographic values for grid locations for which there were no data values. The theory was further developed by the French mathematician Georges Matheron. As with Optimum Interpolation discussed previously, Kriging belongs to the family of linear least-squares estimation algorithms. Again, the goal is to estimate the values of an unknown real-valued function from observations at known, but irregularly, spaced locations. There are several types of Kriging, including Ordinary Kriging and Detrended Kriging (Van Beers and Kleijnen, 2002). In the latter, the Kriging algorithm uses linear regression analysis to "detrend" the data.

In general, spatial interpolation methods such as linear, spline, inverse distance, and triangular interpolation estimate values at given locations as the weighted sum of data values that surround a specific location. Almost all of these methods assign decreasing weights to the data with increasing separation distance from the interpolation point. Similar to Optimum Interpolation, Kriging assigns weights according to a moderately data-dependent weighting function rather than an arbitrary function. Kriging is, however, still an interpolation method. As such, it is subject to many of the basic limitations of other interpolation methods, including the propensity to underestimate the highs and overestimate the lows compared to the actual property distribution. In addition, interpolation methods are highly data-driven, whereby a dense distribution of observations will yield a better representation of reality than a sparse data distribution. Any interpolation method will perform well in data clusters but perform poorly in the gaps between these clusters.

There are some advantages to Kriging compared to other methods, such as the fact that it provides an estimation error (Kriging variance) for the estimate of a given variable. The availability of this estimation error can be used in the assimilation of interpolated fields into numerical simulation models. It also helps to compensate for the effects of data clustering by assigning lower weights for clustered values than for isolated data points.

4.5.1 Mathematical Formulation

All Kriging estimators are variants of the basic linear regression estimator, $Z^*(u)$, defined for a given location, i, as (Bohling, 2005)

$$Z^*(u)_i - m(u) = \sum_{\alpha=1}^{n(u)} \lambda_\alpha [Z(u_\alpha) - m(u_\alpha)] \tag{4.15}$$

where the u are location vectors corresponding to an estimation point (u) and one of the neighboring data points, u_α, indexed by α. The value $n(u)$ is the number of data points in a neighborhood of u that are used for the estimation of $Z^*(u)$, while $m(u)$ and $m(u_\alpha)$ are the expected values of $Z(u)$ and $Z(u_\alpha)$, respectively. The values $\lambda_\alpha(u)$ are the Kriging weights assigned to data values $Z(u_\alpha)$ used to estimate $Z(u)$.

We treat $Z(u)$ as a random field with a trend component, $m(u)$, and a residual component, $R(u) = Z(u) - m(u)$. The Kriging method estimates the residual at u as a weighted sum of residuals at surrounding data points. Kriging weights are derived from the semivariogram, which is another form of the covariance function. The use of distinct functions is one of the fundamental differences between the Kriging and Optimum Interpolation methodologies.

The semivariogram and covariance function quantify the correlation structure of a spatial field and provide measures of how well variables that are spaced closer together correlate compared to variables that are spaced further apart. Both measure the degree of statistical correlation as a function of separation distance, s. The semivariogram is defined as

$$\gamma(s_i - s_j) = \frac{1}{2} \text{var} [Z(s_i) - Z(s_j)] \tag{4.16}$$

where var is the variance of Z. If two locations, s_i and s_j, are close in terms of the distance measure $d(s_i, s_j)$, they are expected to have similar values so that the difference $Z(s_i) - Z(s_j)$ should be small. As locations s_i and s_j become farther apart, they become less similar and the difference $Z(s_i) - Z(s_j)$ is expected to increase (Figure 4.15). Semivariogram functions can also be thought of as dissimilarity functions. As illustrated in Figure 4.15, the height that the semivariogram reaches when it levels out is called the "sill" while the height discontinuity (and y-offset) at zero distance is called the "nugget effect". The nugget effect can itself be divided into a measurement error and a micro-scale variation, either of which can be zero. The distance at which the semivariogram reaches its sill height is called the "range".

There is a direct relationship between the semivariogram and the covariance function, C, whereby

$$\gamma(s_i, s_j) = \text{sill} - C(s_i, s_j) \tag{4.17}$$

where the variable "sill" is shown graphically in Figure 4.15. This equivalence makes it possible to use either the semivariogram or the covariance function to perform the interpolation. One important criterion is that the Kriging interpolated

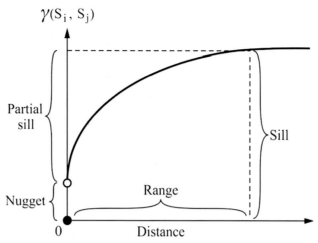

$\gamma(S_i, S_j)$

Partial
sill

Sill

Nugget

Range

0   Distance

FIGURE 4.15 A typical semivariogram showing an increase in the semivariogram with separation distance.

values have nonnegative Kriging standard errors, which means that only some functions can be used as the semivariogram or covariance functions. In those cases where the data-driven semivariogram has negative values, a non-negative function is fit to the semivariogram. This is similar to the step in objective analysis where a non-negative function is fit to the covariance function.

The goal of Kriging is to determine the weights λ_α that minimize the variance of the Kriging estimator

$$\sigma_E^2(u) = \text{var}[Z^*(u) - Z(u)] \tag{4.18}$$

under the unbiased constraint that the expected value E

$$E[Z^*(u) - Z(u)] = 0 \tag{4.19}$$

The random field $Z(u)$ is decomposed into a residual, R, and trend, m, component such that $Z(u) = R(u) + m(u)$, where the residual component is treated as a random field with a stationary mean of zero (0) and a stationary and isotropic covariance function, which is a function of the separation distance, h, but not of the position, u. Thus, we can write

$$E[R(u)] = 0 \tag{4.20a}$$

$$\text{Cov}[R(u), R(u+h)] = E[R(u), R(u+h)] = C_R(h) \tag{4.20b}$$

where $C_R(h)$ is the residual covariance function that is generally derived from the input semi-variogram model, $C_R(h) = C_R(0) - \lambda(h) = \text{sill} - \lambda(h)$. Thus, the semivariogram entered into a Kriging program should represent the residual component of the variable.

There are three different types of Kriging: simple, ordinary and Kriging with a trend. These all differ in their treatments of the trend component, $m(u)$. For simple Kriging, we assume that the trend component is a known constant such that, $m(u) = m$, whereby

$$Z_{SK}^*(u) = m + \sum_{\alpha=1}^{n(u)} \lambda_\alpha^{SK}(u)[Z(u_\alpha) - m] \tag{4.21}$$

This estimate is automatically unbiased since $E[Z(u_\alpha) - m] = 0$, so that $E[Z_{SK}^*(u)] = m = E[Z(u)]$. The estimation error $Z_{SK}^*(u) - Z(u)$ is a linear combination of a random variable representing residuals at the data points, u_α, and the estimation point, u:

$$Z_{SK}^*(u) - Z(u) = \lfloor Z_{SK}^*(u) - m \rfloor - [Z(u) - m] = \sum_{\alpha=1}^{n(u)} \lambda_\alpha^{SK}(u)R(u_\alpha) - R(u) = R_{SK}^*(u) - R(u) \tag{4.22}$$

Using rules for the variance of a linear combination of random variables, the error variance is then given by

$$\sigma_E^2(u) = \operatorname{var}\left[R_{SK}^*(u)\right] + \operatorname{var}[R_{SK}(u)] - 2\operatorname{Cov}\left[R_{SK}^*(u), R_{SK}(u)\right]$$

$$= \sum_{\alpha=1}^{n(u)} \sum_{\beta=1}^{n(u)} \left[\lambda_\alpha^{SK}(u)\lambda_\beta^{SK}(u)C_R(u_\alpha - u_\beta)\right] + C_R(0) - 2\sum_{\alpha=1}^{n(u)} \lambda_\alpha^{SK}(u)C_R(u_\alpha - u) \tag{4.23}$$

To solve for the Kriging weights, which is clearly a principal step in the entire process, we minimize the error variance in (4.23) by taking the derivative of this equation with respect to the Kriging weights and then setting each expression to zero. This leads to the following system of equations:

$$\sum_{\beta=1}^{n(u)} \left[\lambda_\beta^{SK}(u)C_R(u_\alpha - u_\beta)\right] = C_R(u_\alpha - u) \quad \alpha = 1, \ldots, n(u) \tag{4.24}$$

Because of the constant mean, the covariance function for $Z(u)$ is the same as that for the residual component $C(h) = C_R(h)$, so that we can write the simple Kriging systems of equations in terms of $C(h)$:

$$\sum_{\beta=1}^{n(u)} \left[\lambda_\beta^{SK}(u)C(u_\alpha - u_\beta)\right] = C(u_\alpha - u) \quad \alpha = 1, \ldots, n(u) \tag{4.25}$$

or

$$\mathbf{K}\lambda_{SK}(u) = \mathbf{k} \tag{4.26}$$

where $\mathbf{K}$ is the covariance matrix between the measured data points, with elements $K_{i,j} = C(u_i - u_j)$, and $\mathbf{k}$ is the vector of covariances between the data points and the estimation point whose elements are given by $k_i = C(u_i - u)$. Here, it is assumed that the covariance vector is simply a function of the distance between the data point and the estimation point. The vector λ_{SK} represents the weights for simple Kriging for the surrounding data points.

Assuming a positive definite covariance matrix, we can solve for the Kriging weights as:

$$\lambda_{SK}(u) = \mathbf{K}^{-1}\mathbf{k} \tag{4.27}$$

With these weights, we can compute both the Kriging estimate and the Kriging variance from (4.23), which now reduces to

$$\sigma_{SK}^2(u) = C(0) - \lambda_{SK}^T(u) = C(0) - \sum_{\alpha=1}^{n(u)} \lambda_\alpha^{SK}(u)C(u_\alpha - u) \tag{4.28}$$

The result is a set of weights for estimating the value of the variable at the location of interest based on the measured values from a set of neighboring data points. The weighting for each data point generally decreases with increasing separation distance between a specific point and the interpolation grid location.

Bohling (2005) applies simple Kriging to a porosity field (Figure 4.16) where the x and y axes correspond to the eastward and northward directions (eastings and northings), respectively, and are measured in meters. From this field, Bohling derives a spherical semivariogram (Figure 4.17), which has a zero nugget (i.e., it intersects the origin) and a sill of 0.78, with a range of 4,141 m. Since the study uses a spherical semivariogram, the covariance function is given by

$$C(h) = C(0) - \gamma(h) = 0.78\left[1 - 1.5(h/4,141) + 0.5(h/4,141)^3\right] \tag{4.29}$$

for separation distances h up to 4,141 m, and then 0 beyond that range. For the small sample of 6 points marked in Figure 4.16, the matrix of distances between pairs of data points is given in Table 4.1.

Based on the set of data in Table 4.1, the covariance matrix is

$$\mathbf{K} = \begin{bmatrix} 0.78 & 0.28 & 0.06 & 0.17 & 0.40 & 0.43 \\ 0.28 & 0.78 & 0.43 & 0.39 & 0.27 & 0.20 \\ 0.06 & 0.43 & 0.78 & 0.37 & 0.11 & 0.06 \\ 0.17 & 0.39 & 0.37 & 0.78 & 0.37 & 0.27 \\ 0.40 & 0.27 & 0.11 & 0.37 & 0.78 & 0.65 \\ 0.43 & 0.20 & 0.06 & 0.27 & 0.65 & 0.78 \end{bmatrix}$$

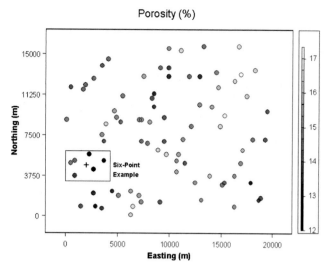

FIGURE 4.16 Point values of porosity (in %) mapped in eastward and northward (easting and northing) geographical coordinates. The plus sign (+) in the box denotes an interpolation grid-point to be determined using the surrounding six observed values. The gray-scale porosity code (%) is shown on the right. *After Bohling (2007).*

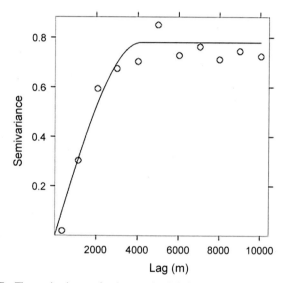

FIGURE 4.17 The semivariogram for the porosity field in Figure 4.16. *After Bohling (2007).*

TABLE 4.1 Matrix of separation distances (in meters) between pairs of points for the distribution of six points presented in Figure 4.16.

	Point 1	Point 2	Point 3	Point 4	Point 5	Point 6
Point 1	0	1,897	3,130	2,441	1,400	1,265
Point 2	1,897	0	1,281	1,456	1,970	2,280
Point 3	3,130	1,281	0	1,525	2,800	3,206
Point 4	2,441	1,456	1,523	0	1,523	1,970
Point 5	1,400	1,970	2,800	1,523	0	447
Point 6	1,265	2,280	3,206	1,970	447	0

where the matrix elements have been rounded off to two decimal places. Note the high correlation between points 5 and 6, which are separated by only 447 m. The corresponding vector of Kriging weights is

$$
\begin{bmatrix} \lambda_1 \\ \lambda_2 \\ \lambda_3 \\ \lambda_4 \\ \lambda_5 \\ \lambda_6 \end{bmatrix} = \mathbf{K}^{-1}\mathbf{k} = \begin{bmatrix} 0.1475 \\ 0.4564 \\ -0.0205 \\ 0.2709 \\ 0.2534 \\ -0.0266 \end{bmatrix}
$$

We note that data point 6 is assigned a very small weight relative to data point 1 even though they are both about the same distance from the interpolation point (+) and have about the same variance. This is because data point 6 is "shielded" by nearby data point 5, with which it is very strongly correlated, and which already has a very strong influence on the estimation point. The covariances and, hence the Kriging weights, are determined entirely by the data configuration and the covariance model, and not by the actual data values themselves.

Using the simple Kriging procedure outlined above and adding back in a mean field to the interpolated porosity values, yields the estimated porosity distribution (in %) presented in Figure 4.18. The standard deviation of the interpolated field is presented in Figure 4.19, which clearly lines up along bands determined by the data distribution, as one might expect from this type of interpolation method.

The preceding results illustrate several important characteristics of the Kriging method (Bohling, 2005). Specifically: (a) kriged surfaces, as with most interpolated surfaces, are smooth, and often much smoother than the actual surface that would be generated if there were more data available; (b) "bull's-eyes", which are caused by local extremes in the data values, are inevitable; and (c) the error map for the interpolated surfaces (as illustrated by the standard deviation in Figure 4.19) is driven primarily by the data distribution rather than by the actual data values at the different locations.

In the case of "ordinary Kriging", the mean data field is no longer assumed to be constant (spatially uniform) over the entire domain but instead is assumed constant in the local neighborhood of each estimation point. In simple Kriging, we equate $C_R(h)$ with $C(h)$, which is the covariance matrix of the data itself due to the presence of a constant mean. While this equality doesn't hold for ordinary Kriging, the practice is to assume it is true on the assumption that the semivariogram, from which $C(h)$ is derived, effectively filters out the influence of large-scale trends in the mean. Both ordinary Kriging and simple Kriging are interpolation methods that follow naturally from a semivariogram analysis and both procedures tend to filter out trends in the mean.

FIGURE 4.18   The result of applying simple Kriging to the porosity field in Figure 4.16. Porosity values (%) are denoted by the color bar on the right. The circles mark to locations of data values.

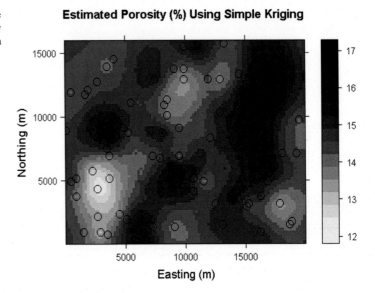

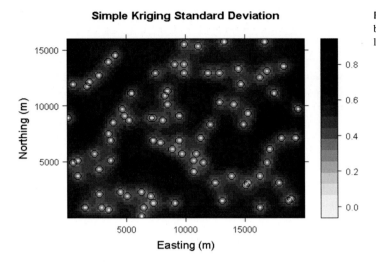

Simple Kriging Standard Deviation

FIGURE 4.19 Standard deviation of the porosity field (in %; see bar on the right) interpolated using simple Kriging. Values are lowest in the vicinity of the data points used in the Kriging.

When applied to the porosity data in Figure 4.16, simple Kriging yields the interpolation field shown in Figure 4.18; the associated standard deviation field is shown in Figure 4.19. The resultant field obtained using ordinary Kriging is similar to that for simple Kriging in Figure 4.18 and normally differs only for datasets having highly detailed features. The standard deviation fields also agree closely, again primarily because they are a function of the data locations more than they are a function of the actual data values. Kriging with a trend (also known as "universal Kriging") resembles ordinary Kriging except that, instead of fitting a local mean in the neighborhood of the estimation point, a linear or higher-order trend is fitted to the (x, y) coordinates of the data points. Including this form of model in the Kriging system, is akin to an extension to a local mean for ordinary Kriging. In fact, ordinary Kriging can be thought of as universal Kriging with a zeroth-order trend model.

Other specialized Kriging methods, such as "indicator Kriging", uses indicator functions rather than a data field for the interpolation process. "Co-Kriging" is a form of multivariate Kriging that uses the covariance between two different variables. "Disjunctive Kriging" is a nonlinear generalization of Kriging, and "lognormal Kriging" interpolates positive data, which has been transformed through the use of logarithms. Mathematically, Kriging is closely related to regression analysis but involves the interpolation of a single variable, while regression is the relationship between multiple data sets. It can also be shown that Kriging results in the best linear unbiased estimate. It is linear in the sense that all estimated values are weighted linear averages of the data values, unbiased because the mean error is zero, and best because it is designed to minimize the variance of the estimation errors.

Kriging that uses a semivariogram with a pronounced nugget will create discontinuities, with the interpolated surface oscillating up or down as it tries to "grab" any data point that happens to coincide with a grid node (estimation point). In such cases, it is possible to use "factorial Kriging", where the nugget is filtered out. Another option is to fit a semi-variogram that does not have a nugget (cf. Figure 4.15).

The method for selecting the appropriate neighboring data points can have at least as much influence on the estimates as the interpolation algorithm itself. The simplest approach is the nearest neighbor criterion; more complex methods may involve quadrant and octant searches that look for data points within a certain distance in each quadrant or octant surrounding the estimation point.

4.6 EMPIRICAL ORTHOGONAL FUNCTIONS

The previous section dealt with the optimal smoothing of irregularly spaced data onto a gridded map. In other studies of oceanic variability, we may be presented with a large data set from a grid of time-series stations, which we wish to compress into a smaller number of independent pieces of information. For example, in studies of climate change, it is necessary to deal with time series of spatial maps, such as sea surface temperature. A useful obvious choice would involve a linear combination of orthogonal spatial "predictors", or modes, whose net response as a function of time would account for the combined variance in all of the observations. The signals we wish to examine may all consist of the same variable, such as temperature, or they may be a mixture of variables such as temperature and wind velocity, or current and sea level. The data may be in the form of concurrent time-series records from a grid (regular or irregular) of stations $x_i(t)$, $y_i(t)$ on a horizontal plane or time-series records at a selection of depths on an $x_i(t)$, $z_i(t)$ cross-section. Examples of time series from cross-sectional data include those from a string of current meters on a single mooring or from moorings of upward-looking bottom-mounted ADCPs strung across a channel.

A useful technique for compressing the variability in this type of time-series data is *principal component analysis* (PCA). In oceanography, the method is commonly known as *empirical orthogonal function* (EOF) analysis. The EOF procedure is one of a larger class of inverse techniques and is equivalent to a data reduction method widely used in the social sciences known as *factor analysis*. The first reference we could find to the application of EOF analysis to geophysical fluid dynamics is a report by Edward Lorenz (1956) in which he develops the technique for statistical weather prediction and coins the term "EOF".

As discussed in Preisendorfer (1988), one of the essential aspects of the PCA method was developed by an Italian geometer, Beltrami, in 1873. He formulated a modern form of the resolution of a general square matrix into its singular value decomposition (SVD), which stands at the core of PCA. This same discovery was made independently by the French algebraist, Jordan, in 1874. PCA appears to have made its first appearance in the United States as an exercise in abstract algebra when Sylvester (1889) considered the problem of the reduction of a square matrix into its singular value decomposition. A decade later, Pearson (1901) recast linear regression analysis into a new form to avoid the common asymmetrical relationship between "dependent" and "independent" variables. In his paper, Pearson introduced a clear geometric visualization of PCA in Euclidean space. The first application of PCA to meteorology appears to have been made at the Massachusetts Institute of Technology (MIT) by G.P. Wadsworth and his colleagues in 1948. The goal of their study was to develop a short-term prediction method for sea level atmospheric pressure over the northern hemisphere. In test calculations over the North Atlantic, Wadsworth was faced in 1944 with the daunting task of hand-calculating the 91 eigenvalues of a 91×91 matrix. Confronted with this unmanageable numerical task, Wadsworth dropped the PCA approach and went on to use theoretical orthogonal functions (Tschebyscheff polynomials) to complete the project. It is interesting to note that, about the same time, a completely independent use of PCA in meteorology was being carried out by Fukuoka (1951).

When the Whirlwind general-purpose computer became available at MIT in the 1950s, E.N. Lorenz, starting with the work of Wadsworth and colleagues, undertook prediction studies of the 500 mb height anomaly for January (1947−52) for a grid of 64 points covering the mainland United States, Southern Canada and portions of the surrounding oceans. Lorenz (1956) is now a classic in the field of statistical-dynamical approaches to weather prediction. The Statistical Forecasting Project at MIT under Lorenz's direction produced some outstanding early applications of PCA to short-range forecasting. Applications of PCA to oceanographic data sets began to appear about a decade after Lorenz's work. Trenberth (1975) related southern hemisphere atmospheric oscillations to sea surface temperature observations. PCA studies based on sea surface temperatures in the Pacific by Barnett and Davis also appeared in the 1970s along with similar work by Weare et al. (1976). An interesting idea involving the use of extended EOFs for moving pattern detection in tropical Pacific Ocean temperatures is explored in Weare and Nasstrom (1982).

The advantage of EOF analysis is that it provides a compact description of the spatial and temporal variability of data series in terms of orthogonal functions, or statistical "modes." Usually, most of the variance of a spatially distributed series is in the first few orthogonal functions, whose patterns may then be linked to possible dynamical mechanisms. It should be emphasized that no direct physical or mathematical relationship necessarily exists between the statistical EOFs and any related dynamical modes. Dynamical modes conform to physical constraints through the governing equations and associated boundary conditions (LeBlond and Mysak, 1979); empirical orthogonal functions are simply a method for partitioning the variance of a spatially distributed group of concurrent time series. They are called "empirical" to reflect the fact that they are defined by the covariance structure of the specific data set being analyzed (as shown below).

In oceanography and meteorology, EOF analysis has found wide application in both the time and frequency domains. Conventional EOF analysis can be used to detect standing oscillations only. To study propagating wave phenomena, we need to use the lagged covariance matrix (Weare and Nasstrom, 1982), or complex principal component analysis in the frequency domain (Wallace and Dickinson, 1972; Horel, 1984). Our discussion, in this section, will focus on space/time domain applications. Readers seeking more detailed descriptions of both the procedural aspects and their applications are referred to Lorenz (1956), Davis (1976), and Preisendorfer (1988).

The best analogy to describe the advantages of EOF analysis is the classical vibrating drum problem. Using mathematical concepts presented in most undergraduate texts, it is well known that we can describe the eigenmodes of drumhead oscillations through a series of two-dimensional orthogonal patterns. These modes are defined by the eigenvectors and eigenfunctions of the drumhead. Generally, the lowest modes have the largest spatial scales and represent the most dominant (most prevalent) modes of variability. Typically, the drumhead has as its largest mode an oscillation in which the whole drumhead moves up and down, with the greatest amplitude in the center and zero motion at the rim where the drum is clamped. The next highest mode has the drumhead separated in the center with one side $180°$ out of phase with other side (one side is up when the other is down). Higher modes have more complex patterns with additional maxima and minima. Now, suppose we had no mathematical theory, and were required to describe the drumhead oscillations in terms of a set of observations. We would look for the kinds of eigenvalues in our data that we obtain from our mathematical analysis. Instead of the analytical or dynamical solutions that can be derived for the drum, we wish to examine "empirical" solutions based strictly on a measured data set. Because we are ignorant of the actual dynamical analysis, we call the resulting modes of oscillation, empirical orthogonal functions.

EOFs can be used in both the time and frequency domains. For now, we will restrict ourselves to the spatial domain application and consider a series of N maps at times $t = t_i$ ($1 \leq i, N$), each map consisting of scalar variables $\psi_m(t)$ collected at M locations, $\mathbf{x}_m$($1 \leq m \leq M$). One could think of N weather maps available every 6 h over a total period of $6N$ h, with each map showing the sea surface pressure $\psi_m(t) = P_m(t)$ ($1 \leq m \leq M$) recorded at M weather buoys located at mooring sites $\mathbf{x}_m = (x_m, y_m)$. The subscript m refers to the spatial grid locations in each map. Alternatively, the N maps might consist of pressure data $P(t)$ from $M - K$ weather buoys plus velocity component records $u(t)$, $v(t)$ from $K/2$ current meter sites. Or, again, the time series could be from $M/2$ current meters on a moored string. Any combination of scalars is permitted (this is a statistical analysis not a dynamical analysis). The goal of this procedure is to write the data series $\psi_m(t)$ at any given location $\mathbf{x}_m$ as the sum of M orthogonal spatial functions $\phi_i(\mathbf{x}_m) = \phi_{im}$ such that

$$\psi(\mathbf{x}_m, t) = \psi_m(t) = \sum_{i=1}^{M} [a_i(t)\phi_{im}] \tag{4.30}$$

where $a_i(t)$ is the amplitude of the ith orthogonal mode at time $t = t_n$ ($1 \leq n \leq N$). Simply put, Eqn (4.30) states that the time variation of the dependent scalar variable $\psi(\mathbf{x}_m, t)$ at each location $\mathbf{x}_m$ results from the linear combination of M spatial functions, ϕ_i, whose amplitudes are weighted by M time-dependent coefficients, $a_i(t)$ ($1 \leq i \leq M$). The weights $a_i(t)$ tell us how the spatial modes ϕ_{im} vary with time. There are as many (M) basis functions as there are stations for which we have data. Put another way, we need as many modes as we have time-series stations so that we can account for the combined variance in the original time series at each time, t. If we wanted, we could also formulate the problem as M temporal functions whose amplitudes are weighted by M spatially variable coefficients. Whether we partition the data as spatial or temporal orthogonal functions, the results should be identical.

Because we want the spatial functions $\phi_i(\mathbf{x}_m)$ to be orthogonal, so that they form a set of basis functions, we require that

$$\sum_{m=1}^{M} [\phi_{im}\phi_{jm}] = \delta_{ij} (\text{orthogonality condition}) \tag{4.31}$$

where the summation is over all observation locations and δ_{ij} is the Kronecker delta

$$\delta_{ij} = \begin{cases} 1, & j = i \\ 0, & j \neq i \end{cases} \tag{4.32}$$

It is worth remarking that two functions are said to be orthogonal when the sum (or integral) of their product over a certain defined space (or time) is zero. Orthogonality in Eqn (4.31) does not mean $\phi_{im}\phi_{jm} = 0$ for each m. For example, in the case of continuous sines and cosines, $\int \sin\theta \cos\theta \, d\theta = 0$ when the integral is over a complete phase cycle, $0 \leq \theta \leq 2\pi$. By itself, the product $\sin\theta \cdot \cos\theta = 0$ only if the sine or cosine term happens to be zero.

There is a multitude of basis functions, ϕ_i, that can satisfy Eqns (4.30) and (4.31). Familiar examples are sine, cosine, and Bessel functions. The EOFs are determined uniquely among the many possible choices by the constraint that the time amplitudes $a_i(t)$ are uncorrelated over the sample data. This requirement means that the time-averaged covariance of the amplitudes satisfies

$$\overline{a_i(t)a_j(t)} = \lambda_i \delta_{ij} \text{ (uncorrelated time variability)} \tag{4.33}$$

in which the overbar denotes the time-averaged value and

$$\lambda_i = \overline{a_i(t)^2} = \frac{1}{N} \sum_{n=1}^{N} [a_i(t_n)^2] \tag{4.34}$$

is the variance in each orthogonal mode. If we then form the covariance matrix $\psi_m(t)\psi_k(t)$ for the known data and use (4.33), we find

$$\overline{\psi_m(t)\psi_k(t)} = \sum_{i=1}^{M}\sum_{j=1}^{M}\left[\overline{a_i(t)a_j(t)}\phi_{im}\phi_{jk}\right] = \sum_{i=1}^{M}[\lambda_i\phi_{im}\phi_{ik}] \tag{4.35}$$

Multiplying both sides of (4.35) by ϕ_{ik}, summing over all k and using the orthogonality condition (4.31), yields

$$\sum_{k=1}^{M}\overline{\psi_m(t)\psi_k(t)}\phi_{ik} = \lambda_i\phi_{im}\,(i\text{th mode at the }m\text{th location}; m=1,...,M) \tag{4.36}$$

Eqn (4.36) is the canonical form for the *eigenvalue problem*. Here, the EOFs, ϕ_{im}, are the *i*th *eigenvectors* at locations $\mathbf{x}_m$, and the mean-square time amplitudes

$$\lambda_i = \overline{a_i(t)}^2$$

are the corresponding *eigenvalues* of the mean product, $\mathbf{R}$, which has elements

$$R_{mk} = \overline{\psi_m(t)\psi_k(t)}$$

The mean product, $\mathbf{R}$, is equal to the covariance matrix, $\mathbf{C}$, if the mean values of the time series $\psi_m(t)$ have been removed at each site $\mathbf{x}_m$. The total of M empirical orthogonal functions corresponding to the M eigenvalues of (4.36) forms a complete basis set of linearly independent (orthogonal) functions such that the EOFs are uncorrelated modes of variability. Assuming that the record means $\overline{\psi_m(t)}$ have been removed from each of the M time series, Eqn (4.36) can be written more concisely in matrix notation as

$$\mathbf{C}\boldsymbol{\phi} - \lambda\mathbf{I}\boldsymbol{\phi} = 0 \tag{4.37}$$

where the covariance matrix, $\mathbf{C}$, consists of M data series of length N with elements

$$C_{mk} = \overline{\psi_m(t)\psi_k(t)}$$

$\mathbf{I}$ is the unity matrix, and $\boldsymbol{\phi}$ are the EOFs. Expanding (4.37) yields the eigenvalue problem

$$\begin{pmatrix} \overline{\psi_1(t)\psi_1(t)} & \overline{\psi_1(t)\psi_2(t)} & \cdots & \overline{\psi_1(t)\psi_M(t)} \\ \overline{\psi_2(t)\psi_1(t)} & \overline{\psi_2(t)\psi_2(t)} & \cdots & \overline{\psi_2(t)\psi_M(t)} \\ \cdots & \cdots & \cdots & \cdots \\ \overline{\psi_M(t)\psi_1(t)} & \overline{\psi_M(t)\psi_2(t)} & \cdots & \overline{\psi_M(t)\psi_M(t)} \end{pmatrix} \begin{pmatrix} \phi_1 \\ \phi_2 \\ \cdots \\ \phi_M \end{pmatrix} = \begin{pmatrix} \lambda & 0...0 \\ 0 & \lambda...0 \\ \cdots \\ 0...\lambda \end{pmatrix} \begin{pmatrix} \phi_1 \\ \phi_2 \\ \cdots \\ \phi_M \end{pmatrix} \tag{4.38a}$$

corresponding to the series of linear system of equations

$$\begin{aligned} \left[\overline{\psi_1(t)\psi_1(t)} - \lambda\right]\phi_1 + \overline{\psi_1(t)\psi_2(t)}\phi_2 + ... + \overline{\psi_1(t)\psi_M(t)}\phi_M &= 0 \\ \overline{\psi_2(t)\psi_1(t)}\phi_1 + \left[\overline{\psi_2(t)\psi_2(t)} - \lambda\right]\phi_2 + ... + \overline{\psi_2(t)\psi_M(t)}\phi_M &= 0 \\ \cdots \\ \overline{\psi_M(t)\psi_1(t)}\phi_1 + \overline{\psi_M(t)\psi_2(t)}\phi_2 + ... + \left[\overline{\psi_M(t)\psi_M(t)} - \lambda\right]\phi_M &= 0 \end{aligned} \tag{4.38b}$$

The eigenvalue problem involves diagonalization of a matrix, which in turn amounts to finding an axis orientation in M-space for which there are no off-diagonal terms in the matrix. When this occurs, the different modes of the system are orthogonal. Because each $\mathbf{C}$ is a real symmetric matrix, the eigenvalues λ_i are real. Similarly, the eigenvectors (EOFs) of a real symmetric matrix are real. Because $\overline{C(x_m, x_k)}$ is positive, the real eigenvalues are all positive.

If Eqn (4.37) is to have a nontrivial solution, the determinant of the coefficients must vanish; that is

$$\det\begin{vmatrix} C_{11} - \lambda & C_{12} & \cdots & C_{1M} \\ C_{21} & C_{22} - \lambda & \cdots & \cdots \\ \cdots & \cdots & \cdots & \cdots \\ C_{M1} & \cdots & \cdots & C_{MM} - \lambda \end{vmatrix} = 0$$

which yields an Mth order polynomial, $\lambda^M + \alpha\lambda^{M-1} + ...$, whose M eigenvalues satisfy

$$\lambda_1 > \lambda_2 > ... > \lambda_M \tag{4.39}$$

Thus, the "energy" (more specifically, the variance) associated with each statistical mode is ordered according to its corresponding eigenvector. The first mode contains the highest percentage of the total variance, λ_1; of the remaining variance, the greatest percentage is in the second mode, λ_2, and so on. If we add up the total variance in all the time series, we obtain

$$\sum_{m=1}^{M} \left\{ \frac{1}{N} \sum_{n=1}^{N} [\psi_m(t_n)]^2 \right\} = \sum_{j=1}^{M} \lambda_j \tag{4.40}$$

Sum of variance in the data = sum of variance in the eigenvalues

The total variance in the M time series equals the total variance contained in the M statistical modes. The final piece of the puzzle is to derive the time-dependent *amplitudes* of the ith statistical mode

$$a_i(t) = \sum_{m=1}^{M} \psi_m(t)\phi_{im} \tag{4.41}$$

Eqn (4.36) provides a computational procedure for finding the EOFs. By computing the mean product matrix, $\overline{\psi_m(t)\psi_k(t)}(m, k = 1,...,M)$ or "scatter matrix" $\mathbf{S}$ in the terminology of Preisendorfer (1988), the eigenvalues and eigenvectors can be determined using standard computer algorithms. From these, we obtain the variance associated with each mode, λ_j, and its time-dependent variability, $a_i(t)$.

As outlined by Davis (1976), two advantages of a statistical EOF description of the data are: (1) the EOFs provide the most efficient method of compressing the data; and (2) the EOFs may be regarded as uncorrelated (i.e., orthogonal) modes of variability of the data field. The EOFs are the most efficient data representation in the sense that, for a fixed number of functions (trigonometric or other), no other approximate expansion of the data field in terms of $K < M$ functions

$$\widehat{\psi}_m(t) = \sum_{m=1}^{K} a_i(t)\widehat{\phi}_{im} \tag{4.42}$$

can produce a lower total mean-square error

$$\sum_{m=1}^{K} \left[\psi_m(t) - \widehat{\psi}_m(t) \right]^2 \tag{4.43}$$

than would be obtained when the $\widehat{\phi}_i$ are the EOFs. A proof of this is given in Davis (1976). Also, as discussed later in this section, we could just as easily have written our data $\psi(\mathbf{x}_m, t)$ as a combination of orthogonal temporal modes $\phi_i(t)$ whose amplitudes vary spatially as $a_i(\mathbf{x}_m)$. Because this is a statistical technique, it doesn't matter whether we use time or space to form the basis functions. However, it might be easier to think in terms of spatial orthogonal modes that oscillate with time rather than temporal orthogonal modes that oscillate in space.

As noted above, EOFs are ordered by decreasing eigenvalue so that, among the EOFs, the first mode, having the largest eigenvalue, typically accounts for a considerable fraction of the variance of the data. Thus, with the inherent efficiency of this statistical description, only a few empirical modes generally are needed to describe the fundamental variability in a very large data set. Often it may prove useful to employ the EOFs as a filter to eliminate unwanted scales of variability. A limited number of the first few EOFs (those with the largest eigenvalues) can be used to reconstruct the data field, thereby eliminating those scales of variability not coherent over the data grid and therefore less energetic in their contribution to the data variance. An EOF analysis can then be made of the filtered data set to provide a new apportionment of the variance for those scales associated with most of the variability in the original data set. In this application, EOF analysis is much like standard Fourier analysis used to filter out scales of unwanted variability. In fact, for homogeneous time series sampled at evenly spaced increments, it can be shown that the EOFs are Fourier trigonometric functions.

The computation of the eigenfunctions $a_i(t)$ in Eqn (4.41) requires the data values $\psi_m(t)$ for all of the time series. Often these time series contain gaps, which make it impossible to compute $a_i(t)$ at those times for which the data are missing. One solution to this problem is to fill the gaps in the original data records using one of the procedures discussed in the previous chapter on interpolation. Most consistent with the present approach is to use optimum interpolation as discussed in the preceding section. While this will provide an interpolation consistent with the covariance of the subject data set, these optimally estimated values of $\psi_m(t)$ often result in large, expected errors if the gaps are large or the scales of coherent variability are small.

An alternative method, suggested by Davis (1976), that can lead to a smaller expected errors is to estimate the EOF amplitude at time, t, directly from the existing values of $\psi_m(t)$, thus eliminating the need for the interpolation of the original data. Conditions for this procedure are that the available number of sample data pairs is reasonably large (gaps do not dominate) and that the data time series are stationary. Under these conditions, the mean product matrix $\overline{\psi_m(t) - \psi_k(t)}(m, k = 1, \ldots, M)$ will be approximately the same as it would have been for a data set without gaps. For times when none of the $\psi_m(t)$ values are missing, the coefficients $a_i(t)$ can be computed from Eqn (4.41). For times t when data values are missing, $a_i(t)$ can be estimated from the available values of $\psi_m(t)$

$$\widehat{a}_i(t) = b_i(t) \sum_{j=1}^{M'} \psi_j(t)\phi_{ij} \tag{4.44}$$

where the summation over j includes only the available data points, $M' \leq M$. From Eqns. (4.37), (4.42), and (4.44), the expected square error of this estimate is

$$\overline{[a_i(t) - \widehat{a}_i(t)]^2} = b_i^2(t) \sum_{j=1}^{M'} \left(\lambda_j \gamma_{ij}^2\right) + \lambda_i[1 + b_i(t)(\gamma_{ii}-1)]^2 \tag{4.45}$$

where

$$\gamma_{ij} = \sum_k \phi_j(k)\phi_i(k) \tag{4.46}$$

and the summation over k applies only to those variables with missing data. Taking the derivative of the right-hand side of (4.45) with respect to b_i, we find that the expected square error is minimized when

$$b_i(t) = (1 - \gamma_{ii})\lambda_i \left/ \left[(1 - \gamma_{ii})^2\lambda_i + \sum_j \lambda_j \gamma_{ji}^2 \right] \right. \tag{4.47}$$

Applications of this procedure (Davis, 1976, 1978; Chelton and Davis, 1982; Chelton et al., 1982) have shown that the expected errors are surprisingly small even when the number of missing data is relatively large. This is because the dominant EOFs in geophysical systems generally exhibit large spatial scales of variability, leading to a high coherence between grid values. As a consequence, contributions to the spatial pattern from the most dominant EOFs at any particular time, t, can be reliably estimated using a relatively small number of sample grid points.

4.6.1 Principal Axes of a Single Vector Time Series (Scatter Plot)

A common technique for improving the EOF analysis for a set of vector time series is to first rotate each data series along its own customized principal axes. In this new coordinate system, most of the variance is associated with the major axis and the remaining variance with the minor axis. The technique also provides a useful application of principal component analysis. The problem consists of finding the principal axes of variance along which the variance in the observed velocity fluctuations $\mathbf{u}'(t) = \left(u_1'(t), u_2'(t)\right)$—or, as commonly written, $\mathbf{u}'(t) = \left(u'(t), v'(t)\right)$—is maximized for a given location; here u_1' and u_2' (or, $u'(t)$ and $v'(t)$), are the respective east-west and north-south components of the wind or current velocity obtained by removing the respective means $\overline{u_1}$ and $\overline{u_2}$ from each record, i.e., $u_1' = u_1 - \overline{u_1}, u_2' = u_2 - \overline{u_2}$. The bulk of the data "scatter" is mainly distributed parallel to the major axis, with lower scatter parallel to the minor axis (Figure 4.20). We also note that principal axes are defined in such a way that the velocity components along the two principal axes are uncorrelated.

The eigenvalue problem (4.37) for a two-dimensional scatter plot has the form

$$\begin{vmatrix} C_{11} & C_{21} \\ C_{12} & C_{22} \end{vmatrix} \begin{vmatrix} \phi_1 \\ \phi_2 \end{vmatrix} = \begin{vmatrix} \lambda & 0 \\ 0 & \lambda \end{vmatrix} \begin{vmatrix} \phi_1 \\ \phi_2 \end{vmatrix} \tag{4.48}$$

where the C_{ij}, are components of the covariance matrix, $\mathbf{C}$, derived from the current velocity data, and (ϕ_1, ϕ_2) are the eigenvectors associated with the two possible values of the eigenvalues, λ. To find the principal axes for the scatter plot of u_2' versus u_1', we set the determinant of the covariance matrix equation derived from Eqn (4.48) to zero, specifically

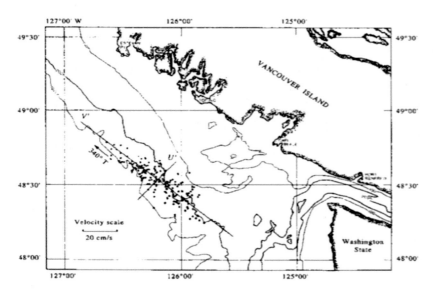

FIGURE 4.20 The principal component axes for daily-averaged velocity components u, v ($\equiv u_1, u_2$) measured by a current meter moored at 175 m depth on the west coast of Canada. Here, the north-south component of velocity, $v(t)$, is plotted as a scatter diagram against the east-west component of current velocity, $u(t)$. Data cover the period October 21, 1992–May 25, 1993. The major axis pointing toward 340° T can be used to define the positive longshore direction, v'.

$$\det|\mathbf{C} - \lambda \mathbf{I}| = \det\begin{vmatrix} C_{11} - \lambda & C_{12} \\ C_{21} & C_{22} - \lambda \end{vmatrix} = \det\begin{vmatrix} \overline{u'^2_1} - \lambda & \overline{u'_1 u'_2} \\ \overline{u'_2 u'_1} & \overline{u'^2_2} - \lambda \end{vmatrix} = 0 \tag{4.49a}$$

where (for $i = 1, 2$) the elements of the determinant are given by

$$C_{ii} = \overline{u'^2_i} = \frac{1}{N} \sum_{n=1}^{N} \left[u'_i(t_n) \right]^2 \tag{4.49b}$$

$$C_{ij} = \overline{u'_i u'_j} = \frac{1}{N} \sum_{n=1}^{N} \left[u'_i u'_j(t_n) \right] \tag{4.49c}$$

Solution of (4.49) yields the quadratic equation

$$\lambda^2 - \left\lfloor \overline{u'^2_1} + \overline{u'^2_2} \right\rfloor \lambda + \overline{u'^2_1 u'^2_2} - \left(\overline{u'_1 u'_2} \right)^2 = 0 \tag{4.50}$$

whose two roots $\lambda_1 > \lambda_2$ are the eigenvalues, corresponding to the variances of the velocity fluctuations along the major and minor principal axes, respectively. The orientations of the two axes differ by 90° and the principal angles θ_p (those along which the sum of the squares of the normal distances to the data points u'_1, u'_2 are extrema) are found from the transcendental relation

$$\tan(2\theta_p) = \frac{2\overline{u'_1 u'_2}}{\overline{u'^2_1} - \overline{u'^2_2}} \tag{4.51a}$$

$$\theta_p = \frac{1}{2} \tan^{-1}\left[\frac{2\overline{u'_1 u'_2}}{\overline{u'^2_1} - \overline{u'^2_2}} \right] \tag{4.51b}$$

where the principal angle is defined for the range $-\pi/2 \leq \theta_p \leq \pi/2$ (Freeland et al., 1975; Kundu and Allen, 1976; Preisendorfer, 1988). As usual, the multiple $n\pi/2$ ambiguities in the angle that one obtains from the arctangent ($\tan^{-1}$) function must be addressed by considering the quadrants of the numerator and denominator in Eqns (4.51a) and (4.51b). Preisendorfer (1988; Figure 2.3) outlines the nine different possible cases. Proof of (4.51a) and (4.51b) is given in Section 4.6.5.

The principal variances (λ_1, λ_2) of the data set are found from the determinant relations (4.49a) and (4.50) as

$$\left.\begin{array}{c}\lambda_1 \\ \lambda_2\end{array}\right\} = \frac{1}{2}\left\{\left(\overline{u_1'^2} + \overline{u_2'^2}\right) \pm \left[\left(\overline{u_1'^2} - \overline{u_2'^2}\right)^2 + 4\left(\overline{u_1'u_2'}\right)^2\right]^{1/2}\right\} \tag{4.52}$$

in which the $+$ sign is used for λ_1 and the $-$ sign for λ_2. In the case of current velocity records, λ_1 gives the variance of the flow along the major axis and λ_2 the variance along the minor axis. The slope, $s_1 = \phi_2/\phi_1$, of the eigenvector associated with the variance λ_1 (i.e., the slope of the eigenvector in the north-south, east-west Cartesian coordinate system) is found from the matrix relation

$$\left| \begin{array}{cc} \left(\overline{u_1'^2} - \lambda_1\right) & \overline{u_1'u_2'} \\ \overline{u_2'u_1'} & \overline{u_2'^2} - \lambda \end{array} \right| \left| \begin{array}{c} \phi_1 \\ \phi_2 \end{array} \right| = 0 \tag{4.53a}$$

Solving (4.53a) for $\lambda = \lambda_1$, gives

$$\left(\overline{u_1'^2} - \lambda_1\right)\phi_1 + \left(\overline{u_1'u_2'}\right)\phi_2 = 0$$
$$\left(\overline{u_2'u_1'}\right)\phi_1 + \left(\overline{u_2'^2} - \lambda_1\right)\phi_2 = 0 \tag{4.53b}$$

so that

$$s_1 = \left|\lambda_1 - \overline{u_1'^2}\right| / \overline{u_1'u_2'} \tag{4.53c}$$

with a similar expression for the slope s_2 associated with the variance $\lambda = \lambda_2$; the products of the slopes $s_1 \cdot s_2 = -1$. If $\lambda_1 \gg \lambda_2$, then $\lambda_1 \approx \overline{u_1'^2} + \overline{u_2'^2}$ and $s_1 \approx \overline{u_2'^2}/\overline{u_1'u_2'}$. The usefulness of principal component analysis is that it can be used to find the main orientation of fluid flow at any current meter or anemometer site, or within a "box" containing velocity variances derived from Lagrangian drifter trajectories (Figure 4.21). Since the mean and low frequency currents in relatively shallow waters are generally "steered" parallel to the coastline or local bottom contours, the major principal axis is often used to define the "alongshore" direction while the minor axis defines the "cross-shore" direction of the flow. In the case of prevailing coastal winds, the major axis usually parallels the mean orientation of the coastline or coastal mountain range that steers the surface winds. Although defining the cross-shore direction is vital to the estimation of cross-shore fluxes, reliable estimates are often difficult to obtain. This is especially true in regions where the alongshore component of flow is strong

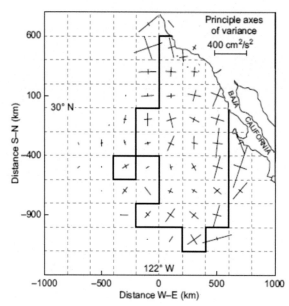

FIGURE 4.21   Principal axes of current velocity variance (kinetic energy) obtained from surface satellite tracked drifter measurements off the coast of southern California during 1985–86. For this analysis, data have been binned into $200 \times 200 \text{ km}^2$ boxes solid border denotes the region for which there were more than 50 drifter days and more than two different drifter tracks. *From Poulain and Niiler (1989).*

and highly variable. In such cases, small "errors" in the specified orientation of the axes can lead to marked relative changes in the cross-shore flux estimates.

4.6.2 EOF Computation Using the Scatter Matrix Method

There are two primary methods for computing the EOFs for a grid of time series of observations. These are: (1) The scatter matrix method which uses a "brute force" computational technique to obtain a symmetric covariance matrix $\mathbf{C}$ which is then decomposed into eigenvalues and eigenvectors using standard computer algorithms (Preisendorfer, 1988); and (2) the computationally efficient singular value decomposition (SVD) method which derives all the components of the EOF analysis (eigenvectors, eigenvalues, *and* time-varying amplitudes) without computation of the covariance matrix (Kelly, 1988). The EOFs determined by the two methods are identical. The differences are mainly the greater degree of sophistication, computational speed, and computational stability of the SVD approach.

Details of the covariance matrix approach can be found in Preisendorfer (1988). This recipe, which is only one of several possible procedures that can be applied, involves the preparation of the data and the solution of Eqn (4.37) as follows:

1. Ensure that the start and end times for all M time series of length N are identical. Typically, $N > M$.
2. Remove the record mean and linear trend from each time-series record $\psi_m(t), 1 \leq m \leq M$, such that the fluctuations of $\psi_m(t)$ are given by $\psi'_m(t) = \psi_m(t) - \left[\overline{\psi_m(t)} + b_m(t - \bar{t})\right]$ where b_m is the slope of the least-squares regression line for each location. Other types of trends can also be removed.
3. Normalize each de-meaned, de-trended time series by dividing each data series by its standard deviation $s = \left[1/(N-1) \sum (\psi_{m'})^2\right]^{1/2}$ where the summation is over all time, t (t_n: $1 \leq n \leq N$). This ensures that the variance from no one station dominates the analysis (all stations have an equal chance to contribute). The M normalized time-series fluctuations, ψ'_m, are the data series to be used for the EOF analysis. The total variance for each of the M eigenvalues $= 1$; thus, the total variance for all modes, $\sum \lambda_i = M$.
4. Rotate any vector time series to its principal axes. Although this operation is not imperative, it helps maximize the signal-to-noise ratio for the preferred direction. For future reference, keep track of the means, trends and standard deviations derived from the M time series records.
5. Construct the $M \times N$ data matrix, $\mathbf{D}$, using the M rows (locations $\mathbf{x}_m$) and N columns (times t_n) of the normalized data series

$$\mathbf{D} = \begin{pmatrix} \psi'_1(t_1) & \psi'_1(t_2) & \cdots & \psi'_1(t_N) \\ \psi'_2(t_1) & \psi'_2(t_2) & \cdots & \psi'_2(t_N) \\ \cdots & \cdots & \cdots & \cdots \\ \psi'_M(t_1) & \psi'_M(t_2) & \cdots & \psi'_M(t_N) \end{pmatrix} \text{Location} \downarrow \quad (4.54)$$

$$\text{Time} \rightarrow$$

and from this derive the symmetric covariance matrix, $\mathbf{C}$, by multiplying $\mathbf{D}$ by its transpose $\mathbf{D}^T$

$$\mathbf{C} = \frac{1}{N-1}\mathbf{D}\mathbf{D}^T \quad (4.55)$$

where $\mathbf{S} = (N - 1) \mathbf{C}$ is the scatter matrix defined by Preisendorfer (1988), and

$$\mathbf{C} = \begin{pmatrix} C_{11} & C_{12} & \cdots & C_{1M} \\ C_{21} & C_{22} & \cdots & C_{2M} \\ \cdots & \cdots & \cdots & \cdots \\ C_{M1} & \cdots & \cdots & C_{MM} \end{pmatrix} \quad (4.56)$$

The elements of the real symmetric matrix $\mathbf{C}$ are

$$C_{ij} = C_{ji} = \frac{1}{N-1} \sum_{n=1}^{N} \left[\psi_i'(t_n) \psi_j'(t_n) \right] \tag{4.57}$$

The eigenvalue problem then becomes

$$\mathbf{C}\boldsymbol{\phi} = \lambda\boldsymbol{\phi} \tag{4.58}$$

where scalar values λ are the eigenvalues and $\boldsymbol{\phi}$ the eigenvectors. Here, $\boldsymbol{\lambda} = \lambda\mathbf{I}$, where $\mathbf{I}$ is the unity matrix.

At this point, we remark that we have formulated the EOF decomposition in terms of an $M \times M$ "spatial" covariance matrix whose time-averaged elements are given by the product $(N-1)^{-1}\mathbf{DD}^T$ (4.55). We could just as easily have formed an $N \times N$ "temporal" covariance matrix whose spatially averaged elements are given by the product $(M-1)^{-1}\mathbf{D}^T\mathbf{D}$. The mean values we remove in preparing the two data sets are slightly different since the preparation of $\mathbf{D}$ involves time averages while the preparation of $\mathbf{D}^T$ involves spatial averages. However, in principle, the two problems are identical and the percentage of the total time-series variance in each mode depends on whether one computes the spatial EOFs or temporal EOFs. As we further point out in the following section, another difference between the two problems is how the singular values are grouped and which is identified with the spatial function and which with the temporal function (Kelly, 1988). The designation of one set of orthogonal vectors as EOFs and the other as amplitudes is quite arbitrary.

Once the matrix $\mathbf{C}$ has been calculated from the data, the problem can be solved using "canned" programs from one of the standard statistical or mathematical computer libraries for the eigenvalues and eigenvectors of a real symmetric matrix. In deriving the values listed in Tables 4.2–4.7, we have used the double-precision program DEVLSF of the International Math and Science Library (IMSL). The program outputs the eigenvalues λ in increasing order. The time varying amplitudes $a_i(t) = \sum_{m=1}^{M} \psi_m(t)\phi_{im}$ (see 4.41) follow from the relationship (4.37); the principal axes are derived (Table 4.3) but are not used to rotate the coordinate system as we had recommended in Step 4 above. Moreover, we used the statistics derived for the Raw Data in Table 4.3 to obtain the results in Table 4.4, but used the Trend Removed data in Table 4.3 to calculate the results in Tables 4.5–4.7. Thus, the principal components (Table 4.4) are based on the non-detrended data because we argue that this is more representative of the actual flow orientation. To obtain λ in decreasing order of importance, we have had to invert the eigenvalue output. For each eigenvector or mode, the program normalizes all values to the maximum value for that mode. The amplitude of the maximum value is unity (=1). Since there are M eigenvalues, the data normalization process gives a total EOF variance of $M(\sum \lambda_i = M)$. The canned programs also allow for calculation of a "performance index" (PI), which measures the error of the eigenvalue problem (4.48) relative to the various components of the problem and the machine precision. The performance of the eigenvalue routine is considered "excellent" if PI < 1, "good" if $1 \leq$ PI ≤ 100, and "poor" if PI > 100. As a final analysis, we can conduct an *orthogonality*

TABLE 4.2 Data matrix $\mathbf{D}^T$ components of velocity (cm/s) at three different sites at 1,700-m depth in the northeast Pacific.

Time (Days)	Site 1 (u_1)	Site 1 (v_1)	Site 2 (u_2)	Site 2 (v_2)	Site 3 (u_3)	Site 3 (v_3)
1	−0.3	0.0	0.4	−0.4	−0.8	−1.4
2	−0.1	0.3	0.4	−0.3	−1.1	0.0
3	−0.1	−0.4	0.0	−0.5	0.0	−2.5
4	0.2	0.6	0.0	−0.6	−0.7	0.4
5	0.3	−0.1	−0.6	−0.3	0.0	−0.3
6	0.5	0.0	0.9	−0.6	0.6	0.3
7	0.2	0.2	−0.1	−0.7	1.2	−2.8
8	−0.5	−0.9	0.0	−0.6	0.0	−1.8

Records start September 29, 1985 and are located near 48°N, 129°W. For each of the three stations we list the east–west (u) and north–south component (v). The means and trends have not yet been removed.

TABLE 4.3 Means, standard deviations, and linear trends for each of the time series components for each of the three current meter sites listed in Table 4.2.

| Component | Mean (cm/s) | Standard deviation (cm/s) | | Trend (cm/s/day) |
		Raw data	Trend removed	
u_1 (east−west)	0.025	0.333	0.328	0.024
v_1 (north−south)	−0.037	0.457	0.418	−0.075
u_2 (east−west)	0.125	0.443	0.433	−0.038
v_2 (north−south)	−0.500	0.151	0.114	−0.040
u_3 (east−west)	−0.100	0.762	0.503	0.233
v_3 (north−south)	−1.012	1.278	1.250	−0.108

Means have been removed from the time series prior to the calculation of the standard deviations. The standard deviations have been calculated in two ways: with no trend removal (raw data) and with the linear trend removed.

TABLE 4.4 Principal axes for the current velocity at each site in Table 4.2.

Station ID	Angle θ (°)	Major axis (cm/s)	Minor axis (cm/s)
Site 1	59.9	0.226	0.054
Site 2	−3.3	0.172	0.020
Site 3	−69.7	1.574	0.362

The angle θ is measured counterclockwise from east. Axes lengths are in cm/s. Values have been derived using (4.51) and (4.52) without removal of the trends. Results can also be derived using eigenvalue analysis but will differ slightly from those in the table since EOF calculations generally involve removal of a linear trend. It is preferable to use the original data series when determining the principal axes.

TABLE 4.5 Eigenvalues and percentage of variance in each statistical mode derived from the data in Table 4.2.

Eigenvalue No.	Eigenvalue	Percentage
1	2.2218	37.0
2	1.7495	29.2
3	1.1787	19.6
4	0.6953	11.6
5	0.1498	2.5
6	0.0048	0.1
Total	6.0000	100.0

check on the EOFs by using the relation (4.31). Here we look for significant departures from zero in the products of different modes; if any of the products

$$\sum_{m=1}^{M} \left[\phi_{im}\phi_{jm} \right]$$

are significantly different from zero for $i \neq j$, then the EOFs are not orthogonal and there are errors in the computation. A computational example is given in Section 4.6.4.

TABLE 4.6 Eigenvectors (EOFs) ϕ_i for the data matrix in Table 4.2.

Station ID	Mode 1	Mode 2	Mode 3	Mode 4	Mode 5	Mode 6
Site 1 u_1	1.000	−0.032	−0.430	0.479	−0.599	−0.969
Site 1 v_1	0.958	−0.078	−0.162	−0.966	1.000	0.085
Site 2 u_2	0.405	0.230	1.000	0.910	0.517	−0.295
Site 2 v_2	−0.329	−0.898	−0.525	1.000	0.784	−0.111
Site 3 u_3	0.349	1.000	−0.474	0.812	0.124	0.907
Site 3 v_3	0.654	−0.964	0.263	0.190	−0.539	1.000

Modes are normalized to the maximum value for each mode.

TABLE 4.7 Time series of the amplitudes, $a_i(t)$ for each of the statistical modes.

Time	Mode 1	Mode 2	Mode 3	Mode 4	Mode 5	Mode 6
Day 1	0.798	−0.773	0.488	0.089	0.091	0.124
Day 2	−0.076	1.258	0.402	0.126	0.595	−0.089
Day 3	1.153	−1.582	−0.458	0.275	−0.492	−0.094
Day 4	−1.531	0.759	0.363	−1.585	−0.382	0.000
Day 5	0.097	1.647	−2.099	0.509	−0.128	0.039
Day 6	−2.169	−0.142	1.084	1.296	−0.171	0.008
Day 7	−0.721	−1.921	−0.866	−0.534	0.503	0.004
Day 8	2.450	0.754	1.085	−0.176	−0.017	0.008

4.6.3 EOF Computation Using Singular Value Decomposition

The above method of computing EOFs requires use of covariance matrix, $\mathbf{C}$. This becomes computationally impractical for large, regularly spaced data fields such as a sequence of infrared satellite images (Kelly, 1988). In this case, for a data matrix $\mathbf{D}$ over N time periods (N satellite images, for example), the covariance or mean product matrix is given by (4.55)

$$\mathbf{C} = \frac{1}{N-1}\mathbf{D}\mathbf{D}^T \tag{4.59}$$

where $\mathbf{D}^T$ is the transpose of the data matrix $\mathbf{D}$. If we assume that all of the spatial data fields (i.e. satellite images) are independent samples, then the mean product matrix is the covariance matrix and the EOFs are again found by solving the eigenvalue problem

$$\mathbf{C}\boldsymbol{\phi} = \boldsymbol{\phi}\boldsymbol{\Lambda} \tag{4.60}$$

where $\boldsymbol{\phi}$ is the square matrix, whose columns are eigenvectors and $\boldsymbol{\Lambda}$ is the diagonal matrix of eigenvalues. For satellite images, there may be $M = 5,000$ spatial points sampled $N = 50$ times, making the covariance matrix a $5,000 \times 50$ matrix. Solving the eigenvalue problem for $\boldsymbol{\phi}$ would take $\max\{O(M^3), O(MN^2)\}$ operations. As pointed out by Kelly (1988), the operation count for the SVD method is $O(MN^2)$, which represents a considerable savings in computations over the traditional EOF approach if M is large. This is primarily true for those cases where M, the number of locations in the spatial data matrix, $\mathbf{D}$, is far greater than the number of temporal samples (i.e., images).

There are two computational reasons for using the singular value decomposition method instead of the covariance matrix approach (Kelly, 1988): (1) The SVD formulation provides a one-step method for computing the various components of the eigenvalue problem; and (2) it is not necessary to compute or store a covariance matrix or other intermediate quantities. This greatly simplifies the computational requirements and provides for the use of canned analysis programs for the EOFs. Our analysis is based on the double-precision program DLSVRR in the IMSL. The SVD method is based on the concept in linear algebra (Press et al., 1992) that any $M \times N$ matrix, **D**, whose number of rows M is greater than or equal to its number of columns, N, can be written as the product of three matrices: an $M \times N$ column-orthogonal matrix, **U**, an $N \times N$ diagonal matrix, **S**, with positive or zero elements, and the transpose $(\mathbf{V}^T)$ of an $N \times N$ orthogonal matrix, **V**. In matrix notation, the SVD matrix becomes:

$$\mathbf{D} = \mathbf{U} \begin{pmatrix} s_1 & & & \\ & s_2 & & \\ & & \ldots & \\ & & & s_N \end{pmatrix} \mathbf{V}^T \tag{4.61}$$

For oceanographic applications, the data matrix, **D**, consists of M rows (spatial points) and N columns (temporal samples). The scalars $s_1 \geq s_2 \geq \ldots \geq s_N \geq 0$ of the matrix **S**, called the *singular values* of **D**, appear in descending order of magnitude in the first N positions of the matrix. The columns of the matrix **V** are called the left singular vectors of **D** and the columns of the matrix **U** are called the right singular vectors of **D**. The matrix **S** has a diagonal upper $N \times N$ part, **S**′, and a lower part of all zeros in the case when $M > N$. We can express these aspects of **D** in matrix notation by rewriting Eqn (4.61) in the form

$$\mathbf{D} = [\mathbf{U}|\mathbf{0}] \left| \frac{\mathbf{S}'}{\mathbf{0}} \right| \mathbf{V}^T \tag{4.62}$$

where $[\mathbf{U}|\mathbf{0}]$ denotes a left singular matrix and **S**′ denotes the nonzero part of **S** which has zeros in the lower part of the matrix (Kelly, 1988).

The matrix **U** is orthogonal, and the matrix **V** has only N significant columns which are mutually orthogonal such that,

$$\begin{aligned} \mathbf{V}^T \mathbf{V} &= \mathbf{I} \\ \mathbf{U}^T \mathbf{U} &= \mathbf{I} \end{aligned} \tag{4.63}$$

Returning to Eqn (4.61), we can compute the eigenvectors, eigenvalues and eigenfunctions of the principal component analysis in one single step. To do this, we prepare the data as before following steps 1−5 in Section 4.6.2. We then use commercially available programs such as the double-precision program DLSVRR in the IMSL. The elements of matrix **U** are the eigenvectors while those of matrix **S** are related to the eigenvalues $s_1 \geq s_2 \geq \ldots \geq s_N \geq 0$. To obtain the time-dependent amplitudes (eigenfunctions), we require a matrix **A** such that

$$\mathbf{D} = \mathbf{U}\mathbf{A}^T \tag{4.64}$$

which, by comparison with Eqn (4.61), requires $\mathbf{A}^T = \mathbf{S}\mathbf{V}^T$ whereby

$$\mathbf{A} = \left(\mathbf{A}^T\right)^T = \mathbf{V}\mathbf{S}^T \tag{4.65}$$

Hence, the amplitudes are simply the eigenvectors of the transposed problem multiplied by the transpose of the singular values, **S**. Solutions of (4.61) are identical (within round-off errors) to those obtained using the covariance matrix of the data, **C**. We again remark that the only difference between the matrices **U** and **V** is how the singular values are grouped and which is identified with the spatial function and which with the temporal function. The designation of **U** as EOFs and **V** as amplitudes is quite arbitrary.

The decomposition of the data matrix **D** through singular value decomposition is possible since we can write it as a linear combination of functions $F_i(x)$, $i = 1, \ldots, M$ so that

$$\mathbf{D} = \mathbf{F}\alpha \tag{4.66a}$$

or

$$\begin{pmatrix} D(x_1, t_j) \\ D(x_2, t_j) \\ \ldots \\ \ldots \\ D(x_N, t_j) \end{pmatrix} = \begin{pmatrix} F_1(x_1) \ldots F_N(x_1) \\ F_1(x_2) \ldots F_N(x_2) \\ \ldots \\ \ldots \\ F_1(x_N) \ldots F_N(x_N) \end{pmatrix} \begin{pmatrix} \alpha_1(t_j) \\ \alpha_2(t_j) \\ \ldots \\ \ldots \\ \alpha_N(t_j) \end{pmatrix} \tag{4.66b}$$

where the α_i are functions of time only. The functions F are chosen to satisfy the orthogonality relationship

$$\mathbf{F}\mathbf{F}^T = \mathbf{I} \tag{4.67}$$

so that the data matrix $\mathbf{D}$ is divided into orthogonal modes

$$\mathbf{D}\mathbf{D}^T = \mathbf{F}\mathbf{a}\mathbf{a}^T\mathbf{F}^T = \mathbf{F}\mathbf{L}\mathbf{F}^T \tag{4.68}$$

where $\mathbf{L} = \mathbf{a}\mathbf{a}^T$ is a diagonal matrix. The separation of the modes arises from the diagonality of the $\mathbf{L}$ matrix, which occurs because $\mathbf{D}\mathbf{D}^T$ is a real and symmetric matrix and $\mathbf{F}$ a unitary matrix. To reduce sampling noise in the data matrix $\mathbf{D}$, one would like to describe it with fewer than M functions. If $\mathbf{D}$ is approximated by $\widetilde{\mathbf{D}}$, which uses only K functions ($K < M$), then the K functions which best describe the $\mathbf{D}$ matrix…in the sense that

$$\left(\widetilde{\mathbf{D}} - \mathbf{D}\right)^T \left(\widetilde{\mathbf{D}} - \mathbf{D}\right)$$

is a minimum … are the empirical orthogonal functions, which correspond to the largest valued elements of the traditional EOFs, found earlier.

4.6.4 An Example: Deep Currents Near a Mid-Ocean Ridge

As an example of the different concepts presented in this section, we again consider the 8 days of daily averaged currents ($N = 8$) at three deep current meter sites in the northeast Pacific near the Juan de Fuca Ridge (Table 4.2). Because each site has two components of velocity, $M = 6$. The data all start on the same day and have the same number of records. Following the five steps outlined in Section 4.6.2, we first removed the average value from each time series. We then calculated the standard deviation for each time series and used this to normalize the time series so that each normalized series has a variance of unity. For convenience, we write the transpose of the data matrix, $\mathbf{D}^T$, where columns are the pairs of components of velocity (u, v) and rows are the time in days.

Time-series plots of the first three eigenmodes are presented in Figure 4.22. The performance index (PI) for the scatter matrix method was 0.026, which suggests that the matrix inversion in the eigenvalue solutions was well defined. A check on the orthogonality of the eigenvectors suggests that the singular value decomposition gave vectors, which were slightly more orthogonal than the scatter matrix approach. For each combination (i, j) of the orthogonality condition (4.31), the products $\Sigma_{i,j}[\phi_{im}\ \phi_{jm}]$ were typically of order 10^{-7} for the SVD method and 10^{-6} for the scatter matrix method. Similar results apply to the orthogonality of the eigenmodes given by Eqn (4.33).

Before closing this section, we remark that we also could have performed the above analysis using complex EOFs of the form

$$\psi_m(t) = u_m(t) + iv_m(t)$$

(where $i = \sqrt{-1}$) in which case $M = 3$. This formulation not only allows the EOF vectors to change amplitude with time, as in our previous decomposition using $2M$ real EOFs, but also to rotate in time.

4.6.5 Interpretation and Examples of EOFs

In interpreting the meaning of EOFs, it is worth keeping in mind that, while EOFs offer the most efficient statistical compression of the data field, empirical modes do not necessarily correspond to true dynamical modes or modes of physical behavior. Often, a single physical process may be spread over more than one EOF. In other cases, more than one physical process may be contributing to the variance contained in a single EOF. The statistical construct derived from this procedure must be considered in light of accepted physical mechanisms rather than as physical modes themselves. It often is likely that the strong variability associated with the dominant modes is attributable to several identifiable physical mechanisms. Another possible clue to the physical mechanisms associated with the EOF patterns can be found in the time-series coefficients $a_i(t)$. The temporal variability of certain processes might resemble the time series of the EOF coefficients, which would then suggest a causal relationship not readily apparent in the spatial structure of the EOF.

One way to interpret EOFs is to imagine that we have displayed the data as a scatter diagram in an effort to discover if there is any inherent correlation among the values. For example, consider two parameters such as sea surface temperature (SST) and sea-level pressure (SLP) measured at a number of points over the North Pacific. This is the problem studied by Davis (1976) where he analyzed sets of monthly SST and SLP over a period of 30 years for a grid in the North Pacific. If we plot $x =$ SST against $y =$ SLP in a scatter diagram, any correlation between the two would appear as an elliptical cluster of points. A more common example is that of Figure 4.20 where we plotted the north–south (y) component of daily mean

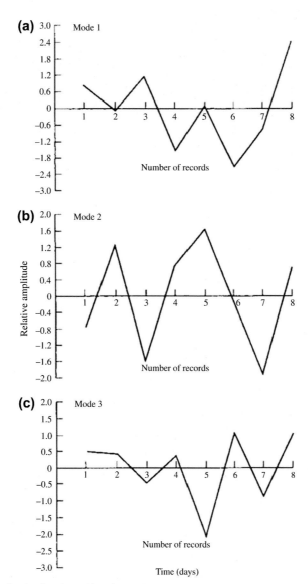

FIGURE 4.22 Eight-day time series for the first three EOFs for current meter data collected simultaneously at three sites at 1,700-m depth in the northeast Pacific in the vicinity of Juan de Fuca Ridge, 1985. Modes 1, 2, and 3 presented in (a), (b) and (c), respectively, account for 37.0%, 29.2%, and 19.6% of the variance, respectively.

current, v, against the corresponding east–west (x) component, u, for a continental shelf region. Here, the mean flow tends to parallel the coastline, so that the scatter plot again has an elliptical distribution. To take this distribution into account, we redefine our coordinate system by rotating x (u) and y (v) through the counterclockwise angle θ to the principal axes representation x', y' (u', v') discussed in Section 4.6.2. These transformations are given by

$$x' = x \cos\theta + y \sin\theta; \quad y' = -x \sin\theta + y \cos\theta \tag{4.69}$$

where (u', v') are found by simply replacing x with u and v with y in the above equations. Note that, in the case of currents, θ is measured in the counterclockwise direction from east in this coordinate system (which confuses things somewhat since east is 90° in terms of True compass bearing; north is 0° ≡ 360° T). What we have done in this rotation is to formulate a new set of axes that explains most of the variance, subject to the assumption that the variance does not change with time.

Since the axes are orthogonal, the total variance will not change with rotation. Let $V = \overline{x'^2} = N^{-1}\sum x'^2$ be the particular variance we want to maximize (as usual, the summation is over all N values of the time series). Note that we have focused on x' whereas the total variance is actually determined by r^2, where r is the distance of each point from the origin. However, we can expand $r^2 = x^2 + y^2$ and associate the variance with a given coordinate. In other words, if we maximize the variance associated with x', we will minimize the variance associated with y'. Using our summation convention, we can write

$$V = \overline{x'^2} = \overline{x^2} \cos^2 \theta + 2\overline{xy} \sin \theta \cos \theta + \overline{y^2} \sin^2 \theta \tag{4.70}$$

and

$$\frac{\partial V}{\partial \theta} = 2\left(\overline{y^2} - \overline{x^2}\right) \sin \theta \cos \theta + 2\overline{xy} \cos 2\theta \tag{4.71}$$

We maximize (4.71) by setting $\partial V/\partial \theta = 0$, giving (4.51), which we previously quoted without proof

$$\tan\left(2\theta_p\right) = \frac{2\overline{xy}}{\overline{x^2} - \overline{y^2}} \tag{4.72}$$

From (4.72), we see that if

$$\overline{xy} \ll \max\left(\overline{x^2}, \overline{y^2}\right)$$

then $\tan(2\theta_p) \to 0$ and $\theta_p = 0$, or $\pm 90°$, and we are left with the original axes. If $\overline{x^2} = \overline{y^2}$ and $\overline{xy} \neq 0$, then $\tan(2\theta_p) \to \pm \infty$ and the new axes are rotated $\pm 45°$ from the original axes.

We now find the expression for V. Since $\sec^2(2\theta) = 1 + \tan^2(2\theta)$

$$\cos(2\theta) = \left(\overline{x^2} - \overline{y^2}\right)/\pm D \tag{4.73a}$$

$$\sin(2\theta) = \left[1 - \cos^2(2\theta)\right]^{1/2} = 2\overline{xy}/\pm D \tag{4.73b}$$

where

$$D = \left[\left(\overline{x^2} - \overline{y^2}\right)^2 + 4\overline{xy}^2\right]^{1/2} \tag{4.74}$$

Then, using the identities

$$\cos^2 \theta = \frac{1}{2}(1 + \cos 2\theta), \ \sin^2 \theta = \frac{1}{2}(1 - \cos 2\theta) \tag{4.75}$$

we can write the variance as

$$V = \overline{x^2}\frac{(1 + \cos 2\theta_p)}{2} + \overline{y^2}\frac{(1 - \cos 2\theta_p)}{2} + \overline{xy} \sin 2\theta_p = \frac{1}{2}\left\{\left(\overline{x^2} + \overline{y^2}\right) \pm \left[\left(\overline{x^2} - \overline{y^2}\right)^2 + 4\overline{xy}^2\right]^{1/2}\right\} \tag{4.76}$$

The two roots of this equation correspond to a maximum and a minimum of V. For a new axis for which $\overline{x'^2}$ is a maximum, we will find $\overline{y'^2}$ is a minimum. This follows automatically from the fact that the total variance is conserved. However, we can confirm this mathematically by computing $\partial^2 V/\partial \theta^2 = 0$. From Eqn (4.71), and using (4.73) and (4.74), we find maximum (minimum) values from

$$\partial^2 V/\partial \theta_P^2 = 2\left(\overline{y^2} - \overline{x^2}\right) \cos 2\theta_P - 4\overline{xy} \sin 2\theta_P = -2\left[\left(\overline{x^2} - \overline{y^2}\right)^2 + 4\overline{xy}^2\right]/\pm D = \pm 2D = 0 \tag{4.77}$$

The positive sign in Eqn (4.77) corresponds to a maximum (because (4.76) is negative); the negative sign corresponds to a minimum. Solving (4.77) using (4.74), yields the relationship between the variances in the x and y variables. It so happens that the variance solutions given by (4.77) are also the eigenvalues of the covariance matrix. Thus, we can return to our previous methods where we used the covariance matrix to compute the EOFs.

A published example of EOF analysis is presented by Davis (1976) who examined monthly maps of SST and SLP for the years 1947–74. The SLP data were originally obtained from the Long-Range Prediction Group of the U.S. National Meteorological Center (NMC) as 1-month averages on a 5° diamond-shaped grid (i.e., 20°N–140°W, 20°N–150°W, ..., 25°N–145°W, 25°N–155°W, etc.). The data were transferred to a regular 5°-square grid using linear interpolation from

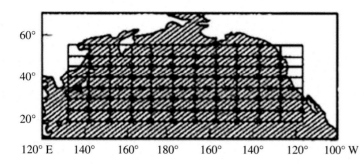

FIGURE 4.23 The grid of sea surface temperature (SST) and sea-level pressure (SLP). The 10° longitude by 5° latitude SLP averages are centered at grid intersections and SST averages are centered at crosses. *From Davis (1976).*

the four nearest diamond grid points to fill in the square grid. The SST data were obtained from the U.S. National Marine Fisheries Service in the form of monthly averages over 2° squares. Because this grid spacing is not a submultiple of 5°, and because sometimes data were missing, the following data analysis scheme was employed. The 2° data were subjectively analyzed to produce maps contoured with a 1°F contour interval. During this stage, missing values were filled in where feasible. The corrected values were then linearly interpolated onto a 1° grid and 25 values were averaged to formulate area averages on the chosen 5° grid coincident with the SLP data. The ship data originated as ship injection temperatures and are subject to all of the problems discussed earlier in the section on SST.

Before carrying out the EOF analysis, the SST and SLP data sets were further averaged onto a grid with a 5° latitude spacing and a 10° longitude spacing (Figure 4.23). In those cases where some SST values were missing, the available observations were used to compute the grid average. Even then there were some 5° × 10° regions with missing data in the SST fields. Both fields were then converted to anomalies using the mean of the 28-year data set as the reference field. Thus, each of the individual monthly maps was transformed into an anomaly map, corresponding to the deviation of local values from the long-term mean.

The standard deviations of both the SLP and SST anomaly fields are shown in Figure 4.24. It is interesting to note some of the basic differences between the variability of these two fields. The SLP field has its primary variability in the central northern part of the field just off the tip of the Aleutian Islands. Here, the Aleutian Low dominates the pressure field in winter and becomes the source of the main variability in the SLP data. In contrast, the SST field has near-uniform variance levels except in the Kuroshio Extension region off of northeast Japan where a maximum associated with advection from the Kuroshio is clearly evident.

To compute the EOFs from the anomaly fields, Davis (1976) used the covariance (scatter) matrix method presented in Section 4.6.2. The fraction of total variance accounted for by the EOFs for both the SST and SLP data is presented in Figure 4.25 as a function of the number of EOFs. The steep slope of the SLP curve means that fewer SLP EOFs are needed

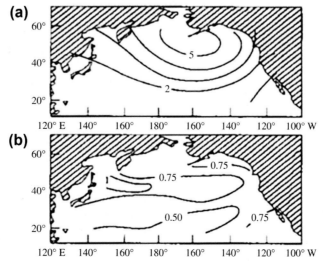

FIGURE 4.24 Standard deviation of: (a) Sea level pressure anomaly (mb); and (b) Sea surface temperature anomaly (°C) for the North Pacific. The anomalies are departures from monthly normal values. Variances are averaged over all months of the 28-year record (1947–74). *From Davis (1976).*

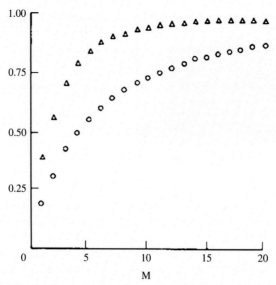

FIGURE 4.25 The fraction of total sea surface temperature (circles, o) and sea-level pressure (triangles, Δ) anomaly variance accounted for by the first M empirical orthogonal functions. *From Davis (1976).*

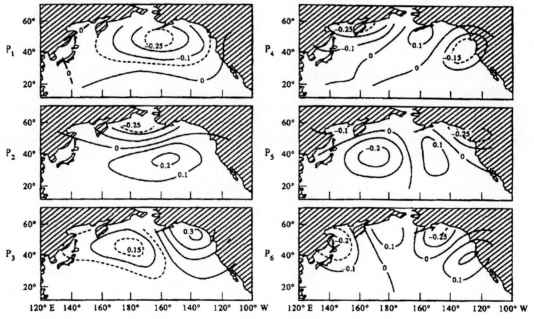

FIGURE 4.26 The six principal empirical orthogonal functions P1—P6 describing the sea level pressure anomalies. Function numbers are written to the left of each panel. *From Davis (1976).*

to express the variance. The slope of the SST EOF curve is consistently below that for the SLP EOF series. As a consequence, Davis presented only the first six SLP EOFs (labeled P_1–P_6 in Figure 4.26) but felt it necessary to present the first eight SST EOFs (labeled T_1–T_8 in Figure 4.27). The SLP EOFs exhibited fairly simple, large-scale patterns with P_1 having the same basic shape as the SLP standard deviation (Figure 4.24). The structural sequence for the first three SLP EOFs was: For P_1, a single maximum; for P_2, two meridionally separated maxima; and for P_3, two zonally separated maxima. Higher modes appear to be combinations of these first three with an increasing number of smaller maxima.

The SST maps obtained by Davis were considerably more complicated than the SLP maps, with large-scale patterns dominating only the first three modes of the temperature field. As with the SLP modes, the sequence seems to be from a central maximum (T_1), to meridionally separated maxima (T_2), and then to zonally separated maxima (T_3). The higher-order EOFs have a number of smaller maxima with no simple structures. The overall scales are much shorter than

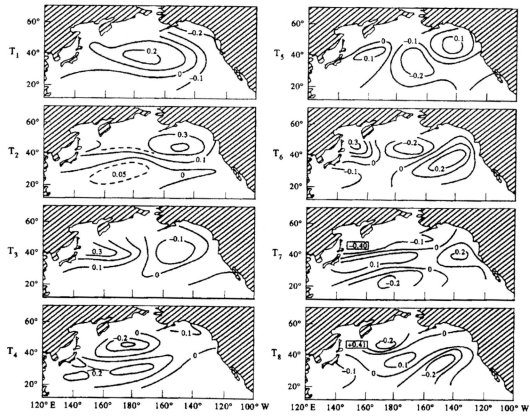

FIGURE 4.27   The eight principal empirical orthogonal functions T1–T8 describing the sea level pressure anomalies. Function numbers are written to the left of each panel. *From Davis (1976).*

those for the SLP EOFs. This turns out to be true for the time scales of the EOFs, with the SLP time scales being much shorter than those computed for the SST EOFs.

The goal of the EOF analysis by Davis (1976) was to determine if there is some direct statistical connection between the SLP and SST anomaly fields. By using the EOF procedure he was able to present the primary modes of variability for both fields in the most compact form possible. This is the real advantage of the EOF procedure. In terms of the two anomaly fields, Davis found that there were connections between the variables. First, he found that SST anomalies could be predicted from earlier SST anomaly fields. This is a consequence of the persistence of individual SST patterns as well as the fact that some patterns appear to evolve from earlier patterns through advective processes. Davis also concluded that it was possible to specify the SLP anomaly on the basis of the coincident SST anomaly field. Finally, it was not possible to statistically predict the SST field from the simultaneous SLP field. These conclusions, would have been difficult to arrive at without using the EOF procedure, are consistent with the much greater heat capacity and persistence ("memory") of SST anomalies compared to SLP anomalies.

4.6.6 Variations on Conventional EOF Analysis

Conventional principal component (EOF) analysis is limited by a number of factors, including the dependence of the solution on the domain of analysis, the requirement for orthogonal spatial modes, and the lumping together of variability over all frequency bands. In addition, the method can detect standing waves but not progressive waves. Over the years, several authors have developed what might be called "variations" on the standard EOF theme. For the most part, the methods differ in the types of variances they insert into the algorithms used to determine the empirical orthogonal functions (principal components). Given that EOF analysis is a strictly statistical method, it is irrelevant how the variance is derived, provided that the type of variance used in the analysis is the same for all spatial locations. All that is required is that the matrix **D**, derived from statistical averages (such as the covariance, correlation and cross-covariance functions) of the gridded time series is a Hermitian matrix.

Departure from standard EOF analysis can have numerous forms. For example, one may choose to work in the frequency domain instead of the time domain by using spectral analysis to calculate the spectral "energy" density for specific frequency bands. In this case, the matrix **D** is complex, consisting of the cross-spectra between the gridded time series over a specific frequency band. The spectral densities represent the data variances, which are used to determine the empirical orthogonal functions. Thus, the method is equally at home with variances obtained in the time or frequency domains. Regardless of variance-type, principal component methods are simply techniques for compressing the variability of the data set into the fewest possible number of modes.

Returning to the time domain, suppose that we are examining the statistical structure of alongshore wind and current fluctuations over the continental shelf and that we have reason to believe that current response to wind forcing is delayed by one, or more, time steps in the combined data series. A delay of half a pendulum day (≈ 12 h at mid-latitudes) is not unreasonable. From a causal point of view, the best way to examine the EOF modes for the combined wind and current data is to first create new time series in which the wind records are lagged (shifted forward in time) relative to the current records. Suppose we want a delay of one time step. Then, alongshore wind velocity values $V_k(t_j)$ at site k at times t_j ($j = 2$, 3, ...) get replaced with the earlier records at times t_{j-1}. That is, $V_k(t_j) \rightarrow V_k(t_{j-1}) = V_k^*(t_j)$, while the current velocity record remains unchanged, $v_k(t_j) = v_k^*(t_j)$. In this case, the asterisk (*) denotes the new time series. Optimal empirical modes are those for which the wind and current records are properly "tuned" with the correct time lags. For large spatial regions with variable wind response times, this can get a little tricky, so caution is advised.

4.7 EXTENDED EMPIRICAL ORTHOGONAL FUNCTIONS (EEOFS)

Extended EOFs are an extension of the traditional spatial EOFs and are formulated to deal not only with spatial but also with temporal correlations in the space-time data sets. This departure from conventional EOF analysis was presented by Kundu and Allen (1976) who combined the zonal (u) and meridional (v) time series of currents into complex time series $U = u + iv$, where each scalar series is defined for times t_j and locations x_k. The method was applied to current data collected during the Coastal Upwelling Experiment (CUE-II) off the Oregon coast in the summer of 1973. The complex covariance matrix obtained from these time series were then decomposed into complex eigenvectors by solving a standard complex eigenvalue problem. Unlike the scalar approach to the problem, this complex EOF technique can be used to describe rotary current variability within selected frequency bands.

A further variation on conventional EOF analysis, which is related to complex EOF analysis, was provided by Denbo and Allen (1984). Using a technique we describe in Chapter 5, the current fluctuations in each of the time series (u, v) records collected during CUE-II were decomposed into clockwise (S^+) and counterclockwise (S^-) rotary spectra. The spectra (corresponding to the variance per unit frequency range) for the dominant spectral components, which is typically S^- in the ocean, were then decomposed into empirical orthogonal functions by solving the standard complex eigenvalue problem. Known as *rotary empirical orthogonal function analysis*, the method is best suited to flows with strong rotary signals such as continental shelf waves and near-inertial motions but is not well suited to highly rectilinear flows such as those in tidal channels for which S^+ and S^- are of comparable amplitude (see Hsieh, 1986; Denbo and Allen, 1986).

The first use of *complex empirical orthogonal functions* in the frequency domain was described by Wallace and Dickinson (1972) and subsequently used by Wallace (1972) to study long-wave propagation in the tropical atmosphere. Early oceanographic applications are provided by Hogg (1977) for long waves trapped along a continental rise and by Wang and Mooers (1977) for long, coastal-trapped waves along a continental margin. In this approach, complex eigenvectors are computed from the cross-spectral matrices for specified frequency bands. This is the most general technique for studying propagating wave phenomena. As noted by Horel (1984), however, EOF analysis in the frequency domain can be cumbersome if applied to time series in which the power of a principal component is spread over a wide range of frequencies as a result of nonstationarity in the data. Horel presents a version of complex EOF analysis in the time domain in which complex time series of a scalar variable are formed from the original time series and their Hilbert transforms. The complex eigenvectors are then determined from the cross-correlation or cross-covariance matrices derived from the complex time series. The Hilbert transform $u_m^H(t)$ of the original time series $u_m(t)$ represents a filtering operation in which the amplitude of each spectral component remains unchanged, but the phase of each component is shifted by $\pi/2$. Because of this 90° shift in phase, the Hilbert transform is also known as the quadrature function. Expanding the scalar time series

$$u_m(t) = \sum_{\omega}[a_m(\omega)\cos(\omega t) + b_m(\omega)\sin(\omega t)] \tag{4.78}$$

as a Fourier series over all frequencies, ω, the Hilbert transform $u_m^H(t)$ is

$$u_m^H(t) = \sum_\omega [b_m(\omega)\cos(\omega t) - a_m(\omega)\sin(\omega t)] \tag{4.79}$$

In practice, the Hilbert transform can be derived directly from the coefficients of the Fourier transform of $u_m(t)$, although with the usual problems caused by aliasing and truncations effects. The complex covariance matrix $r_{mk} = \overline{U_m(t)U_k(t)^*}$ obtained for the series $U_m(t) = u_m(t) + iv_m(t)$ and its complex conjugate, $U_k(t)^*$, are shown to be useful for identifying traveling and standing wave modes; here, (u, v) are the zonal and meridional components of velocity. In the extreme case where the data set is dominated by a single frequency, the frequency domain EOF technique and complex time domain EOF technique are identical. According to Merrifield and Guza (1990), the Hilbert transform complex EOF only makes sense if the frequency distribution in the original time series $(u_m(t), v_m(t))$ is narrow band.

Extended EOFs (EEOFs) were used in metrological studies by Weare and Nasstrom (1982), and later applied to find propagating features in the upper atmosphere (Kimoto et al., 1997; Plaut and Vautard, 1994). In EEOF analysis, the state vector at time t, used in the traditional EOFs is "extended" to include temporal variations as (Hannachi, 2004)

$$\mathbf{X}_t = \left(X_{t,1}, \ldots, X_{t+M-1}; X_{t,2}, \ldots, X_{t+M-2,2}; \ldots; X_{t,p}, \ldots, X_{t-1,p}\right) \tag{4.80}$$

where $t = 1 \ldots, n - M + 1$. The parameter M is known as the window-length or delay parameter, and p refers to the number of eigenvalues or spatial EOFs in the data set. The data matrix becomes

$$\mathbf{X} = \begin{pmatrix} \mathbf{x}_1 \\ \mathbf{x}_2 \\ \ldots \\ \mathbf{x}_{n-M+1} \end{pmatrix} \tag{4.81}$$

From Eqn (4.79), we can see that time is now included with the spatial dimension. We can now write the data matrix as

$$\mathbf{X} = \begin{pmatrix} \mathbf{x}_1^1 & \mathbf{x}_1^2 & \mathbf{x}_1^p \\ \mathbf{x}_2^1 & \cdot & \cdot \\ \cdot & \cdot & \cdot \\ \mathbf{x}_{n-M+1}^1 & \mathbf{x}_{n-M+1}^2 & \mathbf{x}_{n-M+1}^p \end{pmatrix} \tag{4.82}$$

which is similar to the data matrix in Eqn (4.38a) except that now the elements of the data matrix are vectors rather than scalars. This new data matrix is of the order $(n - M + 1)pM$. The covariance matrix of (4.82) is

$$\mathbf{C} = \frac{1}{n-M+1}\mathbf{X}^T\mathbf{X} = \begin{pmatrix} C_{11} & C_{12} & \cdots & C_{1M} \\ C_{21} & C_{22} & \cdots & \cdots \\ \cdots & \cdots & \cdots & \cdots \\ C_{M1} & \cdots & \cdots & C_{MM} \end{pmatrix} \tag{4.83}$$

where each C_{ij} ($1 \leq i, j \leq M$) is a lagged covariance matrix between grid point i and grid point j given by

$$C_{ij} = \frac{1}{n-M+1} \sum_{k=1}^{n-M+1} \mathbf{X}_k^{i\,T}\mathbf{X}_k^j \tag{4.84}$$

For large values of n (as compared to the window length M) the covariance matrix is approximately a diagonal-constant matrix where each descending diagonal from left to right is constant. This is generally the case when we deal with daily observations or even monthly averages derived from more frequently sampled data.

EEOFs are the EOFs for extended versions of the data matrix given by (4.82), corresponding to the eigenvectors of the covariance matrix given by (4.83). The EEOFs can be computed directly by computing the eigenvalues/eigenvectors of (4.82) using the singular-value-decomposition method. With this formulation we can write

$$\mathbf{X} = \mathbf{V}\mathbf{A}\mathbf{U}^T \tag{4.85}$$

where $\mathbf{U} = (u_{ij}) = (\mathbf{u}_1, \mathbf{u}_2, \ldots, \mathbf{u}_d)$ represents the matrix of the d EEOFs, or the left singular vectors, of $\mathbf{X}$, and $d = Mp$ is the number of new variables represented by the number of columns in the data matrix. The diagonal matrix $\mathbf{A}$ contains the singular values $a_1, \ldots, a_d$ of **X** and **V** = $(\mathbf{v_1}, \mathbf{v_2} \ldots, \mathbf{v_d})$ is the matrix of the right singular vectors, or the extended principal components.

In summary, conventional EOF analysis in the time domain works best when the variance is dominated by standing waves and spread over a wide range of frequencies and wavenumbers. Frequency domain EOF analysis should be used when the dominant variability within the data set is concentrated into narrow frequency bands. Rotary spectral EOF analysis is best used for data sets in which the variance is in narrow frequency bands and dominated by either the clockwise or counterclockwise rotating component of velocity. Complex time domain principal component analysis allows for the detection of propagating wave features (if the process has a narrow frequency band) and the identification of these motions in terms of their spatial and temporal behavior. However, regardless of which method is applied, the best test of a method's validity is whether the results make sense physically and whether the variability is readily visible in the raw time series.

4.7.1 Applications of EEOFs

In a study of tropical disturbances using data from the Tropical Ocean Global Atmosphere—Coupled Ocean-Atmosphere Response Experiment (TOGA COARE), Fraedrich et al. (1997) used windowed, vertical time delay Extended EOFs (EEOFs) to examine connections between different modes of vertical and temporal variability in the tropical atmosphere. At the same time, the authors wanted to determine the dominant modes of variability for the wind and diabatic heating in the equatorial Western Pacific during TOGA-COARE. Focus was on observations from the TOGA-COARE Intensive Observing Period (IOP), which consisted of a set of rawinsonde soundings at seven stations in the equatorial Pacific (Figure 4.28). The data were extracted from the Australian Bureau of Meteorology Tropical Analysis and Prediction System (TAPS). Additional ship and island rawinsondes were collected during the IOP and were added to the dataset. Some of the data were acquired from the National Center for Atmospheric Research (NCAR) along with a large number of surface reports, mostly from National Climate Center of the Bureau of Meteorology.

All soundings were quality controlled to remove highly suspicious values. In order to remove the effects of surface winds measured at island stations, surface wind speeds were estimated as 72% of the wind speeds at the 850 hPa pressure level and all data averaged over 24 h to produce daily mean fields from the 6 hourly sonde launches. Only data from the 11 standard levels (surface, 1,000, 850, 700, 500, 400, 300, 250, 200, 150, and 100 hPa) were used. Missing daily values (roughly 5% of the data set) were filled using a horizontal, least-squares fit (Frank, 1979). This procedure produced a complete dataset at all stations at the 11 standard levels for all 120 days of the IOP.

The four variables used in this study were daily averages of zonal and meridional winds, along with heat and moisture fluxes in and out of the atmosphere. Because the heat and moisture budgets were based on daily averages over a fairly large area, they are representative of the net effects of many of the clouds and mesoscale connective systems that were incorporated into the averaged calculations. Extended EOF analysis was applied to the data, which was able to delineate the temporal evolution of the spatial patterns in these datasets. Consider a system described by a vector of K components ordered according to height and evolving with time. The vector series is weighted using a sliding window of length W. This generates a new vector series whose components constitute a height-time section, KXW, which evolves in time and

FIGURE 4.28 The OSA-S rawinsonde array (solid line) of TOGA COARE stations over which the heating, drying, and wind time series are calculated. *From Frank et al. (1996).*

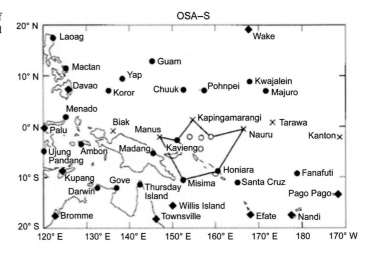

represents the states of the system in KXW-dimensional phase space. Fraedrich et al. (1997) note that the following should be considered when applying this method:

1. Use of a sliding window results in some smoothing. The degree of smoothing depends on the window length, whereby a longer window provides a smoother reconstruction. An optimal choice of the window length (*W*) is the half-period of the longest signal. This resolves the signal with sufficient detail while providing sufficient data smoothing to increase the signal-to-noise ratio.
2. Wavelike oscillations are represented by a pair of EEOFs with similar eigenvalues, with the associated EEOF patterns shifted by quarter of a wavelength. In this way, EEOFs are able to properly represent propagating waves, whereas standard EOFs can only represent standing waves.
3. The advantage of EEOFs is their ability to: (a) represent dominant internally coherent patterns in both space and time; and (b) to distinguish oscillatory components with a variety of frequencies. These oscillatory components are characterized by pairs of EEOFs in quadrature and which are associated with eigenvalues of similar magnitude.

The first two EEOFs of the covariance matrix of the wind (*U, V*) height-time series are presented in the top frames of Figure 4.29. The two EEOFs explain 48.3% of the variance in the wind (Table 4.8). Here, we note that, for each mode, the variations in *U* are much greater than those in *V*.

In Figure 4.29, the wind EEOFs are contoured and the corresponding *U,V* velocity vectors are plotted over these contour fields. EEOF1 shows a vertical wavenumber-one structure with the zonal wind varying most near the tropopause and near 800 hPa. EEOF2 is similar to EEOF1 but its phase is in quadrature with EEOF1. This suggests that this pair of EEOFs represent a single mode of variation. As indicated in Figure 4.29, this variation is approximately twice the window length, or about 40 days, and likely dictated by the 40−60 day Madden-Julian oscillation known to dominate these time-space series.

The first two modes of the heat-flux (*Q₁, Q₂*) height-time series are shown in the lower panels of Figure 4.29 and their contributions to the variance given in Table 4.8. As this table shows, the first two EEOFs explain 41% of the variance in the time-space series of heat flux. As with the wind EEOFs, the two heat-flux modes are similar to each other and shifted in phase by roughly 90°. Again, the time scale is approximately twice the window size, or about 40 days.

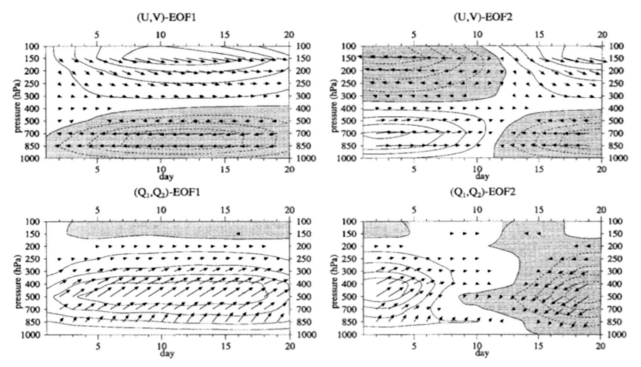

FIGURE 4.29 The first EOF pairs of the (*U, V*) and (*Q₁, Q₂*) height-time series with a 20-day window. Top panels: the vectors represent wind direction and strength. The contours are *U*-component (negative values are shaded); units are in meters per second. Bottom panels: The vectors represent (*Q₁, Q₂*) components with positive *Q₁* upward and positive *Q₂* to the right. The contours are *Q₁* component (negative values are shaded); the units are in degrees per day.

TABLE 4.8 Percent variance (left) and the running total (right) for each of the two pairs of columns contributed by the first six EOFs in the height–time delay analysis of the total (U, V) and (Q_1, Q_2) variances.

EOF	(U, V)(%)	Wind (total %)	(Q_1, Q_2)(%)	Heating (total %)
1	27.3	27.3	28.4	28.4
2	21.0	48.3	12.2	40.6
3	7.7	56.0	7.9	48.5
4	7.1	63.1	7.6	56.1
5	6.8	69.9	4.7	60.8
6	3.4	73.3	3.2	64.0

Variations in Q_1 and, to a lesser extent, in Q_2 are dominated by variability in the vertical velocity field; the vertical structure of these quantities corresponds to an internal wavenumber-one structure in divergence. A similar pattern persists for the high-order EEOFs (not shown) suggesting that there is no significant difference in vertical heating as a function of wavenumber.

One unique aspect of the EEOF analysis is the ability to view the phase relationship between the EEOF modes. As noted, earlier EEOF1 and EEOF2 appear to be similar but phase shifted. This can be seen more clearly in Figure 4.30 where EEOF1 is plotted against EEOF2. In the top-left panel it is clear that EEOF structure is similar, and the phase shift results in an almost circular rotation between the paired EEOF modes for the wind components. Comparison with the lower left panel reveals that this relationship is relatively independent of window size as the only change here is an increase in window size from 20 to 30 days. It is also interesting that the heating EEOFs are similar for the two window sizes and while they are noisier than for the wind, they also exhibit this rotational behavior consistent with a shift in phase between the two EEOF modes. The period of the oscillations is the time required for a complete revolution in these diagrams. For the wind (Figure 4.30) the vertical axis is crossed at days 40 and 80, while the horizontal axis is crossed at days 49 and 88 resulting in an overall period of 39–40 days. Similarly, the heating diagrams yield a period of about 41 days.

In this study the extended EOF analysis has revealed a dominant pattern of space-time variation in both wind and heating-drying with a period of approximately 40 days. This result is very consistent with the temporal period discussed for TOGA-COARE by Gutzler et al. (1994) and McBride et al. (1995). All of these authors associate this variation with the Madden-Julian oscillation. The structure of this oscillation as seen in EEOFs 1 and 2 is dominated by zonal wind variations with a vertical structure of the first internal mode. This zonal wind pattern is consistent with an equatorially trapped Kelvin wave, but the node is not totally a Kelvin wave.

In their analysis of Pacific Ocean sea surface temperature (SST) anomalies, Weare and Nasstrom (1982) present the first and/or most important EEOF at three different times (Figure 4.31). The features are consistent with the standard EOF fundamental modes, which have been associated with El Niño warming of the eastern equatorial Pacific (Weare et al., 1981). The features in Figure 4.31 show remarkable persistence over time, which also agrees with previous studies of equatorial Pacific behavior. Along with this strong persistence, Figure 4.31 suggests a shift of the maximum variability along the equator westward from the South American coast during the 6-month period examined in this study, which also agrees with previous studies of areal averaged data in this region (Weare, 1982).

The second most important EEOF of Pacific SST anomalies (Figure 4.32) shows greater differences in the 6-month sequence than are apparent in the first mode in Figure 4.31. As time processes, there is an extension westward along the equator of the positive values near Peru at a speed of about 0.25 m/s. This speed is in good agreement with estimates of the ocean current speeds in this region (Wyrtki, 1977). During this same period, a large negative feature appears to recede northward at a somewhat slower speed. Figure 4.32 suggests that SST anomalies associated with the Peru-South Equatorial Current system are often opposite in sign to changes in the California-North Equatorial Current system. Since both of these systems advect relatively cold water, this $180°$ (out-of-phase) relationship suggests that the southern currents weaken while the northern currents strengthen, and vice versa.

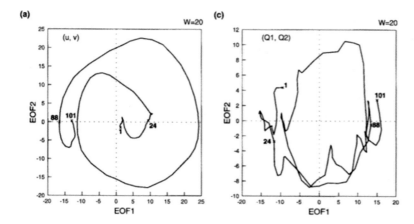

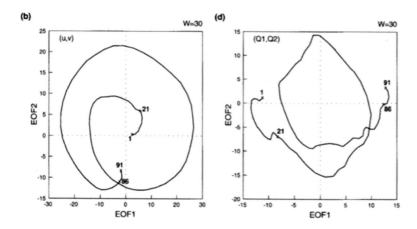

FIGURE 4.30 Dial of the principal components of the first height-time delay eigenvector pair (EOF 1 and EOF 2) for the two window lengths ($W = 20$, 30 days): (a), (b) the U, V winds and (c), (d) the (Q_1, Q_2) heating. Particular day numbers indicated are discussed in the text.

4.8 CYCLOSTATIONARY EMPIRICAL ORTHOGONAL FUNCTIONS (CSEOFS)

In EOF analysis, a set of orthogonal eigenfunctions is found from a spatial covariance function. Data are decomposed into the sum of a set of individual modes composed of a single spatial pattern and a corresponding amplitude time series, referred to as the "loading vectors" (LVs) and "principal component time series" (PCTS), respectively. These empirically derived basis functions can provide useful insight into the physical processes behind the data and serve as a useful tool for developing statistical methods. The underlying assumption in EOF analysis is that the data being analyzed are stationarity so that the covariance function of the data does not depend on time. By definition, the spatial patterns represented by the EOF LVs are time independent (stationary) so that only the amplitudes of these stationary patterns vary in time, as described by the PC time series. However, geophysical variables, including climate-related variables, are rarely stationary even after the removal of cyclic components like the annual and semi-annual cycles. Subsequently, physical inferences based on EOFs of climate signals can be misleading and potentially erroneous. The spatial patterns of many phenomena in geophysics and climate science show the presence of seemingly random fluctuations in addition to a deterministic component such as the annual cycle. Such signals change in time with well-defined periods (deterministic components) in addition to fluctuating at longer timescales and are thus best described by time-dependent covariance functions. These signals are said to be periodically correlated or cyclostationary. Because physical systems are generally not stationary but evolve and change over time, a suitable representation of this time-dependent response is important for the extraction of physically meaningful modes and their space-time evolutions from the data.

The decomposition in terms of a set of basis functions is often useful in understanding the complicated response of a physical system. When decomposed into simpler, basic patterns, insight can be gained into the nature of variability of a given system. While theoretical basis functions have been studied extensively, exact theoretical basis functions are very

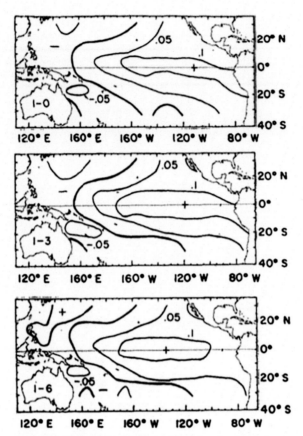

FIGURE 4.31 The first (and most important) "extended" empirical orthogonal function of monthly departures of tropical Pacific Ocean surface temperature for times t, $t + 3$, and $t + 6$ months, reading from top to bottom.

difficult to find and in general, computational basis functions are sought instead. Perhaps the simplest and most common computational basis functions are empirical orthogonal functions. Consider a simple system defined by:

$$T(x,t) = B(x,t)S(t) \tag{4.86}$$

where $B(x,t)$ is a deterministic physical process that is modulated by a stochastic time series process, $S(t)$. It follows that the mean and the space–time covariance function are given by:

$$\mu(x,t) = \langle T(x,t) \rangle = B(x,t)\langle S(t) \rangle = B(x,t)\mu_s \tag{4.87}$$

$$C(x,t;x',t') = \langle T(x,t)T(x',t') \rangle = B(x,t)B(x',t')R_s(\tau) \tag{4.88}$$

where, and μ_s and R_s are the mean and autocovariance function of the stochastic component, $S(t)$, respectively. Thus, the first two moment statistics are time dependent in the presence of a time-dependent physical process $B(x,t)$. This time dependence of the statistics is due, in theory, to the physical component $B(x,t)$ and not to the stochastic component $S(t)$, which is assumed to be stationary over the time scales of interest.

There are many observational examples that suggest geophysical processes and the corresponding statistics are time dependent. In EOF analysis, the response characteristics of a physical process are assumed to be stationary and therefore not dependent on time. In light of the evidence from geophysical observations, the assumption of stationarity restricts the investigator's ability to interpret certain physical signals. The question arises, then, how can time-dependent characteristics be properly accounted for when attempting to compute basis functions that are assumed to be representative of the variability of a physical signal? If the covariance function is time dependent, computational eigenfunctions can be given by the solution of the Karhunen-Loeve equation (Loeve, 1978):

$$\int_D \int_T C(x,t;x',t')B_n(x',t')\,dt'\,dx' = \lambda_n B_n(x,t) \tag{4.89}$$

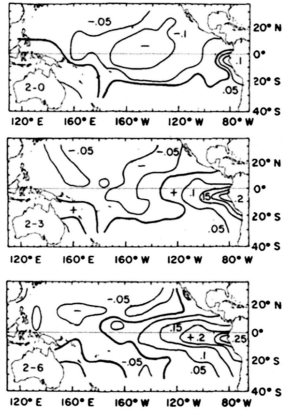

FIGURE 4.32   As in Figure 4.31 but for the second most important function.

where D and T are space and time domains, respectively. Unfortunately, the solution to (4.89) is computationally intensive and not practical to derive. To address this problem, a simplification can be introduced known as the assumption of cyclo-stationarity. Eqn (4.86) can be re-written under the assumption that the response characteristics of the physical process are periodic in time. That is:

$$B(x,t) = B(x,t+d) \tag{4.90}$$

where d is the "nested" periodicity of the process. The periodicity assumption is valid in many cases since many observed physical processes oscillate with a well-defined period. The two moment statistics can then be shown to be periodic:

$$\mu(x,t) = \langle T(x,t) \rangle = \langle T(x,t+d) \rangle = \mu(x,t+d) \tag{4.91}$$

$$C\left(x,t;x',t'\right) = \left\langle T(x,t+d)T\left(x',t'+d\right) \right\rangle = C\left(x,t+d;x',t'+d\right) \tag{4.92}$$

Derivations of the above moment statistics requires that the stochastic component, $S(t)$, in (4.86) be stationary. A physical process that satisfies Eqns (4.91) and (4.92) is said to be "cyclostationary". The stationary case, for which $d = 1 \cdot \Delta t$ shows that stationarity is a special case within the cyclostationary framework. With the assumption of cyclo-stationarity, finding eigenfunctions as solutions to Eqn (4.88) becomes computationally tractable. The resulting eigen-functions are also periodic in time with the same period as the corresponding statistics, leading to the definition of cyclostationary empirical orthogonal functions (CSEOFs) (Kim et al., 1996; Kim and North, 1997). In CSEOF analysis, space-time data are written as:

$$T(x,t) = \sum_n B_n(x,t)P_n(t) \tag{4.93}$$

$$B_n(x,t) = B_n(x,t+d) \tag{4.94}$$

where $B(x,t)$ are CSEOF loading vectors (LVs) and $P(t)$ are corresponding principal component time series (PCTS). Each eigenfunction represents not just one spatial pattern but also multiple spatial patterns, which repeat in time. For example, when using monthly data and a nested period (d) of 1 year, the resulting LVs will be composed of twelve separate spatial patterns, one for each month of the year. In contrast to EOF analysis, the temporal variation of the data in CSEOF analysis has two distinct components: the time-dependent physical process, $B(x,t)$, and the stochastic undulation of the physical processes, $P(t)$. For example, when considering the annual seasonal signal, $B(x,t)$ represents the spatial pattern varying over the course of the year, while $P(t)$ represents the interannual variability in the amplitude of the seasonal signal.

While the assumption of periodic statistics is reasonable for many geophysical variables, it can be difficult to prove this periodicity and subsequently choose the nested period, d, for the CSEOF decomposition. The nested period must be determined based on *a priori* physical understanding of the process being investigated. In many cases, there exists an obvious choice for the nested period. For instance, if one is studying the annual seasonal cycle, the nested period would obviously be 1 year. Sometimes, however, the period of the physical process of interest is not as obvious. For example, the El Niño-Southern Oscillation (ENSO) signal in the Equatorial Pacific does not have a well-defined period, but instead, has cyclicity of somewhere in the approximate range of two to 5 years. There is also the problem of selecting the nested period if one is studying several different geophysical signals, with a range of periods. In general, the nested period should be selected as the least common multiple of the periods of signals of interest. For instance, if there is a dataset in which semi-annual, annual and biennial periodic signals are all present, the nested period should be set at 2 years. The semi-annual cycle LVs would simply repeat four times, while the annual cycle LVs would repeat twice.

The concept of CSEOF analysis was first developed as a way to describe climatic time series with well-defined periods but unpredictable amplitude fluctuations. One of the first published applications involved performing a CSEOF decomposition of the globally averaged surface air temperature field (Kim et al., 1996). More recently, CSEOFs have been used to extract the annual cycle from the sea surface temperature field of the tropical Pacific (Kim and Chung, 2001). This analysis was able to accurately explain the detailed structure and temporal modulation of the annual cycle. Similar work has been completed on the satellite altimetry sea level record, demonstrating the ability to use CSEOFs to extract not only the modulated annual cycle but also the ENSO signal that is present in sea level data (Hamlington et al., 2011).

In the study by Hamlington et al. (2011), the CSEOF technique was applied to the quarter-degree resolution AVISO gridded multiple-altimeter global dataset composed of sea level measurements spanning the years 1993−2008. Using CSEOF analysis, it was possible to extract the time-variant (modulated) annual cycle in the sea level data. In order to distinguish the variability associated with the annual cycle, a nested period of 12 months was used in the CSEOF analysis. The annual cycle is described by the first CSEOF mode as it is the dominant signal in the sea level data. The top panels in Figure 4.33 show the time-dependent LVs associated with the annual cycle, while the bottom panel shows the principal component time series (PCTS) of the CSEOF mode. The PCTS does not exhibit the annual cycle period of 12 months, which is instead described by the LVs that are required to have a period of 1 year. The PCTS describes the longer timescale fluctuations of the annual cycle. The fluctuations of the annual PCTS reflect variations of the strength of the annual cycle about some mean amplitude. The amplitude of the annual cycle varies within 20% of the mean, with values that are less than the mean, indicating weaker than normal annual cycles and values greater than the mean, indicating stronger than normal annual cycles.

The PC time series shows interannual variability that can be related to ENSO. The El Niño phase of ENSO tends to weaken the annual cycle and a weak negative correlation is observed between the PC time series and an ENSO index, such as the multivariate ENSO index (MEI) (Wolter and Timlin, 1998). The MEI can be understood as a weighted average of the main ENSO features contained in six different variables and serves as a tool for monitoring ENSO events. In addition to its relationship with ENSO, the CSEOF mode also contains oscillations with the period of 6 months, which are likely due to the semiannual cycle present in sea level time series. By combining the loading vectors and the principal component time series and then averaging, a global mean sea level (GMSL) time series associated with the annual cycle can be formed (Figure 4.33b). The 12-month periodicity is clear from this figure, as well as the time-varying amplitude of the GMSL annual cycle that is produced by the CSEOF analysis.

In addition to the ability to extract a modulated annual cycle, another advantage of the CSEOF technique comes from the potential to extract physically interpretable modes with less, albeit still significant, variability. The second CSEOF mode is shown in Figure 4.34. The top panels show the temporally varying loading vectors while the bottom panel shows the associated PCTS. After plotting the MEI with the global mode associated with the second CSEOF mode as seen in Figure 4.34, the significance of this mode becomes clear. With a correlation of 0.80 between the MEI and global mean time series of the CSEOF mode and by looking at the spatial patterns of the loading vectors, it is clear that the second CSEOF represents the ENSO variability in the dataset. The ability to extract a mode directly related to the ENSO signal represents one of the strengths of the CSEOF method.

In summary, the usual assumption of stationarity when applying EOF analysis is often not justifiable. Cyclostationarity may be a better assumption for a wide range of geophysical processes, including climate signals such as those discussed in

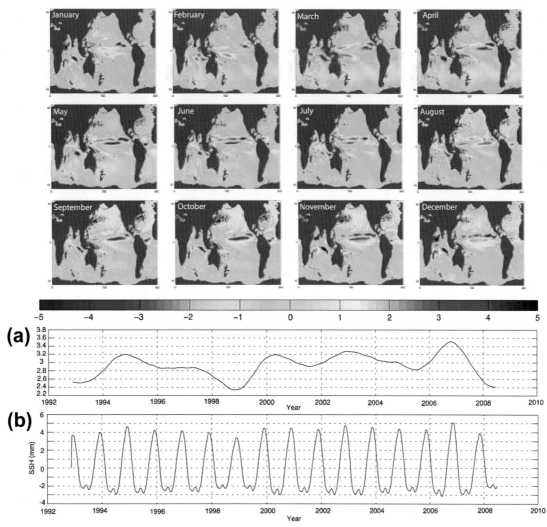

FIGURE 4.33 CSEOF mode 1 representing the modulated annual cycle (MAC) from the AVISO satellite altimetry data set. The color panels show the monthly time-dependent CSEOF LVs, (a) the principal component time series (PCTS), and (b) the reconstructed mode's contribution to global mean sea level in mm (GMSL). *From Hamlington et al. (2011).*

the above example using sea level data. CSEOF analysis finds computational modes of a cyclostationary process. By accounting for the time-dependent response of geophysical signals, clearer and more interpretable information can be extracted regarding the underlying physical processes of a dataset. As with any other analysis technique, CSEOF is based on underlying assumptions, which lead to limitations on the capabilities of the analysis. However, when used in the right context, the CSEOF technique can be a significant improvement over commonly used techniques founded in the assumption of stationarity.

4.9 FACTOR ANALYSIS

As discussed in Preisendorfer (1988), Factor Analysis (FA) can be considered as the generalization of Principal Component Analysis (PCA) and linear regression analysis (LRA). For any given $n \times p$ dataset $\mathbf{Z}$, we can perform either a PCA or an FA. The PCA is the simpler of the two analyses while FA is the conceptually more complex of the two analysis methods. FA is a form of a linear statistical model that can hypothesize about phenomena underlying the dataset $\mathbf{Z}$ while PCA makes no hypotheses about the linearity of the underlying phenomena, and it is basically statistical in character. In spite of these fundamental differences between PCA and FA, their algebraic forms look alike. They are, however, methodologies for reaching quite different goals. If the analyst is interested in isolating the sources of the

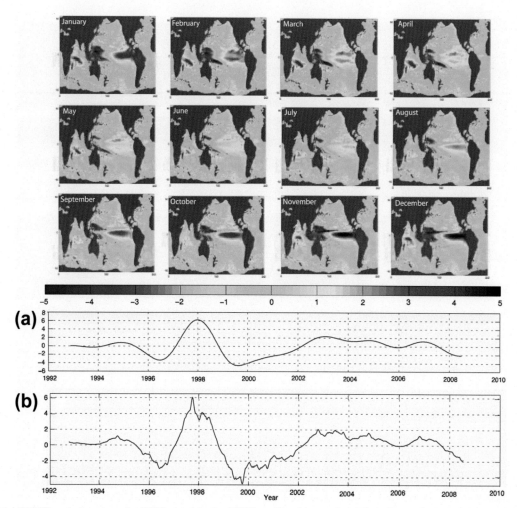

FIGURE 4.34 CSEOF mode 2 captures the ENSO signal in the AVISO satellite altimetry data. The color panels show the monthly time-dependent CSEOF LVs, (a) the PC time series; and (b) and the reconstructed mode's contribution to global mean sea level in mm (GMSL). *From Hamlington et al. (2011).*

data's variability, he or she will use PCA. If instead, the investigator wished to study the sources of data covariability, he or she will use FA.

Factor Analysis itself is a statistical method used to describe variability among observed correlated variables in terms of a potentially lower number of unobserved variables, called *factors*. Factor Analysis searches for potential joint variations. The observed variables are modeled as linear combinations of the potential factors plus errors. The information gained about the interdependencies between observed variables can be used to reduce the set of variables of a dataset. Factor analysis originated in psychometrics and is used largely in the behavioral and social sciences, marketing, product management, operations research and other applied sciences that deal with large quantities of data.

4.10 NORMAL MODE ANALYSIS

In the previous sections, we were concerned with the partition of data variance into an ordered set of spatial and temporal statistical modes and maps. The eigenvalue problem associated with EOF modes was solved with little consideration given to the underlying physics of the oceanic system. In contrast, normal mode decomposition takes into account the physics and associated boundary conditions of the fluid motion. A common approach is to separate the vertical and horizontal components of the motion and to isolate the forced component of the response from the freely propagating response. As illustrations of these techniques, we consider two basic types of normal mode, eigenvalue problem:

(1) The calculation of vertical normal modes (eigenfunctions), $\psi_k(z)$, for a stratified, hydrostatic fluid with specified top and bottom boundary conditions; and

(2) the derivation of the cross-shore orthogonal modes (eigenfunctions), $\phi_k(x, z)$, for coastal-trapped waves over a variable depth, stratified ocean with or without a coastal boundary.

The first problem can be solved without including the earth's rotation, f, while the second problem requires specification of f. Both eigenvalue problems yield solutions only for certain eigenvalues, λ_k, of the parameter, λ.

4.10.1 Vertical Normal Modes

A common oceanographic problem is to find the amplitudes (a_k) and phases (θ_k) of a set of K orthogonal basis functions, or modes, by fitting them to a profile of M ($>K$) observed values of amplitude and phase. For instance, one might have observations from $M = 5$ depths and want to find the modal parameters (a_k, θ_k) for the first three theoretical modes, $k = 1$, 2, 3, derived from an analysis of the equations of motion. Once the set of theoretical modes are derived, they can be fitted using a least-squares regression technique to observations of the along-channel (or cross-channel) current amplitude and phase. This yields the required estimates, (a_k, θ_k), for $k = 1$, 2, 3.

To obtain the vertical normal modes for a nonrotating fluid ($f = 0$), we assume that the pressure, p, density, ρ, and horizontal and vertical components of velocity (u, v) and w, respectively, can be separated into vertical and horizontal variables. This separation of variables has the form

$$[u(\mathbf{x}, t), v(\mathbf{x}, t), p(\mathbf{x}, t)/\rho_0] = \sum_{k=0}^{\infty} P_k(x, y, t)\psi_k(z)$$

$$w = \sum_{k=0}^{\infty} \left[w_k \int_{-H}^{z} \psi_k(z)dz \right] \tag{4.95b}$$

$$\rho = \sum_{k=0}^{\infty} \rho_k \frac{d\psi_k(z)}{dz} \tag{4.95c}$$

where $k = 0$, 1, 2, ... is the vertical mode number, P_k corresponds to one of the variables listed on the left side of the equation, and the variables without subscripts are functions of $(\mathbf{x}, t) = (x, y, t)$. Substituting these expressions into the usual equations of motion (see LeBlond and Mysak, 1979; Kundu, 1990), we obtain the *Sturm-Liouville equation*

$$\frac{d}{dz}\left(\frac{1}{N^2}\frac{d\psi_k}{dz}\right) + \frac{1}{c_k^2}\psi_k = 0 \tag{4.96}$$

where $N(z) = [-(g/\rho)d\rho/dz]^{1/2}$ is the Brunt-Väisälä frequency, c_k^2 are the separation constants and the $1/c_k^2$ are the eigenvalues, λ_k. (Note: as there is no confusion with N standing for the number of values in a sequence, we use the common notation, $N = N(z)$, for the Brunt-Väisälä frequency).

In the case of a rotating fluid (i.e., $f \neq 0$), we assume $N(z)$ is uniform with depth and, following some reorganization of terms, replace N^2/c_k^2 in Eqn (4.96) as follows:

$$N^2 / c_k^2 \rightarrow (N^2 - \omega^2)/gh_k, \quad k = 1, 2, ... \tag{4.97a}$$

where h_k is an "equivalent depth", ω is the wave frequency

$$gh_k = (\omega^2 - f^2)/l^2 + q^2 = c_k^2 - f^2/l^2 \tag{4.97b}$$

and (l, q) are the wavenumbers in the horizontal (x, y) directions. Wave-like solutions are possible provided that $f^2 < \omega^2 < N^2$. For a rectangular channel of width L, the cross-channel wavenumber $q \rightarrow q_m = m\pi/L$ and solutions must be considered for both k, $m = 1$, 2, ... (Thomson and Huggett, 1980). For both the rotating and nonrotating case, solutions to the eigenvalue problem (4.96) are subject to specified boundary conditions at the seafloor ($z = -H$) and the upper free surface ($z = 0$) of the fluid. These end-point boundary conditions are:

$$\frac{d\psi_k}{dz} = 0 \quad (i.e., w = 0) \text{ at } z = -H \tag{4.98a}$$

$$\frac{d\psi_k}{dz} + \frac{N^2}{g}\psi_k = 0 \left(i.e., \frac{\partial p}{\partial t} = \rho g w \right) \text{ at } z = 0 \tag{4.98b}$$

Modal analysis of the type described by (4.96)–(4.98) is valid only for an inviscid hydrostatic fluid in which oscillations occur at frequencies much lower than the local buoyancy frequency, N, and for which the vertical length scale is much smaller than the horizontal length scale. In addition, the ocean must be of uniform depth and have no mean current shear. (For sloping bottoms, the horizontal cross-slope velocity component, u, is linked to the vertical boundary, w, through the bottom boundary condition $u = -w \, dH/dx$ and separation of variables is not possible). The method can be applied to an ocean with zero rotation or with rotation that changes linearly with latitude, y. Solutions to (4.96) are obtained for specified values of $N(z)$ subject to the surface and bottom boundary conditions. Although the individual orthogonal modes propagate horizontally, the sum of a group of modes can propagate vertically if some of the modes are out of phase.

Analytical solutions: Simple analytical solutions to the Sturm-Liouville Eqn (4.96) are obtained with and without rotation when $N = $ constant (density gradient uniform with depth). Assuming the rigid lid condition (i.e., no surface gravity waves so that $w = 0$ at $z = 0$), the vertical shapes of the orthogonal eigenfunctions $\psi_k(z)$ in (4.96) are given by

$$\psi_k(z) = \cos(k\pi z/H), \quad k = 0, 1, 2, \ldots \tag{4.99}$$

where $k = 0$ is the depth-independent barotropic mode, and $k = 1, 2, \ldots$ are the depth-dependent baroclinic modes. The kth mode has k zero crossings over the depth range $-H \leq z \leq 0$ and satisfies the boundary conditions $w = 0$ (cf. 4.98a). Phase speeds (eigenvalues) of the modes are given by

$$c_o = (gH)^{1/2}, k = 0 \text{ (barotropic mode)} \tag{4.100a}$$

$$c_k = NH / k\pi, k = 1, 2, \ldots \text{(barotropic mode)} \tag{4.100b}$$

In general, $N(z)$ is nonuniform with depth and, for a given k, the solutions will have the form

$$c_k = (gh_k)^{1/2} \tag{4.101}$$

where the "equivalent depth" h_k is used in analogy with H in (4.100a). For an ocean of depth $H \approx 2{,}500$ m and buoyancy frequency $N \approx 2 \times 10^{-3}$/s, the eigenvalue for the first baroclinic mode has a phase speed $c_1 \approx 1.6$ m/s and the equivalent depth $h_k = c_1^2/g \approx 0.26$ m. For a 400-m deep tidal channel, we find $N \approx 5 \times 10^{-3}$ m/s, $c_1 \approx 0.8$ m/s and $h_k \approx 0.06$ m.

General solutions: To solve the general eigenvalue problem (4.96)–(4.98) for variable buoyancy frequency, $N(z)$, we resort to numerical integration techniques for ordinary differential equations with two-point boundary conditions. That is, given the start and end values of the function $\psi_k(z)$, and variable coefficient $N(z)$ we seek values at all points within the domain ($-H \leq z \leq 0$). Fortunately, there exist numerous packaged programs for finding the eigenvectors and eigenvalues of the Sturm-Liouville equation for specified boundary conditions. For example, the NAG routine D02KEF (Nag Library Routines, 1986) and MATLAB finds the eigenvalues and eigenfunctions (and their derivatives) of a regular singular second-order Sturm-Liouville system of the form

$$\frac{d}{dz}\left[F(z)\frac{d\psi_k}{dz}\right] + G(z; \lambda)\psi_k = 0 \tag{4.102}$$

together with boundary conditions

$$z_{a2}\psi_k(z_a) = z_{a1}F(z_a)d\psi_k(z_a) / dz \tag{4.103a}$$

$$z_{b2}\psi_k(z_b) = z_{b1}F(z_b)d\psi_k(z_b) / dz \tag{4.103b}$$

for real-valued functional coefficients F and G on a finite or infinite range, $z_a < z < z_b$. Provision is made for discontinuities in F and G and their derivatives. The following conditions hold on the function coefficients:

(1) The function $F(z)$, which equals $1/N^2(z)$ in the case of (4.96), must be nonzero and of one sign throughout the closed interval $z_a < z < z_b$. This is certainly true in a stable oceanic environment where $N^2 > 0$; for $N^2 < 0$, the fluid is gravitationally unstable and vertical modes are not possible.

(2) $\partial G/\partial \lambda$ must be of constant sign and nonzero throughout the interval $z_a < z < z_b$ for all relevant values λ, and must not be identically zero as z varies for any relevant value of λ.

Numerical solutions to the Sturm-Liouville equation are obtained through a Prüfer transformation of the differential equations and a shooting method. (The shooting method and relaxation methods for the solution of two-point boundary value problems are described in *Numerical Methods* (Press et al., 1992)). The computed eigenvalues are correct to within a certain error tolerance specified by the user. Eigenfunctions $\psi_k(z)$ for the problem have increasing numbers of inflection points and zero crossings within the domain $z_a < z < z_b$ as the eigenvalue increases. When the final estimate of λ_k is found by the shooting method, the routine D02KEF integrates the differential equation once more using that value of λ_k and with initial conditions chosen such that the integral

$$I_k = \int\limits_{z_a}^{z_b} [\psi_k(z)]^2 \partial G / \partial \lambda(z; \lambda) \, dz \qquad (4.104)$$

is roughly unity. When $G(z; \lambda)$ is of the form $\lambda w(z) + \psi(z)$, which is the most common case, I_k represents the square of the norm of ψ_k induced by the inner product

$$\overline{\psi_k(z)\psi_m(z)} = \int\limits_{z_a}^{z_b} \psi_k(z)\psi_m(z)w(z) \, dz \qquad (4.105)$$

with respect to which the eigenfunctions are mutually orthogonal if $k \neq m$. This normalization of ψ for $k = m$ is only approximate but typically differs from unity by only a few percent.

If one is working with observed density (σ_t) profiles for the region of interest, a useful approach is to solve the Sturm-Liouville equation using an analytical expression for $N(z)$ by fitting a curve of the type $\sigma_t(z) = [\rho(z) - 1]10^3 = \sigma_o \exp[a/(z + b)]$ or other exponential form, to the data. The eigen (modal) analysis is fairly insensitive to small changes in density so that, even though changes in $N(z)$ are large in the upper oceanic layer, it is usually possible to get by with a simple analytical curve fit. Alternatively, we can specify the actual density on a numerical grid for which modes are to be calculated. Once $N(z)$ is available, we can use numerical methods to solve (4.96) subject to the boundary conditions (4.98), allowing for specified error bounds or degree of convergence on the final boundary estimate. Based on the analytical solutions (4.99), the investigator can expect solutions ψ_k to resemble cosine functions whose vertical structure has been distorted by the nonuniform distribution of density along the vertical profile. There is a direct analogy here with the modes of oscillation of a taut string clamped at either end and having a nonuniform mass distribution along its length.

The normal modes are normalized relative to their maximum value and then fitted to the data in a least-squares sense (Table 4.9). If there are M current meters on a mooring string, the maximum possible number of normal baroclinic modes is $M - 1$. By comparing the normal modes with the data, we can derive the absolute values of the barotropic mode and a maximum of $M - 1$ baroclinic modes. Solutions to the least-squares fitting are described in (Press et al., 1992).

4.10.2 An Example: Normal Modes of Semidiurnal Frequency

Suppose that the along-axis semidiurnal currents, v, in a tidal channel have the form $v_m = a_m \cos(\omega t + \theta_m)$, where t is the time, and a_m, θ_m ($m = 1, ..., M$) are the observed current amplitude and phase, respectively. In terms of tidal current ellipses, we can think of v as the major axis of the current ellipse for each current meter on the mooring line. The oscillations have frequency $\omega = \omega_{M_2}$ corresponding to M_2 semidiurnal tidal currents and the phase, θ, is referenced to

TABLE 4.9 Model amplitudes (cm/s) and phases (degrees relative to 120°W longitude) for Johnstone Strait M_2 tidal currents computed from nine-day current meter records.

Site	M	K_v (cm²/s)	Before (a_0, θ_0)	After (a_0, θ_0)	Before (a_1, θ_1)	After (a_1, θ_1)	Before (a_2, θ_2)	After (a_2, θ_2)
CM13	3	15	42, 55°	42, 55°	12, 172°	25, 171°	–	19, −10°
CM14	3	8	35, 51°	35, 51°	11, 169°	15, 171°	–	8, −4°
CM15	3	13	32, 35°	32, 36°	18, 175°	12, 166°	–	7, −31°
CM02	5	0	36, 42°	NC	21, 220°	NC	9, 13°	NC
CM04	4	7	29, 45°	50, 24°	13, 215°	79, 174°	2, −34°	70, 0°

Column two gives the number of current meters (CM) on the string. The first column or the barotropic mode (a_0, θ_0) and each of the two baroclinic modes (a_k, θ_k), $k = 1, 2$, gives the amplitude and phase (a, θ) before and after the bottom current meters is included in the analysis. The bottom current is included after its amplitude and phase are corrected for bottom boundary layer friction. The vertical eddy viscosity K_v is that value which gives the minimum ratio between the first and second baroclinic modes when the frictionally corrected bottom current meter is included. NC means "no change", implying perfect modal fit for all depths with and without bottom current meter record. At CM04, no near-surface current meter was deployed and the records were only 5 days long and therefore suspect.

some specific time zone or meridian of longitude so that we can intercompare values for different current meters and for the astronomical surface tide. The values a_m, θ_m for the different current meter records can be determined using harmonic analysis techniques (Foreman, 1977; Pawlowicz et al., 2002) provided the measured data are at hourly (or other equally spaced) intervals over a period of 7 days or longer so that the M_2 and K_1 constituents are separable. We next rewrite the above expression for v in the usual way as $v_m = A_m \cos(\omega t) + B_m \sin(\omega t)$, where $\tan\theta_m = A_m/B_m$ and $a_m^2 = (A_m^2 + B_m^2)$. This allows us to examine the sine and cosine components separately. The observed magnitudes A_m and B_m at each current meter depth z_m, $m = 1, \ldots, M$ are then used to compute the amplitudes and phases of the basis functions $\psi_k(z_m)$, for a maximum of K different modes ($K < M$). At best, we can obtain the amplitudes and phases of the barotropic mode ($k = 0$) and up to $M - 1$ baroclinic modes.

Details of the modal analysis at semidiurnal frequency using current meter data from a tidal channel are presented by Thomson and Huggett (1980). The first step is to obtain an exponential functional fit (Figure 4.35a) to the observed mean density structure, $N(z)$. This structure is then used with the local water depth H (assuming a flat bottom), the Coriolis parameter, f, and the wave frequency, ω, to calculate the theoretical dynamic modes (Figure 4.35b). A finite sum of these theoretical modes $\Sigma\psi_k(z)$ is then least-squares fitted to the observed cosine component $A_m(z)$ to obtain estimates of the contributions A_k from each mode, k. This operation is repeated for the sine component B_k. (Recall that the maximum total of barotropic plus baroclinic modes allowed in the summation is fewer than the number of current meter records per mooring string and that the vertical structure of each mode is found through the products $(A_k, B_k)\psi_k(z)$ where the coefficients are constant). Using the relationships $\tan\theta_k = A_k/B_k$ and $a_k^2 = (A_k^2 + B_k^2)$, we obtain the amplitudes and phases of the various modes. In their analysis, Thomson and Huggett (1980) typically had only three reliable current meter records per mooring string. Normally, this would be enough to obtain the first two baroclinic modes. However, the bottom current meter in most instances was within a few meters of the bottom and therefore strongly affected by benthic boundary layer effects. To include a mode-2 solution in the estimates, the observed phase and amplitude of the bottom current meter record had to be

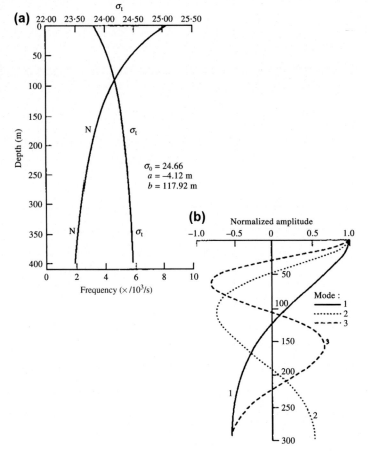

FIGURE 4.35 Baroclinic modes for semidiurnal frequency (ω_{M_2}) in a uniformly rotating, uniform depth channel (a) the mean density structure (σ_t) and corresponding buoyancy frequency $N(z)$ used to calculate the eigenvalues; (b) eigenvectors for the first three baroclinic modes. The barotropic mode (not plotted) has a magnitude of unity at all depths. Phase speeds for the modes fitted to the current meter data are $c_1 = 34$ cm/s; $c_2 = 20$ cm/s. *From Thomson and Huggett (1980).*

adjusted for frictional effects via the added term $\exp(-z')\cos(\omega t + \theta - z')$, where $z' = (z + H)/\delta$, and $\delta \approx (2K_v/\omega)^{1/2}$ is the boundary layer thickness for eddy viscosity K_v.

Because K_v is not known *a priori*, the final solution required finding that value of K_v which minimized the ratio formed by the first mode calculated with and without the bottom current meter included in the analysis (Table 4.9). In the case where five current meters were available, Thomson and Huggett found that there was no difference in the value of the second mode estimate with and without inclusion of the bottom current meter record in the analysis, suggesting that the three-mode decomposition was representative of the actual current variability with depth.

4.10.3 Coastal-Trapped Waves (CTWs)

Stratified or non-stratified oceanic regions characterized by abrupt bottom topography adjacent to deeper regions of uniform depth support the propagation of trapped ocean waves with frequencies, ω, which must be lower than the local inertial frequency, f. Trapped sub-inertial motions ($\omega < f$) typically are found along continental margins where the coastal boundary is bordered by a marked change in water depth consisting of a shallow (<200 m) continental shelf, a steep continental slope, and a deep ($>2,000$ m) weakly sloping continental rise. The alongshore wavelengths vary from tens to thousands of kilometers while the cross-shore trapping scale is determined by the density structure and length scale for the cross-shore topography. For baroclinic waves, the *internal deformation radius*, $r = NH/f$, provides an estimate of the cross-shelf trapping scale while the *stratification parameter*, $S = (N^2_{max}H^2_{max})/f^2L^2$, characterizes the importance of stratification for a shelf-slope region of width L. For a mid-latitude ocean of depth $H \approx 2,500$ m and buoyancy frequency $N \approx 2 \times 10^{-3}$/s, we find $r \approx 50$ km. For wide shelves ($L > 100$ km), the motions are confined mainly to the continental slope, while for narrower shelf regions, the motions extend to the coast where they "lean" up against the coastal boundary. For $S \gg 1$ the CTWs are strongly baroclinic, while for $S \ll 1$, they are mainly barotropic (Chapman, 1983). The case $S \approx 1$ corresponds to barotropic shelf waves modified by stratification.

In addition to continental shelf regions, coastal-trapped waves can occur along mid-ocean ridges and in oceanic trenches (where they are known as *trench waves*; cf., Mysak et al., 1979), as well as around isolated seamounts and islands (Pizarro and Shaffer, 1998). Phase propagation, in all cases, is with the coastal boundary to the right of the direction of propagation in the Northern Hemisphere and to the left of the direction of propagation in the Southern Hemisphere. For strongly baroclinic waves, energy propagation is always in the direction of phase propagation; for barotropic motions, short waves can propagate energy in the opposite direction to phase propagation.

The general coastal-trapped wave solution consists of a Kelvin wave mode ($k = 0$), for which the cross-shore velocity component in the case of a flat-bottom ocean is identically zero at the coast ($U \equiv 0$ at $x = 0$), together with a hierarchy of higher mode shelf waves ($k = 1, 2, ...$) whose cross-shore velocity structures have increasing numbers of zero crossings (sign changes) normal to the coast. (The condition of no normal flow through the coastal boundary requires $U = 0$ at $x = 0$ for the shelf wave component). The first shelf wave mode will have one zero crossing in sea surface elevation ζ over the continental margin, the second mode will have two crossings, and so on. For the current component, U, the first mode shelf wave will have no zero crossing, the second mode will have one crossing, and so on.

Computer programs that calculate the frequencies and cross-shore modal structure of coastal-trapped waves of specified wavelength are available in reports written by Brink and Chapman (1987) and Wilkin (1987). We confine ourselves to a general outline of the programs for the interested reader. Practical difficulties with the numerical solutions to the equations are provided in these comprehensive reports. The programs of Brink and Chapman use linear wave dynamics in which the water depth, $h(x)$, is assumed to be a function of the cross-shore coordinate, x, alone. Similarly, the buoyancy frequency, $N(z)$, is a function of depth alone. The one profile that can be used in the analysis is best obtained by least-squares fitting a function (such as a polynomial or exponential) to a series of observed profiles. The wave parameters such as velocity, pressure and density are assumed to be sinusoidal in time (t) and alongshore direction (y) such that for any particular wave parameter, ξ, we have

$$\xi(x, y, t) = \xi_o(x)\exp[i(\omega t + ly)] \tag{4.106}$$

where ω is the wave frequency and l is the alongshore wavenumber. This gives rise to a two-dimensional eigenvalue problem in (ω, l) of the form

$$L[\xi_o(x; \omega, l)] = 0 \tag{4.107}$$

where L is a linear operator. The problem is solved for arbitrary forcing and a fixed l. In particular, for a given wavenumber, l, the frequency ω is varied until the algorithm finds the free-wave mode resonance. Resonance is defined as the frequency at which the square of the spatially integrated wave variable

$$I_o = \int\limits_0^\infty \xi_0^2 dx, \quad \text{or} \quad I_p = \int\limits_0^\infty \int\limits_{-h}^0 \left(p^2 dz\right) dx \tag{4.108}$$

is at a maximum. The suite of programs tackles the following problems for which the user provides the bottom profile $h(x)$, a mean flow profile (if needed) and a selection of boundary conditions:

(1) The program BTCSW yields the dispersion curves $\omega = \omega(l)$ (the frequency as a function of wavenumber), the cross-shore modal structure for velocity $\mathbf{U}(x) = (U(x), V(x))$ and/or sea surface elevation $\zeta(x)$, and wind coupling coefficients for barotropic coastal-trapped waves—including continental shelf waves and trench waves—for arbitrary topography and mean alongshore current. Options for the long-wave and rigid-lid approximations are included in the program. The user can specify one of two geometries corresponding to topography with and without a coastal boundary. The outer boundary at $x = x_{max}$ is set as $-2L$, where L is the width of the typographically varying domain in the cross-shore direction. Thus, about half the domain has a flat bottom. The outer boundary condition is specified as $\partial U/\partial x = 0$. To obtain solutions for both ζ and $\mathbf{U}$, the depth at the coast should be given a nonzero value $h(0) \geq 1$ m.

(2) For wave frequencies $\omega \leq 0.9f$, the program BIGLOAD2 yields dispersion curves $\omega = \omega(l)$, the horizontal modal structure, and wind-coupling coefficients for an ocean with continuous, horizontally uniform stratification and arbitrary topography. Density in the model has the form $\rho^*(x, y, z, t) = \rho_o(z) + \rho(x, y, z, t)$, where ρ_o is background density and ρ is the density perturbation. Because $\rho \ll \rho_o$, the Boussinesq approximation is assumed throughout (i.e., the small density perturbations are ignored in momentum terms involving the fluid inertia and Coriolis acceleration but are retained in vertical buoyancy terms where they multiply, g, the acceleration due to gravity; cf. LeBlond and Mysak, 1978). The program allows for the component of the β-effect normal to the coast and for both the free surface and rigid lid boundary conditions at the ocean surface. Solutions are obtained using the coordinate transformation $\theta = z/h(x)$ and assuming a linear bottom friction drag. A total of 17 vertical and 25 horizontal grids (rectangles) are generated so that the vertical resolution is much better near shore than in deep water. Problems with singularities are avoided by setting $h(x) \geq 1$ m at the coast, $x = 0$. The program does not work well when the shelf-slope width (or width of a trench at the base of the shelf) is small relative to the internal deformation radius for the first mode in the deep ocean. Spurious features appear in unexpected places and force the user to increase the density of horizontal grids over regions of rapidly varying topography. In addition, a spurious mode occurs in the pressure equation for $\beta = 0$ at the local inertial frequency $\omega = f$, making the overall solution suspect. As noted by the authors, the user will have difficulty finding the barotropic Kelvin wave parameters.

(3) The program CROSS is used to find baroclinic coastal-trapped modes for $\omega \leq f$ for arbitrary stratification and uniform depth.

(4) The program BIGDRV2 is used to obtain the velocity, pressure, and density fluctuations over a continental shelf-slope region of arbitrary depth, stratification, and bottom friction and is driven by an alongshore wind stress of the form $\tau(x) = \tau_o \exp[i(\omega t + ly)]$. Specification of a linear friction coefficient of zero ($r = 0$) results in a divide-by-zero error. As a result, inviscid solutions should not be attempted. As with (2), solutions are obtained on a 25×17 stretched grid. In practice, it is generally best to start a study of coastally trapped waves using BTCSW since it gives first-order insight into the type of modal structure one can expect. However, if the barotropic dispersion curves do not fit the data (e.g., observations reveal strong diurnal-period shelf waves, but the first-mode dispersion curves consistently remain below the diurnal frequency band for realistic topography), then density and mean currents should be introduced using BIGLOAD2 and CROSS.

The Brink and Chapman programs have been used by Crawford and Thomson (1984) to examine free wave propagation along the west coast of Canada and by Church et al. (1986) and Freeland et al. (1986) to examine wind-forced coastal-trapped waves along the southeast coast of Australia (Figure 4.36). In all cases, model results are compared with alongshore sea-level records and current meter observations from cross-shore mooring lines. The cross-shore depth profiles $h(x)$ and associated buoyancy frequencies $N^2(z)$ used in the Australian model are presented in Figures 4.37a and b. From these input parameters, the program was used to generate eigenvalues and eigenfunctions for the first three CTW wave modes (Figure 4.38) and the theoretical dispersion curves (Figure 4.39) relating wave frequency, ω, to alongshore wavenumber, l. The slopes of the (ω, l) curves give the phase speeds c_k for the given modes ($k = 1, 2, 3$) listed on the figure.

Wilkin (1987) presents a series of FORTRAN programs for computing the frequencies and cross-shore modal structure of free coastal-trapped waves in a stratified, rotating channel with arbitrary bottom topography. The

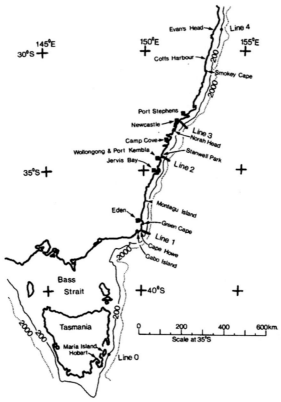

FIGURE 4.36 Southwest coast of Australia showing the locations of the tide gauge stations (-) and current meter lines (0, 1, 2, 3) occupied during the Australian Coastal Experiment (ACE). *From Freeland et al. (1986).*

programs solve the linearized, inviscid, hydrostatic equations of motion using the Boussinesq approximation. The Brunt-Väisälä frequency $N(z)$ is a function of the vertical coordinate only. As with Brink and Chapman (1987), the eigenvalue problem is solved using resonance iteration and finite difference equations. The cross-shore perturbation fields returned by the model include velocity, pressure, and density. The difference with Wilkin's model is that it uses a staggered horizontal (Arakawa "C") grid for which the usual horizontal Cartesian coordinates (x, y) have been mapped to orthogonal curvilinear coordinates (ξ, η). Instead of using finite differencing, the vertical structures of the modes are determined through modified sigma coordinates with expansion of the field variables in terms of Chebyshev polynomials of the first kind. The program has the option of specifying wavenumber, l, and searching for the corresponding free wave frequency, $\omega(l)$, as in Brink and Chapman, or specifying ω and searching for l. For reasons explained by Wilkin, the model is designed to be compatible with the primitive equation ocean circulation model developed by Haidvogel et al. (1991).

In the curvilinear coordinate system, a line element of length ds in the Wilkin model satisfies

$$ds^2 = dx^2 + dy^2 = d\xi^2/dm^2 + d\eta^2/dn^2 \tag{4.109}$$

and the metric coefficients m, n are defined by

$$m = \left[(\partial x/\partial \xi)^2 + (\partial y/\partial \xi)^2 \right]^{-1/2} \tag{4.110a}$$

$$n = \left[(\partial x/\partial \eta)^2 + (\partial y/\partial \eta)^2 \right]^{-1/2} \tag{4.110b}$$

The velocity perturbations for time-dependent solutions of the form $\exp(-i\omega t)$ are then

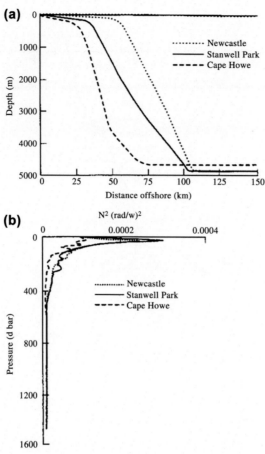

FIGURE 4.37   Parameters used in determining the coastal-trapped wave eigenfunctions along the Cape Howe, Stanwell Park and Newcastle lines; (a) the cross-shore depth profiles $h(x)$; (b) the $N(z)^2$ profiles. Below 600 dbar (≈ 590 m) all curves are similar so that only is drawn. *From Church et al. (1986).*

$$U = \frac{1}{f^2 - \omega^2}\left(i\omega m\frac{\partial \phi}{\partial \xi} - fn\frac{\partial \phi}{\partial \eta}\right) \tag{4.111a}$$

$$V = \frac{1}{f^2 - \omega^2}\left(i\omega n\frac{\partial \phi}{\partial \eta} - fm\frac{\partial \phi}{\partial \xi}\right) \tag{4.111b}$$

$$w = \frac{i\omega}{N^2}\frac{\partial \phi}{\partial z} \tag{4.111c}$$

where (U, V, w) are the usual velocity components and $\phi = p/\rho_o$ is the perturbation pressure. Solutions are then sought for the resulting pressure equation

$$mn\frac{\partial}{\partial \eta}\left(\frac{n}{m}\frac{\partial \phi}{\partial \eta}\right) + (f^2 - \omega^2)\frac{\partial}{\partial z}\left(\frac{1}{N^2}\frac{\partial \phi}{\partial z}\right) + mn\frac{\partial}{\partial \xi}\left(\frac{m}{n}\frac{\partial \phi}{\partial \xi}\right) = 0 \tag{4.112}$$

For a straight coastline, $m\partial/\partial\xi = \partial/\partial x$ and we arrive at the usual solutions for alongshore (x-direction) propagation of progressive waves of the form $F(y)\exp[i(lx - \omega t)]$.

The Wilkin model is less general than the Brink and Chapman model in that application of the rigid-lid approximation does not allow for the barotropic (long wave) Kelvin wave solution and a "slippery" solid wall is placed at the offshore boundary. The new vertical coordinate variable, σ, is defined by

$$\sigma = 1 + 2z/h(\eta) \tag{4.113}$$

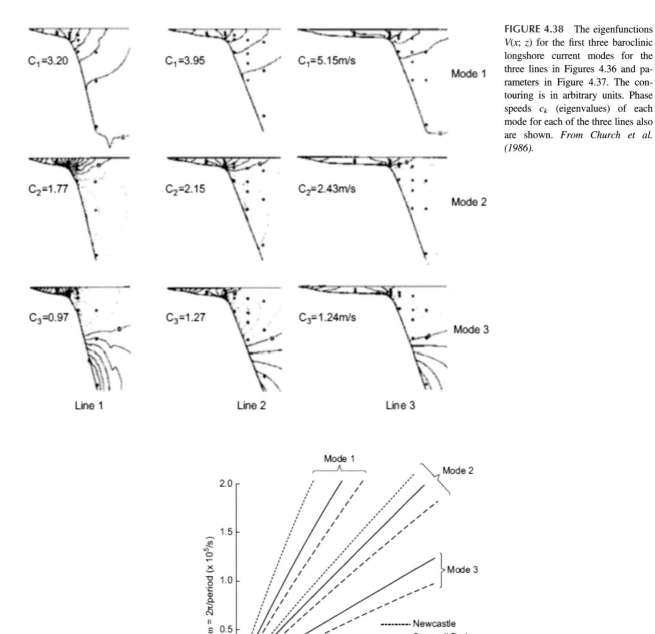

FIGURE 4.38 The eigenfunctions $V(x; z)$ for the first three baroclinic longshore current modes for the three lines in Figures 4.36 and parameters in Figure 4.37. The contouring is in arbitrary units. Phase speeds c_k (eigenvalues) of each mode for each of the three lines also are shown. *From Church et al. (1986).*

FIGURE 4.39 The theoretical dispersion curves $\omega = \omega(l)$ relating the longshore wavenumber, l, to the wave frequency, ω (here λ is the wavelength). Curves correspond to the first three baroclinic modes for each mooring location. For mode 3, the dispersion curve at Stanwell Park and Newcastle are almost identical. The slopes of the lines are the theoretical phase speeds, c_k. *From Church et al. (1986).*

so that the ocean surface is located at $\sigma = 1$ and the (now flattened) seafloor at $\sigma = -1$. Application of this model to the west coast of New Zealand (South Island) is presented by Cahill et al. (1991). Modes 1 and 2 of the alongshore current for the northern portion of this region based on Wilkin's program CTWEIG are reproduced in Figure 4.40. Similar results for the southern region are presented in Figure 4.41. Notice that the coastal-trapped waves are nearly barotropic over the shallow shelf immediately seaward of the coast in both sections but are more baroclinic in the offshore region off the southwest coast.

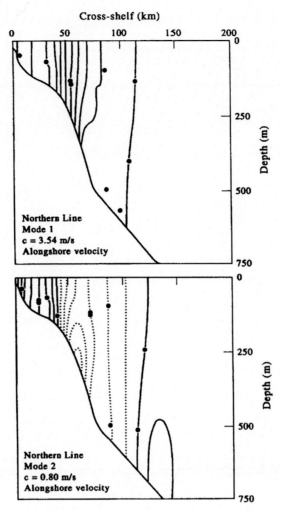

FIGURE 4.40 The alongshore velocity structure of coastal-trapped waves for the northwestern shelf-slope region of South Island, New Zealand, *Top*: Mode 1; *Bottom*: Mode 2. Contour lines when multiplied by 10^{-7} correspond to the alongshore velocities in m/s for unit energy flux in watts. Negative values are dashed. Current meter locations are given by the dots. Here, c is the phase speed of the mode. *From Cahill et al. (1991).*

4.11 SELF-ORGANIZING MAPS

The Self-Organizing Map (SOM) provides a method for extracting spatial patterns from high dimensional data sets by clustering the data into much lower dimensional arrays of orderly and smoothly connected mapping units. Arrays are commonly one- or two-dimensional. Examples of high dimensional data sets include daily time series of sea surface temperature (SST) measured at grid locations within a coastal upwelling region and daily time series of current velocity measured at mooring sites strung across the continental shelf. In the first case, a simple two-map system might have one map showing near-uniform SST values over the entire domain (such as might occur during calm conditions in the middle of winter) and a second map showing cold water near the coast and warmer water offshore (indicative of strong wind-driven upwelling in summer). In the second example, one map unit might consist of a near spatially uniform southeast-ward flow characteristic of upwelling favorable wind conditions while the second map would consist of near unidirectional northwestward flow typical of downwelling favorable wind conditions. The incorporation of intervening map units within the array makes it possible to examine the transitions between the two "end member" states. Self-Organizing Maps of this kind were used successfully to study sea-surface temperature patterns (Liu et al., 2006) and current velocity structure (Liu and Weisberg, 2005) on the West Florida Shelf. Figure 4.42 shows the 12 map units obtained by Liu and Weisberg for the velocity structure on the shelf; Figure 4.43 gives the percentage of time that a particular map unit was "selected" as best matching the input data.

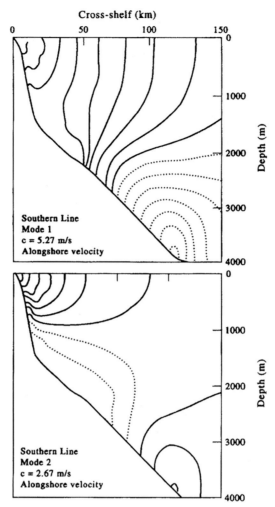

FIGURE 4.41 As for Figure 4.40 but for the shelf-slope region off the southwestern tip of South Island. Note the change in depth and offshore distance scale in the two figures. This line is roughly 500 km to the south of the line in Figure 4.40.

The SOM is an ordered, nonlinear, artificial neural network (ANN) technique with unsupervised learning (no instructor or teacher required) used for mapping high-dimensional spatially and temporally varying input data into a much smaller number of map units (also called, elements, nodes, archetypes, or neurons) of a regular low-dimensional array (Kohonen, 1982, 2001). For oceanographic applications, SOMs provide a pattern recognition and classification tool that clusters input data into an array of gradually changing maps, typically two-dimensional, that can reveal the most commonly occurring structural features embedded in the large-scale data sets. As illustrated by Figure 4.42, adjoining units in the array share similar structures and features but this similarity diminishes with increased separation between the map units.

As with other artificial neural networks, the SOM consists of numerical algorithms that simulate the processing capability of the brain, in which a network of interconnected units (cells, neurons) process information or input data in parallel rather than sequentially. During unsupervised training, the networks learn to generate their own classifications of the training (input) data without external help. This assumes that class membership is broadly defined by the input patterns sharing common features, and that the network will be able to identify those features across the range of input patterns (Bullinaria, 2004). SOMs can deal with noisy and gappy data and require no prior knowledge or requirements about the data, such as the need for normal distribution or equality of parameter variances. The learning is unsupervised in that the training of the network is entirely data-driven and no prior instructions (other than the dimensionality of the output maps) are given to the algorithms on how they are to reach their mapping goal. In supervised learning, target results for the input data vectors are provided to "train" the algorithm (e.g., Bayesian spam email filtering, Support Vector Machines). Self-

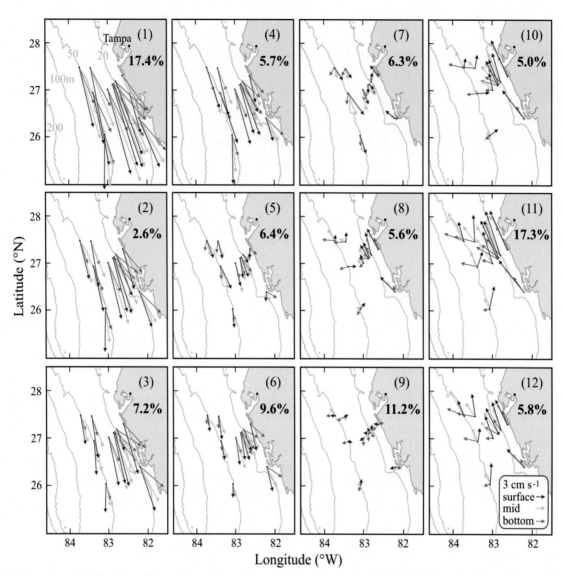

FIGURE 4.42   A 3 × 4 Self-Organizing Map of 2-day low-pass filtered velocity data at three depth levels (see legend) from October 1998 through September 2001 on the West Florida Shelf. Adjoining map units show similar features, but similarity diminishes with separation between maps. The relative frequency of occurrence of each pattern (map unit) is shown in the right corner of each map. *Adapted from Liu and Weisberg (2005).*

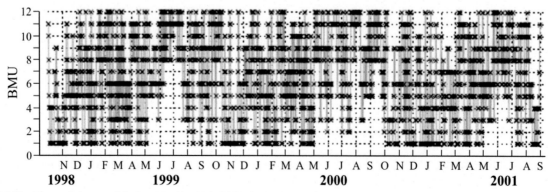

FIGURE 4.43   Temporal changes of the Best Matching Unit (BMU) for the 3×4 Self-Organizing Map in Figure 4.42. The tick marks (x) on the vertical axis range from 1 to 12 and correspond to the pattern numbers in the SOM. The light gray lines connect adjoining tick marks to form a time series. *From Liu and Weisberg (2005).*

organizing map algorithms were developed in the early 1980s with first applications in image analysis and biological cluster analysis (Kohonen, 1982, 2001). The technique has now been used over a wide range of disciplines and was first applied to climate variability by Hewitson and Crane (1994). More recent climate applications can be found in Hsu et al. (2002) and Reusch et al. (2007). Oceanic examples of SOMs include applications to large-scale surface fields such as sea-level pressure (Hewitson and Crane, 2002), sea-surface temperature (Lui and Weisberg, 2006), QuikSCAT winds (Risien et al., 2004), and NCEP/NCAR reanalyses data (Tennant, 2004). SOMs have also been applied to analyses of coastal ocean current patterns derived from moored ADCPs (Liu and Weisberg, 2005) and estuarine circulation as defined by a single ADCP (Cheng and Wilson, 2006). Richardson et al. (2002, 2003) used the SOM to identify characteristic chlorophyll-a profiles in vertical measurements obtained in the Benguela upwelling system. The first public-domain general-purpose SOM software package SOM_PAK was released in 1990 by the Laboratory of Computer and Information Science of the Helsinki University of Technology. The package was then implemented by the same group in MATLAB as a Toolbox (Vesanto et al., 2000). Software for SOM analysis is currently available as part of the Neural Network Toolbox in MATLAB and as "kohonen" in the R programming language. (The reader is referred to chapters 7 and 8 of this book for a more extensive discussion of neural networks and machine learning methods.)

4.11.1 Basic Formulation

The SOM consists of J map units or elements that are typically arranged in a two-dimensional grid with a specified weight vector, $\mathbf{w}_j$, assigned to each map unit (some authors use $\mathbf{m}_j$ for the weights). SOM requires the analyst to specify the number of units in the array and to provide an initial guess for the weight vectors to be used at the first step of the clustering. In their study of currents on the West Florida Shelf, Liu and Weisberg (2005) specified a 3×4 ($J = 12$) SOM consisting of 3 units down by 4 units across (Figure 4.42). The weights can be initialized randomly (as was done for the Florida Shelf study) or the analyst can use prior knowledge, such as output from Principal Component Analysis or the record mean values for each site, to specify the initial weights. Initializing using random weights means that the algorithm will take a bit longer to learn how to cluster the data. After their initial specification, the weights change following a learning rule (presented below) that specifies how to calculate new weights at each step in the clustering process. At a given time step, the jth weight vector for the jth map unit will have the form

$$\left(w_{j,1}, w_{j,2}, \ldots, w_{j,n} \right) \tag{4.114}$$

where n is also the number of elements in the input vector created from the data. The input vector is also referred to as the "training vector" from ANN terminology.

Following specification of the SOM size (but see note * at the end of this paragraph) and initial weighting values, the incremental self-organizing (sequential training) algorithms are ready for the input of the p data vectors, $\mathbf{x}$, constructed from the high-dimensional data set. These input vectors are of length n and have the form

$$\left.\begin{array}{l} \left(x_{1,1}, x_{1,2}, \ldots, x_{1,n} \right) \\ \left(x_{2,1}, x_{2,2}, \ldots, x_{2,n} \right) \\ \cdots\cdots\cdots\cdots\cdots\cdots\cdots \\ \left(x_{i,1}, x_{i,2}, \ldots, x_{i,n} \right) \\ \cdots\cdots\cdots\cdots\cdots\cdots\cdots \\ \left(x_{p,1}, x_{p,2}, \ldots, x_{p,n} \right) \end{array}\right\} p \text{ distinct input (training) vectors} \tag{4.115}$$

where vector elements are real numbers. The data (training) vectors are input into the SOM algorithm and the *activation* of each unit for the specific input vector is calculated. The activation function is typically a pre-selected function of the Euclidian distance, $D^2(t)$, between the input vector and the weight vector for that particular unit (the squared differences between the vectors on a component-by-component basis). For each map unit, j, the SOM algorithm calculates $D^2(t)$ as

$$D_j^2(t) = \sum_{k=1}^{n} \left(x_{i,k} - w_{j,k}(t) \right)^2 \quad j = 1, \ldots m; \ i = 1, \ldots, p \tag{4.116}$$

where the weights and the elements, $x_{i,k}$, of the input vectors have the forms (4.414) and (4.115), respectively. The unit whose weight vector shows the highest activation (i.e., minimum Euclidean distance, D^2, also written as argmin$\|\mathbf{x}_k - \mathbf{w}_i\|$) for the particular input vector is selected as the "winner" or "best matching unit" (BMU) for the particular SOM map unit. In effect, the winner (BMU), determined from the minimum of (4.116), is a measure of how closely a given input data vector, $\mathbf{x}_k$, matches the weight vector $\mathbf{w}_i$ for each of the map units in the array. From a neurological point of view, this class of unsupervised system is a type of competitive learning, whereby the neurons compete amongst themselves to be activated, with the result that only one neuron is activated at any one time. This activated neuron is called a "winner-takes-all neuron" or simply the "winning neuron". Such competition can be induced/implemented by having lateral inhibition connections (negative feedback paths) between the neurons. The result is that the neurons are forced to organize themselves. (* The subjective aspect of selecting the number of map units has led to the formulation of Growing Hierarchical Self-Organizing maps, GHSOM, whereby the optimal number of map units is determined through a more objective SOM procedure; see Section 4.11.5 below.)

The next step after specifying the initial weights and undertaking the first sequential mapping step, is to modify the "winner" weight, $\mathbf{w}_j(t)$, so that it more closely resembles the input data that was presented to it during the previous time step. Specifically, the weight vector of the winner $\mathbf{w}_j(t+\Delta t)$ at $t = t+1 \cdot \Delta t$ is moved toward the presented vector by a fraction of the Euclidian distance as determined by the time-diminishing learning rate, $\alpha(t)$, and the neighborhood function, $\varepsilon_{qi}(t)$. That is,

$$\mathbf{w}_j(t+\Delta t) = \mathbf{w}_j(t) + \alpha(t) \cdot \varepsilon(t) \left[\mathbf{x}_i(t) - \mathbf{w}_j(t) \right], i = 1, \ldots, p \tag{4.117}$$

where the learning rate has one of the following forms (Liu and Weisberg, 2005):

$$\alpha(t) = \begin{cases} \alpha_o(1 - t/T), & \text{linear} \\ \alpha_o(0.05/\alpha)^{t/T}, & \text{power} \\ \alpha_o/(1 + 100t/T), & \text{inverse} \end{cases} \tag{4.118}$$

Here, α_o is the initial learning rate and T is the training duration. In the SOM MATLAB Toolbox, the defaults are a linear function and $\alpha_o = 0.5$ for an initial training session and 0.05 for further fine tuning (Liu and Weisberg, 2005). Users can also specify the initial and final values of α or specify other time decreasing functions. The winner's activation will be even higher the next time the same input vector is presented to the algorithm. In addition to the learning rate, the weight vectors of units in the neighborhood of the winner are modified according to the decreasing spatial-temporal neighborhood function, $\varepsilon_{qi}(t)$, where

$$\varepsilon_{qi}(t) = \begin{cases} H(r_i - D_{qi}) & \text{bubble} \\ \exp\left(- D_{qi}^2/2r_i^2 \right) & \text{Gaussian} \\ \exp\left(- D_{qi}^2/2r_i^2 \right)\delta(r_i - D_{qi}) & \text{cut} - \text{Gaussian} \\ \max\left(0, 1 - (r_i - D_{qi})^2\right) & \text{Epanechikov} \end{cases} \tag{4.119}$$

Here, $\delta(a)$ is the Dirac delta function and $H(a)$ is the Heaviside step function defined as: $H(a) = 0$ if $a < 0$, and $= 1$ if $a \geq 0$. As an example, the initial weight vector, $\mathbf{w}_j(t_{\text{start}})$, at the start of a three-unit SOM array might look something like

$$\left. \begin{array}{l} (w_{1,1}, w_{1,2}, \ldots, w_{1,n}) \\ (w_{2,1}, w_{2,2}, \ldots, w_{2,n}) \\ (w_{3,1}, w_{3,2}, \ldots, w_{3,n}) \end{array} \right\} = \left\{ \begin{array}{l} (0.496, 0.877, \ldots, 0.317) \\ (0.169, 0.714, \ldots, 0.843) \\ (0.522, 0.069, \ldots, 0.589) \end{array} \right. \tag{4.120}$$

where the elements $w_{i,k}$ are random number between 0 and 1 that we selected using the RANDM function on a Hewlett Packard 30S hand calculator. Note that our choice of values could easily represent randomly selected current speeds in meters per second. During the analysis, each of the p vectors in the input data will fall into one of the m clusters or map units corresponding to the output vector, $\mathbf{y} = (y_1, y_2, \ldots, y_m)$ of length m (Figure 4.44). Here, the number of best matching units, m, can be smaller than, greater than, or equal to, the length, n, of the input vector for each time step, p. There is one weight vector of length n associated with each output unit, $\mathbf{y}$.

Two measures commonly used to assess the quality of the SOM analysis, are the *quantization error* (QE) and the *topological error* (TE) (Vesanto et al., 2000). The quantization error is the Euclidean distance between the Best Matching units (winners) and the corresponding input data. The average quantization error, along with its standard deviation, can be used to indicate how well the map units fit the data. The topological error is the ratio of occurrences of map units after all input data have been applied, and where the second-best matching unit is not a direct topological neighbor of the best matching unit. Thus, it gives an indication of how well the map units are topologically ordered.

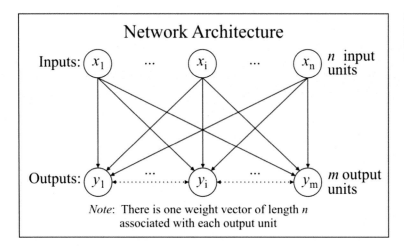

FIGURE 4.44 Network architecture showing input **x** (of length n) and outputs **y** (of lengths m). Arrows denote the interconnectivity between input and output; *Adapted from http://genome.tugraz.at/MedicalInformatics2/SOM.pdf.*

4.11.2 SOM Versus Principal Component Analysis

SOM analysis is often compared to, or used in conjunction with, empirical orthogonal functions (EOFs) or principal component analysis (see Section 4.6). Both methods effectively reduce the dimensionality of the original data and both search for spatial patterns that aid in the interpretation of the structure and dynamical features embedded in the data. In the analysis of a two-dimensional scalar data field, such as sea surface temperature or along-channel velocity field from moored current meters, both methods can be used to isolate a set of representative spatial fields.

The EOFs describe a set of modes of variability in the data, which may or may not be related to the actual physical modes of variability. Often, only the first few modes and their respective Principal Component time series are sufficient to give a reasonable reconstruction of the original detrended data. The EOFs are found by maximizing the variability explained by each EOF while enforcing the integrated orthogonality condition among the EOFs. This results in "orthogonal ordering", whereby the dominant component of the flow, such as the estuarine circulation, determines the first mode. All subsequent modes are orthogonal to this first mode, which can be problematic in that there may be no particular reason why the physical processes should adhere to orthonormal conditions. Thus, the physical dynamics of the system, other than perhaps the one described by the dominant first mode, can be lost in combinations of the higher modes. Techniques have been developed to circumvent some of these shortcomings (see Sections 4.5–4.7), such as rotating the eigenvectors which make up the EOFs, performing EOF analysis after removing certain modes, or using a non-orthogonal (oblique) basis function (Richman, 1986; von Storch and Zwiers, 1999). However, all methods introduce a degree of subjectivity and can lead to non-zero correlation among the principal component time series.

Liu et al. (2006) compare EOF analysis to SOM analysis using an artificial time series made up of a progressive sinusoidal wave plus noise. The EOF analysis can pick out the sinusoidal signal. However, for a more complex signal made up of an admixture of sine, step, sawtooth, and cosine waves, the leading mode was not representative of any of the signals but of a complex hybrid signal. Furthermore, even though the sine and cosine signals are orthogonal functions, they did not represent any of the higher modes in the time series.

4.11.3 The Self-Organizing Map (SOM)

As noted previously, the SOM is a nonlinear, feed-forward neural network with the ability to objectively downscale high dimensional input data into a set of neurons (map units) ordered in a user-selected output space. The user-selected output space in which the map units (also, cells, nodes, archetypes) are organized can be rectangular, cylindrical, toroidal or any other two (or higher) dimensional structure. The purpose of the output space is to provide a graphical view of the connections among the map units. Although SOMs have been used as a cluster analyzing tool, the map units obtained using the SOM approach are topologically ordered, unlike in traditional clustering. Thus, at each stage of the processing, each piece of incoming information is kept in its proper context or neighborhood, and map units (neurons) dealing with closely related pieces of information are kept close together so that they can interact via short synaptic connections. Thus, similar map units are assembled within a region of output space whereas dissimilar map units are forced apart in output space. Because of the connections among map units and the topological ordering, SOM yields a continuum of states representing the original data. In SOM analysis, the map units and their time series are built using the concept of the Best Matching Unit (BMU). The BMU is determined by comparing the map units with the original input data, measuring the similarity (by Euclidean distance) and constructing a time series of the ranking of the map units with respect to their similarity. In this way, SOM and PC analysis can result in similar analysis products.

Topological ordering in SOM occurs because of a neighborhood function. However, to effect topological ordering, each neighbor of the winning map unit is also nudged closer to the input data by some smaller degree determined by a neighborhood function, $\varepsilon_{qi}(t)$, given by (4.119). With only two map units, which we consider as part of the discussion in the section that follows, the complexities in determining neighbors, such as the shape of the map unit array and the lattice of connections between the neurons within that array, do not have to be considered. Furthermore, because the topological ordering has no meaning in the two-map case, the effect of the neighborhood function is to "smooth" or minimize the separation polarization between the two map units. Because the smoothing tends to move the map units closer together, it also increases the quantization error of the mapping, or how well each map unit fits the data it is representing. To achieve the best representation of the input data by two map units, it is best to set the neighborhood function to zero. This allows the SOM software to produce results much like a cluster analysis and is better defined as unsupervised vector quantization.

4.11.4 Application to Estuarine Circulation in Juan de Fuca Strait

Juan de Fuca Strait (Figure 4.45) is a partially mixed tidal channel connecting the freshwater catchment basins of the Strait of Georgia and Puget Sound to the Pacific coast of British Columbia (Canada) and Washington State (USA). The channel is 160 km long, 25–40 km wide, and has a maximum depth of 200 m. Estuarine circulation in the strait prevails 90% of the time in summer and 55% of the time in winter while "transient" wind-forced regimes occur roughly 10% of the time in summer and 45% of the time in winter (Thomson et al., 2007). If we consider the estuarine and transient flow regimes as the end states of a bimodal system, we can use the unsupervised learning capacity of SOM to delineate the "archetypical"

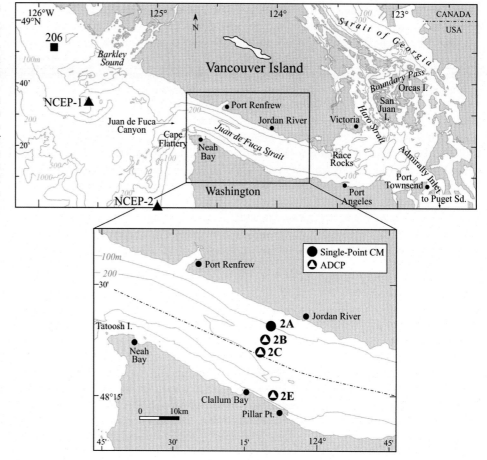

FIGURE 4.45 Map of the Juan de Fuca Strait and adjoining regions. The solid square in the upper panel marks the position of a meteorological buoy (C46206) and the solid triangle marks the National Centers for Environmental Prediction/National Center for Atmospheric Research (NCEP/NCAR) site "126W49NG". Solid circles in the bottom panel show locations of single-point current meter (CM) moorings; circles with triangles denote ADCP or ADCP+CM moorings. *Modified after Thomson et al. (2007).*

structure of the two states and to determine those times in a current meter data set when one or the other of the two states dominates the circulation in the channel. The simple case of two neurons or map units is degenerate with respect topological ordering in SOM analysis. Whether the initial state is positioned to the left or right, upper, or lower panel has no physical meaning and is solely determined by the initialization and the order that input data are presented to the map units. After examining the two-map system, we can progress to a system of multiple map units that characterize the transition between the two basic states.

4.11.4.1 Observations

Following Mihály and Thomson (unpublished report), we use a subset of the current vectors, $\mathbf{x}(t)$, obtained from arrays of ADCPs and single-point current meters moored across central Juan de Fuca Strait from 1998 to 2005 (Figure 4.45). The two criteria used in the selection of the subsets were: (1) sufficient coverage to resolve along-channel currents at the scale of the internal deformation, $r_D \sim 10$ km, in the cross-channel direction; and (2) the availability of continuous high-quality data throughout the selected analysis period. High resolution cross-channel coverage is essential for delineating the variability of the highly horizontally and vertically sheared flows in the strait. Similarly, pattern-recognition algorithms require high quality, long duration records to identify archetypical flow structures. Because the SOM analysis is sensitive to changes in cross-channel coverage and data quality, close attention was paid to variability in data reliability and spatial coverage when interpreting results in terms of dynamical processes.

We selected observations from five ADCP deployments periods covering the period May 2002 to May 2005 (Figure 4.46; Table 4.10). The data span two winter periods, two summer periods, and a separate continuous period encompassing both winter and summer months. All cross-channel configurations typically include upward-looking 150 or 300 kHz ADCPs at stations 2B, 2C, and 2E augmented by a single-point current meter mooring at station 2A with instruments at nominal depths of 25 and 75 m. An additional current meter was sometimes deployed at a nominal depth of 145 m below the ADCP at station 2E on the US side of the strait. The ADCPs were able to achieve maximum vertical ranges in summer but not in winter when there are fewer suitably sized zooplankton scatterers in the strait. This seasonal reduction in scatterers, combined with normal diel migration, resulted in a lowering of backscatter energy from the surface bins during daytime and the subsequent loss of reliable velocity estimates within the upper 2 to 3 ADCP bins. Because of

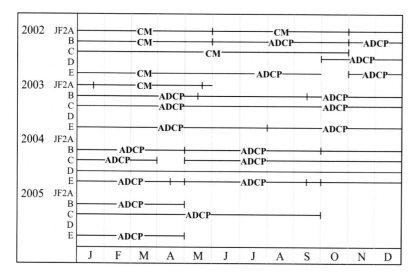

FIGURE 4.46 Yearly timelines for moorings in Figure 4.45 for years 2002–05. ADCP refers to moorings having an upward-looking (bottom- or mid-depth mounted) acoustic Doppler current profiler (ADCP); CM refers to moorings having either Aanderaa RCM4 or InterOcean S4 single-point current meters only. The mid-depth ADCPs often had single-point CMs moored below the ADCP. Gaps denote periods of mooring recovery and servicing, times of instrument damage or data loss, or no moored instrumentation. *Modified after Thomson et al. (2007).*

TABLE 4.10 Times of the five ADCP-single point current meter deployments periods used in the development of Self-Organizing Maps for alongshore currents in central Juan de Fuca Strait (Figure 4.45).

Summer 1	Winter 1	Summer/Winter	Summer 2	Winter 2
May 2002–Sep. 2002	Nov. 2002–May 2003	July 2003–Jan. 2004	May 2004–Sep. 2004	Sep. 2004–April 2005

instrument malfunction and other availability factors, mooring configurations changed slightly from deployment to deployment. For example, for the first series used in this study (summer of 2002), the 150 kHz ADCP at the central site 2C was replaced with three single-point current meters at depths of 28, 78 and 153 m to delineate flow in the three distinct depth ranges normally covered by the ADCP. During the following deployment (winter of 2002/03), the single-point current meters were replaced with an upward-looking 75 kHz ADCP. In the winter deployment of 2004–05, the 75 kHz ADCP at the central mooring was replaced with a 300 kHz ADCP so that its configuration matched that at nearshore stations 2E and 2B.

The ADCPs typically sampled every 15 min, whereas the single-point current meters had sample intervals of 30 or 60 min. All of the time series were converted to hourly samples following low pass filtering and resampling at hourly intervals. Because one of the goals of the SOM analysis was to understand the wind-forced circulation, a low-pass 30-h Kaiser-Bessel filter was applied to the hourly records to remove the tides (see Chapter 5 regarding filters). The filtered data were then resampled at 12:00 h UTC to provide values for the daily mean circulation. The along-channel direction was chosen to be along the first principal component direction as determined for each current vector time series. The along-channel data from each mooring of the array was then interpolated vertically using a one-dimensional Hermite cubic spline. In order to represent a partial slip bottom boundary condition, the interpolating spline was forced to zero at a depth that was arbitrarily chosen as 10% greater than the bottom depth. Near the surface, the velocity at the top-most bin or top-most current meter was extrapolated to the surface. These velocity estimates were then interpolated in two dimensions onto a 10 m vertical by 100 m horizontal cross-channel grid. The grid domain was made approximately 10% wider than the channel width and deeper than the chart depth to allow for slippage at the boundary and resulting (we assume) in a more realistic bottom boundary layer.

4.11.4.2 Archetypical Flows for All Data

The analysis is begun by deriving a two-element (two map unit) SOM representing the archetypical flow conditions for the along-channel current velocity in central Juan de Fuca Strait for all five deployment periods listed in Table 4.10. The left panel in Figure 4.47 is representative of the cross-channel structure during a fully established estuarine circulation regime in summer; the right panel is representative of the flow structure during a major wind-forced transient event in winter, when strong southerly winds drive an intense (~ 1 m/s) eastward flow called the "Olympic Peninsula Countercurrent" on the US side of the strait (Thomson et al., 2007). Analyses conducted using a neighborhood weight function for the two-map array indicates a modest improvement of the average quantization error of about 10% compared to the case when no neighborhood function is used. Quantization errors for the individual deployment periods presented in Table 4.10, as well as the

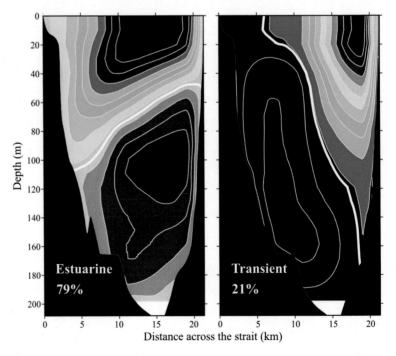

FIGURE 4.47   A two-map unit SOM representing the archetypical flow conditions for the along-channel current velocity, u, in central Juan de Fuca Strait for all five deployment periods listed in Table 4.10. Outflow toward the ocean is colored blue and inflow from the ocean is colored red. The light white line denotes $u = 0$. The left panel is representative of the cross-channel structure during a fully established estuarine circulation regime in summer (peak surface outflow is 14.8 cm/s, peak near bottom inflow is 7.8 cm/s; the right panel is representative of the flow structure during a major wind-forced transient event in winter, when there is a moderately intense eastward flowing Olympic Peninsula Countercurrent on the US side of the channel (peak near-bottom outflow is 8.3 cm/s, peak surface inflow is 29.8 cm/s). *Courtesy, Steve Mihály, Ocean Networks Canada.*

Euclidean distance between the map units, show that the map units for the summer periods are closely spaced (i.e., they share many topological similarities), whereas those for winter deployments—including the "all data" vectors for the period in column 5 of Table 4.10—the map units are spaced much farther apart, indicating greater topological differences. For the linear learning rate $\alpha(t) = \alpha_o(1 - t/T)$ from (4.118), the Epanechnikov neighborhood function (4.119) based on the default initial radii also results in the same map units, indicating that use of the Epanechnikov function for small map unit arrays is equivalent to using no neighborhood function. Applying a decreasing Gaussian neighborhood weight function (4.117), beginning at 1 and proceeding to 0 with sequentially changing weights, also results in the same set of map units when there are sufficient iterations. This implies that, with judicial selection of the neighborhood function parameters in the weight vectors, it is possible to balance the benefits of topological ordering against minimization of the quantization error.

According to the analysis, days for which the estuarine flow regime was the Best Matching Unit occurred 79% of the time; days when transient flow regimes were dominant accounted for the remaining 21% of the time. This objective finding is very close to a subjective analysis of the same period reported in Thomson et al. (2007) who found, by visually inspecting each daily mean flow period, that 78% were representative of an estuarine flow regime and 22% were representative of a transient flow regime. The timeline for the frequency of occurrence of a specific Best Matching Unit (Figure 4.48) shows a clear demarcation between winter and summer. The SOM analysis based on seasons indicates that estuarine (transient) regime prevails 93% (7%) to 97% (3%) of the time in the summer and 56% (44%) to 74% (26%) of the time in the winter. Corresponding values obtained from a subjective inspection of the daily observations over roughly the same periods (Thomson et al., 2007) are 92% (8%) to 93% (7%) over the three summers and 59% (41%) to 70% (30%) for the three winters. Both methods indicate that the winter of 2004−05 had a greater portion of time with estuarine flow conditions, with the SOM and subjective analyses yielding 74% and 70% estuarine conditions, respectively.

The bimodal SOM map structure can be used to estimate the duration of the transient flow events. Over the entire 3-year data set, there were 60 occasions of contiguous transient flows with a maximum duration of 9 days. If we assume it takes 1 day to return to estuarine conditions, there were two transition events with 14-day duration and one each of 10- and 11-day durations. The longer durations tended occur earlier in the winter season (October through December), but otherwise the occurrences of transient events did not reveal any other identifiable pattern during the winter season.

4.11.4.3 One Dimensional Linear SOM Analysis

To capture the transition between the two fundamental archetypical flows (from estuarine to transient and back to estuarine), the bimodal SOM outlined above was expanded along a single dimension to form a linear array. To achieve this, the authors first examined the mean quantization error and the polar (Euclidian) range and distance for arrays of two to ten linear map units for the overall data and each of the five time periods presented in Table 4.10. The parameters used in the SOM analysis to obtain the error metrics are the same as those used for the initial two map unit array described above. The errors fall into two groupings: the smaller grouping of mean quantization error ($MQE = \langle QE \rangle$) and Euclidean distances represent periods which only span the summer when there is very little transient flow activity and a smaller range of

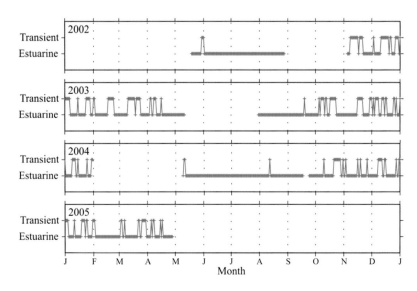

FIGURE 4.48 Frequency of occurrence of Best Matching Units for the two-map unit SOM in Figure 4.47 for observation years 2002−05. *Courtesy, Steve Mihály, Ocean Networks Canada.*

expected flow structures; the larger grouping of errors coincides with times when there are sufficient numbers of transient events to require many more map units to delineate the continuum of features in the data. As expected, the *MQE* diminishes with additional map units. From three map units and upward, the polar distance (the sum of the inter-map unit distances) has a linear increasing trend, whereas the polar range (the distance between the poles), typically begins to level off after five or six map units. This deficit between the polar range and polar distance is a measure of how well the linear array is representing the features in the data. The final metric, the topological error (*TE*), shows that, in general, there is no error until about six map units, and the behavior of the topological error is like that of the polar deficit in that it increases with increasing numbers of map units. Both metrics help define the ability of a linear array to represent the input data and, therefore, are an indication of the linearity of the input data in multidimensional space.

The six-unit linear array (Figure 4.49) gives a reasonable representation of the daily mean along-channel flow in Juan de Fuca Strait and how the flow transitions between the two archetypical flow regimes consisting of the "pure" estuarine mode on the extreme left and the "pure" transient mode on the extreme right. Here, outflow to the ocean is colored blue and inflow from the ocean is colored red. Contiguous pairs of map units have varying degrees of similarity. The values in the individual map units indicate: (1) the number of times that a particular map unit is the best matching unit (BMU); (2) the percentage of input data falling into the map unit; and (3) the ratio of maximum inflow (eastward current, positive) to the maximum outflow (westward current, negative). The strong inflow along the US side of the channel in the right-hand panels indicates the presence of the surface intensified "Olympic Peninsula Countercurrent" that forms during major southerly wind events (Thomson et al., 2007). Note that, while we attribute the transitional maps to a shift in the fundamental flow regime, some of the difference may actually be due to changes in instrumentation and data quality during the different deployment periods.

As in the case for Figure 4.49, the judicial selection of the parameters of the neighborhood function can result in near perfect topological ordering and a minimization of the quantization error (error in Euclidean distance). Large radii, r_i in

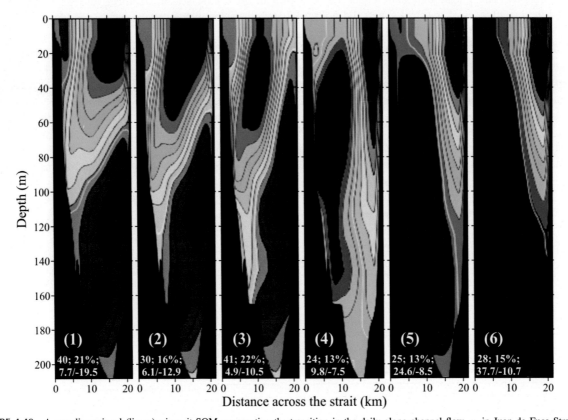

FIGURE 4.49 A one-dimensional (linear), six-unit SOM representing the transition in the daily along-channel flow, *u*, in Juan de Fuca Strait from runoff-driven estuarine circulation (1) to wind-driven transient circulation (6). Outflow toward the ocean is colored blue and inflow away from the ocean is colored red. The light white line denotes $u = 0$; *u* is positive for inflow. Numbers in each panel denote the number of times a particular map is the BMU; the fraction of the input data that is clustered into a particular map; and the peak values of the outflow (cm/s)/inflow (cm/s). *Courtesy, Steve Mihály, Ocean Networks Canada.*

(4.119) result in stiffer arrays, which then result in much larger quantization errors. Conversely, minimum values of the quantization error (D_j) can be reached by allowing the final radius to go to zero or by using no ordering at all. Allowing the radii to diminish to zero, or using no neighborhood function, results in a minimum achievable average D_j of around 289 m/s when applied to the complete data set of along-channel flow in Juan de Fuca. However, topological errors remain unacceptably large in both cases, and the map units do not appear to span a reasonable continuum of data patterns. It is found that using a decreasing radius beginning at 2 and ending at 0.9 within a Gaussian shaped neighborhood function results in topological errors of less than 1% and an average quantization error of 310 cm/s.

4.11.4.4 Comments on Computations

SOM analysis generally begins with specification of the number of map units as well as the space in which they reside and their interconnection. In the oceanographic and meteorological references, we have cited, space has been exclusively limited to a one- or two-dimensional rectangular grid, with the map dimensions chosen subjectively. As in spatial EOF modes, the individual map units have the same dimensions as the input data. After the array of map units have been initialized to some value, each input data field is compared to the array of map units. As in EOF analysis, both the input data and the map units are converted to vector form and the Euclidean distance measured between the input data and each map unit. The closest map unit is identified as the winner and is moved closer to the input data vector by a factor defined by a learning rate, which is specified by the user. At this point, the steps resemble a traditional cluster algorithm and are a form of unsupervised vector quantization. In SOM analysis not only is the winning map unit vector moved closer to the presented input vector but the map units which surround the winning map unit are also incrementally adjusted toward the input vector in inverse proportion to their distances from the winning map unit by use of a neighborhood function. The result is that the SOM array becomes topologically ordered, whereby map units that resemble one another are moved closer together and map units that differ from one another are pushed further apart. This relationship between the winning map units and its neighbors also forces the SOM to produce more map units in regions where there is high input data density as well as forcing a continuum of map units over the entire data set.

To begin a SOM analysis, the map units are first initialized to a set of values. For the SOM Toolbox in MATLAB 5 (Vesanto et al., 2000) used in the Juan de Fuca Strait study, there are two options, a random initialization based on the span of the input data and initialization with the leading EOFs. For the EOF initialization, in the case of the simplest one-dimensional SOM, the map units are initialized as the leading EOF. If the map unit array had been two-dimensional, the first two modes would have been used. Extensive testing of these two initializations indicated that both initializations resulted in the same set of map units; the randomly initialized map units took substantively longer to converge. Tests on larger (10×10) two-dimensional arrays with a small neighborhood function took as many as 9,600 iterations for map units initialized randomly to reach a stable solution. In contrast, for an EOF initialized set of map units with the same neighborhood function, 480 iterations were more than sufficient. However, it should be noted that the number of iterations is strongly affected by the parameters of the neighborhood function; when the neighborhood function is strong, many fewer iterations are needed, and the smoothing brings the SOM to stability much faster. Furthermore, if random initialization is used without a neighborhood function for larger arrays, depending on the algorithm, a few map units will remain random and not converge. In this case, unsupervised vector quantization does not necessarily converge to a stable solution.

Within the MATLAB software packages, two algorithms are available to perform the analysis. Sequential incremental SOM (the traditional method), and batch SOM. In the sequential algorithm, each input vector is presented to the map units and, using a learning rate, the map units are moved incrementally toward the input data. In the batch method, Voronoi tessellation is used to move the map units closer to the data, and no explicit learning rate needs to be specified. Testing the two algorithms using the alongshore flow data indicated that the batch SOM was faster by an order of magnitude. Furthermore, the batch algorithm with a specified set of parameters always resulted in the same solution, whereas the sequential incremental SOM algorithm did tend toward the batch algorithm solution upon increased iterations but repeated sequential algorithms did not always result in the same solution. The main benefit of the sequential algorithm is that it is amenable to a limited set of mathematical analyses, and hence has been studied to examine the behavior of the SOM as a neural network (Kohonen, 2001). Because of its speed and repeatability, Mihály and Thomson (unpublished) chose to use the batch algorithm for all the SOM analyses.

As with the bimodal analysis, Mihály and Thomson used the faster batch algorithm to derive the six-map unit presented in Figure 4.49. This approach provides sufficient iterations so that the map units reach an absorbing state after which there is no change with each iteration. For the global topology, "sheet" is initially chosen (see Vesanto et al., 2000). This places the map units in a two-dimensional array where the interconnections between the map units can be made either with a hexagonal or rectangular lattice. In the limiting case of a linear array, the choice of lattice does not change the interconnections between the map units. Using the sheet topology, the map units at the ends of the one-dimensional array are only connected in varying degrees to map units toward the center of the array. This has the tendency to enhance the polarization between the two end map units. Circular topologies can be chosen so that the two outer end map units of a sheet array are connected to each other (cylindrical) or both the sides and the top and bottom can be connected (toroidal). In addition to these parameters, the number of map units and the nature of interconnection between the map units through a neighborhood function need to be chosen. These provide a strong subjective input into the analysis that cannot be avoided.

The six map units in Figure 4.49 describe a continuum of along-channel flow states in the daily current velocity time series. To quantify the degree of similarity between map unit neighbors, we can use the Euclidean distance metric. To gauge the global similarity of the map units in the array along the linear continuum, some interpretation is necessary. Since each map unit can be represented as a point in multi-dimensional space (defined by a position vector equal in length to the number of grid points of the map unit), the distance between any two map units is easily determined. However, since the magnitude of vector addition is only equal to scalar addition if the vectors have the same "direction" in multidimensional space, it is very unlikely that the distances between map units will sum to the distance between the first and last map units. The vector defining the distance between the "polar" (the two outside) map units is a straight line, but the positions of the map units between the two pole map units will describe a curve in multi-dimensional space. The difference between these two distances is a measure of the linearity of the trajectory of map units transitioning from one polar mode to the other from a particular SOM analysis. Therefore, in addition to quantization and topological error metrics, we can define two other measures specific to a one-dimensional map array to assess SOM analysis. We define the polar range as the Euclidean distance between the first and last map unit, and we define the polar distance as the sum of the Euclidean distances between contiguous map units from the first map unit to the last map unit in the array. With these four metrics—mean quantization error, topological error, polar range, and distance—we can add some insight and objectivity in selecting the number of map units in a linear array as well as assessing the linear arrays adequacy in describing the evolution of the underlying data.

4.11.4.5 Summary

The stages of the SOM algorithm can be summarized as follows:

1. **Initialization** — Choose random values for the initial weight vectors, $\mathbf{w}_j$;
2. **Sampling** — Draw a sample training input vector, $\mathbf{x}$, from the input space.
3. **Matching** — Find the winning neuron, $J(\mathbf{x})$, with weight vector closest to input vector.
4. **Updating** — Apply the weight update equation $\mathbf{w}_j(t+\Delta t) = \mathbf{w}_j(t) + \alpha(t) \cdot \varepsilon(t)[\mathbf{x}_i(t) - \mathbf{w}_j(t)]$;
5. **Continuation** — Keep returning to step 2 until the feature maps stop changing.

4.11.5 Growing Hierarchical Self-Organizing Maps

Growing hierarchical self-organizing maps (GHSOM) were designed to remove some of the subjectivity of choosing the SOM topology. Rules are made so that an initial SOM can be grown by row and column, as well as hierarchically, such that a single map unit can spawn new layers of SOM. According to Liu et al. (2006), the GHSOM improves the basic SOM by (1) providing an incrementally growing version of the SOM, which eliminates the need for the user to directly specify the initial size of the map beforehand; and (2) by enabling the SOM to adapt to hierarchical structures in the input data. Briefly, the steps are:

Step 1

Prior to the training process, a single map (therefore, by definition, not a self-organizing map) is created. The weight vector for this single map is the mean of all input vectors (e.g., the mean flow or the mean SST). This one-unit map is deemed "layer 0", and has a mean quantization error, $\langle QE \rangle_0$, given by

$$\langle QE \rangle_0 = \sum \|\mathbf{x}_k - \mathbf{w}_i\| \tag{4.121}$$

where the double vertical lines denote the Euclidean distance. The only reason for creating this map is to obtain the above quantization error, which is then used in the next step.

Step 2

Below layer 0, a new 2×2 SOM, "layer 1" (Figure 4.50), is created. The mean QE of these 4 maps, $\langle QE \rangle_1$, is compared with $\langle QE \rangle_0$ in (4.121) with the requirement that

$$\langle QE \rangle_1 > \tau_1 \langle QE \rangle_0 \tag{4.122}$$

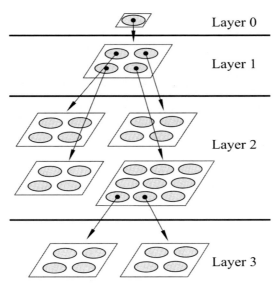

FIGURE 4.50 An example of the hierarchical structure of the growing hierarchical self-organizing maps (GHSOM). The single map from "layer 0" is expanded in the 2×2 map units. All of the four units in the first-layer SOM are expanded in the second layer. Only two units in one of the second layer SOMs are further expanded in the third layer.

where τ_1 is a number between 0 and 1. If τ_1 is set to 1, the above inequality will never be fulfilled and the "growing" will stop at this point, i.e., the result will be layer 1, the 2×2 map unit. If the inequality is fulfilled, the map unit with the greatest mean quantization error $\langle QE \rangle$ is found and defined as the error unit. Next, the most dissimilar adjacent neighbor of the error unit is identified, and a new row or column of map units is inserted between the error unit and the most dissimilar neighbor. In this way, the layer begins to grow in breadth; τ_1 is therefore called the "breadth controlling parameter". If τ_1 is set to zero, the map would grow to infinity, so the idea is to choose a reasonable value for τ_1. The default value in the MATLAB toolbox is 0.3, which indicates that the mean or cumulative QE must be 70% greater than that for the preceding QE for the map to stop growing. Mihály and Thomson (unpublished) found that a higher value (0.6) improved the analysis.

Step 3

If the inequality (4.122) is still not satisfied, the strategy is to pick map units, which might be expanded further into a new hierarchical layer. A second inequality is used:

$$QE_2 > \tau_2 \langle QE \rangle_0 \tag{4.123}$$

where τ_2 is smaller than τ_1 $(0 < \tau_2 \leq \tau_1 < 1)$; here, τ_2 is called the "depth controlling parameter" and has a MATLAB default value of 0.03 (Mihály and Thomson used a lower value of 0.005). In this process, the quantization error of each individual map is compared to the quantization error of the "layer 0" map unit. Therefore, each single map unit that satisfies the above inequality is expanded in a subsequent layer. In general, decreasing the parameter τ_1, grows SOM arrays with greater breadth, while decreasing the parameter τ_2, grows SOMs with more layers in the GHSOM hierarchy. Choosing the appropriate parameters re-introduces a strong degree of subjectivity back into the analysis. Using GHSOM for a single layer illustrates the hierarchical aspect of the process while preserving the objectiveness of the approach.

4.12 KALMAN FILTERS

In 1960, Rudolf E. Kalman published his now famous paper describing a recursive solution to the discrete-data linear filtering problem of trying to estimate the state y, which is governed by the linear stochastic difference equation,

$$y_k = Ay_{k-1} + Bu_{k-1} + \text{random (white)noise} \tag{4.124}$$

The matrix A relates the state at the present step, k, with that at the previous step, $k - 1$, and the matrix B relates the optional control input (u, such as the Navier-Stokes equations of motion) to the state y. (Here, we follow "tradition" in this topic by writing matrices in italics rather than in bold font. To avoid ambiguity, the word "matrix" is added before letters representing matrices".) Kalman's original interest was in determining the orbits of planets from limited Earth observations, but the filter soon found extensive use in autonomous and assisted navigation, with both civilian and military applications. The Kalman Filter is an efficient optimal estimator (a set of mathematical equations) that provides a recursive computational methodology for estimating the state of a discrete data-controlled process from measurements that are typically noisy, while providing an estimate of the uncertainty of the estimates. Because the filter is recursive, new measurements can be processed as they are received from external sensors. Unlike other recursive filters, the method doesn't need to store all previous measurements nor reprocess all data at each time step. If the noise associated with all processes and data contributing to the state have Gaussian (normal) distributions, then the Kalman filter obtains solutions by minimizing the mean square error of the estimated parameters. If the Probability Density Functions of the variables are not known, and only the mean and standard deviation of the noise are available, the Kalman filter is still the best linear estimator; some non-linear estimators may be better than the linear estimators.

Kalman filters have a number of advantages. Specifically, the filter provides reliable practical results due to its optimality, it is designed for real time processing, and equations involving the measurements do not need to be inverted as part of the solution. Moreover, the method is easy to formulate and implement, given a basic understanding of the underlying mathematics. As with other filters, the purpose of the Kalman filter is to determine the "best" estimate of state parameters, Y, from noisy input data. The Kalman filter not only determines values of the parameters within the constraints of the measurement uncertainty, but it also takes into account the error of the measurements relative to the error of the predictions. In essence, Kalman filters fuse prediction and measurement based on a weighted difference between an actual measurement and a measurement prediction. The method proceeds from an *a priori* estimate, $\widehat{y}_k^-$, of the state y_k at step k that is based on information of the process prior to step k (hence the minus sign superscript), to an *a posteriori* estimate of the state, $\widehat{y}_k$, at step k that incorporates newly acquired measurements, z_k (Welch and Bishop, 2006). Specifically,

$$\widehat{y}_k = \widehat{y}_k^- + K\left(z_k - H\widehat{y}_k^-\right) \tag{4.125}$$

where the new measurement

$$z_k = Hy_k + v_k \tag{4.126}$$

is given in terms of the actual state of the process, y_k, plus some uncertainty characterized by a random (white noise) variable, v_k, having a Gaussian probability distribution. Here, H is a matrix linking the state, $\widehat{y}_k$, to the measurement, matrix K is *gain* or *blending factor*, and the difference $z_k - H\widehat{y}_k^-$ is the measurement *residual* or *innovation*. The errors of the *a priori* and *a posteriori* estimates are then, respectively,

$$e_k^- = y_k - \widehat{y}_k^- \tag{4.126a}$$

$$e_k = y_k - \widehat{y}_k \tag{4.126b}$$

with corresponding error covariance functions, E, (traditionally written as P rather than C as we did earlier), where

$$P_k^- = E\left[e_k^- e_k^{-T}\right] \tag{4.127a}$$

$$P_k = E\left[e_k e_k^T\right] \tag{4.127b}$$

The key to solving (4.125) is to find that matrix K that minimizes, in the usual least-squares sense, the *a posteriori* error covariance in (4.127b). This is accomplished by substituting (4.125) into (4.126b), which is then substituted into (4.127b), deriving the indicated expected values, taking the derivative of the trace of the result with respect to K, setting the result to zero (corresponding to minimization), and then solving for K. One expression for K that minimizes (4.127b) is (Welch and Bishop, 2006)

$$K_k = \frac{P_k^- H^T}{\left(HP_k^- H^T + R\right)} \tag{4.128}$$

where R is the measurement error covariance. Note that as R approaches zero, the gain $K_k \to 1/H$ and $K\left(z_k - H\widehat{y}_k^-\right) \to \left(y_k - \widehat{y}_k^-\right)$ in (4.125). As a consequence, K, favorably weights the data in determining the *a posteriori* estimate of the state, $\widehat{y}_k$. In contrast, as the *a priori* estimate error covariance P_k^- approaches zero, $K_k \to 0$ so that the gain K weights the new estimate more favorably than the new data. As an example, consider the problem of determining the precise location of a sailboat participating in a trans-oceanic race. We assume that the boat is equipped with a GPS unit that provides an estimate of the vessel's position within a few meters but that the calculated positions are noisy, with values that "dance" around while remaining within a few meters of the actual position. [Early in our careers, we depended on Loran-C navigation for positioning in offshore regions, for which fixed positions could jump around by as much as several "cables" over periods of several minutes; a cable is 1/10 of a nautical mile]. Because the heading and speed of the boat are known, the boat's location can be computed from the GPS information provided the influence of the surface current and wind on the boat are known or can be estimated. This "dead-reckoning" method—which, in the not-so-distant past, was one of the only ways sailing ships could determine their position at sea—yields smooth estimates of the vessel's position but which drift over time. As in other applications, the Kalman filter can be thought of as operating in two distinct phases of dead-reckoning: *predict* and *update*. During the prediction phase, the boat's position can be modified by the external forces acting on the boat (the dynamic or "state transition" model). In addition to calculating a new estimate of the vessel's position, a new covariance function can also be calculated. Next, in the update phase, a measurement of the boat's position is obtained from the GPS, a measurement that, once again, has a degree of uncertainty. The covariance between the newly predicted position and that of the prediction from the previous phase determines how much the new measurement will affect the updated prediction. Ideally, if the dead-reckoning estimates tend to drift away from the real position, the GPS measurement will help "nudge" the position estimate back toward the real position, while not perturbing the system to the point that the predictions become rapidly changing and noisy. We have summarized the above procedure in Figure 4.51, where the first prediction phase is based on previously available data and precedes the collection of new data in the measurement phase. The variances of the two steps are related.

In Figure 4.51, the conditions for the state and its variance at the previous data step are $\widehat{y}_{k-1}$ and σ_{k-1}, respectively; the prediction for the state and the variance at the next step are $\widehat{y}_k^-$ and σ_k^-. We use the dynamical model (i.e., physical system), together with the initial conditions, to make this prediction. We then take the measurement z_k and compute the corrected state, $\widehat{y}_k$ (with variance σ_k), by blending the prediction and the residual, which is always a case of merging two Gaussian variables. The result is an optimal estimate with a smaller variance.

The blending factor, K, is controlled by knowledge gained from the measurements or from the prediction. As noted earlier, if the prediction is more certain, then the prediction error covariance P_k decreases to zero, K also approaches zero and the filter weights the prediction more heavily than the residual, $z_k - H\widehat{y}_k^-$. In contrast, if the measurements are more certain that the prediction, then the measurement covariance (R) approaches zero as K approaches H^{-1} and the filter weights the residual more heavily than the prediction. The computational basis for the Kalman filter equations is outlined in Figure 4.52, which shows both the prediction and measurement phases of the Kalman filter.

In Figure 4.52, the recursive nature of the filter equations is indicated by the arrows at the top and bottom of the boxes. The boxes represent the prediction and measurement steps of the Kalman filter. The basic assumptions behind the filter are:

a. The model used to predict the "state" needs to be a linear function of the measurements.
b. The model error and the measurement error (measurement noise) must both be Gaussian with zero means.

Even if the noise is not Gaussian, and we know only the mean and standard deviation of the noise and not its probability distribution, the Kalman filter is still the best linear estimator. For highly non-Gaussian distributions, non-linear estimators may be preferable.

It worth emphasizing that the Kalman filter combines a system's dynamics (i.e., the physical laws of motion) known to control the inputs to that system with multiple sequential measurements to form an estimate of the systems varying quantities (its state) that is better than the estimate obtained by using any one measurement alone. Noisy sensor data, approximations to the physical equations that describe how a system changes with time, and unaccounted for external factors, introduce uncertainty in the system's state. The Kalman filter averages a prediction of the system's state with a newly acquired measurement whose contribution is determined by a series of weights. These weights, which also take into account the "trustworthiness" of the data relative to the prediction, are calculated from the covariance function, a measure of the estimated uncertainty of the prediction of the system's state. As a result, the new state estimate lies between the

---

The Kalman Equations

Make prediction based on previous data: $\widehat{y}^-, \sigma^-$

↓

Take measurement: z_k, σ_z

↓

Optimal estimate $(\widehat{y})$ = Prediction + (Kalman Gain) * (Measurement − Prediction)

Variance of estimate = Variance of prediction * (1 − Kalman Gain)

FIGURE 4.51 The sequence of the Kalman equations for estimation of the variable y.

FIGURE 4.52   The Kalman filter equations (see text for details and functions).

The set of Kalman Filtering Equations in detail

Prediction (Time Update)	**Correction (Measurement Update)**
(1) Project the state ahead	(1) Compute the Kalman Gain
$\hat{y}_k^- = A y_{k-1} + B u_k$	$K = P_k^- H^{\mathrm{T}} (H P_k^- H^{\mathrm{T}} + R)^{-1}$
(2) Project the error covariance ahead	(2) Update estimate with measurement z_k
$P_k^- = A P_{k-1} A^{\mathrm{T}} + Q$	$\hat{y}_k = \hat{y}_k^- + K(z_k - H\hat{y}_k^-)$
	(3) Update Error Covariance
	$P_k = (I - KH) P_k^-$

predicted and the measured state and has a smaller estimated uncertainty than either of the two alone. This process is repeated for every time step, with the new estimate and its covariance forming the prediction used in the subsequent iteration. Hence, the recursive character of the Kalman filter, which requires only the last "best guess" rather, then the entire history of a system's state to calculate the new state.

Because the certainty of the measurements is often difficult to specify precisely, it is common to discuss the filter's behavior in terms of its gain. The Kalman gain depends on the relative uncertainty of the measurements and the current state estimate, and it can be adjusted to achieve a particular desired performance of the filter. For a high gain, the filter places more weight on the measurements and follows them more closely; for a low gain, the filter conforms more to the model predictions smoothing out noise and decreasing the responsiveness of the system. At the extremes, a gain of zero causes the measurements to be ignored completely while a gain of unity causes the estimate of the state to be ignored entirely.

Implementation of the Kalman filter is often difficult to achieve in practice due to the requirement for a good estimate of the noise covariance matrices. A study by Furrer and Bengston (2007) used Monte Carlo methods to estimate the Kalman filter variants. Another promising approach is the Autocovariance Least-Squares (ALS) technique that uses the autocovariance of the data to estimate the noise covariance. A study by Odelson et al. (2006) demonstrates that the noise covariances estimated in this way are unbiased and converge to the true values with increasing sample size. They also add positive semi definiteness constraints to these covariances. Abdel-Hafez et al. (2008) uses the same technique to estimate the Global Positioning System (GPS) measurement noise-covariance matrix.

The Kalman filter is known to be optimal because: (a) the model perfectly matches the real system; (b) the measurement noise is white; and (c) the covariances of the noise are exactly known. We have already discussed methods to estimate the error covariances. Once these are known, it is useful to estimate the performance of the Kalman filter itself, i.e., to determine whether it is possible to further improve the state estimation quality. We also know that if the Kalman filter works optimally, the output prediction error is also white noise. This white noise character properly reflects the state estimation quality. To evaluate the performance of the Kalman filter, it is necessary to inspect the "whiteness" of the predictions.

An interesting example of Kalman filter application is given by www.cs.cornell.edu/Courses/cs4758/2012sp/.../MI63slides.pdf. This example concerns tracking the fluid level in a tank being filled. For this problem, we can write the analysis sequence as:

Predict:

$$\hat{y}_{t|t-1} = A_t \hat{y}_{t-1|t-1} + B_t u_t \tag{4.129a}$$

$$P_{t|t-1} = A_t P_{t-1|t-1} A_t^T + Q_t \tag{4.129b}$$

where: $\hat{y}$ is the estimated state; A is the state transition matrix (i.e., the matrix describing the transition between states); u represents the optional control variables on the state; B is the control matrix (i.e., the matrix mapping the control variables to the state variables); P is the state error covariance matrix (i.e., error of the estimation); and Q is the process covariance error matrix (i.e., the error due to the process). Subscripts are $t|t=$ the current time, $t-1|t-1=$ the previous time, and $t|t-1$ is time at intermediate steps. The Kalman filter removes noise from the system by assuming a pre-defined model of the system. This model should be defined by the following:

1. Understand the situation: Examine the problem and break it down in to the mathematical basics.
2. Model the state process: Start with a basic model, which may not be perfect at first, but can be refined later.
3. Model the measurement process: Analyze how to measure the process. The measurement space may not be in the same space as that of the state (e.g., using an electrical diode to measure weight, an electrical reading does not directly translate to weight).
4. Model the noise: This needs to be done both for the state and the measurements processes. The basic Kalman filter assumes Gaussian white noise so that the variance and the covariance (error) functions are meaningful (i.e., make sure that the error you model is suitable for the situation).
5. Test the filter: This step is often overlooked. Use synthetic data if needed. See if the filter is behaving as it should.
6. Refine filter: Try to change the noise parameters (filter), as these are the easiest to change. If necessary, go back further and rethink the situation.

As an example, consider the water in a tank (Figure 4.53) and assume there is a measurement parameter that gives the water level of the tank as it fills or empties. The goal is to estimate the unknown level of water in this tank. In this example, the water level measurements are provided by a float in the tank. The tank could be: (a) Filling, emptying or static (level is increasing, decreasing or constant); or (b) sloshing around or static (level changes from side to side or is flat and constant). We first consider the most basic model in which the fluid in the tank is level (horizontal) and the surface is constant ($L = c$). Using Eqns (4.129a) and (4.129b), we can write the state variable as a scalar so that $\hat{y} = y$, where y is now the estimate of L. Since we are assuming a constant model, there is no time variation and $y_{t+1} = y_t$, so that $A = 1$ for any $t \geq 0$. Both B and $u = 0$.

We now need to model the measurement process, z, which for simplicity we assume is precisely the water level in the tank. Finally, we need to model the noise. Again, for simplicity, we assume that there is noise in the measurement, and that $R = r$ in the expression (4.126) for the Kalman gain. The process is a scalar so we can assume that $P = p$; since the process is not well defined, we adjust the noise so that $Q = q$. To test the filter, we apply Eqns (4.129a) and (4.129b) whereby

$$y_{t|t-1} = y_{t-1|t-1} \tag{4.130a}$$

$$P_{t|t-1} = P_{t-1|t-1} + q_t \tag{4.130b}$$

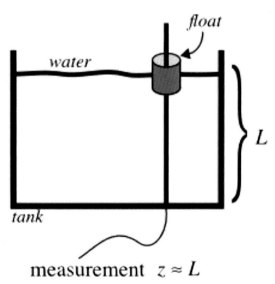

FIGURE 4.53 Water level in a tank. Measurements are provided by the float.

The update is

$$y_{t|t} = y_{t|t-1} + K_t\left(z_t - y_{t|t-1}\right) \tag{4.131a}$$

$$K_t = p_{t|t-1}\left(p_{t|t-1} + r\right)^{-1} \tag{4.131b}$$

$$p_{t|t} = (1 - K_t)p_{t|t-1} \tag{4.131c}$$

where, the new data value is z_t. The filter is now completely defined. To put some numbers into this model, we first assume that the true level is $L = 1$. We initialize the state with an arbitrary number, with an extremely high variance, as it is completely unknown. Specifically, $y_0 = 0$ and $p_0 = 1,000$. If one initializes with a more meaningful value, the filter solution will convergence faster. We choose the system noise as $q = 0.001$ since we believe that we have a realistic and accurate model. Thus,

Predict 1:

$y_{1|0} = 0$
$p_{1|0} = 1,000 + 0.001$

The hypothetical measurement we obtain from the float is $z_t = z_1 = 0.9$, which differs from the true water level value of $L = 1.0$ because of noise. Assuming a measurement noise of $r = 0.1$, we can write:

Update 1:

$K_1 = (1,000 + 0.001) \cdot (1,000 + 0.1)^{-1} = 0.9999$
$y_{1|1} = 0 + 0.9999(0.9 - 0) = 0.8999$
$p_{1|1} = (1 - 0.9999)\,(1,000 + 0.001) = 0.1$

As indicted by this first update step, the initialization value $y_{1|0} = 0$ has been brought close to the true value, $L = 1$, of the system. In addition, the error variance, $p_{1|1}$, has diminished to a more reasonable value.

If we do another step, we have:

Predict 2:

$y_{2|1} = 0.8999$
$p_{2|1} = 0.100 + 0.001 = 0.1001$

and, assuming a second measurement $z_2 = 0.8$ (again differing from unity because of noise), we find:

Update 2:

$K_2 = 0.1001(0.1001 + 0.1)^{-1} = 0.5002$
$y_{2|2} = 0.8999 + 0.5002(0.8 - 0.8999) = 0.8499$
$p_{2|2} = (1 - 0.5002)\,0.1001 = 0.0500$

If we continue this process, we eventually obtain the results in Table 4.11:

Based on the results in Table 4.11, the model works successfully. After stabilization at about time step 4, the estimated state is within 0.05 of the true value, whereas the specified measurements are between 0.8 and 1.2 (i.e., only within 0.2 of the true value). The results are plotted in Figure 4.54, where the purple line represents the values estimated by the Kalman filter; the yellow line is the true value (a constant), and the dark blue line are the input measurements.

As this example shows, the Kalman filter provides an optimal estimate of the true value even when the measurements are very noisy (a 20% error in measurements resulted in only a 5% inaccuracy in the Kalman filter estimate). Hence, the Kalman filter has achieved its purpose.

We now examine a more realistic case in which the tank fills at a constant rate, f, such that the water level $L_t = L_{t-1} + f$. We assume that $f = 0.1$ per unit time and start with $L_0 = 0$. We will also assume that the measurement and the process noise are constant with time (i.e., $q_t = 0.001$ and $r_t = 0.1$). Table 4.11 now takes the form shown in Table 4.12.

We see that, over time, the estimated state stabilizes (i.e., the variance becomes very small). While the estimate reduces the noise, it dramatically underestimates the true value, L, which is much closer to the measured values, z (Figure 4.56).

It is clear from Figure 4.55 that the estimates systematically underestimate the true and even the measured values. This is not a very attractive result even though the noise level has been greatly reduced. There are two possible sources for this problem: (a) the model we have chosen; and/or (b) the reliability of our process model (our chosen q value). The easiest

TABLE 4.11 Kalman filter applied to the water tank problem (Figure 4.53). The predict and update parameters denote the predictions of the water level, y, and error estimation, p, followed by the measured values derived from the updated observations.

Time (t)	Predict		Update			
	$y_{t\|t-1}$	$p_{t\|t-1}$	z_t	K_t	$y_{t\|t}$	$p_{t\|t}$
3	0.8499	0.0501	1.1	0.3339	0.9334	0.0334
4	0.9334	0.0335	1	0.2509	0.9501	0.0251
5	0.9501	0.0252	0.95	0.2012	0.9501	0.0201
6	0.9501	0.0202	1.05	0.1682	0.9669	0.0168
7	0.9669	0.0169	1.2	0.1447	1.0006	0.0145
8	1.0006	0.0146	0.9	0.1272	0.9878	0.0127
9	0.9878	0.0128	0.85	0.1136	0.9722	0.0114
10	0.9722	0.0115	1.15	0.1028	0.9905	0.0103

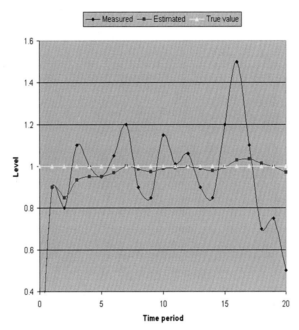

FIGURE 4.54 Kalman filter estimates (purple line) of a constant water level (yellow line) in a tank. The measured values (z) input to the filter are shown in dark blue.

corrective approach is to change the q value. A valid question is "Why did we chose $q = 0.001$ to begin with?" The answer is that we thought our model was a good estimation of the true process and that the error level would be quite small. Apparently, our model was not as good as we had anticipated so we need to relax this requirement. Specifically, we now assume there is a greater error with our process model and set $q = 0.01$. Application of the new q-value to the previous model results in the values plotted in Figure 4.56.

As indicated by Figure 4.56, changing q greatly improved the accuracy of the estimate, although the estimates are still significantly below the true and measured values. If we increase q once again by setting $q = 0.1$, we obtain the results plotted in Figure 4.57, which shows an even greater improvement. The Kalman filter estimate begins to track the noisy measurements, an intentional outcome of increasing q, but still has a bit less noise than the measured points. The filter estimates now approach the true value more effectively than the estimates using smaller noise values.

It is evident that further increases in the process noise level will cause the estimated values to increasingly match the measured values, giving no benefit to using the Kalman filter. The lesson here is that a poorly defined model will not

TABLE 4.12 Kalman filter applied to the water tank problem (Figure 4.53) for the specific values presented in the text (specifically, $f = 0.1$ per unit time and initial water level $L_0 = 0$). The last column gives the actual water level.

	Predict		Measurement and update				Actual
Time (t)	$y_{t\|t-1}$	$P_{t\|t-1}$	z_t	K_t	$y_{t\|t}$	$P_{t\|t}$	L
0	—	—	—	—	0	1,000	0
1	0.000	$1{,}000 \times 0.0001$	0.11	0.9999	0.1175	0.100	0.1
2	0.1175	0.1001	0.29	0.5002	0.2048	0.0500	0.2
3	0.2048	0.0501	0.32	0.3339	0.2452	0.0334	0.3
4	0.2452	0.0335	0.50	0.2509	0.3096	0.0251	0.4
5	0.3096	0.0252	0.58	0.2012	0.3642	0.0201	0.5
6	0.3642	0.0202	0.54	0.1682	0.3945	0.0168	0.6

FIGURE 4.55 Kalman filter estimates (purple line) of a water tank filling at a constant rate $f = 0.1$ per unit time with constant error $q = 0.001$ per unit time. The yellow line shows the true water level, L, and the dark blue line the water level (z_t) measured by the float.

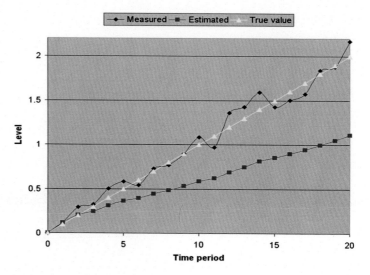

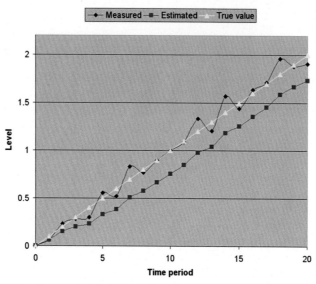

FIGURE 4.56   As with Figure 4.55, but with a much smaller constant error $q = 0.01$ for the process model.

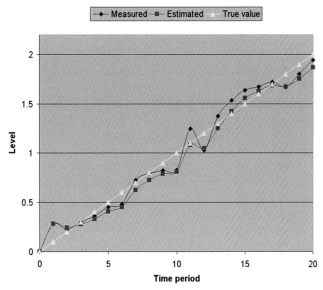

FIGURE 4.57   As with Figures 4.55 and 4.56, but with an even much smaller constant error $q = 0.1$ for the process model.

provide a good filter estimate. However, increasing the estimated error will allow the Kalman filter estimate to rely more on the measurement values, while still allowing some noise removal.

The primary use of Kalman filtering in physical oceanography is for the assimilation of oceanographic data into numerical models (Pham et al., 1998). This application uses the Extended Kalman Filter (EKF), which applies to non-linear state systems or measurement equations. Here, a Kalman filter is a linearized version of these equations, which continues to yield an optimal solution. The key to this form of data assimilation is to approximate the error covariance of the data to be assimilated into the numerical model. This procedure amounts to making no correction in those directions for which the error is the most attenuated by the system. This has the added benefit of improving the filter stability. These "directions of correction" evolve with time according to the model evolution, which is a primary feature of this filter that distinguishes it from other sequential assimilation methods. Pham et al. (1998) suggest a method for initializing the filter based on empirical orthogonal functions (EOFs) discussed earlier in this chapter. They examine assimilation of wind stress forcing into a simple quasi-geostrophic (QG) model for a square ocean domain. Although this is an unrealistic test case, the results of this assimilation method are very encouraging.

In a study of coastal ocean problems, Chen et al. (2009) compared the reduced rank Kalman filter (RRKF) with the ensemble Kalman filter (EnKF) and the ensemble square-root Kalman filter (ENSKF) in three idealized regimes: (a) A flat bottom circular shelf driven by tidal forcing at the open boundary; (b) a linear slope continental shelf with river discharge; and (c) a rectangular estuary with tidal flushing intertidal zones and freshwater discharge. They used the unstructured grid Finite-Volume Coastal Ocean Model (FVCOM). Model run comparisons showed that the success of the data assimilation method depends on sampling location, assimilation methods (univariate or multivariate covariance approaches), and the nature of the dynamical system. In general, for these applications, the EnKF and ENSKF work better than RRKF, particularly for time dependent cases with large perturbations. In EnKF and ENSKF, multivariate covariance methods should be used to avoid the appearance of unrealistic numerical oscillations. Because the coastal ocean features multiscale dynamics in both time and space, an individual case-by-case approach should be used to determine the most efficient and reliable data assimilation approach for different dynamical systems.

4.13 MIXED LAYER DEPTH ESTIMATION

Much of this chapter has focused on methods for constructing smoothed oceanic fields from large-scale spatial and temporal data. Some methods take into account the dynamics of the physical system, while others take a strictly statistical approach. Little has been said about methods designed to delineate physical "breaks" in oceanic distributions. Here, we gave a brief overview of techniques used to define the depth of the surface mixed layer—the top layer of the ocean characterized by uniform to nearly uniform vertical water property structure—which is of interest to a broad variety of oceanic studies including upper ocean productivity, air-sea exchange processes, and climate variability (cf. Curry and Roy,

1989; Robinson et al., 1993; Wijesekera and Gregg, 1996; Kara et al., 2000a,b). Methods for determining temporal "regime shifts" in the ocean are presented in Chapter 5.

Surface wind stress, convective cooling, breaking waves, current shear, and other turbulent processes in the upper ocean generate a surface layer characterized by uniform to near-uniform density, active vertical mixing, and high turbulent dissipation. The depth of this "mixing layer" is ultimately determined by a balance between the destabilizing effects of mechanical mixing and the stabilizing effects of surface buoyancy flux. As a result of temporal variations in these opposing affects, the mixing layer may be imbedded in a deeper "mixed layer" of almost identical density to the mixing layer and representing the time-integrated response to previous mixing events. Mixed layer depth (MLD) can vary by tens of meters over a diurnal cycle and by over 100 m over an annual cycle (cf. Large et al., 1994).

In the absence of direct turbulent dissipation measurements, mixed layer depth is derived from oceanic profile data using a variety of proxy variables. Methods for estimating the MLD from CTDs and other profiling instrumentation data fall into five broad categories: (1) Threshold methods, which find a pre-defined step in the surface profile (Price et al., 1986; Lukas and Lindstrom, 1991; Peters et al., 1988; Giunta and Ward, 2022) or find a critical gradient for the upper layer (Lukas and Lindstrom, 1991); (2) least-squares regression methods, which fit two or more line segments to near-surface profiles (Papadakis, 1981, 1985); (3) integral methods, which calculate a depth-scale for the upper layer based on integral properties of the water column such as conservation of mass (Ladd and Stabeno, 2012; Freeland, 2013; Freeland et al., 1997) or the amount of potential energy available in the upper water column to fully homogenize it to uniform potential density (Reichl et al., 2022); (4) a split-and-merge algorithm that fits straight line segments based on a specified error minimization (Thomson and Fine, 2003, 2009); and (5) hybrid methods that combine versions of the above approaches with regional clustering and pattern recognition algorithms (Holte and Talley, 2009; Gu et al., 2022). The methods have fundamentally different approaches. Methods 1 and 4 consider the mixed layer to be a physically distinct entity whose depth can be determined from the observed density structure independently of any integral conservation constraints. In contrast, method 3 views the mixed layer as the upper component of a two-layer approximation to a continuous density profile who derivation must satisfy certain conservation requirements such as conservation of total mass. Method 5 involves the development of algorithms that combine threshold criterion, the shape of each profile and the gradient criterion using density and temperature profiles. After assembling a suite of possible MLD values, the method then analyzes the patterns in the suite to select a final MLD estimate.

4.13.1 Threshold Methods

In air-sea interaction studies (e.g., Wijesekera and Gregg, 1996; Smyth et al., 1996a,b), the depth of the surface mixed layer, D, is defined as that depth, z, at which the potential density difference $\Delta\sigma_\theta(z) = \sigma_\theta(z) - \sigma_\theta(z_o)$ in the upper ocean exceeds a specified threshold value, typically $0.01 \, kg \cdot m^{-3}$ (Figure 4.58a); here, z_o is a reference depth (generally in the range $z_o = 0$, the ocean surface, to 10 m depth) and $\sigma_\theta(z) = \rho_\theta(z) - 1,000 \, kg \cdot m^{-3}$ is the density anomaly for measured potential density, ρ_θ. Because the threshold method is comparatively simple—depth estimates can be made by hand without analytical computations—and because the threshold difference of $0.01 \, kg \cdot m^{-3}$ generally yields diurnal depth estimates similar to those from turbulent dissipation measurements, the threshold difference method with the threshold $0.01 \, kg \cdot m^{-3}$ has become the *de facto* standard for many mixed layer depth studies. In ocean-climate studies, where focus is on the variability in MLD averaged over periods of months and longer, threshold values in excess of $0.125 \, kg \cdot m^{-3}$ are more commonly used for monthly mean data (cf. Table 1 in Kara et al., 2000b). A drawback to the $0.01 \, kg \cdot m^{-3}$ threshold method is that it often neglects the underlying water of near-identical density encompassing high chlorophyll, nutrient, and particle concentrations (e.g. Robinson et al., 1993; Washburn et al., 1998) and ignores the fact that "… the retreat of turbulent mixing to shallower depths proceeds faster than the erosion of the stratification at the base…" (Kara et al., 2000b). Giunta and Ward (2022) further note that MLD values based on density thresholds typically over-estimate the MLD when compared to that based on turbulent dissipation profiles.

Because oceanographers have access to more temperature profile data than salinity (and hence, density) profile data, and because salinity measurements tend to be noisier than temperature measurements, the mixed layer depth is commonly linked to a step-like change in water temperature, with specified steps in the range $0.01-0.5°C$ (e.g., Levitus, 1982; Weller and Plueddemann, 1996; Kara et al., 2000a,b). However, where possible, it is better to use sigma theta (σ_θ) as an estimator for MLD, primarily because it is the density structure, which directly affects the stability and degree of turbulent mixing in the water column. *In situ* density (sigma-t, σ_t) is less reliable because over-turning turbulence can result in adiabatic changes of $0.04 \, kg \cdot m^{-3}$ over depths of 10 m (Schneider and Müller, 1990).

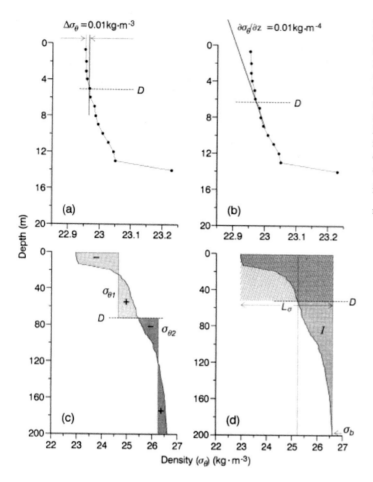

FIGURE 4.58 Estimation of mixed layer depth (D) for different methods for the density profile obtained for a CTD station LH07 off the west coast of Vancouver Island, July 19, 1997 (see Figure 4.60 for station location). (a) Threshold difference method with standard density step $\Delta\sigma_\theta(z) = \sigma_\theta(z) - \sigma_\theta(0) = 0.01$ kg·m^{-3}; (b) threshold gradient method for $\partial\sigma_\theta/\partial z = 0.01$ kg·m^{-4}; (c) two-segment least-squares method for $z_a = 10$ m, $z_b = 200$ m, constant $\sigma_{\theta 1}$ and $\sigma_{\theta 2}$, and $\sigma(D) = (\sigma_{\theta 1} + \sigma_{\theta 2})/2$; and (d) integral depth-scale, D, with $z_a = 10$ m and $z_b = 200$. In (c), each of the paired shaded regions have equal positive ($+$) and negative ($-$) areas. In (d), the darkly shaded rectangle has sides of length D and $L_\sigma = \sigma_\theta(z_b) - \sigma_\theta(z_a)$, with the area $D \times L$ of the rectangle equal to the vertical integral $I = \int_{z_b}^{z}[\sigma_\theta(z) - \sigma_{\theta b}]dz$ denoted by the shaded region to the right of the density profile. *From Thomson and Fine (2003).*

The less frequently used threshold gradient method defines mixed layer depth as the depth at which the density gradient, $\partial\sigma_\theta/\partial z$, first exceeds 0.01 kg·m^{-4} (Figure 4.58b) or other specified level. The density gradient method is considered a less consistent estimator of MLD than the density difference approach (Schneider and Müller, 1990).

4.13.2 Step-Function Least-Squares Regression Method

Problems with profile approximations have led to the formulation of customized curve-fitting algorithms for oceanic profiles, including the use of least-squares linear approximations (Papadakis, 1981; Freeland et al., 1997) and "form oscillators" (Papadakis, 1985). Papadakis (1981) uses a three-segment linear fit and the Newtonian approximation method to find a minimum variance solution to the general mixed layer depth problem. Freeland et al. (1997) uses a two-segment least-squares approach to obtain a time series of winter mixed layer depth at Ocean Station "P" (50°N, 145°W) in the northeast Pacific. The two-segment case can be solved analytically and is useful for pre-CTD data. The three-segment case requires special techniques (e.g., Papadakis, 1985) and solutions can be unstable.

The two-segment approach (Figure 4.58c) seeks a step-like least-squares approximation to a continuous water density profile, $\sigma_\theta(z)$, with z positive downward from the surface such that

$$\sigma_\theta(z) = \begin{cases} \sigma_{\theta_1} & 0 \le z_a < z < D \\ \sigma_{\theta_2} & D < z < z_b \end{cases} \tag{4.132}$$

where z_a is a near-surface depth, D is the estimated mixed layer depth (more realistically, the pycnocline depth rather than the base of the uniformly mixed surface layer), $z_b = 200-500$ m is an arbitrary depth below the depth of seasonal mixing, and $\sigma_{\theta_1}, \sigma_{\theta_2}$ are constant potential densities for the mixed layer and intermediate layer, respectively. Minimizing the integral

$$\Phi = \int_{z_a}^{D} [\sigma_\theta(z) - \sigma_{\theta_1}]^2 dz + \int_{D}^{z_b} [\sigma_\theta(z) - \sigma_{\theta_2}]^2 dz \tag{4.133}$$

with respect to $\sigma_{\theta_1}, \sigma_{\theta_2}$ and D leads to the solution

$$F(D) = \sigma_\theta(D) - \frac{1}{2}\left[\frac{\int_{z_a}^{D} \sigma_\theta(z)dz}{D - z_a} + \frac{\int_{D}^{z_b} \sigma_\theta(z)dz}{z_b - D}\right] = 0, \tag{4.134}$$

which can be solved numerically. Two-segment approximations are computationally stable. Increasing the number of segments or "steps" can improve the approximation to the profile data but typically leads to greater complexity. The quality of the least-squares fit varies from profile to profile depending on the structure of the underlying layering.

4.13.3 Integral Depth-Scale Method

A simple estimate of the mixed layer depth, D, is the integral depth-scale (Figure 4.58d), also called the "trapping depth" (Price et al., 1986), derived from

$$D = \frac{\int_0^{z_b} z N_b^2(z) dz}{\int_0^{z_b} N_b^2(z) dz} = \frac{\int_{z_a}^{z_b} (\sigma_{\theta_b} - \sigma_\theta) dz}{\sigma_{\theta_b} - \sigma_{\theta_a}} \tag{4.135}$$

where z_a and z_b are a near-surface depth and an arbitrary reference depth (e.g., $z_a = 0$ and $z_b \sim 250$ m) and $\sigma_{\theta b} = \sigma_\theta(z_b)$, $\sigma_{\theta a} = \sigma_\theta(z_a)$ (cf. Freeland et al., 1997). Here,

$$N(z) = \left(-\frac{g}{\rho_o}\frac{d\rho_\theta}{dz}\right)^{1/2} \tag{4.136}$$

is the buoyancy (Brunt-Väisälä) frequency, g is the acceleration of gravity, and ρ_o is a reference density. Unlike the threshold methods, which often require only the upper portion of a density profile, the step-function and integral-depth approaches require specification of a deep reference density that is much deeper than the mixed layer depth.

A second integral approach to fitting a two-layer function to a continuous profile finds the mixed layer depth (in effect, the pycnocline depth) by requiring that a parameter, Ψ_{fit}, proportional to the potential energy of the two-layer system, be equal to a corresponding value, Ψ_o, for the original continuous profile of the water column. Once again, we let $\sigma_{\theta 1}, \sigma_{\theta 2}$ be the constant densities of the two layers (mixed and reference layers, respectively) and consider estimates made relative to a deep reference level, $z_b = H = 250$ db (~ 250 m). Following Ladd and Stabeno (2012), we can write

$$\Psi_o = \int_0^H [\sigma_\theta(z) - \overline{\sigma_\theta}] z dz \tag{4.137a}$$

where

$$\overline{\sigma_\theta} = \frac{1}{H}\int_0^H \sigma_\theta(z)dz \tag{4.137b}$$

is the depth-averaged density. Setting the integrals in (4.137a) and (4.137b) equal to the corresponding values obtained for a two-layer representation, specifically,

$$\Psi_{fit} = \frac{1}{2}\left[D^2(\sigma_{\theta_1} - \overline{\sigma_\theta}) + (H^2 - D^2)(\sigma_{\theta_2} - \overline{\sigma_\theta})\right] \tag{4.138a}$$

and

$$\overline{\sigma_\theta} = \frac{\lfloor D\sigma_{\theta_1} + (H-D)\sigma_{\theta_2} \rfloor}{H} \tag{4.138b}$$

yields an estimate for the mixed layer depth,

$$D = \frac{2\Psi_o}{H(\overline{\sigma_\theta} - \sigma_{\theta_1})} \tag{4.139}$$

Solution to (4.139) requires that we specify a value for the upper layer density, σ_{θ_1}. In his study of winter mixed layer depth in the northeast Pacific, Freeland (2013) arbitrarily sets σ_{θ_1} equal to the mean density of the well-mixed upper layer averaged over the top 60 m. This is considered to be the maximum depth that can be chosen without approaching the shallowest mixed layer depth (~ 90 m) ever observed in winter in this region. The method does work for summer months when solar heating and reduced wind mixing can cause the mixed layer depth in the northeast Pacific to shoal to 20 m or less. The method was used to examine temporal variability in the winter mixed layer depth and pseudo potential energy of the water column, Ψ, at Ocean Station P in the northeast Pacific.

4.13.4 The Split-and-Merge Algorithm

The threshold difference method often yields instances where the estimated mixed layer depth differs markedly from the visually estimated mixed layer depth (i.e., the depth to the top of the first pycnocline estimated from density plots). This, and the fact that depth estimates from the step-function and integral-scale methods are more representative of the main pycnocline depth than the surface mixed layer depth, led to formulation of the split-and-merge algorithm for estimating mixed layer depth (Thomson and Fine, 2003, 2009). First developed by Pavlidis and Horowitz (1974) to estimate the optimal decomposition of plane curves and waveforms, the method can also be used to approximate other structural features in the water column.

The split-and-merge algorithm approximates a specified curve using piecewise polynomial functions in which the breakpoints in the fitted curve (locations of subset boundaries and changes in slope) are adjusted to fit the available data (Figure 4.59). The algorithm provides profile decomposition by defining the locations of the breakpoints separating the different segments, the piecewise approximation parameters, and the error of the approximation. In general, the fitted segments are disjointed but can be made to satisfy the continuity requirement by a retrospective local adjustment of the approximating curves. The method addresses the problem of fitting a series of segments to a profile, $\phi(z)$, where z is depth and ϕ represents temperature, salinity, density, fluorescence, or other profile variable. Given a set of points $S = \{z_i, \phi_i\}$, $i = 1, 2, ..., N$, we seek the minimum number, n, such that S is divided into n subsets $S_1, S_2, ..., S_n$ in which the data points for each subset are approximated by a polynomial of order at most $m-1$ with an error norm less than some specified quantity, ε. The algorithm merges adjacent fitted segments with similar approximating coefficients and splits those segments with unacceptable error norms. A least-squares method, or similar approach, is used to fit the curve to the specified data set. Because fitting higher order polynomials has its own set of problems, the method is generally restricted to piecewise linear approximations, for which $m = 2$. The uppermost segment (the mixed layer) is assumed to have a near-uniform vertical distribution so that $m = 1$ for this segment.

The split-and-merge method removes the need for careful *a priori* choice of the number of segments, n, as well as the need to specify the initial segmentation. In this way, it is straightforward to obtain segmentation where the error norm on each segment (or over all segments) does not exceed a specified bound. Computational experience indicates that the local minima found by the split-and-merge algorithm are close to the global minima for computational time of order N. For a specified error norm, the mixed layer depth, D, is equated with the uppermost segment of the piecewise-linear fit. To avoid scaling problems for profile variables, including the need for a different error norm for each variable, the variables z and $\phi(z)$ are normalized such that

$$z^* = \frac{z - z_{\min}}{z_{\max} - z_{\min}}; \; \phi^*(z) = \frac{\phi(z) - \phi_{\min}}{\phi_{\max} - \phi_{\min}} \tag{4.140}$$

where subscripts denote the maximum and minimum values of the given variable over the depth range of interest and $0 \le (z^*, \phi^*(z)) \le 1$. Following other methods (see Table 4.13), the split-and-merge algorithm uses a non-zero starting depth (e.g., $z_{\min} = 2.5$ m), to avoid problems associated with prop-wash or turbulent flow past the hull of the ship during station keeping, and $z_{\max} \ge 150$ m to ensure capture of the mixed layer depth regardless of season. As with the threshold method, the split-and-merge algorithm requires a predefined error norm. To determine the sensitivity of MLD estimates to the specified error norm, Thomson and Fine (2003) examined MLD estimates over a wide range of error values, from 0.001 to

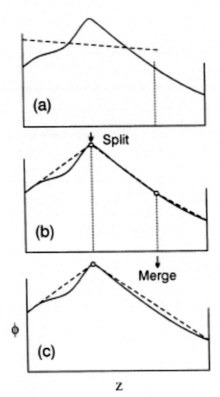

FIGURE 4.59   Illustration of the split-and-merge algorithm for which an optimum segmentation can be found in one iteration. The initial fit (dashed line) to the left-hand segment of the curve $f(z)$ is split at the breakpoint to form three separate segments. The two right-hand segments are then merged to form one segment. *From Thomson and Fine (2003).*

0.03, and found that results for the norm $\varepsilon = 0.01$ typically gave values that were closest to "visual" MLD estimates for CTD profiles collected in July 1997 along ship survey lines off the west coast of Vancouver Island, Canada (Figure 4.60). For the density profile data examined, this error norm yielded the same mean mixed layer depth, $\overline{D}$, as the threshold method for $\Delta\sigma_\theta = 0.03$ kg m^{-3}; the error norm $\varepsilon = 0.003$ yielded the same $\overline{D}$ as the "standard" threshold method for which $\Delta\sigma_\theta = 0.01$ kg m^{-3}. Because the mean MLD estimates determined by the threshold and split-and-merge methods are nearly identical, a second-order statistic (the standard error) was used as a measure of the relative "performance" or predictive skill of a given method.

4.13.5 Hybrid Methods

Hybrid methods combine individual approaches, such as those described above, into a single algorithmic approach in estimating the mixed layer depth. Algorithm methods, such as that formulated by Holte and Talley (2009), analyze patterns in water column profiles and apply logical decision trees to determine whether threshold, gradient, or other variables are optimal choices for the mixed layer depth (MLD) for a given oceanic regime. Because the original focus was on Argo drifter data from the Southern Ocean and South Atlantic, where the surface mixed layer blends into the underlying water column structure in winter, separate decision trees were developed for summer and winter conditions. Thus, the algorithm begins by determining whether a given CTD profile resembles a summer or winter profile. Separate algorithms are provided for temperature (T) and potential density (σ_θ).

For temperature profiles, the five MLD parameters estimated by the Holte and Talley algorithm are: (1) MLFIT $\equiv$ the intersection of straightline fit to the data in the well-mixed surface layer with the straightline fits to the observations in the underlying thermocline (where it occurs); (2) TM $\equiv$ the maximum temperature in the profile (to take into account possible intrusions beneath the mixed layer); (3) DTM $\equiv$ the abrupt change in the temperature gradient below the mixed layer that indicates that top of the thermocline; (4) TDTM $\equiv$ co-located maxima in the temperature and temperature gradient; and (5) TTMLD $\equiv$ the chosen temperature threshold estimate. Similar variables are determined for potential density (σ_θ). A summary of the procedure adopted by the Holte and Talley (2009) MLD algorithm is as follows:

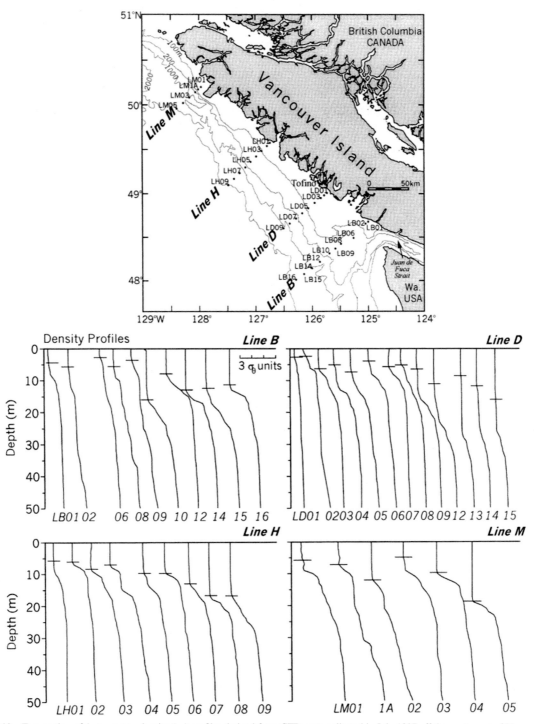

FIGURE 4.60   Top portion of 1-m average density (σ_θ) profiles derived from CTD casts collected in July 1997 off the west coast of Vancouver Island (see map). Horizontal bars denote mixed layer depth derived using the split-and-merge algorithm. Profiles of σ_θ typically range from 22 to 25 kg$\cdot$m^{-3}. A scale of 3σ_θ units is shown in the Line B profile plots. *From Thomson and Fine (2003)*.

(1) The temperature and potential density algorithms begin with (5) and look for the pressure, p, for which

$$p(T = T_o \pm 0.2°C); p(\sigma_\theta = \sigma_o + 0.003 \text{ kg/m}^3) \qquad (4.141)$$

in which T_o and σ_o are surface reference values (typically near, but not at, the surface);

(2) Next applied are the gradient algorithms (DTM for temperature) that search for the pressure at which

$$p\left(\left|\frac{\partial T}{\partial p}\right| > 0.005°C/\text{dbar}\right); p\left(\left|\frac{\partial \sigma_\theta}{\partial p}\right| > 0.0005 \left(\frac{\text{kg}}{\text{m}^3}\right)/\text{dbar}\right) \qquad (4.142)$$

with certain caveats if the criteria are not met;

(3) The algorithm next finds the pressure (depth) of the temperature maximum (TM) and salinity and density minima, which for temperature is

$$p(T_{max}) \qquad (4.143)$$

(4) For the fourth estimate of the MLD, the algorithm sets the MLD to

$$p\left(\left(\frac{\partial T}{\partial p}\right)_{max}, T_{max}\right) \text{ if } \left|p\left[\left(\frac{\partial T}{\partial p}\right)_{max}\right] - p(T_{max})\right| \leq \Delta D \qquad (4.144a)$$

but sets MLD = 0 if

$$\left|p\left[\left(\frac{\partial T}{\partial p}\right)_{max}\right] - p(T_{max})\right| > \Delta D \qquad (4.144b)$$

where the parameter ΔD ($\equiv$ distance from the surface) is set by the user. The algorithm takes the shallowest of the two values;

(5) The final possible estimate is designed to capture the mixed layer depth in profiles having homogeneous mixed layers near the surface and strong thermoclines (pycnoclines) below. In this case

$$p(T_{ML-fit} = T_{Therm-fit}) \qquad (4.145)$$

where T_{ML-fit} is the mixed layer fit and $T_{Therm-fit}$ is the seasonal thermocline fit. Here, MLTFIT = 0 if the two straightline fits do not intersect. Salinity and density use their own straightline fits. This MLD metric works well in summer but can fail in winter when the seasonal thermocline is weak.

Following a description of the various MLD products, Holte and Talley (2009) devote the rest of the study to explaining how the temperature and potential density algorithms pick the appropriate MLD and to providing examples of the method for parts of the World Ocean. The flow diagram for the summer temperature algorithm is shown in Figure 4.61. A more complicated flow diagram is provided for the winter temperature algorithm. The reader is directed to the original publication for further details.

4.13.6 Comparison of Methods

All methods used to estimate the mixed layer depth have drawbacks and all require the specification of parametric values (such as temperature or density thresholds) by the investigator. Clearly, the accuracy of any MLD estimate also depends on

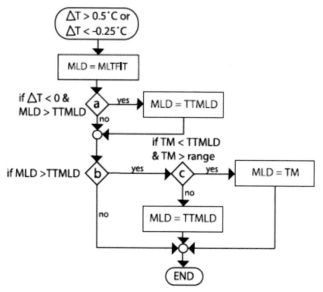

FIGURE 4.61 Flow diagram for the summer temperature algorithm. The metric ΔT (the temperature change across the thermocline) is used to decide if the profile is summer or winter like; for summer profiles, the MLD often equals MLTFIT corresponding to the depth where the straight-line fit to the upper mixed layer intersects that of the thermocline. The search for this intersection is similar to that using the split and merge approach. *From Holte and Talley (2009).*

the quality and vertical resolution of the profile data being used. A comparison among the various mixed layer depth estimators (the split-and-merge method, two threshold-type methods, the two- and three-step regression methods, and the integral-scale method) by Thomson and Fine (2003) found that, in the absence of turbulent dissipation measurements, only the threshold and split-and-merge methods provided accurate "visual" estimates of mixed layer depth. Estimates from the regression and integral methods were more representative of the permanent pycnocline depth than the surface mixed layer depth. The threshold and split-and-merge methods give nearly identical MLD results over a wide range of threshold and error norm values. For the threshold method, MLD estimates increase with increased threshold value, while for the split-and-merge method, the MLD estimate visually captures the mixed layer depth and remains unchanged over wide range of norm values. Where the upper layer has a weak density gradient, the ability of the threshold method to estimate the MLD is strongly dependent on the threshold value. In contrast, the split-and-merge algorithm readily finds a MLD that corresponds closely to the visual estimate. When the upper layer has a "significant" density gradient, both methods have difficulty estimating a mixed layer depth; there is no well-defined mixed layer and depth estimates invariably depend on the specified error values. In such cases, integral methods give a better approximation to a "two-layer" vertical structure.

Table 4.14 presents a statistical summary of the mean depths and standard deviations for five methods for all density profiles collected over 2 decades off the west coast of Vancouver Island. Results for the threshold difference method ("TH") are based on the standard threshold $\Delta\sigma_\theta = 0.01 \text{ kg} \cdot \text{m}^{-3}$ and those for the threshold gradient method ("GR") on a gradient $\Delta\sigma_\theta/\Delta z = 0.006 \text{ kg} \cdot \text{m}^{-4}$. The specified error norm $\varepsilon = 0.003$ for the split-and-merge method coincides with the threshold difference method value of $\Delta\sigma_\theta = 0.01 \text{ kg} \cdot \text{m}^{-3}$. As the tabulated results indicate, the mean mixed layer depths obtained using the threshold and split-and-merge methods are markedly different from those based on the regression and integral methods. Mean depths from the threshold and split-and-merge methods (mean MLD $\approx$ 12 m) are a factor of five smaller than those for the regression and integral methods (mean MLD $\approx$ 60 m), and most accurately match "by-eye" estimates of the uniform-density surface mixed layer.

The algorithm formulated by Holte and Talley (2009) yields a globally applicable MLD diagnostic that considers spatial and seasonal patterns of the surface layer. However, as noted by Reichl et al. (2022), the algorithm uses a multi-step decision tree that is computationally intensive when evaluating many profiles, such as typically occurs with model diagnostics. Furthermore, the branches of the algorithm rely on critical thresholds that are subjectively chosen. The classification scheme also creates discontinuities related to discrete boundaries between classification methods, where a continuous field is desirable.

TABLE 4.13 Selected definitions for mixed layer depth and other upper ocean features. Here, T is temperature, S is salinity, σ_θ is potential density. The "trapping depth", D_T (Price et al., 1986) is analogous to the integral depth scale.

Source	Name	Definition
Price et al. (1986)	Trapping depth, D_T	$D_T = \Delta T^{-1} \int_{z_i}^{z} T dz$
Peters et al. (1989)	Mixed layer depth, MLD	$\Delta\sigma_\theta = 0.01$ kg·m^{-3} (relative to $z = 0$ m)
Schneider and Müller (1990)	Mixed layer depth, MLD	$\Delta\sigma_\theta = 0.01$ kg·m^{-3} (relative to $z = 2.5$ m) $\Delta\sigma_\theta = 0.03$ kg·m^{-3} (relative to $z = 2.5$ m)
Wijffels (1993)	Mixed layer depth, MLD Top of thermocline, TTC	$\Delta\sigma_\theta = 0.01$ kg·m^{-3} (relative to $z = 2.5$ m) $\partial\sigma_\theta/\partial z = 0.01$ kg·m^{-4}
Brainerd and Gregg (1995)	Mixed layer depth, MLD	$\Delta\sigma_\theta = 0.005-0.5$ kg·m^{-3} $\partial\sigma_\theta/\partial z = 0.0005-0.05$ kg·m^{-4}
Smyth et al. (1996a,b)	Diurnal mixed layer, DML Upper ocean layer, UOL	$\Delta\sigma_\theta = 0.01$ kg·m^{-3} $\sigma_\theta < 22$ kg·m^{-3} (top of pycnocline)
Weller and Plueddemann (1996)	Mixed layer depth, MLD Isopycnal layer depth, ILD Seasonal thermocline depth, STD	$\Delta T = 0.01$°C (relative to $z = 2.25$ m) $\Delta\sigma_\theta = 0.03$ kg·m^{-3} (relative to $z = 10$ m) $\Delta\sigma_\theta = 0.15$ kg·m^{-3} (relative to $z = 10$ m)
Wijesekera and Gregg (1996)	MLD MLD_1 MLD_2 MLD_3	$\Delta\sigma_\theta = 0.01$ kg·m^{-3} (relative to $z = 0$ m) $\partial\sigma_\theta/\partial z = 0.01$ kg·m^{-4} $\partial\sigma_\theta/\partial z = 0.025$ kg·m^{-4} $\partial\sigma_\theta/\partial z = 0.01$ psu·m^{-1}
Skyllingstad et al. (1999)	MLD	$\Delta\sigma_\theta = 0.01$ kg·m^{-3} (relative to $z = 0$ m)
Thomson and Fine (2003, 2009)	Split-and-merge	Error norm $\varepsilon = 0.01$

Modified after Thomson and Fine (2003).

TABLE 4.14 The mean mixed layer depth (MLD_{sub}) and standard error (standard deviation/$\sqrt{N}$, $N =$ number of samples) for different estimation methods (depths in meters) based on two decades of measurements off the west coast of Vancouver Island, British Columbia. Subscripts "S&M", "TH", and "GR" denote the split-and-merge, threshold difference, and threshold gradient methods, respectively. D_σ is the integral depth scale, D_2 is the MLD from the two-step function, and D_3 for the three-step-function approach.

	$MLD_{S\&M}$	MLD_{TH}	MLD_{GR}	D_σ	D_2	D_3
Mean (m)	(12.51±0.32)	(12.43±0.33)	(12.57±0.38)	(60.4±0.6)	(60.9±0.7)	(35.60±0.46)

4.14 INVERSE METHODS

4.14.1 General Inverse Theory

General inverse methods have become a sophisticated analysis tool in the earth sciences. For example, in the field of geophysics, a goal of this technique is to infer the internal structure of the earth from the measurement of seismic waves. The essence of the geophysical *inverse problem* is to find an earth structure model, which could have generated the observed acoustic travel-time data. This is in contrast to the *forward problem* which uses a known input and an understood physical system to predict the output. In the inverse problem, the input and output are known, and the result is the *model* required to translate one set of data into the other.

In oceanography, inverse methods are used for a variety of applications, including the inference of absolute ocean currents using known tracer distributions and geostrophic flow dynamics (Wunsch, 1978, 1988). Other applications include the use of underwater acoustic travel times to determine the average temperature of the global ocean for long-term climate studies (Worchester et al., 1988) or the use of tsunami observations from bottom pressure recorders such as DART, cabled observatories, or coastal tide gauges, to help define the seismic source regions for major trans-oceanic tsunamis (Satake, 1995; Fine et al., 2005; Hayashi et al., 2011). Studies by Mackas et al. (1987) and Masson (2006) used inverse techniques to determine the origins and mixing of water masses for the coast of British Columbia. In these oceanographic applications, the "solutions" are what we previously called the "models" in the geophysical problem. The kernel functions are formulated from the physics of the problem in question and the result is found by matching the "solution" to the input data. A cursory look at the problem is provided in this section. The interested reader is referred to Bennett (1992) for detailed insight into the theory and application of inverse methods in oceanography.

In general, the inverse problem takes the form

$$e(t) = \int_a^b C(t, \xi) m(\xi) d\xi \tag{4.146}$$

where $e(t)$ are the input data, $m(\xi)$ is the model and $C(t, \xi)$ is the kernel function for the variable ξ. The kernel functions are determined from the relevant physical equations for the problem and are assumed to be known (Oldenburg, 1984). It is the judicious selection of these kernel functions that makes the inverse problem a complex exercise requiring physical insight from the oceanographer. In order to extract information about the model, $m(\xi)$, we will restrict our consideration of inverse theory to linear inverse methods applied to a set of observations. This is referred to as "finite dimensional inverse theory" by Bennett (1992). In his discussion of this form of inverse theory, Bennett suggests that it applies to:

(1) An incomplete ocean model, based on physical laws but possessing multiple solutions.
(2) Measurements of quantities not included in the original model but related to the model by additional physical laws.
(3) Inequality constraints on the model fields or the data.
(4) Prior estimates of errors in the physical laws and the data.
(5) Analysis of the level of information in the system of physical laws, measurements, and inequalities.

Eqn (4.146) is a *Fredholm equation* of the first kind. Inverse theory is centered on solving this equation in such a way as to extract information about the model, $m(\xi)$, when information is available for the data, $e(t)$. It is important to realize that the inverse problem cannot be solved unless the physics and the geometry of the problem are known (i.e., Eqn (4.146) has been set up). It is, therefore, impossible to consider a solution to the inverse problem unless the forward problem can be solved. The physics of the forward problem may be ill-posed, in which case not all of the solutions will match or, if they do, it is a coincidence and not a solution to (4.146). Thus, the basic questions to ask regarding a solution of the inverse problem are: (1) Does a solution exist? In other words, is there an $m(\xi)$ which produces $e(t)$? (2) How does one construct a solution? (3) Is the solution unique?; and (4) How is the nonuniqueness appraised?

The answers to the above questions will depend on the data, $e(t)$. In theory, there exist three types of data:

(1) An infinite amount of accurate data.
(2) A finite amount of accurate data.
(3) A finite amount of inaccurate data.

In reality, only option (3) occurs as we are forced to work with observations, which contain a variety of measurement and sampling errors. While perfect data are limited to the realm of the mathematical, it is often instructive to consider analytic "inverses". For example, the analytical inverse to

$$x(f) = \int_{-\infty}^{\infty} x(t) e^{-2\pi f t} dt \tag{4.147a}$$

is

$$x(t) = \frac{1}{2\pi} \int_{-\infty}^{\infty} x(f) e^{2\pi f t} dt \tag{4.147b}$$

Similarly, the inverse of

$$\phi(x) = 2/\lambda \int_x^a \left[r\varepsilon(r) / \left(r^2 - x^2 \right)^{1/2} \right] dr \tag{4.148a}$$

is

$$\varepsilon(r) = -\lambda / \pi \int\limits_{r}^{a} \left[(d\phi / dx)/(x^2 - r^2)^{1/2} \right] dx \qquad (4.148b)$$

In the second case, we require knowledge of $d\phi/dx$ to find $\varepsilon(r)$, which is easy to do for ideal continuous data (Figure 4.62a), or even for a finite sample of accurate data (Figure 4.62b). If, however, we have a finite sample of inaccurate data (Figure 4.62c), we have difficulty estimating $d\phi/dx$.

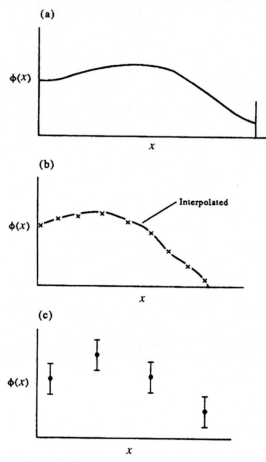

FIGURE 4.62 Three examples of the function $\phi(x)$ required for the inverse solution, $\varepsilon(r)$, of Eqn (4.148b). Analytical (a) and digital (b) versions of $\phi(x)$ for which inversion is readily possible, (c) a typical "observed" version of $\phi(x)$, consisting of four mean values (plus standard deviations) for which inversion is considerably less accurate.

The problem of dealing with a limited sample of inaccurate measurements is the most common obstacle to the application of inverse methods. Usually, these inaccuracies can be treated as additive noise superimposed on the true data and, therefore, can be handled with statistical techniques. These additive errors have the effect of "blurring" or distorting our picture of the solution (model). Unfortunately, one cannot conclude that if the error noise is small that the model distortions also will be small. The reason for this is that most geophysical kernel functions act to smooth the model, thus changing the length scale of the response for both the forward and inverse problems. In other words, the solution obtained with inaccurate data using the inverse procedure may be very different from the model, which actually generated the data. In addition, particular solutions to the model are not unique and a wide variety of solutions is equally possible.

In most oceanographic applications of inverse methods, we are primarily interested in finding a model, which reproduces the observations. Here, the fundamental problem is the nonuniqueness of any inverse solution, which is one of infinitely many functions that can reproduce a finite number of observations. This nonuniqueness becomes more severe when the data are inaccurate, as they must be in any practical oceanographic application. The key to the application of inverse methods in oceanography is to select the "correct" (by which we mean the most probable or the most reasonable) inverse model-solution.

Inverse construction in oceanography may take the form of parametric modeling. In this case, we write our model as $m = f(a_1, a_2, ..., a_N)$ and a numerical scheme is sought to find appropriate values of the parameters, a_i $(i = 1, ..., N)$. Parameterization is justified when the physical system actually has this form and depends on a number of input parameters. The model is solved by collecting more than N data points and finding the parameters through a least-squares minimization of

$$\phi = \sum_{i=1}^{N} \left(e_i - e_i'\right)^2 \tag{4.149a}$$

where

$$e_i' = f(a_1, a_2, ..., a_N; \varepsilon_i) \tag{4.149b}$$

In (4.149b), ε_i is the ith kernel function.

4.14.2 Inverse Theory and Absolute Currents

As reviewed by Bennett (1992), an important application of inverse theory to ocean processes was the computation of absolute currents for large-scale ocean circulation. In the 1970s, two different approaches to this problem were proposed. The first by Stommel and Schott (1977) was called the "beta spiral" technique, which demonstrated that the vertical structure of large-scale, open-ocean velocity fields could be explained using simple equations expressing geostrophy and continuity (conservation of mass). The second method, introduced by Wunsch (1977), showed that reference velocities could be estimated simultaneously around a closed path in the ocean. The resultant absolute velocities were consistent with geostrophy and the conservation of heat and salt at various levels. As a guide to oceanographic applications of inverse techniques, we provide succinct reviews of both applications.

4.14.2.1 The Beta Spiral Method

Insightful reviews of the Stommel and Schott (1977) beta spiral method are provided by Olbers et al. (1985) and Bennett (1992). The basic equations for this application are the usual linearized beta (β)-plane equations for horizontal geostrophic flow (u, v) in a Boussinesq fluid

$$-\rho_o f v = -\partial p/\partial x \tag{4.150a}$$

$$\rho_o f u = -\partial p/\partial y \tag{4.150b}$$

the hydrostatic equation

$$0 = -\partial p/\partial z - \rho g \tag{4.151}$$

which relate pressure perturbations, $p(\mathbf{x}, t)$, to density fluctuations, $p(z, t)$, and the conservation of mass (or continuity) relation

$$\nabla \cdot \mathbf{u} + \partial w/\partial z = 0 \tag{4.152}$$

In these equations, f is the Coriolis parameter, u, v, and w are, respectively, the eastward (x), northward (y) and upward (z) components of current velocity, and $\rho = \rho(x, y, z)$ is the density perturbation about the mean density $\rho_o = \rho_o(z)$. Following Bennett (1992), we will reserve vector notation for horizontal fields and operators ($\mathbf{x} = (x, y)$, $\mathbf{u} = (u, v)$, etc.).

Using the above equations, we can derive the well-known "thermal wind" relation, whose vertically integrated velocity components are

$$u(\mathbf{x}, z) = u_o(\mathbf{x}) + (g/f\rho_o) \int_{z_o}^{z} \rho_y(x, \zeta) d\zeta \tag{4.153a}$$

$$v(\mathbf{x}, z) = v_o(\mathbf{x}) - (g/f\rho_o) \int_{z_o}^{z} \rho_x(x, \zeta) d\zeta \tag{4.153b}$$

where subscripts x, y refer to partial differentiation and $u_o(\mathbf{x})$, $v_o(\mathbf{x})$ are the velocity components at some reference depth. Eqns (4.150a, 4.150b)−(4.152) also give rise to the well-known Sverdrup interior vorticity balance

$$w_z = \beta v / f \tag{4.154}$$

where β is the northward (y) gradient of the Coriolis parameter, and $f = f(y) = f_o + \beta y$ is the beta-plane approximation. These equations cannot be used alone to determine the full absolute velocity field ($\mathbf{u}$, w), even if the density field ρ were known. However, to resolve this indeterminacy, all we need is the velocity field at a particular depth where $\mathbf{u} = \mathbf{u}(\mathbf{x}, z_o)$ and $w = w(\mathbf{x}, z_o)$. Stommel and Schott (1977) demonstrated that these unknown reference values may be estimated by assuming the availability of measurements of some conservative tracer ϕ which satisfy the steady-state conservation law

$$\mathbf{u} \cdot \nabla \phi + w \phi_z = 0 \tag{4.155}$$

This tracer might be salinity (S) or potential temperature (θ), or some function of both S and θ. Combining the vertical derivative of Eqn (4.155) with Eqns (4.153a, 4.153b) and (4.154) yields

$$\left(\mathbf{u} \cdot \nabla + w \frac{\partial}{\partial z} \right) (f \phi_z) = (g / \rho_o) J \tag{4.156}$$

where J is the Jacobian $J(\rho, \phi) = \rho_x \phi_y - \rho_y \phi_x$. In Eqn (4.156), $f \phi_z$ represents the potential vorticity, which would be conserved if density ρ were itself conserved. The tracer equation can be used again to eliminate the vertical velocity w

$$\mathbf{u} \cdot \mathbf{a} = (g / \rho_o) J(\rho, \phi) \tag{4.157}$$

where the vector $\mathbf{a}$ is given by

$$\mathbf{a}(\mathbf{x}, z) = \nabla (f \phi_z) - \frac{\nabla \phi}{\phi_z} f \phi_{2z} \tag{4.158}$$

Using the integrated thermal wind Eqns (4.153a) and (4.153b) yields

$$\mathbf{u_o} \cdot \mathbf{a} = c \tag{4.159}$$

where $\mathbf{u}_o$ is the horizontal velocity at depth z_o and c is given by

$$c(\mathbf{x}, z) = -\mathbf{u}' \cdot \mathbf{a} + (g / \rho_o) J(\rho, \phi) \tag{4.160}$$

In Eqn (4.160), the $\mathbf{u}'$ is that part of the horizontal velocity in the thermal wind relation that depends on the density field.

Because $\mathbf{a}$ and c depend on g, ρ, f, $\nabla \rho$, ∇f, ϕ_z and ϕ_{zz}, they can be determined using closely spaced hydrographic stations through measurements of $T(z)$ and $S(z)$. Thus, from (4.159), we can calculate $\mathbf{u}_o$ using the hydrographic data. Eqn (4.159) holds at all levels so that two different levels can be used to specify u_o and v_o. We can then calculate the vertical velocity w from (4.154). The full velocity solution should be independent of the levels chosen for these computations. In reality, (4.159) is not an exact relation as it was derived from approximate dynamical laws and computed from data that contain measurement and sampling errors. As a consequence, our estimate of $\mathbf{u}_o$ from (4.159) should be done as a best fit to the data from the two levels chosen.

Suppose that N levels are chosen from the hydrographic data ($N \geq 2$). Let $c_n = x(\mathbf{x}, z_n)$ and $\mathbf{a}_n = \mathbf{a}(\mathbf{x}, z_n)$ for $1 \leq n \leq N$. The simple least-squares best fit minimizes

$$R^2 = \sum_{n=1}^{N} R_n^2 = \sum_{n=1}^{N} (c_n - \mathbf{u}_o \cdot \mathbf{a}_n)^2 \tag{4.161}$$

where R_n is the residual at level n and R is the root-mean-square (RMS) total error. R^2 is a minimum if $\mathbf{u}_o$ satisfies a simple linear system

$$\mathbf{M} \mathbf{u}_o = \mathbf{d} \tag{4.162}$$

where the 2×2 systematic, nonnegative matrix $\mathbf{M}$ depends on the components of $\mathbf{a}_n$, while $\mathbf{d}$ depends on $\mathbf{a}_n$ and c. If $\mathbf{a}$ or c varies with depth, Eqn (4.157) implies that the total velocity vector $\mathbf{u}$ must also depend on depth. For the β-spiral problem, we find that the large-scale ocean currents constitute a spiral with depth at each station. The β-spiral in Figure 4.63 is from the study by Stommel and Schott (1977) who used hydrographic data from the North Atlantic to estimate $\mathbf{u}_o$ for a reference

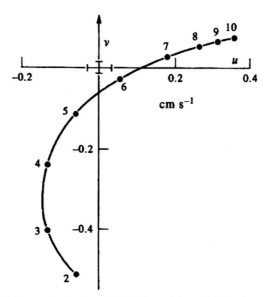

FIGURE 4.63 The β-spiral in horizontal velocity $\mathbf{u} = (u(z), v(z))$ at 28°N, 36°W, with depths in hundreds of meters. Error bars for the two components of velocity are given at the origin. *After Stommel and Schott (1977).*

level of $z_o = 1,000$ m depth. In this application they found, $u_o = 0.0034 \pm 0.00030$ m/s and $v_o = 0.0060 \pm 0.00013$ m/s at 28°N, 36°W.

The β-spiral problem includes two of the basic concepts common to inverse methods. First, we deal with an incomplete set of physical laws (4.150–4.152), or their rearrangement, as in the case of the thermal wind Eqns (4.153a) and (4.153b), which includes the unknown reference velocity. Second, we often resort to the indirect measurement of an additional quantity, which, in the case of the present example, is a conservative tracer. This application could have benefited from the inclusion of prior estimates of the errors in the dynamical equations and in the hydrographic data.

4.14.2.2 Wunsch's Method

In a parallel development to the β-spiral technique, Wunsch (1977) used inverse methods to estimate reference velocities simultaneously around a closed path in the ocean (Bennett, 1992). As discussed by Davis (1978), Wunsch's method and the β-spiral method are closely related. Both approaches assume the vertically integrated thermal wind Eqns (4.153a) and (4.153b) and both provide estimates for the reference velocity $\mathbf{u}_o$. In Wunsch's method, the thermal wind velocity, $\mathbf{u}'$, is assumed to be zero at the reference level z_o, which in general may be a function of position [$z_o = z_o(\mathbf{x})$]. Wunsch chose the reference level to be the ocean bottom at $z_o(\mathbf{x}) = H(\mathbf{x})$, with $\mathbf{u}_o(\mathbf{x})$ defined to be the bottom velocity. He then divided the water column into a number of layers defined by temperature ranges. This is consistent with the general water mass structure of the North Atlantic as defined by Worthington (1976). These layers need not be uniform in depth at each hydrographic station. Together with the coastline of the U.S., the hydrographic stations formed a closed path in the western North Atlantic (Figure 4.64).

We now let v denote the outward component of velocity across the closed triangle formed by the lines of hydrographic stations in Figure 4.64. That is, $v = \mathbf{u} \cdot \mathbf{n}$ where $\mathbf{n}$ is the outward unit normal to the sections. We can further let $v' = \mathbf{u}' \cdot \mathbf{n}$ be the outward thermal wind velocity and $b = \mathbf{u}_o \cdot \mathbf{n}$ be the outward horizontal velocity at the seafloor. Let $v'_n(z)$ and b_n denote the thermal wind velocity estimate and unknown bottom velocity midway between the nth station pair, where $1 \leq n \leq N$, and let v'_{mn} denote the average value of v'_n in the mth layer of the water column, where $1 \leq m \leq M$. Wunsch chose the Mth layer to be the total water column, thus the Mth tracer is the total mass of the water column. The assumption of tracer conservation within each layer can be written as

$$\sum_{i=1}^{N} \left(v'_{mn} + b_n \right) \Delta z_{mn} \Delta x_n = 0, 1 \leq m \leq M \tag{4.163}$$

where Δz_{mn} is the thickness of the mth layer at the nth station pair, and Δz_{mn} is the separation distance between the nth station pair. This system of M equations for N unknowns b_n, $1 \leq n \leq N$, may be written in matrix notation as

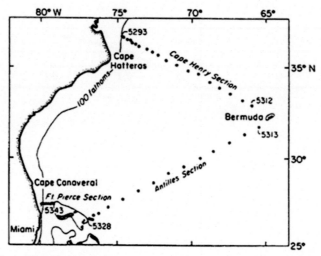

FIGURE 4.64 The locations of hydrographic stations in the North Atlantic used by Wunsch to obtain absolute current estimates using inverse theory. *After Wunsch (1977).*

$$\mathbf{Ab} = \mathbf{c} \tag{4.164a}$$

where $\mathbf{A}$ is an $M \times N$ matrix and $\mathbf{c}$ is a column vector of length M with elements

$$A_{mn} = \Delta z_{mn} \Delta x_n \tag{4.164b}$$

$$c_m = -\sum_{i=1}^{N} \overline{v'_{mn}} A_{mn} \tag{4.164c}$$

Wunsch used $M = 5$ layers as defined by the ranges 12–17°C, 4–7°C, 2.5–4°C, and the entire water column (total mass). The hydrographic data were from $N = 43$ station pairs. For this problem, the matrix Eqns (4.164a–c) represents five equations for 43 unknown velocities, so that the system is underdetermined and has many different solutions.

As reported by Bennett (1992) and Wunsch (1977) somewhat arbitrarily selected the vector $\mathbf{b}$ with the shortest length. This was found by minimizing

$$t_1 = \mathbf{b}^T\mathbf{b} + 2\mathbf{I}^T(\mathbf{Ab} - \mathbf{c}) \tag{4.165}$$

where the superscript T denotes the transpose of the matrix, and $\mathbf{I}$ is an unknown Lagrange multiplier consisting of a column vector of length M. It can be shown that t_1 is a minimum when

$$\mathbf{b} + \mathbf{A}^T\mathbf{I} = \mathbf{0} \tag{4.166a}$$

which gives the minimum solution

$$\mathbf{b} = \mathbf{A}^T\left(\mathbf{AA}^T\right)^{-1}\mathbf{c} \tag{4.166b}$$

which satisfies (4.164a). The symmetric matrix $\mathbf{AA}^T$ has dimensions $M \times M$ and is nonnegative (Bennett, 1992). However, $\mathbf{AA}^T$ may be singular. These singularities may be overcome by allowing errors in the hydrographic data and conservation laws; that is, by not seeking exact solutions of (4.164a). We can instead write (4.164a) in a quadratic form adding weights to each term. It can be shown that for positive weights, we are able to define an exact solution of the problem. This transfers the problem to the selection of these weights.

This cursory presentation of Wunsch's method for computing reference velocities demonstrates, once again, some of the basic elements of inverse methods: A system of incomplete physical laws and inexact measurements of related fields. It is necessary to admit errors into the equations and data values in order to stabilize the solution and to derive a unique solution. In his review, Davis (1978) concluded that both the underdetermined problem of Wunsch's method and the over determined problem of the β-spiral method are consequences of tacit assumptions made about noise levels and fundamental

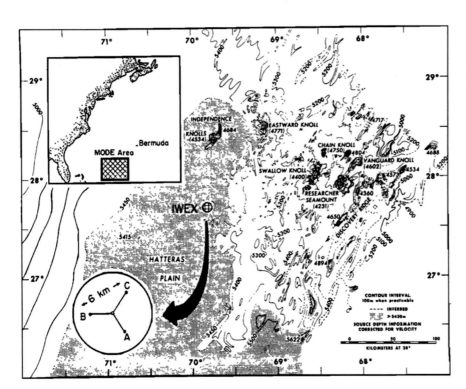

FIGURE 4.65 Location of the IWEX study area showing the positions of the three current meter moorings on the Hatteras Plain in the western North Atlantic. *From Briscoe (1975).*

scales of motion. Davis suggested that a more orderly approach would be based on Gauss-Markov smoothing (Bennett, 1992) which should be an improvement, assuming explicit and quantitative estimates of the noise and its structure.

4.14.3 The IWEX Internal Wave Problem

Another oceanographic example of the inverse method is found in Olbers et al. (1976) and Willebrand et al. (1977). Here, inverse theory is used to determine the three-dimensional internal wave spectrum from an array of moored current meters (Figure 4.65). In this example, the Fredholm Eqn (4.146) is written in matrix form and becomes

$$y_i = A_{ij}x_j; \ 1 \le i \le N; \ 1 \le j \le K \tag{4.167}$$

where y_i are N observed velocity cross-spectra (the data), A_{ij} are the kernel functions (for matrix **A**) representing the physical relations from internal wave theory and x_j are the K internal wave parameters to be determined by the inverse method. The inverse problem is to find the K parameters of the theoretical internal wave energy density cross-spectra using the N observed cross-spectra from the current meter array. We achieve this by using the least-squares method to minimize

$$\varepsilon^2(a) = [\hat{y} - y(a)]\mathbf{W}[\hat{y} - y(a)]^* \tag{4.168}$$

where a represents a set of trial values used to find the minimum and the asterisk (*) denotes the complex conjugate. In Eqn (4.168), **W** is a weighting matrix used to scale the problem and to produce statistical independence (Jackson, 1972).

It is common to expand the kernel function matrix **A** into eigenvectors (Jackson, 1972). Thus, we write

$$\mathbf{A}V_j = \lambda_j u_j, \ \mathbf{A}^T u_j = \lambda_i V_i \tag{4.169}$$

Following singular value decomposition, we conducted in the EOF analysis (Section 4.6.2), we can factor the matrix **A** as

$$\mathbf{A} = \mathbf{UBV}^T \tag{4.170}$$

where **U** is an $N \times P$ matrix whose columns are the eigenvectors u_i, $i = 1, ..., P$; **V** is the $M \times P$ matrix whose columns are the eigenvectors v_i, $i = 1, ..., P$, and **B** is the diagonal matrix of eigenvalues. After **U** and **V** are formed from the eigenvectors corresponding to the P nonzero eigenvalues of $\mathbf{A}$, there remain $(N - P)$ eigenvectors U_j and $(K - P)$ eigenvectors V_j, which correspond to zero eigenvalues. If we assemble these into columns of matrices, we have $\mathbf{U}_o$ (an $N \times (N - P)$ matrix) and $\mathbf{V}_o$ (a $K \times (K - P)$ matrix). This is called *annihilator space* and reveals that our model is composed of both real model space (which corresponds to the data) and annihilator space, which is linked to zeros in the data field. When we perform an inverse calculation, we usually recover a solution, which lies in real model space. We must remember, however, that any function in space a can be added to the solution and still produce a solution that fits the data. With the kernel functions transformed into an orthogonal framework (expanded into eigenvectors) we construct the "smallest" or minimum energy model-solution.

When $P = N$, there is a solution to Eqn. (4.168) and $P = M$ guarantees that a solution, if it exists, is unique. For $P < N$, the system is said to be over constrained, while if $P < M$, the system is both over constrained and underdetermined. In the latter case, an exact solution may not exist but there will be an infinite number of solutions satisfying the least-squares criterion. This is the case for the present internal wave example, which is both over constrained and underdetermined.

Returning to our internal wave problem, we find $\mathbf{W}$ in Eqn (4.168) using the least-squares method which produces the maximum likelihood estimator for a Gaussian distribution. This estimator is defined to be the inverse of the data covariance matrix. From the current meter array, 60 time series were divided into 25 overlapping segments. For each segment, cross-spectral estimates were computed for each of 600 equidistant frequencies. Averaging over segments and frequency bands to increase statistical significance resulted in 3,660 cross-spectra. The resultant $3,660 \times 3,660$ covariance matrix is difficult to invert. The diagonal of the weight matrix was selected to be

$$W = \text{diag}[1/\text{var}(\mathbf{y}_i)] \tag{4.171}$$

which reproduces the main features of the maximum likelihood weight matrix (Olbers et al., 1976). We note that, again for this problem, there are many more data points than parameters so that the system is over constrained.

The least-squares solution procedure for this internal wave example is as follows:

(a) first find a parameter estimate $\hat{a}$ (the best guess).
(b) linearize at the value $a = \hat{a}$, such that

$$\widehat{y}(a) = \widehat{y}(\widehat{a}) + \mathbf{D}(a - \widehat{a}) + \ldots \tag{4.172}$$

where

$$\mathbf{D} = \left\{ \delta\widehat{y}_i / \delta a_j \right\}\big|_{a=\hat{a}} \tag{4.173}$$

(c) improve the parameter estimate by using

$$a - \widehat{a} = \mathbf{H}[\widehat{y}(a) - \widehat{y}(\widehat{a})] \tag{4.174}$$

where the $N \times K$ matrix $\mathbf{H}$ is the generalized inverse of $\mathbf{D}$ derived from the linear terms of (4.172). If the matrix $\mathbf{D} < \mathbf{TWD}$ is nonsingular and well-conditioned, then

$$\mathbf{H} = (\mathbf{D}^T\mathbf{W}\mathbf{D})^{-1}\mathbf{D}^T\mathbf{W} \tag{4.175}$$

and Eqn (4.170) becomes the least-squares solution of (4.169). Since $\mathbf{D}^T\mathbf{W}\mathbf{D}$ is an $K \times K$ matrix, it can be easily inverted using standard diagonalization routines.

Having now arrived at a solution, $\mathbf{A} = \{A_{ij}\}$ of the problem in (4.167), we are left with two additional questions: (1) How well are the data reproduced by our solution? and (2) How accurately do we know our parameters α_{min}? Since our data are subject to random errors, we can treat y as a statistical quantity and test the hypothesis that y and the model estimate $\hat{y}(\alpha_{\text{min}})$ are the same with a 95% probability (inverse estimate must be within the 95% confidence interval of our data point). Using the central limit theorem for our segment and frequency-averaged spectral values, we can approximate the 95% confidence interval on y as

$$\varepsilon_{95\%}^2 = \overline{\delta y W \delta y}\left[1 + O\left(L^{-1}\right)\right] = L \tag{4.176}$$

where $\delta y = y - \hat{y}$, and $O(\cdot)$ indicates the order of magnitude. Now if

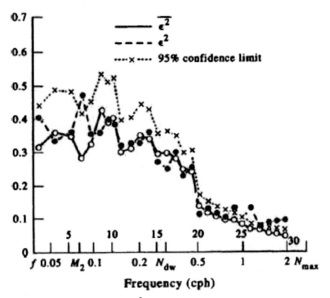

FIGURE 4.66 Consistency for the IWEX study. The error estimate ε^2 is the squared difference between the observed data and the modeled data obtained by inverse methods. Except for motions in the M_2 tidal band and at frequencies great than about 1 cph, the results are within the 95% confidence level. N_{max} and N_{dw} are the maximum Nyquist frequency and the Nyquist frequency for the deep water, respectively. *From Briscoe (1975).*

$$\varepsilon^2(a_{\min}) \leq \varepsilon^2_{95\%} \tag{4.177}$$

the model is a statistically consistent representation of the data. The consistency of the IWEX model is provided by the results in Figure 4.66, where we have plotted the measured, $\varepsilon^2(a)$, and expected, ε^2, values of the parameter ε^2. In this case, all values have been normalized so that magnitudes provide some indication of the percentage to which the observed and estimated (modeled) values of the data, y, coincide. For the most part, the measured values of ε^2 are scattered about the expected values of this parameter. Except at the M_2 tidal frequency and for frequencies greater than 1 cph, the hybrid IWEX model gives a consistent description of the IWEX data set to the 95% level.

Our second question regarding the accuracy of the parameter solution $a_{\min}$, can be answered by calculating the covariance matrix of the parameters. Using Eqn (4.170), we obtain the $K \times K$ covariance matrix of the parameters,

$$\overline{\delta a \delta a} = \mathbf{H} \overline{\delta y \delta y} \mathbf{H}^T \tag{4.178}$$

from the data covariance matrix $\overline{\delta y \delta y}$. As usual, there is a reciprocal relation between the variance and the resolution of the parameters. Statistically uncorrected parameters can be found by diagonalizing the matrix in (4.178).

4.14.4 Summary of Inverse Methods

In this section, we have presented the basic concepts of the general inverse problem and have set up the solution system for two different applications in physical oceanography. Our treatment is by no means comprehensive and is intended to serve only as a guide to understanding the process of forming linear inverse solutions to fit observed oceanographic data.

The first example we treated is the computation of absolute geostrophic velocity by specifying an unknown reference velocity. Both the β-spiral (Stommel and Schott, 1977) and Wunsch's (1977) method are discussed. The dynamics are restricted to geostrophy and the conservation of mass. The second example was the specification of parameters in theoretical internal wave cross-spectra to reproduce the velocity cross-spectra of an array of moored current meters. The statistical nature of both the data and the model are considered and the accuracy of the results are expressed in probabilistic terms. Readers interested in further discussion of these, and other related applications of inverse methods are referred to Bennett (1992). This book contains a complete review of inverse methods along with discussion of most of the popular applications of inverse techniques in physical oceanography. We also direct the interested reader to a recent paper by Egbert et al. (1994) in which a generalized inverse method is used to determine the four principal tidal constituents (M_2, S_2, K_1, O_1) for open ocean tides. The tides are constrained (in a least-squares sense) by the hydrodynamic equations and by observational data. In the first example, solutions are obtained using inversion of the harmonic constants from a set of 80 open ocean tide gauges. The second example uses cross-over data from TOPEX/POSEIDON satellite altimetry. According to the authors, "The inverse solution yields tidal fields which are simultaneously smoother, and in better agreement with altimetric and ground truth data, than previously proposed tidal models." In recent years, the acquisition of high resolution (better than 0.1 mm accuracy), rapidly sampled tsunami wave records have made it possible to augment the delineation of the seismic source regions for the tsunamigenic earthquakes. The inverse travel time estimates of tsunami source regions provided by seismologists provide critical information on the failure regions, which can then be input into high resolution tsunami wave propagation models. Once the tsunami data have been analyzed, these data can be input into the models and the models then run in reverse to more accurately define the boundaries of the seismic sources.

Glossary

ADCP Acoustic Doppler Current Profiler

ANN Artificial Neural Network

AVISO Archiving, Validation and Interpretation of Satellite Oceanographic data (France)

BIGDRV2 Program used to obtain the velocity, pressure and density fluctuations over a continental shelf-slope region of arbitrary depth, stratification and bottom friction, and driven by an alongshore wind stress.

BIGLOAD2 Program that gives frequency dispersion curves for waves in an ocean with continuous, horizontally uniform stratification and arbitrary topography.

BMU Best matching unit

BTCSW Program gives the dispersion curves of coastally trapped waves

COARE Coupled Ocean Atmosphere Response Experiment

CROSS Program that gives baroclinic coastal-trapped modes for arbitrary stratification and uniform depth

CSEOF Cyclostationary EOF

CTD Conductivity-Temperature-Depth profiler

CTW Coastally Trapped Wave

CTWEIG Wilkin (1987) program to compute the frequencies and cross-shore model structure of free coastal-trapped waves in a stratified, rotating channel with arbitrary bottom topography.

db dBar = decibar

DTM the abrupt change in the temperature gradient below the mixed layer that indicates the top of the thermocline.

EEOF Extended EOF

EnKF Ensemble Kalman Filter

ENSO El Niño Southern Oscillation

EOFs Empirical Orthogonal Functions

FA Factor Analysis

FVCOM Finite-Volume Coastal Ocean Model

GHSOM Growing Hierarchical Self-Organizing Maps

GPS Global Positioning System

IOP Intensive Observing Period

IWEX Inverse Wavefield EXtrapolation

LRA Linear Regression Analysis

LV Loading Vectors

MBT Mechanical Bathythermograph

MIT Massachusetts Institute of Technology (USA)

MLD Mixed Layer Depth

MLFIT Intersection of straight-line fit to the data in the well-mixed surface layer with the straight-line fits to the observations in the underlying thermocline.

NCAR National Center for Atmospheric Research (USA)

NCEP National Center for Environmental Prediction (USA)

NMC National Meteorological Center (USA)

OI Optimum Interpolation

PCA Principal Component Analysis

PCTS Principal Component Time Series

QE Quantization Error

QG Quasi-geostrophic

RANDM Random number generator on a Hewlett-Packard 30S hand calculator

RRKF Reduced Rank Kalman Filter

SLP Sea Level Pressure

SOM Self-Organizing Maps

SOM_PAK Public domain software package from the Laboratory of Computer and Information Science of the Helsinki University of Technology.

SST Sea Surface Temperature

SVD Single Value Decomposition

TAPS Tropical Analysis and Prediction System (Australia)

TDTM Co-located maxima in the temperature and temperature gradient

TE Topographical Error

TM The maximum temperature in the profile

TOGA Tropical Ocean Global Atmosphere

TTMLD Chosen temperature threshold estimate

XBT Expendable Bathythermograph

Chapter 5

TIME SERIES ANALYSIS METHODS

5.1 INTRODUCTION

The advent of ocean observing satellites, long-term mooring capability, and cabled marine observatories, coupled with high-density storage devices and advanced satellite communication systems, is enabling oceanographers to collect long, high resolution time series of oceanic and meteorological data. Similarly, the use of rapid-response sensors on moving platforms such as Argo drifters, saildrones and autonomous underwater vehicles (AUVs) has made it possible to generate snapshots of spatial structure over extensive distances. Time series data are collected from moored instrument arrays or by repeated measurements at the same location using ships, satellites, or other instrumented packages. Quasi-synoptic spatial data are obtained from ships, manned-submersibles, remotely operated vehicles, AUVs, satellites, Argo drifters, and satellite-tracked surface drifters. Satellite imaging also produces densely sampled spatial data whose two-dimensional coverage can repeat in time, yielding a three-dimensional data set that can be analyzed in one, two, or all three of these dimensions.

As discussed in Chapters 3 and 4, the first stage of analysis following data verification and editing usually involves estimates of arithmetic means, variances, correlation coefficients, and other sample-derived statistical quantities. These quantities tell the investigator how well the sensors are performing and help characterize the observed oceanographic variability. However, general statistical quantities provide little insight into the different types of signals that are blended together to make the recorded data. The purpose of this chapter is to present methodologies that examine data series in terms of their frequency and/or wavenumber content. With the availability of modern high-speed computers, frequency-domain analysis has become much more central to our ability to decipher the cause and effect of oceanic change. The introduction of fast Fourier transform (FFT) techniques in the 1960s further aided the application of frequency-domain analysis methods in oceanography. Such analyses were not practical prior to the advent of modern digital processors.

5.1.1 Basic Concepts

For historical reasons, the analysis of sequential data is known as *time series analysis*. As a form of data manipulation, it has been richly developed for a wide assortment of applications. While we present some of the latest techniques, the emphasis of this chapter will be on those "tried and proven" methods most widely accepted by the general oceanographic community. Even these established methods are commonly misunderstood and incorrectly applied. Where appropriate, references to other texts will be given for those interested in a more thorough description of analysis techniques. As with previous texts, the term "time series" will be applied to both temporal and spatial data series; methods that apply in the time domain also apply in the space domain. Similarly, the terms *frequency domain* and *wavenumber domain* (the formal transforms of time and spatial series, respectively) are used interchangeably. Wavenumber is the appropriate unit when applying these methods to spatial series and it is poor grammar to refer to wavenumber as "spatial frequency" since frequency only applies to actual time series.

A basic purpose of time series analysis methods is to define the variability of a data series in terms of dominant periodic functions. We also want to know the "shape" of the spectrum. Of all oceanic phenomena, the barotropic astronomically forced tides most closely exhibit deterministic and stationary periodic behavior, making them the most readily predictable motions in the sea. In coastal waters, tidal observations over a period as short as 1 month can be used to predict local tidal elevations with a high degree of accuracy. Where accurate specification of the boundary conditions is possible, a reasonably good hydrodynamic numerical model that has been calibrated against observations can reproduce the regional tide heights to an accuracy of a few centimeters. Tidal currents are much less easily predicted because of the complexities introduced by stratification, seafloor topography, basin boundaries, and nonlinear interactions. For example, although baroclinic (internal) tides generated over abrupt topography in a stratified ocean contribute little to surface elevations, they can lead to strong baroclinic currents. These baroclinic currents typically have both deterministic and nondeterministic (i.e., stochastic) components, and hence are only fully predictable in a statistical sense.

Surface gravity waves are periodic and quasi-linear oceanic features but are generally only predictable in a stochastic sense due to inadequate knowledge of the surface wind fields, the air—sea momentum transfer, and oceanic boundary conditions. Refraction induced by wave-current interactions can be important but difficult to determine. Other oceanic phenomena such as coastal-trapped waves and near-inertial oscillations have marked periodic signatures but are intermittent because of the vagaries of the forcing mechanisms and changes in oceanic and topographic conditions along the direction of propagation. Other less obvious regular behavior can be found in observed time and space records. For instance, oceanic variability at the low-frequency end of the spectrum is dominated by fluctuations at the annual to decadal periods, consistent with baroclinic Rossby waves and short-term climate change, while the spectrum at ultra-low frequencies is dominated by ice-age climate scale variations possibly associated with highly amplified feedback responses to weak Milankovitch-type forcing processes (changes in the caloric summer insolation at the top of the atmosphere arising from changes in the earth's orbital eccentricity, and tilt and precision of its rotation axis).

Common sense should always be a key element in any time series analysis. Attempts to use analytical techniques to find "hidden" signals in a time series often are not very convincing, especially if the expected signal is buried in the measurement noise. Because noise is always present in real data, it should be clear that, for accurate resolution of periodic behavior, data series should span at least a few repeat cycles of the timescale of interest, even for stationary processes. Thus, a daylong record of hourly values will not fully describe the diurnal cycle in the tide nor will a 12-month series of monthly values fully define the annual cycle of sea surface temperature (SST). For these short records, modern spectral analysis methods can help pinpoint the peak frequencies. As we noted in Chapter 1, a fundamental limitation to resolving time series fluctuations is given by the "sampling theorem", which states that the highest detectable frequency or wavenumber (the Nyquist frequency or wavenumber) is determined by the interval between the data points. For example, the highest frequency that we can hope to resolve by an hourly time series is one cycle per 2 h, or one cycle per $2\Delta t$, where Δt is the interval of time between points in the series.

For the most part, we fit series of well-known functions to the data in order to transform from the time domain to the frequency domain. As with the coefficients of the sine and cosine functions used in Fourier analysis, we generally assume that the functions have slowly varying amplitudes and phases, where "slowly" means that coefficients change little over the length of the record. Other linear combinations of orthogonal functions with similar limitations on the coefficients can be used to describe the series. However, the trigonometric functions are unique in that uniformly spaced samples covering an integer number of periods of the function form orthogonal sequences. Arbitrary orthogonal functions, with a similar sampling scheme, do not necessarily form orthogonal sequences. (Note that a basis function $\phi_k(t)$ is orthogonal if the ensemble average and associated integral over the data set are such that $\langle \phi_k(t), \phi_l(t) \rangle = \int \phi_k(t)\phi_l(t)dt = 0, k \neq l$). Another advantage of using common functions in any analysis is that the behavior of these functions is well understood and can be used to simplify the description of the data series in the frequency or wavenumber domain. In this chapter, we consider time series to consist of periodic and aperiodic components superimposed on a secular (long-term) trend and uncorrelated random noise. Fourier analysis and spectral analysis are among the tools used to characterize oceanic processes. Determination of the Fourier components of a time series can be used to generate a *periodogram*, which can then be used to define the spectral power density (*spectrum*) of the time series. However, the periodogram is not the only way to get at the spectral energy density. For example, prior to the introduction of the FFT, the common method for calculating spectra was through the Fourier transform of the autocorrelation function. More modern spectral analysis methods involve autoregressive spectral analysis (including the use of maximum entropy techniques), wavelet transforms, and fractal analysis.

5.2 STOCHASTIC PROCESSES AND STATIONARITY

A common goal of most time series analysis is to separate deterministic periodic oscillations in the data from random and aperiodic fluctuations associated with unresolved background noise (unwanted geophysical variability) or with instrumental effects. It is worth recalling that time series analyses are typically statistical procedures in which data series are regarded as subsets of a stochastic process. A simple example of a stochastic process is one generated by a linear operation on a purely random variable. For example, the function $x(t_i) = 0.5x(t_{i-1}) + \varepsilon(t_i)$ $(i = 1, 2, ...)$ for which $x(t_0) = 0$, say, is a linear random process provided that the fluctuations $\varepsilon(t_i)$ are statistically independent. Stochastic processes are classified as either discrete or continuous. A continuous process is defined for all time steps while a discrete process is defined only at a finite number of points. The data series can be scalar (univariate series) or a series of vectors (multivariate series). While we will deal with discrete data, we assume that the underlying process is continuous.

If we regard each data series as a realization of a stochastic process, each series contains an infinite ensemble of data having the same basic physical properties. Because a particular data series is a sample of a stochastic process, we can apply the same kind of statistical arguments to our data series as we did to individual random variables. Thus, we will be making statistical probability statements about the results of frequency transformations of data series. This fact is important to remember since there is a great temptation to regard transformed values as inherently independent data points. Since many data collected in time or space are highly correlated because of the presence of low frequency, nearly deterministic components, such as long-period tides and the seasonal cycle, standard statistical methods do not really apply. Contrary to the requirements of stochastic theory, the values are not statistically independent. "What constitutes the ensemble of a possible time series in any given situation is dictated by good scientific judgment and not by purely statistical matters" (Jenkins and Watts, 1968). A good example of this problem is presented by Chelton (1982) who showed that the high correlation between the integrated transport through Drake Passage in the Southern Ocean and the circumpolar-averaged zonal wind stress "may largely be due to the presence of a strong semiannual signal in both time series". A strong statistical correlation does not necessarily mean there is a cause-and-effect relationship between the variables.

As implied by the previous sections, the properties of a stochastic process generally are time dependent and the value $y(t)$ at any time, t, depends on the time elapsed since the process started. A simplifying assumption is that the series has reached a steady state or equilibrium in the sense that the statistical properties of the series are independent of absolute time. A minimum requirement for this condition is that the probability density function (PDF) is independent of time. Therefore, a stationary time series has constant mean, μ, and variance, σ^2. Another consequence of this equilibrium state is that the joint PDF depends only on the time difference $t_1 - t_2 = \tau$ and not on absolute times, t_1 and t_2. The term *ergodic* is commonly used in association with stochastic processes for which time (or space) averages can be used in place of ensemble averages. That is, we can average over "chunks" of a time series to get the mean, standard deviation, and other statistical quantities rather than having to produce repeated realizations of the time series. Any formalism involving ensemble averaging is of little value as the analyst rarely has an ensemble at his or her disposal and typically must deal with a single realization. We need the ergodic theorem to enable us to use time averages in place of ensemble averages. (See Chapter 3 for further details on statistics and statistical analysis).

5.3 CORRELATION FUNCTIONS

Discrete or continuous random time series, $y(t)$, have a number of fundamental statistical properties that help characterize the variability of the series and make it easy to compare one time series against another. However, these statistical measures also contain less information than the original time series and, except in special cases, knowledge of these properties is insufficient to reconstruct the time series.

5.3.1 Mean and Variance

If y is a stochastic time series consisting of N values $y(t_i) = y_i$ measured at discrete times t_i $\{t_1, t_2, ..., t_N\}$, the true mean value μ for the series can be estimated by

$$\mu \equiv E[y(t)] = \frac{1}{N} \sum_{i=1}^{N} y_i = \bar{y} \tag{5.1}$$

where $E[y(t)]$ is the expected value, $E[|y(t)|] < \infty$ for all t and $\bar{y} = \frac{1}{N} \sum_{i=1}^{N} y_i$ is the sample mean. The estimated mean value is not necessarily constant in time; different segments of a time series can have different mean values if the series is nonstationary. If $E[y^2(t)] < \infty$ for all t, an estimate of the true variance function is given by

$$\sigma^2 \equiv E\left[\{y(t) - \mu\}^2\right] = \frac{1}{N} \sum_{i=1}^{N} (y_i - \bar{y})^2. \tag{5.2}$$

The positive square root of the variance is the standard deviation, σ, or root-mean-square (RMS) value.

5.3.2 Covariance and Correlation Functions

The covariance and correlation functions are used to describe the covariability of given time series as functions of two different times, $t_1 = t$ and $t_2 = t + \tau$, where τ is the lag time. If the process is *stationary* (unchanging statistically with time) as we normally assume, then absolute time is irrelevant and the covariance functions depend only on τ.

Although the terms "covariance function" and "correlation function" are often used interchangeably in the literature, there is a fundamental difference between them. Specifically, covariance functions are derived from data series following removal of the true mean value, μ, which we typically approximate using the sample mean, $\bar{y}(t)$. Correlation functions use the "raw" data series before removal of the mean. The confusion arises because most analysts automatically remove the mean from any time series with which they are dealing. To further add to the confusion, many oceanographers define correlation as the covariance normalized by the variance, σ^2.

For a stationary process, the *autocovariance function*, C_{yy}, which is based on lagged correlation of a function (or variable) with itself, is estimated as

$$C_{yy}(\tau) \equiv E[\{y(t) - \mu\}\{y(t + \tau) - \mu\}]$$

$$= \frac{1}{N-k} \sum_{i=1}^{N-k} (y_i - \bar{y})(y_{i+k} - \bar{y}) \tag{5.3}$$

where $\tau_k = k\Delta t$ ($k = 0, \ldots, M$) is the lag time, τ, for k sampling time increments, Δt, and $M \ll N$. The corresponding expression for the *autocorrelation function*, R_{yy}, is

$$R_{yy}(\tau) \equiv E[y(t)y(t + \tau)]$$

$$= \frac{1}{N-k} \sum_{i=1}^{N-k} (y_i y_{i+k}) \tag{5.4}$$

At zero lag ($\tau = 0$)

$$C_{yy}(0) = \sigma^2 = R_{yy}(0) - \mu^2 \tag{5.5}$$

where σ^2 is defined by Eqn (5.2) in terms of the normalization factor $1/N$ rather than $1/(N-1)$ (see Chapter 3). From the above definitions, we find

$$C_{yy}(\tau) = C_{yy}(-\tau); \; R_{yy}(\tau) = R_{yy}(-\tau) \tag{5.6}$$

showing that the autocovariance and autocorrelation functions are symmetric with respect to the time lag, τ.

The autocovariance function can be normalized using the variance (Eqn 5.2) to yield the normalized autocovariance function

$$\rho_{yy}(\tau) = \frac{C_{yy}(\tau)}{\sigma^2} \tag{5.7}$$

(Note: as with other oceanographers, we will often, for convenience, refer to Eqn (5.7) as the autocorrelation function). The basic properties of the normalized autocovariance function are:

1. $\rho_{yy}(\tau) = 1$, for $\tau = 0$;
2. $\rho_{yy}(\tau) = \rho_{yy}(-\tau)$, for all τ;
3. $|\rho_{yy}(\tau)| \leq 1$, for all τ;
4. If the stochastic process is continuous, then $\rho_{yy}(\tau)$, must be a continuous function of τ.

If we now replace one of the $y(t)$ functions in the above relations with another function, $x(t)$, we obtain the *cross-covariance function*

$$C_{xy}(\tau) \equiv E[\{y(t) - \mu_y\}\{x(t + \tau) - \mu_x\}]$$

$$= \frac{1}{N-k} \sum_{i=1}^{N-k} [y_i - \bar{y}][x_{i+k} - \bar{x}] \tag{5.8}$$

and the *cross-correlation function*

$$R_{xy}(\tau) \equiv E[y(t)x(t+\tau)]$$

$$= \frac{1}{N-k} \sum_{i=1}^{N-k} y_i x_{i+k} \tag{5.9}$$

The normalized cross-covariance function (or *correlation coefficient function*) for a stationary process is

$$\rho_{xy} \equiv \frac{C_{xy}(\tau)}{\sigma_x \sigma_y} \tag{5.10}$$

Here, $y(t)$ could be the alongshore component of daily mean wind stress and $x(t)$ the daily mean sea-level elevation at the coast. As a result of the Coriolis force, the alongshore current generated by the wind causes a sea-level setup along the coast. Typically, the sea level set up at mid to high latitudes lags the alongshore wind stress by about 1 day (or, roughly one inertial period).

Care should be taken in interpreting covariance and correlation estimates for large lags. Problems arise if low-frequency components are present in the data since the averaging inherent in these functions becomes based on fewer and fewer samples and loses its statistical reliability as the lag increases. For example, at lag $\tau = 0.1T$ (i.e., 10% of the length of the time series) there are roughly 10 independent cycles of any variability on a timescale, $T_{0.1} = 0.1T$, while at lags of $0.5T$ there are only two independent estimates of the timescale $T_{0.5}$. In many cases, low-frequency components in geophysical time series make it pointless to push the lag times beyond 40—80% of the data series (see Section 3.15 of Chapter 3 for further details). Some authors argue that division by N rather than by $N-k$ reduces the bias at large lags. Although this is certainly true ($N-k << N$ at large lags, k), it does not mean that the results are a better representation of statistical reality. In essence, neither of these estimators is optimal. Ideally one should write down the likelihood function of the observed time series, if it exists. Differentiation of this likelihood function would then give a set of equations for the maximum likelihood estimates of the autocovariance function. Unfortunately, the derivatives are in general untraceable and one must work with estimators given above. Results for this section are summarized as follows:

1. Estimators with divisors $T = N\Delta t$ usually have smaller mean square errors (MSEs) than those based on $T - \tau$; also, those based on $1/T$ are positive definite while those based on $1/(T - \tau)$ may not be.
2. Some form of correction for low-frequency trends is required. In simple cases, one can simply remove a mean value while in others the trend can be removed. Trend removal must be done carefully so that erroneous data are not introduced into the time series during the subtraction of the trend.
3. There will be strong correlations between values neighboring values in the autocorrelation function if the correlation in the original series is moderately strong; the autocorrelation function, which can be regarded as a new time series derived from $y(t)$, will, in general, be more strongly correlated than the original series.
4. Due to the correlation in (3), the autocorrelation function may fail to dampen according to expectations; this will increase the basic length scale in the function.
5. Correlation is a relative measure only.

In addition to its direct application to time series analysis, the autocorrelation function was critical to the development of early spectral analysis techniques. Although modern methods typically calculate spectral density distributions directly from the Fourier transforms of the data series, earlier methods determined spectral estimates from the Fourier transform of the autocorrelation function. An important milestone in time series analysis was the proof by N. Wiener and A. Khinchin in the 1930s that the correlation function is related to the spectral density function through a Fourier transform relationship. According to the Wiener—Khinchin relations, the autospectrum of a time series is the Fourier transform of its autocorrelation function.

5.3.3 Analytical Correlation/Covariance Functions

The autocorrelation function of a zero-mean random process $\varepsilon(t)$ ("white noise") can be written as

$$R_{\varepsilon\varepsilon}(\tau) = \sigma_\varepsilon^2 \rho_{\varepsilon\varepsilon}(\tau) = \sigma_\varepsilon^2 \delta(\tau) \tag{5.11}$$

where $\delta(\tau)$ is the Dirac delta function. In this example, σ_ε^2 is the variance of the data series. Another useful function is the cross-correlation between the time-lagged stationary signal $y(t) = \alpha x(t - \tau) + \varepsilon$ and the original signal $x(t)$. For constant α

$$R_{xy}(\tau) = \alpha R_{xx}(\tau - \tau_o) + \sigma_\varepsilon^2 \tag{5.12a}$$

which, for low noise, has a peak value

$$R_{xy}(\tau_o) = \alpha R_{xx}(0) = \alpha \sigma_x^2 \qquad (5.12b)$$

Functions of the type Eqns (5.12a) and (5.12b) have direct use in ocean acoustics where the time lag, τ_o, at the peak of the zero-mean autocorrelation function can be related to the phase speed c and distance of travel d of the transmitted signal $x(t)$ through the relation $\tau_o = d/c$. It is through calculations of this type that modern acoustic Doppler current meters (ADCMs) and scintillation flow meters determine oceanic currents. In the case of ADCMs, knowing τ_o and d gives the speed c and hence the change of the acoustic signal by the currents during the two-way travel time of the signal. Scintillation meters measure the delay τ_o for acoustic signals sent between a transmitter—receiver pair along two parallel acoustic paths separated by a distance d. The relation $\tau_o = d/v$ then gives the mean flow speed, v, normal to the direction of the acoustic path. Sending the signals both ways in the transmitter—receiver pairs circumvents the problem of not knowing the sound speed, c, in detail.

Although the calculation of autocorrelation and autocovariance functions is fairly straightforward, care is needed in interpreting the resulting values. For example, a stochastic process is said to be Gaussian (or normal) if the multivariate PDF is normal. Then the process is completely described by its mean, variance, and autocovariance function. However, there is a class of non-Gaussian processes that have the same normalized autocovariance function, ρ, as a given normal process. Consider the linear system

$$\tau_o \frac{dy}{dt} + y(t) = z(t) \qquad (5.13)$$

where $z(t)$ is a white-noise input and $y(t)$ is the output. Here, $y(t)$ is called a "first-order autoregressive process", which has the normalized autocorrelation function

$$\rho_{yy}(\tau) = e^{-|\tau|/\tau_o} \qquad (5.14)$$

Thus, if the input to the first-order system has a normal distribution, then by an extension of the central limit theorem it may be shown that the output is normally distributed and is completely specified by the autocorrelation function.

Another process with an exponential autocorrelation function, which differs greatly from the normal process, is called the *random telegraph signal* (Figure 5.1). Alpha particles from a radioactive source are used to trigger a flip-flop between $+1$ and -1. Assuming the process was started at $t = -\infty$, we can derive the normalized autocorrelation function as

$$\rho_{yy}(\tau) = e^{-2\lambda|\tau|} \qquad (5.15)$$

If $\lambda = 1/(2\tau_o)$, then this is the same as the autocorrelation function of a normal process, which is characteristically different from the flip-flop time series. Again, one must be careful when interpreting autocorrelation functions. As with any correlation between two variables, the autocorrelation function only indicates how the time series vary together and says nothing about the magnitudes of their variations. A detailed examination of the data requires the analysis of the magnitudes of the values in the series and not just their correlation.

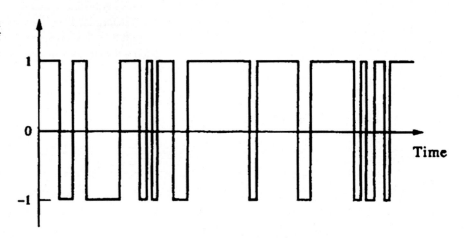

FIGURE 5.1   A realization of a random telegraph signal with digital amplitudes of ± 1 as a function of time.

5.3.4 Observed Covariance Functions

To examine what autocorrelation functions look like in practice, and to emphasize the fact that the methodology applies to spatial as well as temporal data, consider the acoustic profile data in Table 5.1. Here, we have tabulated the calibrated acoustic backscatter anomaly measured over 5-m depth increments in the upper ocean using a 150 kHz acoustic Doppler current profiler (ADCP) lowered from a ship. These spatial data are from the first bin of adjacent Beams 1 and 2 of a four-beam ADCP, and represent the backscatter intensity anomaly (in decibels) from zooplankton ensonified at a distance of 5 m from the instrument transducers. Because each of the transducers is tilted at an angle of 30° to the vertical, the two 5-m increment profiles are separated horizontally by only 3.9 m and so the autocorrelations for the two series should be nearly identical at all lags. In this case, we use the normalized covariance Eqn (5.7) derived from Eqn (5.3) in which the sum is divided by the number of lag values, $N - k$, for lag $\tau = k\Delta z$, where $\Delta z = 5$ m. As indicated by the autocorrelation functions in Figure 5.2, the functions are similar at small lags, where statistical reliability is large, but diverge significantly at higher lags with the decrease in the number of independent covariance estimates.

TABLE 5.1 Acoustic backscatter anomaly (decibels) measured in bin#1 (depth, m) from two adjacent transducers (Beams 1 and 2) on a 4-beam 150 kHz ADCP lowered from a ship in the Northeast Pacific.

Beam	75 m	80	85	90	95	100	105	110	115	120
1	11.56	0.67	−8.33	−9.82	−13.91	−18.00	3.67	−2.00	−12.29	−13.71
2	14.67	3.00	−5.67	−9.64	−12.82	−16.00	−8.50	−11.00	−15.29	−16.71
125 m	**130**	**135**	**140**	**145**	**150**	**155**	**160**	**165**	**170**	**175**
−11.33	−8.00	24.14	38.13	40.00	35.00	29.63	24.00	26.50	28.75	30.63
−10.33	−2.00	23.71	36.63	41.00	33.14	24.38	15.00	20.63	26.25	31.88
180 m	**185**	**190**	**195**	**200**	**205**	**210**	**215**	**220**	**225**	**230**
30.50	31.00	36.00	31.63	21.00	12.25	3.00	−7.00	−4.43	−0.50	0.75
31.00	29.13	29.75	24.75	16.00	7.25	3.25	6.38	11.57	12.25	5.38

The data cover a depth range of 75−230 m at increments of 5 m ($N = 32$ Values). The two vertical profiles are separated horizontally by a distance of roughly 3.9 m. The first line in each set gives the depth of the observations, the next two lines the anomaly values for Beam 1 then Beam 2. The record means have not been removed from the data.

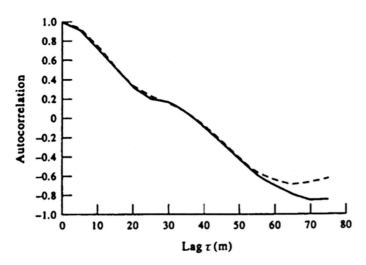

FIGURE 5.2 Autocorrelation functions of the acoustic backscatter data in Table 5.1. The thick line is for acoustic Beam 1, the dashed line for acoustic Beam 2.

5.3.5 Integral Timescales

The integral timescale, T, is defined as the sum of the normalized autocorrelation function (Eqn 5.7) over the length $L = N\Delta\tau$ of the time series correlation for N lag steps, $\Delta\tau$. Specifically, the estimate

$$T = \Delta\tau \sum_{i=1}^{N} \left\{ 1 + 2\left(\frac{N-i}{N}\right)[\rho(\tau_i)] \right\}$$

$$= \frac{\Delta\tau}{\sigma^2} \sum_{i=1}^{N} \left\{ \sigma^2 + 2\left(\frac{N-i}{N}\right)[C(\tau_i)] \right\} \tag{5.16}$$

gives a measure of the dominant correlation timescale for determining the mean statistics of a data series. For lag times $\tau_i = i\Delta\tau$ longer than $N^*\Delta\tau$ (in which N^* is the effective number of degrees of freedom) the data are expected to become decorrelated. (The reader is directed to Chapter 3.15 for a detailed discussion on the derivation of N^* and the associated time scales appropriate to a given statistical property.) In reality, the summation typically is limited to $N^* < N$ since low frequency components within the time series prevent the summation from converging to a constant value over the finite length of the record. In general, one should continue the summation until it reaches a near-constant value, which we take as the value for $T = T^*$. If no plateau is reached within a reasonable number of lags, no integral timescale exists. In that case, the integral timescale can be approximated by integrating only to the first zero crossing of the autocorrelation function (cf. Poulain and Niiler, 1989). (Again, we emphasize that the reader should see Chapter 3.15 for a detailed discussion on calculating the effective degrees of freedom, N^*).

5.3.6 Correlation Analysis Versus Linear Regression

Geophysical data are typically obtained from random temporal sequences or spatial fields that cannot be regarded as mutually independent. Because the data series depend on time and/or spatial coordinates, the use of linear regression to study relationships between data series may lead to incomplete or erroneous conclusions. As an example, consider two time series: A white-noise series, consisting of identically distributed and mutually independent random variables, and the same series but with a time shift. As the values of the time series are statistically independent, the cross-correlation coefficient for the two series will be zero at zero lag, even though the time series are strictly linearly related. Regression analysis would show no relationship between the two series. However, cross-correlation analysis would reveal the linear relationship (a coefficient of unity) for a lag equal to the time shift. Correlation analysis, which takes into lagged offsets between two series, is often a better way to study relations among time series than traditional regression analysis.

5.4 FOURIER ANALYSIS

For many applications, including spectral analysis discussed in Section 5.6, we can view time series as linear combinations of periodic or quasi-periodic components that are superimposed on a long-term trend and random noise. The periodic components are assumed to have fixed, or slowly varying, amplitudes and phases over the length of the record. The trends might include a slow drift in the sensor characteristics or a long-term component of variability that cannot be resolved because of the limited duration of the data series. "Noise" includes random contributions from the instrument sensors and electronics, as well as frequency components that are outside the immediate range of interest and cannot be resolved by the measurement system (e.g., small-scale turbulence). A goal of time series analysis in the frequency domain is to reliably separate periodic oscillations from the random and aperiodic fluctuations. Fourier analysis is one of the most commonly used methods for identifying periodic components in near-stationary time series oceanographic data. If the time series are strongly nonstationary, more localized transforms such as the Hilbert and Wavelet transforms should be used.

The fundamentals of Fourier analysis were formalized in 1807 by the French mathematician Joseph Fourier (1768−1830) during his service as an administrator under Napoleon. Fourier developed his technique to solve the problem of heat conduction in a solid with specific application to heat dissipation in blocks of metal being turned into cannons because during his military service he experienced the fact that French cannons melted much earlier than he expected. Fourier's basic premise was that any finite length, infinitely repeated time series, $y(t)$, defined over the principal interval [0, T] can be reproduced using a linear summation of cosines and sines, or *Fourier series*, of the form

$$y(t) = \overline{y(t)} + \sum_{p} [A_p \cos(\omega_p t) + B_p \sin(\omega_p t)] \tag{5.17}$$

in which $\bar{y}$ is the mean value of the record, A_p, B_p are constants (the Fourier coefficients), and the specified angular frequencies, ω_p, are integer ($p = 1, 2, \ldots$) multiples of the fundamental frequency, $\omega_1 = 2\pi f_1 = 2\pi/T$, where T is the total length of the time series. Provided enough of these Fourier components are used, each value of the series can be accurately reconstructed over the principal interval. By the same token, the relative contribution a given component makes to the total variance of the time series is a measure of the importance of that particular frequency component in the observed signal. This concept is central to spectral analysis techniques. Specifically, the collection of Fourier coefficients having amplitudes A_p, B_p form a *periodogram*, which then defines the contribution that each oscillatory component, ω_p, makes to the total "energy" of the observed oceanic signal. Thus, we can use the Fourier components to estimate the power spectrum (energy per unit frequency bandwidth) of a time series, as described in Section 5.6. Since both A_p, B_p must be specified, there are two degrees of freedom (DoF) per spectral estimate derived from the "raw" or unsmoothed periodogram.

5.4.1 Mathematical Formulation

Let $y(t)$ denote a continuous, finite-amplitude time series of finite duration. Examples include hourly sea-level records from a coastal tide gauge station or temperature records from a moored thermistor chain. If y is periodic, there is a period T_o such that $y(t) = y(t + T_o)$ for all t. In the language of Fourier analysis, the periodic functions are sines and cosines, which have the important properties that:

1. A finite number of Fourier coefficients provides the minimum mean square error (MSE) between the original data and a functional fit to the data series;
2. The functions are orthogonal so that coefficients for a given frequency can be determined independently.

Suppose that the time series is specified only at discrete times by subsampling the continuous series, $y(t)$, at a sample spacing of Δt (Figure 5.3). Since the series has a duration T, there are a total of $N = T/\Delta t$ sample intervals and $N + 1$ sample points located at times $y(t_n) = y(n\Delta t) = y_n$ ($n = 0, 1, \ldots, N$). Using Fourier analysis, it is possible to reproduce the original signal as a sum of sine or cosine waves of different amplitudes and phases. In Figure 5.3, we show a time series, $y(n\Delta t)$, of 41 data points. The time series is followed by plots of the first, second, and sixth harmonics that were summed to create the time series. The frequencies of these harmonics are $f = 1/T$, $2/T$, and $6/T$, respectively, and each harmonic has the form $y_k(n\Delta t) = C_k \cos[(2\pi kn/N) + \phi_k]$, where ($C_k$, ϕ_k) are, respectively, the amplitudes and phases of the harmonics for $k = 1, 2$, and 6. Here, $T = 40\Delta t$ and we have arbitrarily chosen (C_1, ϕ_1) = (2, $\pi/4$), (C_2, ϕ_2) = (0.75, $\pi/2$), and (C_6, ϕ_6) = (1.0, $\pi/6$) in order to generate the time series for this example. The $N/2$ harmonic, which is the highest frequency component that can be resolved by this sampling, has a frequency, $f_{Nq} = (N/2)/N\Delta t = 1/2\Delta t$ cycles per unit time and a period of $2\Delta t$. Called the *sampling* or *Nyquist* frequency, f_{Nq} represents the highest frequency resolved by the sample series in question. (As noted in previous chapters, we use the subscript Nq to denote the Nyquist frequency, which should not be confused with the integer N, as in $n = 1, 2, \ldots, N$, or the depth-dependent buoyancy frequency, $N(z)$).

The fundamental frequency, $f_1 = 1/T$, is used to construct $y(t)$ through the infinite *Fourier series*

$$y(t) = \frac{1}{2} A_0 + \sum_{p=1}^{\infty} \left[A_p \cos\left(\omega_p t\right) + B_p \sin\left(\omega_p t\right) \right] \qquad (5.18)$$

in which

$$\omega_p = 2\pi f_p = 2\pi p f_1 = 2\pi p/T; p = 1, 2, \ldots \qquad (5.19)$$

is the frequency of the pth constituent in radians per unit time (f_p is the corresponding frequency in cycles per unit time) and $A_0/2$ is the mean, or "DC" offset, of the time series. The factor of ½ multiplying A_0 is for mathematical convenience. The length of the data record, T, defines both the lowest frequency, f_1, resolvable by the data series and the maximum frequency resolution, $\Delta f = f_1 = 1/T$, that one can obtain from discretely sampled data.

To obtain the coefficients A_p, we simply multiply Eqn (5.18) by $\cos(\omega_p t)$, then integrate over all possible frequencies. The coefficients B_p, are obtained in the same way by multiplying by $\sin(\omega_p t)$. Using the orthogonality condition for the product of trigonometric functions (which requires that the trigonometric arguments cover an exact integer number of 2π cycles over the interval $(0, T)$), we find

$$A_p = \frac{2}{T} \int_0^T y(t)\cos\left(\omega_p t\right) dt, p = 0, 1, 2, \ldots \qquad (5.20a)$$

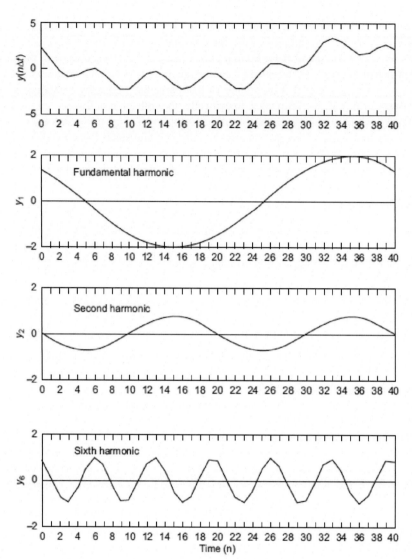

FIGURE 5.3 Discrete subsampling of a continuous signal, $y(t)$. The sampling interval is $\Delta t = 1$ time unit and the fundamental frequency is $f_1 = 1/T$, where $T = N\Delta t$ is the total record length and $N = 40$. The signal $y(t)$ is the sum of the first, second, and sixth harmonics which have the form $y_k(n\Delta t) = C_k \cos[(2\pi kn/N) + \phi_k]$; $k = 1, 2, 6$; $n = 0, 1, ..., 40$.

$$B_p = \frac{2}{T} \int\limits_0^T y(t)\sin\left(\omega_p t\right)dt, p = 1, 2, \dots \tag{5.20b}$$

where the integral for $p = 0$ in (5.20a) yields $A_0 = 2\bar{y}$, twice the mean value of $y(t)$ for the entire record. Since each pair of coefficients, (A_p, B_p), is associated with a frequency ω_p (or f_p), the amplitudes of the coefficients provide a measure of the relative importance of each frequency component to the overall signal variability. For example, if $\left(A_6^2 + B_6^2\right)^{1/2} \gg \left(A_2^2 + B_2^2\right)^{1/2}$ we expect there is much more "spectral energy" at frequency, ω_6 than at frequency, ω_2. Here, spectral energy refers to the amplitudes squared of the Fourier coefficients, which represent the variance, and therefore the energy, for that portion of the time series.

We can also express the Fourier series as amplitude and phase functions in the compact Fourier series form

$$y(t) = \frac{1}{2} C_0 + \sum_{p=1}^{\infty} C_p \cos\left(\omega_p t - \theta_p\right) \tag{5.21}$$

in which the amplitude of the pth component is

$$C_p = \left(A_p^2 + B_p^2\right)^{1/2}, \; p = 0, 1, 2, \ldots \tag{5.22}$$

where $C_0 = A_0 \; (B_0 = 0)$ is twice the mean value and

$$\theta_p = \tan^{-1}\left[B_p / A_p\right], \; p = 1, 2, \ldots \tag{5.23}$$

is the phase angle of the constituent at time $t = 0$. The phase angle gives the relative "lag" of the component in radians (or degrees) measured counterclockwise from the real axis ($B_p = 0$, $A_p > 0$). The corresponding time lag for the pth component is then $\tau_p = \theta_p/(2\pi f_p)$, in which θ_p is measured in radians.

The discrimination of the signal amplitude as a function of frequency given by Eqns (5.18) and (5.21) provides us with the beginnings of spectral analysis. Notice that neither of these expressions allows for a trend in the data. If any trend is not first removed from the record, the analysis will erroneously blend the variance from the trend into the lower frequency components of the Fourier expansion. Moreover, we now see the need for the factor of 1/2 in the leading terms of Eqns (5.18) and (5.21). Without it, the lead $p = 0$ components (A_0, C_0) would equal twice the mean component. Instead, the lead term is $\bar{y} = \frac{1}{2} A_0 = \frac{1}{2} C_0$.

Up to now we have assumed that $y(t)$ is a scalar quantity. We can also expand the time series of a vector property, $\mathbf{u}(t)$. Included in this category are time series of current velocity from moored current meter arrays and wind velocity from moored weather buoys. Expressing vector time series in complex notation, we can write

$$\mathbf{u}(t) = u(t) + iv(t) \tag{5.24}$$

where, for example, u and v might be the north–south and east–west components of current or wind velocity in Cartesian coordinates. An individual vector can be expressed as

$$\mathbf{u}(t) = \overline{\mathbf{u}(t)} + \sum_{p=1}^{\infty} \left[A_p \cos\left(\omega_p t + \alpha_p\right) + iB_p \sin\left(\omega_p t + \beta_p\right)\right] \tag{5.25}$$

Here, $\overline{\mathbf{u}(t)}$ is the mean (time averaged) vector, $\overline{\mathbf{u}} = \bar{u} + i\bar{v}$, and (α_p, β_p) are phase lags or relative phase differences for the separate velocity components.

Vector quantities also can be defined through expressions of the form

$$\mathbf{u}(t) = \overline{\mathbf{u}(t)} + \sum_{p=1}^{\infty} \exp\left[\frac{i\left(\varepsilon_p^+ + \varepsilon_p^-\right)}{2}\right]\left\{\left(A_p^+ + A_p^-\right)\cos\left[\omega_p t + \frac{\varepsilon_p^+ - \varepsilon_p^-}{2}\right] + i\left(A_p^+ - A_p^-\right)\sin\left[\omega_p t + \frac{\varepsilon_p^+ - \varepsilon_p^-}{2}\right]\right\} \tag{5.26}$$

in which A_p^+ and A_p^- are, respectively, the lengths of the counterclockwise ($+$) and clockwise ($-$) rotary components of the velocity vector, and ε_p^+ and ε_p^- are the angles that these vectors make with the real axis at $t = 0$. The resultant time series is an ellipse with major axis of length, $L_M = A_p^+ + A_p^-$ and minor axis of length, $L_m = |A_p^+ - A_p^-|$. The major axis is oriented at angle $\theta_p = \frac{1}{2}\left(\varepsilon_p^+ + \varepsilon_p^-\right)$ from the u-axis and the current rotates counterclockwise when $A_p^+ > A_p^-$ and clockwise when $A_p^+ < A_p^-$. The velocity vector is aligned with the major axis direction, θ_p at a time, t, when $\omega_p t = -\frac{1}{2}\left(\varepsilon_p^+ - \varepsilon_p^-\right)$. Motions are said to be *linearly polarized* (rectilinear) if the two oppositely rotating components are of the same magnitude and *circularly polarized* if one of the two components is zero. In the northern (southern) hemisphere, motions are predominantly clockwise (counterclockwise) rotary. Further details on rotary decomposition are presented in Sections 5.6 and 5.8.

5.4.2 Discrete Time Series

Most oceanographic time or space series, whether they were collected in analog or digital form, are eventually converted to digital data, which may then be expressed as series expansions of the form of Eqn (5.18) or (5.21). These expansions are then used to compute the Fourier transform (or periodogram) of the data series. The basis for this transform is Parseval's theorem, which states that the mean square (or average) energy of a time series, $y(t)$, can be separated into contributions from individual harmonic components to make up the time series. For example, if $\bar{y}$ is the sample mean value of the time series, y_n is the contribution from the nth data value, and N is the total number of data values in the time series, then the mean square value of the series about its mean is (i.e., the variance of the time series)

$$\sigma^2 = \frac{1}{N-1} \sum_{n=1}^{N} [y_n - \bar{y}]^2 \tag{5.27}$$

which provides a measure of the total energy in the time series. The variance in Eqn (5.27) also can be obtained by summing the contributions from the individual Fourier harmonics. This kind of decomposition of discrete time series into specific harmonics leads to the concept of a Fourier line spectrum (Figure 5.4).

To determine the energy distribution within a time series, $y(t)$, we need to find its Fourier transform. That is, we need to determine the coefficients, A_p, B_p in the Fourier series Eqn (5.18) or, equivalently, the amplitudes and phase lags, C_p, θ_p in the Fourier series Eqn (5.21). Suppose that we have first removed any trend from the data record. For any time, t_n, the Fourier series for a finite length, detrended digital record having N (even) values at times $t_n = t_1, t_2, \ldots, t_N$, is

$$y(t_n) = \frac{1}{2} A_0 + \sum_{p=1}^{N/2} [A_p \cos(\omega_p t_n) + B_p \sin(\omega_p t_n)] \tag{5.28}$$

where the angular frequency, $\omega_p = 2\pi f_p = 2\pi p/T$. Using $t_n = n \cdot \Delta t$ together with (5.22) and (5.23), the final form for the discrete, finite Fourier series becomes

$$y(t_n) = \frac{1}{2} A_0 + \sum_{p=1}^{N/2} [A_p \cos(2\pi pn/N) + B_p \sin(2\pi pn/N)]$$

$$= \frac{1}{2} C_0 + \sum_{p=1}^{N/2} C_p \cos[(2\pi pn/N) - \theta_p] \tag{5.29}$$

where the leading terms, $\frac{1}{2} A_0$ and $\frac{1}{2} C_0$, are the mean values ($\bar{y}$) of the record. The coefficients are again determined using the orthogonality condition for the trigonometric functions. In fact, the main difference between the discrete case and the continuous case formulated in the last section (aside from the fact we can no longer have an infinite number of Fourier components) is that the coefficients are now defined through the summations rather than through the integrals

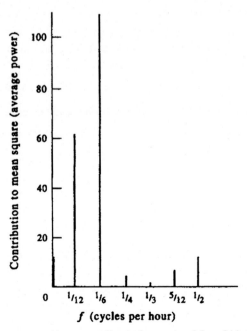

FIGURE 5.4 An example of a Fourier line spectrum with power at discrete frequencies, f, for a 24-h duration record with 1-h sampling increment.

$$A_p = \frac{2}{N} \sum_{n=1}^{N} y_n \cos(2\pi pn/N), p = 0, 1, 2, ..., N/2$$

$$A_0 = \frac{2}{N} \sum_{n=1}^{N} y_n, B_0 = 0$$

$$A_{N/2} = \frac{1}{N} \sum_{n=1}^{N} y_n \cos(n\pi), B_{N/2} = 0 \qquad (5.30)$$

$$B_p = \frac{2}{N} \sum_{n=1}^{N} y_n \sin(2\pi pn/N), p = 1, 2, ..., (N/2)-1$$

Notice that the summations in Eqn (5.30) consist of multiplying the data record by sine and cosine functions that "pick out" from the record those frequency components specific to their trigonometric arguments. Remember, the orthogonality condition requires that the arguments in the trigonometric functions be integer multiples of the total record length, $T = N\Delta t$, as they are in Eqn (5.30). If they are not, the sines and cosines do not form an orthonormal set of basis functions for the Fourier expansion and the original signal cannot be correctly replicated.

The arguments $2\pi pn/N$ in the above equations are based on a hierarchy of equally spaced frequencies, $\omega_p = 2\pi p/(N\Delta t)$, and time increment, "$n$". The summation goes to $N/2$, which is the limit of coefficients we can determine; for $p > N/2$ the trigonometric functions simply begin to cause repetition of coefficients already obtained for the interval, $p \leq N/2$. Furthermore, it should be obvious that because there are as many coefficients as data points and because the trigonometric functions form an orthogonal basis set, the summation over the $2(N/2) = N$ discrete coefficients provides an exact replication of the time series, $y(t)$. Small differences between the original data and the Fourier series representation arise because of roundoff errors accumulated during the arithmetic calculations (see Chapter 3).

The steps in computing the Fourier coefficients are as follows. Step 1: Calculate the arguments, $\Phi_{pn} = 2\pi pn/N$, for each integer p and n. Step 2: For each $n = 1, 2, ..., N$, evaluate the corresponding values of $\cos\Phi_{pn}$ and $\sin\Phi_{pn}$, and collect sums of $y_n \cdot \cos\Phi_{pn}$ and $y_n \cdot \sin\Phi_{pn}$. Step 3: Increment p and repeat steps 1 and 2. The procedure requires roughly N^2 real multiply–add operations. For any real data sequence, roundoff errors plus errors associated with truncation of the total allowable number of desired Fourier components (maximum $f_p < f_{N/2}$) will give rise to a less than perfect fit to the data. The residual $\Delta y(t) = y(t) - y_{FS}(t)$ between the observations, $y(t)$, and the calculated Fourier series, $y_{FS}(t)$, will diminish with increased computational precision and increased numbers of allowable terms used in the series expansion. When computing the phases $\theta_p = \tan^{-1}[B_p/A_p]$ in the formulation (5.29), one must take care to examine in which quadrants A_p and B_p are situated. For example, $\tan^{-1}(0.2/0.7)$ differs from $\tan^{-1}(-0.2/-0.7)$ by 180°. The familiar four-quadrant inverse tangent function ATAN2 in FORTRAN and atan2 in MATLAB are especially designed to take care of this problem.

5.4.3 A Computational Example

The best way to demonstrate the computational procedure for Fourier analysis is with an example. Consider the 2-year segment of monthly mean SSTs measured at the Amphitrite light station off the southwest coast of Vancouver Island (Table 5.2). Each monthly value is calculated from the average of daily surface thermometer observations collected around

TABLE 5.2 Monthly mean sea surface temperature (SST) (°C) at Amphitrite Point (48°55.16′ N, 125°32.17′ W) on the west coast of Canada during January 1982 through December 1983.

Year 1982

N	1	2	3	4	5	6	7	8	9	10	11	12
SST	7.6	7.4	8.2	9.2	10.2	11.5	12.4	13.4	13.7	11.8	10.1	9.0

Year 1983

N	13	14	15	16	17	18	19	20	21	22	23	24
SST	8.9	9.5	10.6	11.4	12.9	12.7	13.9	14.2	13.5	11.4	10.9	8.1

noon local time and tabulated to the nearest 0.1°C. These data are known to contain a strong seasonal cycle of warming and cooling, which is modified by local effects of runoff, tidal stirring, and wind mixing.

The data in Table 5.2 are in the form $y(t_n)$, where $n = 1, 2, ..., N$ ($N = 24$). To calculate the coefficients A_p and B_p for these data, we use the summations Eqn (5.30) for each successive integer p, up to $p = N/2$. These coefficients are then used in Eqn (5.30) to calculate the magnitude $C_p = \left(A_p^2 + B_p^2 \right)^{1/2}$ for each frequency component, $f_p = p/T$. Since C_p^2 is proportional to the variance at the specified frequency, the C_p enables us to rate the order of importance of each frequency component in the data series.

The mean value, $\bar{y}(t) = \frac{1}{2} A_0$, and the 12 pairs of Fourier coefficients obtainable from the temperature record are listed in Table 5.3 together with the magnitude, C_p. Values have been rounded to the nearest 0.01°C. The Nyquist frequency, f_{Nq}, is 0.50 cycles per month (cpmo, $p = 12$) and the fundamental frequency, f_1, is 0.042 cpmo ($p = 1$). As we would anticipate from a visual inspection of the time series, the record is dominated by the annual cycle (period = 12 months) followed by weaker contributions from the biannual cycle (24 months) and semiannual cycle (6 months). For periods shorter than 6 months, the coefficients, C_p have similar amplitudes and likely represent the roundoff errors and background "noise" in the data series. This suggests that we can reconstruct the original time series to a high degree of accuracy using only the mean value ($p = 0$) and the first three Fourier coefficients ($p = 1, 2, 3$).

Figure 5.5 is a plot of the original SST time series and the reconstructed Fourier fit to this series using only the first three Fourier components from Table 5.3. Comparison of these two time series shows that the reconstructed series does not adequately reproduce the skewed crest of the first year nor the high-frequency "ripples" in the second year of the data record. There is also a slight mismatch in the maxima and minima between the series. Differences between the two curves are typically around a few tenths of a degree. In contrast, if we use all 12 components in Table 5.3, corresponding to 24 DoF, we get an exact replica of the original time series to within machine accuracy.

5.4.4 Fourier Analysis for Specified Frequencies

Analysis of time series for specific frequencies is a special case of Fourier analysis that involves adjustment of the record length to match the periods of the desired Fourier components. As we illustrate in the following sections, analysis for specific frequency components (such as the known tidal constituents) is best conducted using LS (Least-Squares) fitting methods rather than Fourier analysis. LS analysis requires that there be many fewer constituents than data values, which is

TABLE 5.3 Fourier coefficients and frequencies for the Amphitrite Point monthly mean temperature data.

p	Frequency (cpmo)	Period (month)	Coefficient A_p (°C)	Coefficient B_p (°C)	Coefficient C_p (°C)	Phase θ_p (degrees)
0	0	–	21.89	0	21.89	0
1	0.042	24	−0.55	−0.90	1.05	−121.4
2	0.083	12	−1.77	−1.99	2.67	−131.7
3	0.125	8	0.22	−0.04	0.23	−10.3
4	0.167	6	−0.44	−0.06	0.45	−172.2
5	0.208	4.8	0.09	−0.07	0.11	−37.9
6	0.250	4	0.08	−0.04	0.09	−26.6
7	0.292	3.4	0.01	−0.16	0.16	−58.0
8	0.333	3	−0.03	−0.16	0.16	−100.6
9	0.375	2.7	−0.14	0.05	0.15	160.3
10	0.417	2.4	−0.09	−0.07	0.11	−142.1
11	0.458	2.2	−0.08	−0.12	0.14	−123.7
12	0.500	2	−0.15	0	0.15	0

Frequency is in cycles per month (ccpmo). $A_0/2$ is the mean temperature and θ_p is the phase lag for the pth component taken counterclockwise from the positive A_p axis.

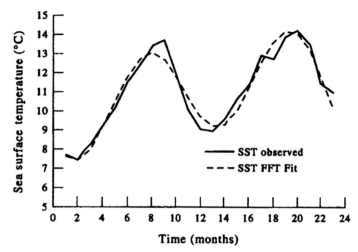

FIGURE 5.5 Monthly mean sea surface temperature (SST) record for Amphitrite Point on the west coast of Vancouver Island (see Table 5.2). The bold line is the original 24-month series; the dashed line is the SST time series generated using the first three Fourier components, f_p, $p = 0, 1, 2$, corresponding to the mean, 24-month, and 12-month cycles (Fourier components appear in Table 5.3). FFT, fast Fourier transform (see Section 5.4.5).

usually the case for tidal analysis at the well-defined frequencies of the tide-generating potential. Problems arise if there are too few data values. For example, suppose that we have a few days of hourly water level measurements and we want to use Fourier analysis to determine the amplitudes and phases of the daily tidal constituents, f_k. To do this, we need to satisfy the orthogonality condition for the trigonometric basis functions for which terms like $\int \cos(2\pi f_j t) \cos(2\pi f_k t) dt = 0$ except where $f_j = f_k$ (the integral is over the entire length of the record, T). The approach is only acceptable when the length of the data set is an integer multiple of all the harmonic frequencies we are seeking. That is, the specified tidal frequencies, f_k, must be integer multiples of the fundamental frequency, $f_1 = 1/T$, such that $f_k \cdot T = 1, 2, ..., N$. If this holds, we can use Fourier analysis to find the constituent amplitudes and phases at the specified frequencies. In fact, this integer constraint on $f_k \cdot T$ is a principal reason why oceanographers often prefer to use record lengths of 14, 29, 180, or 355 days when performing analyses of tides. Because the periods of most of the major tidal constituents (K_1, M_2, etc.) are integer multiples of the fundamental tidal periods (1 lunar day, 1 lunar month $\approx$ 29 days, 1 year, 8.8 years, 18.6 years, etc.) of the above record lengths, the analysis is aided by the orthogonality of the trigonometric functions.

A note for those unfamiliar with tidal analysis terminology: Letters of tidal harmonics identify the different types ("species") of tide in each frequency band. Harmonic components of the tide-producing force that undergo one cycle per lunar day (≈ 25 h) have a subscript 1 (e.g., K_1), those with two cycles per lunar day have subscript 2 (e.g., M_2), and so on. Constituents having one cycle per day are called diurnal constituents, those with two cycles per day, semidiurnal constituents. The main daily tidal component, the K_1 constituent, has a frequency of 0.0418 cph (corresponding to an angular speed of 15.041° per mean solar hour) and is associated with the cyclic changes in the luni-solar declination. The main semidiurnal tidal constituent, the M_2 constituent, has a frequency of 0.0805 cph (corresponding to an angular speed of 28.984° per mean solar hour) and is associated with cyclic changes in the lunar position relative to the earth. Other major daily constituents are the O_1, P_1, S_2, N_2, and K_2 constituents. In terms of the tidal potential, the hierarchy of tidal constituents is M_2, K_1, S_2, O_1, P_1, N_2, K_2, and so on. Other important tidal harmonics are the lunar fortnightly constituent, M_f, the lunar monthly constituent, M_m, and the solar annual constituent, S_a. For further details the reader is referred to Thomson (1981), Foreman (1977, 1978), Pugh (1987), and Pawlowicz et al. (2002).

Returning to our discussion concerning Fourier analysis at specified frequencies, consider the 32-h tide gauge record for Tofino, British Columbia, presented in Figures 5.6 and 5.7. In the case of Figure 5.6, we have used Fourier analysis for the mean and first three Fourier components ($p = 0, 1, 2, 3$) for the 32-hour record to reconstruct the time series. However, as we show in Section 5.5, least-squares (LS) harmonic analysis can be used to reproduce this short record much more accurately using only the K_1 constituent and the M_2 constituent. These are the dominant tidal constituents in all regions of the ocean except near amphidromic points. Because the record is 32-h long, the diurnal and semidiurnal frequencies are not integer multiples of the fundamental frequency, $f_1 = 1/T = 0.031$ cph, and are not among the sequence of 16 possible frequencies generated from the Fourier analysis. In order to have frequency components centered more exactly at the K_1 and M_2 frequencies, we would need to shorten the record to 24 h or pad the existing record to 48 h using zeroes. In either

FIGURE 5.6   Hourly sea-level height (SLH) recorded at Tofino on the west coast of Vancouver Island (see Table 5.8). The bold line is the original 32-h series; the dotted line is the SLH series generated using the mean ($p = 0$) plus the next three Fourier components, f_p, $p = 1, 2, 3$ having nontidal periods, T_p, of 32, 16, and 8 h, respectively.

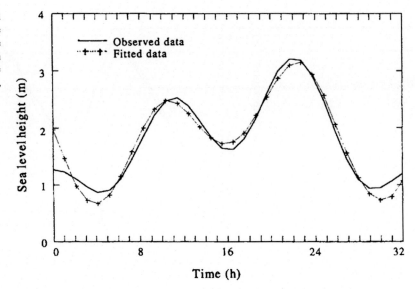

FIGURE 5.7   Hourly sea-level height (SLH) recorded at Tofino on the west coast of Vancouver Island (see Table 5.8; compare to Figure 5.6). The solid line is the original 32-h series; the dotted line is the SLH series obtained from a least-squares fit of the main diurnal (K_1, 0.042 cph) and main semidiurnal (M_2, 0.081 cph) tidal frequencies to the mean-removed data (see Table 5.9).

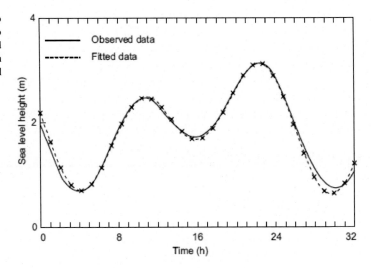

case, the $f_k \cdot T$ for the tides would then be close to integers and a standard Fourier analysis would give an accurate fit to the observed time series. If we stick with the 32-h series, we find that the tidal energy in the diurnal and semidiurnal bands is partitioned among the first three Fourier components at frequencies, $f_1 = 0.031$, $f_2 = 0.062$, and $f_3 = 0.093$ cph. These frequencies are only vaguely close to those of the diurnal and semidiurnal constituents but do span the energy-containing frequency bands. As a result, the time series generated from the record mean combined with the first three Fourier components ($p = 1, 2, 3$) closely approximates the time series obtained using the true tidal frequencies (Figure 5.7). This would not have been possible if the frequencies of the Fourier components had not happened to approximate those of the tides.

5.4.5 The Fast Fourier Transform

One of the main problems with both the autocovariance and the direct Fourier methods of spectral estimation that we discuss in detail in Section 5.6 is low computational speed. The Fourier method requires the expansion into series of sine and cosine terms—a time-consuming procedure. The Fast Fourier Transform (FFT) is a way to speed up this computation, while retaining the accuracy of the direct Fourier method. This makes the Fourier method computationally more attractive than the autocovariance approach when determining spectra.

To illustrate the improved efficiency of the FFT method, consider a series of N values for which $N = 2^p$ (p is a positive integer). The DFT of this series would require N^2 operations whereas the FFT method requires only $8N\log_2 N$ operations. The savings in computer time can be substantial. For example, if $N = 8{,}192$, $N^2 = 67{,}108{,}864$ while $8N\log_2 N = 851{,}968$. Computers are much faster now than when the FFT method was introduced but the relative savings in computational efficiency remains the same. Bendat and Piersol (1986) define the speed ratio between the FFT and discrete Fourier method as $N/4p$. This becomes increasingly more important as the number of terms increases since the direct method computational time is $O(N^2)$, while for the FFT method it is $O(N)$. If one is seeking a smoothed power spectrum, it is often more efficient to compute the spectrum using the FFT technique and then smooth it in spectral space by averaging over adjoining frequency bands rather than smoothing with an autocovariance lag window in the time domain.

To understand the FFT algorithm, we follow the derivation of Danielson and Lanczos (1942) who first helped pioneer the method. Consider a time series of x_k, where $k = 1, 2, \ldots, N$. We want to find the Fourier transform $X_m = X(m/N\Delta t)$, where $m = 0, 1, \ldots, N-1$. To do this, we first partition x_k into two half-series, y_k and z_k, where $y_k = x_{2k-1}$, $z_k = x_{2k}$, $k = 1$, $2, \ldots, N/2$. The series y_k contains values at the odd number times ($x_1, x_3, \ldots$) while the function z_k contains values at the even number times ($x_2, x_4, \ldots$). Both functions have $N/2$ values and their Fourier transforms are

$$Y_m^{(N/2)} = \frac{2}{N} \sum_{k=1}^{N/2} y_k \exp\left[\frac{(-i4\pi km)}{N}\right] \tag{5.31a}$$

$$Z_m^{(N/2)} = \frac{2}{N} \sum_{k=1}^{N/2} z_k \exp\left[\frac{(-i4\pi km)}{N}\right] \tag{5.31b}$$

where the superscript ($N/2$) is used to denote the number of terms used in the expansion. But $X_m^{(N)}$, $Y_m^{(N/2)}$, and $Z_m^{(N/2)}$ are related since

$$
\begin{aligned}
X_m^{(N)} &= \frac{2}{N} \sum_{k=1}^{N/2} x_k \exp\left[\frac{-i4\pi km}{N}\right] \\
&= \frac{1}{N} \sum_{k=1}^{N/2} \left\{ y_k \exp\left[\frac{-i4\pi km}{N}(2k-1)\right] + z_k \exp\left[\frac{-i4\pi km}{N}(2k)\right] \right\} \\
&= \frac{1}{2} \exp\left[\frac{(i2\pi m)}{N}\right] Y_m^{(N/2)} + \frac{1}{2} Z_m^{(N/2)}, \quad 0 \le m \le (N/2)-1
\end{aligned}
\tag{5.32}
$$

Also

$$
\begin{aligned}
Y_{m+N/2}^{(N/2)} &= Y_m^{(N/2)}; \quad 0 \le m \le (N/2)-1 \\
Z_{m+N/2}^{(N/2)} &= Z_m^{(N/2)}; \quad 0 \le m \le (N/2)-1
\end{aligned}
\tag{5.33}
$$

so that

$$
\begin{aligned}
X_{m+N/2}^{(N)} &= \frac{1}{2} \exp\left[i\left(\frac{2\pi}{N}\right)\left(m+\frac{N}{2}\right)\right] Y_m^{(N/2)} + \frac{1}{2} Z_m^{(N/2)} \\
&= -\frac{1}{2} \exp\left(i\frac{2\pi m}{N}\right) Y_m^{(N/2)} + \frac{1}{2} Z_m^{(N/2)}, \quad 0 \le m \le (N/2)-1
\end{aligned}
\tag{5.34}
$$

Thus,

$$X_m^{(N)} = \frac{1}{2} \exp\left[i\frac{2\pi m}{N}\right] Y_m^{(N/2)} + \frac{1}{2} Z_m^{(N/2)}, \quad 0 \le m \le (N/2)-1 \tag{5.35}$$

and

$$X_{m-N/2}^{(N)} = -\frac{1}{2}\exp\left[i\frac{2\pi m}{N}\right]Y_m^{(N/2)} + \frac{1}{2}Z_m^{(N/2)}, 0 \leq m \leq (N/2)-1 \tag{5.36}$$

Thus, the Fourier transform for the series, x_k is found from the Fourier series of the half series, y_k and z_k. Since $N/2$ is even, this can be repeated. If the length of the data is not a power of two, it should be padded with zeros up to the next power of two. For a series of length $N = 2p$ (p a positive integer), the procedure is followed until partitions consist of only one term whose Fourier transform equals itself, or the procedure is followed until N becomes a prime number, i.e., $N = 3$. The Fourier transform is then found directly for the remaining short series.

5.4.5.1 High Resolution of Daily Tides

Provided a tide gauge record is of sufficient duration, sampling interval and quality, with few gaps, it is possible to resolve the fine structure in the astronomical tidal bands using Fast Fourier transform methods. In Figure 5.8, we show the variance per unit frequency (spectra) for well resolved tidal constituents derived using roughly 100-year long, hourly time series of water levels recorded at (a) Victoria and (b) Vancouver in southwestern British Columbia. The middle and bottom panels for each site show the "line spectra" for all tidal constituents in the diurnal and semidiurnal tidal bands, respectively. Prior to the Fourier transform, each tidal record was smoothed using a Kaiser-Bessel window with parameter $\alpha = 3.0$, window length $T = 524288$ ($= 2^{19}$) h — corresponding to a frequency resolution of $1/T = 1.90735 \times 10^{-6}$ cph $= 4.57764 \times 10^{-5}$ cpd (cycles per day) — and overlaps of 50% between data segments. The sharp roll-off of the Kaiser-Bessel window with frequency on either side of the central lobe (see Figure 5.25) means that there was negligible contamination from the side lobes, so that each windowed portion of the time series was nearly statistically independent, yielding a total of 3.6 (≈ 4) DoF per spectral estimate.

In addition to the diurnal and semidiurnal bands highlighted in blue in Figure 5.8, the top panels in each panel reveals well-defined line spectra in the subharmonic tidal bands of 3, 4, 5, 6…. cpd.

5.5 HARMONIC ANALYSIS

Standard Fourier analysis involves the computation of signal amplitudes at equally spaced frequency intervals determined as integer multiples of the fundamental frequency, f_1. That is, for frequencies, $f_1, 2f_1, 3f_1, \ldots, f_{Nq}$ ($f_{Nq} \equiv$ Nyquist frequency). However, as we have shown in the previous section, standard Fourier analysis has major limitations when it comes to the analysis of data series in terms of predetermined frequencies. In the case of tidal motions, for example, it would be impractical to use any frequencies except those linked to the astronomical tidal forces. Equally importantly, we want to determine the amplitudes and phases of as many frequency components as possible using as short a time series as possible. Because there are typically many more data values than prescribed frequencies, we have to deal with an over-determined problem. This leads to a form of signal demodulation known as *harmonic analysis* in which the user specifies the frequencies to be examined and applies LS techniques to solve for the constituents. Harmonic analysis in ocean sciences was originally designed for the analysis of tidal variability but applies equally to analysis at the annual and semiannual periods or any other well-defined cyclic oscillation. The familiar hierarchy of "harmonic" tidal constituents is dominated by diurnal and semidiurnal motions, followed by fortnightly, monthly, semiannual, and annual variability. In this section, we present a general discussion of harmonic analysis. The important subject of harmonic analysis of tides and tidal currents is treated separately in Section 5.5.3.

The harmonic analysis approach yields the required amplitudes and phase lags of the tidal coefficients or any other constituents we may wish to specify. For example, in studies of interannual variability, we may want to first define the *canonical seasonal cycle* as comprised of the amplitudes and phases of the 12-month (or 360 day) cycle plus the first and second subharmonics of 6 months (180 days) and 3 months (90 days). Once these coefficients have been determined, we can subtract the canonical cycle, enabling us to then examine year-to-year variations in the original time series. In the case of tidal motions, subtraction of the reconstructed tidal signal from the original record yields a time series generally termed the *detided* or *residual* component of the time series. In many cases, it is the "detided" signal that is of primary interest. If we break the original time series into adjoining or overlapping segments, we can apply harmonic analysis to the segments to obtain a sequence of estimates for the amplitudes and phase lags of the various frequencies of interest. This leads to the notion of signal *demodulation*.

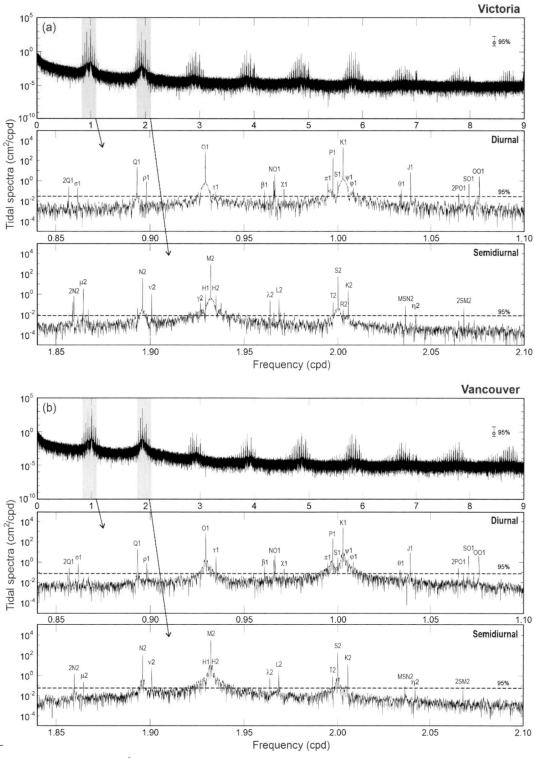

FIGURE 5.8 High resolution spectra (cm²/cph) of water levels derived using FFT transforms on long, hourly time series recorded by tide gauges in (a) Victoria and (b) Vancouver in southwestern British Columbia. Series lengths are: Victoria: 953,578 h (108.89 years) and Vancouver: 809,868 h (92.45 years). The top panel for each site spans the daily tidal frequencies in the bands of 1, 2, 3, 4, … cycles per day (cpd); the middle and bottom panels detail the suite of tidal "constituents" in the diurnal and semidiurnal bands. Prior to the Fourier transform, each record was smoothed using a Kaiser-Bessel window with $\alpha = 3.0$, length $T = 524{,}288 \, (= 2^{19})$ h, and data segment overlaps of 50%. The frequency resolution $1/T = 0.0000457763671875$ cpd. The vertical bar denotes the 95% confidence interval. *Courtesy Igor Medvedev, P.P. Shirshov Institute of Oceanology.*

5.5.1 The Least-Squares (LS) Method

Suppose we wish to determine the harmonic constituents, A_q and B_q, for M-specified frequencies, which, in general, will differ from the Fourier frequencies defined by Eqn (5.19). In this case, $q = 0, 1, ..., M$ and $B_0 = 0$ so that there are a total of $2M + 1$ harmonic coefficients. Assume that there are many more observations, N, than specified coefficients (i.e., that $2M + 1 \ll N$). The problem of fitting M harmonic curves to the digital time series is then overdetermined and must be solved using an optimization technique. Specifically, we estimate the amplitudes and phases of the various components by minimizing the squared difference (i.e., the LS) between the original data series and our fit to that series. The coefficients for each of the M resolvable constituents are found through solution of an $(M + 1) \times (M + 1)$ matrix equation.

For M possible harmonic constituents, the time series, $x(t_n)$, $n = 1, ..., N$ can be expanded as

$$x(t_n) = \bar{x} + \sum_{q=1}^{M} \left[C_q \cos\left(2\pi f_q t_n - \phi_q\right) \right] + x_r(t_n) \tag{5.37}$$

in which $\bar{x} = \overline{x(t)}$ is the mean value of the record; x_r is the residual portion of the time series (which may contain other kinds of harmonic constituents); $t_n = n\Delta t$; and C_q, f_q, and ϕ_q are, respectively, the constant amplitude, frequency, and phase of the qth constituent that we have specified. In the present configuration, we assume that the specified frequencies have the form $f_q = q/(N\Delta t)$ so that the argument, $2\pi f_q t_n = 2\pi q n/N$. Reformulation of Eqn (5.37) as

$$x(t_n) = \bar{x} + \sum_{q=1}^{M} \left[A_q \cos\left(2\pi f_q t_n\right) + B_q \sin\left(2\pi f_q t_n\right) \right] + x_r(t_n) \tag{5.38}$$

yields a representation in terms of the unknown coefficients, A_q, B_q, where

$$C_q = \left(A_q^2 + B_q^2 \right)^{1/2}, \text{(frequency component amplitude)}$$

$$\phi_q = \tan^{-1}\left(B_q/A_q \right), \text{(frequency component phase lag)} \tag{5.39}$$

for $q = 0, ..., M$. To reduce roundoff errors (Section 3.16.3), the mean value, $\bar{x}$, should be subtracted from the record prior to the computation of the Fourier coefficients.

The objective of the LS analysis is to minimize the variance, e^2, of the residual time series, $x_r(t_n)$, in Eqn (5.38), where

$$e^2 = \sum_{n=1}^{N} x_r^2(t_n) = \sum_{n=1}^{N} \left\{ x(t_n) - \left[\bar{x} + \sum_{q=1}^{M} M(t_n) \right] \right\}^2 \tag{5.40}$$

and where, for convenience, we define $\sum M$ as

$$\sum_{q=1}^{M} M(t_n) = \sum_{q=1}^{M} \left[A_q \cos\left(2\pi f_q t_n\right) + B_q \sin\left(2\pi f_q t_n\right) \right]$$

$$= \sum_{q=1}^{M} \left[A_q \cos(2\pi q n/N) + B_q \sin(2\pi q n/N) \right] \tag{5.41}$$

Taking the partial derivatives of Eqn (5.40) with respect to the unknown coefficients, A_q and B_q, and setting the results to zero (the standard method for finding the extrema of a variable), yields $2M + 1$ simultaneous equations for the $M + 1$ constituents

$$\frac{\partial e^2}{\partial A_q} = 0 = 2 \sum_{n=1}^{N} \left\{ \left[x_n - \left(\bar{x} + \sum M \right) \right] \left[-\cos(2\pi q n/N) \right] \right\}, k = 0, ..., M$$

$$\frac{\partial e^2}{\partial B_q} = 0 = 2 \sum_{n=1}^{N} \left\{ \left[x_n - \left(\bar{x} + \sum M \right) \right] \left[-\sin(2\pi q n/N) \right] \right\}, k = 0, ..., M \tag{5.42}$$

Derivation of the coefficients in Eqn (5.42) requires solution of a matrix equation of the form $\mathbf{D}\mathbf{z} = \mathbf{y}$ in which $\mathbf{D}$ is an $(M + 1) \times (M + 1)$ matrix involving sine and cosine summation terms, $\mathbf{y}$ is a vector (column matrix) incorporating summations over the data series, and **z** is a column matrix containing the required coefficients, A_q and B_q. Gaps in the data are still permitted at this stage since the observation times, t_n, used in the LS method are not required to be evenly spaced.

Details on the matrix inversion and related problems can be found in Foreman (1977). To simplify the summations in Eqn (5.42), trigonometric identities are often used. This requires that the data be evenly spaced and that the matrix terms be calculated over segments of the time series with no gaps. The resultant matrix, **D**, is symmetric so that only the upper triangle consisting of $2M + 3M + 1$ elements needs to be stored during the computations. We then seek solutions, **z** through the matrix equation

$$\mathbf{z} = \mathbf{D}^{-1}\mathbf{y} \tag{5.43}$$

where $\mathbf{D}^{-1}$ is the inverse of the matrix:

$$\mathbf{D} = \begin{pmatrix} N & c_1 & c_2 & \cdots & c_M & s_1 & s_2 & \cdots & s_M \\ c_1 & cc_{11} & cc_{12} & \cdots & cc_{1M} & cs_{11} & cs_{12} & \cdots & cs_{1M} \\ c_2 & cc_{21} & cc_{22} & \cdots & cc_{2M} & cs_{21} & cs_{22} & \cdots & cs_{2M} \\ \cdots & \cdots & \cdots & \cdots & \cdots & \cdots & \cdots & \cdots & \cdots \\ \cdots & \cdots & \cdots & \cdots & \cdots & \cdots & \cdots & \cdots & \cdots \\ c_M & cc_{M1} & cc_{M2} & \cdots & cc_{MM} & cs_{M1} & cs_{M2} & \cdots & cs_{MM} \\ \cdots & \cdots & \cdots & \cdots & \cdots & \cdots & \cdots & \cdots & \cdots \\ s_1 & sc_{11} & sc_{12} & \cdots & sc_{1M} & ss_{11} & ss_{12} & \cdots & ss_{1M} \\ s_2 & sc_{21} & sc_{22} & \cdots & sc_{2M} & ss_{21} & ss_{22} & \cdots & ss_{2M} \\ \cdots & \cdots & \cdots & \cdots & \cdots & \cdots & \cdots & \cdots & \cdots \\ s_M & sc_{M1} & sc_{M2} & \cdots & sc_{MM} & ss_{M1} & ss_{M2} & \cdots & ss_{MM} \end{pmatrix} \tag{5.44}$$

and **y** and **z** are column vectors.

$$\mathbf{y} = \begin{pmatrix} yc_o \\ yc_1 \\ yc_2 \\ \cdots \\ \cdots \\ yc_M \\ ys_1 \\ \cdots \\ ys_M \end{pmatrix} \quad \text{and} \quad \mathbf{z} = \begin{pmatrix} A_0 \\ A_1 \\ A_2 \\ \cdots \\ \cdots \\ A_M \\ B_1 \\ \cdots \\ B_M \end{pmatrix} \tag{5.45}$$

The elements of **z** yield the required coefficients, A_q, B_q, for each specified harmonic constituent. To find these solutions, we substitute the elements of **D** for times, $t_n = n\Delta t$ and, using $\alpha_k = f_k T$, $\alpha_j = f_j T$, where f_k and f_j are frequency units of $(\Delta t)^{-1}$ and $T = N\Delta t$ is the record length

$$c_k = \sum_{n=1}^{N} \cos(2\pi\alpha_k n/N), \ s_k = \sum_{n=1}^{N} \sin(2\pi\alpha_k n/N)$$

$$cc_{kj} = cc_{jk} = \sum_{n=1}^{N} \left[\cos(2\pi\alpha_k n/N)\cos(2\pi\alpha_j n/N)\right]$$

$$ss_{kj} = ss_{jk} = \sum_{n=1}^{N} \left[\sin(2\pi\alpha_k n/N)\sin(2\pi\alpha_j n/N)\right] \tag{5.46}$$

$$cs_{kj} = sc_{jk} = \sum_{n=1}^{N} \left[\cos(2\pi\alpha_k n/N)\sin(2\pi\alpha_j n/N)\right]$$

where $\alpha_k n / N = \left(\frac{\alpha_k}{N\Delta t}\right)(n\Delta t)$, and the elements of **y** are given by

$$yc_k = \sum_{n=1}^{N} x_n \cos(2\pi\alpha_k n / N), \quad ys_k = \sum_{n=1}^{N} x_n \sin(2\pi\alpha_k n / N) \tag{5.47}$$

5.5.2 A Computational Example

We can illustrate the power of the LS method by again using the monthly mean SST record of Table 5.2. Our purpose is to estimate the amplitudes and phases of the dominant annual and semiannual constituents in the Amphitrite Point temperature record and compare the results with those we obtained using Fourier analysis in Section 5.4.3. This is also the approach we would use if we wanted to subtract these particular components from the original data record, as we might want to do prior to consideration of less dominant higher frequency variability or before cross-correlation with another data set. We let $f_1 = 1/12$ month ($= 0.0833$ cycles per month cpmo) and $f_2 = 1/6$ month ($= 0.1667$ cpmo) represent the frequencies of interest. From (5.44) and (5.46), we find for $\alpha_1 = f_1 T = 1/12 \times 24 = 2$, and $\alpha_2 = f_2 T = 1/6 \times 24 = 4$ that

$$\mathbf{D} = \begin{pmatrix} N & c_1 & c_2 & s_1 & s_2 \\ c_1 & cc_{11} & cc_{12} & cs_{11} & cs_{12} \\ c_2 & cc_{21} & cc_{22} & cs_{21} & cs_{22} \\ s_1 & sc_{11} & sc_{12} & ss_{11} & ss_{12} \\ s_2 & sc_{21} & sc_{22} & ss_{21} & ss_{22} \end{pmatrix} \tag{5.48}$$

$$= \begin{pmatrix} 24 & 0 & 0 & 0 & 0 \\ 0 & 12 & 0 & 0 & 0 \\ 0 & 0 & 0 & 0 & 0 \\ 0 & 0 & 0 & 12 & 0 \\ 0 & 0 & 0 & 0 & 12 \end{pmatrix} \tag{5.49}$$

and from Eqns (5.45) and (5.47)

$$\mathbf{y} = \begin{pmatrix} yc_0 \\ yc_1 \\ yc_2 \\ ys_1 \\ ys_2 \end{pmatrix} = \begin{pmatrix} 262.70 \\ -21.30 \\ -5.30 \\ -23.87 \\ -0.69 \end{pmatrix} \tag{5.50}$$

where the elements of **y** have units of °C. The solution $\mathbf{z} = \mathbf{D}^{-1}\mathbf{y}$ is the vector

$$\mathbf{z} = \begin{pmatrix} A_0 \\ A_1 \\ A_2 \\ B_1 \\ B_2 \end{pmatrix} = \begin{pmatrix} 10.95 \\ -1.77 \\ -0.44 \\ -1.99 \\ -0.06 \end{pmatrix} \tag{5.51}$$

with units of °C. The results are summarized in Table 5.4. As required, the amplitudes and phases of the annual and semiannual constituents are identical to those obtained using Fourier analysis (see Table 5.3). A plot of the original temperature record and the LS fitted curve using the annual and semiannual constituents is presented in Figure 5.9. The standard deviation for the original record is 2.08°C, while that for the fitted record is 1.91°C. For this short segment of the data record, the two constituents account for 91.7% of the total signal variance.

TABLE 5.4 Coefficients for the annual and semiannual frequencies from a least-squares analysis of the Amphitrite Point monthly mean temperature series (Table 5.2).

q	Frequency (cpmo)	Period (months)	A_q (°C)	B_q (°C)	C_q (°C)
0	–	–	10.95	0.0	10.95
2	0.083	12	−1.77	−1.99	2.67
4	0.167	6	−0.44	−0.06	0.45

Frequency units are cycles per month (cpmo). $q = 0$ gives the mean value for the 24-month record. Other coefficients are defined through Eqn (5.39).

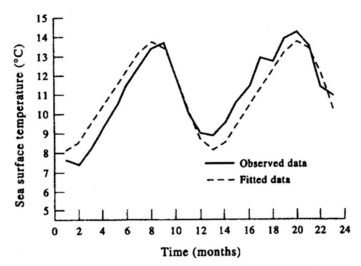

FIGURE 5.9 Monthly mean sea surface temperature (SST) record for Amphitrite Point on the west coast of Vancouver Island (see Table 5.2). The bold line is the original 24-month series. The dashed line is the SST time series obtained from a least-squares fit of the annual (12 month) and semiannual (6 month) cycles to the mean-removed data (see Table 5.4, compare with Figure 5.5).

5.5.3 Harmonic Analysis of Tides

Harmonic analysis is most useful for the analysis and prediction of tide heights and tidal currents. The use of this technique for tides appears to have originated with Lord Kelvin around 1867. Lord Kelvin (Sir William Thomson, 1824−1907) is also credited with inventing the first tide-predicting machine, although the first practical use of such a device was not until several years later. A discussion of tidal harmonic analysis can be found in the *Admiralty Manual of Tides* (Doodson and Warburg, 1941; Godin, 1972; Pugh, 1987). Definitive reports on the least-squares (LS) analysis of current and tide-height data were presented by Foreman (1977, 1978). A highly recommended MATLAB version for the harmonic analysis of tides is provided by Pawlowicz et al. (2002).

The LS harmonic analysis method has a variety of attractive features. It permits resolution of several hundred tidal constituents of which 45 are typically astronomical in origin and identified with a specific frequency in the tidal potential. The remaining constituents include shallow-water constituents that are associated with bottom frictional effects and nonlinear terms in the equations of motion, as well as radiational constituents originating with atmospheric effects. Both scalar and vector time series can be analyzed, with processing of vector series such as current velocity considerably more complex than processing of scalar time series such as sea level and water temperature. If the record is not sufficiently long to permit the direct resolution of neighboring components in the diurnal and semidiurnal frequency bands, the analysis makes provision for the "inference" and subsequent inclusion of these components in the analysis. For example, in the case of the diurnal constituent, P_1, associated with the sun's declination, the phase and amplitude are obtained by lowering the resolution criterion (called the *Rayleigh criterion*) for the separation of frequencies until P_1 is just resolved. The amplitude ratio (amp P_1/amp K_1) and phase difference (phase P_1 − phase K_1) relative to the readily resolved diurnal constituent, K_1, can then be calculated and used to calculate the P_1 constituent for the original record. Equally important, the method allows for gaps in the time series by ignoring those times for which there are no data. Major features of the LS optimization procedure for tidal analysis are outlined below.

A primary aim of LS analysis is to estimate the tidal harmonic constituent amplitudes and phases that can then be used for long-term tidal predictions. Modern applications also include the determination of the tides so that they can be subtracted from a time series to expose weaker underlying signals, such as those from minor tsunamis or infra-gravity waves. The commonly used sampling interval for tidal analysis is 1 h, so that even data collected at shorter time intervals are usually averaged to 1 h intervals for standard analysis packages. (Many modern tide gauges are now mainly used for tsunami detection and research so that sampling intervals have been shortened to 1 min or less). Records must have a minimum length of 13 h in order that they incorporate at least one cycle of the M_2 tidal frequency (period, 12.42 h). The mean component, Z_0, is also included. As the length of the record is increased, additional constituents can be added to the analysis. (As noted in Chapter 1, our ability to resolve adjacent frequencies improves with the length of the time series). Aside from the degree of noise in the data, the main factor limiting the number of derived tidal constituents is the length of the record. For example, the K_1 constituent (period, 23.93 h) can be adequately determined for tidal elevation once the record length exceeds 24 h, although less reliable estimates can be made for shorter record lengths. The criteria for deciding which constituents can be included are discussed in the next section. In essence, inclusion requires that the difference in frequency, Δf, between a given constituent and its so-called *Rayleigh reference constituent* be greater than the fundamental frequency for the record; i.e., $\Delta f \geq f_1 = 1/T$ (see the discussion that follows).

5.5.4 Choice of Constituents

The LS method can be applied to any combination of tidal frequencies. However, the rational approach is to pick the allowable frequencies on the basis of two factors: (1) their relative contribution to the tide-generating potential; and (2) their resolvability in relation to a neighboring principal tidal constituent. In other words, the constituent should be one that makes a significant contribution to the tide-generating force and the record should be of sufficient duration to permit accurate separation of neighboring frequencies. Consideration should also be given to the required computational time, which increases roughly as the square of the number of constituents used in the analysis. Due to noise limitations, the amplitudes of many constituents are too small to be adequately resolved by most oceanic data sets. The high degree of constituent resolution displayed by Figure 5.8 is only possible for very long (~ 100 year) records.

To determine whether a specific constituent should be included in the tidal analysis, the frequency, f_m, of the constituent is compared to the frequency of the neighboring Rayleigh comparison constituent, f_R. The constituent can be included provided

$$|f_m - f_R|T = |\Delta f|T > R \qquad (5.52)$$

where T is the record length and R is typically equal to unity (depending on background noise). In effect, Eqn (5.52) states that f_m should be included if f_R is an included frequency *and* the ratio of the frequency difference, Δf, to the fundamental frequency, $f_1 = 1/T$, is greater than unity. This implies that the fundamental frequency, which corresponds to the best resolution (separation) achievable on the frequency axis, is less than the frequency separation between constituents. Values of $R < 1$ are permitted in the LS program to allow for estimates of neighboring tidal frequencies for record lengths, T, shorter than $1/\Delta f$. Obviously, the longer the record, the more constituents are permitted.

The choice of f_R is determined by the hierarchy of constituents within the tidal band of interest and level of noise in the observations. The hierarchy is in turn based on the contribution a particular constituent makes to the equilibrium tide, with the largest contribution usually coming from the M_2 tidal constituent (Cartwright and Edden, 1973). For the major contributors to the equilibrium tide, the magnitude ratios relative to M_2 in descending order are: $K_1/M_2 = 0.584$, $S_2/M_2 = 0.465$, and $O_1/M_2 = 0.415$. Depending on the level of noise in the observations, the principal semidiurnal constituent, M_2 (0.0805 cph), and the record mean, Z_0, can be determined for records longer than about 13-h duration, while the principal diurnal component, K_1 (0.0418 cph), can be determined for records longer than about 24 h. As a rough guide, separation of the next most significant semidiurnal constituent, S_2 (0.0833 cph), from the principal component M_2 requires a record length, $T > 1/|f(M_2) - f(S_2)| = 355$ h (14.7 days). Similarly, separation of the next most significant diurnal constituent, O_1 (0.0387 cph), from the principal component, K_1, requires an approximate record length, $T > 1/|f(K_1) - f(O_1)| = 328$ h (13.7 days). The frequencies, $f(K_1)$ and $f(O_1)$, then become the Rayleigh comparison frequencies for other neighboring tidal constituents in the diurnal band while the frequencies, $f(M_2)$ and $f(S_2)$, become the comparison frequencies for neighboring frequencies in the semidiurnal band. Extension of this procedure to longer and longer records eventually encompasses all the significant tidal constituents within the diurnal and semidiurnal bands.

The first long-term constituent to be included in the analysis is the lunar–solar fortnightly cycle, M_{sf} (0.00282 cph), requiring an approximate record duration, $T > 14.8$ days, followed by the lunar monthly constituent, M_m (0.00151 cph), duration, $T > 31.8$ days, and the lunar fortnightly cycle, M_f (0.00305 cph), $T > 182.6$ days. These record length requirements are based on stochastic processes; shorter records can be used for deterministic processes, such as tides, provided that noise levels are low. Thus, in all cases, shorter record lengths can be used if the data are highly noise free. By the same token, longer records are often needed to resolve the longer-period tides because of contamination from atmospheric effects, including atmospheric tides.

A summary of the required record lengths for inclusion of the more important constituents is provided in Tables 5.5–5.7 together with a comparison of a given constituent's tidal potential magnitude relative to that of the principal component in the frequency band. Where possible, a candidate constituent is compared to the particular neighboring constituent, which has already been selected and is nearest in frequency.

5.5.5 A Computational Example for Tides

As a simple example of the LS method of harmonic tidal analysis, consider the 32-hourly sea-level heights measured at Tofino, British Columbia during September 10–11, 1986 (Table 5.8). As indicated by Tables 5.5 and 5.6, we can at most

TABLE 5.5 Record lengths (in hours) needed to resolve the main tidal constituents in the semidiurnal tidal band assuming a Rayleigh coefficient, $R = 1$. Magnitude ratios are based on the tidal potential relative to that of the M_2 constituent.

Tidal constituent	Frequency (cph)	Comparison constituent	Magnitude ratio	Record length needed (h)
M_2 (principal lunar)	0.0805	–	1	13
S_2 (principal solar)	0.0833	M_2	0.465	355
N_2 (larger lunar elliptic)	0.0790	M_2	0.192	662
K_2 (luni-solar)	0.0836	S_2	0.029	4,383

TABLE 5.6 Record lengths (in hours) needed to resolve the main tidal constituents in the diurnal tidal band assuming a Rayleigh coefficient, $R = 1$. Magnitude ratios are based on the tidal potential relative to that of the M_2 constituent.

Tidal constituent	Frequency (cph)	Comparison constituent	Magnitude ratio	Record length needed (h)
K_1 (luni-solar)	0.0418	–	0.584	24
O_1 (principal lunar)	0.0387	K_1	0.415	328
P_1 (principal solar)	0.0416	K_1	0.193	4,383
Q_1	0.0372	O_1	0.079	662

TABLE 5.7 Record lengths (in hours) needed to resolve the main tidal constituents in the long-period tidal band assuming a Rayleigh coefficient, $R = 1$. Magnitude ratios are based on the tidal potential relative to that of the M_2 constituent.

Tidal constituent	Frequency (cph)	Comparison constituent	Magnitude ratio	Record length needed (h)
M_{sf} (mixed solar fortnightly)	0.002822	M_f	0.015	355
M_f (lunar fortnightly)	0.003050	–	0.172	4,383
M_m (lunar monthly)	0.001512	M_{sm}	0.091	764
M_{sm} (solar monthly)	0.001310	–	0.017	4,942
S_{sa} (solar semiannual)	0.000228	S_a	0.080	4,383
S_a (solar annual)	0.000114	–	0.013	8,766

TABLE 5.8 Hourly values of sea-level height (SLH) measured at Tofino, British Columbia (49°09.0′ N, 125°54.0′ W), on the west coast of Canada starting September 10, 1986.

N	1	2	3	4	2	6	7	8	9	10	11
SLH	1.97	1.46	0.98	0.73	0.67	0.82	1.15	1.58	2.00	2.33	2.48
N	12	13	14	15	16	17	18	19	20	21	22
SLH	2.43	2.25	2.02	1.82	1.72	1.75	1.91	2.22	2.54	2.87	3.10
N	23	24	25	26	27	28	29	30	31	32	
SLH	3.15	2.94	2.57	2.06	1.56	1.13	0.84	0.73	0.79	1.07	

Heights are in meters above the local chart datum.

resolve the K_1 and M_2 constituents. This problem is similar to that considered in Section 5.5, where we used the LS technique to fit the annual and semiannual components to a 24-month record of SST. Following the analysis in that section, the various matrices are written in terms of a mean component plus the contributions from the K_1 and M_2 frequencies, $f(K_1) = 0.0418$ cph and $f(M_2) = 0.0805$ cph, respectively. From Eqns (5.44) and (5.45), we find

$$\mathbf{D} = \begin{pmatrix} N & c_1 & c_2 & s & s_2 \\ c_1 & cc_{11} & cc_{12} & cs_{11} & cs_{12} \\ c_2 & cc_{21} & cc_{22} & cs_{21} & cs_{22} \\ s_1 & sc_{11} & sc_{12} & ss_{11} & ss_{12} \\ s_2 & sc_{21} & sc_{22} & ss_{21} & ss_{22} \end{pmatrix} \tag{5.53}$$

$$= \begin{pmatrix} 32 & 2.476 & -1.836 & 6.183 & 3.420 \\ 2.476 & 14.809 & 1.450 & 1.136 & 2.117 \\ -1.836 & 1.450 & 16.263 & -2.197 & 0.397 \\ 6.183 & 1.136 & -2.1 & 17.191 & 2.163 \\ 3.420 & 2.117 & 0.397 & 2.163 & 15.737 \end{pmatrix} \tag{5.54}$$

and from Eqns (5.45) and (5.47)

$$\mathbf{y} = \begin{pmatrix} yc_0 \\ yc_1 \\ yc_2 \\ ys_1 \\ ys_2 \end{pmatrix} = \begin{pmatrix} 57.640 \\ 6.514 \\ 6.138 \\ -0.199 \\ -3.335 \end{pmatrix} \tag{5.55}$$

where the elements of $\mathbf{D}$ and $\mathbf{y}$ have units of meters. The solution $\mathbf{z} = \mathbf{D}^{-1}\mathbf{y}$ is the vector

$$\mathbf{z} = \begin{pmatrix} A_0 \\ A_1 \\ A_2 \\ B_1 \\ B_2 \end{pmatrix} = \begin{pmatrix} 1.992 \text{ m} \\ 0.186 \text{ m} \\ 0.523 \text{ m} \\ -0.574 \text{ m} \\ -0.604 \text{ m} \end{pmatrix} \tag{5.56}$$

The results are summarized in Table 5.9. A plot of the original sea-level data and the fitted sea-level curve is presented in Figure 5.7. The standard deviation for the original record is 0.741 m while that for the fitted record is 0.736 m. For this short segment of the data record, the sum of the two tidal constituents accounts for over 99% of the total variance in the record. As a comparison, we have used the full analysis package without inference to analyze 29 days of the Tofino sea-level record beginning at 20:00 h on September 10, 1986. The program finds a total of 30 constituents, including the mean, Z_0, with the sum of the tidal constituents accounting for 98% of the original variance in the signal. The record mean sea

TABLE 5.9 Least-squares estimates of the amplitude (Fourier coefficients) for the K_1 and M_2 tidal constituents for the 32-h Tofino sea level starting at 2000, September 10, 1986.

q	Frequency (cph)	Period (h)	A_q (m)	B_q (m)	C_q (m)	C_q' (m)
0	–	–	3.984	0	3.984	4.100
1	0.042	24	0.186	−0.574	0.365	0.286
2	0.081	12	0.523	−0.604	0.638	0.986

The mean is $\frac{1}{2}A_0$ and q denotes the number of cycles per day ($q = 0$ is for the mean value). The last column, C_q' (m) gives the constituent amplitudes for a more extensive analysis that used a 29-day (685 h) data segment that had the same start time as the 32-h segment used to derive C_q.

level for the month is 2.05 m, and the K_1 and M_2 constituents have amplitudes of 0.286 and 0.986 m, respectively. As expected, these are quite different to the values derived based on only 32 h of data (Table 5.9). Phases for the two constituents for the 29-day records are 122.0 and 12.5° compared with 107.9 and 130.9° for the same two constituents based on the 32-h records. Here, phase angles are, by convention, measured relative to a local (or, alternatively, the Greenwich) meridian of longitude. For example, off the west coast of Canada, it is common to use time in degrees relative to 120° W longitude as the reference time.

5.5.6 Removal of Tides and Atmospheric Effects from Tsunami Records

One of the most important applications of tidal removal ("detiding") is in the study of tsunamis, including those generated by seismic events, landslides (both from submarine and subaerial slope failures) and atmospheric processes ("meteotsunamis"). The first step is to correct any errors in the tide gauge or bottom pressure record, fill in gaps and remove spikes. Least-squares harmonic analysis is then used to reconstruct the tides for the time of the tsunami using either previously derived tidal constituents for the recording station or, if these are not available, by using results from a harmonic analysis of the record for the period prior to the event. The reconstructed tidal variations are then subtracted from the original bottom pressure or water level record on a point-by-point basis to obtain the "residual" values of the data series (e.g., Rabinovich et al., 2006, 2011; Rabinovich and Eblé, 2015; Zaytsev et al., 2017, 2021). To suppress low-frequency pressure or sea-level fluctuations in the residual record arising from synoptic atmospheric processes, and to simplify isolation of the tsunami signal, the records are next high-pass filtered using, for example, a Kaiser-Bessel, Butterworth or Hamming window with a cutoff period, $T_c(= 1/f_c)$, of 3 or 4 h (see also Chapter 6 for further details on filtering). To eliminate possible contamination from high-frequency infra-gravity (IG) waves (especially for recording sites directly exposed to the open ocean), the high-pass filtered record can be smoothed with a low-pass filter with a cutoff period of 6 min (Rabinovich et al., 2017). The final filtered series represents a robust construct of the tsunami motions for plotting and further analyses.

Examples of tsunami waves extracted from two coastal tide gauge records using the above multi-step procedure are presented in Figure 5.10 for observations following the 15 January 2022 Hunga Tonga−Hunga Ha'apai ("Tonga-Hunga") submarine volcanic eruption in the central Pacific. Time series for the two selected sites—North Spit (40°46′ N, 124°13′ W) on the coast of California and King Edward Point (54°17′ S, 36°30′ W) in South Georgia and the South Sandwich Islands in the South Atlantic—span the period of roughly 1 day before and 1 day after arrival of the oceanic tsunami originating from the source region. Arrival of the oceanic tsunami (labeled "O" in the figure) was preceded by meteotsunami waves (labeled "A") that were forced by atmospheric pressure pulses (Lamb waves) that circumvented the globe several times at the speed of sound (~ 10 spherical degrees/hour) for several days after the eruption (Amores et al., 2022; Carvajal et al., 2022). The isochrons in the top panels of Figure 5.10 show the propagation times in hours from the start of the eruption (E) for the leading waves of the oceanic tsunami (top left) and the atmospheric pressure waves (top right). At the final stage of the filtering sequence (bottom panels in the figure), intra-gravity waves, such as those clearly present in the California record before, during and after the tsunami, were attenuated by application of a 6-min low-pass Kaiser-Bessel (KB) window. This sequential filtering of the residual record with a high-pass KB filter (cutoff period = 3 h) followed by filtering of the resulting high-pass record with a low-pass KB filter (cutoff period = 6 min) amounts to filtering the residual time series with a band-pass KB filter with cutoff frequencies $f_{c1} = (1/3)$ cph and $f_{c2} = (1/6)$ cpmin. We further note that the high pass records in the third row of each panel were generated by filtering the residual series with a 3-h low pass window and then subtracting the resultant low-pass record from the residual record.

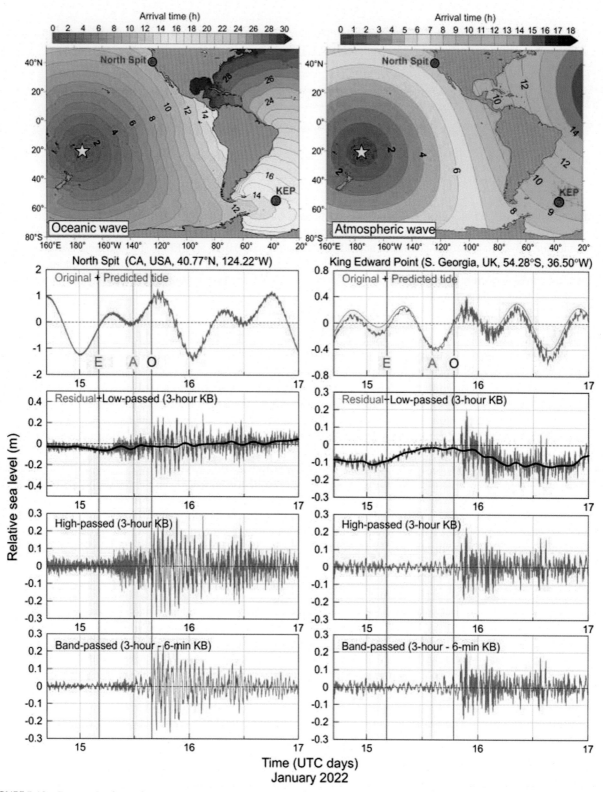

FIGURE 5.10 *Top row*: Isochrons showing the propagation times in hours of the leading waves of the oceanic tsunami (O) and the atmospheric pressure waves (A) generated by the eruption of the Tonga-Hunga volcano on 15 January 2022 (E) in the central Pacific Ocean (the source region is denoted by a star); *Second row*: observed 1-min sea level variations (blue lines) superimposed on the predicted astronomical tide (red line) for North Spit (California, eastern Pacific) and King Edward Point (South Georgia and South Sandwich Islands; South Atlantic) in mid-January 2022 (see station locations in the upper panels); *Third row*: the residual time series obtained by subtracting the predicted tide from the observations; *Fourth row*: time series after processing the residual data series in the second row using a high-pass Kaiser-Bessel filter with a 3-h cutoff period; *Bottom row*: the "true" tsunami record after filtering the record in the third row with a low-pass Kaiser-Bessel filter with a cutoff period of 6 min. *Analysis and plots courtesy of Igor Medvedev and Alexander Rabinovich, P.P. Shirshov Institute of Oceanology.*

The purpose of the above procedure is to isolate tsunami waves from the normally strong tidal variations without distorting the tsunami signal. As with other geophysical time series, we cannot expect to extract "pure" tsunami variations, but we can take analytical steps to eliminate the effect of processes unrelated to the tsunami, thereby enhancing the signal-to-noise ratio for the event. Not only does this analysis minimize distortion of the derived statistical properties of the tsunami waves but it also gives us confidence that we can use the extracted tsunami record to help reconstruct the earthquake source region that created the event in the first place (cf. Fine et al., 2015; Gusman et al., 2016).

To summarize, the stages of the processing are:

1. *Detrending*: In some situations, it is important to remove any trend in the observations caused by instrumental or background drift;
2. *Detiding*. For many sites, the tidal constituents are known and can be used to predict what the water level or bottom pressure would have been in the absence of the tsunami and then subtract the prediction from the original record to obtain the *residual* data series. If the tide is unknown, it can be estimated using the least-squares method from data prior to the event. The investigator need not strictly follow the Rayleigh criterion, but should be reasonable in the choice of tidal constituents. For long data series, we can make use of many harmonics but for short records the selection is limited (e.g., we normally can use 16–17 tidal harmonics for data series of a week or longer, but only the daily M_2 and K_1 constituents for 1-day segments). The calculation of tides also provides some quality control, as the derived constituents must have "reasonable" amplitudes and phases relative to nearby sites. If not, there may be problems with original data series arising from a vertical shift in the gauge, missing values, or instrumental malfunction. Note that it is not wise to use a low-pass filter to suppress the tides when deriving the residual series as application of a strong filter will always distort the tsunami signal.
3. *High-pass filtering*. The purpose of this step is to eliminate low-frequency variations associated with atmospheric activity or other processes. This stage requires careful selection of the correct (optimum) window length. If care is not taken, the investigator may either cut out part of the tsunami signal or retain background noise that will obscure the tsunami signal. For major tsunamis, like those generated by the large source Chile (1960), Alaska (1964) and Japan (2011) earthquakes, a 4-h window length is acceptable, while for weak (small-source earthquake) tsunamis, a shorter (e.g., 2-h) window length is recommended. Note that some recording stations may have anomalous tsunami response characteristics that we should be kept in mind. For example, tsunami waves recorded at Port Alberni on the west coast of Vancouver Island invariably induce significant seiches at the fundamental eigen period of Alberni Inlet of ~ 100 min (Fine et al., 2009). So, if we want to detect a relatively weak tsunami in Alberni Inlet, we need to suppress these oscillations by using (for example) a 1-h high pass filter.
4. *Low-pass filtering*. Some recording sites are susceptible to high-frequency noise associated with storm induced infragravity (IG) waves. As the typical period for IG-waves is from 30 s to 3 min, they can be strongly attenuated using a low pass filter with a 6-min window. However, if this noise is very strong, the investigator may need to use a longer window. For example, such a situation occurred for the Chilean island stations examined by Rabinovich et al. (2017). On the west coast of North America, one of the gauge stations most affected by IG waves was located on Langara Island on the remote northwest coast of Haida Gwaii. Wave-induced contamination of the water level records at the site was so severe that the station had to be closed after many decades of operation. Despite the possible effects from IG waves, we do not recommend using the low-pass filtering stage for weak-source tsunamis. Because such events typically have dominant periods of only a few minutes, low-pass filtering can actually eliminate much of the tsunami signal. As noted earlier, the combination of high-pass filtering with a cut-off frequency of $f_{c1} = (1/3)$ cph followed by low-pass filtering with $f_{c2} = (1/6)$ cpmin is equivalent to applying a band-pass filter with these cutoff frequencies.

5.5.7 Complex Demodulation

In many applications, we seek to determine how the signal characteristics at a specific frequency, ω, change throughout the duration of a time series. For example, we might ask how the amplitude, phase, and orientation of the semidiurnal tidal current ellipses at different depths at a mooring location change with time. Wave packets associated with passing internal tides would be revealed through rapid changes in ellipse parameters at the M_2 and/or S_2 frequencies. The method for determining the temporal change of a particular frequency component for a scalar or vector time series is called *complex demodulation*.

A common technique for finding the demodulated signal is to fit the desired parameters to sequential segments of the data series using LS algorithms. The analysis requires that there be many more data points than frequency components and each segment must span at least one cycle of the frequency of interest. As with any LS analysis, the observations do not have to be at regular time intervals. Inputs to complex demodulation algorithms require specification of the start time of the first segment, the length of each segment, and the time between computation interval start times. Computation intervals may overlap, be end-to-end, or be interspersed with unused data. Following the LS analysis described under the section on harmonic analysis, the time increment between each estimate can be as short as one time step, Δt, thereby providing the maximum number of estimates for a given segment length, or as long as the entire record, thereby yielding a single estimate of the signal parameters.

For each segment of current velocity data, the fluctuating component of velocity at frequency, ω can be expressed as

$$
\begin{aligned}
\mathbf{u}(t) - \overline{\mathbf{u}(t)} &= [u(t) - \overline{u(t)}] + i[v(t) - \overline{v(t)}] \\
&= A^+ \exp[i(\omega t + \varepsilon^+)] + A^- \exp[-i(\omega t + \varepsilon^-)]
\end{aligned}
\tag{5.57}
$$

where $(\overline{u(t)}, \overline{v(t)})$ are the means of the velocity components, and where (A^+, A^-) are the amplitudes and $(\varepsilon^+, \varepsilon^-)$ are the phases of the counterclockwise $(+)$ and clockwise $(-)$ rotating components, respectively. Data are at times, t_k, $(k = 1, ..., N)$ and solutions are found from the matrix Eqn (5.43)

$$
\mathbf{z} = \mathbf{D}^{-1}\mathbf{y}
$$

where

$$
\mathbf{y} = \begin{pmatrix} u(t_1) \\ u(t_2) \\ ... \\ u(t_n) \\ v(t_1) \\ ... \\ v(t_n) \end{pmatrix} ; \mathbf{z} = \begin{pmatrix} A^+ \cos(\varepsilon^+) \\ A^+ \sin(\varepsilon^+) \\ A^- \cos(\varepsilon^-) \\ A^- \sin(\varepsilon^-) \end{pmatrix} \equiv \begin{pmatrix} ACP \\ ASP \\ ACM \\ ASM \end{pmatrix}
\tag{5.58a}
$$

and

$$
\mathbf{D} = \begin{pmatrix}
\cos(\omega t_1) & -\sin(\omega t_1) & \cos(\omega t_1) & \sin(\omega t_1) \\
\cos(\omega t_2) & -\sin(\omega t_2) & \cos(\omega t_2) & \sin(\omega t_2) \\
... & ... & ... & ... \\
\cos(\omega t_n) & -\sin(\omega t_n) & \cos(\omega t_n) & \sin(\omega t_n) \\
\sin(\omega t_1) & \cos(\omega t_1) & -\sin(\omega t_1) & \cos(\omega t_1) \\
... & ... & ... & ... \\
\sin(\omega t_n) & \cos(\omega t_n) & -\sin(\omega t_n) & \cos(\omega t_n)
\end{pmatrix}
\tag{5.58b}
$$

Once the elements of $\mathbf{z}$ are found from the LS solution to the matrix equation (for example, using IMSL routine LLSQAR or specific MATLAB routines), we can find the various ellipse parameters from

$$
A^+ = \left(ASP^2 + ACP^2\right)^{1/2}; A^- = \left(ASM^2 + ACM^2\right)^{1/2}
\tag{5.59a}
$$

$$
\tan(\varepsilon^+) = \frac{ASP}{ACP}; \tan(\varepsilon^-) = \frac{ASM}{ACM}
\tag{5.59b}
$$

For example, we could obtain the demodulated current amplitude and phase for near-inertial motions observed at a midlatitude mooring by setting $\omega = 2\Omega \sin\theta$ and obtaining LS solutions for a series of adjoining 24-h segments with no overlap (here, Ω is the angular earth rotation rate and θ is latitude). For the LS technique to be applicable, data would need to be sampled at roughly hourly intervals so that there were more data points per segment than parameters being estimated. Equatorward of $\theta = \pm 30°$, the period of inertial motions exceeds 24 h and the lengths of individual segments must be increased accordingly. Complex demodulation also can be used to examine inertial motions in Lagrangian-type data. In Figure 5.11a, we have plotted the original and demodulated positions of a satellite-tracked drifter launched in the Canadian Arctic in the fall of 1988. The time series covers 60 days and was analyzed using overlapping 24-h subsections with the assumption that displacements occurred at the inertial period of 12.73 h for 70° N latitude. Figure 5.11b shows a detailed analysis of the trajectory record for the 20 days ending October 11 when the buoy became trapped in growing sea ice. Note

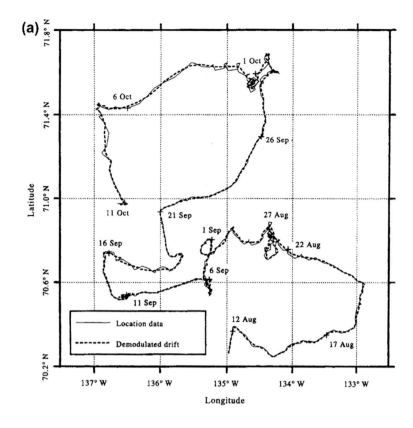

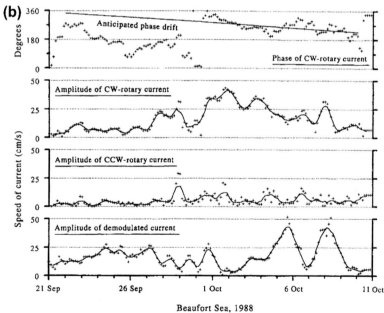

Beaufort Sea, 1988

FIGURE 5.11 Complex demodulation at the inertial period of 12.73 h for the trajectory of a satellite-tracked drifter deployed in the Beaufort Sea in August 1988. (a) Original (solid line) and demodulated version (dashed line) of the drifter track. (b) Parameters of the demodulation over a 20-day period of strong inertial motions. Top panel: phase of the clockwise (CW) rotary component (degrees). Remaining panels: amplitudes of the CW rotary, counterclockwise (CCW) rotary, and speed of the demodulated current. *Courtesy of Humfrey Melling, Institute of Ocean Sciences.*

the intense inertial currents starting on September 30, the prevalence of the clockwise component of rotation, and the roughly $-6.4°$ per day drift in phase of the clockwise component of the current due to the changing latitude of the drifter relative to the reference latitude of $70°$ N.

5.6 SPECTRAL ANALYSIS

Spectral analysis is used to partition the variance of a time series as a function of frequency. For stochastic time series such as wind waves, contributions from the different frequency components are measured in terms of the *power spectral density* (PSD). For deterministic waveforms such as surface tides, either the PSD or the *energy spectral density* (ESD) can be used. Here, power is defined as energy per unit time. The need for two different spectral definitions lies in the boundedness of the integral of signal variance for increasing record length. In practice, the term *spectrum* is applied to all spectral functions including commonly used terms such as autospectrum and power spectrum. The term *cross-spectrum* is reserved for the "shared" power between two coincident time series. We also distinguish between *nonparametric* and *parametric* spectral methods. Nonparametric methods, which are based on conventional Fourier transforms, are not data-specific, while parametric techniques are data-specific and assign a predetermined model to the time series. In general, we use parametric methods for short time series (those having a few cycles of the oscillations of interest) and nonparametric methods for long time series (many cycles of the oscillations of interest).

The word spectrum is a carryover from optics. The "colors" red, white, and blue of the electromagnetic spectrum are often used to describe the frequency distribution of oceanographic spectra. A spectrum whose spectral density decreases with increasing frequency is called a "red" spectrum, by analogy to visible light where red corresponds to longer wavelengths (lower frequencies). Similarly, a spectrum whose magnitude increases with frequency is called a "blue" spectrum. A "white" spectrum is one in which the spectral constituents have near-equal amplitudes throughout the frequency range. In the ocean, long-period variability, such as that associated with climate related processes, tends to have red spectra, while that linked to instrumental noise tends to have white spectra. Blue spectra are confined to certain frequency bands such as the low-frequency portion of wind—wave spectra and within the "weather band" (2 < period < 10 days) for deep atmospherically-generated currents. Spectra of wind-generated inertial currents in the deep ocean are often "blue-shifted" to frequencies a few percent higher than the local inertial frequency (Fu, 1981; Thomson et al., 1990).

In the days before modern computers, it was customary to compute the spectrum of discrete oceanic data from the Fourier transform of the autocorrelation function using a small number of lag intervals, or "lags". First formalized by Blackman and Tukey (1958), the autocorrelation method lacks the wide range of optional improvements to the computations and generalized "tinkering" permitted by more modern techniques. From a historical perspective, the autocorrelation approach has importance for the direct mathematical link it provides through the Wiener—Khinchin relations that connect variance functions in the time domain to those in the frequency domain. Today, it is the spectral *periodogram* generated using the Fast Fourier Transform (FFT) or the Singleton Fourier transform that is most commonly used to estimate oceanic spectra. (We assume that the reader has a basic understanding of Fourier analysis and FFTs provided in the previous sections. Those unfamiliar with these concepts can turn to Section 5.4 where the topics are discussed in considerable detail).

Other methods have been developed over the years as a result of fundamental performance limitations with the periodogram method. These limitations are: (1) restricted frequency resolution when distinguishing between two or more signals, with frequency resolution dictated by the available record length, independent of the characteristics of the data or its signal-to-noise ratio (SNR); (2) energy "leakage" between the main lobe of a spectral estimate and adjacent side-lobes, with a resulting distortion and smearing of the spectral estimates, suppression of weak signals, and the need to use smoothing windows; (3) an inability to adequately determine the spectral content of short time series; and (4) an inability to adjust to rapid changes in signal amplitude or phase. Other techniques such as the maximum entropy method (MEM, best suited to short time series) and the wavelet transform (best suited to event-like signals and other non-stationary processes whose frequency content changes over time) are addressed in this chapter.

Fundamental concepts: Several basic concepts are woven into the fabric of this chapter. First of all, the sample data we collect are subsets of either stochastic or deterministic processes. Deterministic processes are predictable, stochastic ones are not. Secondly, the very act of sampling to generate a time series of finite duration is analogous to viewing an infinitely long time series through a narrow "window" in the shape of a rectangular box-car function (Figure 5.12a). The characteristics of this window in the frequency domain can severely distort the frequency content of the original data series from which the sample has been drawn. As illustrated by Figure 5.12b, the sampling process results in spectral energy being "rippled" away from one frequency (the central lobe of the response function) to a wide number range of adjacent

(a)

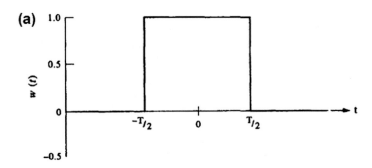

(b)

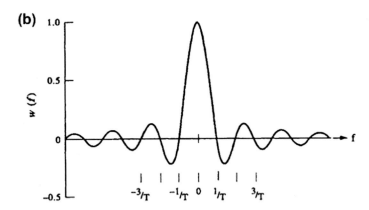

FIGURE 5.12 The box-car (rectangular) window, which creates a sample time series from a "long" time series. (a) The box-car window in the time (t) domain. Here, $w(t) = 1$, $-T/2 \leq t \leq T/2$, and $w = 0$ otherwise. (b) Frequency $W(f)$ response of the box-car window in (a). The central lobe straddles each spectral (frequency) component within the time series and has a width, $\Delta f = 2/T$. Zeros occur at $f = \pm m/T$, where $m = 1, 2, \ldots$.

frequencies. The large side-lobes of the rectangular window—produced by Gibbs phenomenon—are responsible for the leakage of spectral energy from the central frequency to nearby frequencies.

A third point is that the spectra of random processes are themselves random processes. Therefore, if we are to determine the frequency content of a data series with some degree of statistical reliability (i.e., to be able to put confidence intervals on spectral peaks), we need to precondition the time series and average the raw periodogram estimates. Averaging can be done in the time domain by using specially designed windows or in the frequency domain by averaging together adjacent spectral estimates. Windows (which are discussed in detail in Section 5.6.6) suppress Gibbs' phenomenon associated with finite length data series and enable us to increase the number of *degrees of freedom* (DoF) used in each spectral estimate. Here, the term "degrees of freedom" refers to the number of statistically independent variables or values used in a particular estimate. We note that, because they are linked to the signal variance, spectral values are chi-square variables and that the DoF now apply to that particular form of Probability Density Function (PDF). (See Chapter 3 for a discussion of chi-square variables and PDFs). We can also improve spectral estimates by partitioning a time series into a series of segments and then conducting spectral analysis on the separate pieces. Spectral values for each frequency band for each segment of a data record are then averaged together to improve statistical reliability. This is similar to averaging adjacent spectral values in the periodogram, which will give a similar increase in the DoF of the resulting spectral estimate. The penalty for using shorter data segments is a loss of frequency resolution. The alternative—calculating a single periodogram and then smoothing in the frequency domain—suffers the same loss of frequency resolution for a smoothing that gives the same number of degrees of freedom.

Regardless of which averaging approach is chosen, the results will be tantamount to viewing the data through a window in the frequency domain. Any smoothing window used to improve the reliability of the spectral estimates will distort the results and impose structure on the data, such as periodic behavior, when no such structure may exist in the original time series. In addition, conventional methods make the implicit assumption that the unobserved data or correlation lag values situated outside the measurement interval are zero, which is generally not the case. The smoothing window results in smeared spectral estimates. Parametric methods allow us to make more realistic assumptions about the nature of the process outside the measurement interval, other than to assume it is zero or cyclic. This eliminates the need for window functions. The improvement over conventional FFT spectral estimates can be quite dramatic, especially for short records. However, even then, there remain pitfalls, which have tended to detract from the usefulness of these methods to oceanography. Each new method has its own advantages and disadvantages that must be weighed in context of the particular data set and the way it has been collected. Parametric methods are able to identify spectral peaks that standard FFT spectra may not. Unlike the standard FFT approach, however, parametric methods are generally not able to define the statistical significance of these peaks. For time series with low SNR, most parametric methods are no better than the conventional FFT approach.

Means and trends: Prior to spectral analysis, the record mean and trend are generally removed from any time series (Figure 5.13). Unless stated otherwise, we will assume that the time series $y(t)$ we wish to process has the form $y'(t) = y(t) - \overline{y(t)}$, where $\overline{y(t)} = y_o + \alpha t$ is the mean value and αt is the linear trend (y_o and α are constants). If the mean and trend are not removed prior to spectral analysis, they can distort the low-frequency components of the spectrum (see Section 5.6.12). Packaged spectral programs often include record mean and linear trend removal as part of the data preconditioning. Nonlinear trends are more difficult to remove, especially because a single function may not be appropriate for the entire data domain. The latter may apply also to linear trends.

The mean value removed from a record is not always the average for the entire record. For example, to examine interannual variability in the monthly time series of sea-level height, alongshore current velocity, or any other scalar, $\eta(t_m)$, we first calculate the mean monthly values $\overline{\eta}(t_m)$ for each month separately over the entire record (e.g., the individual means for all Januaries, all Februaries, etc.). These mean monthly values for $m = 1, 2, ..., 12$, rather than the simple average value for all values over the entire record, are then subtracted from the original data for the appropriate month to obtain monthly anomalies, $\eta'(t_m) = \eta(t_m) - \overline{\eta(t_m)}$. As with other averaging processes, the user will need to determine how many missing data values will be permitted for a given month before the monthly value is considered "missing" or not available. Trend removal can then be applied to the monthly anomalies to obtain the final anomaly record. As a final comment, we note that certain records, such as those from moored bottom pressure recorders and near-surface transmissometers or dissolved oxygen sensors, will contain long-term nonlinear trends that should be removed from the data record prior to spectral analysis. Trend removal must be done cautiously. Unless one has a justified physical model for a particular trend (including a linear trend), removal of the trend may itself add spurious frequency components to the detrended signal.

FIGURE 5.13   Mean and trend removal for an artificial time series $y(t)$. Here, $y_o = -1.0$, the trend, $\alpha = 0.025$, and the fluctuating component, $y'(t)$, was obtained using a uniformly distributed random number generator. (a) Original time series, showing the linear trend; (b) time series with the mean and linear trend removed.

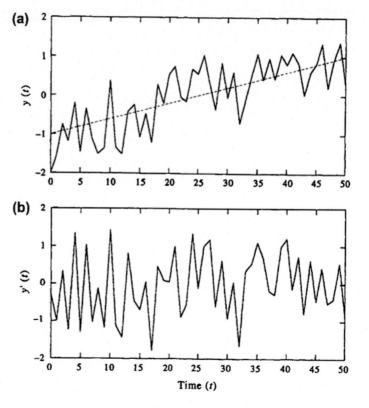

5.6.1 Spectra of Deterministic and Stochastic Processes

Time series data can originate with deterministic or stochastic processes, or a mixture of the two. Turbulence arising from eddy-like motions generated by strong tidal currents in a narrow coastal channel provides an example of mixed deterministic and stochastic processes. To see the difference between the two types of processes in terms of conventional spectral estimation, consider the case of a continuous *deterministic* signal, $y(t)$. If the total signal energy, E, is finite

$$E = \int_{-\infty}^{\infty} |y(t)|^2 dt < \infty \tag{5.60}$$

then $y(t)$ is absolute-integrable over the entire domain and the Fourier transform $Y(f)$ of $y(t)$ exists. This leads to the standard transform pair

$$Y(f) = \int_{-\infty}^{\infty} y(t)e^{-i2\pi ft} dt \tag{5.61a}$$

$$y(t) = \int_{-\infty}^{\infty} Y(f)e^{i2\pi ft} df = \frac{1}{2\pi} \int_{-\infty}^{\infty} Y(\omega)e^{i\omega t} d\omega \tag{5.61b}$$

where $e^{\pm i2\pi ft} = \cos(2\pi ft) \pm i\sin(2\pi ft)$, f is the frequency in cycles per unit time, and $\omega = 2\pi f$ is the angular frequency in radians per unit time. The square of the modulus of the Fourier transform for all frequencies

$$S_E(f) = Y(f)Y^*(f) = |Y(f)|^2 \tag{5.62}$$

is then the Energy Spectral Density (ESD), $S_E(f)$, of $y(t)$. (Here, the asterisk denotes the complex conjugate). To show that Eqn (5.62) is an energy density, we use Parseval's theorem

$$\int_{-\infty}^{\infty} |y(t)|^2 dt = \int_{-\infty}^{\infty} |Y(f)|^2 df \tag{5.63}$$

which states that the total energy, E, of the signal in the time domain is equal to the total energy, $\int S_E(f)df$, of the signal in the frequency domain. Thus, $S_E(f)$, is an energy density (energy per unit frequency) which, when multiplied by df, yields a measure of the total signal energy in the frequency band centered near frequency f. The "power" of a deterministic signal, E/T, is zero in the limit of very long time series ($T \to \infty$).

Now, suppose that $y(t)$ is a stationary *random* process rather than a deterministic waveform. Unlike the case for the finite-energy deterministic signal, the total energy in the stochastic process is unbounded (the characteristics of the process remain unchanged over time) and functions of the form (Eqns 5.61a and 5.61b) do not exist. In other words, the Fourier transform method introduced earlier fails in the sense that the total energy, as defined by Eqn (5.60), does not decrease as the length of the time series increases without bound. To get around this problem, we must deal with the frequency distribution of the signal *power* (the time average of energy or energy per unit time, E/T), which is a bounded function. The basis for spectral analysis of random processes is the autocorrelation function $R_{yy}(\tau) = E[y(t)y(t+\tau)]$ (where E stands for the expected value). Using the Wiener–Khinchin relation, the PSD, $S(f)$, becomes

$$S(f) = \int_{-\infty}^{\infty} R_{yy}(\tau)e^{-i2\pi f\tau} d\tau \tag{5.64a}$$

For an ergodic random process, for which ensemble averages can be replaced by time (or space) averages, R_{yy} has the from

$$R_{yy}(\tau) = \lim_{T \to \infty} \frac{1}{T} \int_{-T/2}^{T/2} [y(t)y^*(t+\tau)]dt \tag{5.64b}$$

By definition, the energy and PSD functions quantify the signal variance per unit frequency. For example, in the case of a stationary random process, integration of $S(f)$ gives the relation

$$s^2 = \int\limits_{f-\Delta f/2}^{f+\Delta f/2} S(f)df \tag{5.65}$$

where s^2 is the integrated signal variance in the narrow frequency range $\Delta f = [f - \tfrac{1}{2}\Delta f, f + \tfrac{1}{2}\Delta f]$. If we assume that the spectrum is nearly uniform over this frequency range, we find

$$S(f) \approx \frac{s^2}{\Delta f} \tag{5.66}$$

which defines the spectrum for a stochastic process in terms of a power density, or variance per unit frequency. The product $S(f)\cdot\Delta f$ is the total signal variance within the frequency band Δf centered at frequency f. At this point, there are several other basic concepts worth mentioning. First of all, a waveform whose autocorrelation function $R(\tau)$ attenuates slowly with time lag, τ, will have a narrow spectral distribution (Figure 5.14a) indicating that there are relatively few frequency components to destructively interfere with one another as τ increases from zero. In the limiting case of only one frequency component, f_a, we find $R(\tau) \approx \cos(2\pi f_a \tau)$ and Fourier *line spectra* appear at frequencies $\pm f_a$ (Figure 5.14b). Because they consist of near monotone signals, tidal motions are highly autocorrelated and produce sharp spectral lines. In contrast, a rapidly decaying autocorrelation function implies a broad spectral distribution (Figure 5.15a) and a large number of frequency components in the original waveform. In the limit $R(\tau) \rightarrow \delta(\tau)$ (Figure 5.15b), there is an infinite number of equal-amplitude frequency components in the waveform and the spectrum $S(f) \rightarrow$ constant (white spectrum).

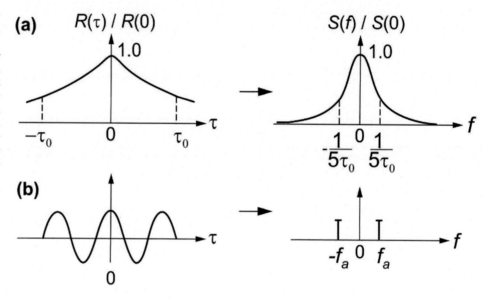

FIGURE 5.14 Examples of slowly decaying autocorrelation functions, $R(\tau)$, as a function of time lag, τ. Functions are normalized by their peak values. (a) The correlation function for a highly correlated signal leads to a relatively narrow power spectra density distribution, $S(f)$; (b) the case for autocorrelation, $R(\tau) \approx \cos(2\pi f_a \Delta t)$ for a single frequency component, f_a, and corresponding line spectra at frequencies $\pm f_a$. *From Konyaev (1990).*

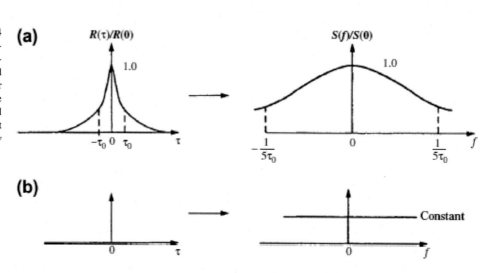

FIGURE 5.15 As for Figure 5.14 but for rapidly decaying autocorrelation functions, $R(\tau)$. (a) Correlation function for a weakly correlated signal leading to a broad power spectra density distribution. (b) The limiting case, $R(\tau) \approx (2\pi f\alpha\Delta t)$, and the related spectrum, $S(f) = $ constant (a white spectrum). *From Konyaev (1990).*

Figure 5.16 provides an example of time series data generated by the relation $y(k) = A\cos(2\pi nk/N) + \varepsilon(k)$, where $k = 0$, ..., N is time in units of $\Delta t = 1$, $n/(N\Delta t) = 0.25$ is the frequency in units of Δt^{-1}, and $\varepsilon(k)$ is a random number between -1 and $+1$. (We will often use this type of generic example rather than a specific example from the oceanographic literature. That way, readers can directly compare their computational results with ours). In the present case, if we set $\Delta t = 1$ day, then the time series $y(k)$ could represent east—west current velocity oscillations of a synoptic (3- to 10-day) period

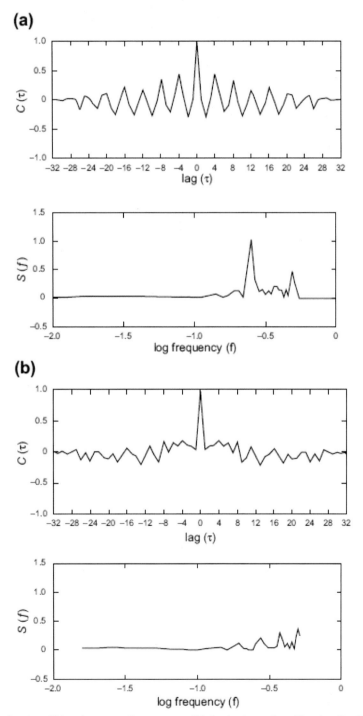

FIGURE 5.16   Autocovariance function, $C(\tau)$, and corresponding spectrum, $S(f)$, for the time series, $y(k) = A\cos(2\pi nk/N) + \varepsilon(k)$; $k = 0$, ..., N, $\Delta t = 1$, $n/N = 0.25$ is the frequency, and $\varepsilon(k)$ is a random number between -1 and $+1$. (a) $C(\tau)$ and $S(f)$ for $A = 1$ and $\varepsilon \neq 0$ (mostly deterministic data); and (b) for $A = 0$ (purely random data). Records have been padded with zeros up to time $k = 2N = 32$.

associated with atmospherically-forced motions (cf. Cannon and Thomson, 1996) or the 5-day non-isostatic bottom pressure fluctuations generated globally by the westward propagating Rossby-Haurwitz surface air pressure wave (e.g., Ponte and Schindelegger, 2022). Here, we set $A = 1$ and $\varepsilon(k) \neq 0$ for mostly deterministic data (Figure 5.16a) and $A = 0$ for random data (Figure 5.16b). In the analysis, the record has been padded with zeroes up to time $k = 2N$. For the mostly deterministic case, the noise causes partial decorrelation of the signal with lag, but the spectral peak remains prominent. For the purely random case, the spectrum resembles white noise but with isolated spectral peaks that one might mistake as originating with some physical process. The latter result is a good example of why we need to attach confidence limits to the peaks of spectral estimates (see Section 5.6.8). It is disconcerting to see the number of papers that are published in reputable journals that present spectra without including confidence intervals. At the same time, one must be cognizant of the meaning of the confidence intervals being presented. Use of a very low significance level might suggest a high degree of confidence in the results. However, this confidence can be rendered meaningless if a low significance level is selected. Significance levels of 95 or 99% are commonly accepted as "meaningful".

5.6.2 Spectra of Discrete Series

Consider an infinitely long time series $y(t_n) = y_n$ sampled at equally spaced time increments $t_n = n\Delta t$, where Δt is the sampling interval and n is an integer, $-\infty < n < \infty$. From sampling theory, we know that a continuous representation of the discrete times series $y_s(t)$, can be determined as the product of the continuous time series $y(t)$ with an infinite set of delta functions, $\delta(t)$, such that

$$y_s(t) = y(t) \sum_{n=-\infty}^{\infty} \delta(t - n\Delta t)$$

$$= y(t) \frac{\Xi(t/\Delta t)}{\Delta t} \tag{5.67a}$$

where Ξ is the "sampling function". The Fourier transform of (5.67a) is

$$Y(f) = \int_{-\infty}^{\infty} \left[\sum_{n=-\infty}^{\infty} y(t)\delta(t - n\Delta t)\Delta t \right] e^{-i2\pi ft} dt$$

$$= \Delta t \sum_{n=-\infty}^{\infty} y_n e^{-i2\pi ft} \tag{5.67b}$$

In effect, the original time series is multiplied by a "picket fence" of delta functions $\Xi(t/\Delta t) \approx \sum_{n=-\infty}^{\infty} \delta(t - n\Delta t)$, which are zero everywhere except for the infinitesimal rectangular region occupied by each delta function (Figure 5.17a and b). Comparison of the above expression with Eqns (5.61a) and (5.61b) shows that retention of the time step Δt ensures the conservation of the rectangular area in the two expressions as $\Delta t \to 0$. Provided that the time series $y(t)$ has a limited number of frequencies (i.e., is band-limited), whereby all frequencies are contained in the Nyquist interval

$$-f_{Nq} \leq f_k \leq f_{Nq} \tag{5.68}$$

in which $f_{Nq} \equiv f_{Nyquist} = 1/(2\Delta t)$ is the Nyquist frequency, the ESD

$$S_E(f) = |Y(f)|^2 \tag{5.69}$$

is identical to that for a continuous function. Conversely, if $Y(f) \neq 0$ for $|f| > f_{Nq}$ then the sampled and original times series do not have the same spectrum for $|f| < f_{Nq}$. The spectrum Eqn (5.69) obtained by Fourier analysis of discrete time series is called a *periodogram* spectral estimate, a term first coined by Schuster (1898) in a study of sunspot cycles. (Note that we always use f_{Nq} for the Nyquist frequency; the subscript Nq for Nyquist should never be confused with the subscript N used in summations or the N used for degrees of freedom (DoF)). The Nyquist frequency is the highest frequency that can be resolved by the series and corresponds to $f_{Nq} = 1/(2\Delta t)$.

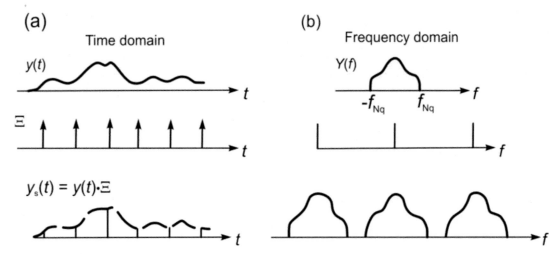

FIGURE 5.17 (a) A "picket fence" of delta functions, $\Xi = \delta(t - n\Delta t)$, used to generate a discrete data series $y_s(t)$ from a continuous time series, $y(t)$. (b) The Fourier transform, $Y(f)$ (schematic only) of the different functions.

Real oceanographic time series data are discrete and have finite duration, $T = N\Delta t$. Returning to Eqns (5.67a) and (5.67b), this means that the summation is over a limited range $n = 1$ to N, and the spectral amplitude for the sample must be defined in terms of the discrete Fourier transform (DFT)

$$Y_k = \Delta t \sum_{n=1}^{N} y_n e^{-i2\pi f_k n \Delta t}$$

$$= \Delta t \sum_{n=1}^{N} y_n e^{-i2\pi kn/N}; f_k = k/N\Delta t, k = 0, \ldots, N \tag{5.70}$$

The frequencies f_k are confined to the Nyquist interval, with positive frequencies, $0 \le f_k \le f_{Nq}$, corresponding to the range $k = 0, \ldots, N/2$ and negative frequencies, $-f_{Nq} \le f_k \le 0$, to the range $k = N/2, \ldots, N$. Because $f_{N-k} = f_k$, only the first $N/2$ Fourier transform values are unique. Specifically, $Y_k = Y_{N-k}$ so that we will generally confine our attention to the positive interval only. The inverse Fourier transform (IFT) is defined as

$$y_n = \frac{1}{N\Delta t} \sum_{k=0}^{N-1} Y_k e^{i2\pi kn/N}, n = 1, \ldots, N \tag{5.71}$$

As indicated by Eqn (5.70), the Fourier transforms, Y_k, are specified for the discretized frequencies f_k, where $f_k = kf_1$ and $f_1 = 1/(N\Delta t) = 1/T$ characterizes both the fundamental frequency, the lowest frequency that can be resolved by the series, and the bandwidth, Δf, for the time series. The energy spectral density (ESD) for a discrete, finite-duration time series is then

$$S_E(f_k) = |Y_k|^2, k = 0, \ldots, N - 1 \tag{5.72}$$

and Parseval's energy conservation theorem (5.20) becomes

$$\Delta t \sum_{n=1}^{N} |y_n|^2 = \Delta f \sum_{k=0}^{N-1} |Y_k|^2$$

where we have used $\Delta f = 1/(N\Delta t)$. A plot of $|Y_k|^2$ versus frequency, f_k, gives the discrete form of the periodogram spectral estimate.

Any geophysical data set that we collect is subject to discrete sampling and windowing. As noted earlier, a time series of geophysical data, $y(t_n)$, sampled at time steps Δt can be considered the product of an infinitely long time series with a rectangular window that spans the duration ($T = N\Delta t$) of the measured data. The discrete spectrum $S(f_k)$ is then the *convolution* of the true spectrum, $S(f)$, with the Fourier transform of the rectangular window (Figure 5.12b). Because the window allows us to see only a segment of the infinite time series, the spectrum $S(f_k)$ provides a distorted picture of the actual underlying spectrum. This distortion, created during the Fourier transform of the rectangular window, consists of a broadening of the central lobe and leakage of power from the central lobe into the side lobes. (The "ripples" on either side of the central lobe in Figure 5.12b are side lobes). A further problem is that the function Y_k and its Fourier transform now become periodic with period N, although the original infinite time series $y(t)$, of which our sample data are a subset, may have been nonperiodic.

As noted in the previous section, the convergence of $|Y(f)|^2$ to $S(f)$ is smooth for deterministic functions in that the function $|Y'(f)|^2$, obtained by increasing the sample record length from T to T' would be a smoother version of $|Y(f)|^2$. For stochastic signals, the function $|Y'(f)|^2$ obtained from the longer time series (T') is just as erratic as the function for the shorter series. The sample spectra of a stochastic process do not converge in any statistical sense to a limiting value as T tends to infinity. Thus, the sample spectrum is not a consistent estimator in the sense that its PDF does not tend to cluster more closely about the true spectrum as the sample size increases. To show what we mean, consider the spectrum of a process consisting of $N = 400$ random, normally distributed deviates (Gaussian white noise) sampled at 1 s intervals. (True white noise is a mathematical construct and is as physically impossible as the spike of an impulse function). The highest frequency we can hope to measure with these data is the Nyquist frequency, $f_{Nq} = 0.5$ cps (cycles per second). The spectra computed from 50 and then from 100 values of the fully white-noise signal are presented in Figure 5.18a. Also shown is the theoretical sample spectrum, corresponding to a uniform amplitude of 1.0. The shorter the sample length used for the discrete spectral estimates, the greater the amplitude spikes in the power spectrum. This same tendency is also apparent in Table 5.10, which lists the means, variances, and MSEs (Mean Square Errors) computed from various subsamples of the white-noise signal. Here, MSE is defined as the variance plus bias of an estimator $\hat{y}(t)$ of the true signal $y(t)$; that is

$$\text{MSE} = E\big[(\hat{y} - y)^2\big] = V[\hat{y}] + B^2 \tag{5.73}$$

where $B = E[\hat{y}] - y$ is the bias of the estimator. The mean is lower in both the $N = 50$ and $N = 400$ cases, while it is greater in the case where $N = 100$ and is exactly 1.0 for $N = 200$. The variance increases as N increases, as does the MSE. However, if this were a purely random discrete process (discrete white noise), the sample spectral estimator of the variance would be independent of the number of observations. Spectral analysis has clearly modified the original signal.

Now consider the spectrum of a second-order autoregressive process for a sample of $N = 400$ measured at 1 s increments (Figure 5.18b). (An autoregressive process of order p is one in which the present value of y depends on a linear combination of the previous p values of y. See Section 5.7.2). The Nyquist frequency is again $f_{Nq} = 0.5$ cps and the maximum bandwidth of the spectral resolution, $\Delta f = 1/(N\Delta t)$, is equal to 0.0025 cps. At higher frequencies, the sample spectrum appears to be a good estimator of the theoretical spectrum (the smooth solid line), while for the lower frequencies there are large spikes in the sample spectrum that are not characteristic of the true spectrum. This misleading appearance is largely a consequence of the fact that the theoretical spectrum has most of its energy at the lower frequencies. In reality, the computed raw spectrum (i.e., with no smoothing) can fluctuate by 100% about the mean spectrum. The fluctuations are much smaller at higher frequencies simply because the actual spectral level is correspondingly smaller.

The basic reason that Fourier analysis breaks down when applied to real time series is that it is based on the assumption of fixed (stationary) amplitudes, frequencies, and phases. Stochastic series are instead characterized by random changes in frequency, amplitude, and phase. Thus, our treatment must be a statistical approach that makes it possible to accommodate these types of changes in our computation of the power spectrum.

5.6.3 Conventional Spectral Methods

The two spectral estimation techniques founded on Fourier transform operations are: (a) the indirect autocorrelation approach popularized by Blackman and Tukey in the 1950s; and (b) the direct periodogram approach presently favored by the oceanographic community. The FFT is the most common algorithm for determining the periodogram. The autocorrelation approach is mainly discussed for completeness. These methods fall into the category of nonparametric techniques, which are defined independently of any specific time series. Parametric techniques, described later in this chapter, make assumptions about the functional variability of the time series and rely on the series for parameter determination.

The following sections first describe the two conventional spectral analysis methods without providing details on how to improve spectral estimates. We wish to first outline the procedures for calculating spectra before describing how to improve the statistical reliability of the spectral estimates. Once this is done, we give a thorough description of windowing, frequency band averaging, and other spectral improvement techniques.

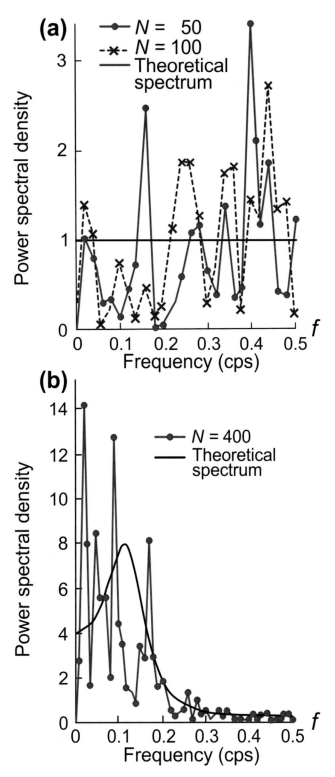

FIGURE 5.18 Power spectra of discrete signals and their theoretical values. Frequency, f, in cycles per second (cps); spectra are in units of amplitude-squared/cps. (a) Power spectrum for the first half ($N = 50$) and full ($N = 100$) realization of a discrete normal white-noise process measured at 1-s intervals. (b) Power spectrum for one realization of a second-order autoregressive process of $N = 400$ values measured at 1-s increments. $f_{Nq} = 0.5$ cps is the Nyquist frequency and the maximum bandwidth of the spectral resolution, $\Delta f = 1/(N \Delta t) = 0.0025/s$. *From Jenkins and Watts (1968).*

TABLE 5.10 Behavior of the statistics of samples drawn from a white noise signal as the record length, N, is increased.

Record length (N)	50	100	200	400
Mean	0.85	1.07	1.00	0.95
Variance	0.630	0.777	0.886	0.826
Mean square error	0.652	0.782	0.886	0.828

Units are arbitrary.
After Jenkins and Watts (1968).

5.6.3.1 The Autocorrelation Method

In the Blackman–Tukey method, the autocovariance function, $C_{yy}(\tau)$ (which equals the autocorrelation function, $R_{yy}(\tau)$, if the record mean has been removed), is first computed as a function of lag, τ, and the Fourier transform of $C_{yy}(\tau)$ is used to obtain the PSD as a function of frequency. An unbiased estimator for the autocovariance function for a data set consisting of N equally spaced values $\{y_1, y_2, ..., y_N\}$ is

$$C_{yy}(\tau_m; N - m) = \frac{1}{N - m} \sum_{n=1}^{N-m} y_n y_{n+m} \tag{5.74a}$$

where $m = 0, ..., M$ is the number of lags ($\tau_m = m\Delta t$) and $M < N$. In place of this estimator, some authors (cf. Kay and Marple, 1981) argue for the use of

$$C_{yy}(\tau_m; N) = \frac{1}{N} \sum_{n=1}^{N-m} y_n y_{n+m} \tag{5.74b}$$

which typically has a lower MSE than $C_{yy}(\tau_m; N - m)$ for most finite data sets. Because $E[C_{yy}(\tau_m; N)] = [(N - m)/N] \times C_{yy}(\tau_m; N - m)$, the function $C_{yy}(\tau; N)$ is a biased estimator for the autocovariance function. Despite this, we will often use the relation Eqn (5.74b) for the autocovariance function since it yields a PSD that is equivalent to the PSD obtained from the direct application of the FFT, as discussed in the next section. However, because the weighting $(N - m)/N$ acts like a triangular (Bartlett) smoothing window to help reduce spectral leakage, we will use Eqn (5.74a) when we want a "stand-alone" unbiased estimator of the covariance function, keeping in mind that this formulation gives largest weight to the most poorly determined components in the Fourier analysis, which is often not desirable despite the reduction in bias.

The one-sided PSD, G_k, for an autocovariance function with a total of M lags is found from the Fourier transform of the autocovariance function

$$G_k = 2\Delta t \sum_{m=0}^{M} C_{yy}(\tau_m) e^{-i2\pi km/M}; k = 0, ..., \frac{M}{2} \tag{5.75a}$$

where $\tau_m = m\Delta t$ and $2\Delta t = 1/f_{Nq}$. As $C_{yy}(\tau_m)$ is an even function, the spectrum of $\{y_n\}$ can be calculated from the cosine transform

$$G_k = 2\Delta t \left[C_{yy}(0) + 2\sum_{m=1}^{M} C_{yy}(\tau_m)\cos\left(\frac{2\pi km}{N}\right) \right]; k = 0, ..., \frac{M}{2} \tag{5.75b}$$

where $G_k = 2S_k$ is centered at positive frequencies $f_k = k/N\Delta t$ and the Nyquist interval $0 \leq f_k \leq f_{Nq}$ is divided into $N/2$ segments (N is even). For the two-sided spectrum, S_k, the first $(N/2) + 1$ frequencies are identical to those for the one-sided spectrum and correspond to positive frequencies in the range $0 \leq f_k \leq f_{Nq}$. The last $(N/2) - 1$ spectral values for the two-sided spectral density, defined for $k = (N/2) + 1, (N/2) + 2 ..., N - 1$, correspond to spectral density estimates for negative frequencies in the range $-f_{Nq} \leq f_k \leq 0$.

The solid line in Figure 5.19 shows the spectrum of monthly mean SSTs for Amphitrite Point on the west coast of Vancouver Island (Table 5.11) derived from the cosine transform using the Blackman–Tukey autocorrelation method for the version (Eqn 5.74b) of the autocovariance function. The temperature data span the 36-month period from January 1982 to December 1984. Since we wish to compare the Blackman–Tukey spectrum in Figure 5.19 with that derived from the

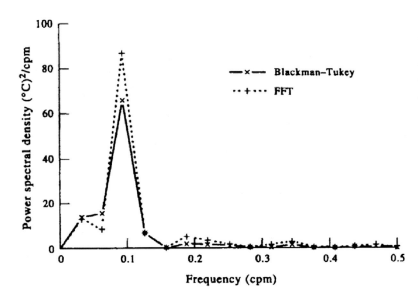

FIGURE 5.19 Spectra (°C)^2/cpm (cpm = cycles per month) versus frequency (cycles per month) for monthly mean sea surface temperatures collected at a coastal station in the northeast Pacific for the period January 1982—December 1984 (cf. Table 5.11). The solid line is the unsmoothed spectrum from the Blackman—Tukey autocorrelation method (the cosine transform of the autocovariance function Eqn (5.74b)); dashed line is the unsmoothed spectrum from the fast Fourier transform (FFT) method based on the first 2^5 (= 32) data values. Spectral peaks span the annual period ($f = 0.083$/month).

TABLE 5.11 Monthly mean sea surface temperatures SST (°C) at Amphitrite Point Lightstation (48°55.16′ N, 125°32.17′ W) on the west coast of Canada for January 1982 through December 1984.

Year 1982

n	1	2	3	4	5	6	7	8	9	10	11	12
SST	7.6	7.4	8.2	9.2	10.2	11.5	12.4	13.4	13.7	11.8	10.1	9.0

Year 1983

n	13	14	15	16	17	18	19	20	21	22	23	24
SST	8.9	9.5	10.6	11.4	12.9	12.7	13.9	14.2	13.5	11.4	10.9	8.1

Year 1984

n	25	26	27	28	29	30	31	32	33	34	35	36
SST	7.9	8.4	9.3	9.9	11.0	11.1	12.6	14.0	13.0	11.7	9.8	8.0

data series using a packaged FFT routine (the dashed line in Figure 5.19), the lags used to generate the Blackman—Tukey were computed for the first 32 (2^5) points only, which is four fewer points than normally would be used in the Blackman—Tukey approach (see Section 5.6.5 for a discussion of FFTs). In this case, artificially extending the lag correlation beyond 40—80% of the data discussed in connection with the effective degrees of freedom in Chapter 3.15, is a necessity if we are to obtain reasonable estimates of the spectrum using the autocorrelation method. As expected, results reveal a strong spectral peak centered near, but not at, the annual frequency ($f = 1.0$ cycles per year $= 0.083$ cycles per month). There are too few data to enable us to accurately resolve the location of the frequency peak. In the present example, all spectral estimates are positive. However, the autocorrelation method can yield erroneous negative spectra for weak frequency components when there are gaps in the data record.

We emphasize that the spectra in Figure 5.19 have been constructed without any averaging or windowing. This means that each spectral estimate has the minimum possible two DoF (corresponding to the orthogonal sine and cosine components obtained from the Fourier transform) so that the error in each estimate is equal to the value of the estimate itself. Some form of averaging is needed if we are to place confidence limits on the spectra (see Sections 5.6.6 and 5.67). The two spectra are slightly different because the record used for the FFT method is shorter than that used for the autocovariance method.

5.6.3.2 The Periodogram (Welch's Method)

The preferred method for estimating the PSD of a discrete sample $\{y_1, y_2, ..., y_N\}$ is the direct or periodogram method (also, Welch's Method). Instead of first calculating the autocorrelation function, the data are transformed directly to obtain the Fourier components $Y(f)$ using Eqn (5.70). To help avoid end effects (Gibbs' phenomenon) and wrap-around problems, the original time series can be padded with $K \leq N$ zeroes after the mean has been removed from the time series. The padding will also increase the frequency resolution of the periodogram (see Section 5.6.9). Although use of $K = N$ zeroes is not recommended for computational reasons, it has one advantage: the N-lag covariance function obtained from the IFT of the $2N$-point PSD is identical to the N-lag covariance function (Eqn 5.74b), as noted previously in Section 5.6.3.1. As with the autocorrelation method, improvements in the statistical reliability of the spectral estimates would be attained by "windowing" the time series prior to spectral estimation or by averaging the raw periodogram estimates over several adjacent frequency bands (see Sections 5.6.6 and 5.6.7).

The two-sided PSD (or autospectral density), for which half the spectral energy is at positive frequencies and half at negative frequencies, for frequency f in the Nyquist interval $-1/(2\Delta t) \leq f \leq 1/(2\Delta t)$ (i.e., $-f_{Nq} \leq f \leq f_{Nq}$) and a padding of K zeroes is

$$S_{yy}(f) = \frac{1}{(N+K)\Delta t}\left|\Delta t \sum_{n=0}^{N+K-1} y_n e^{-i2\pi fn\Delta t}\right|^2$$

$$= \frac{1}{(N+K)\Delta t}|Y(f)|^2 \tag{5.76a}$$

while the one-sided PSD for the positive frequency interval only, $0 \leq f \leq f_{Nq} = 1/(2\Delta t)$, is

$$G_{yy}(f) = 2S_{yy}(f) = \frac{2}{(N+K)\Delta t}|Y(f)|^2 \tag{5.76b}$$

Division by Δt transforms the ESD of Eqn (5.72) into a PSD, $S_{yy}(f)$.

Evaluation of Eqn (5.76a) using the FFT defines $Y(f)$ in terms of the DFT estimates, $Y(f_k) = Y_k$, where the f_k forms a discrete set of $(N+K)/2$ equally spaced frequencies $f_k = \pm k/[(N+K)\Delta t]$, $k = 0, 1,..., [(N+K)/2] - 1$ in the Nyquist interval, $-1/(2\Delta t) \leq f_k \leq 1/(2\Delta t)$ $(-f_{Nq} \leq f_k \leq f_{Nq})$. The case $k = 0$ represents the mean component. The two-sided PSD for N data values and K padded zeroes is then

$$S_{yy}(0) = \frac{1}{(N+K)\Delta t}|Y_0|^2, k = 0$$

$$S_{yy}(f_k) = \frac{1}{(N+K)\Delta t}\left[|Y_k|^2 + |Y_{N+K-k}|^2\right], k = 1,..., \frac{(N+K)}{2} - 1$$

$$S_{yy}(f_{Nq}) = S_{yy}\left(f_{(N+K)/2-k}\right) = \frac{1}{(N+K)\Delta t}|Y_{(N+K)/2}|^2, k = \frac{(N+K)}{2} \tag{5.77a}$$

and the one-sided PSD is

$$G_{yy}(0) = \frac{1}{(N+K)\Delta t}|Y_0|^2, k = 0$$

$$G_{yy}(f_k) = \frac{2}{(N+K)\Delta t}|Y_k|^2, k = 1,..., \frac{(N+K)}{2} - 1$$

$$G_{yy}(f_{Nq}) = G_{yy}\left(f_{(N+K)/2-k}\right) = \frac{1}{(N+K)\Delta t}|Y_{(N+K)/2}|^2, k = \frac{(N+K)}{2} \tag{5.77b}$$

Multiplication of $S_{yy}(f) \equiv S_k$ (or G_k) by the bandwidth of the signal $\Delta f = 1/[(N+K)\Delta t]$ gives the estimated signal variance, σ_k^2, in the kth frequency band; i.e., $\sigma_k^2 = S_k' = S_k\Delta f$. The summation

$$\sum_{n=0}^{N+K-1} S_k' = \sum_{n=0}^{N+K-1} S_k\Delta f \tag{5.78}$$

gives the variance and total power of the signal. The quantity

$$S'_k = \frac{1}{[(N+K)\Delta t]^2} \left[|Y_k|^2 + |Y_{N+K-k}|^2 \right]$$

$$= \frac{1}{(N+K)^2} \sum_{n=0}^{N+K-1} \left| y_n e^{-i2\pi f n \Delta t} \right|^2 \tag{5.79}$$

is often computed as the periodogram. However, this is not correctly scaled as a PSD but represents the "peak" in the spectral plot rather than the "area" under the plot of S_k versus Δf. The representation Eqn (5.79) is sometimes useful although most oceanographers are more familiar with the PSD form of the periodogram.

It bears repeating that the use of Fourier transforms assumes periodic structure within the sampled data when no periodic structure may actually exist in the time series. That is, the FFT of a finite length data record is equivalent to assuming that the record consists of periodic components. We again note that autospectral functions are always real so that $S'_{yy}(f_k) = S'_{yy}(2f_N - f_k)$, and the one-sided autospectral periodogram estimate becomes

$$G'_{yy}(f_k) = 2S'_k = \frac{2}{[(N+K)\Delta t]^2} |Y(f_k)|^2 \tag{5.80}$$

Until the 1960s, the direct transform method first used by Schuster (1898) to study "hidden periodicities" in measured sunspot numbers was seldom used due to difficulties with statistical reliability and extensive computational time. The introduction of the first practical FFT algorithms for spectral analysis (Cooley and Tukey, 1965) greatly reduced the computational time by taking advantage of patterns in DFT functions (see Section 5.10). Problems with the statistical reliability of the spectral estimates are resolved through appropriate windowing and averaging techniques, which we discuss in Sections 5.6.6 and 5.6.7. Figure 5.19 compares the unsmoothed periodogram spectral estimate for the monthly mean SST data at Amphitrite Point (Table 5.11) with the corresponding spectrum obtained from the Blackman–Tukey autocorrelation method. As mentioned earlier, the FFT requires data lengths equal to powers of two so that we have shortened the series to $2^5 = 32$ months. As we found with the Blackman–Tukey autocorrelation method, the FFT spectrum of coastal temperatures has a strong peak near the annual period, albeit with a slightly different spectral amplitude.

5.6.3.3 The PSD for Periodic Data

For a strictly periodic digital time series $y(t)$ having an exact integer number of oscillations over the interval $[0, T]$, we can use the Fourier series expansion Eqn (5.28) and write

$$y(t) = \frac{1}{2} A_0 + \sum_{n=1}^{N} [A_n \cos(\omega_n t) + B_n \sin(\omega_n t)]$$

$$= \frac{1}{2} C_0 + \sum_{n=1}^{N} [C_n \cos(\omega_n t + \phi_n)] \tag{5.81}$$

in which the constants A_n, B_n are given by Eqn (5.30) and in which

$$C_n = \left(A_n^2 + B_n^2 \right)^{1/2}$$

$$\phi_n = \tan^{-1}(B_n/A_n) \equiv \arctan(B_n/A_n) \tag{5.82}$$

are the amplitude and phase of the complex Fourier coefficient for the nth frequency component, $\omega_n = 2\pi f_n$. Because the data record contains periodic components only, a plot of $2|C_n|^2$ against n ($n = 0, \ldots, N-1$) yields a series of distinct "spikes" or line spectra, S_n, with the variance divided equally between negative and positive frequencies

$$S_n = \frac{(\Delta t)^2}{T} \left[|C_n|^2 + |C_{N-n}|^2 \right]$$

$$= \frac{2\Delta t}{N} |C_n|^2 \tag{5.83}$$

where the record mean value C_0 has been subtracted from the record $y(t)$. Here, we have assumed that $y(t)$ is a real function. The squared Fourier components $|C_n|^2$ give the contribution of the nth frequency component to the total variance and the various frequency components contribute additively to the total power of the time series. The contribution from each component is assumed to be independent of that from all other components.

5.6.3.4 Variance-Preserving Spectra

Because the PSD, $S_{yy}(f)$, and frequency, f, of a time series often range over several orders of magnitude, spectral distributions are usually plotted as the logarithm of $S_{yy}(f)$ versus the logarithm of frequency, i.e., $\log[S_{yy}(f)]$ versus $\log(f)$. This format allows the user to provide a compact representation of the spectral distribution. The latter is also useful where a spectrum has a power law dependence of the form $S_{yy}(f) \sim f^{-p}$. In this case, the slope of the spectrum is given as $p = -\log[S_{yy}(f)]/\log(f)$. An example of a more narrowly focused format of $\log[S_{yy}(f)]$ versus f (a log—linear plot) is presented in Figure 5.20a where we have used time series data generated by the relation $y(k) = A\cos(2\pi nk/N) + \varepsilon(k)$ from Section 5.6.1 (Figure 5.16). Spectral density has units of energy/frequency for the same units used for f. For example, the PSD of a current velocity record are typically in units of $(\text{cm/s})^2/\text{cph}$ or $(\text{cm/s})^2/\text{cpd}$ plotted against $\log(\text{frequency})$ or frequency in cph (cycles per hour) or cpd (cycles per day), respectively. (Sometimes, m/s is used in place of cm/s, and vice versa).

In the log—linear format, the integration proceeds over frequency bands of width Δf centered at frequency f_c (where the "c", in this case, stands for center of the frequency band), so that the area under each small rectangular segment of the spectral curve is equal to a pseudo-variance

$$\sigma_*^2(f_c) = \int\limits_{f_c-\Delta f/2}^{f_c+\Delta f/2} \log\big[S_{yy}(f)\big]\,df \tag{5.84}$$

FIGURE 5.20 Two common types of spectral plot derived for the time series $y(k) = A\cos(2\pi nk/N) + \varepsilon(k)$ (see Figure 5.16). (a) A plot of log power spectral density, $\log[S_{yy}(f)]$, versus frequency, f; (b) a variance-preserving spectral plot in which $f \cdot S_{yy}(f)$ is plotted against $\log(f)$.

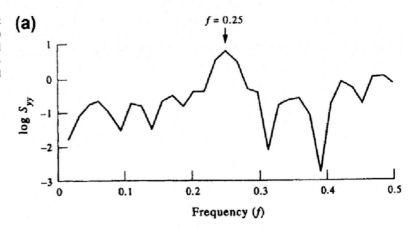

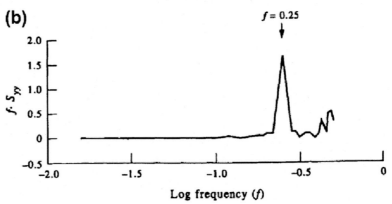

Although log spectra plots have an appealing shape, the integral Eqn (5.84) is certainly not variance preserving. To preserve the signal variance, $\sigma^2(f_c)$, under the spectral curve, we need to plot $fS_{yy}(f)$ versus $\log(f)$, as in Figure 5.20b. Replacing df in Eqn (5.84) with $d[\log(f)]$, the true *variance-preserving* form of the spectrum becomes

$$\sigma^2(f_c) = \int\limits_{f_c-\frac{\Delta f}{2}}^{f_c+\frac{\Delta f}{2}} fS_{yy}(f)d[\log(f)]$$

$$= \int\limits_{f_c-\Delta f/2}^{f_c+\Delta f/2} S_{yy}(f)df \tag{5.85}$$

where we have used the fact that $d[\log(f)] = df/f$. Eqn (5.85) gives the true signal variance within the band Δf. In particular, if $S_{yy}(f) \approx S_c$ is nearly constant over the frequency increment Δf, then $\sigma^2(f_c) \approx S_c\Delta f$ is the signal variance in band Δf centered at frequency f_c. In this format, there is a clear spectral peak at $f = 0.25$ cycles per unit time that is associated with the term $\cos(2\pi nk/N)$ in the original analytical expression.

5.6.3.5 The Chi-Squared Property of Spectral Estimators

Throughout this chapter, we have claimed that each spectral estimate for maximum frequency resolution, $1/T$, obtained from Fourier transforms of stochastic time series has two DoF. We now present a more formal justification for that claim for discrete spectral estimators by showing that each estimate is a stochastic chi-square (pronounced "ki-square") variable with two DoF (i.e., there are two independent squares entering the expression for the chi-square variable). Consider any stochastic white-noise process $\eta(t)$, for which $E[\eta(t)] = 0$. The Fourier components are

$$A(f) = \sum_{n=-N}^{N-1} \eta(n\Delta t)\cos(2\pi f n\Delta t)$$

$$B(f) = \sum_{n=-N}^{N-1} \eta(n\Delta t)\sin(2\pi f n\Delta t) \tag{5.86}$$

where as usual, $-1/(2\Delta t) \le f \le 1/(2\Delta t)$ is the Nyquist interval, and it follows that $E[A(f)] = 0 = E[B(f)]$. Thus, at the harmonic frequencies $f_k = k/N\Delta t$, the variance is

$$V[A(f_k)] = E[A^2(f_k)] = \sigma_\eta^2 \sum_{n=-N}^{N-1} \cos^2(2\pi f_k n\Delta t)$$

$$= \frac{1}{2}N\sigma_\eta^2, k = \pm 1, \pm 2, ..., \pm(N-1)$$

$$= N\sigma_\eta^2, k = 0, -N \tag{5.87a}$$

Similarly

$$V[B(f_k)] = \frac{1}{2}N\sigma_\eta^2, k = \pm 1, \pm 2, ..., \pm(N-1)$$

$$= 0, k = 0, -N \tag{5.87b}$$

When $k \ne j$, the covariance is

$$C[A(f_k), A(f_j)] = \sigma_\eta^2 \sum_{n=-N}^{N-1} \cos(2\pi f_k n\Delta t)\cos(2\pi f_j n\Delta t) = 0 \tag{5.88a}$$

and

$$C[A(f_k), B(f_j)] = 0 \text{ (orthogonality condition)} \tag{5.88b}$$

Because $A(f_k)$ and $B(f_k)$ are linear functions of normal random variables, $A(f_k)$ and $B(f_k)$ are also distributed normally. Hence, the random variables

$$\frac{A^2(f_k)}{V[A(f_k)]} = \frac{2A^2(f_k)}{N\sigma_\eta^2}$$

$$\frac{B^2(f_k)}{V[B(f_k)]} = \frac{2B^2(f_k)}{N\sigma_\eta^2}$$

$$(5.89)$$

are each distributed as χ_1^2, which is a chi-square variable with one DoF.

Because the normal distributions $A(f_k)$ and $B(f_k)$ are independent random variables, the sum of their squares

$$\frac{2}{\sigma_\eta^2}\left[A^2(f_k) + B^2(f_k)\right] = \frac{2}{\Delta t \sigma_\eta^2} S_{yy}(f_k)$$

$$(5.90)$$

is distributed as χ_2^2, which is a chi-square variable with two DoF. Here, $S_{yy}(f_k)$ is the sample spectrum. Thus

$$\frac{E\left[2S_{yy}(f_k)\right]}{\Delta t \sigma_\eta^2} = 2$$

$$(5.91)$$

and

$$E\left[S_{yy}(f_k)\right] = \sigma_\eta^2 \Delta t$$

$$(5.92)$$

which is the spectrum. At the Fourier component frequencies (set by the record length and sampling interval), the sample spectrum is an unbiased estimator of the white-noise spectrum of $\eta(t)$. Also, at these frequencies, the variance of the estimate is constant and independent of sample size. This explains the failure of the sample estimates of the variance to decrease with increasing sample size. We remark further that, even if $\eta(t)$ is not normally distributed, the random variables $A(f_k)$ and $B(f_k)$ are very nearly normally distributed by the central limit theorem. Hence, the distribution of the $S_{yy}(f)$ will be very nearly distributed as χ_2^2 regardless of the PDF of the $\eta(t)$ process.

5.6.4 Spectra of Vector Series

To calculate the spectra of vector time series such as current and wind velocity, we first need to resolve the data into orthogonal components. Spectral analysis is then applied to the combined series of components and the results stored as a complex quantity in the computer. Raw data are recorded as speed and direction by rotor-type meters and as orthogonal components by acoustic and electromagnetic meters. The usual procedure is to convert recorded time series to an earth-referenced Cartesian coordinate system consisting of two orthogonal horizontal components and a vertical component. In the open ocean, horizontal velocities, $\mathbf{u} = (u, v)$, typically are resolved into components of eastward (zonal; u) and northward (meridional; v) time series, whereas in the coastal ocean it is preferable to resolve the vector components into cross-shore (u') and longshore (v') components through the rotation

$$\begin{pmatrix} u' \\ v' \end{pmatrix} = \begin{pmatrix} \cos\theta & \sin\theta \\ -\sin\theta & \cos\theta \end{pmatrix} \begin{pmatrix} u \\ v \end{pmatrix}$$

$$(5.93a)$$

$$u' = u\cos\theta + v\sin\theta$$
$$v' = -u\sin\theta + v\cos\theta$$

$$(5.93b)$$

where the angle θ is the orientation of the coastline (or the local bottom contours) measured counterclockwise from the eastward direction (Figure 5.21). Thus, in the case where the coastline is rotated counterclockwise to lie along a parallel of latitude (i.e., $\theta = \pi/2$), we find $u' = v$ and $v' = -u$. Alternatively, one can let the current velocity observations define θ as the direction of the major axis obtained from principal component analysis; that is, the axis which maximizes the variance in a scatter plot of u versus v (see Figure 4.20).

In coastal regions, the principal axis is usually closely parallel to the coastline. For studies of highly circularly polarized motions, such as inertial currents and tidal currents, resolution into clockwise and counterclockwise rotary components is often more useful. The choice of representation depends on the preference of the investigator and the type of process being investigated. More is said on this subject in Section 5.6.4.2.

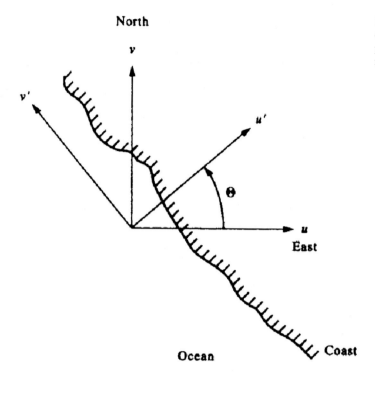

North

v

v'

u'

θ

u

East

Ocean

Coast

FIGURE 5.21 Cross-shore (u') and longshore (v') velocity components in a Cartesian coordinate system rotated through a positive (counterclockwise) angle from the eastward (u) and northward (v) directions.

5.6.4.1 Cartesian Component Rotary Spectra

The horizontal velocity vector can be represented in Cartesian coordinates as a complex function $w(t)$ whose real part, $u(t)$, is the projection of the vector on the zonal (or cross-shelf) axis and whose imaginary part, $v(t)$, is the projection of the vector on the meridional (or long-shelf) axis (Figure 5.22)

$$w(t) = u(t) + iv(t) \tag{5.94}$$

(The use of vector $w(t)$ follows the convention of Gonella (1972), Mooers (1973) and others in their discussion of rotary spectral analysis and is not to be confused with the weights $w(t)$, generally written as $w(t_n)$, used in the sections on data windowing, or the vertical component of velocity, w. Gonella (1972) used u_1 and u_2 for the two horizontal velocity components). A complete description of the time variability of a three-dimensional vector at a single point consists of six functions of frequency: three autospectra for the three velocity components (u, v, w) and three cross-spectra. For the two-dimensional vectors considered in this section, there are two autospectra and one cross-spectrum. The discrete Fourier Transform (DFT), $W(f_k) = U(f_k) + iV(f_k)$, (where $f_k = k/N\Delta t$, $k = 1, ..., N$ and $k = 0$ corresponds to the mean flow) is

$$W(f_k) = \Delta t \sum_{n=0}^{N-1} w(t) e^{-i2\pi kn/N}$$

$$= \Delta t \sum_{n=0}^{N-1} [u(t) + iv(t)] e^{-i2\pi kn/N} \tag{5.95}$$

where $U(f_k)$ and $V(f_k)$ are the Fourier transforms of $u(t)$ and $v(t)$, respectively. If the original record is separated into M blocks of length N', where $N = MN'$ is the total record length if no overlapping of segments is used, the spectral density function is given in terms of the number of segments used to form the block-averaged, one-sided autospectrum $\left(0 \leq f_k' < \infty\right)$

FIGURE 5.22 Horizontal velocity represented as a complex vector, $w = u + iv$, with components (u, v) along the real and imaginary axes, respectively.

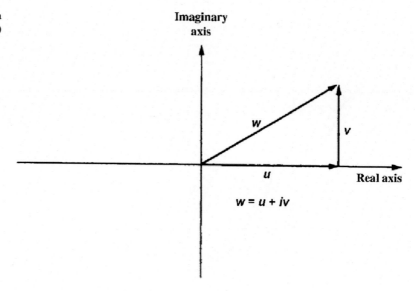

$$G_{ww}\left(f_k'\right) = \frac{2}{N\Delta t} \sum_{m=1}^{M} \left|W_m\left(f_k'\right)\right|^2$$

$$= \frac{2}{N\Delta t} \sum_{m=1}^{M} \left\{\left[W_{Rm}\left(f_k'\right)\right]^2 + \left[W_{Im}\left(f_k'\right)\right]^2\right\} \qquad (5.96)$$

$$= \frac{2}{N\Delta t} \sum_{m=1}^{M} \left\{\left[U_{Rm}\left(f_k'\right) - V_{Im}\left(f_k'\right)\right]^2 + \left[U_{Im}\left(f_k'\right) + V_{Rm}\left(f_k'\right)\right]^2\right\}$$

where $f_k' = \frac{k}{N'\Delta t}$, $k = 0, 1, ..., N'/2$ ($k = 0$ is the mean flow) and for FFT analysis, $N' = 2p$ (positive integer p), and where the subscripts R and I stand for the real and imaginary parts of the given Fourier components.

5.6.4.2 Rotary Component Spectra

Rotary analysis of currents involves the separation of the velocity vector for a specified frequency, ω, into clockwise and counterclockwise rotating circular components with amplitudes A^-, A^+, and relative phases θ^-, θ^+, respectively. Thus, instead of dealing with two Cartesian components (u, v) we deal with two circular components $(A^-, \theta^-; A^+, \theta^+)$. Several reasons can be given for using this approach: (1) The separation of a velocity vector into oppositely rotating components can reveal important aspects of the wave field at the specified frequencies. The method has proven especially useful for investigating currents over abrupt topography, wind-generated inertial motions, diurnal frequency continental shelf waves, and other forms of narrow-band oscillatory flow; (2) in many cases, one of the rotary components (typically, the clockwise component in the northern hemisphere and counterclockwise component in the southern hemisphere) dominates the currents so that we need to deal with one scalar quantity rather than two. Inertial motions, for example, are almost entirely clockwise (counterclockwise) rotary in the northern (southern) hemisphere so that the counterclockwise (clockwise) component can be ignored for most applications; and (3) many of the rotary properties, such as spectral energy $S^-(\omega)$ and $S^+(\omega)$ and rotary coefficient, $r(\omega)$, are invariant under coordinate rotation so that local steering of the currents by bottom topography or the coastline are not factors affecting the analysis.

The vector addition of the two oppositely rotating circular vectors (Figure 5.23a and b) causes the tip of the combined vector (Figure 5.23c) to trace out an ellipse over one complete cycle. The eccentricity, e, of the ellipse is determined by the relative amplitudes of the two rotary components. Motions at frequency ω are circularly polarized if one of the two components is zero; motions are rectilinear (back-and-forth along the same line) if both circularly polarized components have the same magnitude. In rotary spectral format, the current vector $w(t)$ can be written as the Fourier series

(a)

(b)

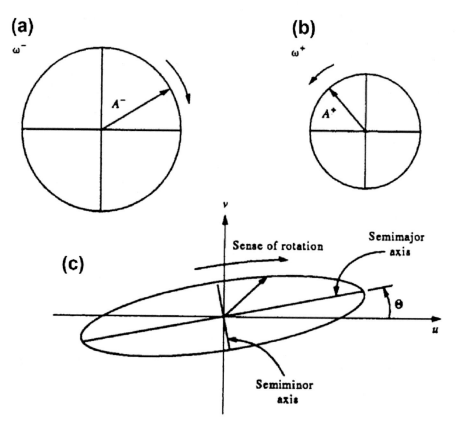

(c)

FIGURE 5.23 Current ellipses formed by the vector addition of two oppositely rotating vectors. (a) Clockwise component (ω^-) and (b) counterclockwise component (ω^+) with amplitudes, A^- and A^+, respectively. (c) General case of elliptical motion with major axis tilted at an angle θ counterclockwise from east. ε^- and ε^+ (not shown) are the angles of the two circular components at time $t = 0$.

$$
\begin{aligned}
w(t) &= \overline{u(t)} + \sum_{k=1}^{N} U_k \cos(\omega_k t - \phi_k) + i\left[\overline{v(t)} + \sum_{k=1}^{N} V_k \cos(\omega_k t - \theta_k)\right] \\
&= [\overline{u(t)} + i\overline{v(t)}] + \sum_{k=1}^{N} [U_k \cos(\omega_k t - \phi_k) + iV_k \cos(\omega_k t - \theta_k)]
\end{aligned}
\tag{5.97}
$$

in which $\overline{u(t)} + i\overline{v(t)}$ is the mean velocity, $\omega_k = 2\pi f_k = 2\pi k/(N\Delta t)$ is the angular frequency, $t\ (= n\Delta t)$ is the time, and (U_k, V_k) and (ϕ_k, θ_k) are the amplitudes and phases, respectively, of the Fourier constituents for each frequency for the real and imaginary components. Subtracting the mean velocity and expanding the trigonometric functions, we find

$$
\begin{aligned}
w'(t) &= w(t) - [\overline{u(t)} + i\overline{v(t)}] \\
&= \sum_{k=1}^{N} \{U_{1k} \cos(\omega_k t) + U_{2k} \sin(\omega_k t) + i[V_{1k} \cos(\omega_k t) + V_{2k} \sin(\omega_k t)]\}
\end{aligned}
\tag{5.98}
$$

in which we have defined the even (U_{1k}, V_{1k}) and odd $(U_{2k}, V_{2k}.)$ functions as

$$
U_{1k} = U_k \cos \phi_k, U_{2k} = U_k \sin \phi_k
\tag{5.99a}
$$

$$
V_{1k} = V_k \cos \phi_k, V_{2k} = V_k \sin \theta_k
\tag{5.99b}
$$

Dropping the prime notation for $w'(t)$ and following some reorganization, we can write the kth frequency component of the series as the sum of counterclockwise (+) and clockwise (−) components

$$
\begin{aligned}
w_k(t) &= w_k^+(t) + w_k^-(t) = A_k^+ \exp(i\varepsilon_k^+)\exp(i\omega_k t) + A_k^- \exp(i\varepsilon_k^-)\exp(-i\omega_k t) \\
&= \exp\left[\frac{i(\varepsilon_k^+ + \varepsilon_k^-)}{2}\right]\left\{[A_k^+ + A_k^-]\cos\left[\frac{\varepsilon_k^+ - \varepsilon_k^-}{2} + \omega_k t\right] + i[A_k^+ - A_k^-]\sin\left[\frac{\varepsilon_k^+ - \varepsilon_k^-}{2} + \omega_k t\right]\right\}
\end{aligned}
\tag{5.100}
$$

where the counterclockwise and clockwise rotary component amplitudes are given by

$$A_k^+ = \frac{1}{2}\left\{[(U_{1k} + V_{2k})]^2 + [(U_{2k} - V_{1k})]^2\right\}^{1/2} \tag{5.101a}$$

$$A_k^- = \frac{1}{2}\left\{[(U_{1k} - V_{2k})]^2 + [(U_{2k} + V_{1k})]^2\right\}^{1/2} \tag{5.101b}$$

and the corresponding phase angles for time $t = 0$, by

$$\varepsilon_k^+ = \tan^{-1}[(V_{1k} - U_{2k}) / (U_{1k} + V_{2k})] \tag{5.102a}$$

$$\varepsilon_k^- = \tan^{-1}[(U_{2k} + V_{1k}) / (U_{1k} - V_{2k})] \tag{5.102b}$$

Each of the constituents contributing to Eqn (5.98) has the form of an ellipse with major semiaxis of length $L_M = \left(A_k^+ + A_k^-\right)$ and minor semiaxis of length $\left|A_k^+ - A_k^-\right|$ (Figure 5.22c). The ellipse is tilted at an angle of $\theta = \frac{1}{2}\left(\varepsilon_k^+ + \varepsilon_k^-\right)$ from the u-axis and the vector is along the major axis of the ellipse at time $t = \left(\varepsilon_k^+ - \varepsilon_k^-\right)/(4\pi f_k)$. The one-sided spectra $\left(G_k^+, G_k^-\right) = \left(S_k^+, S_k^-\right)$ for the two oppositely rotating components for frequencies $f_k = \omega_k/2\pi$ are

$$S\!\left(f_k^+\right) = S_k^+ = \frac{\left(A_k^+\right)^2}{N\Delta t}, f_k = 0,\dots 1/(2\Delta t) \tag{5.103a}$$

$$S\!\left(f_k^-\right) = S_k^- = \frac{\left(A_k^-\right)^2}{N\Delta t}, f_k = -1/(2\Delta t),\dots,0 \tag{5.103b}$$

Plots of rotary spectra are generally presented in two ways. In Figure 5.24a, both S^- and S^+ are plotted as functions of frequency magnitude, $|f| \geq 0$, with solid and dashed lines used for the clockwise and counterclockwise spectra, respectively. In Figure 5.24b, we use the fact that clockwise spectra are defined for negative frequencies and counterclockwise spectra for positive frequencies. The spectra $S\!\left(f_k^+\right)$ and $S\!\left(f_k^-\right)$ used in Figure 5.24a are then plotted on opposite sides of zero frequency. In these spectra, peak energy occurs at the diurnal and semidiurnal periods. The predominantly clockwise rotary motions at semidiurnal periods suggest a combination of tidal and near-inertial motions (at the latitude of these particular observations, the inertial period is close to the semidiurnal tidal period).

Another useful property is the rotary coefficient

$$r(\omega) = \frac{S_k^+ - S_k^-}{S_k^+ + S_k^-} \tag{5.104}$$

which ranges from $r = -1$ for clockwise motion, to $r = 0$ for unidirectional flow, to $r = +1$ for counterclockwise motion. The rotary nature of the flow can change considerably with position, depth, and time. As indicated by Figure 5.25, the observed diurnal tidal currents over Endeavour Ridge in the northeast Pacific change from moderately positive to strongly negative rotation with depth. In contrast, the semidiurnal currents change from strongly negative near the surface to strongly rectilinear at depth. (Data, in this case, are from a string of current meters moored for a period of 9 months). We remark that the definition Eqn (5.104) differs in sign from that of Gonella (1972), who used $S_k^- - S_k^+$ rather than $S_k^+ - S_k^-$ in the numerator. Because many types of oceanic flow are predominantly clockwise rotary in the northern hemisphere, Gonella's definition has the advantage that clockwise rotating currents have positive rotary coefficients. However, we find Gonella's definition a bit awkward since clockwise motions, which are linked to *negative* frequencies, then have *positive* rotary coefficients. We prefer that motions linked to *negative* frequencies have *negative* rotary coefficients.

5.6.4.3 Rotary Spectra (via Cartesian Components)

Gonella (1972) and Mooers (1973) present the rotary spectra in terms of their Cartesian counterparts and provide a number of rotational invariants for analyzing current and wind vectors at specified frequencies. Specifically, the one-sided auto-spectra for the counterclockwise (CCW) and clockwise (CW) rotary components of the vector $w(t) = u(t) + iv(t)$ are, in terms of their Cartesian components

$$G\!\left(f_k^+\right) = \frac{1}{2}[G_{uu}(f_k) + G_{vv}(f_k) + Q_{uv}(f_k)], f_k \geq 0 \text{ (CCW component)} \tag{5.105a}$$

$$G\!\left(f_k^-\right) = \frac{1}{2}[G_{uu}(f_k) + G_{vv}(f_k) - Q_{uv}(f_k)], f_k \leq 0 \text{ (CW component)} \tag{5.105b}$$

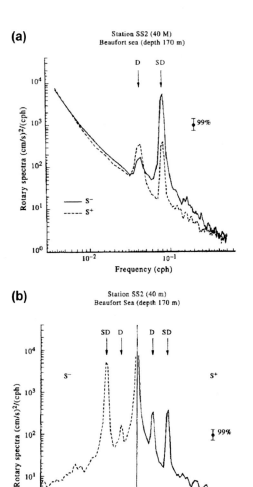

FIGURE 5.24 Rotary current spectra for hourly currents measured at 40-m depth in the Beaufort Sea, Arctic Ocean (water depth = 170 m). Peaks are at the diurnal (D) and semidiurnal (SD) tidal frequencies. Frequency resolution is 0.0005 cph and there are 112 degrees of freedom (DoF) per spectral band. Vertical bar gives the 99% level of confidence, (a) one-sided rotary spectra, $S^-(f)$ and $S^+(f)$, versus f for positive frequency, f; (b) two-sided rotary spectra, $S(f_k^+) = S^+$ and $S(f_k^-) = S^-$ versus log f for positive and negative frequencies, $f_{\pm k}$. *Courtesy E. Carmack, A. Rabinovich, and E. Kulikov.*

where $G_{uu}(f_k)$ and $G_{vv}(f_k)$ are the one-sided autospectra of the u and v Cartesian components of velocity and $Q_{uv}(f_k)$ is the quadrature spectrum between the two components, where

$$Q_{uv}(f_k) = -Q_{uv}(-f_k) = (U_{1k}V_{2k} - V_{1k}U_{2k}) \tag{5.106}$$

As defined in Section 5.8, the spectrum can be written in terms of cospectrum (real part) and quadrature spectrum (imaginary part)

$$G_{uv}(f_k) = C_{uv}(f_k) - iQ_{uv}(f_k) \tag{5.107}$$

5.6.5 Effect of Sampling on Spectral Estimates

Spectral estimates derived by conventional techniques are limited by two fundamental problems: (1) the finite length, T, of the time series; and (2) he discretization associated with the sampling interval, Δt. The first problem is inherent to all real data sets, while the second is associated with finite instrument response times and/or the need to digitize the time series for the purposes of analysis.

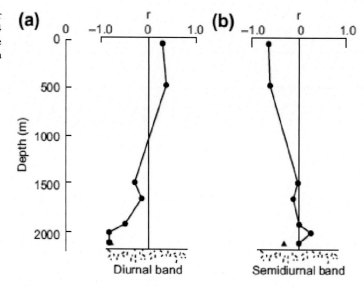

FIGURE 5.25 Rotary coefficient, $r(\omega)$, as a function of depth for current oscillations in (a) the diurnal frequency band ($\omega/2\pi \approx 0.04$ cph) and (b) the semidiurnal band ($\omega/2\pi \approx 0.08$ cph). The *triangles* denote values for a single current meter moored at a nearby site. *From Allen and Thomson (1993).*

Irrespective of the method used to calculate the power spectrum of a waveform, the record duration, $T = N\Delta t$, and sampling increment, Δt, impose severe limitations on the information that can be extracted. Ideally, we would like to have sensors that can sample rapidly enough (small Δt) that no significant frequency component goes unresolved. This also eliminates aliasing problems in which unresolved spectral energy at frequencies higher than the Nyquist frequency is folded back into lower frequencies. At the same time, we wish to record for a sufficiently long period (large N) that we capture many cycles of the lowest frequency of interest. Long-term sampling also enables us to better resolve frequencies that are close together and to improve the statistics (confidence intervals) for spectral estimates. In reality, most data series are a compromise based on the frequencies of interest, the response limitations of the sensor, and cost. The choices of the sampling rate and the record duration are tailored to best meet the task at hand.

5.6.5.1 Effect of Finite Record Length

As noted earlier, we can think of a data sample $\{y(t)\}$ of duration $T = N\Delta t$ as the output from an infinite physical process $\{y'(t)\}$ viewed through a finite length window (Figure 5.12). The window has the shape of a "box-car" function, $w(t_n) = w_n = w(n\Delta t)$, which has unit amplitude and causes zero phase lag over the duration of the data sequence but is zero elsewhere. That is $y(t_n) = w(t_n)\cdot y'(t_n)$ where

$$w_n = 1, n = 0, ..., N - 1$$
$$w_n = 0, \text{ for } n \geq N, n < 0 \tag{5.108}$$

Because it is truncated, the data set has endpoint discontinuities, which lead to Gibbs' phenomena ("ringing") and ripple effects in the frequency domain. The DFT $Y(f)$ of the truncated series $y_n = y(n\Delta t)$ is

$$Y(f) = \sum_{n=-\infty}^{\infty} w_n y'_n e^{-i2\pi f n\Delta t} \tag{5.109}$$

In frequency space, $Y(f)$ is the convolution (written as *) of the Fourier transform of the infinite data set, $Y'(f)$, with the Fourier transform $W(f)$ of the function $w(t)$. That is

$$Y(f) = \int_{-\infty}^{\infty} Y'\left(f'\right) W'\left(f - f'\right) df'$$

$$\equiv Y'(f) * W(f) \tag{5.110}$$

where for a box-car function

$$W(f) = T \exp(i\pi fT) \frac{\sin(\pi fN\Delta t)}{(\pi fN\Delta t)}$$

$$\equiv T \exp(i\pi fT) \sin c(\pi fN\Delta t) \tag{5.111}$$

and $\sin c(x) \equiv \sin(x)/x$. It is the large side-lobes or ripples of the $\sin c$ function (Figure 5.26) that are responsible for the leakage of spectral power from the main frequency components into neighboring frequency bands. In particular, $Y(f)$ for a specific frequency $f = f_o$ is spread to other frequencies according to the phase and amplitude weighting of the window function. Leakage has the effect of both reducing the spectral power in the central frequency component and contaminating it with spectral energy from immediately adjacent frequency bands. Those familiar with the various mathematical forms for the Dirac delta function, $\delta(f)$, will recognize the formulation

$$\delta(f) = \lim_{f \to 0} \left[\frac{\sin(\pi f\Delta t)}{\pi f\Delta t} \right] = \lim_{f \to 0} [\sin c(\pi f\Delta t)]$$

Thus, as the frequency resolution increases (i.e., $f \to 0$), $Y(f) \to Y'(f)$ in (5.110).

In addition to distorting the spectrum, the box-car window limits the frequency resolution of the periodogram, independently of the data. The convolution $Y'(f)*W(f)$ means that the narrowest spectral response of the resultant transform is confined to the main-lobe width of the window transform. For a given window, the main-lobe width (the width between the -3 dB $= 10\log(\frac{1}{2})$ levels on either side of the main lobe) determines the frequency resolution, Δf, of a particular window. For most windows, including the box-car window, this resolution is roughly the inverse of the observation time; i.e., $\Delta f \approx 1/T = 1/(N\Delta t)$.

5.6.5.2 Aliasing

Poor discretization of time series data due to limitations in the response time of the sensor, limitations in the recording and data storage rates, or through postprocessing methods, may cause *aliasing* of certain frequency components in the original waveform (Figure 5.27a). An aliased frequency is one that masquerades as another frequency. In Figure 5.27b, for example, the considerable tidal energy at diurnal and semidiurnal periods (1 and 2 cpd), which is well resolved by the hourly sampled record, gets folded back into the spectra at lower frequencies of roughly 0.065, 0.10, and 0.15 cpd (periods of 14.8, 9.6, and 7.4 days, respectively) when we use a daily ($\Delta t = 24$ h) subsampled version of the original sea-level record. The aliased signals are nowhere near the original higher frequency tidal signals. If we knew nothing about the true spectrum, and were presented only with the aliased spectrum in Figure 5.27b, we would be hard pressed to provide a physical explanation for the strong fortnightly and weather-band cycles in the sea-level time series.

As illustrated by Figure 5.27, it becomes impossible, for a specific sampling interval, to tell with certainty which frequency out of a large number of possible aliases is actually contributing to the signal variability. This leads to differences in the spectra between the continuous and discrete time series. Since we use the spectra of the discrete series to

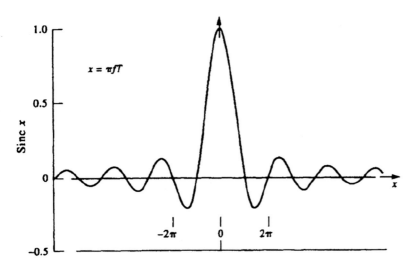

FIGURE 5.26 The function $\sin c(x) = \sin(x)/x$ showing the large side-lobes, which are responsible for leakage of spectral power from a given frequency to adjacent frequencies.

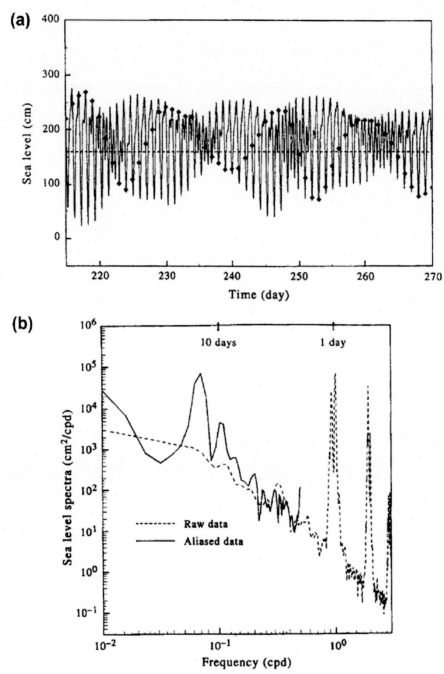

FIGURE 5.27   The origin of aliasing. (a) The solid line is the tide height recorded at Victoria, British Columbia over a 60-day period from July 29 to September 27, 1975 (time in Julian days). The diamonds are the sea-level values one would obtain by only sampling once per day. (b) The power spectrum obtained from the two data series in (a). In this case, the high frequency energy (clearly apparent in the dashed curve) gets folded back into the spectrum at lower (aliased) frequencies (solid curve).

estimate the spectrum of the continuous series, the sampling interval must be properly selected to minimize the effect of the aliasing. If we know from previous analysis that there is little likelihood of significant energy at the disguised frequencies, then aliasing is not a problem. Otherwise, a degree of smoothing may be required to ensure that higher frequencies do not contaminate the lower frequencies. This smoothing must be performed prior to sampling or digitizing since aliased contributions cannot be recognized once they are present in the discrete data series.

The aliasing problem can be illustrated in a number of ways. To begin with, we note that for discrete data at equally spaced intervals, Δt, we can measure only those frequency components lying within the principal frequency ranges

$$-\omega_{Nq} \leq \omega \leq -\omega_o, \omega_o \leq \omega \leq \omega_{Nq}, \omega_{Nq} \geq 0 \qquad (5.112a)$$

$$-f_{Nq} \leq f \leq -f_o, f_o \leq f \leq f_{Nq}, f_{Nq} \geq 0 \qquad (5.112b)$$

in which $\omega_{Nq} = \pi/\Delta t$ and $f_{Nq} = 1/(2\Delta t)$ are the usual Nyquist frequencies in radians and cycles per unit time, respectively, and $\omega_o = 2\pi/T$ and $f_o = 1/T$ are corresponding fundamental frequencies for a time series of duration T. The Nyquist frequency is the highest frequency that can be extracted from a time series having a sampling rate of Δt. Clearly, if the original time series has spectral power at frequencies for which $|f| \geq f_{Nq}$, these spectral contributions are unresolved and will contaminate power associated with frequencies within the principal range (Figure 5.28). The unresolved variance becomes lumped together with other frequency components. Familiar examples of aliasing are the slow reverse rotation of stagecoach wheels in classic western movies due to the under-sampling by the frame rate of the movie camera. Even in modern TV commercials or movies, distinguishable features on moving automobile tires or wheel frames often can be seen to rotate rapidly backwards, slow to a stop, then turn forward at the correct rotation speed as the vehicle gradually comes to a stop. Automobile commercials can avoid this problem by equipping the wheels with featureless hubcaps, spokes, and tires, or by digitizing to a higher resolution.

If $\omega, f \geq 0$ are frequencies inside the principal intervals (Eqns 5.112a and 5.112b), the frequencies outside the interval, which form aliases with these frequencies, are (in sequence)

$$2\omega_{Nq} \pm \omega, 4\omega_{Nq} \pm \omega, ..., 2p\omega_{Nq} \pm \omega \qquad (5.113a)$$

$$2f_{Nq} \pm f, 4f_{Nq} \pm f, ..., 2pf_{Nq} \pm f \qquad (5.113b)$$

where p is a positive integer. These results lead to the alternate term *folding* frequency for the Nyquist frequency since spectral power outside the principal range is folded back, accordion style, into the principal interval. As illustrated by Figure 5.28, folding the power spectrum about f_{Nq} produces aliasing of frequencies $2f_{Nq} - f$ with frequencies f; folding the spectrum at $2f_{Nq}$ produces aliasing of frequencies $2f_{Nq} + f$ with frequencies $2f_{Nq} - f$, which are then folded back about f_N into frequency f, and so forth. For example, if $f_{Nq} = 5$ rad/h, the observations at 2 rad/h are aliased with spectral contributions having frequencies of 8 and 12 rad/h, 18 and 22 rad/h, and so on.

We can verify that oscillations of frequency $2p\omega_{Nq} \pm \omega$ (or $2pf_{Nq} \pm f$) are indistinguishable from frequency ω (or f) by considering the data series $x_\omega(t)$ created by the single frequency component $x_\omega(t) = \cos(\omega t)$. Using the transformation $\omega \to 2p\omega_{Nq} \pm \omega$, together with $t_n = n\Delta t$ and $\omega_{Nq} = \pi/\Delta t$, yields

$$x_\omega(t_n) = \cos\left[(2p\omega_{Nq} \pm \omega)t_n\right] = \text{Re}\left\{\exp\left[i(2p\omega_{Nq} \pm \omega)t_n\right]\right\}$$

$$= \text{Re}\left\{\exp\left[i2p\omega_{Nq}t_n\right]\exp[\pm i\omega t_n]\right\} \qquad (5.114)$$

$$= (+1)^{pn}\text{Re}[\exp(\pm i\omega t_n)] = \cos(\omega t_n) = x_\omega(t_n)$$

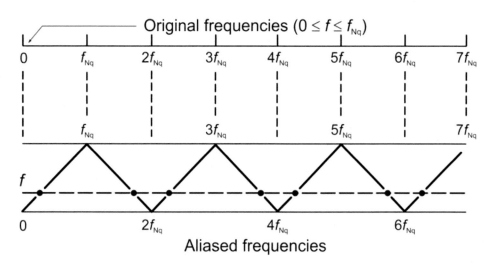

FIGURE 5.28 The spectral energies of all frequencies, $f = \omega/2\pi$, at the nodes (•) located along the dotted line are folded back to the left, accordion style, into the spectral estimate for the spectrum, $S(f)$ for the principal range, $0 \leq f \leq f_{Nq}$ $(0 \leq \omega \leq \omega_{Nq})$, of the original frequencies, as indicated by the arrow in the top panel. *Modified after Bendat and Piersol (1986).*

Original frequencies $(0 \leq f \leq f_{Nq})$

0   f_{Nq} $2f_{Nq}$ $3f_{Nq}$ $4f_{Nq}$ $5f_{Nq}$ $6f_{Nq}$ $7f_{Nq}$

f_{Nq} $3f_{Nq}$ $5f_{Nq}$ $7f_{Nq}$

f

0   $2f_{Nq}$ $4f_{Nq}$ $6f_{Nq}$

Aliased frequencies

In other words, the spectrum of $x(t)$ at frequency ω will be a superposition of spectral contributions from frequencies ω, $2p\omega_{Nq} \pm \omega$, $4p\omega_{Nq} \pm \omega$, and so forth. More specifically, it can be shown that the aliased spectrum $S_a(\omega)$ for discrete data is given by

$$S_a(\omega) = \sum_{n=-\infty}^{\infty} S(\omega + 2n\omega_{Nq}) \tag{5.115a}$$

$$= S(\omega) + \sum_{n=1}^{\infty} \left[S(2n\omega_{Nq} - \omega) + S(2n\omega_{Nq} + \omega) \right] \tag{5.115b}$$

The true spectrum, S, gives rise to the distorted spectrum, S_a, caused by the summation of overlapping copies of measured spectra in the principal interval. Only if the original record is devoid of spectral power at frequencies outside the principal frequency range will the spectrum of the observed record equal that of the actual oceanic variability. To avoid aliasing problems, one has no choice but to sample the data as frequently as justifiably possible (i.e., up to frequencies beyond which energy levels become small) or to filter the sampled data before they are recorded (as in the case of a stilling well used to eliminate gravity waves from a tidal record). A further example of spectral contamination by aliased frequencies is illustrated in Figure 5.29a and b. In Figure 5.29b, we have assumed that the wave recorder was inadvertently programmed to record at 0.13 Hz, corresponding to a limiting wave period of 7.69 s. The energy from the shorter-period waves was not measured but contaminate the energy of the longer-period waves when folded back about the Nyquist frequency.

5.6.5.3 Nyquist Frequency Sampling

Sampling time series that have significant variability at the Nyquist frequency, ω_{Nq}, affords its own set of problems. Suppose we wish to represent $y(t)$ through the usual Fourier relation

$$y(t) = \int_{-\omega_{Nq}}^{\omega_{Nq}} Y(\omega) e^{i\omega t} d\omega \tag{5.116}$$

where we have assumed that $Y(\omega) = 0$ for $|\omega| > \omega_{Nq}$. In this case, there is no aliasing problem because there is no power at frequencies greater than ω_{Nq}. The function $y(t)$ can be constructed from frequency components strictly in the interval $(-\omega_{Nq}, \omega_{Nq})$. In discrete form for infinite length data

$$y(t) = \frac{1}{2\omega_{Nq}} \sum_{n=-\infty}^{\infty} \left[y_n \int_{-\omega_{Nq}}^{\omega_{Nq}} e^{i\omega(t - n\Delta t)} d\omega \right] \tag{5.117a}$$

where the integral has the form of a sinc function such that

$$y(t) = \sum_{n=-\infty}^{\infty} \left[y_n \frac{\sin\left[\omega_{Nq}(t - n\Delta t)\right]}{\omega_{Nq}(t - n\Delta t)} \right] \tag{5.117b}$$

Given the data $\{y_n\}$, we can construct $y(t)$. However, suppose that $y(t)$ fluctuates with the Nyquist frequency ω_{Nq} such that

$$y(t) = y_o \cos(\omega_{Nq} t + \theta) \tag{5.118}$$

where, for the sake of generality, the phase angle is arbitrary, $0 \leq \theta \leq 2\pi$. Then, using $\sin(n\pi) = 0$ for all n (an integer)

$$y_n = y(n\Delta t) = y_o \cos(n\pi + \theta) = y_o[\cos(n\pi)\cos(\theta)]$$
$$= y_o(-1)^n \cos(\theta) \tag{5.119}$$

This leads to a component with amplitude $y_n = y_o(-1)^n \cos(\theta)$, which fluctuates in sign because of the term $(-1)^n$, $-\infty \leq n \leq \infty$. If θ is unknown, the function $y(t)$ cannot be constructed. If $\theta = k\pi/2$, where k is an integer, so that $\cos(\omega_{Nq}t + \theta) = \sin(\omega_{Nq}t)$, the observer will find no signal at all. In general, $0 \leq |\cos\theta| \leq 1$ and the magnitude will always be less than y_o, resulting in biased data.

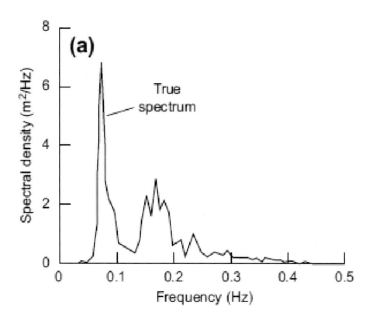

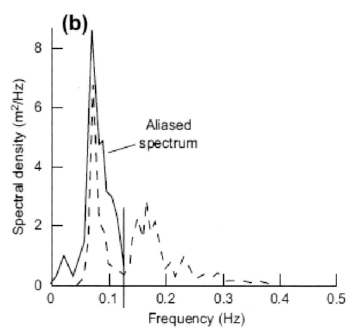

FIGURE 5.29   Example of an aliased autospectrum. (a) The true spectrum, $S(f)$ (m²/cps), of wind-generated waves as a function of frequency (Hz = cycles per second); (b) aliased spectrum, $S_a(f)$, that would arise from folding about a hypothetical Nyquist frequency, $f_{Nq} = 0.13$ Hz.

According to the above analysis, we should sample slightly more frequently than Δt if we are to fully resolve oscillations at the maximum frequency of interest (assumed to be the Nyquist frequency). A sampling rate of 2.5 samples per cycle of the frequency of interest appears to be acceptable whereby $\Delta t = 1/(2.5 f_{Nq}) = (2/5) (1/f_{Nq}) = (4/5)\pi/\omega_{Nq}$.

5.6.5.4 Frequency Resolution

The need to resolve spectral estimates in neighboring frequency bands is an important requirement of time series analysis. Without sufficient resolution, it is not possible to determine whether a given spectral peak is associated with a single frequency, or is a smeared response containing a number of separate spectral peaks. A good example of this for tides is presented by Munk and Cartwright (1966), who show that for long records, the main constituents in the diurnal and semidiurnal frequency bands can be resolved into a multitude of other tidal frequencies (see also Figure 5.8). How well the peaks can be resolved depends on the frequency differences, Δf, between the peaks and the length, T, of the data set used in the analysis. For an unsmoothed periodogram, the frequency resolution in hertz is roughly the reciprocal of the time duration in seconds of the data.

The distinction between well-resolved and poorly resolved spectral estimates is somewhat subjective and depends on how we wish to define "resolution." As illustrated by diffraction patterns in classical optics, we can follow the "Rayleigh criterion" for the separation of spectral peaks (Jenkins and White, 1957). Recall that the diffraction pattern for a given frequency, f, of light varies as $\mathrm{sinc}(\phi) = \sin(\phi - \phi_f)/(\phi - \phi_f)$, where ϕ is the angle of the incident light beam to the grating and ϕ_f is the angle for light of frequency f. This also is the functional form for the spectral peak of a truncated time series (see *windowing* in the next section). Two spectral lines are said to be "well resolved" if the separation between peaks exceeds the difference in frequency between the center frequency to the maximum at the first side-lobe and "just resolved" if the spectral peak of one pattern coincides with the first zero of the second pattern (Figure 5.30a–c). Here, the separation in frequency is equal to the difference in frequency between the peak of one spectrum and the first zero of the function $\sin(\phi)/\phi$ of the second (where $\phi = \omega T/2$). The spectral peaks are "not resolved" if this separation is less than that between the center frequency and the first zero of the $\sin(\phi)/\phi$ functions (Figure 5.30d).

Consider an oceanic record consisting of two sinusoidal components, both having amplitude y_o and constant phase lags such that

$$y(t) = y_o[\cos(\omega_1 t + \theta_1) + \cos(\omega_2 t + \theta_2)], \; -T/2 \le t \le T/2 \tag{5.120}$$

where as usual $\omega = 2\pi f$. The one-sided, unsmoothed PSD, $S(\omega)$, for these data is then found from the Fourier transform

$$S(\omega) = \frac{1}{2} T y_o^2 \left\{ \frac{\sin\left[\frac{1}{2}T(\omega - \omega_1)\right]}{\left[\frac{1}{2}T(\omega - \omega_1)\right]} + \frac{\sin\left[\frac{1}{2}T(\omega - \omega_2)\right]}{\left[\frac{1}{2}T(\omega - \omega_2)\right]} \right\}$$

The power spectrum consists of two terms of the form $\sin(\phi)/\phi$ centered at frequencies ω_1 and ω_2. Using the Rayleigh criterion, we can just resolve the two peaks (i.e., determine if there is one or two sinusoids contributing to the spectrum) provided that the frequency separation $\Delta\omega = |\omega_1 - \omega_2|$ ($\Delta f = |f_1 - f_2|$) is equal to the frequency difference for the peak of one frequency and the first zero of $\mathrm{sinc}(\phi) \equiv \sin(\phi)/\phi$ for the other frequency. Since zeroes of $\sin(\phi)/\phi$ occur at frequencies f equal to $\pm 1/T, \pm 2/T, \ldots, \pm p/T$, the frequencies are just resolved when

$$\Delta\omega = \frac{2\pi}{T}; \Delta f = \frac{1}{T} \tag{5.121a}$$

and well resolved for

$$\Delta\omega > \frac{3\pi}{T}; \Delta f > \frac{3}{2T} \tag{5.121b}$$

FIGURE 5.30 Resolution of spectral lines. (a, b) Well resolved; (c) just resolved; and (d) not resolved. *Based on Jenkins and White (1957).*

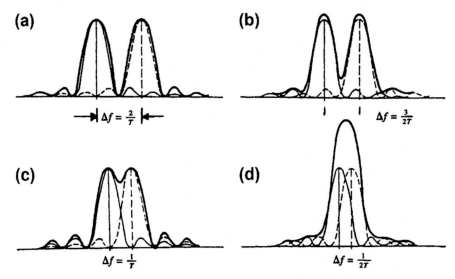

In summary, resolution of two frequencies f_k and $f_{k+1}(= f_k \pm \Delta f)$ using an unsmoothed periodogram or equivalently a rectangular window, requires a record of length T, where $\Delta f = 1/T$ frequency units. Note also that $1/T$ is equal to the fundamental frequency, f_1, which is the lowest frequency that we can calculate for the record. For some nonrectangular windows, the length of the data set must be increased to about $2T = 2/\Delta f$ to achieve the same frequency separation.

In a related study, Munk and Hasselmann (1964) discuss the "super-resolution" of tidal frequency variability. The fact that time series of tidal heights vary at precise frequencies and have relatively large SNRs suggests that the traditional requirement (that a minimum record length T is required to separate tidal constituents separated by frequency difference $\Delta f = 1/T$) is "grossly incomplete." The modified resolvable frequency difference is

$$\Delta f = \frac{1}{rT}; \Delta \omega = \frac{2\pi}{rT} \tag{5.122}$$

in which $r \equiv \sqrt{\text{signal level}/\text{noise level}}$. On this basis, the Rayleigh criterion must be considered a conservative measure of the resolution requirement for deterministic processes.

5.6.6 Smoothing Spectral Estimates (Windowing)

The need for statistical reliability of spectral estimates brings us to the topic of spectral averaging or smoothing. As we have seen, DFTs (Discrete Fourier Transforms) provide an elegant method for decomposing a data sequence into a set of discrete spectral estimates. For a data sequence of N values, the periodogram estimate of the spectrum can have a maximum of $N/2$ Fourier components. If we use all $N/2$ components to generate the periodogram, there are only two DoF per spectral estimate, corresponding to the coefficients A_n, B_n of the sine and cosine functions for each Fourier component (see Sections 5.6.3.1 and 5.6.3.5) or, alternatively, to the magnitude and phase of each Fourier component (see Section 5.6.3.3). Based on the assumption that data are drawn from a normally distributed random sample, we can define the confidence limits for the spectrum in terms of a chi-squared distribution, χ_n^2, where for n DoF

$$E\left[\chi_n^2\right] = \mu = n, E\left[\chi_n^2 - \mu^2\right] = \sigma^2 = 2n \tag{5.123}$$

Substituting $n = 2$ into these expressions, we find that the standard deviation, σ, is equal to the mean, μ, of the estimate, indicating that results based on two DoF are not statistically reliable. It is for this reason that some sort of ensemble averaging or smoothing of spectral estimates is required. The smoothing can be (1) applied directly to the time series through convolution with a sliding averaging function or by (2) averaging adjacent spectral estimates. A one-shot smoothing applied to the entire data record marginally increases the number of DoF per spectral estimate. In most practical applications, the full time series is broken into a series of short overlapping segments and the spectrum for each segment is determined. The analyst then ensemble averages the spectra from each segment to increase the number of DoF per spectral estimate. Because the spectra of the individual segments, as well as that of the ensemble averaged spectra, may be of interest, smoothing can also be applied to each of the overlapping segments prior to averaging the spectra of the individual segments. The greater the smoothing, the greater the number of DoF per spectral band, the narrower the confidence limits, and the greater the reliability of any observed spectral peaks. The trade-off is a longer processing time and a loss of spectral resolution that can remove smaller peaks that may or may not be indicative of real processes (see Figure 5.31).

A window is a smoothing function applied to finite observations or their Fourier transforms to minimize "leakage" in the spectral domain and to increase the statistical reliability of the spectral estimates. (The tradeoff is that the spectral peaks are broadened, making them less well defined). Convolution in the time domain and multiplication in the frequency domain are adjoint Fourier functions (see Appendix H regarding convolution). A practical window is one which allows little of the energy in the main spectral lobe to leak into the side-lobes, where it can obscure and distort other spectral estimates that are present. In fact, weak signal spectral responses can be masked by higher side-lobes from stronger spectral responses. Skillful selection of tapered data windows can reduce the side-lobe leakage, although always at the expense of reduced resolution. Thus, we want a window that minimizes the side-lobes and maximizes (concentrates) the energy near the frequency of interest in the main lobe. These two performance requirements are rather troublesome when analyzing short data records. Short data occur in practice because many measured processes are event-like (of short duration) or have slowly time-varying spectra that may be considered constant over only short record segments. The window is applied to data to reduce the order of the discontinuity at the boundary of the periodic extension since few harmonics will fit exactly into the length of the time series.

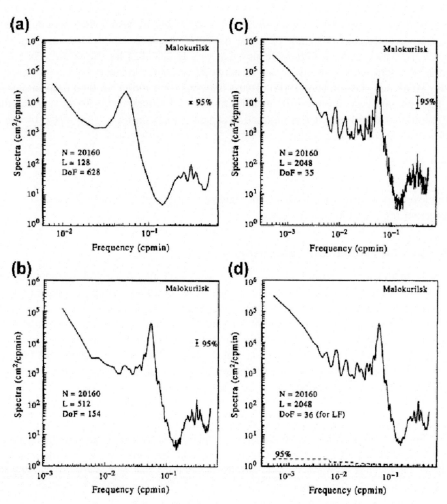

FIGURE 5.31 Spectra of sea-level oscillations recorded by a bottom-pressure gauge in Malokurilsk Bay on the west coast of Shikotan Island, Russia. Time series length, $T = N\Delta t$, where $N = 20,160$ and $\Delta t = 1$ min. Segment lengths are $L = M\Delta t$, $M \ll N$. Each time series segment has been smoothed with a Kaiser-Bessel window (see Section 5.6.6.4) with 50% overlap between segments. Block averaging has been used to smooth the spectral estimates. (a) Highly smoothed spectrum with $M = 128$ (2^7), degrees of freedom (DoF) = 628; (b) moderately smoothed spectrum with $M = 512$ (2^9), DoF = 154; (c) weakly smoothed spectrum with $M = 2048$ (2^{11}), DoF = 36; (d) same as (c) except that DoF = 36 applies to the lowest frequency range only. For $f \geq 6 \times 10^{-2}$ cycles/min, the number of spectral estimates averaged together increases as 3×36, 5×36, and 7×36, for each of the next three frequency ranges. *Courtesy of Alexander Rabinovich, Institute of Ocean Sciences and P.P. Shirshov Institute of Oceanology.*

Signals with frequencies other than those of the basis set are not periodic in the observation window. The periodic extension of a signal, not commensurate with its natural period, exhibits discontinuities at the boundaries of the observational period. Such discontinuities are responsible for spectral contributions or leakage over the entire basis set. In the time domain, the windows are applied to the data as a multiplicative weighting (*convolution*) to reduce the order of the discontinuities at the boundary of the periodic extensions. The windowed data are brought to zero smoothly at the boundaries so that the periodic extensions of the data are continuous in many orders of the derivatives. The value of $Y(f)$ at a particular frequency f, say f_o, is the sum of all the spectral contributions at each f weighted by the window centered at f_o and measured at f

$$Y(f) = Y'(f) * W(f) \tag{5.124}$$

where the asterisk (*) denotes convolution.

There exist a multitude of data windows or tapers with different shapes and characteristics ranging from the rectangular (box-car) window discussed in the previous section, to the classic Hanning and Hamming windows, to more sophisticated windows such as the Dolph–Chebyshev window. The type of window used for a given application depends on the required degree of side-lobe suppression, the allowable widening of the central lobe, and the amount of computing one is willing to endure. We will briefly discuss several of the conventional windows plus the Kaiser-Bessel window recommended by Harris (1978). Additional details on the Kaiser-Bessel window and filter are provided in Section 6.6.6.4.

5.6.6.1 Desired Window Qualities

Windows affect the attributes of a given spectral analysis method, including its ability to detect and resolve periodic waveforms, its dynamic range, confidence intervals, and ease of implementation. Spectral estimates are affected not only by the broadband noise spectrum of the data but also by narrow-band signals that fall within the bandwidth of the window. Leakage of spectral power from a narrow-band spectral component, f_o, to another frequency component, f_a, produces a bias in the amplitude and position of a spectral estimate. This bias is especially disruptive for the detection of weak signals in the presence of nearby strong signals. To reduce the bias, we need a "good" window. Although there are no universal standards for a good window, we would like it to possess the following characteristics in Fourier transform space:

1. The central main lobe of the window (which is centered on the frequency of interest) should be as narrow as possible to improve the frequency resolution of adjacent spectral peaks in the data set, and the first side-lobes should be greatly attenuated relative to the main lobe to avoid contamination from other frequency components. Here, the narrowness of the central lobe is measured by the positions of the -3 dB (half amplitude points, $= 10\log\frac{1}{2}$) on either side of the lobe. Retention of a narrow central lobe, while suppressing the side-lobes, is not as easy as it sounds since suppression of the side-lobes invariably leads to a broadening of the central lobe;
2. The window should suppress the amplitudes of side-lobes at frequencies far removed from the central lobe. That is, the side-lobes should have a rapid asymptotic fall-off rate with frequency so that there is relatively little spreading of energy into the spectral estimate at the central lobe (i.e., into the frequency of interest);
3. The coefficients of the window should be easy to generate for multiplication in the time domain and convolution in the Fourier transform domain.

A good performance indicator (PI) for the time domain window $w(t)$ can be defined as the difference between the equivalent noise bandwidth (ENBW) and the bandwidth (BW), located between the -3 dB levels of the central lobe (Harris, 1978), as

$$\text{PI} = \frac{\text{ENBW} - \text{BW}}{\text{BW}} = \frac{\dfrac{1}{\text{BW}}\sum_n w^2(n\Delta t)}{\left[\sum_n w(n\Delta t)\right]^2} - 1 \tag{5.125}$$

where we have normalized by the BW. The lower the value, the better the performance of the filter; windows that perform well have values for this ratio ($\times$ 100%) of between 4.0% and 5.5%. A summary of the figures of merit for several well-known windows is presented in Table 5.12. PI values are obtained using columns four and five. For example, for the weakly performing box-car (rectangular) window, PI $= 0.124$ (12.4%), while for the strongly performing Kaiser window, PI $= 0.049$ (4.9%). The choice of window can be daunting; Harris lists more than 44 windows for smoothing spectral estimates.

5.6.6.2 Rectangular (Box-Car) and Triangular Windows

As discussed at the beginning of this section, a rectangular window has an amplitude of unity throughout the observation interval of duration $T = N\Delta t$, with the weighting given by

$$w(n\Delta t) = 1, n = 0, 1, ..., N - 1 \ (\text{or}, -N/2 \le n \le N/2)$$

$$= 0, \text{elsewhere} \tag{5.126}$$

(top panel of Figure 5.32). Using the relation $\omega T = N\theta$, where $\theta = \omega\Delta t$ and $T = N\Delta t$, the spectral window obtained from the DFT is

$$W(\theta) = T \exp[-i(N-1)\theta / 2]\frac{\sin(N\theta/2)}{(N\theta/2)} \tag{5.127a}$$

TABLE 5.12 Windows, figures of merit and performance indicator (PI).

Window	Highest side-lobe level (dB)	Side-lobe attenuation (dB/octave)	ENBW (bins)	3 dB BW (bins)	PI	Overlap correlation 75%	Overlap correlation 50%
Rectangle	−13	−6	1.00	0.89	0.124	0.750	0.500
Triangle	−27	−12	1.33	1.28	0.031	0.719	0.250
Hanning	−32	−18	1.50	1.44	0.042	0.659	0.167
Hamming	−43	−6	1.36	1.30	0.046	0.707	0.235
Parzen	−21	−12	1.20	1.16	0.035	0.765	0.344
Tukey $\alpha = 0.5$	−15	−18	1.22	1.15	0.061	0.727	0.364
Kaiser-Bessel $\alpha = 2.0$	−46	−6	1.50	1.43	0.049	0.657	0.169
$\alpha = 2.5$	−57	−6	1.65	1.57	0.051	0.595	0.112
$\alpha = 3.0$	−69	−6	1.80	1.71	0.052	0.539	0.074
$\alpha = 3.5$	−82	−6	1.93	1.83	0.054	0.488	0.048

The last column gives the correlation between adjacent data segments for the specified percentage segment overlap. For completeness, we include the Tukey and Parzen Windows. Squared attenuation is in dB $= 20\log(x)$, where $x =$ (sidelobe amplitude/central peak amplitude). Adapted from Harris (1978).

FIGURE 5.32 A box-car (rectangular) window for $N = 41$ weights. *Top panel*: Weights, $w(n) = 1.0$ in the time domain ($-20 \leq n \leq 20$). *Bottom panel*: Fourier transform of the weights, $W(\theta)$, plotted as $20\log|W(\theta)|$, where $\theta = \omega\Delta t/N = 40\pi/N$ is the frequency span of the window.

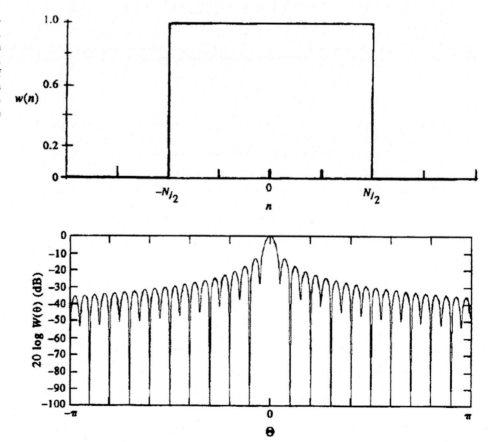

$$|W(\theta)|^2 = T^2 \left[\frac{\sin(N\theta/2)}{(N\theta/2)}\right]^2 \tag{5.127b}$$

(bottom panel of Figure 5.32). The exponential term in Eqn (5.127a) gives the phase shift of the window as a function of the frequency $\omega = \theta/\Delta t$. As indicated by Figure 5.31, the function W, known as the Dirichlet kernel, has strong side-lobes, with the power of the first side-lobe down by only -13 dB (a factor of 0.22) from the main lobe. The remaining side-lobes fall off weakly at -6 dB per octave, which is the functional rate for a discontinuity (an "octave" corresponds to a factor of 2 change in frequency). Zeros of $W(\theta)$ occur at integer multiples of the frequency resolution, $f_1 = 1/T$, for which $N\theta/2 = \omega T/2 = \pm p\pi$. That is, where $f = \pm p/T$ ($\pm 1/T$, $\pm 2/T$, ...).

The triangular (*Bartlett*) window

$$w(n\Delta t) = \begin{cases} = \dfrac{n}{(N/2)}, & n = 0, 1, ..., N/2 \\[2mm] = \dfrac{N-n}{(N/2)}, & n = N/2, ..., N-1 \\[2mm] = \dfrac{N/2 - |n|}{(N/2)}, & 0 \leq |n| \leq N/2 \end{cases} \tag{5.128}$$

is shown in the top panel of Figure 5.32, with the DFT

$$W(\theta) = \frac{2T}{N} \exp[-i(N-1)\theta/2] \left[\frac{\sin(N\theta/2)}{(N\theta/2)}\right]^2 \tag{5.129a}$$

$$|W(\theta)|^2 = \frac{4T^2}{N^2} \left[\frac{\sin(N\theta/2)}{(N\theta/2)}\right]^4 \tag{5.129b}$$

plotted in bottom panel of Figure 5.33, which we recognize as the square of the sinc function for the rectangular window. The main lobe between zero crossings has twice the width of the rectangular window but the level of the first side-lobe is down by -26 dB, twice that of the rectangular window. Despite the improvement over the box-car window, the side-lobes of the triangular window are still extensive and use of this window is not recommended if other windows are available.

The *Parzen* window

$$w(n\Delta t) = 1 - [n/(N/2)]^2, \quad 0 \leq |n| \leq N/2 \tag{5.130}$$

is a squared version of the Bartlett window. This is the simplest of the continuous polynomial windows, with side-lobes that fall off with frequency as $1/\omega^2$. The first side-lobes are down by -22 dB.

5.6.6.3 Hanning and Hamming Windows (50% Overlap)

The Hann window, or *Hanning window* as it is most commonly known, is named after the Austrian meteorologist Julius von Hann and is part of a family of trigonometric windows having the generic form $\cos^\alpha(n)$, where the exponent, α, is typically an integer from 1 through 4. The case $\alpha = 1$ leads to the *Tukey* (or *cosine-tapered*) *window* (Harris, 1978). As α becomes larger, the window becomes smoother, the side-lobes fall off faster, and the main lobe widens. The Hanning window ($\alpha = 2$), also known as the *raised cosine* and *sine-squared* window, is defined in the time domain as

$$w(n\Delta t) = \sin^2\left(\frac{\pi n}{N}\right) = \frac{1}{2}[1 - \cos(2\pi n/N)], \quad n = 0, 1, ..., N-1 \tag{5.131a}$$

$$w(n\Delta t) = \sin^2[\pi(n+N/2)/N] = \frac{1}{2}[1 - \cos[2\pi(n+N/2)/N]], \quad n = -N/2, ..., N/2 \tag{5.131b}$$

(top panel of Figure 5.34), which is a continuous function with a continuous first derivative. The DFT of this weighting function is

$$W(\theta) = \frac{1}{2}D(\theta) + \frac{1}{4}[D(\theta - \theta_1) + D(\theta + \theta_1)] \tag{5.132}$$

FIGURE 5.33 The triangular (Bartlett) window for $N = 51$ weights. *Top panel*: Weights, $w(n)$ in the time domain ($-20 \leq n \leq 20$). *Bottom panel*: Fourier transform of the weights, $W(\theta)$, plotted as $20\log|W(\theta)|$ (cf. Figure 5.32).

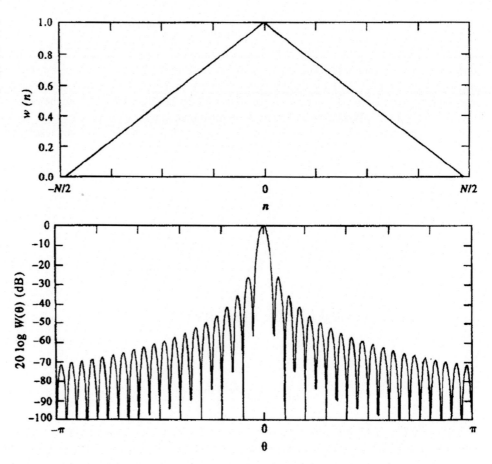

(center panel of Figure 5.34), where $\theta_1 = 2\pi/N$ and

$$D(\theta) = Te^{i\theta/2}\left(\frac{\sin(N\theta/2)}{N\theta/2}\right) \tag{5.133}$$

is the standard function (*Dirichlet kernel*) obtained for the rectangular and triangular windows. Thus, the Hanning window consists of the summation of three sinc functions (bottom panel of Figure 5.34), one centered at the origin, $\theta = 0$, plus two other translated Dirichlet kernels having half the amplitude of the main kernel and offset by $\theta = \pm 2\pi/N$ from the central lobe. There are several important features of the window response $W(\theta)$. First of all, the functions, D, are discrete and defined only at points that are multiples of $2\pi/N$, which also correspond to the zero crossings of the central function, $D(\theta)$. Secondly, for all the zero crossings except those at $\theta_{\pm 1} = \pm 2\pi/N$, the translated functions also have zero crossings at multiples of $2\pi/N$. As a result, only values at $-2\pi/N$, 0, and $+2\pi/N$ contribute to the window response. It is the widening of the main lobes of the translated functions that causes them to be nonzero at the first zero crossings of the central function. Lastly, because the translated functions are out of phase with the central function, they tend to cancel the side-lobe structure. The first side-lobe of the Hanning window is down by -32 dB (factor of 0.025) from the main lobe. The remaining side-lobes diminish as $1/\omega^3$, or at about -18 dB per octave.

An attractive aspect of the Hanning window is that smoothing in the frequency domain can be accomplished using only three convolution terms corresponding to θ_o, $\theta_{\pm 1}$. The Hanning-windowed Fourier transform, $Y_H(f_k)$, representing the spectrum for the frequency, f_k, is then obtained from the raw spectrum Y for the frequencies, f_k and the two adjoining frequencies, f_{k-1} and f_{k+1}; that is

$$Y_H(f_k) = \frac{1}{2}\left\{Y(f_k) - \frac{1}{2}\left[Y(f_{k-1}) + Y(f_{k+1})\right]\right\} \tag{5.134}$$

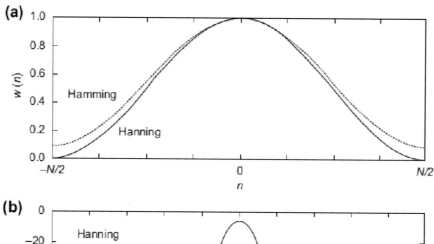

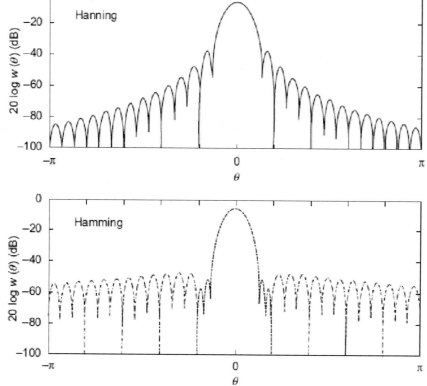

FIGURE 5.34 The Hanning and Hamming windows for $N = 41$ weights. *Top panel*: Weights, $w(n)$, $(-20 \leq n \leq 20)$; *Center panel*: Fourier transform of the weights, $W(\theta)$, of the Hanning window plotted as $20\log|W(\theta)|$ (cf. Figure 5.32); *Bottom panel*: As with the center panel but for the Hamming window. The response functions have not been re-scaled.

The transform $Y(f_k)$ has already been rectangular-windowed by the very act of collecting the data but is "raw" in the sense that no additional smoothing has been applied. Other processing advantages of the Hanning window are discussed in Harris (1978). Because the squares of the weighting terms are $(1/2)^2 + (1/4)^2 + (1/4)^2 = 3/8$, the total energy will be reduced following the application of the Hanning window. To compensate, the amplitudes of the Fourier transforms, $Y_H(f)$ should be multiplied by $\sqrt{8/3}$ prior to computation of the spectrum. Specifically

$$Y_H(f_k) = \Delta t (8/3)^{1/2} \sum_{n=0}^{N-1} y_n [1 - \cos(2\pi n / N)] e^{-i2\pi kn/N} \qquad (5.135)$$

where $f_k = k/(N\Delta t)$.

The *Hamming window* is a variation on the Hanning window designed to cancel the first side-lobes. To accomplish this, the relative sizes of the three Dirichlet kernels are adjusted through a parameter, γ, where

$$w(n\Delta t) = \gamma + (1 - \gamma)[\cos(2\pi n / N)], n = -N/2, ..., N/2 \qquad (5.136a)$$

$$W(\theta) = \gamma D(\theta) + \frac{1}{2}(1 - \gamma)[D(\theta - 2\pi / N) + D(\theta + 2\pi / N)] \tag{5.136b}$$

Perfect cancellation of the first side-lobes (located at $\theta_1 = 2.5\pi/N$) occurs when $\gamma = 25/46 \approx 0.543478$. Taking $\gamma = 0.54$ leads to near-perfect cancellation at $\theta_1 = 2.6\pi/N$ and a marked improvement in side-lobe level. The Hamming window is defined as

$$w(n\Delta t) = 0.54 + 0.46\cos(2\pi n / N), n = -N/2, ..., N/2 \tag{5.137}$$

and has a spectral distribution similar to that of the Hanning window with more "efficient" side-lobe attenuation. The highest side-lobe levels of the Hanning window occur at the first side-lobes and are down by -32 dB from the main lobe. For the Hamming window, the first side-lobe is highly attenuated and the highest side-lobe level (the third side-lobe) is down by -43 dB. To compensate for the filter attenuation, the amplitudes of the Fourier transform $Y_{\text{Ham}}(f)$ should be multiplied by $\sqrt{5/2}$ prior to computation of the spectrum. On a similar note, anyone using any of the windows in this section to calculate running mean time series should make sure each estimated value is divided by the sum of the weights used, $\sum_N (w_n)$.

5.6.6.4 Kaiser-Bessel Window

Harris (1978) identifies the Kaiser-Bessel window as the "top performer" among the many different types of windows he considered. Among other factors, the coefficients of the window are easy to generate and the filter has a high equivalent noise bandwidth (ENBW), one of the criteria used to separate good and bad windows. The trade-off is an increased main-lobe width for reduced side-lobe levels. In the time domain, the filter is defined in terms of a zeroth-order modified Bessel function of the first kind, I_o,

$$w(n\Delta t) = \frac{I_o(\pi\alpha\Omega)}{I_o(\pi\alpha)}, 0 \le |n| \le N/2 \tag{5.138}$$

where the argument $\Omega = [1 - (2n/N)^2]^{1/2}$ and for any variable, x,

$$I_o(x) = \sum_{k=0}^{\infty} \left(\frac{(x/2)^k}{k!} \right)^2 \tag{5.139}$$

The parameter $\pi\alpha$ is half of the time-bandwidth product, with α typically having values 2.0, 2.5, 3.0, and 3.5. The transform is approximated by (5.140)

$$W(\theta) \approx [N / I_o(\pi\alpha)] \frac{\sinh\left\{ [\pi^2\alpha^2 - (N\theta/2)^2]^{1/2} \right\}}{\left\{ [\pi^2\alpha^2 - (N\theta/2)^2]^{1/2} \right\}} \tag{5.140}$$

Plots of the weighting function w and the DFT for W are presented in Figure 5.35 for two values of the parameter α (= 2.0, 3.0). The modified Bessel function I_o is defined as follows:

For $|x| \le 3.75$

$$I_o(x) = 1.0 + 3.5156229\,Z + 3.0899424\,Z^2 + 1.2067492\,Z^3 + 2.659732 \times 10^{-1}Z^4$$
$$+ 3.60768 \times 10^{-2}Z^5 + 4.5813 \times 10^{-3}Z^6 \tag{5.141a}$$

where, for real x

$$Z = (x/3.75)^2 \tag{5.141b}$$

For $|x| > 3.75$, we expand the terms to obtain

$$I_o(x) = \left(\exp(x)/|x|^{\frac{1}{2}} \right) \{ 3.9894228 \times 10^{-1} + 1.328592 \times 10^{-2}Z$$

$$+ 2.25319 \times 10^{-3}Z^2 - 1.57565 \times 10^{-3}Z^3 + 9.16281 \times 10^{-3}Z^4 - 2.057706 \times 10^{-2}Z^5$$

$$+ 2.635537 \times 10^{-2}Z^6 - 1.647633 \times 10^{-2}Z^7 + 3.92377 \times 10^{-3}Z^8 \} \tag{5.141c}$$

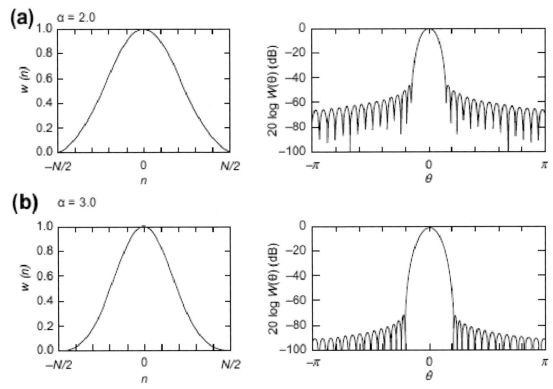

FIGURE 5.35 The Kaiser-Bessel window for $N = 41$ weights and $\alpha = 2.0$ and 3.0. (a) Weights, $w(n)$, $(-20 \leq n \leq 20)$. (b) Fourier transform of the weights, $W(\theta)$, plotted as $20\log|W(\theta)|$ (cf. Figure 5.32). *From Harris (1978).*

where

$$Z = 3.75 / |x| \tag{5.141d}$$

The utility of the Kaiser-Bessel window is nicely illustrated by Figure 5.36. Here, we compare the average spectra (in cm^2/cpd) obtained from a year-long record of hourly coastal sea level following application of a rectangular window (the worst possible window), a Kaiser-Bessel window (the best possible window) and a triangular Bartlett window (nearly as good as the Kaiser-Bessel window) to a series of overlapping data segments. In each case, the window length is 42.7 days and there is 50% overlap between adjoining time series segments. The overlaps give $K = 32$ DoF per spectral estimate for the Kaiser-Bessel and Bartlett windows (corresponding to roughly 16 statistically separate spectral estimates based on 50% data overlaps) but only 16 DoF for the rectangular window for which zero overlap is needed for statistical independence between adjoining data segments. All three windows preserve the strong spectral peaks within the tidal frequency bands centered at 1, 2, and 3 cpd. However, unlike the rectangular window, application of the Kaiser-Bessel and Bartlett windows results in little energy leakage from the tidal bands to adjacent frequency bands. The high spectral levels at periods shorter than about 2 days ($f > 0.5$ cpd) in the nontidal portion of the rectangular-windowed spectrum is an artifact of the window. The slightly better ability of the rectangular window to resolve frequency components within the various tidal bands is outweighed by the high contamination of the spectrum at nontidal frequencies.

5.6.7 Smoothing Spectra in the Frequency Domain

As we noted earlier, each spectral estimate for a random process is a chi-squared function with only two DoF. Because of this minimal number of DoF, some sort of smoothing or filtering is needed to increase the statistical significance of a given spectral estimate. The windowing approach described in the previous section, in which we partitioned the time series into a series of shorter overlapping segments, is one of a number of computational methods used to smooth (average) spectral estimates.

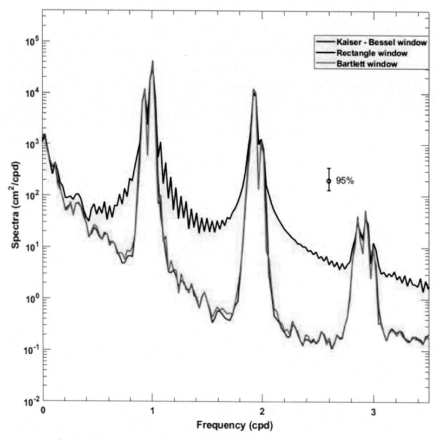

FIGURE 5.36   Spectral power (cm²/cpd) of the hourly coastal sea-level height recorded at Victoria, British Columbia following windowing (number of hourly samples, $N = 8750$); log power versus linear frequency. *Blue line*: application of a rectangular window; *Orange and Green lines*: applications of a Kaiser-Bessel (KB) window (with $\alpha = 3$) and a Bartlett window, respectively. All three windows have a length of 1,024 h ($= 42.67$ days). There are 32 degrees of freedom (DoF) for the Bartlett and KB windows and 16 DoF for the rectangular window, based on a total of $16 \times 50\%$ overlapping data segments (for the rectangular window, adjoining 50% segments are not statistically independent). The tidal peak centered at 3 cpd results from nonlinear interactions between the semidiurnal and diurnal frequency bands. The vertical line is the 95% level of confidence, which appears compact because of the vertical log scale. *Courtesy: Isaac Fine, Institute of Ocean Sciences.*

5.6.7.1 Band Averaging

For a time series consisting of N data points, one of the simplest forms of smoothing is to use the discrete Fourier transform (DFT) or the fast Fourier transform (FFT) to calculate individual spectral estimates for the maximum number of frequency bands ($N/2$) and then average together adjacent spectral estimates. The resultant spectral estimate is assigned to the midpoint of the average. Thus, we could average bands 1, 2, and 3, to form a single spectral estimate centered at band 2, then bands 4, 5, and 6 to form an estimate centered at band 5, and so on. It is often useful in this type of *frequency band averaging* to use an odd-numbered smoother so that the center point is easily defined. In particular, if we were to average groups of three adjacent (and different) bands to form each estimate, the number of DoF per estimate would increase from two to six. In the case of the Blackman–Tukey autocovariance method, the equivalent procedure would be to use larger lag steps in the computation of the autocovariance function before its transform is taken. This is functionally equivalent to smoothing by averaging together the individual spectral estimates.'

5.6.7.2 Block Averaging

As we remarked previously, a common smoothing technique is to segment the time series (of length N) into a series of shorter, equal-length segments of length N_s (where $N_s = N/K$, and K is a positive integer). Spectra are then computed for each of the K segments and the spectral values for each frequency band then *block averaged* to form the final spectral estimates for each frequency band. If there is no overlap between segments, the resulting DoF for the spectral estimates of the composite spectrum will be $2K$. This assumes that the individual sample spectra have not been windowed and that each spectral estimate is a chi-squared variable with two DoF. Since the frequency resolution of a time series is inversely proportional to its length, the major difficulty with this approach is that the shorter time series have fewer spectral values than the original record over the same Nyquist frequency range. In other words, the maximum resolvable frequency $f_{Nq} = 1/(2\Delta t)$ remains the same since Δt is unchanged, but the frequency spacing between adjacent spectral estimates is increased for the short segments because of the reduced record lengths. In addition, the fundamental frequency is higher due to the decreased length of the segments.

However, by not overlapping adjacent segments, we could be overly conservative in our estimate of the number of degrees of freedom (DoF). For that reason, most analysts overlap adjacent segments by 30–50% so that more uniform weighting is given to individual data points. The need for overlapping segments is necessary when a window is applied to each individual segment prior to calculation of the spectra. The effect of the window is to reduce the effective length of each segment in the time domain so that, for some sharply defined windows such as the Kaiser-Bessel window, even adjoining segments with 50% overlap can be considered independent time series for spectral analysis. As in Figure 5.36, the DoF of the periodograms averaged together is $4K$, rather than $2K$ for the nonoverlapping segments (except for the rectangular window for which it is still $2K$). Consideration must be given to the correlation among individual estimates (the greater the overlap the higher the correlation). Nuttall and Carter (1980) report that 92% of the maximum number of equivalent degrees of freedom (EDoF) can be achieved for a Hanning window, which uses 50% overlap. Clearly, we must sacrifice something to gain improved statistical reliability. That "something" is a loss of frequency resolution due to the broad central lobe that accompanies windows with negligible side-lobes.

As an example, consider the spectrum of a 1-min sampled time series $y(t) = A\cos(2\pi f t) + \varepsilon(t)$ of length 512 min composed of Gaussian white noise $\varepsilon(t)$ ($|\varepsilon| \leq 1$) and a single cosine component of amplitude, A, and frequency $f = 0.23$ cpmin (period $T = 1/f = 4.3$ min). The magnitude of the deterministic component, A, is five times the standard deviation of the white-noise signal and the expected variance, $V[\varepsilon] = \left(\frac{1}{\sqrt{2}}\right)$ cm^2. The raw periodogram (Figure 5.37a) reveals a large

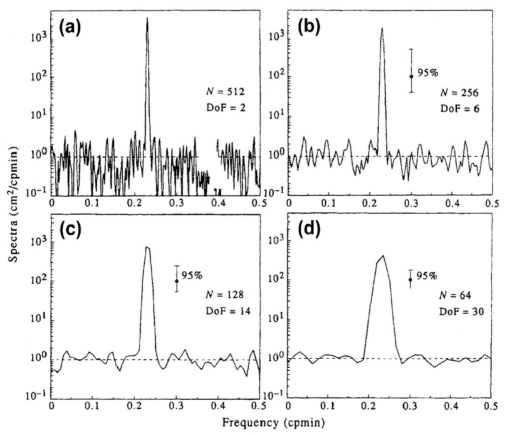

FIGURE 5.37 Periodogram power spectral estimates for a time series composed of Gaussian white noise and a single cosine constituent with a frequency of 0.23 cpmin and amplitude five times that of the white-noise component. N = number of spectral bands and vertical lines are the 95% confidence intervals, which appear compact because of the vertical log scale. (a) Raw (unsmoothed) periodogram, with degrees of freedom (DoF) = 2; (b) smoothed periodogram, by averaging three adjacent spectral estimates such that DoF = 6; (c) As with (b) but for seven frequency bands, and DoF = 14; (d) As with (b) but for 15 frequency bands, DoF = 30.

narrow peak at the frequency (0.23 cpmin) of the single cosine term plus a large number of smaller peaks associated with the white-noise oscillations, where the white noise is indicated by the flat line. The smaller peaks are all caused by leakage of the dominant peak into adjacent frequencies, In this case, there has been no spectral smoothing and the resultant spectral estimates are chi-squared functions with two DoF. The variances of the spectral peaks are as large as the peaks themselves.

If we average together three adjacent spectral components (Figure 5.37b), we obtain a much smoother spectrum, $S(f)$. Here, $S_i = S(f_i)$ is defined by $S_i = 1/3[S(f_{i-1}) + S(f_i) + S(f_{i+1})]$, $S_{i+3} = 1/3[S(f_{i+2}) + S(f_{i+3}) + S(f_{i+4})]$, where the many smaller peaks have been suppressed and the dominant peak has widened. Each of the new spectral estimates now have six DoF instead of only two. The bottom two panels in this figure show what happens if we increase the number of frequency bands averaged together to seven (Figure 5.37c) and then to 15 (Figure 5.37d). Note that, with increasing DoF, our confidence in the existence of a spectral peak increases but delineation of the peak frequency decreases as the peak widens. With increasing DoF, there is increased smoothing of all spectral peaks (see also Figure 5.31). The same effect can be achieved by operating on the autocovariance function rather than on the Fourier spectral estimates. In particular, a spectrum similar to Figure 5.37a is obtained using the autocovariance transform method on the time series $y(t)$ for a time lag of 1 min (the sampling interval). If we apply a lag of 3 min in computing the autocovariance transform, we obtain a spectrum similar to Figure 5.37b, and so on. Any differences between the two methods will be due to computational uncertainties.

To determine the number of DoF for any block averaging, we define the normalized standard error $\varepsilon(G)$ of the one-sided spectrum, $\widetilde{G}_{yy}(f)$, of the time series $y(t)$ of finite length $T = N\Delta t$, as

$$\varepsilon\left[\widetilde{G}_{yy}(f)\right] = \frac{V\left[\widetilde{G}_{yy}(f)\right]^{1/2}}{G_{yy}(f)} \tag{5.142}$$

where $G_{yy} = 2S_{yy}(f)$ is the true one-sided spectrum, $V[\widetilde{G}]$ is the variance of $\widetilde{G}$, the tilde ($\sim$) denotes the raw estimate of the observed time series, and

$$\widetilde{G}_{yy}(f) / G_{yy}(f) = \chi_2^2 / 2 \tag{5.143}$$

is a chi-square variable with $n = 2$ DoF. For the narrowest possible resolution $\Delta f = 1/T$, we have

$$\varepsilon\left[\widetilde{G}_{yy}(f)\right] = \frac{(2n)^{1/2}}{n} = (2/n)^{1/2} \tag{5.144}$$

For maximum resolution, $n = 2$ and so $\varepsilon(\widetilde{G}) = 1$, giving the not-so-useful result that the standard deviation of the estimate is as large as the estimate itself. If, on the other hand, we average the spectral estimates for each frequency for the maximum resolution spectra using a total of N_s separate and independent record segments of length T_s (where $T = N_s \cdot T_s$) we find

$$G_{yy}(f) = \frac{2}{N_s T_s} \sum_{i=1}^{N} |Y_i(f_i; T_s)|^2 \tag{5.145}$$

so that

$$\varepsilon\left[\widetilde{G}_{yy}(f)\right] = (2/2N_s)^{1/2} = (1/N_s)^{1/2} \tag{5.146}$$

The resolution (effective) bandwidth is $b_e = N_s/T = 1/T_s$. Because the first estimate, Eqn (5.144), gives two DoF per spectral band, the spectral averaging expressed by Eqn (5.146) gives $2N_s$ DoF per frequency band.

5.6.8 Confidence Intervals on Spectral Estimates

We can generalize Eqn (5.144) by noting that the ratio of the estimated spectrum and the expected values of the true spectrum

$$\frac{\nu \widetilde{G}_{yy}(f)}{G_{yy}(f)} = \chi_\nu^2 \tag{5.147}$$

is distributed as a chi-square variable with ν DoF. It then follows that

$$P\left[\chi_{\alpha/2,\nu}^2 < \frac{\nu \widetilde{G}_{yy}(f)}{G_{yy}(f)} < \chi_{1-\alpha/2,\nu}^2\right] = 1 - \alpha \tag{5.148}$$

where

$$P\left[\chi_\nu^2 \leq \chi_{\alpha/2,\nu}^2\right] = \alpha / 2 \tag{5.149}$$

Thus, the true spectrum, $G_{yy}(f)$, is expected to fall into the interval

$$\frac{\nu \widetilde{G}_{yy}(f)}{\chi_{1-\alpha/2,\nu}^2} < G_{yy}(f) < \frac{\nu \widetilde{G}_{yy}(f)}{\chi_{\alpha/2,\nu}^2} \tag{5.150}$$

with $(1 - \alpha)100\%$ confidence. In this form, the confidence limit applies only to the frequency f and not to other spectral estimates. We further point out that the DoF, ν, in the above expressions are different for windowed and nonwindowed time series. For windowed time series, we need to use the "equivalent" degrees of freedom (DoF), as presented in Table 5.13 for some of the more commonly used windows. [Note: the term "equivalent degrees of freedom" should not be equated with the "effective degrees of freedom", N^*, that was discussed in some detail in Chapter 3. Here, we are assuming that we are dealing with a random process with N DoF and that we are only correcting for a modest change in the DoF caused by application of the window.]

Another way to view these arguments is to equate $\widetilde{G}_{yy}(f)$ with the measured standard deviation, $s^2(f)$, of the spectrum and $G_{yy}(f)$ with the true variance, $\sigma^2(f)$. Then

$$\frac{(\nu - 1)s^2(f)}{\chi_{1-\alpha/2,\nu}^2} < \sigma^2(f) < \frac{(\nu - 1)s^2(f)}{\chi_{\alpha/2,\nu}^2} \tag{5.151}$$

If spectral peaks fall outside the range Eqn (5.151) then to the $(1 - \alpha)100\%$ confidence level they are unlikely to have occurred by chance. The confidence levels are found by looking up the values for $\chi_{1-\alpha/2,\nu}^2$ and $\chi_{\alpha/2,\nu}^2$ in a chi-square table, then calculating the intervals based on the observed standard deviation, s. It should be noted that, unlike previous symmetric confidence limits that we have discussed (such as those based on the Gaussian distribution), the chi-square distribution is not symmetrical and, as a consequence, the confidence intervals are also non-symmetrical. (Confidence limits on spectral coherency functions are given in Section 5.8.6.1).

5.6.8.1 Confidence Intervals on a Logarithmic Scale

The confidence intervals derived above apply only to individual frequencies, f. This results from the fact that the confidence interval is determined by the value $\widetilde{G}_{yy(f)}$ of the one-sided spectral estimate and will be different for each spectral estimate. It would be convenient if we could have a single confidence interval that applies to all of the spectral values at all frequencies. To obtain such a confidence interval, we transform the spectrum using the $\log_{10}$ function. Transforming the above confidence limits we have

$$\log\left[\widetilde{G}_{yy}(f)\right] + \log\left[\nu / \chi_{1-\alpha/2,\nu}^2\right] \leq \log\left[G_{yy}(f)\right] \leq \log\left[\widetilde{G}_{yy}(f)\right] + \log\left[\nu / \chi_{\alpha/2,\nu}^2\right] \tag{5.152}$$

or

$$\log\left[\nu / \chi_{1-\alpha/2,\nu}^2\right] \leq \log\left[G_{yy}(f)\right] - \log\left[\widetilde{G}_{yy}(f)\right] \leq \log\left[\nu / \chi_{\alpha/2,\nu}^2\right] \tag{5.153a}$$

TABLE 5.13 Equivalent degrees of freedom for spectra calculated using different windows.

Type of window	Equivalent degrees of freedom
Truncated periodogram	(N/M)
Bartlett window	$3(N/M)$
Daniell window	$2(N/M)$
Parzen window	$3.708614(N/M)$
Hanning window	$(8/3)(N/M)$
Hamming window	$2.5164(N/M)$

N is the number of data points in the time series and M is the half-width of the window in the time (or spatial) domain. $N \neq M$ for the truncated periodogram.
From Priestley (1981).

$$\log\left[\nu\,/\,\chi^2_{1-\alpha/2,\nu}\right] \le \log\left[G_{yy}(f)\,/\,\widetilde{G}_{yy}(f)\right] \le \log\left[\nu\,/\,\chi^2_{\alpha/2,\nu}\right] \tag{5.153b}$$

where $\log\left[G_{yy}(f)\,/\,\widetilde{G}_{yy}(f)\right] \to 0$ as the estimated spectrum approaches the real spectrum; i.e., $\widetilde{G}_{yy}(f) \to G_{yy}(f)$. When the estimated spectrum is plotted on a log scale, a single vertical confidence interval is determined for all frequencies by the upper and lower bounds in the above expressions (Figure 5.38a). The spectral estimate $\widetilde{G}_{yy}(f)$ itself is no longer a part of the confidence interval. This aspect, together with the fact that most spectral amplitudes span many orders of magnitude, is a principal reason for presenting spectra as log values. If larger numbers of spectral estimates are averaged together at higher frequencies (i.e., ν is increased), the confidence interval narrows with increasing frequency (Figure 5.38b). Note that, because of the log scaling, the length of the confidence interval is longer above the central point than below.

5.6.8.2 Fidelity and Stability

The general objective of all spectral analysis is to estimate the function $G_{yy}(f)$ (= $2S_{yy}(f)$) as accurately as possible. This involves two basic requirements:

1. The mean smoothed spectrum, $\widetilde{G}_{yy}(f)$, be as close as possible to the actual spectrum $G_{yy}(f)$. That is, the bias

$$B(f) = G_{yy}(f) - \widetilde{G}_{yy}(f) \tag{5.154}$$

should be small. If this is true for all frequencies, then $\widetilde{G}_{yy}(f)$ is said to reproduce $G_{yy}(f)$ with high *fidelity*. 2. For a time series of length T that has been segmented into M pieces for spectral estimation, the variance of the smoothed spectral estimator for bandwidth b_1 is

$$V\left[\widetilde{G}_{yy}(f)\right] \approx \frac{(M/b_1)}{T}\left[G_{yy}(f)\right]^2 \tag{5.155}$$

and should be small. If this is true, the spectral estimator is said to have high *stability*.

FIGURE 5.38 Confidence intervals for current velocity spectra at 50-m depth for three locations (B, C, and 62) on the northeast Gulf of Alaska shelf (59.5° N, 142.2° W), March 15–April 15, 1976. (a) 95% interval for the low-pass filtered currents. The single vertical bar applies to all frequencies; (b) 95% interval for unfiltered records. Confidence interval narrows at higher frequencies with the increased number of degrees of freedom (4–36) used in selected frequency ranges. *Adapted from Muench and Schumacher (1979).*

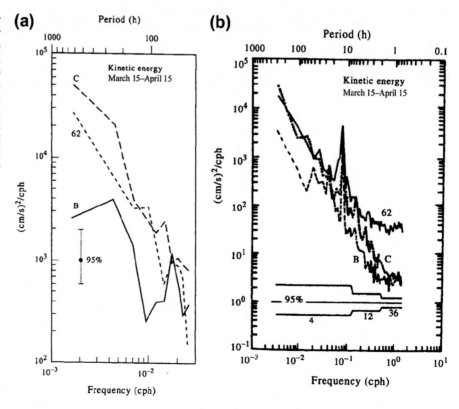

5.6.9 Zero-Padding and Prewhitening

For logistical reasons, many of the time series that oceanographers collect are too short for accurate definition of certain spectral peaks. The frequency resolution $\Delta f = 1/T$ for a record of length T may not be sufficient to resolve closely spaced spectral components. Also, discrete points in the computed spectrum may be too widely spaced to adequately delineate the actual frequency of the spectral peaks. Unfortunately, the first problem—that of trying to distinguish waveforms with nearly the same frequency—can only be solved by collecting a longer time series; i.e., by increasing T to sharpen up the frequency resolution f of the periodogram. However, the second problem—that of locating the frequency of a spectral peak more precisely—can be addressed by padding (extending) the time series with zeros prior to Fourier transforming. Transforming the data with zeros serves to refine the frequency scale through interpolation between PSD estimates within the Nyquist interval $-f_{Nq} \le f \le f_{Nq}$. That is, additional frequency components are added between those that would be obtained with a nonzero-padded transform. Adding zeros helps fill in the shape of the spectrum but in no case is there an improvement in the fundamental frequency resolution. *Zero-padding* is useful for: (1) smoothing the appearance of the periodogram estimates via interpolation; (2) resolving potential ambiguities, where the frequency difference between line spectra is greater than the fundamental frequency resolution; (3) helping define the exact frequency of spectral peaks by reducing the "quantization" accuracy error; and (4) extending the number of samples to an integer power of two for FFT analysis. An example of how zero-padding improves the spectral resolution of a simple digitized data set is provided in Figure 5.39. We again emphasize that increased zero-padding helps locate the frequency of discernible spectral peaks, in this case the peaks of the $\sin x/x$ function, but cannot help distinguish closely spaced frequency components that were unresolved by the original time series prior to padding.

Prewhitening is a filtering or smoothing technique used to improve the statistical reliability of spectral estimates by reducing the leakage from the most intense spectral components and low-frequency components of the time series that are poorly resolved. To reduce the biasing of these components, the data are smoothed by a window whose spectrum is inversely proportional to the unknown spectrum being considered. Within certain frequency bands, the spectrum becomes more uniformly distributed and approaches that of white noise. Information on the form of the window necessary to construct the white spectrum must be available prior to the application of the smoothing. In effect, the time series, $y(n\Delta t)$ is filtered with the weighting function, $w(n\Delta t)$ such that the output is

$$y'(n\Delta t) = w(n\Delta t) \cdot (n\Delta t) \tag{5.156}$$

(a)

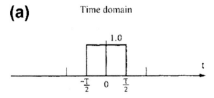

Rectangular data window of length T

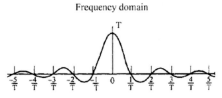

Continuous fourier transform of record 3a

(b)

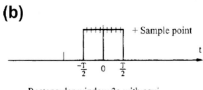

Rectangular window 3a with equi spaced samples

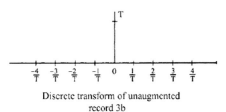

Discrete transform of unaugmented record 3b

(c)

Record 3b augmented with equal number of zeros

Discrete transform of record augmented to length 41

FIGURE 5.39 Use of zero-padding to improve the delineation of spectral peaks. (a) A continuous box-car window of length, T, and its continuous Fourier transform; (b) a discrete sample of (a) at equally spaced sampling intervals and its discrete Fourier transform; (c) same as (b) but with zero-padding of T data points. Note that the middle panel on the right is not a misprint. Transform values lie on the horizontal axis at the points $-4/T$, $-3/T$, and so on. *From Henry and Graefe (1971).*

has a nearly white spectrum. Once the two-sided spectrum $S'_{yy}(\omega)$ is determined, the desired spectrum is derived directly as

$$S_{yy}(\omega) = \frac{S'_{yy}(\omega)}{|W(\omega)|^2} \qquad (5.157)$$

The best aspects of the parametric and nonparametric spectral techniques can be combined if a parametric model is used to prewhiten the time series prior to the application of a smoothed periodogram analysis. In most prewhitening situations, one is limited to using the first-difference filter in which the current data value has subtracted from it the next value multiplied by some weighting coefficient, $0 \leq \alpha \leq 1$. That is $y'(t) = y(t) - \alpha y(t + \Delta t)$. The weighting coefficient can be taken as equal to the correlation coefficient of the initial data series with a shift of one (1) time step, Δt. The filter suppresses low frequencies and stresses high frequencies and has a frequency response

$$W(f) = \left[1 - \alpha e^{-i2\pi f \Delta t}\right]^2 = 1 - 2\alpha \cos(2\pi f \Delta t) + \alpha^2 \qquad (5.158)$$

Prewhitening reduces leakage and increases the effectiveness of frequency averaging of the spectral estimate (reduces the random error). The reduced leakage gives rise to a greater dynamic range of the analysis and allows us to examine weak spectral components. Notice that, if $Y(f)$ is the Fourier transform of $y(t)$, then the Fourier transform of $y'(t)$ is

$$Y'(\omega) = \int_t y'(t)e^{-i\omega t}dt \approx \omega Y(\omega) \qquad (5.159)$$

so that *first differencing* is like a linear high-pass filter with amplitude $|W(\omega)| = |\omega|$. This effect shows up quite well in the processing of satellite-tracked drifter data. Spectra of the drifter positions [longitude, $x(t)$; latitude, $y(t)$] as functions of time, t, are generally "red" whereas the spectra of the corresponding drifter velocities (zonal, $u = \Delta x/\Delta t$; meridional, $v = \Delta y/\Delta t$) are considerably "whiter" (Figure 5.40).

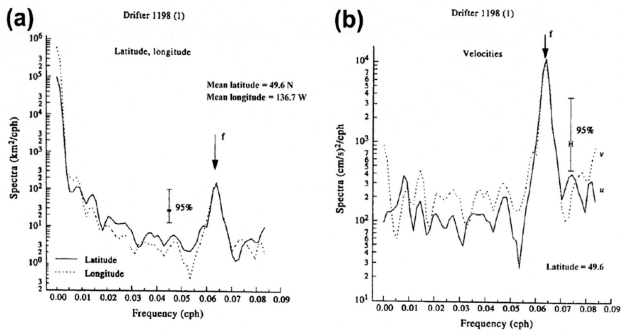

FIGURE 5.40 Reduction in the low frequency content of spectral distributions of time series upon application of a first-difference (high-pass) filter. (a) Spectra of longitude (Δx) and latitude (Δy) displacements of a satellite-tracked drifter launched in the northeast Pacific in September 1990 ($\Delta t = 3$ h; duration, $T = 90$ days); (b) as with (a) but for the zonal ($u = \Delta x/\Delta t$) and meridional velocity ($v = \Delta y/\Delta t$). Mean position of the drifter was 49.6° N, 136.7° W; f denotes the mean inertial frequency; vertical line is the 95% confidence interval.

5.6.10 Spectral Analysis of Unevenly Spaced Time Series

Most discrete oceanographic time series data are recorded at equally spaced time increments. However, some situations arise where the recorded data are spaced unevenly in time or space. For example, positional data obtained from satellite-tracked drifters are sampled at irregular time intervals due to the eastward progression in the swaths of polar-orbiting satellites and to the advection of the drifters by surface currents. Repeated time series oceanic transects are typically spaced at irregular intervals due to the vagaries of station sampling durations, ship scheduling and weather. In addition, instrumental problems and data dropouts generally lead to "gappy," irregularly spaced time series.

As noted in Section 3.17 of Chapter 3, a common technique for dealing with irregularly sampled or gappy data is to interpolate data values to a regular grid. This works well as long as there are not too many gaps and the gaps are of short duration relative to the signals of interest. Long data gaps can lead to the creation of erroneous low-frequency oscillations in the data at periods comparable to the gap lengths. Only for the least-squares (LS) method for harmonic analysis described in Section 5.7 is unevenly sampled data perfectly acceptable. Vaníček (1971), Lomb (1976) and others have devised an LS spectral analysis method for unevenly spaced time series. The Lomb method described by Press et al. (1992) evaluates data, and associated sines and cosines, at the times, t_n, that the data are measured. For the N data values $x(t_n) = x_n$, $i = 1, ..., N$, the Lomb-normalized periodogram is defined as

$$P(\omega) = \frac{1}{2\sigma^2} \left\{ \frac{\left[\sum\limits_{n=1}^{N} (x_n - \bar{x})\cos[\omega(t_n - \tau)] \right]^2}{\sum\limits_{n=1}^{N} \cos^2[\omega(t_n - \tau)]} + \frac{\left[\sum\limits_{n=1}^{N} (x_n - \bar{x})\sin[\omega(t_n - \tau)] \right]^2}{\sum\limits_{n=1}^{N} \sin^2[\omega(t_n - \tau)]} \right\} \tag{5.160}$$

where as usual

$$\bar{x} = \frac{1}{N} \sum_{n=1}^{N} x_n; \ \sigma^2 = \frac{1}{N-1} \sum_{n=1}^{N} (x_n - \bar{x})^2 \tag{5.161}$$

are the mean and standard deviation of the time series, and the time offset, τ, is defined by

$$\tan(2\omega\tau) = \frac{\sum\limits_{n=1}^{N} \sin(2\omega t_n)}{\sum\limits_{n=1}^{N} \cos(2\omega t_n)} \tag{5.162}$$

The offset, τ, renders Eqn (5.160) identical to the equation we would derive if we attempted to estimate the harmonic content of a data set at frequency ω using the linear LS model

$$x(t) = A\cos(\omega t) + B\sin(\omega t) \tag{5.163}$$

In fact, Vaníček's founding paper on the technique refers to it as an LS spectral analysis method. The method, which gives superior results to FFT methods, weights the data on a per point basis rather than on a time-interval basis. By not using weights that span a constant time interval, the method reduces errors introduced by unevenly sampled data. For further details on the Lomb periodogram, including the introduction of significance testing of spectral peaks, the reader is referred to Press et al. (1992; pp. 569−577).

5.6.11 General Spectral Bandwidth and Q of the System

Once the PSD, $S(\omega)$, has been computed, the general spectral bandwidth BW may be determined from the three moments, m_k, of the spectra

$$m_k = \int_0^\infty \omega^k S(\omega) d\omega, \ k = 0, 1, 2$$

$$= \sum_{i=0}^{N/2} \omega_i^k S(\omega_i) \Delta\omega \tag{5.164}$$

where $N/2$ is the number of spectral estimates and $\Delta\omega$ is the frequency resolution of the spectral estimates (cf. Masson, 1996). In particular

$$BW = \left[\left(m_2 m_0 / m_1^2\right) - 1\right]^{1/2} \tag{5.165}$$

The bandwidth, $\Delta\omega_{BW}$, of a particular spectral peak within an oscillatory system can be used to estimate the dissipation of the system at the peak (resonant) frequency, ω_r. Specifically, the "Q" or *Quality factor* of the system measures the amount of energy, E, stored in a linear oscillator compared to the amount of energy lost per cycle through frictional dissipation, $\omega^{-1} dE/dt$ (Rabinovich, 2009). The Q-factor characterizes the sharpness of the resonant frequency and is commonly used as a direct measure of tidal dissipation in the ocean. Suppose that the energy of a simple linear system passes through a maximum at resonance frequency and that the energy of the system falls to 50% of its maximum value at frequencies $\omega \approx \omega_r \pm \Delta\omega_{BW/2}$. The Q of the system is then given by

$$Q = \frac{\omega E}{dE/dt} = \frac{\omega_r}{\Delta\omega_{BW}} = \alpha^{-1} \tag{5.166}$$

where $E = E_o e^{-\alpha\omega t}$ is the system energy as it decays from an initial value E_o with a dimensionless damping coefficient, α. For example, Wunsch (1972) finds $Q \approx 3.3$ for an apparent resonant period of 14.8 h for the North Atlantic Ocean, while Garrett and Munk (1971) obtain a global-wide lower bound of 25 for normal modes near the semidiurnal frequency and Garrett (1972, 1984) gives a value of 5 for the Bay of Fundy. Monserrat et al. (2006) and Rabinovich (2009) provide detailed discussions of the role of the Q-factor in the amplification of long waves (e.g., atmospherically generated meteot-sunamis) arriving in a harbor from the open ocean and Medvedev et al. (2020) discuss Q values for tidal resonance in the Adriatic Sea.

5.6.12 Summary of the Standard Spectral Analysis Approach

In summary, power spectral density (PSD) estimates for time series $y(t)$ can be obtained as follows using the standard autocorrelation and periodogram approaches:

1. Remove the mean and trend from the time series. Failure to remove the trend can lead to spurious energy (power) at low frequencies. Remove *obvious* "spikes" caused by errant sensor responses or other forms of recording glitches, and also try to adjust the data series for discontinuities caused by internal offsets in the instrument or to sudden changes in sensor position or depth (Figure 5.41a). Removing spikes and adjusting for offsets is not as easy as it sounds. However, if not taken into account in the original time series, spikes and offsets can lead to erroneous spectral distributions (Figure 5.41b).

2. If block averaging is to be used to improve the statistical reliability of the spectral estimates (i.e., to increase the number of DoF), divide the data series into M sequential blocks of N' data values each, where $N' = N/M$ (see Section 5.6.7). Depending on which type of window is to be applied, the sequential blocks can have up to 50% overlap.

3. To partially reduce end effects (Gibbs phenomenon) or to increase the series length to a power of two for FFT analysis, pad the data with $K \leq N$ zeroes. Also pad the record with zeroes if you wish to broaden the frequency range or center spectral estimates in specific frequency bands. To further reduce end effects and side-lobe leakage, taper the time series using a Hanning (raised cosine) window, Kaiser-Bessel window, or other appropriate window (see Section 5.6.6).

4. Compute the Fourier transforms, $Y(f_k)$, $k = 0, 1, 2, ..., N - 1$, for the time series (for convenience, we have taken the number of padded values as $K = 0$). For block-segmented data, calculate the Fourier transforms, $Y_m(f_k)$, for each of the M blocks ($m = 1, ..., M$) where $k = 0, 1, ..., N' - 1$ and $N' < N$. To reduce the variance associated with the tapering in step 3, the transforms can be computed for overlapping segments.

5. Rescale the spectra to account for the loss of "energy" during application of the window. That is, adjust the scale factor of $Y(f_k)$ (or $Y_m(f_k)$ in the case of smaller block size partitioning to account for the reduction in spectral energy due to the tapering in step 3. For the Hanning window, multiply the amplitudes of the Fourier transforms by $\sqrt{8/3}$. The rescaling factors for other windows are listed in the right-hand column of Table 5.13.

6. Compute the raw PSD for the time series (or for each block) where for the two-sided spectral density estimates:

$$S_{yy}(f_k) = \frac{1}{N\Delta t}[Y(f_k)Y^*(f_k)], k = 0, 1, 2, ..., N - 1$$

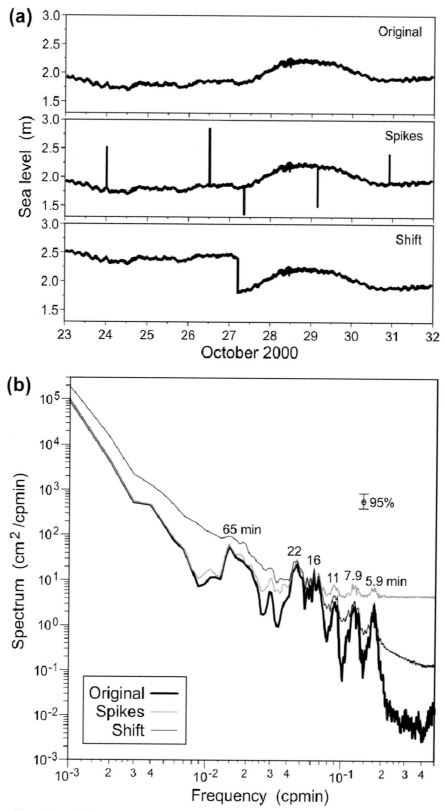

FIGURE 5.41 Effects of data spikes and offsets on spectral estimates. (a) The top line shows a 1-min sampled sea-level times series for Victoria, British Columbia. The middle panel is the same series but in which five data values have been converted into "spikes" (single data points with anomalously high values). In the bottom panel, we have inserted a single negative offset of 0.5 m midway through the original time series; (b) spectra of the three (3) time series in (a). There is considerable loss of high frequency information compared to the original time series and the addition of erroneous low frequency energy in the case of the offset time series. Numbers denote periods in minutes of selected spectral peaks. *Courtesy of Alexander Rabinovich, Institute of Ocean Sciences and P.P. Shirshov Institute of Oceanology*

(no block averaging)

$$S_{yy}(f_k; m) = \frac{1}{N'\Delta t}\left[Y_m(f_k)Y_m^*(f_k)\right], k = 0, 1, 2, ..., N'-1 \tag{5.167a}$$

(block averaging). For the one-sided spectral density estimates

$$G_{yy}(f_k) = \frac{2}{N\Delta t}[Y(f_k)Y^*(f_k)], k = 0, 1, 2, ..., N/2$$

(no block averaging)

$$G_{yy}(f_k; m) = \frac{2}{N'\Delta t}\left[Y_m(f_k)Y_m^*(f_k)\right], k = 0, 1, 2, ..., N'/2 \tag{5.167b}$$

(block averaging).

7. In the case of the block-segmented data, average the raw spectral density estimates from the M blocks of data, frequency-band by frequency-band, to obtain the smoothed periodogram for $S_{yy}(f_k)$ or $G_{yy}(f_k)$. Remember, the trade-off for increased smoothing (more DoF) is a decrease in frequency resolution.
8. Incorporate 80%, 90%, and/or 95% confidence limits in spectral plots to indicate the statistical reliability of spectral peaks. Most authors use the 95% confidence interval.

We can illustrate some additional points in the above summary using the log–log spectra of sea-level oscillations presented previously in Figure 5.31, recorded over 14 days (20,160 min) in 1991 at Malokurilsk Bay on the west coast of Shikotan Island in the western Pacific. All spectra have been obtained using segmented versions of the 14-day time series. Each time series segment has been smoothed using a Kaiser-Bessel window with 50% overlap between segments, and each segment has been treated as an independent time series. An FFT algorithm was used to calculate the spectrum for each segment. The main spectral peak is centered at a period of 18.6 min and corresponds to a wind-generated seiche amplitude of about 25 cm (Rabinovich and Levyant, 1992). The smoothest spectrum (Figure 5.31a) is based on block averaged spectral estimates from roughly 157 overlapping segments ($\sim 20,160$ min/128 ($= 2^7$) min), the moderately smooth spectrum (Figure 5.31b) from the average of 39 overlapping segments, and the noisiest spectrum (Figure 5.31c) from the average of 10 overlapping segments. Taking into account the 50% overlap between segments and the fact that there are two DoF per raw spectral estimate, there are 628 ($= 157 \times 4$), 154, and 36 DoF for the three spectra, respectively. The smoothed spectrum in Figure 5.31d is derived using a slightly different approach. Although the segment lengths are the same as those in Figure 5.31c (i.e., 2,048 min), the number of DoF is increased with increasing frequency, ω. In this sliding scale, the lowest frequency range uses 36 DoF (as with Figure 5.31c), the next frequency band averages together the spectra for three adjacent frequencies to give 108 DoF, the next averages together the spectra for five adjacent frequencies to give 180 DoF, and so on.

As indicated by Figure 5.31, increasing the number of frequency bands averaged in each spectral estimate enhances the overall smoothness of the spectrum and improves the statistical reliability for specific spectral peaks. The number of degrees of freedom (DoF) increases and the confidence interval narrows. The penalty we pay for improved statistical confidence is reduced resolution of the spectral peaks. As in Figure 5.31a, too much smoothing diminishes our ability to specify the frequency of spectral peaks and washes out peaks linked to some of the weaker seiches. Because each time series segment is so short, we also lose definition at the low-frequency end of the spectrum. As indicated by Figure 5.31c, too little smoothing leads to a noisy spectrum for which few spectral peaks are associated with any physical processes. The sliding DoF scale in Figure 5.31d is a useful compromise.

One last point. Up until now, we have assumed that the sensors being used to collect the data have the sensitivity to record all of the variations of interest. If this is not the case, then no form of spectral analysis can extract information from the signal, regardless of the temporal resolution. Consider, for example, Figure 5.42a, which shows a 9-day time series of bottom pressure (sea-level height) collected at 1-min intervals in Saanich Inlet on Vancouver Island, British Columbia. The top line shows the raw bottom pressure data sampled at 0.001 m (1 mm) equivalent vertical resolution. This is followed by time series generated by rounding off the 1-min data values to vertical resolutions that are factors of 10 and 100 lower than that of the original record. The impact of the lower vertical resolution is clearly displayed by the spectra in Figure 5.42b. As would be the case for inadequate sensor resolution, the spectra of vertical displacements become increasingly degraded at higher frequencies. Although the sampling interval is the same for all of the time series, the spectral details of the sea-level signal are lost, including the background roll-off as a function of frequency.

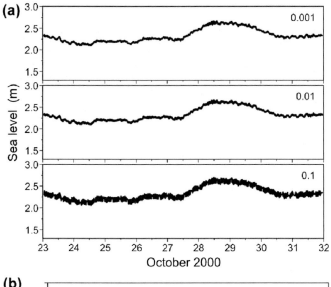

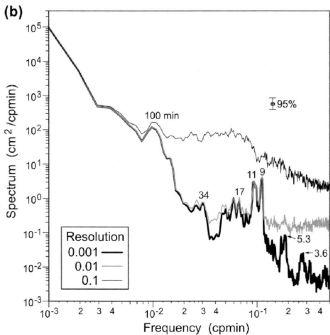

FIGURE 5.42 The importance of sensor resolution to the detection of physical signals using spectral analysis. (a) 1-min sea-level record collected by a modern pressure gauge at Patricia Bay, Saanich Inlet, British Columbia. The top panel shows with original time series at 0.001 m (1 mm) vertical resolution, followed by time series formed by degrading (using decimal runoff) the original series to 0.01 and 0.1 m vertical resolution; (b) Spectra for the three (3) time series showing the loss of information with increased degradation in vertical resolution. Numbers refer to spectral peaks in periods of minutes. The sampling rate is the same in all cases. *Courtesy of Alexander Rabinovich, Institute of Ocean Sciences and P.P. Shirshov Institute of Oceanology.*

Covariance function: Because the covariance function, $C_{yy}(\tau)$, and the autospectrum are Fourier transform pairs, the above analysis can be used to obtain a smoothed or unsmoothed estimate of the covariance function. To do this, first calculate the Fourier transform, $Y(f)$, of the time series, and determine the product $S_{yy}(f) = (1/N)\Delta t[Y(f)Y^*(f)]$. Then take the inverse Fourier transform (IFT) of the autospectrum, $S_{yy}(f)$, to obtain the covariance function, $C_{yy}(\tau)$. If the spectrum is unsmoothed prior to the IFT (or inverse fast Fourier transform (IFFT) if the FFT was used), we obtain the raw covariance function. If, on the other hand, the autospectrum is smoothed prior to the above integral using one of the spectral windows, such as the Hanning window, the covariance function also will be a smoothed function.

A word of caution: Although everyone agrees on the basic formulation for the DFT and the inverse discrete Fourier transform (IDFT), there are several ways to normalize the relations using the number of records, N. In our definitions, Eqn (5.71), N appears in the denominator of the IDFT. Some authors normalize using $1/N$ in the DFT only, while others insist on symmetry by using $1/\sqrt{N}$ in both DFT and its inverse. When using "canned" programs to obtain DFTs and IDFTs, ensure that you know how the transforms are defined and adapt your analysis to fit the appropriate processing routines.

5.7 SPECTRAL ANALYSIS (PARAMETRIC METHODS)

If the analytical model for a time series was known exactly, a sensible spectral estimation method would be to fit the model spectrum to the observed spectrum and determine any unknown parameters. In general, however, oceanic variability is too complex to admit simple analytical models and parametric spectral estimates over the full frequency range of the data series. In addition, the imposition of an overly simplified spectral model could seriously degrade any estimation. On the other hand, it is reasonable that relatively simple spectral models might adequately reflect the system dynamics over limited frequency bands. Under some very general conditions, any stationary series can be represented in closed form by a statistical model in which the corresponding spectrum is a rational function of frequency (i.e., a ratio of two polynomials in ω).

If the time series under investigation is long relative to the timescales of interest, and if the spectrum is not overly complicated and does not have too large of a dynamic range, the simple smoothed periodogram technique will probably yield adequate results. At a minimum, it will identify the major features in the spectrum. For shorter time series or in studies of fine spectral structure, other techniques may be more applicable. One such spectral analysis technique was developed by Burg (1967, 1972), who showed that it was possible to obtain the power spectrum by requiring the spectral estimate to be the most random (i.e., to have the maximum entropy) of any power spectrum, which is consistent with the measured data. This leads to a spectral estimate with a high frequency resolution since the method uses the available lags in the autocovariance function without modification and makes a nonzero estimate (prediction) of the autocorrelation function beyond those, which are routinely calculated from the data. Because the spectral values are computed using a maximum entropy condition, the resulting spectral estimates are not accurate in terms of spectral amplitude.

As we remarked earlier in this Chapter, parametric methods are able to identify spectral peaks that standard FFT spectra may not. The most popular of the "modern" parametric techniques is the *autoregressive power spectral density* (AR PSD) model whose origins are in economic time series forecasting and statistical estimation. Autoregressive estimation was introduced to the earth sciences in the 1960s when it was originally applied to geophysical time series data under the name Maximum Entropy Method (MEM). The duality between AR and MEM estimation has been thoroughly explored by Ulrych and Bishop (1975). Autoregressive spectral estimation is attractive because it has superior frequency resolution compared to conventional FFT techniques. As an example of the frequency resolution capability, consider the 14-year time series of average monthly air temperature for New York City (Figure 5.43a). The unsmoothed periodogram and three smoothed periodograms reveal a broad spectral peak centered at a period of 1 year (Figure 5.43b). This compares to the much sharper annual peak obtained via AR estimation (Figure 5.43c). The results reveal another important difference between the two methods. With the nonparametric periodogram approach discussed in the previous sections, we can determine confidence limits for the spectral peaks, while for the parametric method the significance levels for the peaks are unknown. For example, the MEM is good for finding the location of spectral peaks but is not reliable for computing the correct spectral energy at those peaks. (The periodogram smoothing in Figure 5.43b was performed using a Parzen window with truncation values $N = 16$, 32, and 64; the weights for these windows are $w(n) = 1 - |2n/N|^2$, with $0 \leq |n| \leq \frac{1}{2}N$).

In general, autoregressive and maximum entropy PSD estimations are not as widely used in oceanography as traditional spectral analysis methods. The former finds its greatest application in analytical climate modeling and in wavenumber spectral estimation. Modern parametric techniques are good as long as the model is applicable. On the other hand, if the model is false, the resulting spectrum estimate can be highly misleading. It follows that, if one has no reason for believing a specific model, it is better to use a nonparametric model. For this reason, we limit our presentation to the essential elements of the two methods. The reader is directed to Marple (1987) for a thorough discussion of the topic, including an introduction to Fourier transform methods of spectral analysis.

5.7.1 Basic Concepts of Parametric Methods

Many deterministic and stochastic discrete-time series processes encountered in oceanography are closely approximated by a rational transfer model in which the input sequence $\{x_n\}$ and the output sequence $\{y_n\}$, which is meant to model the input data, are related by the linear difference relation

$$y_n = \sum_{k=0}^{q} b_k x_{n-k} - \sum_{m=1}^{p} a_m y_{n-m} \tag{5.168}$$

Here, y_n is shorthand notation for $y(n\Delta t)$, also written as $y(n)$. In its most general form, the linear model Eqn (5.168) is termed an *autoregressive moving average* (ARMA) model. The PSD of the ARMA output process is

$$P_{\text{ARMA}}(f) = \sigma^2 \Delta t [A(f)/B(f)]^2 \tag{5.169}$$

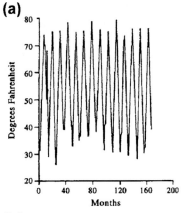

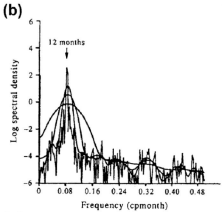

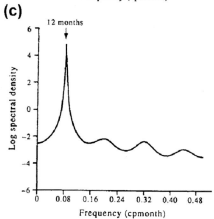

FIGURE 5.43   (a) Time series of monthly average air temperature for New York City (1946−59); (b) the unsmoothed (raw) periodogram and three smoothed periodograms for Parzen windows with truncation lengths of 16, 32, and 64 months; and (c) an autoregressive spectral estimate of (a) showing the sharp peak at 12-month period. *From Pagano (1978). Reprinted with the permission of the Institute of Mathematical Statistics.*

where σ^2 is the variance of the applied white-noise driving mechanism and $\sigma^2 \Delta t$ is the PSD of the noise for the Nyquist interval $-1/(2\Delta t) < f < 1/(2\Delta t)$. Here

$$A(f) = \alpha[\exp(i2\pi f \Delta t)], \; B(f) = \beta[\exp(i2\pi f \Delta t)] \tag{5.170}$$

where the coefficients α, β are defined in terms of the *z-transform*, $X(z)$, of the variable $z = \exp(i2\pi f \Delta t)$ [$= \exp(i2\pi k/N)$ in discrete form], where $k, n = 0, 1, ..., N - 1$

$$X(z) = \sum_{n=0}^{N-1} x_n z^{-n} \tag{5.171}$$

which maps a real-valued sequence into a complex plane. Note that Eqn (5.171) is defined through negative powers of z, the convention used in electrical engineering. Geophysicists expand in positive powers of z (z^{+n}) but define $z = \exp(-iz\pi f \cdot \Delta t)$ so the results are the same. The z-transform of the autoregressive branch is

$$\alpha(z) = \sum_n a_n z^{-n} \tag{5.172a}$$

while that of the moving average branch is

$$\beta(z) = \sum_n b_n z^{-n} \tag{5.172b}$$

Specification of the parameters $\{a_k\}$, termed the autoregressive coefficients, the parameters $\{b_k\}$, termed the moving-average coefficients, and the variance, σ^2, is equivalent to specifying the spectrum of the process $\{y_n\}$. Without loss of generality, one can assume $a_0 = 1$ and $b_0 = 1$ since any gain of the system (5.168) can be incorporated into σ^2. If all the $\{a_k\}$ terms except $a_0 = 1$ vanish then

$$y_n = \sum_{k=0}^{q} b_k x_{n-k} \tag{5.173}$$

and the process is simply a moving average of order q, and

$$P_{MA}(f) = \sigma^2 \Delta t |A(f)|^2 \tag{5.174}$$

This model is sometimes called an *all-zero model* since spectral peaks and valleys are formed through zeroes of the function $A(f)$. If all the $\{b_k\}$ terms except $b_0 = 1$ vanish, then

$$y_n = \sum_{m=1}^{p} a_m y_{n-m} + \varepsilon_n \tag{5.175}$$

and the process is strictly an autoregressive model of order p. The process is called AR in the sense that the sequence y_n is a linear regression on itself with ε_n representing the error. With this model, the present value y_n is expressed as a weighted sum of past values plus a noise term. The PSD is

$$P_{AR}(f) = \frac{\sigma^2 \Delta t}{|B(f)|^2} \tag{5.176}$$

In the engineering literature, this model is sometimes called an *all-pole model* since narrow spectral peaks can be sharply delineated through zeroes in the denominator.

5.7.2 Autoregressive Power Spectral Estimation

The discrete form of an autoregressive model $y(t)$ of order p is represented by the relationship

$$y(n) = a_1 y(n-1) + a_2 y(n-2) + \ldots + a_p y(n-p) + \varepsilon(n) \tag{5.177}$$

where time $t = n\Delta t$, the a_k $(k = 1, \ldots, p)$ are constant coefficients, and $\varepsilon(t)$ is a white-noise series (usually called the "innovation" of the AR process) with zero mean and variance σ^2. Another interpretation of the AR process is one that links $y(t)$ with a value that is predicted from the previous $p - 1$ values of the process with a prediction error equal to $\varepsilon(t)$. Thus, the a_k $(k = 1 \ldots, p)$ represent a p-point prediction filter. If $Y(z)$ is the z-transform of $y(n)$ then

$$Y(z) = \sum_{n=0}^{p} y(n) z^n \tag{5.178}$$

and

$$Y(z) - Y(z)\left(a_1 z + a_2 z^2 + \ldots + a_p z^p\right) = D(z) \tag{5.179}$$

so that

$$|Y(z)|^2 = \frac{|D(z)|^2}{\left|1 - a_1 z - a_2 z^2 \ldots - a_p z^p\right|^2} \tag{5.180}$$

Substituting $z = \exp(-i2\pi f \Delta t)$, we obtain half of the true power spectrum. If the autoregression is a reasonable model for the data, then the AR PSD estimate based on Eqn (5.176) is

$$P_{AR}(f) = \frac{\sigma^2 \Delta t}{\left| 1 - \sum\limits_{k=1}^{p} a_k \exp(-i2\pi f k \Delta t) \right|^2} \tag{5.181}$$

To find the PSD we need to estimate only three things: (1) the autoregressive parameters $\{a_1, a_2, ..., a_p\}$; (2) the variance, σ^2, of the white-noise process that is assumed to be driving the system; and (3) the order, p, of the process. The limitations of the AR model are the degrading effect of observational noise, spurious peaks, and some anomalous effects that occur when the data are dominated by sinusoidal components. Unlike conventional Fourier spectral estimates, the peak amplitudes in AR spectral estimates are not linearly proportional to the power when the input process consists of sinusoids in noise. For high SNRs, the peak is proportional to the square of the power with the area under the peak proportional to power.

5.7.2.1 Autoregressive Parameter Estimation

Yule–Walker (YW) equations: If the autocorrelation function, $R_{yy}(k)$, is known exactly, we can find the $\{a_k\}$ by the YW equations. This method relates the AR parameters to the known (or estimated) autocorrelation function of $y(n)$

$$R_{yy}(k) = \frac{1}{N} \sum_{n=1}^{N-k} \{ [x(n-k) - \bar{x}][x(n) - \bar{x}] \}; \bar{x} = \frac{1}{n} \sum_{n=1}^{N} x(n) \tag{5.182}$$

There are other methods of estimating R_{yy}, but this estimator has the attractive property that its mean-squared error is generally smaller than that of other estimators (Jenkins and Watts, 1968). Since it is generally assumed that the mean $\bar{x}$ has been removed from the data, the autocovariance and autocorrelation functions are equal. To obtain the AR parameters, one need only to choose p equations from the YW equations for $k > 0$, solve for $\{a_1, a_2, ..., a_p\}$, and then find σ^2 from Eqn (5.182) for $k = 0$. The matrix equation to derive the a_i values and σ^2 is

$$\begin{vmatrix} R_{yy}(0) & R_{yy}(-1) & ... & R_{yy}(-p) \\ R_{yy}(1) & R_{yy}(0) & ... & R_{yy}[-(p-1)] \\ ... & ... & ... & ... \\ ... & ... & ... & ... \\ R_{yy}(p) & R_{yy}(p-1) & ... & R_{yy}(0) \end{vmatrix} \begin{vmatrix} 1 \\ a_1 \\ ... \\ ... \\ a_p \end{vmatrix} = \begin{vmatrix} \sigma^2 \\ 0 \\ ... \\ ... \\ 0 \end{vmatrix} \tag{5.183}$$

Thus, to determine the AR parameters and the variance, σ^2, one must solve (5.183) using the $p + 1$ autocorrelation lags, $R_{yy}(0), ..., R_{yy}(p)$, where $R_{yy}(-k) = R^*_{yy}(k)$.

Solutions to the YW matrix equation can be found via the computationally efficient Levinson–Durbin algorithm which proceeds recursively to compute the parameter sets $\{a_{11}, \sigma^2_1\}$, $\{a_{21}, a_{22}, \sigma^2_2\}$, ..., $\{a_{p1}, a_{p2}, ..., a_{pp}, \sigma^2_p\}$. The final set at order p (the first subscript) is the desired solution. The algorithm requires p^2 operations as opposed to the $O(p^3)$ operations of Gaussian elimination. More specifically, the recursion algorithm gives

$$a_{11} = \frac{-R_{yy}(1)}{R_{yy}(0)} \tag{5.184a}$$

$$\sigma^2_1 = \left(1 - |a_{11}|^2\right) R_{yy}(0) \tag{5.184b}$$

with the recursion for $k = 2, 3, ..., p$ given by

$$a_{kk} = \frac{-1}{\sigma^2_1} \left[R_{yy}(k) + \sum_{j=1}^{k-1} a_{k-1,j} R^{(k-j)}_{yy} \right] \tag{5.185a}$$

$$a_{ki} = -a_{k-1,i} + a_{kk} (a_{k-1,k-i})^* \tag{5.185b}$$

$$\sigma^2_k = \left(1 - |a_{kk}|^2\right) \sigma^2_{k-1} \tag{5.185c}$$

Burg algorithm: Box and Jenkins (1970) point out that the YW estimates of the AR coefficients are very sensitive to rounding errors, particularly when the AR process is close to becoming nonstationary. The assumption that $y(k) = 0$, for $|k| > p$ leads to a discontinuity in the autocorrelation function and a smearing of the estimated PSD. For this reason, the most popular method for determining the AR parameters (prediction error filter coefficients) is the Burg algorithm. This algorithm works directly on the data rather than on the autocorrelation function and is subject to the Levinson recursion Eqn (5.185b). As an illustration of the differences in the YW and the Burg estimates, the respective values of a_{11} for the series $y(t_k) = y(k)$ are

$$a_{11} = \frac{\sum\limits_{k=2}^{p} y(k)y(k-1)}{\sum\limits_{k=1}^{p} y(k)^2}, \text{ for the Yule-Walker estimate}$$

(5.186)

$$a_{11} = \frac{\sum\limits_{k=2}^{p} y(k)y(k-1)}{\frac{1}{2}x_1^2 + \sum\limits_{k=1}^{p} y(k)^2 + \frac{1}{2}x_p^2}, \text{ for the Burg estimate}$$

Detailed formulation of the Burg algorithm is provided by Kay and Marple (1981; p. 1392). Again, there are limitations to the Burg algorithm, including spectral line splitting and biases in the frequency estimate due to contamination by rounding errors. Spectral line splitting occurs when the spectral estimate exhibits two closely spaced peaks, falsely indicating a second sinusoid in the data.

Least squares (LS) estimators: Several LS estimation procedures exist that operate directly on the data to yield improved AR parameter estimates and spectra compared with the YW or Burg approaches. The two most common methods use forward linear prediction for the estimate, while a second employs a combination of forward and backward linear prediction. Ulrych and Bishop (1975) and Nuttall (1976) independently suggested this LS procedure for forward and backward prediction in which the Levison recursion constraint imposed by Burg is removed. The LS algorithm is almost as computationally efficient as the Burg algorithm requiring about 20 more computations. The improvement by the LS approach over the Burg algorithm is well worth the added computation time. Improvements include less bias in the frequency estimates, and absence of observed spectral line splitting for short sample sinusoidal data.

Barrodale and Erickson (1978) provide a FORTRAN program for an "optimal" LS solution to the linear prediction problem. The algorithm solves the underlying LS problem directly without forcing a Toeplitz structure on the model. Their algorithm can be used to determine the parameters of the AR model associated with the MEM and for estimating the order of the model to be used. As illustrated by the spectra in Figure 5.44, this approach leads to a more accurate frequency

FIGURE 5.44 Maximum entropy method spectra obtained using (a) the Burg and (b) the Barrodale and Erickson algorithms. Signal consists of a combined 0.2 and 0.03 Hz (cps) sine wave. Spectra are plotted for increasing numbers of coefficients, *p*. *From Barrodale and Erickson (1978).*

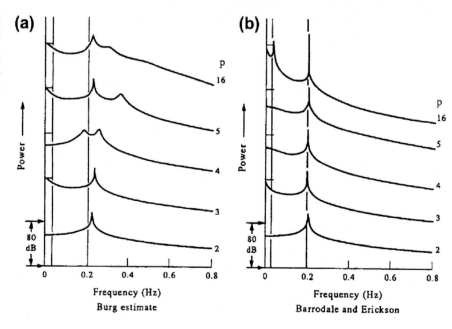

resolution for short sample harmonic processes. In this case, the test data were formed by summing 0.03 and 0.2 Hz sine waves generated in single precision and sampled 10 times per second. The reader is also referred to Kay and Marple (1981; p. 1393) for additional details.

5.7.2.2 Order of the Autoregressive Process

The order p of the autoregressive filter is generally not known *a priori* and is acknowledged as one of the most difficult tasks in time series modeling by parametric methods. The choice is to postulate several model orders then compute some error criterion that indicates which model order to pick. Too low a guess for the model order results in a highly smoothed spectral estimate. Too high an order introduces spurious detail into the spectrum. One intuitive approach would be to construct AR models with increasing order until the computed prediction error power, σ^2_k, reaches a minimum. Thus, if a process is actually an AR process of order p, then $a_{p+1,k} = a_{pk}$ for $k = 1, 2, ..., p$. The point at which a_{pk} does not change would appear to be a good indicator of the correct model order. Unfortunately, both the YW equations and Burg algorithm involve prediction error powers

$$\sigma^2_k = \sigma^2_{k-1}\left[1 - |a_{kk}|^2\right]$$ (5.187a)

that decrease monotonically with increasing order p, so that as long as $|a_{kk}|^2$ is nonzero (it must be ≤ 1) the prediction error power decreases. Thus, the prediction error power is not sufficient to indicate when to terminate the search. Alternative approaches (Kay and Marple, 1981) have been proposed by Akaike (termed the final prediction error, FPE, and the Akaike information criterion, AIC), and by Parzen (termed the criterion autoregressive transfer function). The AIC determines the model order by minimizing an information theoretic function. If the process has Gaussian statistics, the AIC is

$$\text{AIC}(p) = \ln\left(\sigma^2_p\right) + 2(p+1)/N$$ (5.187b)

where σ^2_p is the prediction error power and N is the number of data samples. The second term represents the penalty for the use of extra autoregressive coefficients that do not result in a substantial reduction in the prediction error power. The order p is the one that minimizes the AIC.

5.7.2.2.1 Summary of Algorithms

Method	Model applied	Advantages	Disadvantages
Periodogram method using FFT or discrete Fourier transform	Sum of harmonics (sines and cosines). No specific model needed.	1. Uses harmonic least-squares fit to the data; 2. Output $S(f)$ directly proportional to power; 3. Most computationally efficient; 4. Well-established methodology; 5. Confidence intervals easily computed; 6. Integral of $S(f)$ over frequency band Δf is equal to the variance of the signal in that band; 7. Easily generalized to cross-spectra and rotary spectra analyses.	1. Frequency resolution $\Delta f \approx 1/T$ dependent only on record length, T; 2. Poor performance for short data records; 3. Side-lobe leakage distorts spectra if appropriate windowing not done; windowing reduces frequency resolution, Δf; 4. Must average spectral estimates to improve statistical reliability.
Autoregressive, Yule–Walker algorithm.	Autoregressive (all-pole) process. Specific model.	1. Improved spectral resolution over Fourier transform methods; 2. Sharp spectral peaks; 3. No side-lobe leakage problems; 4. Minimum phase (stable) linear prediction filter guaranteed if biased lag estimates computed; 5. Related to linear prediction analysis and adaptive filtering.	1. AR model order, p, must be specified; 2. Spectral line splitting occurs; 3. Implied windowing distorts spectra; 4. Confidence intervals not readily computed.

| Autoregressive, Burg algorithm. | Autoregressive (all-pole) process. Specific model. | 1. Improved resolution over Fourier transform methods. Uses a constrained recursive least-squares approach;
2. No side-lobe leakage problems;
3. High resolution for low noise signals;
4. Good spectral fidelity for short data series;
5. No windowing implied;
6. Stable linear prediction filter guaranteed. | 1. AR model order, p, must be specified;
2. Spectral line splitting can occur;
3. Confidence intervals not readily computed. |
| Autoregressive, least-squares method. | Autoregressive (all-pole) process. Specific model. | 1. Sharper spectra than for other AR methods;
2. No side-lobes;
3. Good spectral fidelity for short data series;
4. No windowing;
5. No line splitting;
6. Uses exact recursive least-squares solution with no constraint. | 1. AR model order must be specified;
2. Stable linear prediction filter not guaranteed, though stable filter results in most cases. |

5.7.2.3 Maximum Entropy Method (MEM)

The only constraint on the AR method is that the data yield the known autocorrelation function, $R_{yy}(k)$ for the interval $0 < k < p$. The assumption that $y(k) = 0$, for $|k| > p$ leads to a discontinuity in the autocorrelation function and a smearing of the estimated PSD. The MEM was designed, independently of autoregressive estimation, to eliminate the distortion of the spectrum caused by the truncated $R_{yy}(k)$. By adding a second constraint to improve the spectral estimation, the method gets away from the problems with the YW algorithm. In essence, the MEM is a way of extrapolating the known autocorrelation function to lags $k > p$, which are not known. In words, we assume that $\{R_{yy}(0), ..., R_{yy}(p)\}$ are known and find a logical way to extend to lags $\{R_{yy}(p + 1),...\}$. As it turns out, the power spectral estimate for the MEM approach is equivalent to the power spectral estimate for the AR process.

In general, there exists an infinite number of possible extrapolations. Burg (1968) argued that preferred extrapolation should do two things: (1) yield the known R_{yy} for $0 \leq k \leq p$; and (2) generate an extrapolated R_{yy} for $k > p$ that causes the time series to have maximum entropy under the constraint (1). The time series that results is the most random one, which adheres to the known R_{yy} for the first $p + 1$ lags. Alternatively, we can say that the PSD is the one with whitest noise (flattest spectrum) of all possible spectra for which $\{R_{yy}(0), ..., R_{yy}(p)\}$ is known. The reason for choosing the maximum entropy criterion is that it imposes the fewest constraints on the unknown time series by maximizing its randomness thereby causing minimum bias and operator intervention. For a Gaussian process, the entropy per sample is proportional to

$$\int_{-1/2\Delta t}^{1/2\Delta t} \ln\left[P_y(f)\right] df \tag{5.188}$$

where $P_y(f)$ is the PSD of y_n. The spectrum is found by maximizing Eqn (5.188) subject to the constraint that the $p + 1$ known lags satisfy the Wiener–Khinchin relation

$$\int_{-1/2\Delta t}^{1/2\Delta t} P_y(f)e^{-i2\pi fn\Delta t} df = R_{yy}(n), n = 0, 1, ..., p \tag{5.189}$$

The solution is found using the Lagrange multiplier technique (see Ulrych and Bishop, 1975) as

$$P_y(f) = \frac{\sigma_p^2 \Delta t}{\left|1 + \sum_{k=1}^{p} a_{pk} \exp(-i2\pi fk\Delta t)\right|^2} \tag{5.190}$$

where $\{ a_{p1}, ..., a_{pp}\}$ and σ_p^2 are just the order-p predictor parameters and prediction error power, respectively. With knowledge of $\{R_{yy}(0), R_{yy}(1), ..., R_{yy}(p)\}$ the PSD of the MEM is equivalent to the PSD of the autoregressive method. That is, the MEM spectral analysis is equivalent to fitting an AR model to the random process. It is indeed interesting that the representation of a stochastic process by an AR model is that representation that exhibits maximum entropy. The duality of the AR model and MEM has enabled workers to apply the large body of literature on AR time series analysis to overcome shortcomings of the MEM.

The estimation of the MEM spectral density requires knowledge of the order of the AR process that we use to model the data. The importance of correctly estimating the order p is illustrated using the following AR process $y_n \equiv y(t_n)$ at times $t_n = n\Delta t$:

$$y_n = 0.75y_{n-1} - 0.5y_{n-2} + \varepsilon_n \tag{5.191}$$

with noise variance $\sigma_\varepsilon^2 = 1$ (Figure 5.45a). Here $E[y(t)\varepsilon(t)] = \sigma_\varepsilon^2$, but $E[y(t)\varepsilon'(t)] = 0$ for any other additive noise, ε'. As indicated by Figure 5.45b, which compares the theoretical power of a specified second-order AR process with the PSD computed from a realization of this process using $p = 2$ and $p = 11$ (Ulrych and Bishop, 1975), the correct choice of p is vital in obtaining a meaningful estimate of the power spectrum of the process. The peak value and the width of the spectral line of the MEM PSD estimate also may have considerable variance in the MEM estimates.

Although the MEM has numerous advantages over traditional nonparametric spectral techniques, especially for short data series, the usefulness of the approach is diminished by the lack of a straightforward criterion for choosing the length (order) of the prediction model. Too short a length results in a highly smoothed spectrum obviating the resolution advantages of the MEM, whereas an excessive length introduces spurious detail into the spectrum.

Confidence intervals: A major shortcoming of the MEM is the lack of a mathematically consistent variance estimator (confidence interval) for the spectral density, which is characteristic of the parametric spectral methods. One approach is to approximate the confidence bounds in the same way that we compute the bounds in traditional multivariate spectral analysis (i.e., using a chi-square variable with v DoF) under the assumption that the equivalent number of DoF is given by $v = N/p$, where N is the number of data points in the time series and p is the order of the model (Privalsky and Jensen, 1993, 1994). The order p should be chosen on the basis of objective criteria such as AIC, Parzen's criterion, and so on (see Lütkepohl, 1985).

5.7.2.4 An Autoregressive Model of Global Temperatures

One way to determine the effect of initial conditions and random noise on the global temperature predictions of computer-simulated general circulation models (GCMs) is to obtain a control realization, modify the initial conditions and noise, obtain a second realization, and compare results. Since this could take several weeks to months of supercomputing time, a more practical approach is to employ a model of the global air temperature series, $T(t)$, derived by Jones (1988) (Figure 5.46). If we assume that the sensitivity of GCMs to changing conditions is similar to that of a stationary autoregressive model, then marked changes in the AR model that result from slight changes in the initial conditions or inherent noise are evidence that GCMs are too sensitive to these parameters to be reliable.

If $Z_n \equiv Z(t_n)$ represents the temperature deviation (departure from the long-term mean) at year t_n, then the maximum likelihood fourth order AR model for the temperature data in Figure 5.46 is

$$Z_n = 0.669Z_{n-1} - 0.095Z_{n-2} + 0.104Z_{n-3} + 0.247Z_{n-4} + \varepsilon_n \tag{5.192}$$

where $Z_n = T_n - \overline{T}$, and ε_n is an uncorrelated white-noise series with zero mean and variance equal to $0.0115°C^2$ (Tsonis, 1991; Gray and Woodward, 1992). In general, we can state that for any AR process, the initial values will have little effect on forecasts if the sample size is large relative to the order of the process. For this reason, AR processes are often known as "short memory" processes. In the above model, the correlation between $Z(t)$ and $Z(t + m\Delta t)$ is $0.9(0.96)^m$, for values of m greater than about five. For example, the correlation coefficient between $Z(t)$ and $Z(t + 30\Delta t)$ is 0.27, while that between $Z(t)$ and $Z(t + 50\Delta t)$ is 0.14. These correlations imply that, even if we started the model with the same initial values $Z_1, ..., Z_4$, different realizations of the model would typically have low cross-correlation after 30 years and possess very little similarity beyond 50 years (Figure 5.47a). The dissimilarity is associated with the stochastic nature of the noise, $\varepsilon(t)$, which quickly decorrelates the present value of the model from its past values. The fact that the two series take on similar levels near $t = 100$ years is not an indication that they are merging since extending these realizations to even longer times shows them departing from one another.

To show that initial conditions are much less important than noise, Gray and Woodward generated two samples with different starting values but with the same noise sequence. This was intended to mimic a specified set of random conditions driving the weather but having different starting values. As revealed by Figure 5.47b, the realizations begin to merge by 30

FIGURE 5.45   Maximum entropy spectra. (a) Time series for the second-order AR process, $y_n = 0.75y_{n-1} - 0.5y_{n-2} + \varepsilon_n$ (Eqn 5.191). (b) Spectral computation for the AR process. Solid line: the true power spectrum. Dot–dash line: maximum entropy method (MEM) estimate, with 3-point ($p = 2$) prediction error filter. Dashed line: MEM estimate with 12-point ($p = 11$) error filter. *From Ulrych and Bishop (1975).*

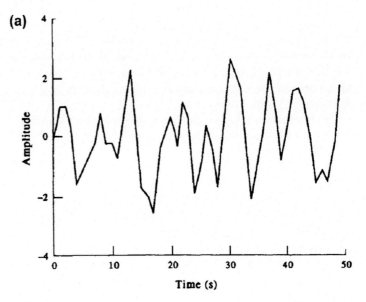

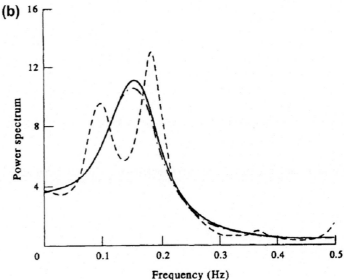

years, demonstrating their insensitivity to the initial conditions. A further point is that for stationary AR processes, the forecast function is only a function of the sample mean and the last four observations. Because the starting values are independent of the last four observations, and small changes in the starting conditions have little effect on the sample mean for a long time series, the forecasts from such a model will be insensitive to changes in initial conditions. In closing their article, Gray and Woodward (1992) note that conventional ARMA modeling methodology indicates that the temperature time series should first be differentiated. Application of a variety of techniques suggests an order 10 (AR(10)) model as the "optimum" model for the differentiated data, which gives rise to an AR(11) model for the original time series, not an AR(4) model used in the analysis. Lastly, Tsonis (1992) points out that it is not appropriate to change the noise of the signal without also changing the initial conditions.

5.7.3 Maximum Likelihood Spectral Estimation

As first demonstrated by Capon (1969), spectra can be defined using the maximum likelihood procedure. Instead of using a fixed window to operate on the autocorrelation function, the window shape is changed as a function of wavenumber or frequency. The window is designed to reject all frequency components in an optimal way, except for the one frequency component, which is desired.

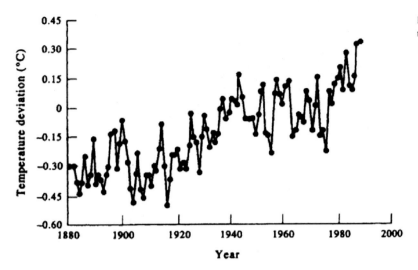

FIGURE 5.46 The annual global mean air temperatures from 1881 to 1988 as deviations (°C) from the 1951−70 average. *From Gray and Woodward (1992).*

(a)

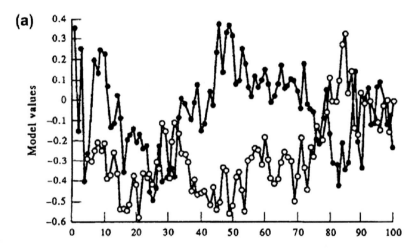

FIGURE 5.47 Two simulated realizations from the AR(4) model given by Eqn (5.192). (a) Same starting values but different and independently derived noise sequence; (b) different starting values but the same noise sequence. *From Gray and Woodward (1992).*

(b)

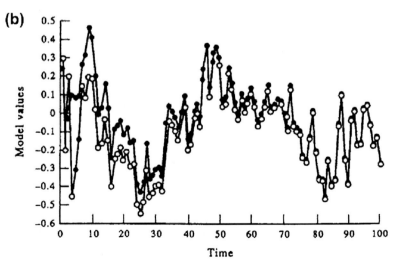

Rather than go through the details of defining the procedure for the maximum likelihood spectrum, we offer here comparisons between the traditional method (in this case, represented by a spectrum computed using a Bartlett window), a maximum likelihood spectrum, and a spectrum computed using the maximum entropy procedure (Figure 5.48). As the figure illustrates, the maximum entropy spectrum has narrow peaks, while both the Bartlett window and maximum likelihood method yield much broader spectral peaks. Note also that, except for the maximum spectral values, the maximum entropy spectrum significantly underestimates the spectral estimates for the 0.15 Hz signal and white noise. The maximum entropy spectrum also has small side-lobe energy that is dramatically less than the off-peak energy in either of the other two spectra. The maximum likelihood spectral values are also systematically lower than those using the standard method with a Bartlett window. A similar comparison is shown in Figure 5.49, which first shows a time series of a 1 Hz (1 cps) sinusoid with 10% white noise added to it (Figure 5.49a). The power spectrum computed as the square of the Fourier coefficients is displayed in Figure 5.49b. This can be compared with the narrow-peaked maximum entropy spectrum in Figure 5.49c. The peaks are located at the same frequency representative of the 1 Hz, but the maximum entropy spectrum is extremely narrow while the Fourier and maximum likelihood power spectra have very wide peaks. It is easy to see that the MEM seriously underestimates the spectral values at frequencies other than the main peak.

5.8 CROSS-SPECTRAL ANALYSIS

Estimation of autospectral density functions deals only with the frequency characteristics of a single scalar or components of a vector time series, $x(t)$. Estimation of cross-spectral density functions performs a similar analysis but for two separate time series, $x_1(t)$ and $x_2(t)$, spanning concurrent times, $0 \leq t \leq T$. Although we often use time series from similar distributions, such as the velocity records from nearby moorings, cross-spectra may also be computed for two completely different quantities. In that sense, we *can* mix apples and oranges. For example, the cross-spectrum formed from the time-varying velocity fluctuations, $x_1(t) = u'(t)$, and the temperature fluctuations, $x_2(t) = T'(t)$, measured over the same time span at the same location gives an estimate of the local eddy heat flux, $q' = \rho C\rho u'T'(t)$, as a function of frequency (ρ is the density and $C\rho$ is the specific heat of seawater). Because autospectra involve terms like $x_1 x_1^*$, where the asterisk denotes the complex conjugate, the spectra are real-valued and all phase information in the original signal is lost. Cross-

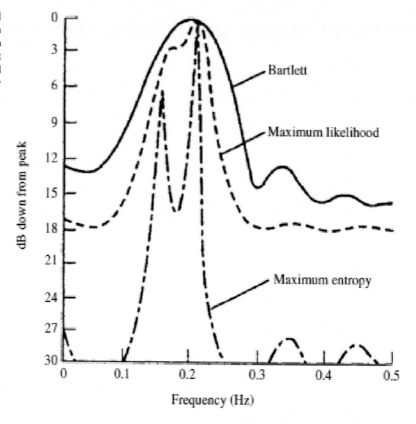

FIGURE 5.48 Power spectral estimates for a signal consisting of white noise plus two sine waves with frequencies 0.15 and 0.2 Hz (cps). Solid line: spectrum using the autocovariance method with a Bartlett smoothing window. Dashed line: maximum likelihood spectral estimate. Dash−dot line: maximum entropy spectrum. *From Lacoss (1971).*

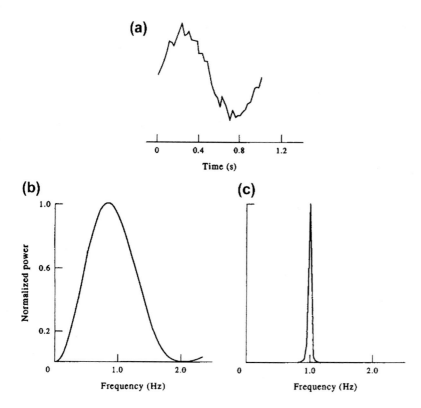

FIGURE 5.49 Comparison of spectra from the periodogram method and the maximum entropy method (MEM). (a) A sinusoid with 10% white noise and truncated with a 1 s window; (b) the power spectrum of (a) computed as the square of the modulus of the Fourier transform; (c) the MEM power spectrum of (a). Frequency in Hz (cps). *From Ulrych (1972).*

spectra, on the other hand, involve terms like $x_1 x_2^*$ and are generally complex quantities whose real and imaginary parts take into account the correlated portions of both the amplitudes and relative phases of the two signals.

There are two ways to quantify the real and imaginary parts of cross-spectra. One approach is to write the cross-spectrum as the product of an amplitude function, called the *cross-amplitude spectrum*, and a phase function called the *phase spectrum*. The sample cross-amplitude spectrum gives the distribution of co-amplitudes with frequency while the sample phase spectrum indicates the angle (or time) by which one series leads or lags the other series as a function of frequency. Alternatively, the cross-spectrum can be decomposed into a *coincident spectral density function (or co-spectrum)*, which defines the degree of co-oscillation for those frequency constituents of the two time series that fluctuate in phase, and a *quadrature spectral density function (or quad-spectrum)*, which defines the degree of co-oscillation for frequency constituents of the two series that co-oscillate but are out of phase by ±90°. Statistical confidence intervals can be provided for normalized versions of the cross-spectral estimates.

5.8.1 Cross-Correlation Functions

In Section 5.6.3.1, we showed that the autocovariance function, $C_{xx}(\tau)$, and the autospectrum, $S_{xx}(f)$, are Fourier transform pairs. Similarly, for separate time series $x_1(t)$ and $x_2(t)$, the cross-covariance function, $C_{x_1 x_2}(\tau)$, and the cross-spectrum, $S_{x_1 x_2}(f)$, are transform pairs. Thus, we can take the Fourier transform of the lagged cross-covariance function to obtain the cross-spectrum or we can take the IFT of the cross-spectrum to obtain the cross-covariance function. As a prelude to cross-spectral analysis, it is worth presenting a brief summary of cross-correlation functions commonly used in oceanography for scalar and vector time series. The cross-correlation functions tell us how closely two records are "related" in the time (space) domain, whereas the cross-spectrum tells us how oscillations within specific frequency bands are related in the frequency (wavenumber) domain.

Using the abbreviation, $C_{12}(\tau)$ for the more awkward notation, $C_{x_1 x_2}(\tau)$, the unbiased *cross-covariance function* is defined as

$$C_{12}(\tau) = \frac{1}{N-m} \sum_{n=0}^{N-m} x_1(n\Delta t) x_2(n\Delta t + \tau) \tag{5.193}$$

where $\tau = m\Delta t$ is the lag time for $m = 0, 1, ..., M, M \ll N$ (see also Section 5.6.3.1). Division of Eqn (5.193) by the product $C_{11}(0)C_{22}(0)$, corresponding to the autocovariance functions for each series at zero lag, gives the *cross-correlation coefficient function* for the data samples

$$\rho_{12}(\tau) = \frac{C_{12}(\tau)}{[C_{11}(0)C_{22}(0)]^{\frac{1}{2}}} \tag{5.194}$$

The time series $x_1(t)$ and $x_2(t)$ represent any two quantities we wish to compare. They also may represent quantities measured at different depths or locations for the same time period. For example, Kundu and Allen (1976) used the lagged covariance function

$$\rho(\mathbf{x}_1, \mathbf{x}_2; \tau) = \frac{\overline{v'(\mathbf{x}_1; t)v'(\mathbf{x}_2; t + \tau)}}{\left[\left(v'(\mathbf{x}_1; t)\right)^2 \left(v'(\mathbf{x}_2; t)\right)^2\right]^{\frac{1}{2}}}$$

$$= \frac{\frac{1}{N-m}\sum_{n=1}^{N-m} v'(\mathbf{x}_1; n)v'(\mathbf{x}_2; n + m)}{\frac{1}{N}\left[\sum_{n=1}^{N}\left(v'(\mathbf{x}_1; n)\right)^2\left(v'(\mathbf{x}_2; n)\right)^2\right]^{1/2}}, \quad m = 0, 1, ..., M \ll N \tag{5.195}$$

to examine the correlation between the alongshore (v) components of current for different coastal sites separated by a distance $d = |\mathbf{x}_1 - \mathbf{x}_2|$. Moreover, if τ_{max} is the lag that gives the maximum correlation, then the speed of propagation, c, of the coherent signal in the direction $\mathbf{d} = \mathbf{x}_1 - \mathbf{x}_2$ is $c = |\mathbf{d}|/\tau_{max}$, the direction of propagation is determined from the sign of τ_{max} (Figure 5.50). In Figure 5.50, the lagged correlations between time series of low-pass filtered alongshore currents, $v(\mathbf{x}; t)$, at different sites along the continental shelf are used to examine the poleward propagation of low-frequency coastal-

FIGURE 5.50 The lag time of maximum correlation of the longshore component of current at 60-m depth versus the distance of separation for the Oregon coast for 1973. Results indicate a mean northward signal propagation of 120 km/day. *From Kundu and Allen (1976).*

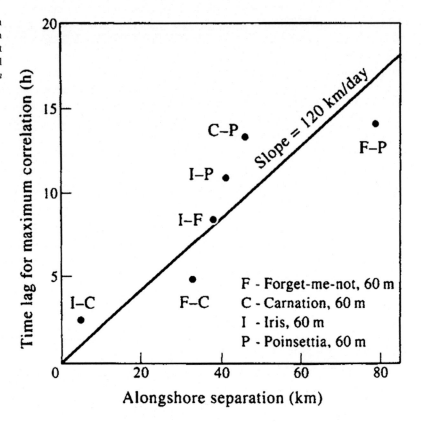

trapped waves. Results in the figure are based on currents at 60-m depth. Letters refer to pairs of stations used; e.g., C − P is the lag between the Carnation and Poinsettia stations.

A generalization of Eqn (5.195) is given by Kundu (1976). If $w = u + iv$ is the complex velocity for horizontal velocity components, (u, v), then the correlation between two rotating velocity vectors is given by the complex correlation coefficient

$$\rho(\mathbf{x}_1, \mathbf{x}_2; \tau) = \frac{\overline{w_1^*(t)w_2(t+\tau)}}{\left[\overline{\left(w_1^*(t)w_1(t)\right)^{\frac{1}{2}}\left(w_2^*(t)w_2(t)\right)^{\frac{1}{2}}}\right]}$$

(5.196)

where subscripts denote locations 1 and 2, and the overbars denote the time or ensemble average. The correlation, ρ, which is independent of the choice of coordinate systems, is a complex quantity whose magnitude gives the overall measure of correlation and whose phase gives the average counterclockwise angle of the second vector with respect to the first.

5.8.2 Cross-Covariance Method

Following the Blackman−Tukey procedure for autospectral density estimation, the Fourier transform of the cross-covariance function, $C_{12}(\tau)$, can be used to find the cross-spectrum, $S_{12}(f)$. Although the cross-covariance method is straightforward to apply, the sample cross-covariance function, $C_{12}(\tau)$, suffers from the same disadvantage as the sample autocovariance function, $C_{11}(\tau)$, in that neighboring values tend to be highly correlated, thereby reducing the effective number of DoF. Moreover, the statistical significance falls off rapidly with increasing lag, τ, so that the number of lags, M, is much shorter than the record length ($M \ll N$). Calculation of cross-spectra is best performed using the discrete Fourier transform method. In fact, it is common practice these days to use the IFT of the cross-spectrum to obtain the cross-covariance function.

5.8.3 Fourier Transform Method

As with autospectral analysis, estimates of cross-spectral density functions are most commonly derived using Fourier transforms. The steps in calculating the cross-spectrum using standard Fourier transforms or FFTs are similar to those discussed in Section 5.6.12 for spectra (see also Bendat and Piersol, 1986):

1. Ensure that the two time series, $x_1(t)$ and $x_2(t)$, span the same period of time, t_n, where $n = 0, 1, ..., N − 1$ and $T = N\Delta t$ is the length of each record. Remove their respective means and trends, and if needed, despike the data series. If block averaging is to be used to improve the statistical reliability of the spectral estimates, divide the available data for each pair of time series into m sequential blocks of N' data values each, where $N' = N/m$. Note that in Figure 5.50, the time series x_1 and x_2 denote the alongshore currents, $v(t)$, at spatial locations $\mathbf{x}_1$ and $\mathbf{x}_2$.
2. To reduce side-lobe leakage, taper the time series, $x_1(t)$ and $x_2(t)$, using a Hanning (raised-cosine) window, Kaiser-Bessel window, or other appropriate taper.
3. Compute the Fourier transforms, $X_1(f_k)$, $X_2(f_k)$, $k = 0, 1, 2, ..., N − 1$, for the two time series, $x_1(t)$ and $x_2(t)$. For block-segmented data, calculate the Fourier transforms, $X_{1m}(f_k)$ and $X_{2m}(f_k)$, for each of the m blocks, where $k = 0, 1, ..., N' − 1$. To reduce the variance associated with the tapering in step 2, the transforms can be computed for overlapping segments.
4. Rescale the spectra to account for the loss of "energy" during application of the window (see Table 5.13). That is, adjust the scale factor of $X_1(f_k)$ and $X_2(f_k)$ [or $X_{1m}(f_k)$, $X_{2m}(f_k)$] for the reduction in spectral energy due to the tapering in step 2. For the Hanning window, multiply the amplitudes of the Fourier transforms by $\sqrt{8/3}$.
5. Compute the raw cross-spectral power density estimates for each pair of time series (or each pair of blocks), where for the two-sided spectral density estimate for no block averaging

$$S_{12}(f_k) = \frac{1}{N\Delta t}\left[X_1^*(f_k)X_2(f_k)\right], k = 0, 1, 2, ..., N − 1$$

or, for block averaging,

$$S_{12}(f_k; m) = \frac{1}{N'\Delta t}\left[X_{1m}^*(f_k)X_{2m}(f_k)\right], k = 0, 1, 2, ..., N' − 1$$

(5.197a)

and for the one-sided spectral density estimates for no block averaging

$$G_{12}(f_k) = \frac{2}{N\Delta t}\left[X_1^*(f_k)X_2(f_k)\right], k = 0, 1, 2, ..., N/2$$

or, for block averaging

$$G_{12}(f_k; m) = \frac{2}{N'\Delta t}\left[X_{1m}^*(f_k)X_{2m}(f_k)\right], k = 0, 1, 2, ..., N'/2 \tag{5.197b}$$

6. In the case of the block-segmented data, average the raw cross-spectral density estimates from the m blocks of data to obtain the smoothed periodogram for $S_{12}(f_k)$, the two-sided cross-spectrum, or $G_{12}(f_k)$, the one-sided cross-spectrum.

Cross-covariance function: Because the cross-covariance function, $C_{12}(\tau)$ $(= R_{12}(\tau)$, the cross-correlation function, if the mean is removed from the record), and the cross-spectrum are Fourier transform pairs, Eqns (5.197a) and (5.197b) can be used to obtain a smoothed or unsmoothed estimate of the cross-covariance function. To do this, we first calculate the Fourier transforms $X_1(f)$ and $X_2(f)$ of the individual time series, and then determine the product $S_{12}(f) = \frac{1}{N\Delta t}\left[X_1^*(f_k)X_2(f_k)\right]$. We then take the IFT of the cross-spectrum, $S_{12}(f)$, to obtain the cross-covariance function

$$C_{12}(\tau) = \int_{-\infty}^{\infty} S_{12}(f)e^{i2\pi f\tau}df \tag{5.198}$$

If the spectrum is unsmoothed prior to the inverse Fourier Transform (IFT), or IFFT if the number of spectral estimates is a power of two, we obtain the raw cross-covariance function. If, on the other hand, the cross-spectrum is smoothed prior to Eqn (5.198) using one of the spectral windows, such as the Hanning window, the cross-covariance function also will be a smoothed function.

We can use the acoustic backscatter data in Table 5.1 to illustrate the direct and indirect methods for calculating the cross-covariance function. In Table 5.14, we present the normalized, unsmoothed cross-covariance function, $\rho_{12}(\tau) = \frac{C_{12}(\tau)}{[C_{11}(0)C_{22}(0)]^{1/2}}$, obtained directly from the definition Eqn (5.194). In this case, the lag τ is in 5-m depth increments. The indirect approach is based on the Fourier estimates presented in Tables 5.15 and 5.16. Here, we first give the Fourier transforms, $X_1(f)$ and $X_2(f)$, of the two profile series as a function of wavenumber, f (Table 5.15). We next calculate the cross-spectrum, $S_{12}(f) = \frac{1}{N\Delta t}\left[X_1^*(f_k)X_2(f_k)\right]$, and then take the inverse transform of $S_{12}(f)$ to obtain the cross-covariance function, $C_{12}(\tau)$, as a function of lag (Table 5.16). No smoothing was applied to either data set, and the results obtained from the IFT method are identical to those listed in Table 5.14, within roundoff error. The advantage of the transform approach is that it is straightforward to derive a smoothed cross-covariance function by windowing the cross-spectral estimate prior to Fourier inversion.

5.8.4 Phase and Cross-Amplitude Functions

Suppose that the constituents of the bivariate time series $\{x_1(t), x_2(t)\}$ have the same frequency, f_0, but different amplitudes (A_1, A_2) and different phases (ϕ_1, ϕ_2), respectively. In particular, let

$$x_k(t) = A_k\cos(2\pi f_0 t + \phi_k), k = 1, 2 \tag{5.199}$$

TABLE 5.14 Unsmoothed, normalized cross-covariance function, $\rho_{12}(\tau)$, given by Eqn (5.194), as a function of lag τ in increments of 5 m for bin 1 of Beams 1 and 2 of the acoustic backscatter spatial series (profiles) listed in Table 5.1.

Lag τ (m)	0	5	10	15	20	25	30	35
	0.96	0.94	0.85	0.71	0.57	0.48	0.40	0.31
Lag τ (m)	40	45	50	55	60	65	70	75
	0.23	0.14	0.02	−0.19	−0.24	−0.37	−0.46	−0.48

The first acoustic bins of the two beams (bin#1 in each case) are separated by a horizontal distance of roughly 3.9 m.

TABLE 5.15 Complex Fourier transforms of $X_1(f_k)$ and $X_2(f_k)$ for the profiles of acoustic backscatter listed in Table 5.1.

FFT	k = 0	1	2	3	4	5	6	7
$X_1(f_k)$	348.13	−289.32	71.17	15.52	55.16	97.59	−28.66	5.07
	0.00	214.96	−16.35	−117.25	105.57	−16.98	−21.37	4.28
$X_2(f_k)$	339.02	−226.53	119.54	55.84	−5.24	59.55	−36.39	4.22
	0.00	227.88	38.22	−93.12	122.33	−24.13	−6.57	−19.09
k = 8	**9**	**10**	**11**	**12**	**13**	**14**	**15**	**16**
1.13	−6.16	41.11	24.03	−1.79	4.63	3.74	4.09	27.13
6.87	21.29	−2.96	−36.43	−4.60	1.08	3.54	18.45	0.00
11.90	5.68	23.89	13.85	3.96	7.37	11.27	2.34	27.79
−5.35	−4.63	−5.13	−18.72	−1.67	−4.93	−4.47	9.00	0.00

For each wavenumber, f_k, the table lists the real part of the transform (top line of each X pair) followed by the imaginary part (bottom line of each X pair), where $X_j(f_k) = ReX_j(f_k) + iImX_j(f_k)$, $j = 1, 2$. The vertical wavenumber, $f_k = kf$, $k = 0, 1,..., 16$, where the fundamental vertical waveumber, $f = 1/155$ m = 0.00645 cpm (cycles per meter).

TABLE 5.16 The inverse fast Fourier transform (IFFT) of the cross-spectrum $S_{12}(f_k) = (N\Delta t)^{-1}[X_1(f_k)^*X_2(f_k)]$ using the values in Table 5.15. The asterisk denotes the complex conjugate.

	τ = 0	1	2	3	4	5	6	7
$C_{12}(\tau)$	13483.7	12752.4	11151.5	9087.4	6992.3	5436.5	4589.9	3411.7
τ = 8	**9**	**10**	**11**	**12**	**13**	**14**	**15**	**16**
2382.5	1393.8	160.6	−1103.6	−2096.0	−3103.5	−3610.5	−3623.8	−3222.1

The values represent the raw (unnormalized) estimates of the cross-covariance function, $C_{12}(\tau)$, as a function of lag τ ($0 \leq \tau \leq 16$) in increments of 5 m for bin 1 of Beams 1 and 2 of the acoustic backscatter spatial series (profiles) listed in Table 5.1.

The Fourier transform of $x_k(t)$, over $-T/2 \leq t \leq T/2$ is

$$X_k(f) = \frac{A_k}{2}\left\{ e^{i\phi_k}\frac{\{\sin[\pi(f - f_0)T]\}}{\pi(f - f_0)} + e^{-i\phi_k}\frac{\{\sin[\pi(f + f_0)T]\}}{\pi(f + f_0)} \right\}, i = 1, 2 \tag{5.200}$$

Hence, the sample cross-spectra of the two series is

$$S_{12}(f) = \frac{1}{T}\left[X_1^*(f)X_2(f) \right] \tag{5.201}$$

where X_1^* is the complex conjugate of X_1. From this expression, we obtain

$$S_{12}(f) = \frac{A_1A_2}{4T}\left\{ e^{-i\phi_1}\frac{\sin[\pi(f - f_0)T]}{\pi(f - f_0)} + e^{i\phi_1}\frac{\sin[\pi(f + f_0)T]}{\pi(f + f_0)} \right\}$$

$$\times \left\{ e^{i\phi_2}\frac{\sin[\pi(f - f_0)T]}{\pi(f - f_0)} + e^{-i\phi_1}\frac{\sin[\pi(f + f_0)T]}{\pi(f + f_0)} \right\} \tag{5.202}$$

where

$$S_{12}(f)_{T\rightarrow\infty} \rightarrow \frac{A_1A_2}{4}\left[e^{-i(\phi_2 - \phi_1)}\delta(f + f_0) + e^{i(\phi_2 - \phi_1)}\delta(f - f_0) \right] \tag{5.203}$$

The phase difference, $(\phi_2 - \phi_1)$, in the above expressions determines the lead (or lag) of one cosine oscillation relative to the other for given frequency, f. The cross-amplitude, $A_1 A_2$, is the square of the geometric mean amplitude of the co-oscillation for frequency, f. From Eqn (5.203), the sample cross-spectrum is

$$S_{12}(f) = \frac{A_1(f)A_2(f)}{T} \left[e^{i[\phi_2(f)-\phi_1(f)]} \right] \tag{5.204}$$

or

$$S_{12}(f) = A_{12}(f)\left[e^{i\phi_{12}(f)} \right] \tag{5.205}$$

where the sample phase spectrum, $\phi_{12}(f) = \phi_2(f) - \phi_1(f)$, is an odd function of frequency, and the sample cross-amplitude spectrum, $A_{12}(f) = A_1(f)A_2(f)/T$, is a positive even function of f.

5.8.5 Coincident and Quadrature Spectra

An alternative description of the above information involves formulating the cross-spectra in terms of coincident (C) and quadrature (Q) spectra. In this case, we can write

$$S_{12}(f) = C_{12}(f) - iQ_{12}(f) \tag{5.206}$$

where

$$C_{12}(f) = A_{12}(f)\cos[\phi_{12}(f)]; \; Q_{12}(f) = -A_{12}(f)\sin[\phi_{12}(f)] \tag{5.207}$$

and

$$A_{12}^2(f) = C_{12}^2(f) + Q_{12}^2(f); \; \phi_{12}(f) = \tan^{-1}\left[\frac{-Q_{12}(f)}{C_{12}(f)} \right] \tag{5.208}$$

Here, $C_{12}(f)$ is an even function of frequency and $Q_{12}(f)$ is an odd function. The cospectral density function, $C_{12}(f)$, for frequency, f is not to be confused with the covariance function, $C_{12}(\tau)$, at time lag τ. Where confusion may arise, we use the cross-correlation, $R_{12}(\tau)$, in place of $C_{12}(\tau)$). If we consider the bivariate cosine example that we used in Eqn (5.199), we have

$$C_{12}(f) = \frac{A_1 A_2}{4}\cos(\phi_2 - \phi_1)[\delta(f+f_0) + \delta(f-f_0)]$$

$$= \left\{ \frac{A_1 \cos \phi_1 \cdot A_2 \cos \phi_2}{4} + \frac{A_1 \sin \phi_1 \cdot A_2 \sin \phi_2}{4} \right\}[\delta(f+f_0) + \delta(f-f_0)] \tag{5.209}$$

where the dot denotes multiplication. The sample cospectrum, $C_{12}(f)$, measures the covariance between the two cosine components and the two sine components. That is, it measures the contributions to the cross-spectrum from those components of the two time series that are "in phase" (phase differences of 0 or 180°). The sample quadrature spectrum, $Q_{12}(f)$, determines the contributions from those components of the time series that are coherent but "out of phase" (phase difference $\pm 90°$).

5.8.5.1 Relationship of Co- and Quad-Spectra to Cross-Covariance

The inverse transform of the cross-spectrum gives the cross-covariance (cross-correlation)

$$R_{12}(\tau) = \int_{-\infty}^{\infty} [C_{12}(f) - iQ_{12}(f)]e^{i2\pi f\tau}df$$

$$= \int_{-\infty}^{\infty} C_{12}(f)\cos(2\pi f\tau)df + \int_{-\infty}^{\infty} Q_{12}(f)\sin(2\pi f\tau)df \tag{5.210}$$

Because $C_{12}(f)$ is an even function, $R_{12}(0) = \int_{-\infty}^{\infty} C_{12}(f)df$. If we define

$$C_{12}(f) = \int_{-T}^{T} R_{12}^{+}(\tau)\cos(2\pi f\tau)d\tau$$

$$Q_{12}(f) = \int_{-T}^{T} R_{12}^{-}(\tau)\sin(2\pi f\tau)d\tau \qquad (5.211)$$

then

$$R_{12}^{+}(\tau) = \frac{1}{2}[R_{12}(\tau) + R_{12}(-\tau)] \text{ (the even part)}$$

$$R_{12}^{-}(\tau) = \frac{1}{2}[R_{12}(\tau) - R_{12}(-\tau)] \text{ (the odd part)} \qquad (5.222)$$

5.8.6 Coherence Spectrum (Coherency)

The *squared coherency, coherence-squared function,* or *coherence spectrum* between two time series, $x_1(t)$ and $x_2(t)$, is defined for frequencies, f_k, $k = 0, 1, ..., N-1$, as

$$\gamma_{12}^2(f_k) = \frac{|G_{12}(f_k)|^2}{G_{11}(f_k)G_{22}(f_k)}$$

$$= \frac{|S_{12}(f_k)|^2}{S_{11}(f_k)S_{22}(f_k)} \qquad (5.223)$$

$$= \frac{\left[C_{12}^2(f_k) + Q_{12}^2(f_k)\right]^2}{S_{11}(f_k)S_{22}(f_k)}$$

where $G_{11}(f_k)$ is the one-sided spectrum (confined to $f_k \geq 0$), $S_{11}(f_k) = \frac{1}{2}G_{11}(f_k)$ is the two-sided spectrum defined for all frequencies (with half the spectral energy at positive frequencies and half at negative frequencies) and $G_{12}(f_k)$ is the one-sided cross-spectrum. Here

$$0 \leq |\gamma_{12}^2(f_k)| \leq 1 \qquad (5.224)$$

and

$$\gamma_{12}(f) = |\gamma_{12}^2(f_k)|^{1/2} e^{-i\phi_{12}f_k} \qquad (5.225)$$

where $|\gamma_{12}^2(f_k)|^{1/2}$ is the modulus of the coherence function and $\phi_{12}(f_k)$ the phase lag between the two signals at frequency f_k, (Figure 5.51). In the literature, both the squared coherency, γ_{12}^2, and its square root are termed "the coherence" so that there is often a confusion in meaning (Julian, 1975). To avoid any ambiguity, it is best to use squared-coherency when conducting coherence analyses once the sign of the coherence function is determined. This has the added advantage that squared coherency represents the fraction of the variance in x_1 ascribable to x_2 through a linear relationship between x_1 and x_2. Two signals of frequency f_k are considered highly coherent and in phase if $|\gamma_{12}^2(f_k)| \approx 1$ and $\phi_{12}(f_k) \approx 0$, respectively (Figure 5.51). The addition of random noise to the functions, x_1 and x_2, of a linear system decreases the coherence-squared estimate and increases the noisiness of the phase associated with the system parameters. Estimation of $\gamma_{12}^2(f_k)$ is one of the most difficult problems in time series analysis since it is so highly noise dependent. We also point out that phase estimates generally become unreliable where coherency amplitudes fall below the 90–95% confidence levels for a given frequency.

The real part of the coherence function, $\gamma_{12}(f_k)$, lies between -1 and $+1$ while the squared-coherency is between 0 and $+1$. If the noise spectrum, $S_{\varepsilon\varepsilon}(f_k)$, is equal to the output spectrum, then the coherence function is zero. This says that white noise is incoherent, as required. Also, when $S_{\varepsilon\varepsilon}(f_k) = 0$, we have $\gamma_{12}^2(f_k) = 1$; that is the coherence is perfect if there

FIGURE 5.51 Coherence between current vector time series at sites Hook and Bell on the northeast coast of Australia (separation distance ≈ 300 km). (a) Coherence squared; (b) phase lag. Solid line: inner rotary coherence (rotary current components rotating in the same sense). Dashed line: outer rotary coherence (rotary current components rotating in the opposite sense). The increase in inner phase with frequency indicates equatorward phase propagation. Positive phase means that Hook leads Bell. *From Middleton and Cunningham (1984).*

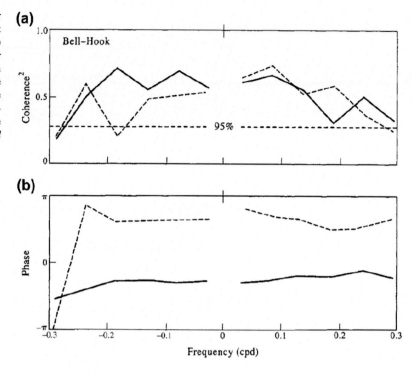

is no spectral noise in the input signal. We note that, if no spectral smoothing is applied, we are assuming that there is no spectral noise. In this case, the coherency spectrum will be unity for all frequencies, which is clearly not physically realistic. Noise can be introduced to the system by smoothing over adjacent frequencies. We also can overcome this problem by a prewhitening step that introduces some acceptable noise into the spectra.

5.8.6.1 Confidence Levels

The final step in any coherence analysis is to specify the confidence limits for the coherence-square estimates. If $1 - \alpha$ is the $(1 - \alpha)100\%$ confidence interval we wish to specify for a particular coherence function, then, for all frequencies, the limiting value for the coherence-square (i.e., the level up to which coherence-square values can occur by chance) is given by

$$\gamma_{1-\alpha}^2 = 1 - \alpha^{[1/(\text{EDoF}-1)]}$$

$$= 1 - \alpha^{[2/(\text{DoF}-2)]} \tag{5.226}$$

where EDoF = DoF/2 is the number of independent cross-spectral realizations in each frequency band (Thompson, 1979). The commonly used confidence intervals of 90%, 95%, and 99% correspond to $\alpha = 0.10, 0.05$, and 0.01, respectively. As an example, suppose that each of our coherence estimates is computed from an average over three adjacent cross-spectral Fourier components, then EDoF = 3 (DoF = 6). The 95% confidence level for the squared coherence would then be $\gamma_{95}^2 = 1 - (0.05)^{0.5} = 0.78$. Alternatively, if the cross-spectrum and spectra were first smoothed using a Hamming window spanning, the entire width of the data series, EDoF = 2.5164 (Table 5.5) and the 95% confidence interval $\gamma_{95}^2 = 1 - (0.05)^{0.6595} = 0.86$. For EDoF = 2, $\gamma_{1-\alpha}^2 = 1 - \alpha$ so that the confidence level for the 95% confidence interval is equal to itself.

A useful reference for coherence significance levels is Thompson (1979). In this paper, the author tests the reliability of significance levels, $\gamma_{1-\alpha}^2$, estimated from Eqn (5.226) with the coherence-square values obtained through the summations

$$\gamma^2(f) = \frac{\left| \sum_{k=1}^{K} X_{1k}(f) X_{2k}^*(f) \right|^2}{\sum_{k=1}^{K} |X_{1k}(f)|^2 \sum_{k=1}^{K} |X_{2k}(f)|^2} \tag{5.227}$$

In this expression, X_{1k} and X_{2k} are the Fourier transforms of the respective random time series, $x_{1k}(t)$ and $x_{2k}(t)$, generated by a Monte Carlo approach, and the asterisk denotes the complex conjugate. The upper limit K corresponds to the value of EDoF in Eqn (5.227). Because $\gamma^2(f)$ is generated using random data, it should reflect the level of squared coherency that can occur by chance. For each value of K, $\gamma^2(f)$ was calculated 1,000 times and the resultant values sorted as 90th, 95th, and 99th percentiles. The operation was repeated 10 times and the means and standard deviations were calculated. This amounts to a total of 20,000 Fourier transforms for each K (EDoF). There is excellent agreement between the significance level derived from Eqn (5.226) and the coherence-square values for a white-noise Monte Carlo process (Table 5.17), lending considerable credibility to the use of Eqn (5.226) for computing coherence significance levels. The comparisons in Table 5.17 are limited to the 90 and 95% confidence intervals for $4 \leq K \leq 30$. Thompson (1979) includes the 99% interval and a wider range of K (EDoF) values.

Confidence intervals for coherence amplitudes, as well as for coherence phase, admittance, and other signal properties (see next section), can be derived using the data itself (Bendat and Piersol, 1986). Let $\widehat{\varphi}$ be an estimator for φ, a continuous, stationary random process, and define the standard error or random error of sample values as

$$\text{random error} = \sigma[\widehat{\varphi}] = \left(E\left[\widehat{\varphi}^2\right] - \{E[\widehat{\varphi}]\}^2\right)^{1/2} \tag{5.228a}$$

and the RMS error as

$$\text{RSM error} = \left(E\left[(\widehat{\varphi} - \varphi)^2\right]\right)^{1/2} = \left(\sigma^2[\widehat{\varphi}] + B^2[\widehat{\varphi}]\right)^{1/2} \tag{5.228b}$$

where B is the bias term $B[\widehat{\varphi}] = E[\widehat{\varphi}] - \varphi$ and $E[x]$ is the expected value of x. If we now divide each error term by the quantity, φ being estimated, we obtain the normalized random error

$$\varepsilon_r = \frac{\sigma[\widehat{\varphi}]}{\varphi} = \frac{\left(E\left[\widehat{\varphi}^2\right] - \{E[\widehat{\varphi}]\}^2\right)^{1/2}}{\varphi} \tag{5.229a}$$

and the normalized RMS error

$$\varepsilon = \frac{\left(E\left[(\widehat{\varphi} - \varphi)^2\right]\right)^{1/2}}{\varphi} = \frac{\left(\sigma^2[\widehat{\varphi}] + B^2[\widehat{\varphi}]\right)^{\frac{1}{2}}}{\varphi} \tag{5.229b}$$

where it is assumed that $\varphi \neq 0$. Provided ε_r is small, the relation

$$\widehat{\varphi}^2 = \varphi^2(1 \pm \varepsilon_r) \tag{5.230}$$

yields

$$\widehat{\varphi} = \varphi(1 \pm \varepsilon_r)^{1/2} \approx \varphi(1 \pm \varepsilon_r/2) \tag{5.231}$$

TABLE 5.17 Monte Carlo estimates, $\gamma^2(f)$, of the significant coherence-squared and prediction of this value using Eqn (5.226) for significance intervals $\alpha = 0.05$ and 0.10 for equivalent degrees of freedom (EDoF) = 4, 5, 6, 8, 10, 20, and 30.

	EDoF = 4	EDoF = 5	EDoF = 6	EDoF = 8	EDoF = 10	EDoF = 20	EDoF = 30
$\alpha = 0.10$							
$\gamma^2(f)$	0.539	0.437	0.371	0.288	0.230	0.114	0.076
$\gamma^2_{0.90}$	0.536	0.438	0.369	0.280	0.226	0.114	0.076
$\alpha = 0.05$							
$\gamma^2(f)$	0.629	0.531	0.452	0.354	0.288	0.144	0.099
$\gamma^2_{0.95}$	0.632	0.527	0.451	0.348	0.283	0.146	0.098

After Thompson (1979).

so that

$$\varepsilon_r\left[\widehat{\varphi}^2\right] \approx 2\varepsilon_r[\widehat{\varphi}] \tag{5.232}$$

Thus, for small ε_r the normalized error for squared estimates $\widehat{\varphi}^2$ is roughly twice the normalized error for unsquared estimates.

When the estimates $\widehat{\varphi}$ have a small bias error, $B[\widehat{\varphi}] \approx 0$, and a small normalized error, e.g., $\varepsilon \leq 0.2$, the probability density for the estimates can be approximated by a Gaussian distribution. The confidence intervals for the unknown true parameter, φ, based on a single estimate, $\widehat{\varphi}$, are then

$$\widehat{\varphi}(1-\varepsilon) \leq \varphi \leq \widehat{\varphi}(1+\varepsilon) \text{ with } 68\% \text{ confidence} \tag{5.233a}$$

$$\widehat{\varphi}(1-2\varepsilon) \leq \varphi \leq \widehat{\varphi}(1+2\varepsilon) \text{ with } 95\% \text{ confidence} \tag{5.233b}$$

$$\widehat{\varphi}(1-3\varepsilon) \leq \varphi \leq \widehat{\varphi}(1+3\varepsilon) \text{ with } 99\% \text{ confidence} \tag{5.233c}$$

5.8.7 Frequency Response of a Linear System

We define the admittance (or transfer) function of a linear system as

$$H_{12}(f_k) = \frac{S_{12}(f_k)}{S_{11}(f_k)} = \frac{G_{12}(f_k)}{G_{11}(f_k)}, f_k = \frac{k}{T}, k = 1, \ldots, N$$

$$= |H_{12}(f_k)|e^{-i\phi_{12}(f_k)} \tag{5.234}$$

where $S_{11}(f_k)$ and $G_{11}(f_k)$ are, respectively, the two-sided and one-sided autospectral estimates for the time series $x_1(t)$ selected here as the input time series. The gain (or admittance amplitude) function, $H(f_k)$, behaves like a spectral regression coefficient at each frequency, f_k. Using the definition $G_{12}(f_k) = C_{12}(f_k) - iQ_{12}(f_k)$, where C is the co-spectrum and Q is the quadrature spectrum, we obtain

$$|H_{12}(f_k)| = \frac{G_{12}(f_k)}{G_{11}(f_k)}$$

$$= \frac{\left|C_{12}^2(f_k) + Q_{12}^2(f_k)\right|^{1/2}}{G_{11}(f_k)} \tag{5.235}$$

and where $\phi_{12}(f_k) = \tan^{-1}[-Q_{12}(f_k)/C_{12}(f_k)]$ from Eqn (5.208). In Figure 5.52 we show the complex admittance for the observed alongshore component of oceanic wind velocity (time series 1) and the alongshore component of wind velocity derived from pressure-derived geostrophic winds (time series 2). As noted in the figure caption, the analysis is based on two separate methods for defining what is meant by the "alongshore component" of wind velocity. For both definitions of "alongshore" (as represented by the solid and dashed lines in the figure), the geostrophic winds closely approximate the amplitude and phase of the actual winds up to a frequency of about 0.05 cph (period = 20 h; log(0.05) = −1.3 (cph)) after which the two signals no longer resemble one another. It is also at this higher frequency that the coherence consistently begins to fall below the 90% confidence level.

5.8.7.1 Multi-Input Systems Cross-Spectral Analysis

Many oceanographic time series are generated through the combined effects of several mutually coherent inputs. For example, low-frequency fluctuations in coastal sea level typically arise through the combined forcing of atmospheric pressure, along- and cross-shore wind stress, and surface buoyancy flux. Coherences between the forcing variables (e.g., pressure, alongshore wind stress, and runoff) are generally quite high. Because of this, it would be physically incorrect to use ordinary cross-spectral analysis, which simply examines the correlation functions, $\gamma_{y:x}^2$, between the output, $y(t)$, and each of the inputs, $x(t)$, individually without taking into account the mutual correlation among all the inputs. If this is not done, the sum of the individual correlation functions can exceed unity. Provided that long-term sea-level fluctuations (the output time series) are linearly related to the individual forcing functions (the input time series), we can use *multi-input systems cross-spectral analysis* to calculate the relative contribution each of the input terms makes to the output. The effective correlation function for the total system will then be less than unity, as required. This concept was pioneered in

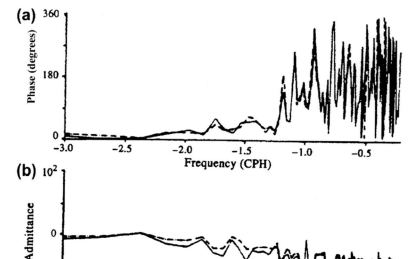

FIGURE 5.52 Complex admittance for observed (series 1) and computed (series 2) alongshore components of oceanic wind velocity for May–September 1980 off the coast of Vancouver Island. (a) Phase; (b) admittance amplitude. Positive phase means that series 1 leads series 2. Solid lines are for "alongshore" defined as parallel to the local shoreline; dashed lines are for "alongshore" derived using principal component analysis. The horizontal axes are log(frequency), where frequency is in cycles per hour (cph). *From Thomson (1983).*

oceanography by Cartwright (1968), Groves and Hannan (1968), and Wunsch (1972). Further advances were made by Sokolova et al. (1992) and Rabinovich et al. (1992). All studies were concerned with sea-level variations.

The purpose of this section is to provide a brief overview of multiple systems analysis. For a thorough generalized presentation, the reader is directed to Bendat and Piersol (1986). Consider K constant-parameter linear systems associated with K stationary and ergodic input time series, $x_k(t)$, $k = 1, 2, ..., K$, a noise function, $\varepsilon(t)$, and a single output, $y(t)$, such that

$$y(t) = \sum_{k=1}^{K} y_k(t) + \varepsilon(t) \tag{5.236}$$

where $y_k(t)$ are the outputs generated by each of the measured inputs, $x_k(t)$. We can only measure the accumulated response, $y(t)$, not the individual responses, $y_k(t)$. In the present context, $y(t)$ represents the measured time series of coastal sea level, $x_k(t)$ the corresponding weather variables, and $\varepsilon(t)$ the deviations from the ideal response due to instrument noise, remotely generated subinertial waves (waves with periods greater than the local inertial frequency), and other physical processes not correlated with the input functions. The Fourier transform of the output, $y(t)$ is

$$Y(f) = \sum_{k=1}^{K} Y_k(f) + E(f)$$

$$= \sum_{k=1}^{K} H_k(f) X_k(f) + E(f) \tag{5.237}$$

where

$$H_k(f) = \frac{Y_k(f)}{X_k(f)}, k = 1, 2, ..., K \tag{5.238}$$

is the admittance (or transfer) function relating the kth input with the kth output at frequency, f. The frequency-domain spectral variables, $X_k(f)$ and $Y(f)$, can be computed from the measured time series, $x_k(t)$ and $y(t)$. Using these variables, we can then determine the functions, $H_k(f)$ and other properties of the system.

Multiplication of both sides of Eqn (5.237) by $X_j^*(f)$, the complex conjugate of $X_j(f)$, for any fixed $j = 1, 2, ..., K$, yields the power spectral relation

$$S_{jy}(f) = \sum_{k=1}^{K} H_k(f)S_{jk}(f) + S_{je}(f), \quad j = 1, 2, ..., K \tag{5.239}$$

in which

$$\begin{aligned} S_{jy}(f) &= \overline{X_j^*(f)Y(f)}, \quad j = 1, 2, ..., K \\ S_{jk}(f) &= \overline{X_j^*(f)X_k(f)}, \quad j = 1, 2, ..., K \end{aligned} \tag{5.240}$$

Here, the overbar denotes the average value, the $S_{jy}(f)$ are the cross-spectra between the K inputs and the single output, $S_{jk}(f)$ are the cross-spectra ($j \neq k$) and spectra ($j = k$) among the input variables, and $S_{je}(f)$ is the cross-spectrum between the input variables and the noise function. If the noise function, $\varepsilon(t)$, is uncorrelated with each input, x_k (as is normally assumed), the cross-spectral terms, $S_{je}(f)$, will be zero and Eqn (5.239) becomes

$$S_{jy}(f) = \sum_{k=1}^{K} H_k(f)S_{jk}(f), \quad j = 1, 2, ..., K \tag{5.241}$$

This expression is a set of K equations in K unknowns—the $H_k(f)$ for $k = 1, 2, ..., K$—where all spectral terms can be computed from the measured records of $y(t)$ and $x_k(t)$. If the model is well defined, matrix techniques can be used to find the $H_k(f)$. Bendat and Piersol (1986) also define the problem in terms of the *multiple and partial coherence functions* for the system. The multiple coherence function is given by

$$\gamma_{y:x}^2 = \frac{S_{vv}(f)}{S_{yy}(f)} = 1 - \frac{S_{\varepsilon\varepsilon}(f)}{S_{yy}(f)} \tag{5.242}$$

where $S_{vv}(f)$ is the multiple coherent output spectrum, $S_{yy}(f)$ is the output spectrum, and $S_{\varepsilon\varepsilon}(f)$ is the noise spectrum. As with any squared coherence function, $0 \leq |\gamma_{y:x}^2| \leq 1$. For any problem with multiple inputs, $\gamma_{y:x}^2$ takes the form of a matrix whose off-diagonal elements take into account the coherent interactions among the different input terms. Expressions (5.241) and (5.242) simplify even further if the inputs themselves are mutually uncorrelated. In that case

$$H_j(f) = \frac{S_{jy}(f)}{S_{jj}(f)}, \quad j = 1, 2, ..., K;$$

$$|H_j(f)|^2 S_{jj}(f) = \gamma_{jy}^2 S_{yy}(f) \tag{5.243}$$

Hence, the contribution of the input variable, $x_j(t)$, to the output variable, $y(t)$, occurs only through the transfer (admittance) function, $H_j(f)$, of that particular input variable. No leakage of $x_j(t)$ takes place through any of the other transfer functions since $x_j(t)$ is uncorrelated with $x_k(t)$ for $k \neq j$.

In general, the output, $y(t)$, is forced not only by the mutually coherent parts of the various inputs but also by the noncoherent portions of the inputs that go directly to the output through their own transfer functions without being affected by other transfer functions. This leads to the need for *partial coherence functions*. If part of one record causes part of, or all of a second record, then turning off the first record will eliminate the correlated parts from the second record and leave only that part of the second record that is not due to the first record. Because we do not want to incorporate the coherent portions of given forcing terms in the partial coherence functions, the partial coherences are found by first subtracting out the coherent parts of the various input signals. Bendat and Piersol (1986) state that, if any correlation between $x_1(t)$ and $x_2(t)$ is due to $x_1(t)$, then the optimum linear effects of $x_1(t)$ to $x_2(t)$ should be found. Denoting this mutual effect as $x_{2:1}(t)$, this should be subtracted from $x_2(t)$ to yield the conditioned (or residual) record, $x_{2:1}(t)$ representing that part of $x_2(t)$ not due to $x_1(t)$.

Multi-input systems cross-spectral analysis takes into account the fact that any input record, $x_k(t)$, with nonzero correlations between other inputs will contribute to variations in the output, $y(t)$, by passage through any of the K linear systems, $H_k(f)$. The conditioned portion of $x_k(t)$ will contribute directly to the output through its own response function only. The problem is to determine what percentage contribution each input function makes to the total variance of $y(t)$ for a specified frequency band. The simplest case is a two-input system consisting of inputs $x_1(t)$ and $x_2(t)$ for which

$$Y(f) = H_1(f)X_1(f) + H_2(f)X_2(f) + E(f) \tag{5.244}$$

and, provided $\gamma_{12}^2 \neq 0$

$$H_1(f) = \frac{S_{1y}(f)\left[1 - \frac{S_{12}(f)S_{2y}(y)}{S_{22}(f)S_{1y}(y)}\right]}{S_{11}(f)[1 - \gamma_{12}^2(f)]} \quad (5.245a)$$

$$H_2(f) = \frac{S_{2y}(f)\left[1 - \frac{S_{21}(f)S_{1y}(y)}{S_{11}(f)S_{2y}(y)}\right]}{S_{22}(f)[1 - \gamma_{12}^2(f)]} \quad (5.245b)$$

What is important to note here is the nonzero coupling between the different input variables when the cross-coherence, $\gamma_{12}^2(f)$, is nonzero. The product, $H_1(f)S_{11}(f)$, in Eqn (5.245a) still represents the ordinary coherent spectrum between the input, x_1, and the output, y. However, when $|\gamma_{12}| \neq 0$, $x_1(t)$ influences $y(t)$ through the transfer function, $H_2(f)$, as well as through its own transfer function, $H_1(f)$. Similarly, $x_2(t)$ influences $y(t)$ through the transfer function $H_1(f)$ as well as through its transfer function $H_2(f)$ (Eqn 5.245b). In general, the sum of $\gamma_{1y}^2(f)$ and $\gamma_{2y}^2(f)$ can be greater than unity when the outputs are correlated. The contributions from the conditioned records of $x_1(t)$ and $x_2(t)$ must also be taken into account when estimating the output response, $y(t)$. Once this is done, it becomes possible to construct reliable forecasting models for y.

Cartwright (1968) used the multiple input method to study tides and storm surges around eastern and northern Britain. He expanded the tide height, ζ, at each of the ports that he studied as a Taylor series of the atmospheric pressure, p, and its horizontal spatial gradients about the port location ($x = 0$, $y = 0$), viz.

$$\zeta(x,y,t) = p_{00}(t) + xp_{10}(t) + yp_{01}(t) + x^2 p_{20}(t) + 2xy p_{11}(t) + y^2 p_{02}(t) + \ldots \quad (5.250)$$

in which the pressure gradient terms $(p_{10}, p_{01}) = (\partial p/\partial x, \partial p/\partial y)$ are proportional to the geostrophic wind stress, the second derivatives $(p_{20}, p_{02}) = (\partial^2 p/\partial^2 x, \partial^2 p/\partial y^2)$ are related to wind stress gradients, and so on for higher order derivatives of the wind stress. As indicated by Table 5.18, the variances in different frequency bands for the sea level at Aberdeen, Scotland are significantly reduced relative to the original values as the pressure, first derivatives, and second derivatives are successively included. Consequently, all of the mutually correlated weather variables are considered relevant to the predictability of sea level. In their study, Sokolova et al. (1992) used the multiple spectral analysis technique to study sea-level oscillations measured from July to September, 1986 at different locations around the perimeter of the Sea of Japan (also, the East Sea). According to their analysis for both the multiple and partial coherences, 46−77% of the sea-level variance was coherent with atmospheric pressure and 5−37% was coherent with the wind stress.

5.8.8 Rotary Cross-Spectral Analysis

As outlined in Section 5.6.4, the decomposition of a complex horizontal velocity vector, $w(t) = u(t) + iv(t)$, into counter-rotating circularly polarized components can aid in the analysis and interpretation of oceanographic time series. (Here, u

TABLE 5.18 Residual variances (cm^2) for different frequency bands for Aberdeen, Scotland sea-level oscillations. SLP denotes sea level pressure and ∇p the horizontal gradient of the pressure field at the point studied.

Variables included	0−0.5 cpd	0.5−0.8 cpd	1.1−1.8 cpd	2.1−2.8 cpd
Original variance	181	16	9.6	4.1
p_{00} (SLP)	88	13	9.1	3.9
p_{00}, p_{10}, p_{01} (∇p)	49	9	7.1	3.6
$p_{00}, p_{10}, p_{01}, \ldots, p_{02}$ ($\nabla^2 p$)	38	6	5.3	3.3

The predictive model explains increasingly more of the variance as additional weather variables are incorporated in the analysis. Periods are in cycles per day (cpd).
Modified after Cartwright (1968).

and v typically represent the eastward and northward or, alternatively, the alongshore and cross-shore, components of the current or wind velocity). Many of the fundamentals of this approach can be found in Fofonoff (1969), Gonella (1972), Mooers (1973), Calman (1978), and Hayashi (1979). In rotary spectral analysis, the different frequency components of the vector, $w(t)$, are represented in terms of clockwise and counterclockwise rotating vectors (Figure 5.23). The counterclockwise component is considered to be rotating with positive angular frequency ($\omega \geq 0$) and the clockwise component with negative angular frequency ($\omega \leq 0$). Depending on which of the two components has the largest magnitude, the vector rotates clockwise or counterclockwise with time, with the tip of the vector tracing out an ellipse. If, for a given frequency, both components are of equal magnitude, the ellipse flattens to a line and the motions are *rectilinear* (back and forth along a straight line). Two one-sided autospectra and two one-sided cross-spectra can be computed for the rotary components. Mooers (1973) formulated these as two two-sided rotary autospectra called, respectively, the *inner* and *outer rotary autospectra*, the terminology originating from the resemblance of the inner and outer rotary autocovariance functions derived from the autospectra to the inner (dot) and outer (cross) products in mathematics. (A note on terminology: Mooers (1973) uses A and C for counterclockwise (+) and clockwise components (−) while Gonella (1972) uses +/− subscripts for these components of the form u_+ and u_-. In this text, we use +/− superscripts where, for example, the amplitudes of the two vector components are written as A^+ and A^-).

To simplify the mathematics, we assume that u and v are continuous, stationary processes with zero means and Fourier integral representations. The velocity vector, $w(t)$, can then be written in terms of its Fourier transform

$$w(t) = u(t) + iv(t) = \sum_p W_p e^{i\omega_p t}$$
$$= \sum_p \left\{ \left[A_{1p} \cos(\omega_p t) + B_{1p} \sin(\omega_p t) \right] + i \left[A_{2p} \cos(\omega_p t) + B_{2p} \sin(\omega_p t) \right] \right\} \tag{5.251}$$

in which the Fourier transform component, W_p, is a complex quantity, A and B are constants, and ω_p is the frequency of the pth Fourier component. As outlined in Section 5.6.4, each Fourier component of frequency $\omega = \omega_p$ can be expressed as a combination of two circularly polarized components having counterclockwise ($\omega \geq 0$) and clockwise ($\omega \leq 0$) rotation. Each of two components has its own amplitude and phase, and the tip of the vector formed by the combination of the two oppositely rotating components traces out an ellipse over a period, $T = 2\pi/\omega$. The semimajor axis of the ellipse has length, $L_M = A^+(\omega) + A^-(\omega)$, and the semi-minor axis has length, $L_m = |A^+(\omega) - A^-(\omega)|$. The angle, θ, of the major axis measured counterclockwise from the eastward direction gives the ellipse orientation.

If we specify $A_1(\omega)$ and $B_1(\omega)$ to be the amplitudes of the cosine and sine terms for the eastward (u) component in Eqn (5.251) and $A_2(\omega)$ and $B_2(\omega)$ to be the corresponding amplitudes for the northward (v) component, the amplitudes of the two counter-rotating vectors for a given frequency are

$$A^+(\omega) = \frac{1}{2} \left\{ [B_2(\omega) + A_1(\omega)]^2 + [A_2(\omega) - B_1(\omega)]^2 \right\}^{1/2} \tag{5.252a}$$

$$A^-(\omega) = \frac{1}{2} \left\{ [B_2(\omega) - A_1(\omega)]^2 + [A_2(\omega) + B_1(\omega)]^2 \right\}^{1/2} \tag{5.252b}$$

and their phases are

$$\tan(\theta^+) = [A_1(\omega) - B_1(\omega)] / [A_1(\omega) + B_2(\omega)] \tag{5.253a}$$

$$\tan(\theta^-) = [B_1(\omega) + A_2(\omega)] / [B_2(\omega) - A_1(\omega)] \tag{5.253b}$$

The eccentricity of the ellipse is

$$\varepsilon(\omega) = 2[A^+(\omega)A^-(\omega)]^{1/2} / [A^+(\omega) + A^-(\omega)] \tag{5.254}$$

where the ellipse traces out an area $\pi[(A^+)^2 - (A^-)^2]$ during one complete cycle of duration, $2\pi/\omega$. The use of rotary components leads to two-sided spectra; i.e., defined for both negative and positive frequencies. If $S^+(\omega)$ and $S^-(\omega)$ are the rotary spectra for the two components, then $A^\pm(\omega) \propto [S^\pm(\omega)]^{1/2}$ can be used to determine the ellipse eccentricity. The sense of rotation of the vector about the ellipse is given by the rotary coefficient (see Section 5.6.4.2)

$$r(\omega) = [S^+(\omega) - S^-(\omega)] / [S^+(\omega) + S^-(\omega)] \tag{5.255}$$

where $-1 \leq r \leq 1$. Values for which $r > 0$ indicate counterclockwise rotation, while values of $r < 0$ indicate clockwise rotation; $r = 0$ is rectilinear motion.

Because u, v are orthogonal Cartesian components of the velocity vector, $w = (u, v)$, the rotary spectra can be expressed as

$$S^+(\omega) = [A^+(\omega)]^2, \omega \geq 0$$

$$= \frac{1}{2}[S_{uu} + S_{vv} + 2Q_{uv}] \tag{5.256a}$$

$$S^-(\omega) = [A^-(\omega)]^2, \omega \leq 0$$

$$= \frac{1}{2}[S_{uu} + S_{vv} - 2Q_{uv}] \tag{5.256b}$$

where S_{uu} and S_{vv} are the autospectra for the u and v components, and Q_{uv} is the quadrature spectrum between the two components. The stability of the ellipse is given by

$$\mu(\omega) = \frac{|\langle (A^-(\omega)A^+(\omega)\exp[i(\theta^+ - \theta^-)])\rangle|^2}{\langle (A^-)^2\rangle\langle (A^+)^2\rangle},$$

$$= \frac{|Y|}{[S^+(\omega)S^-(\omega)]^{1/2}} \quad \omega \geq 0 \tag{5.257}$$

where

$$Y = \frac{1}{2}[S_{uu} - S_{vv} + i2S_{uv}] \tag{5.258}$$

and the ellipse has a mean orientation

$$\phi = \frac{1}{2}\tan^{-1}[2S_{uv} / (S_{uu} - S_{vv})] \tag{5.259}$$

where ϕ is measured counterclockwise from east (the function, ϕ is not coordinate invariant). The brackets $\langle \cdot \rangle$ denote an ensemble average or a band average in frequency space. The ellipse stability, $\mu(\omega)$, resembles the magnitude of a correlation function and is a measure of the confidence one might place in the estimate of the ellipse orientation (Gonella, 1972).

5.8.8.1 Rotary Analysis for a Pair of Time Series

Having summarized the rotary vector analysis for a single location, we now want to consider the coherence and cross-spectral properties for two time series measured simultaneously at two spatial locations. The objective of the rotary spectral analysis is to determine the "similarity" between the two series in terms of their circularly polarized rotary components. For two vector time series, the inner and outer rotary cross-spectra can be computed. As the spectra are complex, they have both amplitude and phase. Hence, coherence and phase spectra can be computed, just as with the cross-spectra of two scalar time series. *Inner* functions describe corotating components (components rotating in the same direction) and *outer* functions describe counter-rotating components (components rotating in opposite directions). We could, of course, use standard Cartesian components for this task. Unfortunately, Cartesian vectors and their derived relationships generally are dependent on the selected orientation of the coordinate system. The advantages of the rotary type of analysis are: (1) the coherence analysis is independent of the coordinate system (i.e., is coordinate invariant); and (2) the results encompass the coherence and phase of oppositely rotating, as well as like-rotating, components for motions that may be highly nonrectilinear. Because the counter-rotating components have circular symmetry, invariance under coordinate rotation follows for coherence.

We consider two vector time series defined by the relations

$$w_1(t) = (u_1, v_1); w_2(t) = (u_2, v_2) \tag{5.260}$$

where, as before, $(u, v) = u + iv$ are complex quantities. If $W_1(\omega)$ and $W_2(\omega)$ are components of the Fourier transforms of these time series, then the transforms can be expressed in the form

$$W(\omega) = \begin{cases} A^+ \exp(-i\theta^+), \omega \geq 0 \\ A^- \exp(-i\theta^-), \omega \leq 0 \end{cases} \tag{5.261}$$

with the same definitions for amplitudes and phases as in the previous subsection. These expressions equate the negative frequency components from the Fourier transform with the clockwise rotary components and the positive frequency components from the transform with the counterclockwise components.

Inner-cross spectrum: The inner cross-spectrum, $S_{w_j w_k}(\omega)$, provides an estimate of the joint energy content of two time series for rotary components rotating in the same direction (e.g., the clockwise component of series 1 with the clockwise component of series 2; Figure 5.53). For all frequencies, $-\omega_{Nq} < \omega < \omega_{Nq}$

$$S_{w_j w_k}(\omega) = \left\langle W_j^*(\omega) W_k(\omega) \right\rangle, \; j, k = 1, 2$$

$$= \begin{cases} A_j^+(\omega) A_k^+(\omega) \exp\left[-i\left(\theta_j^+ - \theta_k^+\right)\right], \omega \geq 0 \\ A_j^-(\omega) A_k^-(\omega) \exp\left[i\left(\theta_j^- - \theta_k^-\right)\right], \omega \leq 0 \end{cases} \tag{5.262}$$

where, as before, $\langle \cdot \rangle$ denotes an ensemble average or a band average in frequency space, and the asterisk denotes the complex conjugate. It follows that the inner-autospectrum for each time series is

$$S_{w_j w_j}(\omega) = \begin{cases} \left[A_j^+(\omega)\right]^2, \omega \geq 0 \\ \left[A_j^-(\omega)\right]^2, \omega \leq 0 \end{cases} \tag{5.263}$$

Thus, $S_{w_j w_j}(\omega)$ ($j = 1, 2$) is the power spectrum of the counterclockwise component of the series j for $\omega \geq 0$, and the power spectrum for the clockwise component for $\omega \leq 0$. The area under the curve of $S_{w_j w_k}(\omega)$ versus frequency equals the sum of the variance of the u and v components. For $\omega \geq 0$, $S_{w_1 w_2}(\omega)$ represents the cross-spectrum for the *counterclockwise* component of series 1 and 2, while for $\omega \leq 0$, $S_{w_1 w_2}(\omega)$ represents the cross-spectrum for the *clockwise rotary* component.

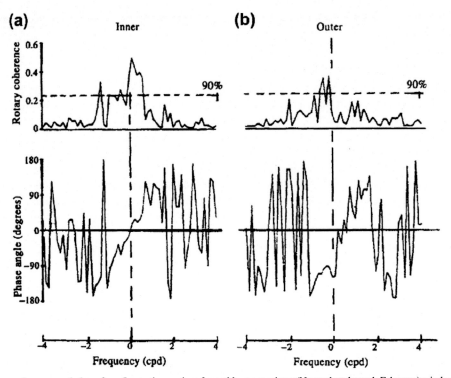

FIGURE 5.53   Rotary coherence and phase for a 5-year time series of monthly mean winter (November through February) wind velocity from two sites off Alaska. (a) Co-rotating (inner) coherence and phase with 90% confidence level; (b) Counterrotating (outer) coherence and phase. *From Livingstone and Royer (1980).*

Inner-coherence squared: The two-sided inner-coherence squared, $\gamma_{12}^2(\omega)$, between the two time series at frequency, ω, is defined in the usual manner. Specifically, using the previous definitions for the rotary components, we find

$$\gamma_{12}^2(\omega) = \begin{cases} \dfrac{\left[\left\langle A_1^+(\omega)A_2^+(\omega)\cos\left(\theta_1^+ - \theta_2^+\right)\right\rangle^2 + \left\langle A_1^+(\omega)A_2^+(\omega)\sin\left(\theta_1^+ - \theta_2^+\right)\right\rangle^2\right]}{\left\langle\left[A_1^+(\omega)\right]^2\right\rangle\left\langle\left[A_2^+(\omega)\right]^2\right\rangle}, & \omega \geq 0 \\[3em] \dfrac{\left[\left\langle A_1^-(\omega)A_2^-(\omega)\cos\left(\theta_1^- - \theta_2^-\right)\right\rangle^2 + \left\langle A_1^-(\omega)A_2^-(\omega)\sin\left(\theta_1^- - \theta_2^-\right)\right\rangle^2\right]}{\left\langle\left[A_1^-(\omega)\right]^2\right\rangle\left\langle\left[A_2^-(\omega)\right]^2\right\rangle}, & \omega < 0 \end{cases} \tag{5.264}$$

where $0 \leq |\gamma_{12}^2| \leq 1$. A coherence of near zero indicates a negligible relationship between the two like-rotating series while a coherence near unity indicates a high degree of variability between the series. The inner-phase lag, ϕ_{12}, between the two vectors is

$$\phi_{12}(\omega) = \tan^{-1}\left[-\operatorname{Im}(S_{w_1w_2}) / \operatorname{Re}(S_{w_1w_2})\right] \tag{5.265}$$

or, in terms of the clockwise and counterclockwise components

$$\tan(\phi_{12}) = \begin{cases} \dfrac{\left\langle A_1^+(\omega)A_2^+(\omega)\sin\left(\theta_1^+ - \theta_2^+\right)\right\rangle}{\left\langle A_1^+(\omega)A_2^+(\omega)\cos\left(\theta_1^+ - \theta_2^+\right)\right\rangle}, & \omega \geq 0 \\[3em] \dfrac{\left\langle -A_1^-(\omega)A_2^-(\omega)\sin\left(\theta_1^- - \theta_2^-\right)\right\rangle}{\left\langle A_1^-(\omega)A_2^-(\omega)\cos\left(\theta_1^- - \theta_2^-\right)\right\rangle}, & \omega < 0 \end{cases} \tag{5.266}$$

The phase, which is the same for both the inner cross-spectrum and the inner coherence, is a measure of the phase lead of the rotary component of time series 1 with respect to that of time series 2. Figure 5.53a shows the inner rotary coherence and phase for 5 years of monthly winter (November–February) wind data measured off Alaska at Middleton Island (59.4° N, 146.3° W) and Environmental Weather Buoy EB03 (56.0° N, 148.0° W). Corotating wind vectors were generally coherent above the 90% confidence level for frequencies $-1 < f < 1$ cpd, with greater coherence at positive frequencies (Livingstone and Royer, 1980). The inner phase was nearly a straight line in the frequency range $-1 < f < 0$ cpd, increasing by 120° over this range.

Outer-cross spectrum: The outer cross-spectrum, $Y_{w_jw_k}(\omega)$, provides an estimate of the joint energy content between rotary components rotating in opposite directions (e.g., between the clockwise component of time series 1 and the counterclockwise component of time series 2). For frequencies in the Nyquist frequency range, $-\omega_{Nq} < \omega < \omega_{Nq}$

$$Y_{w_jw_k}(\omega) = \left\langle W_j(-\omega)W_k(\omega)\right\rangle, \quad j, k = 1, 2$$

$$= A_j^-(\omega)A_k^+(\omega)\exp\left[i\left(\theta_k^+ - \theta_j^-\right)\right], \quad \omega \geq 0 \tag{5.267}$$

$$= A_j^+(\omega)A_k^-(\omega)\exp\left[-i\left(\theta_k^- - \theta_j^+\right)\right], \quad \omega \leq 0$$

(Middleton, 1982). These relations resemble those for the inner-cross spectra but involve a combination of oppositely rotating vector amplitudes and phases. For the case of a single series, j, the outer rotary autospectrum is then

$$Y_{w_jw_j}(\omega) = A_j^-(\omega)A_j^+(\omega)\exp\left[i\left(\theta_j^+ - \theta_j^-\right)\right], \quad \omega \geq 0 \tag{5.268}$$

and is symmetric about $\omega = 0$, and so is defined for only $\omega \geq 0$. Hence, $Y_{w_jw_j}(\omega)$ is an even function of frequency; i.e., $Y_{w_jw_j}(\omega) = Y_{w_jw_j}(-\omega)$. As noted by Mooers, $Y_{w_jw_j}(\omega)$ is not a power spectrum in the ordinary physical sense because it is complex valued. Rather it is related to the spectrum of the *uv*-Reynolds stress.

Outer-coherence squared: After first performing the ensemble or band averages in the brackets $\langle \cdot \rangle$, the outer-rotary coherence squared between series j and k is expressed in terms of the Fourier coefficients as

$$\lambda_{jk}^2(\omega) = \begin{cases} \dfrac{\left\langle A_j^-(\omega)A_k^+(\omega)\right\rangle^2 \left[\left\langle\cos\left(\theta_k^+ - \theta_j^-\right)\right\rangle^2 + \left\langle\sin\left(\theta_k^+ - \theta_j^-\right)\right\rangle^2\right]}{\left\langle\left[A_k^+(\omega)\right]^2\right\rangle\left\langle\left[A_j^-[(\omega)]^2\right]\right\rangle}, \omega \geq 0 \\[20pt] \dfrac{\left\langle A_j^+(\omega)A_k^-(\omega)\right\rangle^2 \left[\left\langle\cos\left(\theta_j^+ - \theta_k^-\right)\right\rangle^2 + \left\langle\sin\left(\theta_j^+ - \theta_k^-\right)\right\rangle^2\right]}{\left\langle\left[A_j^+(\omega)\right]^2\right\rangle\left\langle\left[A_k^-(\omega)\right]^2\right\rangle}, \omega < 0 \end{cases} \tag{5.269}$$

The phase lag, $\psi_{jk}(\omega)$, between the two oppositely rotating components of the time series is then the same for the coherence and the cross-spectrum and is given by

$$\tan(\psi_{jk}) = \begin{cases} \dfrac{\left\langle A_j^-(\omega)A_k^+(\omega)\sin\left(\theta_j^- - \theta_k^+\right)\right\rangle}{\left\langle A_j^-(\omega)A_k^+(\omega)\cos\left(\theta_j^- - \theta_k^+\right)\right\rangle}, \omega \geq 0 \\[20pt] \dfrac{\left\langle A_j^+(\omega)A_k^-(\omega)\sin\left(\theta_k^- - \theta_j^+\right)\right\rangle}{\left\langle A_j^+(\omega)A_k^-(\omega)\cos\left(\theta_k^- - \theta_j^+\right)\right\rangle}, \omega < 0 \end{cases} \tag{5.270}$$

If the values of $A_j^- A_k^+$ and $A_j^+ A_k^-$ change little over the averaging interval covered by the angular brackets, then the phase lag becomes

$$\psi_{jk}(\omega) = \begin{cases} \theta_j^- - \theta_k^+, \omega \geq 0 \\[8pt] \theta_k^- - \theta_j^+, \omega < 0 \end{cases} \tag{5.271}$$

In Figure 5.53b we present the outer rotary coherence and phase for 5-year records of winter winds off the coast of Alaska. Counter-rotating vectors were coherent at negative frequencies in the range $-1 < f < 0$ cpd and exhibited little coherence at positive frequencies. In this portion of the frequency band, the linear phase gradient was similar to that for the corotating vectors (Figure 5.53a).

Complex admittance function: If we think of the wind vector at location 1 as the source (or input) function and the current at location 2 as the response (or output) function, we can compute the complex inner admittance, Z_{12}, between two corotating vectors as

$$Z_{12}(\omega) = S_{w_1 w_2}(\omega) / S_{w_1 w_1}(\omega), \quad -\omega_{Nq} < \omega < \omega_{Nq} \tag{5.272}$$

The amplitude and phase of this function are, respectively

$$|Z_{12}(\omega)| = |S_{w_1 w_2}(\omega)| / S_{w_1 w_1}(\omega) \tag{5.273a}$$

$$\Phi_{12}(\omega) = \tan^{-1}\{\text{Im}[S_{w_1 w_2}(\omega)] / \text{Re}[S_{w_1 w_2}(\omega)]\} \tag{5.273b}$$

For frequency ω, the absolute value of $Z_{12}(\omega)$ determines the amplitude of the clockwise (counterclockwise) rotating response one can expect at location 2 to a given clockwise (counterclockwise) rotating input at location 1. The phase, $\Phi_{12}(\omega)$, determines the lag of the response vector to the input vector.

The corresponding expressions for the complex outer admittance, Z_{12}, between two opposite-rotating vectors are

$$Z_{12}(\omega) = Y_{w_1 w_2}(\omega) / S_{w_1 w_1}(\omega), \quad -\omega_{Nq} < \omega < \omega_{Nq} \tag{5.274}$$

with amplitude and phase

$$|Z_{12}(\omega)| = |Y_{w_1 w_2}(\omega)| / S_{w_1 w_1}(\omega) \tag{5.275a}$$

$$\Phi_{12}(\omega) = \tan^{-1}\{\text{Im}[Y_{w_1 w_2}(\omega)] / \text{Re}[Y_{w_1 w_2}(\omega)]\} \tag{5.275b}$$

For frequency ω, the absolute value of $Z_{12}(\omega)$ yields the amplitude of the clockwise (counterclockwise) rotating response one can expect at location 2 to a given counterclockwise (clockwise) rotating input at location 1. The phase, $\Phi_{12}(\omega)$, determines the lag of the response vector to the input vector.

5.9 WAVELET ANALYSIS

The terms "wavelet transform" and "wavelet analysis" are two recent additions to the lexicon of time series analysis. First introduced in the 1980s for processing seismic data (cf. Goupillaud et al., 1984), the technique has begun to attract attention in meteorology and oceanography, where it has been applied to time series measurements of turbulence (Farge, 1992; Shen and Mei, 1993), surface gravity waves (Shen et al., 1994), low-level cold fronts (Gamage and Blumen, 1993), equatorial Yanai waves (Meyers et al., 1993), storm surge (Rabinovich et al., 2022) and tsunamis (Heidarzadeh et al., 2019; Rabinovich et al., 2019; Fine et al., 2020; Zaytsev et al., 2021).

As frequently noted in the literature, Fourier analysis fairs relatively poorly when dealing with signals of the form $\varphi(t) = A(t)\cos(\omega t)$, where the parameter, A, varies on the slow timescale, τ. Wavelet analysis has a number of advantages over Fourier analysis that are particularly attractive. Unlike the Fourier transform, which generates record-averaged values of amplitude and phase for each frequency component or harmonic, ω, the wavelet transform yields a localized, "instantaneous" estimate for the amplitude and phase of each spectral component in the data set. This gives wavelet analysis an advantage in the analysis of nonstationary data series in which the amplitudes and phases of the constituents may be changing in time or space. Where a Fourier transform of the nonstationary time series would smear out any detailed information on the changing processes, the wavelet analysis attempts to track the evolution of the signal characteristics through the data set. As with other transform techniques, problems can develop at the ends of the time series, and steps must be taken to mitigate these effects. Similar to other transform techniques involving finite length data, steps also must be taken to minimize the distortion of the transformed data caused by the nonperiodic behavior at the ends of the time series. Lastly, we note that increasing the temporal resolution, Δt, of the wavelet analysis decreases the frequency resolution, Δf, and vice versa, such that $\Delta t \Delta f < (1/4)\pi$, reminiscent of the Heisenberg uncertainty relation. The more accurately we want to resolve the frequency components of a time series, the less accurately we can resolve the changes in these frequency components with time and vice versa.

5.9.1 The Wavelet Transform

As noted above, wavelet analysis enables us to study nonstationary signals, in which the amplitudes of the frequency components of the signal are changing with time. This contrasts with traditional spectral methods, which assume that the frequency components of the time series are stationary, thus allowing for a direct transfer into the frequency domain. Wavelet analysis provides a windowing technique with variable-sized windows, which permits customization of the frequency domain analysis by allowing for long time intervals, where more precise low-frequency information is wanted, and also for shorter time intervals, where high-frequency information is wanted. A wavelet is a "small wave" that grows and decays over a limited time span. The oldest and simplest wavelet is the Haar Wavelet, traditionally written as, $\psi(t)$, and introduced at the beginning of the twentieth century (Harr, 1910), where

$$\psi(t) \equiv \begin{cases} \dfrac{-1}{\sqrt{2}} & -1 < t \le 0 \\[2mm] \dfrac{1}{\sqrt{2}} & 0 < t \le 1 \\[2mm] 0 & \text{otherwise} \end{cases} \tag{5.276}$$

From Eqn (5.276), it can be shown that the translations of $\psi(t)$ are orthonormal to both its translated versions (shift in time) and its dilated versions (change in amplitude), i.e.,

$$\int_{-\infty}^{\infty} \psi_{m,n}(t)\psi_{m,n}^*(t)dt = 2^m \int_{-\infty}^{\infty} \psi(2^m t - n)\psi^*\left(2^m t - n'\right)dt$$

$$= \int_{-\infty}^{\infty} \psi(t - n)\psi^*\left(t - n'\right)dt = \delta\left(n - n'\right) \tag{5.277}$$

where the asterisk (*) denotes the complex conjugate. As with the sines and cosines used in spectral analysis, the function, $\psi(t)$, therefore forms an orthonormal basis for the analysis of frequency-dependent variations.

Modern wavelet analysis involves the convolution of a real time-series, $x(t)$, with a set of functions, $g_{a\tau}(t) = g(t: \tau, \alpha)$, that are derived from a "mother wavelet" or analyzing wavelet, $g(t)$, which is generally complex. In particular

$$g_{a\tau}(t) = \frac{1}{\sqrt{a}} g[(t - \tau) / a] \tag{5.278}$$

where τ (real) is the *translation* parameter corresponding to the central point of the wavelet in the time series and a (real and positive) is the *scale dilation* parameter corresponding to the width of the wavelet. For the Gaussian-shaped Morlet wavelet (Figure 5.54) described in detail later in this section, the dilation parameter can be related to a corresponding Fourier frequency (or wavenumber).

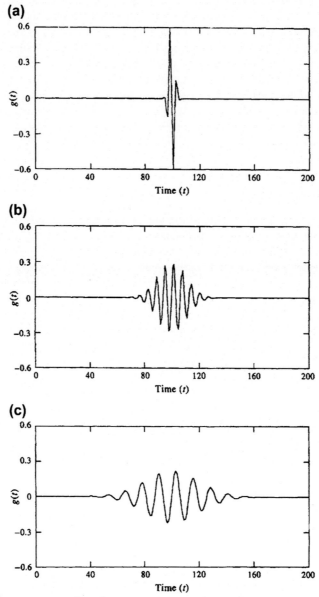

FIGURE 5.54   The Morlet wavelet, $g(t) = (1/\sqrt{a})e^{[(t-\tau)/a]^2}\{2\sin[c(t-\tau)/a]\}$, where t is time in arbitrary units ($t = t_n$; $n = 1, ..., 200$). The example is for $c = 10$ and time lag $\tau = 100$ so that the wavelet is seen midway through the time series. (a) $a = 2$; (b) $a = 10$; (c) $a = 20$.

The continuous wavelet transform, $X(t)$, of the time series with respect to the analyzing wavelet, $g(t)$, is defined through the convolution integral

$$X_g[\tau, a] = \frac{1}{\sqrt{a}} \int_{-\infty}^{\infty} g^*[(t - \tau)/a] x(t) dt \tag{5.279}$$

in which g^* is the complex conjugate of g and variables τ, a are allowed to vary continuously through the domain $(-\infty, \infty)$. Wavelet analysis provides a two-dimensional unraveling of a one-dimensional time series into position, τ, and amplitude scale, a, as new independent variables. The wavelet transformation (Eqn 5.279) is a sort of mathematical microscope, with magnification $1/a$, position τ, and optics given by the choice of the specific wavelet, $g(t)$ (Shen et al., 1994). Whereas Fourier analysis provides an average amplitude over the entire time series, wavelet analysis yields a measure of the localized amplitudes, a, as the wavelet moves through the time series with increasing values of τ. Although wavelets have a definite scale, they typically do not bear any resemblance to the sines and cosines of Fourier modes. Nevertheless, a correspondence between wavelength and scale, a, can sometimes be achieved.

To qualify for mother wavelet status, the function, $g(t)$, must satisfy several properties (Meyers et al., 1993):

1. Its amplitude $|g(t)|$ must decay rapidly to zero in the limit $|t| \to \infty$. It is this feature that produces the localized aspect of wavelet analysis since the transformed values, $X_g[\tau, a]$, are generated only by the signal in the cone of influence about $t = \tau$. In most instances, the wavelet, $g[(t - \tau)/a]$, is assumed to have an insignificant effect at some time $|t| = \tau_c$.

2. $g(t)$ must have a zero mean. Known as the *admissibility condition*, this ensures the invertability of the wavelet transform. The original signal can then be obtained from the wavelet coefficients through the inverse transform

$$x(t) = \frac{1}{C} \int_{-\infty}^{\infty} \int_{-\infty}^{\infty} \{X_g[\tau, a] a^{-2} g_{a\tau}\} d\tau da$$

where

$$C^{-1} = \int_{-\infty}^{\infty} \left(\omega^{-1} |G(\omega)|^2 \right) d\omega \tag{5.280}$$

in which $G(\omega)$ is the Fourier transform of $g(t)$. For $1/C$ to remain finite, $G(0) = 0$.

3. Wavelets are often regular functions, such that $G(\omega < 0) = 0$. These are also called *progressive* wavelets. Elimination of negative frequencies means that wavelets need only be described in terms of positive frequencies.

4. Higher-order moments (such as variance and skewness) should vanish allowing the investigation of higher-order variations in the data. This requirement can be relaxed, depending on the application.

One of most extensively used wavelets is the standard (admissible and progressive) Morlet wavelet

$$g(t) = e^{-t^2/2} e^{+i\omega t} \tag{5.281}$$

consisting of a plane wave of frequency ω (or wavenumber k in the spatial domain), which is modulated by a Gaussian envelope of unit width. Another possible wavelet, which is applicable to a signal with two frequencies, ω_1 and ω_2, is

$$g(t) = e^{-t^2/2} e^{i\omega_1 t} e^{i\omega_2 t} \tag{5.282}$$

while the wavelet

$$g(t) = e^{-t^2/2} e^{i\omega t} e^{ikt^2/2} \tag{5.283}$$

is applicable to short data segments with linearly increasing frequency ("chirps").

5.9.2 Wavelet Algorithms

The choice of $g(t)$ is dictated by the analytical requirements. More specifically, the wavelet should have the same pattern or signal characteristic as the pattern being sought in the time series. Large values of the transform $X_g[\tau, a]$ will then indicate where the time series $x(t)$ has the desired form. The simplest—and most time consuming—method for obtaining the wavelet transform is to compute the transform at arbitrary points in parameter (τ, a) space using the discrete form of Eqn (5.279) for known values of $x(t)$ and $g(t)$. If one integrates from $0 < a \leq M$ and $0 < \tau \leq N$, the integration time goes as MN^2. An alternate method is to use the convolution theorem and then obtain the wavelet transform in spectral space

$$X_g[\tau, a] = \frac{1}{\sqrt{a}} \int_{-\infty}^{\infty} e^{i\tau\omega} G^*(a\omega) X(\omega) d\omega \tag{5.284}$$

where $G(\omega)$ and $X(\omega)$ are the Fourier transforms of $g(t)$ and $x(t)$, respectively. Since FFT transforms can now be exploited, the analysis time drops to $MN\log_2 N$. To use this method, $G(\omega)$ should be known analytically and the data must be preprocessed to avoid errors from the FFT algorithms. For example, if $x(t)$ is aperiodic, the discrete form of Eqn (5.283) will generate an artificial periodicity in the wavelet transform that greatly distorts the results for the end regions. Methods have been devised to work around this problem. Aliasing and bias in FFT routines must also be taken into account.

Meyers et al. (1993) used the standard Morlet wavelet Eqn (5.281), for which $g(t) = e^{-t^2/2}e^{+i\omega t}$, to examine a signal that changes frequency halfway through the measurement. Here, we have broken with tradition and used ω instead of c for frequency. After considerable attempts (including use of raw data, cosine-weighted data, and other variations), the authors decided that the best approach was to taper or buffer the original time series with added data points that attenuate smoothly to zero past the ends of the time series. "The region of the transform corresponding to these points is then discarded after the transform. Without this buffering, a signal whose properties are different near its ends will result in a wavelet transform that has been forced to periodicity at all scales through a distortion (in some cases severe) of the end regions. The greater the aperiodicity of the signal, the greater the distortion."

For the Morlet wavelet, the dilation parameter, a, giving the maximum correlation between the wavelet and a plane Fourier component of frequency, ω_o (i.e., a wave of the form $e^{i\omega_o t}$) is

$$a_o = \frac{\left[\omega + (2 + \omega^2)^{1/2}\right]}{4\pi} T_o \tag{5.285}$$

where $T_o = 2\pi/\omega_o$ is the Fourier period. (In wavenumber space, T_o is replaced by wavelength λ_o and ω_o by k_o). We note that any linear superposition of periodic components results in separate local maxima. Consequently, the wavelet transform of any function $x(t) = \sum A_j e^{ik_j t}$ will have modulus maxima at $a_j = [\omega + (2 + \omega^2)^{1/2}]/(2k_j)$.

5.9.3 Oceanographic Examples

In this section, we will consider two oceanographic wavelet examples (surface gravity wave heights and zonal velocity from a satellite-tracked drifter) using the standard Morlet wavelet

$$g(t) \to g[(t - \tau)/a] = \frac{1}{\sqrt{a}} e^{-1/2[(t-\tau)/a]^2} \sin[\omega(t - \tau)/a] \tag{5.286}$$

In this real expression, the Gaussian function determines the envelope of the wavelet while the sine function determines the wavelengths that will be preferentially weighted by the wavelet. The wavelet function progresses through the time series with increasing τ, its "cone of influence" centered at times $t = \tau$. As a increases, the width of the Gaussian spreads in time from its center value (Figure 5.54a–c). Increasing ω increases the number of oscillations over the span of the function. The processing procedure is as follows: (1) read in the time series $x(n)$ ($n = 0, ..., N - 1$) to be analyzed, where $N = 2^m$ (m is an integer). To reduce ringing, extend each end of the time series by adding a trigonometric taper, tap = $1 - \sin\phi$, where tap = 1.0 at the end values $x(0)$ and $x(N - 1)$. The total length of the buffered time series must remain a power of two; (2) remove the mean of the new record and then take the FFT of the time series to obtain $X(\omega)$; (3) take the Fourier transform of the wavelet, $g(t)$, at given length scales, a, to obtain $G(a\omega)$; (4) calculate the integral Eqn (5.284) by convolving the product $G^*(a\omega)X(\omega)$ in Fourier space; (5) take the inverse FFT of the result to obtain $\sqrt{a}X_g(\tau, a)$ as a function of time dilation, τ, and amplitude, a.

In Figure 5.55a, we have plotted a 300 s record of surface gravity wave heights measured off the west coast of Vancouver Island in the winter of 1993. Maximum wave amplitudes of around 3 m occurred midway through the time series. The Morlet wavelet transform of the record yields an estimate of the wave amplitude (Figure 5.55b) and phase (Figure 5.55c) as functions of the wave period (T) and time (t). Also plotted is the value of the wave period (T = scale a) at peak energy (Figure 5.55d). Comparison of Figure 5.55b and d reveals that the larger peaks near times of 410, 460, and 570 s all have about the same wavelet scale, a, corresponding to a peak wave period of around 8 s. Also, as one would expect, the 2π changes in phase between crests (Figure 5.55c) increase with increasing wave period (scale, a).

In our second example, we have applied a standard Morlet wavelet transform to a 90-day segment of 3-hourly sampled east—west (u) current velocity (Figure 5.56a) obtained from a satellite-tracked drifter launched in the northeast Pacific in August 1990 as part of the World Ocean Circulation Experiment. The drifter was drogued at 15-m depth and its motion was indicative of currents in the surface Ekman layer. The 90-day velocity record has been generated from positional data using a cubic spline interpolation algorithm. We focus our attention on the high-frequency end of the spectrum, $0 < a < 1.5$ days. As indicated by Figure 5.56b and c, the first 30 days of the record, from Julian day (JD) 240—270, were dominated by weak semidiurnal tidal currents with periods of 0.5 days. Beginning on JD 270, strong wind-generated inertial motions with periods around 16 h ($f \approx 1.5$ cpd) dominated the spectrum. These energetic motions persisted through the record, except for a short hiatus near JD 295. A blow-up of the segment from JD 240 to 270 shows a rapid change in the signal phase associated with the shift from semidiurnal tidal currents to near-inertial motions. The contribution from the beat frequency between the M_2 tidal signal and the inertial oscillations, $fM_2 = 0.0805 + 0.0621$ cph = 0.1426 cph can also be seen in the transformed data at the period $T \approx 0.29$ days. Examination of the longer period motions in Figure 5.57 (for $2 < a < 30$ days) suggests the presence of a long-period modulation of the high-frequency motions associated with the near-inertial wave events.

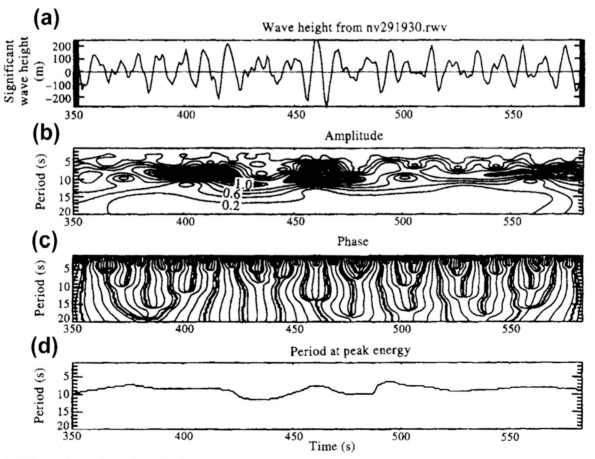

FIGURE 5.55 Morlet wavelet transform of surface gravity waves measured from a waverider buoy moored off the west coast of Vancouver Island. (a) Original 5-min time series of significant wave height for the winter of 1993; (b) wave amplitude (m) and (c) phase (degrees) as function of time; (d) the value of a (wave period) at peak wave amplitude. *Courtesy, Diane Masson, Institute of Ocean Sciences*

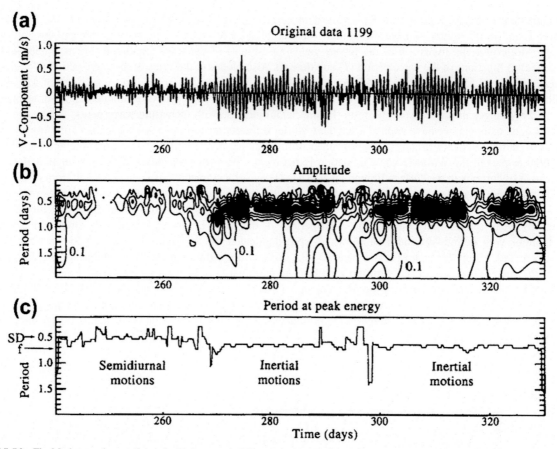

FIGURE 5.56 The Morlet wavelet transform of a 90-day record of the east–west velocity component from the trajectory of a satellite-tracked drifter in the northeast Pacific, September 1990. (a) Original 3-hourly time series of drifter #1199; (b) amplitude (cm/s) versus time as a function of period, T, in the range, $0 < T < 2.0$ days; (c) period (days) of the current oscillations at peak amplitude. *Courtesy, Jane Eert; Institute of Ocean Sciences.*

Our final wavelet example is presented in Figure 5.58. The upper panel shows an 8-month time series of the hourly alongshore components of current velocity, v, for two mooring sites located roughly 100 km apart along the 100 m depth isobath on the continental shelf off the west coast of Canada. Brooks Peninsula is to the north of Estevan Point. The lower panel presents a Morlet wavelet analysis, showing how the coherence amplitude and phase lag between the two time series varies as a function of time and frequency. The plot covers motions with periods in the range of 6 h to about 2 months and is cut off at the bottom where the coherence function for the longest period motions is not resolvable (the longer the period of the motions, the shorter the time period over which the signal coherence can be determined). The coherence amplitudes are color-coded, with red and blue denoting strong and weak coherence amplitudes, respectively. The phase lag of the two signals for a given time and frequency is determined using the vectors and the circular scale at the bottom right of the figure. The phase scale goes from 0 to 180° in the counterclockwise and clockwise direction. If the phase vector in the figure has an upward component, the Brooks series *leads* the Estevan series, whereas if the phase vector has a downward component, the Brooks series *lags* the Estevan series (i.e., the Estevan series leads the Brooks series). Among other aspects of the flow, results show that there are persistent diurnal current motions associated with coastally trapped diurnal shelf waves (Crawford and Thomson, 1982, 1984; Cummins et al., 2000) and more intermittent semidiurnal currents associated with internal tides (Drakopoulos and Marsden, 1993; Cummins and Oey, 1997). The diurnal motions are highly coherent whereas the semidiurnal motions are much less coherent. Highly coherent coastally trapped waves with periods of around 10 days are also sometimes present in the record (Yao et al., 1984).

5.9.4 The S-Transformation

Wavelet transforms are not the only method for dealing with nonstationary oscillations with time-varying amplitudes and phases. The S-transformation (Stockwell et al., 1994) is an extension of the wavelet transform that has been used by Chu

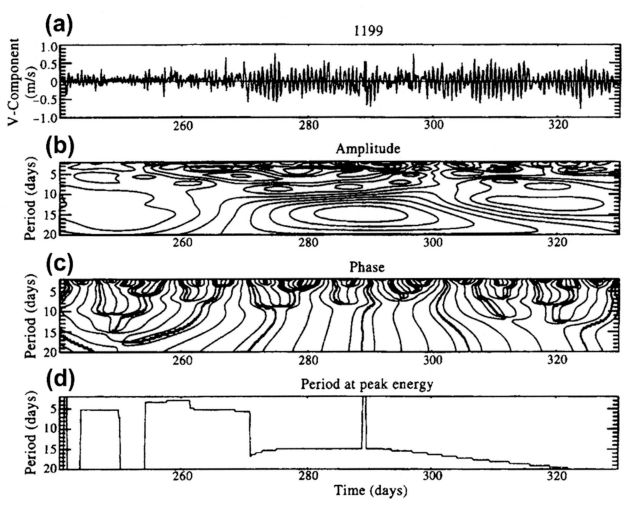

FIGURE 5.57   As for Figure 5.56 but for a larger range of periods. (a) Original 3-hourly velocity time series of drifter #1199; (b) amplitude (cm/s) and (c) phase (degrees) versus time as a function of period, T, in the range, $2 < T \leq 20$ days; (d) period (days) of the current oscillations at peak amplitude.

(1994) to examine the localized spectrum of sea level in the TOGA (Tropical Ocean Global Atmosphere) data sets. For this particular transform, the relationship between the S-transform, $S(\omega, \tau)$, and the data, $x(t)$, is given by

$$S(\omega, \tau) = \int_{-\infty}^{\infty} H(\omega + \alpha)e^{-\left(2\pi^2\alpha^2/\omega^2\right)}e^{i2\pi\alpha\tau}d\alpha \tag{5.287}$$

$$x(t) = \int_{-\infty}^{\infty}\int_{-\infty}^{\infty} S(\omega, \tau)e^{i2\pi\alpha\tau}d\omega d\tau \tag{5.288}$$

where

$$H(\omega + \alpha) = \int_{-\infty}^{\infty} x(t)e^{-i2\pi(\omega+\alpha)\tau}dt \tag{5.289a}$$

$$= \int_{-\infty}^{\infty} S(\omega + \alpha, \tau)d\tau \tag{5.289b}$$

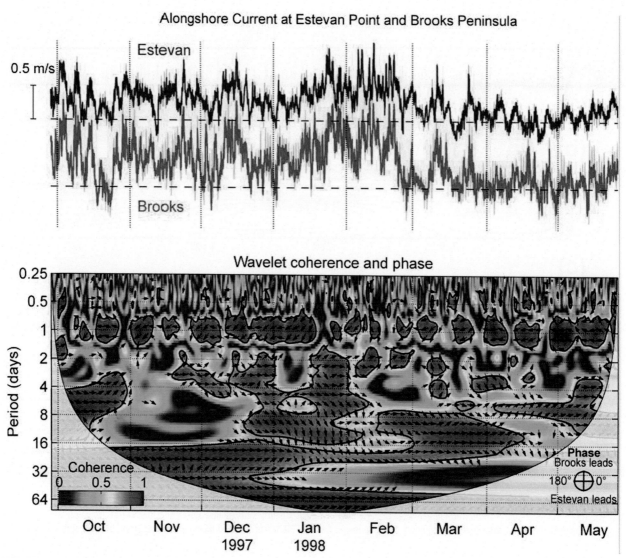

FIGURE 5.58 Morlet wavelet analysis of the coherence amplitude and phase of the alongshore component of current velocity, v, at 35 m depth for current meter mooring sites in 100 m of water off Brooks Peninsula and Estevan Point on the west coast of Vancouver Island, British Columbia. The top panel shows the hourly current velocity in m/s (see scale bar) for the two sites for September 28, 1997 to May 26, 1998. Instruments are separated alongshore by roughly 100 km. Coherence amplitude and phase values over the same time period are given by the scales on the lower left and right, respectively. See the text for a further explanation. *Courtesy of Maxim Krassovski, Institute of Ocean Sciences.*

is the standard Fourier transform of the input time series data. As indicated by Eqn (5.289b), the Fourier transform is the time average of the S-transform, such that $|H(\omega)|^2$ provides a record-averaged value of the localized spectra, $|S(\omega)|^2$, derived from the S-transform. Eqn (5.288) can also be viewed as the decomposition of a time series, $x(t)$, into sinusoidal oscillations, which have time-varying amplitudes $S(\omega, \tau)$.

The discrete version of the S-transformation can be obtained as follows. As usual, let $x(t_n) = x(n\Delta t)$, $n = 0, 1, ..., N-1$ be a discrete time series of total duration $T = N\Delta t$. The discrete version of Eqn (5.287) is then

$$S\left(0, \tau_q\right) = \frac{1}{N} \sum_{m=0}^{N-1} x(m/T), p = 0 \tag{5.290a}$$

$$S\left(\omega_p, \tau_q\right) = \sum_{m=0}^{N-1} \left\{ H[(m+p)/T]e^{-\left(2\pi^2 m^2/p^2\right)} e^{i2\pi mq/N} \right\}, p \neq 0 \tag{5.290b}$$

where $S(0, \tau_q)$ is the mean value for the time series, $\omega_p = p/N\Delta t$ is the discrete frequency of the signal, and $\tau_q = q\Delta t$ is the time lag. The DFT is given by

$$H(p/T) = \frac{1}{N} \sum_{k=0}^{N-1} x(k/T)e^{-i2\pi pk/N} \tag{5.291}$$

The S-transform is a complex function of frequency ω_p and time τ_q, with amplitude and phase defined by

$$A(\omega_p, \tau_q) = |S(\omega_p, \tau_q)| \tag{5.292a}$$

$$\Phi(\omega_p, \tau_q) = \tan^{-1}\left\{ \mathrm{Im}\left[S(\omega_p, \tau_q)\right] / \mathrm{Re}\left[S(\omega_p, \tau_q)\right] \right\} \tag{5.292b}$$

For a sinusoidal function of the form

$$X(\omega_p, \tau) = A(\omega_p, \tau)\cos\left[2\pi\omega_p\tau + \Phi(\omega_p, \tau)\right] \tag{5.293}$$

the function X at frequency ω_p is called the "voice."

Chu (1994) applied the S-transform to the nondimensionalized sea-level records, $x(t)$, collected at Nauru ($0°32'$ S, $166°54'$ W) in the western equatorial Pacific and La Libertad ($2°12'$ S, $80°55'$ W) in the eastern equatorial Pacific. Here

$$x(t) = \frac{[\eta(t) - \overline{\eta}]}{\overline{\eta}} \tag{5.294}$$

and $\overline{\eta}(t)$ represents the mean value of the sea level, $\eta(t)$. A Fourier spectral analysis of the time series revealed a strong annual sea-level oscillation in the western Pacific and a weak annual oscillation in the eastern Pacific. Both stations had strong quasi-biennial oscillations with periods of 24–30 months. The S-transformation was then used to examine the temporal variability in these components throughout the 16- and 18-year time series. For example, the voices for the annual oscillation ($\omega_{16} = 16/T$; $T = 192$ months) were similar at the two locations with higher amplitudes in the late 1970s than in the late 1980s (Figure 5.59). At La Libertad, the annual cycle became weak after 1979. The temporally varying quasi-biennial oscillations ($\omega_8 = 8/T$) were out of phase between the western and eastern Pacific (Figure 5.60).

5.9.5 The Multiple Filter Technique

The multiple filter technique is a form of signal demodulation that uses a set of narrow-band digital filters (windows) to examine variations in the amplitude and phase of dispersive signals as functions of time, t, and frequency, ω (or f). Originally designed to resolve complex transient seismic signals composed of several dominant frequencies (Dziewonski et al., 1969), the technique has recently been modified for the analysis of clockwise and counterclockwise rotary velocity components (Thomson et al., 1997) and in investigations of tsunami frequency content (Rabinovich et al., 2011a, b, 2019) and tsunami wave dispersion (Gonzalez and Kulikov, 1993).

The multiple filter technique relies on a series of band-pass filters centered on a range of narrow frequency bands to calculate the instantaneous signal amplitude or phase. Dziewonski et al. (1969) filter in the frequency domain rather than the time domain, although the results are equivalent to within small processing errors. The filtering algorithm generates a matrix (grid) of amplitudes or phases with columns representing time and rows representing frequency (or period). The gridded values can then be contoured to give a three-dimensional plot of the demodulated signal amplitude (or phase) as a function of time and frequency. González and Kulikov (1993) and Kulikov and González (1997) used the technique to examine the evolution of tsunami waves generated by an undersea earthquake in the Gulf of Alaska on March 6, 1987 (Figure 5.61). Sea-level heights measured by two bottom-pressure recorders deployed in the deep ocean to the south of Kodiak Island show that the tsunami waves were highly dispersive (low frequencies propagated faster than high frequencies) and that the arrival times of the waves closely followed the theoretical predictions for shallow-water wave motions. Peak spectral amplitudes were centered around a period of roughly 5 min, and the signal duration was about 40 min.

5.9.5.1 Theoretical Considerations

Because the technique is used to examine signal energy as a function of time and frequency, it is desirable that the filtering function has good resolution in the immediate vicinity of each center frequency and time value of the "f–t" diagram. The Gaussian function was chosen to meet these requirements since the frequency–time resolution is greater for this function than any other type of nonband-limited function. A system of Gaussian filters with constant relative response leads to a

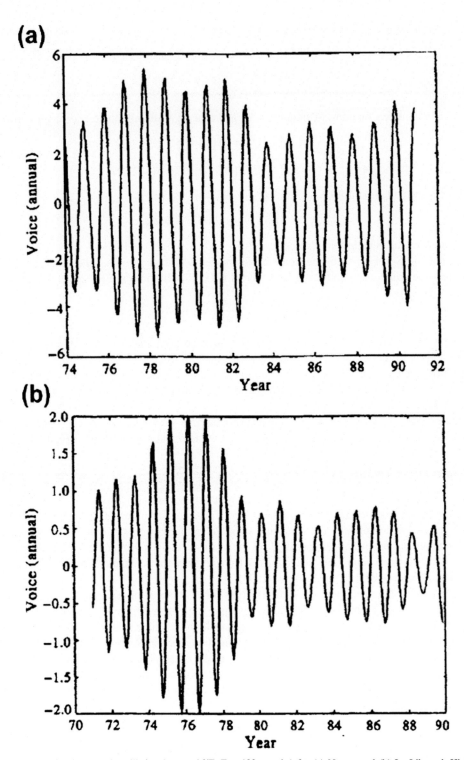

FIGURE 5.59   The "voices" for the annual oscillation ($\omega_{16} = 16/T$; $T = 192$ months) for (a) Nauru; and (b) La Libertad. Higher amplitudes were recorded in the late 1970s than in the late 1980s (Chu, 1994).

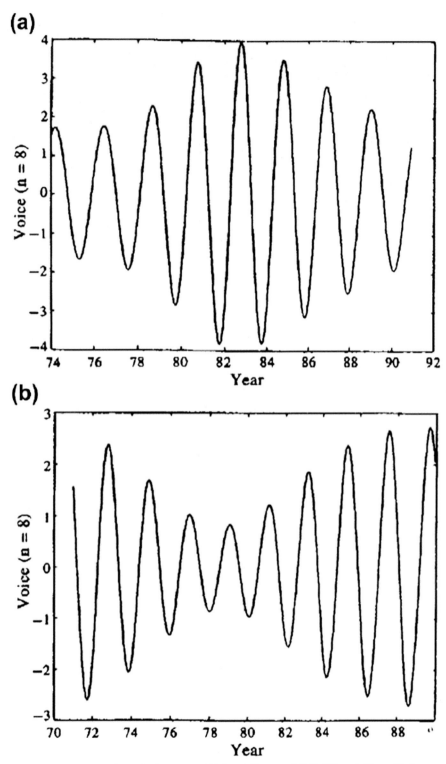

FIGURE 5.60 The "voices" for the quasi-biennial oscillations ($\omega_8 = 8/T$) for (a) Nauru; and (b) La Libertad. The oscillations were out of phase between the western and eastern Pacific (Chu, 1994).

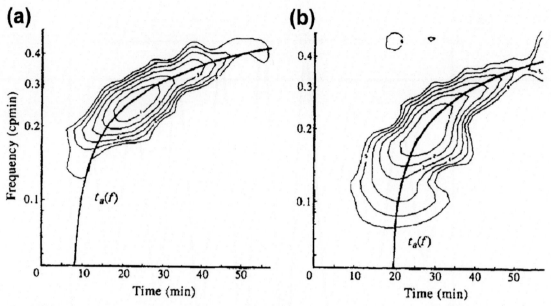

FIGURE 5.61 Multiple filter technique applied to tsunami sea-level heights measured at bottom pressure stations AK1 and AK2 in 5 km of water near 53° N, 156° W in the Gulf of Alaska. (a) November 30, 1987 and (b) 6 March 1988. Amplitude contours in the $f–t$ diagram are normalized by the maximum value and drawn with a step of 1 dB. Solid curve denotes the theoretical arrival time, $t_a(f)$, as a function of frequency for these highly dispersive waves. *From Kulikov and González (1997); see also González and Kulikov (1993).*

constant resolution on a $\log(\omega)$ scale. If $\omega_n = 2\pi f_n$ denotes the center frequency of the nth row, the Gaussian window function can be written as

$$H_n(\omega) = \exp\{-\alpha[(\omega - \omega_n)/\omega_n]^2\} \tag{5.295}$$

The Fourier transform of H_n, which bears a close resemblance to the Morlet wavelet Eqn (5.286), is

$$h_n(t) = \frac{\sqrt{\pi}}{2\alpha}\omega_n \exp\left[-\left(\omega_n^2 t^2 / 4\alpha\right)\right]\cos(\omega_n t) \tag{5.296}$$

The resolution is controlled by the parameter, α. The value of α that we choose depends on the dispersion characteristics in the original signal and, as the user of this method will soon discover, improved resolution in time means reduced resolution in frequency, and vice versa. We also need to truncate the filtering process. Dziewonski et al. (1969) used a filter cut-off where the filter amplitude was down 30 dB from the maximum.

If we let BAND be the relative bandwidth, then the respective lower and upper limits of the symmetrical filter, denoted $\omega_{L,n}$ and $\omega_{U,n}$, are

$$\omega_{L,n} = (1 - \text{BAND})\omega_n \tag{5.297a}$$

$$\omega_{U,n} = (1 + \text{BAND})\omega_n \tag{5.297b}$$

The parameter α in Eqns (5.295) and (5.296) is expressed in terms of the bandwidth and the function β, where

$$\alpha = \beta/\text{BAND}^2 \tag{5.298}$$

and

$$\beta = \ln\left[\frac{H_n(\omega_n)}{H_n(\omega_{L,n})}\right] = \ln\left[\frac{H_n(\omega_n)}{H_n(\omega_{U,n})}\right] \tag{5.299}$$

describe the decay of the window function, $H_n(\omega)$. The window function then takes the form

$$H_n(\omega) = \begin{cases} 0 \text{ for } \omega < (1 - \text{BAND})\omega_n \\ \exp\{-\alpha[(\omega - \omega_n)/\omega_n]^2\} \text{ for } (1 - \text{BAND})\omega_n \leq \omega \leq (1 + \text{BAND})\omega_n \\ 0 \text{ for } \omega > (1 + \text{BAND})\omega_n \end{cases} \tag{5.300}$$

In their analysis of seismic waves, Dziewonski et al. (1969) used BAND = 0.25, $\beta = 3.15$, and $\alpha = \beta/\text{BAND}^2 = 50.3$.

The f–t diagram for the Alaskan tsunamis (Figure 5.61) was obtained by windowing in the frequency domain with the truncated Gaussian function Eqn (5.300). In the time domain, the traces represent the convolution of the original data series with the Gaussian weighting function. The authors first set $\alpha = 25$ and chose $\beta = 1$, so that BAND = 0.20. The choice of β in Eqn (5.299) is arbitrary and can be set to unity, whereupon the bandwidth is determined by the e^{-1} values of the Gaussian function. For $\alpha = 25$ but $\beta = 2$, we have BAND = 0.28, and so on.

The flowchart for the analysis (Figure 5.62) that produced the distributions in Figure 5.61 is as follows:

1. Remove the mean and trend (linear or other obvious functional trend) from the digital time series, $y(t)$.
2. Fourier transform the time series. If an FFT algorithm is to be used for this purpose, augment the time series with zeroes to the nearest power of two.
3. Evaluate the center frequencies, $\omega_n = \omega_{n-1}/\text{BAND}$, for the array of narrow-band filters. The filters have a constant relative bandwidth, BAND, with the total width of each filter occupying the same number of rows in the log(frequency) scale. As noted on numerous occasions in the text, it is the length of the time series and the sampling rate, which determine the frequency of the Fourier components. Because it is often difficult to get the frequencies obtained from the Fourier analysis to line up exactly with the center frequencies of the filters, select those components of the Fourier analysis, which are closest to each member of the array and use these as the center frequencies.
4. Select equally spaced times (columns) for calculation of amplitude or phase, focusing mainly on the times following the arrival of the waves.
5. Filter the wave spectrum (sine and cosine functions of the Fourier transform) in the frequency domain with the Gaussian filter, $H_n(\omega)$. This filter is symmetric about the center frequencies, ω_n.
6. Take the IFT of the spectrum using the same Fourier transform used in step 2. Since the IFT for the wave spectrum as windowed by the function $H_n(\omega)$ yields only the in-phase component of the filtered signal for each ω_n, knowledge of the quadrature spectrum is also required for evaluation of the instantaneous spectral amplitudes and phases. The quadrature spectrum is found from the in-phase spectrum using

$$Q_n(\omega) = H_n(\omega)e^{i\pi/2} \tag{5.301}$$

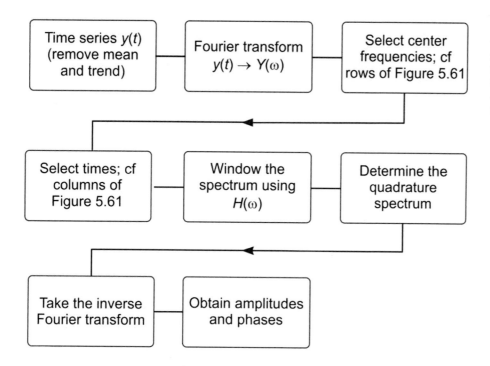

FIGURE 5.62 Flowchart for application of the multiple-filter technique. The results of this method lead to frequency versus time (f–t) distributions similar to those plotted in Figure 5.61. *Adapted from Dziewonski et al. (1969).*

The amplitude and phase of the signal for each center frequency for each time are derived from the IFTs of the spectrum and quadrature spectrum.

7. Instantaneous spectral amplitudes and phases are computed for each time step. The procedures (5)–(7) are repeated for each center frequency.

The multiple filter technique can be used to examine rotary components of current velocity fields. In this case, the input is not a real variable, as it is for scalar time series, but a complex input, $w(t) = u(t) + iv(t)$. We obtain Figure 5.63 from the analysis of a 90-day time series of surface currents measured by a 15-m drogued satellite-tracked drifter launched off the Kuril Islands in the western North Pacific on September 4, 1993 (Thomson et al., 1997). The 3-hourly sampling interval used for this time series was made possible by the roughly eight position fixes per day by the satellite-tracking system. Plots show the variation in spectral amplitude of the clockwise and counterclockwise rotary velocity components as functions of time and frequency. For illustrative purposes, we have focused separately on the high and low frequency ends of the spectrum (periods shorter and longer than 2 days). Several interesting features quickly emerge from these f–t diagrams. For example, the motions are entirely dominated by the clockwise rotary component except within the narrow channel (Friza Strait) between the southern Kuril Islands, where the motions become more rectilinear. The burst of

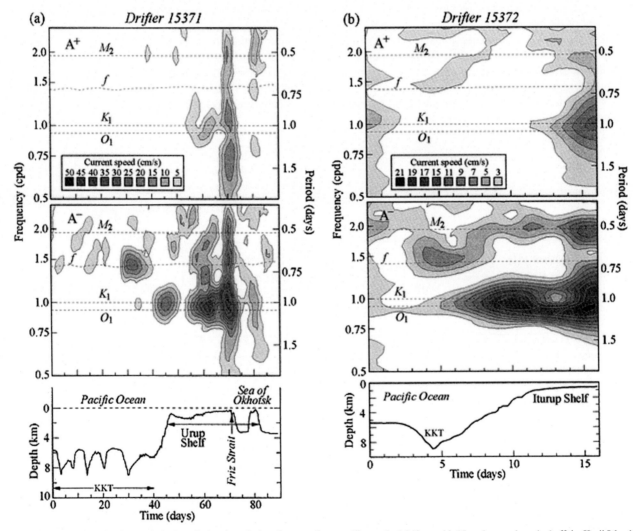

FIGURE 5.63 Multiple-filter technique applied to the velocity of near-surface, satellite-tracked drifters with 15-m drogues launched off the Kuril Islands in the western North Pacific in 1993. (a) A^- denotes the spectral amplitude (cm/s) of the clockwise rotary component versus frequency (cpd) and time (day); (b) A^+ denotes the spectral amplitude of the counterclockwise component'; the bottom panels in (a) and (b) show the water depth and location along each drifter track. Letters and dotted lines denote the frequencies and periods of the main diurnal and semidiurnal tidal constituents; f is the local Coriolis frequency (period) and KKT denotes the Kuril-Kamchatka Trench. *From Thomson et al. (1997).*

clockwise rotary flow encountered by the drifter over the Kuril−Kamchatka Trench starting on day 28 was associated with wind-generated inertial waves, whereas he strong clockwise rotary diurnal currents first encountered on day 40 and again on day 55 were associated with diurnal-period continental shelf waves propagating along the steep continental slope of the Kuril Islands.

5.10 REGIME SHIFT DETECTION

A regime can be broadly defined as an extended period of time over which a given variable or natural system assumes a statistically stable and stationary state. The rapid transition between states is termed a "regime shift." We can associate regimes and regime shifts with a wide variety of natural systems, including national governments, seasonal weather patterns, marine ecosystems, large-scale current patterns, and global climate. If we think of the seasons as distinct earth-ocean regimes, then the transition from winter to summer that occurs each spring (the "Spring Transition" (ST)) and the return to winter in fall (the "Fall Transition" (FT)) can be considered regime shifts. A square wave oscillator, in which the variable jumps abruptly from one stable "flat line" state to another, is a classic example of regimes and regime shifts. Note that "flat" in this context does not necessarily imply constant since there can be large fluctuations about the mean level. In marine and atmospheric sciences, regimes are generally data-determined and identified as periods of time having differing mean levels about which there can be pronounced fluctuations. Investigators might be interested in changes ranging from seasonal conditions during given years (cf., Agapitos and Gajewski, 2012) or in changes in average climate values over interannual timescales (cf., Rodionov and Overland, 2005). Although less common, regime shifts can also be identified through abrupt changes in the signal variance. For example, density-driven bottom water renewal events in coastal inlets can cause the bottom waters to transition from a quiescent state dominated by weak fluctuations in temperature, salinity, dissolved oxygen, and other oceanic variables, to a noisy, highly variable state characterized by pronounced fluctuations in these variables due to spatial gradients introduced by the event. Another example is the effect of a tsunami on water levels; the mean level over days or longer does not change but shorter-term variations may be an order of magnitude greater than average, increasing the variance over a day or so. In ecosystems analysis, step changes in the variance of species abundance in adjacent trophic levels may signal cascading reorganization within the environment. Daskalov et al. (2007) found that a decrease or increase in species abundance variance in the Black Sea since the 1950s was related to a strengthening or weakening of top-down forcing of predatory fish linked through plankton to dissolved oxygen levels in the water column.

This section presents three nonparametric methods for defining regime shifts (statistical transitions) in the ocean. These are: "sequential t-test analysis of regime shifts" (STARS), "adaptive Kolmogorov−Zurbenko filters" (KZA), and "cumulative upwelling index" (CUI). A fourth approach—the use of subjective interpretation in which the investigator uses his or her expertise and knowledge of the data to partition specific data sets into regimes and regime shifts—should not be dismissed out of hand. As pointed out by Overland et al. (2006), "…the main difficulty in its (regime shifts) application is the inability to infer a unique underlying system structure from relatively short time series records." Moreover, all regime-shift algorithms require specification of certain parameters, thereby introducing a degree of subjectivity into the process. In our experience, it is impossible to design a purely objective method and that some degree of subjectivity (call it investigators "expertise" and "experience") is invariably required.

5.10.1 Sequential t-Test Analysis of Regime Shifts (STARS)

STARS was originally designed for climatological time series (Rodionov, 2004, 2006) but has since found applications to other time series such as those for large-scale ecosystems in the Bering Sea (Rodionov and Overland, 2005). STARS was first written in FORTRAN but is now also available for Excel and MATLAB (available at http://www.climatelogic.com; earlier versions of the software are available at http://www.beringclimate.noaa.gov/regimes/). The program uses an algorithm for a sequential t-test (see Appendix Table D.3 for t-test values) in which deviations from the mean value are calculated and compared to a critical value. As each new observation is added to the analysis, a new t-test is performed. The test then determines the validity of the null hypothesis (H_0) that the mean values of two regimes (the original and a possible new regime, which is being considered based on newly added data values) are equal. In essence, the analysis is asking whether the new sequence of data values is part of a new regime or just a minor variation on the existing regime. The strength of a regime is quantified by the Regime Shift Index (RSI), which is dependent on a chosen cutoff length (L) and the probability level (p) of the t-test. Only regimes longer than the assigned cutoff length are detected, unless their RSI values are high enough to be detected for data segments shorter than L (specifically, at the beginning or end of a data series). As shown in Figure 5.64, the duration of regimes detected by STARS becomes shorter as the cutoff length, L, is reduced. The method also requires specification of the Huber weight parameter, H, which determines how outliers are to be

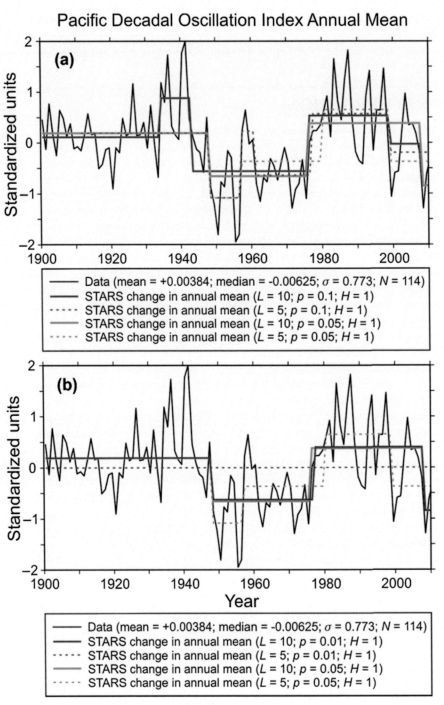

FIGURE 5.64 The effect of changing cutoff length, L, and probability level, p, on the statistically significant difference between regimes for the time series of annual means of the Pacific Decadal Oscillation (PDO) (the first spatial mode of sea surface temperature empirical orthogonal functions). Decreasing L and/or increasing p, increases the number of regimes found in a time series by decreasing the magnitude of the shifts to be detected. (a) Regimes detected in the annual mean of the PDO for $L = 5$ and 10, for $p = 0.1$ and 0.05 ($H = 1$); (b) same as (a) but for and $p = 0.05$ and 0.01. *Courtesy of Roy Hourston, Institute of Ocean Sciences.*

weighted. All values within H standard deviations of the expected mean value of the new regime (default, $H = 1$) have a weight of one, and otherwise are weighted inversely proportional to their distance from the expected mean value of the new regime. In Figure 5.65, we show that the appropriate specification of the Huber weight parameter can be important. In this case, the identification of regimes (top panel) may be confounded by the presence of a single outlier (middle panel), unless H is adjusted appropriately (bottom panel).

Following Rodionov and Overland (2005), we let $x_1, x_2, \ldots, x_i, \ldots$ be a preexisting time series or, alternatively, a data series that has new data being added on a regular basis. The first step is to determine an initial critical regime by calculating the mean and standard deviation over the first L values, $x_1, x_2, \ldots, x_L$. Once this regime is calculated, the program then returns to the start of the time series and begins to recalculate a mean value starting with the mean of the first and second points, x_1 and x_2 (x_1 is assumed part of the initial regime but x_2 may be part of a new regime). A check is performed to determine if the mean of these two values has a statistically significant deviation from the mean value of the current regime. If so, that value is marked as a potential change point, i_c, and subsequent observations are incorporated in the running mean to confirm or reject the null hypothesis. This approach allows detection of regime shifts near the beginning of time series for $i < L$. The hypothesis is tested throughout the data set using the RSI, which is calculated for each i_c as

$$\text{RSI} = \frac{\sum_{i=i_c}^{i_c+m} \hat{x}_i}{L\sigma_L} \qquad (5.302)$$

where $m = 0, \ldots, L-1$ is the number of values since the start of a new regime, σ_L is the average standard deviation for all the regime intervals in the time series to present, and $\hat{x}_i = x_i - \bar{x}_{new}$ is the difference between the data value x_i, and the

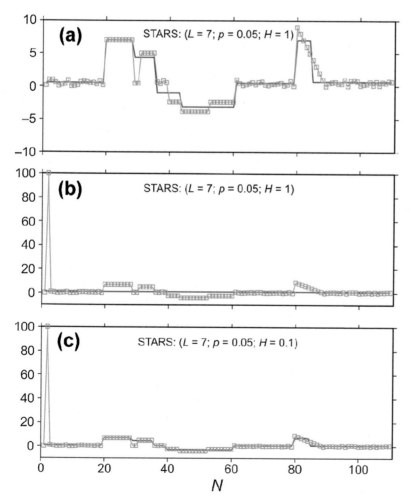

FIGURE 5.65 The importance of considering the Huber weight parameter, H, when using sequential t-test analysis of regime shifts (STARS). The data in the three panels are identical except for the second data value, x_2, which has been changed from $x_2 \sim 0$ in the upper panel to $x_2 = 100$ in the middle and lower panels. The regimes identified by the STARS algorithm in the top panel (a) can no longer be identified in the middle panel (b) due to the addition of the outlier. When the Huber weight parameter is reduced from 1 to 0.1 in the bottom panel (c), the weighting of the outlier is reduced, and the original regimes are identified. *Courtesy of Roy Hourston, Institute of Ocean Sciences.*

possible mean level for the new regime, $\bar{x}_{new}$ that is being tested. For $\bar{x}_{new}$ to be considered as the mean value for a new regime, the difference from the mean level for the current regime $\bar{x}_{cur}$ must be statistically significant according to a studen's t-test specified by

$$\text{difference} = \bar{x}_{new} - \bar{x}_{cur} = t\sqrt{2\sigma_L^2/L} \tag{5.303}$$

where t is the value of the t-distribution with $2L-1$ DoF at the given probability level, p. If, at any time from the start of the new regime, RSI becomes negative, the test fails and a zero value is assigned. If RSI remains positive throughout the range $(0, L-1)$, then i_c is selected to be the time of a regime shift at the selected level $\leq p$. The search for the next regime shift starts at $i_c + L$ so as to ensure that its timing is detected correctly even if the actual duration of the new regime is short relative to the expected durations of actual regimes. Table 5.19 is a shortened version of that presented in Rodionov and Overland (2005) for pollock stock recruitment in the Bering Sea. Data are available from 1963 but RSI values are only listed for the start of the first regime shift in 1978. Only for 3 years (1978, 1985, 1989) did RSI values remain positive up to $m = L-1$.

The average value for the current regime, $\bar{x}_{cur}$, is calculated for the period (i_c-L, i_c). If the transition from one regime to another is gradual, the program might not detect the change because $\bar{x}_{cur}$ is also changing as the window slides along the time axis. Specifically, the difference between the new arriving observations and $\bar{x}_{cur}$ may not be sufficiently statistically significant to become a change point and trigger the calculation of RSI. In the version presented by Rodionov and Overland (2005), $\bar{x}_{cur}$ is calculated for the period that begins from the start of the previous regime shift to the point immediately before the current point in time. As a result, a stepwise function of regimes is produced in most cases. (In a previous version of the method presented by Rodionov (2004), the program could only detect abrupt changes in regimes). To improve the performance at the beginning of the time series, testing for a new regime starts at x_2 rather than at x_{L+1} as the previous version. The average value, $\bar{x}_{cur}$, is still calculated for the entire initial period $(1, L)$ but, if a regime shift occurred prior to $i = L$, it is now detected.

The STARS method may also be used to identify regime shifts based on changes in variance, in addition to changes in the mean. In this approach, shifts in the mean are identified first and then subtracted from the original data series. Changes in the variance of these residuals are then examined similarly to changes in the mean, but using an F-test of the ratio of variances rather than a t-test of the difference of means (Daskalov et al., 2007). Regime identification based on changes in variance may be viewed as supplementary and in support of regimes identified based on changes in the mean.

TABLE 5.19 Truncated version of regime shift index (RSI) table for pollock recruitment ($L = 5$ years, $p = 0.1$; Huber weight parameter = 1).

Year ($i = i_c$)	$m = 0$	$m = 1$	$m = 2$	$m = 3$	$m = 4$
1978	**0.51**	**0.45**	**0.44**	**0.21**	**0.57**
1979	0	0	0	0	0
1980	0	0	0	0	0
1981	0.1	0	0	0	0
1982	0	0	0	0	0
1983	0.15	0	0	0	0
1984	0	0	0	0	0
1985	**0.1**	**0.29**	**0.52**	**0.67**	**0.12**
1986	0	0	0	0	0
1987	0	0	0	0	0
1988	0	0	0	0	0
1989	**0.34**	**0.24**	**0.05**	**0.37**	**0.09**
1990	0	0	0	0	0

The nonzero values of i_c sometimes triggered the calculation of RSI; however, only for the years highlighted in bold did the values remain positive and a regime shift declared.
From Rodionov and Overland (2005).

5.10.2 Adaptive Kolmogorov–Zurbenko (KZA) Filters

The KZA algorithm is based on an iterative moving average Kolmogorov–Zurbenko (KZ) filter. In the KZ filter, a moving average over two times the half-window length plus one ($2q + 1$) is utilized over several iterations, in which the previously averaged values are used as input for the next successive iteration. The time rate of change of the moving averages is computed, and where large, the window length is shortened and moving averages are recomputed. This adaptive aspect of the filtering process defines the KZA filter. The significance of discontinuities may be estimated from the sample variances over the averaging windows. As an example, Figure 5.66 provides a comparison of the regime shift detection capabilities of STARS and KZA for the winter (DJF) time series of the Arctic Oscillation. Both are adjustable and necessitate specifying a window-width, or length of regime one wishes to detect. Both appear to perform reasonably well in detecting sharp discontinuities, or regime shifts, in the time series.

As mentioned in the introductory summary, the nonparametric KZA filter is a low-pass moving average KZ filter that dynamically adjusts the length of the filter according to the rate of change of the process being investigated. As the rate of change of the smoothed data set increases, the length of the filter decreases in order to better resolve the changes. Since the filter depends on an iterative moving average, it filters out high-frequency variations in the data set. Following the notation of Rodionov and Overland (2005), the simple moving average KZ filter is computed according to

$$y_i = \frac{1}{2q + 1} \sum_{j=-q}^{q} x_{i+j} \qquad (5.304)$$

where x_i is the original data, y_i is the filtered data, and $2q + 1$ is the length (number of data values) of the filter window. The filter is iterative, in that the filter Eqn (5.304) is applied, not once, but k-times. After the first pass ($k = 1$), the low-pass filtered values, $y_i = y_{i,1}$, become the new x_i values and the filtering is repeated to produce an even smoother version, $y_i = y_{i,2}$, of the original data and so on until $y_i = y_{i,k}$. The KZ filter is an efficient low-pass linear filter (Zurbenko, 1986) that can be defined as

$$Z(t) = KZ_{q,k}[X(t)] \qquad (5.305)$$

where $X(t)$ is the original time series, q is the half-length of the filter, and k is the number of iterations (successive applications) of the filter to generate an increasingly smooth data series, $y_{i,k}$.

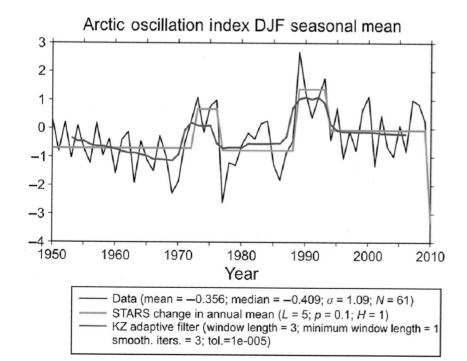

FIGURE 5.66 Comparison of regime shift detection using sequential t-test analysis of regime shifts (STARS) and adaptive Kolmogorov–Zurbenko filters (KZA) for the average winter (January–February–March) Arctic Oscillation Index. For KZA a window length of 3 appears to capture the regime changes better than a window length of 5 (not shown). *Courtesy of Roy Hourston, Institute of Ocean Sciences.*

The absolute value of the differentiated $Z(t)$ time series is defined by

$$D(t) = |Z(t+q) - Z(t-q)| \tag{5.306}$$

while the localized time rate of change of $D(t)$ is given by

$$dD(t)/dt = D(t)' = D(t+1) - D(t) \tag{5.307}$$

When a data point is located in a region of increasing $D(t)$, the half-length of the moving average in the tail region before the data point (q_{T-q}, q_T) is kept equal to the original half-length, q (as in Eqn 5.305), while the half-length ahead of the data point (q_H, q_{H+q}) is shortened as a function of $D(t)$. In the shortened region of $D(t)$, only the half-length behind the data point (the tail region) will be reduced. In the vicinity of the break point, the filter length is reduced, thus sharpening the regime shift resolution of the moving average. Modified after Eqn (5.304), the adaptive filter is defined by

$$Y_t = \frac{1}{q_H(t) + q_T(t)} \sum_{i=-q_T(t)}^{q_H(t)} X_{t+i} \tag{5.308}$$

where

$$q_H(t) = \begin{cases} q, & \text{if } D'(t) < 0 \\ f(D(t))q, & \text{if } D'(t) \geq 0 \end{cases} \tag{5.309a}$$

$$q_T(t) = \begin{cases} q, & \text{if } D'(t) > 0 \\ f(D(t))q, & \text{if } D'(t) \leq 0 \end{cases} \tag{5.309b}$$

and q is the half-length of the filter in the initial $KZ_{q,k}$ filter. The function, $f(D(t))$ is defined as

$$f(D(t)) = 1 - \frac{D(t)}{\max[D(t)]} \tag{5.310}$$

Note that $D(t) = 0$ and $f = 1$ if the two ends of the filtered record $Z(t)$ are equal over the averaging interval, which remains as $(-q, q)$. However, $f = 0$ if the function, $D(t)$ reaches its maximum value $\max[D(t)]$ in the interval being considered. Because $\max[D(t)]$ is the largest change, the latter coincides with the break point.

Plots of the filtered data series, Y_t, reveal times of discontinuities (possible regime shifts) in the time series (Figure 5.67). This qualitative evidence for discontinuities can be placed on a more quantitative basis using the sample variances, σ_t^2, of Y_t defined by

$$\sigma_t^2 = \frac{\sum\limits_{i=q_T}^{q_H} [Y_i - \overline{Y}_t]^2}{q_T + q_H} \tag{5.311}$$

where $\overline{Y}_t$ is the record average. Zurbenko et al. (1996) compare the KZA with the parametric Schwarz criterion for identifying regime shifts in simulated and geophysical time series containing seasonal patterns and trends. Although both methods were successful in locating relatively large discontinuities, the KZA approach was rated more highly for two reasons: (1) it was more accurate in the case of simulated seasonal cycles; and (2) the Schwarz criterion depends on independent and trend-free data, conditions rarely satisfied in geophysical data.

5.10.3 Cumulative Upwelling Index (CUI)

The CUI is a time-integrated version of the upwelling index (UI) devised by Bakun (1973) to define the transition (shift) to upwelling favorable winds within eastern boundary regions of the World Ocean. Examples of eastern boundary upwelling regions are the coasts of northwest and southwest Africa, Peru, Western Australia, and western North America. The timing of the transition from downwelling favorable winds in winter to upwelling favorable winds in summer (known as the Spring Transition, ST) is of particular importance to the functioning of large-scale marine ecosystems (Schwing et al., 1996; Bograd et al., 2009). Although developed for detecting the seasonal transition in the coastal ocean, the idea behind

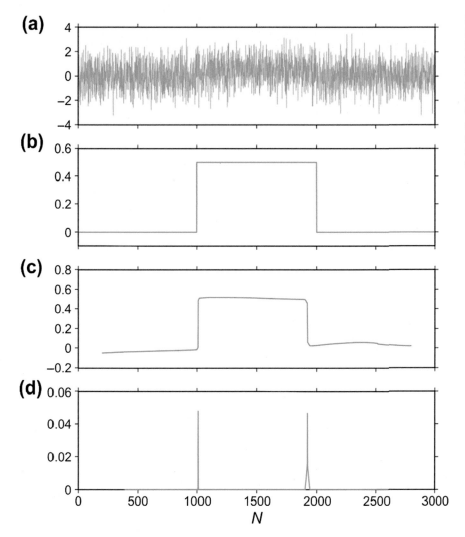

(a)

(b)

(c)

(d)

N

FIGURE 5.67 The results of the adaptive Kolmogorov−Zurbenko filters (KZA) filter for $q = 100$, $k = 3$ applied to 3,000 synthetic standard normal random numbers with breaks of amplitude 0.5σ at times $t = 1,000$ and $2,000$, where σ is the standard deviation of the original time series. (a) Original data; (b) base line box-car function upon which the random data have been superimposed; (c) the KZA filtered data; and (d) the variance σ^2 of the filtered time series. *Adapted from Zurbenko et al. (1996).*

the CUI has possible applications to other time series experiencing seasonal to decadal or even longer-scale regime shifts, providing the parameter in question shifts between periods of positive and negative values.

Despite their name, "upwelling indices" are not based on measurements of upwelling (vertical displacements of isopycnals by currents). In fact, most indices are not even based on oceanic measurements but rather on the alongshore (y-direction) component of coastal wind stress, τ_y, averaged over some time period (typically a day or a month) that is longer than the local inertial period ($2\pi/f$), the time required for geostrophy to modify the circulation. Here, the wind stress serves as a proxy for the cross-shore (x-direction) surface Ekman layer transport, ME_x, generated along eastern boundary current regions. The longest UI series begins in 1967 and is computed from the 6-hourly, $1 \times 1°$ resolution atmospheric sea-level pressure fields generated by the U.S. Navy Fleet Numerical Meteorology and Oceanography Center. In a steady state, the cross-shore transport in the wind-driven surface Ekman layer must be balanced by an oppositely directed cross-transport, MI_x, in the interior region below the Ekman layer of depth, z_E (i.e., $MI_x = -ME_x$). If we ignore the coastal boundaries and assume a constant vertical eddy viscosity, A_z, and a water depth much greater than the surface Ekman layer depth, $z_E = \pi(2A_z/|f|)^{1/2}$ (typically <100 m at mid to high latitudes), the linearized, steady state, alongshore momentum balance for uniform Coriolis parameter, f, is

$$-fu = \frac{1}{\rho} \frac{\partial \tau_y}{\partial z} \qquad (5.312)$$

where u is the cross-shore component of current velocity, ρ is the water density, and the depth, z, is measured vertically downward. Integrating Eqn (5.312) over the depth of the Ekman layer yields

$$ME_x = \int_0^{z_E} \rho u \, dz = \frac{\tau_y}{f} \tag{5.313a}$$

$$MI_x = \int_{z_E}^{z_I} \rho u \, dz = -\frac{\tau_y}{f} \tag{5.313b}$$

where τ_y is positive (negative) in the poleward (equatorward) direction in both hemispheres, and $f \geq 0$ (northern hemisphere) and $f \leq 0$ (southern hemisphere). Thus, ME_x is negative away from the coast in the Northern Hemisphere when the wind is blowing toward the equator, whereby the opposing interior flow, MI_x, is positive (toward the coast, to the left of the wind direction). This is classic wind-driven upwelling. In the Southern Hemisphere, an equatorward wind is also negative, but for a right-hand coordinate system, the x-direction is positive in the offshore direction. Once again, an equatorward wind leads to positive ME_x (away from the coast, to the right of the wind direction) and negative MI_x (toward the coast).

The CUI (Schwing et al., 2006; Pierce et al., 2006; Bograd et al., 2009) is generated by summing the daily mean upwelling indices derived using Eqn (5.313b) at coastal wind-grid locations starting on January 1 of a given year and continuing to the end of the year. As illustrated by Figure 5.68, the CUI can be used to define a variety of seasonal regime shift parameters: (1) the Julian date of the Spring Transition (ST) when the CUI reaches its minimum value at the end of the winter downwelling season; (2) the Julian date of the end of the upwelling season (which we can call the Fall Transition, FT) when the CUI reaches its seasonal maximum (termed END by Bograd et al., 2009); (3) the length of the upwelling season, equal to the total number of days between ST and FT; and (4) the intensity of the upwelling season, Total Upwelling Magnitude Index (TUMI), defined as $\mathrm{TUMI} = \int_{ST}^{FT} \mathrm{CUI}(t)$. Similarly, the Total Downwelling Magnitude Index (TDMI), a measure of the intensity of downwelling during the winter, is the total CUI integrated from the observed FT date to the date of the ST in the following year. Both TUMI and TDMI have units of m^3/s per unit length of coastline (e.g., m^3/s per 100 m). Results for the California Current System from 33 to 48° N (Bograd et al., 2009) show that the upwelling season diminishes from roughly 357 days at 33° N to 151 days at 48° N, with the greatest intensity and variance off northern California (36–42° N). In the northern region of the California Current system, there has been a significant trend of 1 day/year ($r = 0.42$, $p = 0.0083$) toward a later ST, with an accompanying trend toward a shorter upwelling season at 48° N of −1.1 days/year. Bograd et al. (2009) find considerable interannual and decadal variability in the various

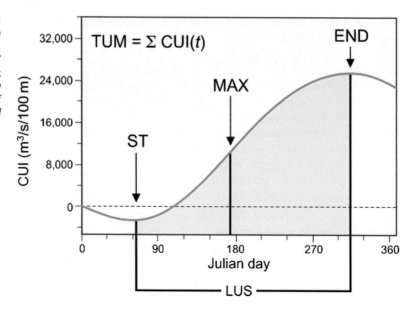

FIGURE 5.68 Schematic of the climatological annual cycle of cumulative upwelling index (CUI) ($m^3/s/100$ m) at 39° N in the California Current system. Times correspond to the following variables: ST (Spring Transition date); LUS (length of upwelling season); and TUM (total upwelling magnitude). END corresponds to the end of the upwelling season (start of the downwelling season, or Fall Transition). *Adapted from Bograd et al. (2009).*

upwelling indices, with El Niño events having a major impact on the upwelling phenology. The upwelling season began late and was of anomalously short duration in strong ENSO years (e.g., 1982–83 and 1997–98). Similar impacts are observed through changes in the currents in the California Current system arising from local and remotely wind-forced coastal trapped waves (Connolly et al., 2014).

In their study of upwelling in the California Current System, Thomson et al. (2014) compared upwelling conditions along the Oregon-Washington-British Columbia coast derived using the cumulative coastal upwelling index with conditions derived from land-based microseismic activity. The high correlation observed between microseismic intensity at ∼0.2 Hz recorded at a broadband seismological station operated by the Pacific Geosciences Centre (PGC) in southwestern British Columbia and an offshore bottom pressure station (Figure 5.69) confirms a direct link between seismic activity and regional wind-wave generation. A comparison of the spring and fall transition times obtained from coincident 20-year records of microseismic intensity and alongshore wind stress from 1993 to 2002 indicates that high-frequency seismic activity may be more representative of coastal upwelling than variations derived using traditional methods. In addition to yearly differences, the study found that the mean wind-derived spring transition dates for southern British Columbia, central Washington, and northern Oregon (mean ± standard deviation of 4 April ± 27 d, 6 April ± 26 d, and 7 April ± 29 d, respectively) are earlier than the mean seismically derived date of 10 April ± 17 d and have twice the standard deviation (Figure 5.70). The seismically defined spring transition series also had a negative trend from 1993 to 2002, which is not present in the wind-derived series. As with the spring transition, the wind-derived fall transition series are similar for the British Columbia, Washington, and Oregon coasts but have much greater interannual variability than the seismically derived fall series. The mean fall transition times derived from the wind data for the British Columbia, Washington and Oregon sites (17 October ± 20 d, 16 October ± 17 d, and 20 October ± 20 d, respectively) are considerably later than the seismically derived date of 7 October ± 11 d and have twice the standard deviation. Similar differences are found between the wind-derived and seismically derived summer duration time series obtained by subtracting the fall and spring transition dates. The mean summer duration for the seismic record (180 ± 22 d) is markedly shorter than the summer duration for the wind stress records (∼195 ± 30 d).

In an attempt to link coastal upwelling more directly with the coastal currents, Thomson and Ware (1996) developed a *current velocity index*, $I_v \equiv \alpha |v'_s \partial v / \partial z| \partial v / \partial z$, that uses daily time series of the longshore component of residual (detided) current velocity, $v(z,t)$, measured at several depths at a single location on the continental slope. Here, z is depth, t is time,

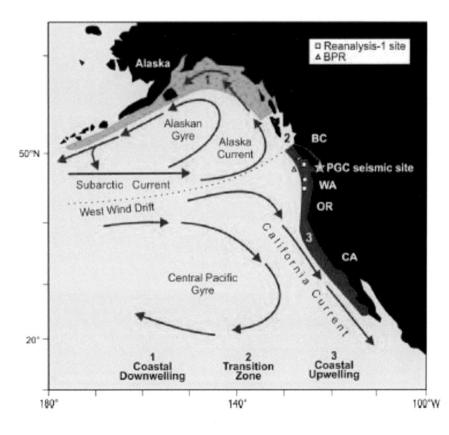

FIGURE 5.69 Upwelling domains and prevailing currents off the west coast of North America. Also shown are locations of the PGC seismic station (star at 48°39′ N; 123°27′ W), the three NCEP/NCAR Reanalysis-1grid points off the British Columbia, Washington and Oregon coasts (open squares at 48°34.2′ N, 125°37.5′ W; 46°40.0′ N 125°37.5′ W; and 44°45.7′ N, 125°37.5′ W, respectively) and the bottom pressure recorders (triangles) at ODP borehole CORK 1026 (47°45.8′ N, 127°45.6′ W) and Ocean Networks Canada (ONC) location 889 (48°40.2′ N, 126°50.9′ W). The Barkley Canyon bottom pressure recorder is near the northern Reanalysis site. *From Thomson et al. (2014).*

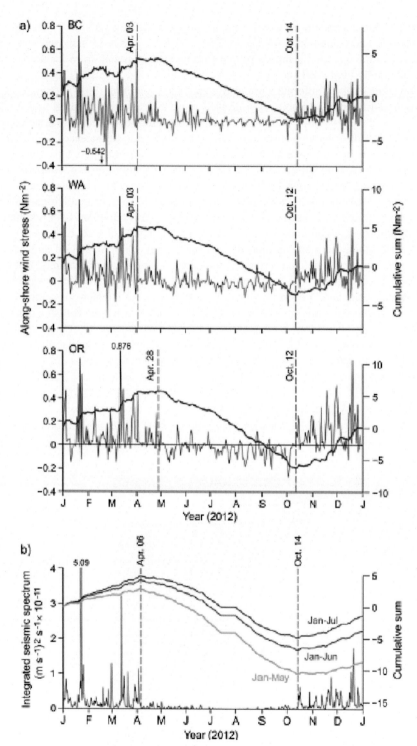

FIGURE 5.70 Time series of input data (thin lines) and resulting cumulative upwelling indices (thick lines) for southern British Columbia and the Pacific Northwest of the United States for 2012. Vertical lines give the derived dates of the spring and fall transitions. Peak input values outside the plotting range have been labeled. (a) Cumulative upwelling index, CUI_τ, based on 6-hourly alongshore wind stress for the three Reanalysis-1 sites (Figure 5.69); and (b) cumulative upwelling index, CUI_S, calculated using the 6-hourly spectra of vertical ground velocity at the PGC site. CUI_S values for the averaging period January-June (in red) are considered most representative of the annual upwelling cycle for the study region. *From Thomson et al. (2014).*

and $\alpha = -\text{sgn}(f)\left[\left(f^2 + \delta^2\right)\sigma_v\right]^{-1}$ is a normalization factor constructed from the local Coriolis parameter, f, a friction parameter, δ, and the standard deviation, σ_v, of the near-surface current velocity, $v_s = \bar{v}_s + v'_s$ where v'_s is the fluctuation about the time series mean, $\bar{v}_s$; $\text{sgn}(f) = +1$ for the Northern Hemisphere and -1 for the Southern Hemisphere. The time series of vertical shear, $\partial v/\partial z \approx \Delta v/\Delta z$, is derived from the velocity differences, Δv, between simultaneous current meter records separated by vertical differences, Δz, within the upper half of the water column. The index characterizes low-frequency baroclinic variability over the continental margin and is linked, through the thermal wind relation, to instability processes that effect the transfer of potential energy stored in the mean cross-slope density field to the kinetic energy of mesoscale meanders and eddies. For eastern boundary regions, the index ranges from large positive values during summer upwelling conditions to large negative values during winter downwelling conditions. The "transition" seasons, spring and fall, are characterized by $I_v \approx 0$. The index was applied to the continental margin of southwest Vancouver Island using current velocity data collected from October 1989 to March 1995 at depths of 35, 100, 175 and 400 m at a long-term mooring site. Results made it possible to specify the start and end times of the oceanic seasons and to quantify the summer upwelling and winter downwelling intensities for the region. Gaps in the current meter time series forced the authors to examine records of coastal sea level, surface temperature, sea surface salinity, and "Bakun" upwelling index (the wind-driven cross-shore component of surface Ekman transport) as possible surrogates for the velocity index. None of these time series was acceptable. Although the Bakun index is a logical substitute for the velocity index, differences between the two indices are significant and suggest that current velocity data are needed for accurate delineation of seasonal oceanic variability along continental margins.

5.11 VECTOR REGRESSION

Oceanographers often use time series from a more readily available variable as a surrogate for the time series of a less easily measured variable. For example, observations of tidal elevation can serve as a surrogate for the tidal currents while the alongshore component of wind stress from meteorological records is used as a surrogate for wind-induced upwelling (see previous section). In such instances, coastal stations, moored offshore platforms, and satellites provide oceanic sea-level data and wind fields whereas direct measurements of ocean currents and coastal upwelling are much more difficult to obtain. For vector processes, the problem is further complicated by the fact that the relationship can be direction-dependent rather than spatially isotropic. This complexity is probably best illustrated by the two-dimensional (vector) anisotropic response of ice-drift and ocean current velocity ("surface drift velocity") to wind forcing in the presence of coastal boundaries.

In this section, we outline regression methods that can be used to examine possible dynamical relationships between vector time series, which, in turn, help to determine if one variable can be used as a surrogate for the other variable. To summarize, we present a basic approach for using wind velocity observations to generate current or ice-drift velocities. The approach consists of estimating the "response matrixes" and corresponding "response ellipses," where the latter defines the drift or current velocity response to a unit wind velocity forcing. For each direction, φ, of the unit-magnitude wind velocity vector there is a corresponding oceanic response consisting of a "wind factor" $\alpha(\varphi)$ and "turning angle" $\theta(\varphi)$. These two factors represent (1) the speed of the current (or ice-drift) relative to the wind speed and (2) the angle that the drift velocity makes with the wind vector. The major ellipse axis corresponds to the direction of the "effective wind" ($\varphi = \varphi_{\max}$) and the minor axis to the direction of the "noneffective" wind. The eigenvectors of the response matrix are along wind directions that are the same as the wind-induced drift velocity directions. Six analytical cases are possible, depending on the water depth, distance from the coast, and other factors. Results range from rectilinear response ellipses near the coast (where the orientation of the shoreline is prominent) to purely circular response ellipses in the open ocean, far from the influence of the coast. Responses derived from the 4-parameter vector-regression method are less constrained and therefore more representative of wind-induced surface motions than those derived using the traditional 2-parameter complex transfer function approach, also discussed in this section. Our presentation closely follows Rabinovich et al. (2007), who used the model to examine ice-drift along the western shelf of Sakhalin Island in the Sea of Okhotsk.

5.11.1 The 2-Parameter Complex Functional Approach

Estimates of ice-drift and currents for shelf and coastal regions are important for a wide variety of marine requirements including navigation, oil and gas exploration, and climate investigations. Seafloor topographic variations, wave-trapping effects, and the formation of land-fast ice (for freezing areas) are among the factors that make continental margins among the most challenging areas for coastal ocean prediction research (Wang et al., 2003). Wind stress is one of the major factors affecting oceanic motions along continental margins (Gill, 1982; Wadhams, 2000). Continental margins are also areas most likely to be instrumented with extensive, near real-time observing systems, thereby allowing for effective application of regression models for both diagnostic and forecasting purposes (cf. Thorndike and Colony, 1982; Fissel and Tang, 1991; Rabinovich et al., 2007).

Regression analysis of two-vector time series is commonly based on a functional relationship between input and output vector series expressed as,

$$\mathbf{u} = \alpha \mathbf{V} \tag{5.314}$$

where $\mathbf{V} = (U, V)$ is the input vector series (e.g., wind velocity), $\mathbf{u} = (u, v)$ is the output vector series (e.g., ice-drift or current velocity; herein, "drift velocity"), and $\alpha = a + ib$ is a complex coefficient determined using an LS regression fit based on coincident wind and drift velocity observations. In the standard Cartesian coordinate system, u, U are positive to the east and v, V are positive to the north. The above formulation is widely used in studies of ice motion (Thorndike and Colony, 1982; Fissel and Tang, 1991; Greenan and Prinsenberg, 1998) and ocean currents (Cherniawsky et al., 2005). Eqn (5.314) can also be written as

$$\mathbf{u} = \alpha_0 \exp(-i\theta_0)\mathbf{V} \tag{5.315}$$

where $\alpha_0 = |\alpha| = \sqrt{a^2 + b^2}$ is the wind factor and where $\theta_0 = -\arctan(b/a)$, the turning angle of the drift velocity direction relative to the wind direction, is measured clockwise (counterclockwise) to the wind in the Northern (Southern) Hemisphere. The values, $\alpha_0 = 0.02$ (a wind of 1 m/s generates a 0.02 m/s ice-drift) and $\theta_0 = 28°$, which describe the relationship between wind and free ice-drift in the open ocean, denote the "Nansen–Ekman ice-drift law" (Thorndike, 1986; Wadhams, 2000). Similar values ($\alpha_0 = 0.01-0.04$, $\theta_0 = 10-40°$) have been obtained for surface currents.

A major limitation of the above equations is that they describe an *isotropic* response of ocean to the wind, in which the parameters α_0 and θ_0 are invariant with respect to the wind direction, φ. This limitation contrasts with Overland and Pease (1988), who observed a markedly *anisotropic* response of ice-drift to the wind near coastal boundaries. Similarly, Fissel and Tang (1991) report that proximity to the coast and the direction of the wind relative to the coastline are major factors affecting the wind factor, $\alpha(\varphi)$, and turning angle, $\theta(\varphi)$. To account for these effects, we need to apply a two-dimensional (matrix) regression model based on regional winds.

5.11.2 The Vector Regression Model

The influence of a nearby coast on wind-driven motions can be quantified using a two-dimensional vector model. This model relates the current or free ice-drift, $\mathbf{u}(t)$, to the wind, $\mathbf{V}(t)$, through the two-dimensional regression equation

$$\mathbf{u} = \mathbf{AV} + \mathbf{E} \tag{5.316a}$$

(Cooley and Lohnes, 1971; Maxwell, 1977) where

$$\mathbf{u} = \begin{pmatrix} u \\ v \end{pmatrix}, \mathbf{V} = \begin{pmatrix} U \\ V \end{pmatrix},$$

$$\mathbf{A} = \begin{pmatrix} a_{11} & a_{12} \\ a_{21} & a_{22} \end{pmatrix} \text{ and } \mathbf{E} = \begin{pmatrix} \varepsilon_u \\ \varepsilon_v \end{pmatrix}. \tag{5.316b}$$

Here, a_{ij} are the regression (response) coefficients that link the cross-shore (u) and alongshore (v) components of drift velocity with the corresponding components (U, V) of the wind velocity, and coefficients (ε_u, ε_v) denote random noise. Without loss of generality, we have assumed that the drift velocity and wind have zero mean speed. As a consequence, Eqn (5.316a) does not require a term representing the mean drift velocity.

The coefficients, a_{ij}, are obtained through an LS method (cf., Section 5.5.1) that minimizes the summations over the duration ($t = t_o$, ..., T) of the applied data set,

$$\sum_{t=t_o}^{T} [a_{11}U(t) + a_{12}V(t) - u(t)]^2; \tag{5.317a}$$

$$\sum_{t=t_o}^{T} [a_{21}U(t) + a_{22}V(t) - v(t)]^2 \tag{5.317b}$$

From Eqns (5.317a) and (5.317b) we obtain the matrix relation (Cooley and Lohnes, 1971)

$$\mathbf{AD} = \mathbf{R},\qquad(5.318a)$$

where $\mathbf{A}$ is defined by expression Eqn (5.316b), $\mathbf{D}$ is the auto-correlation matrix of the input variable (wind), and $\mathbf{R}$ is the cross-correlation matrix between input and output variables; specifically

$$\mathbf{D} = \begin{pmatrix} r_{UU}^2 & r_{UV}^2 \\ r_{UV}^2 & r_{VV}^2 \end{pmatrix} = \begin{pmatrix} \langle UU \rangle & \langle UV \rangle \\ \langle UV \rangle & \langle VV \rangle \end{pmatrix}; \mathbf{R} = \begin{pmatrix} r_{uU}^2 & r_{uV}^2 \\ r_{vU}^2 & r_{vV}^2 \end{pmatrix} = \begin{pmatrix} \langle uU \rangle & \langle uV \rangle \\ \langle vU \rangle & \langle vV \rangle \end{pmatrix}, \qquad(5.318b)$$

where $\langle \cdot \rangle$ denotes a time average. Terms involving random noise average to zero. From (5.304), it follows that

$$\mathbf{A} = \mathbf{RD}^{-1}\qquad(5.319)$$

which yields the four response coefficients, a_{11}, a_{12}, a_{21}, and a_{22}. Here, $\mathbf{D}^{-1}$ is the inverse of the symmetric matrix, $\mathbf{D}$. Because the matrix, $\mathbf{R}$ is generally nonsymmetric, the matrix, $\mathbf{A}$ is also generally nonsymmetric, leading to more complicated wind−ice and wind−current relationships than in Section 5.11.1.

5.11.2.1 Response Ellipses

The regression coefficients, a_{ij}, in Eqns (5.316a) and (5.316b) define a response ellipse, corresponding to the curve traced out by the tip of the output response vector $\boldsymbol{\alpha} = (\alpha_u, \alpha_v)$ of the drift velocity through one complete rotation of a unit amplitude input wind vector, $(U_0, V_0) = (\sin\varphi, \cos\varphi)$, where φ is the angle of the magnitude $= 1$ wind vector measured clockwise from north in the Northern Hemisphere. Specifically,

$$\alpha_u(\varphi) = a_{11} \sin \varphi + a_{12} \cos \varphi \qquad(5.320a)$$

$$\alpha_v(\varphi) = a_{21} \sin \varphi + a_{22} \cos \varphi \qquad(5.320b)$$

so that for each direction of the wind vector, φ, there is a corresponding "wind factor" $\boldsymbol{\alpha}$, having relative drift speed, $\alpha = |\boldsymbol{\alpha}(\varphi)|$, and direction, $\phi = \phi(\varphi)$. The angle between the drift velocity and wind vectors is the turning angle, $\theta = \phi - \varphi$. In the case of an isotropic response, $\alpha_u = \alpha_v = $ constant, $\theta = $ constant, so that the response ellipse is a circle.

In general, a response ellipse is described by four invariant parameters: (1, 2) the semimajor ($\alpha = A_{max}$) and semiminor ($\alpha = A_{min}$) axes corresponding to the maximum and minimum drift responses, respectively; (3) the orientation of the semimajor axis ($\phi = \phi_{max}$); and (4) the direction of the "effective wind" ($\varphi = \varphi_{max}$), corresponding to the wind direction that generates the maximum ice-drift or current response. Specifically, the *semimajor and semiminor axes* are

$$A_{max} = \left[(a_{11} \sin \varphi_{max} + a_{12} \cos \varphi_{max})^2 + (a_{21} \sin \varphi_{max} + a_{22} \cos \varphi_{max})^2 \right]^{1/2} \qquad(5.321a)$$

and

$$A_{min} = \left[(a_{11} \sin \varphi_{min} + a_{12} \cos \varphi_{min})^2 + (a_{21} \sin \varphi_{min} + a_{22} \cos \varphi_{min})^2 \right]^{1/2} \qquad(5.321b)$$

the *orientation of the semimajor axis* is

$$\phi_{max} = \arctan \left(\frac{a_{11} \sin \varphi_{max} + a_{12} \cos \varphi_{max}}{a_{21} \sin \varphi_{max} + a_{22} \cos \varphi_{max}} \right) \qquad(5.321c)$$

(the orientation of the semiminor axis is then given by $\phi_{min} = \phi_{max} \pm 90°$), and the *direction of the effective wind, $\varphi_{max}°$,* the wind direction angle producing the maximum drift response, is

$$\varphi_{max} = \frac{1}{2} \arctan \left[\frac{2(a_{11}a_{12} + a_{21}a_{22})}{(-a_{11}^2 + a_{12}^2 - a_{21}^2 + a_{22}^2)} \right] \qquad(5.321d)$$

where $\varphi_{max}°$ is measured clockwise from north in the Northern Hemisphere. The direction, φ_{min}, of the *noneffective* wind (the direction giving the minimum wind response) is also measured clockwise from north and is related to φ_{max} by $\varphi_{min} = \varphi_{max} \pm 90°$.

5.11.2.2 Eigenvectors of Matrix A

If $\mathbf{A}$ is a square matrix and $\mathbf{V}$ is a column vector such that

$$\mathbf{AV} - \lambda \mathbf{V} = (\mathbf{A} - \lambda \mathbf{I})\mathbf{V} = 0 \qquad(5.322)$$

where

$$\mathbf{I} = \begin{pmatrix} 1 & 0 \\ 0 & 1 \end{pmatrix}$$

is the identity matrix, then $\mathbf{V}$ is said to be an *eigenvector* (*latent vector*) of the matrix $\mathbf{A}$ with scalar eigenvalues, λ (Maxwell, 1977). Each eigenvector is associated with a corresponding eigenvalue. The eigenvectors of the matrix $\mathbf{A}$ determine the transformation from the vector $\mathbf{V}$ to the vector $\mathbf{u}$. Because the directions of the eigenvectors are unchanged by this transformation (i.e., $\varphi_\lambda = \phi_\lambda$), the eigenvectors give the direction along which the wind direction coincides with that of the wind-generated drift velocity. The eigenvalues, λ, are found from the characteristic equation (Maxwell, 1977)

$$(\mathbf{A} - \lambda\mathbf{I}) = \begin{pmatrix} (a_{11} - \lambda) & a_{12} \\ a_{21} & (a_{22} - \lambda) \end{pmatrix} = 0 \tag{5.323}$$

from which we obtain the quadratic equation

$$\lambda^2 - (a_{11} + a_{22})\lambda + (a_{11}a_{22} - a_{12}a_{21}) = 0. \tag{5.324}$$

Eqn (5.324) is used to define properties of the matrix $\mathbf{A}$.

There are three *invariants* of the matrix $\mathbf{A}$ determining the main properties of the transformation from wind to drift velocity (cf. Belyshev et al., 1983):

$$J_1 = a_{11} + a_{22}\,(\text{the trace}); \tag{5.325a}$$

$$J_2 = |a_{ij}| = a_{11}a_{22} - a_{12}a_{21}\,(\text{the determinant}); \tag{5.325b}$$

$$J_3 = a_{12} - a_{21} \ (\text{the index of asymmetry or the turn indicator}). \tag{5.325c}$$

The characteristic Eqn (5.324) has two roots (eigenvalues):

$$\lambda_{1,2} = \frac{1}{2}\left(J_1 \pm \sqrt{J_1^2 - 4J_2}\right) \tag{5.326a}$$

whereby

$$J_1 = \lambda_1 + \lambda_2; J_2 = \lambda_1\lambda_2. \tag{5.326b}$$

The directions of the eigenvectors are found from (5.320) from which,

$$\lambda_j \sin\varphi_j = a_{11}\sin\varphi_j + a_{12}\cos\varphi_j, \ j = 1, 2. \tag{5.327a}$$

whereby,

$$\tan\varphi_j = a_{12} / (\lambda_j - a_{11}), \ j = 1, 2. \tag{5.327b}$$

When $\mathbf{A}$ is a symmetric matrix, $J_3 = 0$, whereby, for given eigenvalues λ_1 and λ_2, the eigenvectors $\mathbf{V}_1$ and $\mathbf{V}_2$ are orthogonal and correspond to the principal ellipse axes associated with the maximum and minimum response (amplification) of the output series (drift velocity) relative to the input series (wind). However, if $J_3 \neq 0$ then the eigenvectors of the response ellipses are nonorthogonal and are rotated relative to the principal ellipse axes. Depending on the sign of J_3, the turning angles, θ, have mainly positive or negative values. The angle between two eigenvectors, which is equal to $90°$ if $J_3 = 0$, becomes smaller with increasing $|J_3|$. In the case $\lambda_1 = \lambda_2$, there is only one eigenvector, and the turning angle always has the same sign except for the two zero-value points coincident with the eigenvector. The one-eigenvector condition ($\lambda_1 = \lambda_2$) can be presented as

$$D = J_1^2 - 4J_2 = (a_{11} - a_{22})^2 + 4a_{12}a_{21} = 0, \tag{5.328}$$

where D is the discriminant of Eqn (5.310). Case (5.328) is possible only if a_{12} and a_{21} have opposite signs. Moreover, if $J_1^2 < 4J_2$, then

$$(a_{11} - a_{22})^2 + 4a_{12}a_{21} < 0 \tag{5.329}$$

so that Eqn (5.326a) does not have real roots and the turning angle always has the same sign.

5.11.2.3 Asymptotic Cases

There are two limiting cases for the ellipse structure: the flat ellipse (corresponding to rectilinear, one-dimensional motions) and the circle (corresponding to isotropic motions). The *flat ellipse* case occurs when the matrix **A** is *singular*, that is, when the determinant

$$J_2 = |a_{ij}| = a_{11}a_{22} - a_{12}a_{21} = 0. \tag{5.330}$$

For a two-dimensional matrix, the determinant is zero if one row (or column) is proportional to the other row (or column), such that

$$\mathbf{A} = \begin{pmatrix} a_{11} & a_{12} \\ ka_{11} & ka_{12} \end{pmatrix} \tag{5.331}$$

where k is a constant. This situation occurs when the wind-induced motions are rectilinear regardless of the wind direction, whereby the ice (or water) moves back and forth along one direction only. Motions of this type are possible in a narrow channel or near the coast, where the wind-generated drift is restricted by the presence of a boundary. The magnitude and sign of k determine the direction of these motions. For example, if $k = 0$ then the direction is west–east; if $k = 1$, then the direction is southwest–northeast (i.e., at 45° to the Cartesian coordinate system).

Using Eqns (5.321a) and (5.330), and several simple transformations, the directions of the most and least effective wind (those wind directions that produce the maximum and minimum drift response, respectively) are found from

$$(a_{11} \sin \varphi + a_{12} \cos \varphi)(a_{11} \cos \varphi - a_{12} \sin \varphi) = 0. \tag{5.332}$$

From Eqns (5.320a) and (5.330), it is clear that when the first term in the brackets of Eqn (5.332) is equal to zero, there is no oceanic response to the wind. In this case,

$$\varphi_{\min} = \varphi_0 = \arctan(-a_{12} / a_{11}), \tag{5.333}$$

so that winds in the direction, $\varphi = \varphi_0 \pm 180°$ produce no oceanic response; for any other direction, there is always a nonzero response. The direction of the maximum response may be found by setting the second term of Eqn (5.332) to zero, yielding

$$\varphi_{\max} = \arctan(a_{11} / a_{12}) = \varphi_0 \pm 90°. \tag{5.334}$$

from which we obtain,

$$\sin\varphi_{\max} = \frac{a_{11}}{\sqrt{a_{11}^2 + a_{12}^2}}; \cos\varphi_{\max} = \frac{a_{12}}{\sqrt{a_{11}^2 + a_{12}^2}} \tag{5.335}$$

From Eqns (5.321a) and (5.335) it follows that the magnitude of the maximum response has the simple form,

$$A_{\max} = \left(1 + k^2\right)^{1/2} \left(a_{11}^2 + a_{12}^2\right)^{1/2} \tag{5.336}$$

and the corresponding direction of the maximum drift velocity,

$$\phi_{\max} = \arctan\left(\frac{1}{k}\right) \equiv \tan^{-1}\left(\frac{1}{k}\right) \tag{5.337}$$

depends only on the coefficient k. Consequently, any drift motions, including the maximum response, are in the direction, $\phi = \pm\arctan(1/k)$. According to Eqns (5.327a) and (5.327b), the matrix Eqns (5.317a) and (5.317b) for this case has two eigenvectors:

$$\lambda_1 = J_1; \varphi_1 = \arctan(1 / k); \tag{5.338a}$$

$$\lambda_2 = 0; \varphi_2 = \arctan(-a_{12} / a_{11}) = \varphi_0. \tag{5.338b}$$

Note that, contrary to the more general case, the case of rectilinear motion is described by three, not four, independent parameters: a_{11}, a_{12}, and k (or $\varphi_{\max}$, $A_{\max}$, and $\phi_{\max}$).

The second limiting case, that of a *circular ellipse*, corresponds to an isotropic response, for which the drift response ellipse is a circle. This occurs when the matrix **A** is antisymmetric ($a_{21} = -a_{12}$) and the main diagonal coefficients are equal ($a_{22} = a_{11}$) so that,

$$\mathbf{A} = \begin{pmatrix} a_{11} & a_{12} \\ -a_{12} & a_{11} \end{pmatrix}. \tag{5.339}$$

This case is equivalent to case Eqn (5.314) with $\mathbf{A} = \boldsymbol{\alpha}$, whereby matrix Eqn (5.339) takes the form

$$\mathbf{A} = \alpha_0 \begin{pmatrix} \cos\theta & \sin\theta \\ -\sin\theta & \cos\theta \end{pmatrix}, \tag{5.340}$$

where $\alpha_0 = \sqrt{a_{11}^2 + a_{12}^2}$ and $\theta = \tan^{-1}(a_{12}/a_{11})$. Matrix Eqn (5.340) describes a combination of stretching and rotation, so that the isotropic response is described by only two independent parameters, a_{11} and a_{12} (or α_0 and θ as in the 2-parameter outlined in Section 5.11.1). In general, matrix Eqn (5.339) has no eigenvectors. The one exception is the case of an isotropic response without any turning ($\theta = 0$), which occurs when $a_{12} = 0$. In this case, wind vectors and response vectors always have the same directions and all vectors are the eigenvectors. The maximum turning angle, $\theta = 90°$ (which is the same for all wind directions) corresponds to the case when the main diagonal elements of the matrix Eqn (5.339) are equal to zero ($a_{11} = 0$).

5.11.3 Wind Versus Surface Drift: The Six Characteristic Cases

In this section, we consider matrices $\mathbf{A}$, whose parameters (the regression coefficients, a_{ij}) are representative of the six possible cases of ice or current response to wind forcing discussed in Section 5.11.2. These six test cases are:

$$(1): \mathbf{A} = \begin{pmatrix} 1.0 & 1.25 \\ 1.6 & 2.0 \end{pmatrix}; \quad (2): \mathbf{A} = \begin{pmatrix} 1.0 & 0.5 \\ 0.5 & 2.0 \end{pmatrix}; \quad (3): \mathbf{A} = \begin{pmatrix} 1.0 & 1.0 \\ 0.5 & 2.0 \end{pmatrix};$$

$$(4): \mathbf{A} = \begin{pmatrix} 1.0 & 0.5 \\ -0.5 & 2.0 \end{pmatrix}; \quad (5): \mathbf{A} = \begin{pmatrix} 1.0 & 1.0 \\ -0.5 & 2.0 \end{pmatrix}; \quad (6): \mathbf{A} = \begin{pmatrix} 2.0 & 1.0 \\ -1.0 & 2.0 \end{pmatrix}.$$

The coefficients, a_{ij} in the above matrices are expressed as the ratio of drift speed in CGS units to the wind speed in MKS units (i.e., cm/s and m/s, respectively), which also corresponds to a percentage. Note that the matrix elements in pairs (2) and (4) and (3) and (5) are identical except for the change in sign of a_{12}. Table 5.20 presents the derived matrix invariants, ellipse parameters, and eigenvector parameters for the six cases.

Case 1: The determinant of matrix $\mathbf{A}$ is equal to zero ($J_2 = 0$), corresponding to a rectilinear (one-dimensional) response of the form Eqn (5.331) with $k = 1.6$. In near-shore regions, the orientation for this flat ellipse response ($\phi_{max} = 32°$ for the present case; Figure 5.71) would typically coincide with the orientation of the coastline, indicating that,

TABLE 5.20 Response ellipse parameters and eigen vectors for the six test examples.

	Matrix invariants				Ellipse parameters				Eigenvectors			
Case	J_1 (%)	J_2 (%)^2	J_3 (%)	D (%)^2	A_{max} (%)	A_{min} (%)	φ_{max} (°)	ϕ_{max} (°)	λ_1 (%)	φ_1 (°)	λ_2 (%)	φ_2 (°)
C1	3.00	0.00	−0.35	9.00	3.020	0.000	38.66	32.00	3.000	32.00	0.00	128.7
C2	3.00	1.75	0.00	2.00	2.207	0.793	22.50	22.50	2.207	22.50	0.793	112.5
C3	3.00	1.50	0.50	3.00	2.422	0.619	23.42	32.89	2.366	36.21	0.634	110.1
C4	3.00	2.25	1.00	0.00	2.081	1.081	−9.22	9.22	1.500	45.00	−	−
C5	3.00	2.50	1.50	−1.00	2.236	1.118	0.00	26.57	−	−	−	−
C6	4.00	5.00	2.00	−4.00	2.236	2.236	−	−	−	−	−	−

From Rabinovich et al. (2007).

(a)

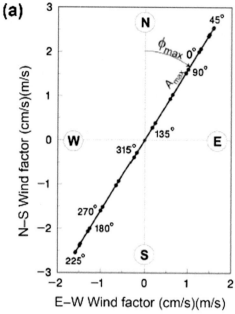

(b)

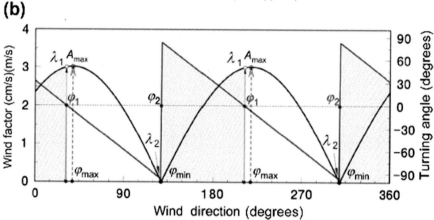

FIGURE 5.71 Case 1. (a) Response ellipse describing rectilinear (one-dimensional) response of drift velocity (ice-drift or ocean current "drift") to the wind. Letters "W," "N," "E," and "S" give the direction of the drift velocity toward the west, north, east, and south, respectively. Numbers 0°, 45°, ..., 315° indicate the direction of the wind; A_{max} and ϕ_{max} denote the magnitude of the maximum response and its direction, respectively. The terms (cm/s)(m/s) in the figure axes denote the ratio of the current response in cm/s to the wind speed forcing in m/s. The axes values can also be considered as the current speed (in m/s) as a percentage of the wind speed (in m/s); (b) variations of the drift velocity response (wind factor) and turning angle as functions of the wind direction. λ_1 and λ_2 denote the eigenvalues; φ_1 and φ_2 are the corresponding wind directions. Shaded areas denote zones of positive turning angles. *From Rabinovich et al. (2007).*

regardless of which way the wind is blowing, only motions in the alongshore direction are possible. This means that the ocean response does not generally have a symmetric directional response to the wind. Moreover, because of Earth's rotation, the turning angle in the Northern (Southern) Hemisphere is expected to be mainly directed clockwise (counterclockwise) relative to the wind vector.

Case 2: The matrix **A** is symmetric ($J_3 = 0$), so that the eigenvectors, $\mathbf{V}_1$ and $\mathbf{V}_2$ for given eigenvalues, λ_1 and λ_2 are orthogonal and correspond to the principal ellipse axes, given here as $\lambda_1 = A_{max} = 2.21$, $\varphi_1 = \phi_{max} = 22.5°$; and $\lambda_2 = A_{min} = 0.79$, $\varphi_2 = \phi_{min} = 112.5°$ (Figure 5.72). The angle between the two eigenvectors is 90°. Because of Earth's rotation, this symmetric response case is possible only near the equator, where the Coriolis parameter, $f \approx 0$. Away from the equator, a solution requires that the response angle switch signs.

Case 3: Here, $J_3 \neq 0$ and the matrix **A** is asymmetric. This means that the eigenvectors are nonorthogonal and do not correspond to the principal ellipse axes (Figure 5.73). Because J_3 is positive in our example ($J_3 = 0.5$), the turning angles, θ, are mainly positive. The angle between the two eigenvectors is equal to 73.9°. For real oceanic conditions, the turning angle is expected to become increasingly smaller with increasing offshore distance.

Case 4: In this case, the discriminant Eqn (5.328) is equal to zero ($D = 0$), so there is only one eigenvector ($\lambda = \lambda_1 = \lambda_2 = 1.50$). The turning angle, θ, is always positive except for two zero-value points corresponding to the eigenvector (Figure 5.74). This case is observed in confined regions near a coastline.

FIGURE 5.72 Case 2. As in Figure 5.71, but describing the case of a symmetric response of the drift velocity to the wind. Eigenvectors, $\mathbf{V}_1$ and $\mathbf{V}_2$ are orthogonal with eigenvalues, λ_1 and λ_2. A_{max} and A_{min} denote the magnitude of the maximum and minimum responses; φ_{max} and φ_{min} indicate the corresponding directions of the wind. Shaded areas denote zones of positive turning angles. The terms (cm/s)(m/s) in the figure axes denote the ratio of the current response in cm/s to the wind speed forcing in m/s. The axes values can also be considered as the current speed (in m/s) as a percentage of the wind speed (in m/s). *From Rabinovich et al. (2007).*

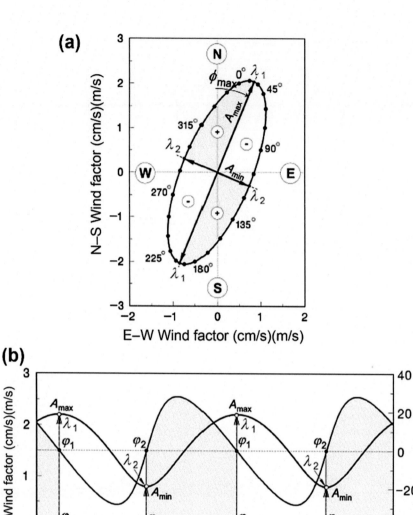

Case 5: For this case, the discriminant is negative ($D = -1.0$), so that Eqns (5.326a) and (5.326b) does not have real roots and the turning angle is always positive (Figure 5.75). This is representative of wind-driven drift motions for offshore regions in the Northern Hemisphere; for the Southern Hemisphere, the turning angle will be negative.

Case 6: In this case, the matrix $\mathbf{A}$ is antisymmetric ($a_{21} = -a_{12} = 1.0$) and the diagonal coefficients are equal ($a_{22} = a_{11} = 2.0$). In this isotropic response example, both the wind factor, $\alpha_0 = 2.24$, and turning angle, $\theta = 26.6°$, are uniform (Figure 5.76). This case corresponds to open-ocean regions, where coastal influence is negligible and the current-ice-drift response to the wind has the same turning angle and response magnitude regardless of the wind direction.

The six cases characterize the way in which the surface drift currents respond to changes in the wind as a function of increasing distance from the shore, ranging from purely rectilinear (alongshore) wind-induced motions near the coast (case 1) to almost circular responses in the open ocean (case 6). The one remaining case (case 2) is only applicable to equatorial regions. In general, ice-drift response changes in a similar way to the ocean currents. However, ice-drift response is also dependent on ice concentration (Shevchenko et al., 2004). Higher ice concentration strengthens the internal ice stress, leading to marked attenuation in ice-motions, especially in the cross-shore direction. In contrast, reduced ice concentration leads to intensification of cross-shore motions, analogous to the effect of increased offshore distance. The effect of coastlines on the directionality of wind-generated currents is clearly demonstrated in Cummins et al. (2022), who used the vector regression model to examine radar-measured surface currents in Queen Charlotte Sound on the west coast of Canada.

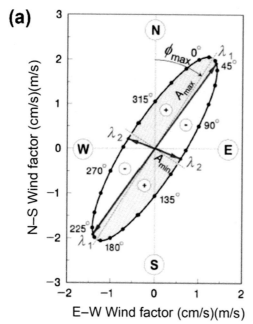

(a)

FIGURE 5.73 Case 3. As in Figure 5.70, but describing the case of a nonsymmetric response of the drift velocity to the wind. The terms (cm/s)(m/s) in the figure axes denote the ratio of the current response in cm/s to the wind speed forcing in m/s. The axes values can also be considered as the current speed (in m/s) as a percentage of the wind speed (in m/s). *From Rabinovich et al. (2007).*

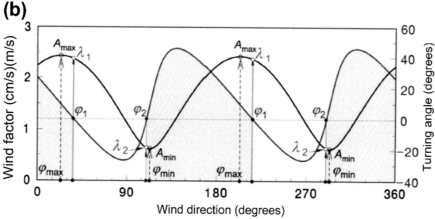

(b)

5.11.3.1 Ice-Drift on the Sakhalin Island Shelf

A study of ice-drift and its vector regression relationship to the wind as a function of offshore distance and ice concentration on the northeast coast of Sakhalin Island, Sea of Okhotsk, is presented in Rabinovich et al. (2007). To quantify the influence of ice concentration on the ice-drift response to wind, the ice-drift data for 1993 were divided into four sequential 18-day segments characterized by distinctly different ice types and concentrations: (1) the period March 12−30 consisted of ice concentrations of approximately 80−90%, with the ice field made up of large and small broken floes; (2) March 31−April 17 had the highest ice concentration (95−100%), consisting of large ice fields; (3) April 18−May 6 had reduced ice concentrations (60−80%) and diminished floe sizes; and (4) May 7−25 consisted of intensive melting with ice concentrations reduced by ∼40−50%. For each of the four segments of time, two observational sites (S1 and S4) were examined, and the matrix **A** computed for eight different cases. Location S1 (located 4 km from shore) yielded the following matrices for the four time segments:

$$(1)\ \mathbf{A} = \begin{pmatrix} 0.90 & -0.33 \\ -2.33 & 2.34 \end{pmatrix};\ (2)\ \mathbf{A} = \begin{pmatrix} 1.08 & -0.52 \\ -2.17 & 1.59 \end{pmatrix};\ (3)\ \mathbf{A} = \begin{pmatrix} 1.20 & -0.38 \\ -2.24 & 2.18 \end{pmatrix};$$

$$(4)\ \mathbf{A} = \begin{pmatrix} 2.10 & 0.31 \\ -1.13 & 2.48 \end{pmatrix};$$

FIGURE 5.74 Case 4. As in Figure 5.71, but for the case for which there is only one eigenvector, $\lambda_1 = \lambda_2 = 1.50$. In this case, the turning angle is always positive, except for two zero-value points of the eigenvector. The terms (cm/s)(m/s) in the figure axes denote the ratio of the current response in cm/s to the wind speed forcing in m/s. The axes values can also be considered as the current speed (in m/s) as a percentage of the wind speed (in m/s). *From Rabinovich et al. (2007).*

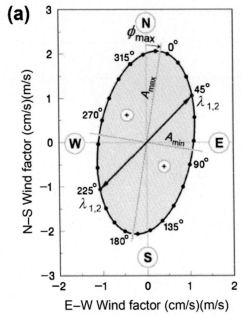

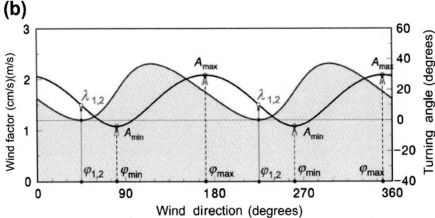

and location S4 (16 km for shore) yielded

$$(1)\ \mathbf{A} = \begin{pmatrix} 0.69 & -0.21 \\ -3.01 & 2.57 \end{pmatrix}; (2)\ \mathbf{A} = \begin{pmatrix} 1.06 & -0.54 \\ -2.30 & 1.49 \end{pmatrix}; (3)\ \mathbf{A} = \begin{pmatrix} 1.18 & -0.33 \\ -2.19 & 2.07 \end{pmatrix};$$

$$(4)\ \mathbf{A} = \begin{pmatrix} 2.04 & 0.25 \\ -1.09 & 2.54 \end{pmatrix}$$

where the matrix coefficients, a_{ij}, give the ratios of the current response speed (in cm/s) to the wind forcing speed (in m/s), or the current speed (in m/s) as a percentage of the wind speed (in m/s). The pronounced flatness of the response ellipses for cases S4-2, S4-3, and S4-4 indicated that the wind-induced ice motions were strongly anisotropic, with the ice response in the alongshore direction much more pronounced than in the cross-shore direction (2.9–6.1% versus 0.2–1.9%, respectively). The alongshore values (2.6–5.4%) of the response coefficients (the wind factor) were similar to those obtained by Fissel and Tang (1991) for the Newfoundland shelf. Response coefficients for the more remote offshore observational area (S4) were greater than for the areas closest to *shore* (S1) by about 15–20%.

The results reveal marked temporal changes in the ice-drift response to the wind, apparently due to changes in ice properties. During the period of the highest ice concentration (period 2), the response ellipses are almost flat, indicating that the ice-drift response was rectilinear (alongshore). There are two eigenvalues, but the turning angles are mainly positive. In general, the ice-response ellipses resemble those for case 1 in Section 5.11.3. For the early spring (period 1), and especially

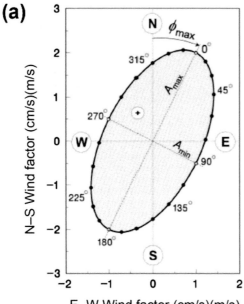

(a)

FIGURE 5.75 Case 5. As in Figure 5.70, but for the case for which there are no eigenvalues and the turning angle is always positive. The terms (cm/s)(m/s) in the figure axes denote the ratio of the current response in cm/s to the wind speed forcing in m/s. The axes values can also be considered as the current speed (in m/s) as a percentage of the wind speed (in m/s). *From Rabinovich et al. (2007).*

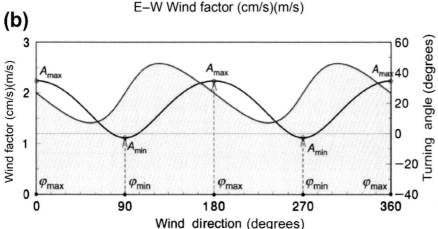

(b)

during the late spring (period 3), the response ellipses have larger magnitudes and are more circular, indicating more intense cross-shore ice motions. Similarly, for the second period, the S1 and S4 matrixes have two eigenvalues and a prevalence of positive turning angles. The response ellipses were of type 3. Finally, during the late spring (period 4), the response ellipses changed from flat to oval, similar to case 5. For period 4, the matrixes for both S1-4 and S4-4 had no eigenvalues and all turning angles were positive. According to this analysis, the last period was a time of free ice-drift, while the three other periods were times of high internal ice stress and influence of the coast.

As illustrated by the above study, the traditional 2-parameter approach for relating drift velocity to the wind is isotropic and, therefore, unrealistic for coastal regions where factors such as the orientation of the coastline and the regional bottom topography are important. The assumption of an isotropic response is likely invalid near the coast. When searching for dynamical relationships and associated surrogate variables, it is best to apply a two-dimensional (vector) regression model. In this model, the relationship between the wind and drift velocity (ice-drift or current velocity) is described by four independent regression (response) coefficients, a_{ij}, linking the cross-shore (u) and alongshore (v) components of the drift to the corresponding components (U, V) of the wind velocity. For each direction of the wind vector, φ, the method prescribes a "wind factor," $\alpha(\varphi)$, (relative drift speed) and "turning angle," $\theta(\varphi)$, (the angle between the drift velocity and wind vector).

Because of its greater number of free coefficients, the 4-parameter vector model should yield a smaller residual variance than the traditional 2-parameter model. However, the number of DoF in the data set being analyzed decreases with an increase in the number of coefficients. As a consequence, calculation of the vector-regression coefficients to same level of

FIGURE 5.76   Case 6. As in Figure 5.71, but for the case describing an isotropic response of the drift velocity to the wind. Both the wind factor, $\alpha_0 = 2.24$ and the turning angle, $\theta = 26.6°$ are spatially uniform. The terms (cm/s)(m/s) in the figure axes denote the ratio of the current response in cm/s to the wind speed forcing in m/s. The axes values can also be considered as the current speed (in m/s) as a percentage of the wind speed (in m/s). *From Rabinovich et al. (2007).*

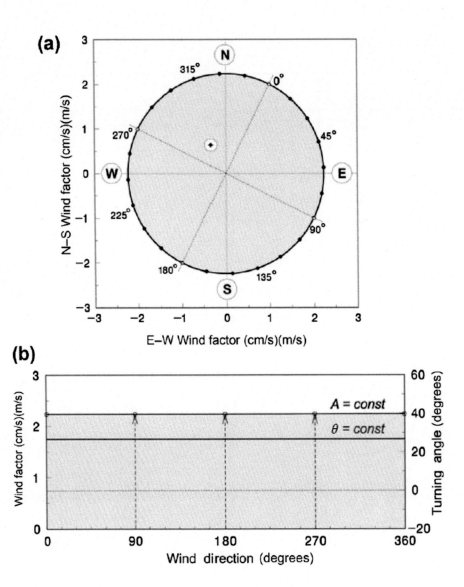

confidence as the traditional model coefficients requires a longer time series. The stability of the response ellipse parameters (relative to small changes in the parameters of the input functions) is the main criterion for determining the reliability of the results. It is also important to note that the structure of these ellipses has physical meaning in the sense that it accounts for the significant difference in ice-drift response to alongshore and cross-shore winds. The results reveal that an anisotropic, vector-regression model is superior to an isotropic, 2-parameter model for examining wind-ice and wind-current processes in coastal zone regions. Moreover, the vector regression model is more likely to capture surface dynamical features of the wind response than the traditional model, and therefore is more reliable in the application of the wind velocity as a surrogate for surface drift velocity.

5.12 FRACTALS

The term "fractal" was coined by Mandelbrot (1967) to describe the bumpiness of geometrical curves and surfaces. Regardless of how closely we examine a fractal object, it fails to become smooth and its degree of jaggedness remains unchanged. Fractal objects are uneven at all scales and possess no characteristic length scales. Fractals are ubiquitous features whose presence has been reported in a wide variety of fluid dynamical settings including the mixing of turbulent flows (Sreenivasan et al., 1989), the trajectories of oceanic drifters (Osborne et al., 1989; Sanderson et al., 1990), and the paths of atmospheric cyclones (Fraedrich et al., 1990). More everyday examples involve the fractal dimensionality of coastlines, the shapes of clouds, and the forms of lightning strikes. The fractal curve in Figure 5.77a, called a *Koch curve*, resembles a coastline or the outline of a snowflake that would be mapped at ever-increasing spatial resolution. In this case, one begins with an equilateral triangle of side-length, L, and then successively attaches smaller and smaller equilateral triangles of size $L/3$, $L/3^2$, and so on to the middle of every straight-line segment. After N iterations, the perimeter consists of N segments of length, r, where $r = L/3^N$ and

$$N = \alpha(L/r)^D \tag{5.341}$$

where $\alpha = 3$ and $D = \log 4/\log 3 \approx 1.262$ is called the fractal dimension. This dimension lies between $D = 1$ for a true one-dimensional curve and $D = 2$ for a true surface area. Figure 5.77b is an example of an area fractal called the *Sierpinski gasket*, which finds use in studies of sediment porosity. Again, one begins with a triangle of side, L, but then cuts out successively smaller triangles of lengths $L/2$, $L/2^2$, and so on. After N iterations, the "pore" space between the sides of the triangles consists only of triangles of size, $r = L/2^N$. The number of such triangles is given by Eqn (5.341), but with $\alpha = 1$ and $D = \log 3/\log 2 \approx 1.585$.

The study of fractal geometry is related to the problem of predictability and propagation of order in nonequilibrium, frictionally dependent dynamical systems, such as turbulent flow in real fluids. In fluid systems, predictability is related to the rate at which initially close fluid particles diverge and the sensitivity of this divergence to initial conditions. Since low

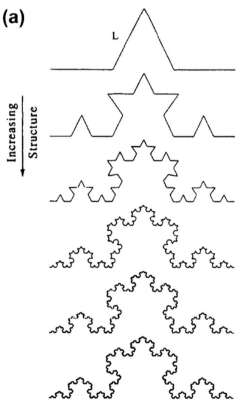

(a)

Increasing Structure

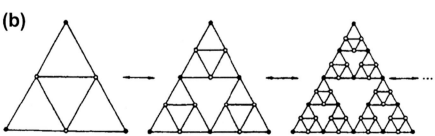

(b)

FIGURE 5.77 Examples of common fractals. (a) Generation of the Koch curve fractal by successive attachment of equilateral triangles; $D = 1.262$; (b) generation of the Sierpinski gasket fractal by successive removal of smaller triangles; $D = 1.585$.

predictability implies a highly irregular dynamical system with sensitive dependence on initial conditions, the dispersion of tagged fluid parcels is related to the ultimate skill that can be achieved by deterministic numerical prediction models.

The fractal (or Hausdorff) dimension, D, provides a measure of the roughness of a geometrical object. For example, drifter trajectories confined to a horizontal plane can have a fractal dimension somewhere between that of a topological curve ($D = 1$) and that of random Brownian motion ($D = 2$). The case $D = 1$ is for a smooth differentiable curve whose length remains constant regardless of how the measurements are made. For fractal curves ($D > 1$), the length of the curve increases without bound for decreasing segment length. In the absence of a stationary mean flow, the track of a fluid parcel undergoing Brownian (random walk) motion will eventually occupy the entire horizontal plane available to it, whereas a parcel displaying fractal Brownian motion will not. The case $D < 2$ implies that the motion has inherent "memory" in the sense that a given incremental displacement in the fluid path is not independent of previous displacements. In terms of dynamical systems, this means that there are a finite number of variables required to explain the dynamics of the fluid motions.

Osborne et al. (1989) examined the scaling properties of drifter trajectories for the upper ocean using yearlong tracks of three satellite-tracked drifters deployed within the Kuroshio Extension region in 1977. Based on results from four fundamentally different fractal analysis methods, the Lagrangian trajectories were found to exhibit fractal behavior with dimension, $D = 1.27 \pm 0.11$, over spatial scales of 20–150 km and temporal scales of 1.5 days to 1 week. These scales are thought to be representative of two-dimensional geophysical fluid dynamical turbulence within the inertial subrange—the eddy cascade region of self-similar turbulence, which separates short-period current motions (daily tidal oscillations and inertial currents) from long-period oscillations such as Rossby waves and mean flows. Sanderson et al. (1990) have reported fractal dimensions at spatial scales of 0.1–4 km for clusters of drifters deployed in Lake Erie, the Atlantic Equatorial Undercurrent, and in coastal waters off the south shore of Long Island. In a related study, the degree of chaotic behavior and predictability of the atmosphere has been studied using tropical and midlatitude maritime cyclone tracks (Fraedrich and Leslie, 1989; Fraedrich et al., 1990). Results suggest that the atmosphere has an e-folding error growth rate of about 24 h and an ultimate predictability of 8–14 days.

In this section, we provide several methods for determining the fractal characteristics of oceanic variability using particle track motions.

5.12.1 The Scaling Exponent Method

Consider a particle track sampled at times (t) along the path, $\mathbf{x}(t) = (x(t), y(t))$ in longitude–latitude (x–y) coordinates. Displacements along each of the two orthogonal horizontal axes are assumed to be independent self-affine (self-scaling) scalar functions. The scaling exponent, H (which may be different for the two axes) is positive, less than or equal to unity, and related to the fractal dimension of the function by $D = \min[1/H, 2]$. Brownian motions have scaling exponent, $H = 1/2$ ($D = 2$) while monofractal scalar displacements exhibit fractional Brownian motions with $H > 1/2$ ($D < 2$). If the scalar series are sampled at equal time intervals, the exponents, H_x, H_y, are given by the *structure functions*

$$\overline{[x(t + \alpha\Delta t) - x(t)]^2} = \overline{[\Delta x(\alpha\Delta t)]^2}$$

$$= \alpha^{2H_x} \overline{[\Delta x(\Delta t)]^2}$$

$$= \alpha^{2H_x} \overline{[x(t + \Delta t) - x(t)]^2} \quad (5.342a)$$

$$\overline{[y(t + \alpha\Delta t) - y(t)]^2} = \overline{[\Delta y(\alpha\Delta t)]^2}$$

$$= \alpha^{2H_y} \overline{[\Delta y(\Delta t)]^2} \quad (5.342b)$$

$$= \alpha^{2H_y} \overline{[y(t + \Delta t) - y(t)]^2}$$

where overbars denote averages over time and the α are assigned integer values. The scaling exponents also can be found using the absolute value of the above functions (Osborne et al., 1989)

$$\overline{|x(t + \alpha\Delta t) - x(t)|} = \alpha^{2H_x} \overline{|x(t + \Delta t) - x(t)|} \quad (5.342c)$$

$$\overline{|y(t + \alpha\Delta t) - y(t)|} = \alpha^{2H_y}\overline{|y(t + \Delta t) - y(t)|} \tag{5.342d}$$

The curves in Figure 5.78 provide examples of the scaling exponents, H_y, derived from Eqns (5.342b) and (5.342d) using 1-year time series of 6-hourly meridional displacements of 120-m-drogued satellite-tracked drifters launched in the northeast Pacific in 1987. Part (a) of the figure is the log of the structure function

$$\overline{\{[y(t + \alpha\Delta t) - y(t)]^2\}}^{1/2}$$

versus $\log(\alpha)$. The slopes of these curves, H_y, are presented in part (b). Figure 5.79 is the same as Figure 5.78, except that it uses artificial drifter tracks generated from a Brownian motion (random-walk) algorithm. For the real drifter data, all four tracks had a constant fractal dimension, $D_y = 1/H_y \approx 1.18 \pm 0.07$, over timescales of about 0.5–10 days. At longer timescales, motions were strongly affected by mesoscale eddies (cf. Thomson et al., 1990) and fractal analysis is no longer valid. For the pseudo-drifters, $D_y \approx 2$, which is what we would expect for a random-walk regime in which the drifters can occupy the entire two-dimensional space available to them.

Although confined to monofractal functions, the scaling dimension approach is attractive because it is computationally fast and defined in terms of simple scaling properties. The principal drawback is that irregularly sampled particle trajectories, such as those of satellite-tracked drifters, must be converted to equally spaced data using a spline or other interpolation scheme. For isotropic monofractal trajectories, a single fractal dimension is sufficient to define the overall scaling properties of the motions including scaling properties of the mean, variance, and higher moments. Anisotropy in the drifter motions may lead to significantly different values for the scaling exponents, H_x, H_y, and associated fractal dimensions. Where these differences are small, fractal dimensions can be expressed through a mean scaling exponent, $\overline{H} = \frac{1}{2}(H_x + H_y)$.

5.12.2 The Yardstick Method

The fractal dimension of a drifter trajectory of length, $L(\Delta)$, can be measured in the usual sense using a ruler (or *yardstick*) with variable length, Δ. As the length of the ruler is decreased and the yardstick estimation of the total length becomes more precise, the length of the trajectory will follow a power-law dependence

$$L(\Delta) \approx \Delta^{1 - D_L}; \lim \Delta \to 0 \tag{5.343}$$

The divider dimension, D_L, which closely approximates the fractal dimension, D, is found from the slope of log-transformed $L(\Delta)$ for small length scales, Δ (Figure 5.79). The case, $D_L = 1$ is the topological dimension for a smooth differential curve. For fractal dimensions, $D > 1$ and the length of the curve increases without bound for decreasing segment length.

A problem with applying Eqn (5.329) to irregularly sampled drifter records is that the data are unequally sampled both in time and space. Although it makes sense to use a spline-interpolation scheme to generate scalar coordinate data with equally spaced time increments, it is less meaningful to generate coordinate series with equally sampled positional increments. The reason is simple enough: Time is single-valued whereas location is not. Drifters often loop back on themselves. If the data are not equally spaced, we cannot define a sequence of fixed-length yardsticks but must measure the curve, $L(\Delta)$, as a function of the average yardstick length, Δ_{av}. This averaging is valid provided the errors introduced by the averaging process are no worse than those arising from other sources (cf. Osborne et al., 1989). Another problem with the yardstick method is that it is based on the slope of Eqn (5.343) for small spatial scales. The measurement of these scales is often difficult in practice due to limitations in the response and/or positioning of the drifters, cyclone, or other Lagrangian particle.

5.12.3 Box-Counting Method

In this method, one counts the numbers, $N_m(L)$, of boxes of length, L in m-dimensional space that are needed to cover a "cloud" or set of points in the space. The Hausdorff–Besicovitch dimension, D, of this set can be estimated by determining the number of cubes needed to cover the set in the limit as $L \to 0$. For a fractal curve, the number of boxes increases without bound as $L \to 0$. That is

$$N_m(L) \to L^{-D}, L \to 0 \tag{5.344}$$

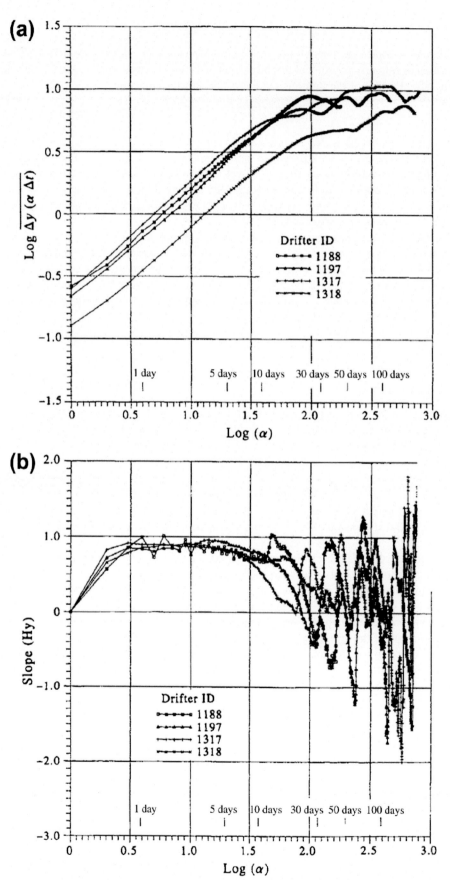

FIGURE 5.78 Structure functions and scaling exponents for trajectories of four 6-hourly sampled, 120-m-drogued satellite-tracked drifters launched in the northeast Pacific in 1987. (a) Absolute values of the structure functions versus the scaling factor, α, plotted on a log–log scale. (b) Slopes, H_y, of the curves in (a) versus scaling factor. Slopes were roughly equal and constant over timescales of 1–10 days.

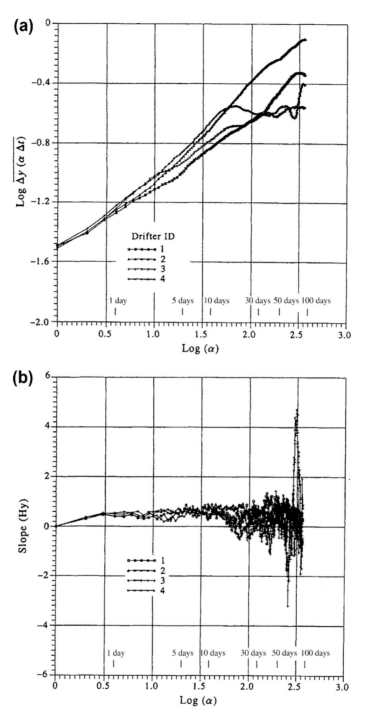

FIGURE 5.79   As in Figure 5.78 except for pseudo-drifter tracks generated using a random number generator. In this case, $H_y \approx 0.5$ and drifters perform a nonfractal random walk with dimension, $D \approx 2$.

If the original series is random, then $D = n$ for any dimension, n (a random process embedded in an n-dimensional space always fills that space). If, however, the value of D becomes independent of n (i.e., reaches a saturation value, D_0, say), it means that the system represented by the time series has some structure and should possess an attractor whose Hausdorff–Besicovitch dimension is equal to D_0. Once saturation is reached, extra dimensions are not needed to explain the dynamics of the system.

As an example, if we were to measure the area of surfaces embedded in three-dimensional space, we would count the number, $N_3(L)$, of cubic boxes of size, L required to cover the surface. The area, S, is then of order

$$S \approx N_3(L)L^2 \tag{5.345}$$

For a nonfractal surface, the area asymptotes to a constant value independent of L, which is the true area of the surface. In general,

$$N_3(L) \approx L^{-D}, S \approx L^{2-D} \tag{5.346}$$

5.12.4 Correlation Dimension

An important method for determining the self-similarity of monofractal curves has been proposed by Grassberger and Procaccia (1983). The technique also has found widespread use in studies of chaos and the dimensionality of strange attractors. Specifically, one determines the number of times that the computed distances, d_{ij}, between points in a time series, $x(t_i)$, (or pair of time series, $x_i(t)$ and $x_j(t)$) are less than a prescribed length scale, ε. That is, one finds what fraction of the total number of possible estimates of the distance, $d_{ij} = |x(t_i) - x(t_j)|$ that are less than ε. For a single discrete vector time series, the Grassberger–Procaccia correlation function is defined as

$$C(\varepsilon) = \frac{1}{M(M-1)} \sum_{ij}^{M} H\big[\varepsilon - |x(t_i) - x(t_j)|\big], M \to \infty \tag{5.347}$$

where $H(\varepsilon, r_{ij})$ is the Heavyside step function (= 0 for $\varepsilon < r$; = 1 for $\varepsilon > r$) and M is the number of points in the time series. In (5.347), the vertical bars denote the norm of the vector, $d_{ij} = [(x(t_i) - x_j)^2 + (y(t_i) - y_j)^2]^{1/2}$. The fractal dimension for a self-affine curve is then obtained as the correlation dimension defined by

$$C(\varepsilon) \approx \varepsilon^v, \varepsilon \to 0 \tag{5.348}$$

The fractal dimension, v, is obtained from the log-transformed version of this equation (plots of C versus length scale are presented in Figure 5.80). According to Osborne et al. (1989), the correlation method gives the least uncertainty in the estimate of the fractal dimension, whereas largest errors are associated with the exponent scaling method.

5.12.5 Dimensions of Multifractal Functions

The various techniques discussed above will (within statistical error) give the same fractal dimension provided that the series being investigated exhibits self-similar monofractal behavior. However, because the techniques rely on different assumptions and measure different scaling properties of the series, the calculated dimensions will be different if the series has a multifractal structure. Multifractal properties are related to multiplicative random processes and are associated with different scaling properties at different scales.

A form of box counting can be used to study the multifractal properties of ocean drifters (Osborne et al., 1989). Given a fractal curve on a plane, the plane is covered with adjacent square boxes of size, Δ and the probability, $p_i(\Delta)$, is computed that the ith box contains a piece of the fractal curve

$$p_i(\Delta) = \frac{n_i(\Delta)}{N} \tag{5.349}$$

where n_i is the number of data points falling in the ith box and N is the total number of points in the time series. For fractal curves for small Δ

$$\sum_i [p_i(\Delta)]^q \approx \Delta^{(q-1)D} \tag{5.350}$$

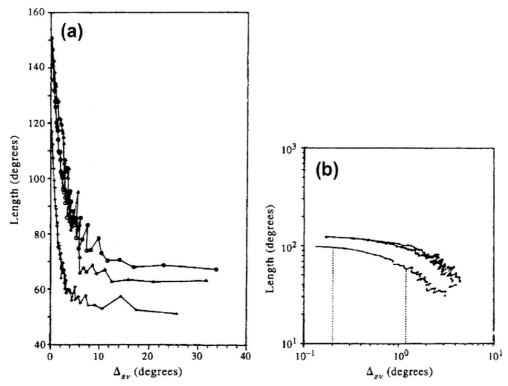

FIGURE 5.80 Yardstick length, $L(\Delta)$, measured using a ruler with variable average yardstick length, Δ_{av} (in degrees of latitude), for three drifters launched in the Kuroshio Extension in 1977. (a) Linear coordinates; and (b) log–log coordinates. Note the divergence of the lengths for small Δ. *From Osborne et al. (1989).*

where the sum is extended over all nonempty boxes. The quantities, $D = D_q$, are the generalized fractal dimensions. A fundamental difference between monofractals and multifractals is that for monofractals, D_q is the same for all q while for multifractals the different generalized dimensions are not equal. In general, $D_q < D_q'$ for $q > q'$.

5.12.6 Predictability

A box-counting method can be used to investigate the degree of chaotic behavior associated with the Lagrangian motions such as those of drifters and tropical cyclones. In this method, one counts the number, $N_n(\Delta)$, of boxes of dimension, Δ in n-dimensional space needed to cover a "cloud" or set of points in the space in the limit $\Delta \rightarrow 0$. In practice, the box-counting method is difficult to apply. Estimates of the predictability of drifter trajectories are more readily obtained using the correlation integral technique of Grassberger and Procaccia (1983). In this case, the degree of predictability is found from the dimension of the attractor derived from an embedded phase space created from all possible pairs of "drifters." The phase space serves, in turn, as a substitute for the state space needed to study the dynamics of a system (Tsonis and Elsner, 1990).

The analysis takes the following steps: (1) We first consider a pair of independent tracks of length, $m\Delta t$, where m is the embedding dimension and Δt is the sampling increment. Specifically, consider the cyclone tracks for Australia for July 1982 and 1983 (Figure 5.81a) examined by Fraedrich et al. (1990). For convenience, the start times and positions of the tracks are reinitialized so that they begin at the same time and location. Fraedrich and Leslie (1989) found that the errors introduced by reinitializing are less than those from other sources; (2) We next examine the divergence of the paths by calculating the multiple track correlation function (or correlation integral), $C_m(\varepsilon)$, for the particular embedding dimension, m and path separation scale, ε. To this end, we count the number of tracks, $N_m(\varepsilon)$ of length, $m\Delta t$ for which the track length remains less than the great circle distance, ε, for all the segments in the track. For $m = 1$, each individual data point forms a unit-length segment of the drifter track. One then counts the number of times, $N_1(\varepsilon)$, that the distance between the drifter positions is less than ε for the $N = m$ possible drifter tracks. The distance between each drifter pair is considered; hence, for 10 drifters or cyclone tracks there would be $10 \times 10 = 100$ pairs. This process is repeated for all values of m to create a

FIGURE 5.81 Correlation functions, $C(\varepsilon)$, for three drifters launched in the Kuroshio Extension in 1977. The slope of the function in log–log coordinates is a measure of the correlation dimension of the signal. The two vertical lines indicate the approximate limits of the scaling range. *From Osborne et al. (1989).*

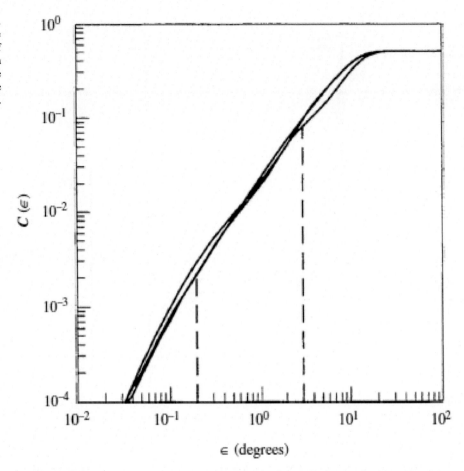

cloud of points in m-dimensional space, which then approximates the dynamics of the system from which the observations, $x(t)$, are drawn. The correlation integral is defined by

$$C_m(\varepsilon) = \frac{N_m(\varepsilon)}{[N_m-1]^2} \tag{5.351}$$

where $N_m(\varepsilon)$ is the number of pairs of trajectories of dimension m that remain less than a distance, ε from one another. Note that the numerator in the above expression is a squared quantity since it is based on the number of drifter pairs; (3) We then plot $\ln[C_m(\varepsilon)]$ versus $\ln(\varepsilon)$ to find the slope, D_2, of the curve

$$C_m(\varepsilon) \approx \varepsilon_2^D, \varepsilon \to 0 \tag{5.352}$$

The subscript "2" indicates that pairs of points are used to create the phase space.

If both original time series are random, then $D_2 = 2m$. A random process embedded in a $2m$-dimensional space always fills that space. On the other hand, if D_2 becomes independent of m at some saturation value, D_0, it means that the system represented by the time series has some structure (i.e., predictability) and should possess an attractor whose Hausdorff–Besicovitch dimension is equal to D_0 (Figure 5.81b). The need to calculate D_0 from the observations arises because we do not know the value of m a priori. We, therefore, calculate D_0 for increasing m until we approach a structure that becomes invariant at higher embedding dimensions, an indication that extra variables are not needed to account for the dynamics of the system. The attractor can be a topological structure such as a point, limit cycle or torus, or a non-topological submanifold with fractal structure. For a random-walk regime, D_2 approaches $2m$ so that there is no corresponding limiting value, D_0.

The independent segments of the paired drifter trajectories of sufficiently long duration embed the attractor in a substitute phase space spanned by the time-lagged coordinates provided by the data. The correlation dimension, D_2, measures the spatial correlation of the points that lie on the attractor. For a random time series, there will be no such spatial correlation in any embedding dimension and thus no saturation will be observed in the exponent, D_2. We note that the dimensionality of an attractor, whether fractal or nonfractal, indicates the minimum number of variables present in the evolution of the corresponding dynamical system. In other words, the attractor must be embedded in a state space of at least its dimension. Therefore, the determination of the Hausdorff dimension of an attractor sets a number of constraints that should be satisfied by any numerical or analytical model used to predict the evolution of the system. The main concern is that we do not extend the interpretation when going from a densely populated low-dimensional space to a sparsely occupied high-dimensional space. We cannot go beyond the critical embedding dimension above which the scaling region cannot be accurately determined (Essex et al., 1987; Tsonis and Elsner, 1990).

Glossary

ADCP Acoustic Doppler Current Profiler
AR Autoregressive
Argo Autonomous Profiling Drifting Buoys
AUV Autonomous Underwater Vehicles
CCW Counterclockwise (applied to rotary spectra)
Chi-squared Probability Density Function (used to define Confidence Intervals for Spectral Estimates)
CORK Circulation Observation Retrofit Kit (applied to Ocean Drilling Program, ODP)
cph Cycles per hour
CUI Cumulative Upwelling Index
CW Clockwise (applied to rotary spectra)
DoF Degrees of Freedom.
EDoF Effective Degrees of Freedom
END When the CUI reaches its seasonal maximum
ENSO El Niño Southern Oscillation index
Ergodic hypothesis Ensemble averages can be replaced by time or space averages
ESD Energy Spectral Density
FFT Fast Fourier Transform
FT Fall Transition
IFFT Inverse Fast Fourier Transform
IFT Inverse Fourier Transform
KZ Kolmogorov-Zurbenko filters
KZA Iterative moving-average Kolmogorov-Zurbenko filters
LS Least-Squares
LUS Length of upwelling season
MEM Maximum Entropy Method
MSE Mean Square Error
NCAR National Center for Atmospheric Research (USA)
NCEP National Center for Environmental Prediction (USA)
ODP Ocean Drilling Program
ONC Ocean Networks Canada
PDF Probability Density Function
Periodogram Unsmoothed FT Spectrum
PGC Pacific Geosciences Centre (Institute of Ocean Sciences, Canada)
PSD Power Spectral Density
RSI Regime Shift Index
SNR Signal to Noise Ratio
SST Sea Surface Temperature
ST Spring Transition
STARS Sequential t-test analysis of regime-shifts
TDMI Total Downwelling Magnitude Index
TUM Total downwelling magnitude
TUMI Total Upwelling Magnitude Index
Wavelet transform Method to analyze a time series using wavelets
Wavenumber The reciprocal of wavelength
White noise Fully random noise with a uniform horizontal line spectrum
YW Yule-Walker equations
Zero padding Procedure to increase the length of a record by adding zeros

Chapter 6

DIGITAL FILTERS

6.1 INTRODUCTION

Digital filtering is often an important step in the processing of oceanographic time series data. Applications involve the use of a series of specifically designed weights for smoothing and decimation of time series, removal of fluctuations in selected frequency bands, and the alteration of signal phase. The term "decimation" technically means the removal of every tenth point but is now commonly used for values other than 10. Digital filtering facilitates data processing by preconditioning the frequency content of the record. For example, filters can be used in studies of inertial waves to isolate current variability centered near the local Coriolis frequency, to remove the tides and atmospherically-forced sea-level fluctuations in investigations of tsunamis, and to eliminate tidal frequency fluctuations in studies of low-frequency current oscillations (Figure 6.1). The terms "detided" or "residual" time series are commonly used to describe time series that have been filtered to remove tidal components. Filters also provide algorithms for data interpolation, for integration and differentiation of recorded signals, and for linear prediction models. Kalman filters used for estimation of the state of a data-controlled process are discussed separately in Chapter 4.

There is no single type of digital filter for general oceanographic use. Selection of an appropriate filter depends on a variety of factors, including the frequency content of the data and the kind of analysis to be performed on the filtered record. Personal preference and familiarity with one type of filter also can be deciding factors. Many oceanographers have their favorite filters and would not consider switching. However, in certain instances, one type of filter may be superior to another for a specific task, and proper filter selection involves some forethought. Often the type of filter must be tailored to the job at hand. For example, some of the so-called "tide-elimination" or "tide-killer" filters once used extensively in oceanography were designed for regions dominated by semidiurnal tides and are, therefore, inadequate for time series with marked diurnal period variability (Walters and Heston, 1982; Thompson, 1983). These filters permit leakage of unwanted diurnal tidal energy into the nontidal (residual) frequency bands of the filtered record. Elimination of this problem is possible through proper filter selection.

This chapter begins with a brief outline of basic filtering concepts and then proceeds to descriptions of some of the more useful digital filters presently used in marine research. We use the term "filter" to cover any linear operation on the data. In *optimal estimation* applications, the term applies specifically to an optimal estimate of the last measurement point. *Smoothing* is reserved for estimates spanned by the observations. Much of the emphasis in this chapter is on the design of

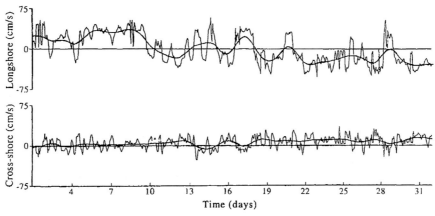

FIGURE 6.1 Time series of hourly alongshore (top) and across-shore (bottom) components of current velocity at 53 m depth on the continental shelf of northern Vancouver Island during March 1980. Thin line, original hourly data; thick line, hourly data filtered with a low-pass Godin tide-elimination filter, $A_{25}^2 A_{24}/(25^2 24)$. *From Huggett et al. (1987).*

Data Analysis Methods in Physical Oceanography. https://doi.org/10.1016/B978-0-323-91723-0.00001-5

low-pass digital filters that remove high-frequency oscillations from a given oceanographic time series. These filters can then be used to construct other types of filters, including high-pass versions of a given type of filter. The running-mean filter, the Lanczos-window cosine (or Lanczos–cosine) filter, and the Butterworth filter are among those commonly used in oceanography. The Kaiser–Bessel filter, which is one of our preferences, is discussed in Section 6.9. As with other time series analysis methods discussed in this book, all filters that we present in this chapter can also be applied to the spatial domain provided that the user takes care to ensure that the filter algorithms are properly formulated.

6.2 BASIC CONCEPTS

From a practical standpoint, a good low-pass filter should have five essential qualities: it should (1) have a sharp cutoff, so that unwanted high-frequency components are effectively removed; (2) have a comparatively flat pass-band that leaves the low-frequency components unchanged; (3) produce a clean transient response, so that rapid changes in the signal do not result in spurious oscillations or "ringing" within the filtered record; (4) cause no phase shift in the data series; and (5) have an acceptable computation time. As a rule, many of these desirable features are mutually exclusive and there are severe limitations to achieving the desired filter. We are invariably faced with a trade-off between the ability of the filter to produce the required results and the amount of filter-induced data loss we can afford to tolerate. For example, improved statistical reliability (increased degrees of freedom) for specified frequency bands decreases the frequency resolution of a filter, while more sharply defined frequency cutoffs lead to greater ringing and associated data loss.

Consider a time series consisting of the sequence

$$x(t_n) = x_n, n = 0, 1, ..., N-1 \tag{6.1}$$

with observations at discrete times $t_n = t_o + n\Delta t$ in which t_o marks the start time of the record and Δt is the sampling increment. A digital filter is an algebraic process by which a sequential combination of the input $\{x_n\}$ is systematically converted into a sequential output $\{y_n\}$. In the case of linear filters, for which the output is linearly related to the input, the time-domain transformation is accomplished through convolution (or "blending") of the input with the weighting function of the filter. Filters having the general form

$$y_n = \sum_{k=-M}^{M} w_k x_{n-k} + \sum_{j=-L}^{L} g_j y_{n-j}, n = 0, 1, ..., N-1, \tag{6.2}$$

in which M, L are integers and w_k, g_j are nonzero weighting functions or "weights" are classified as *recursive* filters because they generate the output by making use of a feedback loop specified by the second summation term. Such filters "remember" the past in the sense that all past output values contribute to all future output values. Filters based on the input data only (weights $g_j = 0$), are classified as *nonrecursive* filters. Any filter for which k lies in the range $-M \leq k \leq M$ is said to be physically unrealizable (in the sense of generating output in real-time) because both past and future data (which is not available in real time) are needed to calculate the output. Filters of this type have widespread application in the analysis of prerecorded data for which all digital values are available beforehand. Filters that use only past and incoming data are said to be physically realizable or causal, and are used in real-time data acquisition and in forecasting procedures.

Impulse response: The output $\{y_n\}$ of a nonrecursive linear filter is obtained through the convolution

$$y_n = \sum_{k=-M}^{M} w_k x_{n-k} = \sum_{k=-M}^{M} w_{n-k} x_k, n = 0, 1, ..., N-1 \tag{6.3}$$

where w_k are the time invariant weights and there are N data values, $x_o, x_1, ..., x_{N-1}$. For a symmetric filter, the time-domain convolution becomes

$$y_n = \sum_{k=0}^{M} w_k (x_{n-k} + x_{n+k}), n = 0, 1, ..., N-1 \tag{6.4}$$

in which $w_k = w_{-k}$. The set of weights $\{w_k\}$ is known as the *impulse response* function (IRF) and is the response of the filter to a spikelike impulse. To see this, we set $x_n = \delta_{0,n}$ where $\delta_{m,n}$ is the Kronecker delta function

$$\delta_{m,n} = 0, m \neq n$$
$$= 1, m = n \tag{6.5}$$

Eqn (6.3) then becomes

$$y_n = \sum_{k=-M}^{M} w_k \delta_{0,n-k} = w_n \tag{6.6}$$

The summations in Eqns (6.3) and (6.4) are based on a total of $2M + 1$ specified weights with individual values of w_k labeled for subscripts $k = -M, -(M + 1), ..., M$. To make practical sense, the number of weights is limited to $M \ll N/2$ where $(N - 1)\Delta t$ is the record length. In reality, it is not possible to use Eqn (6.3) to calculate an output value y_n for each time t_n. Because the response function spans a finite time, equal to $(2M - 1)\Delta t$, difficulties arise near the ends of the data record and we are forced to accept the fact that there are always fewer output data values than input values. There are three options: (1) we can make do with $2M$ fewer estimates of y_n (resulting from time losses of $M\Delta t$ at each end of the record), (2) we can create values of $x(t_n)$ for times outside the observed range $0 \leq t < (N - 1)\Delta t$ of the time series, or (3) we can progressively decrease the filter length, M, in accordance with the number of remaining input values. In the first approach, x_n is defined for $n = 0, 1, ..., N - 1$, whereas y_n is defined for the shortened range $n = M, M + 1, ..., N - (M + 1)$. In the second approach, the appended estimates of x_n should qualitatively resemble the data at either end of the record. For example, we could use the "mirror images" of the data reflected at the end points of the original time series. In the third approach, the values y_{M-1} and $y_{N-(M-1)}$ are based on $(M - 1)$ weights, the values y_{M-2} and $y_{N-(M-2)}$ on $(M - 2)$ weights, and so on.

Frequency response: The Fourier transform of $y(t_n)$ in Eqn (6.3) is

$$Y(\omega) = \sum_{n=-M}^{M} y_n e^{-i\omega n \Delta t}$$

$$= \left(\sum_{k=-M}^{M} w_k e^{-i\omega k \Delta t} \right) \left(\sum_{n=-M}^{M} x_{n-k} e^{-i\omega(n-k)\Delta t} \right)$$

$$= W(\omega)X(\omega) \tag{6.7}$$

so that convolution in the time domain corresponds to multiplication in the frequency domain. The function

$$W(\omega) = \frac{Y(\omega)}{X(\omega)} = \sum_{k=-M}^{M} w_k e^{-i\omega k \Delta t}; \tag{6.8a}$$

$$\omega \equiv \omega_n = 2\pi n/(N\Delta t) \tag{6.8b}$$

$n = 0, ..., N/2$ is known as the *frequency response*, or *transfer function* because it determines how a specific Fourier component $X(\omega)$ is modified as it is transformed from input to output. For the symmetric filter (Eqn 6.4), the transfer function reduces to

$$W(\omega) = w_o + 2\sum_{k=1}^{M} w_k \cos(\omega k \Delta t) \tag{6.9}$$

Once $W(\omega)$ is specified, the weights w_k are found through the inverse Fourier transform

$$w_k = \sum_{n=-N/2}^{N/2} W(\omega) e^{i\omega_n k \Delta t} \tag{6.10}$$

in which ω_n is given by (6.8b).

In general, the frequency response (transfer function), $W(\omega)$, is a complex function that can written in the form

$$W(\omega) = |W(\omega)| e^{i\phi(\omega)} \tag{6.11}$$

where the magnitude $|W(\omega)|$ is called the *gain* of the filter (a term originating with electrical circuitry) and $\phi(\omega)$ is the *phase lag* of the filter. The power $P(\omega)$ of the transfer function is given by

$$P(\omega) = W(\omega)W(-\omega) = W(\omega)W^*(\omega) = |W(\omega)|^2 \tag{6.12}$$

in which the asterisk denotes the complex conjugate.

6.3 IDEAL FILTERS

An ideal filter is one that has unity gain, so that $|W(\omega)| = 1$ at all frequencies within the specified *pass-band(s)*, and zero gain at frequencies within the *stop-band(s)* (Figure 6.2). When processing recorded oceanographic data, it is generally advantageous to have $\phi(\omega) = 0$ for all ω so that the filter produces no alteration in the phase of the frequency components. As we discuss in conjunction with recursive filters, zero phase shift can be guaranteed by first passing the input forward then backward (after inversion of the filtered values from the forward step) through the same set of weights. In the case of nonrecursive filters, zero phase is accomplished using symmetric filters (i.e., those with no imaginary components).

Digital filters commonly used in processing oceanographic data can be classified under the general headings of *low-pass, high-pass,* or *band-pass* filters. Although impossible to achieve in reality, we would like the amplitudes of our ideal filters to satisfy the following relations (see Figure 6.2):

$$\text{Low-pass:} |W(\omega)| = 1 \text{ for} |\omega| \le \omega_c$$

$$= 0 \text{ for } \omega_c < |\omega| \tag{6.13a}$$

$$\text{High-pass:} |W(\omega)| = 0 \text{ for } |\omega| < \omega_c$$

$$= 1 \text{ for } \omega_c \le |\omega| \tag{6.13b}$$

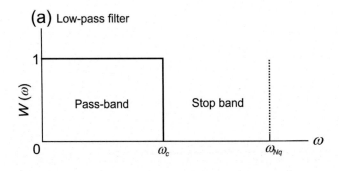

(a) Low-pass filter

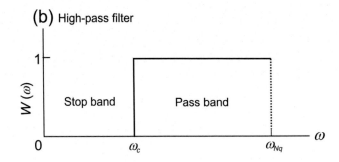

(b) High-pass filter

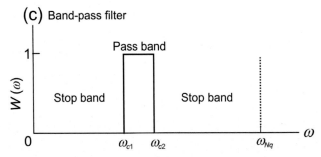

(c) Band-pass filter

FIGURE 6.2   Amplitude of the frequency response (transfer) function, $W(\omega)$, for ideal filters. (a) Low-pass, (b) low-pass, and (c) band-pass. The band-pass filter, with frequency pass band between ω_{c1} and ω_{c2}, has been constructed by combining the low-pass and high-pass filters; $\omega_{Nq}(= 2\pi f_{Nq})$ and $\omega_c(= 2\pi f_c)$ are the Nyquist and cutoff frequencies, respectively.

$$\text{Band-pass}: |W(\omega)| = 1 \text{ for } \omega_{c1} \le |\omega| \le \omega_{c2}$$

$$= 0 \text{ otherwise} \tag{6.13c}$$

The *cutoff frequency*, ω_c ($= 2\pi f_c$), marks the transition from the pass-band to the stop-band. For ideal filters, the transition is steplike, while for practical filters, the transition has a finite width. In the case of real filters, ω_c is defined as the frequency at which the mean filter amplitude in the pass-band is decreased by a factor of $\sqrt{2}$ and should roughly coincide with spectral minima for the time series being analyzed; the power of the filter is down by a factor of 2 (i.e., by $10\log(\frac{1}{2}) = -3$ dB) at the cutoff frequency. As its name implies, a low-pass filter lets through (or is "transparent" to) low-frequency signals but strongly attenuates (removes) high-frequency signals (cf. Figure 6.3a and b). High-pass filters let through the high-frequency components and strongly attenuate (remove) the low-frequency components (cf. Figure 6.3a and c). Band-pass filters permit only frequencies in a limited range (or band) to pass unattenuated.

Low-pass filters are among the most commonly used filters in geophysical data analysis. It is through these filters that low-frequency, long-term signal variability is determined. The running-mean filter, which involves a moving average over an odd number of values, is the simplest form of low-pass filter. More complex filters with better frequency responses, such as the low-pass Kaiser–Bessel window used in Figure 6.3b, also are commonly used (see also Section 6.9). *High-pass filtered* data are readily obtained by subtracting the low-pass filtered data from the original record from which the low-

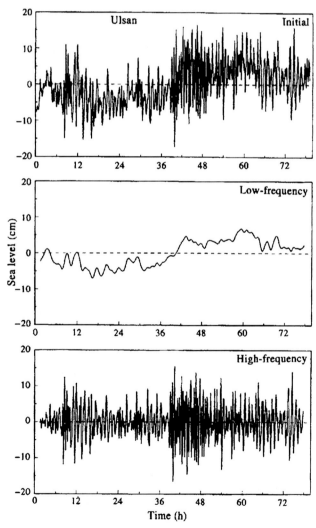

FIGURE 6.3 Filtering of a tide gauge record for Ulsan, Korea, using low- and high-pass Kaiser–Bessel filters (windows) with length $T/27 = 3$ h; $T = 81$ h is the record length and $\Delta t = 0.5$ min the sampling increment. (Top) Original record; (Middle) low-pass filtered record; (Bottom) high-pass filtered record. *Courtesy, Alexander Rabinovich, Institute of Ocean Sciences and P.P. Shirshov Institute of Oceanology.*

pass data were derived; there is no need to create a separate high-pass filter. Similarly, band-pass filters can be formed by an appropriate combination of low-pass and high-pass filters. In the ocean, seawater acts as a form of natural low-pass filter, attenuating high-frequency wave or acoustic energy at a much more rapid rate than low-frequency energy. Acoustic waves of a few hertz (cycles per second) can propagate thousands of kilometers in the ocean, whereas acoustic waves of hundreds of kilohertz or more are strongly attenuated over distances of a few hundred meters.

Applications of high-pass filters include the delineation of high-frequency, high-wave number fluctuations in the internal wave band, roughly $2\omega_{in} < \omega < \omega_{Nq}$ $\left(2f_{in} < f < f_{Nq}\right)$, and in the isolation of seiche or tsunami motions (here, ω_{in} is the local inertial frequency and ω_{Nq} is the Brunt–Väisälä frequency for a given vertical density structure). Band-pass filters are used to isolate variability in relatively narrow frequency ranges such as the near-inertial frequency band or, in North America, the electronic-induced 60-cycle noise in high-frequency oceanic data caused by AC power supplies.

The maximum range of frequencies that can be covered by a digital filter is determined at the high-frequency end by the Nyquist frequency, $\omega_{Nq} = \pi/\Delta t$ (radians/unit time), and at the low-frequency end by the fundamental frequency, $\omega_1 = 2\pi/T$, where $T = N\Delta t$ is the length of the record. The corresponding range in cycles/unit time is determined by $f_{Nq} = 1/(2\Delta t)$ and $f_1 = 1/T$. Provided that the cutoff frequencies are sufficiently far removed from the ends of the intervals, digital filters can be applied throughout the range, $\omega_1 < |\omega| < \omega_{Nq}$ ($f_1 < |f| < f_{Nq}$).

As we discuss in Section 6.3.2, the filter response coefficients for the ideal filters described by Eqns (6.13a)–(6.13c), where the desired filter amplitude is equal to 1 for all the pass-band frequencies and equal to 0 for all the stop-band frequencies, are obtained by taking the discrete Fourier transform (DFT) of the ideal frequency response. The problem is that this leads to infinitely long filter responses since the filters have to reproduce the infinitely steep discontinuities in the ideal frequency response at the edges of the frequency bands. To create a finite impulse response filter, the number of time-domain filter coefficients must be restricted by multiplying them by a finite width window function. The simplest window function is the rectangular window, which corresponds to truncating the sequence after a certain number of terms. In order to suppress the side lobes and make the filter frequency response more closely approximate an ideal filter, the width of the window must be increased and the window function tapered down to zero at the ends. This increases the width of the transition region between the pass- and stop-bands.

6.3.1 Bandwidth

The difference in frequency between the two ends of a pass-band defines an important property known as the *bandwidth* of the filter. To illustrate the relevance of this property, we consider an ideal band-pass filter with constant gain, linear phase, and cutoff frequencies ω_{c1}, ω_{c2} such that

$$W(\omega) = W_o \exp(-i\omega t_o), \omega_{c1} \leq |\omega| < \omega_{c2}$$

$$= 0, \text{otherwise} \tag{6.14}$$

From Eqn (6.13c), the impulse response is $w_k = \dfrac{1}{2\pi}W_0 \left(\displaystyle\int_{\omega_{c1}}^{\omega_{c2}} e^{-i\omega t_o} e^{i\omega k\Delta t} d\omega + \displaystyle\int_{\omega_{c1}}^{\omega_{c2}} e^{i\omega t_o} e^{-i\omega k\Delta t} d\omega \right)$

$$= \left\{ \frac{2W_o}{\pi}\Delta\omega \cdot \cos[\Omega(k\Delta t - t_o)] \frac{\sin[\Delta\omega(k\Delta t - t_o)]}{\Delta\omega(k\Delta t - t_o)} \right\} \tag{6.15}$$

in which $\Omega = \frac{1}{2}(\omega_{c1} + \omega_{c2})$ is the center frequency and $\Delta\omega = \omega_{c2} - \omega_{c1}$ is the bandwidth.

Using the fact that $\sin(p)/p$ ($\equiv \text{sinc}(p)$) $\to 1$ as $p \to 0$, we find that the peak amplitude response of the filter (Eqn 6.15) is directly proportional to the bandwidth $\Delta\omega$ as $\Delta\omega(k\Delta t - t_o) \to 0$. Note also that a narrow-band filter (a filter for which $\Delta\omega \to 0$) will oscillate longer (i.e., persist to higher values of k) than a broadband filter when subjected to a transient loading. Put another way, the persistence of the ringing that follows the application of the filter to a data set increases as the bandwidth decreases. From a practical point of view, this means that the ability of a filter to resolve sequential transient events is inversely proportional to the bandwidth. The narrower the bandwidth (i.e., the finer the resolution in frequency), the longer the time series needed to resolve individual events. For example, if we use a band-pass filter to isolate inertial frequency motions in the range 0.050–0.070 cph, the bandwidth $\Delta f = \Delta\omega/2\pi = 0.020$ cph and the filter could accurately resolve inertial events that occurred about $1/\Delta f = 50$ h apart. If we now reduce the bandwidth to 0.010 cph, the filter is only capable of resolving transient motions that occur more than 100 h apart. (The need to have long records to resolve closely spaced frequencies is exactly the problem we faced in Chapter 5 regarding the Rayleigh criterion for tidal analysis).

Another way of stating the above relationship is that the uncertainty in frequency, Δf (or $\Delta \omega$), is inversely proportional to the length of time T over which the signal oscillates (i.e., $\Delta f \approx 1/T$) so that $T\Delta f \approx 1$ for a given filter. If we wish to use a filter with a very narrow bandwidth, we need to analyze long time series records in which the signals of interest, such as the tides, have a high degree of persistence. In terms of observed data, the measured bandwidth of an oscillation in current speed, sea-level elevation, or other oceanic parameter is directly related to the persistence of the signal. For example, a wind-generated clockwise rotary inertial current having an observed bandwidth $\Delta f \approx 0.10$ cpd implies that the burst of inertial energy had a duration $T \approx 1/\Delta f = 10$ days.

6.3.2 Gibbs Phenomenon

In practice, steplike transfer functions such as described by Eqns (6.13a)–(6.13c) are not possible. Digital filters invariably possess finite slope transition zones between the stop- and pass-bands. To illustrate some of the fundamental impediments to creating ideal filters, consider the steplike transfer function

$$W(\omega) = 1; \ \ 0 \leq \omega \leq \omega_{Nq}$$

$$= 0; \ \ -\omega_{Nq} \leq \omega < 0 \tag{6.16}$$

(see Figure 6.2a) where, for convenience, we specify a cutoff frequency $\omega_c = 0$. Assuming that $W(\omega)$ is repeated over multiples of the basic Nyquist frequency interval $(-\omega_{Nq}, \omega_{Nq})$, the appropriate Fourier series expansion for Eqn (6.16) is given in the usual manner by

$$W(\omega) = \frac{1}{2} a_o + \sum_{n=1}^{\infty} [a_n \cos(\omega n \Delta t) + b_n \sin(\omega n \Delta t)] \tag{6.17}$$

with coefficients

$$a_n = \frac{1}{\omega_{Nq}} \int_{-\omega_{Nq}}^{\omega_{Nq}} W(\omega) \cos(\omega n \Delta t) d\omega \tag{6.18a}$$

$$b_n = \frac{1}{\omega_{Nq}} \int_{-\omega_{Nq}}^{\omega_{Nq}} W(\omega) \sin(\omega n \Delta t) d\omega \tag{6.18b}$$

The fact that $a_n = 1$, for all n, including $n = 0$, suggests reformulation of the problem in terms of the function

$$W_c(\omega) = W(\omega) - \frac{1}{2} \tag{6.19}$$

centered about $W(\omega) = 1/2$, the mean functional value at the discontinuity. Because $W(\omega)$ is then an odd function, the cosine terms in Eqn (6.17) can be eliminated immediately. Moreover, W_c is symmetric about $\omega = \pm\frac{1}{2}\omega_{Nq} = \pm\pi/(2\Delta t)$ so that there are no even sine terms. For odd n, Eqn (6.18b) yields $b_n = 2/n\pi$ and Eqn (6.17) becomes

$$W(\omega) = \frac{1}{2} + \frac{2}{\pi} \left[\sin(\omega \Delta t) + \frac{\sin(3\omega \Delta t)}{3} + \frac{\sin(5\omega \Delta t)}{5} + \ldots \right] \tag{6.20}$$

which must be truncated after a finite number of terms.

Successive approximations to the series (Eqn 6.20), and hence to the function Eqn (6.16), are not convergent near discontinuities such as that for the steplike transition region of the ideal high-pass filter shown in Figure 6.4. In this example, the filter amplitude $|W(\omega)|$ is zero for $\omega < \omega_c$ (the stop-band) and unity for $\omega_c < \omega < \omega_{Nq}$ (the pass-band). The succession of overshoot ripples, or ringing, is known as *Gibbs phenomenon*. The ripple period, $T_p = p\pi\Delta t$ (p is an integer), is fixed but increasing the number of terms in the Fourier series for $W(\omega)$ decreases the distortion due to the overshoot effects. However, even in the limit of infinitely many terms, Gibbs phenomenon persists as the amplitude of the first

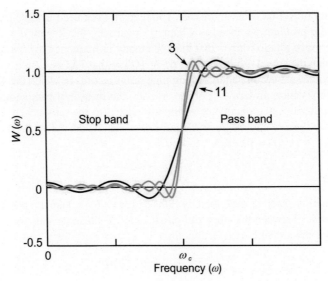

FIGURE 6.4   Gibbs phenomenon (overshoot ripples) arising from successive approximations to the steplike function $W(\omega) = 1$, $\omega_c < \omega \leq \omega_{Nq}$, and zero otherwise; $\omega_c = 2\pi f_c$ is the cutoff frequency. Curves are derived from Eqn (6.20) using $M = 3$, 7, and 11 terms.

overshoot diminishes asymptotically to about 0.18 or about 9% of the pass-band amplitude. The first minimum decreases asymptotically to about 5% of the pass-band amplitude. In the limit of large $N \rightarrow \infty$, it can be shown (Godin, 1972; Hamming, 1977) that

$$W_\infty(0) \rightarrow \frac{1}{\pi} \int\limits_0^\pi \left(\frac{\sin u}{u}\right) du \tag{6.21}$$

The values of $W_\infty(0)$ can be found in tables of the sine integral function. In the case of Figure 6.4, the value for the first maximum is 1.08949 (= 1.0 + 0.08949), while that for the first undershoot is 0.9514 (= 1.0 − 0.04858).

Gibbs phenomenon has considerable importance in that it occurs whenever a function has a discontinuity. For example, suppose that we want to use Eqn (6.20) to remove spectral components near a cutoff frequency, ω_c. Unless the spectral components in the stop- and pass-bands are well separated relative to the width of the transition zone, the finite ripples will cause leakage of unwanted energy into the filtered record. Noise from the stop-band will not be completely removed and certain frequencies in the pass-band will be distorted. A critical aspect of filter design is the attenuation of the overshoot ripples using smoothing or tapering functions (windows). As discussed in Chapter 5, windows are important in reducing side-lobe leakage in spectral estimates.

Further difficulties arise when we apply the weights $\{w_k\}$ of an ideal filter in the time domain. Consider the non-recursive, low-pass filter (positive frequency only)

$$W(\omega) = 1, 0 \leq \omega \leq \omega_c$$

$$= 0, \text{otherwise} \tag{6.22}$$

for which the impulse function is, for $k = -N, ..., N$

$$w(t_k) = w_k = \frac{1}{\omega_{Nq}} \sum_{\omega=0}^{\omega_c} \cos(\omega k\Delta t)\Delta\omega$$

$$= \frac{\sin(\omega_c k\Delta t)}{\omega_{Nq} k\Delta t}$$

$$= \frac{f_c}{f_{Nq}} \frac{\sin(2\pi f_c k\Delta t)}{2\pi f_c k\Delta t} \tag{6.23}$$

whereby, $w_o = f_c/f_{Nq}$. The weights w_k attenuate slowly, as $1/k$, so that a large number of terms are needed if the filter frequency response $W(\omega)$ is to be effectively carried over to the time domain. In addition to being computationally inefficient, filters constructed from a large number of weights lead to considerable loss of information at the ends of the data sequence. Practical considerations force us to truncate the set of weights thereby enhancing the overshoot problem associated with Gibbs phenomenon in the frequency domain. Moreover, if we truncate the length of the data set (Eqn 6.1), we are unable to accurately replicate Eqn (6.22) in the frequency domain. This leads to a finite slope between the stop- and pass-bands of the filter.

The situation is similar for high-pass filters

$$W(\omega) = 0, 0 \leq \omega \leq \omega_c$$

$$= 1, \text{otherwise} \qquad (6.24a)$$

In this case

$$w_k = \frac{1}{\omega_{Nq}} \sum_{\omega=\omega_c}^{\omega_{Nq}} \cos(\omega k \Delta t)\Delta\omega$$

$$= -\frac{f_c}{f_{Nq}} \frac{\sin(2\pi f_c k\Delta t)}{2\pi f_c k\Delta t}, k = -N, ..., N \qquad (6.24b)$$

whereby, $w_o = 1 - f_c/f_o$. Notice that, except for the central term, w_o, the weights w_k of the high-pass filter (Eqn 6.24b) are equal to minus the weights, w_k, of the low-pass filter (Eqn 6.23). The center value, w_o of the high-pass filter is found from w_o of the low-pass filter by: w_o *(high pass)* $= 1 - w_o$ *(low pass)*.

The difficulties that arise with Gibbs phenomenon are somewhat alleviated by applying smoothing functions that attenuate the overshoot ripples. As usual, the price we pay for improved decay of the weighting terms is a broadening of the main lobe centered at the frequency being filtered. As we remarked earlier, the fact that the transition from the pass-to the stop-band takes place over a finite range of frequencies necessitates a working definition for the cutoff frequency, ω_c. Here, ω_c is defined as the frequency at which the power $|W(\omega)|^2$ of the filter is attenuated by a factor of 2 (-3 dB) from its mean pass-band value (power in decibels $= 20\log(A/A_o)$ where A_o is a reference level for the signal amplitude, A, having power proportional to A^2). Alternatively, the cutoff frequency marks the frequency at which the amplitude $|W(\omega)|$ of the filter is reduced by a factor of $\sqrt{2}$ of the pass-band amplitude (amplitude in decibels $= 10\log(A/A_o)$).

6.3.3 Recoloring

The transfer function amplitude $|W(\omega)|$ defines the effectiveness of a particular filter in transmitting or blocking power within specific frequency bands. Since no filter is perfect, in the sense that its transfer function is exactly unity throughout the pass-band(s) and zero in the stop-band(s), it is often necessary to "recolor" (rescale) the output $Y(\omega)$ so that the total variance in the pass-band spectral estimates equals the total variance of the input data for that frequency range. The need to recolor stems from practical considerations involving the choice of filter, cutoff frequency, and filter steepness through the transition band. For a pass-band of width $\Delta\omega$, multiplication of the filter output $|Y(\omega)|$ by a frequency-independent correction factor γ given by

$$\gamma(\omega) = \frac{\text{input variance within the bandwidth } \Delta\omega}{\text{output variance within the bandwidth } \Delta\omega}$$

ensures that the output power is adequately rescaled.

We can illustrate the recoloring process using the Hanning (von Hann) and Hamming windows. If $x(t)$ is any scalar time series of length N, and $y(t)$ is the filtered output of this series following application of one of these windows, then the Fourier transform of the output, $Y(f_k)$, for discrete frequencies $f_k = (k/T)$, $k = 0, 1, ..., (N/2)$ is given by

$$Y(f_k) = 0.50X(f_k) - 0.25X(f_{k-1}) - 0.25X(f_{k+1}) \text{ (Hanning)} \qquad (6.25a)$$

$$Y(f_k) = 0.54X(f_k) - 0.23X(f_{k-1}) - 0.23X(f_{k+1}) \text{ (Hamming)} \qquad (6.25b)$$

where $X(f_k)$ is the Fourier transform of the original time series. The corresponding expected values for $|Y(f_k)|^2$ in Eqns (6.25a) and (6.25b) are

$$E\left[|Y(f_k)|^2\right] = (0.50)^2 + (0.25)^2 + (0.25)^2 = 0.3750 \tag{6.25c}$$

$$E\left[|Y(f_k)|^2\right] = (0.54)^2 + (0.23)^2 + (0.23)^2 = 0.3974 \tag{6.25d}$$

so that the spectral density estimates $S(f_k) \approx |Y(f_k)|^2$ for each frequency component of a time series smoothed by a Hanning window should be rescaled by the exact factor $(0.375)^{-1} = 8/3$ to correct for the loss of power due to the filter. For the Hamming window, the factor is roughly $(0.397)^{-1} \approx 5/2$. Note that, according to Eqns (6.25a) and (6.25b), we can easily obtain spectral estimates $S(f_k)$ for each windowed time series by summing up the squared amplitudes $|X(f)|^2$ of three adjacent Fourier components of the original time series

$$S(f_k) = C_o|X(f_k)|^2 + C_{-1}|X(f_{k-1})|^2 + C_{+1}|X(f_{k+1})|^2 \tag{6.26}$$

where $C_o = 0.50$ and $C_{-1} = C_{+1} = 0.25$ for the Hanning window and $C_o = 0.54$ and $C_{-1} = C_{+1} = 0.23$ for the Hamming window.

6.4 DESIGN OF OCEANOGRAPHIC FILTERS

The isolation of signal variability within specific frequency bands requires filters with well-defined frequency characteristics. The design of application-specific filters can proceed in two basic ways. The first approach is to assemble a combination of simple filters, such as moving averages of variable length, and from them construct a filter with the required characteristics. This is referred to as *cascading* since the output from the lead-off filter is used as input to the second filter, output from the second filter is used as input to the third, and so on. Filter cascading is used in the design of Godin's (1972) tide-elimination filters and the squared Butterworth filters described later in this section. The second approach is to specify the desired characteristics of the filter precisely and then use poles and zeroes of mathematical functions to design a filter that meets these requirements as closely as possible. As an example, we might wish to eliminate the annual cycle from a long time series of upper ocean variability, such as sea surface temperature, so that weaker fluctuations are no longer overwhelmed by the dominant seasonal changes. The filter properties are then directly tailored to the processing requirements and to the data specific to the region of interest. (In this example, we could also use least squares analysis to determine the annual cycle and then subtract this cycle from the original data.)

Regardless of which approach is taken, it is important that the impulse and frequency response functions (FRFs) of the filter have a number of fundamental properties: (1) The FRF should have reasonably sharp transitions between adjacent stop- and pass-bands, especially if the data do not have wide "spectral gaps" between dominant frequencies within the two bands. At the same time, the transition should not be so steep as to introduce large side-lobe effects or cause the filter output to become unstable; (2) the transfer function should have nearly constant amplitude and zero phase (even symmetry) within the pass- and stop-bands so that corrections to amplitude and phase are easily applied. Linear phase change as a function of frequency is acceptable but requires corrective work at the end of the processing; and (3) the impulse response should have as short a span as possible to both minimize the number of points lost (or that are need to be appended at the ends of the data) and reduce the amount of computation.

6.4.1 Frequency Versus Time Domain Filtering

In most instances, filters are designed to precondition the frequency content of the data prior to further analysis. This immediately suggests that the design of a filter begins with specification of the transfer function, $W(\omega)$. Once $W(\omega)$ has been determined there are two ways to proceed. The standard time-domain approach (e.g., Hamming, 1977) is to inverse Fourier transform $W(\omega)$ to obtain the time domain filter weights, w_k, which are then used in the convolution Eqn (6.3) to determine the output $\{y_n\}$. The output is subsequently Fourier transformed to determine $Y(\omega)$. The frequency-domain approach (e.g., Walters and Heston, 1982; Middleton, 1983) makes use of the fact that $Y(\omega) = W(\omega)X(\omega)$, where $X(\omega)$ is the Fourier transform of the data $\{x(t)\}$. In this approach, the data are Fourier transformed to obtain $X(\omega_i)$, $i = 1, 2, \ldots, N/2$, where $X(\omega)$ consists of a set of $N/2$ frequency-dependent amplitudes and phases $(A(\omega_i), \phi(\omega_i))$ at discrete frequencies. The filtered record is obtained by multiplying $X(\omega)$ by $W(\omega)$. The time domain series $\{y_n\}$ can be derived from the inverse Fourier transform of $Y(\omega)$.

There are pros and cons for both approaches. The time-domain approach uses the actual recorded data and filtering consists of simple sums and products. Moreover, the filtered series $\{y_n\}$ can be immediately plotted against the original input $\{x_n\}$ to see directly the effectiveness of the filter. Discontinuities in the time series, which lead to transient filter ringing effects, can be dealt with on the spot. However, if the calculation of $Y(\omega)$ and its associated spectral estimate $|Y(\omega)|^2$ are the ultimate goals, the time-domain approach requires application of two Fourier transforms: First, we use $W(\omega)$ to define the filter weights $\{w_k\}$ and then transform $y_n \rightarrow Y(\omega)$ to obtain the Fourier components. This can lead to roundoff and computational errors.

In the frequency-domain analysis, only one Fourier transform, $x_n \rightarrow X(\omega)$, is required. On this basis, it seems preferable to use the Fourier transform method and just set to zero all those frequency components outside the range of interest. The filtered data $\{y_n\}$ are then found through an inverse transform of the modified Fourier components, $Y(\omega) = W(\omega)X(\omega)$. One obvious difficulty with this procedure is that the discrete frequencies of the Fourier estimates may not be properly positioned relative to the required cutoff frequency of the filter, that is, the cutoff frequency may fall midway between two discrete Fourier components. Walters and Heston (1982) also pointed out that the sharp cutoff associated with this process causes ringing through the entire data set (Figure 6.5). For this reason, the Fourier coefficients must be reduced gradually to zero over a range of frequencies. For example, Nowlin et al. (1986) used a trapezoidal-shaped band-pass filter to study inertial oscillations in data collected in Drake Passage. In this particular instance, "Fourier coefficients within 0.03 cpd of the local inertial frequency were retained undiminished, and this central portion was flanked by two tapered sections 0.06 cpd wide in which the coefficients were reduced linearly to zero". The smooth filter transition results in a substantial reduction in ringing in the filtered data but is certainly reminiscent of data tapering required in the time-domain analysis. A more detailed discussion of frequency-domain filtering is presented in Section 6.10.

6.4.2 Filter Cascades

In some instances, a desired filter $W(\omega)$ can be constructed from a series or *cascade* of basis filters $W_j(\omega)$ such that

$$W(\omega) = W_1(\omega) \times W_2(\omega) \times \cdots \times W_q(\omega) \tag{6.27}$$

where "$\times$" denotes successive applications of individual transfer functions, beginning with W_1. That is, the data are first processed with $W_1(\omega)$ and the output from this filter passed through $W_2(\omega)$; the output from $W_2(\omega)$ is then passed through $W_3(\omega)$, and so on until the last filter, $W_q(\omega)$. The final output from $W_q(\omega)$ corresponds to the sought-after output from $W(\omega)$.

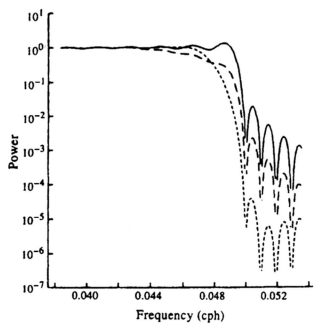

FIGURE 6.5 Squared frequency response functions (with magnitudes in terms of power) for low-pass filters with different transition bands. *Solid line*, a steplike transition band; *long-dashed line*, a nine-point cosine-tapered transition band; *short-dashed line*, a three-point optimally designed transition band. The cutoff associated with each filter causes ringing through the entire data set. *From Elgar (1988).*

Although the technique is straightforward and helps to minimize roundoff error, it has a number of major drawbacks, including the need for extended computations and the possibility of repeated ringing as one filter after another is applied in succession.

A high-pass filter $W_H(\omega)$ is obtained from its low-pass counterpart $W_L(\omega)$ by the relation $W_H(\omega) = 1 - W_L(\omega)$ where, in theory, the combined output from the two filters simply recreates the original data, since $W_L(\omega) + W_H(\omega) = 1$. This has advantages in situations where $W_L(\omega)$ is easily derived or is already available. In the time domain, the high-pass filtered record $\{y'_n\}$ is obtained by subtracting the output $\{y_n\}$ from the low-pass filter from the input time series $\{x_n\}$. Care is needed to ensure that the times of y_n and x_n are properly aligned so that $y'_n = x_n - y_n$ and $n = M, M+1, \ldots, N-2M$.

A band-pass filter can be constructed from an appropriate high- and low-pass filter using the method illustrated in Figure 6.2c. Here, the cutoff frequency of the low-pass filter becomes the high-frequency cutoff of the band-pass filter; similarly, the cutoff frequency of the high-pass filter becomes the low-frequency cutoff of the band-pass filter. The cascade then has the form $W_B(\omega) = W_L(\omega) \times W_H(\omega)$.

Because nonrecursive filters are symmetric ($W(\omega)$ is a real function), there is no shift in phase between the input and output signals. This feature of the filters, as well as their general mathematical simplicity, has contributed to their popularity in oceanography. Recursive filters, on the other hand, are typically asymmetrical. This introduces a frequency-dependent phase shift between the input and output variables and adds to the complexity of these filters for oceanic applications. Despite these difficulties, recursive filters are useful additions to any processing repertoire. Note that, regardless of which type of filter is used, we can remove phase shifts introduced through the "forward" application of the filter by reversing the process and passing the data "backward" through the filter. In performing the latter step, we must be careful to invert the order of the record values between the forward and backward passes. Specifically, if the recursive filter introduces a phase shift $\phi(\omega)$ at frequency ω (or equivalently, a time shift $\phi/\omega = \phi/2\pi f$), it will introduce a compensating shift, $-\phi(\omega)$, when the data are passed in the reverse order through the filter. To show this sequence, let $x_1, x_2, \ldots, x_n$ be the original data sequence used as input to a given filter with nonzero phase characteristics, and $y_1, y_2, \ldots, y_n$ the output from the filter (Figure 6.6). If we now invert the order of the output and pass the inverted signal through the filter again, we obtain a new output $z_1, z_2, \ldots, z_n$. The order of the z-output is then inverted to form $z_n, z_{n-1}, \ldots, z_1$, which returns us to the proper time sequence. For simplicity we can rewrite this later sequence as $y'_1, y'_2, \ldots, y'_n$. The act of applying the filter a second time cancels any phase change from the first pass through the filter. Note that this corresponds to squaring the transfer function so that the final transfer function for the recursive filter is $|W(\omega)|^2$.

As an example of a phase-dependent recursive filter, consider the high-pass *quasi-difference filter*

$$y(n\Delta t) = x(n\Delta t) - \alpha x[(n-1)\Delta t] \qquad (6.28a)$$

where α is a parameter in the range $0 < \alpha \leq 1$; $\alpha = 1$ corresponds to the simple difference filter (Koopmans, 1974). The frequency response (transfer function) for this filter is:

$$W(\omega) = 1 - \alpha e^{-i\omega\Delta t} \qquad (6.28b)$$

FIGURE 6.6 The data processing sequence for a nonsymmetrical recursive filter $W(\omega)$ that removes phase changes $\phi(\omega)$ introduced to the data sequence x_i ($i = 1, \ldots, n$) by the filter. This cascade produces a symmetric squared-filter response $|W(\omega)|^2$.

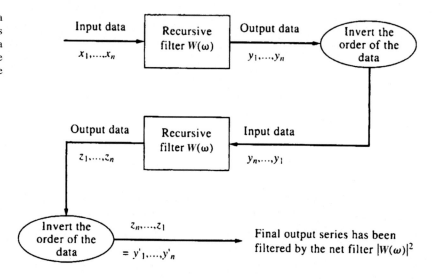

and the phase function is

$$\phi(\omega) = \tan^{-1}[\alpha \sin(\omega \Delta t) / (1 - \alpha \cos(\omega \Delta t))] \tag{6.28c}$$

Reversing the order of the output from the first pass of the data through the filter and then running the time-inverted record through the filter again is tantamount to passing the data through a second filter $W^*(\omega)$. This introduces a phase change $-\phi(\omega)$ which cancels the phase change $\phi(\omega)$ from the first filter (Figure 6.7). The symmetric filter obtained from this cascade is then

$$
\begin{aligned}
|W(\omega)|^2 &= W(\omega) \times W^*(\omega) \\
&= \left(1 - \alpha e^{-i\omega \Delta t}\right)\left(1 - \alpha e^{+i\omega \Delta t}\right) = 1 - 2\alpha \cos(\omega \Delta t) + \alpha^2
\end{aligned} \tag{6.28d}
$$

6.5 RUNNING-MEAN FILTERS

The *running-mean* or *moving average filter* is the simplest and one of the most commonly used low-pass filters in physical oceanography. In a typical application, the filter (which is simply a moving rectangular window) consists of an odd number of $2M + 1$ equal weights, w_k, $k = 0, \pm 1, \ldots, \pm M$, having constant values

$$w_k = \frac{1}{2M+1} \tag{6.29a}$$

where w_k resembles a uniform probability density function in which all occurrences are equally likely. The running-mean filter produces zero phase alteration since it is symmetric about $k = 0$, it satisfies the normalization requirement

$$\sum_{k=-M}^{M} w_k = 1 \tag{6.29b}$$

and is straightforward to apply. To obtain the output sequence $\{y_m\}$ for input sequence $\{x_n\}$, the first $2M + 1$ values of x_n (namely $x_0, x_1, \ldots, x_{2M}$) are summed and then divided by $2M + 1$, yielding the first filtered value $y_M = y(2M\Delta t/2) = y(M\Delta t)$. The subscript M reminds us that the filtered value replaces the original data record x_M at the appropriate location in the time series. The next value, y_{M+1}, is obtained by advancing the filter weights one time step Δt and repeating the process over the data sequence $x_1, x_2, \ldots, x_{2M+1}$ and so on up to $N - 2M$ output values.

The $\{y_m\}$ consists of a "smoothed" data sequence with the degree of smoothing, and associated loss of information from the ends of the input, depending on the number of filter weights. Mathematically

$$y_{M+i} = \frac{1}{2M+1} \sum_{j=0}^{2M} x_{i+j}, \, i = 0, \ldots, N-2M \tag{6.30}$$

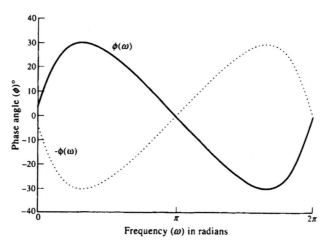

FIGURE 6.7 The phase change $\phi(\omega)$ for a quasi-difference filter (with $\alpha = 0.5$ as a function of frequency, ω).

A high-pass running-mean filter can be generated by subtracting the output $\{y_m\}$ from the original data. The output $\{y_m'\}$ for the high-pass filter is

$$y_m' = x_m - y_{m,}\ m = M, M+1, ..., N-2M \tag{6.31}$$

where we make certain that we subtract data values for the correct times. This technique of obtaining a high-pass filtered record from a low-pass filtered record will also be applied to other types of filters.

The frequency response $W(\omega)$ for the running-mean filter is given by Eqns (6.8a) and (6.8b). Using Eqn (6.29a) and the fact that $\Delta t = \pi/\omega_{Nq}$, we find that

$$W(\omega) = \frac{1}{2M+1}\left\{\frac{1 + 2\sin\left[(\pi/2M)(\omega/\omega_{Nq})\right]\cos\left[(\pi/2(M+1))(\omega/\omega_{Nq})\right]}{\sin\left[(\pi/2)(\omega/\omega_{Nq})\right]}\right\} \tag{6.32a}$$

$$= \frac{1}{2M+1}\frac{\sin\left[(\pi/2(2M+1))(\omega/\omega_{Nq})\right]}{\sin\left[(\pi/2)(\omega/\omega_{Nq})\right]} \tag{6.32b}$$

where $W(\omega) \to 1$ as $\omega/\omega_{Nq} \to 0$. As M increases, the central lobe of the transfer function narrows (Figure 6.8) and the cutoff frequency (at which $|W(\omega)| = e^{-1}|W(0)|$) moves closer to zero frequency. The filter increasingly isolates the true mean of the signal. Unfortunately, the filter has considerable contamination in the stop-band due to the large, slowly attenuating side lobes. Reduction of these side-lobe effects requires a long filter which means severe loss of data at either end of the time series. The running-mean filter should therefore only be used with long data sets ("long" compared with the length of the filter). Accurate filtering requires use of more sophisticated filters.

For the three-point weighted average, $w_k = 1/3$ and Eqn (6.32b) yields

$$W(\omega; 3) = \frac{1}{3}\left[1 + 2\cos\left(\pi\omega/\omega_{Nq}\right)\right]$$

$$= \frac{1}{3}\frac{\sin\left[(3\pi/2)(\omega/\omega_{Nq})\right]}{\sin\left[(\pi/2)(\omega/\omega_{Nq})\right]} \tag{6.33}$$

while for five-point weighted average, $w_k = 1/5$ and

$$W(\omega; 5) = \frac{1}{5}\frac{\sin\left[(5\pi/2)(\omega/\omega_{Nq})\right]}{\sin\left[(\pi/2)(\omega/\omega_{Nq})\right]} \tag{6.34}$$

(Figure 6.8). Numerous examples of running-mean filters appear in the oceanographic literature. A common use of running-mean filters is to convert data sampled at times t to an integer multiple of this time increment for use in standard analysis packages. Data collected at intervals Δt of 5, 10, 15, 20, or 30 min are usually converted to hourly data for use in tidal harmonic programs, although the least squares algorithms used in these programs also work with unequally spaced time series data (e.g., Foreman, 1977, 1978; revised 2004). Running-mean filters are also commonly used to create weekly, monthly, or annual time series (Figure 6.9).

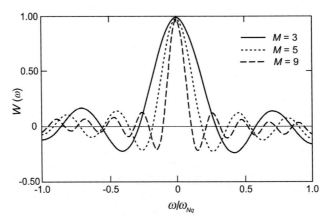

FIGURE 6.8   The amplitude of the frequency response functions, $W(\omega)$, for running-mean (weighted average, rectangular) filters for $M = 3, 5, 9$. $\omega_{Nq} =$ Nyquist frequency.

(a)

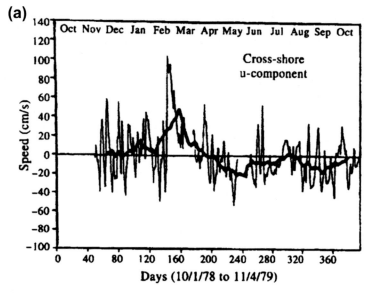

(b)

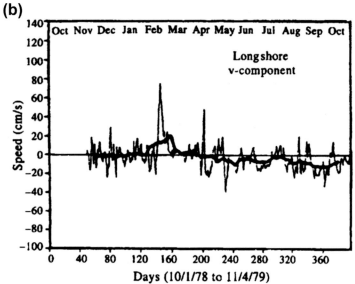

FIGURE 6.9   Daily mean time series of across-shelf (a) and alongshelf (b) near-surface currents off Cape Romain in the South Atlantic Bight for the period January 10, 1979, to April 11, 1979. *Thin line*, daily average data; *thick line*, 30-day running-mean values. *From McClain et al. (1988).*

6.6 GODIN-TYPE FILTERS

For the low-pass filtering of sub-hourly sampled tidal records prior to decimation to "standard" hourly values, Godin (1972) recommends the use of cascaded running-mean filters with response functions of the form

$$\frac{A_n^2 A_{n+1}}{n^2(n+1)}, \frac{A_n A_{n+1}^2}{n(n+1)^2} \tag{6.35}$$

Here, A_n and A_{n+1} are the average values of n and $n+1$ consecutive data points, respectively. Each filter smooths the data three times. In the first version in Eqn (6.35), the smoothing is performed twice using the $\{n\}$-point average and once using the $\{n+1\}$-point average. The alternative version uses the $\{n+1\}$-point average twice and the n-point average once.

Following the filter operation, the smoothed records can then be sub sampled at hourly intervals without concern for aliasing by higher frequency components. For the second version in Eqn (6.35), the response function is

$$W(\omega) = \frac{1}{n^2(n+1)} \frac{\sin^2\left[\left(\frac{\pi}{2n}\right)\left(\frac{\omega}{\omega_{Nq}}\right)\right] \sin\left[\left(\frac{\pi}{2(n+1)}\right)\left(\frac{\omega}{\omega_{Nq}}\right)\right]}{\sin^3\left[\left(\frac{\pi}{2}\right)\left(\frac{\omega}{\omega_{Nq}}\right)\right]} \tag{6.36}$$

Godin filters $(A^2{}_{12}A_{14})/(12^2 14)$ are used routinely to smooth oceanographic time series sampled at multiples of 5-min increments prior to their use in tidal analysis programs. On the other hand, 30-min data would first be smoothed using the filter $(A_2^2 A_3)/(2^2 3)$ (Figure 6.10) and then decimated to hourly data. For example, the conversion of 30-min data collected by the early Aanderaa RCM4 mechanical current meters to hourly data requires such a three-stage running-average filter. The filter is needed to convert the instantaneous directions and average speeds from the current meter to quantities more closely resembling vector-averaged currents. Application of the moving low-pass filter (Eqn 6.36) removes high-frequency components and helps avoid the aliasing errors that would occur if the raw data were simply decimated to hourly values without any form of prior smoothing. Simply picking out a value each hour is, of course, akin to not having recorded the higher frequency variability in the first place. Some care is required in that the smoothing process reduces the amplitude of various Fourier components outside the tidal band. As a result, amplitudes of Fourier components derived after application of the filter must be corrected (recolored) in inverse proportion to the amplitude of the filter at the particular frequency. Phases of the Fourier components are unaltered by this symmetric filter.

The formulation (Eqn 6.35) also can be used to generate low-pass filters to remove diurnal, semidiurnal, and shorter period fluctuations from the hourly records. Although these filters have been criticized in recent years because of their slow transition through the high-frequency end of the "weather band" (periods longer than 2 days), they are easy to apply, have good response in the daily tidal band, and consume relatively little data from the ends of the time series. The most commonly used version of the low-pass Godin filter is $(A^2{}_{24}A_{25})/(24^2 25)$ in which the hourly data are smoothed twice using the 24-point (24-h) average and once using the 25-point average. The filter frequency response is

$$W(\omega) = \frac{1}{24^2 25} \sin^2\left[24(\pi/2)(\omega/\omega_{Nq})\right] \frac{\sin\left[25(\pi/2)(\omega/\omega_{Nq})\right]}{\sin^3\left[(\pi/2)(\omega/\omega_{Nq})\right]}$$

$$= \frac{1}{24^2 25} \sin^2(24\pi f \Delta t) \frac{\sin(25\pi f \Delta t)}{\sin^3(\pi f \Delta t)} \tag{6.37}$$

where, as before, $\omega = 2\pi f$ (f is in cycles per hour), $\omega_{Nq} = \pi/\Delta t$ and $\Delta t = 1$ h. Note that a total of 35 data points (i.e., 35 h) are lost from each end of the time series and that the filter has a half-amplitude point near 67 h (Figure 6.11). The weights of this symmetric 71-h-length filter are (Thompson, 1983)

$$w_k = \frac{1/2}{24^2 25}[1,200 - (12-k)(13-k) - (12+k)(13+k)], 0 \leq k \leq 11$$

$$= \frac{1/2}{24^2 25}(36-k)(37-k), 12 \leq k \leq 35 \tag{6.38}$$

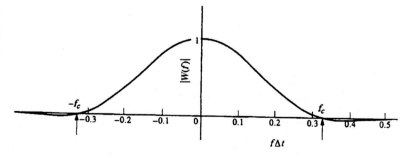

FIGURE 6.10 The frequency response function, $W(f)$, for the Godin-type filter $A_2^2 A_3/(2^2 3)$ used to smooth 30-min data to hourly values. The horizontal axis has units $f\Delta t$, with $f_{Nq}\Delta t = 0.5$; $f_c = \omega_c/(2\pi)$ is the cutoff frequency. *Based on Godin (1972).*

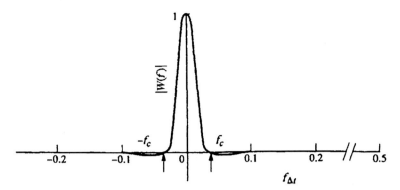

FIGURE 6.11 Same as Figure 6.10 but for the Godin-type low-pass filter $A_{25}^2 A_{24}/(25^2 24)$ used to eliminate tidal oscillations in hourly data. *Based on Godin (1972).*

The Godin low-pass filter (Eqn 6.38) effectively removes all daily tidal period energy except for slight leakage in the diurnal frequency band. More precisely, the filter eliminates variability due to the principal mixed diurnal constituent, K_1, for which the amplitude is down by 3.2×10^{-3}, and is only slightly less effective in removing variability due to the declinational diurnal constituent, O_1. The filter represents a marked improvement over the simple A_{24} and A_{25} running-mean filters and Doodson filter once commonly used earlier for tidal analysis (cf. Groves, 1955). The principal failing of the Godin filter is its relatively slow transition between the pass- and stop-bands which leads to significant attenuation of nontidal variability in the range of 2−3 days. This shortcoming of the filter has inspired a number of authors to investigate more efficient techniques for removing the high-frequency portion of oceanographic signals. The cosine-Lanczos filter, the transform filter, and the Butterworth filter are often preferred to the Godin filter, or earlier Doodson filter, because of their superior ability to remove tidal period variability from oceanic signals.

6.7 LANCZOS-WINDOW COSINE FILTERS

As mentioned in Section 6.3.2, transfer functions for ideal (rectangular) filters are formulated in terms of truncated Fourier series. This leads to overshoot ripples (Gibbs' phenomenon) near the cutoff frequency with subsequent leakage of unwanted signal energy into the pass-band. *Lanczos-window cosine filters* are reformulated rectangular filters which incorporate a multiplicative factor (the *Lanczos window*) in rectangular filters to ensure more rapid attenuation of the overshoot ripples. A variety of other windows can also be used. The terms *Lanczos−cosine filter* and *cosine−Lanczos filter* are commonly used names for a family of filters using windows to reduce the side-lobe ripples. Owing to their simplicity and favorable characteristics, these filters have gained considerable popularity among physical oceanographers over the years (Mooers and Smith, 1967; Bryden, 1979; Freeland et al., 1986).

6.7.1 Cosine Filters

We start with an ideal, low-pass filter with transfer function

$$W(\omega) = 1, 0 \leq |\omega| \leq \omega_c$$

$$= 0, \text{elsewhere} \tag{6.39}$$

and assume that the function $W(\omega)$ is periodic over multiples of the Nyquist frequency domain $(-\omega_{Nq}, \omega_{Nq})$. Written as Fourier series, the response function is

$$W(\omega) = \frac{a_o}{2} + \sum_{k=1}^{M} [a_k \cos(\omega k \Delta t) + b_k \sin(\omega k \Delta t)] \tag{6.40}$$

where we have truncated the series at $M \ll N$; as usual, N is the number of data points to be processed by the filter. To eliminate any frequency-dependent phase shift, we require that $W(\omega) = W(-\omega)$, whereby $b_k = 0$. The resulting *cosine filter* has the frequency response

$$W(\omega) = w_o + \sum_{k=1}^{M} w_k \cos(\pi k \omega / \omega_{Nq}) \tag{6.41}$$

where coefficients $w_k \left(= \frac{1}{2} a_k \right)$ are given by

$$w_k = \frac{1}{\omega_{Nq}} \int\limits_{o}^{\omega_{Nq}} W(\omega) \cos\left(\pi k \omega / \omega_{Nq} \right) d\omega \tag{6.42}$$

with $k = 0, 1, ..., M$. The weighting terms w_k are those which determine the output series $\{y_n\}$ for given $\{x_n\}$. We assume that M is sufficiently large that $W(\omega)$ is close to unity in the pass-band and near zero in the stop-band.

For a low-pass cosine filter, $0 \leq |\omega| \leq \omega_c$ defines the bounds of the integral (Eqn 6.42) and the weights are given by.

$$w_k = \frac{\omega_c}{\omega_{Nq}} \frac{\sin\left(\pi k \omega_c / \omega_{Nq} \right)}{\pi k \omega_c / \omega_{Nq}}, k = 0, \pm 1, ..., \pm M \tag{6.43}$$

for which $w_0 = \omega_c / \omega_{Nq}$. The corresponding weights for a high-pass filter, $|\omega| > \omega_c$, are

$$w_o = 1 - \omega_c / \omega_{Nq}, k = 0 \tag{6.44}$$

$$w_k = \frac{-\omega_c}{\omega_{Nq}} \frac{\sin\left(\pi k \omega_c / \omega_{Nq} \right)}{\pi k \omega_c / \omega_{Nq}}, k = \pm 1, ..., \pm M \tag{6.45}$$

That is, w_o (*high pass*) $= 1 - w_o$ (*low pass*), while for $k \neq 0$, the coefficients w_k are simply of opposite sign. The functions (Eqns 6.43 and 6.45) are identical to those discussed in the context of Gibbs phenomenon. Thus, the cosine filter is a poor choice for accurately modifying the frequency content of a given record based on preselected stop- and pass-bands. As an example of the response of this filter, Figure 6.12 presents the transfer function

$$W(\omega) = 0.4 + 2 \sum_{k=1}^{9} [\sin(0.4k\pi) / k\pi] \cos(k\omega) $$

for a low-pass cosine filter with $\omega_c / \omega_{Nq} = 0.4$ and $M = 10$ terms. This filter response is compared to the ideal low-pass filter response and to the modified cosine filter using the Lanczos window (with sigma factors) discussed in the next section.

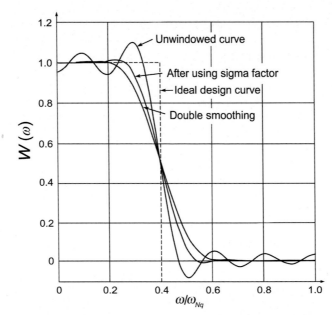

FIGURE 6.12 Approximations to the frequency response, $W(\omega)$, of an ideal low-pass filter (dashed line). Solid curves give the amplitudes of the frequency response for an unwindowed cosine filter, a Lanczos−cosine filter that uses sigma factors, and the response after double application of the Lanczos−cosine filter. Filters use $M = 10$ Fourier terms and $\omega_c = 0.4 \; \omega_{Nq}$; $\omega_{Nq} =$ Nyquist frequency. Gibbs effect is reduced by the sigma factors of the Lanczos window. *After Hamming (1977).*

6.7.2 The Lanczos Window

Lanczos (1956) showed that the unwanted side-lobe oscillations of the form $\sin(p)/p$ in Eqns (6.43) and (6.45) could be made to attenuate more rapidly through use of a smoothing function or window. The window consists of a set of weights that successively average the (constant period) side-lobe fluctuations over one cycle, with the averaging period determined by the last term kept or the first term ignored in the Fourier expansion (Eqn 6.45). In essence, the window acts as a low-pass filter of the weights of the cosine filter. The Lanczos window is defined in terms of the so-called *sigma factors* (cf. Hamming, 1977)

$$\sigma(M, k) = \frac{\sin(\pi k/M)}{\pi k/M} \tag{6.46}$$

in which M is the number of distinct filter coefficients, w_k, $k = 1, ..., M$, and $\omega_M = (M - 1)/M$ is the frequency of the last term kept in the Fourier expansion. Multiplication of the weights of the cosine filter by the sigma factors yields the desired weights of the Lanczos-window cosine filter. Thus, the weights of the low-pass cosine—Lanczos filter become, using $\sigma(M, 0) = 1$

$$w_o = \frac{\omega_c}{\omega_M}, \text{ for } k = 0 \tag{6.47a}$$

$$w_k = \frac{\omega_c}{\omega_{Nq}} \frac{\sin(\pi k \omega_c/\omega_{Nq})}{\pi k \omega_c/\omega_{Nq}} \sigma(M, k) \tag{6.47b}$$

for $k = \pm 1, ..., \pm M$ and $M \ll N$. The corresponding weights for the high-pass Lanczos—cosine filter are

$$w_o = 1 - \frac{\omega_c}{\omega_{Nq}}, \text{ for } k = 0 \tag{6.48a}$$

$$w_k = -\frac{\omega_c}{\omega_{Nq}} \frac{\sin(\pi k \omega_c/\omega_{Nq})}{(\pi k \omega_c/\omega_{Nq})} \sigma(M, k) \tag{6.48b}$$

for $k = \pm 1, ..., \pm M$ and $M \ll N$. The transfer function Eqn (6.47b) for a low-pass cosine—Lanczos filter is then

$$W_L(\omega) = \frac{\omega_c}{\omega_{Nq}} \left[1 + 2 \sum_{k=1}^{M-1} \sigma(M, k) \frac{\sin(\pi k \omega_c/\omega_{Nq})}{\pi k \omega_c/\omega_{Nq}} \cos(\pi k \omega / \omega_{Nq}) \right] \tag{6.49}$$

while for the high-pass cosine—Lanczos filter

$$W_H(\omega) = 1 - W_L(\omega) \tag{6.50}$$

Examination of the transfer functions in Figure 6.12 reveals that the side-lobe ripples are considerably reduced by the sigma factors of the Lanczos window. Again, the trade-off is a broadened central lobe, so that, although there is much less contamination from frequencies within the stop-band, the transition of the filter amplitude at the pass-band is less steep than that for the cosine filter. The effect of this smoothing, which represents a long period modulation of the weighting terms w_k in (6.43), can be illustrated numerically by taking a record length $N = 25$ and calculating the filter response $W(\omega_c/\omega_{Nq})$ with and without the sigma factors. This exercise is instructive in other ways in that it emphasizes the effect of truncation errors during the calculations and indicates what happens if ω_c/ω_{Nq} is too near to the ends of the principal interval $0 \leq \omega/\omega_{Nq} \leq 1$. Consider the case $\omega_c/\omega_{Nq} = 0.022$, $N = 25$, and filter truncation at the fourth decimal place. For a high-pass cosine-type filter with no Lanczos window (which we want to have zero amplitude near zero frequency), we find $W(0) = 0.0740$, whereas use of the sigma factors (Lanczos window) yields $W(0) = 0.4015$. With the cutoff frequency so close to the end of the frequency range, the sigma factors clearly degrade the usefulness of the filter. Increasing the record length to $N = 50$ for the same cutoff frequency improves matters considerably; in this case, $W(0) = 0.0527$ and $W(1) = 0.9997$ using the sigma factors.

6.7.3 Practical Filter Design

Design of a low- or high-pass cosine—Lanczos filter begins with specification of: (1) the cutoff frequency and (2) the number M of weighting terms required to achieve the desired roll-off between the stop- and pass-bands. The cutoff frequency is then normalized by the Nyquist frequency, ω_{Nq}, obtained from the sampling interval Δt of the time series. As with other types of filters, it is advantageous to keep the normalized cutoff frequency away from the ends of the principal interval

$$0 \leq \omega / \omega_{Nq} \leq 1 \tag{6.51}$$

The weights w_k are then derived via Eqns (6.47a, 6.47b) and (6.48a, 6.48b).

Using Eqns (6.47a, 6.47b) and (6.49), and assuming an input $\{x_n\}$, $n = 0, 1,..., N - 1$, the output for a low-pass cosine−Lanczos filter with $M + 1$ weights is

$$y_n = \frac{2\omega_c}{\omega_{Nq}} \left[x_n + \sum_{k=1}^{M} F(k)(x_{n-k} + x_{n+k}) \right] \tag{6.52a}$$

in which

$$F(k) = \frac{1}{2} \frac{\sin(\pi k/M)}{\pi k/M} \frac{\sin\left(\pi k\omega_c/\omega_{Nq}\right)}{\pi k\omega_c/\omega_{Nq}} \tag{6.52b}$$

The output time series begins with $y_M = y(M\Delta t)$ corresponding to the first calculable value for the given filter length, M, and the assumption that the input data begin at $x_n = x_o$. That is

$$y_M = \frac{2\omega_c}{\omega_{Nq}} \left[x_M + \frac{1}{2}F(1)(x_{M-1} + x_{M+1}) + \frac{1}{2}F(2)(x_{M-2} + x_{M+2}) + ... + \frac{1}{2}F(M)(x_o + x_{2M}) \right] \tag{6.53}$$

The chosen number of filter coefficients, M, is always a compromise between the desired roll-off of the filter at the cutoff frequency and the acceptable number of data points ($= 2M$) that are lost from the two ends of the record. The greater the number M, the sharper the filter cutoff and the greater the data loss. Repeated (q times) processing of a given record by the same filter generates an increasingly sharper cascade filter response $(W(\omega/\omega_q))^q$ with a corresponding greater loss (qM) of data values from each end of the record. For a high-pass filter, M should be large enough that, in the time domain, the $2M$ weights for the corresponding low-pass filter span "many" periods of the higher frequency oscillations one is attempting to isolate using the filter.

The sum S of the weights w_k in Eqns (6.47a, 6.47b) and (6.48a, 6.48b)

$$S = \sum_{k=0}^{M} w_k \tag{6.54}$$

gives a qualitative measure of the filter performance. An ideal low-pass filter (i.e., one with no truncation or numerical roundoff effects) should give $S = 1$, while an ideal high-pass filter would have $S = 0$. Close proximity to these values indicates a numerically reliable filter.

6.7.4 The Hanning (von Hann) Window

A variety of cosine-type filters are presented in the recent oceanographic literature under the general term of Lanczos−cosine or cosine−Lanczos filters. A popular formulation having widespread application is the 5-day low-pass filter proposed by Mooers and Smith (1967) in a study of continental shelf waves off Oregon. In this study, a Hanning or raised cosine window defined by

$$w_k = \frac{1}{2}[1 + \cos(\pi k/M)], |k| < M$$

$$= 0, |k| > M \tag{6.55}$$

replaces the sigma factors in Eqn (6.48b).

Let x_n, where $n = 1, 2, ..., N$, denote an hourly digital time series and $2M + 1 = 120$ be the total number of weights spanning a period of 120 h (5 days). The hourly output $\{y_n\}$ from the filter is then

$$y_n = \frac{1}{A} \left[x_n + \sum_{k=1}^{60} F(k)(x_{n-k} + x_{n+k}) \right] \tag{6.56a}$$

where

$$F(k) = \frac{1}{2}[1 + \cos(\pi k/60)]\frac{\sin(p\pi k/12)}{(p\pi k/12)} \tag{6.56b}$$

and

$$A = 1 + 2\sum_{k=1}^{60} F(k) \tag{6.56c}$$

is the normalization factor. Once the number of filter weights k is specified (here, $k = 60$), the transfer function $W_L(\omega)$ is determined by the parameter, p, the half-amplitude frequency of the filter in cycles per day (cpd). Specifically, we find

$$W_L(\omega) = \frac{1}{A}\left[1 + 2\sum_{k=1}^{60} F(k)\cos\left(\frac{\pi k\omega}{\omega_{Nq}}\right)\right] \tag{6.57}$$

in which F and A are given by Eqns (6.56b) and (6.56c).

Comparison of Eqn (6.56b) with Eqn (6.52b) shows that

$$p = 12(\omega_c / \omega_{Nq}) = 24f_c \text{ (in cpd)} \tag{6.58}$$

where $f_c = \omega_c/2\pi$ is the cutoff frequency in cycles per hour (cph) and where we have used the Nyquist frequency $f_{Nq} = 0.5$ cph for the hourly sampled data. The arguments of the angles in Eqns (6.52b) and (6.56b) are, therefore, identical. Where the filters differ is in the use of the sigma factors. Whereas the oscillations of $(1 + \cos(\pi k/M))$ are uniform with k, those of $\sin(\pi k/M)/(\pi k/M)$ decay with increasing k, similar to the way we have seen the term, $\sin(\pi k\omega_c/\omega_{Nq})/(\pi k\omega_c/\omega_{Nq})$, decay in amplitude (e.g., see the ripples in Figure 6.12). In this regard, the raised cosine window provides a more severe weighting of the truncated Fourier series than the sigma factors.

The value $p = 0.7$ cpd, corresponding to a cutoff period of 34.29 h, has been commonly used in the design of low-pass Lanczos–cosine filters (cf. Bryden, 1979). Although this produces an acceptable filter response for periods of 2 days and longer (where 2 days is generally the central period of the oceanic "spectral gap"), it has been shown to pass an unacceptable amount of high-frequency energy from the diurnal band, particularly from the O_1 and Q_1 tidal constituents (Walters and Heston, 1982). In an attempt to further reduce the leakage from the diurnal band, Mooers and Smith (1967) applied a separate filter to the low-pass filtered data from the $p = 0.7$ cpd filter or "Lancz7" filter (Thompson, 1983; Figure 6.13). Walters and Heston (1982) passed the data twice through the filter to produce the 10-day (Lancz7) filter. This not only results in a significantly improved filter amplitude throughout the diurnal band but also doubles the amount of data lost from the ends of the time series. Thompson (1983) suggested the use of a Lanczos–cosine filter with $p = 0.6$ cpd (the "Lancz6" filter) which equates to a cutoff period of 40 h. The Lancz6 filter essentially removes the leakage from the diurnal band but simultaneously shifts the low-pass portion of the filtered record to periods somewhat in excess of 2 days. The

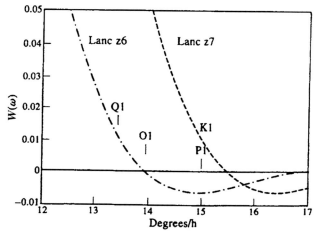

FIGURE 6.13 Expanded views of the filter responses, $W(\omega)$, for two tide-elimination filters for the diurnal frequency band. The Lancz6 and Lancz7 filters are low-pass Lanczos–cosine filters. $15°/h = 1.0$ cpd. *Modified from Thompson (1983).*

difference in the filters is quite subtle. For the Lanczos—cosine filter with $p = 0.7$ (Lancz7 filter), the first zero of the transfer function occurs at 15.4°/h (at 0.0428 cph), which is past the diurnal band (Figure 6.13); for the Lancz6 filter, the first zero is shifted to 14°/h (at 0.0389 cph) near the O_1 frequency of 13.9°/h.

6.8 BUTTERWORTH FILTERS

The windowed cosine filters described in the previous section attempt to approximate an ideal rectangular transfer function using truncated Fourier cosine series. For nonrecursive filters, the output is a simple linear combination of the data and the role of the window is to attenuate the overshoot ripples created by truncation in the time domain (Gibbs phenomenon). We now turn to a specific type of recursive filter for which the transfer function is created using a rational function in sines and cosines. Because this is a recursive filter, the output consists of both input data and past values of the output.

Let $\xi = \xi(\omega)$ be a monotonically increasing rational function of sines and cosines in the frequency, ω. The monotonic function

$$\left| W_L(\omega) \right|^2 = 1 \, / \, \left[1 + (\xi/\xi_c)^{2q} \right] \tag{6.59}$$

(see Figure 6.14) generates a particularly useful approximation to the squared gain of an ideal low-pass recursive filter with frequency cutoff ω_c. The filter design will eventually require $\xi(0) = 0$ so that the final version of $W_L(\omega)$ will closely resemble Eqn (6.59). (Note that we use the variable $\xi(\omega)$ instead of the usual variable notation, $w(\omega)$, to avoid any confusion with filter weights w).

Butterworth filters of the form (Eqn 6.59) have a number of desirable features (Roberts and Roberts, 1978). Unlike the transfer function of a linear nonrecursive filter constructed from a truncated Fourier series, the transfer function of a Butterworth filter is monotonically flat within the pass- and stop-bands, and has high tangency at both the origin ($\omega = 0$) and the Nyquist frequency, ω_{Nq}. The attenuation rate of $W_L(\omega)$ can be increased by increasing the *filter order, q*. However, once again we remark that too steep a transition from the stop-band to the pass-band can lead to ringing effects in the output due to Gibbs phenomenon. Because it has a squared response, the Butterworth filter produces zero phase shift and its amplitude is attenuated by a factor of two at the cutoff frequency, for which $\xi/\xi_c = 1$ for all q. In contrast to nonrecursive filters, such as the Lanczos—cosine filter discussed in the previous section, there is no loss of output data from the ends of

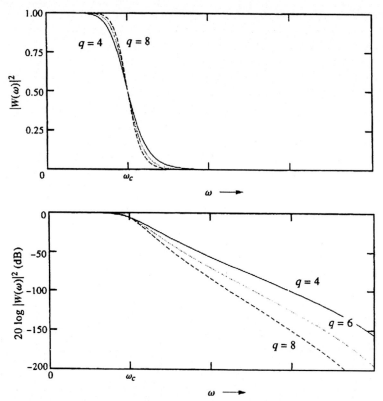

FIGURE 6.14   The frequency response functions $|W(\omega)|^2$ ($= |W_L(\omega)|^2$) for an ideal squared, low-pass Butterworth filter for filter orders $q = 4, 6, 8$. Bottom panel gives response in decibels (dB). Power = 0.5 at the cutoff frequency, ω_c.

the record; N input values yield N output values. However, we cannot expect to get something for nothing. The problem is that ringing distorts the data at the ends of the filtered output. As a consequence, we are forced to ignore output values near the ends of the filtered record, in analogy with the loss of data associated with nonrecursive filters. In effect, the loss is comparable to that from a nonrecursive filter of similar smoothing performance. A subjective decision is usually needed to determine where, at the two ends of the filtered record, the "bad" data end and the "good" data begin.

Butterworth filters fall into the category of physically realizable recursive filters having the time-domain formulation Eqn (6.2) with $k = 0, \ldots, M$. They may also be classified as infinite impulse response filters since the effects of a single impulse input can be predicted to an arbitrary time into the future. To see why we expect $\xi(\omega)$ to be a rational function in sines and cosines, we use Eqn (6.2) and the fact that $W(\omega)$ is the ratio of the output to the input. We can then write

$$W(\omega) = \frac{\text{output}}{\text{input}} = \frac{\sum_{k=0}^{M} w_k e^{-i\omega k \Delta t}}{1 - \sum_{k=1}^{L} g_k e^{-i\omega k \Delta t}} \tag{6.60}$$

where the summations in the numerator and denominator involve polynomials in powers of the form $\exp(-i\omega k \Delta t)$ which can in turn be expressed through the variable ξ. The substitution $z = \exp(i\omega k \Delta t)$ leads to expression of the filter response $W(\omega)$ in terms of the z-transform and zeroes of poles.

6.8.1 High-Pass and Band-Pass Filters

High-pass and band-pass Butterworth filters can be constructed from the low-pass filter (Eqn 6.59). For example, to construct a high-pass filter with cutoff frequency, ω_c, we use the transformation $\xi/\xi_c \to -(\xi/\xi_c)^{-1}$ in Eqn (6.59). The square transfer function of the high-pass filter is then

$$\left| W_H(\omega) \right|^2 = (\xi/\xi_c)^{2q} / \left[1 + (\xi/\xi_c)^{2q} \right] \tag{6.61}$$

where, as required

$$\left| W_H(\omega) \right|^2 = 1 - \left| W_L(\omega) \right|^2 \tag{6.62}$$

Band-pass Butterworth filters (and their counterparts, *stop-band* Butterworth filters) are constructed from a combination of low-pass and high-pass filters. For instance, the appropriate substitution in Eqn (6.59) for a band-pass filter is $\xi/\xi_c = \xi*/\xi_c - (\xi*/\xi_c)^{-1}$ which leads to the quadratic equation

$$(\xi^*/\xi_c)^2 - (\xi/\xi_c)(\xi^*/\xi_c) - 1 = 0 \tag{6.63a}$$

with roots

$$\xi_{1,2}/\xi_c = (\xi/\xi_c)/2 \pm \left[(\xi/\xi_c)^2/4 + 1 \right]^{1/2}. \tag{6.63b}$$

Substitution of $\xi/\xi_c = \pm 1$ (the cut-off points of the low-pass filter) yields the normalized cutoff functions $\xi_1/\xi_c = 0.618$ and $\xi_2/\xi_c = 1.618$ of the band-pass filter based on the cutoff frequency $\pm\omega_c$ of the associated low-pass filter. The corresponding band-pass cutoff functions for the cutoff frequency $-\omega_c$ of the low-pass filter are $\xi_1/\xi_c = -1.618$ and $\xi_2/\xi_c = -0.618$. Specification of the low-pass cutoff determines ξ_1/ξ_2 of the band-pass filter. The bandwidth $\Delta\xi/\xi_c = -(\xi_1 - \xi_2)/\xi_c = 1$ and the product $(\xi_1/\xi_c)(\xi_2/\xi_c) = 1$. Note that specification of ξ_1 and ξ_2 gives the associated function ξ_c of the low-pass filter

$$\xi_1 \xi_2 = \xi_c^2. \tag{6.64}$$

6.8.2 Digital Formulation

The transfer functions (Eqns 6.59−6.62) involve the continuous variable ξ whose structure is determined by sines and cosines of the frequency, ω. To determine a form for $\xi(\omega)$ applicable to digital data, we seek a rational expression with constant coefficients a to d such that the component $\exp(i\omega\Delta t)$ in Eqn (6.60) takes the form

$$\exp(i\omega\Delta t) = \frac{a\xi + b}{c\xi + d} \tag{6.65}$$

(Here, we have replaced $-i\omega\Delta t$ with $+i\omega\Delta t$ without loss of generality.) As discussed by Hamming (1977), the constants are obtained by requiring that $\omega = 0$ corresponds to $\xi = 0$ and that $\omega \rightarrow \pi/\Delta t$ corresponds to $\xi \rightarrow \pm\infty$. Constants b and d (one of which is arbitrary) are set equal to unity. The final "scale" of the transformation is determined by setting $(\omega/2\pi)\Delta t = 1/4$ for $\xi = 1$, which yields

$$\exp(i\omega\Delta t) = \frac{1 + i\xi}{1 - i\xi} \tag{6.66}$$

or, equating real and imaginary parts

$$\begin{aligned} \xi &= \frac{2}{\Delta t}\left[\tan\left(\frac{1}{2}\omega\Delta t\right)\right] \\ &= \frac{2}{\Delta t}[\tan(\pi\omega/\omega_s)], \quad -\omega_{Nq} < \omega < \omega_{Nq} \end{aligned} \tag{6.67}$$

where $\omega_s/(2\pi) = f_s$ is the sampling frequency ($f_s = 1/\Delta t$). We note that the derivation of Eqn (6.67) is equivalent to the conformal mapping

$$\xi = i\frac{2}{\Delta t}\frac{1 - z}{1 + z} \tag{6.68a}$$

where

$$z = e^{2\pi i f\Delta t} = e^{i\omega\Delta t} \tag{6.68b}$$

is the standard z-transform.

The transfer function of the (discrete) low-pass Butterworth filter is then (Rabiner and Gold, 1975)

$$|W_L(\omega)|^2 = \frac{1}{1 + [\tan(\pi\omega/\omega_s)/\tan(\pi\omega_c/\omega_s)]^{2q}} \tag{6.69a}$$

and that of the high-pass Butterworth filter is

$$|W_H(\omega)|^2 = \frac{[\tan(\pi\omega/\omega_s)/\tan(\pi\omega_c/\omega_s)]^{2q}}{1 + [\tan(\pi\omega/\omega_s)/\tan(\pi\omega_c/\omega_s)]^{2q}} \tag{6.69b}$$

The sampling and cutoff frequencies in these expressions are given by $\omega_s = 2\pi/\Delta t$ and $\omega_c = 2\pi/T_c$ in which $T_c = 1/f_c$ is the period of the cyclic cutoff frequency f_c. Plots of Eqn (6.69a) for various cutoff frequencies and filter order q are presented in Figure 6.15.

Use of the bilinear z-transform, $i(1 - z)/(1 + z)$, in Eqn (6.68a) eliminates aliasing errors that arise when the standard z-transform is used to derive the transfer function, these errors being large if the digitizing interval is large. Mathematically, the bilinear z-transform maps the inside of the unit circle ($|z| < 1$, for stability) into the upper half plane. A thorough discussion of the derivation of pole and zeroes of Butterworth filters is presented in Kanasewich (1975) and Rabiner and Gold (1975).

We note that the above relationships define the square of the response of the filter $W(\omega)$ formed by multiplying the transfer function by its complex conjugate, $W^*(\omega) = W(-\omega)$. (In this instance, $W^*(\omega)$ and $W(-\omega)$ are equivalent because $i = \sqrt{-1}$ always occurs in conjunction with ω, as in $i\omega$). The product $W(\omega)W(-\omega)$ eliminates any frequency-dependent phase shift caused by the individual filters and produces a squared and, therefore, sharper, frequency response than produced by $W(\omega)$ alone. The sharpness of the filter (as determined by the parameter q) is limited by filter ringing and stability problems. When q becomes too large, the filter begins to act like a step and Gibbs phenomenon rapidly ensues.

Eqns (6.68a) and (6.68b) are used to design the filter in the frequency domain. In the time domain, we first determine the filter coefficients w_k and g_j for the low-pass filter (Eqn 6.2) and then manipulate the output from the transfer function $W(\omega)$ to generate the output $|W(\omega)|^2$. To obtain the output for a high-pass Butterworth filter, $|W_H(\omega)|^2$, the output from the corresponding low-pass filter, $|W_L(\omega)|^2$, is first obtained and the resulting data values are subtracted from the original input values on a data-point-by-data-point basis.

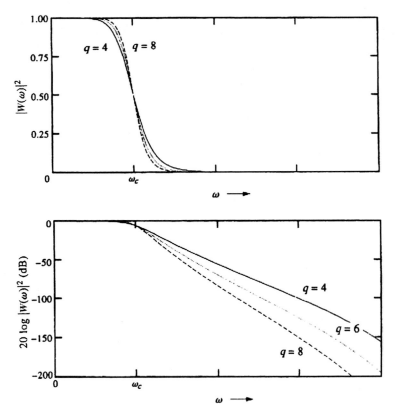

6.8.3 Tangent Versus Sine Filters

Eqns (6.69a) and (6.69b) define the transfer functions of *tangent* Butterworth low-pass filters. The corresponding transfer functions for *sine* Butterworth low-pass filters are given by

$$|W_L(\omega)|^2 = \frac{1}{1 + [\sin(\pi\omega/\omega_s)/\sin(\pi\omega_c/\omega_s)]^{2q}} \tag{6.70}$$

where we have simply replaced tan*x* with sin*x* in Eqns (6.69a) and (6.69b). Although this book deals only with the tangent version of the filter, there are situations where the sine version may be preferable (Otnes and Enochson, 1972). The tangent filter has "superior" attenuation within the stop-band but at a cost of doubled algebraic computation (the sine version has only recursive terms, while the tangent version has both recursive and nonrecursive terms).

6.8.4 Filter Design

The design of Butterworth filters is discussed in Hamming (1977). Our approach is slightly different but uses the same general concepts. We begin by specifying the sampling frequency $\omega_s = 2\pi f_s = 2\pi/\Delta t$ based on the sampling interval Δt for which

$$0 < \omega/\omega_s < 0.5 \tag{6.71}$$

and where the upper limit of 0.5 denotes the normalized Nyquist frequency, ω_{Nq}/ω_s. We next specify the desired cutoff frequency ω_c at the half-power point of the filter. For best results, the normalized cutoff frequency of the filter, ω_c/ω_s, should be such that the transition band of the filter does not overlap to any significant degree with the ends of the sampling domain (Eqn 6.71). Once the normalized cutoff frequency (or frequencies) is (are) known, specification of the filter order q fully determines the characteristics of the filter response. Our experience suggests that the parameter q should be less than 10 and probably not larger than eight. Despite the use of double precision throughout the calculations, roundoff errors and ringing effects can distort the filter response for large q and render the filter impractical.

There are two approaches for Butterworth filter design once the cutoff frequency is specified. The first is to specify q so that the attenuation levels in the pass- and stop-bands are automatically determined. The second is to calculate q based on a required attenuation at a given frequency, taking advantage of the fact that we are working with strictly monotonic functions. Suppose we want an attenuation of $-D$ decibels at frequency ω_a in the stop-band of a low-pass filter having a cutoff frequency $\omega_c < \omega_a$. Using the definition for decibels and Eqn (6.59), we find that

$$q = 0.5 \frac{\log(10^{D/10}-1)}{\log(\xi_a/\xi_c)}$$

$$\approx \frac{\dfrac{D}{20}}{\log\left(\dfrac{\xi_a}{\xi_c}\right)}, \text{ for } D > 10 \tag{6.72}$$

where D is a positive number measuring the decrease in filter amplitude in decibels and ξ is defined by Eqn (6.67). The nearest integer value can then be taken for the filter order provided that the various parameters (ω_a, D) have been correctly specified and q is less than 10. If the latter is not followed, the imposed constraints are too severe and new parameters need to be specified. The above calculations apply equally to specification of q based on the attenuation $-D$ at frequency $\omega_a < \omega_c$ in the stop-band of a high-pass filter, except that $\log(\xi_a/\xi_c)$ in Eqn (6.72) is replaced by $\log(\xi_c/\xi_a)$. Because $\log(x) = -\log(1/x)$, we can simply apply Eqn (6.72) to the high-pass filter, ignoring the minus sign in front of $\log(1/x)$.

6.8.5 Filter Coefficients

Once the characteristics of a transfer response have been specified, we need to derive the filter coefficients to be applied to the data in the time domain. We assume that the transfer function $W_L(\omega; q)$ of the low-pass filter can be constructed as a product, or cascade, of second-order ($q = 2$) Butterworth filters $W_L(\omega; 2)$ and, if necessary, one first-order ($q = 1$) Butterworth filter $W_L(\omega; 1)$. For example, suppose we required a filter of order $q = 5$. The transfer function would then be constructed via the cascade

$$W_L(\omega; 5) = W_L(\omega; 1) \times W_{L,1}(\omega; 2) \times W_{L,2}(\omega; 2) \tag{6.73}$$

in which the two second-order filters, $W_{L,1}$ and $W_{L,2}$, have different algebraic structure. Use of the cascade technique allows for variable order in the computer code for Butterworth filter programs without the necessity of computing a separate transfer function $W_L(\omega; q)$ each time. This eliminates a considerable amount of algebra and reduces the roundoff error that would arise in the "brute-force calculation" of a separate version of W_L for each order.

The second-order transfer functions for a specified filter order q are given by

$$W_L(\omega; 2) = \frac{\left[\xi_c^2(z^2+2z+1)\right]}{a_k z^2 - 2z(\xi_c^2-1) + \left\{1 - 2\xi_c \sin\left[\dfrac{\pi(2k+1)}{2q}\right] + \xi_c^2\right\}} \tag{6.74a}$$

where ξ and z are defined by Eqns (6.67) and (6.68b)

$$a_k = 1 - 2\xi_c \sin[\pi(2k+1)/2q] + \xi_c^2 \tag{6.74b}$$

and k is an integer that takes on values in the range

$$0 \le k < 0.5(q-1) \tag{6.74c}$$

When q is an odd number, the first-order filter $W_L(\omega; 1)$ must also be used where

$$W_L(\omega; 1) = \left(\frac{\xi_c}{1+\xi_c}\right)\left(\frac{z+1}{z - \left(\dfrac{1-\xi_c}{1+\xi_c}\right)}\right) \tag{6.75}$$

Again, suppose that $q = 5$. The transfer function W_L is then composed of the lead filter $W_L(\omega; 1)$ given by Eqn (6.74a) and two second-order filters, for which k takes the values $k = 0$ and 1 in Eqns (6.74a), (6.74b), and (6.74c). Note that we have strictly adhered to the inequality in Eqn (6.74c). The first second-order filter is obtained by setting $k = 0$ in Eqn

(6.74a); the second second-order filter is obtained by setting $k = 1$. For $q = 7$, a third second-order for $k = 2$ would be required, and so on.

The next step is to recognize that the first-order function (Eqn 6.75) has the general form

$$W_L(\omega) = \frac{d_o z + d_1}{z - e_1} \tag{6.76}$$

and that the second-order function (Eqn 6.74a) has the general form

$$W_L(\omega) = \frac{c_o z^2 + c_1 z + c_2}{z^2 - b_1 z - b_2} \tag{6.77}$$

where the sine terms in the coefficients of Eqn (6.74a) change with filter order q. The coefficients d, e in Eqn (6.76) are obtained by direct comparison with Eqn (6.75), while the coefficients b, c in Eqn (6.77) are obtained through comparison with Eqn (6.74a).

The recursive digital filters (Eqn 6.2), whose time-domain algorithms have the transfer functions Eqns (6.76) and (6.77) are, respectively

$$y_n = d_o x_n + d_1 x_{n-1} + e_1 y_{n-1} \tag{6.78}$$

and

$$y_n = c_o x_n + c_1 x_{n-1} + c_2 x_{n-2} + b_1 y_{n-1} + b_2 y_{n-2} \tag{6.79}$$

Direct comparison of Eqn (6.76) with Eqn (6.75) yields the time-domain coefficients for the first-order filter; comparison of Eqn (6.77) with Eqn (6.74a) yields the corresponding coefficients for the second-order filters for each value of k beginning with $k = 0$. In particular, we find, for the first-order filter

$$d_0 = d_1 = \frac{\xi_c}{1 + \xi_c}; e_1 = \frac{1 - \xi_c}{1 + \xi_c} \tag{6.80}$$

and for the second-order filter

$$b_1 = 2(\xi_c^2 - 1)/a_k; b_2 = [(1 + \xi_c^2) - a_k]/a_k$$
$$c_o = \xi_c^2/a_k; c_1 = 2c_o; c_2 = c_o \tag{6.81}$$

where the coefficients in Eqn (6.81) change with the parameter k according to the number of second-order filters needed to create the filter of order q.

To apply the $q = 5$ filter, we process the input data x_n ($n = 0, 1, ..., N$) by the first-order filter (Eqn 6.78). We then take the output from the first-order filter and process it by the first of the second-order filters (Eqn 6.79) with $k = 0$. The resultant output is then processed by the next second-order filter (Eqn 6.79) with $k = 1$. The sequence y_n' ($n = 0, 1, ...$) derived from the three filter applications is the low-pass output for the fifth-order Butterworth filter $W_L(\omega; 5)$, as indicated by Eqn (6.73).

The task is only half complete since our ultimate goal is to remove any filter-induced phase shift by smoothing the data with the squared response of the filter $|W_L|^2$, given by Eqn (6.73). The sequence we require is: $\{x_n\}$ yields $\{y_n'\}$ as the output from $W_L(\omega)$ and $\{y_n'\}$ yields $\{y_n\}$ as the output from $|W_L(\omega)|^2$. To obtain the output $\{y_n\}$ for the square response of the filter, $|W_L(\omega)|^2$, we need to process the output $\{y_n'\}$, obtained from $W_L(\omega)$, with the filter $W_L(-\omega)$. There are three options: (1) We can separately design $W(-\omega)$, a relatively straightforward task involving some sign changes in Eqns (6.74a) and (6.75); (2) we can invert the order of the calculations such that the output $\{y_n'\}$ from $W_L(\omega)$ is passed through the inverted version of this filter. That is, the data from $W_L(\omega)$ are first run through the second-order filter ($k = 1$ for $q = 5$), with the output from this filter passed through the second-order filter ($k = 0$) and finally through the first-order filter; and (3) we can simply invert the chronological order of the data $\{y_n'\}$ and pass the inverted sequence through the original filter $W_L(\omega)$. Since all the data are recorded beforehand, we recommend approach (3). The one caution is that the sequence of the final output must be inverted to regain the original chronological order of the data. In all cases, passing the inverted version of $\{y_n'\}$ through the filter cascade removes any phase shift associated with the first pass which produced $\{y_n'\}$ from $\{y_n\}$. A phase shift $\phi(\omega)$ caused by the first sequence of filters $W_1(\omega) \times W_2(\omega) \times ...$ is canceled by the phase shift $-\phi(\omega)$ due to the second sequence of filters, $W_L(-\omega)$.

Computer programs designed to carry out the Butterworth filter operations should assign the output $\{y'_n\}$ from each filter as the new input $\{x_n\}$ to the next filter in the cascade until the output corresponding to the filter $|W_L(\omega)|^2$ is achieved. The last set of output is then chronologically inverted and rerun through the same filter. Following the final set of calculations, the output sequence is inverted to ensure correct ordering in time.

To obtain the results for a *high-pass* Butterworth filter, one further operation is required. The final output $\{y_n\}$ ($n = 0, 1, \ldots$) from the low-pass filter is subtracted point for point from the original input $\{x_n\}$ ($n = 0, 1, \ldots$) to create the high-pass filtered data $y_n^* = x_n - y_n$. The procedure to obtain the low- and high-pass Butterworth filters is illustrated schematically in Figure 6.16.

FIGURE 6.16 The procedure for filtering data series with low- and high-pass Butterworth filters.

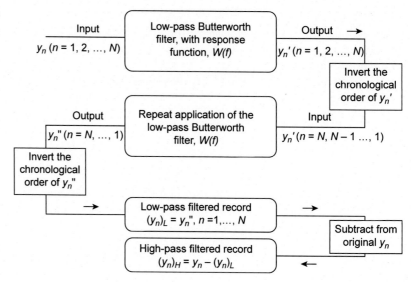

6.9 KAISER–BESSEL FILTERS

As noted in Chapter 5, we consider the Kaiser–Bessel filter among the best filters for processing and analyzing digital oceanographic data. Designed by James Kaiser at Bell Laboratories (Kaiser, 1966), the Kaiser–Bessel filter requires specification of a single parameter, α, and has easy to generate coefficients with high equivalent noise bandwidth, a primary criterion for good digital filter design (note that the parameter $\beta = \pi\alpha$ is sometimes used in place of α to define the filter shape). In the time domain, the M values of the filter weights, $w(m)$, are defined as

$$w(m) = \begin{cases} \dfrac{I_o(\pi\alpha\Omega)}{I_o(\pi\alpha)}, & -\dfrac{M-1}{2} \leq m \leq +\dfrac{M-1}{2} \\ 0, & \text{otherwise} \end{cases} \tag{6.82}$$

where I_o is the zeroth-order modified Bessel function of the first kind, $\alpha > 0$ is an arbitrary real number that determines the shape of the filter, $(M - 1)\Delta t$ is the width of the filter in the time domain for data sampling rate Δt, and

$$\Omega = \left[1 - \left(\frac{2m}{M-1}\right)^2\right]^{1/2} \tag{6.83}$$

The zeroth-order modified Bessel function

$$I_o(x) = \sum_{k=0}^{\infty} \left[\frac{(x/2)^k}{k!}\right]^2 \tag{6.84}$$

has a maximum value of unity at the origin ($I_o(0) = 1$) and oscillates much like the cosine function with a decay rate proportional to $1/\sqrt{x}$, although the roots are not generally periodic, except asymptotically for large x. Insertion of Eqns (6.83) and (6.84) in Eqn (6.82) shows that the filter impulse response peaks at $m = 0$, where $w(0) = 1$, and decays to either side of the central peak. The frequency response of the filter, $W(\omega)$, is obtained from the discrete Fourier transform (DFT) of Eqn (6.82) and is approximated by

$$W(\omega) \approx \frac{(M-1)\Delta t}{I_o(\pi\alpha)} \frac{\sinh\left\{\pi\left[(\alpha)^2 - ((M-1)\omega\Delta t/2\pi)^2\right]^{\frac{1}{2}}\right\}}{\pi\left[(\alpha)^2 - ((M-1)\omega\Delta t/2\pi)^2\right]^{\frac{1}{2}}} \tag{6.85}$$

where $\omega = 2\pi f$, is the angular frequency and the filter length $(M - 1)\Delta t = 1/f_o$ is the inverse of the fundamental frequency, f_o, of the filter. Thus, the product $(M - 1)\omega\Delta t/2\pi = f/f_o$ gives the normalized frequencies for the filter; here, $((M - 1)\Delta t)f$ is referred to as the "DFT bin length". For the purposes of filter design, the modified Bessel function I_o is defined in terms of a Taylor series expansion about $x = 0$ as follows:

$$\text{For}\,|x| \le 3.75 : I_o(x) = 1.0 + 3.5156229\,Z + 3.0899424 Z^2 + 1.2067492\,Z^3$$

$$+ 2.659732 \times 10^{-1} Z^4 + 3.60768 \times 10^{-2} Z^5 + 4.5813 \times 10^{-3} Z^6 \tag{6.86a}$$

where for real x,

$$Z = (x/3.75)^2. \tag{6.86b}$$

$$\text{For}\,|x| > 3.75 : I_o(x) = \left(\exp(|x|)/|x|^{\frac{1}{2}}\right)\left\{3.9894228 \times 10^{-1} + 1.328592 \times 10^{-2} Z\right.$$

$$+ 2.25319 \times 10^{-3} Z^2 - 1.57565 \times 10^{-3} Z^3 + 9.16281 \times 10^{-3} Z^4 - 2.057706 \times 10^{-2} Z^5$$

$$+ 2.635537 \times 10^{-2} Z^6 - 1.647633 \times 10^{-2} Z^7 + 3.92377 \times 10^{-3} Z^8\left.\right\} \tag{6.87a}$$

where

$$Z = 3.75/|x| \tag{6.87b}$$

Examples of the filter weights and their corresponding DFT are presented in Figure 6.17 for two widely separated values of the parameter α.

Recall that the fundamental goal in filter design is to reproduce, as best as possible, an ideal steplike filter whose frequency response is equal to 1 throughout the pass-band and equal to 0 throughout the stop-band. The ideal filter impulse response is then derived by taking the DFT of the ideal frequency response. Because the DFT of a step function leads to an infinite number of filter weights, truncation of the filter is needed, leading to a trade-off between filter width and side-lobe attenuation. As illustrated by Figure 6.17, varying the parameter α permits trade-offs between the width of the main lobe of the pass-band and the amplitudes of the side lobes within the stop-band of the filter response. As α increases, the main lobe increases in width, while the side lobes decrease in amplitude, with the filter taking on a Gaussian shape for large α in both the time and frequency domains. Other filters (such as the running-mean or rectangular filter) have much steeper and confined central lobes but weakly attenuated side lobes compared to the Kaiser−Bessel filter. Because of the highly reduced side-lobe contamination of the Kaiser−Bessel filter, spectral estimates obtained using data segment with 50% overlaps retain near-statistical independence. The Kaiser−Bessel window can be made to approximate other windows by varying the α parameter (Table 6.1). A comparison between the filter weights, w_m, and the corresponding frequency response, $W(f)$, of the Kaiser−Bessel filter and several common filters (windows) is presented in Figure 6.18.

6.9.1 A Low-Pass Kaiser−Bessel Filter

Many oceanic studies require daily mean time series from which the effects of diurnal and semidiurnal tides, inertial oscillations, internal waves, and other "high"-frequency motions have been removed. Except at latitudes less than 30° (the

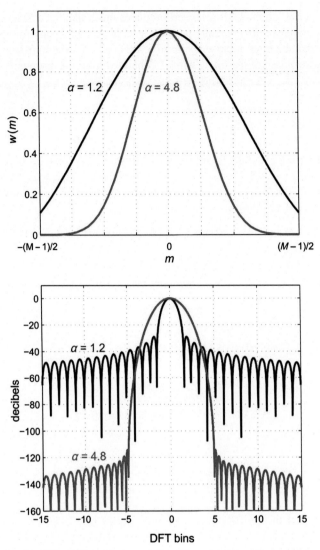

FIGURE 6.17 Frequency response function $W(f) \equiv W(m)$ and corresponding filter weights (impulse response) w_m for a Kaiser–Bessel filter defined by Eqn (6.82) for two values of α (= 1.2 and 4.8) and $M = 31$.

TABLE 6.1 Shapes of Kaiser–Bessel filters relative to other types of filters for different values of the Kaiser-Bessel filter parameters α and $\beta = \pi\alpha$.

Alpha (α)	Beta (β)	Type of window
0	0	Similar to Rectangular
1.6	5	Similar to Hamming
1.9	6	Similar to Hanning
2.7	8.6	Similar to Blackman

Windows

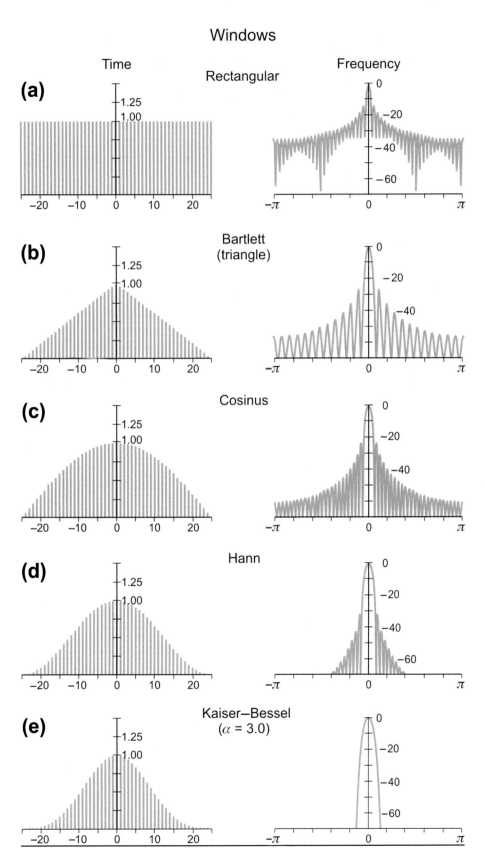

FIGURE 6.18 Comparison of the filter weights (impulse response) w_m and corresponding frequency response function $W(f) \equiv W(m)$ for a Kaiser–Bessel filter (window) and several commonly used filters. All filters have a width of 50 units and are scaled to a maximum $w_0 = 1$. *Courtesy, Alexander Rabinovich, Institute of Ocean Sciences and P.P. Shirshov Institute of Oceanology.*

so-called "turning latitude"), where the inertial period exceeds the diurnal tidal period, removal of the above high-frequency constituents requires low-pass filters with cutoff periods capable of eliminating variations with periods less than roughly 25 h and specifically for diurnal K_1 and O_1 tidal motions with periods of 23.934 and 25.819 h, respectively (frequencies $f = \omega/2\pi = 1.00276$ and 0.92955 cpd, respectively). For this application, we recommend the use of a low-pass Kaiser–Bessel filter with cutoff periods between 25 and 50 h. Such low-pass filters ensure highly suppressed side lobes and, therefore, negligible contamination from higher frequency motions. We can illustrate how well low-pass Kaiser–Bessel filters reproduce an ideal low-pass filter whose stop-band is meant to remove daily tidal variations by examining the characteristics of the Kaiser–Bessel filter impulse amplitude and frequency response for commonly used values of α of 2.0, 2.5, 3.0, and 3.5 and filter lengths M of 25, 31, 37, and 49 hourly values for data series sampled at $\Delta t = 1$ h. Plots of the Kaiser–Bessel filter impulse response amplitudes, $w_m = w(m\Delta t)$, for $-(M-1)/2 \leq m \leq (M-1)/2$ and the corresponding frequency response, $W(f)$, are presented in Figure 6.19. The filter characteristics are listed in Table 6.2 and the filter weights, in Table 6.3. The normalization coefficient, γ, is the value that multiplies each of the filter weights in order that the sum of the weights is equal to unity, viz.

$$\sum_{m=-(M-1)/2}^{(M-1)/2} \gamma w_m = 1 \tag{6.88}$$

Several general trade-off factors emerge from the above results: (1) all the filters achieve high side-lobe attenuation, with the first side lobe diminished by over -45 dB (a factor of $10^{-4.5} \cong 1/32{,}000$); (2) increasing α for a given filter length M increases the width of the filter pass-band (allowing greater leakage from signals in the ideal stop-band lying close to the cutoff period of 25 h, frequency $f = 0.96$ cpd), while at the same time enhancing attenuation of the side lobes (thereby greatly reducing leakage of higher frequency components located in the ideal stop-band); and (3) increasing the filter length, M, causes the filter cutoff frequency (defined as the 1/2 amplitude point or -3-dB level of the frequency response) to shift slightly into the pass-band away from the ideal cutoff period of 0.96 cpd (a negative effect), while greatly attenuating the diurnal K_1 and O_1 tidal signals within the stop-band (a positive effect). Semidiurnal motions with periods well into the stop-band are highly suppressed by all filters. Following low-pass filtering, the hourly record can be decimated to 24-h samples to obtain the daily mean time series.

6.10 FREQUENCY-DOMAIN (TRANSFORM) FILTERING

The type of digital filtering discussed in the previous sections involves convolution of the time series data with weighting functions called *impulse response functions* that eliminate selected ranges of frequencies from the data. In the case of Fourier transform filtering, the weights are defined in terms of a Fourier transform window (frequency response function, FRF), $W(\omega)$, and filtering involves: (1) taking the FFT (Fast Fourier Transform) of the original data set, (2) multiplying the FFT output by the appropriate form of $W(\omega)$ that lets through the frequencies of interest and blocks all the others, and (3) taking the inverse FFT of the result to get back a filtered data set in the time domain. These steps are shown schematically in Figure 6.20. As an example, $W(\omega)$ might be a low-pass filter designed to eliminate frequency components with periods $2\pi/\omega$ that are longer than 40 h. Alternatively, $W(\omega)$ could be a "notch" filter used to isolate oscillations centered near the local Coriolis frequency, or a two-notch filter designed to remove energy in the diurnal and semidiurnal tidal bands. Transform methods have been discussed from an oceanographic perspective by Walters and Heston (1982), Evans (1985), and Forbes (1988). As these papers indicate, the choice of an "appropriate" form for $W(\omega)$ is critical to the success of the method. Filtering in the frequency domain is attractive because of its simplicity compared to convolution in the time domain and because it is conceptually more in accord with our objective in filtering, namely, to remove specific periodicities in the data, while retaining those of interest. Perhaps contrary to expectation, multiplication of the Fourier transform by a window is not always more computationally efficient than convolution of filter weights with the data (Evans, 1985).

We can outline use of the Fourier transform filtering as follows. Suppose we have a time series $\{x(t)\}$ with discrete values $x(n\Delta t) = x_n$, where n is an integer in the range $-N < n \leq N$. The Fourier transform of this time series is

$$X_k = \frac{1}{T} \sum_{n=-N+1}^{N} x_n \exp(-i\omega_k n\Delta t) \tag{6.89}$$

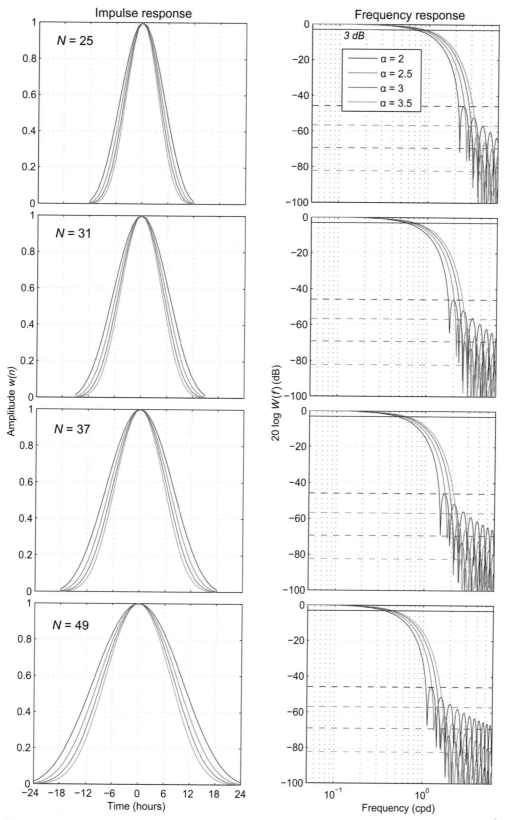

FIGURE 6.19 Frequency response function $W(f)$ (right-hand panels) and corresponding filter weights (impulse response) w_m (left-hand panels) for a low-pass Kaiser–Bessel filter designed for detiding hourly time series. Response functions are shown for four values of α (= 2, 2.5, 3, and 3.5) and four values of the filter length M (= 25, 31, 37, and 49 h). Because it is symmetrical about $f = 0$, only the positive frequencies are shown for $W(f)$. *Courtesy, Alexander Rabinovich, Institute of Ocean Sciences and P.P. Shirshov Institute of Oceanology.*

TABLE 6.2 Low-pass Kaiser–Bessel filter characteristics.

Filter length, M (h)	Alpha (α)	Cutoff (h)	Attenuation (dB)	Attenuation period (h)	Roll-off (h)
25	2	33.0	−46	11.0	16.5
	2.5	30.1	−57	9.1	13.1
	3	27.7	−69	7.7	10.7
	3.5	25.6	−82	6.6	9.0
31	2	41.0	−46	13.8	20.9
	2.5	37.9	−57	11.4	16.3
	3	34.1	−69	9.7	13.5
	3.5	32.0	−82	8.3	11.3
37	2	48.8	−46	16.5	25.0
	2.5	44.5	−57	13.7	19.7
	3	41.0	−69	11.5	16.0
	3.5	37.9	−82	9.9	13.5
49	2	64.0	−46	21.8	33.0
	2.5	60.2	−57	18.3	26.3
	3	53.9	−69	15.3	21.3
	3.5	51.2	−82	13.3	18.0

Columns are as follows: (1) filter length, M (number of hourly values used in the filter); (2) alpha (α) is the filter shape parameter; (3) cutoff is the period at which the frequency response $W(\omega)$ is attenuated by a factor of 2 (the −3-dB level of the maximum response amplitude); (4) attenuation is the amplitude decrease of the first side lobe in the stop-band; (5) attenuation period is the period (inverse frequency) at which the associated attenuation value is achieved; and (6) roll-off is the difference between the lowest frequency at which the "attenuation" is achieved and the cutoff frequency. Cutoff and roll-off are expressed in hours rather than frequency.
Courtesy, Maxim Krassovski, Institute of Ocean Sciences.

where $T = 2N\Delta t$ is the record length and the Fourier frequencies are

$$\omega_k = 2\pi f_k = \frac{2\pi k}{T}, -N < k \leq N. \tag{6.90}$$

Let $w(r\Delta t) = w_r$, $-s \leq r \leq s$, represent a set of filter weights whose sum is unity to preserve the series mean and whose distribution is symmetric about $r = 0$ to preserve the phase information in the data. The number of weights, $S = 2s + 1$, is called the "span" of the filter. Because s points are lost from each end of the input data series, the filtered output series

$$y_n = \sum_{r=-s}^{s} w_r x_{n-r} = \sum_{r=-s}^{s} w_{n-r} x_r \tag{6.91}$$

is shorter than the original series by $2s$ values. The effect of the convolution is to smear the signal $x(t)$ according to the weighting imposed by the impulse response function (IRF), $w(t)$. The FRF or transfer function

$$W(\omega) = \sum_{r=-s}^{s} w_r \exp(-i\omega r\Delta t) = |W(\omega)|\exp(-i\phi(\omega)) \tag{6.92}$$

gives the effect of the IRF on the transform of a sinusoid of unit amplitude and frequency ω ($= 2\pi f$). As stated earlier, the absolute value $|W(\omega)|$ is the *gain factor* of the system and the associated phase angle, $\phi(\omega)$, the *phase factor* of the system. If a linear system is subjected to a sinusoidal input with a frequency ω and produces a sinusoidal output at the same frequency, then $|W(\omega)|$ is the ratio of the output amplitude to the input amplitude and $\phi(\omega)$ is the phase shift between the output and input. The frequency response function is viewed as a window or transfer function that lets through some

TABLE 6.3 Filter weights w(m) for different filter lengths (M, hours) and values of alpha (α) for low pass tide-removing Kaiser–Bessel filter.

Filter length, M (hours)

	25				31				37				49			
Alpha	2	2.5	3	3.5	2	2.5	3	3.5	2	2.5	3	3.5	2	2.5	3	3.5
Normalization coefficient																
	0.085	0.095	0.104	0.112	0.068	0.076	0.083	0.089	0.057	0.063	0.069	0.074	0.043	0.047	0.052	0.056
m																
−24													0.011	0.003	0.001	0.000
−23													0.023	0.007	0.002	0.001
−22													0.038	0.015	0.006	0.002
−21													0.058	0.026	0.011	0.005
−20													0.083	0.041	0.020	0.010
−19													0.113	0.061	0.033	0.018
−18										0.011	0.003	0.001	0.000	0.149	0.087	0.051	0.030
−17									0.027	0.009	0.003	0.001	0.190	0.119	0.075	0.047
−16									0.051	0.021	0.009	0.004	0.236	0.158	0.106	0.071
−15					0.011	0.003	0.001	0.000	0.083	0.041	0.020	0.010	0.287	0.203	0.144	0.102
−14					0.031	0.011	0.004	0.001	0.124	0.069	0.038	0.021	0.343	0.255	0.190	0.141
−13					0.063	0.028	0.013	0.006	0.175	0.108	0.066	0.041	0.403	0.313	0.243	0.189
−12	0.011	0.003	0.001	0.000	0.106	0.057	0.030	0.016	0.236	0.158	0.106	0.071	0.465	0.376	0.305	0.247
−11	0.038	0.015	0.006	0.002	0.164	0.099	0.060	0.036	0.305	0.220	0.158	0.114	0.529	0.444	0.373	0.313
−10	0.083	0.041	0.020	0.010	0.236	0.158	0.106	0.071	0.382	0.293	0.225	0.172	0.594	0.514	0.446	0.386
−9	0.149	0.087	0.051	0.030	0.320	0.234	0.170	0.124	0.465	0.376	0.305	0.247	0.658	0.586	0.523	0.466
−8	0.236	0.158	0.106	0.071	0.415	0.325	0.255	0.200	0.551	0.467	0.396	0.336	0.720	0.658	0.601	0.550
−7	0.343	0.255	0.190	0.141	0.516	0.430	0.359	0.299	0.637	0.562	0.497	0.439	0.779	0.727	0.679	0.634
−6	0.465	0.376	0.305	0.247	0.620	0.543	0.476	0.418	0.720	0.658	0.601	0.550	0.833	0.792	0.754	0.717
−5	0.594	0.514	0.446	0.386	0.720	0.658	0.601	0.550	0.798	0.750	0.705	0.662	0.881	0.851	0.823	0.795
−4	0.720	0.658	0.601	0.550	0.812	0.767	0.724	0.684	0.866	0.833	0.800	0.770	0.923	0.903	0.883	0.864
−3	0.833	0.792	0.754	0.717	0.890	0.862	0.835	0.809	0.923	0.903	0.883	0.864	0.956	0.944	0.933	0.921
−2	0.923	0.903	0.883	0.864	0.950	0.937	0.924	0.911	0.965	0.956	0.946	0.937	0.980	0.975	0.969	0.964

Continued

TABLE 6.3 Filter weights $w(m)$ for different filter lengths (M, hours) and values of alpha (α) for low pass tide-removing Kaiser–Bessel filter.—cont'd

m	0.980	0.975	0.969	0.964	0.987	0.984	0.980	0.977	0.991	0.989	0.986	0.984	0.995	0.994	0.992	0.991
−1	0.980	0.975	0.969	0.964	0.987	0.984	0.980	0.977	0.991	0.989	0.986	0.984	0.995	0.994	0.992	0.991
0	1.000	1.000	1.000	1.000	1.000	1.000	1.000	1.000	1.000	1.000	1.000	1.000	1.000	1.000	1.000	1.000
1	0.980	0.975	0.969	0.964	0.987	0.984	0.980	0.977	0.991	0.989	0.986	0.984	0.995	0.994	0.992	0.991
2	0.923	0.903	0.883	0.864	0.950	0.937	0.924	0.911	0.965	0.956	0.946	0.937	0.980	0.975	0.969	0.964
3	0.833	0.792	0.754	0.717	0.890	0.862	0.835	0.809	0.923	0.903	0.883	0.864	0.956	0.944	0.933	0.921
4	0.720	0.658	0.601	0.550	0.812	0.767	0.724	0.684	0.866	0.833	0.800	0.770	0.923	0.903	0.883	0.864
5	0.594	0.514	0.446	0.386	0.720	0.658	0.601	0.550	0.798	0.750	0.705	0.662	0.881	0.851	0.823	0.795
6	0.465	0.376	0.305	0.247	0.620	0.543	0.476	0.418	0.720	0.658	0.601	0.550	0.833	0.792	0.754	0.717
7	0.343	0.255	0.190	0.141	0.516	0.430	0.359	0.299	0.637	0.562	0.497	0.439	0.779	0.727	0.679	0.634
8	0.236	0.158	0.106	0.071	0.415	0.325	0.255	0.200	0.551	0.467	0.396	0.336	0.720	0.658	0.601	0.550
9	0.149	0.087	0.051	0.030	0.320	0.234	0.170	0.124	0.465	0.376	0.305	0.247	0.658	0.586	0.523	0.466
10	0.083	0.041	0.020	0.010	0.236	0.158	0.106	0.071	0.382	0.293	0.225	0.172	0.594	0.514	0.446	0.386
11	0.038	0.015	0.006	0.002	0.164	0.099	0.060	0.036	0.305	0.220	0.158	0.114	0.529	0.444	0.373	0.313
12	0.011	0.003	0.001	0.000	0.106	0.057	0.030	0.016	0.236	0.158	0.106	0.071	0.465	0.376	0.305	0.247
13					0.063	0.028	0.013	0.006	0.175	0.108	0.066	0.041	0.403	0.313	0.243	0.189
14					0.031	0.011	0.004	0.001	0.124	0.069	0.038	0.021	0.343	0.255	0.190	0.141
15					0.011	0.003	0.001	0.000	0.083	0.041	0.020	0.010	0.287	0.203	0.144	0.102
16									0.051	0.021	0.009	0.004	0.236	0.158	0.106	0.071
17									0.027	0.009	0.003	0.001	0.190	0.119	0.075	0.047
18									0.011	0.003	0.001	0.000	0.149	0.087	0.051	0.030
19													0.113	0.061	0.033	0.018
20													0.083	0.041	0.020	0.010
21													0.058	0.026	0.011	0.005
22													0.038	0.015	0.006	0.002
23													0.023	0.007	0.002	0.001
24													0.011	0.003	0.001	0.000

The normalization factor is the value that multiplies each column of filter weights to ensure that the sum of the weights adds up to unity (i.e., Sum weights (w) = 1). M is the number of filter weights as measured in hours.
Courtesy, Maxim Krassovski, Institute of Ocean Sciences.

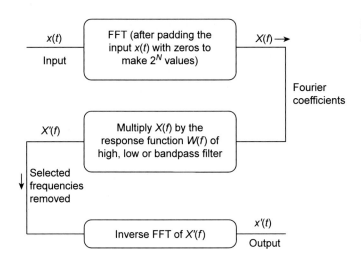

FIGURE 6.20 The procedure for applying discrete Fourier transform (DFT) filters for application in the frequency domain.

frequencies and stops others. Note that W is defined at all frequencies such that $-\pi/\Delta t < \omega \le \pi/\Delta t$, and not just at the Fourier frequencies, ω_k.

The key to Fourier transform filtering is that, for a constant-parameter linear system, the Fourier transform of the filtered data, $Y(\omega)$, is related to the Fourier transform of the input data, $X(\omega)$, through the product

$$Y(\omega) = W(\omega)X(\omega) \tag{6.93}$$

In other words, convolution in the time domain, defined by Eqn (6.91), translates to multiplication in the frequency domain. The merits of a filter are judged by its FRF (frequency domain) and IRF (time domain). We would like the magnitude of the FRF to be near unity in the frequency bands to be passed by the filter and near zero in the bands to be stopped, i.e., $|W(\omega)| \approx 1$ and 0, respectively. The transition band between the stop- and pass-bands should be as narrow as possible since a broad transition band results in a filtered time series whose frequency content may be contaminated by unwanted frequencies. Similarly, the span of the IRF should be short so that the magnitude of weights decays to zero rapidly as r increases toward $\pm s$. If convolution is used, short filters are computationally more efficient and, moreover, result in less data loss. Unfortunately, the two criteria are at odds with one another. In general, the narrower the transition band in the frequency domain, the slower is the decay rate of the IRF in the time domain. Also, the steeper the maximum slope of the transition band, the larger are the side initial side lobes of the IRF that arise from the well-known Gibbs phenomenon. In the limit of a step function-type FRF, in which the transition zone has zero width, the resulting IRF decays very slowly and has large side lobes (ringing). Thus, one must always compromise in specifying an FRF.

In all time-domain filtering (convolution), data are lost from each end of the original digital time series. For example, in the case of nonrecursive filters, in which the output is based on input time series alone, a known segment of the record of length $T/2$ is lost from either end of the time series ($T = (M − 1)\Delta t$ is the filter length). The same applies to recursive filters in which the present output from the filter is based on the original data series as well as previous values of the output. Here, the difficulty is that the amount of data we must discard from either end is not well defined because of ringing effects associated with the convolution and abrupt data discontinuities at the ends of the record. Transform windowing typically results in exactly the same amount of data loss as the equivalent time-domain filter (Walters and Heston, 1982). The Fourier transform treats the data outside the record as if it were zero, so that the ringing at the ends is introduced by the abrupt changes in the series from nonzero to zero and to the circular convolution of the window's IRF with the data. Ringing (Gibbs phenomenon) occurs throughout the entire time series and becomes evident when the filtered FFT data are inverted to recover the desired filtered time series data. The effects of Gibbs phenomenon are mitigated by tapering the frequency-domain filter using a linear or cosine function.

According to Thompson (1983), careful construction of weighting functions in the time domain can more effectively remove tidal components than Fourier transform filtering. This is because tidal frequencies do not generally coincide with Fourier frequencies of the record length. Design of IRF weights to minimize the squared deviation from some specified norm (least squares filter design) offers more control over the FRF at particular non-Fourier frequencies. On the other hand, broadband signals are best served by the FRF approach. Evans (1985) suggests that the ratio of convolution cost to windowing cost is $E = S/(2\log_2(N))$, where S is the filter span. If $E > 1$, then windowing in the frequency domain is a more efficient method. Forbes (1988) addressed the problem of removing tidal signals from the data while retaining the near-inertial signal and argues that Fourier transform filtering is effective provided that careful consideration is given to the filter bandwidth and the amount of tapering of the sides of the filter. Note that, in trying to remove strong tidal signals from a data series, it is sometimes beneficial to first calculate the tidal constituents and then subtract the harmonically predicted tidal signal from the data prior to filtering. This is time consuming and not an advantage if the filter is properly designed.

Figure 6.21a shows the energy-preserving power spectrum for a middepth current meter record from a Cape Howe mooring site (37°35′ S, 150°25′ E) off the coast of New South Wales. To remove the strong tidal motions from this record, Forbes first used an untapered DFT with 12 and 17 adjacent Fourier coefficients set to zero in the diurnal and semidiurnal bands, respectively (Figure 6.21b). However, the greatest improvement in the Fourier transform filtering came from setting only three Fourier terms to zero and tapering the filter with a nine-point cosine taper in the frequency domain at the diurnal and semidiurnal frequencies (Figure 6.21c). Tapering the time series, rather than widening the filter by using more zero frequencies, was found to be a better way to improve filter characteristics. Perhaps, the most important conclusion from Forbes' work is that DFT filters are effective if the number of Fourier coefficients set to zero is sufficient to cover the unwanted frequency band and if the filter is cosine-tapered in the frequency domain to ensure a smooth transition to nonzero Fourier coefficients. In the nonintegral single-frequency case presented here (Forbes was looking at near-inertial motions), this amounted to a three-point filter with a nine-point cosine taper. The widths of the filter and taper must be determined for each application by a careful examination of the spectrum for leakage into adjacent frequencies, but once this is done, the technique is fast and simple to apply.

To summarize the use of Fourier transform filtering:

1. Remove any linear trend (or nonlinear trend if it is well defined) from the data prior to filtering but do not be too concerned with cosine tapering the first and last 10% of the data. Then, Fast Fourier transform the data.
2. Define the Fourier transform filter $W(\omega)$ for both positive and negative frequencies with the extreme frequencies given by $\pm 1/(2\Delta t)$ ($\pm f_{Nq}$, the Nyquist frequency).
3. If the measured data are real, and the filtered output is to be real, the filter should obey $W(-\omega) = W^*(\omega)$, where the asterisk denotes complex conjugate. The easiest way to satisfy this condition is to pick $W(\omega)$ real and symmetric in frequency.
4. If $W(\omega)$ has sharp vertical edges then the impulse response of the filter (the response arising from a short impulse as input) will have damped ringing at frequencies corresponding to these edges. If this occurs, pick a smoother $W(\omega)$. Take the fast Fourier transform (FFT) inverse of $W(\omega)$ to see the impulse response of the filter. The more points used in the smoothing, the more rapid the falloff of the impulse response.
5. Multiply the transformed data series $X(\omega)$ by $W(\omega)$ and invert the resultant data series, $Y(\omega)$, to obtain the filtered data in the time domain. To eliminate ringing effects, discard $T/2$ data points from either end of the filtered time series, where T is the span of the impulse response function (IRF) for the transform filter.

6.10.1 Truncation Effects

For all commonly used digital filters, a percentage of the end values from the filtered record must be omitted prior to further analysis. This loss of information from the ends of the output is linked to ringing effects associated with discontinuities at the ends of the input and to the nonexistence of integrable data prior to the start of the record. The ringing decays toward the interior of the data sequence after the end effects have been smoothed by a sufficient number of filter integrations (Figure 6.22). In the case of the squared Butterworth filter, both ends of the data are affected twice since the data are passed forward and backward through the filter. One approach is to assume that 10% of output data at each end of the filter output is contaminated and remove these points from the final output. However, each case is different and data elimination should be based on a trial and error approach using visual inspection to estimate the extent of the data removal. Padding the ends of the input with zeroes appears to serve no useful purpose. In some cases, the ringing effect can be substantially reduced by using the zero cross-over points (for input centered about the mean record value) as the first record of the input.

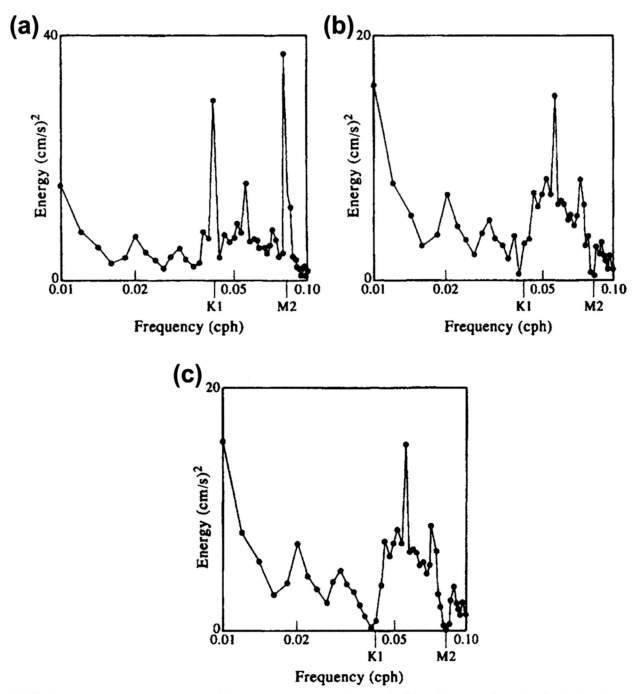

FIGURE 6.21   Energy-preserving spectra for a 4,000-h current meter record at 720 m depth off Cape Howe, Australia. (a) Raw hourly data; (b) after applying a discrete Fourier transform (DFT) filter with 12 and 17 adjacent Fourier coefficients set to zero in the diurnal and semidiurnal bands (no tapering); and (c) after applying a DFT filter with three Fourier coefficients set to zero and nine Fourier coefficients cosine-tapered on each side of the zero coefficients. *From Forbes (1988).*

FIGURE 6.22 Ringing effects following application of different discrete Fourier transform (DFT) filters to an artificial time series with frequency $f = 0.05$ cph and then inverting the transform. (a) Single Fourier coefficient at 0.05 cph set to zero, (b) three Fourier coefficients set to zero, (c) five Fourier coefficients set to zero, and (d) 21 coefficients set to zero. *From Forbes (1988).*

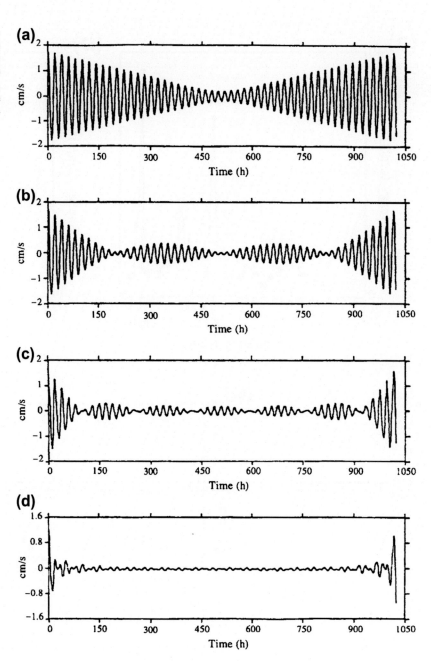

Glossary

cpd Cycles per day
DFT Discrete Fourier Transform
FFT Fast Fourier Transform
FRF Frequency Response Function
IRF Impulse Response Function

Chapter 7

MACHINE LEARNING METHODS

7.1 INTRODUCTION

Machine Learning (ML) applies to a branch of artificial intelligence (AI) whose goal is to create intelligent devices or systems that can automatically and independently assimilate knowledge from data without help from a human analyst. In reality, most ML methods benefit from some guidance from human analyses, but then the ML procedure takes over and performs the bulk of the analysis without further human intervention. These methods range from the basic systems derived from previously discussed statistical estimation methods (such as Maximum Likelihood Estimation) to much more sophisticated systems, such as those that have given rise to "deep learning", an approach where neural networks (NN) are trained to solve a wide variety of scientific and engineering problems. Classical machine learning involves models that can learn from the data presented to it, but which may initially need some human assistance. Even with human guidance (called "supervised learning"), the ML methods are designed to "learn" from application to the data. Neural networks, for example, have nested layers that pass information through hierarchies of these layers, enabling the networks to eventually learn from their mistakes (https://dataconomy.com/2022/09/neural-network-vs-machine-learning/). Thus, neural networks are a form of AI under the umbrella of machine learning that have been made it possible to analyze enormous data sets such as high-resolution satellite images using the ability of modern computers to perform a large number of complicated tasks.

While NNs have progressed from simple single hidden layer systems to complex thousand-layer networks, the biggest recent change has been the introduction of Convolutional Neural Networks (CNNs) that have progressed as a way to handle complex 3D patterns and multi-dimensional satellite images. At the same time, many of the more traditional machine learning methods, including Statistical Trees and Maximum Likelihood Estimation, have been built on and improved, whereby tools such as Random Forest and Support Vector Machines have emerged as techniques to compete with Neural Networks.

In this chapter, we review a number of the statistical estimation methods and discuss several publications that have used the methods in physical oceanography. Chapter 8 then provides a separate discussion on neural networks and convolutional neural networks, and outlines how they have been used in oceanographic studies. This chapter and Chapter 8 are meant to serve as introductions to machine learning methods and neural networks. Readers wanting to pursue these fields in greater depth are encouraged to search the literature for these methods and learn more about how they function.

7.2 MAXIMUM LIKELIHOOD ANALYSES

Machine learning methods are closely related to the Maximum Likelihood Estimation (MLE) procedure discussed in Section 3.11.2. The decision-making criterion is to maximize the likelihood or probability of something occurring. The estimation application amounts to making a probability statement and then maximizing it, usually by taking derivatives and setting them to zero. In other marine science applications, MLE is used to select parameters of model functions in order to solve a given oceanographic problem.

7.2.1 Directional Wave Spectra

Wyatt et al. (1997) uses the MLE method to extract parameters of two-parameter models of the directional spreading of short (0.53 Hz) wind waves derived from the power spectrum of high-frequency (HF) radar backscatter data. These wind-waves have a wavelength that is half the radio wavelength, which gives them their frequency of 0.53 Hz. The parameters that need to be estimated are the short-wave direction, which can be taken as the wind direction, and the directional spreading angle. The spreading angle is model dependent. For the data analyzed in Wyatt et al. (1997), the results indicate that the directional spreading model of Donelan et al. (1985) gives a better description of the spreading than does the $\cos^S$ model, where S is the ocean wave directional spectrum at the Bragg-matched oceanic wavenumber. The results from the HF radar were compared with a similar analysis carried out on wave-rider buoy measurements (Figure 7.1).

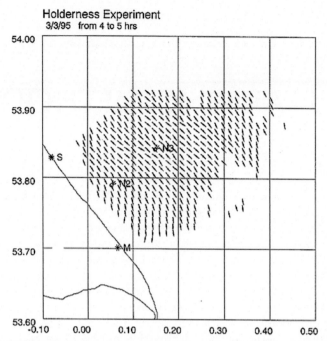

FIGURE 7.1 Map showing location of the Holderness experiment on the northeast coast of England and the radar coverage during the second phase of the experiment in March 1995. The two radar sites are shown: M = master site, S = slave site. The land shown in the lower left of the figure is the Humberside coast north of the Humber estuary, which is at the lower left-hand corner. One of two possible short-wave directions at each cell measured by the master radar is shown with an arrow. The alternative directions are the reflection of those shown about the radar beam direction (the line from M to the point) at each point. The positions N2 and N3 of the directional wave buoys are also shown. *From Wyatt et al. (1997).*

In their study, Wyatt et al. (1997) found that, at some locations and under some meteorological conditions, there are significant differences between the single-radar direction estimates using the MLE model. These differences are not apparent in Figure 7.1, but this is one of the best cases; in other cases, the spatial agreement was not as close. The authors conclude that, if the two radars are indeed measuring the same features, then the differences suggest that the model is not a good description of the directional wave spectrum at these frequencies. There is some evidence that side-lobes corrupt the peaks, but this is not likely the major cause of this discrepancy. Other sources for the discrepancy, such as a difference in signal-to-noise of the two radars are possible, but again this does not seem to be the major cause of this disagreement in surface vectors.

The MLE method used in Wyatt et al. (1997) was that described in Cox and Hinkley (1974), which takes into account the statistical characteristics of the Bragg ratio measurement $r' \approx r F_{\nu,\nu}$, where r' is the observed Bragg ratio, r is the true Bragg ratio (which is assumed to be the ratio obtained using the model), and $F_{\nu,\nu}$ is the F-distribution with ν known degrees of freedom for the Doppler spectral estimates (see also Barrick (1980) and Sova (1995)). The probability density function (PDF) of the F-distribution for two independent random variables with each variable having ν degrees of freedom is given by

$$f\left(\frac{r'}{r}; \nu, \nu\right) = \frac{\Gamma(\nu)}{\left[\Gamma\left(\frac{\nu}{2}\right)\right]^2} \frac{F^{(\nu/2)-1}}{(1+F)^\nu} \tag{7.1a}$$

from which we obtain the PDF for r' as

$$f(r'; \nu, \nu) = \left(\frac{1}{r}\right) \frac{\Gamma(\nu)}{\left[\Gamma\left(\frac{\nu}{2}\right)\right]^2 \left(1+\frac{r'}{r}\right)^\nu} \left(\frac{r'}{r}\right)^{(\nu/2)-1} \tag{7.1b}$$

where Γ is the Gamma function.

In the maximum likelihood method, the goal is to maximize the likelihood of a set of observations with respect to the parameters of a given model. Thus, we seek to maximize the expression $\sum_{i=1}^{N} \ln(f(r'))$, where the subscript, i, denotes one of the N measurements to be included, each of which has a different beam direction, ϕ_i. Using (7.1b), the MLE operation we maximize is the likelihood L, where

$$L = \sum_{i=1}^{N} \left\{ \nu \ln\left(1 + \frac{r_i'}{r}\right) - \left(\frac{\nu}{2} - 1\right) \ln\left(\frac{r_i'}{r}\right) - \ln[\Gamma(\nu)] + 2\ln\left[\Gamma\left(\frac{\nu}{2}\right)\right] + \ln(r_i') \right\} \tag{7.2}$$

Using this method, Wyatt et al. (1997) provides a "dual radar analysis" with the standard cosine model (Figure 7.2).

Wyatt et al. (1997) conclude that the MLE method applied to the HF radar data yielded excellent wave directions that agree well with measurements from a single wave rider buoy. Directional spreading results showed some agreement but were overall less convincing than wave direction, with concern that some of the problem may lie with the buoy measurements. Overall, the Donelan et al. (1985) directional spreading model was found to provide a better description of the wave spreading than the standard cosine model.

7.2.2 Maximum Likelihood Estimators for Physical Oceanographic Equations

The study by Piterbarg and Rozovskii (1996) examined parametric models of the upper ocean variability described by a certain class of stochastic partial differential equations. They used the Maximum Likelihood (MLE) method to solve for the unknown model parameters using both discrete and continuous observations. The study's findings demonstrate that the diffusivity estimate in the advection-diffusion equation is always consistent and that the consistency of the feedback parameter estimate depends upon the spatial dimension. In addition, the MLE estimate of the velocity is consistent if and only if the diffusivity is zero. The authors discuss their estimate of the horizontal diffusivity and bottom friction for the linearized vorticity equation to demonstrate their results.

The Piterbarg and Rozovskii (1996) study was cast as an approach to the general problem of "data assimilation", whereby oceanographers fit a model to observations in some optimal way in order to estimate some unknown model characteristics. In their study, the authors use the MLE approach to derive solutions for their model parameters. They concentrate on some commonly used equations in physical oceanography such as the heat transport equation and the linearized equation for quasigeostrophic motion in the upper ocean. Both equations can be written in the form

$$\frac{\partial u}{\partial t} + A_\theta u = S(t; \mathbf{x}). \tag{7.3}$$

Here, $u = u(t; \mathbf{x})$ is the observed quantity (e.g., temperature, sea level height, etc.) at point $\mathbf{x}$ and time t, and A_θ is a linear operator that governs advection, diffusion, and other dissipation terms. The latter depends on an unknown vector

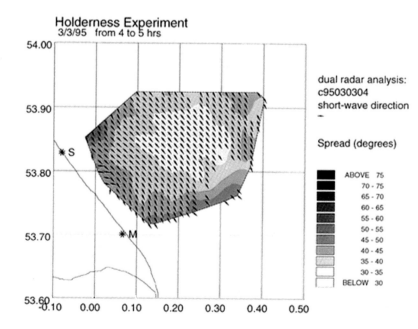

FIGURE 7.2   A dual radar analysis using the standard cosine model. The vectors represent the wave directions and the contours correspond to the spreading angle in degrees, as indicated by the legend. Figure 7.3 shows the improvement resulting from application of the Donelan et al. (1985) model using the same set of data. *From Wyatt et al. (1997).*

FIGURE 7.3 As in Figure 7.2 except using the Donelan et al. (1985) wave model. *From Wyatt et al. (1997).*

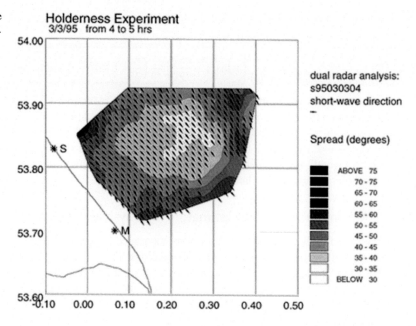

parameter $\theta = (\theta_1, ..., \theta_k)$, which are quantities such as velocity, the diffusion coefficient and the friction coefficients. Finally, $S(t;\mathbf{x})$ is a random field that represents the distributed sources (e.g. heat flux and wind stress), or model noise.

The authors then explore the spatial structure of their solution, assuming an arbitrary orthogonal basis $[\varphi_m(\mathbf{x})]_1^\infty$. They also assume that only the first M amplitudes are observed over the time period, T, so that

$$u_1(t), ..., u_M(t), \qquad t \in [0, T] \tag{7.4}$$

and introduce the spatial Fourier expansion (with separation of variables)

$$u(t; \mathbf{x}) = \sum_m u_m(t) \varphi_m(\mathbf{x}) \tag{7.5}$$

In these equations, Piterbarg and Rozovskii (1996) include data that are both continuous and discrete in time. They use the classical maximum likelihood approach to solve for their unknown quantities. They then discuss how these solutions could be extended to infinite-dimensional systems like stochastic Partial Differential Equations (PDEs).

Before introducing their specific methodology, the authors felt they needed to pose several questions that they intended to answer in their study:

1. Why can/should we view the equations of physical oceanography as stochastic PDEs?
2. What is the utility of the spectral representation of oceanographic data?
3. What is a good estimator and why are MLE estimators "good"?

The equations of physical oceanography are phenomenological equations representing an approximate macroscopic description of ocean dynamics. An analysis of the deeper levels to understand ocean dynamics is much more difficult than is intended in the present analysis. Therefore, to account for the neglected microscopic nature of the physical system, together with measurement error and other fine structure, one can consider appropriate perturbations of the macroscopic equations. One class of such models is represented by random perturbations of the forcing terms, as well as the initial and boundary conditions, which naturally lead to stochastic PDE models.

A number of studies (Frankignoul and Reynolds, 1983; Frankignoul et al., 1993; Fiadeiro and Veronis, 1984; Kelly, 1989; Ghil and Malanotte-Rizzoli, 1991; Ostrovskii and Piterbarg, 1985; Herterich and Hasselman, 1987) have applied similar methods to mapping sea surface temperature and ocean surface velocity. Despite differences in the approach to the equations, the results were considered quite similar. In the northeast Pacific, the analysis revealed circulations that were very similar to those inferred from ship-drift observations. In contrast, in the northwest Pacific, the equation solutions did not exhibit advection patterns consistent with the observed circulation in the subtropical gyre and the Kuroshio extension. The reason for this lack of agreement was due to the poor spatial resolution of the temperature data used for their analysis. In both Ostrovskii and Piterbarg (1985) and Herterich and Hasselman (1987), the Namias sea surface temperature product was used, with its coarse $5° \times 5°$, spatial resolution based on ship measurements (Namias and Cayan, 1981).

Piterbarg and Rozovskii (1996) state that the general problem of parameter estimation can be formulated in a straightforward manner: given the chosen model and the observed data, find the optimal values of the unknown parameters that satisfy the imposed restrictions. While there is a multitude of possible approaches, the problem reduces to the minimization of some misfit functional of the data and the parameters with respect to the latter. The misfit functional is a measure of the discrepancy between the model and the data. It may be chosen in a variety of ways. Some examples of misfit functionals popular in the oceanographic literature were mentioned above, while others can be found in Kelly (1989), Tsiperman and Thacker (1989), and Malanotte-Rizzoli (1996). Most misfit functionals lead to some form of least squares procedure.

Piterbarg and Rozovskii (1996) further state that, while the task of minimization of a particular misfit functional might be technically very difficult, a more difficult (if at all solvable) problem is the comparison between different "optimal" estimates. In many cases, an optimal estimate simply does not exist (at least for some fixed sample sizes). In these cases, an asymptotic approach to the statistics of the solutions gives a convenient framework for comparison and classification of estimators. This asymptotic approach only becomes reasonable for relatively large samples. Fortunately, this is not a critical limitation as modern satellite data sets allow for simultaneous measurements of sea surface temperature (SST) and sea surface height (SSH) at thousands of grid points.

The Piterbarg and Rozovskii (1996) paper concentrates on the MLE approach to parameter estimation. This approach has a twofold advantage: firstly, the MLE estimates are usually simple and computationally effective; and secondly, they often provide explicit criteria for consistency, asymptotic normality and asymptotic efficiency. As shown by (7.4), the model of observations consists of two parameters related to the volume of the sample, the number of observed modes, M, and the duration, T, of the observations. Here, we focus on the case where M goes to infinity, while T remains fixed, which states that the sample volume increases due to an indefinitely improving spatial resolution. This is typical of satellite data fields. The case when T goes to infinity and M is small and fixed is characteristic of climate time series. When T tends to infinity, under general condition many popular estimators, such as MLE estimators, least squares estimators, and so on, are consistent and asymptotically normal (Loges, 1984). In the present case ($M \rightarrow$ infinity, T fixed), the answer is much more complex.

The Piterbarg and Rozovskii (1996) method considers the problem of estimating a vector parameter $\theta = (\theta_1, ..., \theta_k)$ in the equation

$$\frac{\partial u}{\partial t} + (A_0 + \theta_1 A_1 + \cdots + \theta_k A_k)u = S(t; \mathbf{x}). \tag{7.6}$$

where $u = u(t; \mathbf{x})$ is the observed random field, $\mathbf{x} \in G$, G is a bounded domain in the d-dimensional Euclidean space R^d, A_k, $k = 0,..., K$ are linear self-adjoint operators defined on a convenient function space and $S(t; \mathbf{x})$ is a zero mean Gaussian "white noise" process in t that does not contain unknown parameters.

The authors construct the MLE estimator for θ using an expansion of the observed field in terms of an orthogonal basis and find necessary and sufficient conditions under which the estimator based on the first M expansion terms ("modes") are consistent, asymptotically Gaussian, and asymptotically efficient when M goes to infinity. Piterbarg and Rozovskii (1996) also discuss a number of cases where this solution is subject to different boundary and initial conditions.

7.3 DECISION TREES, RANDOM FORESTS AND GRADIENT BOOSTING

7.3.1 Decision Trees

A decision tree is a type of supervised machine learning algorithm that can use a variety of indicators for its decision (https://www.analyticsvidhya.com/blog/2016/04/tree-based-algorithms-complete-tutorial-scratch-in-python/). The method is mainly used for classification, but it can work with what the statistical literature calls "categorical" data (what we call discrete data) and also with continuous data, both as input and output variables. The goal of the method is to divide the input data into two or more homogeneous groups (Figure 7.4). The criteria upon which this homogeneity is judged can be defined in a number of ways.

The top of the decision tree starts with the root, known as a node, which essentially means that the "tree" is turned upside down (Figure 7.5).

Each node comprises an attribute (feature) that becomes the root cause of further splitting in the downward direction. The basic questions that the tree algorithms need to address are:

1. How to decide which feature should be located at the root node;
2. How to pick the most accurate feature to serve as internal nodes or leaf nodes;
3. How to divide the tree;
4. How to measure the accuracy of splitting the tree many more times.

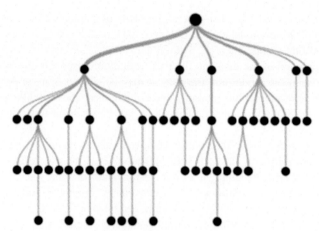

FIGURE 7.4   Schematic of tree-based modeling. *From https://www.analyticsvidhya.com/blog/2016/04/tree-based-algorithms-complete-tutorial-scratch-in-python.*

FIGURE 7.5   An upside-down tree representing a decision tree. *From https://www.analyticsvidhya.com/blog/2016/04/tree-based-algorithms-complete-tutorial-scratch-in-python.*

There is some important terminology when working with tree-based algorithms (see Figure 7.6):

1. Root Node: This represents entire populations or samples that the algorithm needs to divide into two or more homogenous sets;
2. Splitting: This process consists of dividing a node into two or more sub-nodes;
3. Decision Node: When a sub-node splits into further sub-nodes, the node is called a decision node;
4. Leaf/Terminal Node: Nodes that do not split any further are called Leaf or Terminal nodes;
5. Pruning: The removal of sub-nodes of a decision node is called pruning. It is the opposite of splitting;
6. Branch/Sub-Tree: A sub-section of an entire tree is called a branch of the sub-tree;
7. Parent and Child Node: A node that is divided into sub-nodes is called a parent node of the sub-nodes who are therefore the children of the parent nodes.

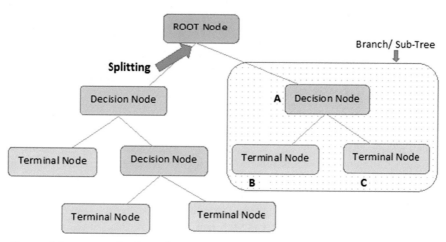

Note:- A is parent node of B and C.

FIGURE 7.6 Splitting of a root node into decision nodes to obtain terminal nodes. *From https://www.analyticsvidhya.com/blog/2016/04/tree-based-algorithms-complete-tutorial-scratch-in-python/.*

There are definite advantages and disadvantages to using decision trees for classification or other statistical estimation procedures. These are:

Advantages

1. Easy to Understand: A decision tree output is easy to understand even for people with a non-analytical background. It is very intuitive and does not require any statistical knowledge or training to read and interpret the output. Users can easily relate the output to their hypothesis.
2. Useful in Data Exploration: A decision tree is one of the fastest ways to identify the most significant variables and the relationships between two or more variables. With the help of decision trees, the user can create new variables/features that have a better ability to predict the target variables. Decision trees can also be useful in the data exploration stage to help identify the most significant features.
3. Less Data Cleaning Required: Decision trees require less data cleaning when compared with other modeling methods because they are not influenced by outliers and missing values.
4. Data Type is Not a Constraint: Decision trees can handle both discrete and continuous data.
5. Non-parametric Method: A decision tree is considered a non-parametric method, which means that the method requires no assumptions concerning the spatial distribution and the classifier structure.

Disadvantages

1. Over Fitting: The main problem with decision trees is their tendency to overfit the model to the data. This has led to the need for pruning methods, as defined earlier and to be discussed in more detail later.
2. Lost Information when Working with Continuous Variables: When working with continuous variables, a decision tree loses information when it places variables into different categories.

7.3.1.1 Deciding Where to Split

The essence of tree-based methods is knowing where to split and create new classification groups. Decision trees use multiple algorithms to decide where to split, with the goal at all times being to increase the homogeneity of the spawned populations. That is, to increase the purity of the target variable in the node. There are four most commonly used algorithms in making the split decision.

7.3.1.2 Gini

The Gini Index (also, Geni Impurity, or simply "Gini") is a probabilistic measure of the impurity of a node. Thus, a node with multiple classes after a split is said to be "impure", whereas a node with only a single class after the spit is considered to be "pure". The goal of the decision tree is to increase the purity of the node after the split. (Developed by Italian statistician Corrado Gini in 1912, the index provides a measure of economic inequality, as measured by income distribution or, less commonly, wealth distribution, across a population and among nations.) Mathematically, Gini ranges between

0 and 0.5, with higher values indicating a higher probability that the node is impure and that a specific feature will be misclassified upon being randomly sampled. If we choose two samples randomly, they must be of the same class if the population is pure and the probability of drawing that class equals 1. The higher the value of the Gini Index, the higher the inhomogeneity and the greater the impurity in the given class.

To calculate the Gini Index (Impurity), G, for sub-nodes, the method subtracts the sum of the squared probabilities of each class, p_i, from unity

$$G = \sum_{i=1}^{n} p_i(1 - p_i) = 1 - \sum_{i=1}^{n} (p_i)^2 \tag{7.7}$$

where n is the number of classes of the i elements. A value $G = 0.5$ means that there is a perfectly equal distribution of elements among the classes. An index of $G = 0$ means that the node has achieved purity in the randomly drawn elements Thus, in designing the decision tree, the features possessing the lowest value of the Gini Index is preferred.

In economics, the Gini coefficient is defined as the area (A) between the perfect equality line and the Lorenz curve (Figure 7.7) divided by the sum of this area and the area beneath the Lorenz curve ($A + B$). The Gini coefficient has been called a measure of inequality and, as noted earlier, is used to rank the income disparity of different nations over a range from 0 to 1. (The range of the Gini index is half that of the Geni coefficient.)

7.3.1.3 Chi-Square Algorithm

This algorithm determines the statistical significance in the differences between sub-nodes and parent nodes. It is measured as the sum of squares of standardized differences between observed and expected frequencies of occurrence of the target variable. The higher the value of Chi-Square, the higher the statistical significance of the difference between a sub-node and a Parent node. Chi-Square (χ^2) is computed as

$$\chi^2 = \left[(\text{Actual} - \text{Expected})^2 / \text{Expected} \right]^{1/2} \tag{7.8}$$

To calculate a split based on Chi-Square, we carry out the following steps (https://www.kdnuggets.com/2020/01/decision-tree-algorithm-explained.html):

1, Calculate Chi-Square for an individual node by calculating the deviations for both success and failure.
2. Calculate Chi-Square of the split using the sum of all Chi-Square value of success and failure of each node of the split.

7.3.1.4 Entropy

One of the most widely used algorithms for splitting nodes is "entropy", which, by analogy to thermodynamic entropy, is a measure of disorder or randomness in the data points. A high order of disorder means a low level of purity. Entropy, E, varies between 0 and 1, where $E = 1$ is the highest level of disorder and $E = 0$ is the lowest level of disorder. Thus, if the

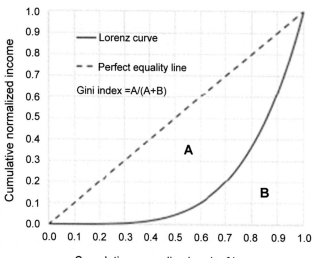

FIGURE 7.7 Definition of the Gini coefficient in terms of the Lorenz curve. In this format, the index can attain a maximum value $G = 1$ (i.e., when $B = 0$).

sample has a uniform class distribution its entropy is zero, whereas if the sample is evenly mixed with all elements, its entropy is 1. This contrasts with the Gini Index, which closely resembles entropy, but is limited to a maximum value of 0.5 (not to be confused with the Gini coefficient which can attain a maximum value of 1).

Entropy is calculated as:

$$\text{Entropy} = -\sum_{i=1}^{n} p_i \log_2 p_i \tag{7.9}$$

where n is the number of class elements and p_i and $q_i = 1 - p_i$ are the probabilities of success and failure, respectively, for a given node. Entropy is used in a split by comparison with the parent node. The lower the entropy, the better the split.

The steps in using entropy for a split:

1. Calculate the entropy of the parent node.
2. Calculate the entropy of each individual node of a split and calculate the weighted average of all sub-nodes available in the split.

7.3.1.5 Reduction of Variance

The last algorithm we review for splitting nodes is the reduction of variance. The previously reviewed decision methods are most appropriate for discrete data, while the reduction in variance method is used for continuous target variables. This algorithm employs the standard formula for variance to choose the best split at a node. The split with the lower variance is selected as the criterion to split the population, with

$$\text{Variance} = \frac{\sum_{i=1}^{n} \left(X_i - \overline{X}_i \right)^2}{n} \tag{7.10}$$

where $\overline{X}_i$ is the mean of the values, X_i is the actual value for element, i, and n is the number of elements. The process is:

1. Calculate the variance for each node.
2. Calculate the variance for each split as the weighted average of each node variance.

7.3.1.6 Pruning

Before presenting examples of applying decision trees to oceanographic data, we need to address the problem with decision trees involving the overfitting of a model to the data. One of the main approaches to reducing this overfitting is known as "pruning." With pruning, the user looks a few steps ahead before making a choice. Thus, the choices are no longer "greedy" in the sense that they are conditioned only by the immediate situation. So, how do we perform pruning?

1. We first make a decision tree with a large depth.
2. We then start at the bottom and start removing those leaves that are giving us negative returns when compared from the top.

Suppose a split is giving us a gain of -10 (i.e., a loss of 10) and then the next split gives a gain of 20. A simple decision tree will stop at the first step but one that has pruning will decide that the overall gain is $+10$ and will keep both leaves.

7.3.1.7 Ensemble Methods in Tree Based Algorithms

Ensemble methods involve groups of predictive models to achieve a better accuracy and model stability. Ensemble methods are known to impart supreme boosts to tree-based models. Like every other predictive model, a tree-based algorithm suffers from the plague of bias and variance. Bias indicates the degree to which the predicted values are different, on average, from the actual values; variance indicates how different will be the predictions of the model at the same point if samples are taken from the same population.

7.3.2 Alternating Decision Trees for Cloud Masking in MODIS and VIIRS NASA Sea Surface Temperature Products

Our first example of an oceanographic application of the decision tree method involves the identification of clouds in MODIS and VIIRS satellite data in order to improve the mapping of sea surface temperature (SST) from these instruments (Kilpatrick et al., 2019). In general, clouds are identified using uniformity tests over a small number of satellite pixels, brightness temperature range tests, and comparisons with low-resolution gap-free reference fields. Together, these methods are adequate at identifying large, upper-level, very cold cumulus clouds and moderately sized patchy cumulus clouds. These cloud detection methods are less effective at cloud edges, for thin cirrus clouds, and for the lower, more uniform stratus clouds, for which cloud top temperatures can be similar to those of the SST. In their paper, Kilpatrick et al. (2019) present the results for an ensemble cloud classifier based on boosted alternating decision trees applied to NASA MODIS and VIIRS SST imagery.

A small decision tree yields a model with low variance and a high bias. Clearly, there is a need to have a balance between bias and variance. As the complexity of a model is increased, there will be a reduction in prediction error due to lower model bias. As the model becomes more complex, it will end up overfitting the data and start encountering high variance. A successful model should maintain a balance between these two types of errors. This is known as trade-off management of bias-variance errors. Ensemble learning is one way to carry out this trade off. Some of the commonly used ensemble methods include Bagging, Boosting and Stacking. We discuss boosting.

Boosting refers to a family of algorithms that converts weak learners to strong learners. To convert weak learners to strong learners, we combine the predictions of each weak learner by:

a. Using averages or weighted averages;
b. Considering predictions that have a higher "vote".

As an example, consider the problem of spam email identification. To classify an email as Spam or Not Spam, we would use the following criteria:

1. Email has only one image file (promotional image), it's Spam;
2. Email has only link(s), it's Spam;
3. Email body consists of sentences like "You won prize money of $...", it's Spam;
4. Email is from the user's official domain, it's not Spam
5. Email is from a known source, it's not Spam

As these rules are individually strong enough to successfully classify an email at a better than average rate, they are called "weak learners". To convert these weak learners to "strong learners" that can successfully classify the email regardless of source, we combine the prediction of each weak learner using the methods in (a) and (b). Boosting places higher focus on examples that are misclassified or have higher errors in the classifications by the preceding weak rules. There are also algorithms such as Gradient Boosting Machines (GBMs) and XGboost (an optimized distributed gradient boosting machine learning library) that impart an additional boost to a model's accuracy.

In the Kilpatrick et al. (2019) SST study, the goal was to use a boosted, alternating decision tree to better identify clouds in MODIS and VIIRS imagery in order to remove the clouds when computing SST. As the authors note, the versions of R.2014 (and earlier) for the MODIS SST products and VIIRS Pathfinder SST, the cloud mask was developed using recursive binary decision trees (BDtrees; Kilpatrick et al., 2001, 2015). These earlier trees were, in turn, based on the classification algorithm of Breiman et al. (1984), a statistical classifier that uses known characteristics of an object to place that object into one or more classes. The performance of classification algorithms, such as the boosted decision tree (BDtree), is often presented as a "confusion matrix". A confusion matrix is used in supervised learning to visualize the performance of an algorithm. The rows represent the actual (or observed) classes, while the columns are the model predicted values. The name "confusion matrix" denotes the fact that it is easy to see whether or not the model is confusing the classes.

For a binary classification, as is the case for cloud contamination (either cloud or not cloud), there are four possible outcomes represented by the confusion matrix: (a) The four cells in the matrix contain the number of records for each combination of observed and predicted classes; (b) two cells in the matrix report the numbers of records correctly classified as cloud-contaminated or clear; (c) the remaining two cells correspond to classification errors: (c1) clear pixels identified as cloudy (false cloud) and (c2) cloud-contaminated pixels incorrectly identified as clear (false clear). The confusion matrix allows the calculation of various performance metrics such as sensitivity (the proportion of true clouds correctly identified as a cloud) and specify (the proportion of truly clear pixels that are correctly identified as clear). These metrics enhance one's ability to evaluate the performance of a classification algorithm rather than a single metric such as accuracy (the overall proportion of records correctly classified).

These two different classification errors have different implications for satellite-derived SST. Misclassification of a cloud-contaminated pixel as "clear" introduces an error in the retrieved SST value. Unidentified cloud presence in an SST pixel introduces a negative bias in the SST. An overly conservative cloud mask (one in which considerable numbers of clear pixels are misclassified as cloud-contaminated) can introduce significant sampling errors as many actual SST pixels will be eliminated from the SST field, resulting in reduced SST coverage. This will be particularly problematic in areas of persistent cloud cover such as the polar regions. The result is a failure to capture the true geographical variability of the SST.

In their study, Kilpatrick et al. (2019) compared the results of the two machine learning (ML) classification algorithms discussed below to distinguish cloudy and clear-sky pixels in MODIS and VIIRS SST images.

7.3.2.1 Boosted Decision Trees (BDtrees) and Alternative Decision Trees (ADtrees)

Both BDtrees and ADtrees represent an ensemble of Boolean functions, f, that are easy to interpret. Here, $f:\{0,1\}^n \to \{0,1\}$ maps each length-n binary vector into a single binary value. The ADtree classification algorithm was selected for the Kilpatrick et al. study because of its three advantages. First, like Bdtrees, ADtrees are easy to interpret and straightforward to implement. Both trees have low computational costs. Second, an ADtree not only predicts class membership of a pixel, but also provides an estimate of the confidence in that prediction. Finally, an ADtree prediction represents a collective vote from an ensemble of both strong and weak classifiers, rather than a decision from a single terminal node. It is this collective weighted majority vote that is the key advantage and power of an ADtree.

The data Kilpatrick et al. (2019) used for their study were a subset of the SST matchup databases (MUDBs) described in Kilpatrick et al. (2001, 2015), that are publicly available for VIIRS and MODIS from the NASA Ocean Biology distributed active archive system (OB.DAAC) SeaBASS validation system (https://seabass.gsfc.nasa.gov/archive/ SSTVAL). The VIIRS and MODIS MUDBs each contain several million records that are temporally (± 30 min) and spatially (± 10 km) coincident with *in situ* SST measurements. For the MUDBs, the *in situ* SSTs were taken from the NOAA/NCEI/STAR *in situ* Quality Control Monitor (Xu and Ignatov, 2014) and records eligible for inclusion in the training dataset were required to be from drifting or moored buoys and located within the VIIRS or MODIS pixel.

Subsets of MODIS and VIIRS were randomly selected from the MUDB for use in training and cross-validation. Of these, records were assigned to the "cloud-contaminated" class if the difference between the retrieved skin SST and the *in situ* SST exceeded -1.5 K (after correcting for the median skin-subsurface SST temperature difference of -0.17 K), to produce a zero bias with respect to the subsurface buoy SST. The cloud threshold of -1.5 K was chosen as it is ~ 3 standard deviations from the typically reported uncertainty of 0.5 K for clear-sky, best-quality IR SST retrievals.

One of the challenges in decision tree classification is classification under an imbalanced data distribution, where one class has many more records than the others. Oceans are significantly cloudier than clear, with a cloud fraction of about 72%, and only small seasonal variability (Eastman et al., 2011; King et al., 2013). Consistent with this fraction, the initial ratio of cloudy versus cloud-free records in the selected subset from MUDB was roughly 3 to 1 in favor of cloudy. This bias can seriously affect the decision tree classification. Kilpatrick et al. (2019) addressed this imbalance by resampling the training subset with under sampling, that is, randomly removing cloudy instances (the majority) to produce smaller datasets with approximately equal numbers of cloud-contaminated and clear-sky instances. This class-balanced dataset was further split into four subsets for classifier training.

Classification models were built for four different conditions: (1) nighttime; (2) daytime and no sun glint contamination ≤ 0.005 (cf. glint coefficient Cox and Munk (1954)); (3) daytime moderate glint (glint coefficient between 0.005 and 0.01); and (4) day severe glint, when red ($\lambda = 678$ nm) reflectance > 0.065 and the glint coefficient > 0.01. A series of labels were designed based on these conditions. The probability density functions and the bar charts in Figure 7.8 present the spatial and temporal characteristics of the buoy SSTs used in the training set by latitude and month. The instances used for tree fitting for moderate and high glint conditions were identical, but the resulting models are different. In high glint conditions, many of the reflective bands saturate and could not be used, reducing the training set to the same as that used at night.

A variety of open-source software packages were evaluated for the Kilpatrick et al. (2019) study, and the authors settled on a "workbench" developed and maintained by the University of Waikato Environment for Knowledge and Analysis in New Zealand. The classifiers BDtrees and ADtrees were evaluated for each of the satellite data sets (MODIS on Terra, MODIS on Aqua and VIIRS on SNPP) and all four conditions. All classifiers were built and validated using tenfold cross-validation. The decision tree algorithms they applied selected attributes based on ranking the information gain ratio. This ratio provides a measure of the relative entropy or homogeneity of the class outcome. During training, the split that results in the most homogeneous daughter node (lowest relative entropy) is selected until the node was either pure or the information gain is zero (0).

For the BDtree classifier, Kilpatrick et al. (2019) used a "best first tree approach" that selects attributes to split the datasets based on their contribution to the global entropy loss and not just the entropy loss of a particular split. The ADtree algorithm doesn't start with a single root node but rather is built as an ensemble of decision nodes and predictions. In this method, each boosting interaction concentrates on misclassified prior instances, adding another layer of decision and prediction to the ADtree at each iteration. The weaker learners are then "boosted in importance" and those associations with the greatest entropy loss are added to the ensemble of prior strong classifiers. The method produces a decision tree with alternating layers of predictions and splitters. The basic rule at each branch is a condition, or set of conditions, and two prediction nodes, each with a score. The score is a signed real value that is updated after each iteration by an amount proportional to the gradient decent in the mean squared error of the prediction in that layer. As depicted in Figure 7.9, a pixel's final classification is determined by the sign of the weighted majority vote from the cumulative sum of all the layers of the true prediction nodes.

The optimum number of boosting operations was 15, found by evaluating the decrease in the log loss-function after each iteration. For more than 15 iterations, there was little gain in classifier accuracy, for either class.

The Alternative Decision (ADtree) classifiers outperformed the Boosted Decision Tree (BDtree) classifiers when applied to VIIRS and MODIS SST matchups. Under all conditions, the ADtrees showed a slightly higher percentage of overall correctly classified records and a reduction in the rate of false positives (clear pixels identified as cloudy), particularly in glint regions. For VIIRS there was a 6—10% reduction in false positives for cloud. The validated values of ADtrees are several percentage points higher than for BDtrees, with the improvement occurring during daylight under moderate to high glint conditions. The difference in the false positive rate between sensors may be due to the higher spatial resolution of VIIRS.

An examination of the SST images processed with the above methods indicates that the ADtrees improved the discrimination of clouds near ocean thermal fronts, such as the edge of the Gulf Stream (Figure 7.10). The bottom panels, derived using the operational BDtree method, show the strong thermal gradients along the edge of the Gulf Steam as cloudy due to the dominant influence of strict spatial homogeneity tests. In contrast, the ADtree processed fields exhibited a marked reduction in the false classification as cloudy pixels that are located at the strong SST front, particularly at night.

Global SST images classified using the operational BDtrees (Figure 7.11, left) and the ADtrees (Figure 7.11, right) indicate that the better performance of the ADtree, as well as the gain in daytime clear-sky SSTs for VIIRS and MODIS, which occurs primarily in higher latitudes for both day and night. Globally, individual cloud sizes (data gaps) appear slightly smaller, and the overall clear pixel density is greater in many areas using the ADtrees. For the daytime images, there is a significant ($\sim$335%) increase in the count of cloud free grid cells, with a quality level of good and best.

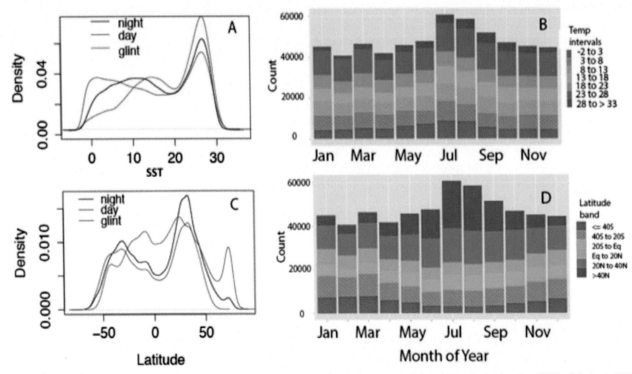

FIGURE 7.8 Spatial and temporal characteristics of buoy SST in the decision tree training set: (a) Probability density function (PDF) of the buoy SST, °C; (b) count of instances by month for each 2°C in temperature; (c) PDF by latitude; and (d) count by month and latitude. *From Kilpatrick et al. (2019).*

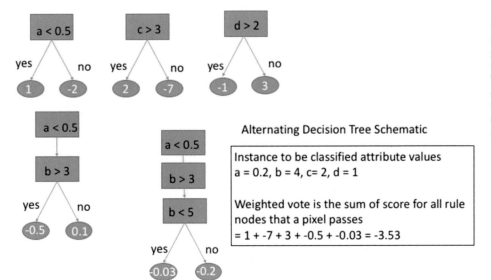

FIGURE 7.9 Schematic of boosted alternating decision tree classifier. Boxes represent the decision rule in each layer and ovals are the weighted score for the prediction. The sign of the score represents the class. The magnitude of the sum of true rules represents the overall confidence in the outcome. *From Kilpatrick et al. (2019).*

Alternating Decision Tree Schematic

Instance to be classified attribute values
a = 0.2, b = 4, c= 2, d = 1

Weighted vote is the sum of score for all rule nodes that a pixel passes
= 1 + -7 + 3 + -0.5 + -0.03 = -3.53

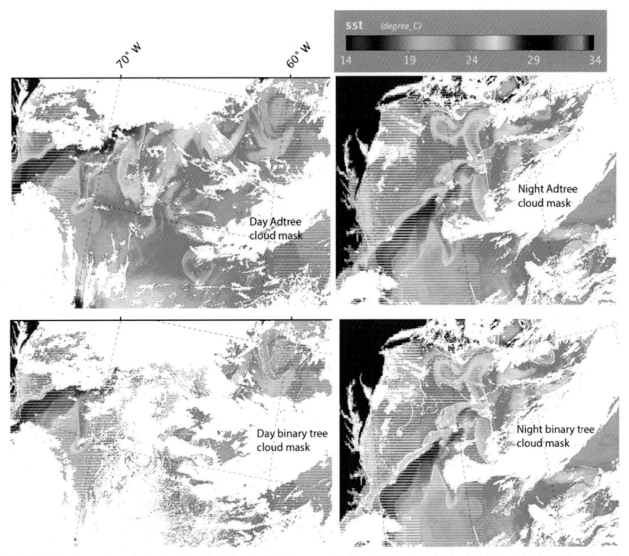

FIGURE 7.10 Comparison of cloud mask methods for the Gulf Stream on 19 June 2014 daytime (left) and nighttime (right) VIIRS SST$_{skin}$ (the nighttime SST is computed with the NSST$_{triple}$). White areas indicate pixels identified as cloudy; black indicates land. The ADtree classifiers (top) are more compact and there is improved retention of clear pixels at the high gradient edges of the Gulf Stream compared to the BDtree processed images below. *From Kilpatrick et al. (2019).*

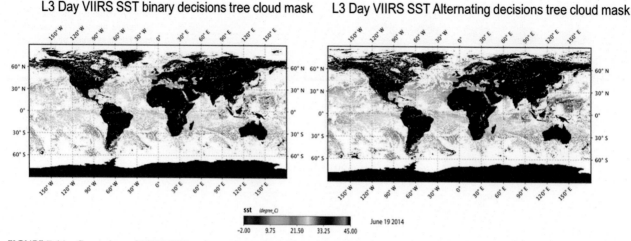

FIGURE 7.11 Comparison of VIIRS SST$_{skin}$ data coverage for daytime retrievals, 4-km maps for 19, June 2014 good or better quality, using different cloud mask models: (left) cloud identification based on a Boosted Decision Tree (BDtree) model and (right) cloud identification based on an ensemble of Alternative Decision Tree (ADtree) models. White indicates areas identified as cloudy. The use of an ADtree algorithm significantly increases the number of valid daytime retrievals everywhere, with the largest gains occurring in the mid- to high latitudes between 30° and 60°. *From Kilpatrick et al. (2019).*

The increases obtained using the ADtree classification models are supported by cross-validation results based on independent sets of data. The ADtree classifier outperformed the BDtree on the same training set by several percentage points in glint conditions compared to non-glint or at night. Glint conditions can cover up to 1/3 of daytime pixels depending on the pixel's geographic distance from the center of the sun glint. The BDtree, with its higher rate of false positives, significantly impacts the SST field by reducing the number of correctly identified clear-sky pixels.

7.3.3 Random Forest

Many data analysts consider the "random forest" method as the panacea for all data science problems. It is a highly versatile, non-parametric machine learning technique that can be used for both regression and classification tasks. It also reduces the dimension of the problem, treats missing values, outliers and other essential steps in data examination, and does moderately well handling all of these factors. Random Forest is a type of ensemble learning method where a group of weak learner decision tree models combine to form a more powerful model. Random Forest models combine the simplicity of Decision Trees with the flexibility and power of an ensemble model.

Unlike the single tree used in standard methods, Random Forest models grow multiple trees. To classify a new object, based on its attributes, each tree gives a classification, and the tree is said to "vote" for that class. The forest chooses the classification having the most votes (over all the trees in the forest) and, in the case of regression, takes the average of the outputs from different trees. The analysis has the following steps (Figure 7.12):

1. Assume the number of cases in the training set is N. Then, a sample of N cases is taken at random but with replacement. This sample will be the training set for growing the tree.
2. If there are M input variables, a number $m < M$ is specified such that at each node, m variables are selected at random out of the M. The best split on these m is used to split the node. The value of m is held constant while we grow the forest.
3. Allow each tree to grow to the largest extent possible, with no pruning.
4. Predict new data by aggregating the predictions of the N trees (i.e., use the majority votes for classification, the average for regression).

Random Forest models combine the simplicity of Decision Trees with the flexibility and power of an ensemble model. In a forest of trees, the model can ignore the high variance of a specific tree and is less concerned about each individual element. The model can, therefore, grow larger trees that have more predictive power than a pruned one. While Random Forest models don't offer as much interpretation ability as a single tree, their fitting performance is better, and the investigator need not worry about perfectly tuning the parameters of the forest, as in the case with individual trees. As discussed in the following subsections, building a random forest has three main phases.

7.3.3.1 Creating a Bootstrapped Data Set for Each Tree in the Forest

To build an individual decision tree, we use a training data set and all of the observations (input). Not surprisingly, the tree can adjust very well to the one training data set but might not work well with a new set of observations. To avoid this problem, one stops the tree from growing very large, usually at the cost of reducing its performance.

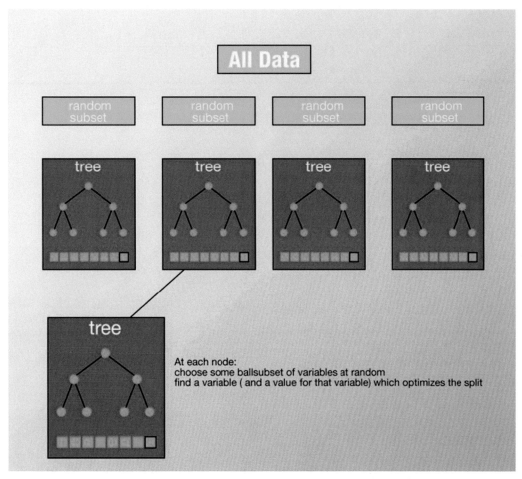

At each node:
choose some ballsubset of variables at random
find a variable (and a value for that variable) which optimizes the split

FIGURE 7.12 Processing steps in a random forest classifier. *From https://www.analyticsvidhya.com/blog/2016/04/tree-based-algorithms-complete-tutorial-scratch-in-python/.*

Building a Random Forest requires the training of *N* decision trees. However, not all of the same data are used to train all of the trees and we do not use all of the training data. This is where the first random feature enters the process. To train each individual tree, the investigator picks a random sample of the entire data set (Figure 7.13).

As illustrated by Figure 7.13, the size of the dataset needed to train an individual tree is not the size of the whole dataset. Also, individual data values may be present more than once in the data used to train a single tree (as in tree number n° 2 in Figure 7.13). This is called "Sampling with Replacement" or Bootstrapping (see the discussion on Bootstrapping in Chapter 3). Here, each data point is picked randomly from the whole dataset, so that a data point can be picked more than once. By choosing the training data randomly, the investigator avoids one of the problems in individual decision trees, namely the tree's affinity to the selected training data. In the case of a random forest, each decision tree is trained with different data, and the problem is mitigated. This makes it possible to grow larger trees, as we have avoided the tendency of the tree to overfit. If we restrict our training sets to small portions of the entire data set, we increase the randomness of the forest (reducing over-fitting), but usually at the cost of lower performance.

7.3.3.2 Training a Forest of Trees Using Random Data Sets and Adding a Little More Randomness with the Feature Selection

To build an individual decision tree, we evaluate a certain metric at each node using the Gini Index, entropy, or other metric and then pick the feature or variable of the data to go in the node that minimized/maximized this metric. This procedure works well when training only one tree but not a whole forest. To train a forest, we use an ensemble model. Ensemble models, like the Random Forest, work best if the individual models (individual trees in this case) are uncorrelated. In a Random Forest, this is achieved by randomly selecting certain features to evaluate at each node (Figure 7.14).

At each node, only a subset of all of the initial features is evaluated. For the root node, we consider E, A and F (and F wins). At Node 1, we consider C, G and D (and G wins). Finally, in Node 2, we take into account only A, B and G (and A wins). We would carry on with this until we have built the entire tree. The procedure avoids including features that have

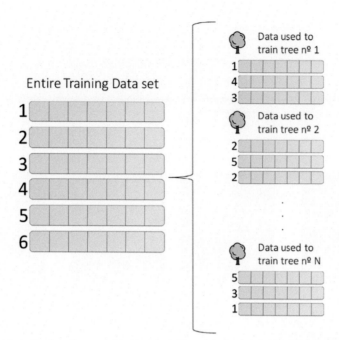

FIGURE 7.13 Training sequence for a Random Forest model. *From https://www.analyticsvidhya.com/blog/2016/04/tree-based-algorithms-complete-tutorial-scratch-in-python/.*

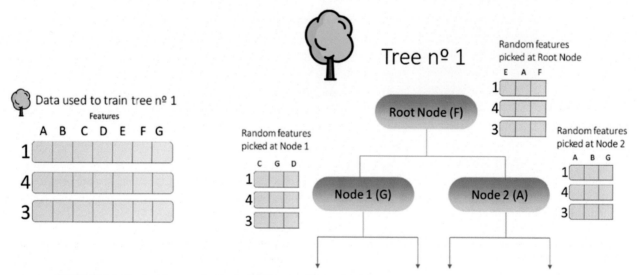

FIGURE 7.14 Random feature selection for building tree nº 1 in a Random Forest. *Designed by Freepik (www.freepik.com).*

very high predictive power in every tree, while creating many uncorrelated trees. This introduces the second type of randomness in that we introduce random feature in addition to the random training data.

7.3.3.3 Repeat the Previous Procedure for the N Trees to Create the Forest

The previous section outlined how to build a single decision tree that will be an element in a forest. This procedure is repeated for *N* trees by randomly selecting each node of each of the *N* trees, and randomly selecting from each node which variables enter the contest for being picked as the feature to split on.

In summary the Random Forest procedure is:

1. Create a bootstrapped data set for each tree.
2. Create a decision tree using the corresponding data set but at each node use a random sub-sample of variables or features to split on.

3. Repeat all steps hundreds of times to build a massive forest with a wide variety of trees.

This variety is what makes a Random Forest much more preferable to the use of a single decision tree.

7.3.3.4 Making Predictions Using a Random Forest

Once the forest is designed and trained, making predictions is fairly straightforward. The model simply takes the individual trees, passes the new observation (for which the user wants prediction made) through the trees to obtain a prediction from each tree (*N* predictions) and then computes an overall, aggregated prediction. Bootstrapping the data and then using an aggregate to make a prediction is called "bagging." For continuous regression problems, the aggregate decision is the average of the decisions of the individual trees.

7.3.3.5 Multivariable Integration Method for Estimating Sea Surface Salinity in Coastal Waters from In Situ Data and Remotely Sensed Data Using Random Forest Algorithm

As an example of using a random forest algorithm, we consider the estimates of sea surface salinity (SSS) in the Hong Kong Sea, combining *in situ* direct measurements with optical remotely sensed data from China's HJ-1 satellite (Liu et al., 2015). The random forest predictive model of salinity was trained using *in situ* measurements of sea surface temperature (SST), pH, total inorganic nitrogen (TIN) and Chl-*a*, all of which are strongly related to sea surface salinity according to a Pearson's correlation analysis. An earlier study (Urquhart et al., 2012) used eight different statistical models (such as artificial neural network, ANN and random forest) to predict SSS in Chesapeake Bay based on MODIS-Aqua satellite data. In their study, Liu et al. (2015) felt that it was adequate to estimate SSS from satellite data using a Random Forest (RF) model. The RF method is able to model non-linear relationships and can handle both discrete and continuous data. Compared to ANN, the RF method is less affected by noise in the data variables.

The study area chosen was the Hong Kong Sea (Figure 7.15), which is a Special Administrative Region of China. The region has a warm and humid climate, and the Pearl River flows through large catchment basins before entering the area on its way to the South China Sea. The *in situ* data used in the study were collected at the 76 stations shown in Figure 7.15 on a monthly basis between 2003 and 2011. Data from 10 water control zones were also available for the study. The measurements collected at each station included sea surface salinity (SSS), total inorganic nitrogen (TIN), total suspended particles (TSP), dissolved oxygen (DO), sea surface temperature (SST), total volatile solids (TVS), chlorophyll-*a* (Chl-*a*), pH and total nitrogen (TN).

The remotely sensed data consisted of HJ-1 images from a charge-coupled device (CCD) camera and a hyperspectral imager (HIS) or infrared camera (IRS). This satellite was launched on 6 September 2008, with the plan to monitor the

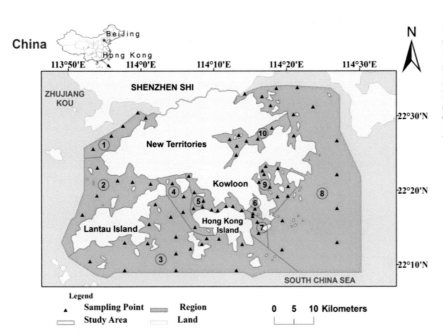

FIGURE 7.15 The Hong Kong study area and sampling points. The numbers 1—10 in the figure represent the ten water control zones, namely Deep Bay (DB), North-Western (NW), South (S), Western Buffer (WB), Victoria Harbour (VH), Junk Bay (JB), Eastern Buffer (EB), Mirs Bay (MB), Port Shelter (PS) and Tolo Harbour and Channel (TH), respectively. *From Liu et al. (2015).*

environment and natural disasters. The Hj-1 camera has four spectral bands (0.43–0.52 μm, 0.52–0.60 μm, 0.63–0.69 μm and 0.76–0.90 μm), with a 30 m × 30 m spatial resolution. The two identical cameras in Hj-1-A and HJ-1-B have ground swaths of 700 km and revisit cycles of 2 days. Preprocessing of the HJ-1 imagery includes atmospheric and geometric corrections, concluding with a resampling down to a 120 m × 120 m resolution. Images with significant cloud cover were excluded from the study.

To construct the Random Forest (RF) model, Liu et al. (2015) selected four cloudless images in March, June, and September 2010 and January 2011 to represent the four seasons. The whole set of *in situ* samples (76 × 4 = 304) was split into two subsets. After eliminating the outliers, approximately 2/3 of the samples were then used for training ($n = 200$ samples) and the rest ($n = 98$) for external validation. The overall process for assessing the spatial distribution of SSS is shown in Figure 7.16.

An important step in the RF model development is the selection of the appropriate predictor input variables. In their study, Liu et al. (2015) used a Pearson correlation analysis to determine which variables (SST, TN, TSP, TIN, Chl-*a* DO, TVS and pH) control, or at least markedly affect, SSS in coastal waters. A high Pearson's correlation coefficient indicates that variables were significantly correlated with each other. According to the study, DO and pH were positively correlated with SSS, while TIN, SST, Chl-*a* and TN were strongly negatively correlated with SSS. The authors decided that TN could be dropped from their analysis because it was very strongly related to TIN. Thus, they used TIN, Chl-*a*, SST and pH for the SSS estimation model. Ordinary Kriging was used to interpolate the SST and pH data onto a 120 m × 120 m spatial grid to provide the large area coverage needed to match the remotely sensed data.

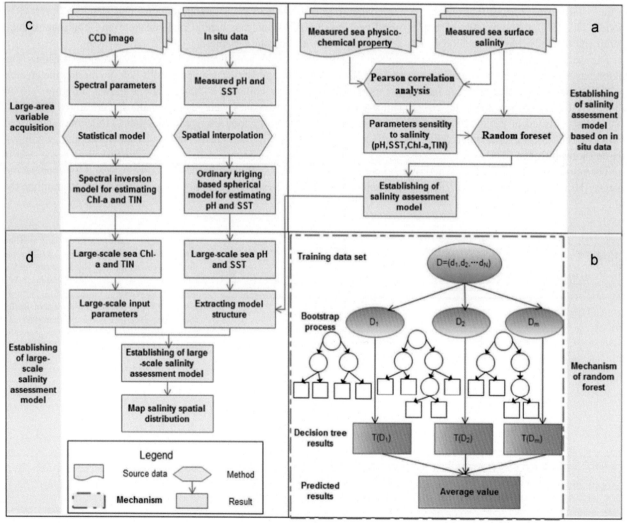

FIGURE 7.16 Flow diagram for the procedure to predict sea surface salinity (SSS) using the Random Forest Classification Method. Predictor input variables are total inorganic nitrogen (TIN), sea surface temperature (SST), chlorophyll-*a* (Chl-*a*), and pH (Liu et al., 2015).

Areal maps of the input variables Chl-*a* and TIN were derived from optical remote sensing data. Liu et al. (2015) then constructed retrieval models for Chl-*a* and TIN using the CCD data from HJ-1 together with their *in situ* measurements. The authors then turned to the RF model to retrieve SSS values. As part of the training procedure, each of the multiple trees was based on a bootstrap sample of the original training data. Numerous trees were generated and finally aggregated to give a single estimation of SSS, which is the average of the individual tree outputs (Figure 7.16). The processes depicted in Figure 7.16 are:

1. Identify the input and output to develop (train) the RF model.
2. Draw a bootstrap sample from the available dataset. The RF model is made up of hundreds of decision trees and, in this study, each decision tree is built from a bootstrap sample of the original data set. In general, 2/3 of the samples will be included in a bootstrap sample and 1/3 will be left out (called the "out of bag" samples).
3. Adjust the essential parameters in the RF algorithm: the number of trees in the forest, the minimum number of data points in each terminal node and the number of features tried at each node. In this study node size = 4.
4. Evaluate the estimation errors and accuracy. The RF was facilitated using two important evaluation parameters: the first is the mean-square-error (MSE) and the second is the variance explained by the RF model. The greater the increase in the node "purity," which dictates how the data are split at each node, the greater the importance of that particular variable.
5. Calculate the final output. Each tree acts as a regression function on its own and the final output is taken as the average of the individual tree outputs.

After finishing the training of the RF algorithm to establish the relationship between the above four parameters and SSS, Liu et al. (2015) used an independent dataset ($n = 98$) to validate the predictive performance of the RF regression model. Input variable importance showed that different input variables have different influences on the model (Figure 7.17).

From these results, it is clear that SST is the most important variable for SSS estimation, closely followed by TIN and then Chl-*a*. This finding agreed with the study's earlier results, where the correlation coefficient was highest between SSS and SST, followed by TIN and Chl-*a*. The results of the RF algorithm gave an excellent correlation with the training data (Figure 7.18). A correlation of 0.92 between the RF model and SSS is very high and the root mean square error of 1.68 psu (or, g/kg) is quite small.

To validate the RF algorithm, an independent dataset of 98 independent samples of SST, TIN, Chl-*a* and pH was used as input variables (Figure 7.19). The correlation of 0.86 is fairly close to the 0.92 in the original training data, verifying that the RF model has been adequately trained to predict SSS as a function of these input variables. The RMSE is again very similar to that in Figure 7.18, while the APE is smaller.

In conclusion, Liu et al. (2014) demonstrated that the RF salinity model based on four input variables was useful for assessing the temporal and spatial changes of SSS in a coastal area. Their RF model was able to estimate SSS for a period of 2 years with an RMSE of less than 2.0 psu (g/kg), and an APE lower than 5%. They checked their results against other statistical models and found that their model had lower estimation errors.

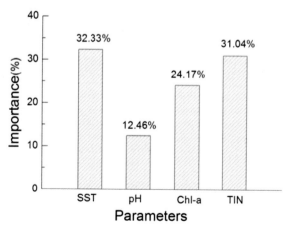

FIGURE 7.17 Variable influence of different input parameters (variables) in the RF SSS estimation model. Input variables are sea surface temperature (SST), pH, chlorophyll-*a* (Chl-*a*), and total inorganic nitrogen (TIN). *From Liu et al. (2015).*

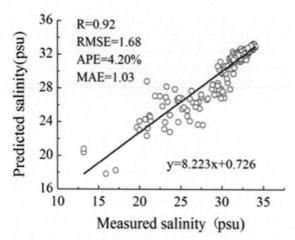

FIGURE 7.18 Correlation of sea surface salinity (SSS) in practical salinity units (psu) against the measured training dataset. Here, *y* denotes the predicted Random Forest (RF) regression estimate of surface salinity against *x* (the observed surface salinity). R, RMSE, APE and MAE denote the correlation coefficient, root-mean-square error (psu (or, g/kg)), the absolute percent error and the mean absolute error, respectively. *From Liu et al. (2015).*

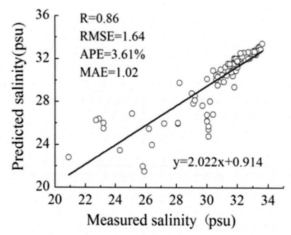

FIGURE 7.19 As in Figure 7.18 but for the correlation of SSS against the validation data. *From Liu et al. (2015).*

7.3.4 Naive Bayes

This method is considered one of the most popular and simple machine learning classification algorithms. It is based on Bayes Theorem for calculating probabilities and conditional probabilities. The method works with both discrete and continuous data. It is considered as naïve because it assumes that all of the data are independent of each other; the presence of any one feature does not depend on the presence of any other feature. As a classification method it also works much faster than many other classification methods.

In Bayesian classification, the user is interested in finding the probability of a label (*L*) given some observed features, which we can write as $P(L|features)$. Bayes's theorem tells us how to express this in terms of quantities that we can compute directly:

$$P(L|features) = \frac{P(features|L)P(L)}{P(features)} \tag{7.11}$$

Suppose we are trying to decide between two labels, L_1 and L_2. One way to make this decision is to compute the ratio of the posterior probabilities for each label:

$$\frac{P(L_1|features)}{P(L_2|features)} = \frac{P(features|L_1)P(L_1)}{P(features|L_2)P(L_2)} \tag{7.12}$$

All that is needed now is a model by which we can compute $P(L_i|features)$ for each label, i. Such a model is called a "generative model" because it specifies the hypothetical random process that generates the data. Specifying this generative model for each label in the main part of training such a Bayesian classifier.

A general version of this training step is very difficult, but we can make some simplifying assumptions, and this is where the label "naïve" comes in, since we make some very naïve assumptions about the generative model. Different types of naïve Bayes classifiers use very different naïve assumptions regarding the data.

7.3.4.1 Gaussian Naïve Bayes

Perhaps the easiest naïve Bayes classifier to understand is the Gaussian naïve Bayes, where it is assumed that data from each label are drawn from a Gaussian (normal) distribution. We illustrate this with a simple example (https://jakevdp. github.io/PythonDataScienceHandbook/05.05-naive-bayes.html). Assume that we want to classify the data shown in Figure 7.20. Further assume that the data are best described by a Gaussian distribution with no covariance between dimensions. This model can be fit by simply finding the mean and standard deviation of the points within each label. The results of this naïve Gaussian assumption are shown here in Figure 7.21. Here, the ellipses represent the Gaussian generative model for each label, with the probability increasing toward the center of the ellipses. With this generative model in place for each class, we have a simple recipe to compute the likelihood $P(features|L_1)$ for any data point, and therefore we can easily determine which label is most probable for a given point.

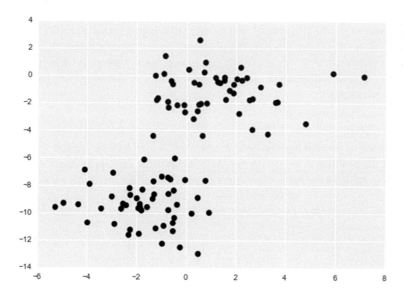

FIGURE 7.20 Data to be classified with a Gaussian naïve Bayes classifier. The $x-y$ coordinates give the value for each point. *Credit: VanderPlas, 2022. Python Data Science Handbook, second ed., O'Reilly Media, Inc.*

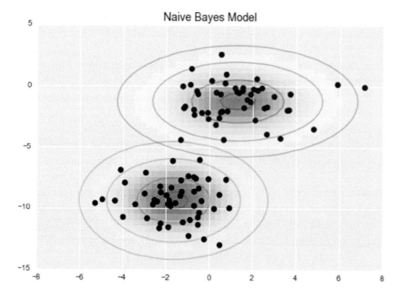

FIGURE 7.21 Naïve Bayes Gaussian model fit to the data in Figure 7.20. Ellipses correspond to the probability distribution from the model fits. The x-y coordinates give the value for each point *Credit: VanderPlas, 2022. Python Data Science Handbook, second ed., O'Reilly Media, Inc.*

7.3.4.2 Multinomial Naïve Bayes

Another useful assumption that can be used with the Naïve Bayes model is the multinomial naïve Bayes, where features are assumed to be generated from a simple multinomial distribution. The multinomial distribution describes the probability of observing counts among a number of categories, and thus multinomial naïve Bayes is most appropriate for features that represent counts or discrete values. The procedure is the same as before except we now model the data distribution with a best-fit multinomial distribution.

7.3.4.3 Advantages and Disadvantages of Naïve Bayes

The naïve Bayes classifier has definite advantages and disadvantages.

Advantages

1. The classifiers are computationally extremely fast for both training and prediction.
2. They provide straightforward probabilistic predictions.
3. They are often very easy to interpret.
4. They have few (if any) tunable parameters.
5. When an assumption such as statistical independence holds, a Naïve Bayes classifier performs better than other models such as logistic regression and requires less training data.

Disadvantages

1. If a discrete data set has a value that was not present in the training dataset, the model will assign a zero probability and will be unable to make a prediction. To solve this, a smoothing function is used to fill in the missing values.
2. Some researchers consider a naïve Bayes to be a poor estimator, so the probability outputs are not taken too seriously.
3. A serious limitation of Naïve Bayes is the assumption of independent predictors. In reality, it is almost impossible to have a set of predictors that are completely independent.

7.3.4.4 Some Naïve Bayes Applications

Naïve Bayes is a quick learning classifier and is attractive for making real time predictions. It also has the capability for providing multi-class predictions with discrete data. In addition, it is popular for text classification and spam filtering. In text classification, this classifier has a higher success rate compared to other algorithms and is therefore widely used in Spam filtering (to identify spam e-mail) and Sentiment Analysis (in social media analysis, to identify positive and negative customer sentiments.) Finally, it is popular as a "recommendation system" that uses machine learning and data mining techniques to filter unseen information.

7.3.4.5 Algal Bloom Prediction Based on Naïve Bayesian Model and Satellite Images

In their study of algal blooms in lakes, Mu et al. (2021) used a Naïve Bayes classifier applied to satellite images to detect and map the algal blooms. Predicting algal blooms is important for resource management and protection. Because the mechanisms of algal bloom formation are not well understood, a predictive model needs to be based on existing data. Consequently, a Naïve Bayes classifier was used. The desired outcome was the ability to predict bloom occurrence probabilities 1–7 days in advance under different weather conditions in Dianchi Lake, a shallow lake near Kunming City, China (Figure 7.22).

The proposed model used data from MODIS satellite images, the floating algae index (FAI) for the previous 7 days and five meteorological variables: mean wind speed; air pressure; relative humidity on the prediction day; accumulated sunshine hours in the previous 3 days; and accumulated air temperature in the previous 7 days. The probabilities for each pixel were calculated on a monthly basis to highlight the algal bloom's temporal-spatial differences, while the 1–7-day posterior probabilities were calculated by combining the prior and conditional probabilities.

Mu et al. (2021) sought to establish a predictive model for algal bloom occurrences in Dianchi Lake that considers not only the meteorological factors but also the spatial heterogeneity of the algae accumulation in the water before the prediction day.

The lake is about 40.4 km long and 7 km wide, with a mean depth of 4.4 m. It is located in a humid monsoon climate zone and its climate is mainly affected by the southwest monsoon and tropical continental air masses. The annual average air temperature is $16.19 \pm 4.87°C$ and the lake is in a windy zone with daily mean wind speeds of 2.99 ± 0.55 m/s during the dry season, and 2.21 ± 0.40 m/s during the wet monsoon season. Compared to the dry seasons, the rainy seasons have more precipitation and lower wind speeds, which are more conducive for algal blooms.

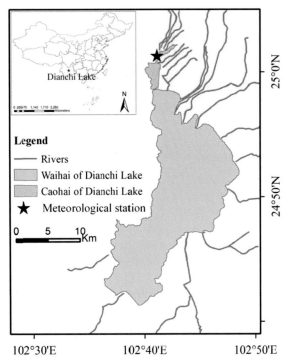

FIGURE 7.22   Location of Dianchi Lake in southern China. The area studied is in the southern, much larger portion of the lake known as "Waihai". Also shown is the closed northern part of the lake and the many rivers that feed into the lake. *From Mu et al. (2021).*

The satellite data used in this study were MODIS images with a 1-day revisit frequency. Approximately 6,000 images from 2002 to 2018 were obtained from NASA's Goddard Space Flight Center website (http://oceancolor.gsfc.nasa.gov). These images were filtered for clouds, cloud shadows and thick aerosols, leaving a total of 872 clear images for the study. The images were further corrected for Rayleigh scattering using the SeaWiFS Data System (SeaDAS 7.5). The spatial resolution of the resulting MODIS dataset was 250 m. In their study, Mu et al. (2021) determined the bloom or non-bloom state on a pixel scale using the floating *algae* index (FAI) value (Hu, 2009), defined as the difference between the reflectance at 859 nm and the linear baseline between the red (645 nm) and the shortwave infrared (1,240 nm) bands of the MODIS instrument. When FAI $\geq$ 0, the class variable was "1", indicating a bloom; when FAI < 0, the class variable was 0, indicating the absence of a bloom.

The meteorological data were from the Chinese Meteorological Administration (CMA) Station #56778 and included daily resolved air temperature (°C), wind speed (m/s), atmospheric pressure (hPa), relative humidity (%), and sunshine hours (h). These data were all available from the CMA website (http://data.cma.cn).

All of the data were used with a Naïve Bayes analysis to predict algal bloom occurrence probabilities. This consisted of two parts: (1) *The calculation of bloom occurrence posterior probability.* This was done by calculating the FAI obtained from the satellite images and meteorological data. Then, the conditional probability was combined with the prior probability to obtain posterior probability of bloom occurrence 1−7 days in advance; and (2) *the prediction of bloom occurrence probability in the future.* By combining the posterior probability lookup tables with the satellite images and meteorological parameters, the proposed model can predict and map the bloom occurrence probabilities 1−7 days in advance (Figure 7.23).

In their analysis, Mu et al. (2021) recognized the problem of the model giving false results when new values were not present in the training set. They also recognized that these problems can be solved by Laplacian smoothing. The main element of each pixel observation is the meteorological variables, which are closely related to the algal floatation and ultimately to algal bloom events. Previous studies have shown that meteorological conditions have a significant influence on bloom occurrence. Mu et al. (2021) found that sunshine hours for the previous 3 days, the atmospheric pressure on the prediction day, the sum of the air temperatures in the previous 7 days and the relative humidity on the prediction day, had importance values of 0.623, 0.642, 0.619 and 0.688 respectively. These variables, themselves, are sufficiently different that they should satisfy the independence assumption of the Naïve Bayes method and could therefore be used in training the model.

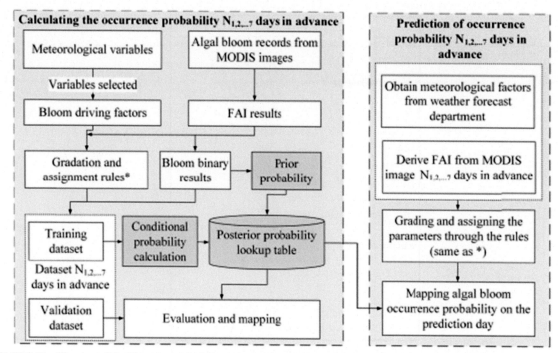

FIGURE 7.23 Flowchart of the Naïve Bayes algal bloom predictions scheme used by Mu et al. (2021) for Dianchi Lake in southern China.

The prior probability is an important part of the Naïve Bayes model. The traditional method of defining prior information is from ground observation records; however, blooms can occur suddenly and in patches scattered over a large area, making it difficult to obtain information using traditional methods. Therefore, the authors used satellite images to determine the bloom prior probability, which can be calculated using the following equation for the two possible states (0, 1) for a bloom:

$$P_m(c_1) = \frac{number\ of\ class\ c_1 + 1}{N + 2} \tag{7.13}$$

where m indicates the pixel number (here, $m = 1, 2, \ldots, 3{,}889$), c_1 is the bloom state and $N = 872$ is the number of observations, which, in this study, was the number of satellite images. In general, the higher the prior probability, the higher the algal bloom occurrence probability. However, in different seasons and different parts of the lake, there are different prior probabilities for the lake and, hence, different bloom occurrence probabilities (Figure 7.24).

Prior to the Naïve Bayes probability calculations, the MODIS images were divided into two sets (training and validation) at a 9:1 ratio. The larger set was used to train a set of Naïve Bayes models to predict the probability of algal blooms within the next 1−7 days. Each data set in this training consisted of the bloom state variable (0 or 1), the selected meteorological variables and FAI at n days before the prediction day. The final probability was calculated as the joint probability of all the Naïve Bayes models. The models were tested at the pixel scale using the 87 images held out for the validation data set (Figure 7.25).

A majority of the pixels' CCI values were very high. Lower CCI values were mainly located in the southern and middle portions of the lake; algae migration in these areas may lead to higher uncertainty in the predictive mode, especially under high wind conditions. Mu et al. (2021) find that, within the limits of the observations, their NB predictive models are satisfactory for providing 1−7-day predictions of algal bloom probability in Dianchi Lake.

7.4 K-MEANS CLUSTERING

A method of vector quantization, K-means, was originally developed for signal processing. The aim of the method is to partition n observations into k clusters in which each observation belongs to the cluster, with the nearest mean (cluster centers or cluster centroids) serving as a prototype of the cluster (https://medium.com/analytics-vidhya/the-most-comprehensive-guide-to-k-means-clustering-youll-ever-need-2a570ff2c0a3). This is termed partitioning of the data

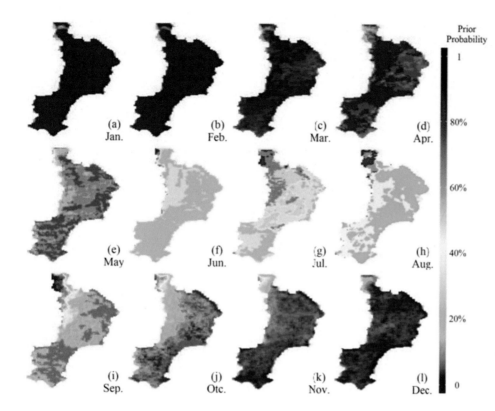

FIGURE 7.24 Monthly prior probability of algal bloom occurrences for each pixel derived from the 872 MODIS images for Dianchi Lake in southern China. *From Mu et al. (2021).*

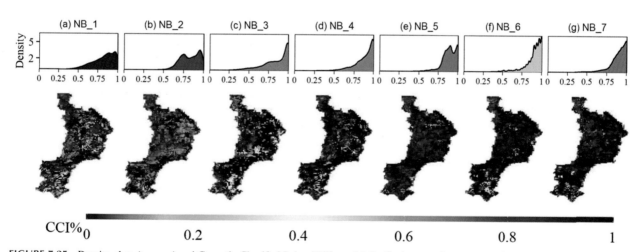

FIGURE 7.25 Density plots (top row) and Correctly Classified Index (CCI) spatial distribution map (bottom row) of the various Naïve Bayes (NB) prediction models using the validation data set. The distributions NB_1, ..., NB_7 refer to the algal bloom occurrence probability predictions 1—7 days in advance. Density = receiver-operating characteristic curve. *From Mu et al. (2021).*

into "Voroni cells". The method minimizes within-cluster variances (squared Euclidean distances), but not regular Euclidean distances, which would be much more difficult. K-means is a very popular, unsupervised machine learning algorithm to classify data into clusters. To apply the method, one defines a target number, k, which refers to the number of centroids needed in the dataset. A centroid is the imaginary or real location representing the center of the cluster. Every data point is allocated to each of the clusters through reducing the in-cluster sum of the squares. Thus, this algorithm identifies k centroids and then allocates every data point to the nearest cluster, while keeping the number of centroids as small as possible. The "means" refers to the averaging of the data to find the centroid.

The K-means method starts with a first group of randomly selected centroids, which are used as the beginning point for every cluster, and then performs iterative (repetitive) calculations to optimize the positions of the centroids. The operations creating and optimizing the clusters halt when either:

1. The centroids have stabilized, whereby there is no change in their values because the clustering has been successful;
2. The defined number of iterations has been achieved.

In an unsupervised method like K-means, there is no target to predict. The investigator examines the data and tries to group similar observations together to form classes. All of this is done without any training of the clustering method; hence, the method is unsupervised. Three key features of K-means that make it efficient are often considered as the methods' biggest drawbacks:

1. Euclidean distance is used as a metric and variance is used as a measure of cluster scatter;
2. The number of clusters, k, is an input parameter, and an inappropriate choice of k may yield poor results. For this reason, when performing a K-means analysis, it is important to run some early diagnostic checks to estimate the number of clusters in the data set;
3. Convergence to a local minimum may produce counterintuitive (wrong) results.

The K-means method is based on spherical clusters that are separable, making the mean converge toward the cluster center. Clusters are expected to be of similar size, so that assignment to the nearest cluster does not bias the clustering. As with any other clustering algorithm, K-means assumes that the data satisfy certain criteria. As a consequence, the method works well with some data sets and fails with others.

The relaxed solution of K-means clustering, specified by the cluster indicators, is given by principal component analysis (PCA), discussed in Chapter 4 in this book. The intuition is that K-means describe the spherically shaped clusters. If the data have $k = 2$ clusters, the line connecting the two centroids is the best 1-dimensional projection direction, which is also the first PCA direction. Cutting the line in half, separates the clusters (this is the continuous relaxation of the discrete cluster indicator). If the data have three clusters, the 2-dimensional plane containing the three cluster centroids is the best 2D projection. This plane is also defined by the first two PCA dimensions.

There are two important properties of clusters: first all the data points in a cluster should be similar to each other and second the data points from different clusters should be as different as possible. The first requirement leads to the need to minimize the distance between points within a cluster. This is done by minimizing the sum of the distances between the points and their respective cluster centroid. One of the common challenges with K-means is that the sizes of the clusters are different. Another challenge is that the densities of the original points may vary. One solution is to use a higher number of clusters.

Picking the optimum number of clusters is a major challenge in K-means. One procedure is to start with a low number, such as 2, and calculate the clusters. The clusters are then plotted against inertia on a graph, where inertia equates to the sum of the squares between each data point and the centroid. Next, increase the number of clusters and train the model, again plotting the clusters against the inertia. This process is repeated, resulting in a plot, such as that in Figure 7.26. A good model has low inertia and a small number of clusters (k). The tradeoff is that k increases as inertia decreases.

To determine the optimal number of clusters for a dataset, we can use the *Elbow method*, in which we find the point where the rate of decrease in inertia begins to diminish. In Figure 7.26, there is a dramatic drop in inertia between 2 and 4

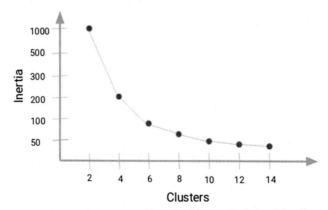

FIGURE 7.26 Inertia as a function of the number of clusters (k) (https://medium.com/analytics-vidhya/the-most-comprehensive-guide-to-k-means-clustering-youll-ever-need-2a570ff2c0a3).

clusters and the inertia levels out about 10 clusters. We could pick any k-value elbow between 6 and 10 for our analysis. One has to also consider the computational cost, which increases as the number of clusters increases.

There is a wide range of free software packages that address K-means. These software packages are evolving rather rapidly so that we have chosen not list them here as the list will likely be out of date by the time this revision is published. It should be easy for the reader to find one or more software packages that will carry out the K-means analysis of interest.

7.4.1 Cluster Analysis of North Atlantic-European Circulation Types and Links with Tropical Pacific Sea Surface Temperatures

In their study of atmospheric circulation over the North Atlantic-European (NAE) region, Fereday et al. (2008) used a K-means cluster analysis applied to daily mean sea level pressure (MSLP) fields to derive a set of circulation types for six 2-month seasons. They found that the K-means algorithm produced clusters of equal size (although there was a tendency toward a broader distribution of sizes as the number of k clusters was increased). Results suggest that rather than finding distinct regimes, the algorithm is merely partitioning a smooth cloud of data into similarly sized volumes (as was also noted by Christiansen, 2007). For almost every choice of k, different solutions can be found for which the sum of within-cluster variance is within 1% of the best estimate of its global minimum value, but whose clusters are substantially different (Figure 7.27). This suggests that the data are not distributed in real clusters. In addition, each cluster possesses a high density of points near to the origin in the figure, with decreasing density away from the origin. This contrasts with the K-means intuition that each cluster consists of a dense kernel of points at its center surrounded by a lower-density of points.

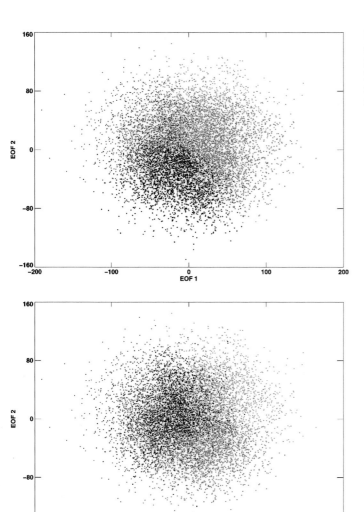

FIGURE 7.27 EOF scatterplot of EOF2 (y-axis) versus EOF1 (x-axis) for two alternative classifications for the January—February (JF) season, with different clusters indicated by different colors. The data are projected onto the plane spanned by the leading pair of EOFs. The totals of within-cluster variance for each classification were within 1% of each other. *From Fereday et al. (2008).*

A study of the cluster stability did not yield a clear indication of the best k value for a particular season. If stability were the sole criterion, $k = 2$ would always be selected, which is clearly wrong as it classifies the minimum number of groups and has a very large variance. Cluster analysis can nevertheless be useful in generating a representative set of circulation patterns for each season. Cluster centroids are averages of similar circulation fields, and therefore correspond to physical circulation patterns. Clusters are preferable to EOFs, which are constrained to be mutually orthogonal, which may not be consistent with the physical circulation patterns.

In their study, Fereday et al. (2008) chose $k = 10$ for each 2-month season. This represents a compromise in that there are enough cluster centroids to span the space of data, but, at the same time, not so many clusters that the similarity between neighboring cluster centroids is too great. The authors emphasize that the choice of 10 is arbitrary and that there is no underlying reason for this choice. Earlier studies (Lund, 1963; Barnston and Livezey, 1987) have used a similar value. The value of $k = 10$ produces a range of circulation types, while keeping cluster centroids reasonably distinct (Figure 7.28). The centroids exhibit a wide range of patterns, including a variety of strong zonal flow and blocking.

The cluster classification in Fereday et al. (2008) was used to explore the influence of global SST on the North Atlantic-European (NAE) circulation. SST relationships were investigated for all seasons by a lagged regression with the frequency of occurrence (days per season) of each cluster, to give one frequency value per cluster per year. The regression was performed over the period of 1870–2002, corresponding to the availability of SST in the Hadley (HadISST) dataset. The study used SST from the month preceding the 2-month season. When looking for the forcing from the atmosphere on SST, Fereday et al. (2008) applied SST following the season. Trends in global SST were removed from the SST data before computing regressions.

Significant relationships between SST and circulation-type frequencies were found for all seasons, although the strengths and significance of these links varied between clusters. In general, a stronger effect was found for mean sea level pressure (MSLP) leading SST, with marked patterns in the North Atlantic consistent with atmospheric forcing of the ocean. SST regression patterns were positively spatially correlated with cluster MSLP over the North Atlantic (high pressure is associated with warm SST) in 57 of the 60 clusters. The correlation coefficients are generally larger in the Northern Hemisphere summer than in winter. The enhanced correlation of high MSLP with warm SST in summer likely results from increased solar insolation under clear skies.

Fereday et al. (2008) concentrate their exploration of links to SST on the influence of the El Niño Southern Oscillation (ENSO) phenomenon. Correlations were very modest. In particular, the correlation between the Niño-3 SST index and lagged cluster frequency emphasize that these links explain only a fraction of the circulation variance. This may explain why ENSO effects on European climate have not been more readily apparent. They also found that there is a marked difference in the effect of ENSO on European climate between early and late winter.

7.4.2 Mapping Near Surface Global Marine Ecosystems Through Cluster Analysis of Environmental Data

Zhao et al. (2019) used cluster analysis to classify ocean surface waters based on long-term averages of 20 oceanic variables. Both principal component analysis (PCA) and K-means clustering were applied. The study focus was on ecosystems and the variables that can identify these ecosystems. The authors found seven distinct areas that fit the definition of ecosystems, which means that these regions have environmental properties that persist and clustered according to both statistical methods. Their strict definition of an ecosystem consisted of a spatially bounded environment, where biological and energy interactions are greater within the area than with other ecosystems.

According to the authors, there have been very few studies that have tried to objectively identify marine ecosystems. Oliver and Irwin (2008) used hierarchical and K-means clustering on SST and solar irradiance to map 81 "provinces", while Sayre et al. (2017) used K-means applied to six variables (SST, surface salinity, oxygen, nitrate, phosphate and silicate) down to a depth of 5,500 m. They identified 37 three-dimensional regions that they named "Ecological Marine Units" (EMUs), of which 22 comprised 99% of the ocean's volume. All of these earlier classifications were limited by the data that were available at the time. Many more data are now available. The study by Zhao et al. (2019) analyzed 20 physical, biochemical and nutrient variables to map global marine ecosystems derived from data acquired from the Global Marine Environmental Datasets (GMED; http://gmed.auckland.ac.nz).

Although there are local scale variations due to tidal changes, as well as diel and seasonal oceanic cycles, these are shorter than the lifespans of most marine species and, therefore, affect their local abundance rather than their geographic distribution. Only a few marine species, typically larger cetaceans and birds, migrate seasonally. Because their aim was to identify the features that endured over time, the use of data averaged over decades for ice cover, wind speed, salinity, oxygen, silicate, phosphate nitrate, PAR and pH, and from 2002 to 2009 (or 2010) for temperature, chlorophyll, and calcite was considered suitable.

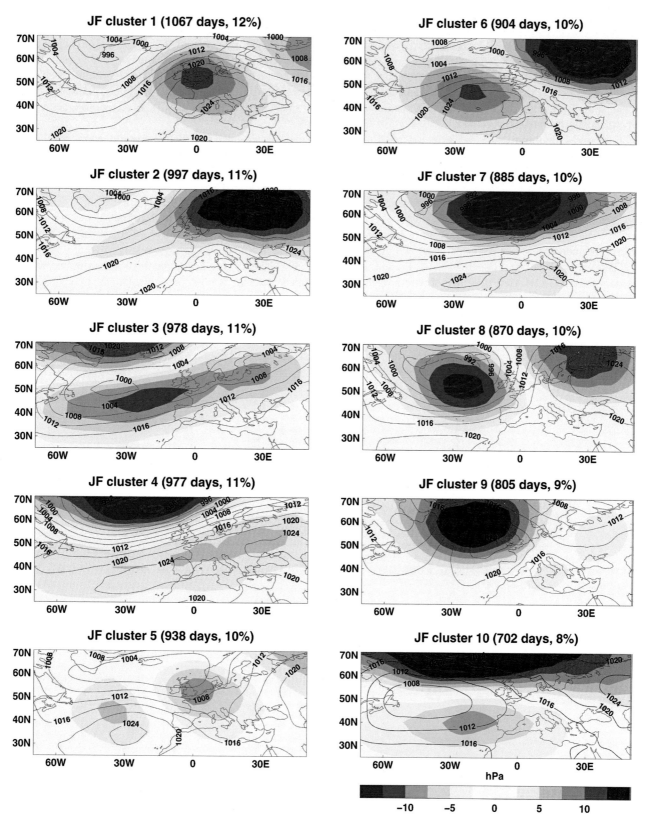

FIGURE 7.28 Cluster centroids for the January–February (JF) season. Colors denote the cluster mean anomaly fields derived from the seasonally varying climatology. Overlaid contours show the cluster mean fields of absolute air pressure. *From Fereday et al. (2008).*

7.5 SUPPORT VECTOR MACHINES

A Support Vector Machine (SVM) is a supervised computer algorithm that learns by example to assign labels to objects and performs supervised learning for classification of data groups (https://www.analyticsvidhya.com/blog/author/sunil-ray/). Although SVMs can be used for regression problems, they are the most widely used for classification algorithms based on a learning theory developed by Vladimir Vapnik in 1979 but first published in 1995 (Vapnik, 1995). An SVM performs classification by finding the hyperplane that maximizes the margin between the two classes. The vectors that define the hyperplane are the support vectors.

In the SVM algorithm, each data point is plotted in n-dimensional space (where n is the number of features and hence the number of possible points) with the value of each feature being the value of a particular coordinate. The algorithm then performs the classification by finding the hyperplane (in 2D, the line) that strongly differentiates the two classes (Figure 7.29).

The Support Vectors are simply the coordinates of the individual observations. The SVM classifier is the frontier that best segregates the two classes (hyperplane/line). The term machine is added as this is an algorithm in the domain of machine learning. The basic question is "how do we identify the right hyperplane?" We explore this by considering a few cases. First, consider a data set made up of two distinct classes (Figure 7.30).

There are three different hyperplanes (A, B and C) in Figure 7.30. The goal is to determine the one that best separates the stars and the circles. Here, the "rule of thumb" is that the hyperplane does the best job of segregating the two classes. In this example, hyperplane B does the best job.

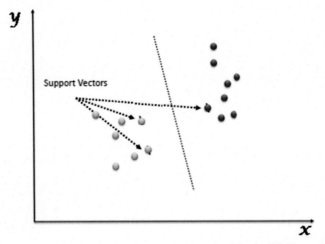

FIGURE 7.29   A data set with point coordinates x, y classified into 2 groups (green and red) by the SVM hyperplane (here, a line). *From https://www.analyticsvidhya.com/blog/author/sunil-ray/.*

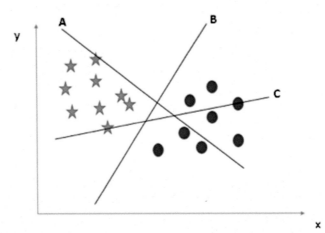

FIGURE 7.30   Case 1: a data set with two distinct classes of points with coordinates (x, y) and three different hyperplanes, A, B and C. *From https://www.analyticsvidhya.com/blog/author/sunil-ray/.*

In the second example (Figure 7.31), there also are three hyperplanes (lines A, B and C) that clearly separate the stars and the circles, so we need a way to identify the best one. In this case, maximizing the distance between the point and the hyperplane gives the best line to separate the two different types of points (Figure 7.32).

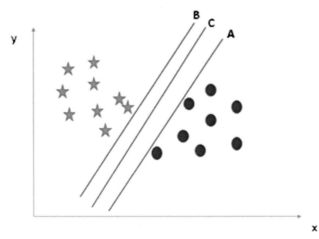

FIGURE 7.31 Case 2: an example population of star and circles with coordinates (x, y) separated by three parallel lines. *From https://www. analyticsvidhya.com/blog/author/sunil-ray/.*

It is easy to see that distance (called the "Margin") in Figure 7.32 is greatest for the best hyperplane separating the triangle and circle elements of the population. In addition, selecting the hyperplane with the highest margin is robust, in the sense that there is a lower probability of miss-classification. There is a case (Figure 7.33) where maximizing the margin alone will not select the best hyperplane.

In the example presented in Figure 7.33, maximizing the margins would lead to the selection of B as the optimum hyperplane but that splits the population of stars because there is a star displaced from the general star population that is much closer to the circles. In this case, hyperplane A has correctly classified the two elements of the total population and it therefore the correct hyperplane to choose for classification.

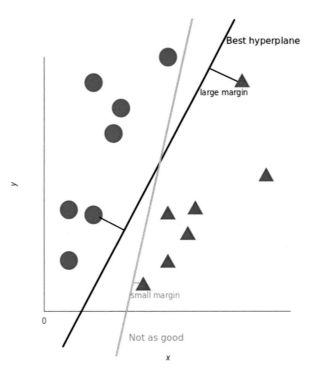

FIGURE 7.32 Population of triangles and circles with coordinates (x, y) showing the "margins" for two hyperplanes. *From https://www.analyticsvidhya. com/blog/author/sunil-ray/.*

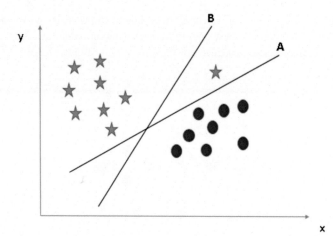

FIGURE 7.33 Two hyperplanes in a population of stars and circles with coordinates (x, y). *From https://www.analyticsvidhya.com/blog/author/sunil-ray/.*

There can be cases where there is no hyperplane that can clearly separate the two elements of the population (Figure 7.34). Here, there is a star in the area of the circles that appears to be an outlier of the star population. The star embedded in the circle elements in Figure 7.34 is an outlier of the star elements. The SVM algorithm has a feature that lets it ignore outliers and find the hyperplane that has the maximum margin (Figure 7.34). Thus, the SVM algorithm is robust with respect to outliers. However, what if the distributions of the sub-elements are highly nonlinear, as they are in Figure 7.35?

In the case of Figure 7.35, we transform this distribution into a new space by introducing $z^2 = x^2 + y^2$ (the equation for a circle with its center at the origin) and plot the new distribution in Figure 7.36. All the values of z are positions because they are derived from the sum of x^2 and y^2. In Figure 7.35, the red circles appear close to the origin of the x and y axes, resulting in lower values of z, with stars all having higher values of z and being father from the origin in Figure 7.35. An examination of Figure 7.36 clearly shows that the SVM algorithm can easily find the best hyperplane in z space. If we then transform the hyperplane (line) in z space back to x, y space, we find that the hyperplane is a circle surrounding the red circles (Figure 7.37).

The transformation to make a nonlinear problem solvable by the SVM algorithm is called the "kernel trick." It is commonly used with nonlinear distributions. We have provided a very simple case, where we could transform all of the support vectors. In some problems, this transform can be computationally expensive, particularly if it is applied to every vector in a multi-dimensional space. It is in such situations that the "kernel trick" can be applied. Because the SVM can use the dot products of the vectors rather than the vectors themselves, we can avoid this burdensome computer processing. Here, the SVM kernel is a function that takes low dimensional input space and transforms it to a higher dimensional space, thus converting a non-separable problem to a separable problem. It is mostly useful in non-linear separation problems. In

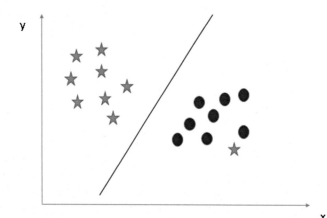

FIGURE 7.34 Population of stars and circles with coordinates (x, y) with a star outlier (https://www.analyticsvidhya.com/blog/author/sunil-ray/).

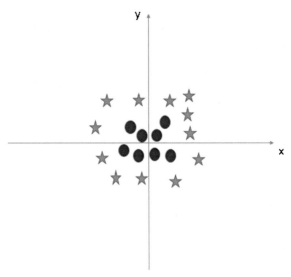

FIGURE 7.35 A nonlinear element distribution of circles and stars with coordinates (x, y). *From https://www.analyticsvidhya.com/blog/author/sunil-ray/.*

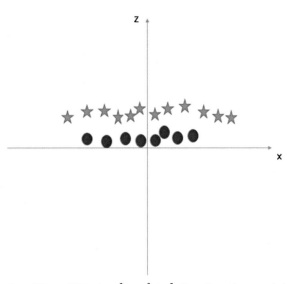

FIGURE 7.36 The transformed version of Figure 7.35 using $z^2 = x^2 + y^2$. *From https://www.analyticsvidhya.com/blog/author/sunil-ray/.*

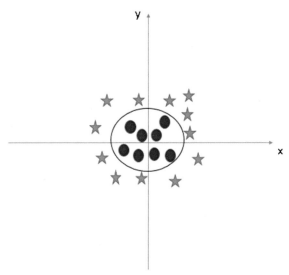

FIGURE 7.37 The SVM hyperplane in the original (x, y) coordinate system (https://www.analyticsvidhya.com/blog/author/sunil-ray/).

other words, this trick does some clever data transformations and then finds the process to separate the data based on the labels that the user defined. In the above problem, we have shown that we could have obtained a nonlinear classifier without transforming the data; we simply change the dot product to that of the space we want and the SVM will give us the hyperplane we are looking for. Note, the kernel trick is not actually part of the SVM and can be used with other linear classifiers such as regression.

Now that we can classify vectors in multidimensional space, we need to determine how to get the data transformed into vectors. The most common answer is to use "frequency of occurrence." Thus, we can vectorize the values by counting the number of times they occur and then normalizing that value by the total number of features. As we will see in later sections, satellite images use the different channels of the satellite sensor and different target land/sea cover values to train the SVM for the classification. It must be remembered that SVM is a supervised classification method, and results will depend on the training set developed by an independent analysis of the input data and the classes that are desired for the output.

Once the data are established as support vectors, a kernel function needs to be chosen for the model. Every problem is different, and the kernel function depends on the properties of the data and the desired outcome. SVM is ideal for 2D analysis but as has been discussed it can be extended to multi-dimensional analyses. Some SVM applications use tens or even hundreds of features. In such cases, a nonlinear kernel will likely end up overfitting the data. It is best to stick to a linear kernel to obtain the best possible performance from an SVM. So, how is the best kernel selected? Unfortunately, the only realistic answer is by trial and error. It is best to start with a simple SVM, and then experiment with a variety of standard kernel functions. In addition to allowing SVMs to handle nonlinearly separable data sets, and to incorporate prior knowledge, the kernel function yields at least two additional benefits. First, kernels can be defined on input that are not vectors. This ability to handle non-vector data is critical in analyzing some types of data. Second, kernels provide a mathematical formalism for combining different types of data (cf. what_is_a_support_vector_machine.pdf).

The most obvious drawback to the SVM algorithm is that it apparently only handles binary classification problems. How can we use SVM with a large number of classes? There exists a variety of ways to extend the SVM to multi-dimensions. One simple approach is to train multiple, one-versus-all classifiers. To recognize three classes, A, B and C, we simply have to train three separate SVMs to answer the binary questions: "is it A?" "is it B?" and/or "is it C?" More sophisticated approaches also exist to generate an SVM to account for multiple classes. For data sets with thousands of examples, solving the SVM optimization problem is quite fast. Empirically, running times of state-of-the-art SVM learning algorithms scale approximately quadratically, which means that an SVM input with twice as much data requires four times as long to run.

What remains is training. The user has to take a subset of the labeled input values and assign them to classes independent of the SVM. These data then train the SVM so that it learns those values belong to a certain class. Keep in mind that classifiers learn and get smarter as one feeds the classifier more training data. At the same time, the more one defines the training data, the less data are available for the SVM to classify. The training data are identified by tagging. At least 2 points are needed to get started. One can always add more tags later. The investigator starts training the SVM by choosing tags for each feature in the data set. After manually tagging some examples, the SVM classifier will start making predictions on its own. If one wants the model to be more accurate, more data will need to be tagged to continue training the model. There is always a trade-off between the training number and the accuracy of the model.

In summary, a SVM allows one to classify data that are linearly separable. If it isn't linearly separable, the user can apply the "kernel trick" to transform the nonlinear data into a linear one or perform the SVM on the dot-product rather than the "support vectors." For many problems, it is best to use a linear kernel.

7.5.1 Sea Level Prediction by Support Vector Machines Combined with Particle Swarm Optimization

Noting the absence of satellite altimetry data in the near coastal region, Moura et al. (2010) analyzed coastal sea level measurements using Support Vector Machines (SVMs) to predict future coastal sea levels arising from the ongoing effects of global warming. The coastline of interest is that of Brazil, which has very few sea level monitoring stations. There are only 7 tide gauge stations along the 7,367 km coastline of the country. Given this paucity of data, it is important to have a model that is able to predict future sea level changes.

In their study, the authors applied the SVM method to the available coastal sea level data in order to provide a model to predict future sea level changes. SVM have been successfully applied to time series data and have produced useful predictive model results even when relatively few data are available. In many of these applications, SVM models have provided superior performance relative to other supervised learning techniques for classification, including Artificial Neural Networks (ANN). The ANN are based on the Empirical Risk Minimization (ERM), which only minimizes the training errors. SVMs use Structural Risk Minimization (SRM), and, through the mathematical application of this principle, it seeks to minimize an upper bound of the generalization error, which appears to be less burdensome as the ERM are designed for dealing with very large sample sizes. In addition, the SRM corresponds to dealing with a convex quadratic optimization for which the Karush-Kuhn-Tucker (KKT) conditions are necessary and sufficient for a global optimum.

Despite these attractive characteristics, SVMs are not widely used in environmental prediction problems (Sapankevych and Sankar, 2009). The SVM performance depends on a set of parameters from the related learning problem, known as training. In order to select the optimal values for the parameters, a Particle Swarm Optimization (PSO) method was used. Thus, the overall method is a combination of PSO and SVM. The adequacy of the SVM for the Brazilian sea level data was evaluated with the Normalized Root-Mean-Square-Error (NRMSE). This indicator will demonstrate the performance and adequacy of the SVM for the sea level analysis.

Non-linear SVMs entail the use of three parameters: ε, C and γ. SVM performance depends strongly on these parameters and their selection is known as the model selection problem. As it is very difficult to select ε, C and γ by trial and error, we need a structured method to perform this task. PSO (Fei et al., 2009) was chosen because it is well-suited to handle real-valued decision variables and does not require derivatives. Also, PSO requires less computational time. For the SVM model selection problem, the available data are divided into training, validation and test sets, denoted by $\ell - r$, r and m examples, respectively. The validation set, by means of the validation error, guides the selection of the parameter values. Once the search is finished, the trained SVM is used to predict the output values (y_i, $i = \ell +1, \ell +2, ..., \ell +m$) from the test data set, in order to assess its generalization ability by the test error. Both the validation and test errors using NRMSE, which uses real (y_i) and predicted ($\widehat{y}_i$) output values:

$$\text{NRMSE} = \sqrt{\frac{\sum_i (y_i - \widehat{y}_i)^2}{\sum_i (y_i)^2}} \tag{7.14}$$

in which $i = \ell - r + 1, \ell - r + 2, ..., \ell$ when validation NRMSE is considered and $i = \ell + 1, \ell + 2, ..., \ell + m$ when test NRMSE is evaluated.

7.5.1.1 Combining PSO and SVM

The real solution to the problem comes when PSO and SVM are combined. PSO is a probabilistic approach based on the behavior of organisms that move in groups (e.g., birds, fishes). It was initially used to solve non-linear optimization problems. For the SVM model selection problem described above the SVM parameters, ε, C and γ become decision variables for the PSO. Thus, the ith particle is described by its current position "1" in the 3-dimension search space $\mathbf{s}_i = (s_{i1}, s_{i2}, s_{i3})$, the best individual position is having occupied $\mathbf{p}_i = (p_{i1}, p_{i2}, p_{i3})$ and its velocity $\mathbf{v}_i = (v_{i1}, v_{i2}, v_{i3})$, where the first, second and third dimensions refer to C, ε, and γ. A particle moves through the search space in order to find an optimal position by using its own information and also from its neighboring particles. Its performance is measured by its fitness that is evaluated by an objective function (NRMSE) computed at its current position. The combination of the PSO and the SVM takes place at the fitness evaluation step. The particle provides the SVM algorithm with its corresponding parameter values. Then, SVM gives a trained machine that is used to predict output values from the validation set. With these predicted values, validation is based on the lowest value of NRMSE and the PSO process continues.

7.5.1.2 Sea Level Prediction with Support Vector Machines

The goal of the Moura et al. (2010) analysis was to predict coastal sea level based on a limited set of observations. The sea level data were provided by the Ponta de Armacåo Station, which is located on the Southeastern Brazilian Coast at 22°53′ S, 43°08′ W and maintained by the Brazilian Navy Hydrographic Center. A total of 214 daily average sea levels from June to December 2004 were used in this analysis. It was assumed that a future sea level value (y_t) is a function of the immediately previous value (y_{t-1}). Thus, the data set is made up of 213 pairs (y_{t-1}, y_t). From these, the first 170 points were used for training, the next 20 were put in the validation data set and the last 23 in the test set.

Because the PSO is a stochastic method, the PSO + SVM procedure was replicated 10 times in order to explore its behavior. The descriptive statistics are in Table 7.1. The values of the machine parameters associated with the best PSO + SVM combination was the one with the smallest test NRMSE value (3.714×10^{-2}) for which the study found $C = 987.7249$, $\varepsilon = 7.1222$ and $\gamma = 3.2865 \times 10^{-2}$. The corresponding NRMSE value was 5.2351×10^{-2}. These are the values

TABLE 7.1 Descriptive statistics for the 10 PSO + SVM runs of sea level predictions for the coast of Brazil. NRMSE denotes the Normalized Root-Mean-Square-Error between the observations and the model predictions.

		Minimum	Maximum	Median	Mean	Std. dev.
Parameter	C	96.1314	1389.2354	270.9468	388.4644	440.6952
	ε	4.4714	7.1222	4.6660	5.0802	1.0753
	γ	3.287×10^{-2}	21.8210	4.636×10^{-2}	2.2224	6.8862
NRMSE	Validation	5.166×10^{-2}	7.109×10^{-2}	5.168×10^{-2}	5.368×10^{-2}	6.121×10^{-3}
	Test	3.714×10^{-2}	3.931×10^{-2}	3.746×10^{-2}	3.761×10^{-2}	6.148×10^{-4}
				Absolute (relative, %) frequency		
Stop criteria	Maximum iterations (6,000)			0 (0)		
	Equal best fitness for 600 iterations			10 (100)		
	Tolerance $\delta = 1 \times 10^{-12}$			0 (0)		
				Metric value		
Performance	Mean time per run (minutes)			13.2868		
	Mean number of trainings			73178.2		
	Mean number of predictions			73178.2		

Modified after Moura et al. (2010).

that the machine used for the sea level Support Vector Regression (SVR) analysis. The resulting sea level predictions for the validation and test runs are compared with the recorded sea levels in Table 7.2.

The graphical presentation of these results in Figure 7.38 demonstrates how well the PSO + SVM model did in predicting sea level. This method appears to provide an excellent way to predict sea level in a region where altimetry is not available and there are relatively few tide gauges in operation.

TABLE 7.2 Observed and predicted SVR sea levels (cm) for data pairs (y_{t-1}, y_t) for the southeast coast of Brazil.

Validation			Test		
Example	Observed	Prediction	Example	Observed	Prediction
171	190.6667	177.9579	191	173.9583	178.4348
172	197.7500	190.1206	192	179.8333	177.7922
173	196.4167	189.3322	193	171.3333	177.8987
174	172.0000	186.7060	194	182.6667	178.9812
175	157.7083	177.7466	195	176.7917	181.2931
176	153.1250	162.5183	196	170.8750	178.9812
177	158.2500	155.6255	197	176.2917	179.3478
178	165.2917	162.4041	198	177.2917	178.1885
179	172.3750	167.5212	199	173.0000	177.8970
180	183.6250	177.5037	200	184.7083	177.4231
181	181.2083	183.7403	201	188.9167	186.5204
182	184.2500	178.8391	202	182.1667	186.6943
183	175.2917	185.4152	203	191.2917	180.2508

TABLE 7.2 Observed and predicted SVR sea levels (cm) for data pairs $(y_{t-1}, y_t$ for the southeast coast of Brazil.—cont'd

Validation			Test		
Example	Observed	Prediction	Example	Observed	Prediction
184	164.2917	178.8391	204	199.5417	192.1807
185	154.1250	163.7561	205	188.5833	197.2796
186	150.6250	156.4856	206	178.5833	186.6691
187	158.2500	162.6228	207	178.5833	177.6865
188	168.7500	162.2368	208	177.5417	177.6865
189	177.8750	181.6912	209	183.8750	177.8297
190	171.4167	177.7566	210	201.1250	184.4201
			211	191.5000	197.1957
			212	180.2917	192.8508
			213	178.7500	178.1065

Modified after Moura et al. (2010).

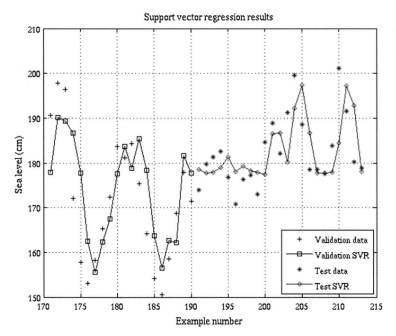

FIGURE 7.38 The modeled PSO + SVM regression results for sea level on the southeast coast of Brazil. Data are from Table 7.2. *From Moura et al. (2010).*

7.5.2 Classification of Sea Ice in the Baltic Sea with TanDEM-X Multiparametric Data Features and Support Vector Machines

Marbouti et al. (2020) used X-band interferometric Synthetic Aperture Radar (SAR) imagery to develop a support vector machine (SVM) for automated classification of sea ice in the Baltic Sea. A bistatic SAR scene acquired by the TanDEM-X mission over the Bothnian Bay in March of 2012 was used in this analysis. Backscatter intensity, interferometric coherence magnitude and interferometric phase were included as informative features in the classification experiments. Various machine learning tools were applied for classification (maximum likelihood and random forest), but we restrict our focus to Support Vector Machines (SVMs) in this example. With the SVM, the authors were able to discriminate between open water and several ice types, including undeformed ice, ridged ice, moderately deformed ice, brash ice, thick level ice, and new ice. The addition of interferometric phase and coherence magnitude to backscatter-intensity resulted in improved overall classification performance compared to using backscatter-intensity only. The best SVM accuracy was 72.91% when the interferometric parameters were added, compared to 63.05% when using backscatter intensity only.

As with all radars, SAR has the benefit that the atmosphere is basically transparent, even when densely cloud covered. Thus, SAR provides excellent coverage in the polar regions, where clouds persist nearly year around. SAR imagery has been used in polar regions to compute sea ice motion, ice concentration, ice-type classification and iceberg detection and monitoring. In the Baltic Sea, the primary application of SAR imagery is winter navigation, when the absence of sunlight makes it difficult to observe the sea surface. Under these conditions, radar can guide operations in an ice-covered ocean. Because the Baltic Sea has a relatively low salt content, sea ice forms more rapidly than in the adjacent Atlantic Ocean.

Operational sea-ice forecasting for the Baltic is performed by the Finnish Meteorological Institute (FMI), which provides daily ice charts primarily from the analysis of imagery from RADARSAT-2 and Sentinel-1 SAR satellite missions. These sensors have sufficient spatial resolution (100 m) for general navigation needs. In some cases, however, in order to distinguish ice ridges, heavily deformed ice and new ice formation in greater detail, it is necessary to use SAR imagery in the X-band, with its higher spatial resolution. The short X-band wave-length means that the SAR will have a greater sensitivity to varying surface sea ice properties such as small-scale surface roughness. Currently, the FMI uses trained experts to classify sea ice images, which is both time consuming and subjective. An objective machine-based analysis system was needed to remove the error possible with the present FMI procedure.

While there have been previous studies focusing on the use of interferometric SAR (InSAR) for sea ice monitoring, it appears that none prior to Marbouti et al. (2020) had utilized machine learning in carrying out the classification activities. In addition, the Marbouti et al. (2020) study combines both SAR interferometry and backscatter intensity for SVM based sea ice classification. The study area is located in the vicinity of Hailuoto Island in the north arm of the Baltic Sea (Figures 7.39a). Sea ice in the brackish Baltic waters begins to form in November, reaching its maximum extent between January and March. Ice breakup starts in April and typically completes by the beginning of June. The average winter ice cover is 40%.

There was a mild winter in 2012 but the northern and eastern basins of the Baltic Sea still froze over. At the end of March, the ice in the Bay of Bothnia was tightly packed into the northeast corner (Figure 7.39b), with the InSAR data covering both landfast ice and drift ice. In the area of satellite coverage, the thickness of the landfast ice was generally 35−60 cm, while the drift ice largely contained deformed ice. As the chart in Figure 7.39b was not detailed enough for the study, the FMI performed an independent high-resolution reference ice chart based on other operational satellite data and ground truth data from icebreakers. The ice types chosen for this study are listed in Table 7.3.

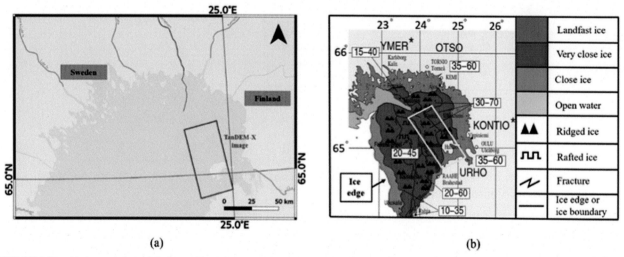

FIGURE 7.39   (a) An overview of the Bay of Bothnia in the Baltic Sea with the TanDEM-X footprint indicated by the red rectangle for an image collected on 30 March 2012; and (b) ice chart over the Bay of Bothnia for 30 March 2012, with the TanDEM-X coverage indicated by the yellow rectangle. *From Marbouti et al. (2020).*

TABLE 7.3 Description of the ice classes.

Ice type	Definition
Open water	A sea surface which is free of ice but may contain some ice fragments, by definition covering less than 1/10 of the surface.
New ice	New ice represents all forms of thin newly formed ice from frazil ice with almost liquid-like attributes to a uniform crust of up to 5 cm in thickness.
Undeformed Ice	Ice thicker than 10 cm, not significantly deformed but including rough surface properties.
Thick Level Ice	Ice thicker than 10 cm with a smooth surface and without any deformation. The ice surface might have been refrozen from flooded water or melted snow.
Ridged ice	Ice thicker than 10 cm with frequent occurrences of deformed ice, both rafted and ridged. Most of this ice type has been broken and piled to pressure ridges by compressive forcing. In this ice type, there is typically a network of ridge lines crisscrossing in a sheet of otherwise level first-year ice but also heavily deformed ice with virtually no level ice present.
Moderately deformed ice	Drift ice thicker than 10 cm, including a mixture of different ice types. The field consists of originally broken drift ice of different stages of development, including both rafted and ridged ice but also patches of level ice.
Brash ice	Ice broken into a fairly homogenous surface of very small ice blocks and typically identified by a high backscatter coefficient. In the scenes studied in this paper, brash ice can mainly be identified along known, fixed shipping lanes. It was chosen into the categorization because of its distinct features and evident detectability. Brash ice also forms at ice edges, broken and compacted by wind and waves into wide and thick zones and causing considerable harm to navigation.

From Marbouti et al. (2020).

Landfast ice covers a large fraction of the satellite image coverage but was not included in the ice classifications. The boundary between landfast ice and drift ice is dynamic, changing with the formation of fractures that depend on the thickness of the ice and the consolidation between ice floes. Thus, landfast ice cannot be easily distinguished from drift ice in a single satellite image. Fracturing processes cause mixtures between drift and landfast ice.

Marbouti et al. (2020) used TanDEM-X reregistered single look slant range complex (CoSSC) image products in their study. The images were acquired on 30 March 2012, in the strip-map mode in the bistatic InSAR configuration, at HH polarization. The bistatic formation makes it possible to avoid temporal decorrelation, a unique benefit of the TanDEM-X mission. The mission's relatively short along track baseline and large perpendicular baseline are also beneficial for ice mapping. As we stated earlier, X-band has advantages over C-band in that it has a higher sensitivity to surface features on the ice.

Several features of the TanDEM-X image are provided in Figure 7.40; specifically, backscatter-intensity (Figure 7.40a), InSAR coherence-magnitude (Figure 7.40b), and InSAR-phase (Figure 7.40c). TanDEM-X image data were orthorectified using ESA SNAP software and the TanDEM-X InSAR-phase ramp was compensated to remove its effect on classification. All features were additionally filtered using a 7×7 boxcar filter (cf. Chapter 6 on filters). Land areas were removed using a land mask. Finally, linear stretching to a dynamic range (0; 255) was applied over each field before running the classification experiments. As noted previously, highly detailed daily ice classification maps were generated for the study area by the Finnish Meteorological Institute (FMI). The charts were based on expert visual interpretation of SAR imagery. Currently, RADARSAT-2 and Sentinel-1 SAR imagery, with their wide swaths, are used along with some X-band imagery from Cosmo-SkyMed and TerraSAR-X. The experts also used visible and thermal infrared imagery from the Moderate Resolution Imaging Spectrometer (MODIS), *in situ* observations, sea reports from icebreakers and sea ice models to generate the high-resolution ice maps. The reference map (Figure 7.41) generated for the study is more highly detailed than the routine FMI ice product.

Because the Support Vector Machine (SVM) is a supervised classifier, it was important to select an appropriate training set. A total of 14,000 pixels (2,000 per each class) were randomly selected based on the feature properties dictated by the sea ice experts. The SVM classifier was implemented using MATLAB. In order to evaluate the benefits of adding InSAR features to backscatter-intensity, several classification experiments were carried out for each TanDEM-X feature and for their various combinations. The SVM is basically a binary classifier but in the case of multiclass problems, such as sea ice

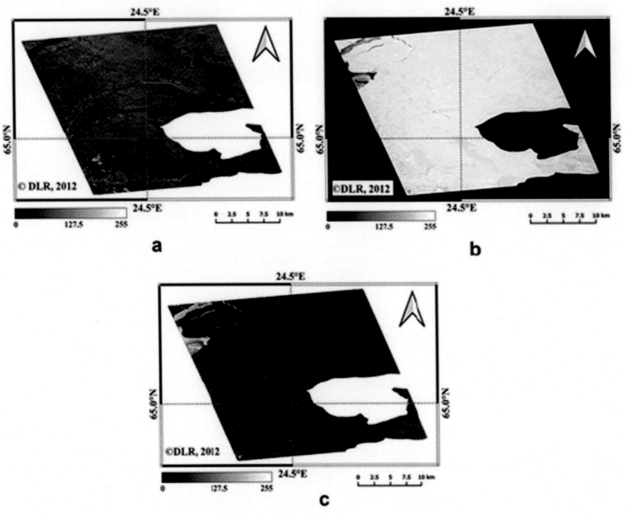

FIGURE 7.40 TanDEM-X image features. (a) Backscatter-intensity; (b) InSAR coherence-magnitude; and (c) InSAR-phase. The image for the Bay of Bothnia (Baltic Sea) was acquired on 30 March 2012. A land mask was applied. The features were linearly stretched to a [0;255] dynamic range. *From Marbouti et al. (2020).*

FIGURE 7.41 Reference ice classification map for 30 March 2012 for the Bay of Bothnia in the Baltic Sea. A description of the ice types associated with the colors is given at the left along with the training data used in the analysis. New ice class with training plots are shown in the upper left corner of the image. *From Marbouti et al. (2020).*

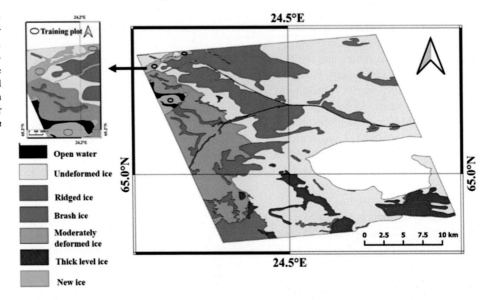

classification, SVM classification must be extended to a number of classes. Two of the common methods to enable this adaptation include the one-versus-one (OVO) and one-versus-all (OVA) approaches. The OVA method represents the earliest and the most common SVM multiclass approach. It involves the division of a K-class dataset into K-two class cases. On the other hand, the OVO approach involves construction of a machine for each pair of classes resulting in $K(K-1)/2$ machines.

In this sea ice problem, an error-correcting output code (ECOC) multiclass model based on SVM binary learners was trained using the MATLAB environment. Thus, an OVO approach with K binary learners is used in the analysis. The number of binary learners depends on the number of classes, which in this case are seven, for which the OVO model has 21 binary learners. As noted earlier, 2,000 pixels were used to train each class.

The accuracy of the SVM classification was assessed using the stratified random sampling approach, keeping the same number of pixels in each class for the accuracy assessment. The confusion matrix (CM) was calculated for ice classes and the following accuracy parameters were used: Overall Accuracy (OA), User Accuracy (UA), Producer Accuracy (PA) and the Kappa coefficient of determination (see the following Glossary for definitions). The parameters are defined as:

OA = (Overall number of correctly identified pixels)/(Total number of pixels);
UA = (Number of correctly identified pixels in a given map class)/(Total number of pixels claimed to be in that map class);
PA = (Number of correctly identified pixels in the reference plots for a given map class)/(Total number of pixels claimed to be in that map class);
Kappa = (Observed accuracy − chance agreement)/(1 − chance agreement).

The OA and Kappa values for the SVM results for the different classifications are listed in Table 7.4. The classified map for the final case with all of the parameters (last row in Table 7.4) is presented in Figure 7.42.

Note the similarities between Figures 7.41 and 7.42 (the reference map made by experts at the FMI). Even the detailed features of the FMI map are well captured by the SVM classification in Figure 7.42. Thus, the combination of all the parameters had the best overall accuracy of 72.91% and the highest Kappa coefficient of 68.39%. The results of the SVM classification are summarized in Figure 7.43, which shows the UA and PA values for the sea ice classes and the various combination of data parameters. Marbouti et al. (2020) present similar plots for the RF and ML classifications.

The computation time for SVM classification was about 30 s per any image feature; the time increased when using combinations of two or three features. The best detected class was open water with UAs of 87.84−97.72%. Rigid ice was the hardest class to discriminate inn four classification features with UAs on the order of 37.39−56.92%. Brash ice in InSAR-phase was not discriminated at all and it was also low in coherence-magnitude + InSAR-phase combination (22.41%). Moderately deformed ice with UAs on the order of 19.77% was also difficult to discriminate in backscatter-intensity. In PA new ice was the best detected class. Brash ice was not discriminated at all with any single classification feature. Other classes particularly hard to discriminate were ridged ice with PAs on the order of 42.87−52.76%, undeformed ice with PAs on the order of 39.18% and moderately deformed ice with PAs of about 27.59%.

TABLE 7.4 Support Vector Machine (SVM) overall accuracies and Kappa coefficients for the sea ice study in the Bay of Bothnia in the Baltic Sea.

Classification features	OA SVM (%)	Kappa SVM (%)
Backscatter-intensity	63.05	56.90
Coherence-magnitude	61.88	55.53
InSAR-phase	52.62	46.16
Backscatter-intensity + coherence-magnitude	72.15	68.38
Backscatter-intensity + InSAR-phase	72.78	68.24
Coherence-magnitude + InSAR-phase	61.89	55.54
Backscatter-intensity + Coherence-magnitude + InSAR-phase	72.91	68.39

From Marbouti et al. (2020).

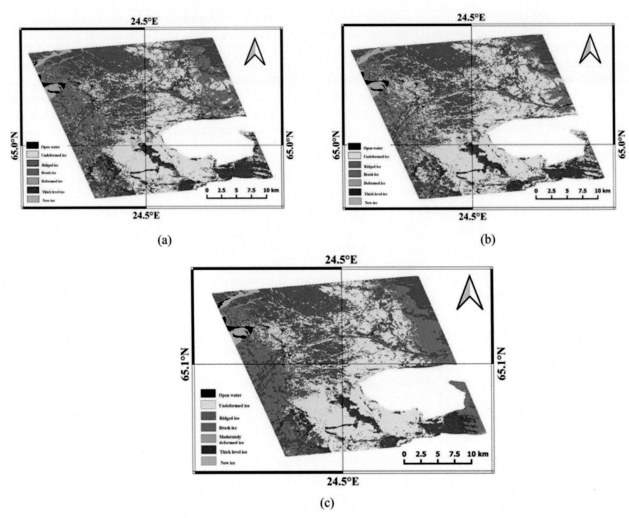

FIGURE 7.42   Final classification maps for water and sea ice classes for the Bay of Bothnia using combined backscatter-intensity + coherence-magnitude + InSAR-phase. (a) Random Forest (RF); (b) Maximum Likelihood Estimation (MLE); and (c) Support Vector Machine (SVM) classifications. *From Marbouti et al. (2020).*

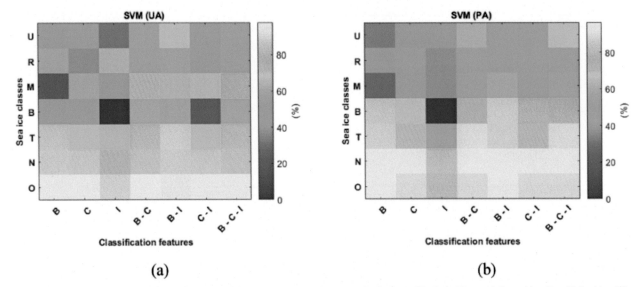

FIGURE 7.43   (a) UA and (b) PA for the SVM classifier for water and sea ice classes in the Bay of Bothnia (U = undeformed ice, R = Ridged ice, M = Moderately deformed ice, B = Brash ice, T = Thick level ice, N = New ice, O = Open water) versus classification features (B = Backscatter-intensity, C = Coherence-magnitude, I = InSAR-phase.). See text for a definition of UA and PA. *From Marbouti et al. (2020).*

7.5.3 Bridging Observations, Theory and Numerical Simulation of the Ocean Using Machine Learning

The study of the ocean provides some unique challenges that machine learning tools can help to investigate. Observational data are spatially sparse, biased to the surface, with a few time series spanning more than a few decades. Important time scales in the ocean range from seconds to millennia, with strong interactions and numerical modeling complicated by details such as coastlines, seafloor topography, baroclinic tides and internal waves. A study by Sonnewald et al. (2001) examines three elements of ocean science: observations, theory and numerical modeling. The authors discuss both the historical treatments and introduce the relevant machine learning tools, with a focus on machine learning analysis of *in situ* observations and satellite observations. Their study shows how Support Vector Machines (SVMs) can best be utilized to classify data in a way that theory and numerical simulations can best be applied to understand the data analyzed with ML tools. In these applications, Sonnewald et al. (2001) address model error, bias corrections and data assimilation.

Glossary

ADtrees Alternate Decision Trees
ANN Artificial Neural Network
Aqua NASA afternoon polar orbiting spacecraft
BDtrees Boosted Decision Trees
CCI Correctly Classified Index
Chl-*a* Chlorophyll-*a* is a form of chlorophyll used in oxygenic photosynthesis
CM Confusion Matrix
CNN Convolutional neural networks
DO Dissolved Oxygen
DT Decision Trees
ECOC Error Correcting Output Code
FAI Floating Algae Index
FMI Finnish Meteorological Institute
GB Gradient Boosting
GMED Global Marine Environmental Datasets
HJ-1A,B Small Chinese environmental satellites
InSAR Interferometric SAR
ML Machine Learning
MLE Maximum Likelihood Estimation
MODIS Moderate Resolution Imaging Spectrometer
MSLP Mean Sea Level Pressure
MUDB Match Up Data Base
NAE North Atlantic-European
NOAA/NCEI/STAR Center for Satellite Applications and Research
NRMSE Normalized Root Mean Square Error
OA Overall Accuracy
OBDAAC Ocean Biology Distributed Active Archive Center
OVA One-versus-all
OVO One-versus-one
PCA Principal Component Analysis
PDE Partial Differential Equation
PDF Probability density function
pH "Potential of hydrogen", a scale to specify the acidity or basicity of a fluid
PSO Particle Swarm Optimization
RF Random Forest
SAR Synthetic Aperture Radar
SNPP Suomi NPOESS Preparatory Platform
SSS Sea Surface Salinity
SST Sea Surface Temperature
SVM Support Vector Machine
TanDEM-X German X-band SAR satellite
Terra NASA morning polar orbiting spacecraft

TIN Total inorganic nitrogen
TN Total Nitrogen
TSP Total Suspended Particles
TVS Total Volatile Solids
UA User Accuracy
VIIRS Visible Infrared Imaging Radiometer Suite

Chapter 8

NEURAL NETWORKS, CONVOLUTIONAL NEURAL NETWORKS AND DEEP LEARNING

8.1 INTRODUCTION

Until recently, most computer code was written by programmers using basic first principles. This provides a way to digitalize the environment into code that the computer could operate with. Neural networks, on the other hand, are codes designed to mimic the processes by which our brain performs its functions. The model of the brain is used to create a processing algorithm that can be carried out by the computer. The advantage of transferring such processes to the computer is that the computer can process a large amount of data very quickly. This is a general goal of machine learning, as discussed in the previous chapter. All machine leaning methods have the objective of teaching the computer to do the processing without further human intervention.

Considerable time could be spent discussing what precisely constitutes data. For now, we restrict our discussion to data with a suitable numerical representation. Each example (or data sample) typically consists of a set of attributes called "features" from which our model makes its predictions. In supervised learning problems, the goal is to predict features that are members of the larger population but identified by the features used to "train" the model. When we are working with image data, each individual image might constitute an example represented by an ordered list of numerical values corresponding to the brightness of each picture element (pixel). A 200 pixel × 200 pixel color image would have $200 \times 200 \times 3 = 120,000$ numerical values, corresponding to the brightness of the red, green and blue values for each spatial location. When every case is characterized by the same number of numerical values, the data are said to consist of fixed-length vectors, and we can ascribe the length of the vectors to the dimensionality of the data. However, not all data can be easily represented by fixed-length vectors. Images from the same instrument are expected to be of fixed length, but images taken by different instruments, even of the same object, will have different dimensions, orientations and shapes. For images, the user could consider cropping them all to a standard size but that might lead to the loss of useful information. Cropping also won't account for differences in shape and orientation.

In general, the more data that are available, the easier the data are to model. With more data, we can train more powerful models and rely less on pre-conceived assumptions. The change from relatively small data sets to today's "big data" has been a major contributor to the development of modern "deep learning." Many of the more important new models do not work without large data sets. However, it is not enough to just have large volumes of data and to process it in an intelligent way; we must also have the "right" data. If the data are riddled with errors, or if they are not related to the target quantity, the learning process is going to fail. Caution must be exercised to not use a faulty dataset to train the model as it will give erroneous results that may appear correct. While deep learning and neural nets are powerful tools, their successful application requires an informed user to dictate their application. Not all input data problems are apparent as obvious errors, and the user must carefully ensure that none of the categories of data are underrepresented in the training data.

8.2 MODELS

Most machine learning applications involve transforming the data in some quantitative way. For example, we may have a group of satellite images from which we want to extract a particular target object, such as an oil slick. In this sense, the model is a computational process which ingests the data so that the model gives us the target we are looking for. This process is called a "prediction" because the model provides information that was not apparent in the input data. (An early introduction to the application of neural network models for the prediction and analysis of meteorological and

Data Analysis Methods in Physical Oceanography. https://doi.org/10.1016/B978-0-323-91723-0.00015-5

oceanographic data is found in Hsieh and Tang (1998).) While simple models can perform this task well, with relatively small datasets, "deep learning" can address rather large datasets and yield a useful answer that would not be possible with a simpler model. Here, deep learning refers to much more powerful models that generally require a large number of "hidden layers" in the neural network. Thus, these models result in many transformations of the input data that are chained together "top to bottom"; hence the name, deep learning.

Deep learning models are considered an improvement on classical predictive models. To judge the actual level of improvement, we need metrics to measure the improvement. These metrics are termed "objective functions". We usually define these metrics such that lower values are considered to be better (more accurate or more representative). However, as this is merely a convention, one can take any function in which higher is better and turn it into a new function that is quantitatively identical but for which lower is better by flipping the sign. When lower is better, the functions are often called "loss functions." The most common loss function is "squared error" (the square of the difference between the prediction and the validation or "ground" truth). For classification cases, the most common objective is to minimize the error rate (i.e., the fraction of examples for which the predictions differ from the validation or ground truth data). Some objectives, such as the squared error, are easy to optimize, while others, like error rate, are more difficult to optimize.

Typically, the loss function is defined relative to the model's parameters, depending on the particular dataset. We define the best values of our model's parameters by minimizing the loss incurred on a set consisting of the examples collected for training. It is important to realize, however, that doing well with the training dataset does not ensure that the model does well with new data. Hence, the investigator normally splits the available data into two parts: the training data for defining the model parameters, and the "test" data, which is held aside for validation and for a measure of how well the model performs. When a model performs well with the training data set but fails to generalize to unseen data, the model is said to be "overfitting".

8.3 SUPERVISED LEARNING

In supervised learning, known examples are used to train a model so that it can "classify" any unknown input data into similar "predicted" labels. The supervision applies to the fact that the user (the supervisor) provides the model with a dataset consisting of known labeled examples, where each known input parameter is matched with a known "ground truth" label. In probabilistic terms, we are interested in estimating the conditional probability of a particular label given a specific input feature. While supervised learning is only one type of machine learning, it accounts for the majority of successful applications, particularly in industry. Even for the simple case of "predicting labels given input features", supervised learning can take a wide variety of forms that all require specific decisions, depending, in part, on the number of inputs and outputs.

Supervised learning can be separated into parts (Figure 8.1): The first part consists of the training step, whereby known examples, having both the input data and corresponding labels, are selected for training; in the second part, the user needs to set aside an equally large group of examples to serve as the validation, or ground truth, data. For both the training data and the validation data, it is important to select (where possible) cases that cover the full range of the data. In an image, one doesn't want to pick all of the training and validation pixels from the same part of the image. The selected training data are input into a supervised learning model algorithm and the model parameters are adjusted to produce labels that match the training inputs. The second set of data is used to validate the model.

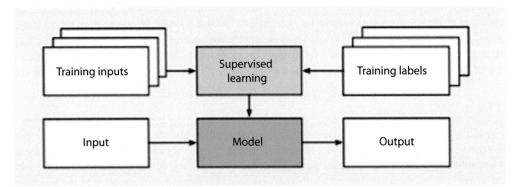

FIGURE 8.1   Schematic for training in Supervised Learning applications. The second set of "input" data are used for validation (ground truthing) of the model.

Once the model has been trained, the user then feeds the previously unseen input data into the trained model to obtain new output predictions. The model accuracy can then be evaluated by feeding the validation data input into the trained model and comparing the output values with the labels of the validation data set. This generates a model that can be used to analyze any amount of data that are similar to the training and validation data sets.

8.4 CLASSIFICATION

There are simple applications of Supervised Learning, such as regression, but we are interested in more general problems that can be extended easily to two-dimensions. In the classification of images, we want a model that can analyze the images to identify features by classifying the pixel values. It is easiest to express the classification model in terms of probabilities. Thus, the desired model assigns a probability to each possible class in the image. Consequently, an output probability of 0.9 means that the model is 90% certain that the pixel in question belongs in a certain class. Classification can be binary, multiclass or even multi-label. In some cases, classification can address hierarchies in which an erroneously misclassified pixel is assigned to a nearby related class rather than to a distant unrelated class.

8.5 UNSUPERVISED AND SELF-SUPERVISED LEARNING

We are sometimes faced with a large volume of data that has no previously known relationships to assist in the choice of training or validation data. Nevertheless, we still want to find possible relationships among these data. This type of approach is called "Unsupervised Learning" and the types and numbers of questions that can be asked are limited only by the investigator's ingenuity. To this end, we can ask:

1. Is it possible to identify a small number of features that accurately summarize the data?
2. Can we identify a small number of parameters that accurately capture the relevant properties of the data?
3. Is there a representation of objects in Euclidean space such that symbolic properties can be well matched?
4. Is there a description of the root source of many of the data characteristics that we observe?

An important recent development in unsupervised learning is the advent of generative adversarial networks.

A form of unsupervised learning known as "self-supervised learning" leverages unlabeled data to provide supervision in training, such as by predicting a withheld part of the data in another part. For images, we may train models to determine the relative position between two cropped portions of the same image.

8.6 FUNDAMENTALS

The basic concepts of neural networks have been around for centuries. Many of the tools used in data science today trace their roots back to developments such as the Gaussian (normal) distribution and the least squares procedure discovered by Carl Friedrich Gauss (1777−1855). Statistics rapidly advanced with the collection and availability of data. One of the leaders, Ronald Fisher (1890−1962), created concepts like "linear discriminant analysis." A major development was the emergence of digital computing and the influence of mathematicians like Claude Shannon (1916−2001), who introduced information theory, and Alan Turing (1912−54), who addressed the idea of whether or not machines can "think." Major contributions can also be found in neuroscience and psychology. One of the oldest algorithms was formulated by Donald Hebb (1903−85) who posited that neurons learn by positive reinforcement. This became known as the Hebbian learning rule, which laid the foundation for many of the stochastic gradient descent algorithms that underpin deep learning today.

Biological inspiration gave neural networks their name. For well over a century, researchers have tried to assemble computational circuits that resemble networks of interacting neurons. Over time, the interpretation of biology has become less literal, but the name has prevailed. There are a few key principles that can be found in most networks today.

1. The alternation of linear and nonlinear processing units, often referred to as "layers".
2. The use of the chain rule (also known as back propagation) for simultaneously adjusting parameters in the entire network.

After initial rapid progress, research in neural networks was somewhat forgotten from 1995 to about 2005. There were several reasons for this. First, training a network is computationally expensive and, at the time, computer power was very limited. Second, datasets were relatively small and did not require the processing ability of deeper neural networks. As a result, more traditional methods such as kernel methods, decision trees and graphical models were computationally adequate. Unlike neural networks, these methods do not require large amounts of computational time.

8.7 THE PATH TO DEEP LEARNING

Two developments set the stage for the advancement of data science into deep learning. One was the emergence of high-speed digital computing and the other the ready availability of large amounts of data. These elements were later brought together with the creation of the Internet that linked data with computing capabilities. While computing power has out-stripped the increase in data volume, the amount of random-access memory has not. As a consequence, statistical models need to become more memory efficient (typically, by adding nonlinearities), while spending more time on opti-mizing these parameters using more computer time. Thus, data science moved from linear models and kernel methods to deep neural networks. This also explains why many of the important elements of deep learning (multilayer perceptrons, convolutional neural networks, long short-term memory and Q-Learning (where "Q" stands for Quality)) have been "rediscovered" in the last decade after laying somewhat dormant for a considerable time. Deep learning frameworks now play a critical role in distributing ideas.

Recent developments in deep learning include:

1. Novel methods for capacity control that have helped to mitigate overfitting.
2. Attention mechanisms that make it possible to increase the memory and complexity of a system without increasing the number of learnable parameters.
3. Multi-state designs and the neural programmer-interpreter that allows modelers to describe iterative approaches to reasoning.
4. The invention of generative adversarial networks, which allow the user to replace the sampler with an arbitrary algo-rithm having differentiable parameters.
5. The ability to create parallel and distributed training algorithms that make it possible to process large amounts of data available for training in cases where a single GPU (graphics processing unit) is not sufficient.
6. The ability to parallelize computations, which has contributed to progress in reinforcement learning.

Before addressing Deep-leaning Neural Networks, we need review some of their basic concepts.

8.8 LINEAR NEURAL NETWORKS

In Chapter 3 of this book, we discuss linear regression (invented by Gauss in 1795), where it is assumed that the rela-tionship between y and x is linear. As discussed earlier, we need to determine a measure of how well the linear model fits the data. For the case where y is regressed on x, the feature to minimize is the squared error of y relative to x, as indicated in Figure 8.2.

As depicted by Figure 8.3, we can think of a linear regression model as a neural network by expressing its components in neural network terms.

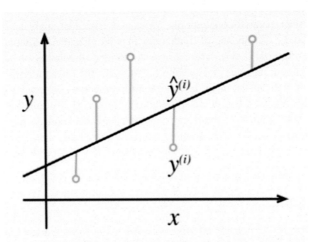

FIGURE 8.2 Linear regression of y on x showing the errors of the y values ($y^{(i)}$) relative to the linear fit denoted by $\hat{y}^{(i)}$.

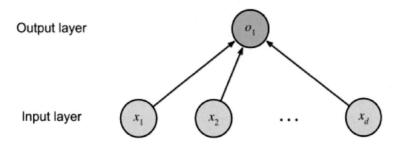

FIGURE 8.3   Linear regression as a single-layer neural network, with a single computed neuron.

Note that, in this depiction, the output consists of a layer containing the y output elements. Also, in this model, there is a single computed neuron, O_1. Because we do not count the input as a separate layer, the network in Figure 8.3 has only one layer. As every input x value is connected to an output y value, this transformation can be considered as a fully-connected, or dense, layer.

While the basics of neural networks were inspired by biology, the modern extensions into deep learning have deviated from the biology of neuroscience. A classic text on Artificial Intelligence by Russell and Norvig (2016) points out that, while flight might have been inspired by birds, ornithology has not been the driver of the development of airplanes. Aeronautics has deviated far from the flight of birds. Similarly, the development of deep learning originates more from mathematics, statistics and computer science than it does from neuroscience.

The most basic neural network (NN) is the "Multilayer Perceptron" (MLP), which overcomes the assumption of linearity in regression models by introducing a hidden layer that can accommodate non-linear solutions (Figure 8.4).

The MLP in Figure 8.4 has 4 inputs, 3 outputs and 5 hidden layer nodes. Because the input layer does not involve any computations, generating the outputs involves only the hidden layer and the output layer. Thus, this MLP has 2 layers. Note that all the nodes are "fully connected" and that each node influences every other node.

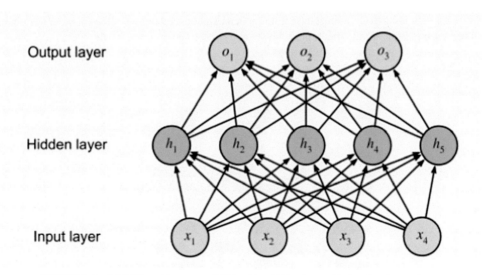

FIGURE 8.4   A Multilayer Perceptron (MLP) with a hidden layer with 5 nodes ($h_1, ..., h_5$).

8.9 THE STRUCTURE OF MORE COMPLEX NEURAL NETS

The basic idea of a neural network is to simulate the densely interconnected brain cells (now nodes in a computer) and to have this network of neurons learn certain patterns that enables it to make future decisions (Woodford, 2021; https://www. explainthatstuff.com/introduction-to-neural-networks.html). As with the human brain, once properly designed, the neural network is able to learn on its own. However, it must be remembered that a NN is not a brain but rather a software simulation that has been created by programming very ordinary computers. In effect, a NN differs from a human brain in much the same way that a computer model of the weather differs from actual weather features such as clouds, snow or sunshine. Computer simulations are just collections of algebraic variables and the mathematical equations that link them. The simulations mean something to the people who programmed the computers but not to the computers themselves. Strictly speaking, these are "artificial" neural networks (ANNs) and they are often referred to as such.

An ANN can have from a few dozen to hundreds, thousands or even millions of artificial neurons called "units", arranged in a series of layers, each of which connects to the layers on either side. Some of these layers are known as input units, which transfer information from outside of the ANN to be used for training the ANN or to exercise the ANN in its decision-making process. On the other side of the ANN, is the output unit, where the results of the ANN model analysis are deposited. Between the input and output units are the hidden units, which together with the other layers, form the artificial brain that makes the decisions that lead to the "predictions" that become the output of the ANN (Figure 8.5). Most ANNs are "fully connected," meaning that each hidden layer is linked to each output layer. Hidden layers are connected to the layers on both sides. The connections between one layer and another are represented by a number called a "weight". Weights can be either positive (if on layer excites another) or negative (if one layer suppresses or inhibits another). The larger the absolute value of the weight, the more influence the layer has on another layer. This is similar to the way that brain cells trigger one another across tiny gaps called synapses. An ANN with more than one hidden layer is called a "deep neural network" (DNN).

A DNN similar to that shown in Figure 8.5 is typically used for analyzing highly complex problems. In theory, a DNN can map any type of input to any kind of output. In reality, these complex networks require a much larger training set than a standard ANN to be able to cover all of the eventualities that the DNN will have to deal with.

For a neural network to learn, there has to be an element of feedback involved, just as we learn from experience as we mature. Everyone learns through feedback as we determine what does or doesn't work, so that the situation can be avoided in the future. Likewise, when something does work, we want to be able to repeat that in the future. A neural network learns in the same way, typically in a feedback process called "back propagation" (Figure 8.6), which involves comparing the output of a NN to the output that was expected in the training data set and using the difference between them to modify the weights of the connections between the layers, working from the output layers through the hidden layers to the input layers. Hence, the name, back propagation. Thus, a training data set must have both input and associated output values in order to train the NN. A DNN needs more training examples due to the many more layers that need to be adjusted by the back propagation procedure.

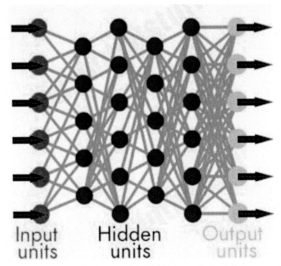

FIGURE 8.5 A fully connected ANN with six input units, 22 hidden units and 6 output units. *From Woodford (2021).*

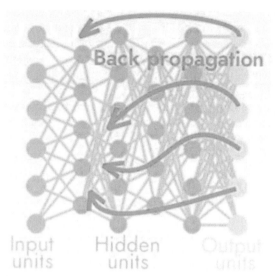

FIGURE 8.6 A neural network (NN) learning by back propagation, where differences between output values linked to the input values are used to adjust the weights in the hidden layers. *From Woodford (2021).*

With time, back propagation enables the network to learn by reducing differences between the actual and predicted outputs up to an assumed limit (close to zero). When trained, a network can be applied to input data with no corresponding output in order to predict the outputs consistent with the training provided to the network.

For example, consider a DNN that has been trained with 50 examples of trucks and 50 examples of cars. Following the training, we can input a bus and determine how the network classifies it. The network will try to identify the object as a truck or a car, and will look to determine if the object has 6 windows or 3, how long it is, how tall it is, how many doors it has, how many seats, and so on. These characteristics represent the nodes in the hidden layers that were tuned in the training process. The weights defined in the training process specify whether a characteristic node is activated or not. Thus, when faced with a new object, like a "bus", these weights will dictate which of the characteristics are compared and which are ignored.

8.10 FEED FORWARD NETWORKS

A primary goal of a neural network is to arrive at the point of least error as quickly as possible. In effect, the system is in a computational "race" in which it continuously repeats its processing steps. Each step involves a "guess," an error measurement and a slight update to the weights, as the network learns to pay attention to the most important features. A collection of weights produces a neural network that we also call a model because it is an effort to model the relationships between the input data and the validation or ground-truth data. In general, models are frequently inadequate but improve over time with additional training (cf. A Beginners Guide to Neural Networks and Deep Learning by Chris Nicholson, Pathmind Inc. 2020).

A neural network is born ignorant. It must learn which particular weights and biases will translate the input to a prediction of the best possible outputs. The network has to start with a guess and then make better guesses sequentially as it learns from its mistakes. In general, these steps are the following: (1) Input * weight = guess; (2) Ground truth − guess = error, and (3) Error * weights contribution to error = adjustment. These three steps are repeated in the training process for the network. When the nodes in a NN are summed, the input is passed through a non-linear function. What the process is trying to create at each node is a switch (like a neuron) that turns on and off, depending on whether (or not) it should let the signal of the input pass through to affect the ultimate decision of the network. The switch has to decide whether the input signal is "enough" or "not-enough" to activate the node.

8.11 GRADIENT DESCENT

One of the more widely used methods for optimizing the NN weights is called the "gradient descent". The gradient represents the slope in the relationship between two variables (x and y). In this case, the slope of concern is the one that describes the relationship between the network's error and a single weight, which informs the user how the error varies as the weight is adjusted. The objective is to determine which weight will produce the least error, or the one which correctly represents the signals in the input data and their transform to the corresponding output data. As a NN learns, it slowly adjusts a large number of weights so that they can map the input signal to the desired output correctly. The difference between the model prediction and the expected output (ground truth) is the network error. This error can be characterized as a derivative of the error (*Error*) with respect to weight (*dError/dWeight*) that measures the degree to which a slight change in a weight will cause a slight change in the output error.

In a DNN, each weight is simply one factor in a network that has many transforms, whereby the user must use the "chain rule of calculus" to return back through the network activations and outputs to arrive at the weight and its relationship to the overall error. Specifically, we can write

$$\frac{dError}{dWeight} = \frac{\partial Error}{\partial Activation} \times \frac{dActivation}{dWeight} \tag{8.1}$$

According to (8.1), the two variables, *Error* and *Weight*, are linked through a third variable, *Activation*. The model can then calculate how a change in *Weight* affects a change in *Error* by first calculating how a change in *Activation* affects a change in *Error*, and how a change in *Weight* influences a change in *Activation*. Activation functions decide whether a neuron (node) should (or should not) be activated by calculating the weighted sum and then adding bias to the sum. Thus, in deep learning, the system needs to adjust the model's weights in response to the error it produces until the process can no longer reduce the error.

The final layer in a deep neural network with many layers has an important role. In the training phase, the network deals with "labeled" data and the output layer classifies each example in its link to the input. This labels the node, which then turns on or off according to the strength of the signal it receives from the previous layer. Each output node produces either 0 or 1 and is driven by whether the input variable deserves a label or not.

Although neural networks themselves deal with binary values, the input variable often consists of continuous data. Thus, the output layer has to convert a range of continuous data into a range between 0 and 1, corresponding to the probability that a given input should be labeled or not. The tool used to convert continuous data into binary values is called "logistic regression," which is somewhat of an unfortunate term in that we are really using the conversion for classification rather than for linear regression (the concept we normally think of when we use the term "regression"). The formula that calculates the probability that a set of inputs matches the label is

$$F(x) = \frac{1}{1 + e^{-x}} \tag{8.2}$$

where *x* represents the model output.

Eqn (8.2) is known as the Sigmoid Function and originates with the earliest work with neural networks in which investigators were interested in modeling biological neurons that either fire or not. Because the goal is to represent continuous inputs as probabilities, the conversion (8.2) can only give positive results; probabilities cannot be negative. Hence, the exponent, *x*, is the output, which forces the result of the neural network to be greater than zero. Because the exponent of *e* in (8.2) is negative, the input variable *x* can only make the *e*-term smaller. In the limit of large *x*, the *e*-term approaches zero and the numerator approaches 1. So, $F(x) \rightarrow 1$, which is the maximum positive probability; any non-zero value of e^{-x} will only decrease the probability from unity. Input that correlates negatively with the corresponding output will have its value switched by the negative sign on the *e*-exponent, and the *e*-value will grow, pushing the entire fraction closer to zero. There are a variety of Sigmoid Functions other than (8.2) that can be used in a NN analysis.

In the simple case outlined above, we considered just one input value for *x*. In the general case, the exponent consists of the sum of the products of all the weights and their corresponding inputs, which represents the total signal passing through the network. This signal is ultimately what is being fed into the "logistic regression" layer at the output layer of the neural network classifier. In this layer, the user can set a decision threshold (in terms of probability) above which an example is labeled and below which it is not. A low threshold will increase the number of false positives, while a higher threshold will increase the number of false negatives. The user must choose on which side they want to err.

8.12 CONVOLUTIONAL NEURAL NETWORKS

One way NN can be used to analyze image data is to discard the spatial structure and flatten the image into one-dimensional vectors that can be analyzed by a fully-connected Multilayer Perceptron (MLP). The vectors contain the location of the image elements or pixels. Convolutional Neural Networks (CNN) were created to handle such two-dimensional (i.e., vector) data in a more correct fashion (cf. A Beginner's Guide to Convolutional Neural Networks

(CNNs), Pathmind.html, 2020). CNN applications are wide-spread in the field of computer vision and have saturated the NN analysis of all types of imagery. Not only do CNN properly represent a 2-dimensional field but they are also computationally efficient because they require fewer parameters than fully-connected NNs and because convolutions are easy to parallelize across Graphical Processor Unit (GPU) cores. As a consequence, analysts frequently use CNN whenever possible. Here, we outline the basic operations in CNN, including the convolutional layers, details on "stride" (the pooling layers used to aggregate information across adjacent spatial regions) and the use of multiple channels at each layer.

Suppose one needs to find a particular item in an image. The user has some characteristics of the target and can scan the image looking for these characteristics. Each location in the image would have a score relative to the target characteristics and the one with the highest score would be the target location. This score represents the likelihood (probability) that the target is located at the position with the highest probability. CNNs provide a procedure of "systematic invariance" that can be used to derive answers with fewer parameters. This approach is fairly simple if the image is black and white. In this case, the data can easily be converted into binary values. The procedure becomes decidedly more complex for color images that are made up of three channels red, green and blue. In this case, an image becomes a tensor, which makes it possible to handle all three channels.

CNNs consider color images as three-dimensional objects rather than flat fields because the images have width and height as well as color (red-green-blue, RGB) encoding to produce the color spectrum that our brain sees. A CNN takes in such images as three separate color levels stacked on top of one another. Hence, a CNN inputs a color image as a rectangular box whose width and height are given by the number of pixels along those dimensions, with a depth that is three levels deep, with one level for each of the RGB components. These levels are referred to as "channels".

As an image moves through a CNN, it can be described in terms of input and output volumes that are expressed mathematically as matrices of multiple dimensions in the form $20 \times 20 \times 3$, where we have assumed a square image in space and RGB color. Between the layers, these dimensions will change. The multiple dimensions of the image volume are the foundations of the linear algebra used to process the images. For each pixel of an image, the intensity of R, G and B is expressed by a number, and that number is an element in one of the three stacked, 2-dimensional matrices that together form the image volume. These numbers are the raw sensory features that are fed into the CNN, which then works to find the numbers that have significant signals that help to classify the images more accurately (in the same sense as we discussed with respect to Feed Forward Networks). Rather than focusing on one pixel at a time, a CNN examines square patches of pixels and passes them through a filter, which is a square matrix equal in size to the patch. The matrix is called a "kernel", similar to what was discussed with respect to Support Vector Machines, whose role is to find patterns in the image pixels.

In the present example, the CNN takes a 30 pixel $\times$ 30 pixel patch of the image and a filter that is 3×3, whereby the filter is 1/100 of the patch's single channel surface area. The system then takes the dot product of the filter with this patch of the single channel image. If the two matrices have high values when the dot product is taken, the dot product will have a high value. If not, the product will be low. Thus, the dot product tells us if the image pixel pattern matches that expressed by the filter. The approach begins in the upper left-hand corner of the image and moves the filter along the top step by step, until it reaches the upper right-hand corner. The size of the step is called the "stride".

At each step, the system takes another dot product, and places the results of the dot products into a third matrix called the "activation map". As a result, the width (number of columns) of the activation map is equal to the number of steps the filter takes to cross the image. Larger strides result in fewer steps and, as a consequence, a large stride will result in a smaller activation map. This is significant because the size of these matrices controls the computational time needed to process the CNN. A larger stride will require less computational time but at the expense of a less detailed pattern recognition. Because images have pixel patterns of different shapes and sizes, the user will want to apply other filters designed to find these other patterns. The dot products of each filter results in another activation map; and there are as many of these maps as there are filters that have been run over the image. It is this rolling of the filter over the image and taking the dot products to form the activation map that constitutes the convolution.

8.13 PREDICTION OF SEA LEVEL VARIATIONS WITH ARTIFICIAL NEURAL NETWORKS

Makarynskyy et al. (2004) applied an artificial neural network (ANN) to predict 24 h, semi-diurnal, daily, 5-day and 10-day sea levels at Hillarys Boat Harbour north of Perth in Western Australia for the period December 1991 to December 2002. Sea level measurements from the tide gauge at the harbor were used to train and to validate the ANN. The ANN was then used to forecast the sea level at the study site.

The three-layer, feed-forward network applied in this study had a log-sigmoid transfer function in the hidden layer and a linear transfer function in the output layer. Training of the networks was through a resilient back propagation algorithm in 200 training epochs. The size of the ANN was determined by a "saliency analysis" derived from the concept that an ANN must continue to function when the input data are incomplete, or if an internal component of the net fails. The saliency analysis estimates the relative importance of the input and processing nodes of an ANN in addition to the corresponding optimization of the ANN when missing neurons are introduced.

The hourly sea level records used for training and validation purposes were obtained from a SEAFRAME (Sea-level Fine Resolution Acoustic Measuring Equipment) tide gauge installed at the harbor (Figure 8.7).

The raw measurements shown in Figure 8.8 exhibit a seasonal variability, with minima sea levels in the Southern Hemisphere summer and maxima in the winter. Sea level ranges over 140 mm, which is small compared to the large tidal ranges that are observed farther north in Western Australia. The tidal variability in the harbor is also influenced by the location of the tide gauge inside this sheltered harbor (Figure 8.7). For the ANN study, the record was broken into three sections: (1) A section for training the ANN; (2) a section for correcting the ANN; and (3) a section for validation of the ANN results (see the time periods denoted by the vertical lines in Figure 8.8).

To obtain the hourly sea level forecasts, the first part of the time series (December 1991–September 1995) was used to train the ANNs, the second part (September 1995–May 1999) for independent training of the "corrected" networks, and the last segment (May 1999–December 2002) for model validation. The performance was evaluated in terms of the correlation coefficient, R, the root-mean-square-error ($RMSE$), and the scatter index SI ($RMSE$/mean). Using the saliency analysis noted above, ANNs with architectures ranging from 72 input neurons, 145 processing neurons and 24 output neurons (the $72 \times 145 \times 24$ network) to a $12 \times 25 \times 24$ network were used to produce 1-day (24 h) predictions of hourly

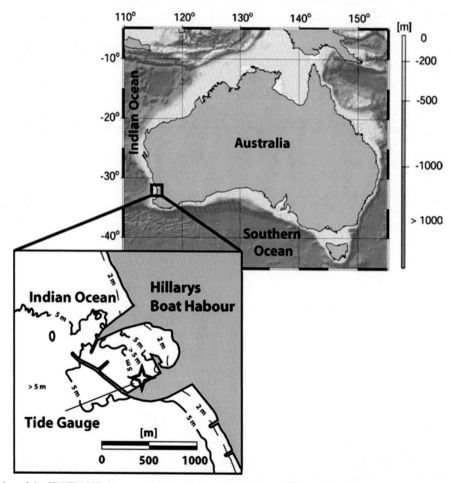

FIGURE 8.7 Location of the SEAFRAME tide gauge (star) at Hillarys Boat Harbour (Western Australia). This tide gauge has been operated since December 1991 by the Australian National Tidal Centre and for this study the entire record was used (see Figure 8.8). *From Makarynskyy et al. (2004).*

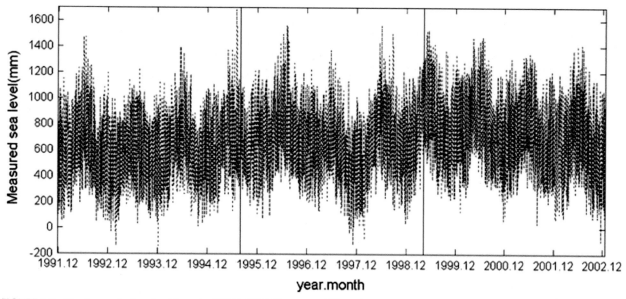

FIGURE 8.8 Hourly record of sea level from the SEAFRAME tide gauge at Hillarys Boat Harbour. The vertical lines mark the boundaries of the three partitions of the record used for training and validation. *From Makarynskyy et al. (2004).*

sea level (Table 8.1). All correlations were about 0.83 regardless of net architecture. The *RMSE* ranged from 111 to 139 and the *SI* from 0.155 to 0.195 (Table 8.1).

The results of the ANN model are presented graphically in Figure 8.9, where all of the hourly predictions are shown for the various input nodes. It is evident from Figure 8.9 and Table 8.1 that all networks perform in a similar way. However, it is also clear that the two networks with the greatest number of neurons in the input and hidden layers (i.e., $72 \times 145 \times 24$ and $60 \times 121 \times 24$) produce predictions with consistently lower accuracy than the three neural nets having a more simplified architecture (i.e., $48 \times 97 \times 24$, $36 \times 73 \times 24$, and $24 \times 49 \times 24$). The network with the best overall performance was the one with the same even number of inputs and outputs, and with the number of hidden nodes (processing neurons) equal to the sum of input and output nodes ($24 \times 49 \times 24$) plus 1. Thus, a large amount of input information combined with an overcomplicated network lowers the quality of the simulations because the more complicated ANNs are unable to find consistent relationships between the large number of input-output nodes.

An added benefit of the simpler ANNs is the fact that a decrease in the number of neurons leads to lower computational costs and a quicker result. While the simplest network gives the best results, Figure 8.9 indicates that even this result is poor for the 5–10-h forecasts. This is most evident in the ensemble averages, which show a steep decrease in correlation (*R*) and slight decreases in ensemble *RMSE* and *SI*. The authors of the study concluded that too few initial data (12-hourly measurements in this case) are not sufficient for an accurate simulation for the variability of sea level over the next 24 h. Thus, using 12-h training segments is not the best approach for longer term predictions. The overall conclusion is that the simple ANN successfully predicts hourly sea levels with lead times from 1 to 24 h.

TABLE 8.1 Number of nodes in the input and hidden layers of the Artificial Neural Networks (ANNs) and the corresponding performance statistics for the Hillarys Boat Harbour (sea level) prediction study.

Input neurons	Hidden neurons	Averaged *R*	Averaged *RMSE* (mm)	Averaged *SI*
72	145	0.816	139	0.195
60	121	0.830	130	0.182
48	97	0.859	118	0.165
36	73	0.859	119	0.166
24	49	0.875	111	0.155
12	25	0.823	122	0.172

From Makarynskyy et al. (2004).

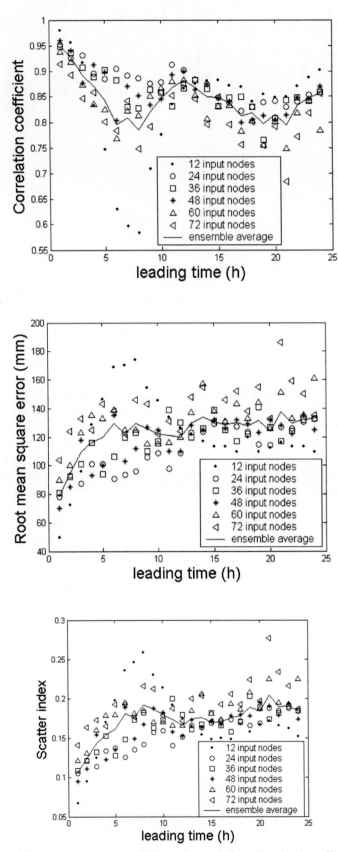

FIGURE 8.9   Correlation coefficients (*R*), root mean square errors (*RMSE*) and scatter indices (*SI*) as functions of lead time for the initial simulations of sea level at Hillarys Boat Harbour. *From Makarynskyy et al. (2004).*

8.14 PREDICTION OF BASIN-SCALE SEA SURFACE TEMPERATURES WITH ARTIFICIAL NEURAL NETWORKS

The prediction of sea surface temperature (SST) using artificial neural networks (ANNs) is considered as complimentary to SST predictions from numerical models. Usually, ANN models are site specific, while numerical models frequently treat entire basins. In their SST study, Patil and Deo (2018) use a very large number of ANNs to treat the entire tropical Indian Ocean (TIO, 30°N–30°S, 30°–120°E). The data set analyzed is from the Hadley Centre Sea Ice and SST (HadISST) data covering the past 140 years. Monthly SST anomalies were predicted at 3,813 nodes in the TIO basin over nine time steps into the future. The 20 million ANN models used in the analysis were found to accurately predict SST in both the eastern and western TIO, with lead times of 4 and 5 months, respectively. The predictive skill of the ANN models for the TIO region was found to be better than that of the physics-based, coupled atmosphere-ocean numerical models. In addition, the ANNs are capable of providing advanced warning of the Indian Ocean dipole and abnormal basin warming.

Knowledge of SST variability is important for understanding local air-sea interaction and long-term climate variability. For example, sea surface temperature is the most important contributor to cyclogenesis over the ocean and the formation and continuation of tropical cyclones. SST also exerts a strong influence on marine biodiversity and, hence, knowledge of the SST patterns can be helpful in locating possible fishing areas. SST is also an important variable in studying coral reefs, which can be severely damaged by anomalously high surface water temperatures.

Most sea surface temperature predictions have been made using complex, physics-based numerical models that rely on a suite of assumptions and observations to make them solvable. Data are also required for model initialization and assimilation. Physics-based simulations predicting SST are typically from coupled atmosphere-ocean models, which also predict a number of other ocean variables at the same time. Different statistical methods have also been used to predict SST, including linear regression, correlation analysis and principal component analysis. All have resulted in somewhat poor predictions of SST, with rather large root-mean-square (RMS) errors.

Neural networks have been applied in the past to predict SST on monthly time scales and found to be more effective than other nonlinear methods. There are many advantages to using ANNs to predict SST since they are statistical constructs, with no fixed physical-mathematical relationships, that can handle nonlinear connections. ANNs are, however, site-specific as they are restricted to the source region for the data used to train the ANN. In the Patil and Deo (2018) study, the authors have tried to extend the coverage area by using a number of ANNs to cover entire basins of the tropical Indian Ocean. The ANN SST predictions were then compared with those from numerical models of the same areas.

The HadISST data used in the Patil and Deo (2018) study were reconstructed from various *in situ* and satellite datasets using a two-stage, reduced-space optimal interpolation method. The SST fields are in the form of monthly values having a 1° spatial resolution and running from January 1870 to the present. In their study, the authors used the data from January 1870 to August 2017 (147 years and 8 months), providing 1,772 observations at each grid point. The SST values were converted to SST anomaly (SSTA) values by subtracting the long-term mean SST. Analyses were conducted via a feed forward, back propagation (FFBP) architecture for the ANN. While there were many other architectures that could have been applied, the FFBP has been widely used and is much simpler than most of the alternatives. It was observed in the previous example for Hillarys Boat Harbour in Western Australia that the simpler networks gave the better sea level predictions. In addition, the FFBP had the much faster processing speed needed to process so much data and over a very large area. As with many earlier studies, Patil and Deo (2018) used a tan-signmoid transfer function in the hidden layer and a linear function in the output layer.

The SST data for the Tropical Indian Ocean (TIO) region extracted on a 1° × 1° grid from the global HadISST dataset yielded a total of 3,813 nodes. At each node, the SST was converted to SST anomalies (SSTA) by subtracting the long-term SST of 30 years between 1961 and 1990. This latter period was taken to include as much satellite data as possible. The training samples were divided into subsets of training, cross validation and testing in portions of 75%, 5% and 20%, respectively. A separate ANN was set up to predict the SSTA for given lead times in months (ranging from 1 to 9 months). Thus, a total of $3,813 \times 9 = 34,317$ nodes needed to be trained. Each ANN model was trained with 60 sets of input combinations. During training for each input combination, 10 different random initial conditions were used to reduce the effect of initial conditions. The training goal was to minimize the mean square error between the predicted and real output values, which dictated which model was retained out of the 10 different initial conditions and 60 different input combinations. Thus, 34,317 best ANNs were retained out of the 20.6 million possible models.

The prediction skill of the ANN for SSTA in the TIO was derived using the 27 years of the "testing data" between January 1990 and December 2016. The spatial distribution of the correlation coefficient (r) is presented as gray shades in Figure 8.10.

Overall, the prediction skill of the ANN SSTA model proved to be good ($r > 0.5$) up to a lead time of about 3 months for most of the TIO. For longer lead times, high r values are restricted to the small area in the southwest corner of the region, as indicated by the square in the images. The r-value exceeds 0.5 in the Bay of Bengal for lead times of up to about 5 months, while the rest of the eastern TIO shows relatively low r values. The area just to the east of Madagascar (marked by the boxes in Figure 8.10) shows persistently high values regardless of lead time.

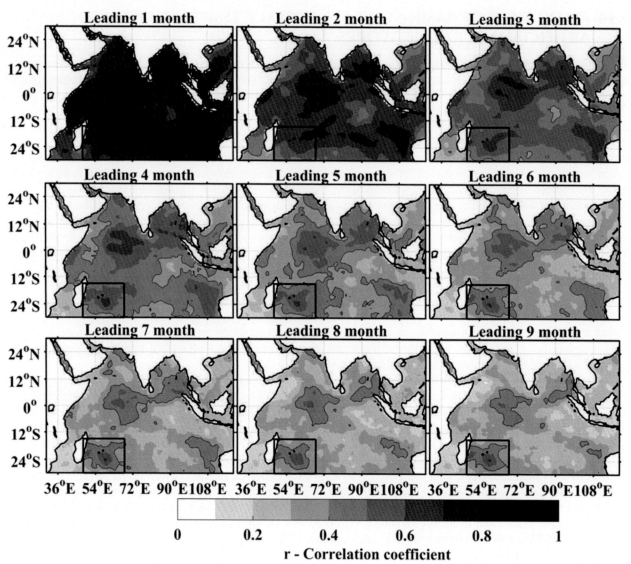

FIGURE 8.10 Spatial maps of the correlation coefficient, r, between the observed and the ANN predicted SSTAs as a function of lead time, ranging from 1 to 9 months, during the ANN testing phase. The rectangles show the regions with the best performance, particularly for long lead times. *From Patil and Deo (2018).*

To evaluate the seasonal prediction skill of the ANNs, the correlations were computed for the western (WIO) and eastern (EIO) Indian Ocean along with the Indian Ocean Dipole (IOD). Results are presented in Figure 8.11 as functions of lead time versus the annual monthly values.

The Western Indian Ocean (WIO) was found to have better predictability in the southern-hemisphere autumn than in spring, where $r > 0.6$ in autumn, even at 6 months lead time. In spring, r exceeds 0.6 only for lead times less than 3 months. The opposite is true for the Eastern Indian Ocean (EIO), where in the spring the model had better skill ($r < -0.6$; $|r| \geq 0.6$) for up to 7 months lead time, which fell to only 2 months in the autumn. The Indian Ocean Dipole (IOD) prediction is better in spring than in autumn and, overall, had longer predictive lead times. Because the IOD usually occurs between September and November, the ANN results are consistent.

The Patil and Deo (2018) study also used the ANN to examine the prediction of the well-known strong IOD events in 1994, 1996 and 1997. The 1997 event was particularly strong. In addition to the fact that the SSTA in the WIO was much greater than that in the EIO, both areas in 1997 had SSTAs that exceeded 1.5°C. This strong IOD profoundly impacted the local climate, resulting in wildfires, the destruction of coral reefs and damage to several ecosystems.

As shown by Figure 8.12, the ANN predicted the SSTA values and captured the warm and cold fronts (0°C contour) well. In addition, an observed warm band extending from the Arabian Sea to the central WIO was well predicted. A warm

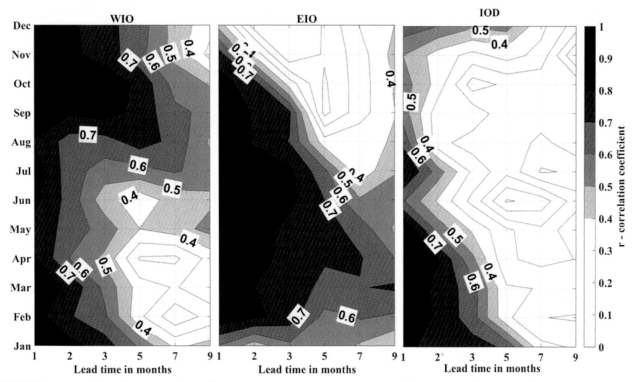

FIGURE 8.11 Variation of the correlation coefficient, *r*, between the observed and predicted SST anomaly (SSTA) values for lead times (months) from January to December for SSTA predictions for the western (WIO), eastern (EIO) and dipole (IOD) regions of the Indian Ocean. *From Patil and Deo (2018).*

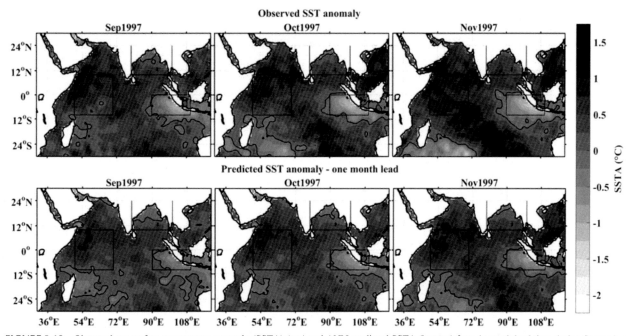

FIGURE 8.12 Observed sea surface temperature anomaly (SSTA) (top) and ANN predicted SSTA (bottom) for a 1-month lead time during September, October, and November for the extreme Indian Ocean Dipole (IOD) of 1997. The rectangles focus on the WIO and EIO regions and the open upper square at the top focuses on an unusual warming in the Bay of Bengal region. The thick black contour is the 0°C contour SSTA. *From Patil and Deo (2018).*

(cold) bias in the WIO (EIO) is seen in both the observed and predicted maps. The EIO cold bias is greater than that of the WIO in both fields. Also, the Bay of Bengal, which is normally cool during an IOD, became unusually warm in 1997 in both the observed and predicted fields.

Another feature well replicated by the ANN models was a warming of the Red Sea, which has adversely impacted biodiversity and slowed down the growth of coral reefs by 30%. The overall magnitude of the warming has been on the order of 0.7°C after 1994. Figure 8.13 reveals the progression of this warming for the years 1990—2016 for both the observed and modeled SSTs. In the early 1990s, the Red Sea was cooler, which is even more pronounced in the SSTA record than in the ANN model SSTA. From 1994 onward, the SSTAs started to increase. This warming continued to a maximum in 2010 and covered all seasons for both the observed and predicted values. This warming trend is also predicted by the ANN model, but is not as strong as in the observations. Most of the warming occurred in the northern part of the Red Sea.

In summary, the ANN model approach yielded results consistent with, and complimentary to, those from physics-based numerical models. While the focus of this study was to examine basin-scale variations, the ANN models appeared to replicate individual features quite well. The extreme IOD events were well forecasted, as was the warming of the Red Sea.

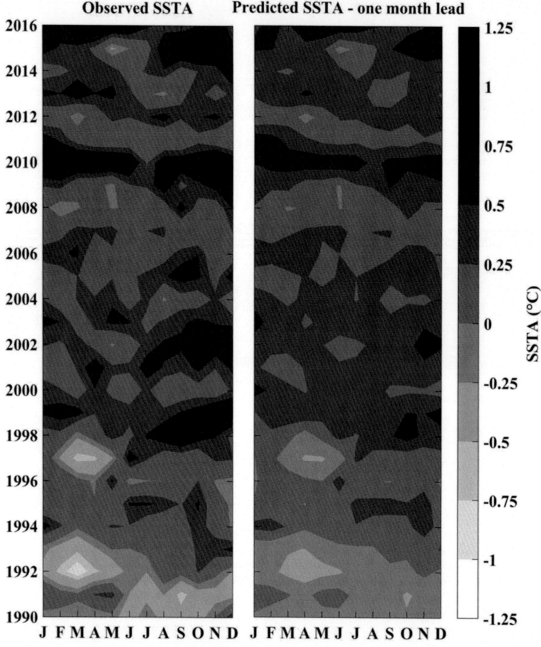

FIGURE 8.13 Evolution of the intraseasonal warming of the Red Sea from 1990 to 2016 as derived from spatial averages of the observed sea surface temperature anomaly (SSTA) (left column) and the ANN modeled SSTA (right column) for a 1-month lead time. *From Patil and Deo (2018).*

8.15 NEURAL NETWORKS FOR OIL SPILL DETECTION USING ERS-SAR DATA

Synthetic Aperture Radar (SAR) imagery has proven useful in detecting the ocean surface signature of oil spills. Oil on the water damps the capillary waves due to the increased viscosity, which dramatically reduces the radar backscatter, making the oil spill an excellent SAR target. The study by Del Frate et al. (2000) seeks a method for the semi-automatic detection and mapping of oil spills using SAR data from the European Research Satellites (ERS). An artificial neural network (ANN) was designed to provide this automatic detection. The training data contain oil spill signatures identified by visual inspection of SAR imagery. A similar data set is used to evaluate the performance of the ANN, which incorporated network "pruning" to improve the performance of the model.

In their application of the ANN, Del Frate et al. (2000) first assumed that a selected area in a SAR image contains a dark image that may be an oil spill. Thus, the ANN does not have to process the entire SAR image but instead can concentrate on the area thought to contain an oil slick. As a result, the method first requires a visual inspection to locate that part of a SAR image that contains a dark object consistent with the signature of an oil spill.

The ANN selected for this study easily accounts for the nonlinearities that are inherent in this type of analysis. An ANN can handle both linear and nonlinear relationships in the classification analysis. The input to the ANN consists of SAR pixels associated with the dark feature in the SAR image, while the output of the ANN yields the probability that the candidate is a real oil spill. The full network is optimized using pruning to define which nodes in the network are most important.

The spatial resolution of ERS SAR data is 25 m × 25 m, but for oil spill detection a lower image resolution of 100 m × 100 m was sufficient, significantly reducing the amount of data to be analyzed. For their study, Del Frate et al. (2000) used "fast delivery images" produced at the Fucino Station in Italy. That way, the results could be quickly available to support the Italian Coast Guard in an oil spill clean-up. The total archive for this study was a set of 600 images collected between 1997 and 1998 over various parts of the Mediterranean Sea. From this archive, all of the images that contained a dark area that might be an oil spill (Figure 8.14), or a natural slick (Figure 8.15), were extracted.

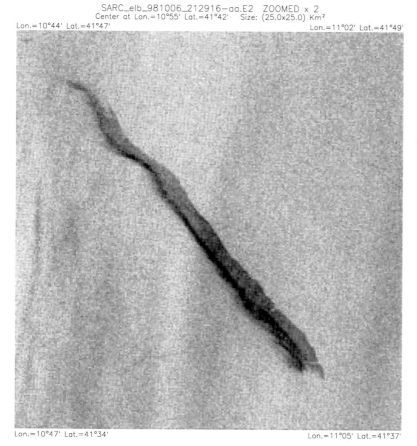

SARC_elb_981006_212916-aa.E2 ZOOMED x 2
Center at Lon.=10°55' Lat.=41°42' Size: (25.0x25.0) Km²
Lon.=10°44' Lat.=41°47'
Lon.=11°02' Lat.=41°49'

Lon.=10°47' Lat.=41°34'
Lon.=11°05' Lat.=41°37'

FIGURE 8.14 SAR image of a verified oil spill taken on 10 June 1998 off Elba Island (Italy), centered at 41°42′ N, 10°55′ W and covering an area of 25 km × 25 km. *From Del Frate et al. (2000).*

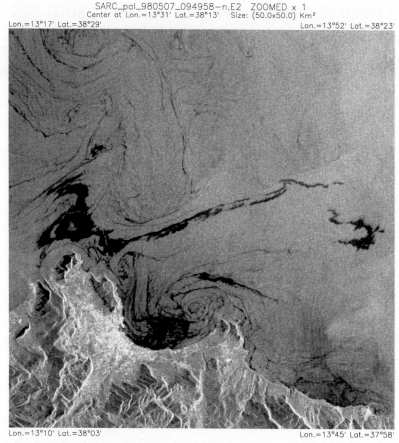

SARC_pal_980507_094958−n.E2 ZOOMED x 1
Center at Lon.=13°31' Lat.=38°13' Size: (50.0x50.0) Km²
Lon.=13°17' Lat.=38°29'   Lon.=13°52' Lat.=38°23'

Lon.=13°10' Lat.=38°03'   Lon.=13°45' Lat.=37°58'

FIGURE 8.15 SAR image of a verified natural surface film taken on 5 July 1998, near the coast of Palermo (Italy), centered at 38°13′ N, 13°31′ W and covering an area of 50 km × 50 km. *From Del Frate et al. (2000).*

Based on a histogram of one of these selected images (Figure 8.16), it is evident that there are two peaks, with the lower peak representing the mean backscattering value of the dark object and the much higher peak representing the background backscatter. Note that the "dark maximum" shows considerable variability whereas the background peak is more uniform. This is consistent with the variations in the dark object in Figure 8.14.

A feed-forward, artificial neural network (ANN) was set up to link all of the input data (Figure 8.17) and to determine from the SAR imagery to a binary output conversion, whether a feature was an oil slick or not.

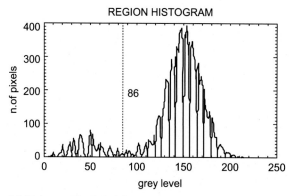

FIGURE 8.16 Histogram of an analyzed SAR image. The dashed line corresponds to the local minimum between the two peaks and indicates the threshold value needed for the edge detection of an oil spill candidate. *From Del Frate et al. (2000).*

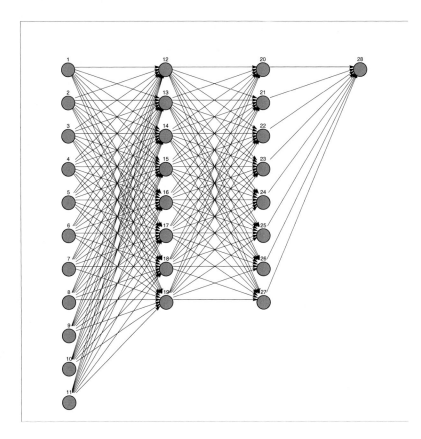

FIGURE 8.17 Feed forward topology progressing from left to right for the ANN to detect oil spills from SAR imagery. *From Del Frate et al. (2000).*

In a feed forward network, the input flows only in one direction—toward the output and each neuron of a layer is connected to all neurons of the successive layer, but with no feedback to neurons in the previous layers. The oil spill detection study used two hidden layers between the inputs and the single binary output. Each neuron is characterized by its activation function, which in this case is the common nonlinear sigmoid function (defined earlier). The input vector contains the image values, while the output contains the classification provided either by previous ground truth information, or by the careful analysis of image interpretation experts. The input values were scaled to the range between 0.01 and 0.99. Del Frate et al. (2000) trained their network using the back propagation method, which uses a gradient search technique and then iteratively adjusts the nodal weights in the network to minimize the difference between the ANN prediction and the actual output. The iterations are stopped when the size of this difference remains nearly constant.

The neural network simulator (SNNS) developed at the University of Stuttgart, Germany was the basic software used for the classification implementation and it proved to be a high-level, flexible and reliable software package. Several different network formulations were tried, and a topology of 11−8−4−1 was finally chosen for the network (Figure 8.17). This topology required 15,000 training cycles for the network to be fully established. Once trained, the ANN was able to accurately classify a SAR image. It gave a 0 if a dark spot in the image is an oil slick and a 1 if it is a look-alike natural film.

The ANN was then applied to a new set of non-training data. For this step, the authors decided to use the "leave-one-out" method. In this method, if there are N data points, the net is trained with $N − 1$ data points and tested against the remaining data point. This process is repeated for each of the N possible choices. For this particular study, $N = 139$ and the authors found that 18% of real oil spills had been misclassified as "look-alikes," and 10% of natural-slicks had been misclassified as oil spills. This gave an overall misclassification rate of roughly 14%, which was considered a very good performance of the ANN. This performance encouraged an optimization of the ANN using the "pruning" procedure, where the network nodes are examined to identify those with the greatest importance as dictated by their weights in the net. Those nodes with the least importance are discarded. Typically, pruning is followed by some additional training of the pruned network. The cycle of pruning and training may be repeated for several cycles. In the Del Frate et al. (2000) study, every time a node was removed, the resulting network was retrained. This process continued until further removal did not change the output error. The final ANN is present in Figure 8.18.

FIGURE 8.18 The ANN used to examine oil slicks in the Mediterranean Sea after pruning was invoked. The input nodes are on the left, separated by two hidden layers to reach the goal on the right. *From Del Frate et al. (2000).*

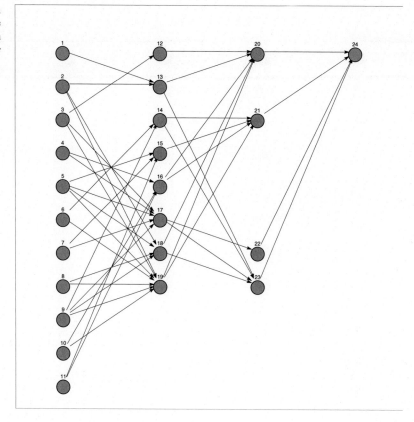

As illustrated by Figure 8.18, the major change is in the second hidden layer, which has gone from 8 nodes to only 4 nodes, resulting in a reduction of connections from 160 to 45. A consequence of this reduction is that the computational time of the pruned network was much lower than that of the original ANN. Still, the pruned ANN gave overall outputs similar to those from the first ANN. It was concluded that the ANN could automatically and correctly detect (and map) an oil spill in SAR imagery. With a classification error of only 14%, the ANN model could discriminate between an oil spill and a natural slick.

Glossary

ANN Artificial Neural Network
CNN Convolutional Neural Network
DNN Deep Neural Network
EIO Eastern Indian Ocean
ERS European Space Agency (remote sensing satellites)
FFBP Feed Forward Back Propagation
GPU Graphical Processor Unit
HadISST Hadley Centre Sea Ice and Sea Surface Temperature data set
IOD Indian Ocean Dipole
MLP Multilayer Perceptron
NN Neural Network
Pixel Picture element (image element)
R (or r) Correlation Coefficient
RMSE Root Mean Square Error
SAR Synthetic Aperture Radar
SI Scatter Index
SST Sea Surface Temperature
SSTA Sea Surface Temperature Anomaly
TIO Tropical Indian Ocean
WIO Western Indian Ocean

Appendix A

Units in Physical Oceanography

Length	1 micrometer (μm; micron) = 10^{-3} millimeter (mm) = 10^{-6} meter (m)
	1 centimeter (cm) = 10 mm = 10^{-2} m = 0.3937 inches (in)
	1 meter (m) = 10^2 cm = 39.37 in = 3.281 feet (ft) = 1.094 yards (yd)
	1 kilometer (km) = 10^3 m = 0.5396 nautical mile (naut mi) = 0.6214 statute mile (mi)
	1 nautical mile = 1 minute latitude = 6080 ft = 1.152 statute mi = 1.8532 km
	1°latitude = 111.19 km; 1°longitude = 111.19 cos|(latitude)| km
	At 45°; 1°longitude = 78.62 km
	1 cable = 0.1 nautical mile = 608 ft = 185.3 m
	1 fathom (fm) = 6 ft = 1.8288 m
	1 league = 3040 fathoms = 3 nautical miles
	1 inch (in) = 2.54 cm; 12 in = 1 ft; 36 in = 1 yard
Area	1 square kilometer (km^2) = 10^6 m^2 = 100 hectares (ha) = 0.386 mi^2
	1 hectare (ha) = 2.471 acres (ac)
	1 square mile (mi^2) = 640 ac = 259 ha
Volume	1 cubic meter (m^3) = 264.2 US gallon (gal) = 220.0 imperial gal = 35.314 ft^3
	1 litre (L) = 10^3 mL = 10^{-3} m^3 = 0.264 US gal = 0.220 imperial gal
	1 barrel (oil; bbl) = 42 US gal = 0.159 m^3 = 158.987 litres (L)
	1 US gal = 0.83 imperial gal
Time	1 hour (h) = 3.6×10^3 seconds (s)
	1 solar day = 24 h = 8.64×10^4 s
	1 sidereal day = 23 h, 56 min, 4 s
Mass	1 gram (g) = 0.03527 ounces (oz) = 0.03215 troy ounces
	1 kilogram (kg) = 10^3 g = 2.205 lb; 1 pound (lb) = 0.4536 kg
	1 metric ton (tonne) = 10^3 kg = 10^6 g = 2205 lb = 1.1025 ton
	1 ton = 2000 lb
Pressure	1 pascal (Pa) = 1 newton/m^2 (N/m^2) = 10^{-5} bar = 10^{-4} decibar (dbar)
	1 atmosphere (atm) = 1.01325×10^5 Pa
	1 bar = 0.98692 atm = 10^5 Pa = 1.02 kg/cm^2
	1 millibar (mb) = 10^{-3} bar = 10^{-1} kPa = 1 hPa
	1 kPa = 10^3 Pa = 10^{-1} dbar = 10 millibar (mb)
	The inverse barometer effect: A 1 mb drop in atmospheric sea surface pressure causes an approximately 1 cm rise in sea level (and vice versa).

(Continued)

(Cont'd)

Stress	$1 \text{ dyn/cm}^2 = 10^{-1} \text{ N/m}^2$
	$1 \text{ kg/cm}^2 = 0.96784 \text{ atm} = 14.2233 \text{ lb/in}^2$
Speed	1 knot (nautical mi/h; kn) = 0.5148 m/s = 51.48 cm/s = 44.48 km/day
	1 meter per second (m/s) = 2.24 statute mi/h = 1.943 knots = 86.4 km/day
	1 (statute) mi/h = 1.609 km/h = 0.868 knots
	Sound speed in water $\approx$ 1482 m/s ($T = 5°C$; $S = 34$ g/kg (psu); depth = 1000 m)
	Sound speed in air $\approx 331.3 + 0.61 \ T_{air}$ m/s (T_{air} in °C)
Temperature	°F (Fahrenheit) = 9/5 × °C (Celsius) +32
	°C (Celsius) = (°F − 32) × 5/9
	K (kelvin) = °C + 273.15 (0 K = absolute zero)
Dissolved O_2	1 milliliter per liter (ml/l) = 1.43 mg/l
	1 ml/l $\approx$ 43.3 μmol/kg (μmol/kg) for $S = 34.7$ g/kg (psu), $T = 3.5°C$, $\sigma_t = 27.96$
	1 mol = a quantity of N_0 atoms or molecules, where $N_0 = 6.022 \times 10^{23}$ is Avogadro's number. Atomic weight is the weight of 1 mole of atoms, and molecular weight is the weight of 1 mole of molecules. For example, the molecular weight of sodium chloride (N_0 molecules of NaCl; i.e., 1 mol of sodium +1 mol of chlorine) = 22.990 + 35.453 = 58.443 g.
Earth's rotation rate	$\Omega = 0.72921 \times 10^{-4}$ rad/s = 0.04178 cycles/hour (cph)
Earth's gravity	g (or g) = 9.81 m/s^2 = 981 cm/s^2 = 32.1722 ft/s^2 (has weak variation with latitude)
Force	1 newton (N) = 1 kg m/s^2 = 10^5 dynes (dyn) = 2.2 lb
Energy and power	1 (thermochemical) calorie (cal) = 4.184 joules (J) = 3.968 × 10^{-3} British thermal units (BTU; Btu; @ 60°F)
	1 J = 1 newton meter (N m) = 10^7 ergs = 0.2390 cal = 2.78 × 10^{-7} kWh
	1 watt (W) = 1 joule per second (J/s) = 1.341 × 10^{-3} horsepower (hp)
	1 kilowatt hour (kWh) = 3.6 × 10^6 J = 3.41 × 10^3 Btu
	1 hp = 7.457 × 10^2 W
Geopotential	1 dynamic centimeter (dyne cm) = 10^3 cm^2/s^2 $\approx$ 1.02 cm (equivalent change in sea surface elevation that this work can produce)
	1 joule/kg (1 J/kg) = 1 m^2/s^2 = 10 dyne cm
	Specific volume anomaly: $\delta = \alpha_{S,T,P} - \alpha_{35,0,P}$ where $\alpha = \rho^{-1}$ m^3/kg
	Geopotential anomaly (or dynamic height anomaly in the older literature): $\Delta\Phi$(or ΔD) $= \Phi_{P_2} - \Phi_{P_1} = -\int_{P_1}^{P_2} \delta dP$ joules/kg $\quad$ Potential energy: $PE = g^{-1} \int_{P_1}^{P_2} \delta \cdot P dP$ joules/m^2
Transport	1 sverdrup (Sv) = 10^6 m^3/s
Specific heat	Liquid water: $c_p = 4.1855$ J/(g·K) = 1 cal/g°C (at 15 °C, 101.325 kPa)
Capacity	Air: $c_p = 1.006$ J/(g·K) (at −50 to 40°C; 101.325 kPa pressure)

Appendix B

Glossary of Statistical Terminology

Alternative hypothesis: Value of a parameter of a population other than the value hypothesized or believed to be true by the investigator.

Asymptotically normal distribution: A distribution of values that is not truly normal, but which approaches a normal distribution as the number of samples becomes very large.

Autocorrelation: In a time series, $x(t)$, the statistical relationship between values of a variable taken at certain times in the series and values of a variable taken at other times (a function of the time lag, τ, between the two series) written as,

$$R_{xx}(\tau) = E[x(t)x(t+\tau)]$$

where E is the expected value. Autocovariance, $C_{xx}(\tau)$, is similar except that the mean of the record is subtracted prior to the analysis.

Biased estimator: An estimator $\hat{x}$ for which the expected value $E[\hat{x}]$ of a sample has a systematic error with respect to the true expected value, μ_x; i.e., $E[\hat{x}] \neq \mu_x$.

Bin interval: A specified arbitrary interval, which partitions a quantity whose number of occurrences are being measured; used for constructing a histogram (frequency of occurrence distribution) of the data set.

Central limit theorem: States that the distributions of sample means taken from large populations approach a normal (Gaussian) distribution regardless of the original distributions of the individual samples.

Chi-squared distribution: The distribution function generated by the random variable,

$$\chi_n^2 = X_1^2 + X_2^2 + X_3^2 + \ldots + X_n^2$$

where $X_1, X_2, \ldots, X_n$ are n independent random variables drawn from a normal population with zero mean. The chi-squared variable χ_n^2 has an expected mean value of n and a variance of $2n$.

Confidence interval: An interval that has a specified probability of containing a given value or characteristic. If the values (e.g., measurements) are normally distributed, the commonly used 95% confidence interval implies that 95% of the values will lie within ± 2 standard deviations of the mean; for the 99% interval of confidence, this range increases to ± 3 standard deviations.

Continuous random variable: A random variable that may be characterized as a continuous function.

Correlation (covariance): A quantitative measure of the interdependence or association between two variables.

Countable: Either finite or denumerable.

Cross correlation: The correlation between corresponding members of two or more different series of the same duration: if $x(t) = (x_1, x_2, \ldots, x_n)$ and $y(t) = (y_1, y_2, \ldots, y_n)$ are two series, the cross correlation is the correlation between $x(t)$ and $y(t)$, or between $x(t)$ and $y(t + \tau)$ for lag τ such that,

$$R_{xy}(\tau) = E[x(t)y(t+\tau)]$$

(see Autocorrelation).

Cumulative distribution function: The integral of the probability density function, $p(x)$, over some specified interval, (x_1, x_2). The integral or sum of $p(x)$ from $-\infty$ to x gives the cumulative total of all values whose value is less than or equal to x.

Degrees of freedom: The number of truly independent samples used to estimate a parameter; when each sample of a series of length N is independent of all other values, the degrees of freedom, $v = N - 1$.

Discrete random variable: A random variable that may be characterized as a discrete function.

Ensemble average: An average over several realizations of a random variable taken over times of equal duration.

Ergodic hypothesis: The hypothesis that replaces statistical replications (ensemble averages) with averages in space or time. Allows us to compute averages in space or time as representative of ensemble averages.

Estimator bias: The difference between an estimate and the true value of the parameter being estimated.

Expected value: For a random variable x with a probability function $f(x)$, this is the integral from $-\infty$ to ∞ of $xf(x)$; also known as the expectation and written as,

$$E(x) = \int_{-\infty}^{\infty} xf(x)\mathrm{d}x = \mu$$

where μ is the mean of the variable.

Gamma distribution: A normal distribution whose probability distribution function involves the gamma function as,

$$f(x) = \frac{x^{\alpha-1}e^{-\frac{x}{\beta}}}{\beta^{\alpha}\Gamma(\alpha)}; \quad \alpha, \beta > 0; \quad 0 \leq x \leq \infty$$

$$= 0 \text{ elsewhere}$$

where α and β are parameters of the distribution and $\Gamma(\alpha)$ is the gamma function.

$$\Gamma(\alpha) = \int_{0}^{\infty} x^{\alpha-1}e^{-x}\mathrm{d}x.$$

Gauss−Markov theorem: An unbiased linear estimator of a parameter has minimum variance (i.e., is the best estimator), when it is determined by the least squares method. Note that implementation of the least squares method requires no knowledge of the distribution from which the observations are drawn for the purpose of the parameter's estimation.

Hypothesis testing: The branch of statistics that considers the problem of choosing between two actions on the basis of the observed value of a random variable whose distribution depends on a parameter, the value of which would indicate the correct choice to make.

Independent random variables: The discrete random variables, $X_1, X_2, \ldots, X_n$, are independent if for arbitrary values x_1, $x_2, \ldots, x_n$ of the variables, the probability that $X_1 = x_1, X_2 = x_2, \ldots$ is equal to the product of the probabilities that $X_i = x_i$, for $i = 1, 2, \ldots, n$; in this case, the random variables are unrelated.

Inference: The act of passing from statistical sample data to generalizations (as when inferring the values of true population parameters) with calculated degrees of certainty (confidence intervals at selected significance levels).

Joint probability density function: The distribution that gives the probability that $X_i = x_i$, for $i = 1, 2, \ldots, n$ for all values x_i of the random variable X_i.

Jointly sufficient statistics: Let $X_1, X_2, \ldots, X_n$ be a random sample from a probability density function $p(X; \theta)$, where θ is an unknown statistical parameter such as the mean or variance; the statistics $S_1, \ldots, S_r$ are defined to be jointly sufficient if and only if the conditional distribution of $X_1, X_2, \ldots, X_n$ given $S_1 = s_1, \ldots, S_r = s_r$ does not depend on θ.

Least squares method: A technique for fitting a line, polynomial, or other curve to a given distribution of points, which minimizes the sum of the squares of the deviations of the given points from the fitted curve.

Likelihood L(x): The likelihood of occurrence of a sample of independent values of $x_1, x_2, \ldots, x_n$, with a probability function $f(x)$, is the product $f(x_1) \cdot f(x_2) \cdot \ldots \cdot f(x_n)$.

Likelihood ratio: The probability of a random drawing of a specified sample from a population, assuming a given hypothesis about the parameter of the population, divided by the probability of a random drawing of the same sample assuming that the parameters of the population are such that this probability is maximized; i.e., $L(x_1)/L(x_2)$ is maximized.

Linear regression: The straight line running through the points of a scatter diagram about which the amount of scatter is a minimum in the least squares sense.

Lognormal distributions: Because many parameters (especially biological quantities such as zooplankton biomass and surface chlorophyll-*a* concentration) change in an exponential manner in response to growth and mortality rates, they can be considered to have a lognormal distribution. The use of log-transformed data (using the natural logarithm), $\ln(x)$ for data values, x, permits use of standard statistical analysis assumptions and associated methodologies.

Maximum likelihood estimation: A method whereby the likelihood distribution is maximized to produce an estimate of the random variable.

Mean square error: A measure of the extent to which a collection of numbers, $x_1, x_2, \ldots, x_n$, is unequal and defined by the expression,

$$\text{mean square error} = E\big[(\widehat{x} - x)^2\big] = E\big[(\widehat{x} - E[\widehat{x}])^2\big] + E\big[(E[\widehat{x}] - x)^2\big]$$

where $E[x]$ is the expected (mean) value and $\widehat{x}$ is the estimator for x.

Method of moments: A procedure for estimating the parameters (such as the mean and variance) of a distribution.

Minimal sufficient statistic: A set of jointly sufficient statistics is defined to be minimal sufficient if and only if it is a function of every other set of sufficient statistics.

Moments: The nth moment of a distribution $f(x)$ about a point x_0 is the expected value $E[f^n(x - x_0)]$; the first moment is the mean of the distribution, while the variance is a function of the first and second moments (this definition only applies to moments about the mean and not the origin).

Moment generating function: Let x be a random variable with probability density function $p(x)$; the expected value of some function $f(rx)$, $E[f(rx)]$ is defined to be the moment generating function $m(r)$ of f if the expected value exists for every value of r in some interval $-\eta < r < \eta$; $\eta > 0$ whereby,

$$m(r) = E[f(rx)] = \int_{-\eta}^{\eta} f(rx)p(x)\mathrm{d}x.$$

Moment of a power spectrum, $S(f)$: Defined as $m_n = \int_0^\infty S(f)f^n df$ where f is the frequency. In the case of surface gravity waves, the *significant wave height* $H_s = (16m_o)^{1/2}$ and the *mean zero crossing period* is $T = (m_0/m_2)^{1/2}$.

Multivariate analysis: The study of random variables, which are multidimensional.

Normal (or Gaussian) probability distribution function: A normally distributed frequency distribution of a random variable x with a mean μ and standard deviation σ given by,

$$p(x) = \frac{1}{\sigma\sqrt{2\pi}}\exp\left[\frac{(x - \mu)^2}{2\sigma^2}\right].$$

Null hypothesis: The hypothesis that there is no validity to the specific claim that two variations (treatments) of the same thing (such as the mean values) can be distinguished by a specific procedure.

Pivotal statistic: The statistic that allows one to compute a confidence interval for a specific estimate.

Population: Any finite or infinite collection of individuals or elements that can be specified or labeled.

Population distribution: The distribution that characterizes a population; may be displayed using a histogram or frequency of occurrence diagram.

Population mean and variance: The population mean, μ, is the arithmetic average of values x_i ($i = 1, \ldots, N$) obtained from all members in a population of size N based on the measurement of some quantity, x, associated with each member; population variance is the arithmetic average of the numbers $(x_i - \mu)^2$.

Population moment: The rth moment associated with a particular population.

Probability: The probability of an event is the ratio of the number of times the event occurs relative to the total number of trials that take place.

Probability density function (PDF): A real-valued function whose integral over any set gives the probability that a random variable has values in this set.

Random variable: A well-defined function that allows us to assign a real number to any outcome of an experiment. Specifically, the random outcome of a particular experiment, indexed by k, can be represented by a real number x_k called the random variable.

Relative frequency distribution: A frequency distribution in which the individual class frequencies are expressed as a fraction of the total frequency range.

Sample: A selection of values from a larger collection of values.

Sample mean and variance: The mean value $\bar{x}$ and variance s of a sample taken from a given data set, X. In general, these differ from the true mean, μ, and variance, σ^2, of the population.

Sample moment: The moment of a sample taken from a given set of samples.

Significance level: The probability of a false rejection of the null hypothesis.

Standard error: A measure of the variability any statistical constant would be expected to show in taking repeated random samples of a given size from the same universe of observations.

Standard normal variable: A normal (Gaussian) distributed random variable with specified mean and standard deviation, which has been transformed to a random variable with a mean of zero ($\overline{X} = 0$) and a standard deviation of unity ($s = 1$).

Stationarity: The property by which the statistics of a random variable do not change with time. For a stationary time series, quantities such as the mean and variance are nearly identical for different segments of the record.

Student's t-distribution: A probability distribution used to test the hypothesis that a random sample of n observations comes from a normal population with a given mean in the case of low sample sizes.

Sufficiency: Condition of an estimator that uses all the information about the population parameter contained in the sample observations.

Sufficient statistics: Let $X = (x_1, x_2, ..., x_n)$ be a random sample from the probability density function $p(x; \theta)$; a statistic $S = s(x_1, x_2, ..., x_n)$ is defined to be a sufficient statistic if and only if the conditional distribution of X given S does not depend on θ for any statistic $R = r(x_1, x_2, ..., x_n)$.

Tchebysheff's (Chebyshev's) theorem: Given a nonnegative random variable $f(x)$, and number $k \geq 1$, the probability $p(f(x) - \mu) \geq k\sigma \leq 1/k^2$, where μ and σ are the mean and standard deviation of $f(x)$.

Unbiased estimator: An estimate $\widehat{\theta}$ for a parameter θ whose expected value is $E[\widehat{\theta}] = \theta$.

Uniform probability density function: The distribution of a random variable in which each value has the same probability of occurrence.

Appendix C

Means, Variances and Moment-Generating Functions for Some Common Continuous Variables

Distribution	Probability Function	Mean	Variance	Moment-Generating Function
Uniform	$f(x) = \dfrac{1}{\theta_2 - \theta_1}; \ \theta_1 \leq x \leq \theta_2$	$\dfrac{\theta_2 + \theta_1}{2}$	$\dfrac{(\theta_2 - \theta_1)^2}{12}$	$\dfrac{e^{t\theta_2} - e^{t\theta_1}}{t(\theta_2 - \theta_1)}$
Normal $N(\mu, \sigma^2)$	$f(x) = \dfrac{1}{\sigma\sqrt{2\pi}}\exp\left[-\left(\dfrac{1}{2\sigma^2}\right)(x-\mu)^2 \right]$	μ	σ^2	$\exp\left(\mu t + \dfrac{t^2\sigma^2}{2}\right)$
Gamma $\Gamma(\alpha, \beta)$	$f(x) = \left[\dfrac{1}{\Gamma(\alpha)\beta^\alpha}\right] x^{\alpha-1} e^{-x/\beta}$	$\alpha\beta$	$\alpha\beta^2$	$(1 - \beta t)^{-\alpha}$
Chi-squared χ_ν^2	$f(\chi^2) = \dfrac{(\chi^2)^{\left(\frac{\nu}{2}\right)-1} e^{-\chi^2/2}}{2^{\nu/2}\,\Gamma\left(\dfrac{\nu}{2}\right)}$	ν	2ν	$(1 - 2t)^{-\nu/2}$
Beta	$f(x) = \left[\dfrac{\Gamma(\alpha+\beta)}{\Gamma(\alpha)\Gamma(\beta)}\right] x^{\alpha-1}(1-x)^{\beta-1}$	$\dfrac{\alpha}{\alpha+\beta}$	$\dfrac{\alpha\beta}{(\alpha+\beta)^2(\alpha+\beta+1)}$	Does not exist in closed form

Appendix D

Statistical Tables

TABLE D.1 Cumulative normal distribution. The area or cumulative distribution, $F(z)$, Under the standardized normal distribution curve for $z \leq z_F$, such that the probability $P(z < z_F) = F(z)$. For example, $P(z < z_F = 1.21) = 0.8869$, and $P(z > z_F = 1.21) = 1 - 0.8869 = 0.1131$

$$F(z) = \int_{-\infty}^{z} \frac{1}{\sqrt{2\pi}} e^{-z^2/2} dz$$

z	0.00	0.01	0.02	0.03	0.04	0.05	0.06	0.07	0.08	0.09
0.0	0.5000	0.5040	0.5080	0.5120	0.5160	0.5199	0.5239	0.5279	0.5319	0.5359
0.1	0.5398	0.5438	0.5478	0.5517	0.5557	0.5596	0.5636	0.5675	0.5714	0.5753
0.2	0.5793	0.5832	0.5871	0.5910	0.5948	0.5987	0.6026	0.6064	0.6103	0.6141
0.3	0.6179	0.6217	0.6255	0.6293	0.6331	0.6368	0.6406	0.6443	0.6480	0.6517
0.4	0.6554	0.6591	0.6628	0.6664	0.6700	0.6736	0.6772	0.6808	0.6844	0.6879
0.5	0.6915	0.6950	0.6985	0.7019	0.7054	0.7088	0.7123	0.7157	0.7190	0.7224
0.6	0.7257	0.7291	0.7324	0.7357	0.7389	0.7422	0.7454	0.7486	0.7517	0.7549
0.7	0.7580	0.7611	0.7642	0.7673	0.7704	0.7734	0.7764	0.7794	0.7823	0.7852
0.8	0.7881	0.7910	0.7939	0.7967	0.7995	0.8023	0.8051	0.8078	0.8106	0.8133
0.9	0.8159	0.8186	0.8212	0.8238	0.8264	0.8289	0.8315	0.8340	0.8365	0.8389
1.0	0.8413	0.8438	0.8461	0.8485	0.8508	0.8531	0.8554	0.8577	0.8599	0.8621
1.1	0.8643	0.8665	0.8686	0.8708	0.8729	0.8749	0.8770	0.8790	0.8810	0.8830
1.2	0.8849	0.8869	0.8888	0.8907	0.8925	0.8944	0.8962	0.8980	0.8997	0.9015
1.3	0.9032	0.9049	0.9066	0.9082	0.9099	0.9115	0.9131	0.9147	0.9162	0.9177
1.4	0.9192	0.9207	0.9222	0.9236	0.9251	0.9265	0.9279	0.9292	0.9306	0.9319
1.5	0.9332	0.9345	0.9357	0.9370	0.9382	0.9394	0.9406	0.9418	0.9429	0.9441

Continued

TABLE D.1 Cumulative normal distribution. The area or cumulative distribution, $F(z)$, Under the standardized normal distribution curve for $z \leq z_F$, such that the probability $P(z < z_F) = F(z)$. For example, $P(z < z_F = 1.21) = 0.8869$, and $P(z > z_F = 1.21) = 1 - 0.8869 = 0.1131$

$$F(z) = \int_{-\infty}^{z} \frac{1}{\sqrt{2\pi}} e^{-z^2/2} dz\text{—cont'd}$$

z	0.00	0.01	0.02	0.03	0.04	0.05	0.06	0.07	0.08	0.09
1.6	0.9452	0.9463	0.9474	0.9484	0.9495	0.9505	0.9515	0.9525	0.9535	0.9545
1.7	0.9554	0.9564	0.9573	0.9582	0.9591	0.9599	0.9608	0.9616	0.9625	0.9633
1.8	0.9641	0.9649	0.9656	0.9664	0.9671	0.9678	0.9686	0.9693	0.9699	0.9706
1.9	0.9713	0.9719	0.9726	0.9732	0.9738	0.9744	0.9750	0.9756	0.9761	0.9767
2.0	0.9772	0.9778	0.9783	0.9788	0.9793	0.9798	0.9803	0.9808	0.9812	0.9817
2.1	0.9821	0.9826	0.9830	0.9834	0.9838	0.9842	0.9846	0.9850	0.9854	0.9857
2.2	0.9861	0.9864	0.9868	0.9871	0.9875	0.9878	0.9881	0.9884	0.9887	0.9890
2.3	0.9893	0.9896	0.9898	0.9901	0.9904	0.9906	0.9909	0.9911	0.9913	0.9916
2.4	0.9918	0.9920	0.9922	0.9925	0.9927	0.9929	0.9931	0.9932	0.9934	0.9936
2.5	0.9938	0.9940	0.9941	0.9943	0.9945	0.9946	0.9948	0.9949	0.9951	0.9952
2.6	0.9953	0.9955	0.9956	0.9957	0.9959	0.9960	0.9961	0.9962	0.9963	0.9964
2.7	0.9965	0.9966	0.9967	0.9968	0.9969	0.9970	0.9971	0.9972	0.9973	0.9974
2.8	0.9974	0.9975	0.9976	0.9977	0.9977	0.9978	0.9979	0.9979	0.9980	0.9981
2.9	0.9981	0.9982	0.9982	0.9983	0.9984	0.9984	0.9985	0.9985	0.9986	0.9986
3.0	0.9987	0.9987	0.9987	0.9988	0.9988	0.9989	0.9989	0.9989	0.9990	0.9990
3.1	0.9990	0.9991	0.9991	0.9991	0.9992	0.9992	0.9992	0.9992	0.9993	0.9993
3.2	0.9993	0.9993	0.9994	0.9994	0.9994	0.9994	0.9994	0.9995	0.9995	0.9995
3.3	0.9995	0.9995	0.9995	0.9996	0.9996	0.9996	0.9996	0.9996	0.9996	0.9997
3.4	0.9997	0.9997	0.9997	0.9997	0.9997	0.9997	0.9997	0.9997	0.9997	0.9998

Adapted from Introductory Statistical Analysis by Harnett, D.L., Murphy, J. L., 1976. Addison-Wesley, 197.

TABLE D.2 Cumulative chi-square distribution. The area or cumulative distribution, $F(\chi^2)$, under the χ^2 distribution curve for different degrees of freedom, v, such that the probability $P(\chi^2_v < \chi^2_{v;F}) = F(\chi^2)$. For example, for $v = 16$, the probability $P(\chi^2_{16} < \chi^2_{16;F} = 26.3) = F(26.3) = 0.950$. Consequently, $P(\chi^2_{16} > \chi^2_{16;F} = 26.3) = 1 - F(26.3) = 0.050$

$$F(\chi^2) = \int_0^{\chi^2} \frac{x^{(v-2)/2} e^{-x/2}}{2^{v/2}[(v-2)/2]!} dx$$

						F							
v	0.005	0.010	0.025	0.050	0.100	0.250	0.500	0.750	0.900	0.950	0.975	0.990	0.995
1	0.0^4393	0.0^3157	0.0^3982	0.0^2393	0.0158	0.102	0.455	1.32	2.71	3.84	5.02	6.63	7.88
2	0.0100	0.0201	0.0506	0.103	0.211	0.575	1.39	2.77	4.61	5.99	7.38	9.21	10.6
3	0.0717	0.115	0.216	0.352	0.584	1.21	2.37	4.11	6.25	7.81	9.35	11.3	12.8
4	0.207	0.297	0.484	0.711	1.06	1.92	3.36	5.39	7.78	9.49	11.1	13.3	14.9
5	0.412	0.554	0.831	1.15	1.61	2.67	4.35	6.63	9.24	11.1	12.8	15.1	16.7
6	0.676	0.872	1.24	1.64	2.20	3.45	5.35	7.84	10.6	12.6	14.4	16.8	18.5
7	0.989	1.24	1.69	2.17	2.83	4.25	6.35	9.04	12.0	14.1	16.0	18.5	20.3
8	1.34	1.65	2.18	2.73	3.49	5.07	7.34	10.2	13.4	15.5	17.5	20.1	22.0
9	1.73	2.09	2.70	3.33	4.17	5.90	8.34	11.4	14.7	16.9	19.0	21.7	23.6
10	2.16	2.56	3.25	3.94	4.87	6.74	9.34	12.5	16.0	18.3	20.5	23.2	25.2
11	2.60	3.05	3.82	4.57	5.58	7.58	10.3	13.7	17.3	19.7	21.9	24.7	26.8
12	3.07	3.57	4.40	5.23	6.30	8.44	11.3	14.8	18.5	21.0	23.3	26.2	28.3
13	3.57	4.11	5.01	5.89	7.04	9.30	12.3	16.0	19.8	22.4	24.7	27.7	29.8
14	4.07	4.66	5.63	6.57	7.79	10.2	13.3	17.1	21.1	23.7	26.1	29.1	31.3
15	4.60	5.23	6.26	7.26	8.55	11.0	14.3	18.2	22.3	25.0	27.5	30.6	32.8
16	5.14	5.81	6.91	7.96	9.31	11.9	15.3	19.4	23.5	26.3	28.8	32.0	34.3
17	5.70	6.41	7.56	8.67	10.1	12.8	16.3	20.5	24.8	27.6	30.2	33.4	35.7
18	6.26	7.01	8.23	9.39	10.9	13.7	17.3	21.6	26.0	28.9	31.5	34.8	37.2
19	6.84	7.63	8.91	10.1	11.7	14.6	18.3	22.7	27.2	30.1	32.9	36.2	38.6
20	7.43	8.26	9.59	10.9	12.4	15.5	19.3	23.8	28.4	31.4	34.2	37.6	40.0
21	8.03	8.90	10.3	11.6	13.2	16.3	20.3	24.9	29.6	32.7	35.5	38.9	41.4
22	8.64	9.54	11.0	12.3	14.0	17.2	21.3	26.0	30.8	33.9	36.8	40.3	42.8
23	9.26	10.2	11.7	13.1	14.8	18.1	22.3	27.1	32.0	35.2	38.1	41.6	44.2
24	9.89	10.9	12.4	13.8	15.7	19.0	23.3	28.2	33.2	36.4	39.4	43.0	45.6
25	10.5	11.5	13.1	14.6	16.5	19.9	24.3	29.3	34.4	37.7	40.6	44.3	46.9
26	11.2	12.2	13.8	15.4	17.3	20.8	25.3	30.4	35.6	38.9	41.9	45.6	48.3
27	11.8	12.9	14.6	16.2	18.1	21.7	26.3	31.5	36.7	40.1	43.2	47.0	49.6
28	12.5	13.6	15.3	16.9	18.9	22.7	27.3	32.6	37.9	41.3	44.5	48.3	51.0
29	13.1	14.3	16.0	17.7	19.8	23.6	28.3	33.7	39.1	42.6	45.7	49.6	52.3
30	13.8	15.0	16.8	18.5	20.6	24.5	29.3	34.8	40.3	43.8	47.0	50.9	53.7

Adapted from Introductory Statistical Analysis by Harnett, D.L., Murphy, J.L., 1976. Addison-Wesley; abridged from Tables of percentage points of the incomplete beta function and of the chi-square distribution Thompson, C.M., Biometrika, 1941, 32.

TABLE D.3A Cumulative *t*-distribution. The area or cumulative distribution, $F(t)$, under the *t*-distribution curve for different degrees of freedom, v, such that the probability $P(t_v < t_{v;F}) = F(t)$. The example here is for $n = 20$. For example, for $v = 9$, the probability $P(t_9 < t_{9;F} = 2.262) = F(2.262) = 0.975$ and $P(t_9 > t_{9;F} = 2.262) = 1 - F(2.262) = 0.025$, corresponding to the 95% confidence interval ($F = F_{0.025}$). Note that $F_{0.100}$, $F_{0.50}$, and $F_{0.005}$ correspond to the 80%, 90%, and 99% levels, respectively.

$$F(t) = \int_{-\infty}^{t} \frac{\left(\frac{v-1}{2}\right)!}{\left(\frac{v-2}{2}\right)! \sqrt{\pi n} \left(1 + \frac{t^2}{v}\right)^{(v+1)/2}} \, dt$$

				F			
v	0.75	0.90	0.95	0.975	0.99	0.995	0.9995
1	1.000	3.078	6.314	12.706	31.821	63.657	636.615
2	0.816	1.886	2.920	4.303	6.965	9.925	31.598
3	0.765	1.638	2.353	3.182	4.541	5.841	12.941
4	0.741	1.533	2.132	2.776	3.747	4.604	8.610
5	0.727	1.476	2.015	2.571	3.365	4.032	6.859
6	0.718	1.440	1.943	2.447	3.143	3.707	5.959
7	0.711	1.415	1.895	2.365	2.998	3.499	5.405
8	0.706	1.397	1.860	2.306	2.896	3.355	5.041
9	0.703	1.383	1.833	2.262	2.821	3.250	4.781
10	0.700	1.372	1.812	2.228	2.764	3.169	4.587
11	0.697	1.363	1.796	2.201	2.718	3.106	4.437
12	0.695	1.356	1.782	2.179	2.681	3.055	4.318
13	0.694	1.350	1.771	2.160	2.650	3.012	4.221
14	0.692	1.345	1.761	2.145	2.624	2.977	4.140
15	0.691	1.341	1.753	2.131	2.602	2.947	4.073
16	0.690	1.337	1.746	2.120	2.583	2.921	4.015
17	0.689	1.333	1.740	2.110	2.567	2.898	3.965
18	0.688	1.330	1.734	2.101	2.552	2.878	3.922
19	0.688	1.328	1.729	2.093	2.539	2.861	3.883
20	0.687	1.325	1.725	2.086	2.528	2.845	3.850
21	0.686	1.323	1.721	2.080	2.518	2.831	3.819
22	0.686	1.321	1.717	2.074	2.508	2.819	3.792
23	0.685	1.319	1.714	2.069	2.500	2.807	3.767
24	0.685	1.318	1.711	2.064	2.492	2.797	3.745
25	0.684	1.316	1.708	2.060	2.485	2.787	3.725
26	0.684	1.315	1.706	2.056	2.479	2.779	3.707
27	0.684	1.314	1.703	2.052	2.473	2.771	3.690
28	0.683	1.313	1.701	2.048	2.467	2.763	3.674
29	0.683	1.311	1.699	2.045	2.462	2.756	3.659
30	0.683	1.310	1.697	2.042	2.457	2.750	3.646
40	0.681	1.303	1.684	2.021	2.423	2.704	3.551
60	0.679	1.296	1.671	2.000	2.390	2.660	3.460
120	0.677	1.289	1.658	1.980	2.358	2.617	3.373
∞	0.674	1.282	1.645	1.960	2.326	2.576	3.291

Adapted from Introductory Statistical Analysis by Harnett, D.L., Murphy, J.L., 1976. Addison-Wesley; abridged from the Statistical tables of Fisher, R.A., Frank Yates, Oliver & Boyd, Edinburgh and London, 1938.

TABLE D.3B Cumulative t-distribution (two-tailed tests). Similar to Table D.3A except that values give cumulative distribution, $F(t)$, under the t-distribution curve for different degrees of freedom, ν, regardless of sign, such that the probability $P(|t_\nu| > |t_{\nu;F}|) = F(t)$. The example here is for $n = 20$. For example, for $\nu = 9$, the probability $P(|t_9| > |t_{9,F}| = 2.262) = F(2.262) = 0.05$ and $P(|t_9| < |t_{9,F}| = 2.262) = 1 - F(2.262) = 0.95$, corresponding to the 95% confidence interval. Note that $F_{0.200}$, $F_{0.100}$, and $F_{0.010}$ correspond to the 80, 90, and 99% levels, respectively.

	F probability of a larger value, sign ignored								
ν	0.500	0.400	0.200	0.100	0.050	0.025	0.010	0.005	0.001
1	1.000	1.376	3.078	6.314	12.706	25.452	63.657		
2	0.816	1.061	1.886	2.920	4.303	6.205	9.925	14.089	31.598
3	0.765	0.978	1.638	2.353	3.182	4.176	5.841	7.453	12.941
4	0.741	0.941	1.533	2.132	2.776	3.495	4.604	5.598	8.610
5	0.727	0.920	1.476	2.015	2.571	3.163	4.032	4.773	6.859
6	0.718	0.906	1.440	1.943	2.447	2.969	3.707	4.317	5.959
7	0.711	0.896	1.415	1.895	2.365	2.841	3.499	4.029	5.405
8	0.706	0.889	1.397	1.860	2.306	2.732	3.355	3.832	5.041
9	0.703	0.883	1.383	1.833	2.262	2.685	3.250	3.690	4.781
10	0.700	0.879	1.372	1.812	2.228	2.634	3.169	3.581	4.587
11	0.697	0.876	1.363	1.796	2.201	2.593	3.106	3.497	4.437
12	0.695	0.873	1.356	1.782	2.179	2.560	3.055	3.428	4.318
13	0.694	0.870	1.350	1.771	2.160	2.533	3.012	3.372	4.221
14	0.692	0.868	1.345	1.761	2.145	2.510	2.977	3.326	4.140
15	0.691	0.866	1.341	1.753	2.131	2.490	2.947	3.286	4.073
16	0.690	0.865	1.337	1.746	2.120	2.473	2.921	3.252	4.015
17	0.689	0.863	1.333	1.740	2.110	2.458	2.898	3.222	3.965
18	0.688	0.862	1.330	1.734	2.101	2.445	2.878	3.197	3.922
19	0.688	0.861	1.328	1.729	2.093	2.433	2.861	3.174	3.883
20	0.687	0.860	1.325	1.725	2.086	2.423	2.845	3.153	3.850
21	0.686	0.859	1.323	1.721	2.080	2.414	2.831	3.135	3.819
22	0.686	0.858	1.321	1.717	2.074	2.406	2.819	3.119	3.792
23	0.685	0.858	1.319	1.714	2.069	2.398	2.807	3.104	3.767
24	0.685	0.857	1.318	1.711	2.064	2.391	2.797	3.090	3.745
25	0.684	0.856	1.316	1.708	2.060	2.385	2.787	3.078	3.725
26	0.684	0.856	1.315	1.706	2.056	2.379	2.779	3.067	3.707
27	0.684	0.855	1.314	1.703	2.052	2.373	2.771	3.056	3.690
28	0.683	0.855	1.313	1.701	2.048	2.368	2.763	3.047	3.674
29	0.683	0.854	1.311	1.699	2.045	2.364	2.756	3.038	3.659
30	0.683	0.854	1.310	1.697	2.042	2.360	2.750	3.030	3.646
35	0.682	0.852	1.306	1.690	2.030	2.342	2.724	2.996	3.591
40	0.681	0.851	1.303	1.684	2.021	2.329	2.704	2.971	3.551
45	0.680	0.850	1.301	1.680	2.014	2.319	2.690	2.952	3.520
50	0.680	0.849	1.299	1.676	2.008	2.310	2.678	2.937	3.496
55	0.679	0.849	1.297	1.673	2.004	2.304	2.669	2.925	3.476

Continued

TABLE D.3B Cumulative t-distribution (two-tailed tests). Similar to Table D.3A except that values give cumulative distribution, $F(t)$, under the t-distribution curve for different degrees of freedom, ν, regardless of sign, such that the probability $P(|t_\nu| > |t_{\nu;F}|) = F(t)$. The example here is for $n = 20$. For example, for $\nu = 9$, the probability $P(|t_9| > |t_{9,F}| = 2.262) = F(2.262) = 0.05$ and $P(|t_9| < |t_{9,F}| = 2.262) = 1 - F(2.262) = 0.95$, corresponding to the 95% confidence interval. Note that $F_{0.200}$, $F_{0.100}$, and $F_{0.010}$ correspond to the 80, 90, and 99% levels, respectively.—cont'd

ν	F probability of a larger value, sign ignored								
	0.500	0.400	0.200	0.100	0.050	0.025	0.010	0.005	0.001
60	0.679	0.848	1.296	1.671	2.000	2.299	2.660	2.915	3.460
70	0.678	0.847	1.294	1.667	1.994	2.290	2.648	2.899	3.435
80	0.678	0.847	1.293	1.665	1.989	2.284	2.638	2.887	3.416
90	0.678	0.846	1.291	1.662	1.986	2.279	2.631	2.878	3.402
100	0.677	0.846	1.290	1.661	1.982	2.276	2.625	2.871	3.390
120	0.677	0.845	1.289	1.658	1.980	2.270	2.617	2.860	3.373
∞	0.6745	0.8416	1.2816	1.6448	1.9600	2.214	2.5758	2.8070	3.2905

TABLE D.4A Critical values of the F-distribution for $\alpha = 0.05$. The distributions represent the area exceeding the value of $F_{0.05,\nu_1,\nu_2}$ and $F_{0.01,\nu_1,\nu_2}$ as shown by the shaded area in the figure for different degrees of freedom, ν. For example, if $\nu_1 = 15$ and $\nu_2 = 20$, the critical value for $\alpha = 0.05$ is 2.20.

Values of $F_{0.05,\nu_1,\nu_2}$

$\nu_1 =$ Degrees of freedom for numerator

ν_2/ν_1	1	2	3	4	5	6	7	8	9	10	12	15	20	24	30	40	60	120	∞
1	161	200	216	225	230	234	237	239	241	242	244	246	248	249	250	251	252	253	254
2	18.5	19.0	19.2	19.2	19.3	19.3	19.4	19.4	19.4	19.4	19.4	19.4	19.4	19.5	19.5	19.5	19.5	19.5	19.5
3	10.1	9.55	9.28	9.12	9.01	8.94	8.89	8.85	8.81	8.79	8.74	8.70	8.66	8.64	8.62	8.59	8.57	8.55	8.53
4	7.71	6.94	6.59	6.39	6.26	6.16	6.09	6.04	6.00	5.96	5.91	5.86	5.80	5.77	5.75	5.72	5.69	5.66	5.63
5	6.61	5.79	5.41	5.19	5.05	4.95	4.88	4.82	4.77	4.74	4.68	4.62	4.56	4.53	4.50	4.46	4.43	4.40	4.37
6	5.99	5.14	4.76	4.53	4.39	4.28	4.21	4.15	4.10	4.06	4.00	3.94	3.87	3.84	3.81	3.77	3.74	3.70	3.67
7	5.59	4.74	4.35	4.12	3.97	3.87	3.79	3.73	3.68	3.64	3.57	3.51	3.44	3.41	3.38	3.34	3.30	3.27	3.23
8	5.32	4.46	4.07	3.84	3.69	3.58	3.50	3.44	3.39	3.35	3.28	3.22	3.15	3.12	3.08	3.04	3.01	2.97	2.93
9	5.12	4.26	3.86	3.63	3.48	3.37	3.29	3.23	3.18	3.14	3.07	3.01	2.94	2.90	2.86	2.83	2.79	2.75	2.71
10	4.96	4.10	3.71	3.48	3.33	3.22	3.14	3.07	3.02	2.98	2.91	2.85	2.77	2.74	2.70	2.66	2.62	2.58	2.54
11	4.84	3.98	3.59	3.36	3.20	3.09	3.01	2.95	2.90	2.85	2.79	2.72	2.65	2.61	2.57	2.53	2.49	2.45	2.40
12	4.75	3.89	3.49	3.26	3.11	3.00	2.91	2.85	2.80	2.75	2.69	2.62	2.54	2.51	2.47	2.43	2.38	2.34	2.30
13	4.67	3.81	3.41	3.18	3.03	2.92	2.83	2.77	2.71	2.67	2.60	2.53	2.46	2.42	2.38	2.34	2.30	2.25	2.21
14	4.60	3.74	3.34	3.11	2.96	2.85	2.76	2.70	2.65	2.60	2.53	2.46	2.39	2.35	2.31	2.27	2.22	2.18	2.13
15	4.54	3.68	3.29	3.06	2.90	2.79	2.71	2.64	2.59	2.54	2.48	2.40	2.33	2.29	2.25	2.20	2.16	2.11	2.07
16	4.49	3.63	3.24	3.01	2.85	2.74	2.66	2.59	2.54	2.49	2.42	2.35	2.28	2.24	2.19	2.15	2.11	2.06	2.01
17	4.45	3.59	3.20	2.96	2.81	2.70	2.61	2.55	2.49	2.45	2.38	2.31	2.23	2.19	2.15	2.10	2.06	2.01	1.96
18	4.41	3.55	3.16	2.93	2.77	2.66	2.58	2.51	2.46	2.41	2.34	2.27	2.19	2.15	2.11	2.06	2.02	1.97	1.92
19	4.38	3.52	3.13	2.90	2.74	2.63	2.54	2.48	2.42	2.38	2.31	2.23	2.16	2.11	2.07	2.03	1.98	1.93	1.88
20	4.35	3.49	3.10	2.87	2.71	2.60	2.51	2.45	2.39	2.35	2.28	2.20	2.12	2.08	2.04	1.99	1.95	1.90	1.84
21	4.32	3.47	3.07	2.84	2.68	2.57	2.49	2.42	2.37	2.32	2.25	2.18	2.10	2.05	2.01	1.96	1.92	1.87	1.81
22	4.30	3.44	3.05	2.82	2.66	2.55	2.46	2.40	2.34	2.30	2.23	2.15	2.07	2.03	1.98	1.94	1.89	1.84	1.78
23	4.28	3.42	3.03	2.80	2.64	2.53	2.44	2.37	2.32	2.27	2.20	2.13	2.05	2.01	1.96	1.91	1.86	1.81	1.76
24	4.26	3.40	3.01	2.78	2.62	2.51	2.42	2.36	2.30	2.25	2.18	2.11	2.03	1.98	1.94	1.89	1.84	1.79	1.73
25	4.24	3.39	2.99	2.76	2.60	2.49	2.40	2.34	2.28	2.24	2.16	2.09	2.01	1.96	1.92	1.87	1.82	1.77	1.71
30	4.17	3.32	2.92	2.69	2.53	2.42	2.33	2.27	2.21	2.16	2.09	2.01	1.93	1.89	1.84	1.79	1.74	1.68	1.62

Continued

TABLE D.4A Critical values of the F-distribution for α = 0.05. The distributions represent the area exceeding the value of $F_{0.05,\nu_1,\nu_2}$, and $F_{0.01,\nu_1,\nu_2}$ as shown by the shaded area in the figure for different degrees of freedom, ν. For example, if $\nu_1 = 15$ and $\nu_2 = 20$, the critical value for α = 0.05 is 2.20.—cont'd

Values of $F_{0.05,\nu_1,\nu_2}$

ν_1 = Degrees of freedom for numerator

ν_2/ν_1	1	2	3	4	5	6	7	8	9	10	12	15	20	24	30	40	60	120	∞
40	4.08	3.23	2.84	2.61	2.45	2.34	2.25	2.18	2.12	2.08	2.00	1.92	1.84	1.79	1.74	1.69	1.64	1.58	1.51
60	4.00	3.15	2.76	2.53	2.37	2.25	2.17	2.10	2.04	1.99	1.92	1.84	1.75	1.70	1.65	1.59	1.53	1.47	1.39
120	3.92	3.07	2.68	2.45	2.29	2.18	2.09	2.02	1.96	1.91	1.83	1.75	1.66	1.61	1.55	1.50	1.43	1.35	1.25
∞	3.84	3.00	2.60	2.37	2.21	2.10	2.01	1.94	1.88	1.83	1.75	1.67	1.57	1.52	1.46	1.39	1.32	1.22	1.00

ν_1 = Degrees of freedom for numerator.
ν_2 = Degrees of freedom for denominator.
$P(F > 2.20) = 0.05$.
$P(F < 2.20) = 0.95$.
Adapted from Introductory Statistical Analysis by Harnett, D.L., and Murphy, J.L., 1976. Addison-Wesley; abridged from tables of percentage points of the inverted beta (F) distribution by Merrington, M., Thompson, C.M., Biometrika, 1943, 33.

TABLE D.4B Critical values of the F-distribution for $\alpha = 0.01$ The distributions represent the area exceeding the value of $F_{0.05,\nu_1,\nu_2}$ and $F_{0.01,\nu_1,\nu_2}$ as shown by the shaded area in the figure for different degrees of freedom, ν. For example, if $\nu_1 = 15$ and $\nu_2 = 20$, then the critical value for $\alpha = 0.01$ is 3.09.

Values of $F_{0.01,\nu_1,\nu_2}$

ν_1 = Degrees of freedom for numerator

ν_1/ν_2	1	2	3	4	5	6	7	8	9	10	12	15	20	24	30	40	60	120	∞
1	4052	5000	5403	5625	5764	5859	5928	5982	6023	6056	6106	6157	6209	6235	6261	6287	6313	6339	6366
2	98.5	99.0	99.2	99.2	99.3	99.3	99.4	99.4	99.4	99.4	99.4	99.4	99.4	99.5	99.5	99.5	99.5	99.5	99.5
3	34.1	30.8	29.5	28.7	28.2	27.9	27.7	27.5	27.3	27.2	27.1	26.9	26.7	26.6	26.5	26.4	26.3	26.2	26.1
4	21.2	18.0	16.7	16.0	15.5	15.2	15.0	14.8	14.7	14.5	14.4	14.2	14.0	13.9	13.8	13.7	13.7	13.6	13.5
5	16.3	13.3	12.1	11.4	11.0	10.7	10.5	10.3	10.2	10.1	9.89	9.72	9.55	9.47	9.38	9.29	9.20	9.11	9.02
6	13.7	10.9	9.78	9.15	8.75	8.47	8.26	8.10	7.98	7.87	7.72	7.56	7.40	7.31	7.23	7.14	7.06	6.97	6.88
7	12.2	9.55	8.45	7.85	7.46	7.19	6.99	6.84	6.72	6.62	6.47	6.31	6.16	6.07	5.99	5.91	5.82	5.74	5.65
8	11.3	8.65	7.59	7.01	6.63	6.37	6.18	6.03	5.91	5.81	5.67	5.52	5.36	5.28	5.20	5.12	5.03	4.95	4.86
9	10.6	8.02	6.99	6.42	6.06	5.80	5.61	5.47	5.35	5.26	5.11	4.96	4.81	4.73	4.65	4.57	4.48	4.40	4.31
10	10.0	7.56	6.55	5.99	5.64	5.39	5.20	5.06	4.94	4.85	4.71	4.56	4.41	4.33	4.25	4.17	4.08	4.00	3.91
11	9.65	7.21	6.22	5.67	5.32	5.07	4.89	4.74	4.63	4.54	4.40	4.25	4.10	4.02	3.94	3.86	3.78	3.69	3.60
12	9.33	6.93	5.95	5.41	5.06	4.82	4.64	4.50	4.39	4.30	4.16	4.01	3.86	3.78	3.70	3.62	3.54	3.45	3.36
13	9.07	6.70	5.74	5.21	4.86	4.62	4.44	4.30	4.19	4.10	3.96	3.82	3.66	3.59	3.51	3.43	3.34	3.25	3.17
14	8.86	6.51	5.56	5.04	4.70	4.46	4.28	4.14	4.03	3.94	3.80	3.66	3.51	3.43	3.35	3.27	3.18	3.09	3.00
15	8.68	6.36	5.42	4.89	4.56	4.32	4.14	4.00	3.89	3.80	3.67	3.52	3.37	3.29	3.21	3.13	3.05	2.96	2.87
16	8.53	6.23	5.29	4.77	4.44	4.20	4.03	3.89	3.78	3.69	3.55	3.41	3.26	3.18	3.10	3.02	2.93	2.84	2.75
17	8.40	6.11	5.19	4.67	4.34	4.10	3.93	3.79	3.68	3.59	3.46	3.31	3.16	3.08	3.00	2.92	2.83	2.75	2.65
18	8.29	6.01	5.09	4.58	4.25	4.01	3.84	3.71	3.60	3.51	3.37	3.23	3.08	3.00	2.92	2.84	2.75	2.66	2.57
19	8.19	5.93	5.01	4.50	4.17	3.94	3.77	3.63	3.52	3.43	3.30	3.15	3.00	2.92	2.84	2.76	2.67	2.58	2.49
20	8.10	5.85	4.94	4.43	4.10	3.87	3.70	3.56	3.46	3.37	3.23	3.09	2.94	2.86	2.78	2.69	2.61	2.52	2.42
21	8.02	5.78	4.87	4.37	4.04	3.81	3.64	3.51	3.40	3.31	3.17	3.03	2.88	2.80	2.72	2.64	2.55	2.46	2.36
22	7.95	5.72	4.82	4.31	3.99	3.76	3.59	3.45	3.35	3.26	3.12	2.98	2.83	2.75	2.67	2.58	2.50	2.40	2.31
23	7.88	5.66	4.76	4.26	3.94	3.71	3.54	3.41	3.30	3.21	3.07	2.93	2.78	2.70	2.62	2.54	2.45	2.35	2.26
24	7.82	5.61	4.72	4.22	3.90	3.67	3.50	3.36	3.26	3.17	3.03	2.89	2.74	2.66	2.58	2.49	2.40	2.31	2.21
25	7.77	5.57	4.68	4.18	3.86	3.63	3.46	3.32	3.22	3.13	2.99	2.85	2.70	2.62	2.53	2.45	2.36	2.27	2.17
30	7.56	5.39	4.51	4.02	3.70	3.47	3.30	3.17	3.07	2.98	2.84	2.70	2.55	2.47	2.39	2.30	2.21	2.11	2.01

Continued

TABLE D.4B Critical values of the F-distribution for $\alpha = 0.01$ The distributions represent the area exceeding the value of $F_{0.05,\nu_1,\nu_2}$ and $F_{0.01,\nu_1,\nu_2}$ as shown by the shaded area in the figure for different degrees of freedom, ν. For example, if $\nu_1 = 15$ and $\nu_2 = 20$, then the critical value for $\alpha = 0.01$ is 3.09.—cont'd

Values of $F_{0.01,\nu_1,\nu_2}$

ν_1 = Degrees of freedom for numerator

ν_1/ν_2	1	2	3	4	5	6	7	8	9	10	12	15	20	24	30	40	60	120	∞
40	7.31	5.18	4.31	3.83	3.51	3.29	3.12	2.99	2.89	2.80	2.66	2.52	2.37	2.29	2.20	2.11	2.02	1.92	1.80
60	7.08	4.98	4.13	3.65	3.34	3.12	2.95	2.82	2.72	2.63	2.50	2.35	2.20	2.12	2.03	1.94	1.84	1.73	1.60
120	6.85	4.79	3.95	3.48	3.17	2.96	2.79	2.66	2.56	2.47	2.34	2.19	2.03	1.95	1.86	1.76	1.66	1.53	1.38
∞	6.63	4.61	3.78	3.32	3.02	2.80	2.64	2.51	2.41	2.32	2.18	2.04	1.88	1.79	1.70	1.59	1.47	1.32	1.00

ν_1 = Degrees of freedom for numerator.
ν_2 = Degrees of freedom for denominator.
$P(F > 3.09) = 0.01$.
$P(F < 3.09) = 0.99$.

Appendix E

Correlation Coefficients at the 5% and 1% Levels of Significance for Various Degrees of Freedom, ν

Degrees of Freedom (ν)	5%	1%
1	0.997	1.000
2	0.950	0.990
3	0.878	0.959
4	0.811	0.917
5	0.754	0.874
6	0.707	0.834
7	0.666	0.798
8	0.632	0.765
9	0.602	0.735
10	0.576	0.708
11	0.553	0.684
12	0.532	0.661
13	0.514	0.641
14	0.497	0.623
15	0.482	0.606
16	0.468	0.590
17	0.456	0.576
18	0.444	0.561
19	0.433	0.549
20	0.423	0.537
21	0.413	0.526
22	0.404	0.515
23	0.396	0.505
24	0.388	0.496
25	0.381	0.487
26	0.374	0.478

(Continued)

—cont'd

27	0.367	0.470
28	0.361	0.463
29	0.355	0.456
30	0.349	0.449
35	0.325	0.418
40	0.304	0.393
45	0.288	0.372
50	0.273	0.354
60	0.250	0.325
70	0.232	0.302
80	0.217	0.283
90	0.205	0.267
100	0.195	0.254
125	0.174	0.228
150	0.159	0.208
200	0.138	0.181
300	0.113	0.148
400	0.098	0.128
500	0.088	0.115
1000	0.062	0.081

Appendix F

Approximations and Nondimensional Numbers in Physical Oceanography

Beta parameter, β^*: A nondimensionalized form of β (the beta parameter) defined as the ratio of the horizontal gradient in relative vorticity, $\nabla_h \zeta$, to the horizontal gradient in planetary vorticity, $\nabla_h f$, such that

$$\beta^* \equiv \frac{\beta L^2}{U}$$

Here, $\zeta = \partial v/\partial x - \partial u/\partial y$ is the relative vorticity (for velocity components u, v in the x, y directions, respectively), f is the local Coriolis parameter, U is a horizontal velocity scale, L is a horizontal length scale (*see also Rhines length*) and β is defined as

$$\beta \equiv \nabla_h f \approx 10^{-11}/\text{ms}$$

where "ms" denotes "meter·seconds". In the *beta-plane approximation*, the curved surface of the earth is approximated by a flat plane tangent to the earth for which $f = f_o + \beta y$, where f_o is a reference value for f and y is the latitude. For this case, $\beta = df/dy$ is a constant.

Boussinesq approximation: Assumes that density changes in the fluid can be neglected except where density, $\rho = \rho_o + \rho'$, is multiplied by the acceleration of gravity, g. That is, the effects of density fluctuations, ρ', can be neglected in terms of the form $\rho F = (\rho_o + \rho')F$ for any variable F except for those involving, g (i.e., $\rho' g$). Here, $\rho_o = \rho_o(z)$ is the mean density and $\rho'/\rho_o \approx 10^{-3}$. At large *Mach* numbers ($U/c > 1$), the compressibility of the fluid becomes important and large density changes can occur. Since the speed of sound in water, $c \approx 1500$ m/s, is almost always large compared to the flow speed, U, the approximation is good for normal oceanic conditions.

Brunt–Väisälä frequency, $N(z)$ (also Väisälä or Buoyancy frequency): The natural frequency of oscillation of a parcel of water displaced vertically (z-direction is positive upward) from its level of equilibrium:

$$N(z) = \sqrt{-\left(\frac{g}{\rho_o}\frac{d\rho}{dz} + \frac{g^2}{c^2}\right)} \simeq \sqrt{-\left(\frac{g}{\rho_o}\frac{d\rho}{dz}\right)}$$

where $g = 9.81$ m/s^2 is the acceleration due to gravity, c is the speed of sound, $d\rho/dz \leq 0$ is the vertical *in situ* density gradient, and ρ_o is a reference density. N marks the maximum intrinsic frequency of oscillation obtainable by internal gravity waves. The compressibility term, $g^2/c^2 \approx 5 \times 10^{-5}$/s, associated with adiabatic displacement of the fluid can generally be ignored in the upper few thousand meters of the ocean where $10^{-4} < N < 10^{-2}$/s (periods of 17.5 h to 10.5 min). However, this is not the case in the deep ocean where $d\rho/dz$ is small and N can be of order 10^{-5}/s. Derivation of N using conductivity-temperature-depth data usually requires low-pass filtering to eliminate erroneously high or negative values of the density gradient (see Figure 1.41).

Burger number, B (also Stratification parameter, S): The squared ratio of the internal (Rossby) radius of deformation, r_i, to a longwave scale, L (such as the wavelength of a Rossby or coastal-trapped wave)

$$B(S) \equiv \frac{N^2 H^2}{f^2 L^2}$$

where N is a characteristic Brunt-Väisälä frequency for the water depth H. For continental shelf-slope regions influenced by coastal-trapped waves, motions are baroclinic when $B \gg 1$ and barotropic when $B \ll 1$. For internal waves over a sloping bottom tilted at an angle ϕ to the horizontal, $H = L \sin\phi$, and

$$B(S) \equiv \frac{N^2 \sin^2\phi}{f^2}$$

Coriolis parameter (also Inertial frequency, Coriolis frequency, Planetary vorticity): The local vertical component of the earth's rate of rotation given by $f = 2\Omega \sin\theta$, where θ is the latitude and $\Omega = 0.72921 \times 10^{-4}$/s is the angular rate of rotation based on a *sidereal day* of 23 h 56 min 4 s. A sidereal day is the time for the earth to complete one rotation relative to an absolute reference point in space. Because of the movement of the earth about the sun, a sidereal day differs slightly from the solar day of 24 h. Latitude θ is positive for the northern hemisphere and negative for the southern hemisphere. At 50° latitude, $|f| = 1.117 \times 10^{-4}$/s (rad/s), corresponding to a cyclic frequency of 0.0640 cph and a period $T_f = 2\pi/|f|$ of 15.6 h; at 10° latitude, $|f| = 0.253 \times 10^{-4}$/s, corresponding to a period of 68.92 h = 2.87 days.

Cox number, C_θ: A relative measure of high vertical wavenumber temperature structure (temperature "noisiness") defined as the ratio of the mean vertical gradient squared to the mean-square vertical gradient for temperature, $T(z)$;

$$C_\theta \equiv \frac{\langle dT/dz \rangle^2}{\langle (dT/dz)^2 \rangle}$$

Ekman number, E: A nondimensional number giving the relative importance of frictional forces at a boundary to the Coriolis force

$$E \equiv \frac{\text{frictional force}}{\text{Coriolis force}} = \frac{\nu}{f D^2}$$

where ν is the turbulent eddy viscosity, f is the Coriolis parameter, and D is a representative depth-scale for the fluid (e.g., the depth of the main pycnocline in the case of the surface Ekman layer). The characteristic thickness, δ, of the Ekman layer is given by

$$\delta = \sqrt{\frac{2\nu_v}{f}}$$

where ν_v is the vertical component of eddy viscosity. For ν_v of order 10^{-2} m²/s, $\delta \approx 20$ m at mid-latitudes ($f \approx 1 \times 10^{-4}$/s).

Froude number, F_r: The square root of the ratio of the inertial force to the gravitational force for barotropic motions with a free surface

$$F_r \equiv \left(\frac{\text{inertial force}}{\text{gravitational force}} \right)^{1/2} = \frac{U}{\sqrt{gH}}$$

where U is the flow velocity and $c = \sqrt{gH}$ is the phase speed of a surface wave in a fluid of depth H. The flow is *supercritical* if $F_r > 1$ and *subcritical* if $F_r < 1$. The Froude number is analogous to the *Mach number* used for compressible fluids, such as air. Hydraulic jumps occur where the fluid speed transitions from supercritical to subcritical flow.

Froude number (internal), F_r': The square root of the ratio of the inertial force to the buoyancy force for baroclinic motions in a stratified fluid

$$F_r' \equiv \left(\frac{\text{inertial force}}{\text{buoyancy force}} \right)^{1/2} = \frac{U}{\sqrt{g' H_n}}$$

where $g' = g(\rho_2 - \rho_1)/\rho_2$ is the reduced gravity and $c_n' = \sqrt{g' H_n}$ is the phase speed of a mode n internal wave in a fluid with an effective depth, H_n (see *Internal Rossby radius*). The flow is *supercritical* if $F_r' > 1$ and *subcritical* if $F_r' < 1$. The internal Froude number is used in studies of density-driven turbidity currents and landslide-generated tsunamis.

Geostrophic approximation: Assumes that the Rossby number is small ($Ro \ll 1$) so that horizontal motions are mainly a balance between the Coriolis force and the horizontal pressure gradient. It takes just over one inertial period, $T_f = 2\pi/|f|$, for a perturbed geostrophic flow to return to near geostrophic balance.

Hydrostatic approximation: Assumes that the vertical velocity, w, can be ignored in the vertical component of the momentum balance and that the vertical pressure gradient $\partial p/\partial z$ is proportional to the density, ρ:

$$\frac{\partial p}{\partial z} = -g\rho$$

Integration from depth z to the ocean surface $z = \eta$ gives, for near-uniform density $\rho \approx \rho_o$,

$$p = p_o + g\rho_o(\eta - z)$$

where p_o is the atmospheric pressure at the ocean surface. The approximation cannot be used to study high-frequency internal wave dynamics.

Inertial period, T_f: The period of oscillation for the Coriolis frequency, f (*see Coriolis frequency*)

$$T_f = \frac{2\pi}{|f|} = \frac{\pi}{\Omega|\sin\theta|}$$

$T_f \approx 68.92$, 15.62, and 12.74 h for inertial motions at latitudes θ of $10°$, $50°$, and $70°$, respectively.

Intrinsic frequency: If ω_o is the frequency of a wave measured at a fixed point and $\boldsymbol{k}$ is the wavenumber vector of the wave, the intrinsic frequency, ω, of the wave as seen by an observer in a coordinate system moving with the mean flow, $\boldsymbol{U}$, is given by

$$\omega = \omega_o - \boldsymbol{k} \cdot \boldsymbol{U}$$

Thus, the frequency, ω_o, of the wave measured at the fixed point is Doppler shifted by the amount $+\boldsymbol{k} \cdot \boldsymbol{U}$ relative to the intrinsic frequency, ω. For most oceanic motions, the Doppler shift measured at fixed point is within a few percent of the intrinsic frequency.

Kolmogorov microscale, η: The length scale at which turbulent motions begin to be damped out by small-scale molecular viscosity, ν

$$\eta \equiv 2\pi\left(\frac{\nu^3}{\varepsilon}\right)^{1/4}$$

in which $\varepsilon = 2\nu\left\langle\left(\partial u_i/\partial x_j\right)^2\right\rangle$ is the mean rate of dissipation of turbulent kinetic energy (*see Ozmidov scale*). In the upper ocean, η is a few centimeters.

Mach number, M: The relative importance of fluid compressibility defined by the relation

$$M \equiv \left(\frac{\text{inertial force}}{\text{compressibility force}}\right)^{1/2} = \frac{U}{c}$$

where c is the speed of sound (≈ 1500 m/s in water) and U is the velocity of the fluid. Flows are subsonic if $M < 1$ and supersonic if $M > 1$. Compressibility effects can be ignored if $M < 0.3$.

Monin−Obukhov length, L_M: The height above a heated boundary at which mechanical (shear) production of turbulent kinetic energy equals the buoyant (convective) destruction of turbulent kinetic energy

$$L_M \equiv \frac{\text{shear production}}{\text{buoyant destruction}} = \frac{u^3}{k\alpha g\overline{wT'}}$$

where u, k, g and α are, respectively, the friction velocity, the von Kármán constant (≈ 0.40), the acceleration of gravity, and the coefficient of thermal expansion, and $\overline{wT'}$ is the mean heat flux for vertical velocity fluctuations w and temperature fluctuations T'.

Ozmidov (buoyancy) scale, η_b (or L_R): The ratio of nonlinear to buoyancy scales in a turbulent fluid; the scale above which eddy-like motions are damped by stratification (as characterized by the buoyancy frequency, N);

$$\eta_b \equiv 2\pi\left(\frac{\varepsilon}{N^3}\right)^{1/2}$$

Here, $\varepsilon = 2\nu\left\langle \left(\rho u_i / \rho x_j\right)^2 \right\rangle$ is the rate of dissipation of turbulent kinetic energy, and u_i is the ith component of velocity in the jth direction, x_j ($i, j = 1, 2, 3$ corresponding to the x, y, z directions, respectively). In the upper ocean, η_b can be up to a few meters.

Péclet Number, Pe: The diffusivity analog to the Reynolds number (*Re*):

$$Pe \equiv \frac{UL}{K} = Pr \cdot Re$$

where K is the diffusivity of heat or salt. In geophysical fluid dynamics, K corresponds to the turbulent eddy diffusivity (see *Prandtl number, Pr* and *Reynolds number, Re*).

Prandtl number, Pr: The ratio of momentum to heat (or salt) diffusivity:

$$Pr \equiv \frac{\text{momentum diffusivity}}{\text{heat diffusivity}} = \frac{\nu}{K_T}$$

For typical values of molecular viscosity $\nu \approx 10^{-2}$ cm^2/s (10^{-6} m^2/s) and molecular heat diffusivity $K_T \approx 10^{-3}$ cm^2/s (10^{-7} m^2/s), $Pr \approx 10$. For salt, $K_S \approx 10^{-5}$ cm^2/s (10^{-9} m^2/s) and $Pr \approx 1000$. A turbulent *Prandtl number* can be defined in terms of the turbulent eddy viscosity and turbulent diffusivities of heat and salt. The *Schmidt number* is similar to the Prandtl number with momentum diffusivity replaced by mass diffusivity.

Rayleigh number, Ra: The ratio of the destabilizing effect of the buoyancy force to the stabilizing effect of the viscous force:

$$Ra \equiv \frac{g\alpha\Gamma d^4}{K_T\nu}$$

where α is the coefficient of thermal expansion, $\Gamma = -d\langle T\rangle/dz$ is the vertical gradient of the background temperature $\langle T\rangle$ (the adiabatic temperature gradient, also known as the "lapse rate" by meteorologists), d is the depth of the layer, K_T is the thermal diffusivity, and ν is the kinematic viscosity. The "lapse rate" is the fastest rate at which the temperature can decrease with height without causing instability.

Reynolds number, Re: The ratio of the inertial (nonlinear) force to the viscous force

$$Re \equiv \frac{\text{inertial force}}{\text{viscous force}} = \frac{UL}{\nu}$$

where U is the flow speed, L is a characteristic length scale, and ν is the kinetic viscosity; $\nu \approx 0.01$ cm^2/s (0.01×10^{-4} m^2/s) for molecular processes. Viscous effects become important at small Reynolds numbers, $Re \ll 1$. In geophysical fluid dynamics, such as in the formation of mesoscale vortex streets, ν appears to correspond to the turbulent eddy viscosity.

Rhines length, l: The scale at which barotropic mesoscale eddies transform from individual features to Rossby wave packets (the scale at which the effect of planetary vorticity, β, becomes comparable to nonlinear turbulent eddy effects). Rossby wave propagation causes anisotropic elongation of the eddies in the zonal (east–west, x) direction and the eddy size in the meridional (north–south, y) direction stops growing at the scale (Rhines, 1975) $l = \sqrt{\dfrac{u}{\beta}}$, where u is the root-mean-square velocity of the eddying motions and β is the north–south gradient of the Coriolis parameter, $f = f_o + \beta y$ (*see Beta parameter*). When the eddy length scale exceeds the Rhines scale, the linear Rossby wave term dominates the nonlinear turbulent term. The corresponding "Rhines effect" consists of an interaction of the Rossby waves with the two-dimensional turbulence that induces alternating zonal flows ("jets"), thereby deforming and eventually destroying the coherent vortices that might exist.

Richardson number, Ri: A measure of the dynamic stability of the water column. In a two-layer fluid with reduced gravity g', mean flow U, and horizontal length scale, L, the *local Richardson number* is defined as

$$Ri \equiv \frac{g'L}{U^2}$$

while for a continuously stratified fluid with buoyancy frequency $N(z)$

$$Ri \equiv \frac{N^2L^2}{U^2}$$

The above expressions also are known as the *bulk Richardson number* since they define the overall stability characteristics of the water column. In both cases, $Ri \propto 1/Fr'^2$, where Fr' is the internal Froude number. For $Ri > 0$, the stratification is stable; for $Ri = 0$ it is neutral; and for $Ri < 0$ it is unstable. The *gradient Richardson number*

$$Ri \equiv \frac{N^2}{(dU/dz)^2}$$

is a measure of the localized stability of the water column in which the stabilizing effect of the density gradient, or buoyancy N, competes with the destabilizing effect of turbulent mixing due to the vertical shear, dU/dz. Shear instability typically can be expected for $Ri \leq 1$ (the often-used $Ri \leq 1/4$ criterion is a necessary, but not sufficient condition for instability). The *flux Richardson number*, which is the ratio of the rate of increase in fluid potential energy due to entrainment (buoyant destruction of turbulent kinetic energy) to the rate of production of turbulent energy associated with the velocity shear, may be defined as

$$Rf \equiv \frac{-g\alpha\overline{wT'}}{-\overline{uw}(dU/dz)} \approx \frac{\nu_v N^2}{\varepsilon}$$

where $g\alpha\overline{wT'}$ is the production of turbulent kinetic energy by the vertical heat flux $\overline{wT}$, $-\overline{uw}(dU/dz)$ is the production of turbulent kinetic energy by the Reynolds stress $\overline{uw}$ working against the mean shear dU/dz, ν_v is the vertical diffusion coefficient, N is the buoyancy frequency, and ε is turbulent energy production.

Rigid-lid approximation: For surface displacement $\eta(t)$, the rigid-lid approximation requires that the vertical velocity $w = \partial\eta/\partial t + \mathbf{u} \cdot \nabla\eta = 0$ at the surface ($z = 0$) and that vertical baroclinic motions within the fluid greatly exceed those at the surface. One implication of the rigid-lid approximation is that the external Rossby radius, r_o, becomes infinite; hence, a measure of the validity of the approximation is that for motions of length scale L, $L/r_o \ll 1$. The rigid-lid approximation allows surface pressure in the ocean to vary spatially but eliminates surface gravity waves. If one could put a rigid cover on top of the ocean, the upward pressure beneath the cover would vary in space but gravity waves would be eliminated. Application of the rigid lid approximation removes barotropic Kelvin waves from the coastal trapped wave problem and simplifies calculation of baroclinic modes.

Rossby number, Ro: The ratio of nonlinear to Coriolis forces, and the ratio of the relative vorticity to the planetary vorticity, defined by

$$Ro \equiv \frac{\text{nonlinear accelerations}}{\text{Coriolis force}} = \frac{U^2/L}{fU} = \frac{U}{fL}$$

For common oceanic scales $U \approx 0.1$ m/s, $L \approx 100$ km, and $f \approx 10^{-4}$/s, we find $Ro \approx 0.01$ so that nonlinear terms are often of second order in the equations of motion.

Rossby radius of deformation (external; barotropic), r_o: The natural e-folding scale for barotropic currents in the sea defined as

$$r_o \equiv \frac{\sqrt{gH}}{f} = \frac{c}{f}$$

where $c = \sqrt{gH}$ is the propagation speed of long gravity waves (e.g., the tide) in water of depth H. For a mid-latitude ocean of depth 1000 m, $r_o \approx 1000$ km.

Rossby radius of deformation (internal; baroclinic), r_i: The natural e-folding scale for baroclinic motions which, for a continuously stratified ocean, is normally written as

$$r_i \equiv \frac{NH}{f}$$

where H is the local water depth and N is a representative value for the local buoyancy frequency. We may also define the baroclinic Rossby radius as

$$\pi r_i \equiv \frac{\sqrt{gH_n}}{f} = \frac{c_n}{f}$$

where H_n is the "equivalent depth"

$$H_n = H^2 N^2 / g n^2 \pi^2$$

and

$$c_n = NH/n\pi, n = 1, 2, \ldots$$

are the horizontal phase speeds of the different vertical wave modes. For first mode ($n = 1$) wave propagation in a mid-latitude region of depth $H \approx 1000$ m, buoyancy frequency $N \approx 3 \times 10^{-3}$/s, and Coriolis frequency $f \approx 10^{-4}$/s, we find $c_1 \approx 1.0$ m/s and $r_i \approx 60$ km. For a two-layer fluid with upper and lower layer densities and thicknesses ρ_1, H_1 and ρ_2, H_2, respectively, we have

$$r_i \equiv f^{-1} \left(\frac{g(\rho_2 - \rho_1)}{\rho_2} \cdot \frac{H_1 H_2}{H_1 + H_2} \right)^{1/2} = f^{-1} \left(g' \frac{H_1 H_2}{H_1 + H_2} \right)^{1/2}$$

Schmidt number, *Sc*: The ratio of viscosity (momentum diffusivity) to mass diffusivity (such as in convection processes) defined by

$$Sc \equiv \frac{\text{viscous diffusion rate}}{\text{molecular (mass) diffusion rate}} = \frac{\nu}{K_m} = \frac{\mu}{\rho K_m}$$

where $\nu = \mu/\rho$ (m²/s) is the kinematic viscosity, μ is the dynamic viscosity (N s/m²), ρ is the density (kg/m³), and K_m is the mass diffusivity (m²/s).

Strouhal number, *S*: The ratio of the boundary-imposed frequency of fluid oscillation from an object, n_s, to the "natural" frequency of oscillation, U/D, based on the flow speed U and length scale D (often the diameter) of an obstacle:

$$S \equiv \frac{n_s D}{U}$$

In the case of an obstacle in a steady flow, n_s is the frequency of vortex shedding of the leeward flow.

Thorpe scale, L_T: An objective measure of the vertical overturning scale in a turbulent stratified fluid. First proposed by Thorpe (1977) to describe overturning structures within turbulent mixing events in a Scottish loch, the scale is obtained by rearranging an observed density profile, which may contain inversions, into a profile in which density increases monotonically with depth. Heat and mass are conserved during the rearrangement process. Consider an observed profile of n density values, ρ_n, sampled at depths z_n. If a given sample with density ρ_n must be moved to a depth z_m in generating the stable profile, then the Thorpe displacement for the sample is $z_m - z_n$. In general, a unique displacement will result from each density sample and n Thorpe displacements will be generated from the original profile. The Thorpe scale, L_T, is the RMS of these displacements (Dillon, 1982; Libe Washburn, personal communication). Typical values are of the order of 1 m.

Turner angle, T_u: The diffusivity of heat, K_T, in the ocean is roughly 100 times that of salt ($K_T \approx 100 K_S$). In regions of the ocean where the vertical gradients of temperature and salinity have the same sign, this differential diffusivity can lead to the formation of sharply defined thermohaline "staircases" through the process of double diffusion (Turner, 1973; Kelley, 1990; Ruddick and Gargett, 2004; Spear and Thomson, 2012). The strength of double diffusion can be characterized by the density gradient ratio $R_\rho = \alpha T_z / \beta S_z$, where $\alpha = -\rho^{-1} \partial \rho / \partial T$ is the thermal expansion coefficient, $\beta = \rho^{-1} \partial \rho / \partial S$ is the haline contraction coefficient, ρ is density, and T_Z, and S_Z are the vertical temperature and salinity gradients, respectively. Both diffusive convection and salt fingering intensify as R_ρ approaches unity. To avoid sign ambiguities associated with R_ρ, Ruddick (1983) proposed the "Turner angle", $T_u = \tan^{-1}(R_\rho) - 45°$, which remains defined as S_Z approaches zero. For T_u between $-45°$ and $-90°$ (R_ρ between 0 and 1), diffusive convection is possible; when T_u lies between $45°$ and $90°$ (R_ρ between 1 and ∞), salt fingering can be expected. For T_u between $-45°$ and $45°$, the water column is doubly stable; for all other values of T_u, the water column is statically unstable. Double diffusion is characterized as strong ($|T_u| \geq 75°$), medium ($75° > |T_u| \geq 60°$), or weak ($60° > |T_u| \geq 45°$).

Appendix G

Convolution

G.1 Convolution and Fourier Transforms

Consider the time-dependent functions $g(t)$ and $h(t)$ and their respective frequency-dependent Fourier transforms $G(f)$ and $H(f)$. The convolution of the two original functions (written $g * h$) is defined as

$$g * h \equiv \int_{-\infty}^{\infty} g(t)h(t - \tau)\mathrm{d}t \tag{G.1}$$

where $g * h$ is a function of the time lag, τ, and $g * h = h * g$. There is a one-to-one relationship between the function $g * h$ and the product of the Fourier transforms of the two functions such that

$$g * h \leftrightarrow G(f) \cdot H(f)^* \tag{G.2}$$

Known as the convolution theorem, Eq. (G.2) states that the Fourier transform of the convolution term on the left is the product of the Fourier transforms of the individual functions on the right side. In other words, convolution in one domain equates to the multiplication in the other domain. We further note that the correlation of g and h [$C(g,h) = \mathrm{corr}(g,h)$; see Chapter 5] is written as

$$C(g, h) \equiv \int_{-\infty}^{\infty} g(t + \tau)h(t)\mathrm{d}t \tag{G.3}$$

which is also a function of the lag τ. As with convolution, we can form the transform pair

$$C(g, h) \leftrightarrow G(f) \cdot H(f)^* \tag{G.4}$$

called the correlation theorem, where $H(f)^*$ is the complex conjugate of $H(f)$ and $H(f)^* = H(-f)$, since we are restricting discussion to the usual case in which g and h are real functions. As this relationship indicates, multiplying the Fourier transform of one function by the complex conjugate of the Fourier transform of the other function yields the Fourier transform of their correlation. The correlation of a function with itself is called its autocorrelation (Chapter 5).

G.2 Convolution of Discrete Data

The analysis of geophysical data commonly involves the convolution of specially designed "data windows" (convolution functions or filters) with time series records in order to smooth the spectral estimates obtained from these data and to improve the statistical reliability of spectral peaks. Good filters are those that minimize unwanted spectral leakage associated with the filter's side lobes in the frequency domain. Consider a filter $h(t_k)$ applied to a discrete data series $g(t_j)$, where the t_j and t_k ($j, k = 0, \ldots$) are discrete times in the data series. The filter will have nonzero values over a short segment of the data to which it is being applied and will be zero elsewhere, yielding a single value for the central time of the filter for that specific piece of the data. The filter $h(t_k)$ typically has a central peak and falls off to zero on either side of the maximum.

The convolution theorem can be extended to discrete time series as follows. Assume that the time series, $g(t_j)$, has duration N and is completely determined by the N values $g(t_0)$, ..., $g(t_{N-1})$. The convolution of this function with the window, $h(t_k)$, is a member of the discrete Fourier transform pair

$$\sum_{k=-N/2+1}^{N/2} g(t_{j-k}) h(t_k) \leftrightarrow G_n H_n \tag{G.5}$$

where G_n $(n = 0, ..., N - 1)$ is the discrete Fourier transform of the time series $g(t_j)$ $(j = 0, ..., N - 1)$, and H_n $(n = 0, ..., N - 1)$ is the discrete Fourier transform of the function $h(t_k)$, $(k = 0, ..., N - 1)$. The values of $h(t_k)$ typically span a small fraction of the full data range $k = -N/2 + 1, ..., N/2$.

In Figure G.1, the original time series, $g(t_j)$—chosen for illustrative purposes to be the normalized monthly values of the Southern Oscillation Index, SOI—is shown in the top panel and the convolution function, $h(t_k)$, used to filter the time series is presented in the middle panel. Here, we have used a simple 5-year long Hamming window [see Chapter 6]. The window (filter) is symmetrical, uses 61 monthly weights (with nonzero first and last weights), and begins with the first month of the time series.

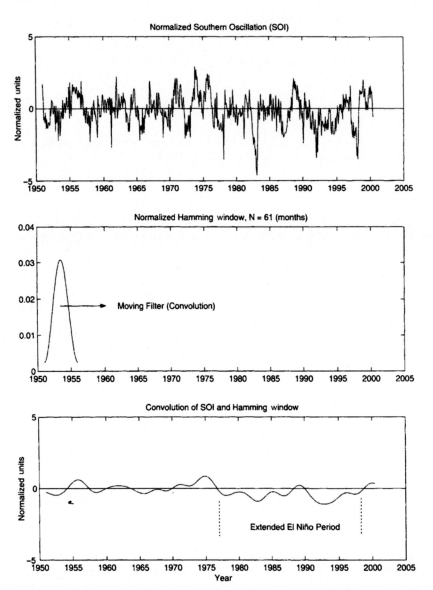

FIGURE G.1 Convolution of the monthly time series of normalized southern oscillation index (see U.S. government web site ftp.necp.noaa.gov/pub/cpc/wd52dg/data/indices/...) using a 61-month Hamming window (filter). Negative (positive) values of the index are associated with El Niño (La Niña) events. The convolution emphasizes the low-frequency variability of the El Niño-La Niña phenomenon in the equatorial Pacific.

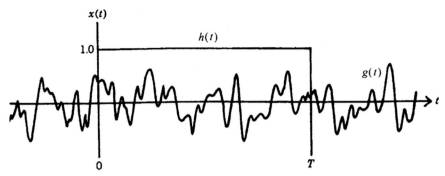

FIGURE G.2 Sampling a time series segment of duration T. The measurement is analogous to application of a rectangular window, $h(t)$, of amplitude 1.0 and duration T to an extensive time series $g(t)$.

The bottom panel in Figure G.1 shows the convolution of $h(t_k)$ with $g(t_j)$. As the filtered result clearly demonstrates, $h(t_k)$ acts as a smoothing function that flattens out the "bumpiness" of $g(t_j)$, reducing sharp year-to-year changes in the normalized SOI. This smoothing depends on the duration and the shape of the window, $h(t_k)$. A more sharply peaked $h(t_k)$ would produce less time series smoothing, leaving more of the large year-to-year variability. The function, $h(t_k)$, has exactly the same purpose as a moving average, except that the weights of the filter (the filter coefficients) are specially designed to reduce side lobe spectral leakage problems. In contrast, for a moving average filter, all weights would be of equal value. The convolved data (bottom panel of Figure G.1) consists of variations longer than 5 years. Note the extended period of El Niño events (negative SIO) in the 1980 and 1990s.

G.3 Convolution as Truncation of an Infinite Time Series

An observed time series, $x(t)$, can be considered a subset of an unlimited duration time series $g(t)$, obtained by convolving $g(t)$ with a rectangular window $h(t)$ of the form

$$h(t) = \begin{cases} 1 & 0 \le t \le T \\ 0 & \text{otherwise} \end{cases} \tag{G.6}$$

As illustrated in Figure G.2, the series $x(t)$ can be defined as

$$x(t) = h(t)g(t) \tag{G.7}$$

It follows that the Fourier transform of $x(t)$ is the convolution of the Fourier transforms of $h(t)$ and $g(t)$, namely

$$X(f) = \int_{-\infty}^{\infty} H(\zeta)G(f - \zeta)\mathrm{d}\zeta \tag{G.8}$$

In this case, multiplication in the time corresponds to convolution in the frequency domain, whereas in the previous case we examined convolution in the time domain (as with a running average) and multiplication in the frequency domain. These concepts are essential for the application of data windows in both the time and frequency domains.

G.4 Deconvolution

Deconvolution is the process of reversing (undoing) the smoothing that took place during application of the "data window", either in the time or frequency domains. It is assumed in this case that the response function is known and the process of deconvolution requires only a reverse of the process described above. Thus, the equation for deconvolution follows from that for convolution presented in Eq. (G.1).

Appendix H

Optimal Interpolation MATLAB Live Scripts

MATLAB Live scripts demonstrating aspects of Optimal Interpolation.

The optimal interpolation methodology follows the material in Chapter 4 on *The Spatial Analysis of Data Fields*. In: Thomson, R.E., Emery, W. J. (Eds.), 2014. Data Analysis Methods in Physical Oceanography, Elsevier. https://doi.org/10. 1016/C2010-0-66362-0.

The MATLAB codes presented here can be downloaded from the GitHub repository https://github.com/johnwilkin/ gda_exercises maintained by John L. Wilkin (jwilkin@rutgers.edu).

From the README.md

There are three exercises using data that depict the occurrence of "Tropical Instability Waves in the eastern equatorial Pacific Ocean". For an overview of the dynamics of Tropical Instability Waves, see: Willett, C.S., Leben, R.R., Lavín, M.F., 2006. Eddies and tropical instability waves in the eastern tropical Pacific: A review. *Prog. Oceanogr.* 69 (2−4), 218−238. https://doi.org/10.1016/j.pocean.2006.03.010.

Exercise 1: Estimation of covariance length scales from pseudo-observations of SST generated by sampling output from the NOAA Global Real-Time Ocean Forecast System (RTOFS). The Live script demonstrates how to compute a binned-lagged covariance function from a sample data set and use this to fit covariance length scales and signal variance appropriate to the data set. These fitted parameters are used in Exercise 2 to perform an optimal interpolation of the data.

Exercise 2: Optimal Interpolation of pseudo-observations generated by sampling output from the NOAA-RTOFS ocean model. This live script allows the user to vary the sampling resolution and the magnitude of the simulated observational error to explore the sensitivity of the interpolated field in comparison to the "truth".

Exercise 3: Optimal Interpolation of Infrared SST observations from Low Earth Orbiting (LEO) satellites. This live script parallels Exercise 2, but uses actual ocean observations acquired by infrared imagers on LEO satellites during November 2022.

The Live Script codes and outputs are presented below, but users are encouraged to download and experiment with these codes to gain experience with how Optimal Interpolation performs in practice.

Exercise 1: Estimating covariance length scales from pseudo-observations of SST in the eastern Equatorial Pacific generated by sampling output from the NOAA Global Real-Time Ocean Forecast System (RTOFS).

This Live script shows how to compute a binned-lagged covariance function from a sample data set and use this information to choose covariance length scales and signal variance appropriate to the data set. These fitted parameters are used in Exercise 2 to perform an optimal interpolation of the data. It should be stressed that, if the covariance functional form of chosen length scale and other parameters do not represent the underlying data well, the interpolation can be far from "optimal".

Load example data set and subsample spatially to create a set of pseudo-observations.

```
% Pseudo data are generated using example outputs from the NOAA RTOFS
% ocean model in the equatorial Pacific showing Tropical Instability
% Waves
load rtofs2.mat
% Plot the full data
pcolor(rt.lon,rt.lat,rt.sst')
shading flat
xtickformat('degrees');
ytickformat('degrees');
set(gca,'tickdir','out')
colorbar
title("RTOFS nowcast SST "+rt.date)
```

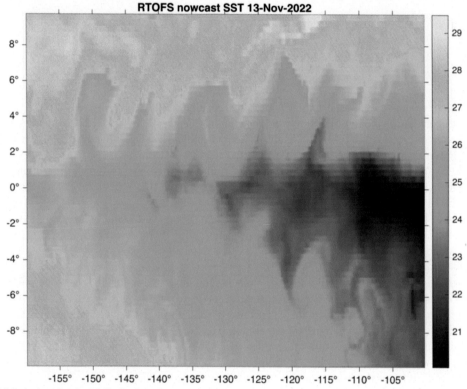

Figure H.1 Spatial (latitude vs longitude) distribution of sea surface temperature (SST) from the NOAA RTOFS nowcast model for 13 November 2022. The temperature scale on the right is in °C.

Estimate covariance length scales from data.

```
% Optimal interpolation is not optimal if the covariance scale used
% poorly represents the true covariance of the quantity being mapped.
% We can use the data themselves to estimate the covariance scale of a
% geophysical signal.
% Before analyzing the data to estimate covariance scales, remove long
% wavelengths by detrending in two dimensions so as to focus on the
% ocean mesoscale
sstanom = detrend(detrend(rt.sst)')';
% Plot spatially detrended anomaly field
pcolor(rt.lon,rt.lat,rt.sstanom')
shading flat
xtickformat('degrees');
ytickformat('degrees');
set(gca,'tickdir','out')
caxis([-2 2])
colorbar
title("RTOFS detrended SST anomaly")
```

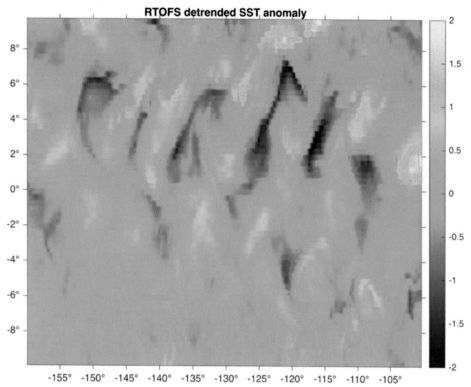

Figure H.2   Spatial (latitude vs longitude) distribution of detrended sea surface temperature anomaly (SSTA) from the NOAA-RTOFS nowcast for 13 November 2022. The temperature scale on the right is in °C.

Subsample/decimate the "truth" to prepare a data set of pseudo-observations.

```
% The data will be subsampled spatially to simulate a sparse observing
% network, and random noise added to simulate observational error.
% In script oi_example.mlx Optimal Interpolation will be used to
% explore how well we can recover the original data ("truth")
% Start with the detrended SST anomaly on the full spatial domain
dataf = sstanom;
disp("Set the fraction of data to retain as pseudo observations")
```

Set the fraction of data to retain as pseudo-observations.
frac = 0.1
frac = 0.1000

```
% Get set of unique random integers spanning the length of the full
% data set and use this to subsample/decimate the true field
N = numel(dataf);
ss = randperm(N,round(frac*N));
data = dataf(ss);
% Get corresponding coordinates of the sample set
[LON,LAT] = ndgrid(rt.lon,rt.lat);
x = LON(ss);
y = LAT(ss);
% Add random noise to simulate observational error
disp("Set the magnitude of pseudo observational error to add to the data")
```

Set the magnitude of pseudo-observational error to add to the data.

e = 0.2 % degrees C e = 0.2000

Sample Covariance function

```matlab
data = data + e*randn(size(data));
% Force column vectors
x = x(:);
y = y(:);
data = data(:);
% Use the simulated data to infer the shape and scales of the
% covariance function. It is recommended before attempting optimal
% interpolation to interrogate your data in this way to get an
% objective sense of the covariance scales that dominate your data.
% Create matrices that repeat the pseudo observation coordinates:
% columns of x, and rows of y. Then X-X' and Y-Y' will be all possible
% distance combinations between pairs of points
X = repmat(x,[1 length(y)]);
Y = repmat(y',[length(x) 1]);
Rdd = sqrt((X-X').^2+(Y-Y').^2);
% The sample data were detrended, so they have a mean of zero. The
% covariance of the sample data is then simply
Cdd = data*data';
% The Rdd distance matrix and C are symmetric because they repeat the
% covariances c(i,j)=c(j,i). We want only unique covariance pairs,
% which can be achieved by keeping only the upper triangle of Cdd.
% Create a mask M = 0 on lower triangle and 1 on the diagonal and
% upper triangle.
% Applying this mask then eliminates repeats c(i,j)=c(j,i) i~=j
M = triu(ones(size(Rdd)));
C = Cdd(M==1);
R = Rdd(M==1);
% Compute the sample covariance for values grouped into separation
% distance bins of size dr, and limit the maximum range considered to
% focus on the ocean mesoscale. Retaining long teleconnections would
% bias the estimate of the covariance length scale.
dr = 0.1;
rfitmax = 6;
C(R>rfitmax) = [];
R(R>rfitmax) = [];
% Find range bin number
B = round(R/dr);
B = B+1;
% deg lon/lat
% deg lon/lat
% unique integer index for each range bin +/-dr
% add 1 because rounding R<0.5*dr is zero
% accumarray is a handy Matlab function to construct an array by
% accumulation
% Find all in arg 2 (range) with a common index in arg 1 (range bin)
% and apply operation arg 4 (take the mean)
rf = accumarray(B,R,[],@mean);
cf = accumarray(B,C,[],@mean);
% Replace covariance for bin r = 0 (actually 0 < r < dr/2) with the
% covariance at precisely zero lag to capture sum of signal variance
% and observation error variance
cf(1) = mean(diag(Cdd));
% Trap the case of bins having no elements
rf(cf==0) = [];
cf(cf==0) = [];
```

Fit functional forms to sample covariance.

```
% Use FMINSEARCH to fit the user defined covariance functional forms
% in MYCOVFUNCTIONS.m to the sample covariance. The 2nd argument to
% FMINSEARCH is a vector [C0 A] of the starting guesses for C0 =
% covariance at r = 0, and length scale, A. The 4th and 5th arguments
% are the sample data which are passed on to MYCOVFUNCTIONS. The 6th
% argument is the functional form (one of Gauss, Markov or Le Traon).
% The help on FMINSEARCH and OPTIMSET explain the nonlinear parameter
% fitting process.
% Starting guess assumes C0 = 1 (half the span of the data range) and
% length scale A = 3 (half the rfitmax of 6). FMINSEARCH is quite
% robust to very bad initial guesses.
X0 = [0.5 3];
% f = a1*exp(-0.5*(r/a2)^2)
a_ga = fminsearch('mycovfunctions',X0,[],rf(2:end),cf(2:end),'gauss');
% f = b1*(1+r/b2)*exp(-r/b2)
a_mk = fminsearch('mycovfunctions',X0,[],rf(2:end),cf(2:end),'markov');
% f = c1*exp(-r/c2)*(1+r/c2+(r^2)/6-(r^3)/6
a_lt = fminsearch('mycovfunctions',X0,[],rf(2:end),cf(2:end),'letra');
% evaluate the final fit at convergence
misfit_a = mycovfunctions(a_ga,rf(2:end),cf(2:end),'gauss');
misfit_b = mycovfunctions(a_mk,rf(2:end),cf(2:end),'marko');
misfit_c = mycovfunctions(a_lt,rf(2:end),cf(2:end),'letra');
% For these test data the best functional fit is consistently Le
% Traon
disp("For these test data the best functional"+newline...
 +"form is consistently Le Traon"+newline+newline...
 +"Le Traon function values fitted for this"+newline...
 +"test data subsample and added noise were"+newline...
 +" length scale "+num2str(a_lt(2))+newline...
 +" variance at zero lag (r=0) "+num2str(a_lt(1))+newline)
```

For these test data, the best functional form is consistently Le Traon. Le Traon function values fitted for this test data (a subsample plus added noise) have a length scale 0.89468 and a variance at zero lag ($r = 0$) of 0.17277.

```
% Plot the sample covariance and the fits for the different functions.
han = plot(rf,cf,'x',...
 rf,a_ga(1)*exp(-0.5*(rf/a_ga(2)).^2),...
 rf,a_mk(1)*(1+rf/(a_mk(2))).*exp(-rf/a_mk(2)),...
 rf,a_lt(1)*(1+rf/a_lt(2)+1/6*(rf/a_lt(2)).^2 -1/6*(rf/a_lt(2)).^3).*exp(-rf/a_lt
legend("Sample covariance",...
 "Gaussian: misfit = "+num2str(misfit_a),...
 "Markov:  misfit = "+num2str(misfit_b),...
 "Le Traon: misfit = "+num2str(misfit_c))
grid
title("Covariance functions estimated from the pseudo-data")
```

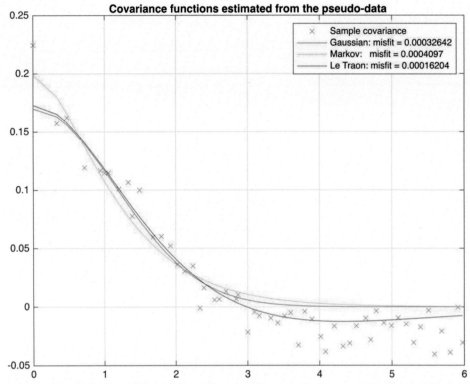

Figure H.3 Covariance functions estimated from the pseudo-data (SST anomaly) presented in the two previous figures. Note that at zero lag, $r = 0$, the sample covariance jumps up dramatically. This is because it includes the variance inherent to the geophysical signal, plus the observation error variance. The effect of observation error does not appear at lags other than zero because the pseudo error of different distinct observations is uncorrelated. This is the reason that fitting a covariance function to actual data should always ignore the covariance of the data with itself. This can be an effective way to objectively estimate what the observation error actually is when the error is uncertain.

Exercise 2: SST patterns of Tropical Instability Waves in the eastern Equatorial Pacific—Pseudo-observations generated by sampling output from the NOAA-RTOFS ocean model.

This live script allows the user to vary the sampling resolution and the magnitude of the simulated observational error to explore the sensitivity of the interpolated field in comparison to the "truth".

Load example data set and subsample spatially to create a set of pseudo-observations.

```
% Pseudo data are generated using example outputs from the NOAA RTOFS
% ocean model showing equatorial Pacific Tropical Instability Waves
load rtofs2.mat
DATAF = rt.sst;
[LON,LAT] = ndgrid(rt.lon,rt.lat);
% Plot the full data set again for comparison
pcolor(LON,LAT,DATAF)
shading flat
caxis([20 29])
colorbar
title("RTOFS SST in eastern Pacific on "+rt.date)
xlabel('longitude');ylabel('latitude')
xtickformat('degrees');ytickformat('degrees')
```

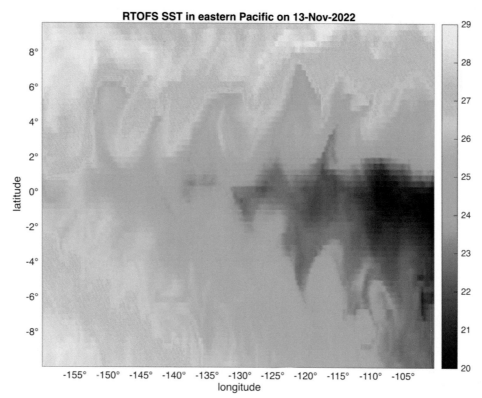

Figure H.4 Spatial (latitude vs longitude) distribution of sea surface temperature (SST) from the NOAA-RTOFS nowcast model for 13 November 2022. The temperature scale on the right is in °C.

```
frac = 0.1;
disp(int2str(100*frac)+"% of the data are retained as pseudo observations")
```

10% of the data are retained as pseudo-observations.

```
% Get set of unique random integers spanning the length of the full data set
% and use this to subsample/decimate the true field
N = numel(DATAF);
ss = randperm(N,round(frac*N));
data = DATAF(ss);
disp("OI will be performed using "+int2str(numel(ss))+" observations")
```

OI will be performed using 1080 observations.

Add random noise to simulate observational error.

The pseudo data simulated observation error is 0.2°C.

```
% Get corresponding coordinates of the sample set
% [LON,LAT] = ndgrid(rt.lon,rt.lat);
x = LON(ss);
y = LAT(ss);
e = 0.2; % degrees C
disp("The pseudo data simulated observation error is "+num2str(e)+" degC")
data = data + e*randn(size(data));
% Force column vectors
x = x(:);
y = y(:);
data = data(:);
% Plot the data
scatter(x,y,10,data,'o','filled')
colorbar
title("Pseudo observations at "+int2str(round(frac*100))+...
  "% of data points")
xlabel('longitude');ylabel('latitude')
xtickformat('degrees');ytickformat('degrees')
```

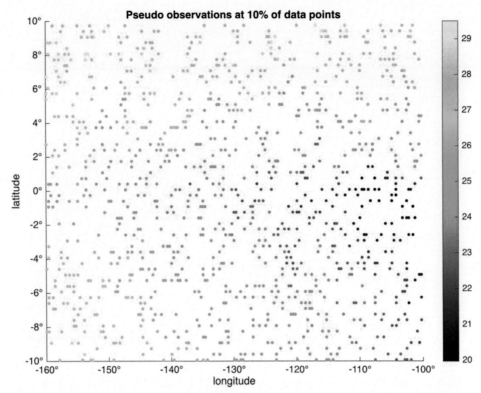

Figure H.5   Spatial distribution of sea surface temperature (SST) for 10% of the pseudo-observations from the NOAA-RTOFS nowcast for 13 November 2022. The temperature scale on the right is in °C.

Prepare for Optimal Interpolation.

```
% save the mean of the data for later
datam = mean(data);
% compute data anomaly from the mean - optimal interpolation works on
% the anomaly
dataa = data - datam;
% Create matrices that repeat the observation coordinates: columns
% of x, and rows of y. X-X' and Y-Y' will be all possible distance
% combinations between pairs of points
Xobs = repmat(x,[1 length(y)]);
Yobs = repmat(y',[length(x) 1]);
Rdd = sqrt((Xobs-Xobs').^2+(Yobs-Yobs').^2);
% The live script oi_cov_estimation.mlx was previously used to analyze
% these data and determine that the Le Traon covariance function
% f = ssq*exp(-r/a)*(1+r/a+((r/a)^2)/6-((r/a)^3)/6 represented well
% the covariance of a sample of these data, where ssq is the signal
% variance at zero lag and the length scale is parameter "a" in
% degress lon/lat.
a = 0.9;
disp("The covariance length scale (in degree lon/lat) is "+num2str(a))
```

The covariance length scale (in degree lon/lat) is 0.9.

The signal standard deviation squared at zero lag is 0.17.

Using Le Traon function for covariance $f = a0*exp(-r/a)*(1 + r/a + (r^2)/6 - (r^3)/6)$ with length scale $a = 0.9$ deg lon/lat or $\sim$80 km and the estimated signal variance at zero lag ($r = 0$) has a value of 0.17.

A noise to signal variance ratio of 0.235 will be added to the diagonal of the data-data covariance matrix.

Build the data-data covariance matrix.

```
ssq = 0.17;
disp("The signal standard deviation squared at zero lag is "+num2str(ssq))
% In OI, the ratio of error variance to signal variance, e^2/ssq, is
% added to the diagonal of the data-data covariance matrix. If, as
% user, you wish only to set this ratio and not explicit values of e
% and ssq separately, then set ssq = 1 here, and above set e^2 to be the assumed normalized observation
error variance.
disp("Using Le Traon function for covariance "+newline...
  +" f = a0*exp(-r/a)*(1+r/a+(r^2)/6-(r^3)/6)"+newline...
  +" with length scale"+newline...
  +" a = "+num2str(a)+" deg lon/lat or ~80 km"+newline...
  +" and estimated signal variance at zero lag (r=0) of "+num2str(ssq,2))
disp("A noise to signal variance ratio of "+num2str(e^2/ssq,3)...
  +" will be added to the diagonal of the data-data covariance matrix")
% The data-data covariance matrix is modeled based on the chosen
% continuous covariance function and the distance between observation
% points. Since the covariance function is isotropic, i.e., the same in
% all directions, we can evaluate the matrix simply using the
% separation distance of every data point from every other point in
% Rdd computed above. Note, to keep this code compact, no adjustments
% are made for spherical geometry to distances expressed in lon/at.
% Strictly speaking, the covariance function should be expressed in
% true geographic distance, but we are at the equator and the effects
% are small. Arguably, there are situations when covariance scales are
% anisotropic and it might be best to actually select different scales
% in lon and lat.
%
% The normalized (C=1 at r=0, instead of ssq) fitted
% data covariance matrix is then
Rddona = Rdd/a;
Cdd0 =  exp(-Rddona).*(1+Rddona+(Rddona.^2)/6-(Rddona.^3)/6);
% Add error variance normalized by signal variance along the diagonal
lambda = e^2/ssq;
Cddf = Cdd0 + lambda*eye(size(Cdd0));
```

Build the model-data covariance function.

```
% Here, "model" is the mapped solution on some other set of target
% coordinates. Typically the target coordinates are a regular grid,
% which can be at any resolution.
%
% In this exercise, we map to the original RTOFS grid to facilitate
% easy comparison to the true field and optimally interpolated field
% (from the subsampled data with added observation noise).
% Get distances between every grid point and every data point
Xgrid = repmat(LON(:),[1 numel(y)]);
Xdata = repmat(x(:),[1 numel(LAT)]);
Ygrid = repmat(LAT(:)',[length(x) 1]);
Ydata = repmat(y',[numel(LAT) 1]);
Rmd = sqrt((Xgrid-Xdata').^2+(Ygrid'-Ydata).^2);
% Evaluate the normalized covariance using the chosen covariance
% function
Rmdona = Rmd/a;
Cmd =  exp(-Rmdona).*(1+Rmdona+(Rmdona.^2)/6-(Rmdona.^3)/6);
```

Perform the Optimal Interpolation.

```
% Formally, OI evaluates Cmd*inv(Cdd)*data, but it is faster to use a
% Matrix Left Divide. See >> help mldivide
Da = Cmd*(Cddf\dataa);
Da = reshape(Da,size(LON));
DOI = datam + Da; % must add back the mean value
% Plot the interpolated field
pcolor(LON,LAT,DOI)
shading flat
caxis([20 29])
colorbar
title("Optimal Interpolation Solution")
xlabel('longitude');ylabel('latitude')
xtickformat('degrees');ytickformat('degrees')
```

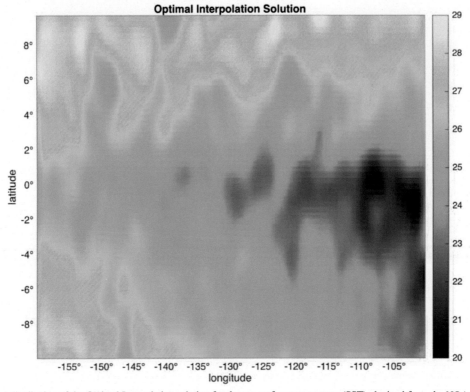

Figure H.6   Spatial distribution of the Optimal Interpolation solution for the sea surface temperature (SST) obtained from the NOAA-RTOFS nowcast for 13 November 2022. The temperature scale on the right is in °C.

```
% Plot the full data set again for comparison
pcolor(LON,LAT,DATAF)
shading flat
caxis([20 29])
colorbar
title("Original RTOFS data")
xlabel('longitude');ylabel('latitude')
xtickformat('degrees');ytickformat('degrees')
```

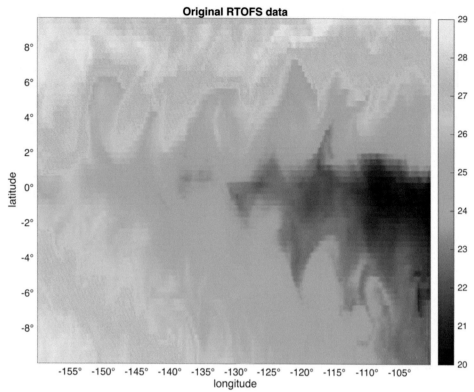

Figure H.7   Spatial distribution of the original sea surface temperature (SST) from the NOAA-RTOFS nowcast for 13 November 2022. The temperature scale on the right is in °C.

```
% Plot the difference of the interpolated and original fields
pcolor(LON,LAT,DOI-DATAF)
shading flat
caxis([-5 5])
colorbar
title("Optimal interpolation minus original RTOFS data"+newline...
  +"and locations of the pseudo-data")
hold on
han = scatter(x,y,20,'w','o','MarkerEdgeColor','w');
han.MarkerEdgeColor = 0.9*[1 1 1];
hold off
disp("Notice that differences between the mapped and true fields"+newline...
  +"are greatest where the observations are sparse")
```

Notice that differences between the mapped and true fields are greatest where the observations are sparse.

```
xlabel('longitude');ylabel('latitude')
xtickformat('degrees');ytickformat('degrees')
```

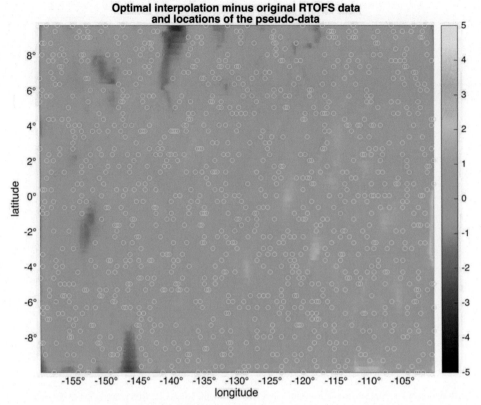

Figure H.8 Difference in the sea surface temperature distributions between the Optimal Interpolation distribution minus the original NOAA-RTOFS nowcast distribution for 13 November 2022. The temperature scale on the right is in °C.

Examine residuals of the interpolated noisy data.

```
% Repeat the optimal interpolation, but map to the pseudo-observation
% locations. In OI, the target "model" coordinates are arbitrary and
% need not be on a regular grid. We can use the OI machinery to map
% the data back to the sample locations so that we can examine the
% residuals of the optimal interpolation fit. If OI has recovered the
% original data well, the variance of the residuals should be close to
% the random observation noise we added when creating the pseudo
% observations.
% Mapping to the data locations is achieved by using Cdd0 in place of
% the Cmd matrix (i.e. the 'model' grid points are now just the data
% coordinate points). Note that we don't use Cddf because Cddf has the
% observation error variance added to the diagonal.
D0 = Cdd0*(Cddf\dataa);
% Plot histogram of residuals
res = D0-dataa; % both have mean(data) removed
histogram(res)
text(min(xlim)+0.05*diff(xlim),max(ylim)-0.1*diff(ylim),...
  "\sigma of residuals = "+num2str(std(res),3)+newline...
  + "obs. error "+num2str(e,3),'FontSize',14)
title("Histogram of residuals of OI minus observations")
```

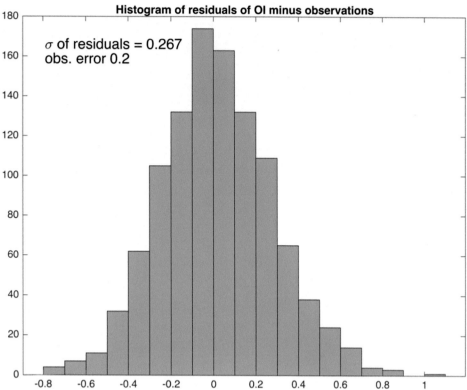

Figure H.9   Histogram showing the difference in sea surface temperature between the Optimal Interpolation values and the original NOAA-RTOFS nowcast values for 13 November 2022. Temperature values are in °C.

Compute expected error estimate.

```
% The expected error variance of OI result is
% E[e^2] = s^2 - Cmd inv(Cdd) Cmd'
% We will plot this normalized by the signal variance s^2 so that the
% map shows the relative expected error.
%
% Where this value is near 1, the expected error is as large as the
% data variance itself and the OI can provide no information about the
% solution that is better than the large scale mean. This occurs when
% the data are several covariance scales distant from the estimation
% point. Conversely, where the data are numerous the expected error
% decreases. However, expected error will never be less than the
% observation error because the OI estimate is limited by the
% precision of the underlying observations.
oierrornorm = diag(1-Cmd*inv(Cddf)*Cmd');
oierrornorm = reshape(oierrornorm,size(LON));
% Plot a map of the OI error
pcolorjw(LON,LAT,oierrornorm); caxis([0 1])
colorbar
hold on
han = scatter(x,y,15,'w','o','MarkerEdgeColor','w');
han.MarkerEdgeColor = 0.9*[1 1 1];

hold off

title("normalized expected error")
xlabel('longitude');ylabel('latitude')
xtickformat('degrees');ytickformat('degrees')
```

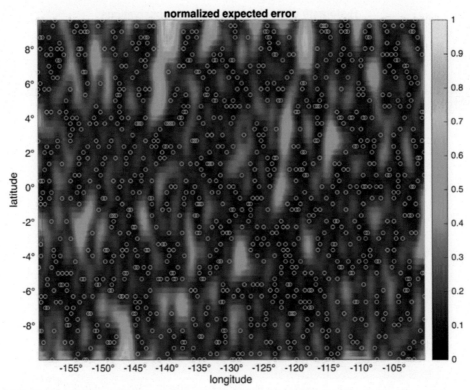

Figure H.10 Spatial distribution of the normalized expected errors in sea surface temperature derived from Optimal Interpolation versus values from the original NOAA-RTOFS nowcast values for 13 November 2022. Values are in °C.

Test how many residuals fall within the expected error estimate.

```
% Calculate expected error at the data locations, but this time
% multiply by signal variance to get dimensional values (deg C^2)
oierror0 = ssq*diag(1-Cdd0*inv(Cddf)*Cdd0');
% Plot OI error in comparison to residuals
plot(1:length(dataa),sqrt(oierror0),'r-',1:length(dataa),abs(D0-dataa),'k.')
legend("expected error","actual abs(data-OI)")
within_limits = find(abs(D0-dataa)<=sqrt(oierror0));
xlabel('sample number')
ylabel('^oC')
titlestr = int2str(round(100*length(within_limits)/length(dataa)))+...
  " % of residuals are within expected error bars";
title(titlestr)
```

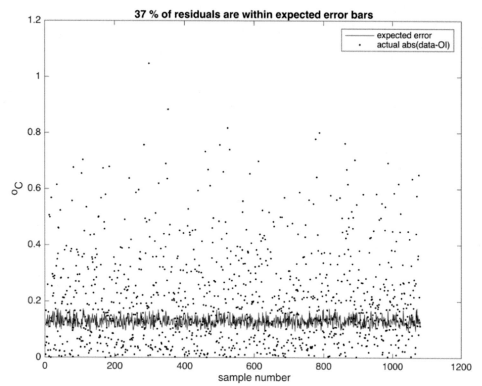

Figure H.11 Expected error in sea surface temperature (in °C) derived from optimal interpretation versus the original SST values obtained from the NOAA-RTOFS nowcast values for 13 November 2022. In this case, 37% of the differences between the OI values and the original data are within the expected error bars.

disp(titlestr+newline+" (expect 68% for normally distributed random variable)") 37 % of residuals are within expected error bars

(expect 68% for normally distributed random variable)

Further exercises.

% Go back in this script and set covariance length scales that are
% larger, or shorter, than the estimated best fit and see what happens
% to the quality of the map compared to the original "truth".
% Go back in the script and change the density of the data
% subsampling, or the assumed observation error, and see what happens
% to the quality of the map compared to the original "truth".

Exercise 3: SST patterns of Tropical Instability Waves in the eastern Equatorial Pacific—Infrared SST observations from Low Earth Orbiting (LEO) satellites

This live script parallels Exercise 2 but uses actual ocean observations acquired by infrared imagers on LEO satellites during November 2022.

Load Level-3 gridded example data set and subsample spatially to create a set of pseudo observations.

```
% Data downloaded on 11 February 2023, from NASA PO.DAAC Dataset
% catalog ID ACSPO-L3S-LEO-AM-v2.80
% The data file contains 10 days of data centered on 2022-11-13
f = 'L3S_LEO_ACSPO_SST_20221113_10day.nc';
% These are very high resolution 1/50 arc degree lon/lat (~2.2 km)
% data. To make the OI manageable, decimate the data on load by factor
% of s in both horizontal coordinates
s = 50;
leo.lon = ncread(f,'lon',1,Inf,s);
leo.lat = ncread(f,'lat',1,Inf,s);
leo.time = roms_get_time(f,'time');
for v = {'sea_surface_temperature','sses_bias','quality_level'}
  vname = char(v);
  leo.(vname) = ncread(f,vname,[1 1 1],[Inf Inf Inf],[s s 1]);

end


[leo.lon,leo.lat] = ndgrid(leo.lon,leo.lat);
% Only accept data that meets quality_level flag = 5 (see NetCDF
% global attributes for more information on quality_level)
leo.sea_surface_temperature(leo.quality_level~=5) = NaN;
% Apply sses_bias correction (see NetCDF global attributes for more
% information on sses_bias)
leo.sea_surface_temperature = leo.sea_surface_temperature-leo.sses_bias;
% Convert F to C

leo.sea_surface_temperature = leo.sea_surface_temperature-273.15;
```

Choose a day or set of days of data.

```
K = 5; % K = 5 is 2022-11-13
% K = [4 5 6]; % 3 days centered on 2022-11-13
data = leo.sea_surface_temperature(:,:,K(1));
valid = ~isnan(data);
x = leo.lon(valid);
y = leo.lat(valid);
data = data(valid);
% if more than one day
for k=2:numel(K)
  next = leo.sea_surface_temperature(:,:,K(k));
  valid = ~isnan(next);
  x = cat(1,x,leo.lon(valid));
  y = cat(1,y,leo.lat(valid));
  data = cat(1,data,next(valid));

end

disp("OI will be performed using "+int2str(numel(data))+" observations")
```

OI will be performed using 1049 observations.

Plot the data.

```
scatter(x,y,10,data,'o','filled')
colorbar
title("LEO satellite SST")
xlabel('longitude');ylabel('latitude')
xtickformat('degrees');ytickformat('degrees')
```

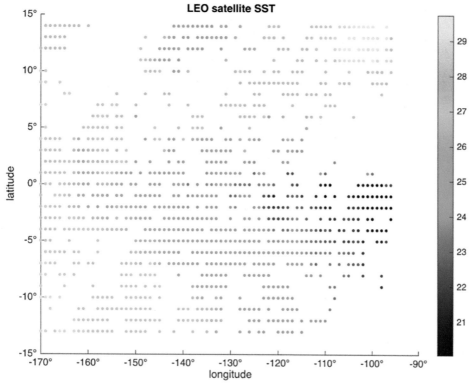

Figure H.12 Spatial distribution of the infrared sea surface temperature obtained from the low earth orbiting (LEO) satellites over 10 days centered on 13 November 2022. The temperature scale on the right is in °C.

Prepare for Optimal Interpolation.

```
% save the mean of the data for later
datam = mean(data);
% compute data anomaly from the mean - optimal interpolation works on
% the anomaly
dataa = data - datam;
% Create matrices that repeat the observation coordinates: columns
% of x, and rows of y. X-X' and Y-Y' will be all possible distance
% combinations between pairs of points
X = repmat(x,[1 length(y)]);
Y = repmat(y',[length(x) 1]);
Rdd = sqrt((X-X').^2+(Y-Y').^2);
% Set the covariance length scale in degrees lon/lat
a = 0.9;

% Set the signal standard deviation squared
ssq = 1;
% Set the observation error standard deviation (IR SST +/- 0.4 degC)
e = 0.4;
```

Build the data-data covariance matrix.

```
Rddona = Rdd/a;
Cdd0 =  exp(-Rddona).*(1+Rddona+(Rddona.^2)/6-(Rddona.^3)/6);
% Add error variance nomalized by signal variance along the diagonal
lambda = e^2/ssq;
Cddf = Cdd0 + lambda*eye(size(Cdd0));
```

Build the model-data covariance function.

```
% Map to same grid as the pseudo data example
if ~exist('LON','var')
  load rtofs2.mat
  [LON,LAT] = ndgrid(rt.lon,rt.lat);

end

% Get distances between every grid point and every data point
Xgrid = repmat(LON(:),[1 numel(y)]);
Xdata = repmat(x(:),[1 numel(LAT)]);
Ygrid = repmat(LAT(:)',[length(x) 1]);
Ydata = repmat(y',[numel(LAT) 1]);
Rmd = sqrt((Xgrid-Xdata').^2+(Ygrid'-Ydata).^2);
% Evaluate the normalized covariance using the chosen covariance
% function
Rmdona = Rmd/a;
Cmd =  exp(-Rmdona).*(1+Rmdona+(Rmdona.^2)/6-(Rmdona.^3)/6);
```

Perform the Optimal Interpolation.

```
% Formally, OI evaluates Cmd*inv(Cdd)*data, but it is faster to use a
% Matrix Left Divide (see Matlab help mldivide)
Da = Cmd*(Cddf\dataa);
Da = reshape(Da,size(LON));
% restore the mean
DOI = datam + Da;
% Plot the interpolated field
pcolor(LON,LAT,DOI)
shading flat
caxis([20 29])
colorbar
title("Optimal Interpolation Solution")
xlabel('longitude');ylabel('latitude')
xtickformat('degrees');ytickformat('degrees')
```

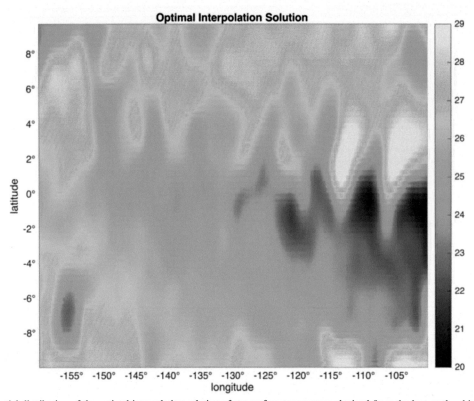

Figure H.13 Spatial distribution of the optimal interpolation solution of sea surface temperature obtained from the low earth orbiting (LEO) infrared SST observations over 10 days centered on 13 November 2022. The temperature scale on the right is in °C.

Compute expected error estimate

```
oierrornorm = diag(1-Cmd*inv(Cddf)*Cmd');
oierrornorm = reshape(oierrornorm,size(LON));
% Plot a map of the normalized OI error
pcolorjw(LON,LAT,oierrornorm); caxis([0 1])
colorbar
hold on
han = scatter(x,y,20,'w','o','MarkerEdgeColor','w');
han.MarkerEdgeColor = 0.1*[1 1 1];
hold off
title("normalized expected error")
xlabel('longitude');ylabel('latitude')
xtickformat('degrees');ytickformat('degrees')
```

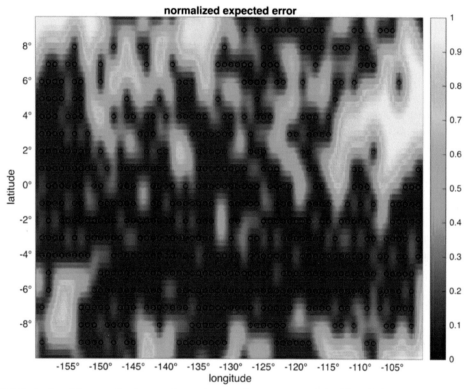

Figure H.14   Spatial distribution of the normalized expected errors in sea surface temperature derived from Optimal Interpolation versus values from the LEO satellite SST values for 13 November 2022. The temperature scale on the right is in °C.

```
% Calculate expected error at the data locations, but this time
% multiply by signal variance to get dimensional values (deg C^2)
oierror0 = ssq*diag(1-Cdd0*inv(Cddf)*Cdd0');
% oierror0 = max(eps,oierror0); % this is a variance - take sqrt to
% get deg C
% Plot OI error in comparison to residuals
D0 = Cdd0*(Cddf\dataa);
plot(1:length(dataa),sqrt(oierror0),'r-',1:length(dataa),abs(D0-dataa),'k.')
legend("expected error","actual abs(data-OI)")
within_limits = find(abs(D0-dataa)<=sqrt(oierror0));
xlabel('sample number')
ylabel('^oC')
titlestr = int2str(round(100*length(within_limits)/length(dataa)))+...
  " % of residuals are within expected error bars";
title(titlestr)
```

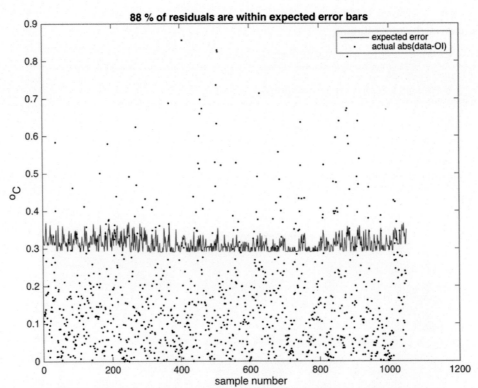

Figure H.15 Expected error in sea surface temperature (in °C) derived from optimal interpretation (red line) versus the original SST values obtained from 10-day low earth orbiting (LEO) satellite data centered on 13 November 2022. In this case, 88% of the differences between the original data and the OI values are within the expected error bars.

disp(titlestr+newline+" (expect 68% for normally distributed random variable)")

In the example presented in Figure H.15, 88% of residuals are within expected error bars (expect 68% for normally distributed random variable).

References

Abdel-Hafez, M.F., Kim, D.J., Lee, E., Chun, S., Lee, Y.J., Kang, T., Sung, S., 2008. Performance improvement of the wald test for GPS RTK with the assistance of INS. Int. J. Cont. Automat. Syst. 6, 534–543.

Agapitos, K., Gajewski, K., 2012. Analysis of the seasonal cycle of the climate record of Ottawa, Ontario. Can. Meteor. Oceanogr. Soc. Bull. 40, 120–126.

Allen, S.E., Thomson, R.E., 1993. Bottom-trapped subinertial motions over midocean ridges in a stratified rotating fluid. J. Phys. Oceanogr. 23, 566–581.

Amores, A., Monserrat, S., Marcos, M., Argueso, D., Villalonga, J., Jorda, G., Gomis, D., 2022. Numerical simulation of atmospheric Lamb waves generated by the 2022 Hunga-Tonga volcanic eruption. Geophys. Res. Lett. 49 (6), e2022GL098240. https://doi.org/10.1029/2022GL098240.

Anderson, D.L., 1993. ^{3}He from the mantle; primordial signal or cosmic dust? Science 261, 170–176.

Aoyama, M., Hamajima, Y., Hult, M., Uematsu, M., Oka, E., Tsumune, D., Kumamoto, Y., 2016. ^{134}Cs and ^{137}Cs in the North Pacific Ocean derived from the March 2011 TEPCO Fukushima Dai-ichi nuclear power plant accident, Japan. Part one: surface pathway and vertical distribution. J. Oceanogr. 72, 53.

Apel, J.R., Holbrook, J.R., Liu, A.K., Tsai, J.J., 1985. The Sulu Sea internal soliton experiment. J. Phys. Oceanogr. 15, 1625–1651. https://doi.org/10.1175/1520-0485(1985)015<1625:TSSISE>2.0.CO;2.

Arnault, S., Menard, Y., Merle, J., 1990. Observing the tropical Atlantic Ocean in 1986–87 from altimetry. J. Geophys. Res. 95, 17921–17946.

Arns, A., Wahl, T., Haigh, I.D., et al., 2013. Estimating extreme water level probabilities: a comparison of the direct methods and recommendations for best practice. Coast. Eng. 81, 51–66.

Arns, A., Dangendorf, S., Jensen, J., Bender, J., Talke, S.A., Pattiaratchi, C., 2017. Sea-level rise induced amplification of coastal protection design heights. Nat. Sci. Rep. 7, 40171. https://doi.org/10.1038/srep40171.

Baker, D.J., 1981. Ocean instruments and experiment design. In: Warren, B.A., Wunsch, C. (Eds.), Evolution of Physical Oceanography. MIT Press, Cambridge, MA, pp. 396–433.

Baker, E.T., Lavelle, J.W., 1984. The effect of particle size on the light attenuation coefficient of natural suspensions. J. Geophys. Res. 89, 8197–8203.

Baker, E.T., Lupton, J.E., 1990. Changes in submarine hydrothermal ^{3}He/heat ratios as an indicator of magmatic/tectonic activity. Nature 346, 556–558.

Baker, E.T., Massoth, G.J., 1986. Hydrothermal plume measurements: a regional perspective. Science 234, 980–982.

Baker, E.T., Massoth, G.J., 1987. Characteristics of hydrothermal plumes from two vent fields on the Juan de Fuca Ridge, Northeast Pacific Ocean. Earth Planet. Sci. Lett. 85, 59–73.

Baker, E.T., Lavelle, J.W., Feely, R.A., J Massoth, G., Walker, S.L., Lupton, J.E., 1989. Episodic venting of hydrothermal fluids from the Juan de Fuca Ridge. J. Geophys. Res. 94, 9237–9250.

Baker, E.T., German, C.R., Elderfield, H., 1995. Hydrothermal plumes over spreading-center axes: global distributions and geological inferences. In: Humphris, S.E., Zierenberg, R.A., Mullineaux, L.S., Thomson, R.E. (Eds.), Seafloor Hydrothermal Systems: Physical, Chemical, Biological and Geological Interactions. AGU, Geophysical Monograph, pp. 47–71, 91.

Bakun, A., 1973. Coastal Upwelling Indices, West Coast of North America, 1946–71. NOAA Technical Report NMFS SSRF-671. US Department of Commerce.

Banks, R., Henry, J.B., 1996. Good computer contouring/modeling can help the geophysical interpreter. Recorder 21 (1), 10.

Bard, E., Arnold, M., Östlund, H.G., Maurice, P., Monfray, P., Duplessy, J.-C., 1988. Penetration of bomb radiocarbon in the tropical Indian Ocean measured by means of accelerator mass spectrometry. Earth Planet. Sci. Lett. 87, 379–389.

Bardone, A., Pennecchi, F., Raiteri, G., Repetto, L., Reseghetti, F., 2020. XBT, ARGO float and ship-based CTD profiles intercompared under strict space-time conditions in the Mediterranean Sea: assessment of metrological comparability. J. Mar. Sci. Eng. 8. https://doi.org/10.3390/jmse8050313.

Barkley, R.A., 1968. Oceanographic Atlas of Pacific Ocean. University of Hawaii Press, Honolulu.

Barnes, C., Bethea, D.M., Brodeur, R.D., Spitz, J., Ridoux, V., Pusineri, D., Chase, D.C., Hunsicker, M.E., Juanes, F., Kellermann, A., Lancaster, J., Menard, F., Bard, F.X., Munk, P., Pinnegar, J.K., Scharf, F.S., Rountree, R.A., Stergeiou, K.I., Sassa, C., Sabates, A., Jennings, S., 2008. Predator and prey body sizes in marine food webs. Ecology 89, 881.

Barnett, T.P., 1983a. Recent changes in sea level and their possible causes. Clim. Change 5, 15–38.

Barnett, T.P., 1983b. Long-term changes in dynamic height. J. Geophys. Res. 88, 9547–9552. https://doi.org/10.1029/JC088iC14p09547.

Barnett, T.P., Patzert, W.C., Webb, S.C., Brown, B.R., 1979. Climatological usefulness of satellite determined sea-surface temperatures in the tropical Pacific. Bull. Am. Meteorol. Soc. 60, 197–205.

Barnston, A.G., Livezey, R.E., 1987. Classification, seasonality and persistence of low-frequency atmospheric circulation patterns. Mon. Wea. Rev. 115. https://doi.org/10.1175/1520-0493(1987)115<1083:CSAPOL>2.0.CO;2.

Barrick, D.E., 1980. Accuracy of parameter extraction from sample-averaged sea-echo Doppler spectra. IEEE Trans. Antennas Propag. 28, 1–11.

Barrodale, I., Erickson, R.E., 1978. Algorithms for Least-Squares Linear Prediction and Maximum Entropy Spectral Analysis. MS Report, DM-142-IR. University of Victoria.

Barth, J.A., Menge, B.A., Lubchenco, J., Chan, F., Bane, J.M., Kirincich, A.R., McManus, M.A., Nielsen, K.J., Pierce, S.D., Washburn, L., 2008. Delayed upwelling alters nearshore coastal ocean ecosystems in the northern California Current. Proc. Natl. Acad. Sci. U.S.A. 104, 3719–3724.

Barth, A., Alvera-Azcárate, A., Troupin, C., Ouberdous, M., Beckers, J.-M., 2010. A web interface for griding arbitrarily distributed in situ data based on Data-Interpolating Variational Analysis (DIVA). Adv. Geosci. 28, 29–37. https://doi.org/10.5194/adgeo-28-29-2010.

Bartlett, M.S., 1946. On the theoretical specification and sampling properties of autocorrelated time-series. Suppl. J. R. Stat. Soc. 8, 27–41.

Barton, I.J., 1995. Satellite-derived sea surface temperatures: current status. J. Geophys. Res. 100, 8777–8790.

Bartz, R., Zaneveld, J., Pak, H., 1978. A transmissometer for profiling and moored observations in water. Soc. Photo-Opt. Instrum. Eng. J. 160 (V), 102–108.

Bashkaryn, S., Lagerloef, G.S.E., Born, G., Emery, W.J., Leben, R.L., 1993. Variability in the Gulf of Alaska from GEOSAT altimetry data. J. Geophys. Res. 98, 16290–16311.

Batchelor, G.K., 1967. An Introduction to Fluid Dynamics. Cambridge University Press, Cambridge.

Bates, J.J., Diaz, H.F., 1991. Evaluation of multichannel sea surface temperature product quality for climate monitoring. J. Geophys. Res. 96, 20613–20622.

Bates, J.J., Smith, W.L., 1985. Sea surface temperatures from geostationary satellites. J. Geophys. Res. 90, 11609–11618.

Bauer, S., Griffa, A., 2002. Eddy means flow decomposition and eddy diffusivity estimates in the tropical Pacific Ocean; 2. Results. J. Geophys. Res. 107. https://doi.org/10.1029/2000JC000613.

Bauer, S., Swenson, M.S., Groffa, A., Mariano, A.J., Owens, K., 1998. Eddy-mean flow decomposition and eddy-diffusivity estimates in the tropical Pacific Ocean: 1. Methodology. J. Geophys. Res. 103, 30855–30871. https://doi.org/10.1029/1998JC900009.

Beardsley, R.C., 1987. A comparison of the vector-averaging current meter and new Edgerton, Germeshausen, and Gier, Inc., vector-measuring current meter on a surface mooring in Coastal Ocean Dynamics Experiment 1. J. Geophys. Res. 92, 1845–1859.

Beardsley, R.C., Boicourt, W.C., 1981. On estuarine and continental-shelf circulation in the Middle Atlantic Bight. In: Warren, B.A., Wunsch, C. (Eds.), Evolution of Physical Oceanography. MIT Press, Cambridge, MA, pp. 198–233.

Beardsley, R.C., Boicourt, W.C., Huff, L.C., McCullough, J.R., Scott, J., 1981. CMICE: a near-surface current meter intercomparison experiment. Deep-Sea Res. 28A, 1577–1603.

Becker, K., Thomson, R.E., Davis, E.E., Villinger, H., Wheat, C.G., 2021. Geothermal heating and episodic cold-seawater intrusions into an isolated ridge-flank basin near the Mid-Atlantic Ridge. Nat. Commun. Earth Environ. 2, 226. https://doi.org/10.1038/s43247-021-00297-2.

Beguería, S., 2005. Uncertainties in partial duration series modelling of extremes related to the choice of the threshold value. J. Hydrol. 303, 215–230. https://doi.org/10.1016/j.jhydrol.2004.07.015.

Belkin, I.M., O'Reilly, J.E., 2009. An algorithm for Oceanic front detection in chlorophyll and SST satellite imagery. J. Mar. Sys. 78, 319–326.

Belyshev, A.P., Klevantsov, Y.P., Rozhkov, V.A., 1983. Probability Analysis of the Sea Currents (In Russian). Gidrometeoizdat, Leningrad, p. 264.

Bendat, J.S., Piersol, A.G., 1986. Random Data: Analysis and Measurement Procedures. John Wiley, New York.

Bennett, A.F., 1976. Poleward heat fluxes in southern hemisphere Oceans. J. Phys. Oceanogr. 4, 785–798.

Bennett, A.F., 1992. Inverse Methods in Physical Oceanography. Cambridge University Press, Cambridge, p. 346.

Bernard, E.N., González, F.I., Meinig, C., Milburn, H.B., 2001. Early detection and real-time reporting of deep-ocean tsunamis. Proc. Int. Tsunami Symp. 2001, 97–108. Seattle, WA.

Bernstein, R.L., 1982. Sea surface temperature estimation using the NOAA-6 satellite advanced very high resolution radiometer. J. Geophys. Res. 87, 9455–9466.

Bernstein, R.L., Chelton, D.B., 1985. Large-scale sea surface temperature variability from satellite and shipboard measurements. J. Geophys. Res. 90, 11619–11630.

Beron-Vera, F.J., Olascoaga, M.J., Lumpkin, R., 2016. Inertia-induced accumulation of flotsam in the subtropical gyres. Geophys. Res. Lett. 43, 12228–12233. https://doi.org/10.1002/2016GL071443.

Berteaux, H.O., 1990. Program SSMOOR: Users' Instructions. Cable Dynamics and Mooring Systems (CDMS), Woods Hole, MA.

Berteaux, H.O., 1991. Coastal and Oceanic Buoy Engineering. H.O. Berteaux, Woods Hole, MA., USA.

Birkhoff, G.D., 1931. Proof of the ergodic theorem. Proc. Natl. Acad. Sci. 17, 656–660.

Blackman, R.B., Tukey, J.W., 1958. The Measurement of Power Spectra. Dover, New York.

Boehlert, G.W., Costa, D.P., Crocker, D.E., Green, P., O'Brien, T., Levitus, S., Le Boef, B.J., 2001. Autonomous pinniped environmental samplers: using instrumented animals as Oceanographic data collectors. J. Atmos. Ocean. Technol. 18, 1882–1893.

Bograd, S.J., Rabinovich, A.B., Thomson, R.E., Eert, A.J., 1999. On sampling strategies and interpolation schemes for satellite-tracked drifters. J. Atmos. Ocean. Technol. 16, 893–904.

Bograd, S.J., Schroeder, I., Sarkar, N., Qiu, X., Sydeman, W.J., Schwing, F.B., 2009. Phenology of coastal upwelling in the California current. Geophys. Res. Lett. 36, L01602. https://doi.org/10.1029/2008GL035933.

Bohling, G., 2005. Introduction to Geostatistics and Variogram Analysis. http://people.ku.edu/wgbohling/Cp-940.

Bohling, G.C., 2007. Introduction to Geostatistics. Kansas Geological Survey Open File Report 2007-26, p. 50.

Boicourt, W.C., 1982. The recent history of Ocean current meter measurement. In: Proceedings of IEEE Second Workshop Conference on Current Measurement, pp. 9–13.

Bommier, E., 2014. Peaks-Over-Threshold Modelling of Environmental Data. Department of Mathematics, Uppsala University, U.U.D.M. Project Report 2014:33, p. 35.

Bordone, A., Pennecchi, F., Ralteri, G., Repetti, L., Reseghetti, F., 2020. XBT ARGO Float and ship based CTD profiles intercompared under strict space-time conditions in the Mediterranean Sea: assessment of meteorological comparibility. J. Mar. Sci. Eng. 8. https://doi.org/10.3390/jmse8050313.

Born, G.H., Richards, M.A., Rosborough, G.W., 1982. An empirical determination of the effects of sea-state bias on the SEASAT altimeter. J. Geophys. Res. 87, 3221–3226.

Born, G.H., Mitchell, J.L., Heyler, G.A., 1987. GEOSAT-ERM mission design. J. Astron. Sci. 35, 119–134.

Bowditch, N., 1977. American Practical Navigator. Defense Mapping Agency Hydrographic Center, DMA No. NVPUB9V1.

Bowen, M., Emery, W.J., Wilkin, J.L., Tildesley, P., Barton, I.J., Knewtson, R., 2002. Extracting multi-year surface currents from sequential thermal imagery using the maximum cross correlation technique. J. Atmos. Ocean. Technol. 19, 1665–1676.

Box, G.E.P., Jenkins, G.M., 1970. Time Series Analysis: Forecasting and Control. Holden-Day, San Francisco, CA.

Boyd, P.W., Trull, T.W., 2007. Understanding the export of biogenic particles in Oceanic waters: is there a consensus? Prog. Oceanogr. 72, 276–312.

Brainerd, K.E., Gregg, M.C., 1995. Effects of monsoon winds and topographical features on the vertical thermohaline and biogeochemical structure in the Gulf of Tadkpiraj (Djibouti). Deep Sea Res. Part A Oceanogr. Res. Paper 42, 1521–1543. https://doi.org/10.1016/0967-0637(95)00068-H.

Breiman, L., Friedman, J.H., Olshen, R.A., Stone, C.J., 1984. Classification and Regression Trees. Routledge, New York, p. 368. https://doi.org/10.1201/9781315139470.

Bretherton, F.P., McWilliams, J.C., 1980. Estimations from irregular arrays. Rev. Geophys. Space Phys. 18, 789–812.

Bretherton, F.P., Davis, R.E., Fandry, C.B., 1976. A technique for objective analysis and design of Oceanographic experiments applied to MODE-73. Deep-Sea Res. 23, 559–582.

Brink, K.H., Chapman, D.C., 1987. Programs for Computing Properties of Coastal-Trapped Waves and Wind-Driven Motions over the Continental Shelf and Slope, second ed. Woods Hole Oceanographic Institution, Technical Report No. WHOI-87–24.

Briscoe, M.G., 1975. Internal waves in the Ocean. Rev. Geophys. Space Res. 13, 591–598.

Broecker, W.S., Peng, T.H., 1982. Tracers in the Sea. Lamont-Doherty Geological Observatory. Eldigio Press, New York.

Broecker, W.S., Peng, T.H., Östlund, G., Stuiver, M., 1985. The distribution of bomb radiocarbon in the Ocean. J. Geophys. Res. 90, 6953–6970.

Broecker, W.S., Peng, T.H., Östlund, G., 1986. The distribution of bomb Tritium in the Ocean. J. Geophys. Res. 91, 14331–14344.

Broecker, W.S., Virgilio, A., Peng, T.-H., 1991. Radiocarbon age of waters in the deep Atlantic revisited. Geophys. Res. Lett. 18, 1–3.

Brooks, C.F., 1926. Observing water-surface temperatures at sea. Mon. Weather Rev. 54, 241–254.

Brower, R.L., Gohrband, G.S., Pichel, W.G., Signore, T.L., Walton, C.C., 1976. Satellite Derived Sea-Surface Temperatures from NOAA Spacecraft. NOAA Technical Memo, NESS No. 79, Washington, DC.

Brown, N.L., 1974. A precision CTD microprofiler. In: Ocean 74 Record, 1974 IEEE Conference on Engineering in the Ocean Environment, vol 2. IEEE Publication 74 CHO873–0 OEC, Institute of Electrical and Electronics Engineers, New York, pp. 270–278.

Brown, R.A., 1983. On a satellite scatterometer as an anemometer. J. Geophys. Res. 88, 1663–1673.

Brown, J.D., 2008. The bonferroni adjustment, JALT testing and evaluation. SIG Newsl. 42 (1), 23–28.

Brown, N.L., Morrison, G.K., 1978. WHOI/Brown Conductivity, Temperature and Depth Profiler. WHOI Report No. 78–23.

Brown, O.B., Brown, J.W., Evans, R.H., 1985. Calibration of advanced very high resolution radiometer infrared observations. J. Geophys. Res. 90, 11667–11678.

Bruland, K.W., Donat, J.R., Hutchins, D.A., 1991. Interactive influences of bioactive trace metals on biological production in Oceanic waters. Limnol. Oceanogr. 36, 1555–1577.

Bryden, H., 1979. Poleward heat flux and conversion of available potential energy in Drake Passage. J. Mar. Res. 37, 1–22.

Buchanan, M.K.M.O., E Kopp, R., 2017. Amplification of flood frequencies with local sea level rise and emerging flood regimes. Environ. Res. Lett. 12, 064009. https://doi.org/10.1088/1748-9326/aa6cb3.

Bullinaria, J.A., 2004. Introduction to Neural Networks – Lecture 1, p. 16. http://www.cs.bham.ac.uk/wjxb/inn.html.

Bullister, J.L., 1989. Chlorofluorocarbons as time-dependent tracers in the Ocean. Oceanography 2 (2), 12–17.

Bullister, J.L., Weiss, R.F., 1988. Determination of CCl_3F and CCl_2F_2 in seawater and air. Deep-Sea Res. 35, 839–853.

Burd, B.J., Thomson, R.E., 1993. Flow volume calculations based on three-dimensional current and net orientation data. Deep-Sea Res. 40, 1141–1153.

Burd, B.J., Thomson, R.E., 1994. Hydrothermal venting at endeavour ridge: effect on zooplankton biomass throughout the water column. Deep-Sea Res. 41, 1407–1423.

Burd, B.J., Thomson, R.E., 2012. Estimating zooplankton biomass distribution in the water column near the endeavour segment of Juan de Fuca Ridge using acoustic backscatter and concurrently towed nets. Oceanography 25, 269–276. https://doi.org/10.5670/oceanog.2012.25.

Burd, B.J., Thomson, R.E., 2015. The importance of hydrothermal venting to water column secondary production in the Northeast Pacific. Deep Sea Res. Part II: Top. Stud. Oceanogr. 121, 85–94.

Burd, B.J., Thomson, R.E., 2019. Seasonal patterns in deep acoustic backscatter layers near vent plumes in the northeast Pacific Ocean. FACETS 4, 183–209. https://doi.org/10.1139/facets-2018-0027.

Burd, B.J., Thomson, R.E., 2022. A review of zooplankton and deep carbon fixation contributions to carbon cycling in the dark Ocean. J. Mar. Syst. 236. https://doi.org/10.1016/j.jmarsys.2022.103800.

Burg, J.P., 1967. Maximum entropy spectral analysis. Paper Presented at the 37th Annual International Meeting. Society of Exploration Geophysicists, Oklahoma City, OK.

Burg, J.P., 1968. A new analysis technique for time series data. In: NATO Advanced Study Institute on Signal Processing Emphasis on Underwater Acoustics. Enschede, The Netherlands.

Burg, J.P., 1972. The relationship between maximum entropy spectra and maximum likelihood spectra. Geophysics 37, 375–376.

Burke, E.J., Perry, R.H.J., Brown, S.J., 2010. An extreme value analysis of UK drought and projections of change in the future. J. Hydrol. 388, 131–143.

Busalacchi, A.J., O'Brien, J.J., 1981. Interannual variability of the equatorial Pacific in the 1960's. J. Geophys. Res. 86, 10901–10907.

Cahill, M.L., Middleton, J.H., Stanton, B.R., 1991. Coastal-trapped waves on the west coast of the South Island, New Zealand. J. Phys. Oceanogr. 21, 541–557.

Calman, J., 1978. On the interpretation of Ocean current spectra. J. Phys. Oceanogr. 8, 627–652.

Caires, S., 2007. Extreme Wave Statistics. Confidence Intervals. WL|Delft Hydraulics Report H4803.30, p. 33.

Calder, M., 1975. Calibration of echo sounders for offshore sounding using temperature and depth. Int. Hydrogr. Rev. LII (2), 13–17.

Camilli, R., Di Ioriob, D., Bowen, A., Reddy, C.M., Techet, A.H., Yoerger, D.R., Whitcomb, L.L., Seewald, J.S., Sylva, S.P., Fenwick, J., 2012. Acoustic measurement of the Deepwater Horizon Macondo well flow rate. Proc. Natl. Acad. Sci. U. S. A. 109 (50), 20235–20239. https://doi.org/10.1073/pnas.1100385108.

Cannon, G.A., Thomson, R.E., 1996. Characteristics of 4-day oscillations trapped by the Juan de Fuca ridge. Geophys. Res. Lett. 23, 1613–1616.

Capon, J., 1969. High resolution frequency-wavenumber spectral analysis. Proc. IEEE 57, 1408–1418.

Caries, S., 2011. Extreme Value Analysis: Still Water Level. JCOMM Technical Report No. 58, p. 29.

Cartwright, D.E., 1968. A unified analysis of tides and surges round North and East Britain. Philos. Trans. R. Soc. London A263, 1–55.

Cartwright, D.E., Edden, A.C., 1973. Corrected tables of tidal harmonics. Geophys. J. R. Astron. Soc. 33, 253–264.

Carvajal, M., Sepúlveda, I., Gubler, A., Garreaud, R., 2022. Worldwide signature of the 2022 Tonga Volcanic tsunami. Geophys. Res. Lett. 49 (6), e2022GL098153. https://doi.org/10.1029/2022GL098153.

Cavanaugh, N.R., Gershunov, A., Panorska, A.K., Kozubowski, T.J., 2015. The probability distribution of intense daily precipitation. Geophys. Res. Lett. 42, 1560–1567. https://doi.org/10.1002/2015GL063238.

Chadwick, W.W., Nooner, S.L., Butterfield, D.A., Lilley, M.D., 2012. Seafloor deformation and forecasts of the April 2011 eruption at axial seamount. Nat. Geosci. 5 (7), 474–477. https://doi.org/10.1038/NGEO1464.

Chadwick, W., Clague, D., Embley, R., Perfit, M., Butterfield, D., Caress, D., Paduan, J., Martin, J., Sasnett, P., Merle, S., Bobbitt, A., 2013. The 1998 eruption of axial seamount: new insights on submarine lava flow emplacement from high-resolution mapping: the 1998 eruption of axial seamount. Geochem. Geophys. Geosyst. 14, 3939–3968. https://doi.org/10.1002/ggge.20202.

Chapman, D.C., 1983. On the influence of stratification and continental shelf and slope topography on the dispersion of subinertial coastally trapped waves. J. Phys. Oceanogr. 13, 1641–1652.

Charnock, H., Lovelock, J.E., Liss, P.S., Whitfield, M., 1988. Tracers in the Ocean. In: Proceedings of a Royal Society Meeting Held on 21 and 22 May, 1987. The Royal Society, London.

Chavanne, C., 2018. Do high frequency radars measure the induced Stokes drift? J. Atmos. Ocean. Technol. 325, 1023–1031.

Chave, A.D., Luther, D.S., 1990. Low-frequency, motionally induced electromagnetic fields in the Ocean. J. Geophys. Res. 95, 7185–7200.

Chave, A.D., Luther, D.S., Filloux, J.H., 1992. The barotropic electromagnetic and pressure experiment. 1. Barotropic current response to atmospheric forcing. J. Geophys. Res. 97, 9565–9593.

Chelton, D.B., 1982. Statistical reliability and the seasonal cycle: comments on "Bottom pressure measurements across the Antarctic Circumpolar Current and their relation to the wind". Deep-Sea Res. 29, 1381–1388.

Chelton, D.B., 1983. Effects of sampling errors in statistical estimation. Deep-Sea Res. 30, 1083–1101.

Chelton, D.B., 1983. Effects of sampling errors in statistical estimation. Deep Sea Res. Part A Oceanogr. Res. Paper 30, 1083–1103.

Chelton, D.B., 1984. Commentary: Short-term climatic variability in the Northeast Pacific Ocean. In: Pearcy, W.G. (Ed.), The Influence of Ocean Conditions on the Production of Salmonids in the North Pacific. Oregon Sea Grant Program, Oregon State Univ., Corvallis. Publ, pp. 87–99. ORESU-W-83-001.

Chelton, D.B., 1988. WOCE/NASA Altimeter Algorithm Workshop. US WOCE Technical Report, No. 2.

Chelton, D.B., Davis, R.E., 1982. Monthly mean sea level variability along the west coast of North America. J. Phys. Oceanogr. 6, 757–784.

Chelton, D.B., Risien, C.M., 2020. A hybrid precipitation index inspired by the SPI, PDSI, and MCDI. Part II: Application to investigate precipitation variability along the West Coast of North America. J. Hydrometeorol. 21, 1977–2002.

Chelton, D.B., Schlax, M.G., 1996. Global observations of Oceanic Rossby waves. Science 272, 234–238.

Chelton, D.B., Bernal, P.A., McGowan, J.A., 1982. Large-scale interannual physical and biological interaction in the California current. J. Mar. Res. 40, 1095–1125.

Chen, Z., Xu, S., Lin, Y., Jia, X., 2009. Modeling the effects of fishery management and marine protected areas on the Beibu Gulf using spatial ecosystem simulation. Fish. Res. 100 (3), 222–229. https://doi.org/10.1016/j.fishres.2009.08.001.

Cheney, R.E., Marsh, J.G., Becley, B.D., 1983. Global mesoscale variability from collinear tracks of SEASAT altimeter data. J. Geophys. Res. 88, 4343–4354.

Cheney, R.E., Emery, W.J., Haines, B.J., Wentz, F., 1991. Improvements in GEOSAT altimeter data. EOS, Trans. Am. Geophys. Union 72 (51), 577. https://doi.org/10.1029/90EO00401.

Cheng, P., Wilson, R., 2006. Temporal variability of vertical nontidal circulation pattern in a partially mixed estuary: comparison of self-organizing map and empirical orthogonal functions. J. Geophys. Res. 111 (C12). https://doi.org/10.1029/2005JC003241.

Cherniawsky, J.Y., Crawford, W.R., 1996. Comparison between weather buoy and comprehensive Ocean-atmosphere data set wind for the west coast of Canada. J. Geophys. Res. 101, 18377–18389.

Cherniawsky, J.Y., Foreman, M.G.G., Crawford, W.R., Henry, R.F., 2001. Ocean tides from TOPEX/Poseidon sea level data. J. Atmos. Ocean. Technol. 18, 649–664. https://doi.org/10.1175/1520-0426(2001)018<0649:OTFTPS>2.0.CO;2.

Cherniawsky, J.Y., Crawford, W.R., Nikitin, O., Carmack, E.C., 2005. Bering Strait transports from satellite altimetry. J. Mar. Res. 63, 887–900.

Chiodi, A.M., Zhang, C., D Cokelet, E., Yang, Q., Mordy, C.W., Gentemann, C.L., Cross, J.N., Lawrence-Slavas, N., Meinig, C., Steele, M., Harrison, D.E., Stabeno, P.J., Tabisola, H.M., Zhang, D., Burger, E.F., O'Briend, K.M., Wang, M., 2021. Exploring the Pacific Arctic seasonal ice zone with saildrone USVs. Front. Mar. Sci. 03. https://doi.org/10.3389/fmars.2021.640697.

Chisholm, S.W., Morel, F.M.M. (Eds.), 1991. What controls phytoplankton production in nutrient-rich areas of the open sea? Limnol. Oceanogr. 36, 1507–1970.

Chiswell, S., October, 1992. Inverted echo sounders at the WOCE deepwater station. WOCE Note. 4 (4), 1–6. US Office, Department of Oceanography. Texas A&M University, College Station.

Chiswell, S.M., Wimbush, M., Luckas, R., 1988. Comparison of dynamic height measurements from an inverted echo sounder and an island tidge gauge in the central Pacific. J. Geophys. Res. 93, 2277–2283.

Christensen, E.J., J Haines, B., Keihm, S.J., Morris, C.S., Norman, R.A., Purcell, G.H., Williams, B.G., Wilson, B.D., Born, G.H., Parke, M.E., Gill, S.K., Shum, C.K., Tapley, B.D., Kolenkiewicz, R., Nerem, R.S., 1994. Calibration of TOPEX/POSEIDON at platform harvest. J. Geophys. Res. 99, 24,465–24,487.

Christiansen, B., 2007. Atmospheric circulation regimes: can cluster analysis provide the number. J. Clim. 20, 2229–2250. https://doi.org/10.1175/JCLI4107.1.

Christiansen, F., Sironi, M., Moore, M.J., Di Martino, M., Ricciardi, M., Warick, H.A., Irschick, D.J., Gutierrez, R., Uhart, M.M., 2019.

Estimating body mass of free-living whales using aerial photogrammetry and 3D volumetrics. Methods Ecolog. Evol. 10 (12), 2034−2044. https://doi.org/10.1111/2041-210X.13298.

Chu, P.C., 1994. Localized TOGA sea level spectra obtained from the S-transformation. TOGA Note. 17, 5−8.

Church, J.A., Freeland, H.J., L Smith, R., 1986. Coastal-trapped waves on the East Australian continental shelf. Part I: propagation of modes. J. Phys. Oceanogr. 16, 1929−1943.

Church, J.A., et al., 2013. Chapter 13: Sea Level change, climate change: the physical science basis. In: Stocker, T.F., Qin, D., Plattner, G.K., Tignor, M., Allen, S.K., Boschung, J., Nauels, A., Xia, Y., Bex, V., Midgley, P. (Eds.), Contribution of Working Group I to the Fifth Assessment Report of the Intergovernmental Panel on Climate Change. Cambridge University Press, pp. 1137−1216. https://doi.org/10.1017/CBO9781107415324.026.

Clayson, C.A., Bogdanoff, A.S., 2013. The effect of Diurnal Sea surface temperature warming on climatological air−sea fluxes. J. Clim. 26, 2546−2556. https://doi.org/10.1175/JCLI-D-12-00062.1.

Clayson, C.A., Emery, W.J., Savage, R., 1993. Wind speed comparisons between GEOSAT, SSM/I, buoy and ECMWF wind speeds. J. Geophys. Res. 19, 558−563.

Coale, K.H., Johnson, K.S., Fitzwater, S.E., Blain, S.P.G., Stanton, T.P., Coley, T.L., 1998. IronEx-I, an *in situ* iron enrichment experiment: experimental design, implementation and results. Deep Sea Res. Part 2 45, 919−945.

Cochran, W.G., Cox, G.M., 1992. Experimental Designs, second ed. John Wiley, New York, p. 640.

Cochrane, G.R., Nasby, N.M., Reid, J.A., Waltenberger, B., Lee, K.M., 2003. Nearshore Benthic Habitat GIS for the Channel Islands National Marine Sanctuary and Southern California State Fisheries Reserves Volume 1. U.S. Geological Survey Open-file Report 03-85.

Coles, S., 2001. An Introduction to Statistical Modeling of Extreme Values, Springer series in statistics. Springer-Verlag, London, p. 208.

Collins, C.A., Giovando, L.F., Abbott-Smith, K.A., 1975. Comparison of Canadian and Japanese merchant ship observations of sea surface temperature in the vicinity of the present Ocean Station P, 1927−53. J. Fish. Res. Board Can. 32, 253−258.

Connolly, S.R., Dornelas, M., Bellwood, D.R., Hughes, T.P., 2009. Testing species abundance models: a new bootstrap approach applied to Indo-Pacific coral reefs. Ecology 90 (11), 3138−3149.

Connolly, T.P., Hickey, B.M., Shulman, T., Thomson, R.E., 2014a. Coastally trapped waves, alongshore pressure gradients, and the California Undercurrent. J. Phys. Oceanogr. 44, 319−342. https://doi.org/10.1175/JPO-D-13-095.1.

Connolly, P.J., McFiggins, G.B., Wood, R., Tsiamis, A., 2014b. Factors determining the most efficient spary distribution for marine cloud brightening. Philos. Trans. R. Soc. A372, 1−15. https://doi.org/10.1098/rsta.2014.0056.

Conway, K.W., Barrie, J.V., Thomson, R.E., 2012. Submarine slope failure and tsunami hazard in coastal British Columbia: Douglas Channel and Kitimat Arm. Geol. Surv. Can. Curr. Res. (Online) No. 2012-10, 17. https://doi.org/10.4095/291732.

Cooley, W.W., Lohnes, P.R., 1971Cooley, J.W., Tukey, J.W., 1965. An algorithm for machine calculation of complex Fourier series. Math. Comput. 19, 297−301.

Cooley, W.W., Lohnes, P.R., 1971. Multivariate Data Analysis. Wiley, New York, p. 364.

Cowles, T., Delaney, J., Orcat, J., Weller, R., 2010. The Ocean observatory initiative: sustained Ocean observing across a range of spatial scales. Mar. Technol. Soc. J. 44, 54−64. https://doi.org/10.4031/MTSJ.44.6.21.

Cox, R.A., 1963. The salinity problem. Prog. Oceanogr. 1, 243−261.

Cox, D.R., Hinkley, D.V., 1974. Properties of the maximum likelihood estimate and bias reduction for logistic regression model. Open Access Lib. J. 4. https://doi.org/10.4236/oalib.1103640.

Cox, C., Munk, W., 1954. Measurements of the roughness of the sea surface from photographs of the sun's glitter. J. Opt. Soc. Amer. 44, 838−850. https://doi.org/10.1364/JOSA.44.000838.

Cox, R.A., Culkin, F., Riley, J.P., 1967. The electrical conductivity/chlorinity relationship in natural sea water. Deep-Sea Res. 14, 203−220.

Craig, H., 1994. Retention of Helium in subducted interplanetary dust particles. Science 265, 1892−1893.

Crawford, A.B., 1969. Sea Surface Temperatures, Some Instruments, Methods and Comparison. Tech. Note No. 103, WMO, pp. 117−129.

Crawford, W.R., Thomson, R.E., 1982. Diurnal-period continental shelf waves along Vancouver Island: a comparison of observations with theoretical models. J. Phys. Oceanogr. 14, 1629−1646.

Crawford, W.R., Thomson, R.E., 1984. Diurnal-period continental shelf waves along Vancouver Island: a comparison of observations with theoretical models. J. Phys. Oceanogr. 14, 1629−1646.

Cresswell, G.R., 1976. A drifting buoy tracked by satellite in the Tasman Sea. Aust. J. Mar. Freshwat. Res. 27, 251−262.

Crocker, R.I., Emery, W.J., Matthews, D., Baldwin, D., 2007. Computing Ocean surface currents from infrared and Ocean color imagery. Trans. Geosci. Remote Sens. 45 (2), 435−447.

Cummins, P.F., Oey, L.Y., 1997. Simulation of barotropic and baroclinic tides off northern British Columbia. J. Phys. Oceano. 27, 762−781. https://doi.org/10.1175/1520-0485(1997)027<0762:sobabt>2.0.co;2.

Cummins, P.F., Masson, D., Foreman, M.G.G., 2000. Stratification and mean flow effects on diurnal tidal currents off Vancouver Island. J. Phys. Oceanogr. 30, 15−30. https://doi.org/10.1175/1520-0485(2000)030<0015:SAMFEO>2.0.CO;2.

Cummins, P.F., Freeland, H.J., Thomson, R.E., 2011. Transport of Japan Tsunami Marine Debris to the Coast of British Columbia: An Updated Review. Science Response 2012-006. Canadian Science Advisory Secretariat (CSAS). http://www.dfo-mpo.gc.ca/csas-sccs/Publications/ScR-RS/2012/2012_006-eng.pdf.

Cummins, P.F., Blanken, H., Hannah, C.G., 2022. HF Radar observations of wintertime surface currents over Hecate Strait, British Columbia. Atmosphere-Ocean 60 (5), 580−599. https://doi.org/10.1080/07055900.2022.2068995.

Cury, P., Roy, C., 1989. Optimal environmental window and pelagic fish recruitment success in upwelling areas. Can. J. Fish. Aquatic Sci. 46, 670−680. https://doi.org/10.1139/F89-086.

Cutter, G.R., Reiss, C.S., Nylund, S., Watters, G.M., 2022. Antarctic krill biomass and flux measured using wideband echosounders and acoustic Doppler current profilers on submerged moorings. Front. Mar. Sci. 9. https://doi.org/10.3389/fmars.2022.784469.

Dahlen, J.M., Chhabra, N.K., 1983. Slippage errors and dynamic response of four drogued buoys measured at sea. Report P-1729, C.C. Draper Laboratory. In: Presented at Marine Technology Society/NOAA Data Buoy Center, 1983 Symposium on Buoy Technology, New Orleans.

Dalbosco, A.L.P., Franco, D., Barletta, R. do Carma, Trevisan, A.B., 2020. Analysis of currents on the continental shelf off the Santa Catarina Island through measured data. RBRH (Braz. J. Water Resour.) 5. https://doi.org/10.1590/2318-0331.252020180175.

Dallimore, A., Thomson, R.E., Bertram, M.A., 2005. Modern to late Holocene deposition in an anoxic fjord on the west coast of Canada: implication for regional Oceanography, climate and paleoseismic history. Mar. Geol. 219, 47–69.

Daniel, T., Manley, J., Trenaman, N., 2011. The Wave Glider: Enabling a New Approach to Persistent Ocean Observation and Research. Liquid Robotics Internal Report.

Danielson, G.C., Lanczos, C., 1942. Some improvements in practical Fourier analysis and their application to X-ray scattering from liquids. J. Franklin Inst. 233, 365–380 and 435–452.

Dantzler, H.L., 1974. Dynamic salinity calibrations of continuous salinity/ temperature/depth data. Deep-Sea Res. 21, 675–682.

Daskalov, G.M., Grishin, A.N., Rodionov, S., Mihneva, V., 2007. Trophic cascades triggered by overfishing reveal possible mechanisms of ecosystem regime shifts. Proc. Natl. Acad. Sci. 104 (25), 10518–10523.

Datawell bv, 1992. Directional Waverider. Datawell bv, Haarlem.

Davis, R.E., 1976. Predictability of sea surface temperature and sea level pressure anomalies over the North Pacific Ocean. J. Phys. Oceanogr. 6, 249–266.

Davis, R.E., 1977. Techniques for statistical analysis and prediction of geophysical fluid systems. Geophys. Astrophys. Fluid Dyn. 8, 245–277.

Davis, R.E., 1978. Predictability of sea level pressure anomalies over the North Pacific Ocean. J. Phys. Oceanogr. 8, 233–246.

Davis, R.E., 1983. Oceanic property transport, Lagrangian particle statistics, and their prediction. J. Mar. Res. 41 (1), 163–194.

Davis, R.E., 1991. Lagrangian Ocean studies. Annu. Rev. Fluid Mech. 23, 43–64.

Davis, R.E., Regier, L., 1977. Methods for estimating directional wave spectra from multielement arrays. J. Mar. Res. 35, 453–477.

Davis, R.E., Webb, D.C., Regier, L.A., Dufour, J., 1992. The autonomous Lagrangian circulation explorer (ALACE). J. Atmos. Ocean. Technol. 9, 264–285.

Davison, A., Huser, R., 2015. Statistics of extremes. Annu. Rev. Stat. Appl. 2, 203–235. https://doi.org/10.1146/annurev-statistics-010814-020133.

Davison, A.C., Smith, R.L., 1990. Models for exceedances over high thresholds. J. R. Stat. Soc. B: Stat. Methodol. 393–442.

De Robertis, A., Levine, M., Lauffenburger, N., Honkalehto, T., Ianelli, J., Monnahan, C.C., Towler, R., Jones, D., Stienessen, S., McKelvey, D., 2021. Uncrewed surface vehicle (USV) survey of walleye pollock, *Gadus Chalcogrammus,* in response to the cancellation of ship-based surveys. ICES J. Mar. Sci. 78, 2797–2808. https://doi.org/10.1093/icesjms/fsab155.

Defant, A., 1936. Schichtung und Zirkulation des Atlantischen Ozeans. Die Troposphare. In: Wissenschaftliche Ergebnisse der Deutschen Altantishcen Expedition auf dem Forschungs- und Vermessungsschiff "Meteor", pp. 1925–1927.

Defant, A., 1937. Stratification and circulation of the Atlantic Ocean. In: Emery, W.J. (Ed.), The Troposphere, Scientific Results of the German Atlantic Expedition of the Research Vessel, Meteor, 1925–27, English Translation. Amerind, New Delhi, 1981.

Defant, A., 1961. Physical Oceanography, VI. Pergamon Press, New York.

Deines, K., 1999. Backscatter estimation using Broadband acoustic Doppler current profilers. In: Proceedings of the IEEE Sixth Working Conference on Current Measurement. San Diego, CA. https://doi.org/10.1109/CCM.1999.755249.

Del Frate, F., Petrocchi, A., Lichtenegger, J., Calabresi, G., 2000. Neural networks for oil spill detection using ERS-SAR data. IEEE Trans. Geosci. Remote Sens. 38 (5), 2282–2287. https://doi.org/10.1109/36.868885.

Denbo, D.W., Allen, J.S., 1984. Rotary empirical orthogonal function analysis of currents near the Oregon Coast. J. Phys. Oceanogr. 14, 35–46.

Denbo, D.W., Allen, J.S., 1986. Reply to "Comments on: rotary empirical orthogonal function analysis of currents near the Oregon Coast". J. Phys. Oceanogr. 16, 793–794.

Denman, K.L., Freeland, H.J., 1985. Correlation scales, objective mapping and a statistical test of geostrophy over the continental shelf. J. Mar. Res. 43, 517–539.

Dhage, L., Strub, P.T., 2016. Intra-seasonal sea level variability along the west coast of India. J. Geophys. Res. Oceans 121, 8172–8188.

Dibarboure, G., Pujol, M.-I., Briol, F., Le Traon, P.Y., Larnicol, G., Picot, N., Mertz, F., Ablain, M., 2011. Jason-2 in DUACS: updated system description, first tandem results and impact on processing and products. Mar. Geod. 34, 214–241.

Di Iorio, D., Lavelle, J.W., Rona, P.A., Bemis, K., Xu, G., Germanovich, L.N., Lowell, R.P., Genc, G., 2012. Measurements and models of heat flux and plumes from hydrothermal discharges near the deep seafloor. Oceanography 25 (1), 168–179. https://doi.org/10.5670/oceanog.2012.14.

Di Lorenzo, E., Mantua, N., 2016. Multi-year persistence of the 2014/15 North Pacific marine heatwave. Nat. Clim. Change 6, 1042–1047. https://doi.org/10.1038/nclimate3082.

Di Lorenzo, E., Miller, A.J., Schneider, N., 2005a. The warming of the California Current System: dynamics and ecosystem implications. J. Phys. Oceanogr. 35 (3), 336–362.

Di Lorenzo, E., Foreman, M.G.G., Crawford, W.R., 2005b. Modelling the generation of Haida Eddies. Deep-Sea Res. II-Top. Stud. Oceanogr. 52 (7–8), 853–873. https://doi.org/10.1016/j.dsr2.2005.02.007.

Di Vittorio, A.V., Emery, W.J., 2002. An automated, dynamic threshold cloud-masking algorithm for daytime AVHRR images over land. IEEE Trans. Geosci. Remote Sens. 40 (8), 1682–1694. https://doi.org/10.1109/TGRS.2002.802455.

Diaconis, P., Efron, B., 1983. Computer-intensive methods in statistics. Sci. Am. 248, 116–130.

Diaconis, P., Skyrms, B., 2018. Ten Great Ideas about Chance. Princeton University Press, Princeton and Oxford, p. 255.

Dickson, A.G., Sabine, C.L., Christian, J.R. (Eds.), 2007. Guide to Best Practices for Ocean CO_2 Measurements. PICES Special Publication 3, p. 191. http://cdiac.ornl.gov/ftp/oceans/Handbook_2007/Guide_all_in_one.pdf.

Dillon, M.T., 1982. Vertical overturns: A comparison of Thorpe and Ozmidov length scales. J. Geophys. Res. 87, 9601–9613.

DiVittorio, A.V., Emery, W.J., 2002. An automated dynamic threshold cloud-masking algorithm for daytime AVHRR images over land. Trans. Geosci. Remote Sens. 40, 1682–1694.

Dodimead, A.J., Favorite, F., Hirano, T., 1963. Salmon of the North Pacific. II: review of Oceanography of the subarctic Pacific region. Int. North Pac. Fish. Comm. Bull. 13, 195.

Dodson, S.I., 1990. Predicting diel vertical migration of zooplankton. Limnol. Oceanogr. 35, 1195—1200.

Doney, S.C., Fabry, V.J., Feely, R.A., Kleypas, J.A., 2009. Ocean acidification: the other CO_2 problem. Ann. Rev. Mar. Sci. 1, 169—192. https://doi.org/10.1146/annurev.marine.010908.163834.

Donelan, M.A., Hamilton, J., Hu, W.H., 1985. Directional spectra of wind-generated waves. Phil. Trans. R. Soc. Lond. A 315, 509—562. https://doi.org/10.1098/rsta.1985.0054.

Doodson, A.T., Warburg, H.D., 1941. Admiralty Manual of Tides. Hydrographic Department, Admiralty, London.

Drakopoulos, P.G., Marsden, R.F., 1993. The internal tide off the west coast of Vancouver Island. J. Phys. Oceanogr. 23, 758—775. https://doi.org/10.1175/1520-0485(1993)023<0758:TITOTW>2.0.CO;2.

Drozdowski, A., Greenan, B.J.W., 2013. An intercomparison of acoustic current meter measurements in low to moderate flow regions. J. Atmos. Ocean Technol. 30 (8), 1924—1939. https://doi.org/10.1175/JTECH-D-12-00198.1.

Druffel, E.R.M., Williams, P.M., 1991. Radiocarbon in seawater and organisms from the Pacific coast of Baja California. Radiocarbon 33, 291—296.

Druffel, E.R.M., Williams, P.M., Druffel, E.R.M., 1989. Decade time scale variability of ventilation in the North Atlantic: high-resolution measurements of bomb radiocarbon in banded corals. J. Geophys. Res. 94, 3271—3285.

Ducet, N., Le Traon, P.Y., Reverdin, G., 2000. Global high-resolution mapping of ocean circulation from TOPEX/Poseidon and ERS-1 and -2. J. Geophys. Res. 105, 19,477—19,498.

Dupuis, D., 1998. Exceedances over high thresholds: a guide to threshold selection. Extremes 261, 251—261.

Dziewonski, A., Bloch, S., Landisman, M., 1969. A technique for the analysis of transient seismic signals. Bull. Seismol. Soc. Am. 59, 421—444.

Eastman, R., Warren, S.G., Hahn, C.J., 2011. Variations in cloud cover and cloud types over the ocean from surface observations, 1954—2008. J. Clim. 24, 5914—5934. https://doi.org/10.1175/2011JCLI3972.1.

Ebisuzaki, W., 1997. A method to estimate the statistical significance of a correlation when the data are serially correlated. J. Clim. 10, 2147—2153.

Edmond, J.M., Measures, C.M., McDuff, R.E., Chan, L.H., Collier, R., Grant, B., Gordon, L.I., Corliss, J.B., 1979. Ridge crest hydrothermal activity and the balance of the major and minor elements in the Ocean: the Galapagos data. Earth Planet. Sci. Lett. 46, 1—18.

Efron, B., Gong, G., 1983. A leisurely look at the bootstrap, the jackknife, and cross-validation. Am. Stat. 37, 36—48.

Egbert, G.D., Bennett, A.F., Foreman, M.G.G., 1994. TOPEX/POSEIDON tides estimated using a global inverse model. J. Geophys. Res. 99, 24,821—24,852.

Eisner, J.B., Tsonis, A.A., 1991. Comparison of observed Northern Hemisphere surface air temperature records. Geophys. Res. Lett. 18, 1229—1232.

Ekman, V.W., 1905. On the Influence of the Earth's Rotation on Ocean Currents. Arkiv für mathematik astronomi cch fysik. 2.

Ekman, V.W., 1908. Die Zusammendruckbarkeit des Meerwassers. Conseil Perm, Intern. p. l'Explor. de la Mer, Pub. de Circonetance, No. 43, p. 47.

Ekman, V.W., 1932. On an improved type of current meter. J. Cons. Int. Explor. Mer. 7, 3—10.

Elgar, S., 1988. Comment on "Fourier transform filtering: a cautionary note" by A.M. Forbes. J. Geophys. Res. 93, 15755—15756.

Eliot, M., 2010. Influence of interannual tidal modulation on coastal flooding along the Western Australian coast. J. Geophys. Res. 115, C11013. https://doi.org/10.1029/2010JC006306.

Emery, W.J., 1983. On the geographical variability of the upper level mean and eddy fields in the North Atlantic and North Pacific. J. Phys. Oceanogr. 12, 269—291.

Emery, W.J., Meincke, J., 1986. Global water masses: summary and review. Oceanogl. Acta 9, 383—391.

Emery, W.J., Schluessel, P., 1989. Global differences between skin and bulk sea surface temperatures. EOS 70, 210—211.

Emery, W.J., Royer, T.C., Reynolds, R.W., 1985. The anomalous tracks of North Pacific drifting buoys 1981—83. Deep-Sea Res. 32, 315—347.

Emery, W.J., Thomas, A.C., Collins, M.J., Crawford, W.R., Mackas, D.L., 1986a. An objective procedure to compute surface advective velocities from sequential infrared satellite images. J. Geophys. Res. 91, 12865—12879.

Emery, W.J., Lee, W.G., Zenk, W., Meincke, J., 1986b. A low-cost digital XBT system and its application to the real-time computation of dynamic height. J. Technol. Oceanogr. Atmos. Sci 3, 75—83.

Emery, W.J., Born, G.H., Baldwin, D.G., Norris, C.L., 1989a. Satellite derived water vapor corrections for GEOSAT altimetry. J Geophys. Res. 95, 2953—2964.

Emery, W.J., Brown, J., Novak, V.P., 1989b. AVHRR image navigation; summary and review. Photogramm. Eng. Remote Sens. 8, 1175—1183.

Emery, W.J., Born, G.H., Baldwin, D.G., Norris, C.L., 1990. Satellite-derived water vapor corrections for GEOSAT altimetry. J. Geophys. Res. 95, 2953—2964.

Emery, W.J., Fowler, C.W., Clayson, C.A., 1992. Satellite image derived Gulf Stream currents. J. Ocean. Atmos. Sci. Technol. 9, 285—304.

Emery, W.J., Yu, Y., Wick, G., Schluessel, P., Reynolds, R.W., 1994a. Improving satellite infrared sea surface temperature estimates by including independent water vapor observations. J. Geophys. Res. 99, 5219—5236.

Emery, W.J., Fowler, C.W., Maslanik, J., 1994b. Arctic sea ice concentrations from special sensor microwave imager and advanced very high resolution radiometer satellite data. J. Geophys. Res. 99, 18329—18342.

Emery, W.J., Kantha, L., Wick, G.A., Schluessel, P., 1994c. The relationship between skin and bulk sea surface temperatures. In: Jones, I. (Ed.), Proceedings of the First Pacific Ocean Remote Sensing Conference, Okinawa, Japan, August 1992, pp. 25—40.

Emery, W.J., Baldwin, D., Matthews, D., 2003. Maximum cross correlation automatic satellite image navigation and attitude corrections for open-ocean image navigation. IEEE Transact. Geosci. Rem. Sens. 41 (1), 33—42. https://doi.org/10.1109/TGRS.2002.808061. Jan. 2003.

Encyclopedia Britannica 2010: fifteenth ed., Benton Foundation and Encyclopædia Britannica, Inc., 32,604.

Enfield, D.B., Cid, L., 1990. Statistical analysis of El Niño/Southern oscillation over the last 500 years. TOGA Notes 1, 1—4.

Espinosa, A.L., 2012. Determination of the Acidification State of Canadian Pacific Coastal Waters Using Empirical Relationships with Hydrographic Data (M.Sc. thesis). School of Earth and Ocean Sciences, University of Victoria, p. 92.

Essex, C., Lookman, T., Nerenberg, M.A.H., 1987. The climate attractor over short timescales. Nature 326, 64–66.

Evans, J.C., 1985. Selection of a numerical filtering method: convolution or transform windowing? J. Geophys. Res. 90, 4991–4994.

Fabry, V.J., Seibel, B.A., Feely, R.A., Orr, J.C., 2008. Impacts of Ocean acidification on marine fauna and ecosystem processes. ICES J. Mar. Sci. 65, 414–432. https://doi.org/10.1093/icesjms/fsn048.

Fallmann, J., Lewis, H., Castillo, J.M., Arnold, A., Ramsdale, S., 2017. Impact of sea surface temperature on stratiform cloud formation over the North Sea. Geophys. Res. Lett. 44, 4296–4303.

Fang, J., Wahl, T., Zhang, Q., Muis, S., Hu, P., Fang, J., Du, S., Dou, T., Shi, P., 2021. Extreme sea levels along coastal China: uncertainties and implications. Stoch. Environ. Res. Risk Assess. 35, 405–418. https://doi.org/10.1007/s00477-020-01964-0.

Farge, M., 1992. Wavelet transforms and their applications to turbulence. Annu. Rev. Fluid Mech. 24, 395–457.

Farland, R.J., November, 1975. Salinity Intercomparison Report, the Oceanographic Subprogramme for the GARP Atlantic Tropical experiment (GATE). Nat. Oceanogr. Instrum. Center, Washington, DC.

Farmer, D.M., Armi, L., 1999. The generation and trapping of solitary waves over topography. Science 283, 188–190. https://doi.org/10.1126/science.283.5399.188.

Farmer, D., Smith, J.D., 1980. Nonlinear internal waves in a fjord. In: Freeland, H.J., Farmer, D.M., Levings, C.D. (Eds.), Fjord Oceanography. Plenum, New York, pp. 465–493.

Farmer, D.M., Clifford, S.F., Verral, J.A., 1987. Scintillation structure of a turbulent tidal flow. J. Geophys. Res. 92, 5369–5382.

Favorite, F., Dodimead, A.J., Nasu, K., 1976. Oceanography of the subarctic Pacific region. Int. North Pac. Fish. Comm. Bull. 33, 187.

Feely, R.A., Gendron, J.F., Baker, E.T., Lebon, G.T., 1994. Hydrothermal plumes along the east Pacific rise, 8°40′ to 11°50′N: particle distribution and composition. Earth. Planet. Sci. Lett. 128, 19–36.

Feely, R.A., Sabine, C.L., Lee, K., Berelson, W., Kleypas, J., Fabry, V.J., Millero, F.J., 2004. Impact of anthropogenic CO_2 on the $CaCO_3$ system in the Oceans. Science 305, 362–366.

Feely, R.A., Sabine, C.L., Byrne, R.H., Millero, F.J., Dickson, A.G., Wanninkhof, R., Murata, A., Miller, L.A., Greely, D., 2012. Decadal changes in the aragonite and calcite saturation state of the Pacific Ocean. Global Biogeochem. Cyc. 25. https://doi.org/10.1029/2011GB004157.

Fei, S.-W., Wang, M.-J., Miao, Y.-B., Tu, J., 2009. Particle swarm optimization-based support vector machine for forecasting gases content in power transformer oil. Energy Convers. Manag. 50 (6), 1604–1609. https://doi.org/10.1016/j.enconman.2009.02.004.

Feistel, R., 2003. A new extended Gibbs thermodynamic potential of seawater. Prog. Ocean. 58, 43–114.

Feistel, R., 2008. A Gibbs function for seawater thermodynamics for −6 to 80° C and salinity up to 120 g/kg. Deep-Sea Res. Part 1 55, 1639–1671. https://doi.org/10.1016/j.dsr.2008.07.004.

Feistel, R., Weinreben, S., Wolf, H., Seitz, S., Spitzer, P., Adel, B., Nausch, G., Schneider, B., Wright, D.G., 2010a. Density and absolute salinity of the Baltic Sea 2006–2009. Ocean Sci. 6, 3–24. https://doi.org/10.5194/os-6-3-2010.

Feistel, R., Marion, G.M., Pawlowicz, R., Wright, D.G., 2010b. Thermophysical property anomalies of Baltic seawater. Ocean Sci. 6, 949–981. https://doi.org/10.5194/os-6-949-2010.

Feng, X., Tsimplis, M.N., 2014. Sea level extremes at the coasts of China. J. Geophys. Res. Ocean. 119 (3), 1593–1608.

Fereday, D.R., Knight, J.R., Scaife, A.A., Folland, C.K., Philipp, A., 2008. Cluster analysis of North Atlantic-European circulation types and links with Tropical Pacific sea surface temperature. J. Clim. 21, 3687–3703. https://doi.org/10.1175/2007JCLI1875.1.

Ferreira, J.A., Guedes Soares, C., 1998. An application of the peaks over threshold method to predict extremes of significant wave height. J. Offshore Mech. Arct. Eng. 120, 165–176.

Fiadeiro, M.E., Veronis, G., 1984. Obtaining velocities from tracer distributions. J. Phys. Oceanogr. 14, 1734–1746. https://doi.org/10.1175/1520-485(1984)014<1734:OVFTD>2.0.CO;2.

Fine, R.A., 1985. Direct evidence using tritium data for throughflow from the Pacific into the Indian Ocean. Nature 315, 478–480.

Fine, I.V., Thomson, R.E., March, 2021. Storm Surge Simulations for the Southern Strait of Georgia and Boundary Bay for Selected Future Sea Level Rise Predictions. Defence Research and Development Canada's Centre for Security Science (DRDC CSS) Program (CSSP), Coastal Flood Mitigation Canada CSSP-2018-CP-2352, p. 38.

Fine, I.V., Rabinovich, A.B., Bornhold, B.D., Thomson, R.E., Kulikov, E.A., 2005. The Grand Banks landslide-generated tsunami of November 18, 1929: preliminary analysis and numerical modeling. Mar. Geol. 215, 45–57. https://doi.org/10.1016/j.margeo.2004.11.007.

Fine, I.V., Cherniawsky, J.Y., Rabinovich, A.B., Stephenson, F., 2009. Numerical modeling and observations of tsunami waves in Alberni Inlet and Barkley Sound, British Columbia. Pure Appl. Geophys. 165, 2019–2044. https://doi.org/10.1007/s00024-008-0414-9.

Fine, I.V., Cherniawsky, J.Y., Thomson, R.E., Rabinovich, A.B., Krassovski, M.V., 2015. Observations and numerical modeling of the 2012 Haida Gwaii tsunami off the coast of British Columbia. Pure Appl. Geophys. 172 (3–4), 699–718.

Fine, I.V., Thomson, R.E., Chadwick, W.W., Fox, C.G., 2020. Towards a universal frequency of occurrence distribution for tsunamis: statistical analysis of a 32-year bottom pressure record at axial seamount. Geophys. Res. Lett. https://doi.org/10.1029/2020GL087372.

Fischer, J., Visbeck, M., 1993. Deep velocity profiling and self-contained ADCPs. J. Atmos. Ocean. Technol. 10, 764–773.

Fissel, D.B., Tang, C.L., 1991. Response of sea ice drift to wind forcing on the northeastern Newfoundland shelf. J. Geophys. Res. 96 (C10), 18,397–18,409.

Flagg, C.N., Smith, S.L., 1989. On the use of the acoustic Doppler current profiler to measure zooplankton abundance. Deep-Sea Res. 36, 455–479.

Flierl, G., Robinson, A.R., 1977. XBT measurements of thermal gradient in the MODE eddy. J. Phys. Oceanogr. 7, 300–302.

Fofonoff, N.P., 1960. Transport Computations for the North Pacific Ocean 1955–1958. Fisheries Research Board of Canada, Manuscript Report Oceanography and Limnology, No. 77–80.

Fofonoff, N.P., 1969. Spectral characteristics of internal waves in the Ocean. Frederick C. Fuglister Sixtieth Anniversary volume. Deep-Sea Res. 76 (Suppl. l), 58–71.

Fofonoff, N.P., 1985. Physical properties of seawater: a new salinity scale and equation of state for seawater. J. Geophys. Res. 90, 3332–3342. https://doi.org/10.1029/JC090iC02p03332.

Fofonoff, N.P., Fischer, C.F., Pacific Oceanographic Group, and Canadian Committee on Oceanography, 1958. Program for Oceanographic Computations and Data 32 Oceanography,Vol. 23, No. 3. In: Processing on the Electronic Digital Computer ALWAC III-E: PSW-I Programs for Properties of Seawater. Manuscript Report Series no. 27, Nanaimo, BC. Pacific Oceanographic Group, p. 39.

Fofonoff, N.P., Froese, C., 1960. Programs for Oceanographic Computations and Data Processing on the Electronic Digital Computer ALWAC III-E. M-l Miscellaneous Programs. Fisheries Research Board of Canada, Manuscript Report Oceanography and Limnology, No. 72.

Fofonoff, N.P., Tabata, S., 1966. Variability of Oceanographic conditions between Ocean Station P and Swiftsure Bank off the Pacific coast of Canada. J. Fish. Res. Board Can. 23, 825–868.

Fofonoff, N.P., Hayes, S.P., Millard, R.C., 1974. WHOI/Brown CTD Microprofiler: Methods of Calibration and Data Handling. WHOI-74–89.

Foppert, A., Donohue, K.A., Watts, D.R., Tracey, K.L., 2017. Eddy heat flux across the Antarctic Circumpolar Current estimated from sea surface height standard deviation. J. Geophys. Res. 122 (8), 6947–6964. https://doi.org/10.1002/2017JC012837.

Forbes, A.M.G., 1988. Fourier transform filtering: a cautionary note. J. Geophys. Res. 93, 6958–6962.

Forch, C., Knudsen, M., Sorensen, S.P.L., 1902. Berichte ueber die Konstantenbestimungen zur Aufstellunt der hydrographischen Tabellen. Kgl. Danske Vedenskab. Selskab Skrifter, 7 Taekke, Naaturvidensk, og Mex.

Foreman, M.G.G., 1977. Manual for Tidal Height Analysis and Prediction. Institute of Ocean Sciences, Sidney, BC. Pacific Mar. Sci. Rep. 77–10.

Foreman, M.G.G., 1977. Manual for tidal height analysis and prediction. Pac. Mar. Sci. Rep. 77 (10) (Institute of Ocean Sciences, Sidney, BC).

Foreman, M.G.G., 1978. Manual for Tidal Currents Analysis and Prediction. Institute of Ocean Sciences, Sidney, BC. Pacific Mar. Sci. Rep. 78–6.

Foreman, M.G.G., Crawford, W.R., Marsden, R.F., 1995. De-tiding: theory and practice. In: Lynch, D.R., Davies, A.M. (Eds.), Quantitative Skill Assessment for Coastal Ocean Models, Coastal and Estuarine Studies, 47, pp. 203–239. https://doi.org/10.1029/CE047p0203.

Foreman, M.G.G., Crawford, W.R., Cherniawsky, J.Y., Gower, J.F.R., Cuypers, L., Ballantyne, V.A., 1998. Tidal correction of TOPEX/poseidon altimetry for seasonal sea surface elevation and current determination off the Pacific coast of Canada. J. Geophys. Res. 103 (C12), 27979–27998.

Foreman, M.G.G., Czajko, P., Stucchi, D.J., Guo, M., 2009. A finite volume model simulation for the Broughton Archipelago, Canada. Ocean Model. 30, 29–47. https://doi.org/10.1016/j.ocemod.2009.05.009.

Introduction to the special issue: in RIDGE to ridge 2000. In: Fornari, D.J., Beaulieu, S.E., Holden, J.F., Mullineaux, L.S., Tolstoy, M. (Eds.), Oceanography 25 (1), 12–17. https://doi.org/10.5670/oceanog.2012.01.

Fornari, D.J., Von Damm, K.L., Bryce, J.G., Cowen, J.P., Ferrini, V., Fundis, A., et al., 2012b. The East Pacific rise between 9°N and 10°N: twenty-five years of integrated, multidisciplinary Oceanic spreading center studies. Oceanography 25, 18–43. https://doi.org/10.5670/oceanog.2012.02.

Fox, C.G., 2016. Processed bottom pressure recorder (BPR) data from uncabled instruments deployed at axial seamount on the Juan de Fuca Ridge. Integr. Earth Data Appl. (IEDA). https://doi.org/10.1594/IEDA/322344.

Fraedrich, K., Leslie, L.M., 1989. Estimates of cyclone track predictability. I: tropical cyclones in the Australian region. Q. J. R. Meteorol. Soc. 115, 79–92.

Fraedrich, K., Grotjanh, R., Leslie, L.M., 1990. Estimates of cyclone track predictability. II: fractal analysis of mid-latitude cyclones. Q. J. R. Meteorol. Soc. 116, 317–335.

Fraedrich, K., Jiang, J., Gerstenbarbe, W., 1997. Multiscale detection of abrupt climate changes: Application to Nile River flood levels. Int. J. Climatol. 17, 1301–1315.

Franco, A.C., Ianson, D., Ross, T., Hamme, R.C., Monahan, A.H., Christian, J.R., Davelaar, M., Johnson, W.K., Miller, L.A., Philippe, M., Tortell, D., 2021. Anthropogenic and climatic contributions to observed carbon system trends in the Northeast Pacific. Global Biogeochem. Cycles 35 (7). https://doi.org/10.1029/2020GB006829.

Frank, W.M., Wang, H., McBride, J.L., 1996. Rawinsonde budget analyses during the TOGA COARE IOP. J. Atmos. Sci. 53, 1761–1780.

Frank, M., 1979. Radiogenic Isotropes: tracers of past ocean circulation and erosional input. Rev. Geophys. 40. https://doi.org/10.1029/2000RG000094.

Frankignoul, C., Scoffier, N., Cane, M.A., 1993. An adaptive inverse method for model tuning and testing statistical methods. In: Phys. Oceanogr., Proceeding Hawaii. Winter Workshop. Univ. of Hawaii at Manoa, pp. 331–350.

Fratantoni, D.M., 2001. North Atlantic surface circulation during the 1990's observed with satellite-tracked drifters. J. Geophys. Res. 106, 22067–22093. https://doi.org/10.1029/2000JC000730.

Freeland, H.J., 2013. Evidence of change in the winter mixed layer in the Northeast Pacific Ocean: a problem revisited. Atmos. Ocean 51, 126–133.

Freeland, H.J., Gould, W.J., 1976. Objective analysis of meso-scale Ocean circulation features. Deep-Sea Res. 23, 915–923.

Freeland, H.J., Rhines, P.B., Rossby, T., 1975. Statistical observations of the trajectories of neutrally buoyant floats in the North Atlantic. J. Mar. Res. 33, 383–404.

Freeland, H., Church, J.A., Smith, R.L., Boland, F.M., 1985. Current Meter Data from the Australian Coastal Experiment: A Data Report. CSIRO Marine Laboratories, Report 169, Hobart, Australia.

Freeland, H.J., Boland, F.M., Church, J.A., Clarke, A.J., Forbes, A.M.G., Huyer, A., Smith, R.L., Thompson, R.O.R.Y., White, N.J., 1986. The Australian coastal experiment: a search for coastal-trapped waves. J. Phys. Oceanogr. 16, 1230–1249.

Freeland, H.J., Denman, K.L., Wong, C.S., Whitney, F., Jacques, R., 1997. Evidence of change in the winter mixed layer in the Northeast Pacific Ocean. Deep-Sea Res. 44, 2117–2129.

Freilich, M.H., Chelton, D.B., 1986. Wavenumber spectra of Pacific winds measured by the Seasat scatterometer. J. Phys. Oceanogr. 16, 741–757. https://doi.org/10.1175/1520-0485(1986)016<0741:WSOPWM>2.0.CO;2.

Freund, Y., Mason, L., 1999. The alternating decision tree learning algorithm. In: Bratko, I., Dzeroski, S. (Eds.), Proceedings of the Sixteenth International Conference on Machine Learning (ICML '99). Morgan Kaufmann Publishers Inc., San Francisco, CA, USA, pp. 124–133.

Friehe, C.A., Pazan, S.E., 1978. Performance of an air-sea interaction buoy. J. Appl. Meteorol. 17, 1488−1497.

Fu, L.-L., 1981. Observations and models of inertial waves in the deep ocean. Rev. Geophys. 19, 141−170.

Fuglister, F.C., 1960. Atlantic Ocean Atlas of Temperature and Salinity Profiles and Data from the International Geophysical Year of 1957−1958. Woods Hole Oceanographic Institution Atlas Series 1.

Fukuoka, A., 1951. A study of 10-day forecast (A synthetic report). Geophys. Mag. 22, 177−208.

Furrer, R., Bengston, T., 2007. Estimation of high-dimensional prior and posterior covariance matrices in Kalman filter variants. J. Multivar. Anal. 98, 227−255.

Gamage, N., Blumen, W., 1993. Comparative anlaysis of low-level cold fronts: wavelet, Fourier, and empirical orthogonal function decompositions. Month. Weather Rev. 121, 2867−2878.

Gammon, R.H., Cline, J., Wisegarver, D., 1982. Chlorofluoromethanes in the Northeast Pacific Ocean: measured vertical distributions and application as transient tracers of upper Ocean mixing. J. Geophys. Res. 87, 9441−9454.

Ganachaud, Wunsch, C., 2000. Improved estimates of global Ocean circulation, heat transport and mixing from hydrographic data. Nature 408, 453−457.

Gandin, L.S., 1965. Objective Analysis of Meteorological Fields. Israel Program for Scientific Translations, Jerusalem.

Garcia-Berdeal, I.G., Hautala, S., Thomas, L.N., Johnson, H.P., 2006. Vertical structure of time-dependent currents in a mid-Ocean ridge axial valley. Deep Sea Res. 53, 367−386. https://doi.org/10.1016/j.dsr.2005.10.004.

Garrett, C.J., 1972. Tidal Resonance in the Bay of Fundy and Gulf of Maine. Nature 238, 441−443.

Gargett, A.E., 1994. Observing turbulence with a modified acoustic Doppler current profiler. J. Atmos. Ocean. Technol. 11, 1592−1610.

Gargett, A.E., Östlund, G., Wong, C.S., 1986. Tritium time series from Ocean Station P. J. Phys. Oceanogr. 16, 1720−1726.

Gargett, A.E., Wells, J., Tejada-Martinez, A.E., Grosch, C.E., 2004. Langmuir supercells: a mechanism for sediment resuspension and transport in shallow seas. Science 306, 1925−1928. https://doi.org/10.1126/science.1100849.

Garrett, C.J.R., Munk, W.H., 1971. The age of the tide and the "Q" of the Oceans. Deep-Sea Res. 18, 493−503.

Garrett, C.J.R., Munk, W.H., 1979. Internal waves in the Ocean. Ann. Rev. Fluid Mech. 11, 339−369.

Garrett, C., Petrie, B., 1981. Dynamical aspects of the flow through the Strait of Belle Isle. J. Phys. Oceanogr. 11, 376−393.

Gaspar, P., 1988. Modeling the seasonal cycle of the upper Ocean. J. Phys. Oceanogr. 18, 161−180. https://doi.org/10.1175/1520-0485(1988)018<0161:MTSCOT>2.0.CO;2.

Geist, E.L., Parsons, T., ten Brink, U.S., Lee, H.J., 2009. Chapter 4: Tsunami probability. In: Bernard, E.N., Robinson, A.R. (Eds.), The Sea, 15, ISBN 978-0-674-03173-9.

Gendron, J.F., Cowen, J.P., Feely, R.A., Baker, E.T., 1993. Age estimate for the 1987 megaplume on the southern Juan de Fuca Ridge using excess radon and manganese partitioning. Deep-Sea Res. 40, 1559−1567.

Gentemann, C.L., Hilburn, K.A., 2015. In situ validation of sea surface temperatures from the GCOM-W1 AMSR2 RSS calibrated brightness temperatures. J. Geophys. Res. 120, 3567−3585. https://doi.org/10.1002/2014JC010574.

Georgi, D.T., Dean, J.P., Chase, J.A., 1980. Temperature calibration of expendable bathythermographs. Ocean Eng. 7, 491−499.

German, C.R., Lin, J., Parson, L.M., 2004. Mid-Ocean Ridges: Hydrothermal Interactions Between the Lithosphere and Oceans, Geophysical Monograph Series. American Geophysical Union, Washington, DC, p. 318.

Geyer, W.R., 1989. Field calibration of mixed-layer drifters. J. Atmos. Oceanic Technol. 6 (2), 333−342. https://doi.org/10.1175/1520-0426(1989)006<0333:FCOMLD>2.0.CO;2.

Ghanbari, M., Arabi, M., Obeysekera, J., Sweet, W., 2019. A coherent statistical model for coastal flood frequency analysis under nonstationary sea level conditions. Earth's Future 7, 162−177. https://doi.org/10.1029/2018EF001089.

Ghil, M., Malanotte-Rizzoli, P., 1991. Data assimilation in meteorology and oceanography. Adv. Geophys. 33, 141−266. https://doi.org/10.1016/S0065-2687(08)60442-2.

Gibbs, J.W., 1873. A method of geometrical representation of the thermodynamic properties of substances by means of surfaces. Trans. Conn. Acad. Arts Sci. 2, 382−404.

Gilbert, J.A., Steele, J.A., Caprasp, J.G., Steinbrück, L., Reeder, J., Temperton, B., Huse, S., McHardy, A.C., Knight, R., Joint, I., Somerfield, P., Fuhrman, J.A., Field, D., 2012. Defining seasonal marine microbial community dynamics. ISME J. 6, 298−308.

Gill, A.E., 1982. Atmosphere-ocean Dynamics. Int. Geophys. Series 30. Academic Press, London, p. 663.

Giunta, V., Ward, B., 2022. Ocean mixed layer depth from dissipation. J. Geophys. Res. Oceans 127 (4), e2021JC017904. https://doi.org/10.1029/2021JC017904.

Gladkova, U., Ignatov, A., Shahriar, F., Kihai, Y., Hilger, D., Petrenko, B., 2016. Improved VIIRS and MODIS SST imagery. Remote Sens. https://doi.org/10.3390/rs8010079.

Godfrey, J.S., Golding, T.J., 1981. The Sverdrup relation in the Indian Ocean and the effect of Pacific-Indian Ocean throughflow on the Indian Ocean circulation and on the East Australia current. J. Phys. Oceanogr. 11, 771−779.

Godin, G., 1972. The Analysis of Tides. University of Toronto Press, Toronto.

Goldstein, R.M., Barnett, T.P., Zebker, H.A., 1989. Remote sensing of Ocean currents. Science 246, 1282−1285.

Gonella, J., 1972. A rotary component method for analyzing meteorological and oceanographic vector time series. Deep-Sea Res. 19, 833−846.

Gonzalez, F.I., Kulikov, Y.A., 1993. Tsunami dispersion observed in the deep Ocean. In: Tsunamis in the World. Kluwer, pp. 7−16.

Gonzalez, F.I., Milburn, H.B., Bernard, E.N., Newman, J., 1998. Deep-ocean Assessment and Reporting of Tsunamis (DART): Brief Overview and Status Report. Pacific Marine Environmental Laboratory, NOAA, Seattle, WA, USA.

González, F.I., Bernard, E.N., Meinig, C., Eble, M., Mofjeld, H.O., Stalin, S., 2005. The NTHMP tsunameter network. Nat. Hazard. 35 (1), 25−39. https://doi.org/10.1007/s11069-004-2402-4 (Special Issue, U.S. National Tsunami Hazard Mitigation Program).

Gooberlet, M.A., Swift, C.T., Wilkerson, J.C., 1990. Ocean surface wind speed measurements of the special sensor microwave imager (SSM/I). IEEE Trans. Geosci. Remote Sens. 28, 823−828.

Gordon, A.L., 1986. InterOcean exchange of thermocline water. J. Geophys. Res. 91, 5037−5046.

Gordon, R.L., 1996. Acoustic Doppler Current Profilers. Principles of Operation: A Practical Primer, second ed. RD Instruments, San Diego, CA.

Gordon, H.R., Wang, M., 1994. Retrieval of water-leaving radiance and aerosol optical thickness of the Oceans with SeaWIFS: a preliminary algorithm. Appl. Opt. 33, 443−452.

Gould, W.J., 1973. Effects of non-linearities of current meter compasses. Deep-Sea Res. 20, 423−427.

Gould, W.J., Sambuco, E., 1975. The effect of mooring type on measured values of Ocean currents. Deep-Sea Res. 22, 55−62.

Goupillaud, P., Grossmann, A., Morlet, J., 1984. Cycle-octave and related transforms in seismic signal analysis. Geoexploration 23, 85−105.

Gouretski, V.V., Koltermann, K.P., 2004. WOCE global hydrographic climatology [CD-ROM]. Ber. Bundesamt Seeschiffahrt Hydrogr. Rep. 35, 52. Bundesamt Seeschiffahrt Hydrogr., Hamburg, Germany.

Graham, F.S., McDougall, T.J., 2013. Quantifying the nonconservative production of conservative temperatures, potential temperature and entropy. J. Phys. Oceanogr. 43, 838−862. https://doi.org/10.1175/JPO-D-11-0188.

Grantham, W.I., Braclente, E.M., Jones, W.L., Johnson, J.W., 1976. The SeaSat-A satellite scatterometer. J. Ocean. Eng OE-2, 199−206.

Grassberger, P., Procaccia, I., 1983. Measuring the strangeness of strange attractors. Physica 9D 189−208.

Grasshoff, K., 1983. Methods of Seawater Analysis. Verlag Chemie, Weinheim.

Grassl, H., 1976. The dependence of the measured cool skin of the Ocean on wind stress and total heat flux. Bound. Layer Meteorol. 10, 465−474.

Gray, H.L., Woodward, W.A., 1992. Autoregressive models not sensitive to initial conditions. EOS Trans. AGU 73 (25), 267−268.

Grayver, A.V., Olsen, N., 2019. The magnetic signatures of the M_2, N_2, and O_1 oceanic tides observed in Swarm and CHAMP satellite magnetic data. Geophys. Res. Lett. 4230−4238. https://doi.org/10.1029/2019GL082400.

Green, A., 1984. Bulk dynamics of the expendable bathythermograph (XBT). Deep-Sea Res. 31, 415−426.

Green, D., 2006. Transitioning NOAA Moored Buoy Systems from research to operations. In: Proceedings of OCEANS'06 MTS/IEEE Conference, 18−21 September 2006, Boston, MA, CD-ROM.

Greenan, B.J.W., Prinsenberg, S.J., 1998. Wind forcing of ice cover in the Labrador shelf marginal ice zone. Atmos. Ocean 36 (2), 71−93.

Grimaldi, C.M., Lowe, R.J., Benthuysen, J.A., Green, R.H., Reyns, J., Kernkamp, H., Gilmour, J., 2022. Wave and tidally driven flow dynamics within a coral reef atoll off Northwestern Australia. J. Geophys. Res. Ocean. 127, e2021JC017583. https://doi.org/10.1029/2021JC017583.

Grinsted, A., Moore, J.C., Jevrejeva, S., 2009. Reconstructing sea level from paleo and projected temperatures 200 to 2100AD. Clim. Dyn. https://doi.org/10.1007/s00382-008-0507-2.

Groves, G.W., 1955. Numerical filters for discrimination against tidal periodicities. Trans. Am. Geophys. Union 36, 1073−1084.

Groves, G.W., Hannan, E.J., 1968. Time series regression of sea level on weather. Rev. Geophys. 6, 129−174.

Gruza, G.V., Ran'kova, E.Y., Rocheva, E.V., 1988. Analysis of global data variations in surface air temperature during instrument observation period. Meteor. Gridr. 16−24.

Gu, C., Qi, J., Zhao, Y., Yin, W., Zhu, S., 2022. Estimation of the mixed layer depth in the Indian Ocean from surface parameters: a clustering-neural network method. Sensors 22, 5600. https://doi.org/10.3390/s22155600.

Guenther, G.C., 2007. Airborne LiDAR bathymetry. In: Maune, D.F. (Ed.), Digital Elevation Model Technologies and Applications: The DEM Users Manual, second ed.

Gumbel, E.J., 1958. Statistics of Extremes. Columbia University Press, New York, p. 375.

Gusman, A.R., Sheehan, A.F., Satake, K., Heidarzadeh, M., Mulia, I.E., Maeda, T., 2016. Tsunami data assimilation of Cascadia seafloor pressure gauge records from the 2012 Haida Gwaii earthquake. Geophys. Res. Lett. 43, 4189−4196. https://doi.org/10.1002/2016GL068368.

Gutenberg, B., Richter, C.F., 1944. Frequency of earthquakes in California. Bull. Seismol. Soc. Am. 34, 185−188.

Gutierrez-Villanueva, M.O., Chereskin, T.K., Sprintall, J., 2020. Upper-ocean eddy heat flux across the Antarctic Circumpolar Current in Drake Passage from observations: time-mean and seasonal variability. J. Phys. Oceanogr. 50 (9), 2507−2527. https://doi.org/10.1175/JPO-D-19-0266.1.

Gutzler, D.S., Kiladis, G.N., Meehl, G.A., Weickmann, K.M., Wheeler, M., 1994. The global climate of December 1992-February 1993. Part II: Large-scale variability across the Tropical Western Pacific during TOGA CORE. J. Clim. 7, 1606−1622.

Guymer, T.H., Businger, J.A., Jones, W.L., Stewart, R.H., 1981. Anomalous wind estimates from SEASAT scatterometer. Nature 294, 735−737.

Haidvogel, D.B., Wilkin, J.L., Young, R., 1991. A semi-spectral primitive equation ocean circulation model using vertical sigma and orthogonal curvilinear horizontal coordinates. J. Computat. Phys. 94 (1), 151−185.

Hall, M., Frank, E., Holmes, G., Pfahringer, B., Reutemann, P., Witten, I.H., 2009. The WEKA data mining software: an update. SIGKDD Explor. Newsl. 11, 10−18. https://doi.org/10.1145/1656274.1656278.

Hallock, Z.R., Teague, W.J., 1992. The fall rate of the T7 XBT. J. Atmos. Ocean. Technol. 9, 470−483.

Halpern, D., 1978. Mooring motion influences on current measurements. In: Proceedings of a Workshop Conference on Current Measurement, Technical Report DEL-SG-3−78. College of Marine Studies of Deleware, Newark.

Halpern, D., Reed, R.K., 1976. Heat budget of the upper Ocean under light winds. J. Phys. Oceanogr. 6, 972−975. https://doi.org/10.1175/1520-0485(1976).

Halpern, D., Weller, R.A., Briscoe, M.G., Davis, R.E., McCullough, J.R., 1981. Intercomparison tests of moored current measurements in the upper Ocean. J. Geophys. Res. 86, 419−428.

Halpern, D., Knauss, W., Brown, O., Wentz, F., 1993. An Atlas of Monthly Mean Distributions of SSMI Surface Wind Speed, ARGOS Buoy Drift, AVHRR/2 Sea Surface Temperature, and ECMWF Surface Wind Components During 1991. JPL Publications 93−10, Jet Propulsion Laboratory, Pasadena.

Halverson, M., Pawlowicz, R., 2008. Estuarine forcing of a river plume by river flow and tides. J. Geophys. Res. 113, 1−15.

Halverson, M., Pawlowicz, R., Chavanne, C., 2017. Dependence of 25-MHz HF radar working range on near-surface conductivity, sea state, and tides. J. Atmos. Ocean. Technol. 34, 447−462. https://doi.org/10.1175/JTECH-D-16-0139.1.

Hamlington, B.D., Leben, R.R., Nerem, R.S., Kim, K.-Y., 2011. The effect of signal-to-noise ratio on the study of sea level trends. J. Clim. 24, 1396−1408.

Hamme, R.C., Webley, P.W., Crawford, W.R., Whitney, F.A., DeGrandpre, M.D., Emerson, S.R., Eriksen, C.C., Giesbrecht, K.E., Gower, J.F.R., Kavanaugh, M.T., Pena, M.A., Sabine, C.L., Batten, S.D., Coogan, L.A., Grundle, D.S., Lockwood, D., 2010. Volcanic ash fuels anomalous plankton bloom in subarctic Northeast Pacific. Geophys. Res. Lett. 37. https://doi.org/10.1029/2010GL044629.

Hamming, R.W., 1977. Digital Filters. Prentice-Hall, Englewood Cliffs, NJ.

Hamon, B.V., 1955. A temperature-salinity-depth recorder. ICES J. Mar. Sci. 21, 72−73. https://doi.org/10.1093/icesjms/21.1.72 (Conseil Permanent International pour le Exploration de la Mer. J. Cons. 21, 22−73).

Hamon, B.V., Brown, N.L., 1958. A temperature-chlorinity-depth recorder for use at sea. J. Sci. Instrum. 35, 452−458.

Hanawa, K., Yasuda, T., 1992. New detection method for XBT depth error and relationship between the depth error and coefficients in the depth-time equation. J. Oceanogr. 48, 221−230.

Hanawa, K., Yoritaka, H., 1987. Detection of systematic errors in XBT data and their correction. J. Oceanogr. Soc. Japan 32, 68−76.

Hanawa, K., Yoshikawa, Y., 1991. Re-examination of depth error in XBT data. J. Atmos. Ocean. Technol. 8, 422−429.

Hannachi, A., 2004. A Primer for EOF Analysis of Climate Data. Department of Meteorology, University of Reading, United Kingdom, p. 33.

Hansen, J., Lebedeff, S., 1987. Global trends of measured surface air temperature. J. Geophys. Res. 92, 13345−13372.

Hanson, S., Nicholls, R., Ranger, N., Hallegatte, S., Corfee-Morlot, J., Herweijer, C., Chateau, J., 2011. A global ranking of port cities with high exposure to climate extremes. Climatic Change 104, 89−111. Springer. https://doi.org/10.1007/s10584-010-9977-4.

Hansson, D., Stigebrandt, A., Liljebladh, B., 2013. Modelling the Orust Fjord system on the Swedish west coast. J. Mar. Syst. 113, 29−41.

Harada, K., Tsunogai, S., 1986. Ra-226 in the Japan Sea and the residence time of the Japan Sea water. Earth Planet. Sci. Lett. 77, 236−244.

Harnett, D.L., Murphy, J.L., 1975. Introductory Statistical Analysis. Addison-Wesley, Reading, MA.

Harr, A., 1910. Zur theorie der orthogonalen Funktionen-Systeme. Math Ann. 69, 331−371.

Harris, F.J., 1978. On the use of windows for harmonic analysis with the discrete Fourier transform. Proc. IEEE 66, 51−83.

Haxby, W.F., 1985. Gravity Field of the World's Oceans, Chart Scale 1:51,400,000. Office of Naval Research, Washington, DC.

Hayashi, Y., 1979. Space-time spectral analysis of rotary vector series. J. Atmos. Sci. 36, 757−766.

Hayne, G.S., Hancock, D.W., 1982. Sea state-related altitude errors in the Seasat altimeter. J. Geophys. Res. 87, 3227−3231.

Heidarzadeh, M., Šepić, J., Rabinovich, A.B., Allahyar, M., Soltanpour, A., Tavakoli, F., 2019. Meteorological tsunami of 19 March 2017 in the Persian Gulf: observations and analyses. Pure Appl. Geophys. 177, 1231−1259. https://doi.org/10.1007/s00024-019-02263-8.

Heinmiller, R.H., 1968. Acoustic Release Systems. WHOI Technical Report 68−48. Woods Hole, MA, USA.

Heinmiller, R.H., Ebbesmeyer, C.C., Taft, B.A., Olson, D.B., Nitkin, G.P., 1983. Systematic errors in expendable bathythermograph (XBT) profiles. Deep-Sea Res. 30, 1185−1197.

Helland-Hansen, B., 1918. Nogen Hydrografiske Metoder. Forh. Skand. Naturforskeres, 16, Kristiania.

Hellerman, S., Rosenstein, M., 1983. Normal monthly wind stress over the World Ocean with error estimates. J. Phys. Oceanogr. 13, 1093−1104.

Hendry, R.F., 1993. Canadian Technical Report of Hydrography and Ocean Sciences. Bedford Institute of Oceanography CTD Trials. BIO, Dartmouth, Nova Scotia, B2Y 4A2. Fisheries and Oceans Canada.

Henry, R.F., Graefe, P.W.U., 1971. Zero Padding as a Means of Improving Definition of Computed Spectra. Manuscript Series Report Series No. 20. Canadian Department of Energy, Mines and Resources, Ottawa.

Henyey, F.S., Hoering, A., 1997. Energetics of borelike internal waves. J. Geophys. Res. 102 (C2), 3323−3330. https://doi.org/10.1029/96JC03558.

Hersbach, H., Bell, B., Berrisford, P., Hirahara, S., Horanyi, A., Munoz-Sabater, J., Nicolas, J., Peubey, C., Radu, R., Schepers, D., Simmons, A., Soci, C., Abdalla, S., Abellan, X., Balsamo, G., Bechtold, P., Beavati, G., Bidot, J., Bonavita, M., De Chiara, G., Dahlgren, P., Dee, D., Diamantakis, M., Dragani, R., Flemming, J., Forbes, R., Fuentes, M., Geer, A., Haimberger, L., Healy, S., Hogan, R.J., Hölm, E., Janiskova, M., Keeley, S., Laloyaux, P., Lopez, P., Lupu, C., Radnoti, G., de Rosnay, P., Rozum, I., Vamborg, F., Villaume, S., Thepaut, J.N., 2020. The ERA5 global analysis. Q. J. R. Meteorol. Soc. 146, 1999−2049.

Herterich, K., Hasselman, K., 1987. Extraction of mixed layer advection velocities, diffusion coefficients, feedback factors and atmospheric forcing parameters from the statistical analysis of North Pacific SST anomaly fields. J. Phys. Oceanogr. 17, 2145−2156. https://doi.org/10.1175/1520-0485(1987)017<2145:EOMLAV>2.0.CO;2.

Hewison, T.J., Wu, X., Yu, F., Tahara, Y., Hu, X., Kim, D., Koenig, M., 2013. GSICS inter-calibration of infrared channels of geostationary imagers using Metop/IASI. IEEE Trans. Geosci. Remote Sens. 51, 1160−1170.

Hewitson, B.C., Crane, R.G., 1994. Neural nets: applications in geography, 29. Springer Link, GeoJournal Library, p. 196.

Hewitson, B.C., Crane, R.G., 2002. Self-organizing maps: applications to synoptic climatology. Clim. Res. 22, 13−26.

Hichman, M.L., 1978. Measurement of Dissolved Oxygen. John Wiley, New York.

Hickey, B.M., Dobbins, E.L., Allen, S.E., 2003. Local and remote forcing of currents and temperature in the central Southern California Bight. J. Geophys. Res. 108, 3081. https://doi.org/10.1029/2000JC000313.

Hill, M.N. (Ed.), 1962. The Sea. Physical Oceanography, vol 1. Interscience, New York.

Hill, P.R., Lintern, D.G., 2022. Turbidity currents on the open slope of the Fraser Delta. Mar. Geol. 445. https://doi.org/10.1016/j.margeo.2022.106738.

Hilland, J.E., Chelton, D.B., Njoku, E.G., 1985. Production of global sea surface temperature fields for the Jet Propulsion Laboratory workshop. J. Geophys. Res. 90, 11642−11650.

Hiller, W., Käse, R.H., 1983. Objective analysis of hydrographic data sets from mesoscale surveys. Ber. Inst. Meereskd. Univ. Kiel 116.

Hine, R., Willcox, S., Hine, G., 2009. The Wave Glider: A Wave-Powered Autonomous Marine Vehicle MTS/IEEE Oceans, pp. 1−6. https://doi.org/10.23919/OCEANS.2009.5422129.

Hirach, E., Koren, I., 2021. Record-breaking aerosol levels explained by smoke injection into the stratosphere. Science 371 (6535), 1269−1274. https://doi.org/10.1126/science.abe1415.

Hiyagon, H., 1994. Retention of solar helium and neon in IDPs in deep sea sediment. Science 263, 1257–1259.

Hoegh-Guldberg, O., Mumby, P., Hooten, A.J., Steneck, R.S., Greenfield, P.F., Gomez, E., Harvell, C.D., Sale, P., Edwards, A.J., Caldeira, K., Knowlton, N., Eakin, C.M., Iglesias-Prieto, R., Muthiga, N., Bradbury, R.H., Dubi, A., Hatziolos, M.H., 2007. Coral reefs under rapid climate change and Ocean acidification. Science 318, 1737–1742. https://doi.org/10.1126/science.1152509.

Hogg, N., 1977. Topographic waves along 70°W on the continental rise. J. Mar. Res. 39, 627–649.

Hogg, N.G., Frye, D.E., 2007. Performance of a new generation of acoustic current meters. J. Phys. Oceanogr. 37, 148–161.

Holgate, S.J., Matthews, A., Woodworth, P.L., Richards, L.J., Tamisiea, M.E., Bradshaw, E., Foden, P.R., Gordon, K.M., Jevrejeva, S., Pugh, J., 2013. New data system and products at the permanent service for mean sea level. J. Coast. Res. 29 (3), 493–504. https://doi.org/10.2112/JCOASTRES-D-12-00175.1.

Holl, M.M., Mendenhall, B.R., 1972. Fields by Information Blending, Sea-Level Pressure Version, Technical Note 72–2. Fleet Numerical Weather Central, Monterey, CA, USA.

Hollinger, J., 1989. DMSP Special Sensor Microwave/Imager Calibration/validation. Final report, vol 1. Navy Research Laboratory, Washington, DC.

Holte, J., Talley, L., 2009. A new algorithm for finding mixed layer depths with applications to Argo data and subantarctic mode water formation. J. Atmos. Ocean. Technol. 26 (9), 1920–1939. https://doi.org/10.1175/2009JTECHO543.1.

Horel, J.D., 1984. Complex principal component analysis: theory and examples. J. Clim. Appl. Meteorol. 23, 1660–1673.

Horne, E.P.W., Toole, J.M., 1980. Sensor response mismatches and lag correction techniques for temperature-salinity profilers. J. Phys. Oceanogr. 10, 1122–1130.

Horsburgh, K.J., Wilson, C., 2007. Tide-surge interaction and its role in the distribution of surge residuals in the North Sea. J. Geophys. Res. Ocean. 112, C08003. https://doi.org/10.1029/2006JC00403.

Hosegood, P., van Haren, H., 2004. Near-bed solibores over the continental slope in the Faeroe-Shetland Channel. Deep-Sea Res. Part II 51, 2943–2971.

Hosking, J.R.M., Wallis, J.R., Wood, E.F., 1985. Estimation of the generalized extreme-value distribution by the method of probability-weighted moments. Technometrics 27 (3), 251–261. https://doi.org/10.2307/1269706.

Hosking, J.R.M., Wallis, J.R., 1987. Parameter and quantile estimation for the generalized Pareto distribution. Technometrics 29 (3), 339–349.

Hosokawa, S., Okura, S., 2022. Long-term observation of current at the mouth of Tokyo Bay. Coast. Eng. J. https://doi.org/10.1080/21664250.2022.2122300.

Hourston, R.A.S., Martens, P.S., Juhász, T., Page, S.J., Blanken, H., 2021. Surface ocean circulation tracking drifter data from the Northeastern Pacific and Western Arctic Oceans, 2014–2020. Can. Data Rep. Hydrogr. Ocean Sci. 215 (vi+), 36.

Howe, B.M., Milsis-Olds, J., Rehm, E., Sagen, J., Worcester, P.F., Haralabus, G., 2019. Observing the Oceans acoustically. Front. Mar. Sci. 26. https://doi.org/10.3389/fmars.2019.00426.

Hsieh, W.W., 1986. Comments on: "Rotary empirical orthogonal function analysis of currents near the Oregon Coast". J. Phys. Oceanogr. 16, 791–792.

Hsieh, W.W., Tang, B., 1998. Applying neural network models to prediction and data analysis in meteorology and Oceanography. Bull. Am. Meteorol. Soc. 79, 1855–1879.

Hsu, Y.L., Allard, R.A., Mettlach, T.R., 2002. Wave Model Validation for the Northern Gulf of Mexico Littoral Initiative (NGLI) Project. NRL/FR/7322-02-10,032.

Hu, C., 2009. A novel ocean color index to detect floating algae in the global oceans. Rem. Sens. Environ. 113, 2118–2129. https://doi.org/10.1016/j.rse.2009.05.012.

Huang, N.E., Leitao, C.D., Parra, C.G., 1978. Large-scale Gulf stream frontal study using Geos 3 radar altimeter data. J. Geophys. Res. 83, 4673–4682.

Huang, H.P., Kaplan, A., Curchitser, E.N., Maximenko, N.A., 2007. The degree of anisotropy for mid-Ocean currents from satellite observations and an eddy permitting model simulation. J. Geophys. Res. 112. https://doi.org/10.1029/2007JC004105.

Huggett, W.S., Crawford, W.R., Thomson, R.E., Woodward, M.V., 1987. Data Record of Current Observations Volume XIX. Coastal Ocean Dynamics Experiment (CODE). Institute of Ocean Sciences, Part 1.

Hughes, T., Kerry, J., Álvarez-Noriega, M., et al., 2017. Global warming and recurrent mass bleaching of corals. Nature 543, 373–377. https://doi.org/10.1038/nature21707.

Hunter, J., 2012. A simple technique for estimating an allowance for uncertain sea-level rise. Clim. Change 113, 239–252.

Hydes, D.J., Hartman, M.C., Kaiser, J., Campbell, J.M., 2009. Measurement of dissolved oxygen using optodes in a Ferry Box system. Estuar. Coast. Shelf. 83, 485–490. https://doi.org/10.1016/j.ecss.2009.04.014.

Ichikawa, H., Sakajiri, H., Nakamura, H., Nishina, A., 2001. Year-to-year variation of sea surface salinity in the East China Sea. In: Proceedings, Extended Abstract Volume, The 11th PAMS/JECSS Workshop, April 11–13, 2001, Cheju, Korea, pp. 73–76.

Igouchi, T., Kozu, T., Meneghni, T., Awaka, J., Okamoto, Ken'ichi, 2000. Rain-profiling algorithm for the TRMM precipitation radar. J. Appl. Meteorol. 39, 2018–2051. https://doi.org/10.1175/1520-0450(2001)040<2038:RPAFTT>2.0.CO;2.

Iler, R.K., 1979. The Chemistry of Silica. John Wiley, New York.

Intergovernmental Oceanographic Commission, 2010. The International Thermodynamic Equation of Seawater. http://www.teos-10.org/pubs/TEOS-10_Manual.pdf.

IOC, SCOR, and IAPSO, 2010. The International Thermodynamic Equation of Seawater—2010: Calculation and Use of Thermodynamic Properties. Intergovernmental Oceanographic Commission, Manuals and Guides No. 56, UNESCO (English), Paris, p. 196. Available online at: http://www.TEOS-10.org/TEOS-10_Manual_21Apr10.pdf. (Accessed 26 May 2010).

IPCC, 2021. Climate change 2021: the physical science basis. In: Masson-Delmotte, V., Zhai, P., Pirani, A., Connors, S.L., Péan, C., Berger, S., Caud, N., Chen, Y., Goldfarb, L., Gomis, M.I., Huang, M., Leitzell, K., Lonnoy, E., Matthews, J.B.R., Maycock, T.K., Waterfield, T., Yelekçi, O., Yu, R., Zhou, B. (Eds.), Contribution of Working Group I to the Sixth Assessment Report of the Intergovernmental Panel on Climate Change. Cambridge University Press, Cambridge, UK and New York, NY, USA, p. 2391. https://doi.org/10.1017/9781009157896.

IPCC, 2007. Climate Change 2007. Synthesis Report. In: Pachauri, R.K., Reisinger, A. (Eds.), Contribution of Working Groups I, II and III to the Fourth Assessment Report of the Intergovernmental Panel on Climate Change. IPCC, Geneva, Switzerland, p. 104.

IPCC, 2013. Summary for Policymakers. In: Stocker, T.F., Qin, D., Plattner, G.-K., Tignor, M., Allen, S.K., Boschung, J., Nauels, A., Xia, Y., Bex, V., Midgley, P.M. (Eds.), Climate Change 2013: The Physical Science Basis. Contribution of Working Group I to the Fifth Assessment Report of the Intergovernmental Panel on Climate Change. Cambridge University Press, Cambridge, United Kingdom and New York, NY, USA.

IPCC, 2014. Climate change 2014: Synthesis report. In: Pachauri, R.K., Meyer, L.A. (Eds.), Contribution of Working Groups I, II and III to the Fifth Assessment Report of the Intergovernmental Panel on Climate Change. IPCC, Geneva, Switzerland, pp. 1–112. https://doi.org/10.1017/CBO9781107415324.

IPCC, 2022. Climate Change 2022: Impacts, Adaptation, and Vulnerability. In: Pörtner, H.-O., Roberts, D.C., Tignor, M., Poloczanska, E.S., Mintenbeck, K., Alegría, A., Craig, M., Langsdorf, S., Löschke, S., Möller, V., Okem, A., Rama, B. (Eds.), Contribution of Working Group II to the Sixth Assessment Report of the Intergovernmental Panel on Climate Change. Cambridge University Press, Cambridge, UK and New York, NY, USA, p. 3056. https://doi.org/10.1017/9781009325844.

Jackett, D.R., McDougal, T.J., 1997. A neutral density variable for the world's Oceans. J. Phys. Oceanogr. 27, 237–263. https://doi.org/10.1175/1520-0485(1997)027<0237:ANDVFT>2.0.CO;2.

Jackson, D.D., 1972. Interpretation of inaccurate, insufficient and inconsistent data. Geophys. J. R. Astron. Soc. 28, 97–109.

Jackson, D.R., Jones, C.D., Rona, P.A., Bemis, K.G., 2003. A method for Doppler acoustic measurement of black smoker flow fields. G-Cubed 4, 1–12. https://doi.org/10.1029/2003G000509, 1095.

Jacobsen, A.W., 1948. An instrument for recording continuously the salinity, temperature and depth of sea water. Trans. Am. Meteorol. Soc. 1057–1070.

Jakobsen, P.K., Ribergaard, M.H., Quadfasel, D., Schmith, T., Hughes, C.W., 2003. Near-surface circulation in the Northern North Atlantic as inferred from Lagrangian drifters: variability from the mesoscale to interannual. J. Geophys. Res. 108. https://doi.org/10.1029/2002JC001554.

James, R.W., Fox, P.T., 1972. Comparative Sea-Surface Temperature Measurements. Report No. 5, Report on Marine Science Affairs. WMO, Geneva, pp. 117–129.

James, T.S., Robin, C., Henton, J.A., Craymer, M., 2021. Relative Sea-Level Projections for Canada Based on the IPCC Fifth Assessment Report and the NAD83v70VG National Crustal Velocity Model. Geological Survey of Canada, Open File 8764, 1. zip file. https://doi.org/10.4095/327878.

Jamous, D., Mémery, L., Andrié, C., Jean-Baptiste, P., Merlivat, L., 1992. The distribution of helium 3 in the deep western and southern Indian Ocean. J. Geophys. Res. 97, 2243–2250.

Janniasch, H.W., Coletti, L.J., Johnson, K.S., Fitzwater, S.E., Needoba, J.A., Plant, J.N., 2008. The Land/Ocean biogeochemical observatory: a robust networked mooring system for continuously monitoring complex biogeochemical cycles in estuaries. Limnol Oceanogr. Methods 6, 263–276.

Jansen, E., Overpeck, J., Briffa, K.R., Duplessy, J.-C., Joos, F., Masson-Delmotte, V., Olago, D., Otto-Bliesner, B., Peltier, W.R., Rahmstorf, S., Ramesh, R., Raynaud, D., Rind, D., Solomina, O., Villalba, R., Zhang, D., 2007. Palaeoclimate. In: Solomon, S., Qin, D.,

Manning, M., Chen, Z., Marquis, M., Averyt, K.B., Tignor, M., Miller, H.L. (Eds.), Climate Change 2007: The Physical Science Basis. Contribution of Working Group I to the Fourth Assessment Report of the Intergovernmental Panel on Climate Change. Cambridge University Press, pp. 433–497.

Jenkins, W.J., 1988. The use of anthropogenic tritium and helium-3 to study subtropical gyre ventilation and circulation. Philos. Trans. R. Soc. London A325, 43–61.

Jenkins, G.M., Watts, D.G., 1968. Spectral Analysis and its Applications. Holden-Day, San Francisco, CA, USA.

Jenkins, F.A., White, H.E., 1957. Fundamentals of Optics. McGraw-Hill, New York.

Jenkins, W.J., Edmond, J.M., Corliss, J.B., 1978. Excess ^{3}He and ^{4}He in Galapagos submarine hydrothermal waters. Nature 272, 156–158.

Jerlov, N.G., 1976. Marine Optics. Elsevier, New York.

Johnson, G.C., McPhaden, M.J., Firing, E., 2001. Equatorial Pacific Ocean horizontal velocity, divergence, and upwelling. J. Phys. Oceanogr. 31, 839–849.

Johnson, K.S., Needoba, J.A., Riser, S.C., Showers, W.J., 2007. Chemical sensor networks for the aquatic environment. Chem. Rev. 107, 623–640.

Johnson, K.S., Riser, S.C., Karl, D.M., 2010. Nitrate supply from deep to near surface waters of the North Pacific subtropical gyre. Nat. Lett 465 (24), 1062–1065.

Johnston, D.T., Poulton, S.W., Fralick, P.W., Wing, B.A., Canfield, D.E., Farquar, J., 2006. Evolution of the Oceanic sulfur cycle at the end of the Paleoproterozoic. Geochim. Cosmochim. Acta. 70, 5723–5739.

Jones, P.D., 1988. Hemispheric surface air temperature variations: recent trends and an update to 1987. J. Clim. 1, 654.

Jones, P.D., Wigley, T.M.L., Wright, P.B., 1986. Global temperature variations between 1861 and 1984. Nature 322, 430–434.

Joos, F., Plattner, G.-K., Stocker, T., Wallace, D.W.R., 2003. Trends in marine dissolved oxygen: implications for Ocean circulation changes and the carbon budget. EOS Trans. Am. Geophys. Union 84. https://doi.org/10.1029/2003EO210001.

Joyce, R.J., Arkin, P.A., 1997. Improved estimates of tropical and subtropical precipitation using the GOES precipitation idex. J. Atmos. Ocean. Technol. 14, 997–1011.

Julian, P.R., 1975. Comments on the determination of significance levels of the coherence statistic. J. Atmos. Sci. 32, 836–837.

Kadko, D., Rosenburg, N.D., Lupton, J.E., Collier, R., Lilley, M., 1990. Chemical reaction rates and entrainment within the Endeavour Ridge hydrothermal plume. Earth Planet. Sci. Lett. 99, 315–335.

Kagan, Y.Y., 2002. Seismic moment distribution revisited: I. Statistical results. Geophys. J. Int. 148, 520–541.

Kaiser, J.F., 1966. Digital Filters. In: Kuo, F.F., Kaiser, J.F. (Eds.), Systems Analysis by Digital Computer. Wiley and Sons, New York. Chapter 7.

Kaiser, J.A.C., 1978. Heat balance of the upper Ocean under light winds. J. Phys. Oceanogr. 8, 1–12. https://doi.org/10.1175/1520-0485(1978).

Kalnay, E., Kanamitsu, M., Kistler, R., Collins, W., Deaven, D., Gandin, L., Iredell, M., Saha, S., White, G., Woollen, J., Zhu, Y., Chelliah, M., Ebisuzaki, W., Higgins, W., Janowiak, J., Mo, K.C., Ropelewski, C., Wang, J., Leetmaa, A., Reynolds, R., Jenne, R., Joseph, D., 1996. The NCEP/NCAR 40-year reanalysis project. Bull. Am. Meteorol. Soc. 77, 437–472.

Kanamitsu, M., Ebisuzaki, W., Woollen, J., Yang, S.-K., Hnilo, J.J., Fiorino, M., Potter, G.L., 2002. NCEP–DOE AMIP-II Reanalysis (R-

2). Bullet. Am. Meteorol. Soc. 83 (1), 1631–1644. https://doi.org/10.1175/BAMS-83-11-1631.

Kanasewich, E.R., 1975. Time Series Analysis in Geophysics. University of Alberta Press, Edmonton.

Kara, A.B., Rochford, P.A., Hurlburt, H.E., 2000a. An optimal definition for ocean mixed layer depth. J. Geophys. Res. 105, 16803–16821.

Kara, A.B., Rochford, P.A., Hurlburt, H.E., 2000b. Mixed layer depth variability and barrier layer formation over the North Pacific Ocean. J. Geophys. Res. 105, 16783–16801.

Kara, A.B., Wallcraft, A.J., Metzger, E.J., Hurlburt, H.E., Fairall, C.W., 2007. Wind stress drag coefficient over the global Ocean. J. Clim. 20, 5856–5864. https://doi.org/10.1175/2007JCLI1825.1.

Katsumata, K., Yoshinari, H., 2010. Uncertainties in global mapping of Argo drift data at the parking level. J. Oceanogr. 66, 553–569.

Kautsky, H., 1939. Quenching of luminescence by oxygen. Trans. Faraday Soc. 35, 216–219.

Kay, S.M., Marple Jr., S.L., 1981. Spectrum analysis-a modern perspective. Proc. IEEE 69, 1380–1417.

Kelley, D.S., Delaney, J.R., Juniper, S.K., 2014. Establishing a new era of submarine volcanic observatories: cabling axial seamount and the Endeavour segment of the Juan de Fuca Ridge. Mar. Geol. 50th Ann. Speci. 352, 426–450. https://doi.org/10.1016/j.margeo.2014.03.010.

Kelly, K.A., 1988. Comment on "Empirical orthogonal function analysis of advanced very high resolution radiometer surface temperature patterns in Santa Barbara Channel" by G.S.E. Lagerloef and R.L. Berstein. J. Geophys. Res. 93, 15753–15754.

Kelley, D.E., 1990. Fluxes through diffusive staircases: a new formulation. J. Geophys. Res. 95, 3365–3371.

Kelly, K.A., 1989. An inverse model for near-surface velocity from infrared images. J. Phys. Oceanogr. 19, 1845–1864. https://doi.org/10.1175/1520-0485(1989)019<1845:AIMFNS>2.0.CO;2.

Kelly, K.A., Strub, P.T., 1992. Comparison of velocity estimates from advanced very high resolution radiometer. J. Geophys. Res. 97, 9653–9668.

Kennedy, J.J., 2014. A review of uncertainty in *in situ* measurements and data sets of sea surface temperature. Rev. Geophys. 52, 1–32. https://doi.org/10.1002/2013RG000434.

Kennedy, J.J., Berry, D., 2008. Assessment of the Marine Observing System (ASMOS): Final Report. NOCS Research and Consultancy Report No. 32. NOC, Southampton, p. 55.

Kennedy, J.J., Smith, R., Rayner, N., 2011. Reassessing biases and other uncertainties in sea surface temperature observations since 1850 Part 1: measurement and sampling errors. J. Geophys. Res. 116. https://doi.org/10.1029/2010JD015218.

Keyte, F.K., 1965. On the formulae for correcting reversing thermometers. Deep-Sea Res. 12, 163–172.

Kiliszek, D., Kroszczyński, K., Araszkiewicz, A., 2022. Analysis of different weighting functions of observations for GPS and Galileo precise point positioning performance. Remote Sens 14 (9), 2223. https://doi.org/10.3390/rs14092223.

Kilpatrick, K.A., Podestá, G.P., Evans, R.H., 2001. Overview of the NOAA/NASA Pathfinder algorithm for sea surface temperature and associated matchup database. J. Geophys. Res. 106, 9179–9198. https://doi.org/10.1029/1999JC000065.

Kilpatrick, K.A., Podestá, G., Walsh, S., Williams, E., Halliwell, V., Szczodrak, M., Brown, O.B., Minnett, P.J., Evans, R., 2015. A decade of sea surface temperature from MODIS. Remote Sens. Environ. 165, 27–41.

Kilpatrick, K.A., Podestá, G., Williams, E., Walsh, S., Minnett, P.J., 2019. Alternating decision trees for cloud masking in MODIS and VIIRS NASA sea surface temperature products. J. Atmos. Ocean. Technol. 36, 387–407. https://doi.org/10.1175/jtech-d-18-0103.1.

Kim, K.Y., Chung, C., 2001. On the evolution of the annual cycle in the tropical Pacific. J. Clim. 14, 991–994. https://doi.org/10.1175/1520-0442(2001)014<0991:OTEOTA>2.0.CO;2.

Kim, K.Y., North, G.R., 1997. EOFs of harmonizable cyclostationary processes. J. Atmos. Sci. 54, 2416–2427.

Kim, K.Y., North, G.R., Huang, J., 1996. EOFs of one-dimensional cyclostationary time series: computation, examples and stochastic modeling. J. Atmos. Sci. 53, 1007–1017.

Kim, S.Y., Terrill, E.J., Cornuelle, B.D., Jones, B., Washburn, L., Moline, M.A., Paduan, J.D., Garfield, N., Largier, J.L., Crawford, G., Kosro, P.M., 2011. Mapping the U.S. west coast surface circulation: a multiyear analysis of high-frequency radar observations. J. Geophys. Res. 116, 2011–2026.

Kimoto, M., Yoshikawa, I., Ishii, M., 1997. An ocean data assimilation system for climate monitoring. J. Meteor. Soc. Japan 75, 471–487.

King, M.D., Platnick, S., Menzel, W.P., Ackerman, S.A., Hubanks, P.A., 2013. Spatial and temporal distribution of clouds observed by MODIS onboard the Terra and Aqua satellites. IEEE Trans. Geosci. Remote Sens. 51, 3826–3852. https://doi.org/10.1109/TGRS.2012.2227333.

Kipphut, G.W., 1990. Glacial meltwater input to the Alaska coastal current: evidence from oxygen isotope measurements. J. Geophys. Res. 95, 5177–5181.

Kirwan, A.D., Chang, M.S., 1976. On the micropolar Ekman problem. Int. J. Eng. Sci. 14, 685–692.

Kirwan, A.D., McNally, G., Chang, M.-S., Molinari, R., 1975. The effect of wind and surface currents on drifters. J. Phys. Oceanogr. 5, 361–368.

Kirwan, A.D., McNally, G.J., Reyna, E., Merrell, W.J., 1978. The near-surface circulation of the eastern North Pacific. J. Phys. Oceanogr. 8, 937–945.

Kirwan, A.D., McNally, G., Pazan, S., Wert, R., 1979. Analysis of surface current response to wind. J. Phys. Oceanogr. 9, 401–412.

Kistler, R., Kalnay, E., Collins, W., Saha, W., White, G., Woolen, J., Chellah, M., Ebisuzaki, W., Kanamitsu, M., Kousky, V., van den Dool, H., Jenne, R., Florino, M., 2001. The NCEP-NCAR 50-year reanalysis: monthly means CD-ROM and documentation. Bull. Am. Meteorol. Soc. 82, 247–268.

Kizu, S., Sukigara1, C., Hanawa, K., 2010. Comparison of the fall rate and structure of recent T-7 XBT manufactured by Sippican and TSK. Ocean Sci. Discuss 7, 1811–1847. https://doi.org/10.5194/osd-7-1811-2010. www.ocean-sci-discuss.net/7/1811/2010/.

Kizu, S., Sukigara, C., Hanawa, K., 2020. Comparison of the fall rate and structure of recent T-7 XBT manufactured by Sippican and TSK. Ocean Sci. Discuss 7, 1811–1947.

Klemas, V., 2011. Remote sensing techniques for studying coastal ecosystems: an overview. Coast. Res. 27, 2–17. https://doi.org/10.2112/JCOASTRES-D-10-00103.1.

Klymak, J.M., Legg, S., Alford, M.H., Buijsman, M., Pinkel, R., Nash, J.D., 2012. The direct breaking of internal waves at steep topography. Oceanography 25, 150–159. https://www.jstor.org/stable/24861352.

Knudsen, M. (Ed.), 1901. Hydrographical Tables. GEC, Copenhagen.

Kohonen, T., 1982. Self-organized information of topologically correct feature maps. Bio. Cyber. 43, 59–69.

Kohonen, T., 2001. Self-organizing maps, third ed. Springer, Information Sciences, Berlin. 30, 501 p.

Kohno, N., Dube, S.K., Entel, M., Fakhruddin, S.H.M., Greenslade, D., Leroux, M.-D., Rhome, J., Thuy, N.B., 2018. Recent progress in storm surge forecasting. Trop. Cyclone Res. Rev. 7 (2), 128–139. https://doi.org/10.6057/2018TCRR02.04.

Konyaev, K.V., 1990. Spectral Analysis of Physical Oceanographic Data. National Science Foundation, Washington, DC.

Koopmans, L.H., 1974. The Spectral Analysis of Time Series. Academic Press, New York.

Koracin, D., Dorman, C., 2001. Marine atmospheric boundary layer divergence and clouds along California in June 1996. Mon. Weather Rev. 129, 2040–2055.

Krauss, W., 1993. Ekman drift in homogenous water. J. Geophys. Res. 98, 20187–20209.

Krauss, W., Kase, R.H., 1984. Mean circulation and eddy kinetic energy in the eastern North Atlantic. J. Geophys. Res. 84, 3407–3415.

Kremling, K., 1972. Comparison of specific gravity in natural sea-water from hydrographical tables and measurements by a new density instrument. Deep-Sea Res. 19, 377–383.

Kuhn, H., Quadfasel, D., Schott, F., Zenk, W., 1980. On simultaneous measurements with rotor, wing and acoustic current meters, moored in shallow water. Dtsch. Hydrogr. Z. 33, 1–18.

Kulikov, E.A., Gonzalez, F.I., 1995. On Reconstruction of the Initial Tsunami Signal from Distant Bottom Pressure Records. Doklady Akademii Nauk.

Kulikov, E.A., González, F.I., 1996. Recovery of the shape of a tsunami signal at the source from measurements of oscillations in the ocean level by a remote hydrostatic pressure sensor. Trans. (Doklady) Russian Acad. Sci., Earth Sci. Sect. 345A, 585.

Kulikov, E.A., González, F.I., 1997. Reconstruction of the initial tsunami signal from distant bottom pressure records. Dokl. Akad. Nauk 344 (6), 814–815 (in Russian).

Kummerow, C., Hong, Y., Olson, W.S., Yang, S., Adler, R.F., Mccollum, J., Ferraro, R., Petty, G., Shin, D.B., Wilheit, T.T., 2001. The evolution of the Goddard Profiling Algorithm (GPROF) for rainfall estimate from passive microwave sensors. J. Appl. Meteorol. 40, 1801–1820. https://doi.org/10.1175/1520-0450(2001)040<1801:TEOTGP>2.0.CO;2.

Kundu, P.K., 1976. An analysis of inertial oscillations observed near the Oregon coast. J. Phys. Oceanogr. 6, 879–893.

Kundu, P.K., 1990. Fluid Mechanics. Academic Press, San Diego, CA, p. 628.

Kundu, P.K., Allen, J.S., 1976. Some three-dimensional characteristics of low-frequency current fluctuations near the Oregon coast. J. Phys. Oceanogr. 6, 181–199.

Kundu, P.K., Allen, J.S., Smith, R.L., 1975. Modal decomposition of the velocity field near the Oregon coast. J. Phys. Oceanogr. 5, 683–704.

Kurihara, Y., Murakami, H., Kachi, M., 2016. Sea surface temperature from the new Japanese geostationary meteorological Himawari-8 satellite. Geophys. Res Lett. 43, 1234–1240.

Kyselý, J., Picek, J., Beranová, R., 2010. Estimating extremes in climate change simulations using the peaks-over-threshold method with a non-stationary threshold. Global Planet. Change 72 (1–2), 55–68. https://doi.org/10.1016/j.gloplacha.2010.03.006.

Labrecque, A.M., Thomson, R.E., Stacey, M.W., Buckley, J.R., 1994. Residual currents in Juan de Fuca Strait. Atmosphere-Ocean 32, 375–394.

Lacoss, R.T., 1971. Data adaptive spectral analysis methods. Geophysics 36, 661–675.

Ladd, C., Bond, N.A., 2002. Evaluation of the NCEP/NCAR reanalysis in the NE Pacific and the Bering Sea. J. Geophys. Res. 107. https://doi.org/10.1029/2001JC001157.

Ladd, C., Stabeno, P.J., 2012. Stratification on the eastern Bering Sea shelf revisited. Deep-Sea Res. Part II 65 (70), 72–83. https://doi.org/10.1016/j.dsr2.2012.02.009.

LaFond, E.C., 1951. Processing Oceanographic Data, HO Publication No. 614. US Hydrographic Office.

Lanczos, C., 1956. Applied Analysis. Prentice-Hall, Englewood Cliffs, NJ (Reprinted in 1988, Dover, New York).

Lane, R.W.P.M., Manuels, M.W., Staal, W., 1984. A procedure for enriching and cleaning up rhodamine B and rhodamine WT in natural waters, using a SEP-PAK C18 cartridge. Water Res. 18, 163.

Langley, R.B., 2021. Navigator Notes: Editorial Highlights from the Editor-in-Chief. Navigation 68 (1), 1–235. https://doi.org/10.1002/navi.414.

Langdon, C., 1984. Dissolved oxygen monitoring system using a pulsed electrode: design, performance and evaluation. Deep-Sea Res. 31, 1357–1367.

Large, W.C., McWilliams, J.C., Doney, S.C., 1994. Oceanic vertical mixing: A review and a model with nonlocal boundary layer parameterization. Rev, Geophys. 32, 363–403.

Laurinda, L.C., Mariano, A.J., Lumpkin, R., 2017. An improved near-surface velocity climatology for the global Ocean from drifter observations. Deep-Sea Res. Part 1 124, 73–92. https://doi.org/10.1016/j.dsr.2017.04.009.

Laxon, S., McAdoo, D., 1994. Arctic Ocean gravity field derived from ERS-1 satellite altimetry. Science 265, 621–625.

Layton, C., Greenan, B.J.W., Hebert, D.E., Kelley, D., 2018. Low-frequency oceanographic variability near Flemish Cap and Sackville Spur. J. Geophys. Res. Oceans 123, 1814–1826. https://doi.org/10.1002/2017JC013289.

L'Ecuyer, T.S., Jiang, J.H., 2010. Touring the atmosphere aboard the A-Train. Phys. Today 63, 36–41.

Le Menn, M., Poli, P., David, A., Sagot, J., Lucas, M., O'Carroll, A., Belbeoch, M., Herklotz, K., 2019. Development of surface drifting buoys for fiducial reference measurements of sea-surface temperature. Front. Mar. Sci. 13. https://doi.org/10.3389/fmars.2019.00578.

LeBlond, P.H., Mysak, L.A., 1978. Waves in the Ocean. Elsevier Sci. Pub. Co., p. 602

LeBlond, P.H., Mysak, L.A., 1979. Waves in the Ocean. Elsevier, Amsterdam, p. 602.

Ledwell, J.R., St. Laurent, L.C., Girton, J.B., Toole, J.M., 2011. Diapycnal mixing in the Antarctic Circumpolar Current. J. Phys. Oceanogr. 41 (1), 241–246. https://doi.org/10.1175/2010JPO4557.1.

Lee, Z.P., Carder, K.L., Mobley, C.D., Steward, R.G., Patch, J.S., 1998. Hyperspectral remote sensing for shallow water: 1. A semi-analytical model. Appl. Opt. 37, 6329–6338.

Lee, Z.P., Carder, K.L., Mobley, C.D., Steward, R.G., Patch, J.S., 1999. Hyperspectral remote sensing for shallow water: 2. Deriving bottom depth and water properties by optimization. Appl. Opt. 38, 3831–3843.

Lee, J.A., Garcia, C.A., Larkin, A.A., Carter, B.R., Martiny, C.A., 2021. Linking a latitudinal gradient in Ocean hydrography and elemental stoichiometry in the Eastern Pacific Ocean. Glob. Biogeochem. Cycles 35. https://doi.org/10.1029/2020GB006622 e2020GB006622.

Legeckis, R., 1975. Application synchronous meteorological satellite data to the study of time dependent sea surface temperature changes along the boundary of the Gulf Steam. Geophys. Res. Lett. 2, 435−438. https://doi.org/10.1029/GL002i010p00435.

Lemon, D.D., Farmer, D.M., April 3−5, 1990. Experience with a multi-depth scintillation flowmeter in the Fraser Estuary. In: Proceedings IEEE Fourth Working Conference on Current Measurement, Clinton, MD, pp. 290−298.

Lemon, D.D., Thomson, R.E., Delaney, J.R., Farmer, D.M., Rowe, F., Chave, R.A.J., 1996. Acoustic Scintillation Velocity Measurement of a Buoyant Hydrothermal Plume. Preliminary Report. ASL Environmental Science, Sidney, BC, Canada.

Lenn, Y.D., Chereskin, T.K., 2009. Observations of Ekman currents in the Southern Ocean. J. Phys. Oceanogr. 39, 768−779.

Lerner, R., Hollinger, J., 1977. Analysis of 1.4 GHz radiometric measurements from Skylab. Remote Sens. Environ. https://doi.org/10.1016/0034-4257(77)90047-5.

Levitus, S., 1982. Climatological Atlas of the World Ocean. NOAA Professional Paper 13. US Department of Commerce, Rockville, MD, USA.

Lewis, E.L., 1980. The practical salinity scale 1978 and its antecedents. J. Ocean. Eng. 5, 3−18.

Lewis, E.L., Perkin, R.G., 1978. Salinity: its definition and calculation. J. Geophys. Res. 83, 466. https://doi.org/10.1029/JC083iC01p00466.

Lewis, E.L., Perkin, R.G., 1981. The practical salinity scale 1978: conversion of existing data. Deep-Sea Res. 28A, 307−328.

Li, W.K.W., McLaughlin, F.A., Lovejoy, C., Carmack, E.C., 2009. Smallest algae thrive as the Arctic Ocean freshens. Science 326. https://doi.org/10.1126/science.1179798.

Li, J., Carlson, B.E., Yung, Y.L., Lv, D., Hansen, J., Penner, J.E., Liao, H., Ramaswamy, V., Kahn, R.A., Zhang, P., Dubovik, O., Ding, A., Lacis, A.A., Zhang, L., Dong, Y., 2022. Scattering and absorbing aerosols in the climate system. Nat. Rev. Earth Environ. 3, 363−379. https://doi.org/10.1038/s43017-022-00296-7.

Lilley, M.D., Feely, R.A., Trefry, J.H., 1995. Chemical and biochemical transformations in hydrothermal plumes. In: Humphris, S.E., Zierenberg, R.A., Mullineaux, L.S., Thomson, R.E. (Eds.), Seafloor Hydrothermal Systems: Physical, Chemical, Biological and Geological Interactions. AGU, Geophysical Monograph 91, pp. 369−391.

Lindsey, R., 2022. Climate Change: Global Sea Level. NOAA Climate.gov report.

Lipps, F.B., Hemler, R.S., 1992. On the downward transfer of tritium to the Ocean by a cloud model. J. Geophys. Res. 97, 12889−12900.

Liu, Y., Minnett, P.J., 2016. Sampling errors in satellite-derived infrared sea-surface temperatures. Part I: global and regional MODIS fields. Remote Sens. Environ. 177, 48−64. https://doi.org/10.1016/j.rse.2016.02.026.

Liu, M., Tanhua, T., 2021. Water masses in the Atlantic Ocean: characteristics and distributions. Ocean Sci. (Europ. Geosci. Union) 17, 463−486. https://doi.org/10.5194/os-17-463-2021, 2021.

Liu, Y., Weisberg, R.H., 2005. Patterns of ocean current variability on the West Florida Shelf using the self-organizing map. J. Geophys. Res. 110. https://doi.org/10.1029/2004JC002786.

Liu, Y., Weisberg, R.H., 2006. Sea surface temperature patterns on the West Florida Shelf using the growing hierarchical self-organizing maps. J. Atm. Ocean Tech. 2, 325−328.

Liu, Y., Weisberg, R.H., Mooers, C.N.K., 2006. Performance evaluation of the self-organizing map for feature extraction. J. Atm. Ocean Tech. 111. https://doi.org/10.1029/2005JC003117.

Liu, C., Zipser, E.J., Nesbitt, S.W., 2007. Warm rain in the tropics: seasonal and regional distribution base on 9 years of TRMM data. J. Clim. 22, 767−779.

Liu, M., Liu, X., Liu, D., Ding, C., 2015. Multivariable integration method for estimating sea surface salinity in coast waters from in situ data and remotely sensed data. Comput. Geosci. 75. https://doi.org/10.1016/j.cageo.2014.10.016.

Liu, Y., Chin, T.M., Minnett, P.J., 2017. Sampling errors in satellite-derived infrared sea- surface temperatures. Part II: sensitivity and parameterization. Remote Sens. Environ. 198, 297−309. https://doi.org/10.1016/j.rse.2017.06.011.

Livingstone, D., Royer, T.C., 1980. Eddy propagation determined from rotary spectra. Deep-Sea Res. 27A, 835−883.

Llewellyn-Jones, D.T., Minett, P.J., Saunders, R.W., Zavody, A.M., 1984. Satellite multichannel infrared measurements of sea surface temperature of the Northeast Atlantic Ocean using AVHRR/2. Q. J. R. Meteorol. Soc. 110, 613−631.

Loeve, M., 1978. Probability Theory II (46). Graduate texts in mathematics, 0-387.

Loges, W., 1984. Girsnov's theorem in Hilbert space and an application to the statistics of Hilbert space valued stochastic differential equations. Stoch. Proc. Appl. 17, 243−263.

Lohrmann, A., October 15, 2001. Monitoring Sediment Concentration with Acoustic Backscattering Instruments. Nortek Tech. Note No. 003.

Lomb, N.R., 1976. Least-squares frequency-analysis of unequally spaced data. Astrophys. Space Sci. 39, 447−462.

Lorenz, E., 1956. Empirical Orthogonal Functions and Statistical Weather Prediction. Scientific Report No. 1. Air Force Cambridge Research Center, Air Research and Development Command, Cambridge, MA.

LPD, 2021. Living Planet Database. www.livingplanetindex.org.

Lueck, R.G., 1990. Thermal inertia of conductivity cells: theory. J. Atmos. Ocean. Technol. 7, 741−755.

Lueck, R.G., Hertzman, O., Osborn, T.R., 1977. The spectral response of thermistors. Deep-Sea Res. 24, 951−970.

Lukas, R., Lindstrom, E., 1991. The mixed layer of the western equatorial Pacific Ocean. J. Geophys. Res. https://doi.org/10.1029/90JC01951.

Lukas, R., 1994. HOT results show interannual variability of Pacific Deep and Bottom waters. WOCE Note. 6 (2), 4.

Lumpkin, R., 2003. Decomposition of surface drifter observations in the Atlantic Ocean. Geophys. Res. Lett. 30. https://doi.org/10.1029/2003GL017519.

Lumpkin, R., Garraffo, Z., 2005. Evaluating the decomposition of tropical Atlantic drifter observations. J. Atmos. Ocean. Technol. 22, 1403−1415. https://doi.org/10.1175/JTECH1793.1.

Lumpkin, R., Johnson, G.C., 2013. Global Ocean surface velocities from drifters: mean, variance, El Nino-Southern Oscillation response, and seasonal cycle. J. Geophys. Res. 118, 2992−3006.

Lumpkin, R., Pazos, M., 2005. Measuring Surface Currents with Surface Velocity Program Drifters: The Instrument, Its Data, and Some Recent Results. NOAA, AOML.

Lumpkin, R., Pazos, M., 2006. Chapter 2: Measuring surface currents with surface velocity program drifters: the instrument, its data, and some recent results. In: Griffa, A., Kirwan, A.D., Mariano, A.J., Ozgokmen, T., Rossby, T. (Eds.), Lagrangian Analysis and Prediction of Coastal and Ocean Dynamics (LAPCOD). Cambridge University Press, pp. 39−67. https://doi.org/10.1017/CBO9780511535901.003.

Lumpkin, R., Treguier, A.M., Speer, K., 2002. Lagrangian eddy scales in the Northern North Atlantic Ocean. J. Phys. Oceanogr. 32, 2425–2440.

Lund, I., 1963. Map-pattern classification by statistical methods. J. Appl. Meteor. 2, 56–65.

Lupton, J.E., Craig, H., 1981. A major helium-3 source at 15°S on the East Pacific Rise. Science 214, 13–18.

Lupton, J.E., Delaney, J.R., Johnson, H.P., Tivey, M.K., 1985. Entrainment and vertical transport of deep-Ocean water by buoyant hydrothermal plumes. Nature 316, 621–623.

Lupton, J.E., Baker, E.T., Massoth, G.J., 1989. Variable ^{3}He/heat ratios in submarine hydrothermal systems: evidence from two plumes over the Juan de Fuca Ridge. Nature 337, 161–164.

Lupton, J.E., Baker, E.T., Mottl, M.J., Sansone, F.J., Wheat, C.G., Resing, J.A., Massoth, G.J., Measures, C.I., Feely, R.A., 1993. Chemical and physical diversity of hydrothermal plumes along the East Pacific Rise, 8°45′N to 11°50′N. Geophys. Res. Lett. 20, 2913–2916.

Luther, D.S., 1982. Evidence of a 4-6 day barotropic, planetary oscillation of the Pacific Ocean. J. Phys. Oceanogr. 12, 644–657.

Lütkepohl, H., 1985. Comparison criteria for estimating the order of a vector autoregressive process. J. Time Ser. Anal. 6, 35–52.

Lynn, R.J., Reid, J.L., 1968. Characteristics and circulation of deep and abyssal waters. Deep-Sea Res. 15, 577–598.

Lynn, R.J., Svejkovsky, J., 1984. Remotely sensed sea surface temperature variability off California during a "Santa Ana" clearing. J. Geophys. Res. 89, 8151–8162.

Macdonald, R.W., McLaughlin, F.A., Wong, C.S., 1986. The storage of reactive silicate samples by freezing. Limnol. Oceanogr. 31, 1139–1142.

MacDonald, A., Scarrott, C.J., Lee, D., Darlow, B., Reale, M., Russell, G., 2011. A flexible extreme value mixture model. Comput. Statistics Data Anal. 55 (6), 2137–2157. https://doi.org/10.1016/j.csda.2011.01.005.

Mackas, D.L., Denman, K.L., Bennett, A.F., 1987. Least squares multiple tracer analysis of water mass composition. J. Geophys. Res. 92, 2907–2918.

Mackas, D.L., Denman, K.L., Bennett, A.F., 1987. Least squares multiple tracer analysis of water mass composition. J. Geophys. Res. https://doi.org/10.1029/JC092iC03p02907.

Mackenzie, K.V., 1981. Nine term equation for sound speed in the Oceans. J. Acoust. Soc. Am. 70, 807–812.

Macklin, S.A., Stabeno, P.J., Schumacher, J.D., 1993. A comparison of gradient and observed over-the-water winds along a mountainous coast. J. Geophys. Res. 98, 16555–16569.

MacPhee, S.B., 1976. Acoustics and Echo Sounding Instrumentation. Canadian Hydrographic Service Technical Report 76–1.

Makarynskyy, O., Makarynska, D., Kuhn, M., Feathersone, W., 2004. Predicting sea level variations with artificial neural networks at Hillarys Boat Harbour, Western Australia, Estuarine. Coastal Shelf Sci. 61 (2), 351–360. https://doi.org/10.1016/j.ecss.2004.06.004.

Malanotte-Rizzoli, P., 1996. Modern Approaches to Data Assimilation in Ocean Modeling, , first ed.61, ISBN 9780080536668.

Mandelbrot, B.B., 1967. How long is the coast of Britain? Statistical self-similarity and fractional dimension. Science 155, 636–638.

Manley, J., Willcox, S., 2010. The wave glider: a new concept for deploying ocean instrumentation. IEEE Instrum. Meas. Mag. 13 (6), 8–13. https://doi.org/10.1109/MIM.2010.5669607.

Mann, M.E., Bradley, R.S., Hughes, M.K., 1998. Global-scale temperature patterns and climate forcing over the past six centuries. Nature 392, 779–787.

Mantyla, A.W., Reid, J.L., 1983. Abyssal characteristics of the World Ocean waters. Deep-Sea Res. 30 (8A), 805–833.

Marbouti, M., Antropov, O., Praks, J., Eriksson, P.B., Arabzadeh, V., Rinne, E., Leppäranta, M., 2020. TanDEM-X multiparametric data features in sea ice classification over the Baltic Sea. Geo-spatial Inform. Sci. 24 (2), 313–332. https://doi.org/10.1080/10095020.2020.1845574.

Marion, G., Millero, F.J., Feistel, R., 2009. Precipitation of solid phase calcium carbonates and their effect on applications of seawater SA-T-P models. Ocean Sci. 5, 285–291.

Marks, K.M., McAdoo, D.C., Smith, W.H.F., 1993. Mapping the southeast Indian ridge with geosat. EOS 74 (8), 81–86.

Marple, S.L., 1987. Digital Spectral Analysis. Prentice-Hall, Englewood Cliffs, NJ.

Marra, G., Fairweather, D.M., Kamalov, B., Gaynor, P., Cantono, M., Mulholland, S., Baptie, B., Castellanos, J.C., Vagenas, G., Gaudron, J.O., Kronjäger, J., Hill, I.R., Schioppo, M., Barbeito Edreira, I., Burrows, K.A., Clivati, C., Caonico, D., Curtis, A., 2022. Optical interferometry-based array of seafloor environmental sensors using a transOceanic submarine cable. Science 376, 874–879. https://doi.org/10.1126/science.abo193.

Marsden, R.F., 1987. A comparison between geostrophic and directly measured surface winds over the Northeast Pacific Ocean. Atmosphere-Ocean 25, 387–401.

Masson, D., 2006. Seasonal water mass analysis for the Straits of Juan de Fuca and Georgia. Atm. Ocean 44, 1–15.

Martin, J.H., Knauer, G.A., 1973. The elemental composition of plankton. Geochem. Cosmochim. Acta 37, 1639–1653.

Martin, M., Talley, L.D., de Szoeke, R.A., May, 1987. Physical, Chemical and CTD Data from Marathon Expedition R/V Thomas Washington 261, 4 May–4 June 1984. Oregon State University, College of Oceanography, Data Report 131, Ref. 87–15.

Martini, K.I., Alford, M.H., Nash, J.D., Kunze, E.L., Merrifield, M.A., 2007. Diagnosing a partly standing internal wave in Mamala Bay, Oahu. Geophys. Res. Lett. 34. https://doi.org/10.1029/2007GL029749.

Marullo, S., Minnett, P.J., Santoleri, R., Tonani, M., 2016. The diurnal cycle of sea-surface temperature and estimation of the heat budget of the Mediterranean Sea. J. Geophys. Res. Ocean. 121. https://doi.org/10.1002/2016jc012192.

Masson, D., 1996a. A case study of wave-current interaction in a strong tidal current. J. Phys. Oceanogr. 26, 359–372.

Masson, D.G., 1996b. Catastrophic collapse of the volcanic island of Hierro 15 ka ago and the history of landslides in the Canary Islands. Geology 24. https://doi.org/10.1130/0091-7613(1996)024<0231:CCOTVI>2.3.CO;2.

Maturi, E., Harris, A., Merchant, C., Mittaz, J., Potash, B., Meng, W., Sapper, J., 2008. NOAA's sea surface temperature products from operational geostationary satellites. Bull. Am. Meteorol. Soc. https://doi.org/10.1175/2008BAMS2528.1.

Maul, G., Bravo, N.J., 1983. Fitting of satellite and in-situ Ocean surface temperatures: results for polymode during the winter of 1977–1978. J. Geophys. Res. 88, 9605–9616.

Maxwell, A.E., 1977. Multivariate Analysis in Behavioral Research. Chapman and Hall, London, p. 164.

Mazzotti, S., Jones, C., Thomson, R.E., 2008. Relative and absolute sea level rise in western Canada and Northwestern United States from a combined tide gauge-GPS analysis. J. Geophys. Res. 113, C11019. https://doi.org/10.1029/2008JC004835.

McBride, R.A., Byrnes, M.R., Hiland, M.W., 1995. Geomorphic response-type model for barrier coastlines: a regional perspective. Mar. Geolog. 126, 143−159.

McClain, E.P., 1981. Multiple atmosphere-window techniques for satellite sea surface temperatures. In: Gower, J.F.R. (Ed.), Oceanography from Space. Plenum, New York, pp. 73−85.

McClain, E.P., Pichel, W.G., Walton, C.C., Ahmad, Z., Sutton, J., 1983. Multi-channel improvements to satellite-derived global sea surface temperatures. Adv. Space Res. 2, 43−47.

McClain, E.P., Pichel, W.G., Walton, C.C., 1985. Comparative performance of AVHRR- based multichannel sea surface temperatures. J. Geophys. Res. 90 (6), 11609−11618. https://doi.org/10.1029/JC090iC06p11609.

McClain, C.R., Yoder, J.A., Atkinson, L.P., Blanton, J.O., Lee, T.N., Singer, J.J., Müller-Karger, F., 1988. Variability of surface pigment concentration in the South Atlantic Bight. J. Geophys. Res. 93, 10675−10697.

McDougall, T.J., 1985a. Double-diffusive interleaving/Part 1: linear stability analysis. J. Phys. Oceanogr. 15, 1532−1541.

McDougall, T.J., 1985b. Double-diffusive interleaving/Part 2: finite amplitude, steady state interleaving. J. Phys. Oceanogr. 15, 1542−1556.

McDougall, T.J., 1987. Neutral surfaces in the Ocean: implications for modeling. Geophys. Res. Lett. 14, 797−800. https://doi.org/10.1029/GL014i008p00797.

McDougall, T.J., Barker, P.M., 2011. Thermodynamic Equation of Seawater TEOS-10. https://www.teos-10.org/index.htm.

McDougall, T.J., Feistel, R., Millero, F.J., Jackett, D.R., Wright, D.G., King, B.A., M Marion, G., Chen, C.T.A., Spitzer, P., 2009. Calculation of the Thermodynamic Properties of Seawater. Global Ship-Based Repeat Hydrography Manual, IOCCP Report No. 14, ICPO Publication Series No. 134, p. 112. Online available at: www.marine.dsiro.au/∼jackett/TEOS-10/.

McDougall, T.J., Jackett, D.R., Millero, F.J., Pawlowicz, R., Barker, P.M., 2012. A global algorithm for estimating absolute salinity. Ocean Sci. 8, 1123−1134. https://doi.org/10.5194/os-8-1123-2012.

McDuff, R.E., 1988. Effects of vent fluid properties on the hydrography of hydrothermal plumes. EOS Trans. AGU 69, 1497.

McDuff, R.E., 1995. Physical dynamics of deep-sea hydrothermal plumes. In: Humphris, S.E., Zierenberg, R.A., Mullineaux, L.S., Thomson, R.E. (Eds.), Seafloor Hydrothermal Systems: Physical, Chemical, Biological, and Geological Interactions. American Geophysical Union, Geophysical Monograph 91, Washington, DC, pp. 357−368.

McLane Research Laboratories, Inc., 2022. 121 Bernard St Jean Dr., E. Falmouth, Massachusetts 02536, USA.

McManus, J., Collier, R.W., Chen, C.-T.A., Dymond, J., 1992. Physical properties of Crater Lake, Oregon: a method for the determination of conductivity- and temperature-dependent expression of salinity. Limnol. Oceanogr. 37, 41−53.

McMillin, L.M., 1975. Estimation of sea surface temperatures from two infrared window measurements with different absorption. J. Geophys. Res. 80, 5113−5117. https://doi.org/10.1029/JC080i036p05113.

McMillin, L.M., Crosby, D.S., 1984. Theory and validation of the multiple window sea surface temperature technique. J. Geophys. Res. 89 (C3), 3655−3661.

McNally, G.J., 1981. Satellite-tracked drift buoy observations of the near-surface flow in the eastern mid-latitude North Pacific. J. Geophys. Res. 86, 8022−8030.

McNally, G.J., White, W.B., 1985. Wind driven flow in the mixed layer observed by drifting buoys during autumn-winter in midlatitude North Pacific. J. Phys. Oceanogr. 15, 684−694.

McNally, G.J., Patzert, W.C., Kirwan, A.D., Vastano, A.C., 1983. The near-surface circulation of the North Pacific using satellite tracked drifting buoys. J. Geophys. Res. 88, 7507−7518.

McWilliams, J.C., 1976. Maps from the mid-Ocean dynamics experiment: part I. Geostrophic streamfunction. J. Phys. Oceanogr. 6, 810−827.

Meckler, A.N., Sigman, D.M., Gibson, K.A., Francois, R., Martinez-Garcia, A., Jaccard, S.L., Röhl, U., Peterson, L.C., Tiedemann, R., Haug, G.H., 2013. Deglacial pulses of deep-Ocean silicate into the subtropical North Atlantic Ocean. Nature 495, 495−498.

Medvedev, I.P., Vilibić, I., Rabinovich, A.B., 2020. Tidal resonance in the Adriatic Sea: observational evidence. J. Geophys. Res. Ocean. 125. https://doi.org/10.1029/2020JC016168.

Meinen, C.S., Watts, D.R., 2000. Vertical structure and transport on a transect across the North Atlantic Current near 42°N: time series and mean. J. Geophys. Res. 105. https://doi.org/10.1029/2000JC900097.

Meinen, C.S., Garzoli, S.L., Johns, W.E., Baringer, M.O., 2004. Transport variability of the deep western boundary current and the Antilles current off Abaco Island, Bahamas. Deep Sea Res.-I 51, 1397−1415. https://doi.org/10.1016/j.dsr.2004.07.007.

Meinig, C., Stalin, S., Nakamura, A.I., Gonsalez, F., Milburn, H.B., 2005. Technology developments in realtime tsunami measuring, monitoring and forecasting. Proc. MTS/IEEE. https://doi.org/10.1109/OCEANS.2005.1639996.

Meisel, D.D., 1978. Fourier transforms of data sampled at unequaled observational intervals. Astron. J. 83, 538−545.

Meisel, D.D., 1979. Fourier transforms of data sampled in unequally spaced segments. Astron. J. 84, 116−126.

Meissner, T., Wentz, F.J., 2012. The emissivity of the Ocean surface between 6 and 90 GHz over a large range of wind speeds and Earth incidence angles. Trans. Geosci. Remote Sens. 50, 3004−3016.

Memery, L., Wunsch, C., 1990. Constraining the North Atlantic circulation with tritium data. J. Geophys. Res. 95, 5239−5256.

Méndez, F.J., Menéndez, M., Luceño, A., Losada, I.J., 2006. Estimation of the long-term variability of extreme significant wave height using a time-dependent Peak Over Threshold (POT) model. J. Geophys. Res. 111, C07024. https://doi.org/10.1029/2005JC003344.

Menéndez, M., L Woodworth, P., 2010. Changes in extreme high water levels based on a quasi-global tide-gauge data set. J. Geophys. Res. 115, c10011. https://doi.org/10.1029/2009jc005997.

Merchant, C.J., Harris, A.R., Maturi, E., MacCallum, S., 2005. Probabilistic physically based cloud screening of satellite infrared imagery for operational sea surface temperature retrieval. Q. J. R. Meteorol. Soc. 131, 2735−2755. https://doi.org/10.1256/qj.05.15.

Merchant, C.J., Embury, O., Rayner, N.A., Berry, D.I., Corlett, G.K., Lean, K., Veal, K.L., Kent, E.C., Llewellyn-Jones, D.T., Remedios, J.J., Saunders, R., 2012. A 20 year in- dependent record of sea surface temperature for climate from along track scanning radiometers. J. Geophys. Res. 117. https://doi.org/10.1029/2012JC008400.

Mero, T.M., 1982. Performance results for the EG7G vector-measuring current meter (VMCM). In: Dursi, M., Woodward, W. (Eds.), Proceedings of IEEE Second Working Conference on Current

Measurement. Institute of Electrical and Electronics Engineers, New York, pp. 159—164.

Merrifield, M.A., Guza, R.T., 1990. Detecting propagating signals with complex empirical orthogonal functions: a cautionary note. J. Phys. Oceanogr. 20, 1628—1633.

Meyers, S.D., Kelly, B.G., O'Brien, J.J., 1993. An introduction to wavelet analysis in Oceanography and meteorology: with application to the dispersion of Yanai waves. Mon.Weather Rev. 121, 2858—2866.

Michaelsen, J., 1987. Cross-validation in statistical climate forecast models. J. Appl. Meteorol. Climatol. 26, 1589—1600.

Middleton, J.H., 1982. Outer rotary cross spectra, coherences and phases. Deep-Sea Res. 29 (10A), 1267—1269.

Middleton, J.H., 1983. Low-frequency trapped waves on a wide, reef-fringed continental shelf. J. Phys. Oceanogr. 13, 1371—1382.

Middleton, J.F., 1985. Drifter spectra and diffusivities. J. Mar. Res. 43, 37—55.

Middleton, J.H., Cunningham, A., 1984. Wind-forced continental shelf waves from geographical origin. Cont. Shelf Res. 3, 215—232.

Miller, L., Cheney, R., 1990. Large-scale meridional transport in the tropical Pacific Ocean during the 1986—1987 El Niño from Geosat. J. Geophys. Res. 95, 17905—17919.

Millero, F.J., 1979. The thermodynamics of the carbonate system in seawater. Geochim. Cosmochim. Acta 41, 1651—1661.

Millero, F.J., 2010. History of the equation of state of seawater. Oceanography 23, 18—33. https://doi.org/10.5670/oceanog.2010.21.

Millero, F.J., Kremling, K., 1976. The densities of Baltic Sea waters. Deep-Sea Res. 23 (1), 1129—1138.

Millero, F.J., Feistel, R., Wright, D.G., McDougall, T.J., 2008. The composition of standard seawater and the definition of the reference-composition salinity scale. Deep Sea Res. Part 1 Oceanogr. Res. Pap. 55 (1), 50—72. https://doi.org/10.1016/j.dsr.2007.10.001.

Minami, T., Toh, H., Ichihara, H., Kawashima, I., 2017. Three-dimensonal time domain simulation of Tsunami-generated electromagnectic fields: application to the 2011 Tohoku Earthquake Tsunami. J. Geophys. Res. 122, 9559—9579. https://doi.org/10.1002/2017J B014839.

Minnett, P.J., 1991. Consequences of sea surface temperature variability on the validation and applications of satellite measurements. J. Geophys. Res. 96, 18409—18475. https://doi.org/10.1029/91JC01816.

Minnett, P.J., Kaiser-Weiss, A.K., 2012. Group for High Resolution Sea-Surface Temperature Discussion Document: Near-Surface Oceanic Temperature Gradients, p. 7. Available at. https://www.ghrsst.org/wp-content/uploads/2016/10/SSTDefinitionsDiscussion.pdf.

Minnett, P.J., Alvera-Azcarate, A., Chiin, T.M., Corlett, G.K., Gentemann, C.L., Karagali, I., Li, X., Marsouin, A., Marullo, S., Maturi, E., Santoleri, R., Saux Picart, S., Steele, M., Vazquez-Cuervo, J., 2019. Hal a century of satellite remote sensing of sea-surface temperature. Remote Sens. Environ. 233. https://doi.org/10.1016/j.rse.2019.111366.

Mitrovica, J.X., Milne, G.A., Davis, J.L., 2001. Glacial isostatic adjustment on a rotating earth. Geophys. J. Int. 147, 562—578. https://doi.org/10.1046/j.1365-246x.2001.01550.x.

Miyake, Y., Saruhashi, K., 1967. A study on the dissolved oxygen in the Ocean. In: Geochemistry Study of the Ocean and the Atmosphere, Yasuo Miyake Seventieth Anniversary. Geochemical Laboratory, Meteorological Research Institute, Tokyo, 1978, pp. 91—98.

Miyakoda, K., Rosati, A., 1982. The variation of sea surface temperature in 1976 and 1977: the data analysis. J. Geophys. Res. 87, 5667—5680.

Mofjeld, H.O., Whitmore, P.M., Elbé, M.C., González, F.I., Newman, J.C., 2001. Seismic- wave contributions to bottom pressure fluctuations in the North Pacific—implications for the DART tsunami array. Proc. Int. Tsunami Symp. 97—108, 2001, Seattle, WA, CD.

Mofjeld, H.O., 2009. Tsunami measurements. In: Bernard, E.N., Robinson, A.R. (Eds.), The Sea, vol 15. Harvard University Press, Cambridge, pp. 201—235.

Monserrat, S., Vilibic, I., Rabinovich, A.B., 2006. Meteotsunamis: atmospherically induced destructive Ocean waves in the tsunami frequency band. Nat. Hazards Earth Sys. Sci. 6, 1035—10551.

Montgomery, R.B., 1938. Circulation in the upper layer of the southern North Atlantic deduced with the aid of isentropic analysis. Pap. Phys. Oceanogr. Meteorol 6 (2), 55.

Montgomery, R.B., 1958. Water characteristics of Atlantic Ocean and of World Ocean. Deep-Sea Res. 5, 134—148.

Mooers, C.N.K., 1973. A technique for the cross-spectrum analysis of pairs of complex-valued time series, with emphasis on properties of polarized components and rotational invariants. Deep-Sea Res. 20, 1129—1141.

Mooers, C.N.K., Smith, R.L., 1967. Continental shelf waves off Oregon. J. Geophys. Res. 73, 549—557.

Mosquera-Vásquez, K., Dewitte, B., Illig, S., Takahashi, K., Garric, G., 2013. The 2002/2003 El Niño: equatorial waves sequence and their impact on sea surface temperature. J. Geophys. Res. Oceans 118, 346—357. https://doi.org/10.1029/2012JC008551.

Moura, M., Veleda, D., Lins, I.D., Droguett, E.L., Araújo, M., 2010. Sea level prediction by support vector machines combined with particle swarm optimization. Conference paper. https://www.researchgate.net/publication/253795544.

Mu, M., Li, Y., Bi, S., Lyu, H., Xu, J., Lei, S., Miao, S., Zeng, S., Zheng, Z., Du, C., 2021. Prediction of algal bloom occurrence based on the naive Bayesian model considering satellite image pixel differences. Ecol. Indicat. 124. https://doi.org/10.1016/j.ecolind.2021.107416.

Muench, R.D., Schumacher, J.D., 1979. Some Observations of Physical Oceanographic Conditions on the Northeast Gulf of Alaska Continental Shelf. NOAA Technical Memorandum ERL PMEL-17, Seattle, WA.

Muis, S., Verlaan, M., Winsemius, H.C., et al., 2016. A global reanalysis of storm surges and extreme sea levels. Nat. Commun. 7 (1), 1—12.

Mungov, G., Eble, M., Bouchard, R., 2013. DART tsunameter retrospective and real-time data: a reflection on 10 years of processing in support of tsunami research and operations. Pure Appl. Geophys. 170, 1369—1384. https://doi.org/10.1007/s00024-012-0477-5.

Munk, W.H., Cartwright, D.E., 1966. Tidal spectroscopy and prediction. Philos. Trans. R. Soc. London A259, 533—581.

Munk, W.H., Hasselman, C., 1964. Upper resolution of tides. In: Studies in Oceanography. Tokyo Geophysical Institute, University of Tokyo, pp. 339—344.

Murty, T.S., 1984. Storm surges: meteorological Ocean tides. Can. Bull. Fish. Aquat. Sci. 212, 897.

Mysak, L.A., LeBlond, P.H., Emery, W.J., 1979. Trench waves. J Phys. Oceanogr. 9, 1001−1013. https://doi.org/10.1175/1520-0485(1979) 009<1001:TW>2.0.CO;2.

Nadstrom, G.D., Gage, K.S., Jasperson, W.H., 1984. Kinetic energy spectrum of large- and mesoscale processes. Nature 310, 36−38.

Nagano, A., Hasegawa, T., Ariyoshi, K., Matsumoto, H., 2022. Interannual bottom-intensified current thickening observed on the continental slope off the southeastern coast of Hokkaido, Japan. Fluids 7. https://doi.org/10.3390/fluids7020084.

NAG Library, 1986. https://nag.com/nag-library/.

Nastrom, G.D., Jasperson, W.H., 1984. Kinetic energy spectrum of large- and mesoscale processes. Nature 310, 36−38.

National Research Council, 2012. Sea-Level Rise for the Coasts of California, Oregon, and Washington: Past, Present, and Future. The National Academies Press, Washington, DC. https://doi.org/10.17226/13389.

Needham, H.F., Keim, B.D., Sathiaraj, D., 2015. A review of tropical cyclone-generated storm surges: global data sources, observations, and impacts. Rev. Geophys. 53, 545−591.

Nemac, A.F.L., Brinkhurst, R.O., 1988. Using the bootstrap to assess statistical significance in the cluster analysis of species abundance data. Can. J. Fish. Aquat. Sci. 45, 965−970.

Nicholoson, D.P., Emerson, S., Eriksen, C.C., 2008. Net community production in the deep euphotic zone of the subtropical North Pacific gyre from glider surveys. Limnol. Oceanogr. 53, 2226−2236.

Nielsen-Englyst, P., Hoyer, J.L., Pedersen, L.T., Gentemann, C.L., Alerskans, E., Block, T., Donlon, C., 2018. Optimal estimation of sea surface temperature from AMSR-E. Remote Sens 10. https://doi.org/10.3390/rs10020229.

Niiler, P., 2001. The world Ocean surface circulation. In: Church, J., Siedler, G., Gould, J. (Eds.), Ocean Circulation and Climate- Observing and Modeling the Global Ocean. Academic, London, pp. 193−204.

Niiler, P.P., Paduan, J.D., 1995. Wind-driven motions in the Northeast Pacific as measured by Lagrangian drifters. J. Phys. Oceanogr. 25, 2819−2830.

Niiler, P.P., Davis, R.E., White, H.J., 1987. Water-following characteristics of a mixed layer drifter. Deep-Sea Res. 34, 1867−1882.

Niiler, P.P., Maximenko, N.A., Panteleev, G.G., Yamagata, T., Olson, D.B., 2003. Near- surface dynamical structure of the Kuroshio Extension. J. Geophys. Res. 108 (C6), 3193. https://doi.org/10.1029/2002JC001461.

Niiler, P.P., Maximenko, N.A., McWilliams, J.C., 2004. Dynamically balanced absolute sea level of the global Ocean derived from near-surface velocity observations. Geophys. Res. Lett. 30, 2164. https://doi.org/10.1029/2003GL018628.

Ninnis, R.N., Emery, W.J., Collins, M.J., 1986. Automated extraction of sea ice motion from AVHRR imagery. J. Geophys. Res. 91, 10725−10734.

Nooner, S.L., Chadwick, W.W., 2016. Inflation-predictable behavior and co-eruption deformation at axial seamount. Science 354 (6318), 1399−1403. https://doi.org/10.1126/science.aah4666.

North American Datum, 1927, 1983. National Geodetic Survey, National Oceanic and Atmospheric Administration (NOAA), Washington, DC.

Nowlin, W.D., Bottero, J.S., Pillsbury, R.D., 1986. Observations of internal and near-inertial oscillations at Drake Passage. J. Phys. Oceanogr. 16, 87−108.

Nowlin, W.D., Clifford, M., 1982. The kinematical and thermohaline zonation of the Antarctic Circumpolar Current at Drake Passage. J. Marine Res. 40, 481−507. https://elischolar.library.yale.edu/journal-of-marine research 1653.

Núñez-Riboni, I., Chelton, D.B., Valentina, M., 2023. The spectral color of natural and anthropogenic time series and its impact on the statistical significance of cross correlation. Sci. Total Environ. 860, 160219. https://doi.org/10.1016/j.scitotenv.2022.160219.

Nuttall, A.H., 1976. Spectral Analysis of a Univariate Process with Bad Data Points, via Maximum Entropy, and Linear Predictive Techniques. Naval Underwater systems Center, Technical Document 5419, New London, CT.

Nuttall, A.H., Carter, G.C., 1980. A generalized framework for power spectral estimation. IEEE Trans. Acoust. Speech Signal Process, ASSP 28, 334−335.

Obeysekera, J., Park, J., Irizarry-Ortiz, M., Barnes, J., Trimble, P., 2013. Probabilistic projection of mean sea level and coastal extremes. J. Waterw. Port Coast. Ocean Eng. 139 (2), 135−141. https://doi.org/10.1061/(ASCE)WW.1943-5460.0000154.

Odelson, B.J., Rajamani, M.R., Rawlings, J.B., 2006. A new autocovariance least squares method for estimating noise covariance. Automatica 42 (2), 303−308. https://doi.org/10.1016/j.automatica.2005.09.006.

Olbers, D.J., Müller, P., Willebrand, J., 1976. Inverse technique analysis of large data sets. Phys. Earth Planet. Inter. 12, 248−252.

Olbers, D.J., Wenzel, M., Willebrand, J., 1985. The inference of North Atlantic circulation patterns from climatological hydrographic data. Rev. Geophys. 23, 313−356.

Oldenburg, D.W., 1984. An introduction to linear inverse theory. IEEE Geosci. Remote Sens. GW-22, 665−674.

Oliver, M.J., Irwin, A.J., 2008. Objective global ocean biogeographic provinces. Geophys. Res. Lett. 35. https://doi.org/10.1029/2008GL034238.

Omura, H., 1969. Black Right Whales in the North Pacific, Scientific Reports of the Whales Research Institute, No. 21. Whales Research Institute, Tokyo, p. 78.

Osborne, A.R., Kirwan, A.D., Provenzale, A., Bergamasco, L., 1989. Fractal drifter trajectories in the Kuroshio extension. Tellus, 41A, 416−435. https://doi.org/10.3402/tellusa.v41i5.11850.

Östlund, H.G., Rooth, C.H., 1990. The North Atlantic tritium and radiocarbon transients 1972−1983. J. Geophys. Res. 95, 20147−20165.

Ostrovskii, A.G., Piterbarg, L.I., 1985. Diagnosis of the seasonal variability of water surface temperature anomalies in the Noth Pacific. Meteorol. Hydrol. 51−57 (in Russian).

Otnes, R.K., Enochson, L., 1972. Digital Time Series Analysis. John Wiley, New York.

Overland, J.E., Pease, C.H., 1988. Modeling ice dynamics of coastal seas. J. Geophys. Res. 93 (C12), 15,619−15,637.

Overland, J.E., Percival, D.B., Mofjeld, H.O., 2006. Regime shifts and red noise in the North Pacific. Deep Sea Res. Pt. I. Oceanogr. Res. Pap. 53 (4), 582−588.

OWID, 2021. Data from "Our World in Data". https://ourworldindata.org.

Paduan, J., Washburn, L., 2013. High-frequency radar observations of Ocean surface currents. Ann. Rev. Mar. Sci. https://doi.org/10.1146/annurev-marine-121211-172315.

Pagano, M., 1978. Some recent advances in autoregressive processes. In: Bril-linger, D.R., Tiao, G.C. (Eds.), Directions in Time Series. Institute for Mechanical Statistics.

Papadakis, J.E., 1981. Determination of the wind mixed layer by an extension of Newton's method. Pac. Mar. Sci. Rep. 81-9, 32. Institute of Ocean Sciences, Sidney, BC. Canada.

Papadakis, J.E., 1985. On a class of form oscillators. Speculations Sci. Technol. 8, 291−303.

Paros, J.M., 1976. Digital pressure transducers. Meas. Data 10 (2), 74−79.

Parsons, T.R., Maita, Y., Lalli, C.M., 1984. A Manual of Chemical and Biological Methods for Seawater Analysis. Pergamon Press, Oxford, UK.

Parsons, R.J., Breitbart, M., Lomas, M.W., Carlson, C.A., 2012. Ocean time-series reveals recurring seasonal patterns of virioplankton dynamics in the Northwestern Sargasso Sea. ISME 6, 273−284.

Patil, K., Deo, M.C., 2018. Basin-scale prediction of sea surface temperature with artificial neural networks. J. Atmos. Oceanic Technol. 35, 1441−1454. https://doi.org/10.1175/JTECH-D-17-0217.1.

Patterson, D., Macdonald, J., Skibo, K., Barnes, D., Guthrie, I., Hills, J., 2007. Reconstructing the summer thermal history for the lower Fraser Rivers, 1941 to 2006, and implication for adult sockeye salmon spawning. Fish. Oceans Can. 2724, 1−5.

Patterson, R.T., Chang, A.S., Prokoph, A., Roe, H.M., Swindles, G.T., 2013. Influence of the Pacific Decadal Oscillation, El Niño-Southern Oscillation and solar forcing on climate and primary productivity changes in the Northeast Pacific. Quat. Int. 310, 124−139.

Paulson, C.A., Simpson, J.J., 1981. The temperature difference across the cool skin of the Ocean. J. Geophys. Res. 86, 11044−11054.

Pavlidis, T., Horowitz, S.L., 1974. Segmentation of plan curves. IEEE Trans. Comput. C-23, 860−870.

Pawlowicz, R., 2010. A model for predicting changes in the electrical conductivity, preactical salinity, and absolute salinity of seawater due to variations in relative chemical composition. Ocean Sci. 6, 361−378. https://doi.org/10.5194/os-6-361-2010.

Pawlowicz, R., 2013. Key physical variables in the Ocean: temperature, salinity, and density. Nat. Educ. Knowl. 4 (4), 13.

Pawlowicz, R., Beardsley, B., Lentz, S., 2002. Classical tidal harmonic analysis including error estimates in MATLAB using T_TIDE. Comput. Geosci. 28, 929−937.

Pawlowicz, R., Riche, O., Halverson, M., 2007. The circulation and residence time of the Strait of Georgia using a simple mixing-box approach. Atmos.-Ocean 45, 173−193.

Pawlowicz, R., Wright, D.G., Millero, F.J., 2011. The effects of biogeochemical processes on Oceanic conductivity/salinity/density relationships and the characterization of real seawater. Ocean Sci. 7, 363−387. https://doi.org/10.5194/os-7-363-2011.

Pawlowicz, R., McDougall, T.J., Feistel, R., Tailleux, R., 2012. An historical perspective on the development of the Thermodynamic Equation of Seawater − 2010. Ocean Sci. 8, 161−174. https://doi.org/10.5194/os-8-161-2012.

Pazan, S.E., Niiler, P.P., 2001. Recovery of near-surface velocity from undrogued drifters. J. Atmos. Ocean. Technol. 18, 476−489. https://doi.org/10.1175/1520-0426(2001)018<0476:RONSVF>2.0.CO;2.

Pearson, K., 1901. On lines and planes of closest fit to systems of points in space. Philos. Mag. 2, 559−572. https://doi.org/10.1080/14786440109462720.

Pearson, C.A., Schumacher, J.D., Muench, R.D., 1981. Effects of wave-iduced mooring noise on tidal and low-frequency current observations. Deep-Sea Res. 28A, 1223−1229.

Pearson, J., Fox-Kemper, B., Barkan, R., Choi, J., Bracco, A., McWilliams, J.C., 2018. Impacts of convergence on Lagrangian statistics in the Gulf of Mexico. J. Phys. Oceanogr. https://doi.org/10.1175/JPO-D-17-0036.1.

Peimani, M., Kharestani, N., Khalilabadi, M.R., 2022. Simulation of the effect of the Holey-Sock Drogue on the drifter performance. J. Oceanogr. 13 (50), 107−116.

Peltier, W.R., 1990. Glacial isostatic adjustment and relative sea-level change. In: Sea-level Change. National Academy Press, Washington, DC, pp. 73−87.

Peltier, W.R., 2004. Global glacial isostasy and the surface of the ice-age Earth: the ICE-5G (VM2) model and GRACE. Ann. Rev. Earth Planet. Sci. 32, 111−149.

Peltier, W.R., 2009. Closure of the budget of global sea level rise over the GRACE era: the importance and magnitudes of the required corrections for global glacial isostatic adjustment. Quat. Sci. Rev. 1−17.

Peng, T.-H., Broecker, W.S., Mathieu, G.G., Li, Y.-H., 1979. Radon invasion rates in the Atlantic and Pacific Oceans as determined during the geosecs program. J. Geophys. Res. 84, 2471−2486.

Peng, S., Qian, Y.-K., Lumpkin, R., Du, Y., Wang, D., Li, P., 2015. Characteristics of the near- surface currents in the Indian Ocean and deduced from satellite-tracked surface drifters. Part 1: Pseudo-Eulerian statistics. J. Phys. Oceanogr. 45. https://doi.org/10.1175/JPO-D-14-0050.1.

Perez, R.C., Hormann, V., Lumpkin, R., Brandt, P., Johns, W.E., Hernandez, F., Schmid, C., Bourlès, B., 2014. Mean meridional currents in the central and eastern equatorial Atlantic. Clim. Dyn. 43, 2943−2962. https://doi.org/10.1007/s00382-013-1968-5.

Permanent Service for Mean Sea Level, 2021. National Oceanography Centre, Joseph Proudman Building. Liverpool, United Kingdom.

Peteherych, S., Appleby, W.S., Woiceshyn, P.M., Spagnol, J.C., Chu, L., 1988. Application of seasat scatterometer wind measurements for operational short-range weather forecasting. Weather Forecast. 3, 89−103.

Petereit, J., Saynisch-Wagner, J., Morschhauser, A., Pick, L., Thomas, M., 2022. On the characterization of tidal Ocean-dynamo signals in coastal magnetic observatories. Earth Planet. Space 74. https://doi.org/10.1186/s40623-022-01610-9.

Peters, H., Gregg, M.C., Toole, J.M., 1988. On the parameterization of equatorial turbulence. J. Geophys. Res. 93, 1199−1218.

Peters, H., Gregg, M.C., Toole, J.M., 1989. Meridional variability of turbulence through the equatorial undercurrent. J. Geophys. Res. 94, 18003−18009.

Peterson, J.I., Fitzgerald, R.V., Buckhold, D.K., 1984. Fiber-optic probe for in vivo measurement of oxygen partial pressure. Anal. Chem. 56, 62−67.

Peterson, W., Robert, M., Bond, N., 2015. The warm Blob continues to dominate the ecosystem of the Northern California Current. North Pac. Mar. Sci. Organ. PICES Press 23, 44−46.

Pettigrew, N.R., Irish, J.D., September, 1983. An evaluation of a bottom mounted Doppler acoustic profiling current meter. In: Proceedings Oceans 83.

Pettigrew, N.R., Beardsley, R.C., Irish, J.D., 1986. Field evaluations of a bottom mounted acoustic Doppler current profiler and conventional current meter moorings. In: Proceedings of the IEEE Third Working Conference on Current Measurement, January 22−24, 1986, Airlie, VA, pp. 153−162.

Pettigrew, N.R., Wallinga, J.P., Neville, F.P., Schlenker, K.R., 2005. Gulf of Maine Ocean observing system: current measurement approaches in a prototype integrated Ocean observing system. In: IEEE Proceedings, Eighth Current Measurement Technology Conference 2005, pp. 127–131.

Pfeffer, W.T., Harper, J.T., O'Neel, S., 2008. Kinematic constraints on glacier contributions to 21st-century sea-level rise. Science 321 (5894), 1340–1343. https://doi.org/10.1126/science.115909.

Pham, D.T., Verron, J., Roubaud, M.C., 1998. A singular evolutive extended Kalman filter for data assimilation in oceanography. J. Mar. Sci. 16, 323–340.

Phillips, O.M., 1966. The Dynamics of the Upper Ocean. Cambridge University Press., p. 270

Phillips, O.M., Gu, D., Donelan, M., 1993. Expected structure of extreme waves in a Gaussian sea. Part I: theory and SWADE buoy measurements. J. Phys. Oceanogr. 23, 992–1000.

Pickard, G.L., Emery, W.J., 1982. Descriptive Physical Oceanography, fourth ed. Pergamon Press. https://doi.org/10.1016/C2013-0-10174-2.

Pickard, G.L., Emery, W.J., 1992. Descriptive Physical Oceanography: An Introduction, fifth ed. Pergamon Press, New York.

Pierce, S.D., Barth, J.A., Thomas, R.E., Fleischer, G.W., 2006. Anomalously warm July 2005 in the northern California Current: historical context and the significance of cumulative wind stress. Geophys. Res. Lett. 33 (22).

Pierson, W.J., 1981. The variability of winds over the Ocean. In: Beal, R., DeLeonibus, P.S., Katz, I. (Eds.), Space-borne Synthetic Aperature Radar for Oceanography, Johns Hopkins Oceanographic Studies, vol 7. Johns Hopkins University Press, Baltimore, MD.

Pillsbury, R.D., Bottero, J.S., Still, R.E., Gilbert, W.E., 1974. A Compilation of Observations from Moored Current Meters. vols. VI and VII. Refs 74–2 and 74–77, School of Oceanography, Oregon State University, Corvallis, OR.

Piola, A.R., Gordon, A.L., 1984. Pacific and Indian Ocean upper-layer salinity budget. J. Phys. Oceanogr. 14, 747–753.

Pitcher, T.J., Clark, M.R., Morato, T., Watson, R., 2010. Seamount fisheries. Oceanography 23, 134–144.

Piterbarg, L.I., Rozovskii, B., 1996. Maximum likelihood estimators in the equations of physical oceanography. In: Conference on Stochastic Modeling in Physical Oceanography. https://doi.org/10.1007/978-1-4612-2430-3_15.

Pizarro, O., Shaffer, G., 1998. Wind-driven, coastal-trapped waves off the Island of Gotland, Baltic Sea. J. Phys. Oceanogr. 28, 2117–2129.

Plaut, G., Vautard, R., 1994. Spells of low-frequency oscillations and weather regimes in the Northern Hemisphere. J. Atm. Sci. 51, 210–236.

Poli, P., Lucas, M., O'Carroll, A., Le Menn, M., David, A., Corlett, G.K., Blouch, P., Meldrum, D., Merchant, C.J., Belbeoch, M., Herklotz, K., 2019. The Copernicus Surface Velocity Platform drifter with Barometer and Reference Sensor for Temperature (SVP-BRST): genesis, design, and initial results. Ocean Sci. 15, 199–214. https://doi.org/10.5194/os-2018-109.

Polster, A., Fabian, M., Villinger, H., 2009. Effective resolution and drift of Paroscientific pressure sensors derived from long-term seafloor measurements. Geochem. Geophys. Geosyst. 10, Q08008. https://doi.org/10.1029/2009GC002532.

Polzin, K.L., Kunze, E., Hummon, J., Firing, E., 2002. The fine-scale response of lowered ADCP velocity profiles. J. Ocean. Atmos. Technol. 19, 205–224.

Polzin, K.L., Wang, B., Wang, Z., Thwaites, F., Williams, A.J., 2021. Moored flux and dissipation estimates from the Northern Deepwater Gulf of Mexico. Fluids 6, 237. https://doi.org/10.3390/fluids6070237.

Pond, S., Pickard, G., 1983. Introductory Dynamical Oceanography, second ed. Pergamon Press, New York, p. 329.

Ponte, R.M., Schindelegger, M., 2022. Global Ocean response to the 5-day Rossby-Haurwitz atmospheric mode seen by GRACE. J. Geophys. Res. Ocean. 127. https://doi.org/10.1029/2021JC018302.

Pope, P.A., Emery, W.J., 1994. Sea surface velocities from visible and infrared multispectral atmospheric mapping sensor (MAMS) imagery. IEEE Trans. Geosci. Remote Sens. 32, 220–222.

Poulain, P.-M., Niiler, P.P., 1989. Statistical analysis of the surface circulation in the California current system using satellite-tracked drifters. J. Phys. Oceanogr. 19, 1588–1603.

Poulain, P.-M., Gerin, R., Mauri, E., Pennel, R., 2009. Wind effects on drogued and undrogued drifters in. the eastern Mediterranean. J. Atmos. Ocean. Technol. 26, 1144–1156.

Pratt, J.H., 1859. See Vogt and Jung (1991).

Pratt, J.H., 1871. See Vogt and Jung (1991).

Preisendorfer, R.W., 1988. Principal Component Analysis in Meteorology and Oceanography. Elsevier Science Ltd, p. 444.

Press, W.H., Teukolsky, S.A., Vetterling, W.T., Flannery, B.P., 1992. Numerical Recipes in Fortran, second ed. Cambridge University Press, Cambridge, p. 963.

Price, J.F., Weller, R.A., Pinkel, R., 1986. Diurnal cycling: observations and models of the upper ocean response to diurnal heating, cooling, and wind-mixing. J. Geophys. Res. 91, 8411–8427.

Priestley, M.B., 1981. Spectral Analysis and Time Series. Academic Press, London.

Privalsky, V.E., Jensen, D.T., 1993. Time Series Analysis Package. Utah Climate Center, Logan.

Privalsky, V.E., Jensen, D.T., 1994. Assessment of the influence of ENSO on annual global air temperatures. Dyn. Atmos. Ocean. 22, 161–178.

Pugh, D., Woodworth, P.L., 2014. Sea-Level Science: Understanding Tides, Surges, Tsunamis and Mean Sea-Level Changes. Cambridge University Press, Cambridge, p. 395.

Pugh, D.T., 1987. Tides, Surges and Mean Sea Level—A Handbook for Engineers and Scientists. John Wiley & Sons, Chichester, UK, p. 486.

Pyper, B.J., Peterman, R.M., 1998a. Comparison of methods to account for autocorrelation in correlation analyses of fish data. Can. J. Fish. Aquat. Sci. 55, 2127–2140.

Pyper, B.J., Peterman, R.M., 1998b. Erratum: Comparison of methods to account for autocorrelation in correlation analyses of fish data. Can. J. Fish. Aquat. Sci. 55, 2710.

Qazi, W.A., Emery, W.J., Fox-Kemperm, B., 2013. Computing Ocean surface currents over the coastal California current system using 30-min-lag sequential SAR images. Trans. Geosci. Remote Sens. 52, 7559–7580.

Quadfasel, D., Schott, F., 1979. Comparison of different methods of current measurements. Dt. Hydrogr. Z. 32, 27–38.

Quay, P.D., Stuiver, M., Broecker, W.S., 1983. Upwelling rates for the equatorial Pacific Ocean derived from the bomb 14C distribution. J. Mar. Res. 41, 769–792.

Quinn, W.H., Neal, V.T., Antunez de Mayolo, S., 1987. El Niño occurrences over the past four and a half centuries. J. Geophys. Res. 92, 14449–14461.

Rabiner, L., Gold, B., 1975. Theory and Application of Digital Signal Processing. Prentice-Hall, Englewood Cliffs, NJ.

Rabinovich, A.B., Eblé, M.C., 2015. Deep-Ocean measurements of tsunami waves. Pure Appl. Geophys. 172, 3281–3312.

Rabinovich, A.B., Levyant, A.S., 1992. Influence of seiche oscillations on the formation of the long-wave spectrum near the coast of the southern Kuriles. Oceanology 32, 17–23.

Rabinovich, A.B., Thomson, R.E., 2001. Evidence of diurnal shelf waves in satellite- tracked drifter trajectories off the Kuril Islands. J. Phys. Oceanogr. 31 (9), 2650–2668.

Rabinovich, A.B., Thomson, R.E., Bograd, S.J., 2002. Drifter observations of anticyclonic eddies off Bussol' Strait, Kuril Islands. J. Oceanogr. 58, 661–671.

Rabinovich, A.B., Thomson, R.E., Stephenson, F.E., 2006. The Sumatra tsunami of 26 December 2004 as observed in the North Pacific and North Atlantic oceans. Surv. Geophys. 27, 647–677. https://doi.org/10.1007/s10712-006-9000-9.

Rabinovich, A.B., Shevchenko, G.V., Thomson, R.E., 2007. Sea ice and current response to the wind: a vector regressional analysis approach. J. Atmos. Oceanic Technol. 24, 1086–1101. https://doi.org/10.1175/JTECH2015.1.

Rabinovich, A.B., Stroker, K., Thomson, R.E., Davis, E.E., 2011. DARTs and CORK in Cascadia Basin: high-resolution observations of the 2004 Sumatra tsunami in the Northeast Pacific. Geophys. Res. Lett. 38. https://doi.org/10.1029/2011GL047026.

Rabinovich, A.B., Thomson, R.E., Fine, I.V., 2012. The 2010 Chilean Tsunami off the West Coast of Canada and the Pacific Northwest coast of the United States. Pure Appl. Geophys. https://doi.org/10.1007/s00024-012-0541-1. Springer Basel AG.

Rabinovich, A.B., Thomson, R.E., Fine, I.V., 2013. The 2010 Chilean tsunami off the west coast of Canada and the Northwest coast of the United States. Pure Appl. Geophys. 170. https://doi.org/10.1007/s00024-012-0541-1.

Rabinovich, A.B., Titov, V.V., Moore, C.W., Eblé, M.C., 2017. The 2004 Sumatra tsunami in the southeastern Pacific Ocean: new global insight from observations and modeling. J Geophys. Res. Ocean. 122, 7992–8019. https://doi.org/10.1002/2017JC013078.

Rabinovich, A.B., Thomson, R.E., Krassovski, M.V., Stephenson, F.E., Sinnott, D.C., 2019. Five great tsunamis of the 20th century as recorded on the coast of British Columbia. Pure Appl. Geophys. 176, 2887–2924. https://doi.org/10.1007/s00024-019-02133-3.

Rabinovich, A.B., Šepić, J., Thomson, R.E., 2023. Strength in numbers: The tail end of typhoon Songda combines with local cyclones to generate extreme sea level oscillations on the British Columbia and Washington coasts during Mid-October 2016. J. Phys. Oceanogr. 53 (1), 131–155. https://doi.org/10.1175/JPO-D-22-0096.1.

Rabinovich, A.B., 2009. Seiches and harbor oscillations. In: Kim, Y.C. (Ed.), Handbook of Coastal and Ocean Engineering. World Scientific Publ., Singapore, pp. 193–236.

Ralph, E., Niiler, P.P., 1999. Wind-driven currents in the tropical Pacific. J. Phys. Oceanogr. 29, 2121–2129.

Ramos, P.L., Louzada, F., Ramos, E., Dey, S., 2019. The Fréchet distribution: estimation and application – an overview. J. Stat. Manag. Syst. https://doi.org/10.1080/09720510.2019.1645400.

Rao, P.K., Smith, W.L., Koffler, R., 1972. Global sea surface temperature distribution determined from an environmental satellite. Mon. Weather Rev. 100, 10–14.

Ray, R.D., 1998. Ocean self-attraction and loading in numerical tidal models. Mar. Geod. 21, 181–192.

Rayner, N.A., Brohan, P., Parker, D.E., Folland, C.K., Kennedy, J.J., Vanicek, M., Ansell, T.J., Tett, S.F.B., 2006. Improved analyses of changes and uncertainties in sea surface temperature measured in situ since the mid-nineteenth century: the HadSST2 dataset. J. Clim. 19, 446–469.

RD Instruments, 1989. (See also Gordon, R.L.) Acoustic Doppler Current Profilers. Principles of Operation: A Practical Primer. RD Instruments, San Diego, CA.

Redfield, A.C., 1958. The biological control of chemical factors in the environment. Am. Sci. 46, 205–221.

Redfield, A.C., Ketchum, B.H., Richards, F.A., 1963. The influence of organisms on the composition of sea-water. In: Hill, M.N. (Ed.), The Sea, vol 2. Interscience, New York, pp. 26–77.

Reichl, B.G., Adcroft, A., Griffies, S.M., Hallberg, R., 2022. A potential energy analysis of Ocean surface mixed layers. J. Geophys. Res. Ocean. 127, e2021JC018140. https://doi.org/10.1029/2021JC018140.

Reid, J.L., 1965. Intermediate Waters of the Pacific Ocean. Johns Hopkins Oceanographic Studies, No. 2.

Reid, J.L., 1982. Evidence of an effect of heat flux from the East Pacific Rise upon the characteristics of the mid-depth waters. Geophys. Res. Lett. 9, 381–384.

Reid, J.L., 1994. On the total geostrophic circulation of the North Atlantic Ocean: flow patterns, tracers and transports. Prog. Oceanogr. 33, 1–92.

Reid, J.L., Lynn, R.J., 1971. On the influence of the Norwegian-Greenland and Weddell Seas upon the bottom waters of the Indian and Pacific Oceans. Deep-Sea Res. 18, 1063–1088.

Reid, J.L., Mantyla, A.W., 1978. On the mid-depth circulation of the North Pacific Ocean. J. Phys. Oceanogr. 8, 946–951.

Reseghetti, F., Cheng, L., Borghini, M., M Yashayaev, I., Raiteri, G., Zhu, J., 2018. Assessment of quality and reliability of measurements with XBT Sippican T5 and T5/20. J. Atmos. Ocean. Technol. 35. https://doi.org/10.1175/JTECH-D-18-0043.1.

Reul, N., Saux-Picart, S., Chapron, B., Vandermark, D., Tournadre, J., Salisbury, J., 2009. Demonstration of Ocean surface salinity microwave measurements from space using AMSR-E data over the Amazon plume. Geophys. Res. Lett. 36. https://doi.org/10.1029/2009GL038860.

Reusch, D.B., Alley, R.B., Hewitson, B.C., 2007. North Atlantic climate variability from a self-organizing map perspective. J. Geophys. Res. 112. https://doi.org/10.1029/2006JD007460.

Reverdin, G., Niiler, P.P., Valdimarsson, H., 2003. North Atlantic Ocean surface currents. Geophys. Res. 108. https://doi.org/10.1029/2001JC001020, 2-1–2-21.

Reynolds, R.W., 1982. A Monthly Averaged Climatology of Sea Surface Temperature. NOAA Technical Report NWS-31. National Oceanic and Atmospheric Administration, Silver Springs, MD.

Reynolds, R.W., 1983. A comparison of sea surface temperature climatologies. J. Glim. Appl. Meteorol. 22, 447−459.

Reynolds, R.W., Smith, T.M., 1994. Improved global sea surface temperature analyses using optimum interpolation. J. Clim. 7, 929−948. https://doi.org/10.1175/1520-0442(1994) 007<0929:IGSSTA>2.0.CO;2.

Reynolds, R.W., Rayner, N.A., Smith, T.M., Stokes, D.C., Wang, W., 2002. An improved in situ and satellite SST analysis. J. Clim. 15, 1609−1625.

Rhines, P., 1975. Waves and turbulence on a β-plane. J. Fluid Mech. 69, 417−443.

Ribatet, M.A., 2006. A User's Guide to the Pot Package (Version 1.4). http://cran.r-project.org/.

Richardson, P.L., 1993. A census of eddies observed in North Atlantic SOFAR float data. Prog. Oceanogr. 31, 1−50.

Richardson, W.S., Stimson, P.B., Wilkins, C.H., 1963. Current measurements from moored buoys. Deep-Sea Res. 10, 369−388.

Richardson, P.L., Price, J.F., Owens, W.B., Schmitz, W.J., 1981. North Atlantic subtropical gyre: SOFAR floats tracked by moored listening stations. Science 213, 435−437.

Richardson, A.J., Pfaff, M.C., Field, J.G., Silulwane, N.F., Shillington, F.A., 2002. Identifying characteristic chlorophyll *a* profiles in the coastal domain using an artificial neural network. J. Plankton Res. 24, 1289−1303.

Richardson, A.J., Risien, C., Shillington, F.A., 2003. Using self-organizing maps to identify patterns in satellite imagery. Prog. Oceanogr. 59, 223−239.

Riche, O., 2011. Time-dependent Inverse Box-Model for the Estuarine Circulation and Primary Productivity in the Strait of Georgia. Ph.D. Thesis. University of British Columbia, Vancouver, British Columbia, p. 228.

Richman, M.B., 1986. Rotation of principal components. J. Clim. 6 (3), 293−335. https://doi.org/10.1002/joc.3370060305.

Rienecker, M.M., Suarez, M.J., Gelaro, R., Todling, R., Bacmeister, J., Liu, E., Bosilovich, M.G., Schubert, S.D., Takacs, L., Kim, G.-K., Bloom, S., Chen, J., Collins, D., Conaty, A., da Dilva, A., Gu, W., Joiner, J., Koster, R.D., Lucchesi, R., Molod, A., Owens, T., Pawson, S., Pegion, P., Redder, C.R., Reichle, R., Robertson, F.R., Ruddick, A.G., Sienkiewicz, M., Woollen, J., 2011. MERRA: NASA's Modern-era retrospective analysis for research and applications. J. Clim. 24, 3624−3648.

Rio, M.-H., Hernandez, F., 2003. High-frequency response of wind-driven currents measured by drifting buoys and altimetry over the world Ocean. J. Geophys. Res. 108. https://doi.org/10.1029/2002jc001655.

Riser, S.C., 1982. The quasi-Lagrangian nature of SOFAR floats. Deep-Sea Res. 29, 1587−1602.

Risien, C.M., Reason, C.J.C., Shilliongton, F.A., 2004. Variability in satellite winds over the Benguela upwelling system during 1999-2000. J. Geophys. Res. 109, C3. https://doi.org/10.1029/2003JC001880.

Ritzema, H.P., 1994. Drainage Principles and Applications, second ed. Compl. Rev. (ILRI publication; No. 16). ILRI https://edepot.wur.nl/149491.

Riva, R.E.M., Bamber, J.L., Lavallée, D.A., Wouters, B., 2010. Sea-level fingerprint of continental water and ice mass change from GRACE. Geophys. Res. Lett. 37. https://doi.org/10.1029/2010GL044770.

Roache, P.J., 1972. Computational Fluid Dynamics. Hermosa, Albuquerque, NM.

Roarty, H., Cook, T., Hazard, L., George, D., Harlan, J., Cosoli, S., Wyatt, L., Alvarez Fanjul, E., Terrill, E., Otero, M., Largier, J., Glenn, S., Ebuchi, N., Whitehouse, B., Bartlett, K., Mader, J., Rubio, A., Corgnati, L., Mantovani, C., Griffa, A., Reyes, E., Lorente, P., Flores-Vidal, X., Johanna Saavedra-Matta, K., Rogowski, P., Prukpitikul, S., Lee, S.-H., Lai, J.-W., Antoine Guerin, C., Sanchez, J., Hansen, B., Grilli, S., 2019. The global high frequency radar network. Front. Mar. Sci. 6. https://doi.org/10.3389/fmars.2019.00164.

Roberts, J., Roberts, T.D., 1978. Use of the Butterworth low-pass filter for Oceanographic data. J. Geophys. Res. 83, 5510−5514.

Robinson, I.S., 1985. Satellite Oceanography. Ellis Horwood, Chichester.

Robinson, A.R., McGillicuddy, D.J., Colman, J., Ducklow, H.W., Fasham, M.J.R., Hoge, F.E., Leslie, W.G., McCarthy, J.J., Podewski, S., Porter, D.L., Saure, G., Yoder, J.A., 1993. Mesoscale and upper ocean variabilities during the 1989 JGOFS bloom study. Deep-Sea Res. Part II 40, 9−35.

Rodionov, S., Overland, J.E., 2005. Application of a sequential regime shift detection method to the Bering Sea ecosystem. ICES J. Marine Sci. 62, 328e332. http://www.beringclimate.noaa.gov/regimes/JMSPublArticle.pdf.

Rodionov, S.N., 2004. A sequential algorithm for testing climate regime shifts. Geophys. Res. Lett. 31, L09204. https://doi.org/10.1029/2004GL019448. https://docs.google.com/file/d/0B8eNwWtdAAJbWGg4dFhRM1RqT28/edit.

Roemmich, D., Cornuelle, B., 1987. Digitization and calibration of the expendable bathythermograph. Deep-Sea Res. 34, 299−307.

Roesler, C.J., Emery, W.J., Kim, S.Y., 2013. Evaluating the use of high-frequency radar coastal currents to correct satellite altimetry. J. Geophys. Res. 118, 3240−3259. https://doi.org/10.1002/jgrc.20220.

Roll, H.U., 1951. Wassertemperaturemessungen an Deck und in Maschinenraum. Ann. Meteorol. 4, 439−443.

Roquet, F., Madec, G., McDougall, T.J., Barker, P.M., 2015. Accurate polynomial expressions for the density and specific volume of seawater using the TEOS-10 standard. Ocean Model. https://doi.org/10.1016/j.ocemod.2015.04.002.

Rørbæk, K., 1994. Comparison of Aanderaa Instruments DCM 12 Doppler Current Meter with RD Instruments Broadband Direct Reading 600 kHz ADCP. Danish Hydraulic Institute, Copenhagen, Denmark.

Rosenberg, N.D., Lupton, J.E., Kadko, D., Collier, R., Lilley, M.D., Pak, H., 1988. Estimation of heat and chemical fluxes from a seafloor hydrothermal vent field using radon measurements. Nature 334, 604−607.

Rossby, H.T., 1969. On monitoring depth variations of the main thermocline acoustically. J. Geophys. Res. 74, 5542−5546.

Rossby, H.T., Webb, D., 1970. Observing abyssal motions by tracking Swallow floats in the SOFAR channel. Deep-Sea Res. 17, 359−365.

Rossby, H.T., Dorson, D., Fontaine, J., 1986. The RAFOS systems. J. Atmos. Ocean. Technol. 3, 672−679.

Royer, T.C., 1981. Baroclinic transport in the Gulf of Alaska. Part II. A fresh water driven coastal current. J. Mar. Res. 39, 251−266.

Rual, P., June, 1991. XBT Depth Correction. Addendum to the Summary Report of the Ad Hoc Meeting of the IGOSS Task Team on Quality Control for Automated Systems, Marion, Mass., USA, IOC/INF-888 Add, pp. 131−144.

Ruddick, B., Gargett, A.E., 2004. Oceanic double-diffusion: introduction. Prog. Oceanogr. 56, 381–393.

Ruddick, B., 1983. A practical indicator of the water column to double-diffusive activity. Deep Sea Res. 30, 1105–1107.

Rueda, A., Camus, P., Mendez, F.J., Tomas, A., Luceno, A., 2016. An extreme value model for maximum wave heights based on weather types. J. Geophys. Res. Oceans 121, 1262–1273. https://doi.org/10.1002/2015JC010952.

Rusby, R.L., 1991. The conversion of thermal reference values to the ITS-90. J. Chem. Thermodyn. 23, 1153–1161. https://doi.org/10.1016/S0021-9614(05)80148-X.

Russell, S.J., Norvig, P., 2016. Artificial Intelligence: A Modern Approach. Pearson Education Limited, Malaysia.

Sabaka, T.J., Tyler, R.H., Olsen, N., 2016. Extracting ocean-generated tidal magnetic signals from Swarm data through satellite gradiometry. Geophys. Res. Lett. 43 (7), 3237–3245. https://doi.org/10.1002/2016GL068180.

Sakazaki, T., Hamilton, K., 2020. An array of ringing global free modes discovered in tropical surface pressure data. J. Atmos. Sci. 77, 2519–2539.

Salas, J.D., Obeysekera, J., 2014. Revisiting the concepts of return period and risk for nonstationary hydrologic extreme events. J. Hydrol. Eng. 19 (3), 554–568. https://doi.org/10.1061/(ASCE)HE.1943-5584.0000820.

Sanderson, B.G., Okubo, A., Goulding, A., 1990. The fractal dimension of relative Lagrangian motion. Tellus 42A, 550–556.

Sandford, T.B., 1971. Motionally induced electric and magnetic fields in the sea. J. Geophys. Res. 76, 3476–3492.

Santos, N.R., M Mata, M., de Azevedo, J.L.L., Cirano, M., 2018. An assessment of the XBT fall-rate equation in the Southern Ocean. J. Atmos. Ocean. Technol. 35. https://doi.org/10.1175/JTECH-D-17-0086.1.

Sapankevych, N.I., Sankar, R., 2009. Time series prediction using support vector machines: a survey. IEEE Computat. Intell 4, 24–38. https://doi.org/10.1109/MCI.2009.932254.

Sarmiento, J.L., Feely, H.W., Moore, W.S., Bainbridge, A.E., Broecker, W.S., 1976. The relationship between vertical eddy diffusion and buoyancy gradient in the deep sea. Earth Planet. Sci. Lett. 32, 357–370.

Sarmiento, J.L., Toggweiller, J.R., Najjar, R., 1988. Ocean carbon cycle dynamics and atmospheric pCO_2. Philos. Trans. R. Soc. A325, 3–21.

Satake, K., 1995. Linear and nonlinear computations for the 1992 Nicaragua earth- quake tsunami. Pure Appl. Geophys. 144, 455–470.

Saunders, P.M., 1976. Near-surface current measurements. Deep-Sea Res. 23, 249–258.

Saunders, P.M., 1980. Overspeeding of a Savonious rotor. Deep-Sea Res. 27A, 755–759.

Saunders, P.M., 1990. Cold outflow from the Faeroe bank channel. J. Phys. Oceanogr. 20, 29–43.

Saur, T., 1963. A study of the quality of sea water temperatures reported in logs of ships' weather observations. J. Appl. Meteorol. 2, 417–425.

Sayles, M.A., Aagaard, K., Coachman, L.K., 1979. Oceanographic Atlas of the Bering Sea Basin. University of Washington Press, Seattle, WA.

Sayre, R.G., Wright, D.J., Breyer, S.P., Butler, K.A., Van Graafeiland, K., Costello, M.J., Harris, P.T., Goodin, K.L., Guinotte, J.M., Basher, A., Kavanaugh, M.T., Halpin, P.N., Monaco, M.E., Cressie, N., Aniello, P., Frye, C.E., Stephens, D., 2017. A three-dimensional mapping of the ocean based on environmental data. Oceanography. https://doi.org/10.5670/oceanog.2017.116.

Scarborough, J.B., 1966. Numerical Mathematical Analysis. Johns Hopkins Press, Baltimore.

Scarlet, R.I., 1975. A data processing method for salinity, temperature, depth profiles. Deep-Sea Res. 27, 509–515.

Schaad, T., 2009. Infrasound Signals Measured with Absolute Nano-Resolution Barometers. Paroscientific In., Doc. No. G8221 Rev. A.

Schlosser, P., Bönisch, G., Rhein, M., Bayer, R., 1991. Reduction of deepwater formation in the Greenland Sea during the 1980s: evidence from tracer data. Science 251, 1054–1056.

Schluessel, P., Emery, W.J., 1989. Atmospheric water vapor over Ocean from SSM/I measurements. Int. J. Remote Sens. 11, 753–766.

Schluessel, P., Shin, H.Y., Emery, W.J., Grassl, H., 1987. Comparison of satellite derived seas surface temperature with in situ skin measurements. J. Geophys. Res. 92, 2859–2874.

Schneider, N., Muller, P., 1990. The meridional and seasonal structures of the mixed-layer depth and its diurnal amplitude observed during the Hawaii-to-Tahiti Shuttle experiment. J. Phys. Oceanogr. 20, 1395–1404.

Schneider, U., Becker, A., Finger, P., Anja, M.-C., Ziese, M., Rudolf, B., 2014. GPCC's new land surface precipitation climatology based on quality-controlled in situ data and its role in quantifying the global water cycle. Theor. Appl. Climatol. 115, 15–40. https://doi.org/10.1007/s00704-013-0860-x.

Schott, F., 1986. Medium-range vertical acoustic Doppler current profiling from submerged buoys. Deep-Sea Res. 33, 1279–1292.

Schott, F., Leaman, K.D., 1991. Observations with moored acoustic Doppler current profilers in the convection regime in the Golfe du Lion. J. Phys. Oceanogr. 21, 558–574.

Schrama, E.J.O., Ray, R.D., 1994. A preliminary tidal analysis of TOPEX/POSEIDON altimetry. J. Geophys. Res. 99, 24799–24808. https://doi.org/10.1029/94JC01432.

Schueler, C.F., Lee, T.F., Miller, S.D., 2013. VIIRS constant spatial-resolution advantages. Int. J. Remote Sens. 34. https://doi.org/10.1080/01431161.2013.796102.

Schumacher, J.D., Reed, R.K., 1986. On the Alaska coastal current in the western Gulf of Alaska. J. Geophys. Res. 91, 9655–9661.

Schuster, A., 1898. On the investigation of hidden periodicities with application to a supposed 26 day period of meteorological phenomena. Terr. Magn. 3, 13–41.

Schwing, F.B., O'Farrell, M., Steger, J.M., Baltz, K., 1996. Coastal Upwelling Indices, West Coast of North America, 1946–1995. NOAA Tech. Memo. NOAA-TMNMFS-SWFSC-231, p. 144.

Schwing, F.B., Bond, N.A., Bograd, S.J., Mitchell, T., Alexander, M.A., Mantua, N., 2006. Delayed coastal upwelling along the US west coast in 2005: a historical perspective. Geophys. Res. Lett. 33 (22).

Seaver, G.A., Kuleshov, S., 1982. Experimental and analytical error of the expendable bathythermograph. J. Phys. Oceanogr. 12, 592–600.

Šepić, J., Orlić, M., Vilibić, I., 2008. The Bakar Bay seiches and their relationship with atmospheric processes. Acta Adriat. 49 (2), 107–123.

Šepić, J., Pasarić, M., Međugorac, I., Vilibić, I., Karlović, M., Mlinar, M., 2022. Climatology and process-oriented analysis of the Adriatic sea level extremes. Prog. Oceanogr. https://doi.org/10.1016/j.pocean.2022.102908.

Shen, Z., Mei, L., 1993. Equilibrium spectra of water waves forced by intermittent wind turbulence. J. Phys. Oceanogr. 23, 505–531.

Shen, Z., Wang, W., Mei, L., 1994. Finestructure of wind waves analyzed with wavelet transform. J. Phys. Oceanogr. 24, 1085–1094.

Shevchenko, G.V., Rabinovich, A.B., Thomson, R.E., 2004. Sea-ice drift on the northeastern shelf of Sakhalin Island. J. Phys. Oceanogr. 34 (11), 2470–2491.

Shibata, A., 2006. Features of Ocean microwave emission changed by wind at 6 GHz. J. Oceanogr. 62, 321–330.

Shum, C.K., Werner, R.A., Sandwell, D.T., Zhang, B.H., Nerem, R.S., Tapley, B.D., 1990. Variations of global mesoscale eddy energy observed from Geosat. J. Geophys. Res. 95, 17865–17876.

Siemens, C.W., 1876. On determining the depth of the sea without the use of a sounding line. Philos. Trans. R. Soc. London 166, 671–692.

Sippican, 1994. MK12 Oceanographic Data Acquisition System, User's Manual 306677-1. Sippican Ocean Systems, Inc., Marion, MA.

Skyllingstad, E.D., Smyth, W.D., Moum, J.N., Wijesekera, H., 1999. Upper-ocean turbulence during a westerly wind burst: A comparison of large-eddy simulation results and microstructure measurements. J. Phys. Oceanogr. 29, 5–28.

Smith, W.L., Rao, P.K., Koffler, R., Curtis, W.R., 1970. The determination of sea-surface temperature from satellite high resolution infrared window radiation measurements. Mon. Weather Rev. 95, 604–611.

Smith, J.N., Rossi, V., Buesseler, K.O., Cullen, J.T., Cornett, J., Nelson, R., Macdonald, A.M., Robert, M., Kellogg, J., 2017. Recent transport history of Fukushima radioactivity in the Northeast Pacific Ocean. Environ. Sci. Technol. 51 (18), 10494–10502.

Smyth, W.D., Hebert, D., Moum, J.N., 1996a. Local ocean response to a multiphase westerly windburst. Part 1: The dynamic response. J. Geophys. Res. 101, 22495–22512.

Smyth, W.D., Hebert, D., Moum, J.N., 1996b. Local ocean response to a multiphase westerly windburst. Part 2: Thermal and freshwater responses. J. Geophys. Res. 101, 22513–22533.

Snedecor, G.W., Cochran, W.G., 1967. Statistical Methods. Iowa State University Press, Ames, IA.

Snodgrass, F.E., 1968. Deep sea instrument capsule. Science 162, 78–87.

Sokolova, S.E., Rabinovich, A.B., Chu, K.S., 1992. On the atmosphere-induced sea level variations along the western coast of the Sea of Japan. La Mer 30, 191–212.

Solari, S., Losada, M.A., 2012. A unified statistical model for hydrological variables including the selection of threshold for the peak over threshold method. Water Resour. Res. 48, W10541. https://doi.org/10.1029/2011WR011475.

Solari, S., Eguen, M., Polo, M.J., Losada, M.A., 2017. Peaks Over Threshold (POT): a methodology for automatic threshold estimation using goodness of fit p-value. Water Resour. Res. 53, 2833–2849. https://doi.org/10.1002/2016WR019426.

Sonnewald, M., Lguensat, R., Jones, D.C., Dueben, P.E., Brakard, J., Balaji, V., 2021. Bridging observations, theory and numerical simulation of the ocean using machine learning. Environ. Res. Lett. 16. https://doi.org/10.48550/arXiv.2104.12506.

Sova, M.S., 1995. The Sampling Variability and the Validation of High Frequency Radar Measurements of the Sea Surface. Ph.D. thesis. School of Mathematics and Statistics, University of Sheffield, UK.

Spear, D.J., Thomson, R.E., 2012. Thermohaline staircases in a British Columbia fjord. Atmos. Ocean 50, 127–133. https://doi.org/10.1080/07055900.2011.649034. First Article, 1-7.

Spencer, R.W., Hood, H.M., Hood, R.E., 1989a. Precipitation retrieval over land and Ocean with the SSM/I: identification and characteristics of the scattering signal. J. Atmos. Ocean. Technol. 6, 254–273.

Spencer, R.W., Hinton, B.B., Olson, W.S., 1989b. Nimbus-7 37 GHz radiances correlated with radar rain rates over the Gulf of Mexico. J. Clim. Appl. Meteorol. 22, 2095–2099.

Spiess, F., 1928. The Meteor Expedition-Research and Experiences During the German Atlantic Expedition 1925–27. Amerind, New Delhi, 1985.

Sprent, P., Dolby, G.A., 1980. The geometric mean functional relationship. Biometrics 36, 547–550.

Sreenivasan, K.R., Ramshankar, R., Meneveau, C., 1989. Mixing, entrainment and fractaldimensions of surfaces in turbulent fluids. Proc. R. Soc. London A421, 79–109.

Stacey, M.W., Pond, S., LeBlond, P.H., 1988. An objective analysis of the low-frequency currents in the Strait of Georgia. Atmosphere-Ocean 26, 1–15.

Stegen, G.R., Delisi, D.P., Von Collins, R.C., 1975. A portable, digital recording, expendable bathythermograph (XBT) system. Deep-Sea Res. 22, 447–453.

Steinhart, J.C., Hart, S.R., 1968. Calibration curves for thermistors. Deep-Sea Res. 15, 497–503.

Stephens, S.A., Bell, R.G., Lawrence, J., 2018. Developing signals to trigger adaptation to sea-level rise. Environ. Res. Lett. 13, 104004. https://doi.org/10.1088/1748-9326/aadf96.

Stockwell, R.G., Mansinha, L., Lowe, R.P., 1994. Localization of the complex spectrum: the S transformation. AGU Trans. 55.

Stommel, H., Schott, F., 1977. The beta spiral and the determination of the absolute velocity field from hydrographic station data. Deep-Sea Res. 24, 325–329.

Stommel, H., Saunders, K., Simmons, W., Simmons, J., 1969. Observations of the diurnal thermocline. Deep Sea Res. 16, 269–284.

Strickland, J.D.H., Parsons, T.R., 1968. A Practical Handbook of Seawater Analysis. Bulletin Fisheries Research Board of Canada.

Strickland, J.D.H., Parsons, T.R., 1972. A Practical Handbook of Seawater Analysis, second ed. Bulletin Fisheries Research Board of Canada.

Strong, A.E., Pritchard, J.A., 1980. Regular monthly mean temperatures of the Earth's Oceans from satellites. Bull. Am. Meteorol. Soc. 61, 553–559.

Stuiver, M.P., Quay, P.D., Östlund, N.D., 1982. Abyssal water carbon-14 distribution and the age of the world Oceans. Science 219, 849–851.

Sturges, W., 1983. On interpolating gappy records for time-series analysis. J. Geophys. Res. 88, 9736–9740.

Suijlen, J.M., Buyse, J.J., 1994. Potentials of photolytic rhodamine WT as a large-scale water tracer assessed in a long-term experiment in the Loosdrecht lakes. Limnol. Oceanogr. 6, 1411–1423.

Sverdrup, H.U., 1947. Wind driven currents in a baroclinic Ocean with applications to the equatorial currents in the eastern Pacific. Proc. Natl. Acad. Sci. U. S. A. 33, 318–336.

Sverdrup, H.U., Johnson, M.W., Fleming, R.H., 1942. The Oceans, Their Physics, Chemistry and General Biology. Prentice-Hall, New York. http://ark.cdlib.org/ark:/13030/kt167nb66r/.

Swallow, J.C., 1955. A neutrally-buoyant float for measuring deep current. Deep-Sea Res. 3, 74–81.

Sweers, H.E., 1971. Sigma-T specific volume anomaly and dynamic height. Marine Technol. J. 5, 7–25.

Sweet, W.V., Horton, R., Kopp, R.E., LeGrande, A.N., Romanou, A., 2017. Sea level rise. In: Wuebbles, D.J., Fahey, D.W., Hibbard, K.A., Dokken, D.J., Stewart, B.C., Maycock, T.K. (Eds.), Climate Science Special Report: Fourth National Climate Assessment, Volume I. U.S. Global Change Research Program, pp. 333–363. https://doi.org/10.7930/J0VM49F2.

Sweet, W.V., Dusek, G., Obeysekera, J., Marra, J.J., February, 2018. Patterns and projections of high tide flooding along the U.S. coastline using a common impact threshold. NOAA Tech. Rep. NOS CO-OPS 44.

Swift, C.T., McIntosh, R.E., 1983. Considerations for microwave remote sensing of Ocean- surface salinity. Trans. Geosci. Remote Sens. GE-21, 480−491. https://doi.org/10.1109/TGRS.1983.350511.

Sybrandy, A.L., Niiler, P.P., 1990. The WOCE/TOGA SVP Lagrangian Drifter Construction Manual. Scripps Institution of Oceanography, University of California, San Diego, SIO. Reference 90−248.

Sybrandy, A.L., Niiler, P.P., Martin, C., Scuba, W., Charpentier, E., Meldrum, D.T., 2009. Global Drifter Program, Barometer Drifter Design Reference, Data Buoy Cooperation Panel, DBCP Report No. 4, Rev. 2.2.

Sylvester, J.J., 1889. On the reduction of a bilinear quantic of the nth order to the form of a sum of n products by a double orthogonal substitution. Messenger Math 19, 42−46.

Tabata, S., 1978a. On the accuracy of sea-surface temperatures and salinities observed in the Northeast Pacific. Atmosphere-Ocean 16, 237−247.

Tabata, S., 1978b. Comparison of observations of sea-surface temperatures at Ocean Station "P" and N.O.A.A. buoy stations and those made by merchant ships travelling in their vicinities, in the Northeast Pacific Ocean. J. Appl. Meteorol. 17, 374−385.

Tabata, S., Stickland, J.A., 1972. Summary of Oceanographic Records Obtained from Moored Instruments in the Strait of Georgia, 1969−70. Current Velocity and Seawater Temperature from Station H-06: Pacific Marine Science Report 72−7. Environment Canada.

Talke, S.A., Kemp, A.C., Woodruff, J., 2018. Relative sea level, tides, and extreme water levels in Boston harbor from 1825 to 2018. J. Geophys. Res. Oceans 123. https://doi.org/10.1029/2017JC013645.

Talley, L.D., Joyce, T.M., 1992. Double silica maximum in the North Pacific. J. Geophys. Res. 97, 5465−5480.

Talley, L.D., Martin, M., Salameth, P., 1988. Trans Pacific Section in the Subpolar Gyre (TPS47): Physical, Chemical, and CTD Data, R/V Thomas Thompson TT190, 4 August 1985−7 September 1985. SIO Ref. 88−9. Scripps Institute of Oceanography, La Jolla, CA, USA.

Talley, L.D., Joyce, T.M., deSzoeke, R.A., 1991. Trans-Pacific sections at 47°N and 152°W: distribution of properties. Deep-Sea Res. 38, 563−582.

Talley, L.D., Pickard, G.L., Emery, W.J., Swift, J.H., 2011. Descriptive Physical Oceanography: An Introduction, sixth ed. Elsevier, Boston, p. 560.

Tapley, B.D., Born, G.H., Park, M.E., 1982. The seastat altimeter data and its accuracy assessment. J. Geophys. Res. 87, 3179−3188.

Tauber, G.M., 1969. The Comparative Measurements of Sea Surface Temperature in the USSR, Technical Note 103, Sea Surface Temperature. WMO, pp. 141−151.

Taylor, K.E., 2001. Summarizing multiple aspects of model performance in a single diagram. J. Geophys. Res. 106, 7183−7192.

Taylor, K.E., 2005. Taylor Diagram Primer. Wikipedia, p. 4.

Tchernia, P., 1980. Descriptive Regional Oceanography. Pergamon Marine Series, vol 3. Pergamon Press, Oxford.

Tebaldi, C., Strauss, B.H., Zervas, C.E., 2012. Modeling sea level rise impacts on storm surges along US coasts. Environ. Res. Lett. 7 (1). https://doi.org/10.1088/1748-9326/7/1/014032.

Teledyne-RD Instruments, 2011. Acoustic Doppler Current Profiler Principles of Operation A Practical Primer, P/N 951-6069-00 (January 2011), p. 56.

Tengberg, A., Hovdenes, J., Andersson, H.J., Brocandel, O., Diaz, R., Hebert, D., Arnerich, T., Huber, C., Körtzinger, A., Khripounoff, A., Rey, F., Rönning, C., Schimanski, J., Stommer, S., Stanglemayer, A., 2006. Evaluation of a lifetime-based optode to measure oxygen in aquatic systems. Limnol. Oceangr. 4, 7−17.

Tennant, W., 2004. Considerations when using the pre-1979 NCEP/NCAR reanalyses in the southern hemisphere. Geophys. Res. Lett. 31 (11). https://doi.org/10.1029/2004GL019751.

Teukolsky, W.H.,S.A., Vetterling, W.T., Flannery, B.P., 1992. Numerical Recipes, second ed. Cambridge University Press, Cambridge.

Thadathil, P., Saran, A.K., Gopalakrishna, V.V., Vethamony, P., Araligidad, N., Bailey, R., 2002. XBT fall rate in waters of extreme temperature: a case study in the Antarctic Ocean. J. Atmos. Ocean. Technol. 19, 391−396. https://doi.org/10.1175/1520-0426-19.3.391.

Thompson, R., 1971. Spectral estimation from irregularly spaced data. IEEE Trans. Geosci. Electron. GE-9, 107−119.

Thompson, R.O.R.Y., 1979. Coherence significance levels. J. Atmos. Sci. 36, 2020−2021.

Thompson, R.O.R.Y., 1983. Low-pass filters to suppress inertial and tidal frequencies. J. Phys. Oceanogr. 13, 1077−1083.

Thompson, T.W., Weissman, D.E., Gonzalez, F.I., 1983. L-band radar backscatter dependence upon surface wind stress: a summary of new Seasat-I and aircraft observations. J. Geophys. Res. 88, 1727−1735.

Thomson, R.E., LeBlond, P.H., Rabinovich, A.B., 1998. Satellite-tracked drifter measurements of inertial and semidiurnal currents in the northeast Pacific. J. Geophys. Res. 103 (1), 1039−1071.

Thompson, P., Cai, Y., Reeve, D.E., Stander, J., 2009. Automated threshold selection methods for extreme wave analysis. Coast. Eng. 56 (10), 1013−1021. https://doi.org/10.1016/j.coastaleng.2009.06.003.

Thomson, R.E., 1977. Currents in Johnstone Strait, British Columbia: supplementary data on the Vancouver Island side. J. Fish. Res. Board Can. 34, 697−703.

Thomson, R.E., 1981. Oceanography of the British Columbia coast. Can. Special Pub. Fish. Aquat. Sci. 56, 291. Ottawa.

Thomson, R.E., 1983. A comparison between computed and measured Oceanic winds near the British Columbia coast. J. Geophys. Res. 88, 2675−2683.

Thomson, R.E., Davis, E.E., 2017. Equatorial Kelvin waves generated in the wester tropical Pacific Ocean trigger mass and heat transport within the Middle America Trench off Costa Rica. J. Geophys. Res. 122, 5850−5869. https://doi.org/10.1002/2017JC012848.

Thomson, R.E., Fine, I.V., 2003. Estimating mixed layer depth from Oceanic profile data. J. Atmos. Ocean. Technol. 20, 319−329. https://doi.org/10.1175/1520-0426(2003)020<0319:EMLDFO>2.0.CO;2.

Thomson, R.E., Fine, I.V., 2009. A diagnostic model for mixed layer depth estimations with application too Ocean Station P in the Northeast Pacific. J. Phys. Oceanogr. 39, 1399−1415. https://doi.org/10.1175/2008JPO3984.1.

Thomson, R.E., Fine, I.V., 2021. Revisiting the Ocean's non-isostatic response to 5-day atmospheric loading: new results based on global bottom pressure records and numerical modeling. J. Phys. Oceanogr. 51, 2845−2859. https://doi.org/10.1175/JPO-D-21-0025.1.

Thomson, R.E., Freeland, H.J., 1999. Lagrangian measurement of mid-depth currents in the eastern tropical Pacific. Geophys. Res. Lett. 26, 3125−3128.

Thomson, R.E., Huggett, W.S., 1980. M2 baroclinic tides in Johnstone Strait, British Columbia. J. Phys. Oceanogr. 10, 1509−1539.

Thomson, R.E., Krassovski, M.V., 2010. Poleward reach of the California undercurrent extension. J. Geophys. Res. 115. https://doi.org/10.1029/2010JC006280.

Thomson, R.E., Krassovski, M.V., 2015. Remote alongshore winds drive variability of the California undercurrent off the British Columbia-Washington coast. J. Geophys. Res. Ocean. 120, 8151−8176. https://doi.org/10.1002/2015JC011306.

Thomson, R.E., Spear, D.J., 2020. Gravity currents facilitate the generation and propagation of internal bores and solitons at the bottom of the Strait of Georgia. J. Geophys. Res. Ocean. 125, e2020JC016589. https://doi.org/10.1029/2020JC016589.

Thomson, R.E., Ware, D.M., 1996. A current velocity index of Ocean variability. J Geophys. Res. 101, 14297−14310.

Thomson, R.E., Tabata, S., Ramsden, D., 1985a. Comparison of sea level variability on the Caribbean and the Pacific coasts of the Panama Canal. In: Time Series of Ocean Measurements, vol. 2. UNESCO, IOC Technical Series 30, pp. 33−37.

Thomson, R.E., Crawford, W.R., Huggett, W.S., 1985b. Low-pass filtered current records for the west coast of Vancouver Island: coastal oceanic dynamics experiment, 1979−81. Can. Data Rep. Hydrogr. Ocean Sci. 40, 102.

Thomson, R.E., Curran, T.A., Hamilton, M.C., McFarlane, R., 1988. Time series measurements from a moored fluorescence-based dissolved oxygen sensor. J. Atmos. Ocean. Technol. 5, 614−624.

Thomson, R.E., LeBlond, P.H., Emery, W.J., 1990. Analysis of deep-drogued satellite-tracked drifter measurements in the Northeast Pacific. Atmosphere-Ocean 28, 409−443.

Thomson, R.E., Gordon, R.L., Dolling, A.G., 1991. An intense acoustic back-scattering layer at the top of a mid-Ocean ridge hydrothermal plume. J. Geophys. Res. 96, 4839−4844.

Thomson, R.E., Burd, B.J., Dolling, A.G., Gordon, R.L., Jamieson, G.S., 1992a. The deep scattering layer associated with the Endeavour Ridge hydrothermal plume. Deep-Sea Res. 39, 55−73.

Thomson, R.E., Delaney, J.R., McDuff, R.E., Janecky, D.R., McLain, J.S., 1992b. Physical characteristics of the endeavour Ridge hydrothermal plume during July 1988. Earth Planet. Sci. Lett. 111, 141−154.

Thomson, R.E., LeBlond, P.H., Rabinovich, A.B., 1997. Oceanic odyssey of a satellite-tracked drifter: North Pacific variability delineated by a single drifter trajectory. J. Oceanogr. 53, 81−87.

Thomson, R.E., Mihály, S.F., Kulikov, E.A., 2007. Estuarine versus transient flow regimes in Juan de Fuca Strait. J. Geophys. Res. Oceans 112, C09022. https://doi.org/10.1029/2006JC003925.

Thomson, R.B., Phillips, R.A., Tuck, G.N., 2009. Modeling the impact of fishery bycatch on wandering and black-browed albatrosses of South Georgia: preliminary results. ICCAT, Collect. Vol. Sci. Pap. 64, 2342−2382.

Thomson, R.E., Davis, E.E., Heesemann, M., Villinger, H., 2010. Observations of long-duration episodic bottom currents in the Middle America Trench: Evidence for tidally initiated turbidity flows. J. Geophys Res.Oceans 115, C10020. https://doi.org/10.1029/2010JC006166.

Thomson, J., Polagye, B., Durgesh, V., Richmond, M.C., 2012. Measurements of turbulence at two tidal energy sites in Puget Sound, WA. IEEE J. Ocean. Eng. 37, 363−374.

Thomson, J., D'Asaro, E.A., Cronin, M.F., Rogers, W.E., Harcourt, R.R., Shcherbina, A., 2013. Waves and the equilibrium range at Ocean Weather Station P. J. Geophys. Res. 118, 5951−5962.

Thomson, R.E., Heesemann, M., Davis, E.E., Hourston, R.A.S., 2014. Continental microseismic intensity delineates Oceanic upwelling timing along the west coast of North America. Geophys. Res. Lett. 41, 6872−6880. https://doi.org/10.1002/2014GL061241.

Thorndike, A.S., Colony, R., 1982. Sea ice motion response to geostrophic winds. J. Geophys. Res. 87 (C8), 5845−5852.

Thorndike, A.S., 1986. Kinematics of the sea ice. In: Untersteiner, N. (Ed.), The Geophysics of Sea Ice. Plenum, New York, pp. 489−549.

Thorpe, S.A., 1977. Turbulence and mixing in a Scottish loch. Phil. Trans. R. Soc. London A286, 125−181.

Thurnherr, A.M., 2011. Vertical velocity from LADCP data. 2011 IEEE/OES Tenth Current, Waves and Turbulence Measurements (CWTM). IEEE 198−204. https://doi.org/10.1109/CWTM.2011.5759552.

Thurnherr, A.M., Ledwell, J., Lavelle, J., Mullineaux, L., 2011. Regional circulation near the crest of the East Pacific Rise between 9° and 10°N. Deep Sea Res. Part I 58, 365−376. https://doi.org/10.1016/j.dsr.2011.01.009.

Thurnherr, A.M., Jacobs, S.S., Dutrieux, P., Giulivi, C.F., 2014. Export and circulation of ice cavity water in Pine Island Bay, West Antarctica. J. Geophys. Res. Ocean. 119 (3), 1754−1764.

Thurnherr, A.M., Kunze, E., Toole, J.M., Laurent, L.C.S., Richards, K.J., Ruiz-Angulo, A., 2015. Vertical kinetic energy and turbulent dissipation in the ocean. Geophys. Res. Lett. 42, 7639−7647. https://doi.org/10.1002/2015GL065043.

Thurnherr, A.M., Goszczko, I., Bahr, F., 2017. Improving LADCP velocity with external heading, pitch, and roll. J. Atmos. Ocean. Technol. 34, 1713−1721. https://doi.org/10.1175/JTECH-D-16-0258.1.

Tichelaar, B.W., Ruff, L.J., 1989. How good are our best models? Jackknifing, bootstrapping, and earthquake depth. EOS 70 (20), 593−605.

Titov, V.V., Rabinovich, A.B., Mofjeld, R.E., Thomson, F.I., González, 2005. The global reach of the 26 December 2004 Sumatra tsunami. Science 309, 2045−2048.

Toggweiler, J.R., Trumbore, S., 1985. Bomb-test 90Sr in Pacific and Indian Ocean surface water as recorded by banded corals. Earth Planet. Sci. Lett. 74, 306−314.

Togneri, M., Jones, D., Neill, S., Lewis, M., Ward, S., Piano, M., Masters, I., 2017. Comparison of 4- and 5-beam acoustic Doppler current profiler configurations for measurement of turbulent kinetic energy. Energy Proc. 125, 260−267. https://doi.org/10.1016/j.egypro.2017.08.170.

Tokamamkian, R., Strub, P.T., McClean-Padman, J., 1990. Evaluation of the maximum cross-correlation method of estimating sea surface velocities from sequential satellite images. J. Atmos. Ocean. Technol. 7, 852−865.

Toole, J.M., Andres, M., Le Bras, I.A., Joyce, T.M., McCartney, M.S., 2017. Moored observations of the Deep Western Boundary current in the NW Atlantic: 2004−2014. J. Geophys. Res. Ocean. 122, 7488−7505. https://doi.org/10.1002/2017JC012984.

Topham, D.R., Perkins, R.G., 1988. CTD sensor characteristics and their matching for salinity calculations. IEEE J. Ocean. Eng. 13, 107−117.

Treasure, A.M., Roquet, F., Ansorge, L.J., Bester, M.N., Boehme, L., Bornemann, H., Charrassin, J.B., Chvallier, D., Costa, D.P., Fedak, M.A., Guinet, C., Hammill, M.O., Harcourt, R.G., Hindell, M.A., Kovacs, K.M., Lea, M.A., Lovell, P., Lowther, A.D., Lydersen, C., McIntyre, T., McMahon, C.R., Muelbert, M.M.C.,

Nicholls, K., Picard, B., Reverdin, G., trites, A.W., Williams, G.D., N de Bruyn, P.J., 2017. Marine mammals exploring the Oceans pole to pole: a review of the MEOP consortium. Oceanography 30, 132–138. https://doi.org/10.5670/oceanog.2017.234.

Trenberth, K.E., 1975. A quasi-biennial standing wave in the Southern Hemisphere and interrelations with sea surface temperature. Quart. J. Royal Metero. Soc. 101 (427), 55–74. https://doi.org/10.1002/qj.49710142706.

Trenberth, K.E., Olson, J.G., 1988. ECMWF Global Analysis 1979–1986: Circulation Statistics and Data Evaluation. NCAR Technical Report Note NCAR/TN-300+STR. National Center for Atmospheric Research, USA.

Trivett, D.A., Terray, E.A., William, A.J., 1991. Error analysis of an acoustic current meter. IEEE J. Ocean. Eng. 16, 329–337.

Trumbore, S.E., Jacobs, S.S., Smethie, W.M., 1991. Chlorofluorocarbon evidence for rapid ventilation of the Ross Sea. Deep-Sea Res. 38, 845–870.

Trump, W., 1983. Effect of ship's roll on the quality of precision CTD data. Deep-Sea Res. 30 (11 A), 1173–1183.

Tsiperman, E., Thacker, W.C., 1989. An optimal-control/adjoint-equations approach to studying the oceanic general circulation. J. Phys. Oceanogr. 1471–1485.

Tsonis, A.A., 1991. Sensitivity of the global climate system to initial conditions. EOS Trans. AGU 72, 313–328.

Tsonis, A.A., 1992. Autoregressive models not sensitive to initial conditions. Reply. EOS Trans. AGU 25, 268.

Tsonis, A.A., Eisner, J.B., 1990. Comments on "Dimension analysis of climatic data". J. Clim. 3, 1502–1505.

Turner, J.S., 1973. Buoyancy Effects in Fluids. Cambridge University Press, Cambridge, p. 367.

Turner, J., Kraus, E., 1967. A one-dimensional model of the seasonal thermocline I. A laboratory experiment and its interpretation. Tellus A 19, 88–97.

Tushingham, A.M., Peltier, W.R., 1992. ICE-3-G: a new global model of late Pleistocene deglaciation based upon geophysical predictions of post glacial relative sea level change. J. Geophys. Res. 96, 4497–4523.

Tyler, R., Maus, S., Lühr, H., 2003. Satellite observations of magnetic fields due to ocean flow. Science 299, 239–241.

Ulrych, T.J., 1972. Maximum entropy spectrum of truncated sinusoids. J. Geophys. Res. 77, 1396–1400.

Ulrych, T.J., Bishop, T.N., 1975. Maximum entropy spectral analysis and autoregressive decomposition. Rev. Geophys. Space Phys. 13, 183–200.

UNESCO, 1966. International Oceanographic Tables. UNESCO, Place de Fontenoy, Paris. National UNESCO Office of Oceanography, Institute of Oceanography, Wormley, UK.

UNESCO, 1981. The Practical Salinity Scale 1978 and the International Equation of State of Seawater 1980, 36. UNESCO Technical Papers in Marine Science, p. 25.

UNESCO, 1985. The International System of Units (SI) in Oceanography, 45. UNESCO Technical Papers in Marine Science, p. 131.

Upstill-Goddard, R.C., Suijlen, J.M., Malin, G., Nightingale, P.D., 2001. The use of photolytic rhodamines WT and sulpho G as conservative tracers of dispersion in surface waters. Limnol. Oceanogr. 46 (4), 927–934. https://doi.org/10.4319/lo.2001.46.4.0927.

Urick, R.J., 1967. Principles of Underwater Sound. McGraw-Hill, New York.

Urquhart, D., Sell, Z., Dhouieb, E., Bell, G., Oliver, S., Black, R., Tallis, M., 2012. Effects of a supervised, outpatient exercise and physiotherapy programme in children with cystic fibrosis. Pediatr. Pulmonol. https://doi.org/10.1002/ppul.22587.

U.S. Geological Survey, 2017. Current Velocity Measurements, U.S. Geological Survey Open-File Report 2007-119. cmgds.marine.usgs.gov/publications/of2007-1194v1/html/velocity.html. (Accessed 6 December 2017).

Vachon, W.A., 1973. Scale Model Testing of Drogues for Free Drifting Buoys. Technical Report. The Charles Stark Draper Laboratory, Inc., Cambridge, MA, USA.

Vachon, P.W., Emery, W.J., 1988. A simulation for spaceborne SAR imagery of a distributed, moving scene. IEEE Trans. G.R.S. 27, 67–78.

Van Beers, W.C.M., Kleijnen, J.P.C., 2002. Kriging interpolation in simulation: a survey. In: Proceedings of the Winter Simulation Conference. https://doi.org/10.1109/WSC.2004.1371308.

VanderPlas, J., 2022. Python Data Science Handbook, second ed. O'Reilly Media, Inc.

van Leer, J., Düing, W., Erath, R., Kennelly, E., Speidel, A., 1974. The cyclesonde: an unattended vertical profiler for scalar and vector quantities in the upper Ocean. Deep-Sea Res. 21, 385–400.

Van Scoy, K.A., Fine, R.A., Östlund, H.G., 1991. Two decades of missing tritium into the North Pacific Ocean. Deep-Sea Res. 38, S191–S219.

Vaniček, P., 1971. Further development and properties of the spectral analysis by least-squares. Astrophys. Space Sci. 12, 10–73.

Vapnik, V., 1995. The Nature of Statistical Learning Theory. Springer Verlag, New York.

Vazquez, J., Zlotnicki, V., Fu, L.-L., 1990. Sea level variability in the gulf stream beween cape hateras and 50°N, a GEOSAT study. J. Geophys. Res. 95, 17957–17964.

Vélez-Belchí, P., Caínzos, V., Romero, E., Casanova-Masjoan, M., Arumí-Planas, C., Santana- Toscano, D., González -Santana, A., Pérez-Hernández, M.D., Hernández-Guerra, A., 2021. The canary intermediate poleward undercurrent: not another poleward undercurrent in an eastern boundary upwelling system. J. Phys. Oceanogr. 51 (9), 2973–2990. https://doi.org/10.1175/JPO-D-20-0130.151.

Vesanto, J., Himberg, J., Alhoniemi, E., Parhandkangas, J., 2000. SOM Toolbox for Matlab 5, report. Helsinkii University of Technology, Helsinki, Finland.

Vilibić, I., Šepić, J., Pasarić, M., Orlić, M., 2017. The Adriatic Sea: a long-standing laboratory for sea level studies. Pure Appl. Geophys. 174 (10), 3765–3811. https://doi.org/10.1007/s00024-017-1625-8.

Vilibić, I., Rabinovich, A.B., Anderson, E.J., 2021. Editorial on Special issue on the global perspective on meteotsunami science. Nat. Hazard. 106, 1087–1104.

Vitousek, S., Barnard, P.L., Fletcher, C.H., et al., 2017. Doubling of coastal flooding frequency within decades due to sea-level rise. Sci. Rep. 7 (1), 1–9.

Vogt, P.R., Jung, W.-Y., 1991. Satellite radar altimetry aids seafloor mapping. EOS 72 (43), 465, 468–469.

Volkov, D.L., Fu, L.-L., Lee, T., 2010. Mechanisms of the meridional heat transport in the Southern Ocean. Ocean Dyn. 60, 791–801.

von Arx, W.S., 1950. An electromagnetic method for measuring the velocities of Ocean currents from a ship under way. Pap. Phys. Oceanogr. Meteorol. 11, 1–61.

Von Storch, H., Zwiers, F.W., 1999. Statistical Analysis in Climate Research. Cambridge University Press. https://doi.org/10.1017/CBO9780511612336.

Wadhams, P., 2000. Ice in the Ocean. Gordon and Breach Sci. Publ., Amsterdam, p. 351.

Wadsworth, J.L., Tawn, J.A., June, 2012. Dependence modelling for spatial extremes. Biometrika 99 (2), 253−272. https://doi.org/10.1093/biomet/asr080.

Wahl, T., Haigh, I.D., Nicholls, R.J., Arns, A., Dangendorf, S., Hinkel, J., Slangen, A.B.A., 2017. Understanding extreme sea levels for broad-scale coastal impact and adaptation analysis. Nat. Commun. https://doi.org/10.1038/ncomms160752.

Walden, H., 1966. Zur messung der Wassertemperatur auf Handelsschiffen. Dtsch Hydro. Z. 19, 21−28.

Waliser, D.E., Gautier, C., 1993. Comparison of buoy and SSM/I-derived wind speeds in the tropical Pacific. TOGA Note. 12, 1−7.

Walker, E.R., Chapman, K.D., 1973. Salinity-Conductivity Formulae Compared. Pacific Marine Science Report 73−5. Institute of Ocean Sciences, Sidney, BC, Canada.

Wallace, J.M., 1972. Empirical orthogonal representation of time series in the frequency domain. Part II: application to the study of tropical wave disturbances. J. Appl. Meteorol. 11, 893−900.

Wallace, J.M., Dickinson, R.E., 1972. Empirical orthogonal representation of time series in the frequency domain. Part I: theoretical considerations. J. Appl. Meteorol. 11, 887−892.

Wallace, D.W.R., Lazier, J.R.N., 1988. Anthropogenic chlorofluoromethanes in newly formed Labrador Sea water. Nature 332, 61−63.

Walters, R.A., Heston, C., 1982. Removing tidal-period variations from time-series data using low-pass digital filters. J. Phys. Oceanogr. 12, 112−115.

Walton, C.C., Pichel, W.G., Sapper, J.F., 1998. The development and operational application of nonlinear algorithms for the measurement of sea surface temperatures with the NOAA polar-orbiting environmental satellites. J. Geophys. Res 103, 27999−28012.

Wang, D.-P., Mooers, C.N.K., 1977. Long coast trapped waves off the west coast of the United States, summer 1973. J. Phys. Oceanogr. 7, 856−864.

Wang, H., Fyke, J., Lenaerts, J., Nusbaumer, J., Singh, H., Noone, D., Rasch, P., 2020. Influence of sea ice anomalies on Antarctic precipitation using source attribution. Cryosphere 14, 429−444. https://doi.org/10.5194/tc-14-429-2020.

Wang, D.-P., Oey, L.-Y., Ezer, T., Hamilton, P., 2003. Near-surface currents in DeSoto Canyon (1997-99): comparison of current meters, satellite observation, and model simulation. J. Phys. Oceanogr 33 (1), 313−326.

Warren, B.A., 1970. General circulation of the South Pacific. In: Wooster, W.S. (Ed.), Scientific Exploration of the South Pacific. National Academy of Sciences, Washington, DC, pp. 33−49.

Warren, B.A., 1983. Why is no deep water formed in the North Pacific? J. Mar. Res. 41, 327−347.

Warren, B.A., Carl, W. (Eds.), 1980. Evolution of Physical Oceanography. MIT Press.

Washburn, L., Emery, B.M., Jones, B.H., Ondercin, D.G., 1998. Eddy stirring and phytoplankton patchiness in the subarctic North Atlantic in late summer. Deep-Sea Res. I 45, 1411−1439.

Watanabe, Y.W., Watanabe, S., Tsunogai, S., 1991. Tritium in the Japan Sea and the renewal time of Japan Sea deep water. Mar. Chem. 34, 97−108.

Waterhouse, A.F., Mackinnon, J.A., Nash, J.D., Alford, M.H., Kunze, E., Simmons, H.L., Polzin, K.L., Laurent, L.C.S., Sun, O.M., Pinkel, R., Talley, L.D., Whalen, C.B., Huussen, T.N., Carter, G.S., Fer, I., Waterman, S., Naveira Garabato, A.C., Sanford, T.B., Lee, C.M., 2014. Global patterns of diapycnal mixing from measurements of the turbulent dissipation rate. J. Phys. Oceanogr. 44, 1854−1872.

Watson, A.J., Ledwell, J.R., 1988. Purposefully released tracers. Philos. Trans. R. Soc. London A325, 189−200.

Watson, A.J., Liddicoat, M.I., 1985. Recent history of atmospheric trace gas concentrations deduced from measurements in the deep sea: application to Sulphur hexafluoride and carbon tetrachloride. Atmos. Environ. 19, 1477−1484.

Watts, D.R., Rossby, H.T., 1977. Measuring dynamic heights with inverted echo sounders: results from MODE. J. Phys. Oceanogr. 7, 345−358.

Weare, B.C., Nasstrom, J.S., 1982. Examples of extended empirical orthogonal function analyses. Mon. Weather Rev. 110, 481−485.

Weare, B.C., Navato, A.R., Newell, R.E., 1976. Empirical orthogonal analysis of Pacific sea surface temperatures. J. Phys. Oceanogr. 6, 671−678.

Weare, B.C., Strub, P.T., Samuel, M.D., 1981. Annual mean surface heat fluxes in the Tropical Pacific Ocean. J. Phys. Oceanogr. 11, 705−717. https://doi.org/10.1175/1520-0485(1981)011<0705:AMSHFI>2.0.CO;2.

Wearn, R.B., Baker, D.J., 1980. Bottom pressure measurements across the Antarctic Circumpolar Current and their relation to the wind. Deep-Sea Res. 21 A, 875−888.

Weaver, A.J., Wiebe, E.C., 2006. Micro Meteorological Network in Greater Victoria Schools. University of Victoria School of Earth and Ocean Sciences Bulletin, p. 10. www.victoriaweather.ca. https://www.islandweather.ca/resources/info/2006CMOSBulletin.pdf.

Webb, A.J., Pond, S., 1986. The propagation of a Kelvin wave around a bend in a channel. J. Fluid Mech. 169, 257−274. https://doi.org/10.1017/S0022112086000629.

Wei, J., Wang, M., Lee, Z., Briceno, H.O., Yu, X., Jiang, L., Garcia, R., Wang, J., Luis, K., 2020. Shallow water bathymetry with multi-spectral satellite Ocean color sensors: leveraging temporal variation in image data. Remote Sens. Environ. 250. https://doi.org/10.1016/j.rse.2020.112035.

Weinreb, M.P., Hamilton, G., Brown, S., Koczor, R.J., 1990. Nonlinearity corrections in calibration of advanced very high resolution radiometer infrared channels. J. Geophys. Res. 95, 7381−7388.

Weiss, W., Roether, W., 1980. The rates of tritium input to the world Oceans. Earth Planet Sci. Lett. 49, 446−453.

Welch, G., Bishop, G., 2006. An Introduction to the Kalman Filter. TR 95-041, Dept. Computer Sci., Univ. of North Carolina at Chapel Hill, Chapel Hill, NC.

Weller, R.A., Davis, R.E., 1980. A vector measuring current meter. Deep-Sea Res. 27, 565−582.

Weller, R.A., Plueddemann, A.J., 1996. Observations of the vertical structure of the oceanic boundary layer. J. Geophys. Res. 101, 8789−8806. https://doi.org/10.1029/96JC00206.

Weller, R.A., Rudnick, D.L., Pennington, N.J., Trask, R.P., Valdes, J.R., 1990. Measuring upper Ocean variability from an array of surface moorings in the subtropical convergence zone. J. Atmos. Ocean. Technol. 7, 68−84. https://doi.org/10.1175/1520-0426(1990)007<0068:Muovfa>2.0.Co;2.

Wenner, F., Smith, E.H., Soule, F.M., 1930. Apparatus for the determination aboard ship of the salinity of sea water by the electrical conductivity method. Bur. Stand. J. Res. 5, 711−732.

Wentz, F.J., Mattox, L.A., Peteherych, W., 1986. New algorithms for microwave measurements of Ocean winds: applications to SEASAT and the special sensor microwave imager. J. Geophys. Res. 91, 2289–2307.

Wentz, F.J., Meissner, T., Gentemann, C., Hilburn, K.A., Scott, J., 2014. Remote Sensing Systems GCOM-W1 AMSR2 Environmental Suite on 0.25_ grid, Version V.8. Remote Sensing Systems, Santa Rosa, CA. Available online at. http://www.remss.com/missions/amsr.

Whitney, F., Robert, M., 2002. Structure of Haida eddies and their transport of nutrients from coastal margins into the NE Pacific Ocean. J. Oceanogr. 58. https://doi.org/10.1023/A:1022850508403.

Wick, G.A., Emery, W.J., Schluessel, P., 1992. A comprehensive comparison between skin and multi-channel sea surface temperatures. J. Geophys. Res. 97, 5569–5596.

Wiggins, S., Manley, J., Brager, E., Woolhiser, B., 2010. Monitoring Marine Mammal Acoustics Using Wave Glider. OCEANS 2010 MTS/IEEE Seattle Washington State Convention and Trade Center Seattle, Washington, USA September 20–23, 2010.

Wüst, G., 1957. Stromgeschwindigkeiten und Strommengen in den Tiefen des Atlantischen Ozeans. Wissenschaftliche Ergebnisse der Deutschen Atlantischen Expedition Meteor 1925–1927 6, 261–420.

Wijesekera, H.W., Gregg, M.C., 1996. Surface layer response to weak winds, westerly bursts, and rain squalls in the western Pacific warm pool. J. Geophys. Res. 101, 977–997. https://doi.org/10.1029/95JC02553.

Wijffels, S.E., 1993. Exchanges Between Hemispheres and Gyres: A Direct Approach to the Mean Circulation of the Equatorial Pacific. Ph.D. thesis. Mass. Inst. of Technol./Woods Hole Oceanographic Institution. Joint Program, Woods Hole, MA.

Wikipedia, 2013, 2022. https://www.wikipedia.org.

Wikipedia 2021, 2023: https://www.wikipedia.org.

Wilkin, J.L., 1987. A Computer Program for Calculating Frequencies and Modal Structures of Free Coastal-Trapped Waves. Woods Hole Oceanographic Institution, Technical Report WHOI-87–53.

Willebrand, J.W., Müller, P., Olbers, D.J., 1977. Inverse Analysis of the Trimoored Internal Wave Experiment (IWEX). Berichte aus dem Institut für Meereskunde, 20a,b.

Williams, A.J., 2004. Principles of Oceanographic Instrument Systems – Sensors and Measurements. MIT. https://ocw.mit.edu/courses/2-693-principles-of-oceanographic-instrument-systems-sensors-and-measurements-13-998-spring-2004/fdba55b169e259099d1754b3fcb4ef4b_lec16.pdf.

Williams, P.J. leB., Quay, P.D., Westberry, T.K., Behrenfeld, M.J., 2013. The oligotrophic Ocean is autotrophic. Ann. Rev. Mar. Sci. 5, 161–1615. https://doi.org/10.1146/annurev-marine-121211-172335.

Wilson, W.D., 1960. Speed of sound in sea water as a function of temperature, pressure and salinity. J. Acoust. Soc. Am. 32, 641–644.

Wimbush, M., 1977. An inexpensive sea-floor precision pressure recorder. Deep-Sea Res. 24, 493–497.

Wimbush, M., Chiswell, S.M., Lukas, R., Donohue, K.A., 1990. Inverted echo sounder measurement of dynamic height through an ENSO cycle in the Central Equatorial Pacific. IEEE J. Ocean. Eng. 15, 380–383.

Witter, D.L., Chelton, D.B., 1988. Temporal variability of sea-state bias in SEASAT altimeter height measurements. In: Chelton, D.B. (Ed.), Proceedings of the WOCE/NASA Altimeter Algorithm Workshop, Oregon 1987. US WOCE Technical Report No. 2, WOCE (World Ocean Circulation Experiment) Implementation Plan, vol. 1. Detailed requirements. 1988. WOCE International Planning Office, Wormley, UK.

Witter, D.L., Chelton, D.B., 1991. A Geosat altimeter wind speed algorithm and a method for altimeter wind speed algorithm development. J. Geophys. Res. 96, 8853–8860.

WOCE, 1988. World Ocean Circulation Implementation Plan, vols. 1 and 2. WOCE International Planning Office, Wormley, UK.

WOCE Science Steering Committee (SSC), 1991. SSC discusses WOCE priorities in Pacific, Indian and Atlantic Oceans. WOCE Note. 3 (3), 4–5, 1.

Wolter, K., Timlin, M.S., 1998. Measuring the strength of ENSO events – how does 1997/98 rank? Weather 53, 315–324.

Woodford, C., 2021 (updated May 12, 2023). https://www.explainthatstuff.com/introduction-to-neural-networks.html.

Woods, J.D., 1985. The world Ocean circulation experiment. Nature 314, 501–511.

Woodward, M.J., Crawford, W.R., August, 1992. Loran-C drifters for coastal Ocean measurements. Sea Technol. 33, 24–27.

Woodward, M.J., Huggett, W.S., Thomson, R.E., 1990. Near-surface Moored Current Meter Intercomparisons. Canadian Technical Report of Hydrography and Ocean Sciences, No. 125. Fisheries and Oceans Canada.

Woodworth, P.L., 1991. The permanent service for mean sea level and the global sea level observing system. J. Coast. Res. 7 (3), 699–710.

Wooster, W.S., Lee, A.J., Dietrich, G., 1969. Redefinition of salinity. Deep-Sea Res. 16, 321–322.

Worcester, P.F., Howe, D.M., Luther, D.S., 1988. Damping and phase advance of the tide in western Hudson Bay by annual ice cover. J. Phys. Oceanogr. 18, 1744–1751.

Worcester, P.F., Cornuelle, B.D., Spindel, R.C., 1991. A review of Ocean acoustic tomography: 1987–1990. Rev. Geophys. Supp. 29, 557–570.

World Climate Research Programme, 1988. Global Surface Velocity Programme (SVP): Workshop Report of WOCE/SVP Planning Committee and TOGA Pan-Pacific Surface Current Study. World Climate Research Programme, Miami, FL.

World Geodetic System, 1984. World Geodetic System — 1984 (WGS-84) Manual. International Civil Aviation Organization, Montreal, Quebec, Canada, p. 135.

Worthington, L.V., 1976. On the North Atlantic circulation. In: The Johns Hopkins Oceanographic Studies. Johns Hopkins University Press, Baltimore, MD.

Worthington, L.V., 1981. Chapter 2: The water masses of the world Ocean: some results of a fine-scale census. In: Warren, B.A., Wunsch, C. (Eds.), Evolution of Physical Oceanography. MIT Press, Cambridge, MA, pp. 42–69.

Wright, D.G., Pawlowicz, R., McDougall, T.J., Feistel, R., Marion, G.M., 2011. Absolute salinity, "density salinity" and the reference-composition salinity scale: present and future use in the seawater standard TEOS-10. Ocean Sci. 7, 1–26. https://doi.org/10.5194/os-7-1-2011.

Wu, Q.X., 1991. Tracking evolving sea surface temperature features. In: Proceedings of 6th New Zealand Image Processing Workshop. DSIR Physical Sciences, Lower Hutt, New Zealand.

Wu, Q.X., 1993. Computing velocity fields from sequential satellite images. In: Jones, I.S.F., Sugimore, Y., Stewart, R.W. (Eds.), Satellite Remote Sensing of the Oceanic Environment. Seibutsu Kenkyusha Co., Tokyo.

Wunsch, C., 1972. Bermuda sea level in relation to tides, weather, and baroclinic fluctuations. Rev. Geophys. Space Phys. 10, 1–49.

Wunsch, C., 1977. Determining the general circulation of the Oceans: a preliminary discussion. Science 196, 871–875.

Wunsch, C., 1978. The North Atlantic general circulation west of 50°W determined by inverse methods. Rev. Geophys. Space Phys. 16, 583–620.

Wunsch, C., 1988. Transient tracers as a problem in control theory. J. Geophys. Res. 93, 8099–8110.

Wüst, G., 1935. Die Sratosphäre. Wissenshaftliche Ergebinesse der Deutschen Atlantischen Expedition Meteor 1925–27.

Wyatt, L.R., Ledgard, L.J., Anderson, C.W., 1997. Maximum-likelihood estimation of the directional distribution of 0.53-HZ ocean waves. J. Atm. Ocean. Tech. 14, 591–603. https://doi.org/10.1175/1520-0426(1997)014<0591:MLEOTD>2.0.CO;2.

Wyrtki, K., 1961. The oxygen minimum in relation to Ocean circulation. Deep-Sea Res. 9, 11–23.

Wyrtki, K., 1962. The oxygen minima in relation to Ocean circulation. Deep-Sea Res. 1, 11–23.

Wyrtki, K., 1971. Oceanographic Atlas of the International Indian Ocean Expedition. NSF, Washington, DC.

Wyrtki, K., 1977. Sea level during the 1972 El Nino. J. Phys. Oceanogr. 7 (6), 779–787.

Wyrtki, K., Meyers, G., 1975. The Trade Wind Field over the Pacific Ocean. Part I, the Mean Field and the Mean Annual Variation. Hawaii Institute of Geophysics Report, HIG-75–1. University of Hawaii, Honolulu.

Xu, G., Di Iorio, D., 2011. The relative effects of particles and turbulence on acoustic scattering from deep-sea hydrothermal vent plumes. J. Acoust. Soc. Am. 130. https://doi.org/10.1121/1.3624816.

Xu, F., Ignatov, A., 2014. In situ SST Quality Monitor (iQuam). J. Atmos. Oceanic Technol. 31, 164–180. https://doi.org/10.1175/JTECH-D-13-00121.1.

Xu, G., Jackson, D.R., Bemis, K.G., Rona, P.A., 2013. Observations of the volume flux of a seafloor hydrothermal plume using an acoustic imaging sonar. Geochem. Geophys. Geosyst. 14, 2369–2382. https://doi.org/10.1002/ggge.20177.

Xu, G., Bemis, K., Jackson, D., Ivakin, A., 2021. Acoustic and in-situ observations of deep seafloor hydrothermal discharge: an OOI cabled array ASHES vent field case study. Earth Space Sci. 8, e2020EA001269. https://doi.org/10.1029/2020EA001269.

Yao, T., Freeland, H.J., Mysak, L.A., 1984. A comparison of low frequency current observations off British Columbia with coastal-trapped wave theory. J. Phys. Oceanogr. 14, 22–34.

Yashayaev, I., 2024. Intensification and shutdown of deep convection in the Labrador Sea were caused by changes in atmospheric and freshwater dynamics. Commun. Earth Environ. 23. https://doi.org/10.1038/s43247-024-01296-9.

Yashayaev, I., Loder, J.W., 2009. Enhanced production of Labrador Sea Water in 2008. Geophys. Res. Lett. 36, L01606. https://doi.org/10.1029/2008GL036162.

Yashayaev, I., Loder, J.W., 2016. Recurrent replenishment of Labrador Sea Water and associated decadal-scale variability. J. Geophys. Res. 121, 8095–8114. https://doi.org/10.1002/2016JC012046.

Yashayaev, I., Loder, J.W., 2017. Further intensification of deep convection in the Labrador Sea in 2016. Geophys. Res. Lett. 44, 1429–1438. https://doi.org/10.1002/2016GL071668.

Yoshikawa, Y., Matsuno, T., Marubayashi, K., Fukudome, K., 2007. A surface velocity spiral observed with ADCP and HF radar in the Tsushima Strait. J. Geophys. Res. 112, C06022. https://doi.org/10.1029/2006JC003625.

Yu, Y., Emery, W.J., Leben, R., 1995. The annual variation of equatorial surface currents in the Western Tropical Pacific computed from satellite altimetry. J. Geophys. Res. 100, 25069–25085.

Yu, X., Naveira Garabato, A.C., Martin, A.P., Buckingham, C.E., Brannigan, L., Su, Z., 2019. An annual cycle of submesoscale vertical flow and restratification in the upper Ocean. J. Phys. Oceanogr. 49 (6), 1439–1461.

Yueh, S.H., West, R., Wilsonn, W.J., Li, F.K., Njoku, E.G., Rahmat-Samii, Y., 2001. Error sources and feasibility for microwave remote sensing of Ocean surface salinity. Trans. Geosci. Remote Sens. 39, 1049–1060. https://doi.org/10.1109/36.921423.

Závody, A.M., Mutlow, C.T., Llewellyn-Jones, D.T., 1995. A radiative transfer model for sea surface temperature retrieval for the along-track scanning radiometer. J. Geophys. Res. Ocean. 100, 937–952.

Zaytsev, O., Rabinovich, A.B., Thomson, R.E., 2017. The 2011 Tohoku Tsunami on the Coast of Mexico: A Case Study, Pure and Applied Geophysics, Topical Issue: Global Tsunami Science, vol II. Springer International Publishing AG. https://doi.org/10.1007/s00024-017-1593-z.

Zaytsev, O., Rabinovich, A.B., Thomson, R.E., 2021. The impact of the Chiapas tsunami of 8 September 2017 on the Coast of Mexico. Part 1: observations, statistics, and energy partitioning. Pure Appl. Geophys. 178, 4291–4323. https://doi.org/10.1007/s00024-021-02893-x.

Zenk, W., Halpern, D., Kase, R., 1980. Influence of mooring configuration and surface waves upon deep-sea near-surface current measurements. Deep-Sea Res. 27, 217–224.

Zhai, L., Greenan, B., Thomson, R., Tinis, S., 2019. Use of Oceanic reanalysis to improve estimation of extreme storm surge. J. Atmos. Ocean. Technol. 36, 2205–2219. https://doi.org/10.1175/JTECH-D-19-0015.1.

Zhan, L., Zhang, J., Ouyang, Z., Lei, R., Xu, S., Qi, D., Gao, Z., Sun, H., Li, Y., Wu, M., Liu, J., Chen, L., 2021. High-resolution distribution pattern of surface water nitrous oxide along a cruise track from the Okhotsk Sea to the western Arctic Ocean. Limnol. Oceanogr. 66, S401–S410.

Zhang, X., Church, J.A., Monselesan, D., McInnes, K.L., 2017. Sea level projections for the Australian region in the 21st century. Geophys. Res. Lett. 44, 8481–8491. https://doi.org/10.1002/2017GL074176.

Zhang, Y., Rueda, C., Kieft, B., Ryan, J.P., Wahl, C., O'Reilly, T.C., Maughan, T., Chavez, F.P., 2019. Autonomous tracking of an Oceanic thermal front by a Wave Glider. J. Field Robot. 36, 940–954. https://doi.org/10.1002/rob.21862.

Zhao, Y., Bao, Z., Wan, Z., Fu, Z., Jin, Y., 2019. Polystyrene microplastic exposure disturbs hepatic glycolipid metabolism at the physiological, biochemical, and transcriptomic levels in adult zebrafish. Sci. Total Environ. https://doi.org/10.1016/j.scitotenv.2019.136279.

Zhurbas, V., Lyzhkov, D., Kuzmina, N., 2014. Drifter-derived estimates of lateral eddy diffusivity in the world ocean with emphasis on the Indian Ocean and problems of parameterization. Deep-Sea Res. 83, 1–11. https://doi.org/10.1016/j.dsr.2013.09.001.

Zurbas, V., Oh, I.M., 2004. Drifter-derived maps of lateral diffusivity in the Pacific and Atlantic Oceans in relation to surface circulation patterns. J. Geophys. Res. 109, C05015. https://doi.org/10.1029/2003JC002241.

Zurbas, V.M., Lyyzhkov, D.A., Kuzima, N.P., 2014. Estimates of the lateral eddy diffusivity in the Indian Ocean as derived from drifter data. Mar. Phys. 54, 281–288.

Zurbenko, I., Porter, P.S., Rao, S.T., Ku, J.Y., Gui, R., Eskridge, R.E., 1996. Detecting discontinuities in time series of upper air data: development and demonstration of an adaptive filter technique. J. Clim. 9, 3548–3560. http://ams.allenpress.com/pdfserv/10.1175%2F1520-0442(1996)009%3C3548:DDITSO%3E2.0.CO%3B2.

Zurbenko, I.G., 1986. The Spectral Analysis of Time Series. North Holland, p. 248.

Index

Note: 'Page numbers followed by *f* indicate figures and *t* indicate tables.'

Printed in the United States
by Baker & Taylor Publisher Services